Second Edition

Economic Geography
An Institutional Approach

Roger Hayter & Jerry Patchell

OXFORD
UNIVERSITY PRESS

OXFORD
UNIVERSITY PRESS

Oxford University Press is a department of the University of Oxford.
It furthers the University's objective of excellence in research, scholarship,
and education by publishing worldwide. Oxford is a registered trade mark of
Oxford University Press in the UK and in certain other countries.

Published in Canada by
Oxford University Press
8 Sampson Mews, Suite 204,
Don Mills, Ontario M3C 0H5 Canada

www.oupcanada.com

Library and Archives Canada Cataloguing in Publication

Hayter, Roger, 1947-, author
Economic geography : an institutional approach / Roger Hayter and
Jerry Patchell. – Second edition.

Includes bibliographical references and index.
ISBN 978-0-19-901328-9 (paperback)

1. Economic geography. I. Patchell, Jerry, 1957-, author II. Title.

HF1025.H39 2016   330.9   C2015-908295-1

Cover image: Neil Emmerson/Getty Images

Oxford University Press is committed to our environment.
Wherever possible, our books are printed on paper which comes from
responsible sources.

Printed and bound in the United States of America

1 2 3 4 — 19 18 17 16

Contents

PART III   LOCATION DYNAMICS OF VALUE CHAINS

Figures, Tables, and Case Studies

Tables

Case Studies

Preface

Institutional concepts and themes have been present in economic geography texts for decades, but for the most part they have been discussed only as parts of an eclectic mix, not as the building blocks of a coherent architecture. This book provides such an architecture and is distinctive, in two main ways; first, by emphasizing markets as the key institution underlying modern economies and, second, by exploring the main institutions that influence markets and the nature of global economic geographies. For the past two centuries markets have driven economic development and increased living standards but have also led to vast income inequalities and environmental degradation. This book offers a comprehensive introduction to the interplay of economic, social, and political institutions that shape local development, global integration, and levels of living. Students will become critically aware of how markets are organized and governed by people and their institutions. As an introductory text, this book equips students with basic principles and concepts of how markets are deemed to organize space and the institutions of governance necessary both to make efficient organization of space possible and improve the outcomes for society.

About this Book

This second edition of our book updates, clarifies, and elaborates our argument while retaining the original overall structure. Thus the book is organized in three main parts. Part 1 ("Markets in Place, Space, and Time") lays the foundations, introducing markets (Chapter 1), the spatial division of labour (Chapter 2), and the role that innovation plays in the evolution of market economies (Chapter 3). Part 2 ("Institutional Pillars of Modern Space Economies") defines the central economic and non-economic institutions (business, labour, government, non-government organizations) and explores how their activities shape economic activities in place and space (Chapters 4–7). Finally, Part 3 ("Location Dynamics of Value Chains") explores how various institutions influence the location dynamics of value chains beginning with the powerful coordinating role of cities (Chapter 8) and is further explored in the resource, manufacturing, and service sectors (Chapters 9–12); examines how transportation and communication networks function to integrate economies (Chapter 13); and culminates with an assessment of the nature and impact of consumption (Chapter 14).

The three parts constitute progressive building blocks, and it is important to read them in sequence (although individual chapters in Parts 2 and 3 may be rearranged somewhat). The book is designed as an introductory text in economic geography, that is, for first- or second-year university and college students. We hope that it will provide a solid foundation for more advanced studies not only in economic geography but in any discipline with an interest in economic affairs and the roles that institutions play in shaping them.

Acknowledgements

The revisions to this book have been greatly helped by a number of people. Indeed, we most gratefully acknowledge the extensive, constructive comments of our reviewers noted below, the numerous insightful comments of Ian MacLachlan and Søren Kerndrup, the many colleagues and friends who prepared exhibits for us (they are acknowledged in the text), the work of John Ng at Simon Fraser who has provided wonderfully patient help in drawing all the illustrations while the research provided by Julia Affolderbach and Klaus Edenhoffer for the first edition continued to be useful. We also thank the editorial staff at Oxford, especially Meg Patterson, Karri Yano, and Lisa Ball who provided prompt, friendly, and excellent advice throughout the process. Similarly Jacquie, Sue, and our families were always understanding, supportive, and often inspirational! Finally, we dedicate this book to our undergraduate students, not incidentally recalling our own undergraduate lives, teachers, mentors, and friends.

Our reviewers who provided both terrific help and much encouragement are:

Jeff Boggs
Brock University

David Edgington
University of British Columbia

Brian Lorch
Lakehead University

Heather Nicol
Trent University

Dan Todd
University of Manitoba

An Institutional Approach to Economic Geography

Human economic life . . . is the affair of organized groups.

Wagner 1960: 63

Why does economic activity vary so tremendously from place to place around the world? Why do some places effectively control the global economy while others are integrated into it through specialized roles and some remain marginalized? What are the consequences for the well-being of countries, regions, companies, households, and individuals? These are some of the basic questions that economic geography tries to answer.

The introduction is divided into two parts: The first part introduces the discipline of economic geography and outlines the basics of institutionalism. The second part offers a preview of our approach to the subject that focuses on the inter-relationships among markets, institutions, technology, and places and spaces, and the organization of economic activities within value chains.

About Economic Geography

Economic geography studies the variations in the location and spatial distribution of economic activities, the geographic differences in the economic functions and well-being of places, and the connections among economic activities across space. Economic geography's mandate is pursued at different geographic scales—from the local to the global—and across different time horizons, from instantaneous snapshots to evolutionary changes across historical periods. The theories and tools of the discipline have implications for government policy and practical applications for private and state-owned companies, non-governmental organizations, and consumers.

Economic geography has four interrelated objectives:

1. To understand how and why places differ in terms of economic activities. Economic geographers examine community, regional, and national patterns of economic growth and decline; why rich and poor nations or regions exist; why standards of living vary among places and what accounts for disparities within places; and the use and abuse of the environment.

2. To understand how and why economic activities are connected across space. Economic geographers examine how the complementarity and competition among places generates massive flows of money, information, labour, goods, and

services within nations and across national borders. Economic geography is at the forefront of explanations of global integration and how it works. That understanding provides feedback to understanding the interdependence of spatially uneven development and what globalization means "on the ground" in particular places.

3.   To advise business, government, and other institutions on how best to locate economic activities and organize patterns of land and resource use, and make connections in place and across space so as to meet social objectives, however these objectives may be defined in terms of efficiency, equity and sustainability criteria. And, finally,

4.   To teach skills and develop perspectives that students can apply in a wide variety of careers and that, ultimately, may help them become better citizens.

Outside academia, "economic geographers" are rarely labelled as such, but demand for their expertise is widespread in the public and private sectors. For example, location-analysis skills can be used to determine the best location for a manufacturer to put together products and still have access to markets, or which piece of real estate is best for which commercial uses. As real estate agents say, it is all about location, location, location. Similarly, a background in economic geography provides a valuable perspective for anyone seeking to advise business or government on the economic, social, and environmental implications of their decisions to, for example, open a new factory (or close an old one) or to build a highway. Economic geography addresses a host of issues directly relevant to local and regional development planning, environment-economy management, transportation planning, etc.

As an introductory text, we will study economic geography with the understanding that the foundation of human society is rooted in institutions: formal and informal structures, routines, conventions, and customs that organize all human relationships. In terms of economic behaviour, the central organizing institution is the "market." But the market is neither an abstract or independent entity and there are many different kinds of markets. They began as *places of exchange* and continue as such, for example, as shopping malls where consumers buy goods in person from stores or suppliers. However, markets increasingly are becoming *spaces of exchange* such as e-bay, electronically conducted between remotely located buyers and sellers. Within these two basic types, markets vary widely depending not only on what is bought and sold, but more importantly, on *how* products are bought and sold, or how products are organized and regulated by particular societies. Our key point is: *markets are institutions interdependent with other institutions and influences*. We will see how markets vary over space and time and examine their various manifestations as business, labour, governments, and non-profit organizations shape and are shaped by them. Together, these institutional forces determine a place's economic efficiency, the degree of social equity it enjoys, its ability to economically and environmentally sustain itself, and its position in a global system.

Historical Thumbnail Sketch of Economic Geography

Economic geography has changed profoundly since its beginnings. Around 1960, economic geography (and geography as a whole) experienced a "paradigmatic" change in its way of thinking from descriptive (ideographic) to theoretical (nomothetic) approaches (Table I.1).

TABLE I.1   The Evolution of Economic Geography

Main Phases	Space–Place Focus and Comment	Texts
Descriptive (ideographic)		
Commercial geography (1600–1900)	During the age of exploration, emphasis is on describing activities in "distant" places around the globe and their potential for trade. Institutions are not considered, although descriptions are organized on a national basis and colonial power is implicit.	Chisholm 1889 (revised by Stamp 1932)
Environmental determinism (1900–1930)	Economic geography emerges as the study of economic activities (especially resource and industrial production) in particular places and the implications of those activities for trade and national economic development. Physical environment is seen as the most important determinant of economic activity in any place.	Huntingdon 1915; Taylor 1937
Areal differentiation 1930–1960	Human geography emphasizes the uniqueness of places through historical accounts and regional syntheses of interactive social, economic, and physical processes. Resource geography is also an important expression of economic geography. The influence of cultural institutions is recognized.	Jones and Darkenwald 1949; Zimmerman 1933; Wagner 1960
Theoretical (nomothetic)		
Quantitative revolution 1960–	Dramatic shift away from the description of unique places towards the explanation of activities across space in terms of general laws and principles drawn from neoclassical economic theory, tested by formal quantitative methods. Strong interest in regional development and planning. Greater emphasis on urban–industrial economies. Markets are recognized, albeit abstractly; otherwise institutions are marginalized.	Morrill 1974; Lloyd and Dicken 1972; Berry et al. 1993 See also Isard 1960
Conceptual pluralism and fragmentation 1970–	Economic geography explores a variety of theoretical approaches that include Marxism and institutionalism. Emphasis on global perspectives and the roles of place and space (local–global dynamics). Institutions—big business, government, labour—become an explicit theme.	Knox and Agnew 1989; de Stouza and Stutz 1994; Coe et al. 2007. Also edited texts, e.g., Daniels and Lever 1996, Barnes and Shepherd 2000

For a more detailed account of these phases see Berry, Conkling, and Ray 1993.

From around 1600, early commercial geographers set out to document the economic activities of the distant places that European powers were discovering. If descriptive, their work nevertheless addressed the nature of goods production among countries and provided heightened awareness of economies and societies around the globe at a time when international trade was primarily commodity-driven. By the late 1800s, commercial geography was describing the attributes of different places in an effort to determine their potential for trade, usually for the benefit of European interests. At that time, (bearing in mind that the newly discovered lands were largely subsistence agricultural economies) textbooks related spatial variations in commodity production with variations in physical factors such as climate, soils, and topography.

In the early twentieth century, variations in development levels were attributed to the features of climate and landscape, a way of thinking labelled environmental determinism. But geographers soon began to recognize that this approach failed to appreciate the many ways in which human populations act as agents of change. By the 1930s geographers began to emphasize the role of human agency to explain how both developed and less developed areas became different from each other. For economic geography, this **areal differentiation** typically meant an historical approach to describe the characteristics of

areal differentiation An approach to human geography that refers to regions as unique places shaped by distinct evolutionary interactions among economic, political, cultural, and environmental forces.

economic activities in particular places, tracing the development of unique patterns to particular combinations of social, political, economic, and environmental factors. Some geographers focused on resource exploitation and other human–environment interactions. These pioneering textbooks were not simply descriptive, but sought to classify economic activities and begin to conceptualize human behaviour in institutional terms. Thus Zimmermann's (1933) resource geography argued for the cultural interpretation of resources and Wagner's (1960) economic geography recognized that "human life . . . is the affair of organized groups."

But areal differentiation approaches were swept aside in the 1950s and 1960s as part of a quantitative revolution. Pioneering economic geographers urged the development of explicit theories, conceptual frameworks, data collection and its quantitative analysis and modelling. This shift in thinking was so profound that it was considered a change in scientific paradigm.

Economic geography's embrace of theory and quantitative methods was driven in part by a desire for scientific respectability, a growing recognition of the potential of systematic data manipulation, the need to educate students with analytical capabilities, and new policy challenges that demanded systematic, problem-solving research. Social problems, such as regional and international income and employment disparities, urban and regional restructuring, and environmental sustainability, drove the research and methodological agendas. The spirit and purpose of the discipline were redefined. Economic geography transformed itself from an "ideographic"/descriptive pursuit concerned primarily with empirical description and unique places to a "nomothetic"/theoretical one focused on discovering and testing generalizations about spatial organization. In practical terms, this transformation was initially driven by the development of theories of location, land use, and regional development that were closely connected with the abstractions of neoclassical economics and its emphasis on the universality of economic behaviour in competitive markets. Explanations in economic geography became similarly abstract, focused solely on economic factors such as access to markets, resources, and labour as explanations for the spatial distribution of economic activities.

Neoclassical thinking remains influential in economic geography. But alternative theoretical perspectives mushroomed after 1970 to the point that at times the discipline seemed to be on the verge of fragmentation. Marxism, rooted in a different, but no less universalist view of economic behaviour, became a significant influence, in direct opposition to neoclassical models. Meanwhile, proponents of various other approaches to economic geography argued against over-emphasis on universal laws and sought to develop theories that would give more emphasis to spatial differences both in human behaviour and organization and in the ways economic and non-economic factors intertwine with one another. If fragmentation is still evident, overlapping themes in this conceptual pluralism carry many institutional concepts that have permeated all approaches to economic geography.

Institutionalism and Economic Geography

The theoretical approach advocated in this book is rooted in the radical or dissenting institutional economics that developed at the beginning of the twentieth century to provide alternative explanations of economic behaviour than those offered by neoclassical economics or Marxism. Thorstein Veblen (1904) and John R. Commons (1893; 1934) were the leaders of this movement. They and their followers employed research methods linked by common emphasis on institutions, the understanding of actual behaviour, and how they evolve over time in specific contexts. This focus on "real" behaviour contrasted to the abstract models used by neoclassical and Marxist economists that provide more general ("universal") interpretations of economic behaviour and systems. Against the prevailing orthodoxy, however, institutionalists took particular issue with three central features of (traditional) neoclassical thought.

First, foundational neoclassical theories start from a number of highly abstract assumptions about human decision-making. In particular, they assume that producers and consumers act purely on the basis of *economic self-interest*; that they are relentlessly rational, invariably choosing to minimize costs, maximize profits, or otherwise optimize some narrowly defined economic objective; and that they have all the information they need to make the correct decision. By contrast, institutional explanations are based on observations of "real world" behaviour that recognize the diverse abilities and motives of decision-makers, and the power of large corporations to shape markets and the nature of economic development.

Second, neoclassical explanations are powerfully rooted in the idea that economic systems tend to seek a stable equilibrium condition, balancing the forces of supply and demand. In contrast, institutionalism stresses that capitalist economies are by nature transformative—continually changing rather than simply moving back and forth around some equilibrium point.

Third, in neoclassical economics, market choices are made by individual consumers and producers whose reasoning is entirely economic (or "economistic"), unaffected by any other considerations. By contrast, the institutionalists argued that economic reasoning cannot be divorced from social and political factors.

Like Marxism, institutionalism recognizes the profound social problems created by markets, but since its beginnings the focus of institutionalism was not limited to class relations or the exploitation of labour. Rather, it addresses the tensions and interdependencies that exist among society's many different stakeholders with their multiple motives and interests. Moreover, while recognizing the conflicts of interest inherent in many of these relations, institutionalism appreciates that humans are extraordinarily creative in building and adapting institutions to address social problems. The analysis of technological change and innovation, and their diverse socio-economic consequences, is a central concern of institutionalism, more so than in Marxist approaches.

During the 1950s and 1960s institutionalists remained outside mainstream economics but gained a public profile through the iconic works of Gunnar Myrdal (1944, 1957) and John Kenneth Galbraith (1958, 1967). Myrdal interpreted poverty, racism, and underdevelopment as deeply engrained "circular and cumulative" problems. Galbraith drew attention to the power of big business in shaping the modern industrial state and affluent society, and his arguments made him the most widely read economist of the twentieth century. The ideas of both these leaders impacted economic geography.

Recently, institutionalism has become a much stronger force in economics, its spirit and purpose reflected in the work of Geoffrey Hodgson (2006), and in other disciplines such as sociology (Granovetter 1985), and economic geography (Martin 1994, Storper 1997; Hayter 2004; Gertler 2010). Indeed, this surging interest has revived appreciation for the classic works of Schumpeter (1943) on the dynamics of capitalist economies as "creative destruction" and for Polanyi's (1944) study of the "great transformation" of market economies. Further, within mainstream economics, the development of the "new" institutional economics and behavioural economics have incorporated institutional thinking and questioned narrow assumptions of economic rationality. Notable economists of these approaches include Coase and Williamson and Simon and Kahneman respectively. Marxist approaches, such as regulation theory, have similarly embraced institutions.

In the geographical context, conventional (neoclassical) theories of location and land use have long recognized the role that markets play in creating global economic geographies through their influence on the spatial distribution of production. *This recognition remains important in Von Thünen's agricultural land use theory (Chapter 10), Christaller's central place theory (Chapters 1 and 12), Weber's industrial location theory (Chapter 11) and urban land theory (Chapter 8).* But they take the market itself as given and rarely examine how a particular market actually works and is organized in a particular place, according to particular rules, regulations, and habits of thought. The institutional approach, by contrast, explicitly recognizes the part that non-economic institutions play in all economic activity, regardless of the nature of the market or the parties to the transaction (business–consumer, business–business, labour–employer, government–business). Among those non-economic institutions are not only governments, unions, and NGOs, but also the gamut of formal and informal political relationships and social conventions that determine the rules of the game in a particular place. An institutional analysis incorporates these multiple perspectives in an integrating framework.

An Institutional Framework for Economic Geography

In this book, economic geography is understood through a focus on the interaction of institutions, markets, and technology in place and space (Figure I.1). These interactions are characterized by three basic principles: embeddedness, evolution, and differentiation.

Economic and non-economic factors and processes are inseparable—so thoroughly integrated that they can be understood only in relation to one another or symbiotically. This is what we mean by "embeddedness." Thus individual markets that set prices for the exchange of goods and services simultaneously involve issues such as trust, conventions, regulation, control, and social influence. More generally economic, political, and social (and environmental) institutions continually interact or feed back on one another.

Market economies are fundamentally transformative, driven by processes of creative destruction that imply growth and crisis as a result of technological innovation, competition, population dynamics, and changes in human–environment relations. In other words, they "evolve" and become qualitatively different over time, sometimes slowly, sometimes rapidly. In this book, we emphasize technological innovation because it is an important stimulus to competition and drives systemic and paradigmatic transformations throughout economic systems that occur in periodic cycles or waves, changing the nature, location, and organization of economic activities. Cumulative causation and path dependency are two related important concepts that recognize that evolution is deeply engrained and self-reinforcing. Thus the economic trajectories of individuals, places, firms, and industries are shaped by their accumulated assets, income, know-how, and mutual influence. Broadly stated, economic evolution is constrained by history. The ability of people to think, innovate, and compete threatens these constraints.

Finally, (geographic) "differentiation" describes the unique nature of the places and spaces that evolve through the distinct interactions of markets, other institutions, and technology. It is the recognition that capitalism appears in many different configurations among and within countries and at different scales. In geography, this differentiation principle is expressed by the use of concepts such as "local models" (Barnes 1987) and "regional worlds of production" (Storper 1997). In both cases, geographic variations in economic activity are seen as resulting from the interplay of the local or regional with more general global forces or models from the rest of the world. Areas or regions at varying scales are typically distinguished by shared political and administrative boundaries, common activities, shared values and attitudes, and/or by strong levels of functional integration. Indeed, regions themselves may be considered as institutions, albeit complex ones, where people co-operate, compete, and resolve problems within particular boundaries.

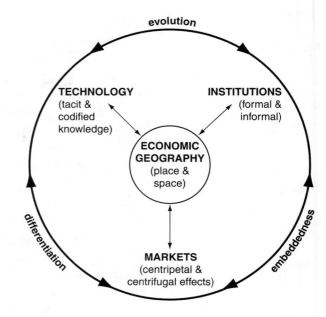

FIGURE I.1   **The Starting Point: An Institutional Perspective for Economic Geography**
See also Storper 1997: 27.

Markets

Our approach to economic geography centres on markets, whose importance, we believe, has been set in the background in our subject—our text books assume markets are there, but don't often discuss them directly. Markets are the key institutions organizing exchanges and flows of goods and services among economic actors, whether within a particular economy or among multiple economies. In neoclassical economics, supply and demand, competition, and economies of scale are central to understanding how market economies function. Economic geography asks why and how markets organize economic activity in places and across space, why and how exchange becomes concentrated in places, and why and how transactions occur over space. These questions are pertinent whether the transactions are an arm's-length buying and selling of goods and services in a store or the more mediated exchange of knowledge, labour, and materials within a firm's production process. In all cases, the essence of markets is their ability to reduce transaction costs. As Chapter 1 explains, these three basic types of exchange correspond to three basic market types: *open markets* (numerous competing buyers and sellers); *relational markets* (between parties with an established relationship of some kind); and *administered or hierarchical markets* (within a single company). In whatever form, reducing transaction costs provides distinct market geographies that are revealed by the roles of cities and value chains.

Thus markets, to reduce transaction costs, act as a centripetal force by drawing firms, labour, and consumers together for competitive and co-operative interaction. Indeed, cities can be seen as the congregation of marketplaces for diverse industrial sectors, labour skills, and consumer demand. Within cities, these market places sort themselves out, in part according to land costs and need for interaction. Simultaneously, cities act as centrifugal forces, stimulating the creation of transportation and communication networks to distribute goods and services. Understanding markets helps understand cities, the economies they offer, the products and services produced in the spaces between cities, and the hierarchies of infrastructure that channel resources, people, and information into them and out of them. Moreover, the centripetal power of markets is increasing as the world becomes increasingly urbanized, urbanization focuses on megalopolises, and expectations grow for the city to function as the engine of economic development.

The value chain is a linked series of markets. Resources, labour, and capital are combined in a linked series of production stages that eventually deliver a final product that is purchased by a consumer, business, or government. To produce and distribute a complex good such as an automobile, thousands of transactions may be required, some of which may span the globe. To accomplish that goal these markets have to be coordinated by some entity, usually a multinational enterprise, but many goods and services are exchanged through spatially differentiated markets based on speculation of further buyers downstream. The global value chain looks at this coordination from the product perspective, and the global value *cycle* considers product coordination all the way through to recovery and recycling, or from an environmental perspective.

Significantly, as Chapter 1 further clarifies, markets do not always operate efficiently or fairly, and markets can fail, sometimes with dire consequences. The near collapse of the global financial system in 2008–9, following the failure of banks in the US and UK, illustrates the savage implications of market failure. The subsequent meeting of the G20 leaders of the world's most powerful countries reflected the need for some form of global orchestration of the global economy (Case Study I.1). The importance of understanding markets as institutions that are socially created and organized by groups of people was graphically underlined by the source of this failure, which stemmed directly from the removal of regulations governing the banking system and from contemporary efforts to resolve the problems. Indeed, the institutional perspective recognizes the importance of regulation and the influential role of institutions other than markets in shaping the economies of capitalist societies.

Institutions

Sympathetic to the Veblen tradition the sense of institutions in this textbook is captured by habits of thought, and how conventions, routines, and rules shape social behaviour.

Case Study I.1

THE G20 LONDON SUMMIT AGREEMENT, 2 APRIL 2009

For two Nobel Prize-winning economists, George Stiglitz and Paul Krugman, the root cause of the global financial crisis was the deregulation of the banking system: specifically, the decision by the institutions responsible for governing the world's banks (in particular, the US Federal Reserve and the Bank of England) to relax the existing restrictions on banks' ability to engage in risky speculative ventures. Alan Greenspan, chairman of the US Federal Reserve from 1987 to 2006, had supported deregulation because he believed that "free markets" and minimum government intervention (i.e., regulation) were necessary to stimulate competition, efficiency, and wealth creation. Deciding how to re-regulate the financial system became the most urgent task of the Obama administration when it took office in 2009, but the crisis was not limited to the United States. US investment banks had sold their financial products around the world, and banks in countries such as the UK, Switzerland, and France had joined the race for paper profits. Finally, in April 2009, the leaders of the G20 countries met to work out a solution.

Only 19 of the 20 world leaders who met for the summit were present for this photo; the missing leader was Canada's Prime Minister Stephen Harper.

The G20 agreement sought to address the global financial crisis in two ways: by re-regulating the global financial markets and by stimulating development in hopes of bringing the global economy out of its deep downturn. With respect to regulation, the agreement proposed to

- regulate (for the first time) hedge funds and derivatives: two relatively new instruments that had come into wide use among financial institutions since the 1980s, largely uncontrolled by supervisory agencies (and largely unfamiliar to the public); generally speaking, hedge funds and derivatives are created when financial institutions lend to one another using other loans (rather than real assets) as collateral;

(Continued)

- publish immediately a list of tax havens and impose sanctions on countries that did meet anti-secrecy regulations;
- tighten regulations on credit agencies to prevent conflicts of interest;
- develop an international body of supervisors to oversee national regulators; and
- reform the International Monetary Fund (IMF) to provide a greater role for recently industrialized countries such as China.
    With respect to development the agreement proposed to
- inject a total of $1 trillion into the global economy, including $500 billion for the IMF, $250 billion specifically for loans to developing countries, and $250 billion for trade assistance;
- provide $50 billion for the world's poorest countries; and
- renew commitment to global development goals.

Whether the plan was implemented, or to what degree, is hard to tell. The G20 does not seem to evaluate its initiatives. There has been some re-regulation of the US/UK banking systems, but the reform of the IMF has not occurred (as of mid-2014), and it is not known if the funds for loans, trade assistance, and help for the world's poorest countries have been made.

The G20 agreement is an appropriate reminder that, in general, "human economic life . . . is the affair of organized groups." In this case, the organized group consisted of the G20 leaders, along with their supporting bureaucracies and advisors. The agreement also underlines the point that markets are institutions that are socially created in specific places and specific times.

According to Hodgson's (2006, p. 2) widely cited definition, institutions are "systems of established and prevalent social rules that structure social interactions." Even more pithily, North (1990, p. 3) refers to them as "the rules of the game in a society," while Allen (2012, p. 4) sees them as "the rules we live by and how life is organized." The institutional approach to economic geography is rooted in the idea that groups of people, both at broad societal and individual organizational levels, are held together by institutions that coordinate their behaviour to some extent. If coordination is less defined at the societal level, institutions shape behaviour and attitudes within nation states, regions, or local communities and have a pervasive and profound impact on the functioning of markets and on economic development. Social institutions may be informal, such as customs, traditions, and norms, for trust and reciprocity or formal as in the legal system, contracts, property rights regimes, work manuals, and committee mandates. Societies actively try to modify their institutions, from core values to government policies. Many societies, for example, have adopted the property and legal norms of other nations in the quest to improve the efficiencies of their markets.

Are organizations also institutions? We think so. Organizations such as businesses, labour unions, government departments, non-government organizations, and social clubs, like social institutions, hold people to certain values, beliefs, and behaviours, and coordinate behaviour to a greater degree. That is especially true of corporations that have the authority to determine employee behaviour as derived from the power of ownership and employees' acceptance of working conditions. Corporations structure or "institutionalize" behaviour around specified routines and rules, supported by common (organizational) ways of thinking and committee structures with specified terms of reference. The rules and routines of corporations are typically legitimized and constrained by laws (contracts, property rights, labour laws, health and safety regulations). Similar conditions prevail in government, NGOs, and so on. However, membership in NGOs, industry associations, social and sports clubs are voluntary in nature, and these organizations cannot impose behavioural expectations; rather common values and reciprocity shape behaviour. As individuals, we are typically governed by multiple institutions imposed by the legal system, places of work, and the various social and recreational groups in which we participate.

Big corporations, governments, and labour unions and associations are so powerful they are able to shape the rules of the game at societal levels. On the other hand, informal rules of the game may evolve regionally, among clusters of firms, and even in particular neighbourhoods. The fact that institutions are animated, passed on, and changed by organizations (and organized groups) makes it virtually impossible to discuss one without the other. Indeed, this book emphasizes the role that organized groups—markets, business, labour, government, NGOs—play in shaping institutions. In general, it is useful to think of institutions as operating at different levels or in different domains, from society as a whole to segments of society (business, consumers, local cultural groupings) or movements (environmentalism) to individual organizations (such as businesses that vary from small logging operations or corner stores to big banks and giant MNCs such as Apple or Toyota). Broad institutional types are also readily classified. In every case, institutions are defined by habitual, durable, and distinctive behaviour.

A principal social function of institutions is to provide the stability and continuity necessary to allow people to live their daily lives. Stable routines and conventions are essential for the efficient, orderly functioning of societies because they establish the regular patterns of behaviour and shared expectations that make it possible for people to coordinate all the individual activities required to produce the goods and services they need. Relatedly, institutions potentially impose penalties or constraints on inappropriate behaviour, whether this behaviour is deemed illegal, opportunistic, or socially unacceptable. Penalties range from fines and jail sentences to ostracism or exclusion from groups or loss of trust.

In addition, by representing diverse social interests, institutions help to maintain social stability and cohesion. At the same time, by interacting with one another—whether competing for influence and power or working together, co-operatively or resolving conflicts—they make it possible for societies to evolve. One reason disputes over the use of land and resources are so hard to resolve is that the competition between rival interests in society is often driven by different, equally legitimate values.

Market economies are inherently dynamic. They are constantly evolving, driven by innovation, population dynamics, and changing tastes, and subject to disruption at any time by economic crises, wars, or natural hazards. From this perspective, institutions serve two contrasting functions: to constrain behaviour and to facilitate change towards new routines. In Allen's (2012) provocative thesis, the transformation from pre-modern (aristocratic) to modern (market) institutions between 1750 and 1850 in England were stimulated by new systems of measurement (of inputs and outputs). In the former, patronage, trust, and venality underlay aristocratic behaviour (and made sense), but the development of abilities to measure (weight, distance, time, levels, position, etc.) stimulated alternative more efficient institutions that in turn facilitated the productive potentials of the new technologies of the Industrial Revolution to be realized. In contemporary times, the "transitional" economies of Eastern Europe and China provide spectacular illustrations of the power of institutional change. China's reform and opening up began in 1978 and has lifted hundreds of millions out of poverty. The policies of *perestroika* and *glasnost* (economic and political liberalization) introduced in the Soviet Union in 1988 heralded a massive institutional transformation as the countries of Eastern Europe sought to catch up with the economically advanced Western countries. At the heart of this transformation was the curtailment of state planning in favour of decision-making by individual enterprises operating in free markets where prices are determined by demand and supply. While the transition to free markets was seen as the ultimate institutional change, the creation of the ancillary institutions required to enable those markets to function has been no less notable.

In both China and Eastern Europe, for example, the transition required some legal recognition of the private property rights on which virtually all market activity depends (the specifics vary from place to place). Similarly, laws restricting inward foreign direct investment (FDI) by multi-national corporations (MNCs) have been eased, albeit rather cautiously. Other institutions are also being established. In China, the state is rushing to put in place social safety nets and environmental regulations required to deal with the consequences of market liberalization. At the same time, it is slowly opening up to individual rights, partly because it recognizes that economic freedom by itself is not enough to drive a sophisticated economy. In fact, in 2008 a UN study found that legal reform—improving access to justice, reducing corruption, and recognizing private property rights—has created even more new wealth in China than FDI has.

Markets for the exchange of goods and services interact in various ways with other institutions. The most direct links are with business and labour, the principal economic institutions of market economies. Less directly, markets are connected with the various layers of government that regulate them (along with business and labour). In recent decades, non-government organizations (NGOs) such as environmental, human rights, and consumer groups have also come to play an important role in shaping business behaviour and market regulations.

Technology

Markets and the institutions that govern them do not drive economic development all by themselves. The force that powers development is progress in technology, defined by Galbraith (1967, p. 12) as "the systematic application of scientific or other organized knowledge to practical tasks." The common view of technology is that of hardware and software: machinery, equipment, infrastructure, operating systems, and applications. But technology also includes the skills and knowledge required to perform practical tasks. Technological knowledge takes two forms: the tacit knowledge that is acquired through experience and the codified knowledge that is written down using standard terms and formulas.

Technological progress drives socio-economic development because it is the key to the increases in productivity that make it possible for a firm—or an economy—to grow. Technologically advanced societies encourage continuing innovation by supporting the necessary institutions, from public schools to universities and research and development (R&D) organizations. Freeman and Louçã's (2001) theory of technological–economic paradigms explains how, since the eighteenth century, scientific and technological innovations have interacted with social and political institutions to shape the evolution of economic activity. The theory also emphasizes the importance of institutional innovation when diverse groups—firms, universities, governments, communities, consumers, each operating in its own particular context—develop distinctive institutions and mechanisms of collaboration and governance. Indeed technological and institutional innovation co-evolve, each stimulating the other (Chapter 3).

Place and Space

Place and space are two sides of the same coin, inseparably connected and dependent on one another, but representing distinctly different perspectives. For economic geographers, *place is the specific location where economic activities occur*—where people live, work, and consume—while *space is the expanse of territory across which connections are made*, whether within a single place or among many places. A place may be any size, from a village to a nation. Every place has its own mixture of habits, conventions, rules, and routines and therefore its own distinctive economic characteristics. Yet diverse places are connected across space by flows of goods, services, investment, ideas, information, and people. These connections are both complementary and competitive.

In this text, the interaction of place and space is reflected in the interaction of local and global (non-local) institutions (Figure I.2). Connections across space reflect the importance of general (global or non-local), if not universal, institutions, but their influence is constantly modified and even resisted by local institutions. As a consequence, different places retain their unique identities. The distinctions between place and space, local and global, are also expressed as **endogenous** and **exogenous forces**. Thus local institutions are endogenous (internal to specific places) while global institutions have space-based exogenous influences acting on them. Geography's traditional distinction between site (absolute location) and situation (relative location) is a related duality.

The span of control by decision-makers is the conventional basis for distinguishing local and global institutions or organizations. An organization that is owned, controlled, and operated within a particular jurisdiction, as most small businesses and local governments are, is clearly local. MNCs, national governments, international bodies such as the World Trade Organization or the United Nations, and international NGOs are global institutions. Their decisions reach across space and influence local development. This distinction is nevertheless blurred as some local firms internationalize their operations while branch plants of MNCs also operate as local institutions, as do agencies of national governments and international NGOs. Establishing a local presence helps MNCs gain legitimacy and attune themselves to local labour and commercial markets; in some cases, it may also be a legal requirement if they are to gain access to local markets.

Places and spaces are intimately connected. The viability of any economic activity depends on both the characteristics of a particular place (e.g., local resource endowments or labour force skills) and its ability to connect with other places across space to gain access to distant markets or sources of supply. Over time, economic activities may be either strengthened or weakened by changes in any number of conditions both at home

endogenous forces In the context of local development are initiatives driven by local, internal, or domestic actors. The birth of new locally based businesses and the local generation of innovations are illustrations.

exogenous forces Forces that drive local development from non-local, external, or global sources, no matter how the local is defined. Export demands, foreign competition, and foreign-based MNCs are illustrations.

and abroad. For example, local resources may become depleted or distant markets may come to prefer alternative (higher quality or less expensive) sources of supply. Indeed, central contributions—virtual axioms—of economic geography are (a) the recognition that local economic (place-based) development needs to be understood in terms of connections across space, and (b) the recognition that spatial connections or flows among economic activities are grounded and often modified in the attributes of particular places.

The contemporary global economy is highly interdependent: local economies at every level (national, regional, intra-regional) are connected by flows of goods, services, information, people, and ideas (Figure I.2). The fundamental idea underlying the concept of globalization is that these local or regional places and territories are becoming more open to the influence of external or foreign institutions, and that their fortunes are increasingly tied into the fortunes of other places. Yet local economies retain considerable power to shape their own future through their capacity to generate and capture value within a globally integrated system. That is why we prefer the term "value chain" to alternatives such as "supply" or "commodity chain," or "production system" or "network." In a value chain, individual places are not simply arranged in a linear sequence: each one is recognized as a complex of institutional forces that may make or break its linkage to the chain.

The idea of the value chain reminds us that the three sectors of any economy—primary (resources), secondary (manufacturing), and tertiary (services), along with transportation and consumption—are interdependent (Figure I.3). Thus there is a flow of goods from the resource to the manufacturing to the service sector, and there are flows of services to the manufacturing and resource sector, and of manufacturing to resources. There are also many flows within each sector. Further, because activities are interdependent, growth in any one activity generates growth (multiplier effects) throughout the global economy via the linkages of the value chains. Value chain models provide awareness of the geographic disposition of these multiplier effects.

In general, whether measured by employment or output, with economic development, a nation's economy shifts in emphasis from resources to manufacturing and then to services. In tandem, economies become increasingly urbanized and "networked" via a variety of transportation and communication linkages that enrich domestic and international connections. There are variations to this trajectory. Nevertheless, developed countries have long reached that third stage in which most jobs are provided by the service sector. In these countries, primary and secondary sectors have become highly efficient, and still play

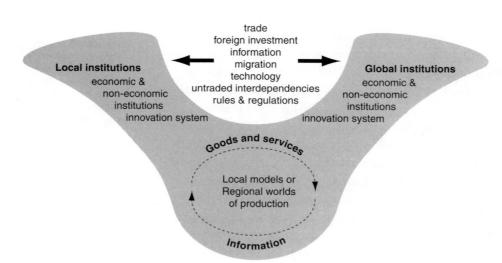

FIGURE I.2   Global–Local Dynamics

vital roles in the economy, with manufacturing becoming increasingly high-value oriented. At the same time, in countries that are still developing, the income growth that is generated by expansion of the manufacturing sector remains a priority. However, the globalization of value chains means that growth (and decline) in one country is rapidly transmitted elsewhere. Chinese industrialization, for example, has led directly to increasing demands for business services and resources around

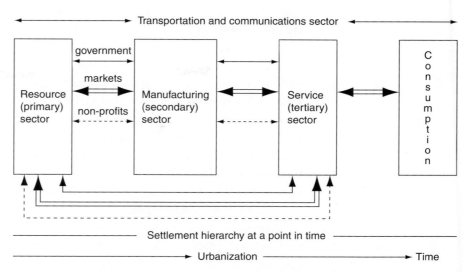

FIGURE I.3   **Value Chains in Aggregate Sectoral Perspective**

the globe (as well as generating multiplier effects within China). Meanwhile, increased incomes for Chinese people have increased demand for imported goods and have globalized other value chains. For example, somewhat unexpectedly China has become the world's largest importer of wine. As a converse example, the US-led financial crisis of 2007–8 and associated collapse in house prices and building reduced demand in a wide range of supplier industries within the US and elsewhere, not least in the Canadian forest sector. As American consumers suffered, so did exporters to the US.

The institutional analysis of value chains is particularly useful for the light it sheds on the economic geography of economic activity and the questions it raises. What are the consequences, both for rich market economies and for developing countries, of extending value chains from the one to the other? How are value-added activities, profits, and employment opportunities distributed throughout value chains? Are alternative arrangements possible? How can the various socio-economic institutions in a given place co-operate to secure a position in a global value chain? And how are environmental costs generated throughout the value chain and subsequently distributed?

The value chain, as conceptualized within and outside economic geography, describes a linear linkage through extraction of resources, production and distribution, segmented in various graduations and set in particular locales. This conceptualization has, for the most part, focused on inputs at each stage, without considering the other waste and emissions outputs otherwise known as externalities. Nor has the value chain considered how the use of a product affected the environment and the need for products to be recycled. A major challenge facing contemporary market economies and economic geography, thus, is to incorporate environmental sustainability into the value chain.

In this book, we take up this challenge by extending the value chain model to a *value cycle*. Essentially, the value cycle incorporates the need to account for the three Rs (reduce, reuse, and recycle) throughout all phases of economic activity from production to final consumption. Similar concepts include green or sustainable value chains, but we believe value cycles more evocatively expresses the challenge of closing the loop. The value cycle of a cellphone, for example, will not only discuss how the parts are put together in different places, but will also consider the environmental impacts from the mining for rare metals and the need to extract the same metals at the end of the phone's life. Control of embodied greenhouse gas emissions or their offsetting adds another layer of complexity to the value cycle. Conversion of value chains to value cycles remains a massive undertaking and a "paradigmatic challenge" for societies around the world and collectively. The value chains of products reach around the world, and can require thousands of parts and transactions while global population and income growth continues apace and offsets three-R gains. The awareness raised by this perspective is sobering for consumers in rich market economies whose own urban environments have become green but their consumer spending, electronic trash, and travel habits might suggest otherwise.

Important challenges facing society, and economic geography, relate to the relationships between economy (development) and environmental goals. Put bluntly, can economy and environment evolve in positive sum relationships, or must they be part of zero sum games in which one must be sacrificed for the other? Marxist analysis, for example as outlined by Lipitz (1992), often points to capitalist culprits of environmental as well as social exploitation, and tends to be pessimistic about the future (under capitalism). Neoclassical approaches are much more bullish but perhaps over-emphasize market solutions. Institutional analyses, rooted in the innovation-driven creative destruction tendencies, offer "reasoned hope" for contemplating development goals that combine income with quality of life and environmental imperatives (Patchell and Hayter 2013). An institutional economic geography can contribute to these reasoned hopes at multiple scales.

Geography has a long tradition of investigating human–environment interdependence and as economic geographers we feel a special responsibility to encourage students to examine the relationship between economic activity and the environment, and how to make economies sustainable. While a separate chapter on the environment would serve that purpose, we believe that it is necessary to see how sustainability has to be integrated throughout economic activities. The value cycle, showing interrelations of environmental impacts from product design through production, consumption, and recycling, demonstrates the need for such integration. The environment, or nature, is of course, not itself an institution. However, the only way the value cycle is going to be reformed at its many stages and its interdependencies coordinated is through the actions of institutions—pressure from environmental NGOs, changes in household behaviour and demand, government regulations, and of course companies developing the capacities to implement changes.

Market institutions have played an instrumental role in improving the conditions of human life since the dawn of civilization; in fact, civilization and market institutions are interdependent. The economic development, population growth, and increasing material prosperity of the last 250 years would not have been possible without the expansion (and increasing sophistication) of those institutions. Yet many problems remain unresolved, most notably the vast inequalities in wealth that persist around the world and the degradation of the environment. Clearly, our economic institutions must continue to evolve. We hope this book will help its readers both to understand and to participate in that evolution, and to deepen their thinking regarding the economic geography of value cycles and development.

Conclusion

This introduction has sketched the nature and scope of economic geography and the institutional approach the book advocates. This approach is theoretically informed and rooted in economic geography's evolving pragmatic traditions that emphasize understandings of actual behaviour and relationships as they evolve in place and across space. In an institutional approach, evidence and observation interact with (and inform) one another. It continually raises judgements as to whether theory requires modification or behaviour needs to be changed, for example, through new policies. Inevitably these judgments involve suggestions as to how people and society should behave. Increasingly, economic geography has shifted its focus from employment, production, and development to incorporate environmental imperatives. In market economies, economic geography's theory, analysis, and policy prescriptions, we argue, need to begin with an understanding of markets themselves. This theme is taken up in Part 1.

Key Terms

areal differentiation 3
endogenous and exogenous forces 13

Recommended Resources

Barnes, T.J. 1999. "Industrial geography, institutional economics and Innis." Pp. 1–22 in T.J. Barnes and M. Gertler, eds. *The New Industrial Geography: Regions, Regulations and Institutions*. London: Routledge.
An interesting review of institutional approaches to economic geography with reference to Harold Innis—a leading Canadian institutionalist who was considered an economist, economic geographer, and economic historian.

Flowerdew, R., ed. 1982. *Institutions and Geographical Patterns*. London: Croom Helm.
An early edited collection of essays that demonstrate an institutional approach to economic geography.

Gertler, M. 2010. "Rules of the game: The place of institutions in regional economic change." *Regional Studies* 44: 1–15.
Recent review of the role of institutions in local development, consistent with this text book's approach.

Granovetter, M. 1985. "Economic action and social structures: The problem of embeddedness." *American Journal of Sociology* 91: 481–510.
A sociologist's interpretation reflecting a cross-disciplinary revival in institutionalism, widely cited in Geography.

Heilbroner, R.L. 1985. *The Nature and Logic of Capitalism*. New York: Norton.
The insightful summary views of a leading economist who is often associated with institutional perspectives.

Hodgson, G.M. 2001. *How Economics Forgot History*. London: Routledge.

An account of the (neglected) importance of history and evolution in conventional economics, and the case for an institutional approach.

———. 1999. *Evolution and Institutions: On Evolutionary Economics and the Evolution of Economics*. Northampton, MA: Edward Elgar.
Broadly based reflections by the one of the leading contemporary theorists on institutional economics.

Hollingsworth, J.R., Miller, K.H., and Hollingsworth, E.J., eds. 2002. *Advancing Socio-economics: An Institutional Perspective*. Lanham: Rowman and Littlefield.
A collection of essays emphasizing the social nature of modern economies and the power of institutions.

McNee, R.B. 1958. "Functional geography of the firm with an illustrative case study from the petroleum industry." *Economic Geography* 34: 321–7.
Introduction to the idea of MNCs as dominant space-organizing institutions.

Pred, A.R. 1965. "Industrialization, initial advantage and American metropolitan growth." *Geographical Review* 55: 158–85.
Classic early elaboration of the principle of circular and cumulative causation to understanding American urbanization.

Scott, A.J. 2000. "Economic geography: The great half century." *Cambridge Journal of Economics* 24: 483–504.
An excellent review of the conceptual development of economic geography since the beginning of the nomothetic phase.

Tickell, A., and Peck, J.A. 1995. "Social regulation after Fordism: Regulation theory, neoliberalism and the global–local nexus." *Economy and Society* 24, 3: 357–86.

An application of Marxist-based regulation theory to contemporary capitalist economies and neoliberalism in which institutions are important.

Williamson, O.E. 1975. *Markets and Hierarchies, Analysis and Anti-trust Implication: A Study in the Economics of Industrial Organization.* New York: Free Press.

An elaboration of the "new institutional economics" perspective from a neoclassical economics tradition based on the theme of transaction costs.

References

Allen, D.W. 2012. *The Institutional Revolution.* Chicago: University of Chicago Press.

Barnes, T. 1987. "*Homo economicus*, physical metaphor, and universal models in economic geography." The Canadian Geographer 31: 299–308.

Berry, B.J.L., Conkling, C., and Ray, D.M. 1993. *Economic Geography.* Englewood Cliffs, NJ: Prentice-Hall.

Chisholm, G.G. 1889. *Handbook of Commercial Geography.* (Many subsequent editions were edited by L.D. Stamp until the early 1950s).

Coe, N.M., Kelly, P.F., and Yeung, H.W.C. 2007. *Economic Geography: A Contemporary Introduction.* Oxford: Blackwell.

Commons, J.R. 1893. *The Distribution of Wealth.* New York: Augustus M. Kelley.

———. 1934. *Institutional Economics.* New York: McGraw-Hill.

Daniels, P.W., and Lever, W.F. 1996. *The Global Economy in Transition.* Edinburgh: Longman.

de Souza, A., and Stutz, F.R. 1989. *The World Economy.* New York: Macmillan.

Freeman, C. and Louçã, F. 2001. *As Time Goes By: From the Industrial Revolution to the Information Revolution.* Oxford: Oxford University Press.

Galbraith, J.K. 1958. *The Affluent Society.* Boston: Houghton Mifflin.

———. 1967. *The New Industrial State.* Boston: Houghton Mifflin.

Gertler, M. 2010. "Rules of the game: The place of institutions in regional economic change." *Regional Studies* 44: 1–15.

Granovetter, M. 1985. "Economic action and social structures: The problem of embeddedness." *American Journal of Sociology* 91: 481–510.

Hayter, R. 2004. "Economic geography as dissenting institutionalism: The embeddedness, evolution and differentiation of regions." *Geografiska Annaler* 86B: 95–115.

Hodgson, G. 2006. "What are institutions?" *Journal of Economic Issues* 40 (1): 1–25.

Huntingdon, E. 1915. *Civilization and Climate.* Princeton: Yale University Press.

Isard, W. 1960. *Methods of Regional Analysis: An Introduction to Regional Science.* New York: John Wiley and Sons.

Jones, C.F., and Darkenwald, G. 1949. *Economic Geography.* New York: Macmillan.

Knox, P., and Agnew, J. 1989. *The Geography of the World Economy.* London: Edward Arnold.

Krugman, P. 2009. *The Return of Depression Economics and the Crisis of 2008.* New York: Norton.

Lipitz, A. 1992. *Green Hopes: The Future of Political Ecology.* London: Malcolm Slater Polity Press.

Lloyd, P., and Dicken, P. 1972. *Location in Space: A Theoretical Approach to Economic Geography.* London: Harper Row.

Martin, R. 1994. "Economic theory and human geography." Pp. 21–53 in D. Gregory, R.L. Martin, and G.E. Smith, eds. *Human Geography: Society, Space and Social Science.* Basingstoke: Macmillan.

———. 2000. "Institutional approaches in economic geography." In E. Sheppard and T.J. Barnes, eds. *A Companion to Economic Geography,* 77–94. Oxford: Blackwell.

Morrill, R.L. 1974. *The Spatial Organization of Society.* North Scituate, RI: Duxbury Press.

Myrdal, G. 1944. *An American Dilemma: The Negro Problem and Modern Democracy.* New York: Harper.

———. 1957. *Economic Theory and Underdeveloped Regions.* London: Gerald Duckworth.

Nelson, R.R., and Winter, S.G. 1982. *An Evolutionary Theory of Economic Change.* Cambridge: Harvard University Press.

North, D.C. 1990. *Institutions, Institutional Change and Economic Performance.* Cambridge: Cambridge University Press.

Patchell, J. and Hayter, R. 2013. Environmental and evolutionary economic geography: Time for EEG? *Geografisca Annaler* Series B 95: 111–130.

Polanyi, K. 1944. *The Great Transformation.* New York: Rinehart.

Schumpeter, J. 1943. *Capitalism, Socialism and Democracy.* New York: Harper.

Shepherd, E., and Barnes, T. 2000. *A Companion to Economic Geography.* Oxford: Blackwell.

Stiglitz, G. 2010. *Freefall: America, Free Markets and the Sinking of the World Economy.* New York: Norton.

Storper, M. 1997. *The Regional World: Territorial Development in a Global Economy.* London: Harvard University Press.

Taylor, G.T. 1937. *Environment, Race and Migration.* Chicago: University of Chicago Press.

Wagner, P. 1960. *The Human Use of the Earth.* Glencoe: Free Press.

Veblen, T. 1904. *The Theory of Business Enterprise.* New York: Charles Scribner's Sons.

Zimmermann, E.W. 1933. *Introduction to World Resources.* New York: Harper Row.

PART I

Markets in Place, Space, and Time

CHAPTER 1

Markets in Place and Space

The term "market" is inherently ambiguous, implying on the one hand abstract sets of economic phenomena and, on the other hand, specific social relationships and institutions within which economic transactions take place.

Bestor 1998: 154–5

The main goal of this chapter is to introduce the relationships between markets, market institutions, and the concepts of place and space. This chapter's main objectives are

- To establish markets as the foundation of economic geography;
- To show markets as exchanges between consumers and suppliers and the influence of their spatial distribution;
- To identify the three basic types of market—open, administered, relational—and explain their coordination in terms of transaction costs;
- To connect these markets with the tendencies towards agglomeration (clustering), dispersal, and value chain formation;
- To introduce the institutions of market governance, formal and informal, and the concept of property rights;
- To explore the nature of market failure, the idea for countervailing powers, and the challenge of sustainable development.

The chapter unfolds in three sections, each one focusing on a key aspect of markets. The first section introduces what economists call the "perfect competition" model which sees markets in purely economic terms as venues where competitive exchange takes place. The second section discusses the coordinating role that markets play in reducing transaction costs and facilitating the exchange of information. Finally, the third section focuses on *market governance*—how markets are regulated. This section looks at why market governance is needed; how it is embedded in economic, political, and social institutions; and how weak market governance can lead to market failure.

Contemplating Markets

As Bestor states in the opening quotation, the term "market" is inherently ambiguous. Thus the way markets actually operate cannot be explained only in terms of narrowly defined price mechanisms ("product exchange" or economic transactions). Markets necessarily integrate economic, political, and social governance into their operations. Indeed, markets are **embedded** in the broader institutional structures of society, both formal and informal, and organized by specific rules and conventions. To help clarify their ambiguous nature, this chapter discusses markets from three perspectives:

embedded A term emphasizing the interrelatedness of economic and non-economic factors or processes.

1. First we will view markets as (invisible) forums of exchange between (spatially distributed) suppliers and consumers that are self-organized by price and competition.

2. Second, we will view the role of markets in coordinating exchange as seeking to minimize **transaction costs** between consumers and suppliers. Transaction costs refer to the costs of using markets, including basic administrative expenses, information search, and the risks associated with incomplete information. This discussion recognizes three main types of *market coordination*. An **open market** has many buyers and sellers; a **relational market** typically has few buyers and many sellers that are involved in long-term relationships with one another; and an **administered market** is one in which the exchanges are internal (where different departments or divisions within the same corporation supply goods or services to one another).

3. The third perspective will focus on market governance and how markets are subject to both formal and informal regulation by governments (local, national, and international), other **countervailing powers** (including rival big businesses), labour unions and NGOs, and by social norms, conventions, and other "rules of the game." Market governance exists to ensure markets work efficiently, fairly, and for the social good. **Market failures**—for example, those arising because social and environmental values have not been considered (or "externalized" by market actors)—underline the need for governance. An important challenge for policy is to better regulate markets to reduce market failure.

Most readers of this book live in a "market economy" of one kind or another—societies that rely on the voluntary price-based production and exchange of goods and services for consumption. Market economies vary greatly as to how and to what extent they allow or promote *intervention* into the market—the freedom they give their markets to direct transactions on their own. Within economies, individual markets also differ in the ways they are embedded, their forms of governance, and the rules they follow. **Market economies** are known by this name because they substantially rely on **markets** to organize the exchange of goods and services that are produced for profit and consumption. Yet market economies are historically recent, emerging (albeit rapidly) from around 1750. Market exchanges had long existed before then, but the material needs of societies, along with land occupancy, crafts, trading rights, and so on, were governed by "non-market" customs rooted in cultural factors, kinship, reciprocity, class, and patronage, while self-subsistence was a widespread reality. Exchange was limited among community members who specialized in particular tasks, often according to gender and age, such as hunting, cooking, clothes making, and tool making. After the mid-eighteenth century, the innovations of the Industrial Revolution stimulated transformations in agriculture, manufacturing, and transportation. In tandem the market system grew rapidly to facilitate exchanges in the escalating and diversifying supply and demand of goods, services, and labour. Increasingly, people supplied less of their own needs in favour of producing specialized goods for a price that could be used to buy the goods of other specialized groups. Karl Polanyi termed these intertwined changes in technology and markets the "Great Transformation" because they began the shift of economies from primarily non-market to primarily market-based forms of social organization. Most countries now depend on markets to organize their economies and human well-being.

transaction costs Generally refer to the costs of establishing and maintaining systems of rules and rights, for example, with respect to markets.

open market A market made up of numerous small, independent buyers and sellers engaged in voluntary interactions.

relational market A type of market characterized by repeated transactions between firms or other partners that are mediated by formal contracts and informal behavioural expectations. Information exchange and collaboration, for example, are important features in such relationships between core firms and suppliers.

administered market Exchange of goods or services that is internal to a particular (usually large) organization; also referred to as hierarchical, integrated, or internal.

countervailing power Refers to major economic and non-economic institutions that constrain the power of "big business."

market failure Refers to situations where markets do not work properly (e.g., when competition does not exist or when market behaviour results in unacceptable environmental or social consequences).

market economies Societies that rely on market institutions to organize the production of goods and services for consumption.

markets Institutions for the price-based exchange of goods and services.

neoclassical economics Refers to mainstream economics that developed in the nineteenth century as a formal, abstract interpretation of economic behaviour that is rational, self-interested, and perfectly informed.

institutions The rules, conventions, habits, and routines that organize human behaviour, formally and informally.

place A particular territory, locality, region, or neighbourhood where people live and work.

space The area over which people and activities are linked; regions, nations, continents, or even the globe itself could all be defined as spaces, depending on the context.

The key mechanism of markets is to exchange goods between suppliers and consumers as mediated by prices. Within **neoclassical economics**, markets are often described in terms of an "invisible hand." In this view, a seemingly natural force brings together suppliers and consumers to set a price and voluntarily exchange goods on that basis. Such markets are powered by the pursuit of individual self-interest and disciplined only by competition. However, markets are real **institutions**—not abstractions. Markets are created and regulated in particular places at particular times with specific purposes in mind. In other words, markets are embedded in social relations which means they are not only disciplined by competition, but also by the institutions of law, convention, values, attitudes, and by the myriad formal organizations that make up the private and public sectors. Many institutions have discretionary powers to influence market exchanges. Moreover, markets are fundamentally shaped by **place** and **space**. Most basically market exchange is affected by the spatial distribution of consumers and suppliers and because market potentials and supply costs vary across space. As *places*, markets are physical locations with distinct functions. They are part of daily life for the people who meet to exchange goods, services, money, investment, and information. As *spaces*, markets are institutions that integrate local and global economies to permit those exchanges. The institutions organizing markets in place and across space vary in kind, practices, and degree of influence.

Case Study 1.1
TOKYO'S TSUKIJI FISH MARKET

© iStockphoto.com/alvarez

The Tsukiji fish market is the biggest food market of any kind in the world. It processes approximately 17 per cent of the world's fish catch. Originally set up when Japan's military ruler established Edo (Tokyo) in the early 1600s, it is now owned

The Tsukiji fish market in Tokyo is an excellent example of market exchanges organized by *specific* social relationships and institutions (Case Study 1.1). The Tsukiji market is a real *place*, organized by particular rules and conventions. Despite its location in a non-descript area of Tokyo, it is part of the city's identity. But its reach is across *space*—drawing on fish stocks around the world and accessing sushi bars, grocery stores, and fish lovers throughout Japan. These interactions can be measured and mapped. The power of Tsukiji is that the prices set there determine the extent of those interactions, locally and globally, and their viability.

Market Exchange

In economics, **perfect competition** is the ideal, foundational model of markets, in which economic exchange is based solely on price as negotiated between suppliers (producers, sellers) and consumers (buyers, demanders). Actual markets rarely conform to the perfect competition model. However, the model introduces the important roles of demand, supply, and price, and more generally promotes the power of markets to stimulate efficiency, competition, and innovation.

perfect competition The ideal market model in neoclassical economic theory in which fair competition between multiple buyers and sellers is self-regulating.

by the Tokyo Metropolitan Government, which together with the Ministry of Agriculture, Forestry and Fisheries controls and licenses the participants. It is located on a 26-hectare site on the Sumida River in central Tokyo and handles approximately 2,400 tonnes of fish, worth approximately US$20 million, every day. One-third of the fish is fresh, another third is frozen, and the remainder is dried or processed. The market employs 14,000 people directly and attracts about 35,000 buyers. Every morning as many as 50,000 people may gather in its 230,836 square-metre space.

The market's primary function is to serve as a venue for the fair pricing of fish and seafood from all over the world. In addition, however, it reduces distribution costs, provides up-to-date information on quality and availability, and improves hygiene control by setting strict standards for cleanliness and traceability of the supply chain. Fish is brought to the market by fishing companies, regional co-operatives, distributors, and importers. Then one of seven licensed primary wholesalers auction the fish to licensed intermediate wholesalers and authorized buyers (e.g., retailers). The intermediate wholesalers sell on to retailers and distributors of various types.

Tsukiji combines the "arm's-length" tradition of an open market with a significant administered component. Only seven primary wholesalers are licensed to auction the fish to about 800 intermediate wholesalers, who then sell it on to retailers. The majority of the wholesalers are small family firms that have been handed down through the generations. On the other hand, Daito Gyorui is both a primary and intermediate wholesaler and part of one of the world's largest fishing and distribution companies. Tsukiji is scheduled to move soon (2016?) to a much larger site, where land costs are lower and its operations will be computerized and verbal auctions will be replaced by electronic trading; the repercussions for the social and economic relations that have developed over four centuries around the site are unknown, although redevelopment plans indicate a smaller market will be retained.

The Perfect Competition Model

Perfect competition assumes a market is perfect or ideal in three respects:

1. *Competition*: Large numbers of rival suppliers bring goods to the market and large numbers of rival consumers purchase them with all actors competing with one another.

2. *Fairness*: Market transactions are voluntary; consumers and suppliers are independent of one another, free of any personal or organizational ties; and all consumers are more or less equal in status and therefore unable to arbitrarily influence prices.

3. *Self-regulation*: Consumers and suppliers have the perfect rationality of **Homo economicus** ("Economic man")—who seeks to maximize profits and minimize costs—that ensures the self-regulation of market exchanges. Suppliers who charge too high a price for their goods will lose customers to competitors who have reduced their costs and charge less. Only the efficient survive.

In the perfect market model, prices are set through the interaction of demand and supply (Figure 1.1). The **demand curve** (or demand schedule) indicates the quantity of a good that consumers will be able and willing to buy at various prices at a given time. The "law" of demand predicts that consumers will be able and willing to buy more as the price falls. The **supply curve** (or supply schedule) indicates the quantity of a good that suppliers will be able and willing to supply at various prices at a given time. The "law" of supply predicts that suppliers will be willing and able to supply more as the price increases. Thus in a highly competitive market, we can imagine that the numerous suppliers and consumers will haggle until they arrive at an "equilibrium price," where the quantity supplied is equal to the quantity demanded. That equilibrium price will then remain in place until something happens that causes either supply or demand to change.

Demand, supply, and the price mechanism are profoundly important forces in market economies. Increases (or declines) in the demand for a good or service puts upwards (or downwards) pressures on prices, and increases (decreases) in the supply of a good or service puts downwards (or upwards) pressure on prices, **ceteris paribus** ("all other things being equal"). Nevertheless, there is a crucial geographic factor that this model of demand–supply interaction fails to take into account: the location where the exchange takes place. The prices negotiated at the Tsukiji fish market are intimately related to the spaces of consumption and production.

Homo economicus "Economic man," the personification of rationality in neoclassical economics; the ideal economic actor, in possession of all the information necessary for decision-making, who seeks to maximize profits and/or minimize costs.

demand curve (schedule) A graph showing the quantity of products consumers are able and willing to buy at different prices.

supply curve (schedule) A graph showing the quantity of products suppliers are able and willing to supply at different prices.

ceteris paribus Latin for "all other things being equal"; a standard phrase in abstract neoclassical theorizing, used to rule our extraneous factors that could affect the outcome of a prediction.

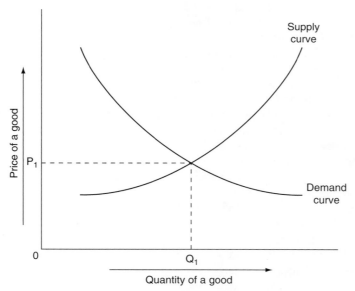

FIGURE 1.1   The Market as Interaction between Supply and Demand

At P_1, the quantity supplied equals the quantity demanded which determines the equilibrium price.

The Geography of a Market

Demand for a Good over Space: Where Consumers Are Located

The real-world locations of markets—as opposed to considering markets in abstract— raise many questions. For example, how does the distribution (location across space) of consumers influence the nature of demand? Or the location of supply? The main customers of the Tsukiji fish market are restaurants located throughout Tokyo's metropolitan area. Restaurants close to the market, in inner Tokyo, can obtain fish more quickly and cheaply than restaurants in the outer suburbs; the farther away from the Tsukiji fish market, the higher the transportation costs and, as a result, the less buyer demand. The increased time needed to travel further affects demand by reducing the freshness of the fish and therefore its value. Indeed, it is significant that the Tsukiji market is located in the Tokyo area, where demand is concentrated in terms of both the numbers and the wealth of consumers. A market located in a remote community in Japan would experience similar overall supply costs (i.e., the cost of procuring the raw fish), and lower land and related costs than in Tokyo, but demand would also be greatly reduced by additional distribution costs.

The theory that the location of suppliers is highly dependent on the spatial distribution of consumers is highlighted in Walter Christaller's **central place theory**. In this theory, perfect competition and *Homo economicus* are assumed, but the location of central place goods are defined and determined by accessibility costs to (or of) consumers. For spatially distributed consumers, the *real* price of a good—the price charged at the market (e.g., a shop) plus the transportation costs incurred by consumers going to the market—increases with distance from the market, while the quantity of the good demanded decreases (Figure 1.2). The **range of a good** (or service) is the maximum distance that consumers are willing and able to travel in order to purchase it (or the maximum distance over which it can be economically distributed to consumers). In turn, the range of a good defines the maximum extent or accessibility of a market's hinterland or the market area over which the good is sold. If consumers are evenly distributed over a homogeneous plain and transportation costs are the same in all directions, the market area is circular and the radius of the circle is the range. The location of a single supplier of a service (say, a barber) serving a population uniformly distributed over a homogeneous plain can be anywhere, so long as that area holds the **threshold population**—the minimum number of customers that must be contained within the range in order for a good to be viable or profitable. Ranges that encompass more than this minimum number of consumers allow suppliers to earn "surplus profits"—profits beyond the minimum necessary (the "normal profit") to

central place theory A theory of the location of suppliers in response to the distribution of consumers; in this theory, minimizing distances to markets is the predominant imperative.

range (of a good or service) The distance that consumers are willing to travel to buy a particular good or service (see Chapter 11).

threshold population The minimum population required to ensure that demand for a good is sufficient to make its production economically viable (see Chapter 11).

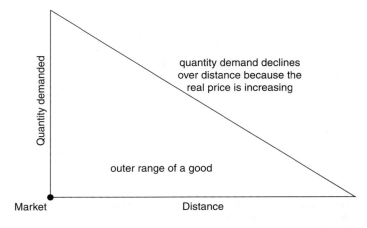

Quantity distance relationship

quantity demand declines over distance because the real price is increasing

outer range of a good

Market   Distance

Quantity demanded

FIGURE 1.2   **Demand Costs over Space for "Central Place" Goods**

make doing business worthwhile. In a competitive market, if the existing barbers cannot serve all the consumers in the region (because consumers exist outside their range), or if surplus profits are made, additional barbershops will be attracted until there are enough to serve all consumers and surplus profits are eliminated.

On Christaller's homogeneous plain, additional barbers that provide the *same* service will disperse themselves in such a way as to ensure that the threshold population exists within the range of the service (Figure 1.3a). Eventually, each barber will serve all consumers within market areas that take the form of hexagons—the geometric shape that most efficiently encloses a space without overlaps (Figure 1.3b).

Different kinds of services are associated with different thresholds and ranges. Barbers, bakeries, and gas stations, for example, are "low threshold" goods with smaller ranges and threshold populations while fashion hairdressers, opera companies, or the huge Tsukiji fish market are "high threshold" goods and services. Between these two types, various "middle orders" exist. On this basis, a nested hierarchy of central places ("market centres") can be loosely identified (Table 1.1). In practice, the threshold and range of goods are important concepts in urban and regional planning (see Chapter 12).

(a) Three suppliers begin to divide market share

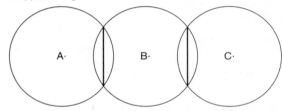

A, B, and C are suppliers providing a similar (and similarly priced) service (e.g., hair cuts) to consumers distributed evenly on a homogeneous plain. If consumers minimize their costs by travelling to the nearest barber, the potential market areas defined by the outer range of the service that overlap are bifurcated.

(b) Multiple suppliers create hexagonal market areas

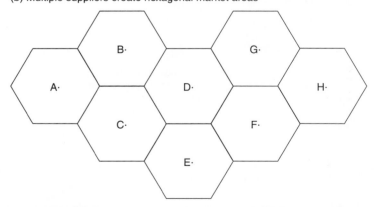

In perfectly competitive markets, the most efficient distribution of services is geometrically defined by hexagonal market areas.

FIGURE 1.3 **Perfect Competition and Hexagonal Market Areas on a Homogeneous Plain for a Single Homogeneous Good**

Supply of a Good across Space: Where Suppliers Are Located

The geographic distribution of suppliers also influences supply costs. The Tsukiji fish market is supplied from inland waters at varying distances from Tokyo and from distant fisheries in the oceans of the world. Normally, fish suppliers at greater distances face higher costs not only for transporting fish to the market, but for other distance-related costs, such as keeping fish fresh. More distant suppliers try to compensate for these higher costs, usually by catching large quantities of fish to achieve economies of scale by reducing average costs in processing and transportation (Chapter 2) or possibly by catching higher-value fish. Thus so-called "in-shore" commercial fishing is typically "labour-intensive" with small boats operated by individuals or small crews that operate relatively close to shore, often returning to port on a daily basis. Deep-sea fishing, on the other hand, is typically capital intensive involving large, well-equipped boats that can spend weeks at sea; to pay for the capital costs and for

TABLE 1.1 The Central Place Hierarchy

Order of Centre	Regional Scale	Urban Scale	Comments
A	Metropolitan area	Central-business district	A-order goods/services with the highest threshold population and largest ranges are purchased here. All B-, C-, D-, and E-order goods and services are also available.
B	City	Suburban centre	B-, C-, D-, and E-order goods are available.
C	Town	Neighbourhood centre	C-order goods, along with D- and E-order goods, and services are available.
D	Village	Local centre	D- and E-order goods are available.
E	Hamlet	Corner store	Only the goods and services with the lowest (E-order) thresholds and ranges are available.

Note: Chapter 11 discusses factors that modify this central place hierarchy in practice.

transportation costs fish catches need to be high to generate sufficient returns. At the Tsukiji fish market, for example, tuna is caught by small boats and delivered from local in-shore fisheries such as the one near Oma (Case Study 1.3) while huge trawlers fish for tuna in larger quantities around the Pacific, but also farther away.

Both in-shore and deep-sea commercial fishing are vulnerable to rising fuel prices that increase the cost of transportation and declining fish stocks that reduce the volume and accessibility (supply) of fish available (and add to costs). In many parts of the world, such as around New Zealand, Newfoundland and Labrador, and Taiwan, capital-intensive trawling has been in conflict with, and has threatened the viability of, smaller-scale in-shore fishing. Mainly since the 1970s many countries have introduced 200-mile fishing zones, in part to protect local fishing operations and in part to sustain stocks.

More distant mineral and energy supplies similarly face higher transportation costs to markets. *Ceteris paribus*, geographic dispersal increases supply costs. More distant mines need to produce in larger quantities to reduce operating and transportation rates and ensure viability. For example, let's imagine that four coal mines (A, B, C, and D) are located at progressively greater distances around a market (Figure 1.4), and that these four mines are part of a larger group operating within a competitive market situation. As we saw in Figure 1.1, in such a market the supply curve would reflect varying degrees of willingness and ability among producers (coal mines) to deliver coal to the market at particular prices. We might expect that the distribution costs of delivering coal to markets would be greater for the

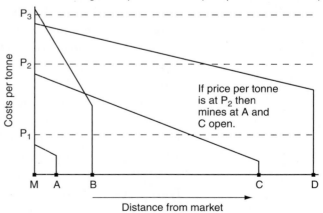

A, B, C, and D are mines, M = Market. Note: The slopes of lines reflect variations in average transportation costs (cost per tonne kilometre)

If price per tonne is at P_2 then mines at A and C open.

FIGURE 1.4 Supply Costs over Space: Mines around a Market

If price per tonne is P_1, then mine A opens; if P_2, then mines A and C open; if P_3, then mines A, C, and D open.

Source: Based on Hay, A. 1976. "A simple location theory for mining" *Geography* 71: 65–76. Sage Publications.

more remote mines than for those located closer to the market. However, the average costs of transporting one tonne of coal one kilometre vary from mine to mine: A and D enjoy low *per capita* transportation costs because they are able to ship their coal by relatively inexpensive water transportation, while B has to pay more to transport its coal overland by truck. Perfectly competitive mines do not care where the coal comes from, all that matters is the price.

The supply situation is further complicated by the fact that the coal mines incur different extraction (production) costs. Coal seams vary in many respects: not only in quality and quantity of the coal itself, but in the ease of working them, the hazards faced, and ease of accessibility. Variations in wage rates, taxation levels, energy costs, toxic content, and so on mean additional variations in production costs. In Figure 1.4, A is the lowest-cost mine in *per capita* terms and D is the highest-cost mine. The market price must cover both production and transportation costs if the mines are to be viable. If the price is P_1, then no mine is viable; at P_2, C can operate, etc. Moreover, all these predictions are subject to change. If a new coal mine opens, the quantity of coal available to the market is likely to increase (and prices decrease); if the increase is dramatic, the supply curve will show a sharp step up rather than a smoothly sloping line.

In general, suppliers at locations farther away from their markets need lower production costs, low transportation costs, and/or the ability to command high prices if they are to be viable.

Market Coordination

The basic model of perfect competition has an interesting omission, namely it does not explain how market exchange mechanisms actually work. Markets are not automatic, nor divined, but require real institutions to coordinate exchanges between buyers and sellers according to a set of regulations and conventions. In the language of the new institutional economics, markets exist to reduce transaction costs.

Market coordination occurs in many different forms, as illustrated by *open, relational*, and *administered* markets. Open markets bring together many independent buyers and sellers in order to physically transfer products and exchange information. In contrast to large corporations, in administered markets exchanges of goods and services occur and are controlled internally within corporations. Relational markets differ again, often involving repeated bilateral exchanges between two firms, for example, between a core firm and its suppliers.

Transaction Costs

Transaction cost analysis provides an economic explanation of how exchanges are coordinated in various market forms. Transaction costs, *not* to be confused with production or transportation costs, generally refer to the costs of establishing and maintaining "any systems of rules and rights" (Allen 2012: 19). From this perspective transaction costs are incurred by both sellers and buyers in order to take part in (the rules and rights of) market exchanges. There are two basic types of transaction cost: coordination costs and opportunism costs. **Coordination costs** are the expenses associated with the administration and organization of market participants, and usually include the operation costs of the actual marketplace where sellers and buyers meet and conduct their business. Coordination costs also include the search costs that market participants incur when seeking the information they need to make fully informed decisions—for example, when consumers do not know how the price charged by one supplier compares with the price charged by others. Collecting the necessary information takes the time, effort, and cost of researching. Consumers have to decide how long to continue searching, for example, for the lowest price since the money saved may not be enough to compensate for those costs. The cost of legal services is another important search cost when the precise nature of legal rights, obligations and liabilities, or encumbrances, must be identified. Real estate transactions, for example, typically require legal verification to ensure that the seller holds clear title to the property.

Opportunism costs arise when one participant in a transaction takes advantage of another participant in some way. A buyer, for instance, may have no intention of paying the supplier, while a supplier may lie about the cost of producing a good or service, or about its capabilities. Collaboration among workers or firms can be strained if some participants decide to "free ride," receiving earnings without putting in their share of the work. To minimize the risk of opportunism, companies monitor the behaviour of transaction partners and take legal action to enforce the conditions of agreements. Both monitoring and enforcement costs can be high.

Moral hazard is a type of opportunism when one party changes its behaviour to take more risks because another party takes on the negative consequences of those risks. The term was first used in the insurance industry to refer to clients who deliberately conceal conditions or behaviour that would disqualify them for coverage; in effect, such people take advantage of the firm's ignorance in order to transfer their risk to it. The term has since come to be applied more broadly, especially in situations where some people are allowed to use, and potentially profit from using, other people's money, at little or no risk of loss to themselves. During the 2009 financial crisis, for example, critics argued that the $500-billion bailout given by the US government to financial institutions would only encourage the latter to make risky investments, since they would stand to profit if the gamble were successful, but any losses would be borne by the taxpayers. Not surprisingly, fear of opportunism breeds distrust and reluctance to invest in a project when there is a risk that one partner in the transaction may capture all the benefits. While laws can be introduced to protect against opportunism, the cultivation of mutual trust offers a lower cost alternative or complement.

Remember that markets are not simply places where goods are exchanged for a price. Since prices depend on the information that sellers and buyers have—both about the products for sale and about one another—markets are also places of information exchange. Of course there are many routine transactions where information is standardized and for which search and evaluation costs have long since been accounted, for example, when consumers buy milk and bread, or metal fabricators buy steel, or house builders buy lumber. The pertinent information, such as quality, brand, and price is often known or taken for granted. The costs associated with incomplete information are minimal in such cases.

coordination costs The expenses of administering marketplaces and the search costs of market participants.

moral hazard A form of opportunism where people take risks they would otherwise avoid because they are insured against losses; such changes in behaviour are usually detrimental to the interests of the insurer.

On the other hand, there are markets in which new information is crucial and incurs high costs to communicate, for example, when a consumer purchases a car or house, or a manufacturing firm seeks a high performance component for a new product. Information is incorporated into a good when a firm collaborates with its suppliers on the design and specifications for a proposed component. The suppliers may have to engage in extensive research and development (R&D) before they are able to manufacture the component. Individuals or firms that need expert advice of some kind (legal, engineering, medical, accounting, etc.) pay fees that are calculated on the basis of hours worked, the personal responsibilities assumed, and the nature of the information required (e.g., its complexity or its implications).

As stated before, markets are organized to reduce transaction costs. There are no universal rules for sorting out the pros and cons of using one type of market over the other. At its simplest, the choice may be understood as a "make or buy" decision: whether to make something internally ("in-house") or to buy it from an external supplier. However, there is a spectrum of markets and interactions among transaction partners. Most people, companies, and governments rely on a combination of them.

Three Types of Market Exchange

The distinctions between *open*, *relational*, and *administered* markets provide insights into the spectrum of market types (Figure 1.5).

Open Markets

Of the three market types, open markets come closest to the perfect competition model. They involve many sellers and buyers, and are unlikely to require formal, legal contracts; instead, the exchange of goods from seller to buyer is governed by weaker forms of "general" contract law, along with established routines and codes of conduct (in the absence of formal contracts, "buyer beware" is an important principle of open markets). Most open markets are (or have evolved from) actual marketplaces where people congregate to sell and buy, compete, and compare. The fact that transaction costs can be reduced by congregating in a particular place for exchange is the centripetal force that has drawn people together for thousands of years. It is the key to understanding not only markets, but civilization.

The Tsukiji fish market represents one type of open market, in which goods are sold to the highest bidder via verbal auctions. Farmers' markets and "flea" markets are similar except that, instead of

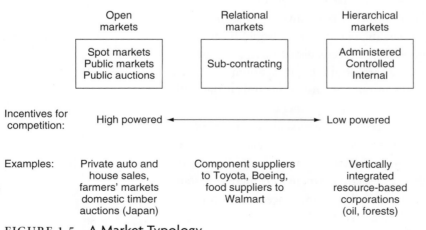

FIGURE 1.5   **A Market Typology**

Source: Adapted from Rosen, Christine Meisner, Janet Bercovitz, and Sara Beckman. 2000. "Environmental Supply-Chain Management in the Computer Industry: A Transaction Cost Economics Perspective." *Journal of Industrial Ecology*. John Wiley and Sons.

placing bids, customers compare similar goods, prices, and qualities, and haggle with the sellers over the final price (Case Study 1.2). The **agglomeration** or clustering of buyers and sellers within a market reduces the transaction costs associated with both search and opportunism by allowing for comparison, competition, and information exchange (Chapter 2). A downtown shopping area or mall containing five shoe stores or 15 clothing stores is similarly effective at reducing transaction costs, although it offers fewer opportunities for haggling. Business-to-business sales are also highly concentrated for the same reason: it's easier to find cheaper, better-quality suppliers if they are all located in the same place and competing for business. Product differentiation is another strong imperative that pushes companies to cluster spatially (Chapter 2).

agglomeration The spatial concentration of related manufacturing and service activities; also increasingly referred to as clusters.

Cyberspace markets such as eBay recreate marketplace conditions over space; online shoppers do the same thing when they search for similar goods at different websites. Consumers no longer have to pay the costs of visiting physical markets when they surf the web to find product information, compare prices, and sometimes even haggle with suppliers. EBay is trusted because it provides money-back guarantees if suppliers do not meet obligations. In addition, databases on global suppliers allow companies to select parts, components, and services over the web. Most evocatively, financial products that once required a trading-room floor on Bay Street or Wall Street now trade electronically around the globe. Indeed, the New York Stock Exchange is no longer an inherently place-based entity because the digitalization of exchange has enabled ownership to be taken over by Intercontinental Exchange, a private electronic futures exchange based in Atlanta. Of course these markets not only comprise many small buyers but many, especially the finance markets, are dominated by large buyers and sellers.

Administered (Hierarchical) Markets

Administered or corporate hierarchical markets occur when corporations exchange goods and services within their own ("affiliated") functions, departments, or operations rather than buying and selling on open-market transactions. This process is also known as the *internalization* of markets. By creating an administered (or "hierarchical" or "vertically or horizontally integrated") market, a corporation gains direct managerial control to replace reliance on independent suppliers or distributors. The corporation seeks to reduce transaction costs by reducing the costs of coordination and guarding against the risks and opportunism associated with open markets. The main disadvantage of internalization is that corporations lose the strong or "high-powered" incentives of open competition. Instead, corporations stress the benefits of administered markets in terms of efficiency, stability, and security of operations.

Administrative (hierarchical) exchange violates the principles of perfect competition, because the transaction partners are not independent, their behaviour is not voluntary, and the good being exchanged remains controlled by the same corporate owners. Yet administered exchange within a corporation is a type of market exchange. Corporations are not closed systems: they depend on *externalized* relations in both open and relational markets. Most obviously, the majority of their sales are external. Moreover, "contracting out" or "outsourcing" by corporations is an option that often replaces an internal activity with one bought on an open or relational market.

Case Study 1.2

FARMERS' MARKETS

Farmers' markets were the foundations on which most of the world's towns and cities were built. They also established the basics of market economics. Selling agricultural produce, crafts, and sundry manufactured goods, traders come together to access a diverse group of customers for their products. A market minimizes the transaction costs of transportation, research, comparison, and selection for farmers and consumers alike. Prices are set through haggling between consumers and sellers, who differentiate themselves by price, quality, selection, display, and salesmanship, and monitor one another in open competition. Competition in such markets is regulated by the repeated interaction among farmers and consumers, and the need to share space and develop common norms of trade. Although farmers' markets have been largely replaced in developed economies by supermarkets with fixed prices, many continue to operate. Today they are often used to revitalize neighbourhoods and link the city to local agriculture.

© Toronto Star Syndicate 2003

The St Lawrence market has been part of Toronto since the early 1800s.

Administered markets directly integrate economic activities over space. Oil, forest, aluminium, and other resource corporations that exploit, process, and distribute resource-based products to retail outlets or final consumers in some instances are examples of vertical integration, which means they control technically linked stages of production and distribution. Oil-based giants such as Shell and Exxon are examples that extract, refine, and distribute oil and oil products (see Chapter 9). To some degree, auto, electronic, and other secondary manufacturing corporations assemble products that utilize components their affiliates have made. Indeed, it is estimated that over half of all international trade in goods occurs within corporations. Moreover, large corporations "horizontally" expand across space by building the same or similar facilities in different locations. These dispersed operations are typically controlled (to varying degrees) by common managerial functions and head-office control. Retail giants such as Walmart are good examples of such "horizontal integration." The same administration minimizes the transaction costs of the many services required to make a product available in the store—transportation, unloading, refrigeration, stocking, recording, checkout. Coordination of corporate activities is facilitated when the flow of information from head office to branch plants is standardized through various corporate routines and readily communicated. Branch plants, offices, or stores are then dispersed across space to places offering other location factors, such as infrastructure, subsidies, or cheap labour.

Administered markets must to some degree focus their activities in particular places, stimulated by various kinds of economies of scale, or reductions in average cost with increasing size (Chapter 2). For example, the "control functions" (accounting, marketing, engineering, design, research, decision-making) performed by head-office operations involve the exchange of sophisticated, complex information requiring daily interaction or the common use of facilities. Often the information exchanged among internal corporate control functions is secret. Agglomerating (clustering) functions in particular places facilitates such interactions. In general, the capacity to coordinate a complex division of labour across space is the main reason for the dominance of administered markets across most sectors of the economy. Even the largest corporations, however, cannot internalize all transactions. As the biggest corporation in the world, Walmart relies on millions of consumers and thousands of suppliers. Even so, it seeks to influence the former by advertising and the latter by its purchasing power and the creation of relational markets.

Relational Markets

Relational markets are a hybrid of open and administered markets. The two parties to an exchange are different companies (or a company and a consumer), but the exchange includes agreements on more than price. These agreements, or "relational contracts," involve matters such as mutual investments, collaboration on research, varying pricing agreements, repeated transactions, or extended service, and are often governed by formal contracts. For example, suppliers enter into relational contracts with client firms when they collaborate on design, set quality standards, determine delivery times, and so on. Large firms rely on independent, often smaller firms as suppliers, while suppliers typically rely heavily on large firms as markets, such as those that occur in auto manufacturing when core firms buy component parts from suppliers (Chapter 11). Relational markets that involve numerous, ever-changing suppliers lean towards the open-market model, while those that are more stable involve fewer suppliers and lean towards the administrative/hierarchical model. Consumers enter into relational contracts when they hire someone to perform a service or purchase a warranty on a car. The key mechanism of relational markets is that they enhance both control and coordination, while retaining the high-powered incentives of the market.

Relational contracts reinforce the agglomeration (clustering) tendencies of open markets when arm's-length relations fail to provide customers with the solutions they desire and face-to-face negotiation and collaboration are required. Firms that need to collaborate closely with one another on the development and production of specialized or sophisticated products are especially likely to congregate in certain regions. Auto assemblers, for instance, require specially designed devices from electronic suppliers for new models; movie production requires sophisticated graphics from computer specialists; and high-end fashion retailers rely on high-end designers. In these cases, product (service) exchange is information-intensive, requires originality, and is facilitated by close personal interaction. Examples of such concentrations can be seen in Hollywood, Silicon Valley, Toyota City, Wall Street, the Pearl River Delta, Baden-Württemberg, or the rue du Faubourg Saint-Honoré in Paris. On the other hand, advances in transportation and communication technologies have tended to undermine agglomeration tendencies for routine purchases or standardized products.

The ever-improving capabilities of electronic communication have reduced the advantages of personal proximity in relational markets. For example, designs for new, highly complex components are co-developed on the web, while committee meetings are held through teleconferencing or non-visual conference calls. Nevertheless, face-to-face communication remains vitally important for certain types of information exchange— when the ideas are difficult to understand, when strategies and plans need to be probed and developed, when secrecy is important, or more generally, when non-routine information is important.

Markets and Value Chains

In reality, most goods and services are not truly single products; rather, they are assembled from various other goods and services bought through several markets and typically several different types of markets. The integration and coordination of these product systems create a **value chain** which links multiple markets operating in different places and over space. A series of linked market exchanges of goods and services progressively adds value to the "final" product or service. The end-consumer may be an individual, firm, or government department.

value chain The complex set of interrelated activities required to produce a product (a good or service) and distribute it to consumers.

The Tsukiji market is part of a value chain composed of several markets, each with its own characteristics, participants, and functions, and each market exchange adds some value, which is reflected in an increase in the price of fish (Figure 1.6). The role of auctions—both the inner and live auctions—underlines the role of open markets, especially evident in the case of tuna, where there is generally competition among fishing firms as well as buyers. Big actors and relational markets also exist, as indicated by "negotiated sales." Further, the fishing fleets that supply Tsukiji—like most international fisheries—treat the ocean as an "open-access resource" from which any number of fishers can take all the fish

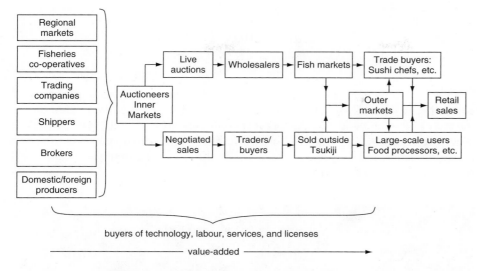

FIGURE 1.6   Idealization of the Fish Value Chain at Tsukiji

The Tsukiji fish market involves multiple markets in which buying and selling activities are intermingled with fishing, transporting, storing, icing, sorting, processing, packing, preparing, and eating activities.

Source: Partially based on Bestor, T.C. 2004. *Tsukiji: The Fish Market at the Center of the World*, Berkeley: University of California Press, 54.

they want without restriction. High market prices may indicate scarcity but don't necessarily encourage sustainability as is sometimes supposed. In the traditional new year auction of 2013, for example, a restaurant owner bought a 222 kilogram tuna for $1.78 million ($7/94 per kg), but such a price may well stimulate rather than reduce fishing. Because everybody competes for fish while the stocks last, the resource becomes depleted. Unfortunately, the international community has not yet mustered the will required to correct this fundamental problem, through regulation or other means. In the town of Oma, Japan, however, in-shore fishermen have created their own rules to govern their fishing areas as "common pool resources" that restrict fishing to members only and to agreed-upon rules (Case Study 1.3).

The other segments of the value chain are equally complex. Many international fish companies not only catch, but also process the fish (on board or shore) and transport it to market themselves; others hire independent firms to handle different stages in the process. The auctioneers perform an important intermediary role. The wholesalers at the market are *both* buyers and sellers of fish, as are the restaurants and other retail operations. Profitability or viability in these segments requires that prices cover both costs and a "normal" or "satisfactory" profit—defined as sufficient incentive for suppliers to stay in business. But definitions of normal or satisfactory profits vary widely, depending on individual factors, such as a desire to stay in the industry, the availability of other job opportunities, the potential for higher returns on investment elsewhere, traditions, lifestyle choice, and so on. The sellers in the Tsukiji fish market, for example, have inherited their skills and entrepreneurial traditions through generations, and may feel committed to staying in the industry even if they face the prospect of losses and a slow demise.

On the distribution side, increasingly more sushi restaurants belong to large firms with set prices, salaried employees, and customers interested in standardized fare who

Case Study 1.3

BRANDING TUNA BY THE TOWN AND NOT BY THE TIN

The town of Oma in northern Japan has long been famous as the source of the choicest tuna for sushi and sashimi. Its fame increased after a TV show focused on the town's fishermen, and now the local economy benefits from millions of dollars brought by tourists. The Tuna Festival in October is especially popular. However, Oma's reputation and income are threatened by the Japanese government's refusal to support sustainable practices of the fishermen.

Branding Tuna by the Town

The Oma fishermen catch the tuna in two-person open boats using hand-held lines. They claim that this method allows them to select the tuna they catch, taking only the large adults and leaving the younger fish to sustain the population. Whether this method is actually the best way to ensure sustainability is unclear, since the biggest fish may in fact be the fittest rather than the oldest. Nevertheless, the local fisher people determine who will be allowed to fish and how much fish can be taken.

Oma and other fishing communities in Japan are among the world's best examples of sustainable governance over common pool resources. However, their sustainable approach is undermined both by the national government, which allows other Japanese fishermen to use trawling and large vessel long-lining in the area, and by fishermen from all nations who help themselves to the fish when they travel outside Japan's territorial waters. In fact, in order not to threaten tuna supplies from the Mediterranean and North Atlantic Ocean, the Japanese government has actively worked against the efforts of other countries to impose controls on fishing of the severely threatened tuna.

Based on Martin Fackler (2009), "Tuna Town in Japan Sees Falloff of Its Fish," *New York Times*. http://www.nytimes.com/2009/09/20/world/asia/20tuna.html (accessed 9 Sept. 2010).

pay no particular allegiance to a restaurant. Still large numbers of neighbourhood *sush-iyasan* remain, frequented by loyal customers and run by owner-chefs who learned their craft as apprentices. Each service link in the value chain adds value: from process-ing and transporting, through sorting and packaging, to distribution. Moreover, each organization in the value chain is also a consumer both of technology in the form of equipment, tools, and furnishings (e.g., fishing boats, cutting knives, platforms, trucks, tables, etc.) and of labour and many services from the private and public sectors (e.g., legal, accounting, inspecting, marketing, R&D, financing).

The value chain is further affected by variations in consumer preference over space. Income powerfully affects people's ability to consume. Even at the same income level, however, tastes and preferences vary across space. Tsukiji would not exist without Japanese consumers' demand for fish products, especially fresh fish, stemming from a traditional diet of seafood that has developed over thousands of years. It is no coinci-dence that the world's largest fish market is in Tokyo, one of the largest high-income metropolitan areas in the world, and one that has a high demand for raw fish. Yet it's a complex push-and-pull process. At Tsukiji, brokers are gatekeepers, assessing value on the basis of colour, fat content, uniformity, and ideas about what the fish themselves have eaten. Japanese and foreign companies are constantly searching for new sources of fish and new ways of selling. And consumption preferences continue to vary within countries and between countries.

What appears to be a simple market exercise—getting fish to the table—is in real-ity the culmination of a remarkably complex set of connected activities constitut-ing a value chain that spans the globe. At each stage, there is a market that sets the value for that activity or service, sometimes in counter-intuitive ways. For example, the fishing costs account for only a minor portion of the price of a piece of sushi, since competition reduces the labour costs of fishing while technological upgrading further lowers the price of caught fish. On the other hand, ambience, location, and restaurant services push up the final cost of the sushi to consumers. Importantly, all these contributions will be compensated only if the overall scale of activity is large enough (see Chapter 2).

Market Governance

No market (or value chain linking several markets) operates simply according to the principles of exchange. Indeed, the definition of transaction costs noted earlier implies the need for rules, rights, and some form of market governance. Thus, mar-kets operate within the framework of the controls of governance provided by formal and informal institutions. In addition to influencing economic transactions, both forms of institutions influence the way firms operate within their broader social, political, and environmental contexts (Figure 1.7). In short, markets are not simply a matter of economic relations; markets are embedded in social and political life, as we have seen in the Tsukiji fish market example. In general, governance in some form is required to establish trust in transactions, to protect buyers and sellers from inappropriate behaviour, and more generally to ensure markets work in society's interests. Diverse forms of governance have evolved to meet these needs. The fact

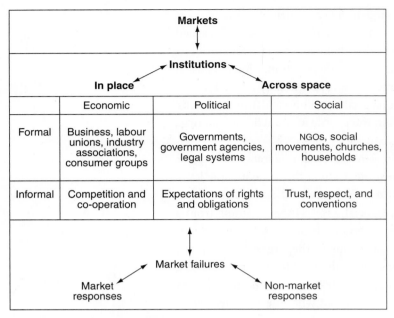

FIGURE 1.7   Market Governance

property rights The rights attached to ownership of any asset, physical or otherwise, and the associated income or benefit.

formal institutions Structured constraints on human behaviour; for example, laws at the societal level and corporate regulations at the organizational level.

informal institutions Tacit, culturally rooted constraints on human behaviour.

that market failures exist, and market economies experience substantial economic and social problems, indicates there are still significant challenges facing market governance.

Markets in advanced economies are embedded in legal systems. For example, the law pertaining to **property rights** is especially important, since market transactions involve the transfer of ownership from seller to buyer (except in administered transactions where property rights are retained under the same ownership). Thus property rights refer not only to buildings and land, but to all the products and services that are sold between legally separate entities; for example, consumers who purchase goods at a retail store acquire the property rights to those goods. As *basic* principles, property rights confer exclusive and secure ownership, control over benefits and income, and the right to transfer ownership. However, in practice, property rights (and contracts law) are complicated because they can include services and intellectual property (such as patents, see Chapter 3), and the sale of goods can involve specific legal contracts or simply be covered by more general laws. Property rights can also be private, held by individuals (including corporations), held in common as in common pool resources (Case Study 1.3) or controlled by the government. Open access resources means property rights are absent.

But markets are also embedded in a much wider range of formal and informal institutions. **Formal institutions** can be classified in three main groups: *economic* (business), *political* (governments, legal systems), and *social* (NGOs, churches), each of which has its own specific, legally mandated organizational forms and is characterized by certain explicit values, conventions, expectations, and norms. **Informal institutions** also have economic, political, and social expressions. The difference is that these expressions are implicit rather than explicit, unwritten rather than written, enforced through values, conventions, expectations, and norms. These institutions coexist within any economy or market, sometimes in harmony, sometimes in conflict. Examination of market failure provides a context for understanding the importance of market governance in its various forms.

Formal Governance

Formal governance of markets operates through the regulations imposed by various economic and non-economic institutions at different spatial scales (from the local to the global), as illustrated in Table 1.2. In practice, the "hierarchy" is much looser than the table suggests. Nevertheless, all institutions have distinct geographic origins (and biases), and their geographies exist at several scales. At the national, regional, and local

TABLE 1.2   Hierarchy of Markets and Institutions

Market Levels	Governing Institutions	Supporting Institutions
International trade conducted by companies selling in foreign markets within multinational companies (MNCs) for their different markets	World Trade Organization (WTO), bilateral and multilateral treaties between countries, NGOs (e.g., Greenpeace), industry associations (e.g., Chemical Manufacturers Association, International Chamber of Commerce), certifying organizations (e.g., International Organization for Standardization [ISO])	World Health Organization (WHO); United Nations Educational, Scientific, and Cultural Organization (UNESCO); Oxfam; Greenpeace; other NGOs; Organization of Economic Co-operation and Development (OECD)
National trade by MNCs and national companies	In addition to the above: national (federal) commerce regulations, national industry associations, NGOs, and certifying organizations	In addition to the above: national departments of health, education, and welfare; various NGOs
Regional trade by MNCs, national companies, and regional-sized firms	In addition to the above: state or provincial commerce regulations, regional industry associations, NGOs, and certifying organizations	In addition to the above: state or provincial departments of health, education, and welfare; various NGOs
Local trade by MNCs, national companies, regional firms, and local firms	In addition to the above: municipal and county commerce regulations, industry associations, NGOs, and certifying organizations	In addition to the above: municipal and county departments of health, education, and welfare; various NGOs

levels, governments have real authority to impose conditions on businesses and consumers alike. National companies control operations in several regions of a national space, while the most localized firms operate as single plants at just one site. Localized firms are often broadly categorized as small- and medium-sized enterprises (SMEs) and typically, though not invariably, serve local markets.

At the global level, trade is dominated by MNCs (or TNCs) that control operations in several countries. These giant firms supply goods and services both to external markets and to their own administered and "affiliated" operations. The governance of international market relations is less clear. The World Trade Organization (WTO) was created in 1995 by the national governments of the world's market economies as the successor to the General Agreement on Tariffs and Trade (GATT) that was formed in 1947. It has two principal responsibilities: (1) it develops the rules that govern trade and investment flow among countries and (2) it monitors government compliance with these rules. However, its real power is to enable aggrieved countries to set their own trade conditions when they are being taken advantage of. Although it cannot enforce its rulings, the WTO can authorize countervailing measures, and the legitimacy this gives to grievances is thought to deter member countries from breaking international trade regulations.

The governance capacities of other global institutions are even more diffuse. The World Bank was established at the end of the Second World War to provide financial advice and support for "problem" countries. The International Standards Organization (ISO) is a quasi-independent institution designed to establish standards for quality control and environmental performance. Labour interests are partially represented by unions at national and local scales, while the World Labour Organization (WLO) seeks to establish international standards of employee welfare. In the education and health sectors, NGOs and international bodies such as the World Health Organization (WHO) provide rudimentary medical services in poor countries that lack the sophisticated national, regional, and local systems of advanced countries. Unfortunately, many of these institutions' efforts to set international standards have failed, largely because success would require the compliance of both big business and national governments. NGOs are independent voluntary organizations that seek to persuade MNCs and governments to change their behaviour.

Informal Governance

Because markets are embedded in social relations, they are underpinned by particular norms, conventions, and habits of behaviour that have become established over time. These informal social characteristics of markets are often discussed in terms of cultural traditions and "untraded interdependencies." While they are intangible, difficult to measure, and often highly localized in nature, they are crucially important. What is essential to all market interactions are the shared expectations and values—especially trust and co-operation. The significance of trust—not only for a civilized society, but also for economic development and productivity—was highlighted by Fukuyama's (1995) distinction between "high trust" and "low trust" societies. Markets with a high level of trust are efficient because trust reduces transaction costs. Trust allows suppliers and buyers to anticipate one another's behaviour and feel confident—with or without formal contracts—that partners in a deal will not behave opportunistically. Trust is an advantage in market exchanges because it reduces the time and costs involved in negotiation, as well as the uncertainty associated with opportunism, and facilitates co-operative behaviour. In contrast, when competition is a zero-sum game, or when social relations are marked by suspicion and alienation, transaction costs rise and markets are inefficient.

In practice, trust is underpinned both by formal laws and regulations and by informal social and moral factors. Japan and Germany are widely regarded as high-trust societies, strongly associated with both formal regulation and an informal social commitment to stable, co-operative behaviour; in both countries, many business transactions take place without formal contracts. The US and Canada may also be considered high-trust societies from the legal perspective; however, their tendencies towards individualism can mean a higher risk of opportunism and can work against the formation of enduring business relationships. North American firms are considerably more likely than their Japanese or German counterparts to switch suppliers whenever a lower-cost alternative appears.

Admittedly, the distinction is blurred if legal systems are deemed corrupt. Fukuyama characterizes China as a low-trust economy because its legal system is not well developed and is subject to corruption. However, he does not give proper consideration to the informal governance provided by family ties. The growth of Chinese corporations and supplier networks, for example, has been largely driven by family-owned businesses that find competitive advantages in their ability to depend on relatives—people they know and trust—to manage geographically dispersed operations and organize supplier networks.

Although trust and shared values are universal concepts, the particular conventions through which these are expressed vary from culture to culture, locally and globally. For example, socializing over a meal is a universal way of enhancing business relationships, but the meaning of such occasions varies from place to place. In North America, a shared meal tends to mark the sealing of a deal. In China and Japan it is an occasion for building trust. Conventions are built within cultural contexts, and a North American hoping to seal a deal in China needs to understand what an invitation to dinner really means. Locally, concentrations of economic activity—whether in the financial districts of London, New York, or Toronto, the high-tech centres of Silicon Valley, Cambridge, or Waterloo, or in auto, steel, and shipbuilding districts—typically feature distinctive forms of mutual trust and the sharing of information that is captured by the term "untraded interdependency." The social nature of localized market relations is often reflected in voluntary exchanges of information between consumers and suppliers (or users and producers). Such exchanges are particularly important in local firm–supplier relations that involve personal contact.

Yet informal governance, even where it is well established, can be undermined. Opportunists on both the supply and demand side can take advantage of situations where trust is the norm. And when conventions are discriminatory—for example, when the job market in a particular place is biased against a certain group—both the efficiency and the morality of the market are diminished. In such cases formal government intervention is required.

Social relations also have a significant effect on demand. As Thorstein Veblen recognized, tastes (and values) are not "given" but socially learned and subject to change (see Chapter 14). Indeed, social relations frequently confound the *Homo economicus* assumptions that consumers are both rational (driven to buy the lowest-cost good) and highly individualistic in their preferences. Thus consumers' choices in clothing are often (though not always!) determined by the choices of colleagues, friends, or trends: many people dress to "fit in," whether the choice is a teen's backwards baseball cap or a financier's pin-striped suit. Sometimes consumers' choices are determined by elites whose expensive choices in clothes or lifestyles have become the models for lower-cost options produced for the masses. Some people will be more likely to buy a higher-priced good than a less expensive alternative, simply because they associate a high price with high status (or quality). Consumer choices are also socialized by ethical concerns that in recent years have emphasized environmentally sustainable products. Indeed, consumers are increasingly willing to pay extra for eco-certified wood products, sustainable fish, and hybrid autos etc., and such purchases are socially motivated. Consumption is a highly social experience, even in Western countries where individualism is assumed to be a widespread characteristic.

Market Failure

Markets exist to facilitate the efficient exchange of goods and services. But markets can fail in that role in various ways to various degrees, and market economies as a whole can experience major disruptions in economic activity. Table 1.3 outlines a number of problems that can be seen as consequences of market failure—though it is important to keep in mind that broad societal problems can rarely be traced to one single cause. In many cases, non-market factors play a contributing role and circumstances vary by geographic context. For present introductory purposes, four basic types of market failure are briefly described below:

TABLE 1.3   Selected Market Failures in Geographic Perspective

Failure	Geographic Perspective
Income disparities: Markets create massive differences in income between rich and poor.	Poverty is widespread in developing countries, especially in rural areas, but it also exists in rich countries in both urban and rural communities.
Busts: Recessions occur when supply exceeds demand and output falls. Depressions are structural crises characterized by long-term unemployment.	Recessionary effects are spatially uneven, often varying with economic structure. Recessions in major economies have global effects.
Inflation: Occurs when demand exceeds supply, causing prices to increase. Hyper-inflation is a major problem for people on fixed incomes, and threatens overall stability.	Inflation typically begins when a boom in a particular region causes a rapid increase in its cost of living, which spills over elsewhere.
Monopolies: Markets concentrate capital in the hands of monopolies and oligopolies; eventually, the dominant firms threaten competition and the rights of consumers and labour.	Regional monopolies can take advantage of consumers. The geographic concentration of corporate head offices in core regions leaves peripheries subject to the control of MNCs.
Public goods: Markets underinvest in public goods that societies would prefer be accessible to all members of the population.	Public goods are less available in poor countries (governments have limited funds) and where strong market ideology exists.
Land-use conflicts: Economic and non-economic interests clash over the use of land.	Land-use conflicts occur around the globe and are especially virulent in resource peripheries.
Unpaid work: Markets do not recognize the value of unpaid activities such as housework, childcare, and volunteer activities.	Concentrated among women in low-income groups and areas, especially poor countries.
Labour: Markets treat labour as a variable cost, hiring and firing workers as demand dictates. Unemployment and underemployment result.	Lack of appropriate work is a big problem in poor countries, and regionally and locally concentrated in rich countries.
Environmental degradation: Markets fail to take environmental values into account and regard nature as a free sink or resource.	Global and local problems, driven by population growth and economic growth.

- *Monopoly and oligopoly.* Markets driven by competition benefit society by encouraging innovation and keeping prices low. Paradoxically, though—as Marx recognized in the mid-nineteenth century—competition can lead to the elimination of competitors, reducing the numbers of firms in the field and increasing the power of the ones that are left. Eventually, a single firm (**monopoly**) or handful of firms (**oligopoly**) will so dominate the industry that the pressure to reduce prices and innovate is seriously reduced. Corporate concentration is problematic when it gives giant firms *disproportionate* influence over prices and production levels or allows them to exploit labour, abuse the environment, gouge consumers, or control government policy. With the rise of the MNC, regional development has come to be increasingly dominated by distant decision-makers and multi-plant firms whose decisions regarding the allocation of investment and jobs reflect corporate rather than local priorities.

monopoly A market controlled by a single supplier.

oligopoly A market dominated by relatively few suppliers; a highly concentrated oligopoly has very few suppliers, while a weak one has many suppliers.

- *Negative externalities.* Externalities are social or "third-party" costs (or benefits) resulting from market activities that are not paid for by the partners in a transaction. A trip in a taxi, for example, means a ride for the passenger and a fare for the driver; but it also comes with social costs in the form of increased air pollution and traffic congestion.

Whether a given externality is an unintended effect, a matter of willful negligence, or something in-between, it can usually be attributed to either the limitations of property rights themselves or to inadequate enforcement of them. Among the best-known examples are manufacturing, mining, and commercial fishing operations that have treated air, land, water, and fish stocks as "free" ("open access" or unregulated) resources, either exploiting them directly for profit without thought to long-term sustainability or using them as sinks for pollutants at no expense to themselves. The situation is complicated because property rights do not automatically guarantee wise use of resources. A related type of market failure involves phenomena that markets simply do not recognize or value. These externalities range from childcare and volunteer work to what have been called ecosystem services—the natural processes essential to life, such as the production of oxygen.

- *Failure to invest in public goods.* For a society to function well, all its members need access to services, such as roads, sanitation systems, transit systems, health care, education, policing, parks, electricity, telephones, even Internet connections. However, because markets operate on the basis of profit, they have no reason to provide access to goods or services for people who cannot afford them. Thus it falls to governments and NGOs to provide them as public goods on a non-profit or subsidized basis. Geographically, the ability to provide public goods and services varies from place to place. In general, richer economies are financially capable of providing them, at least to some level, but willingness to do so varies significantly, even within a single society.

- *Information asymmetries.* An information asymmetry is an imbalance in the information possessed by the two parties to a transaction. A classic example is the problem of opportunism, in which the more-informed partner in a transaction takes advantage of the less-informed partner. Consumers face information asymmetries when they lack the time or the means to research the claims made by the suppliers of goods and services; stockholders in a company face them when they do not know whether the manager is working for them or for him- or herself (principal–agent problems). Information asymmetries are pervasive because of unequal knowledge, and neither party to a transaction can ever be completely certain of the other's capacities or intentions. The risks associated with them can be reduced, however, by legal means (contracts, etc.) or by choosing to do business with people we trust because we share certain values and norms.

The consequences (direct and indirect) of these and other market failures take economic, social, and environmental forms. A major task of democratic governments in market-based economies is to deal with those consequences. In the nineteenth and twentieth centuries, according to Polanyi (1944), capitalist societies saved themselves from the most harmful effects of market forces by developing institutions to govern them and regulations to improve them (e.g., working conditions). Government is the most powerful of those institutions, but others, such as labour unions and NGOs, also play major roles in the broader guidance of society that we refer to as governance.

The Challenges of Sustainable Development

Income inequalities and environmental degradation are the central challenges (market failures) facing global society in the twenty-first century. With roots at the start of market economies in the late eighteenth century, both (sets of) problems have become more pronounced. Global income inequality was formally recognized by the so-called "development decade" of the 1950s while environmental degradation became a truly global problem at that time, awareness sparked by Rachel Carson's (1962) *Silent Spring* and Farley Mowat's (1963) *Never Cry Wolf*. It was made clear in 1987 that income inequality (poverty) and environmental degradation had become urgent when the UN's World Commission on Environment and Development (the Brundtland Commission) delivered its plea for **sustainable development** with the goal of reducing poverty and income inequalities while sustaining environmental values.

sustainable development
Economic growth that can be sustained over long periods of time without harming the environment.

The scope of these challenges becomes clear when we consider global inequalities in income and energy consumption.

Global Income Inequality

gross domestic product (GDP)
The total value of market-related activity in a nation over a period of time, usually a year; gross national product (GNP) consists of GDP plus whatever income was earned by citizens working abroad.

The standard way to measure income is per capita **gross domestic product (GDP)**. GDP is a summary measure of the market value of all the final goods and services produced in an economy in a year. ("Final goods" are those purchased by the ultimate consumer, as opposed to the intermediate goods used to produce them.) In (circa) 2012, the US was the world's biggest economy, with a GDP of US$16,245 trillion; China, Japan, Germany, and France followed, with GDPs of $8,227, $5,959, $3,428, and $2,612 trillion respectively. These five countries accounted for 50.3 per cent of the total GDP ($72,420 trillion) of 190 countries (Canada's GDP was $1,821 trillion). Per capita GDP, measured by dividing national GDP by the nation's total population, provides an index of average income and insights into average standards of living (and poverty). In order to account for national variations in purchasing power (a dollar typically buys more in poor countries than in rich countries), the International Monetary Fund (IMF) calculates GDP at purchasing power parity (PPP).

Average wealth based on these figures is only an estimate; in many countries, it is not easy to measure either population or market activity with any precision, while non-market activity is not taken into account at all. Going back in time, estimates of GDP and population are even harder to make, while per capita averages say nothing about income distribution within countries. Nevertheless, per capita GDP provides a useful "ballpark" reference for assessing variations in income and (by implication) in standards of living. Pioneering work by Angus Maddison (2007) on global wealth levels and distribution indicates that until the sixteenth century, per capita GDP (PPP) was roughly equal around the world. Then Europe—with the Industrial Revolution and colonization—gradually started to become relatively wealthy until the early nineteenth century, when global income disparities started to widen more rapidly. Maddison estimates, for example, that the ratio of the average GDP (PPP) for 12 leading west (and northern) European countries was 1.5 times greater than the global average in 1700, rising to 1.8 in 1820, 2.8 in 1973, and 3.1 in 2007. For the US, GDP (PPP) was about the same as the world average in 1700, was on par with western Europe in 1820, and was over

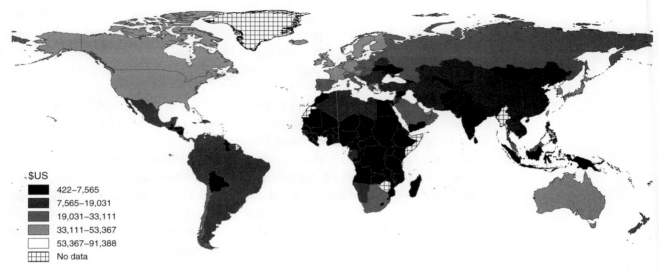

$US

- 422–7,565
- 7,565–19,031
- 19,031–33,111
- 33,111–53,367
- 53,367–91,388
- No data

FIGURE 1.8   Global Distribution of GDP (PPP) Per Capita, Circa 2012

four times the global average in 1973 and almost five times more in 2003. Compared to the poorest region (Africa), the gaps are much wider. In 1820, for example, US and west European per capita income levels (in real purchasing power) were about three times more than the African average and by 2003 the ratios had grown respectively to 18.7 and 12.9 times higher.

The income gaps between rich and poor countries continue to be large (Figure 1.8). Thus in 2012 the world's 32 richest countries, with per capita GDPs (PPP) in excess of US$30,000, included the long-established, rich industrialized countries in Europe and North America, plus Australia, New Zealand, and Japan. In contrast, the bottom 64 countries had per capita GDPs (PPP) of less than US$5,000; in 34 countries it was less than $2,000; and in 8 countries, less than $1,000. These very poor countries are mainly located in Africa with some in Asia such as Bangladesh.

Over the past two decades global income distributions reveal important changes. In terms of absolute size of economies measured by GDP, Brazil (seventh) and India (tenth) joined China (second) in the world's top ten list in 2012. In per capita income terms, Singapore, Hong Kong, South Korea, and probably Taiwan (if statistics could be published) have become rich countries, as have several others, including oil-rich states such as Kuwait and Saudi Arabia (although in these countries incomes are strongly skewed to the super rich). Nevertheless the income gap between rich and poor is daunting. Even if per capita GDP in the US did not increase, the 34 poorest countries would need to grow by a factor of 50 to 100 to catch up. Given their vast populations, China and India are particularly significant in this regard. China's economy has grown spectacularly over the last two decades, while India's growth is gathering momentum. Even so, at US$9,233 (PPP), China's per capita GDP is one-fifth of US levels ($49,965), and India's, at $3,876, is only one-thirteenth. Moreover, evidence shows the income gap between the richest and poorest countries is widening.

Poverty and income inequality are connected. People in poverty are not merely poor: they are without the means to maintain an "adequate" **standard of living**. It

standard of living A measure of people's ability to live a healthy and productive life in accordance with acceptable social norms; often approximated by per capita GDP in international comparisons.

has been estimated that roughly half the world's people (almost three billion) live on less than two dollars a day, 790 million people are chronically undernourished, and unemployment and under-employment are widespread. Children under the age of five are particularly at risk; UNICEF suggests that 30,000 children living in poverty die each day.

Should we be concerned about extreme variations in wealth and poverty within our own countries and across the world? Don't people receive what they are worth, the market might say. On the other hand, income extremes may be regarded as morally repugnant; markets rarely operate fairly; and wealth cumulates from one generation to the next so starting points are scarcely equal. Indeed, there are economic reasons for considering wide income disparities to be a problem. Low incomes and poverty reduce a nation's aggregate purchasing power, productivity, and ability to participate, while exclusion of large segments of the population may contribute to social and political discord. Unregulated markets will not resolve these issues.

Global Energy Consumption

With economic growth, environmental problems have escalated and are now truly global in scope. It is now widely (if not universally) acknowledged that climate change is powerfully influenced by human activity and will drastically alter temperature regimes, water levels and availability, and weather patterns, with global consequences that are likely to be particularly severe in poor countries. Of all the human impacts on the environment, those associated with energy consumption are the most pervasive. Indeed, climate change has become the bell-weather of global environmental concern. Massive increases in the exploitation of fossil fuels from the late eighteenth century, coal, oil, and natural gas along with deforestation, have released carbon dioxide, which acts as a greenhouse gas that (partially) traps heat within the earth's atmosphere and oceans. The scientific community agrees that as a result of anthropogenic (man-made) causes, global temperatures increased by 1°F during the twentieth century and in the next hundred years will likely increase by 5–9°F with many implications (sea-level rise, ice-sheet melting, stormier weather, etc.).

Even today, a very large proportion of the world's energy comes from non-renewable fossil fuels, such as oil and coal. Thus production and per capita energy consumption—such as kilograms per oil equivalent per capita—provides a comparative indicator of environmental impact (Figure 1.9). Typically, the richest western countries; oil exporting countries, such as Kuwait, Bahrain, and the United Arab Emirates; and some small countries, such as Iceland and Gibraltar, occupy the two highest energy consumption categories shown on Figure 1.9 (between 4–7000 and >7000 kg per oil equivalent per capita). In the poorest countries, by contrast, energy consumption is less than 1000 kg per oil equivalent per person, and in 15 countries it is less than half that. Meanwhile, Chinese and Indian energy consumption rates are roughly less than one-third and one-tenth of Canadian/American levels of energy consumption (or about half and a sixth of German levels). Thus the disparities in energy consumption are roughly similar to the disparities in per capita GDP (PPP).

The "straightforward" implication is that in order to develop economically (i.e., increase GDP), energy consumption must also rise—in other words, development

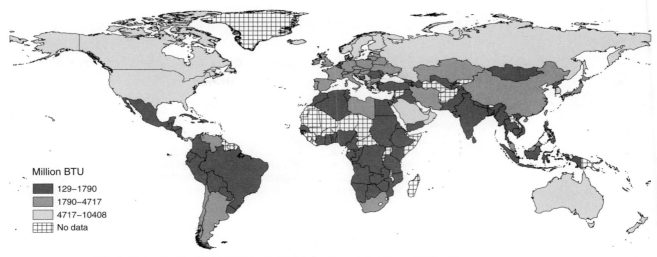

Million BTU

- 129–1790
- 1790–4717
- 4717–10408
- No data

FIGURE 1.9   World Per Capita Total Primary Energy Consumption, 2011–12

and environment are in a trade-off situation. Given that China and India account for approximately half the world's population, the impact of their growth on the environment is potentially huge. However, the income and energy data at a single point in time do not inform about processes of change. Most rich countries have at least initiated improved energy use efficiencies and policies to reduce environmental impacts. Significantly, it is vital to recognize that market economies are highly innovative and that the connections between income and energy can be restructured (Chapter 3).

Countervailing Power

In the perfect competition model, individual producers and consumers have no power to influence either prices or the choices of others (Figures 1.1–1.4). In the real world, however, contemporary markets are characterized by significant imbalances in market power, led by the rise of MNCs. Contrary to the perfect competition model, the markets in which these giant firms operate have relatively few sellers. While monopolies (industries with just one seller) are rare, oligopolies (industries dominated by a few sellers) are a dominating feature of market economies. In oligopolies, rival behaviour is interdependent; decisions regarding matters such as pricing, advertising, investment, and location decisions tend to be strongly influenced by the decisions of their competitors. Because MNCs, size and geographical scope give them influence over public policy and the nature of both national and local development, they cast giant shadows over market economies.

Beginning with John D. Rockefeller's Standard Oil Company, which rose to monopolize the US oil industry in the late 1800s, the rise to dominance of the corporate giants marked a new phase in capitalism and industrialization. Oligopolies have dominated market economies ever since, led by MNCs first in the resource and manufacturing sectors and more recently in the service sector as well. The biggest MNCs are big not only in their market share, but in their revenues (Table 1.4). Thus in 2012 Walmart, Exxon, and General Motors collectively took in more money—and had more to spend—than

TABLE 1.4   Revenues of Four MNCs Compared to GDP of Four Countries, 2012

Country	GDP ($US millions)	Corporation	Revenues ($US millions)
Canada	1,821.4	Exxon Mobil	482.8
Australia	1,532.8	Walmart	443.9
The Netherlands	700.6	Royal Dutch Shell	467.1
Sweden	523.8	BP	375.8

The GDP figures are from The World Bank 2012 Gross Domestic Product 2013, http://siteresources.worldbank.org/DATASTATISTICS/Resources/GDP.pdf (accessed February 2014). The GDP is an index of the overall wealth of countries. The revenue figures are from the Fortune 500 listing of America's largest publicly traded corporations.

invisible hand Adam Smith's description of the way market self-regulation operates under conditions of (preferably perfect) competition.

Canada's federal government, which is responsible for more than 35 million people. As one Japanese economist put it, contemporary markets are dominated by the "face-you-can-see competition" rather than the "**invisible hand**" and the numerous independent suppliers and consumers envisioned by the perfect competition model.

Much of giant corporations' power can be attributed to their sheer size and geographic scope with operations in multiple countries and abilities to scan business trends globally. Thus giant MNCs directly control many markets in terms of the quantity and quality of goods and services sold and the prices charged; these administered relations are not routinely exposed to the competitive influences of open markets. Moreover, their size and geographic diffusion give giant corporations serious bargaining power in relation to governments, labour, and suppliers. Instead of accepting demand and supply as "given" conditions, MNCs seek to modify them to suit their own interests. They do this by shaping consumer preferences through massive spending on advertising while, on the supply side, they reduce competition by acquiring rivals. They reduce costs by both forcing hard bargains on suppliers that lack alternative customers and negotiating lower wages, taxes, energy costs, and other inputs in particular jurisdictions by threatening to pull out and invest elsewhere.

The social implications of big corporate power are not straightforward. The social legitimacy of such corporations derives from their ability to meet the material needs of society. Technological development is extremely expensive, and large-scale operations are vital to reduce costs and cope with the uncertainties of developing new technologies. On the other hand, if corporations use their power to exploit consumers, workers, smaller rivals and suppliers, shareholders, politicians, or the environment, for private advantage, they undermine the interests of society. Whichever way their social impacts are judged, institutional models of market economies recognize that big, powerful corporations are an inherent feature of market economies. Their power is not absolute, but is balanced by "countervailing" powers. When John Kenneth Galbraith advanced this view in the 1960s he argued that the power of Big Business was kept in check by the countervailing powers of Big Government and Big (unionized) Labour, along with rival Big Business (Figure 1.10). He deliberately excluded individual consumers to emphasize their relative lack of influence and to reject prevailing ideas about the importance of consumer sovereignty.

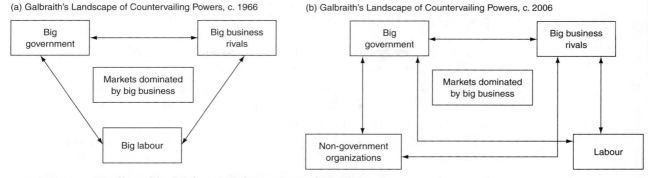

FIGURE 1.10   Galbraith's Landscape of Countervailing Powers

Galbraith's argument remains relevant today. However, "Big Labour" is less power-ful now than it was several decades ago, while NGOs of various kinds have become more important in shaping market activities (Figure 1.10b). While countervailing powers are in conflict with one another, they nevertheless work together in ways that shape the daily routines of all industrial societies. For example, unions and corporations have both differing and shared interests and so work together to gain mutually acceptable benefits. Similarly, environmental NGOs are adversaries of big business, but can cooperate with MNCs, encouraging them to incorporate environmental objectives into their bottom lines. The role of governments in facilitating and regulating the interests of business is similarly complicated.

Conclusion

Markets have established themselves as humanity's most efficient means of allocat-ing resources. One of the main reasons for this efficiency is the geography of markets that bring people together to exchange, not only saving transportation costs, but vastly decreasing the transaction costs of search and governance. The power of markets in promoting economic development is further explored in the next two chapters that examine urbanization and economies of scale (Chapter 2) and innovation (Chapter 3). The risks of market failure are well recognized, and all societies have imposed rules to help markets function better. The case for minimal external regulation is based on two beliefs: that markets will regulate themselves through the pursuit of self-interest by suppliers and consumers, and that the collective pursuit of self-interest will always be in the best interests of society. The case for stronger regulation rejects this simple equation between private and public interest and maintains that external governance is needed for social purposes. But there is inevitably much debate about the nature and extent of that governance. This debate is powerfully influenced by governments, NGOs, labour, and other actors defending their territories and seeking fair treatment within the global economic system. We will see these countervailing forces in action in all the subsequent chapters.

Practice Questions

1. Do you shop on eBay? Explain why or why not. Do you trust this market?
2. Farmers markets have proliferated recently. Why? Visit a local example and find out where the produce is coming from and how it gets there. Why do consumers go to farmers' markets?
3. Define transaction costs and high-order goods. Assess the transaction costs you have incurred in purchasing a high-order good. What did you learn from the experience?
4. Conduct a class tutorial or project on the purchase of organically grown goods. Identify reasons for their purchase or non-purchase.
5. Discuss in class what you observe as examples of market failure in your community.

Key Terms

embeddedness   20
transaction costs   21
open market   21
relational market   21
administered market   21
countervailing power   21
neoclassical economics   22
institutions   22
place   22
space   22
perfect competition   23

market failure   21
market economies   21
markets   25
Homo economicus   24
demand curve   24
supply curve   24
ceteris paribus   24
central place theory   25
range (of a good or service)   25
threshold population   25
coordination costs   29

moral hazard   29
agglomeration   31
value chain   34
property rights   38
formal institutions   38
informal institutions   38
monopoly   42
oligopoly   42
sustainable development   44
gross domestic product (GDP)   44
standard of living   45

Recommended Resources

The classic work on markets is Adam Smith's *The Wealth of Nations*, first published in 1776. Karl Marx (*Capital*, vol. I, 1867) and Thorstein Veblen (*The Theory of the Leisure Class*, 1899) offer very different, and equally important, perspectives. For a comparative overview of these three (and other) pioneering critics of markets, see R.L. Heilbroner's *The Worldly Philosophers: The Lives, Times and Ideas of Great Economic Thinkers* (1972). Heilbroner's study reveals how the ideas of the great economists reflected the times—and the places—in which they lived.

More recently, O.E. Williamson's 1975 *Markets and Hierarchies* resparked interest in the nature of transaction costs (including Allen, below), but this book is for economists.

Fukuyama, F. 1995. *Trust: The Social Virtues and the Creation of Prosperity*. New York: Free Press.

A provocative book that stimulated much debate on the role of trust in economic development.

Granovetter, M. 1985. "Economic action and social structures: The problem of embeddedness." *American Journal of Sociology* 91: 481–510.

The article that stimulated the modern debate on embeddedness.

Reiffenstein, T., and Hayter, R. 2006. "Domestic timber auctions and flexibly specialized forestry in Japan." *The Canadian Geographer* 50: 4.

Discusses the nature of market auctions and emergence, functions, and prospects of timber auctions in Japan that closely resemble open markets, and explains how these markets reduce transaction costs for sellers and buyers.

References

Allen, D.W. 2012. *The Institutional Revolution*. Chicago: University of Chicago Press.

Bestor, T.C. 1998. "Making things clique: Cartels, coalitions, and institutional structure in the Tsukiji wholesale seafood market." Pp. 154–80 in M.W. Fruin, ed. *Networks, Markets, and the Pacific Rim, Studies in Strategy*. Oxford: Oxford University Press.

———. 2004. *The Fish Market at the Center of the World*. Berkeley: University of California Press.

Carson, R. 1962. *Silent Spring*. Boston: Houghton Mifflin.

Fukuyama, F. 1995. *Trust: The Social Virtues and the Creation of Prosperity*. New York: Free Press.

Galbraith, J.K. 1967. *The New Industrial State*. Boston: Houghton Mifflin.

Mowat, F. 1963. *Never Cry Wolf*. Toronto: McClelland and Stewart.

CHAPTER 2

The (Evolving) Spatial Division of Labour

This division of labour, from which so many advantages are derived, is not originally the effect of any human wisdom. . . . It is the necessary, though very slow and gradual, consequence of a certain propensity of human nature . . . to truck, barter and exchange one thing for another.

Smith 1986 [1776]: 117

• • •

Because of . . . circular causation a social process tends to be cumulative and often to gather speed at an accelerating rate.

Myrdal 1957: 13

geographic multipliers The ratio of total economic impacts to the initial economic impact of a new development within a region.

The overall goal of this chapter is to explore the relationships between markets, economies of scale, and the evolving spatial division of labour, especially in relation to urbanization trends and the organization of value chains. More specifically, this chapter's six main objectives are

- To explain the (spatial) division of labour and the relationships between urbanization and economic development;
- To examine internal and external economies and diseconomies of scale;
- To introduce the concept of the **geographic multiplier**;
- To see urbanization and de-industrialization as cumulative, path-dependent processes;
- To explore globalization in relation to the free trade debate and the evolution of value chains.

The chapter is divided into two main parts. The first part defines various types of economies of scale, explaining in broad terms the cumulative implications of their efficiencies (and inefficiencies) in regard to urbanization. The second part explores the implications of scale economies for international trade, the controversies about free trade and globalization, and the main themes in the (global) organization of value chains.

Introducing the Division of Labour

As we discussed in Chapter 1, markets are distinct institutions of exchange operating under particular forms of governance and linked through value chains and trade. As the opening quotation illustrates, Adam Smith recognized in the late eighteenth century that the evolution of markets and the human predisposition to "barter and exchange" is related to the division of labour (or specialization) that generates improvements in productivity. He argued that specialization potentially reduces the average cost of production by increasing the scale of an activity. In turn, the specialization of production depends upon the creation of markets to facilitate exchange (trade). Moreover, this co-evolution of market, specialization, and scale economies has propelled fundamental changes in global economic geographies, most generally in the transformation of rural-based to urban-based economies. Rich western economies have long been highly urbanized, while in contemporary times, this ongoing transformation has featured the globalization of value chains, rapid urbanization in virtually all parts of the world, and huge rural-urban migrations in China and India on unprecedented scales.

The specialization or **division of labour** is the practice whereby people perform a specific task according to their specialized skill, rather than one person performing multiple tasks. The related **spatial division of labour** is the distribution of specific types of work to different *places* and is central to economic geography's mandate to understand the specialized functions of places and their integration into national and global economies. Note also that the allocation of tasks within firms creates an internal division of labour; a **social division of labour** is created when tasks are organized among a population or number of different firms.

The purpose of the division of labour is to take advantage of the efficiencies known collectively as **economies of scale**. In general, increases in productivity, especially reductions in average cost, that arise with increases in the size or scale of individual operations (plants, establishments, or factories) or firms (Case Study 2.1) are known as internal economies of scale. The term external economies of scale refers to the advantages firms derive from the social division of labour and notably includes the broadly based economic advantages and reductions in average cost gained from location in cities (urbanization economies) or more specifically to the economic advantages of agglomeration (clustering) of firms in related activities (localization economies). However, disadvantages to increasing size can also occur as internal or external *diseconomies* of scale. Indeed, economies and diseconomies can both be present, as seen in many metropolitan areas where economic growth is matched with problems of congestion and pollution. Moreover, economies and diseconomies of scale are dynamic and powerfully contribute towards broader, self-reinforcing social processes of urbanization or urban decline that have long been explained with the principle of **circular and cumulative causation** (Figure 2.1). Recently, self-reinforcing processes by firms and in regions, especially in relation to technology choices and labour relations and attitudes, have been described as **path-dependent behaviour**.

From economic geography's perspective, the spatial division of labour is organized in two related ways: (1) by cities in place and (2) by value chains across space that link

division of labour The allocation of specific jobs to different people, and by extension to equipment and places; also referred to as the "specialization" of labour.

spatial division of labour The allocation of specialized tasks among and within places and across space.

social division of labour The distribution of tasks among firms in an economy. In practice, this term is often used when small firms play important roles.

economies of scale A reduction in average costs of production by increasing the scale of a particular activity.

circular and cumulative causation Refers to social processes that are interdependent and self-reinforcing in particular specialized directions for lengthy periods of time.

path-dependent behaviour The tendency for decisions, for example, by firms regarding the development of a new technology, to be constrained by decisions made in the past, even though circumstances may have changed. Often used in relation to "lock-ins" to inefficient practices.

Case Study 2.1
INTERNAL ECONOMIES OF SCALE

There are several main sources of internal economies of scale in the use of capital equipment:

- *Cube-square law.* Increasing the volume of capital equipment (e.g., wine vats, refineries, airplanes) creates economies of scale because area increases as the square of a container's dimensions, while volume increases to the cube. Thus with increasing output, costs that are based on the quantities of materials needed to make containers increase more slowly than revenue, based on volumes generated.
- *Principle of multiples.* In a linked production process, each stage of production may require specialized equipment operating at different rates. To coordinate the flow of production, different numbers of machines may be needed at each stage of the process.
- *Reduced energy costs.* On-site integration of linked operations can save energy costs. If manufactured alone, for instance, wood pulp has to be baled for transport and then re-pulped before it can be used to make paper. Thus combining the pulping and paper-making processes at a single site would save not only the cost of transport to a separate facility but also the cost of the energy required for heating the material and re-pulping.
- *Massing of reserves.* Reserve machines may be kept if different machines in a process fail at different times.
- *Dynamic interdependencies between specialized capital and labour.* For example, a winemaker who invests in new vats that are twice the size of the ones used previously can increase wine output with little extra effort. Similarly, increasing the size of the trucks used to transport goods will reduce costs in several ways, not only because of gas savings, licensing, maintenance, or the cube-square law, but because bigger trucks often only require one driver.
- *Technical changes.* Over the past 30 years technical changes, especially those involving microelectronics, have enhanced quality control and possibilities for product differentiation through economies of scope. In sawmills, for example, at the touch of a button a sawyer can change the cutting process depending upon timber species and market requirements. Such "technical" economies are general throughout an organization, or for that matter a national or regional economy.

Oil tankers, road haulage trucks, iron and steel plants, assembly lines of all kinds, and paper machines have all become bigger and faster as a result of the ongoing

globalization A term with multiple meanings, in economic geography for example, often used as shorthand to refer to the current stage of capitalist development, the intensification of international networks of all kinds; qualitative changes in the relationships between markets and governments; and the increasing role of exogenous forces in national and regional development. But it also implies the expansion of environmental problems to a truly global scale.

the settlement system as a whole. As cities specialize and grow by exploiting economies of scale, value chains reach out to peripheries for resources and labour, as well as to other cities for specialized goods and services and as markets for their goods. With the rapid industrialization in the nineteenth and early twentieth centuries of a relatively few market economies, especially in western Europe and North America, cities typically made strong value chain connections with local hinterlands (Chapter 8). In recent decades, however, the **globalization** of the spatial division of labour has spread urban-industrial growth around the world that has both increased competition and market opportunities, while value chains have become more spatially extensive and complex.

search for economies of scale and scope. For example, a state-of the-art paper machine in 1912 ran at 650 feet per minute to produce 39,000 metric tonnes a year. By 1981, with the same number of employees, new paper machines were running at 3,500 feet per minute to produce 220,000 tonnes a year. In the case of oil tankers, sizes range from coastal tankers that carry about 50,000 deadweight tonnes to ultra-large crude carriers that carry more than 550,000 deadweight tonnes. While big paper machines and tankers are expensive to buy, they reduce average costs (of production and transportation respectively) so long as they are used to full capacity.

At the same time, economies of scale amplify risks. The *Exxon Valdez* was a mid-sized tanker carrying 53.1 million US gallons of oil when it struck a reef in Alaska's Prince William Sound in 1989 and spilled 10 million gallons into the Sound, causing severe environmental damage. The clean-up cost alone amounted to $2 billion, and Exxon was charged an additional $500 million in punitive damages. Perhaps the only beneficial effect of the disaster was that it stirred regulators to require that all oil tankers be built with double hulls. In 2010, the Deepwater Horizon Oil Spill in the Gulf of Mexico became the largest oil spill in history when a BP oil well became uncapped and spilt over 210 million (US) gallons of oil in the Gulf, causing massive environmental and economic damage and 11 deaths.

© Accent Alaska.com/Alamy

Urbanization, Scale Economies, and Cumulative Causation

The search for economies of scale has been a major driving force behind the spatial division of labour and its urbanization around the world. The UK, for example, was a largely rural society when the Industrial Revolution began in the late eighteenth century (Table 2.1). Yet by 1900 more than half the population lived in cities, and by 2011 urbanization in the UK had reached almost 90 per cent. The same association between urbanization and economic development can be seen in other countries: in

TABLE 2.1   Urbanization in the UK, 1800–2011

	1800	1850	1890	1950	1960	1970	1980	1990	2000	2011
United Kingdom	20.3	40.8	50.3	79.0	78.4	77.1	87.9	88.7	88.9	89.2

Note: Figures indicate percentage of population living in cities >10,000 people and for England and Wales in the first three time periods.

Source: De Vries, J. 1984. *European Urbanization 1500–1800*. Cambridge, MA: Harvard University Press (years 1800–1890).

TABLE 2.2   Urbanization in Selected Countries, 1950–2011

	1950	1960	1970	1980	1990	2000	2008–11
Brazil	36.0	44.9	55.8	66.2	74.7	81.1	84.3*
Canada	60.8	68.9	75.7	75.7	76.6	79.4	77.8
China	12.5	16.0	17.4	19.6	27.4	35.8	49.7*
Germany	71.9	76.1	79.6	82.6	85.3	87.5	88.5
India	17.3	18.0	19.8	23.1	25.5	27.7	29.4^
Japan	37.5	63.5	72.1	76.2	77.4	78.3	90.7
Russian Federation	44.7	53.7	62.5	69.8	73.4	73.3	74.4
United Kingdom	79.0	78.4	77.1	87.9	88.7	88.9	89.2
United States	64.2	70.0	73.6	73.7	75.3	79.1	80.7*
World total	29.1	32.9	36.0	39.1	43.0	46.6	48.6

Note: Figures refer to share of population living in urban areas. The figures for China do not include Hong Kong and Macao. Also note that if a higher population threshold is used to define an urban area, urbanization trends remain strongly evident. For the last column * is for 2010, ^ is for 2008; otherwise figure is for 2011. The UK and Germany figures are from census.

Source: De Vries, J. 1984. *European Urbanization 1500–1800*, Cambridge, MA: Harvard University Press (years 1800–1890); United Nations Statistics Division (years 1950–2011).

China, for example, about half its population lived in cities in 2011, compared to just 12.5 per cent in 1950, when its population was much smaller (Table 2.2). Meanwhile, in the established market economies, as many cities have lost manufacturing jobs, both absolutely and relatively, and gained service sector jobs, the spatial division of labour continues to deepen in association with ongoing urbanization. Indeed, cities are deemed vital to national economic development. The economic geographers Florida's (2006) and Scott's (2006) respective notions of creative and resurgent cities underlie this claim.

The powerful relationships between economic development and urbanization in general became starkly evident with the industrial cities that emerged in North America and western Europe during the nineteenth century. These cities evolved around distinct manufacturing specialties that remained defining features of city identity into the 1970s. The steel and metal working specialisms of Sheffield (UK), Pittsburgh (US), Hamilton (Canada), and Dortmund (Germany) illustrate this evolution, each city being part of broader regional "manufacturing belts" of specialized cities. Indeed, the evolution of industrial cities was summarized in the 1960s in terms of

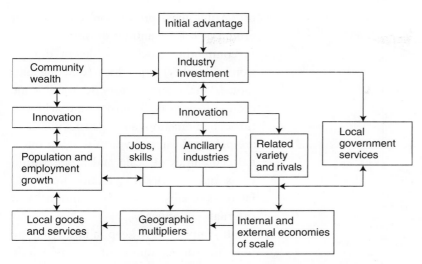

FIGURE 2.1   **Virtuous Circular and Cumulative Urbanization**

Martin (2010) provides an advanced, recent discussion of a related model of path dependent evolution.

Source: Adaptation based on Keeble (1967) and Ped (1964).

a virtuous model of cumulative causation in which some "initial advantages" encouraging industrial investment and innovation activities in particular became the basis for all kinds of cumulative and linked growth (Figure 2.1). Note that Chapter 8 addresses the economic geographies of cities and their inter-connections as urban systems in more detail.

The cumulative causation model does not explain the initial advantages for city growth but it does point to key mechanisms in how urban growth evolves in specialized directions over long periods of time, most notably with respect to investment in new facilities and the realization of internal and external economies of scale, the development of local linkages and local multiplier effects, and innovation. Implicit to this model is the significance of cities as centres of consumption themselves and as vital forums of market exchange and trade. While innovation is discussed in Chapter 3, the nature of scale economies and multipliers are clarified below. The deindustrialization of cities, including the four cities mentioned above, implies the long-term reversal of these inter-related causal factors.

Internal Economies

Economies of scale are achieved when the *average* (or per unit) cost of production (or service delivery) declines with increasing output or scale of production, even though *total* costs increase (Figure 2.2). In most cases, these economies reflect the efficiencies gained in the production of one good or service such as when newsprint is manufactured on larger machines or oil transported in larger tankers and pipelines (Chapter 1). **Economies of scope**, by contrast, are achieved when efficiencies are gained when several products or product types can be supplied by the same workforces, equipment, and technology. Economies of scale and scope can be combined, for example, in forms

economies of scope A reduction in average costs of production achieved by using existing resources to perform tasks over a range of product categories.

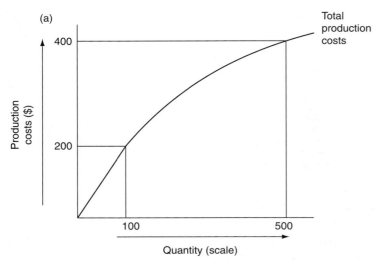

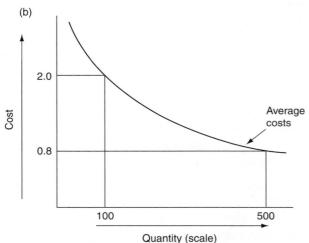

FIGURE 2.2   **Economies of Scale for Widgets**

of "flexible mass production" when a firm produces large quantities of different types of commodity (such as different dimensions of lumber) or products (different auto models) in the same factory. Scale and scope economies can be "external" and achieved through interactions among firms, but there are powerful imperatives to develop internal economies of scale and scope within firms and factories. In practice, firms and factories of all sizes seek both internal and external economies.

It's important to keep in mind that the potential for economies of scale is strictly limited by market size; if there is demand for 100 widgets, there is no point in producing 500. Among the most significant advantages of big, rich economies such as the US, Germany, Japan, the UK, and France have been their huge domestic markets. In the twentieth century, market size in itself played a major role in the growth of MNCs, the majority of which grew out of large domestic firms. MNCs from smaller market economies, such as Sweden, Finland, Switzerland, and the Netherlands, typically had to internationalize as smaller firms. From this perspective, China and India both have enormous potential today. Further, big cities have important market access advantages, for example, for high threshold activities (Chapter 1). Internal economies of scale within factories may also face technological limits at points in time.

Operating-Level Economies of Scale

Operating-level economies of scale occur when equipment, labour, materials, and other requirements can be engaged in the quantities necessary for efficient coordination of the division of labour and specialization. The location imperatives of **operating-level economies of scale** are obvious in huge plants such as refineries, ports, or assembly factories, which require a variety of capital-intensive and specialized equipment in one place. Capital equipment is a major source of internal economies at this level: specialized and/or larger machines can produce more than smaller ones, operate at higher speeds, and so on (Case Study 2.1). The same imperatives are at work in a large office, laboratory, or call centre; and even a small store must maintain minimum levels of stock and equipment.

operating-level or plant-level economies of scale Reductions in average costs of production or operation resulting from increasing the size of individual factories, stores, or other types of operation.

Labour specialization is an important internal economy of scale for large operations such as factories, hospitals, and schools. The specific benefits depend on the nature of the business. In a factory, specialization will mean increases in speed (as workers become more familiar with their tasks) and elimination of "waste time" (because specialist workers don't move from one task to another). In other types of work, specialization allows individuals to focus on the quality of their job performance, to improve their problem-solving abilities, and generally do the work for which they are best suited. Specialist workers are equipped with specialized tools and machines to further increase the scale, speed, and quality of production. Learning or experience-based effects can also increase productivity as specialist workers learn on the job to perform their tasks more efficiently. The principles of specialization are powerful forces throughout society and apply to intellectual and professional work as well as to manual labour.

Labour is generally considered a variable cost in that the quantity required varies with the level or scale of output or production (higher outputs require more labour inputs, and vice versa). Energy and materials are also **variable costs**. By contrast, **fixed costs** (e.g., investments in buildings and machinery) do not change with the scale of production. While total cost includes both variable and fixed costs, average (or per unit) cost is the ratio of total cost to total production. Labour can also be a "quasi-fixed cost" if there are important recruitment and training costs before workers become productive. Average variable costs can be reduced with increasing size by employing more specialized and productive labour. Average fixed costs can be reduced simply by producing more in the same factory with the same equipment. For example, if a factory costs $10 million to build and equip, increasing its production from 10 units a week to 1000 would mean a substantial reduction in average fixed costs. Few processes are completely focused on one good, however, and even in manufacturing there is room for economies of scope.

variable costs Costs that vary with the scale of activity.

fixed costs Costs that remain the same regardless of the scale of activity.

The geographical effect of increasing economies of scale at the factory level is to concentrate production, related work forces, and supporting infrastructure in a smaller number of more accessible places. Factories typically represent enormous **sunk costs** in physical capital that is not geographically mobile. Specialized labour forces may possibly be regarded as sunk costs, since their distinctive skills and talents, gained through experience and training, are not easily reproduced. The immobility of labour is compounded by the investments that members of the work force make in a particular community through (for instance) the purchase of homes and commitment to social infrastructure such as schools.

sunk costs Investments that once made cannot be recovered, and include investments in physical capital such as factories and machinery or operating systems that are (more or less) fixed in place and cannot be easily moved.

Firm-Level Economies of Scale

Whatever a firm's business—exploiting resources, manufacturing products, providing services—it must also undertake many distinct decision-making and related functions. In large and giant firms with multi-plant operations, these firm-level functions are largely performed by the head office for all operating units. They include purchasing "inputs" (especially materials, services, and machinery), budget planning, and long-range planning; financing; establishing operating standards; marketing (including the establishment of distribution and market networks); and research and development

(R&D). Even a small firm must be "big enough"—that is, must generate sufficient revenue—to cover the costs of these firm-level activities while providing sufficient profit to make staying in business worthwhile. The **minimum efficient scale (MES)** is the lowest level of output at which average costs are minimized, and hopefully low enough to make production viable. The survival of new firms depends on how quickly they are able to reach the MES. **Firm-level economies of scale** have played a major role in the evolution of giant MNCs.

Finding economies of scale and scope in **logistics**—both inbound (the coordination of inputs) and outbound (distribution of outputs)—requires investment in capital and labour for warehousing and reception, documentation, transport, and tracking. The greater the quantities requiring coordination, the greater the economies of scale that can be realized on a single product. In addition, scope economies can be achieved when the same resources are used for several different products. Although large-scale purchasing allows firms to drive harder bargains with suppliers, such orders give suppliers themselves opportunities to achieve economies of scale. Economies of scale and scope can be achieved in R&D, financing, and human resources as well. An auto manufacturer that invests $1 billion in developing a new engine will achieve proportionally higher returns if the engine is used not only for large-scale production of one model, but for several models. Clearly, fixed costs of that magnitude are a powerful incentive to increase production. The same is true of marketing activities: the average cost of providing these services decreases as output increases.

The primary driver behind economies of scope is **product differentiation**: the modification of one product to create one or more similar but distinct products to serve different market niches. Many companies have several brands at different price points, and some have different product types under the same corporate brand umbrella (Apple, for example, sells several different computers, phones, and iPods, each with different functions and price points). Product differentiation may mean the loss of some scale efficiencies, but it often requires only minor modifications to the basic product—a simple change in colour or some variation in exterior design—can significantly differentiate model types that are similar if not identical in terms of technology and market. A car such as the Ford Fusion or Toyota Corolla can be offered with the same engine, on the same chassis, in at least three sub-models (2-door, 4-door, hatchback), each of which can be further differentiated with various options. (Toyota's Matrix is another variation). Mass customization—which takes advantage of computer-aided design and manufacturing systems to differentiate products according to consumer demands late in the value chain—allows even greater consumer choice. In this case, any loss in economy of scale is made up for by increased economies of scope in purchasing, distribution, marketing, research, financing, customer service, and so on. Such activities are at the heart of the (now-dominant) service economy, and they represent the majority of costs and income even for manufacturing companies. The focus on scope economies is reflected in the marketing strategies of many well-known brands, including Microsoft (several distinct products in one "Office" package); Toyota (a line of products under the same banner); and Tommy Hilfiger (products from clothing to perfume marketed under the same name).

Although sunk costs may serve to anchor firms in one place, the search for economies of scale and scope also encourages firms to set up new branches or factories to access new markets and/or obtain lower-cost locations. The fact that efficiencies in different activities—R&D, component manufacture, assembly, recycling, financing, design—tend to be found in different places has been a major factor in the globalization of value chains. Moreover, the effects of economies of scale and scope on increasing the size of firms are often reinforced by economies of size or the advantages of being big in negotiating prices with suppliers, for example. In addition, the resources and experience in establishing new facilities can help firms learn in plotting further growth.

It is widely supposed that as industries grow, large corporations typically become dominant with respect to sales and employment. However, corporate consolidation can be offset by consumer demands for specialized, niche products and by the ability of entrepreneurs to establish innovative small-scale production. In the beer brewing

---

Case Study 2.2
THE MICRO-BREWERY EXPLOSION IN THE US

In the US micro-breweries are defined as small, independent manufacturers of beer with production capacities of less than six million barrels of beer per annum. Many "nano" breweries are much smaller. Beer production in the US after Prohibition in the 1930s expanded greatly and became dominated by relatively few firms that manufactured beer in large volumes in a few large breweries. As of 2012, the largest single brewery in the US, located in Colorado, produced 22 million barrels of beer and is owned by a foreign-owned MNC. The evolution of the industry towards corporate domination is an excellent example of the impact of internal economies of scale at the level of the plant (Case Study 2.1) and firm. Yet since around 1980 there has been an explosion of micro-breweries across the US so that there were 2,403 in 2012 (producing around 13 million barrels). Average costs and prices charged are higher for these smaller breweries. While deregulation has been a factor underlying this growth, consumer preferences for more varied beers and possibly for local products and the recognition of market potentials by entrepreneurs have been key. Similar trends have occurred in Canada and elsewhere, including the "real ale" movement in the UK.

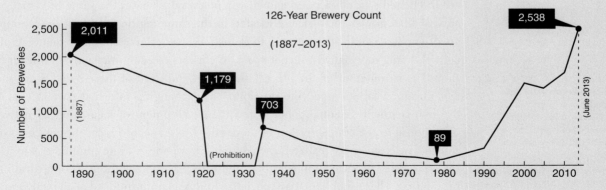

The 126-year Micro-Brewery Count

Source: The Brewers Association

industry, for example, while a classic case of increased corporate and operational consolidation over the mass production of a few related products, there has been a spectacular flowering of micro-breweries in recent years (Case Study 2.2).

External Economies

Firms that seek economies of scale look at every segment of the value chain and decide whether to produce the good (or perform the service) themselves or to purchase it from an external supplier. These "make or buy" decisions depend on whether it is more efficient to produce the service in-house or obtain it through the market. The firm has to decide not only whether it has the capacity to provide the service, but also—equally important—whether its internal economies of scale can match the cost and quality available from firms serving the broader market. In so doing, it also has to take into consideration its internal transaction costs and uncertainties in comparison to the costs of external search, collaboration, contracting, and monitoring. If firms choose to achieve economies of scale externally (and all do to some extent), they become engaged in the broader location dynamics of value chains, whether locally or other places, in terms of the mix of labour, firms and capital, public infrastructure, and governance.

External economies—the efficiencies (reductions in average cost) and some related benefits that a firm may be able to realize by purchasing inputs from external suppliers—may be found in established agglomerations, newly created economic spaces, or a combination of the two. In a sense, providing access to **external economies of scale** is the reason open and relational markets exist: to give firms (and consumers) access to the benefits of better technologies, less costly and/or more skilled labour, and economies of scale developed by other firms.

Agglomeration Economies of Scale

Agglomeration economies of scale are the benefits that come with the concentration of activities in a particular geographic area. They are generally classified as either **localization** or **urbanization economies of scale**. Localization economies of scale are the benefits or efficiencies gained when several firms in the same industry, or in closely related lines of work, are located in the same region. Localization economies are often associated with reference to industrial districts that comprise manufacturing activities and, especially in recent years, to clusters of economic activities in services as well as manufacturing. In general, the most significant of these benefits are access to a common labour pool; the possibility of exchanging information and expertise through frequent personal contact; and the development of local supplier networks in response to the concentration of demand. With respect to local supplier networks, external economies are realized when demand from one or more companies stimulates the emergence of local suppliers that might not be viable without a concentration of production. These suppliers may specialize in inputs not produced at all by the buying companies, or they may complement the purchasing firms' production by providing additional inputs during boom periods. Specialist suppliers of machinery and equipment may develop in the same way. Competition among localized concentrations of suppliers offers the purchasing firms additional flexibility, while competition among purchasing firms offers the same flexibility to suppliers. Access to versatile suppliers who can readily adapt to meet new requirements and challenges may encourage a firm to pursue economies of scope.

Access to a concentration of appropriately skilled labour reduces the costs associated with recruitment, training, and turnover, since firms can advertise positions by word of mouth or public notices; new recruits may already know something about the work, whether through the local schools or family connections; and employees who leave can be readily replaced. A localized pool of skilled workers who circulate from one firm to another in the same industry will require little or no additional training, and may even hone their skills along the way, to the benefit of their employers.

Central to all localization economies is the sharing of information on matters such as market development, technological change, government policy, or labour issues. Some of this information (for example, legal information) is bought ("traded") for a fee. Some, however, is exchanged without direct charge or price. This kind of informal information exchange is one of the "untraded interdependencies" that come with location in an agglomeration of firms in related activities. The difference between untraded and traded interdependencies can be blurred; for example, the cost of negotiations involved in setting up a deal may be incorporated in the eventual price. The basic point, however, is that the localization of market activities facilitates information exchanges of all kinds, helping to develop pools of know-how and opportunities for specialist activities. The opportunity for face-to-face exchanges of information is especially important when the issues at stake involve personal judgment, uncertainty, sensitivities, security, or long-term planning. This kind of information exchange—highly nuanced, interactive, and often sensitive—is not easily done through other channels. Decisions to invest in a foreign country, research and develop a new product, issue shares for public tender, or address charges of malpractice typically involve decision-makers to seek out expert opinions and advice that cannot be easily "googled."

Information exchange is facilitated by chambers of commerce and development agencies, as well as industry associations, conventions, and seminars. Industry associations provide collective inputs that individual members cannot or will not supply themselves: marketing services, labour training, research and development, and political lobbying. They are especially helpful to small firms and are sometimes partially funded from government sources. They can also help to "brand" a place by promoting its reputation as a centre for a particular type of product (as the Wine Council of Ontario and the British Columbia Wine Institute have done in recent decades for wines from Ontario and BC). Chambers of commerce pursue broader representational mandates and provide information services for local business as a whole.

Public institutions also contribute to localization economies. Universities and government research agencies provide long-term research and problem-solving expertise as well as highly trained labour for local industry. Specialized government departments (small business, agriculture, forestry) provide regulations, direction, and representation, while public and local development agencies offer information, subsidies, and planning to support new investments.

In addition, urbanization economies are the advantages that come with location in a large city that provides roads, sewers, electricity, telecommunications, health and education facilities, and general (wide-ranging) labour skills. Even if a smaller town offered some of the same types of infrastructure, amenities, and support services, a larger city would have the advantage of greater quantity, quality, and diversity. Indeed cities are themselves major centres of consumption for a vast array of goods and services and are especially attractive to those activities that seek close access to markets while providing opportunities for entrepreneurs seeking to serve new market niches. Moreover, cities provide key roles in the movement and exchange of goods and services, both locally and with respect to imports and exports, such as in relation to distribution, communication, travel, storing, warehousing, and logistical and insurance operations that in turn stimulate and maintain a diversity of transport and communication networks. More generally, city growth provides opportunities for in-migrants with new and locally limited skills, while urban growth implies an expanding population that adds to local demands in the public and private sector, and new local sources of ideas for investment. Note also that public institutions provide many key urban services. Therefore public policy plays an important part in determining the nature of the urbanization economies available in any given city. Since many of these services are financed indirectly by taxes, taxpayers need to agree that they will generate sufficient benefits to the community. In advanced countries, the concentration of high-tech activities in cities is closely related to the access they offer to highly diversified urbanization economies of scale, for example, large universities, specialized technology institutes, sophisticated amenities, and specialized education programs.

In practice, localization and urbanization economies overlap. Universities, for instance, typically provide expertise to local industries in addition to educating individuals whose knowledge will contribute to overall amenity levels and, perhaps, diversification of the local economy. More generally, the idea of related variety indicates the emergence of new activities in a city that draws upon or "recombines" inputs (such as skills and know-how) derived in part from existing specialties. The range of external economies provided by institutions such as universities, technical colleges, chambers of commerce, local development agencies, and NGOs is sometimes referred to as **institutional thickness**. The supposition (not well tested) is that institutional thickness improves a place's ability to diversify and respond to economic crises.

Localization and urbanization economies have generally been seen as immobile in the sense that their benefits are restricted to firms located within the same agglomeration or region, either because they depend on highly proximate, personal interactions and/or because there are membership rules for local co-operatives and industry associations or conventions. However, it is possible for **immobile economies** to become more mobile—or at least accessible to firms located elsewhere—as a result of improvements in communication technology or relaxation of local membership requirements. An example mentioned by Robinson (1959) is the Liverpool cotton exchange, which was made possible by the existence of hundreds of local textile firms but which, once established, was accessed by firms from India. Another example can be seen in the efforts of a leading forestry association in BC to improve access to the Japanese market on behalf of its member firms; by improving the forest industry's knowledge of the Japanese market and encouraging it to accept common standards, those efforts

institutional thickness The range of external economies provided to local firms by institutions (private and public) such as industrial associations or government planning departments.

immobile economies External economies that are available only when the firms concerned are in close proximity to one another.

indirectly benefited some American firms as well. More generally, increasingly sophisticated forms of communication are contributing to more mobile external economies by enabling virtual forms of direct contact and facilitating the secure transmission of confidential documents and blue-prints.

Local Multipliers

Multipliers refer to how expansion of one activity stimulates growth in other activities, directly and indirectly, and are usually measured in employment or income terms; local multipliers measure these effects in a particular region (Case Study 2.3). Thus an initial increase in demand for a product, commodity, or service stimulates further growth in connected activities throughout the value chain, and the multiplier measures the *ratio of initial to total growth*. In general, links, networks, or interdependencies among economic functions are the channels through which an expanding economy grows, and local multipliers are higher when local economies are highly linked. Local multiplier effects are dampened by non-local linkages that "leak" income (and jobs) from local circulation through, for example, imports, savings, and non-local taxes. On the other hand, new multiplier effects are created by exports, government expenditures, and investment in general.

> **multiplier** How an initial spurt of growth in one activity generates growth in other activities.

Local income and employment-multiplier effects are especially strong in agglomerations where many small firms interact with one another. Such agglomerations offer strong external economies of scale and numerous localized supplier and consumer linkages within value chains. The addition of new suppliers and the creation of related market niches further reinforce localization and urbanization economies of scale and scope. Other less measurable but still important local multipliers involve the transfer of skills, ideas, and expertise from one firm or sector to others through market transactions and untraded interdependencies, such as knowledge gained in conversations over business matters between managers and their suppliers or consumers, or even among rivals. These quality multipliers reflect a region's ability to absorb and build on best practices in organizational routines and technology. In contrast, local income and employment multipliers are low in regions dominated by foreign-owned **branch plants**, which tend to depend on the head office for services such as marketing and R&D, instead of buying them locally. Similarly, branch plants that direct their purchases to affiliated plants elsewhere reduce local multiplier effects. Internal economies of scale are undoubtedly high, but external economies of scale and scope are limited. When large firms use preferred non-local suppliers rather than local sources, the effect is similar to reducing local multipliers. Walmart and other retail giants, along with the fast-food and coffee chains such as Starbucks, typically concentrate their supply chains on imported sources that they can purchase more cheaply than local sources. Some firms, of course, celebrate their preference to buy locally and enhance local multipliers!

> **branch plant** A plant or factory that is owned and controlled by a company whose head office is elsewhere.

Short-term multiplier analyses assume no changes in the structure of a local urban economy. However, in expanding urban economies and cities, the development of local linkages, scale economies, and multiplier effects *reinforce and provide feedback* on one another along the lines suggested by the model of circular and cumulative causation (Figure 2.1).

Case Study 2.3
LOCAL MULTIPLIERS

Employment and income *multipliers* address how the injection of new growth in a particular activity reverberates around the economy through value chain linkages to create additional growth. Declines in growth in particular activities occur as negative multipliers. The multiplier measures the ratio of initial or direct impacts in the value chain to total impacts created throughout the economy. The impacts are usually measured in terms of employment or income. A *local* multiplier (k) for a specific region is measured by:

$$K = \frac{\text{total economic impacts within the region}}{\text{initial economic impact within the region}}$$

One category of local multipliers is based on the impacts of exports on local economies (export-based multipliers). For example, if an increase in auto export sales in a city *directly* generates 100 new manufacturing jobs in car assembly and *indirectly* creates another 50 jobs then k = 150/100 = 1.5. These indirect jobs occur principally among local suppliers of goods and services that in the case of auto firms may include purchases of parts, steel, equipment, and marketing services. These purchases from suppliers help define the value chain of producing autos. In addition, increases in the size of the labour force (and population), along with incomes, in turn lead to more consumer ("household") expenditures on local goods and services that also employ more people. The impacts of such consumer expenditures are known as *induced* multiplier effects. Multipliers help estimate impacts of future exports. Using the above numbers, if new auto exports require an increase of 1000 workers then a total of 1500 total workers can be predicted. Clearly, such predictions assume underlying technologies and their employee requirements remain the same.

In general, the size of local multipliers depends on the size of the original injection of growth and the extent to which value chain purchases stay *within* the region. In multiplier analysis, non-local purchases and other losses to local income (and employment) streams are known as *leakages*. Thus if suppliers of equipment and components to the auto firm are located outside the city then the additional jobs created by the firm's purchases are located—"leaked out"—elsewhere, and the (local) multiplier effects are reduced. In general, payments that are made for imported goods and services, savings that are sent to distant relatives, profits sent to distant headquarters, and taxes to non-local governments represent income (and employment) "leakages" out of the region and reduce the size of local multipliers. On the other hand, export-based multipliers are enhanced if exports are used to promote further local value-adding activity—extending the local value chain towards final consumers.

In this virtuous model, given some "initial advantage" in a particular place for a specialized activity, investments in this activity generate sales from exports and a paid work force generates demands for goods and services, including government services through taxes. These activities then stimulate further investments in related activities; rivals that seek to expand markets including via product differentiation; ancillary or supply industries that supply components, professional services, or machinery; and infrastructure in the form of economic overhead capital (EOC) such as transportation systems, and social overhead capital (SOC), such as education, health, and housing systems. Governments need to evolve with this growth to provide and maintain services, including with respect to SOC and EOC, and all types of investments will require construction activity. All economic activities in the public and private sectors generate employment that support families and add to demands for local goods and services including increasingly high threshold goods (Chapter 1). Over time, the search for economies of scale and scope, and their associated multiplier effects and demands for public services, add to and are reinforced by increases in population and employment; and innovation activities likely focused on the city's specialized activities may lead to diversification as a bigger, more educated population embraces new ideas. This is *virtuous cumulative causation*.

But growth is not guaranteed in market economies, virtuous cumulative causation can have limits, and even large cities are vulnerable, not only to problems of size, but to decline as a result of competition from elsewhere, changing consumer demands, and technological changes that favour alternative locations. The widespread occurrence of deindustrialized cities throughout the manufacturing belts of North America and Europe illustrate the power of downward spiralling cumulative causation and locked-in path-dependent behaviour. Internal and external diseconomies of scale and reverse multipliers help explain these processes.

Diseconomies of Scale and Scope

Internal diseconomies of scale and scope refer to increases in average costs that occur with the increasing size of factories or firms.

Internal diseconomies of scale occur when increases in production cause average costs to increase either for individual operations (**operating-** or **plant-level diseconomies**) or for the business as a whole (firm-level diseconomies). For example, a plant may experience rising average costs as a result of overspecialization. If labour skills are too narrowly defined, opportunities for workers to solve problems themselves may be limited. Then expensive specialist problem-solvers (maintenance workers, engineers) may have to be called in; meanwhile, regular employees may become bored and alienated, with the result that absenteeism, turnover, and mistakes increase, leading to higher costs of control and supervision.

Firm- or organization-level diseconomies of scale and scope can occur in various ways, including by bureaucratization. As the division of labour becomes more complex, more managerial levels and positions are established to control it. For businesses as a whole, diseconomies of scale can occur when these decision-making structures become too unwieldy. Excessive layers of management, communication breakdowns, group thinking, collusion, and slow decision-making processes can all increase a firm's average costs. Lack of competition and the difficulties of monitoring performance compound these problems. Such issues can be exacerbated by a company's geographical reach. For example, head offices remote from operating units may not understand the local conditions facing dispersed operating units. Diseconomies of scope result

operating- or **plant-level economies (diseconomies) of scale** Reductions (increases) in average costs of production resulting from increasing levels of output within a plant or factory.

when companies become too diversified and the advantages of managing several products and product-lines are outweighed by the disadvantages. Firm diseconomies can also arise from overly fast growth. For example, Toyota's 2010 recall crisis for accelerator and other problems has been described as a symptom of overly rapid global growth, compounded by management's desire to retain control in Japan. Especially in North America, Toyota was criticized for failing to provide local management with the authority to establish proper quality control procedures and respond to complaints.

All firms suffer from such diseconomies to some degree and must guard against them; this is one reason why firms frequently reorganize. When diseconomies become overwhelming, however, even the largest companies can be threatened by failure. The local impact of such failure can be particularly severe if firms were the focus for localization economies. These negative consequences can be interpreted in terms of cumulative causation and path dependency. Thus the same localization economies that drew certain firms together can become diseconomies if the path-dependent behaviour of firms becomes "locked-in" to outdated, dysfunctional attitudes and ways of doing things. The situation is worsened if labour relations and local political attitudes are similarly locked-in. For example, the US auto industry, centred in Detroit, built a large agglomeration through the American midwest and southern Ontario. It was once highly competitive, but by the 1950s, General Motors (GM) alone had more than 50 per cent of industry sales in North America. The arrival of Japanese competition in the 1970s made it clear that the US makers, not only GM but also Ford and Chrysler, had become locked into inappropriate managerial practices, supplier relations, R&D planning (e.g., alternative vehicles), distribution systems, labour agreements, and compliant local and state governments. Finally, after decades of seeing its market share decline, GM filed for bankruptcy in 2009 (while Ford and Chrysler experienced extensive rationalization that involved plant closures and workforce reductions). Many of its suppliers and distributors were forced into bankruptcy as a consequence, thousands of employees lost their jobs, and many communities were devastated. GM survived with the help of massive government subsidies from the US and Canada. But such subsidies invite moral hazard, tempting management to continue with past practices, while undermining the viability of more efficient rivals.

Indeed, the closing of large-scale operations can set in motion processes of reverse cumulative causation, locked-in path dependency, and downward multiplier effects as the investments in infrastructure, supporting services, and amenities that create robust urbanization economies decline. The collapse of Detroit, which in 2010 had an unemployment rate of almost 30 per cent and in 2013 declared bankruptcy, is an illustration. As plants are closed, workers laid off, and their incomes reduced, demand for local supplies and consumer goods is also reduced. In tandem, the providers of localized external economies, such as industry associations, become less viable and they too have to close. Moreover, declining incomes and population levels imply a lower tax base and the ability of individuals to pay for even basic public services decreases; in 2014, many citizens of Detroit, for example, still could not pay water taxes and lost water supply to their homes. Meanwhile city governments are unable to invest in **public goods** such as roads, education, and health services and that poses serious problems for people living in poverty.

public goods Goods and services that are available to everyone in a society either at no direct charge or at prices below the actual cost of supplying them (Chapter 6).

Urbanization economies can also become diseconomies when cities grow too large or too rapidly or when they experience decline. Wealthy and vibrant cities, such as London and New York, experience significant **negative externalities** in the forms of overcrowding, traffic congestion, noise and air pollution, housing price inflation, and, in some cases, crime. In these cities, both urban economies and diseconomies of scale can be strong. In cities that are trapped in a vicious circle of decline, businesses and residents alike are repelled by derelict land, run-down housing, declining amenities, and business closures. Services, job opportunities, and investments disappear. Once this cycle has been set in motion, it is difficult to stop, as localized and urban diseconomies of scale interact negatively together. Distressed cities, nevertheless, have local assets available to rejuvenate: existing populations, amenities, and infrastructure. Sometimes urban rejuvenation—attempts to spark new positive cumulative growth cycles—can create new activities that have roots in the old economy. For example, Pittsburgh has developed expertise in environmental technologies and services to solve industrial waste problems of the kind it produced for over 100 years, while Sheffield's metal working expertise is being applied in aerospace. These cases illustrate path interdependence in which a new expanding path is created out of an old declining one and in which local universities are playing important research and training roles. If urban economic rejuvenation towards new specialized and resilient activities is difficult, globalization poses both opportunities and threats.

negative externalities Costs that fall on third parties or the environment that were not taken into account in the decisions made by market actors.

Globalization

In present times, the evolving spatial division of labour is usually framed by reference to "globalization" and the related idea of **free trade**. These terms are controversial, even ambiguous concepts. Indeed, both the proponents and critics of globalization and free trade emphasize the role played by the expansion of markets, the intensification of the division of labour, and the search for scale economies on a global scale. As a term, globalization has only been in popular usage since around 1980 (while free trade has been debated since the mid-nineteenth century). However, the *meaning* of globalization as an idea or political project is contested. Meetings of WTO and G8—now G20—leaders about trade, investment, and related matters generate so much opposition that they are held only in secure, often remote locations. At the heart of these protests are fears of a neoliberal agenda that seeks to privilege market forces and empower business by deregulating and privatizing the public sector, while failing to deal with social and environmental problems. Yet globalization need not be conflated with **neoliberalism**.

free trade Market exchanges without political interference in the form of tariff and non-tariff barriers. The market exchanges refer to goods, services, and investment.

In more neutral terms, globalization refers to social, political, and economic processes that are deepening international integration of all kinds. The (extreme) antithesis of a globalized world is one comprising self-subsistence economies that have no contacts or knowledge of each other. Clearly, international integration is not new and many scholars see contemporary globalization as an evolution of trends that have developed over the past 1000 years. Long-distance, even world-spanning trade and markets have long existed before capitalism (Case Study 2.4). The European Voyages of Discovery from the fifteenth century onwards opened up trade and encouraged colonialization.

neoliberalism The school of thought that emphasizes the power of markets to meet society's needs and argues for minimal government interference in the economy; neoliberals often advocate deregulation and privatization of the services provided by the public sector.

Case Study 2.4
THE SILK ROAD

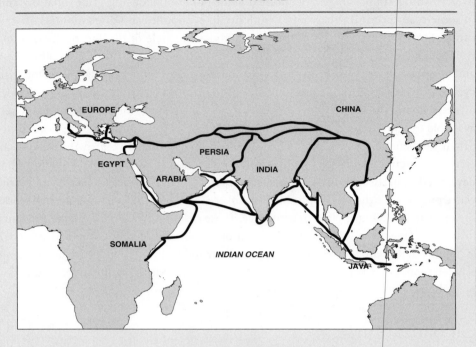

The Silk Road was not an actual single road but several connected sea-based, as well as land-based, trade routes that linked various parts of Asia (China, Java, and India) through western Asia (Persia and Arabia) with the Mediterranean region. The principal northern and southern land routes were approximately 4000 km in length and were already in operation 3000 years ago—some 1500 years before Marco Polo made his legendary journeys. Silk made in China was an important trade item, along with many other products, including satins, perfumes, spices, medicines, and jewels. In addition to merchants, the routes carried pilgrims, missionaries, and armies, and were conduits for information and ideas (about technologies as well as products), as well as disease. In short, the Silk Road (with its related sea-routes) was both a cause and an effect of the rise of Asian and Mediterranean civilizations. Typically, trade along the Silk Road was organized as a whole series of markets in oasis towns involving local consumers and traders moving goods along the route.

At the height of the Silk Road, the world's leading civilizations were in Asia and the Mediterranean. Europe was on the geographic margin. Marco Polo's travels in the late thirteenth century, and his trade connections with Asia, signalled the rise of Europe as a centre of production and trade. With the beginning of the Industrial Revolution, Europe quickly became the world's economic centre (see Chapter 3).

Big trading companies—such as Hudson's Bay and the East India Trading Company that presaged the modern MNC—were international in scope and, in the case of the former, helped shaped the contours of the future Canada. The nineteenth century witnessed massive international migrations, the beginnings of the free trade debate, and the diffusion of industrialization. More recently, the WTO (formed 1995) is a continuation of GATT (formed 1947) while the dominant role of contemporary MNCs was prevalent by the 1960s in resources and manufacturing.

Thus from an economic perspective, globalization is usually related to the intensification of trade, markets, the division of labour, and scale economies on a global scale with important implications for government autonomy. However, globalization can also be interpreted in cultural terms rooted in the intensification of information communication. Further, it is important to recognize that a significant feature of contemporary times has been the globalization of environmental problems, as in global climate change, from the more localized environmental problems associated with early industrialization. The increased scale of environmental problems is a direct consequence of the increased scale of industrial activity, urbanization, and resource exploitation.

Historically, and now, internationalization is based on technological innovations in transportation and communications. In this regard, globalization is distinguished by the development of information and communication technologies (ICT) that have vastly reduced—and for many practical purposes eliminated—the costs and time of exchanging information. These same technologies have exerted massive effects on the location dynamics of production. Globalization is further distinguished by the dominance of market systems achieved by the integration of the former, centrally planned economies of the Soviet Union, eastern Europe, and China. With globalization, market failures have also become more pronounced, especially in relation to global income inequalities and environmental problems, themselves subject to powerful forces of cumulative causation.

Globalization and Local Development

Proponents of the "end of geography" thesis claim that globalization has created a borderless world in which national governments have lost their power to influence flows of goods, services, people, and investment. Simultaneously, MNCs in particular have been empowered by the high mobility of their investments and the ability to locate wherever costs are lowest. There are three basic criticisms of this thesis. First, even as governments' roles have changed they remain influential: national governments are increasingly involved in the development of rules, regulations, and policies that are international in scope and designed to govern markets in "groupings of countries"—such as the European Union (EU)—and globally through agencies such as the WTO. National governments also have roles to play in addressing global market failures of the type identified by the Kyoto Protocol on climate change. Meanwhile, local governments must now deal not only with national regulations, but also with an array of international regulations that both constrain and stimulate economic activity. Second, location conditions are complex, and often highly nuanced. Location choices must provide access to markets and, in many cases, close access. Third, location choices involve capital investments with sunk and fixed costs, and are typically made with long-term commitments in mind.

Nevertheless, globalization has important implications for local development, not least because increased flows of trade, investment, information, and labour across political boundaries would "open up" regional development to external market forces. The data presented in Table 2.3 provide some tentative support for this claim, especially with respect to trade. As you can see in Table 2.3, the export flow to GDP ratio

TABLE 2.3   Global GDP, Inward FDI Flows and Merchandise Exports, Selected Years 1970–2012 ($m)

	1970	1980	1990	2000	2005	2012
GDP	3,244,755	11,903,792	22,274,225	32,370,841	45,849,262	71,435,240
FDI (% GDP)	13,346 (0.41)	54,069 (0.45)	207,362 (0.93)	1,413,169 (4.37)	989,618 (2.16)	1,350,926 (1.89)
Exports (% GDP)	318,019 (9.80)	2,049,407 (17.22)	3,459,585 (15.53)	6,448,969 (19.92)	10,499,521 (22.90)	18,402,184 (25.76)

Source: UNCTAD Handbook of Statistics Online: http://www.unctad.org (accessed 24 Feb. 2012).

grew consistently from 1970 to 2012 in the years selected, apart from 1980. In 1970, for example, exports accounted for 9.8 per cent of global GDP and by 2012 for 25.76 per cent of global GDP. The relative importance of inward foreign direct investment FDI (the flows of FDI received by host countries) compared to global GDP was also noticeably greater in 2000 than in 1970 or 1980 but then declined in 2005 and 2012. However, these simple statistics should be approached with caution: for example, aggregate global trends in exports cannot be assumed to apply for all individual nations or regions, or for all products or industries. To a considerable degree, trade (and investment) has also regionalized within continental regions (North America; the EU; Asia-Pacific). Foreign-owned branch plants can also be financed by retained earnings or from other domestic sources. Yearly data would give a better idea of trends.

The New (and Even Newer) International Division of Labour (NIDL)

Before the 1970s, the development of markets, value chains, and the spatial division of labour, especially in relation to manufactured goods and services, was overwhelmingly concentrated in relatively few countries: specifically, in the rich market economies of the members of the OECD (Organization for Economic Co-operation and Development). To an important degree, value chains were organized along national lines—for example, most rich countries had significant auto industry agglomerations (assemblers and suppliers) with important shares of their domestic markets—although the OECD markets were closely connected to one another both through trade in goods and through direct foreign investment. At the same time, the OECD countries sought access to developing countries, primarily for their resources, and to a much lesser degree as markets for their own products. In other words, the OECD countries constituted the global core, often summarized as "the north," while the developing countries constituted the global periphery, often summarized as "the south."

Beginning in the 1960s, MNCs established branch plants in selected developing countries, especially several newly industrialized countries (NICs) such as Hong Kong, Taiwan, and Singapore, as export platforms for goods that were relatively easy to assemble from components manufactured in the OECD countries. The NICs offered low-cost labour that was readily trained for repetitive factory work, along with basic infrastructure and pro-business governments. This **new international division of labour** (NIDL) was based on the idea that the OECD countries would retain the well-paid, highly skilled operations—R&D, design, services, innovative manufacturing—but lose many of

new international division of labour (NIDL) A term initially used in the 1970s to refer to the establishment of export-based secondary manufacturing in developing countries. The NIDL has evolved considerably since then.

their less-skilled assembly jobs to the NICs. Meanwhile, economic development in the NICs would be limited to low-wage operations controlled by foreign MNCs.

In reality, after two decades of globalization, the "newer" international division of labour has turned out to be more complex. There has been a remarkable expansion of manufacturing outside the OECD: far from being limited to assembly, operations in east and south Asia in particular now include not only component manufacturing but R&D, design, and innovative activity. The service sector has also internationalized. In the twenty-first century, value chains are organized by increasingly complex divisions of labour that link economic activities not only within regions and nations, but within and between trading blocs around the globe. The establishment of a few routine electronics factories in the Asian NICs in the 1960s has evolved into a sophisticated intra-Asian division of labour that produces a wide variety of final consumer products and components. Foreign, especially Japanese, MNCs remain influential but so are Taiwanese and South Korean giants, while the Hsinchu Science Park (Taiwan) and Bangalore (India) have become globally significant R&D clusters serving the industry. In general, Asian industrialization, led by China's emergence as the world's leading exporter of manufactured goods, has contributed towards raising the per capita incomes and the growing significance of Asian markets.

Free Trade and Neoliberalism

International trade is conducted through markets that cross national borders and free-trade policies facilitate the spatial division of labour among countries by removing political or protectionist barriers to trade in the form of **tariff** and **non-tariff barriers**.

Tariff barriers impose taxes on imported goods, such as a percentage "*ad valorem*" tax on the value of goods, while non-tariff barriers refer to deliberately bureaucratic practices, or impediments to trade expressed as regulations that ostensibly refer to safety, health, and security concerns. Nations seek to discourage imports in order to protect existing domestic jobs, ways of life, and give new "infant" industries time to grow to the stage where they will be able to compete with established foreign rivals. Nations also impose tariffs as a source of revenue.

Basically, proponents of free trade have argued that political barriers to trade should be eliminated in order to allow market forces to work efficiently. In this view, unimpeded access to global markets is necessary to allow full realization of the benefits associated with labour specialization and economies of scale. Historically, the **comparative advantage model** explains why a nation (or region) should not rely solely on domestic (or local) production, but should specialize in the activities in which it is most competitive and obtain other goods through trade with the places that are most competitive in them (Case Study 2.5). Although the original nineteenth-century model referred only to the trade in goods, it has been elaborated and extended to support free trade in investment (especially foreign investment) and services as recommended by the GATT and, since 1995, the WTO.

The rationale for free trade is not hard to appreciate. After all, the benefits of labour specialization and economies of scale cannot be realized without access to markets. If specialization is essential for increases in productivity then trade is essential for specialization. In this regard, the debate over free trade is not about whether trade

tariffs Taxes imposed on the goods and services imported into a country.

non-tariff barriers Restrictions on imports using means other than taxation (e.g., quotas, punitive inspection requirements).

comparative advantage model Proposes that a region or nation (or firm or individual, for that matter) should specialize in the goods and services that it can produce more efficiently than others—specifically, with lower opportunity costs.

Case Study 2.5

INTERNATIONAL TRADE AND COMPARATIVE ADVANTAGE:
FREE TRADE OR PROTECTIONISM?

The theory of comparative advantage proposes that trade itself is a stimulus to specialization (and growth), creating mutual benefits for all trading partners. This theory
was contrary to the idea of trade as a zero-sum game in which one nation's gain was
another's loss (often associated with the idea of mercantilism).

To demonstrate mutual gains from trade, the basic comparative advantage model
outlines a simplified (hypothetical) global economy consisting of two regions (A and
B) and assumes that each region produces two classes of products (resources and
manufactured goods) with the same amount of labour. Within each region, labour
is "perfectly mobile" between the two economic sectors, but between regions it is
immobile. The model further assumes perfectly competitive open markets. In the
following scenario, each region has an absolute advantage in one sector. In the pre-
trade scenario (1a) each region splits its labour force in half between the two sectors
to produce certain outputs (e.g., Region A produces 5000 units of resources and
10,000 units of manufactured goods).

Scenario 1a (Pre-trade)			Scenario 1b (Post-trade)		
Each region has an absolute advantage.			Each region concentrates on its absolute advantage.		
	Region A	**Region B**		**Region A**	**Region B**
Resources	5,000	2,000	Resources	10,000	0
Manufacturing	10,000	20,000	Manufacturing	0	40,000

In scenario 1a, total production in region A and B combined is 7,000 units of resources and 30,000 units of manufacturing. But Region A is (absolutely) better at producing resources than Region B, and the latter is better at manufacturing. In scenario
1b, each region has shifted its labour force into its "best" sector—the one in which
it has an absolute advantage. The result is that the total output of resources and

and specialization should exist or not, but over appropriate policies towards trade and
industrial specialisms (and competitive advantage).

Free trade policies have always been controversial; they have often been imposed by
powerful countries on the weak, and **protectionist** sentiments are prevalent. Moreover,
criticisms have been inflamed by the conflation of free trade and globalization with the
rise of neoliberalism. In general, neoliberalism promotes globalization by advocating the
empowerment of market forces and the reduction of political and social regulations that
affect trade and market transactions to the bare minimum, as well as privatization of public-sector activities. Encouraged by this advice, the WTO has expanded the scope of free
trade discussions to encompass services, including services traditionally provided locally
by the public sector, such as education, health care, and utilities (water, electricity, etc.).

protectionism Protecting
domestic industries by imposing
penalties and restrictions on
market exchanges between
countries.

manufacturing is increased, suggesting that both regions would be better off if they specialized and traded.

Specialization involves **opportunity costs**: that is, the potential benefits that are lost when one alternative is chosen and the other is rejected. By specializing in resources, for example, Region A foregoes the opportunity of manufacturing. Indeed, region A "loses" 4 units of manufacturing for every unit of resources, while Region B "loses" 10 units of manufacturing for every unit of resources. A key claim of the comparative advantage model is that trade creates mutual benefits so long as the domestic ratios that define opportunity costs are different, even if a region has no absolute advantage in any product.

Consider the following scenario:

Scenario 2a (Pre-trade)			Scenario 2b (Post-trade)		
Each region has an absolute advantage.			Each region concentrates on its absolute advantage.		
	Region A	**Region B**		**Region A**	**Region B**
Resources	5,000	3,000	Resources		
Manufacturing	20,000	9,000	Manufacturing		

Can specialization and trade improve overall outputs in this case? Region A has absolute advantages in both resources and manufactured goods, producing more efficiently than Region B. However, the opportunity costs of specializing in the two sectors are different. Thus Region B gives up "only" three units of manufacturing for every unit of resources, while Region A gives up four units of manufacturing. In this sense, resources are cheaper in Region B than in Region A: hence Region B has the comparative advantage. As a result, it is possible that specialization and trade would increase production in both sectors. Assume that Region A allocates 3/4 of its labour to manufacturing and 1/4 to resources, while Region B allocates all its labour to resources. Calculate the total units produced for scenario 2b to demonstrate the overall increase in production and the benefit of specialization and trade. In this model, prices are set through the (perfectly) competitive interaction of demand and supply.

opportunity costs The potential benefits that are lost when one alternative is chosen and the other is rejected.

Neoliberalism promotes the "economic freedom" of individuals (whether consumers or corporations) to invest and trade, and therefore argues against the imposition of political and social barriers to international trade based on broader interests, however they may be defined. In this view, competitive markets effectively "self-regulate" and produce the most efficient outcomes, a rationale that also underlay the reductions in regulations governing the borrowing and lending activities by banks, especially in the US and UK, at the turn of the twenty-first century. There are also attempts to push a free trade agenda on the poorest, weakest countries that lack competitive firms, skilled labour, or much infrastructure.

The institutionalist (sometimes called "nationalist") critics of free trade, it should be emphasized, do not seek economic self-sufficiency; they are well aware that economic growth depends on markets, specialization, and economies of scale. Where they differ from the neoliberal advocates of free trade is in their belief that regulations and governments are necessary to offset market failures and that protectionism can have a role to play in economic development. These free-trade critics reject the comparative advantage model that takes endowments such as natural resources and labour skills as "givens" from which opportunities for specialization and trade "automatically" flow. Rather, they emphasize the theory of **competitive advantage** that proposes businesses and governments should actively shape opportunities for specialization and trade. More specific criticisms of free trade are

competitive advantage Proposes that businesses and governments should actively seek out and shape opportunities for specialization and trade.

- Free trade is never actually free. Like all markets, free-trade (international) markets are subject to transaction costs, as well as social and political regulation. Indeed, much trade occurs within MNCs and is influenced by corporate policies.
- Markets are regulated to meet social and political goals—for example, to ensure health and safety in the supply of goods and services. Trade protectionism also has social and political goals: to prevent unemployment and related social problems, or to allow small-scale "infant" industries to grow to an efficient size before they are being exposed to international rivals.
- All major economies have recognized the necessity of protecting "infant" industries as a basis for economic growth. There are no important exceptions. Governments also protect their industries for strategic and security reasons, such as activities that develop advanced technologies for military as well as commercial purposes.
- All major economies, including all those that support free trade, continue to protect or heavily subsidize industries for social and political reasons. The massive subsidies provided to GM and Chrysler in 2009 by the US and Canada are cases in point that could be multiplied.
- Trade conflicts are widespread, even between "free-trade" partners. The mechanisms that have been created to address such conflicts represent another form of regulation, typically designed to address issues of fairness.
- Free-trade arrangements are a matter of policy and the terms of reference are often unfair. For example, in the 2008 round of WTO talks, the strong market economies were willing to open their markets to the agricultural produce of poor countries, but only if the latter opened their countries to foreign investment. At the same time, the rich countries refused to drop their subsidies to domestic farmers, limiting the ability of poor countries to compete.
- Finally, market failures with respect to income equality and environmental protection are significant. A free-trade agenda would likely cement narrow specialization in poor countries, increasing their vulnerability. Meanwhile, a global consensus on tackling environmental problems has proven elusive, and sustainable development has yet to be substantively integrated into trade agreements. Naomi Klein's *This Changes Everything: Capitalism versus the Climate* argues that continuing support for free trade is threatening the environment.

Further, significantly, the reality of international trade is that it is organized by value chains that are dominated by MNCs. Indeed, most international trade is conducted as administered trade within MNCs while many other international transactions involve relational markets and subcontracting arrangements. In other words, international trade does not largely conform to the model of open competition.

Global Value Chains as the Search for Scale Economies

From a value-chain perspective, each business needs to make two basic choices between (a) making a product itself or buying from another firm and (b) making or buying this product from local or non-local sources (Figure 2.3). Thus each business comprises an internal division of labour that provides primary and support activities; the former adds value directly to a good and the latter adds value indirectly. But no firm is self-sufficient and must buy many inputs—both goods (e.g., fuels, materials, parts, components, capital goods, office supplies, buildings) and services (e.g., transportation, distribution, after-sales service)—from other firms (and value chains) within the social division of labour. At the same time, firms need to assess the location advantages and disadvantages of developing local or dispersed networks. In general, the internal value chains of firms are integrated into larger value chains that extend both *upstream* (towards suppliers) and *downstream* (towards markets) within cities and beyond. The concept of the value chain thus emphasizes the linkages and coordination required to make any product or service. Increasingly it extends through the consumption stage to include eventual recycling or disposal. The configuration of the spatial division of labour in any given value chain is determined by how the firms involved find the conditions best suited to achieving internal economies of scale and scope and augmenting their internal capacities with external economies.

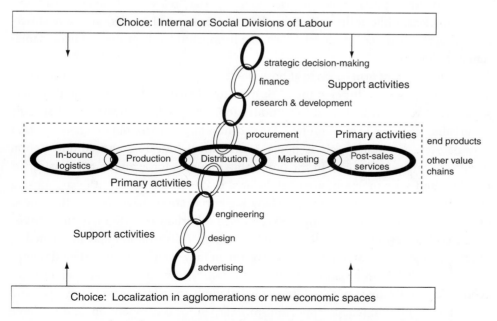

FIGURE 2.3   Value Chains in Geographical and Organizational Perspective

Source: Edington, David W., and Roger Hayter, "International Trade, Production Chains and Corporate Strategies: Japan's Timber Trade with British Columbia," in *Regional Studies* 31.2, © Regional Studies Association, reprinted by permission of Taylor & Francis Ltd, www.tanfonline.com, on behalf of Regional Studies Association.

Value chains exist for all goods and services—coffee, housing, rock concerts, autos, computers, tourism, buildings—and involve a multiplicity of markets that are governed by an array of formal and informal institutions. Typically, giant companies have powerful roles in organizing value chains both in providing functions themselves (internally) and with respect to the (external) contributions of large numbers of small and medium-sized firms. As conceptualized by Porter (1990), the links that make up a value chain *add value* to the final product (Figure 2.3). In Porter's terms, a firm itself can be seen as a value chain in which each link consists of a separate specialized activity. For any particular firm, Porter distinguished between primary and support activities. In the case of a manufacturer, primary activities directly add value to the end product (e.g., procurement of materials, production processes, distribution, marketing, and post-sales service of various kinds) and the supporting activities add value indirectly (e.g., strategic decision-making, finance, R&D, design).

Whether conceptualized at the level of the firm or industry, two broad questions concerning the economic geography of value chains can be immediately raised: (1) How are value chains organized in terms of combinations of open, relational, and hierarchical markets? and (2) How are value chains geographically structured, in other words, how are forces of agglomeration and dispersal sorted out? In terms of the choice between administrative markets (performing functions internally) or relying on relational or open markets (buying functions externally) there are no hard and fast rules. Even firms in the same industry make different choices. The rise of Japanese and German auto firms after 1950 to world prominence was based on developing strong, stable relational supplier-markets while their established North American rivals favoured the stronger control offered by integration within administrative markets. However, stronger relational markets have become the norm in recent years (see Chapter 11). Presently, value chains in the auto industry as a whole comprise both important roles for relational and administered markets, while at the end of the chain consumers and retailers are more closely aligned in an open market model (albeit one in which the role of the giant companies remains important).

Geographically, firms need to choose whether to agglomerate—that is co-locate—a set of functions or disperse them. Bear in mind that agglomeration and dispersal are relative terms that need to be understood in specific contexts and at specific geographic scales. A firm nominally located in a large metropolitan agglomeration may disperse functions such as upper management, research, and administration in different areas of the city. In the publishing industry, for example, administrative, editorial, and sales functions are often located in downtown cores with warehousing and printing in the suburban area. Many big corporations have located their head offices in city cores while locating R&D centres in suburban areas. Similar functions or others such as resource extraction, manufacturing, or distribution may be located in other agglomerations, and the scale of investment may even stimulate urban development. Nevertheless in broad terms, powerful trends towards the agglomerations of economic activity are stimulated in part by attempts to reduce transaction costs and coordinate activities, reduce opportunism, sharpen local competition, and facilitate comparison and information exchange. The fact that so many actual markets continue to thrive in particular places—the Tsukiji fish market, Toronto's Kensington

Market, shopping malls, neighbourhood garage sales, the financial markets of Bay Street and Wall Street—testifies to the power of transaction costs to promote agglomeration. Individual consumers or firms are able to comparison-shop among suppliers to elicit information on the various attributes of alternative goods that are already manufactured, and can do so on a face-to-face basis. Those benefits are magnified for non-routine purchases when the process of searching for, collecting, and evaluating alternatives is not straightforward and repeated interactions may be required. Moreover, over time the search for internal and external economies of scale reinforce these agglomeration tendencies.

Core–Periphery Relations

In tandem, the rise of agglomerations stimulates the dispersal of economic activities to distant peripheries in order to access vital resources and for other reasons such as the availability of low wage labour. The rise of MNCs and administered markets has greatly facilitated this dispersal as centralized services at head offices can readily support and control distant branch plants while the latter focus on operational functions. Indeed, the search for resources and cheap labour has contributed towards core–periphery structures within cities, between cities and rural areas, between regions within countries, and globally.

Broadly speaking, cores are large, diverse, and accessible centres of innovation, production, and consumption. They are also the home bases of the head offices of MNCs (and other countervailing powers) where strategic decisions influencing consumption and supply in other parts of the world are made. Peripheries, by contrast, are areas of specialized resource and commodity production (e.g., agriculture, coal mining) for export markets, and the decisions made there are typically limited in their influence. In terms of power relations, peripheries are clearly subordinate to cores. Cores are home to large, rich institutions with global reach, and large concentrations of people with varied skills. They have the ability to diversify the geography of their economic base and change their ties with peripheries, including replacing adjacent hinterlands with more distant sources of supply and markets for distribution. In contrast, peripheries are vulnerable to market fluctuations, threatened by redundancy from technological change and product substitution, and/or resource depletion. To a great extent, distant decision-makers who can choose where and when to invest control the fate of peripheries.

The need for countervailing powers in such situations is clear. Thus governments in peripheries often seek (though not always successfully) to diversify and develop the region's economic base by, for example, using the wealth generated by resource activities to fund new endeavours (see Chapters 9 and 10). And employees can organize in labour unions to demand better working conditions, salaries, pensions, and so on. In all cases, core–periphery relations are marked by interdependency: the core providing investment, management, and technology in return for the periphery's cheap labour, resources, and land. These relations exist at every scale, from regional to global. Nevertheless, some peripheries are much better off than others.

Core–periphery relations are deeply engrained, but they can change, sometimes profoundly. Several former low-income national peripheries, for example, have evolved to become classified as "semi-peripheries," "developing," or "transitional," and even more "core-like." Some of these countries have used manufacturing or trade, rather than resources or agriculture, as their springboard to development. But there is more than one path to development. Not so long ago, Hong Kong, South Korea, and Singapore were largely "undeveloped" peripheries, then they became cheap labour manufacturing semi-peripheries, but they have now taken on many core functions. Other countries, such as Puerto Rico or the Mexican borderlands, have not had the same success in this regard. Meanwhile, on the global scale, Canada and Australia are often classified as (rich) peripheries, even though their national territories contain urban core regions of global significance. Similarly, British Columbia is perceived as a resource periphery in Canada, but its metropolitan core—Vancouver—has half the province's population, a diversified economy, and global connections. The size and geographical scope of the decision-making mandates of private and public institutions and functional diversity are important indicators of core–periphery status.

Globalization of Value Chains

Within this globalized world, individual firms are continually seeking location advantages in their value chains. BMW, for example, in 2013, began production of its BMW i3, the firm's first mass-produced, zero-emissions urban electric vehicle which is constructed with carbon-fibre reinforced plastic, designed to reduce weight while maintaining structural integrity. Carbon fibre is half the weight of steel and a third lighter than aluminum. The raw material is bought from Japan and further processed and heated to high intensities, in a new $100 million facility at Moses Lake, eastern Washington, drawing on large quantities of cheap power provided by nearby hydro-electric power stations on the Columbia River. Spools of carbon fibre are then shipped to BMW plants in the small Bavarian towns of Wackersdorf and Landshut in Germany to be woven into fabric and then shipped for final assembly in a plant in Leipzig—itself a relatively new site for BMW—that offers lower land and labour costs, lower taxation levels and subsidies, while most of its head office, core production facilities, and main supplier firms are located in Munich and the surrounding region of southern Bavaria.

The above example hints at two key trends shaping contemporary globalization. First, globalization has meant the internationalization of competition on a global stage as, for example, auto rivals from Europe, North America, and Asia have established manufacturing plants in one another's traditional markets. This trend both "standardizes" options available to consumers while offering product variety and differentiation within specific markets. Nevertheless, while globalization is often equated with corporate consolidation and the standardization of technologies and consumer tastes, there are off-setting tendencies in which local preferences stimulate local innovation, as revealed by micro-breweries (Case Study 2.2).

Second, globalization means the increasing significance of environmental considerations for the economic geography of production. The development of electric cars, carbon-fibre auto bodies, and the use of renewable energy in the BMW case cited above are part of broader trends occurring in the auto industry and others to create more sustainable production, and to shift towards value cycles. Resource-based industries are increasingly engaged in R&D to reduce their environmental impacts. Meanwhile end-product firms such as BMW are not committed to using fossil fuel energy sources and have greater freedom to seek more sustainable power sources. In BMW's case, this search has meant the dispersal of production within a more global value chain that includes the selection of small, remote locations such as Moses Lake for production.

More generally, the globalization of value chains underlies processes of contemporary urbanization. Thus the globalization of urbanization has meant the rise of industrial cities in so-called new economic spaces, for example, in the rural areas and small towns of south China where industry had not previously existed except on a small scale. In the rich market economies that have lost much industrial production during globalization, resurgent and creative cities have grown and become part of a global division of labour. Drawing on Scott and Florida, several features of this trend can be noted. First, this growth has been inspired by R&D, design centred, creative, and green activities in services and innovative manufacturing that are providing a variety of high threshold consumer services and inputs to global value chains. Second, the clustering of these activities is driven by external economies of scale and scope that reduce transaction costs by facilitating information exchanges, brain storming, the development of pools of skilled labour and supporting services, and high levels of accessibility both locally and internationally. Third, city regions are coalescing and supplying multiple clusters of activities. Fourth, these resurgent and creative cities are increasingly tied together in a global division of labour and less tied to their local hinterlands for inputs such as resources and labour. Fifth, the process of resurgence is highly uneven within and among city regions. If Pittsburgh has moved on from its steel production base, Detroit struggles to be so resurgent. If Los Angeles and New York illustrate world cities that have vast urban regions and help orchestrate the global economy, they also contain much poverty and crime, with many people scarcely touched by creative or green jobs.

Conclusion

The spatial division of labour is constantly evolving, quantitatively and qualitatively. Established regions have either transformed their economies or declined. While new industrial spaces—especially in Asia—have taken on much of the world's manufacturing, cities in rich market economies have moved on to more sophisticated service activities.

The income generated through the spatial division of labour has massively expanded the amount of wealth in the world. However, for individuals and for places, specialization means that jobs are vulnerable to competition and technological change. Moreover, the distribution of wealth is profoundly unequal and, in recent years, increasingly so (see Figure 1.9). Poor countries have remained reliant on subsistence activities and/or on the less profitable segments of value chains. Moreover, as the scale of economic activity has increased, so has the scale of environmental damage caused by it. The social processes generating income inequalities and environmental degradation are profoundly cumulative and self-reinforcing. Radical efforts are needed to resolve them.

Practice Questions

1. Conduct a survey of a single, locally owned establishment (e.g., grocery store, restaurant, coffee house) on a street near where you live. How does it survive in a world dominated by big chains exploiting economies of scale? Do you think it is very profitable?

2. Discuss the main urban diseconomies of scale in your city. What can be done?

3. Distinguish between different types of internal and external economies of scale. What are the effects of (a) increased internal economies of scale and (b) external economies of scale on Christaller's hierarchical settlement pattern (introduced in Chapter 1)? Discuss with respect to retail functions.

4. Investigate an auto mall and document the various sales establishments and related services. Why have they clustered together? Do "stand-alone" auto sales establishments exist in your community? Explain their ability to remain profitable.

5. How has globalization shaped your city in recent decades?

Key Terms

geographic multipliers   52
division of labour   53
spatial division of labour   53
social division of labour   53
economies of scale   53
circular and cumulative
   causation   53
path-dependent behaviour   53
globalization   54
economies of scope   57
operating-level or plant-level
   economies of scale   58
variable costs   59
fixed costs   59
sunk costs   59

minimum efficient (optimal) scale
   (MES/MOS)   60
firm-level economies
   (diseconomies) of scale   60
logistics   60
product differentiation   60
external economies (diseconomies)
   of scale   62
localization economies
   (diseconomies) of scale   62
urbanization economies of
   scale   62
institutional thickness   64
immobile economies   64
multiplier   66

branch plant   66
operating- or plant-level economies
   (diseconomies) of scale   67
public goods   68
negative externalities   69
free trade   69
neoliberalism   69
New International Division of
Labour (NIDL)   72
tariffs   73
non-tariff barriers   73
comparative advantage model   73
protectionism   74
opportunity costs   75
competitive advantage   76

Recommended Resources

Amin, A., and Thrift, N. 1994. "Globalization, institutional 'thickness' and the local economy." Pp. 91–108 in P. Healey, S. Cameron, S. Davoudi, S. Graham, and A. Madinpour, eds. *The New Urban Context*. Chichester: Wiley.

A pioneering exploration of institutional thickness in terms of institutional presence and arrangements, interactions, governance and representation, and inclusiveness and mobilization.

Boschma, R. and Martin, R. (eds.) 2010. *The Handbook of Evolutionary Economic Geography*. Cheltenham: Edward Elgar.

A collection of essays that explore urban evolution over long periods of time, and explore the concept of path dependency.

Coe, N.M., Hess, M., Yeung, H., Dickin, P. and Henderson, J. 2004. "'Globalising' regional development: A global productions network perspective." *Transactions of the Institute of British Geographers* 29: 468–84.

An elaboration of the central theme of this chapter, focusing on global production networks (our "value chains") and regional development.

Frenken, K., Van Oort, F., and Verburg, T. 2007. "Related variety, unrelated variety and regional economic growth." *Regional Studies* 41: 685–97.

A discussion of related (and unrelated) variety that has become an important theme in recent literature.

Martin, R. and Sunley, P. 2006. "Path dependence and regional economic evolution." *Journal of Economic Geography* 6: 395–437.

Explanation of path dependence from the perspective of regional development.

References

Florida, R. 2012. *The Rise of the Creative Class Revisited*. New York: Basic Books.

Porter, M.E. 1990. *Competitive Advantage*. New York: Free Press.

Klein, N. 2014. *This Changes Everything: Capitalism versus the Climate*. New York: Simon and Schuster.

Myrdal, A.G. 1957. *Economic Theory and Underdeveloped Regions*. London: Gerald Duckworth.

Robinson, E.A. 1959. *The Structure of Competitive Industry*. Chicago: University of Chicago Press.

Scott, A.J. 2006. *Social Economy of the Metropolis*. Oxford: Oxford University Press.

Smith, A. 1986. *The Wealth of Nations Books I–III*. London: Penguin (first published 1776).

United Nations. 2005. *Report on the World Social Situation*. New York: UN.

United Nations Development Report. 1999. *Globalization with a Human Face*. New York: United Nations Development Program.

CHAPTER 3

Innovation, Evolution, and Inequality

[The] process of industrial mutation . . . incessantly revolutionizes the economic structure from within, incessantly destroying the old one, incessantly creating a new one. This process of Creative Destruction is the essential fact about capitalism.

Schumpeter 1942: 83

The overall goal of this chapter is to explain how innovation compels the evolution of the economies of place and space, and conversely, how these geographical contexts shape innovative processes. This chapter's six objectives are

- To explain why innovation arises and persists in economic systems (or fails to do so);
- To describe innovation in terms of basic types, scale of impacts, life cycle dynamics, and as social processes shaped by actors, and diverse R&D organizations and related institutions that form innovation systems;
- To trace the evolution of market economies over the last 250 years using the techno-economic paradigm framework;
- To explain the changing organization and location dynamics of innovation with reference to R&D investments and patent counts;
- To illustrate how innovation drives virtuous and vicious cycles of cumulative development; and
- To identify broad possibilities for technology transfer to developing countries.

The chapter has four parts. The first part analyzes the social processes of innovation, situating innovation in the product life cycle, and then describes the roles of various actors and institutions. The second part examines the concept of the techno-economic paradigm (TEP). The last two sections look at the geography of innovation and the role that innovation plays in global inequality.

Innovation and Markets

Markets do more than set prices on transactions and institutions do more than organize and govern production. The market system promotes constant innovation, compelling both the economy and the society to evolve. Entrepreneurs and companies that seek competitive advantages (Chapter 2) in the market constantly generate new economic activities that destroy existing patterns of production and consumption. Some innovations involve advances in technology that transform either products or production processes, while organizational or institutional innovations change the management of production and the nature of consumption.

Innovation is evolutionary in a Darwinian sense because only the products and companies that the market accepts will survive, and because new products shape, and are shaped by, their external environment. New technologies loom large in this evolution because much of the economic progress made over the last two centuries has been more attributable to **technological innovation** than to changes in trade, management, or advances in economic theory. Yet that progress would not have been possible without a concurrent evolution of market and institutional relations. The Internet is a classic example of the power of a technology to transform human lives in rapid order (Case Study 3.1).

Geography, particularly how people seek to facilitate transactions across space, is a powerful stimulus to innovation. Cyberspace overcomes the limitations of place by allowing communities and business partners to meet across space—online. Long before the Internet, however, many of the most dramatic innovations—trains, planes, and automobiles; telephone, radio, television, and satellite communications—had already gone a long way towards overthrowing the "tyranny" of distance by reducing the costs in money, time, and uncertainty of moving goods, people, and ideas across space. Many other innovations have dramatically changed the geography of production, distribution, and consumption. For example, refrigeration makes it possible to consume fresh produce thousands of kilometres from where it was grown; the use of recycled inputs has encouraged paper and steel factories to locate near markets; and Internet sales sites, such as Amazon and eBay, bring markets to customers virtually anywhere in the world. In general, innovations are spurred by the competition-based evolution of our economies. For that reason, this chapter focuses on *innovation as social process*.

From a long-term perspective, the innovation-driven evolution of market economies has three broad tendencies. First, economic evolution is path dependent and cumulative: development moves in particular directions, shaped by earlier investments in infrastructure, equipment, facilities, and skills, but not determined by them. Second, the evolution of market economies is marked by alternating booms and busts of varying scales and intensities, and is subject to periodic restructuring involving fundamental changes in technology, organization, environmental impact, and location. Third, economic evolution has a differentiating effect, generating variations of wealth and poverty within places and across space. Economic evolution is virtuous to the extent that it promotes increases in knowledge, productivity, and wealth. It is vicious to the extent that it serves to marginalize segments of society, fossilizing obsolescent attitudes, and reinforcing vested interests. Yet change is "guaranteed." The lives of our grandparents and parents were different from ours and our children's lives will be different again.

technological innovation
The first commercial introduction of new methods of production ("processes") or new products that typically involves manufacturing, processing, communication, or transportation activities.

Innovation as a Social Process

Innovation is a process that culminates in the commercial or practical introduction of a new idea or invention, usually by new or existing businesses. The OECD classifies four types of innovation: (1) new products or qualitative change to existing products;

Case Study 3.1
REVOLUTION AND EVOLUTION

Most readers of this textbook were born into an Internet-connected world, but not so the authors. The world of their youth was connected by snail mail, analog phones, faxes, and typewriters. The Internet has dramatically reconfigured human society, enabling instantaneous exchange of information and money across the globe. Indeed we are not sure where the Internet will take us. That said, the socio-economic transformation follows familiar patterns of paradigm shifts.

A Radical Technology

The Internet is a main carrier technology, and at its core is radical innovation in coding and transmitting information. "Packet switching" uses the standard Internet Protocol (IP) to enable communication among computer networks, linked by copper wires, fibre optics, and wireless connections. Basically, packet switching allows different units of information ("traffic") to be transferred together (in contrast to circuit switching) across different physical networks, with the help of queuing techniques. The principle of packet-switching technology hasn't changed since the conception of the Internet, but the various infrastructures for transmission among computers have evolved with incremental advances in fibre optics, wireless transmission, information compression, operating systems, etc. The Internet has gone through four main phases:

1. *DARPA and the ARPANET Age 1960–1989*

Within the US Defense Advance Research Project Agency (DARPA), J.C.R. Licklider envisioned man-computer symbiosis in 1960, and shortly thereafter P. Baran, D. Davies, and L. Kleinrock provided the conceptual and mathematical theories necessary to create packet switching. DARPA funded further development of the technologies, principally by establishing ARPANET (Advanced Research Projects Agency Network) networks among universities. From its first message in 1969, ARPANET grew to 1000 hosts by 1984 and 100,000 in 1990, as similar applications of packet-switching computer communications were advanced in other US departments, Britain, France, and Canada. While universities wanted to keep the Internet uncommercialized, by the late 1980s various companies became involved and soon Internet service providers were surpassing the speed of government networks.

2. *The World Wide Web 1990–1999*

In the late 1980s the diverse packet switching protocols used in the US were refined into the Transmission Control Protocol (TCP) and Internet Protocol (IP) (TCP/IP) by the US National Science Foundation. Shortly thereafter, the European Centre for Nuclear Research (CERN) adopted TCP/IP and the Internet was on its way to world

(2) new production processes; (3) new markets; and (4) new organizational methods and practices within and among firms. Institutional innovations by governments, associations, and related agencies that change the policy framework of a sector or of economies as a whole may be recognized as a fifth type. Growing reference has also

domination. Also at CERN, Tim Berners-Lee transformed the Internet into the World Wide Web (WWW) by creating a system of hyperlinks and URLs (web addresses). Then a team at the University of Champagne-Urbana led by Marc Andreessen produced the first web browser, which evolved into Netscape Navigator. Search engines also provided access to the web: first Yahoo and others tried to provide users with an index, then Google offered the individualized search.

Other industries realized the potential of the Internet to improve services and advance sales. In the late 1990s, entrepreneurs tried to convert all industries to online versions, but after massive run-ups in expectations, most collapsed in the bursting of the dot.com bubble. A few giants like eBay and Amazon emerged, but perhaps the finance industry made the most of the Internet's capacities.

3. *The Mobile Web and Social Media 2000–*

The Japanese first made web access available to mobile phones and for a while the mobile web was an Asian phenomenon. The Canadian firm Research in Motion (RIM) pioneered the web for email in North America and started to offer services available to computers when, in 2007, Apple introduced the iPhone, its iOS, and application store. Various social media such as Facebook, Twitter, and multi-player games were developed to take advantage of these new capacities. The web and mobile web have been wedded with Geographic Information Systems (GIS) to enable people to locate themselves and search for sites with great precision. The Internet grew from 15 million users in 1993 to 2.4 billion in 2012 (and more than two million hosts).

4. *Governance and Socio-economic Change*

The Internet is primarily a government-fostered techno-economic paradigm shift. Originally developed by DARPA, responsibility for its control was shifted to the National Science Foundation (NSF). The US government, through the NSF and Commerce department, still maintains authority over various addressing and registration issues. However, technical evolution is now largely governed by voluntary organizations such as the Internet Society and the Internet Engineering Task Force that have worldwide membership.

The key to the Internet's growth has been increasing productivity and expanding utilities, not only in industry, but also in facilitating social interaction. The Internet epitomizes the economies of scale produced by adding new users to a network (see network economies in Chapter 13). Ways of working, management and socializing, and the organization of leisure have been transformed. However, the Internet critic Jaron Lanier has decried how network owners have profited from making information freely available while depriving the producers of intellectual content of their property rights, and how these network owners demand protection of their network access while freely impinging and profiting from the privacy of their users. The Internet evolution continues, driven by technological development, but whether present governance capacities will ensure maximum social benefits is unclear.

been made to "social innovations" that address some social problems related to health, disability, poverty, or the environment, and are not primarily profit-driven.

Innovation is usually the product of direct and indirect efforts on the part of many social actors, from the inventors and researchers who pursue basic R&D; to the various firms and agencies involved in designing, developing, and transferring technology; to the producers of goods and services; to the marketers who target and obtain feedback from consumers. Most innovations are interdependent within an existing social-technological network of infrastructure, equipment, training, and knowledge necessary to make the innovation workable, while other innovations require new networks to be established. Once introduced commercially at a particular place and time, a successful innovation typically proliferates and spreads over time and space. During this process of diffusion, innovations are either "adopted" with few if any changes or "adapted" to satisfy the needs or preferences of particular customers, which may vary widely with local differences in customs, tastes, regulations, and physical conditions. For example, any innovation to the chassis or software of a car designed for global markets will still require design adjustments for left-hand drive cars in the UK, Japan, and Australia and for right-hand drive cars in other countries.

Product Life Cycles

The sources of innovation are variable. Many are experiential, stemming from practice, observation, and intuition. Managers, engineers, and workers in many activities recognize new, often minor, ways to improve efficiencies ("learning by doing") on the job, while in other cases feedback from consumers ("learning by using") sparks innovation. Many new firms are initiated almost spontaneously when an entrepreneur spots a market opportunity for a new product or service. In contrast, innovation can be a much longer and formal process, involving large-scale, formal **research and development (R&D)** programs or based in data-driven, market-driven market analysis—usually both. Product innovations, such as new consumer electronics, that require long and costly processes of R&D are often labelled "high-tech" and their evolution is illustrated by the S-shape **product life cycle** model (Figure 3.1). Because of development and production problems and small markets, production is usually limited both in scale and in location in the early stages of commercialization.

   With increased consumer acceptance, and as technological and marketing problems are overcome, increasingly rapid growth allows the realization of economies of scale and reductions in average cost and unit price (Chapter 2). If the product is successful, sales will reach the break-even point—when revenues compensate for the accumulated costs of R&D, production, and marketing. **Innovation diffusion** occurs over time and space as new (household or industrial) users buy or accept the new product, service, design, etc. and/or when production moves to new locations. Apple's iOS, for example, is the same everywhere. Innovation adoption occurs when there are no changes made in the diffusion process, while adaptation involves some modification to the innovation to meet particular consumer needs (or site requirements). The Android operating system for example is modified by different phone manufacturers and for different markets. Eventually, in the process of **product maturity**, revenues will stagnate and begin to decline, perhaps because demand is fully met or consumer tastes have changed with the introduction of innovative rival products. Further technological innovation, however, can rejuvenate production and sales through product differentiation (that also illustrates adaptation).

   As production levels and sales increase, economies of scale reduce the per unit price. Subsequently, as the product matures, the firm may be able to realize economies

research and development (R&D) Systematic, learning-based creative work that increases society's stock of knowledge and levels of technological capability.

product life cycle A metaphorical linking of the evolutionary trajectories of products with life-cycle stages, such as birth, growth, maturity, old age, and death.

innovation diffusion The spread of an innovation over time (temporal diffusion) and space (spatial diffusion).

product maturity The stage in the product life cycle when mass production technology has been developed, market potentials are predictable, and low-cost, relatively unskilled labour is the key location condition.

of scope through product differentiation, drawing on its accumulated R&D, production, and marketing expertise to modify the product so as to meet new demands and extend its life cycle. New auto models, for example, are differentiated by colour, style, and performance characteristics, in an effort to maintain market share and recoup fixed costs (R&D, factories, marketing channels, labour training). The sales profiles of many electronic products, such as desktop computers and the laptops derived from them (Figure 3.1), conform to the S-shaped trajectory. Typically, sales expansion reduces per unit costs, although new models with superior performance often command higher prices.

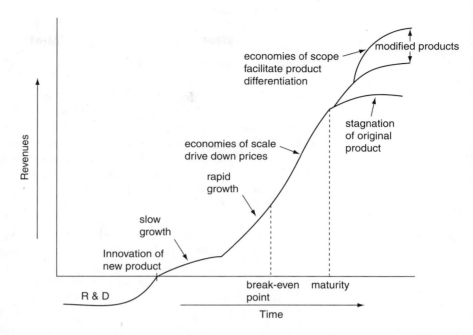

FIGURE 3.1 Idealized S-Shaped (life-cycle) Model of Innovation Diffusion

Actors and Institutions

At the outset of industrialization, all the risks and uncertainties associated with the commercial development of innovative products were borne by individual innovators (often also the inventors) and perhaps others who were willing to invest in them. Certainly this was the case with the innovator-entrepreneurs who transformed the manufacturing, agricultural, engineering, and transportation sectors during the Industrial Revolution; Benjamin Huntsman is an example (Case Study 3.2).

Individuals, working alone or in small firms, remain important sources of innovation today. However, they are now overshadowed by the teams of professional scientists and engineers employed by governments, universities, and corporations to conduct R&D in large-scale laboratories. The professionalization of R&D was an institutional innovation of the late nineteenth century, stimulated by the increasing scale and complexity of industrial activity and made possible by the rise of large corporations capable of organizing and financing large-scale R&D work. The first corporate or "in-house" R&D laboratories were developed in the US and Germany in the electrical engineering and chemical industries respectively. At the same time, the advanced market economies were expanding their school and university systems, not least to supply the educated professionals required by corporate bureaucracies and R&D groups. With the professionalization of R&D came the development of intricate, quasi-formal **innovation systems**: interlocking networks of formal and informal, national and regional institutions that develop and market innovations using scientific and technological knowledge to enhance competitive advantage. At the heart of

innovation system The interlocking network of institutions that creates and markets innovations, applying scientific and technological knowledge to enhance competitive advantage. The core of innovation systems is a "triple helix" of corporate, government, and university R&D.

Case Study 3.2

BENJAMIN HUNTSMAN AND THE INVENTION OF CRUCIBLE STEEL

Crucible steel, developed by Benjamin Huntsman (1704–76) in Sheffield, England, made the city the world's largest steel producer by the 1840s. At its peak, Sheffield produced some 200,000 tonnes of crucible steel from over 14,000 crucible pots every year. While much larger scale production methods replaced crucible steel in the late nineteenth century, local innovations such as stainless steel helped Sheffield to retain a substantial steel industry until the crash of the 1980s.

Born in Lincolnshire to German parents, Huntsman was working as a clockmaker (among other things) in Doncaster when he began experimenting to improve the quality of the soft blister steel used for the springs and pendulums in his clocks. In 1740, he moved to Sheffield, which offered varied resources such as supplies of coal for fuel and clays for pot-making, plus a labour pool already skilled in metal-working. Numerous water wheels dispersed along five main "rivers" (streams) provided power. After years of secret experimentation, Huntsman succeeded in producing the first crucible steel sometime in the 1740s.

Crucible steel-making was a refining process in which blister steel (made from Swedish bar iron) was mixed with lime which was fused with impurities and made it possible to remove them as "slag." The ingredients were mixed in clay pots ("crucibles") and lowered by hand into a coke-fired furnace where they cooked for roughly four hours at temperatures around 1525°C. The molten "cast" steel was then poured into moulds and left to cool. Quality assessment (of the carbon content) was visual.

The pots—each weighing approximately 25 kg—could be used only two or three times, and their manufacture was no less labour-intensive than the steel-making itself. They were produced on site by mixing local fireclays with china clay, coke dust, and water on the "treading floor." The pot-maker used his bare feet to work

innovation systems is what is sometimes described as a "triple helix" of corporate, government, and university R&D, and related association and cooperative endeavours.

Diversity of R&D Organizations

Private- and public-sector institutions play complementary roles with respect to innovation (Table 3.1). Thus in a highly generalized conventional view, universities provide basic research as a public good, accessible to all interested parties. Government R&D is also designed to meet public goals, but varies considerably from highly secretive defence and military-oriented R&D to activities that serve specific sectors (e.g., agriculture) or have specific missions (e.g., cancer research). Industry conducts its own "in-house" or "industry" R&D, often in secret, to meet firm-specific goals. In addition, firms in an industry may jointly fund association R&D for projects of benefit to all or most members, or may cooperate with universities and governments.

Generally speaking, different types of institutions have comparative advantages at different stages along the R&D continuum. Thus basic research, defined as the search for fundamental laws and the study of natural and social phenomena for its own sake, is performed mainly by public institutions, primarily universities, followed by government

the clay, treading it and removing extraneous materials in a process that took more than five hours. The clay was then forced into a flask and "plugged" into the shape of a crucible. Once formed, the pots had to be cured and eventually fired for some eight hours before they could be used. Such "industrial arts" (tacit knowledge) were impossible to learn except through direct experience.

Inside Abbeydale Industrial Hamlet, Sheffield. Powered by a water wheel on the River Sheaf, metal production at Abbeydale Industrial Hamlet may have begun as early as the thirteenth century. A crucible steel furnace, built in the 1830s, is located in the building with the tall chimneystacks. The site was closed in the 1930s and reopened as an industrial museum in 1970.

TABLE 3.1   Comparative Advantages of Alternative Forms of R&D

Organization	Basic Research	Applied Research	Process Development	Product Development	Technology Transfer
University	Strong	Strong	Moderate	Weak	Weak
Government	Moderate	Moderate	Weak	Weak	Weak
Industry (Equipment suppliers)	Weak (Weak)	Moderate (Weak)	Moderate (Strong)	Strong	Strong (Strong)
Association/Cooperative	Weak	Weak	Weak	Weak	Strong

Source: Hayter, R. 1988. *Technology and the Canadian Forest-Product Industries: A Policy Perspective*. Background Study 54, Ottawa: Science Council of Canada, p. 30. Note that references to strong, moderate, and weak are highly generalized; see text for explanation.

laboratories. Association and cooperative laboratories are typically oriented towards applied R&D with a focus on some industry specific process, material, device, or commercial objectives (usually in relation to industry-wide problems). Corporate or in-house laboratories usually focus on developmental R&D, testing, evaluating, and improving practical innovations (including the design, construction, and operation of pilot plants). The process culminates in the transfer of an innovation to a specific site (e.g., a factory) for commercial production to serve specific markets. Government laboratories, with exceptions, are often relatively weak in terms of developing and transferring technology as they lack direct connections with markets and have "public" mandates in that they help firms within an industry as a whole.

These patterns vary, and there are numerous exceptions, perhaps increasingly so. In agriculture, government laboratories have been important at every stage, because most farmers have neither the funds nor the incentive to undertake R&D. In technologically dynamic sectors, by contrast, firms searching for competitive advantage push in-house groups to engage in both applied and basic R&D; IBM, for one, has long employed "pure" scientists. Universities are now becoming much more strongly engaged in the later stages of the R&D process than they once were.

Firms conduct R&D for "firm-specific" purposes—to expand their market share or address particular technological challenges. They have no interest in benefiting their competitors. Therefore they invest in R&D only when they think they can "appropriate" most of the returns on any innovations. Firms also have unique know-how in linking specific market opportunities or supply sources with specific R&D efforts. For example, designers will create new fashions based on a close understanding of the preferences or needs of their consumers, while paper machine manufacturers will use knowledge of the timber species to guide their development of new pulping processes. Typically, firms protect their innovations by keeping their R&D activities as secret as possible.

In contrast, public-sector R&D is strongly (though not completely) motivated to disseminate knowledge as far as possible. Public R&D provides the long-term basic research, and the educated labour pool, that the private sector needs. At the same time it helps to prevent powerful firms from monopolizing knowledge. Association R&D, funded by industry members, sometimes with government support, is a kind of public–private hybrid in that innovations are usually made available to all members (but not to non-members). However, product innovations that are crucial to the competitive advantage of individual firms are not likely to come from either university or association R&D.

In market economies, competition between R&D organizations provides social insurance against the technical and economic uncertainties of the innovation process. It is hard to predict success or failure in advance, and competitive R&D makes it more likely that, if one innovation fails, an alternative will succeed. Competition—with all its implications for market rewards and losses—also stimulates innovation.

Social Supports for Innovation

In addition to R&D, successful innovation requires the support of many interdependent institutions. An education system that provides breadth and depth of opportunities at an affordable cost is paramount. Universal access to basic communications infrastructure, whether telephone or Internet, is also essential, and these services require frequent improvements. Patent and copyright laws must be strong enough to protect inventors without obstructing the flow of ideas or preventing competition. Other laws and policies support or hinder innovation. Also important are trust and willingness to share findings on the part of both scientists and firms; objectivity and fairness in assessing grant applications; recognition of the value of scientific research and willingness on the part of governments, firms, and foundations to fund it; and adequate incentives to pursue such research as a career. In addition, massive R&D infrastructure investments (such as the Large Hadron Collider or supercomputer facilities) require significant public support.

Techno-Economic Paradigms

Innovations may be classified along a continuum of significance. **Incremental innovations** are minor in cost and localized in impact, involving almost imperceptible adjustments (e.g., to a machine by its operator or to work routines in an office). Yet the cumulative impact of incremental improvements to productivity in an industry over time is often substantial.

Other types of innovations are typically more expensive to develop and have wider impacts on productivity, organization, or attitudes. Major innovations tend to have significant implications for particular industries, while **radical innovations** have broader social implications. The personal computer was a radical innovation, as was the Internet, both vital parts of a broader application of microelectronic technologies. So was the birth control pill: when it was introduced in the 1960s, it had a tremendous impact not just on women's lives and prospects but on the entire society. Decades earlier, in the 1930s, the introduction of the shopping mall as a way of re-organizing retail activities had important social consequences, encouraging suburbanization—often to the detriment of city centres.

Paradigmatic innovations or what Lipsey et al. (2005) call "**general purpose technologies**" (GPTs) are even more fundamental and have large-scale, complex social impacts. The introduction of the steam engine, the internal combustion engine, and microelectronics illustrate GPTs that created new industries and were transformative in nature, each in turn helping to stimulate a new **techno-economic paradigm (TEP)**. TEPs are bundles of related technological and institutional innovations that restructure entire economies, creating entirely new industries and exerting a pervasive influence across all sectors, changing not only methods of production and communication but the very nature of work, as well as patterns of leisure, education, and health.

The concept of the TEP builds on the model of "long waves" of economic activity known as **Kondratieff cycles**. According to the Russian Nikolai Kondratieff (1892–1938), the evolution of capitalist economies demonstrates growth, fluctuations in price, and output levels that feature waves or cycles of more or less 50 years that begin with a surge of economic growth, reach a plateau, and then decline, ending in depression or severe recession. In this view, the first such cycle began in the late 1700s and ended with a deep depression in the 1820s. Subsequent booms and busts occurred at roughly 50-year intervals: in the 1880s, 1930s, and 1980s. In their TEP model, Freeman and Perez (1988) relate Kondratieff long waves to deep-seated structural changes that they interpret as innovation and "paradigmatic" transformations in the institutional ("economic") and technological ("techno") bases of market economies as they undergo creative destruction. Thus new TEPs both threaten the viability of existing structures and places, and highlight investment and job growth in different industries, often in new economic spaces. They reveal possibilities for rejuvenation or adjustment of declining industries and places.

incremental innovation Organizational and technological changes with relatively small, almost imperceptible, impacts on productivity.

radical innovation An organizational or technological innovation with large-scale impacts on productivity throughout an industrial sector or social group.

general purpose technologies (GPTs) Major innovations (such as the steam engine) that have a transformative effect on economic life.

techno-economic paradigm (TEP) Strategic, long-term (roughly 50-year) shifts or waves in the economy driven by paradigmatic innovations, and associated with particular key factor industries, productivity principles, leading industries, labour relations, R&D models, and organizational structures.

Kondratieff cycles Long waves of economic growth and decline lasting about 50 years.

TEP transformations are seen in new industrial and institutional structures, new forms of regional and national integration, and new lead economies. With respect to technology, each TEP is associated with distinctive key factor industries, main carrier branches, infrastructures, newly emerging industries, and productivity principles (Table 3.2). **Key factor industries** provide crucial inputs that are cheap, abundant, and applicable across most, if not all, sectors of the economy, reducing average costs everywhere. Main carrier branches are the dominant or lead industries, which are further complemented by newer, smaller, but fast-growing industries, which often become the

TABLE 3.2   Selected Technological Characteristics of Techno-Economic Paradigms

TEP (key factor industries)	Main carrier branches and infrastructure	Other industries growing rapidly from small base	Productivity improvements
1. Early mechanization 1770s–1830s (cotton, pig iron)	Textiles, textile chemicals, textile machinery, iron working and castings, water power, potteries. Canals, turnpike roads.	Steam engines, machinery	Mechanization creates new economies of scale in factories.
2. Steam power and railway 1830s–1880s (coal, transport)	Steam engines, steamships, machine tools, iron, railway equipment. Railways, world shipping.	Heavy engineering	Factory-level scale economies deepened by further mechanization, steam power, and railways.
3. Electrical and heavy engineering 1890s–1930s (steel)	Electrical engineering, electrical machinery, cable and wire heavy engineering, armaments, steel strips, heavy chemicals, synthetic dyestuffs. Electricity supply and distribution.	Autos, aircraft	Factory-level scale economies deepened by use of cheap steel (superior to iron), more flexible electrical-power for belts, pulleys, and machinery, and overhead cranes and power tools, which lead to efficiencies in factory layouts and standardization. Multi-plant and multi-location economies of scale created by improved transportation and communication systems (telegraph and phone).
4. Fordist mass production 1930s–1980s (energy, oil)	Autos, trucks, tractors, tanks, armaments, aircraft, consumer durables, process plants, synthetic materials, petrochemicals. Highways, airports, airlines.	Computers, radar, numerically controlled machine tools, drugs, nuclear weapons and power, missiles, micro-electronics	Assembly line and flow processes greatly deepen factory-level economies of scale. Related standardization of components. MNCs spread rapidly and deepen economies of scale and size.
5. Information and communication 1980s (silicon chips)	Computers, electronic capital goods, software, telecommunications, optical fibres, robotics, flexible manufacturing systems (FMS), ceramics, data banks, information services, digital telecommunications, satellite networks	"Third generation" biotechnology products and processes, space activities, fine chemicals, Strategic Defense Initiative	Flexible manufacturing systems enhance internal and external economies of scope via electronic control systems. Systemization of efficiencies throughout global value chains comprising design, production, and marketing activities.

Note: The TEPs coincide with Kondratieff waves I–V.
Source: Based on Freeman and Louçã (2001).

lead industries of the next TEP. Productivity improvements redefine "engineering common sense," leading to new economies of scale and scope. The national technology leaders are the nations most responsible for developing the innovations that define a TEP. In this regard, the UK's early leadership in manufacturing and world trade ("Pax Britannica" and the Common Wealth) was effectively challenged by the US and Germany by 1900. After World War Two, Fordism gave US domination over global market economies, but during the **information and communication techno-economic paradigm (ICT)** the US has been challenged by Germany, Japan, and now China. A more multi-centred global system has evolved with deepening trends towards both multi-lateral free trade and regional trading blocs.

Crucially, the technological and production developments of each TEP co-evolve or are "matched" with organizational innovations in business structures, labour relations, and R&D. Decision-making structures once dominated by individual entrepreneurs are now overshadowed by giant firms and MNCs. New hierarchical and network based decision-making structures must cope with managing increasingly complex and geographically diffuse operations (Chapter 4). As an institutional innovation, the spread of limited liability laws in the nineteenth century reduced the risks facing investors (to the size of their investment), and allowed firms greater ease in obtaining funds by which to grow. The rise of giant firms has facilitated the realization of economies of scale and scope, including with respect to the costs and risks of R&D that has become increasingly professionalized and based on a division of labour. As an institutional innovation, free trade allows scale economies on a global basis. These and other matching organizational innovations facilitate the productivity improvements created by technological changes.

The TEP model emphasizes that innovation is a key dimension underlying patterns of economic development, and that broader political, economic, scientific, technological, and cultural forces shape innovation. An interesting question in this regard is why the first TEP originated in western Europe with a succession of innovations in the agricultural, transportation, and (above all) manufacturing sectors, beginning around the mid-eighteenth century. According to the pioneering work of Angus Maddison, in the year 1000 both China and Europe were well-developed rural societies with more or less equal populations and standards of living as indicated by measure per capita GDP measures (Table 3.3). Subsequently west European per capita GDP grew relatively faster than in China (and other regions) until around 1800 when income divergence became more rapid. The US, Canada, Australia, New Zealand, and Japan also experienced rapid growth from the late 1800s.

Important innovations originated in Asia; China and Korea, for example, developed the first mechanical printing presses. However, as Maddison argues, the potential of western Europe for innovative change, and more generally for The "Industrial Revolution" or "Great Transformation," was facilitated in prior centuries by reforms in economic, political, social, and religious institutions that stimulated societies to become more democratic, better educated, and interested in material progress. The great European voyages of discovery, sparked by Portuguese and Spanish expeditions in the fifteenth century across the Atlantic Ocean, led to increasing trade and wealth and the growth of rich trading centres, while national rivalries stimulated competition

information and communication techno-economic paradigm (ICT) The (TEP) that has developed since the 1970s, characterized by flexibility in mass production, specialization, and labour, exploiting economies of scope as well as scale.

TABLE 3.3   World Per Capita GDP for Selected World Regions 1–2003 AD

Region	1	1000	1500	1700	1820	1870	1913	1950	1970	2003
West Europe (12 countries)	599	425	798	1,032	1,243	2,087	3,688	5,018	12,517	20,597
East Europe	412	400	496	606	683	937	1,695	2,111	4,988	6,476
Former USSR	400	400	499	610	688	943	1,488	2,841	6,059	5,397
USA	400	400	400	527	1,257	2,445	5,233	9,268	16,179	28,039
Latin America	400	400	416	527	691	676	1,493	2,503	4,513	5,786
Japan	400	425	500	570	669	737	1,387	1,921	11,434	21,218
China	450	450	600	600	600	530	552	448	838	4,803
India	450	450	550	550	533	533	673	619	853	2,160
Africa	472	425	414	421	420	500	637	890	1,410	1,549
World	467	450	566	616	667	873	1,526	2,113	4,091	6,516

Note: The figures are in inflation adjusted 1990 dollars. In absolute levels of GDP China and India were the world's two biggest economies prior to 1800.
Source: Maddison 2007.

and innovation in relation to shipping and navigation (and colonization). As religious controls lessened, inheritance laws favoured families, accumulation, and individualism. Moreover, the university system was significantly expanded across Europe, and the spread of knowledge was facilitated by the diffusion of the printing press in Europe after 1450. A related vital key to industrialization was the formalization of science that first occurred in Europe. In addition to establishing the laws of motion that became the basis of modern mechanics, Isaac Newton (1642–1727) set down the fundamental principles of the scientific method, promoting rigorous hypothesis-testing, systematic application, and a cumulative approach to developing and transferring knowledge. The rise of the modern engineering profession was a closely related development that applied and elaborated scientific knowledge across the economic spectrum. Modern banking systems were evolving through this time period as well. Most fundamentally, by the eighteenth century the advantages of markets as a basis for exchange and decision-making had gained recognition in practice, in law, and in theory, especially in the UK.

Income growth in Europe and other rich market economies became especially strong after 1950, and it has only been in recent decades that other economies, especially in Asia, have been catching up and, not coincidentally, becoming more innovative.

The most recent TEPs, and the possibility of a future **Green TEP**, are now briefly elaborated.

Fordism's Long Boom

Named after Henry Ford, the pioneering car manufacturer who introduced the assembly line and the use of interchangeable parts, **Fordism** became the dominant paradigm of the twentieth century, beginning in the 1920s and culminating in the long boom period that followed the Second World War. Fordism standardized the production of cornflakes, soft drinks, beer, laundry detergent, and much more. The mass production

green TEP A TEP in which economic development would meet the challenge of environmental sustainability as well as economic goals related to jobs and income.

Fordism The TEP that dominated the half-century from the 1920s to the 1970s, characterized by mass production based on internal economies of scale organized by giant firms typically employing unionized labour.

of housing facilitated the process of suburbanization, and retailing was transformed as small general stores were replaced by department stores, supermarkets, and fast-food chains. Along with mass production, Fordism depended on workers receiving higher wages and mass consumption becoming the social norm.

With the quest for economies of scale came the rise of the multinational corporation, dedicated machinery, specialized labour, and unionization. Unions and collective bargaining agreements, especially among big corporations, sanctioned and stabilized the trend towards increasing size and specialization by accepting managerial control over production and structured work rules based on job demarcation and seniority (see Chapter 5). The wage gains achieved through collective bargaining steadily increased the demand for goods. Hierarchical MNCs became defined by layers of management organized around different divisions and groups and reporting to the chief executive officer (CEO) who was located in head offices; corporate networks were held together by formalized routines and mandates (regarding work and communication). National governments sought to promote stable economic growth domestically, through policies mitigating business cycle effects, and internationally, by moving towards free trade in goods and foreign direct investment (see Chapter 6). Governments also took increasing responsibility for expanding access to education, health, and social security (welfare and unemployment insurance), as well as problems of international development. Under Fordism, policy-making in general was strongly "top-down" in nature, dominated by national-level government bureaucracies.

Globally, the economy was divided between the centrally planned economies (CPEs) of the "East" and the market economies of the "West" (with poor developing countries peripherally connected to one or the other). The US—the global leader of the market economies—developed structured rules and approaches to international economic relations, promoting fixed exchange rates among the different national currencies, control over financial movements between countries, and orderly reductions in tariff barriers. However, the key factor input for Fordism was oil, and in the 1970s its price rose dramatically. Energy crises, inflation, rising unemployment, and declining productivity followed, signaling that Fordism's rigid production structures and international economic relations were in trouble. Deindustrialization—the massive loss of manufacturing jobs—in the traditional manufacturing belts of the US and Europe became pronounced in the deep recession of the early 1980s. This marked the demise of Fordism and the need for restructuring. The decline of Fordism in the US and Europe was hastened by competition from Japan and its more flexible and leaner approach to mass production.

The ICT and Flexibility

At the heart of the ICT is an interconnected set of radical innovations in microelectronics—computers, software engineering, control systems, integrated circuits, and telecommunications—that have drastically reduced the cost of storing, processing, communicating, and disseminating information. Computers and robots have revolutionized the processes used by established industries. Microelectronics and computer-assisted design and manufacturing have improved design and product differentiating capabilities in everything from sawmills to cameras. Microelectronics have revolutionized the service sector as well. On airplanes, empty seats and extra meals disappeared as airlines became capable of matching supply and demand precisely. Stock markets were transformed from trading pits to electronic markets open to anyone, anywhere, in real time. Meanwhile, we have all become consumers of a vast array of microelectronic and software products.

In contrast to Fordism, with its emphasis on stability, the ICT era brought managerial and **organizational innovations** that emphasized flexibility. Firms developed the capacity to rapidly alter product design and mix in response to consumer demand and, for many firms, distribution and marketing functions became more important than production itself. Workers and management had to become more flexible, to be willing and able to do more than one kind of work, to shift to new jobs and employers, and to accept varying work schedules, including part-time work. Economies of scope became as important as economies of scale. The imperatives of flexibility supported the neoliberal perspective and provided a rationale for financial deregulation while drawing attention to the bureaucratic rigidities of the CPEs, which could not keep up with the innovation-based competition from the West. Governments changed their approach to the economy, putting less emphasis on top-down policies and more on flexible, bottom-up approaches, attuned to local conditions. Simultaneously, the ICT era witnessed profound changes in the location of economic activities as a result of deindustrialization, the partial rejuvenation of some deindustrialized areas, and the formation of new economic spaces (for example, in south China).

organizational innovation An innovation that changes the way economic activity is governed or organized.

Towards a Green Paradigm?

Industrialization has always had impacts on the environment, but these were rarely considered important by decision-makers, at least until relatively recently. Similarly, theories of economic development, whether neoclassical, Marxist, or institutional (including the TEP model), largely ignored environmental concerns. Yet the growth of mass production and consumption during Fordism, coupled with exponential population growth, multiplied resource exploitation and environmental degradation on a global scale. Environmental problems such as climate change and biodiversity loss now threaten human and natural welfare, and society as a whole. As a result, nations, regions, and cities are seeking to make development sustainable.

The TEP model gives hope that sustainable development can be achieved. These hopes are rooted in the periodic restructurings in capitalist evolution, the recognition of related environmental phases, and innovation shaped by policy. Thus the model recognizes fundamental restructurings in the organization of market economies, and that these restructurings can be extended to embrace distinct "environmental phases" reflected in public and private policy-making and attitudes toward environmental values (Table 3.4). In early TEPs and laissez-faire environmental phases, environmental impacts were basically ignored, apart from a few scattered pioneering policies (e.g., local pollution laws and creation of parks). During Fordism, even as global impacts of pollution, biodiversity loss, and climate warming escalated, public, corporate, and household decision-makers were more interested in the amenity values of the environment. For example, community-land–use plans incorporated parks, recreation areas, gardens, and

TABLE 3.4   Techno-Economic Paradigms and Environmental Phases

Techno-economic paradigm	Environmental phase	Scale of impact	Enviro-policy/industry strategies
Mechanical, 1760s–1820s	Frontierism (uncontrolled use and abuse of resources as if on an isolated frontier)	Local	"Laissez-faire" (belief that markets work best when free of regulation); environment seen as a free "good," pollution as externality. Efforts at social reform, improved housing for workers, etc., are isolated, "utopian" initiatives.
Steam, 1820s–1870s	Frontierism awareness	Regional	As above, but Alkali Act (UK) of 1860s signals government recognition of pollution as public bad.
Electrical, 1870s–1920s	Frontierism awareness	National	Policy initiatives to (a) preserve enviro-values (Yellowstone National Park, 1872), and (b) improve living conditions of workers (e.g., garden cities). Industry attitudes remain unchanged.
Fordism, 1920s–1970s	Amenity and protection	Global	Amenity (e.g., clean air, gardens, parks) is central to urban/regional planning; "environment" is a location factor for firms. But the rapid growth of mass production and consumption creates fears for "limits to growth," and "Silent Spring" draws attention to threat of enviro-catastrophe.
ICT, 1970s–2020s	Resource management	Global	Rapid movement towards legislation to reduce pollution; biodiversity becomes priority; Montreal/Kyoto protocols seek global responses. "Green consumer" appears. Recycling expands, industry develops technologies to reduce enviro-impacts (e.g., electrical vehicles); eco-certification.
Green	Eco-development	Local	Dematerialization of the economy; industry internalizes environmental values (eco-certification, take-back policies, emphasis on selling services rather than goods), and sustainability is defined. Environment R&D prioritized.

Source: Adapted from Hayter (2008: 831–50)

tree-lined streets while communities with high levels of environmental amenity became attractive to footloose activities (e.g., high-tech activities such as software design or new electronic products that rely on know-how and whose location does not depend on access to material inputs) that enjoyed considerable discretion as to where to locate.

During the ICT era the sustainable management of resources has become an important theme, while new computerized technologies and forms of communication have increased abilities to reduce environmental impacts, for example, by reducing waste, using more recycled materials, and allowing more people to work at home thus reducing commuting. Yet the primary goal of the ICT was to increase economic efficiency rather than resolve environmental problems that, in absolute terms, have continued to increase.

In a green paradigm, environmental sustainability will have to be integrated into all economic activities as value chains are redesigned as value cycles, driven by the 3Rs: reduce, reuse, recycle (see Introduction). In this scenario pollution elimination, greenhouse-gas reduction, and biodiversity are designed into economic activities, rather than becoming forms of compensation after development. This was the spirit advocated in the UN's Brundtland Commission of 1987 which came to the conclusion that the best way to save the environment was to solve poverty at the same time. Since then, many international agreements and national policies have not only increased environmental regulations, but also funded and prioritized green technologies. Industry has responded with innovations of all kinds—from green buildings and electric automobiles to green chemistry, and organic agriculture. Environmental non-profits have been part of this paradigm shift, raising awareness and pushing for change (Chapter 7).

Energy use is a central challenge facing a Green TEP, particularly with respect to reducing carbon emissions that are driving climate warming. In this regard, the TEP model recognizes the role of market-based approaches such as a carbon trading system or carbon taxes. Although market-based, these approaches require government policies to put prices on environmental outputs, thereby encouraging lower demands for fossil fuels. But market prices fluctuate and can be deceptive indicators of development priorities. Thus, the TEP model emphasizes the need for innovations—technological, organizational, and institutional—in increasing the efficiency of existing fossil fuel production and to reinforce developments of non-renewable sources, such as solar energy. Indeed, as a result of technological change and increasing demand, the price of solar energy has dropped considerably, and this trend will likely continue. Reductions in the massive subsidies supporting the fossil energy industries and transferral to the R&D of renewable options would further hasten this trend. More environmentally benign energy systems may also shift to more distributed, flexible forms of production, storage, and pricing that involve highly localized, small scale sources of energy and incentives to consumers to both contribute to the electricity grid and not use it at peak times. In addition, the TEP model emphasizes the need for long-term thinking, and a green TEP requires truly comprehensive, effective participation in which local initiatives make global sense, and vice-versa. A green TEP, also referred to as a sustainability transition, is a challenge to globalization, economically and environmentally.

Finally, several overall features of TEPs are worth highlighting:

- The TEP is an intellectual device that summarizes complex evolutionary processes in terms of key technological and organizational innovations. It is recognized that economic development trajectories over long periods of time are complicated, varied, and shaped by distinct mixes of economic, political, social, scientific, technological, and environmental factors.
- Technological and institutional innovations evolve in tandem, each supporting the other (e.g., corporate R&D laboratories, labour and environmental laws, social housing, and unionism).
- The transition from one TEP to another is complex, involving

  o   declining productivity gains from existing production and organizational structures;

  o   severe recessionary crises that undermine traditional activities in existing, high-cost places, and increased awareness of the need for change;

  o   lengthy periods between invention, innovation, and large-scale commercialization;

  o   war and military expenditures that stimulate innovation (for example, the jet fighter planes and supporting tanker aircraft developed during the Second World War that provided technology for the commercial aircraft industry).

- There are relatively few "lead nations." The rich, industrialized countries of 1900 (and earlier) continue to lead the way in innovation (a few of the newly industrializing countries may become leaders in innovation, but their emergence is still relatively recent).
- The evolution of TEPs is path-dependent and open at the same time. Countries, regions, and companies carry the baggage of their existing technologies and institutions, but they also learn, experiment, and observe experiences from elsewhere and generate new policies and institutions.
- New TEPs create "new economic spaces," which then put competitive pressure on old industrial spaces, forcing them either to adapt or to give way.

The Changing Geography of Innovation

The emergence of the "mechanical" TEP in the late eighteenth century marked both a decisive escalation in the rate, scale, and scope of innovation and its concentration in Europe, North America, Japan, and a few other places such as Australia, New Zealand, and South Africa. Continental Asia, Latin America, and Africa, which did not participate in the new surge of innovation, were left behind. As we noted above, this development was facilitated by a variety of cultural and institutional shifts in Europe, including the development of modern scientific approaches to problem-solving. The development of corporate R&D and innovative behaviour has also been massively cumulative, and overwhelmingly concentrated in the leading market economies. This dominance has only recently been challenged by a few countries, especially in Asia. Typically, as TEPs evolved, the geography and organization of innovation and R&D systems has changed. Individual innovator-entrepreneurs, such as Huntsman (Case Study 3.2), often working in secret in (then) relatively isolated places gave way to the professionalization of R&D in formal laboratory settings within the triple helix of institutions. This professionalization was typically urban-centred and, in recent decades, increasing global connections among urban centres have become an important trend, forming more globalized "spatial innovation systems."

Knowledge and Learning

Innovation systems operate at many scales: national, regional, urban, and—increasingly—global. But the mechanisms of innovation are similar at every scale. Innovative societies are continually adding to their knowledge and technological capacity to produce and organize economic activity. Learning takes place in various ways—formal and informal, private and public—and is cumulative and synergistic as individuals, firms, and regions build on and transform existing concepts and methods. In addition, formal learning in education systems (primary, secondary, and post-secondary education) and at work (apprenticeships, learning by doing, periodic off-the-job training) is explicitly based on the progressive accumulation of knowledge.

A distinction is commonly made between **codified** and **tacit knowledge** (see Case Study 3.2): Codified knowledge is that which can be written down (as in a textbook, blueprint, or formula), while tacit knowledge, by contrast, can be learned only through observation, imitation, and hands-on experience. Codified knowledge is conventionally considered to be more easily communicated across space than tacit

codified knowledge Knowledge that can be communicated precisely through language or formulas (e.g., in textbooks or patents).

tacit knowledge Knowledge that is acquired through observation, imitation, and hands-on experience, and is difficult to communicate verbally.

knowledge which relies on personalized and localized interactions. However, codified knowledge and its implications are not necessarily easy to understand. In any event, both types of knowledge are important, and organizational practices often combine them; an illustration is the transfer to Japan's electronic music industry of patented and tacit knowledge developed in the US (Case Study 3.3).

National Innovation Systems

An innovation system is an interactive mix of institutions, policies, and infrastructures that support the creation and diffusion of innovations. A *national* innovation system emphasizes the importance of national policies, regulations, industry specialties, and cultures in shaping innovation systems. Government sponsorship of technological development, particularly for military purposes, has a long history. By the nineteenth century, many governments were systematically promoting innovation in an effort to develop national industrial strengths, protecting domestic industrial firms in new industries, funding the search for and transfer of technological expertise from other lead countries, and establishing comprehensive educational systems. In the twentieth century, to support commercial and military innovation, professional public- and private-sector R&D expanded rapidly, and by the 1960s and 1970s national and local development plans were increasingly explicit in their emphasis on the importance of innovation. At the core of every such system are

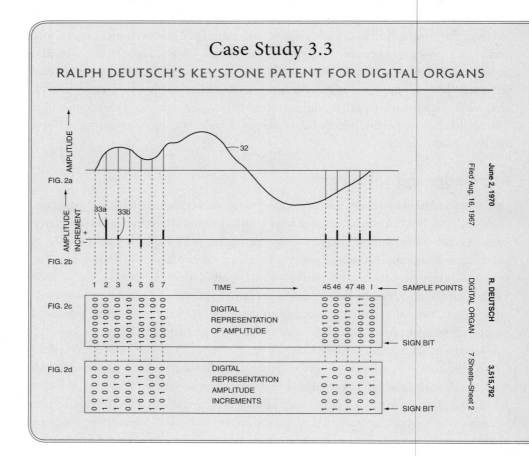

Case Study 3.3
RALPH DEUTSCH'S KEYSTONE PATENT FOR DIGITAL ORGANS

public institutions—schools, universities, government laboratories—and policies designed to increase learning, knowledge, skills, and technological capability and to promote the development of competitive advantages. The in-house R&D programs of private-sector firms are equally important. The objective of a national innovation system is to align the nation's various public and private initiatives so as to enhance the flow of information and technological know-how between them, to provide a supply of qualified workers, and to stimulate technological transfer. The underlying rationale for a national innovation policy is the recognition that product innovation is vital to enhancing productivity and generating high-paying jobs.

The developed nations have had *de facto* innovation systems for well over a century. The nature of those systems varies, however. In the mid-twentieth century, innovation in the Soviet Union and its satellite countries was handled almost exclusively by the public sector (government and universities). Privately owned operations were tolerated only in fields such as agriculture, where operations were small and not research-intensive. The sophistication of the USSR's system became clear in 1961, when Yuri Gagarin became the first human to fly in space. More generally, however, the lack of private-sector R&D was a significant weakness, depriving the CPEs of the market competition that has been such a powerful incentive for innovation in the West.

National innovation systems have taken various forms in market economies. The US, the UK, and France devote a much greater portion of their R&D spending to military

---

Ralph Deutsch was a prolific American engineer who registered more than 140 inventions, among them four of the top ten patents in electronic music. The document above, issued in 1970, was the first patented rendering of a musical tone in binary code, the basis for the digital organ. Yet Deutsch, who was working for American Rockwell, had difficulty convincing most US musical firms of the value of his invention. The Japanese firms Yamaha and Kawai were more prescient. In hiring Deutsch, they acquired both his patented (codified) knowledge and his ability to develop further innovations in what Reiffenstein cites as "a cycling process between tacit and codified domains."

Two other pioneering American engineers, John Chowning (Stanford University) and David Smith (who worked in Silicon Valley), also found Japanese musical firms to be more receptive than their US counterparts. Chowning licensed his key patent, which provided the basis for the world's first all-digital programmable polyphonic synthesizer, to Yamaha. And although Smith never patented his ideas, he passed on his know-how to Yamaha, which subsequently patented it and employed him, leading to the world's first mass-market programmable instrument and the Musical Instrument Digital Interface (MIDI).

Clearly, codified knowledge is not necessarily easy to understand. Also, codified and tacit knowledge interact. The transfer of this know-how from the US to Japan had massive implications for the global geography of the electronic musical industries. Initially concentrated in the US, by the 1980s it was overwhelmingly concentrated in Japan, notably in the industrial agglomeration of Hamamatsu.

Source: From *Canadian Geographer* 50(3), 2006, Tim Reiffenstein, "Codification, patents and the geography of knowledge transfer in the Canadian furniture industry." By permission of the author.

and space-related activities than do Germany and Japan, in part because the treaties that concluded the Second World War expressly restricted the latter countries' ability to invest in their militaries. Consequently, they focused their innovation systems on private-sector firms and commercial applications. These differences create interesting debates. For example, the US and UK have suggested that their defence-related investments in R&D are in the interest of global security and peace, and therefore countries such as Japan and Germany should contribute financially to those efforts. Those opposed to the idea point out that defensive R&D often subsidizes private-sector developments, among other arguments.

There are also more subtle differences in the national innovation systems of market economies. In Canada, for example, the private sector plays a relatively smaller role and the public sector a relatively larger one than is the case in other market economies. Since private-sector innovation is more important even in smaller countries like Sweden and Finland, market size does not appear to be a major factor. It seems more likely that institutional factors—notably Canada's high levels of foreign ownership in key industries, especially in the manufacturing sector—are to blame. MNCs, which are responsible for about half the world's R&D, typically centralize R&D in their home countries, leaving places where they are a major presence, such as Canada, with significantly less such activity than might otherwise be expected (for more on the "truncating" effect of foreign ownership, see Case Study 4.4).

The Japanese government began to organize innovation for economic and military purposes to close the technology gap with the West in the 1880s. After its defeat in the Second World War, it strategically targeted industrial activity for commercial purposes, supporting domestic growth in the directions it perceived to be most desirable. To promote the auto industry, for example, it restricted imports and required suppliers to be local. For their part, Japanese auto firms made considerable efforts to "reverse engineer"—that is, understand and improve on the practices of leading firms in the US and Europe. The apparent success of that strategy encouraged governments in Europe and North America to create science and technology ministries and develop formal innovation policies. Sweden, Denmark, and Finland have been at the forefront of this trend, creating national and regional innovation systems that emphasize the value of clustering and enhancing information flows. The US and Canada have taken a less direct approach, focusing on targeted research funding, improving laws governing university–industry collaboration, and forcing universities to look to industry for funding. In an international context, the US emphasis on defence research and its spin-offs and research funding for distribution by the National Institute for Health stand out in terms of absolute and relative size of funding commitment. In Asia, however, Korea, Taiwan, Singapore, and China have all preferred the Japanese model of development.

R&D Expenditures

A key factor in innovation capacity is the nation's willingness to spend on R&D. The US, Japan, and Germany—the world's biggest market economies—have had the biggest R&D budgets in both absolute and relative (per capita) terms. China has recently begun to invest enormous amounts in government, market, and military R&D. Internationally, corporate R&D tends to be overwhelmingly located in the same countries as the head offices of MNCs, although in recent years many MNCs have established foreign R&D subsidiaries, partly to gain access to a wider pool of scientific and engineering talent and partly to address distinct market opportunities.

A nation's input into innovation is measured in terms of Gross Expenditures on R&D (GERD). Although not a perfect measure—since innovation occurs without R&D, and R&D does not always lead to innovation—GERD is a useful indicator. GERD is both a response and stimulus to wealth creation, and historically concentrated in rich countries. In 2002, the (rich) developed countries, with around 19 per cent of the world's population, accounted for almost 60 per cent of global GDP and fully 77.8 per cent of global GERD (Figure 3.2). The US, Japan, and Germany alone accounted for 56.7 per cent of global totals. In 2009, these rich countries still accounted for 73.0 per cent of global GERD and the three leading countries for 48.5 per cent. Yet interesting changes are occurring. A large group of developing countries that accounted for around 70 per cent of the world's population have growing shares of both the global GDP and GERD. China's investments in R&D have been massive recently, and the country's global share of GERD has increased from 5 per cent in 2002 (already the third largest national share) to 12.1 per cent in 2009, second only to the US. South Korea's performance is also impressive and these two countries combined to account for 15.5 per cent of global GERD in 2009, a reflection of their own economic development and a global shift in industrial power and innovativeness towards Asia. In contrast, the very poorest countries, with 11.1 per cent of the world's population in 2002, had only 1.5 per cent of global GDP and barely registered any GERD then or in 2009.

Patents

If a country's R&D spending is a measure of its national input into innovation, then the number of patents it issues is an output measure of the efficiency of both the spending and of the national system. The social virtue of patents is that they provide incentives for innovation

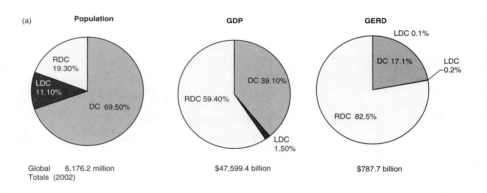

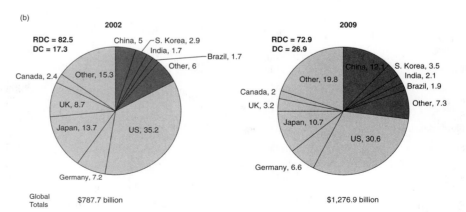

FIGURE 3.2 Global Distribution of GERD by Stage of Development, 2002 and 2009

Notes: GERD: Gross Expenditures on Research and Development; RDC: Rich developed countries; DC: Developing countries; LDC: Less-developed countries. For definitions of these categories see UNESCO's webpage. Basically, the regions are classified according to per capita income.

Source: UNESCO Institute for Statistics http://data.uis.unesco.org/index.aspx?queryname=79 (accessed February 2015.)

by protecting the patent-holder's right to the rewards. In fact, firms (and individuals) can take out multiple related patents, either at one point in time or over time, building on their knowledge and experience to create additional innovations, differentiate existing products, and further exploit the know-how embodied in their goods and services through exports and foreign direct investment or branch plants. On the other hand, the social problem with patents is that, because they sustain institutional monopolies for long periods of time, they can disrupt the efficient flow of information and reduce overall capacities for innovation.

Countries have dealt with these trade-offs in various ways. There are different rules governing the duration of patent protection and some countries award a patent to the actual inventor of a product, while others award a patent to the first person to file for it. Such differences, together with the limitations of national enforcement capacities, often force companies to register the patent for a particular product in several countries. Harmonizing international patent regulations is one of the key goals of the WTO.

The attractions of the US market are such that most companies register their patents there. Table 3.5, showing the national origins of patents issued in the US, clearly illustrates the dominance of the advanced countries. Admittedly, raw numbers of patents provide only a crude indicator of innovativeness. Not all important innovations are patented, while many trivial ones are; and US figures naturally include a preponderance of Americans. Nevertheless, before 1990, the patent counts were overwhelmingly dominated by rich Western countries, especially the US, and also Japan. In 1990, the

TABLE 3.5   US Patents Granted by Country of Origin

Selected Countries	Pre-1990 Total (%)	1990 Total (%)	2003 Total (%)	2013 Total (%)
US Origin	1,182,124 (65.1)	47,391 (52.4)	87,901 (52.0)	133,593 (52.6)
Japan	184,348 (10.2)	19,925 (21.6)	35,517 (21.0)	51,919 (20.5)
Germany	146,010 (8.0)	7,614 (8.4)	11,444 (6.8)	15,498 (6.1)
UK	70,825 (3.9)	2,789 (3.1)	3,627 (2.2)	5,806 (2.3)
France	54,725 (3.0)	2,866 (3.2)	3,869 (2.3)	6,083 (2.2)
Canada	31,144 (1.7)	1,859 (2.1)	3,426 (2.0)	6,547 (2.6)
Taiwan	2,341 (0.1)	732 (0.8)	5,298 (3.1)	11,071 (4.4)
Sweden	19,681 (1.1)	768 (0.9)	1,521 (0.9)	2,271 (0.9)
Netherlands	17,353 (1.0)	960 (1.1)	1,325 (0.8)	2,253 (0.9)
South Korea	599 (0.0)	225 (0.3)	3,944 (2.3)	14,548 (5.7)
Australia	6,402 (0.4)	432 (0.5)	900 (0.5)	1,631 (0.6)
USSR	6,415 (0.35)	174 (0.2)	1 (0.0)	0
South Africa	1,981 (0.1)	114 (0.1)	112 (0.1)	161 (0.1)
China	240 (0.0)	47 (0.1)	297 (0.2)	5,928 (2.3)
India	329 (0.0)	23 (0.0)	341 (0.2)	2,424 (0.1)
Brazil	552 (0.0)	41 (0.1)	130 (0.1)	254 (0.1)
World Total	1,815,531 (100.0)	90,365 (100.0)	169,028 (100.0)	253,835 (100.0)

Source: US Patent and Trademark Office, http://www.uspto.gov/web/offices/ac/ido/oeip/taf/cst_utl.htm

countries that led the way through the first five TEPs (the US, Japan, Germany, and the UK) were still, for the most part, leading in patent counts, although the UK was slightly less important than France. Japan in particular had become more important. The same pattern was true for 2003 and 2013, but significant new trends are emerging. In 2013, the clear patent leaders remained the US, Japan, and Germany, accounting for 79.2 per cent of all patents, and these countries remain the most important global centres of innovation. Fourth and fifth place, however, went to South Korea and Taiwan in 2013, and the former accounted for almost as many patents as Germany, and with twice the totals of the UK, France, or Canada. China and India are also increasing their patent counts as they industrialize.

Local Innovation Systems

National innovation systems are strongly urbanized and regionalized. In the US, for example, over half the corporate R&D budgets and roughly 40 per cent of the patents generated by the leading R&D centres stem from five metropolitan areas (Table 3.6). The San Jose–San Francisco area, which includes Silicon Valley, alone accounts for over 20 per cent of these R&D budgets. Most remaining R&D efforts by leading corporations are located in big city regions; the three leading ones are on the Pacific Coast.

Cities are important to innovation because they are "meta-places of interaction" for actors and institutions, public and private, and because they provide a mix of local and non-local, economic and non-economic influences that are vital for sustaining innovation. Cities bring together the various agents of innovation and facilitate face-to-face meetings in which they can exchange tacit knowledge, solve complex technological problems, and identify research priorities. Moreover, the amenities desired by the "creative class" are focused in cities. In practice, many R&D and related highly innovative activities are located in suburban areas that can offer extensive "pleasant" and highly secure space for buildings and recreation, easy parking or access by public transportation, accessibility to airports, and reasonable proximity to downtown corporate head offices. Such a location facilitates interaction without too much ongoing interference with work that often has long-time horizons. Universities are also urban and suburban centred, and in general, agglomerations facilitate interactions throughout the R&D process that are "loopy"—that is, it features feedback between the different stages, rather than a simply linear process.

TABLE 3.6 Biggest US Corporate R&D Locations, 2011

City region	Reporting Companies	R&D Budgets ($ millions) (%)	Patent Count from Location
San Jose–San Francisco–Oakland	380	23,346 (21.9)	17,596
Seattle–Tacoma–Olympia	76	10,496 (9.9)	4,208
Los Angeles–Long Beach	214	8,797 (8.3)	6,065
New York–Newark	209	8,154 (7.7)	8,996
Detroit–Warren–Ann Arbor	98	7,736 (6.9)	2,972
US Total	2,931	106,440	108,592

Note: These statistics refer only to reporting companies regarding their biggest R&D location.

Source: (US) National Center for Science and Engineering Statistics, InfoBrief August 2014, p. 4.

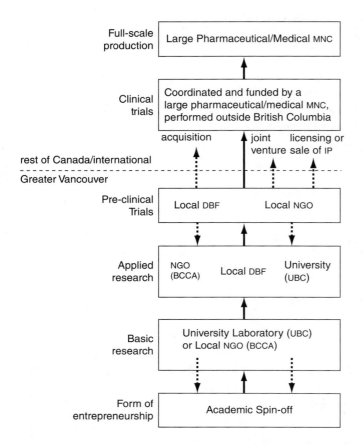

FIGURE 3.3   Network Linkages of Vancouver Medical Biotechnology Firms

Notes: NGO: Non-government organization; UBC: University of British Columbia; DBF: Dedicated biotechnology firms; BCCA: British Columbia Cancer Agency.

Source: Modified figure and text provided by Kevin Rees, May 2009. See K. Rees 2004.

An interesting trend in localized innovation clusters is the growing role played by universities as they move beyond their traditional roles as providers of basic R&D and suppliers of a skilled labour force to become actively involved in the commercialization of technology. This role is not new—universities had connections to the first in-house laboratories in the late nineteenth century—but it has become increasingly important in the ICT era (anticipated by the pioneering efforts of MIT in the 1940s). The Bayh–Dole Act of 1980 gave universities in the US ownership of intellectual property created with federal funding, and with it the right to reap rewards from the commercialization of research. At the same time, the research intensity and complexity of the microelectronic and biotechnology industries, and the corporate desire for rapid product cycles, encouraged firms to access expertise within universities. University–industry research collaboration has deepened considerably; by 1990 there were 1056 co-operative research centres in the US alone. A related, widespread trend is the growth of incubator and entrepreneurial facilities on or near university campuses, designed to promote the commercialization of science and technology by encouraging university faculty to participate in market-driven research and business start-ups themselves.

One illustration of the role universities play in stimulating local high tech activities is the development of the biotechnology industry in Vancouver (Figure 3.3). Although Vancouver lacked a major pharmaceutical firm, since the 1990s the city has developed a major cluster of approximately 90 biotechnology companies, of which the majority specialize in the development of biopharmaceutical and bio-diagnostic products for the global market. Vancouver's biomedical cluster is the seventh largest in North America and the fastest growing in Canada. The origin of the local industry owes much to academic entrepreneurs elaborating on ideas that originated in the basic research they were conducting in local universities, especially the University of British Columbia, which has given rise to approximately 70 per cent of the new firms. These firms specialize in the discovery (upstream) stages of the innovation process and network linkages in the Vancouver cluster and mainly involve corporate collaboration with university labs on applied research, and with not-for-profit research organizations, such as the BC Cancer Agency, on relatively inexpensive pre-clinical trials.

Universities are one component of local innovation systems; whether or not important in directly generating innovations, they continue to play vital roles in basic research and, along with related post-secondary institutions, in developing an educated, creative, problem-solving work force, not only in the natural sciences but also in the humanities and social sciences as well. Moreover, local innovation systems or networks can take a variety of forms.

Globally, Silicon Valley is the well-established icon for agglomerations or clusters of high-tech activities—that is, activities that require high levels of creativity not only in R&D but in design, financing, and marketing. In Silicon Valley, the firms are connected in many ways: through a common labour pool whose members often move from one firm to another; entrepreneurial spin-offs; market exchanges; and use of common services such as venture capital agencies, all of which point to powerful localization economies. The development of Silicon Valley, beginning before the Second World War, was anchored by key institutions such as Stanford University and firms like Hewlett Packard, but essentially it was an organic evolution, albeit with the support of government procurement.

Other famous clusters that have organically transformed old industries and established new ones, as Fordism gave way to the ICT, include the Baden-Württemberg region in Germany, northeastern Italy, and Ota ward in Tokyo. Recognizing the success of these regions, academics and policy-makers encouraged the formation of innovation-led, high-tech, or R&D clusters. Commitments were made to promote education and training to help create innovative, skilled workforces. Governments provided support for R&D and innovation in the form of tax relief and specialized infrastructure and amenities. In many cases governments have sought to "clone" the Silicon Valley experience with varying success. In Japan, for example, the development of Tskuba "science city" was a substantial experiment on these lines without achieving similar global outcomes. In contrast, there have been dramatic successes in developing countries. For example, the Bangalore (officially Bengaluru) R&D centre is India's SiliconValley and has attracted R&D operations of MNCs as well as many domestic public- and private-sector research institutions that have been highly innovative. The Hsinchu R&D centre in Taiwan has been similarly successful and in particular has attracted the R&D and design activities of hundreds of domestic firms in the electronics industry, including globally renowned semi-conductor activities.

Globalization of Local (and National) Innovation Systems

Although the nature of localized networking varies, the clustering of R&D institutions, innovation, and design activities is a well-established pattern. At the same time, local innovation systems and creative links are becoming increasingly globalized and connected to one another as part of **spatial innovation systems**. The emergence of spatial innovation systems has been led by the activities and mobility of highly educated professionals and the growth of the Internet that continues to improve flows of information and allows companies to geographically diversify R&D among new and old clusters. The movement of key ideas between the electronic music industries in the US and Japan, led by American scientists with strong connections to Silicon Valley, and other innovation clusters in the US, are examples (Case Study 3.3). More generally, iconic global innovative clusters such as Silicon Valley have attracted talented individuals from around the world, who in turn have gained qualifications and experience. Many such individuals, termed "argonauts" by Saxenian (2006), have returned

spatial innovation system (SIS)
An international system linking agents around the world involved in the research, development, and transfer of new technologies.

to their home countries and helped establish high-tech clusters in such places as Bangalore, India, and Hsinchu Park, Taiwan. In some instances, such as the Taiwan Semi-conductor Manufacturing Company, argonauts have set up domestic champions with globally networked innovative activities. MNCs have also been more willing to establish important R&D centres in innovative clusters in foreign countries. In part, this strategy has sought to capture the expertise and ideas of these clusters and to diversify innovation potentials within the corporation. This strategy also involves industrializing countries, such as China, which has attracted major R&D investments by foreign MNCs such as Microsoft and Volvo.

Non-local links are often vital to the local viability of emerging innovative activities in peripheral places. In Vancouver's biotechnology cluster, for example, emerging bio-technology firms depend not only on strong contacts with local universities and medical research institutes, but also on access to medical clusters and pharmaceutical giants elsewhere, notably in the US (including Silicon Valley), but also in Europe and Japan (Figure 3.3). Very few drugs developed in Vancouver are actually manufactured there, and local firms must look beyond British Columbia to commercialize their discoveries. A range of commercialization strategies are common within the Vancouver biomedical cluster, including the formation of strategic alliances, the licensing-out of technology in exchange for royalty payments, or the outright sale of technology to another firm. These strategies involve partnering with, or selling technology to, large pharmaceutical firms outside Vancouver and reflect both the substantial cost of performing the necessary clinical trials and the lack of local marketing and distribution capabilities. An alternative strategy is to sell the entire firm to a large pharmaceutical corporation. A local vaccine development company named ID Biomedical, for example, was sold to GlaxoSmithKline in 2005 for $1.7 billion. The survival of this knowledge-based cluster therefore depends on continuing innovation with respect to promising drug therapies and the protection and sale of intellectual property rather than cost-competitive manufacturing.

Innovation and Global Inequality

As Gunnar Myrdal recognized decades ago, the massive disparities in income and standard of living that exist between (and within) rich and poor nations are rooted in powerful forces of circular and cumulative causation. In this view, economic and non-economic processes reinforce one another over very long periods of time to confound any natural tendency towards equilibrium by which, for example, poor regions attract investment by offering low wages and begin to grow, thereby increasing the demand for labour and incomes. Moreover, on a global scale, cumulative patterns of growth and poverty are interwoven with innovation. Simply put, rich advanced societies are typically innovative, while poverty stricken societies are not.

In human history, innovation has always been a source of wealth and power. However, since the Industrial Revolution of the later eighteenth century, investments in R&D, patent activity, and advances in innovation have become more continuous, central to the goals of economic development, and overwhelmingly concentrated in the rich industrialized countries of North America, Europe, and Japan. As stressed by the TEP model and related literatures, widening global-income gaps that have become apparent since 1800 have been significantly driven by innovation. In this (virtuous) view, rich advanced economies are innovative societies, in which innovation stimulates investments, new jobs, skills, and income that progressively increase productivity, wealth, and market demands, driven by increasing scale economies, mutliplier effects, and deepening divisions of labour and levels of knowledge, learning, and know-how. In these societies, markets and businesses are dynamic and highly competitive, and their economic influence is further indicated by the rise of multinational corporate production and marketing networks, and related global institutions, such as patent laws, trade regulations, and the G20. With rising wealth, innovative societies support public service provision and enduring investments in both economic and social infrastructure, and healthier, better-educated workforces that add to the ability and willingness to innovate. In these countries, economic development has been powerfully self-reinforcing, as already illustrated in creating cumulative positive or **virtuous cycles** of self-sustaining urban-industrial growth (Figure 2.1). For present purposes, the key role of innovation can be summarily and broadly underlined (Figure 3.4). Indeed, it has only been since the 1970s and 1980s that other countries—led by South Korea, Taiwan, Hong Kong, Singapore, and China—have begun to catch up with western income levels. It is no

virtuous cycle A cycle of processes that reinforce favourable socio-economic changes (for example, the ability of core regions to attract more people). Investment reinforces their attractiveness for further economic development.

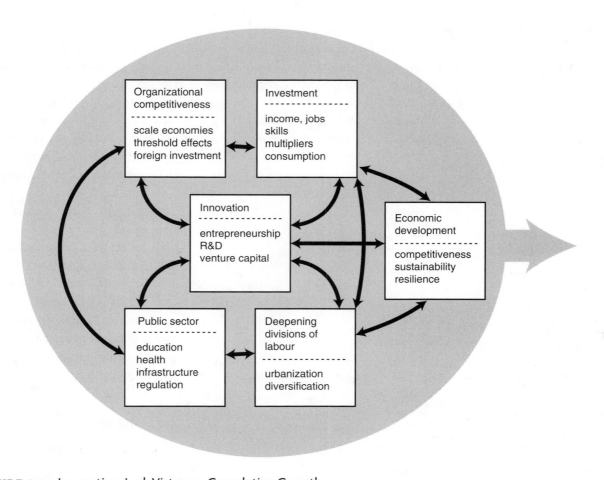

FIGURE 3.4   Innovation-Led, Virtuous, Cumulative Growth
Path-dependent accumulation and commercialization of know-how. Market institutions stimulate and reward innovation. Expansion of exports and foreign direct investment. Know-how protected and elaborated via MNCs, patents, and trade agreements.

coincidence that the level of commitment to innovation of these countries has increased as well. Although policies have varied, all governments in these countries have invested heavily in education, training, and R&D (more so than in Africa and Latin America).

Broadly stated, in virtuous cycles, innovations generate dynamic comparative advantages that lead to increased investment, jobs, income, and taxes, which themselves are reinforced through knowledge spillovers, multiplier effects, and localization economies. This growth generates increasing demand for more goods and services, as well as expansion of the education, training, and medical services that create a smarter, healthier, more productive, and more specialized workforce. At the same time, the division of labour deepens, including with respect to "the creative class," as science, engineering, design, marketing, and advertising activities become increasingly interdependent, professionalized, urbanized, and dedicated to producing further rounds of innovation. While virtuous cycles can wind down, rich economies have enormous resources to reinvigorate themselves (see Figure 2.2 and related discussion).

In contrast, very poor nations and regions left outside of virtuous cycles can experience the multiple self-reinforcing effects of dysfunctional institutions that create **vicious cycles** of poverty or "poverty traps" that lack connections to innovation (Figure 3.5). These problems are deep-seated and involve localized connected problems of lack of education, income, jobs, useful experience, social networks, sources of investment, loss of anomie and self-respect, all of which combine to limit innovative change. In turn, the self-reinforcing nature of poverty can be further perpetuated by unfavourable economic, social, political, and geographical conditions, such as remoteness from markets, racist attitudes, class and caste-systems, foreign control, inappropriate policy choices, and the undermining of local products by imports. Indeed, there is no single "magic bullet" explanation or solution to resolve low incomes and poverty. The challenges posed by the differences in wealth between rich and poor nations become even more daunting when we take into account issues of environmental degradation and justice.

Bearing in mind its cumulative, multi-faceted nature, the "development problem" of stimulating growth and incomes (or more generally standards of living) in poor countries is often summarized as seeking to reduce the "technology gap" (or productivity gap).

vicious cycle A cycle of cumulative self-reinforcing processes that reinforce unfavourable socio-economic changes (in poor regions, for example, outmigration and disinvestment further reduce the prospects for economic development).

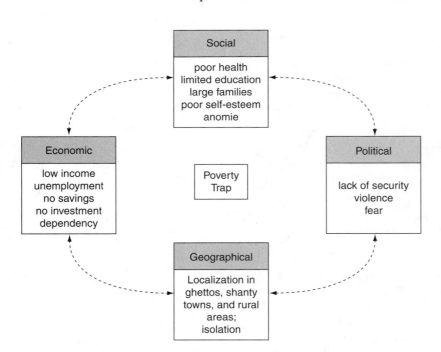

FIGURE 3.5   **Idealization of Vicious (Cumulating) Poverty Cycle**

Breaking the Vicious Cycle of Underdevelopment

Globally, the vast income gaps between rich and poor countries were widely recognized in the 1950s and related to differences in technological capabilities in generating higher levels of productivity. Consequently, efforts to break the vicious cycle of underdevelopment have generally focused on improving the technological capabilities of poor countries and in reducing technology gaps with rich countries. Among western economies, conventional wisdom and policy initiatives in the 1950s emphasized the role of MNCs in transferring technology to catalyze growth in poor countries. Problems with a dependency on MNCs to promote development, however, stimulated alternative approaches to enhancing local **technological capability** related to intermediate technologies and to latecomer firms. The implications of the ICT (and the Internet) for addressing development problems have also been raised.

technological capability The ability (within an economy or firm) to generate commercially profitable innovations.

(1) MNCs as Technology Catalysts

The World Bank was created in 1944 to deal with the post-war reconstruction of Europe, but it soon turned its attention to global poverty. The Bank provided financial support for MNC-led projects in poor countries, but required those countries to show commitment to market (free-trade) principles by reducing barriers to foreign direct investment (FDI). Since 1998, the Bank has recognized the need for developing countries themselves to play a part in formulating long-term plans, but it still looks to MNCs to resolve problems. Underlying the Bank's thinking is the conventional view that MNCs can provide capital for projects in poor countries, offer best-practice technology and supporting expertise, and provide access to global market channels, financing, and internal R&D. From this perspective, MNCs stimulate local technology learning by investing in branch plants that give employees opportunities to enhance their skills, demonstrate best practices to local firms, and introduce new products. In effect, branch plants allow poor countries to "leapfrog" some steps in the process of technological learning.

As the rich countries' preferred approach to resolving global poverty, the involvement of MNCs in development has been criticized from various perspectives. Hirschman (1958) is among the one-time advocates who eventually recognized that foreign-controlled large-scale industrialization was inappropriate to developing countries. One concern is that the technology (machines and equipment) used in branch plants is not readily diffused locally and does not enhance local technological capability. The latest, highly automated technology often requires minimal factory skills to operate, and MNCs often bring in key inputs and expertise from their own facilities rather than buy them from local suppliers. Further, even if large-scale, capital-intensive projects provide a few well-paid jobs and export income, they often disrupt domestic production and traditional practices and damage the local environment, while siphoning any profits away from the country to pay for the foreign investment. Recently, the Nobel-Prize–winning economist and former president of the World Bank, Joseph Stiglitz (2006) has comprehensively criticized the Bank, the International Monetary Fund (IMF), and western governments in general for their inappropriate imposition of market institutions on developing countries. Free-trade rules, for example, may allow subsidized agricultural exports from rich market economies to undermine unsubsidized production by poor farmers, while privatization of land in poor countries has often displaced peasant producers and local crops in favour of specialized export crops organized by MNCs in plantations. The latter may generate export income (and a few high-paying jobs) but the costs can be substantial (high imports of machinery and products like fertilizer, profit loss, unemployment).

(2) Bottom-Up Development and Intermediate Technology

In contrast to the "top-down" imposition of advanced technologies in which know-how is embodied and not easily transferred, **intermediate** or **appropriate technologies** are techniques and organizational practices that are based on local resources and markets and understood and controlled by local people (Case Study 3.4). The term *intermediate* suggests the potential for innovation in traditional societies. Schumacher

intermediate or **appropriate technologies** Techniques and organizational practices that are based on local resources and markets and understood and controlled by local people. The term *intermediate* suggests the potential for innovation in traditional societies.

Case Study 3.4
ALUMINUM POT MAKING IN GHANA

Appropriate technology makes optimum use of the resources in a given environment. The concept is generally discussed in the context of developing countries and is famously associated with Mahatma Gandhi's plea, in the 1920s, for rural Indian communities to develop self-reliance as part of India's struggle to escape British rule. And E.F. Schumacher used the term *intermediate technology* in the 1970s to promote his "small is beautiful" thesis.

In developing countries, appropriate and intermediate technologies are characterized by labour-intensive practices that use local materials and are built, operated, and maintained by local people in a sustainable way. However, appropriate technologies are developed in various ways: by taking technologies that originated elsewhere and adapting them to local circumstances; through R&D aimed at creating them from scratch; and by improving existing indigenous technologies. NGOs and related government agencies such as the Canadian International Development Agency (CIDA) are active in transferring appropriate technologies around the world. Examples include improved toilets designed to reduce disease by providing better sanitation (www.itdg.org/, accessed 25 April 2009) and LED lights designed to provide healthier, more efficient lighting in rural Nepal by the Light Up the World Foundation (http://lutw.org/project_nepal.htm, accessed 25 April 2009). The example of LED lights, a capital intensive, new technology, contrasts with the example of indigenous manufacturing of aluminum pots in the Jamra area of central Ghana.

The aluminum pot industry established in Ghana in the 1960s was economically viable and serving markets throughout the country by the 1980s. But the industry has faced problems, notably shortages of the scrap aluminum used as raw material, the relatively crude nature of the products, and market saturation (the pots are so durable that demand has declined, and many operations have had to leave the business). In response, there has been some product diversification, arrangements have been made with large-scale aluminum plants to supply materials, and the Kumasi Technology Consultancy Centre was established in 1972 to promote appropriate technology. Should the term *appropriate technology* only apply to indigenous technology?

Source: The text on the pot industry was kindly provided by Ben Ofori-Amoah (see Ofori-Amoah 1988).

(1973), who inspired the "small is beautiful" movement, described intermediate technology as a £1 labour-intensive technology and advanced technology as a £1000 capital-intensive technology.

Intermediate technology is not only labour-intensive, but also locally oriented in terms of inputs and outputs, and dispersed in rural as well as urban areas, promising what has been called grassroots or "bottom-up" development (a common concept in community development). Such projects promote skill development in incremental ways that are economically viable. This approach is supported by NGOs and government agencies, such as the Canadian International Development Agency (CIDA). On the other hand, critics of intermediate technology note that harnessing its potential is not easy, that mistakes have been made by "foreign" NGOs and government agencies, and that there is limited potential for learning to further advance technology capabilities. What is appropriate technology, however, has been transformed by the digital age and demands of sustainability. There have been repeated attempts to develop low cost computers and cell phones for use for the poor. The most effective use of digital technology has been the use of cell phones in developing countries to provide market information to farmers and fishermen, to allow electronic transfer of funds, e-banking, and so on. Relatedly, energy for the new electronics has been provided by the use of photovoltaic panels, biogas, and other technologies that combine appropriateness, market efficiency, and sustainability. Micro-finance to initiate small scale entrepreneurship is a complementary financial innovation.

(3) Latecomer Firms as Local Champions

Latecomer firms or **domestic champions** are locally owned businesses established specifically to emulate MNCs in their use of leading-edge practices. Sometimes they even surpass their models in terms of innovation. The rationale behind the latecomer strategy is to overcome the innovation advantages held by MNCs that continue to concentrate R&D in their home countries. The economic history of rich market economies reveals the significance of locally controlled R&D and national innovation systems. In this view, developing advanced technological capability requires big, sophisticated domestic organizations that will pursue R&D and innovative activities at home and act as catalysts within the economy through strong ties with local institutions. Government policy is required to stimulate latecomers by subsidizing R&D and protecting them from MNCs until they are themselves globally competitive and as technically advanced.

Since the 1970s, latecomer firms have been especially associated with Taiwan and South Korea, which themselves have been labelled "latecomer countries" because of their recent industrialization and achievements in technological capability. In South Korea, the latecomer category is dominated by giant, diversified MNCs called "chaebols," such as Samsung, while in Taiwan the leading latecomer firms have tended to be in the electronics industry. Both countries invested considerably in education and introduced various policies to encourage the development of local champions. In Taiwan, for example, the government funded major national research associations, science parks, and related infrastructure; initially encouraged FDI to stimulate development but later emphasized local ownership; and provided substantial tax incentives for local R&D. China is producing many of this type of firm, such as Huawei in telecommunications

latecomer firm A locally owned firm that has emerged in growing industries in latecomer countries—that is, countries that have industrialized successfully relatively recently, including by exporting. Large latecomer firms are sometimes labelled domestic champions.

domestic champions A large, locally (as opposed to foreign-) owned firm that helps to advance strategic industries in terms of innovation, employment, local supply networks, and exports. Some latecomer firms are domestic champions.

equipment and Haier in kitchen appliances. Chinese firms have the added advantage of having a domestic market that is the world's largest within which to develop their economies of scale and innovative capacities. The challenges facing latecomer firms include the identification of appropriate industrial priorities, the development of domestic entrepreneurs, the need for close coordination between government and business, and the necessity of investment in appropriate educational and research infrastructure.

(4) Engaging the ICT

The origins of the ICT reflected and reinforced technological gaps between rich and poor countries. Yet the know-how of the ICT has been successfully transferred to countries such as South Korea and Taiwan by their latecomer firms. Moreover, the ICT has much potential for serving poor people and remote areas by increasing access to banking services and small loans via cellphones, on-line education, and new ways of providing localized sources of power to marginalized populations. The ICT is also stimulating social innovations such as portable sterile surgical packs that can be used in hospitals with limited basic services, and is behind the substantial support of the Bill and Melinda Gates Foundation to provide clean water and effective vaccinations for diseases for the world's poorest people. The global spread of the Internet has also been helpful (Case Study 3.1). If the Internet has many social problems and can be manipulated it also provides ready access to information and networking that can be seen as aiding more meritocratic societies and enhancing the education levels and mobility of individuals.

Conclusion

Markets and institutions alike are pushed by constant innovation. However, innovation involves complex social processes with many actors and networks that operate at and across global, national, and local scales and requires broad institutional support (Figure 3.6).

Thus national innovation capacities are embedded within education and training systems, regulatory and macroeconomic policy contexts (e.g., support for local R&D), communication infrastructures, factor market conditions (e.g., the quality and availability of local labour, risk capital, consumer preferences), and product market conditions (e.g., nature of oligopolies). All of these institutions bear on how knowledge is generated and diffused within and among firms, and related organizations that help create local, national, and global networks. From a policy perspective, the challenge of innovation varies among countries, most obviously at different stages of development and levels of technological capability. Technologically advanced countries seek to maintain global leadership in developing new products, while the poorest countries seek to meet basic needs in terms of food and shelter, and the vicious cycle of underdevelopment. Within countries, innovation challenges and policies also vary. Indeed, virtually all cities and regions in rich countries look to innovation to maintain income levels that reflect distinct local competitive advantages to being creative. At any scale, the very idea of sustainable development depends on innovation. The themes of market, scale, and innovation are present throughout the remainder of this book.

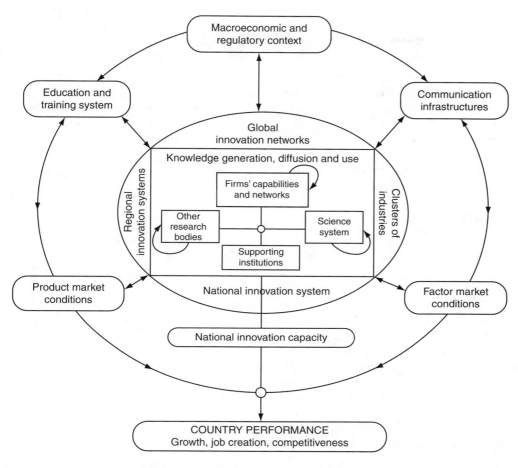

FIGURE 3.6   Actors and Linkages in the Innovation System

OECD (1999), *Managing National Innovation Systems*, OECD Publishing, Paris.

Practice Questions

1. Troll though the business pages of a local city newspaper and identify examples of innovation by small firms. Summarize the nature of these innovations (scale, purpose, jobs created, location, entrepreneurial characteristics).

2. Define "high-tech activity." What kinds of high-tech activity can you identify in your region? Explain any location patterns you observe.

3. Are there engrained pockets of poverty and hardship within your own community or region? Can you explain them in terms of cumulative causation? How can they be integrated into mainstream society?

4. What do you understand by the "Green Paradigm"? Do you agree that public policy should promote such a paradigm? Outline what you think it would look like (key factor industries, main carrier branches, principles of productivity, etc.).

5. How are universities and other institutions of higher learning in your region contributing to the commercialization of innovation?

6. Reference is increasingly made to "green jobs." Is this a useful term? Is your community seeking to promote green jobs?

Key Terms

Recommended Resources

Asheim, B. and Gertler, M.S. 2005. "The geography of innovation: Regional innovation systems." In J. Fagerberg, D. Mowery, and R. Nelson (eds.) *The Oxford Handbook of Innovation*. Oxford: Oxford University Press, 291–317.

Informative review of literature on regional innovation systems.

Bathelt, H., Malmberg, A. and Maskell, P. 2004. "Clusters and knowledge: Local buzz, global pipelines and the process of knowledge creation." *Progress in Human Geography* 28: 31–56.

A conceptual discussion of how clusters or agglomerations shape learning locally and are linked to one another globally.

Bunnell, K., and Coe, N. 2001. "Spatializing knowledge communities: Towards a conceptualization of transnational innovation networks." *Progress in Human Geography* 25: 437–56.

A conceptual discussion of how clusters are globally linked by knowledge flows.

Cooke, P. 1992. "Regional innovation systems: competitive regulation in the new Europe." *Geoforum* 23: 365-82.

Diamond, J. 1997. *Guns, Germs, and Steel*. New York: W.W. Norton.

An insightful effort to understand present patterns of global inequities by tracing them back to the origins of civilization.

Military might (guns), nature (germs), and innovation (steel) are interwoven central themes.

Freeman, C., and Louçã, F. 2001. *As Time Goes By*. Oxford: Oxford University Press.

A lengthy, advanced discussion of the five major TEPs and their central characteristics; elaborates on Freeman and Perez (1988).

Malecki, E.J. 1997. *Technology and Economic Development: The Dynamics of Local, Regional and National Competitiveness*. Harlow: Longman.

An economic geography textbook designed for senior level courses focusing on innovation, technological change, and local development.

Oinas, P. and Malecki, E.J. 2002. "The evolution of technologies in time and space: From national and regional to spatial innovation systems." *International Regional Science Review*, 25: 102–31.

Argues that innovation systems that were once strongly localized have become increasingly globalized and interdependent on one another.

References

Freeman, C., and Perez, C. 1988. "Structural crises of adjustment, business cycles and investment behaviour." Pp. 38–66 in Dosi, G., Freeman, C., Nelson, R., Silverberg, R., and Soete, L. eds. *Technical Change and Economic Theory*. London: Pinter.

Hayter, R. 2008. *Environmental Economic Geography in Institutional Perspective*. Blackwell Compass.

Lipsey, R.G., Carlaw, K.I. and Bekar, C.T. 2005. *Economic Transformations: General Purpose Technologies and Long Term Economic Growth*. Oxford: Oxford University Press.

Maddison, A. 2007. *Contours of the World Economy*. Oxford: Oxford University Press.

Myrdal, G. 1944. *An American Dilemma: The Negro Problem and Modern Democracy*. New York: Harper.

———. 1957. *Economic Theory and Underdeveloped Regions*. London: Gerald Duckworth.

Ofori-Amoah, B. "Ghana's informal aluminium pottery: Another grassroots industrial revolution?" *Appropriate Technology* 14 (4): 17–19.

Rees, K. 2005. "Interregional collaboration and innovation in Vancouver's emerging high-tech cluster." *Tijdschrift voor Economische en Sociale Geografie* 96: 298–312.

Reiffenstein, T. 2006. "Codification, patents and the geography of knowledge transfer in the electronic music industry." *The Canadian Geographer* 50: 298–318.

Saxenian, A. 2006. *The New Argonauts: Regional Advantage in a Global Economy*. Cambridge: Harvard University Press.

Schumacher, E.F. 1973. *Small is Beautiful: A Study of Economics as if People Mattered*. London: Blond and Briggs.

Schumpeter, J. 1943. *Capitalism, Socialism and Democracy*. Harper: New York.

Stiglitz, J. 2006. *Making Globalization Work*. Penguin.

PART 2

Institutional Pillars of Modern Space Economies

CHAPTER 4

Business

The material framework of modern civilization is the industrial system, and the directing force which animates this framework is business enterprise.

Veblen 1904: 7

The goal of this chapter is to establish a foundation to understand and analyse the evolution of businesses (firms) in relation to place and space—how they start up, how they expand, and how they separate and integrate activities over space. This chapter's specific objectives are

- To examine the nature, motivations, and location behaviour of business firms;
- To explain the concept of business segmentation and the different ways in which the activities of SMEs, large firms, and giants are integrated;
- To discuss the locational dynamics of firms, especially with respect to MNCs and foreign direct investment (FDI);
- To explain the location behaviour of local firms and their integration in spatial value chains at various scales;
- To note the greening of value chains.

The chapter has three main parts. The first looks at the nature of business decision-making and integration strategies. The second explores the idea of business segmentation and the characteristics of three types of firm: small firms, large firms, and (giant) MNCs. The last examines the location dynamics of spatial value chains, as well as the influence of environmental imperatives and their implications for local development.

Introducing Business

multinational corporations (MNCs) Firms that control operations (branch plants, mines, offices, etc.) in at least two countries. The transnational corporation (TNC) is a related term that is sometimes used to imply MNCs that no longer identify with a "home" national jurisdiction.

The myriad, ever-changing flows of goods and services that take place within and among market economies, and that constitute "the material framework of modern civilization," are organized mainly by business institutions. For Veblen, writing at the end of the nineteenth century, "business" meant the giant, in some cases rapacious, enterprises—famously represented by John D. Rockefeller's Standard Oil Company—that dominated the US economy in his day. Since that time, giant **multinational corporations (MNCs)** that control operations across international boundaries have become the pre-eminent organizers of global markets. MNCs typically organize and coordinate production and distribution on a global basis, seeking to develop substantial economies of scale and scope among multiple functions, such as production, R&D, finance, purchasing, or marketing, while incorporating the activities of myriad suppliers and distributors. Directly and indirectly, MNCs employ tens of thousands of people and generate sales in the multi-billions.

Corporate giants such as Walmart, General Motors, Exxon, Vodafone, Deutsche Telecom, IBM, Toyota, Coca-Cola, and Nestlé have significant shares of production, sales, investment, and jobs in industries across the economic spectrum, and they enjoy

Case Study 4.1
PROFILE OF WALMART

In 2012, Walmart recorded sales worth more than $489 billion, generated from over 10,000 stores employing 1.8 million people ("associates" in Walmart language) located in 11 countries. Economists claim that Walmart accounted for half the productivity gains achieved in the US between 1995 and 1999, and in 2005 it accounted for 10 per cent of US imports from China. Both Walmart's growth and its geographical diffusion have been spectacular; for many observers, the company is the supreme example of globalization. The spatial division of labour that Walmart has used so successfully combines a supplier network involving many SMEs in competitive open markets with its own big-box store distribution system. All corporate giants such as General Motors, Exxon, Vodafone, Deutsche Telecom, IBM, Toyota, Coca-Cola, Nestlé, and Microsoft develop spatial divisions of labour, but in different configurations.

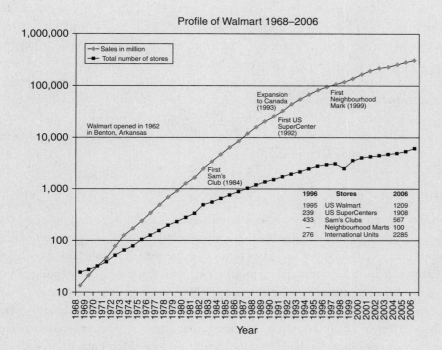

Profile of Walmart 1968–2006

annual revenues that are greater than the GNPs of most nations. In 2012, Walmart was the world's biggest MNC in terms of revenues—the first retail company to achieve that status (see Case Study 4.1). Yet for all their power, the giants make up only a tiny fraction of all business firms. A far greater proportion consists of **small and medium enterprises (SMEs)** with diverse organizational, location, and technological characteristics. In addition, there are many **large firms (LFs)**, which share some of the characteristics of both SMEs and giant MNCs to form a distinctive business segment.

Businesses, firms, corporations, or enterprises are private sector organizations that are inherently based in both place and space. They are place-based in the sense that they require specific locations—offices, factories, stores, warehouses—in which to conduct their economic activities. The location of those activities has ramifications for local employment and spending on goods and services, as well as tax revenue. The fact that firms are embedded in place also influences the character of local development: the nature and organization of economic activities, environmental impacts, and standards of living (as a result of incomes generated directly and indirectly through *multiplier* effects). SMEs that operate in only one location are intimately connected to specific places, typically owned and managed by local people, and often highly dependent on local markets, labour, and sources of supply. But the branch plants of MNCs are place-based, too (in Walmart's case, in the form of "local" retail stores). At the same time businesses are space-based in that their functions are integrated into dispersed *value chains* of decision-making, production, innovation, distribution, and marketing, in which MNCs with operations in multiple national jurisdictions play an especially important role as organizers. Moreover, MNCs are likely to have a pivotal role in transforming *value chains* to *value cycles* driven by the environmental imperatives of reduce, reuse, and recycle.

For statistical purposes, national and international bodies collect data on businesses according to the sector in which they operate. Each of the three basic sectors—primary (resources and agriculture), secondary (manufacturing), and tertiary (services)—is further divided into many other categories: for example, mining (primary), textiles (secondary), and finance (tertiary). However, individual businesses often perform functions that fall into different sectors. The reality of production is a flow of materials, energy, and services within value chains to and from multiple sectors and industries, including not only activities directly related to production but also supporting activities such as finance. These value chains include firms of many different types within and among sectors.

Defining Business

Businesses or firms in capitalist societies are *legal organizations* that supply goods and services to consumers in exchange for remuneration and profit. As *legal* entities, firms are subject to social sanction and have various entitlements and obligations, while as *organizations* firms integrate the activities they control and have some degree of durability. Legal entitlements consist most obviously of the property rights to the goods and services offered for sale and the right to obtain remuneration for the goods and services so provided. Depending on the nature of their activity, firms also incur a variety of legal liabilities or responsibilities related to safety, health, and environmental impact.

small and medium enterprises (SMEs) Relatively small firms, owned and controlled by individual entrepreneurs and employing no more than 500 people (usually fewer). SMEs are usually highly localized, single-facility operations.

large firms (LFs) Firms with roughly 500 to 10,000 employees, typically focused on a limited range of products and serving global markets.

business or firm A legally recognized private-sector organization that employs at least two people to provide a good or service for sale.

Legal definitions of a firm vary from one jurisdiction to another, but three features are standard:

1. A firm is an entity with legal responsibilities and liabilities of its own, separate from those of its managers, and often its owners.
2. A firm owns the property rights (and obligations) attached to the goods and services it offers for sale until the exchange is completed. (A market, by contrast, is the vehicle for the exchange of those rights: it never possesses them itself.)
3. A firm is durable in the sense that it must exist for some time. There is no prescribed minimum period, but to qualify as a firm, a business must be in operation at least long enough to incur legal obligations and participate in a market.

The distinction between privately owned (and controlled) businesses and publicly owned (or "listed" businesses) is important to understanding their behaviour. Scale and distribution of ownership are key factors. Thus most SMEs are owned and controlled by individuals, individual families, a partnership, or some other form of limited ownership. In contrast, larger firms and MNCs are typically listed on public stock exchanges where literally thousands of people can buy shares and become part-owners in the firm but without much, if any, influence over decision-making, which is conducted by hired executives and managers. Two institutional innovations in the late nineteenth century greatly facilitated the growth of giant firms and ultimately MNCs: limited liability and decentralized decision-making. The spread of limited liability laws limited the risks facing shareholders to their investment if businesses failed, and reduced financial restraints on the growth of firms. Meanwhile decentralized decision-making structures removed the "fixed" managerial constraint on growth. Alfred P. Sloan became famous for his development of a "divisional" decision-making structure at General Motors.

The distinction between locally owned and controlled firms and listed firms is often blurred in practice. SME owners can appoint professional managers, and there are giant companies that are privately controlled and owned and others, including Walmart, that are publicly listed but remain privately controlled by a few individuals or families. Joint venture businesses typically involve joint ownership and control over specific projects. Large-scale operations can also be "nationalized" by being moved to public ownership under government control and conversely government-owned operations can be "privatized" by sale on the stock market or by other means. Note that state-owned companies (SOCs) that are professionally managed and mandated to be economically viable are significant actors in the resource sector (see Chapter 9). In general, the nature of ownership and control has important implications for how organizations behave. Theories of the firm and business segmentation explore these implications as we shall see.

Theories of the Firm

While there are numerous theories of the firm, three long-established approaches remain useful starting points for understanding firm behaviour and roles in the economy. These approaches differ in terms of the nature and goals of decision-making and how firms relate to their business environments.

1. Neoclassical Theory of the Firm

neoclassical theory of the firm A theory based on *Homo economicus*, in which firms are assumed to make optimal decisions based on complete information and perfect rationality.

optimal decisions Decisions and behaviour that maximize or minimize some specified objective (e.g., maximizing profit or minimizing cost); associated with neoclassical theory.

In the basic **neoclassical theory of the firm**, firms are abstractions ("mental constructs") that make decisions objectively and automatically. This decision-maker is sometimes personified as *Homo economicus* (a characterization defining the corporation legally as a person). Graced with perfect information and rationality, *Homo economicus* (a firm) makes **optimal decisions**, whether by minimizing costs or maximizing profits (Table 4.1). In this theory, the economic landscape is understood in terms of costs and revenues—for example, how costs of production and transportation vary from place to place. This landscape is taken as a "given" to which firms must respond. Neoclassical theory justifies its abstract approach by arguing that, in competitive markets, only firms that are efficient and rationally located will survive; how they actually make their decisions is irrelevant. In other words, competition weeds out inefficient behaviour and poor choices with regard to matters such as location and land use. This approach is primarily concerned with understanding how cost and revenue "fundamentals" shape aggregate patterns of production, prices, and locations.

2. Behavioural Theory of the Firm

behavioural theory of the firm A theory that sees firms as using their limited information and bounded rationality to reach decisions that are acceptable rather than optimal. See also **satisficers**.

satisficers Decision-makers who make choices intended to meet aspiration levels. The choices are based on the availability of, and ability to interpret, information.

franchise A business owned and operated by an individual or partnership but controlled according to the system of rules and conventions established by a core firm.

In contrast, according to the **behavioural theory of the firm**, decision-makers have limited information about their business environment and limited capacity to understand it (i.e., "bounded rationality"). Unlike the efficiency-optimizing firm envisioned by neoclassical theory, in behavioural approaches decisions are based on how information is found, learned, and evaluated. In this view, decision-making is influenced by the experience, abilities, and personal preferences of decision-makers and the time and costs used in search processes. In behavioural theory, decision-makers are not perfectly well-informed, relentlessly logical, or optimizers. Rather they are **satisficers** who tend to assess alternatives sequentially rather than comparatively and base their decisions, at least in part, on their aspirations and with what they feel comfortable, often choosing the first alternative deemed satisfactory. Why would one popular local restaurant stay local and another decide to open **franchises** across the country (Case Study 4.2)? According to the behaviourists, the answer lies in the information that each operation obtains, the different

TABLE 4.1	Theories of the Firm and Decision-Making Characteristics		
Characteristic	**Neoclassical theory**	**Behavioural theory**	**Institutional theory**
Decision-making	*Homo economicus*	Satisficing	Technostructure
Decision-making assumptions	Perfect information, perfect rationality	Imperfect information, imperfect bounded rationality	Hierarchical structure, strategic
Goals	To minimize costs and/or maximize profits	To meet or better aspiration levels	To grow, achieve security, be profitable
Nature of market relations	Open and relational	Unclear	Administered, relational
Choice process	"Instantaneous," best chosen from alternatives	First viable option chosen on basis of information available	Bargaining with other actors, strategy shapes choice

Case Study 4.2
SATISFICING AND FRANCHISING

Richmond Row in London, Ontario, is well known for its agglomeration of nightlife venues (aka bars). Long the hangout of University of Western Ontario students, this section of Richmond Street underwent gentrification in the late 1970s. One of the institutions that helped to kick off that process was Joe Kool's. Serving massive portions of pizza and nachos with a touch of nostalgia, it remains one of London's most popular spots and could have capitalized on its success by franchising. Instead, its owners chose to open a few different-themed restaurants nearby, satisfied to be royalty in London.

Around the same time, in the suburbs, the Jeffery brothers opened up Kelsey's Roadhouse. This initiative was a great success, and within two years the brothers established a franchise operation called Kelsey's. Economies of scale in advertising, purchasing, training, and management made it possible to expand to over 100 locations in central and western Canada. Kelsey's also gave birth to the Montana's and Outback chains. Eventually, the chains were sold to CARA, the largest operator of full-service restaurants in Canada (Harvey's, Swiss Chalet, Milestones, etc.). Like many franchise chains, Kelsey's stardom eventually faded and many outlets were closed in 2012–13.

aspirations of the decision-makers involved, and the ways in which they evaluate information.

3. Institutional Theory of the Firm

The **institutional theory of the firm** emphasizes how and why firms grow, the nature of corporate strategies and structures, and their bargaining power and organizing role in value chains. Thus an institutional analysis of a restaurant chain such as Kelsey's (Case Study 4.2) would focus on the owners' motivations for expansion, the rationale behind the decision to establish a franchise operation rather than a directly owned and managed chain, and the factors that account for the strategy's success (or failure). This approach recognizes the reality that markets are dominated by giant firms in which major

institutional theory of the firm Focuses on how and why firms grow with emphasis on the structures and strategies of large and giant firms run by technostructures with the power to influence the behaviour of other organizations.

technostructure The top-level management and supporting staff that control large firms and MNCs; a concept associated with institutional theories.

decisions are made by professionalized managerial bureaucracies or **technostructures**. These firms—well illustrated by Walmart—are motivated by a relentless commitment to growth of revenues and market share. Their size allows them to realize economies of scale that minimize the opportunities for new competitors to enter the market, while their bargaining power allows them to influence suppliers, distributors, consumers, and governments. Institutional theorists maintain that, as a result, advanced market economies are typically characterized not by "fair" competition among small firms, but by oligopolies and monopolies (Chapter 1). In this approach, corporations are powerful, and public policy is constantly challenged to ensure corporate behaviour accords with social priorities.

Corporate strategies and structures explain the how and why of corporate growth.

Corporate Strategies

corporate strategy Allocation of investment to meet long-term corporate goals (growth, profit, market share, downsizing, restructuring).

internal (corporate) growth Investment in new plants and facilities.

external (corporate) growth Growth through acquisition of existing plants and facilities.

vertical integration Expansion either upstream, towards sources of supply (backward vertical integration), or downstream, towards markets (forward vertical integration).

Many firms have no desire to grow beyond a "satisfactory" size (Case Study 4.2) that meets the MES (Chapter 2). But for large and giant firms, there are a range of powerful motivations to keep growing, related to increased profits and market share, the realization of economies of scale and scope, the prestige and egotism of managers and owners, and survival. Indeed, growth itself can be an objective; too little growth creates threats from shareholders and acquisitions attempts by rivals while too much growth can over-extend managerial and financial resources. Firms are able to grow for various reasons. For example, firms expand their operations to multiple locations because they have developed core competencies, know-how, and experience they can draw upon, have established marketing networks, and access to financing. Typically, firms grow in particular *related or path-dependent* ways that are expressed in terms of corporate strategies. (Figure 4.1).

Corporate strategies reflect plans for growth in the "long run" and refer to the allocation of investments that take time to implement. They meet basic corporate goals regarding growth, profitability, and market share and relate to changes in markets, competition, and governance. Strategies may be informal, but large corporations need a formalized strategy to plan for consistent, coordinated growth and for diffusion among employees and stockholders. Firms can choose to grow by investment in new facilities—referred to as "**internal**" **growth**—or by acquisition of existing facilities—referred to as "**external**" **growth**. Whether internal or external, business evolution is path dependent along connected or integrated strategies. Two basic types of corporate strategy are vertical and horizontal integration strategies.

Vertical and Horizontal Integration

When businesses expand through **vertical integration** they take control of successive stages or tasks in a particular value chain. Firms can integrate backwards to an earlier stage in the supply

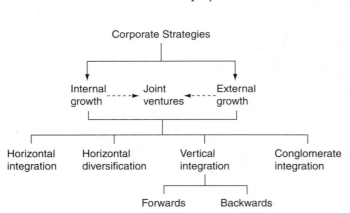

FIGURE 4.1 **Classification of Corporate Strategies**

chain or forwards towards a later one such as distribution or even recycling. Vertical integration can occur through investment in new facilities (internal growth) or through acquisition of former suppliers or customers (external growth). The rational for such strategies is to reduce the transaction costs of using markets (Chapter 1) and to gain control and stability over the quality, quantity, and timing of the flows of goods and services in production processes. Resource companies often exhibit a strong commitment to vertical integration; oil companies, for instance, in addition to drilling for oil, often transport, refine, and distribute as well. Even so, oil firms also rely on many external suppliers of equipment, inputs, and services, as well as many end users, such as car owners.

Horizontal integration occurs when firms expand their existing activities into new locations, as illustrated by the geographical proliferation of Walmart stores (Case Study 4.1). Horizontal diversification extends the concept of vertical integration to include a related but different range of products or services. For example, the Canadian company Bombardier initially developed as a vertically integrated manufacturer of snowmobiles, but later diversified horizontally into the production of trains, airplanes, boats, and ATVs. Horizontal integration allows a firm to exploit the competitive advantages and expertise it has developed in one location in other locations and take advantage of economies of scope by applying the managerial, financial, distribution, and marketing (especially brand) capacities it has already developed for one product to a range of products. The term *horizontal integration* is most often used to refer to firms that expand (by internal investment or external acquisition) in the same or similar product categories. However, horizontal integration also occurs when two firms selling competing products collaborate on activities such as R&D, management, or marketing.

Firm growth typically involves geographical expansion and the creation and integration of multi-plant and even multinational operations. Thus a primary reason for vertical integration is to access distant markets or remote resources, while horizontal integration is strongly associated with expansion into new markets. Geographical diversification may itself become a corporate objective and, for example, associated with the risks and loss of bargaining power of operating in one jurisdiction. Firm growth also requires adjustments to corporate structures, forms of governance, or managerial capacity. In particular, as firms become bigger, more complex, and geographically diversified, they need to invest in larger more specialized technostructures to provide coordination and direction. With growth, firms typically relocate technostructures from original locations, usually adjacent to operating sites, to separate head offices with their own location requirements. Similarly, R&D activities often begin near an operating location and with firm growth are established in their own location. Head offices and R&D activities are the control activities of firms, responsible for their survival, growth, and evolving integration as spatial systems.

Horizontal and vertical integration strategies, and their variations, emphasize business evolution in terms of path-dependent trajectories, or "common threads," that seek to realize value by focused learning processes. These processes develop upon the accumulation of specialized production and marketing know-how, assets, and networks. If rooted in internal abilities and motivations, strategies of value realization, as ultimately measured by sales, profits, growth rates, dividends, and size, also need to take into account external considerations, such as levels of competition, market opportunities,

horizontal integration Expansion in the same or related lines of business. Expansion in related lines of business is sometimes called horizontal diversification.

and resource availability, as an exception. Conglomerate (or holding company) growth strategies involve diversification beyond existing know-how and assets, typically by acquisitions of firms with different expertise. However, if conglomeration is justified, in terms of the spreading of risks and for facilitating the movement of capital from declining to growth activities, it has been criticized for undermining efforts to fully realize value within specialized units.

At the same time, integration strategies of value realization can be relatively diverse, and even involve small firms in distinct activities. Sitka Surfboards, for example, was recently (2011) founded in Victoria, British Columbia, by two university graduates who recognized a market niche for surfboards specifically for cold water surfing. Yet their ability to more than break-even depended on designing a clothing line with sales from their own retail store. Because they have a close friend from New Zealand, a store has been opened there as well. In general, value realization strategies can be hard to predict, and common threads can be complexly interwoven, including among networks of firms of varying sizes.

Business Segmentation

There are a very large number of firms in the world and their size distribution looks like a pyramid, with many SMEs at the bottom and a small number of giant MNCs at the top. Although the precise shape of the pyramid varies from place to place—SMEs are a more important part of the national economy in Japan and Germany, for example, than they are in the US and Canada—the general pattern is similar everywhere. The size distribution of business establishments (e.g., individual factories, mills, workshops, offices, warehouses, retail outlets) also conforms to a population pyramid. Most SMEs operate at just one location, but giant MNCs typically control numerous establishments in several countries. In addition, population pyramids of firms and business establishments have been an enduring characteristic of national economies for a long time. However, their composition is dynamic. Individual firms can change in size and fail; small firms typically exhibit high turnover rates with death rates being especially high in the first few years after formation. But through growth, decline, failure, and acquisition the largest firms change over time too.

As one example of the inverse relationship between the number of establishments and their size by number of employees in Canada in 2012, 75 per cent of the 1,107,540 businesses that employed workers had less than 10 employees, and less than 1 percent employed more than 500 employees (Industry Canada 2014). (In addition there are many "tiny" owner-operated businesses). The distribution pattern is similar when size is defined by sales rather than number of employees. However, large and giant businesses directly account for important shares of jobs, investment, and production, and help organize the activities of many smaller firms within value chains. Moreover, businesses differ, not simply according to size, but as institutions and in their interrelations.

The idea of **business segmentation** recognizes that firms vary according to their institutional characteristics and that it is not easy for firms to move between segments (because of the need to change as institutions). The most obvious institutional differences are between giant MNCs or **transnational corporations (TNCs)** and SMEs.

business segmentation The grouping of businesses by type in terms of routines, structures, and strategies.

transnational corporation (TNC) A term often used as a synonym for MNC. However, TNCs were originally defined as corporations with no loyalty or commitment to any particular home region.

TABLE 4.2   Business Segmentation: Idealized Characteristics of Capitalist Firms

	Giant firms	Large firms	SMEs
Decision-making structures	bureaucratic, techno-structures	entrepreneurial, decentralized	entrepreneurial
Product-mix	diversified	focused	niche-oriented
Markets	global	global	local
Economies of scale	significant	significant	limited
Economies of scope	moderate	significant	significant
Research and development	significant	significant	variable
Innovativeness	moderate	high	variable
Employee relations	structured, often unionized	structured, mainly non-union	unstructured, largely non-union
Ownership	public	public	private

Source: Based on Hayter et al. (1999).

Galbraith (1967) referred to these as "planning system" and "market system" firms respectively, because the former have important powers to shape the direction of the overall economy, while the latter survive by responding to market forces. Further, these two types differ significantly in terms of their routines and conventions regarding decision-making structures, production, marketing strategies, labour relations, innovation, and local impacts. But there are "in-between" firms that cannot be classified into these polar segments—here classified as large firms (LFs)—with their own distinct characteristics. LFs also recognize the possibility of moving from one segment to another, although it is not straightforward. Table 4.2 summarizes the overall characteristics of the three types, while Figure 4.2 shows the place of LFs in the triad segmentation model.

Giants (MNCs and TNCs)

Giant national and multinational corporations are the dominant firms and principal integrators of the global economy. Their ability to bring together multiple parts of their value chains is the result of the economies of scale and scope they have achieved, both in managerial and supporting functions (e.g., finance, accounting, human resources, R&D, health and safety) and in operational functions (e.g., logistics, production, engineering, marketing). A giant need not be superior in all functions—it need not even perform all of them—but it must have outstanding advantages in some of them. Walmart doesn't produce anything, but it has created extremely efficient supply-chain and

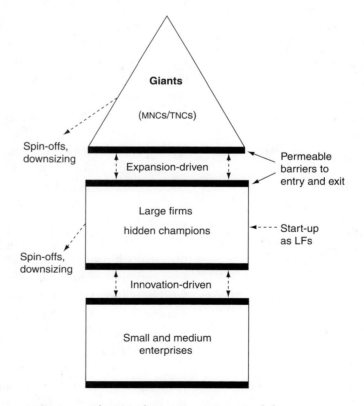

FIGURE 4.2   The Triad Segmentation Model

Source: Based on Hayter et al. 1999: 429

distribution systems that serve a vast network of retail outlets (more than 8,400, under various names) in 15 countries around the world. Sony, on the other hand, concentrates on the production end of the value chain: although it has its own retail shops, they account for a relatively small proportion of the firm's revenue.

Giants are big relatively and absolutely: revenues generated, jobs provided, size and geographic distribution of facilities (e.g., factories, offices, stores). Around the world, there are more people employed by Walmart than there are people with jobs of any kind in British Columbia, and the firm's annual revenues amount to roughly 50 per cent of the annual budget of Canada's federal government. Corporate concentration—the extent to which markets are controlled by relatively few firms—is another indicator of giant presence. Typically, giant firms are part of oligopolies in at least one market and hold significant shares in others. At one time General Motors had 50 per cent of the North American auto market, as well as large shares of the markets in Europe, China, and South America. Walmart's 30 per cent share of the North American market provided the base it needed to expand to Europe, Asia, and South America (although with varying levels of success). Big firms invest in big factories, mines, and service-based functions that are important in local job markets. Dependence on a single dominant employer is a feature of many specialized resource communities, but it can also be observed in bigger cities. In general, the power of giant MNCs is even more apparent in regional and local contexts than it is on the national level.

Organizational Structure

managerial structures How decision-making functions within a firm are coordinated across space and in places.

Giant firms have **managerial structures** that coordinate their activities across space and within particular places. All MNCs are essentially hierarchical, as top management has ultimate authority. However, that authority is delegated to the management of various divisions, departments, teams, or other groupings around the world. Operation units may be organized in various ways: on the basis of functions (e.g., accounting, production, marketing), products, brands, processes, or geographical regions. In turn, decentralized decision-making is controlled by formal and informal chains of command and lines of communication. An important advantage available to MNCs is the ability to enhance the two-way flow of knowledge and resources between "control functions," notably head offices and R&D laboratories, and diverse operating units in the realization of core competencies, or the assets and expertise that are the basis of corporate competitive advantages. Vertical and horizontal integration develop core competencies and structure the growth of firms in particular product and geographical directions. As a firm grows, it typically adapts (decentralizes) its organizational structure to better manage operations of increasing size and complexity. Occasionally, a firm may choose to transform its core competencies in radically different strategic directions, in effect transforming its organizational structure. IBM's transformation from a vertically integrated hardware hierarchy to a horizontally integrated services network is a classic example (Case Study 4.3).

Structure varies with sector and financial strategy. While MNCs such as Federal Express, McDonalds, Mazda, and HSBC each focus on a single sector, more diversified companies such as Honda (motorcycles, robots, cars), Apple (computers, phones, MP3 players), Pepsico (beverages, snack foods, cereals), and Virgin (music sales, airlines, banking, communications) span several sectors. Very often such firms try to establish

Case Study 4.3

INTERNATIONAL BUSINESS MACHINES (IBM): FROM THE VERTICAL TO THE HORIZONTAL

As the developer of key tabulating, typing, mainframe, and personal computing and information storage technologies, IBM is one of the world's most influential companies. For most of its history IBM was vertically integrated, hardware-focused development. When software entered the picture, anti-monopoly regulations required that it be either bundled with the hardware or sold as a separate product.

In the late 1980s, the computer industry began to segment horizontally, with different firms producing individual components (e.g., hard drives, micro-processors, RAM, software). Workstation computer-makers challenged IBM's mainframe business. IBM attempted to transform its organization into independent horizontal divisions, but by the early 1990s was forced to reduce its workforce from 370,000 to 225,000 employees.

In 1993 Louis Gerstner, as the new CEO, developed a strategy focused on generating value by offering computing and information services designed to improve customers' business process efficiency. Since then, IBM has developed a workforce, acquired companies, and created both the hardware and the software necessary to fulfill the strategy. In the 2000s, the firm divested itself of its hard-disk and PC manufacturing businesses, and today hardware is merely a component of its service products. With technical professionals accounting for roughly half of its 400,000 employees, IBM now generates its profits by providing integrated services to companies, governments, and NGOs through local experts around the globe.

IBM's operations are perhaps the most pervasively global of any company. Close to 300,000 of its employees work outside the US, and it has research centres in China, India, Japan, Switzerland, and Israel (its Swiss research centre has produced four Nobel Prize winners). The success of IBM's strategy is such that in 2008–9, HP, Xerox, Cisco, and Dell all adopted similar strategies.

IBM's Income Streams

1994			2012		
Division	$ Millions	Percentage	Division	$ Millions	Percentage
Hardware sales	32,344	50.4	Global technology services	40,236	38.5
Software	11,346	17.7	Global business services	18,566	17.8
Services	9,715	15.1	Software	25,448	24.4
Maintenance	7,222	11.2	Systems and technology	17,667	16.9
Rentals and financing	3,425	5.3	Global financing	2,013	1.9
Total revenue	64,052	100	Total revenue	104,507	100

IBM's income segments have changed significantly from 1994 to 2012, as the table shows.

connections among these sectors, as Apple does with its operating systems and app store and as Virgin has done with its unique approach to customer service. Holding companies (conglomerates), which control shares and provide overall financial control among otherwise separately managed and different types of businesses, capitalize on economies of scale in financial resources and expertise to draw together many companies; examples include Thomson Reuters, GE, Berkshire-Hathaway, and Cheung Kong Holdings.

In most giant firms, ownership and management are formally separated and governance (or decision-making) is the responsibility of the professional management and staff that make up the technostructure. Most individual owners of shares belonging to giant companies listed on stock exchanges have little influence over technostructures. However, some powerful owners (with significant share holdings) have representation on the boards of directors, which choose the chief executive officers (CEOs) of corporations. And technostructures are necessarily disciplined to some extent by the need to provide adequate returns to all shareholders. This discipline is strongest in the countries where the firm's shares are listed on public stock exchanges, but firms still have to pay attention to the concerns of investors in other countries. Companies in which a single family has a controlling interest are less subject to stock-market pressures; even if their holdings are listed on stock markets, the family may take the lead in matters of governance. Stockholder influence is also less influential in Japanese *keiretsu* and other similar groups where cross-shareholding among firms predominates.

Some location decisions are related to vertical integration: for example, when retailers establish their own manufacturing capacities or manufacturers develop retail outlets for their products. Others are necessitated by horizontal expansion: for example, when an auto company builds a plant in a new location to increase its production capacity, or when it acquires a plant that produces a related product.

Large Firms

Giant MNCs play the key role in integrating the various parts of value chains, and many of them gain their competitive advantage from cost leadership and differentiation across a wide range of products. By contrast, large firms focus on a particular product (or set of related products) that fits into a value chain, and they gain their competitive advantage from economies either of scale or of scope. For example, the real value in your cellphone is produced by dozens of component manufacturers that make various processors, memory, screens, etc. on which all the cellphone manufacturers depend. As Table 4.2 indicates, large firms share some characteristics with both SMEs and giants. They are entrepreneurially organized and they employ skilled, flexible labour to produce specialized product mixes for international markets; they have global marketing networks, and often operate branch plants. In addition, large firms are typically global market leaders in their particular market niches, committed to ongoing research, development, and innovation. In short, they share the entrepreneurialism and flexibility of SMEs, but resemble giants in their ability to exploit economies of scope and scale, to reach global markets, and to engage in ongoing R&D. Giants clearly impose powerful barriers to the growth of smaller firms. What allows large firms to thrive despite those barriers is their focus on innovation within an area of specialization not dominated by giants. Driven by innovation, large firms develop and protect their competitive advantages and market niches, legally through patents, through brand names that influence consumer preferences, and through production and marketing networks that they control. Large firms can resist acquisition threats from giants if they prefer independence.

Although not widely known like giants, large firms, also labelled hidden champions and mittelstand (Germany), little giants and gazelles (US), threshold firms (Canada), middle-market companies or brittelstand (UK), and backbone firms (Japan), are significant among market economies. Indeed, a recent (2015) UK report indicated that firms with annual sales of US$50–500 million accounted for 13–17 per cent of GDP in leading EU and US economies. Their head offices can be found in a diverse range of locations, including in relatively small places; they can be "big firms locally" in terms of adding skilled jobs, R&D, and exports; they often offer alternative, often rival products to giants; and they are growth-oriented.

Small and Medium Enterprises (SMEs)

Giant firms cast a giant shadow because of their integrating functions, market power, and ability to take over both rivals and smaller firms. Nevertheless, even as large corporations have extended their dominance throughout the primary, secondary, and tertiary sectors, small firms continue to form the broad base of every market economy. There are many reasons for the persistence of SMEs:

- *Ongoing entry of new SMEs*: New firms are continually created by people wishing to start up their own businesses. Across the service sector, for instance, new restaurants, dental offices, and retail stores open every day.
- *Economies of scale*: There are many cases in which smaller firms are better suited than large ones to realize appropriate economies of scale. For example, dispersed markets may be too small to permit large-scale local production or distant suppliers may face significant transportation costs. SMEs play powerful roles in providing external economies for larger firms.
- *Diseconomies of scale*: Firms will eschew growth in many activities when the returns become negative because of bureaucratic, equipment, or other costs. Here too, opportunities open up for SMEs to serve the needs of MNCs.
- *Innovation*: MNCs do not have a monopoly on knowledge or innovation. SMEs (and individuals) can also generate new technologies, products, and services, and can take out patents on their work. Many new firms are created to serve a new market niche, and represent market-driven innovations (Chapter 3).
- ***Anti-monopoly laws:*** Modern economies have long had laws that prevent monopolies or socially undesirable restraints on competition ("anti-trust law" is the usual term in the US). Such laws often serve to deter giant firms from seeking monopoly positions.

  anti-monopoly laws Laws that prevent excessive corporate concentration in an industry and/or unfair business practices.
- *Personal preference*: Many people—entrepreneurs, managers, and workers alike—would rather work in SMEs than in giant firms and massive factories. Small firms can provide more opportunities to get in on the ground floor of a growing company, and social relations within them are often less hierarchic and more collegial than they are in larger businesses.

Independent SMEs are present in every sector and location in the economy, but they share several characteristics in common. They are typically owner-managed and reliant on private funding and bank loans. Decisions regarding matters such as marketing, production, employee relations, and financing are made by individual people (as opposed to management groups) and involve personal risks. Employees tend not to be unionized; entrepreneurs often see unions as a threat to their decision-making discretion and an additional cost burden. Two factors in the flexibility of SMEs are the speed with which decisions can be made and the ability to exploit economies of scope in human and capital resources. Usually specializing in one particular activity, SMEs tend to operate from very few locations (often just one), to create only a handful of jobs, and to have low levels of sales or output.

SMEs vary greatly in their strategies and competitive strengths. Many are "laggards" that rarely change business practices and have little interest in growth, but others are highly dynamic. Many SMEs are closely tied to giant companies such as suppliers, distributors, or franchisees in highly price-competitive and technologically dependent relationships. Others possess a specialization that ensures market power. Some compete with the giants in local markets; others compete internationally, using cost and differentiation strategies to their advantage. All SMEs are integrated into value chains to some degree. Moreover, SMEs are closely associated with the idea of entrepreneurship and their collective contributions are a vital indicator of local vitality. Thus SMEs contribute towards local decision-making and control; they explore and "test" local market opportunities; they add to employment levels at little cost; they are typically strongly connected to local supplier systems, enhancing multiplier effects; decision-making is less bureaucratic than in giant firms and more flexible; they can operate in a wide range of locations and make use of old buildings and equipment; and they are effective training grounds for future entrepreneurs. The formation of new SMEs can reinforce new directions; for example, the idea of "green entrepreneurship" refers to recent trends among some new firms to ensure their activities are environmentally benign, as well as profitable, from inception.

Location Dynamics

Firms extend their production and sales activities beyond their home bases in order to meet strategic goals (market size, share, profitability, stability, and growth). Geographic expansion can be advantageous in any number of ways: by permitting economies of scale and scope; accessing markets or resources; increasing shares in foreign markets; reducing production costs; securing skilled labour; countering and/or anticipating the strategies of rivals; and avoiding excessive geographic concentration.

Firms develop new locations in two basic ways. First, firms can invest in new facilities in a new or greenfield site or/and (especially) region. Firms sometimes undertake such investments as part of a joint venture with other firms. **Greenfield investments** allow firms to choose preferred locations, hire new labour forces, and design their own buildings. However, they involve the time and uncertainties of choosing the appropriate location, construction, and operating in unfamiliar places. Second, firms can acquire existing activities and locations. This reduces such costs and uncertainties, but may not be an available option or may raise problems such as the need for renovation. In-situ

greenfield investment
Investment in new locations where entirely new facilities can be constructed.

adjustments are also locationally selective in that they involve investments at existing operations rather than elsewhere to expand, modernize, and/or change product-mix. Among multi-plant firms, decisions to close or downsize a particular factory are similarly locationally selective.

Understanding why firms choose particular locations for their operations is not straightforward. The term *location* is also ambiguous in regard to scale. It may refer to a specific site or a particular community, region, or even nation. Location choices at any of these scales, and the associated implications for development, are influenced by industry, history, and geographical context as well as by the organizational and behavioural characteristics of the firm (as depicted in the business segmentation model).

International Expansion of Giant and Large Firms

Giant and large firms gain control of locations across space, usually beginning in their **home economies** and then expanding first inter-regionally and eventually internationally to host economies. The expansion of giant corporations to oligopoly or monopoly control of their national space increased rapidly in most developed countries in the decades after the Second World War. Similar processes are now underway in China, India, Brazil, and other developing countries. Because of the interpenetration of trade among countries, prices are set on global markets and companies must compete with the most efficient producers and distributors wherever they are located. The reality for many firms today is that they must internationalize rapidly.

home economies The countries or regions where MNCs that control FDI in host economies have their head offices.

The world's biggest firms are multinational (or transnational) in scope, and while estimates vary some suggest that the largest 500 firms account for almost half of global GDP and about one-third of world trade. MNCs' share of R&D budgets is also impressive. To cite one example, Apple's R&D budget for 2012 at $3.4 billion was around 80 per cent of Harvard University's operating budget and greater than that of the University of Toronto. Only about 20 or so nations have GDPs bigger than the sales of the largest MNCs, such as Walmart. MNCs often have high levels of foreign as well as domestic assets; in 2014, for example, Walmart employed around 2 million people, 800,000 of whom lived outside the US.

MNCs internationalize by investing directly in branch plants that are part of foreign-based "subsidiaries." According to the OECD, such **foreign direct investment (FDI)**, to be defined as such, must amount to at least 10 per cent of the value of the enterprise, and must come with significant management control. Foreign investors can be individuals, governments, or other entities, as well as businesses. Most foreign investment goes into the acquisition of existing firms. In 2008, FDI around the world amounted to US$1.8 trillion; most of that total involved acquisitions ($1.6 trillion) and most ($1.2 trillion) went to countries in the developed world.

foreign direct investment (FDI) Investment by MNCs in foreign countries in operations that they own and control.

The location entry model provides a useful framework for understanding the economic geography of FDI. Firms internationalize (or inter-regionalize) their operations to exploit internally generated advantages or competencies, whether in production, marketing, or technological know-how. This know-how is a powerful incentive for FDI as it provides extra returns on the related fixed costs in R&D, marketing networks, and other corporate services. In addition, establishing branch plants typically serves

entry advantages Advantages based on accumulated expertise that allow firms to establish operations in new regions.

spatial entry barriers The costs and uncertainties that face firms contemplating investment in a new region or country.

to extend administered markets, thus reducing transaction costs. Such **entry advantages** provide incentives both to expand and to overcome the various **spatial entry barriers** that foreign investors face in unfamiliar environments. These barriers include the managerial costs and uncertainties of trying to understand local conditions and whether firms need to make adjustments. Local firms conducting business in their home territory enjoy several advantages over foreign firms. For example, they "inherit" a great deal of information regarding local laws and legal practices, government policies, social attitudes and customs, language, and religious beliefs, as well as the nature of local suppliers, consumers, and financial organizations, without direct cost. Local firms also have access to networks of friends and contacts with local knowledge of one kind or another. Knowledge of the local physical environment is another component of local entrepreneurial advantage. Furthermore, local firms do not have to communicate and coordinate dispersed operations across national boundaries and can often rely on direct personal contact—the most effective form of information exchange about business matters. The fact that internationally expanding firms enjoy none of these advantages constitutes a significant barrier for them.

Spatial entry barriers are especially important when firms contemplate investing in places where differences in language, taste, customs, and so on define psychological distances. Even when the host community speaks the same language, differences in tax and accountancy laws, labour relations, supplier conventions, and government practices need to be understood. Thus geographically expanding firms must engage in costly, time-consuming research into host economies. Failure to properly assess local institutions and conditions frequently causes problems—sometimes outright failure.

The advantages that allowed Walmart to cross borders stemmed in part from its strengths in procurement and logistics and also in part from the huge volume of the purchases required to stock its hundreds of "big-box" stores. Using these advantages as the basis of its strategy in Latin America, Walmart was able to buy or build hundreds of stores and to impose massive entry barriers to any other firm, existing or new, that tried to compete on its terms. In European markets, such as Germany, however, Walmart faced competitors with similar strategies, and was unable to buy enough good locations from them. Further, its customer practices and labour relations didn't meet national norms or regulations. In 2006 Walmart sold its 85 stores in Germany to a German rival, losing $1 billion in the process. The fact that it has also experienced difficulties in the UK and South Korea underlines the significance of spatial entry barriers, even for giant firms that are already multinational in scope. Target's failed expansion into Canada in 2013 (by acquiring an existing firm), terminated less than two years after entry, cost $1 billion, again reflecting poor planning and recognition of what (Canadian) consumers were expecting.

To establish branch plants in unfamiliar places, internationalizing firms rely on accumulated expertise, know-how, or competence in production, marketing, procurement, and/or technology. The costs involved in the creation of these competencies need not be re-incurred in foreign economies if they are internalized in the branch-plant operations. In effect, the size of the foreign firm's entry advantage is defined by the fixed costs and uncertainties that would face any local firm that tried to acquire the same competencies.

Aggregate Patterns of FDI

Historically, most FDI has occurred among rich ("developed") countries. The most notable exceptions have been poor countries with important natural resources, such as the Middle Eastern countries that attracted investment from major oil companies in the West. In 1990 and 2012, rich countries continued to be the main destinations and origins of FDI (Table 4.3). Significant shifts are nevertheless evident. The share of inward FDI in developed countries declined from 75.3 per cent in 1990 to 62.3 percent in 2012, while their dominance of outward FDI declined from 93.1 to 79.1 per cent between 1990 and 2012.

Among developing countries, Taiwan, Singapore, Hong Kong, and Malaysia were important recipients of FDI in the 1970s and 1980s, while China has dominated inward flows since 1990. Several "transitional" countries in East Europe, such as Poland, have also emerged as important recipients of FDI. Taiwan, Hong Kong, and China have also emerged as significant sources of FDI, with many flows occurring within "Greater China." The US, however, continues to be the dominant source of inward and outward FDI.

FDI continues to be attracted to the markets (and skilled labour) of rich countries. The strong shift in FDI towards China is not only chasing its vast pool of "cheap" labour, but reflects China's growing market potential as well. Overall, the combination of low costs, growing pools of highly skilled labour, market potential, and relative political stability helps to explain the shift of FDI to Asia. (In fact, rapidly growing Asian countries are themselves increasingly important sources of FDI.) Until recently, Africa's limited markets and labour skills and high political instability have discouraged FDI, but now China's demand for natural resources is transforming its economies.

TABLE 4.3 Inward and Outward Stocks of Foreign Direct Investment 1990–2012 ($US millions)

	Inward		Outward	
	1990	2012	1990	2012
Developed countries	1,563,939	14,220,303	1,946,832	18,672,623
USA	539,601	3,931,976	731,762	5,906,169
Canada	112,882	636,972	84,807	715,053
UK	203,905	1,321,352	229,307	1,808,167
Germany	111,231	716,344	151,581	1,547, 185
Japan	9,850	205,361	201,141	1,054,928
Developing countries	514,319	7,744,523	144,664	4,459,356
Africa	60,675	629,632	20,229	144,735
Asia	340,270	4,779,316	67,010	3,159,803
Latin America	111,373	2,310,630	57,357	1,150,092
World	2,078,267	22,812,680	2,091,496	23,592,739

Source: UNCTAD Handbook of Statistics Online (http://www.unctad.org). Accessed 2 March, 2014.

No Place Like Home: New Firms and SMEs

New firms are typically small operations, almost always located in the home communities of their founders. For them, as for most SMEs, location is not the subject of a decision-making process: it's simply a given. The tendency for new firms to be embedded in home places is well-documented and expressed as the **seedbed hypothesis**. Often a new business starts in the founder's basement or garage before larger premises are found nearby if growth is sufficient. In this view, the apparently casual way of choosing a location in the founder's home environment is rational. There are many personal costs and uncertainties involved in establishing a new business as founders must take on multiple tasks related to production, marketing, financing, accounting, and hiring, etc. The struggle to reach a minimum efficient scale of production is too much for many new firms, even in familiar settings. To locate a new business in an unfamiliar place would compound the costs and uncertainties of the initial start-up and search for viability. By locating in familiar territory, new founders capitalize on their "inherited" knowledge of their local institutions, laws, and customs.

Entrepreneurs who are thoroughly familiar with their home locales are also likely to have a good number of contacts there. Potential entrepreneurs are likely to know where to look for premises in which to locate; they may also know about possible workers, and will at least understand the characteristics of local labour. They may have contacts with local financial institutions and knowledge of local markets and suppliers. The entrepreneur's "real" home provides a ready-made head office and even manufacturing space (in garages, spare rooms, etc.), at least for a while. In other words, local entrepreneurs "inherit" considerable knowledge about their home environment as part of their birthright. The very market opportunities that potential entrepreneurs set out to exploit are typically rooted in their experience of their home places. There are important exceptions to seedbed generation of entrepreneurs—places such as Silicon Valley that attract people from around the world because the culture and financial resources encourage business builders. Even so, Silicon Valley is also famous for local kids, such as Steve Jobs, starting a business out of a garage.

New Firm Characteristics

The motivations for starting new businesses vary. Some are economic: to make money, to earn a decent living, to be rewarded for effort. Non-economic motivations involve quality of life and work satisfaction. Quality-of-life goals include the desire to follow a particular lifestyle, achieve peace of mind, or work at one's own pace; being one's own boss, taking on new challenges, and realizing one's potentials are important work satisfaction goals.

Traditionally, SMEs had mostly male owners and managers. In recent decades, however, female entrepreneurship has increased. Many businesses are started by serial entrepreneurs who establish new businesses following earlier failures; others are spun off from existing enterprises, with which they often maintain business relationships.

SMEs themselves are the greatest source of new entrepreneurs. Thus the nature of a region's existing industrial structure strongly influences new business formation there. This path dependency can be both beneficial and detrimental. The fact that SMEs beget SMEs has important implications for regional variations in rates of new firm formation.

seedbed hypothesis Argues that new firms are typically located close to their founders' home place, typically where they have lived for a long time. Many new firms begin in the homes of founders and move to nearby locations when they need more space.

Regions with large populations of SMEs tend to have higher rates of new firm formation than regions that are dominated by large-scale establishments. The reasons for this "industry structure" effect on new firm formation are several: employees in SMEs perform a greater variety of tasks than employees of large plants, are more familiar with owners and managers, and often have opportunities to watch and gain insights into running a business. On the other hand, since they tend to have less job security than employees of large firms, they have less to lose if they leave. In contrast, large firms offer more specialized jobs, are more likely to be unionized, and pay higher wages and benefits (pension schemes, vacation rights and pay, medical insurance, etc.) These benefits give workers powerful incentives to stay.

Value Chains and Inter-firm Relations

Economic activities are organized into value chains in which many different types of firms have roles; giants and LFs provide key internally controlled links, while SMEs are integrated into value chains through the external economies of both localization and urbanization economies (Chapter 2). The two main types of external economy involve integration into either global or urban value chains. SMEs that are integrated into global value chains specialize in a particular function (e.g., parts manufacture, advertising, retail sales, recycling). Most SMEs are organized within a social division of labour, horizontally and vertically. Horizontally, SMEs and LFs are important in enriching product differentiation and rivalry in providing similar goods and services. Vertically, SMEs play important roles in upstream and downstream activities and are often coordinated by LFs and giant firms.

Upstream Integration

In the hierarchical (or vertical) model, core firms buy goods or services from a first or top tier of suppliers or subcontractors, who in turn purchase goods or services from a second tier of suppliers, who in turn purchase from a third tier, and so on. The core firm engages in direct market relationships only with **tier one suppliers** (Figure 4.3); this process is often referred to as **hierarchical subcontracting**. In the **lateral subcontracting** model, core firms negotiate directly with all their suppliers; however, the latter may use subcontractors themselves, in which case the distinction between lateral and hierarchical organization is blurred. In both cases, the key to the relationship is the fact that specialist suppliers with specialized knowledge and expertise are able to produce certain parts or services at lower cost, or of higher quality, than the core firm could. Because specialist suppliers can serve numerous customers, they are able to achieve minimum efficient economies of scale.

tier one suppliers Suppliers of key components to core firms in relational markets at the top of the hierarchy of suppliers.

hierarchical subcontracting Contracting out to a limited number of important suppliers of key components, who in turn engage other suppliers to provide the good or service in question.

lateral subcontracting Contracting out directly to all suppliers.

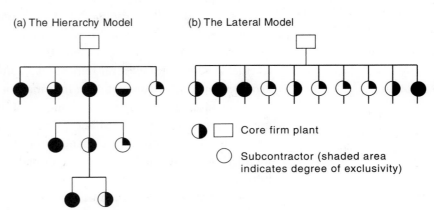

FIGURE 4.3   Two Models of Subcontracting Organization around a Core Firm

The specific mechanisms of both hierarchical and lateral arrangements depend on the industry, region, and cultural context. The suppliers may depend on one core firm for all their revenue or some proportion of it. When one or more core firms are present and there is significant competition and specialization among suppliers, the area can be called a cluster. Large specialist firms may complement and compete with SMEs in the cluster. Also within clusters, capacity suppliers provide a product that the core firm also produces. Core firms will purchase from capacity suppliers during upturns in the business cycle while reducing purchases during downturns. Capacity suppliers reduce core firms' costs by allowing "production smoothing" of employment and equipment operation and by keeping employees' wage demands in check. Core firms and suppliers may also use contractual and part-time workers to deal with rise and falls in demand.

Cluster relationships are particularly deep (vertically) and broad (horizontally) in manufacturing industries, but clusters are powerful economic forces in other sectors and segments as well: natural resources and services, software, tourism, sports and entertainment, advertising, logistics, R&D, and finance. Some of the most famous clusters include finance in New York, London, and Toronto; aerospace in Los Angeles, Toulouse, and Montreal; automobiles in Nagoya-Hamamatsu, Detroit, Baden-Württemberg, and southern Ontario; film in Hollywood and Bollywood (Mumbai); electronics in Silicon Valley, the Pearl River Delta, and Osaka; and oil in Houston. Major Canadian clusters include film in Toronto and Vancouver, aerospace in Montreal, fossil fuels in the oil patch (Calgary–Edmonton), electronics and software in Ottawa–Hull, finance on Bay Street (Toronto), and automobiles in southern Ontario. The integration and development of these areas are greatly influenced by territorial governments, universities, and industry associations that provide R&D and training support, infrastructure, a political voice, and venues for communication.

Downstream Integration

Downstream activities—logistics, warehousing, and retailing—are also organized on a cluster basis, although they are often internally integrated. However, logistics, warehousing/jobbing, and retail firms offer producers significant external economies of scale in terms of financing, distribution, promotion, and contact with buyers and consumers. They also offer various efficiencies because of their experience with customer preferences. SME distribution firms are completely independent of larger distributors, logistics firms, or manufacturers, but others may be influenced by them through contractual or administrative arrangements. It is common for distributors to have exclusive arrangements, either contractual or informal, for specific areas.

Franchises are more detailed arrangements under which the larger firm has much greater control over its brand and pricing while providing guidance and imposing restrictions on the proprietor. Manufacturers and service firms organize franchises for wholesalers, retailers, and service providers. SMEs may also organize themselves into voluntary chains and co-operatives. Global manufacturing and service firms play important roles in local economic activities, especially in small communities.

An increasingly important downstream activity is recycling. Most recycling and waste management programs are organized by municipalities, but overseen by national or provincial regulations. Municipalities are directly in charge of taxing or imposing fees on waste, providing facilities, and organizing collection. In most cases they stipulate what types of waste have to be separated and how each is to be disposed of. However, persuading companies to make their products recyclable—for example, through "take-back" laws requiring sellers to take back goods once they are no longer useable by consumers—usually requires the authority of governments at the national level. National governments in Japan and Europe have instigated the development of national recycling facilities for many products. Large producers, distributors, and recyclers (e.g., Waste Management Ltd in North America) play major roles in the integration of the system. Many SMEs are locally contracted to carry out the collection, separation, recycling, and marketing of recycled materials.

Urban and Community Integration

Every place participates in both local and global value chains. Every place has some companies (or branches of companies) that export services, parts, components, or finished products, and even more companies or branches that import products from other places. On the other hand, most local or regional economies feature local value chains, often dominated by services, from hairdressers to realtors, from construction workers to physicians. Such local value chains are generally self-sustaining, although they depend on multiplier effects resulting from participation in global value chains.

Local specialists in many fields—optometry, dentistry, real estate, restaurants and bars, bicycle repair, acupuncture, and pharmacy—have no need of integration into global value chains. Nevertheless, many such services are integrated into large corporations or franchised in order to achieve economies of scale in functions such as marketing. Another way of achieving integration is to work with others in the same field, as realtors do with multiple-listing services.

Other local specialists rely on local integrators to help them find a niche. General contractors perform that function for specialists such as plumbers, carpenters, surveyors, and architects. The largest local economic institutions, however, are often not businesses at all, but hospitals, schools, and universities. These big institutions are internally integrated to a great extent, and offer a diversity of services to local communities. At the same time, they often rely on the private sector for specialist services: public hospitals and private physicians, for instance, routinely work together. Private-sector services may also compete with them to some extent.

Manufacturers generally have complete control over the particular product they make, and some control over suppliers, distribution channels, and use by consumers. Retailers depend on suppliers for products, but have significant influence over them. Logistics firms, which provide bundles of support services such as transportation, warehousing, inventory management, and freight forwarding, are increasingly important to the organization of value chains.

The most pervasive examples of integration are the various components of the urban infrastructure. While private utilities typically provide the energy infrastructure (electricity, gas, etc.), the rest of the necessary infrastructure (roads, sewage systems, transit, water) is usually provided by government or government corporations. Businesses also tend to rely on government to handle matters such as land use and planning. These arrangements can vary, however. Jurisdictions that outsource some of these functions (e.g., private water supply, private transit) offer great opportunities for local SMEs.

The Greening of Value Chains

Environmental considerations have played an increasingly significant role in shaping corporate strategies over the last fifty years, especially in developed countries. Policy legislation since the 1960s has imposed increasingly strict regulations on air and water pollution, and in recent years policies have sought to reduce wider impacts on climate warming and loss of bio-diversity. The impetus for this change has been the recognition that polluters should pay to eliminate the damages incurred by their activities.

Throughout the history of industrialization, most companies were allowed to externalize the costs of pollution by making society pay for the health and ecological damage incurred by pollution, but society no longer is willing to let companies escape that responsibility. Government regulations at local, regional, and national levels remain the most powerful forms of governance as they impose legal authority on corporate pollution. Regulations are generally classified as either "command and control" or market-based instruments. The first requires that companies meet performance standards (i.e., limit pollution) or adopt specific technologies, while the second allows companies more freedom to devise their own ways to limit pollution while minimizing costs.

Globally, the UN does not have legal authority, but has succeeded on several occasions (e.g., Montreal Protocol on ozone depleting substances) in getting countries to collectively act on environmental regulations. Environmental NGOs (see Chapter 7) also impose governance at the local to global scales. Alongside governance requirements, companies have responded to demands for better environmental performance according to the geography of the impacts incurred by their processes and products. A timeline of the geography of corporate responses can be summarized according to three main response eras.

End-of-Pipe Localized Solutions: 1960s

End-of-pipe solutions were company responses when the effluents they were releasing from pipes into rivers or the emissions they were releasing from smokestacks were forced by legislation to be reduced or eliminated. Companies first figured out how to capture and treat these releases in situ using various waste treatment facilities, precipitators, or scrubbers on smokestack emissions, sealed landfills, etc.

Sometimes the pollution was just sent farther away; for example, a new pulp mill that opened on British Columbia's coast in 1967 was required to build an effluent disposal pipeline to the ocean floor where it was assumed "nature" would take care of the problem; and in Sudbury, Ontario, the nickel smelter INCO reduced the pollution that was devastating the local landscape by building a much taller smokestack to spread the pollution around the region.

Pollution Prevention Extends Changes from Local Processes to Global Value Chains: 1980s

In-process changes are innovations in the actual production process, whether they are in raw material refining, manufacturing, or distribution, in order to eliminate pollution, and solve the environmental and financial costs of end-of-pipe solutions. For example, much pollution from paper-making has been eliminated by substitution of oxygen bleaching for chlorine bleaching, and nickel and copper smelting pollution has been reduced by the use of flash smelting over conventional technologies. Pollution prevention also allowed companies to take the initiative and put in place a market-base response to the governance or the threat of governance.

In so far as pollution-prevention measures typically affected local operations, however, they didn't deal with the causes arising from the makeup of the product itself. Lifecycle analysis (LCA) identified a product's environmental impact throughout the resource extraction, material processing, manufacturing, distribution, consumer use, and disposal/recycling stages of the value chain wherever they occurred. LCA forced companies to recognize the full extent of their responsibilities and many responded by using design for environment (DFE) to change the makeup, inputs, processes, functioning, and recyclability of their products (e.g., in electronics, eliminate heavy metals, brominated flame retardants, and PVCs; improve energy efficiency; reduce packaging; take-back for recycling). Regulations on elimination of hazardous waste, product take-back, energy efficiency, and eco-labelling pushed companies to do LCA and DFE. Often, these laws were drafted to eliminate impacts in a particular location (e.g., hazardous waste into municipal landfills in consuming countries), but forced changes to process and inputs far upstream (e.g., manufacturing in producing countries).

Sustainability Strategies Implemented Globally and Locally: 2000s

MNCs first improved environmental governance in developed home and host countries, and then transferred product and process improvements to developing countries to avoid criticism. Transferral of environmental management practices was spurred by the need to geographically standardize production processes, save costs, and meet NGO pressures. Companies realized that global demands for sustainability would continue to escalate and, as a result, incorporated sustainability into corporate strategy, thus making issues such as environmental design and management priorities. SMEs don't have the power to reconfigure the value chain according to environmental priorities, but manufacturers are compelled by MNC-control of the value chain to improve their performance and distributors (large and small) are increasingly expected to provide products that minimize the environmental burden on the local and global environment.

Local Development Perspective

Local development is closely tied to the nature of business segmentation, interfirm relations and location dynamics. As we have outlined, SMEs, LFs, and MNCs have distinctive characteristics, roles, and locations within value chains (and cycles). Thus there are places that are dominated by highly interactive networks of SMEs, and perhaps some LFs, whose localization economies define the nature of local development, often summarily labelled as "flexible specialization." In this model, SMEs are individually specialized and collectively flexible, able to adjust to new demands and innovative new products. On the other hand, there are enclave economies that are dominated by the branch plants or mines or plantations of MNCs that can bring substantial growth and income to remote places but are vulnerable to the priorities of businesses with multiple options elsewhere. Other small, specialized places are the locations for the head office, R&D, and major operations of an LF, offering a range of high income, stability, and problem-solving skills.

There are also places with highly diverse mixes of SMEs, LFs, and MNCs. In some places, a giant may dominate by organizing local subcontracting and in other places networking may be more lateral (see Chapter 11). Then there are places that have attracted a wide range of functions that are connected to multiple value chains but that are foreign or externally controlled as branch plant economies.

The local development implications of MNCs are especially complex and controversial, given their size and geographical scope. While the investment decisions of MNCs are typically large-scale, offering massive local development impacts, they have options to locate elsewhere. Consequently, local development is not only shaped by competition, technological change, and market and resource access but also by corporate discretionary powers and priorities. FDI has been especially controversial.

Potentially, FDI is a catalyst to local development by adding jobs, transferring best practice technology, improving local labour training, stimulating local supply connections, and providing demonstration effects that occur when local firms are able to observe and learn from innovative larger firms. Such behaviour stimulates local multiplier effects in both quantitative and qualitative terms. Core firms can promote upgrading through the instructions they provide regarding product specifications, operating processes, and management systems. Leading-edge firms provide benchmarks that other firms can either adopt wholesale or adapt to their needs. However, as we noted in Chapter 3, the role of MNCs in transferring technology to poor countries has been criticized. In part, problems arise in these contexts because of a lack of locally based SMEs, LFs, and related institutions that can absorb and develop the technology and demands of MNCs. Even in rich countries such as Canada, it is claimed that high levels of foreign ownership can distort or "truncate" the economy by undermining domestic technological capabilities and possibilities for SMEs and LFs (Case Study 4.4).

Further, MNCs may price the internal exchange of goods and services to and from affiliated operations to minimize tax and royalty payments or extract profits from distant locations. Perhaps even more egregious behaviour occurs when MNCs locate head offices and related functions in tax-free havens, in many cases paying little or no taxes in the places they operate. Apple, Google, Starbucks, and Amazon have been criticized recently in this regard, and so have many others. Such manipulation of transactions raises questions about "disproportionate" business power, corporate social responsibility, and the need for more global regulatory responses. In whatever precise form, tax avoidance and secret funds on a massive scale poses significant problems: they reduce the ability of governments to provide services; the tax burden is transferred to residents; local competitors who pay taxes are undermined; the ability of workers to negotiate wages is reduced; income-gaps are widened; and the power of global financial centres is reinforced.

As dominant players in the global economy, the way MNCs allocate specific functions to particular locations influences the nature of employment opportunities across space. For example, it is argued that MNCs have helped to institutionalize patterns of uneven regional development and inequality in terms of jobs and income within and among countries. In this view, high paying, stable head-office and R&D jobs are concentrated in a few major cities in core regions, while low paying, vulnerable routinized jobs are located in peripheries (Chapter 1). Yet, this view on so classifying peripheries needs to be tempered by the role of SMEs and LFs as job providers, as well as the role of public sector

Case Study 4.4
THE TRUNCATED FIRM

The term "truncated firm" was coined in Canada to refer to branch plants in secondary manufacturing that are controlled by MNCs (mainly American). Such subsidiaries are seen as truncated because they do not perform all the functions that domestic firms in the same line of business would; they do not even take much responsibility for buying the necessary inputs and services. Rather, via administered markets, they rely on the head offices of their parent companies for R&D, strategic decision-making, and services such as accounting, marketing, and procurement. Basically, the truncated branch plant produces a standard product for local markets. Revenues from these sales are used to pay for the services and inputs supplied by the parent company and to provide profits. Because truncated firms do not have the mandate to buy such services and inputs from local sources, local multiplier effects are reduced, as are opportunities for high-paying jobs. Regions with high levels of foreign ownership in the manufacturing sector, such as southern Ontario, can be said to have truncated (industrial) economies.

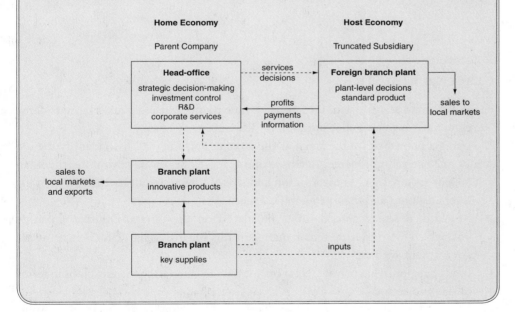

organizations. Moreover, the conditions underlying business segmentation and location dynamics evolve, sometimes profoundly, as implied by the transformation from Fordism to the ICT and the imperative of flexibility and globalization (Chapter 3). This transformation has both threatened existing business practices and led to widespread business failure and job loss while stimulating new forms of business practices and job gains. Meanwhile, places once considered cores have declined and have sought to restructure even as new economic spaces have been generated. The implications of the search for flexibility by business have been especially important for labour (Chapter 5) and have led to changing roles for governments (Chapter 6) and non-profit organizations (Chapter 7).

Conclusion

Business is the most important institution of the spatial economy. It draws people and capital together in one place to create a division of labour that will produce a good or service. Thus in its most basic form, the private firm is inherently centripetal, and the majority of small firms stay focused at this organizational and geographic scale. However, increasing scale and complexity of production requires integrating activities over space and minimizing associated transportation and transaction costs. The segmentation of SMEs, MNCs, and LFs and the coordination among them arises chiefly through the larger firms developing the capacity to organize value chains. Diverse parts, products, and services must be assembled, distributed, serviced, and, eventually, either disposed of or recycled, and the locations where these activities are carried out may span multiple environments, jurisdictions, languages, and cultures. The advantages of clustering need to be weighed against those of dispersal, and today any location decision must also take into account environmental impacts. Spatial integration also creates challenges for labour, governments, communities, and NGOs that seek to modify the behaviour of business in line with their interests.

Practice Questions

1. Why should MNCs' use of tax havens to hide profits and avoid taxes be considered egregious? Is this a policy question for the G20?
2. Using a published source (such as the annual "Fortune 500" list), identify the top 10 to 20 private companies that have one or more operations in your metropolitan area or region. Discuss the location and size of these operations, their roles in the local economy, and their geographical reach.
3. At national scales, should limits be placed on the foreign control of private business? Does industry sector matter in this regard? What if foreign firms are government-owned?
4. In your region, what private sector businesses are generating the most jobs in terms of size and sector?
5. What industrial sector is Apple a part of? What functions of that sector does Apple perform, and where? What other firms perform different functions in the same sector, and where?
6. Identify the largest 10 or so largest private sector businesses by employment and/or sales in your region. What are their functions? Where are their head offices? How dominant are they in your region?
7. Assess the view that FDI is driven by desires to access low cost labour.

Key Terms

Recommended Resources

Dicken, P. 2010. *Global Shift*. London: Sage.

A broad, authoritative review of the impact of MNCs in several sectors.

Iammarino S. and P. McCann. 2013. *Multinationals and Economic Geography*. Cheltenham: Edward Elgar.

A balanced assessment of the economic geography of MNCs.

Patchell, J. 1993. "Composing robot production systems: Japan as a flexible manufacturing system." *Environment and Planning A* 25: 923–44.

An analysis of value chains based on the hierarchical model of relational contracting in Japanese robot manufacturing; notes the ability of the firms in question to create localized production systems in new economic spaces within Japan.

Rice, M.D. and Lyons, D.I. 2009. "Geographies of corporate decision-making and control: Development, applications, and future research in headquarters location research." *Compass* 2/6: 1–15.

Examines location dynamics of corporate headquarters, especially rapidly growing firms of all sizes.

Yeung, H. 2004. *Chinese Capitalism in a Global Era: Towards a Hybrid Capitalism*. London: Routledge.

Examines the role played by ethnic Chinese in the dynamic growth of East and Southeast Asian economies and how Chinese capitalism has changed in response to globalization.

Web Sources

Greenwald, R. *The High Cost of Low Price*, http://www.walmartmovie.com.

A DVD that takes a critical look at Walmart; the website provides a free trailer.

References

Galbraith, J.K. 1967. *The New Industrial State*. Boston: Houghton Mifflin.

Hayter, R., Patchell, J. and Rees, K. 1999. "Business segmentation and location revisited: Innovation and the terra incognita of large firms." *Regional Studies* 33: 425–42.

Industry Canada 2014 SME Research and Statistics. Ottawa: http://www.ic.gc.ca/eic/site/061.nsf/eng/02715.html (accessed 28 February 2014).

UNCTAD. 2005. *Global Statistics* (http://globstat.unctad.org/).

Veblen, T. 1904. *The Theory of Business Enterprise*. New York: Charles Scribner's Sons.

CHAPTER 5

Labour

Fictitious commodities . . . possess qualities that are not expressed in the formal rationality of the market.

Polanyi 1944: 71

The overall goal of this chapter is to explore how labour markets function in place and across space. Its specific objectives are

- To explore the employment relationship as an agreement between workers and employers;
- To examine labour market segmentation as a process that varies over space and time;
- To explain the shift from Fordism to flexible labour markets and its geographical implications;
- To note the geographic mobility of labour and its implications for local development; and
- To identify indicators of labour market performance and the main forms of labour market failure.

The chapter is divided into three main parts. The first part introduces the idea of labour market segmentation and focuses on employment relations, governments, and unions as formal institutional influences on labour markets. It also discusses how segmentation can be anticipated by the career expectations of young people. The second part explores how the structured forms of labour segmentation that were developed during Fordism, especially in North America, have been challenged by more flexible work cultures, especially those developed in new economic spaces. The third part outlines labour market failures of various kinds (e.g., high rates of unemployment, income inequality, and job discrimination).

Labour as Countervailing Power

The labour market resembles other markets in that it involves exchange between businesses and other organizations (the "demanders") that offer remuneration (wages, salaries, fees) to workers (the "suppliers") in return for their labour. It differs from other markets, however, as Karl Polanyi (1944) pointed out, in that labour is a "fictitious commodity." Unlike commodities such as apples or automobiles, labour is embodied in individuals who are not created for sale in markets, who exist outside of markets, and who have ideas, skills, habits, social goals, networks, and so on. Even within markets, labour, unlike other commodities, is able to promote its own interests, most obviously through labour unions but also through various social and political forums. That is, *labour has institutional power.*

Historically, labour has been the most powerful countervailing power to business. For example, back in 1937, a strike by General Motors workers in Flint, Michigan, won significant benefits for workers and their families and marked a turning point in the rise of industrial unionism in the US (Case Study 5.1). More recently, in China, 2010 saw the rise of labour unrest at electronics and auto factories that may represent a turning point in the ongoing effort to shift control of labour unions from the communist state to the workers. Yet labour cannot escape the vagaries of markets. The deep recession that started in 2008 meant massive lay-offs and the loss of many previously won benefits for autoworkers throughout the US Midwest. Meanwhile, in the Pearl River Delta, employers can no longer rely on cheap migrant labour as jobs and pay in China's interior improve.

Labour markets are both place- and space-based. In places, labour markets shape standards of living, social structures, and employment opportunities. For most employees, the workplace is in the same community as the home. Paid work accounts for a substantial part of most people's daily experience and has broad implications for family and social routines. Rooted in different places, the characteristics of labour markets vary across space. The employment relationships worked out in one place are influenced by the conditions and bargains worked out elsewhere—a consequence of corporate and **labour mobility**, as well as trade in goods and services. A defining feature of globalization is the constant relocation of companies to different labour markets, coupled with national and transnational migration of people in search of jobs, and increasing trade among regions and nations.

The relationship between employer and employee is regulated by law and by the nature of the labour market in which the relationship is formed. Indeed, labour markets are highly varied and segmented, characterized by distinct routines and working conditions. In the **Fordist (or dual) labour market model** labour markets in Western market economies are divided into two broad segments. There is a **primary labour market** that itself has two sub-segments: (1) unionized blue-collar workers and (2) non-unionized professionals and white-collar workers, both with relatively high pay, good benefits, and structured working conditions. Then there is a highly diverse **secondary labour market** that includes large populations of non-unionized workers, with relatively poor pay, few (if any) benefits, insecure (often temporary or part-time) jobs, and poor (sometimes hazardous) and less structured working conditions.

With the advent of the information and communication techno-economic (ICT) paradigm, however, the structured arrangements underlying the primary segment of Fordist (dual) labour markets have been challenged by demands for more flexible working conditions. In present times, firms are seeking to adapt quickly to changing circumstances including by the development of flexible workforces. The implications are sweeping and controversial. Thus in the **flexible labour market model**, the key distinction, and basis for segmentation, is between core workers—who are multi-skilled,

labour mobility Voluntary movement of workers from one region and job to another, whether drawn by opportunity or pushed by lack of jobs.

Fordist (or dual) labour market model A model developed to explain labour markets that evolved in the US (and elsewhere) after the 1950s, based on the differences between primary and secondary labour markets.

primary labour market The privileged segment of Fordist labour markets comprising salaried and production workers in large firms who receive higher wages, job security, enhanced health care, pensions, and other benefits and whose work conditions are highly structured.

secondary labour market The less-privileged segment of the Fordist labour market where jobs are typically non-unionized, insecure, unstructured, and poorly paid, and where working conditions are often poor or hazardous. Most employers in this segment are SMEs; disproportionate numbers of employees are women, minorities, and students.

flexible labour market model A workplace model where labour is organized according to flexibility principles that are variously defined in terms of multi-tasking, multi-skilling, team work, financial discretion, job intensification, and ease of hiring and firing.

Case Study 5.1

TURNING POINTS IN THE LABOUR MOVEMENT:
DIFFERENT TIMES, DIFFERENT PLACES

The Great Flint Strike, 1936–7

After three decades of automated assembly work and six years of a severe economic depression, American auto workers organized themselves with enough strength to take on the most powerful companies in the US. The United Auto Workers of America (UAW) targeted Flint, Michigan—a company town dominated by GM and home to 45,000 (poorly paid) employees. Because it was at the heart of GM's empire, the union had to win there if it was to become truly effective. They took full advantage of the fact that Flint's Fisher Body plant was one of only two sources of the dies used to produce the stamps for GM's body parts, and workers at the other plant in Cleveland were already on strike. Although the Flint strikers had to contend with many threats, including spies and attacks by police, they prevailed with the help of the community, which supplied both food and moral support.

Among the key union demands that the company eventually met were the introduction of hourly pay (as opposed to piecework rates), shorter hours, higher wages, and union influence over the speed of work. In addition, the contract recognized the UAW and collective bargaining; a notable feature of its preamble was a provision that the workers would return to their jobs and no longer be threatened by prejudicial acts, such as the use of spies on the factory floor, personal threats to labour leaders, and discrimination by GM. Within weeks, the UAW had 100,000 members. The Ford Motor Company resisted, refusing even to recognize the UAW until 1941, but the Flint strike was the event that set the auto industry as the global standard for collective bargaining.

stable, and well-paid—and peripheral workers—including low-skilled and low-wage contingents, who are hired or contracted as needed.

The Nature of Labour Markets

labour catchment area The geographic area where local labour markets operate, centred on a spatial concentration of workplaces within the same commuting zone.

Labour markets exist within local place-based **labour catchment areas**: the physical spaces around cities or towns where most residents live and work, statistically defined by commuting patterns. Yet, every labour catchment area contains a variety of labour markets distinguished by particular employment relations that specify forms of remuneration, routines, rules, and conditions of work, and by informal work cultures. In other words, *labour markets are segmented*. Thus auto production workers, managers, cleaners, doctors, and teachers each work within their own labour markets, and mobility between them is limited. Indeed, the hiring of workers, both high- and low-waged, from distant places typically occurs within specific labour market segments driven by

> ### Labour Unrest in the Pearl River Delta, 2010
>
> By the turn of the millennium, China had become the workshop of the world, largely on the backs of harshly exploited workers who received no support from the government-controlled trade unions. The first major stirrings of labour organization came in the large foreign-based sector, especially the Pearl River Delta (PRD) where contract electronics manufacturers had established massive facilities. At Shenzhen, for instance, the Taiwanese firm Foxconn employed 400,000 Chinese workers, the vast majority of them young women from rural areas living in dormitories (because the national residency permit system didn't allow permanent residency in urban areas) and working under military-style discipline for pay so low that they had to put in extended overtime hours to make ends meet. Local governments tolerated exploitive conditions at Foxconn and other firms for the sake of tax and other revenues.
>
> In 2008, however, the national government called for improved wages and benefits for workers (partly in order to boost domestic consumption) while regional governments, including that of Shenzhen, raised minimum wages in an effort to push manufacturers into upgrading. Then in 2010, a rash of suicides forced Foxconn's management to double salaries and improve working conditions. The dormitory-based workers, many of whom worked so hard that they didn't even know the names of their bunkmates, remained relatively unorganized. But unrest quickly spread, particularly to Japanese auto subcontractors in the PRD. These workers didn't live in dormitories and were able to use electronic communications to organize themselves and reach out to NGOs and other labour activists for support. Settlements on better wages and conditions were quickly reached, and some progress has been made towards reducing residency restrictions. Time will tell whether the spring of 2010 marked the same kind of turning point for China's labour movement that the Flint strike did for the US movement.
>
> Note: Union contracts providing progressively higher wages and benefits made Flint an increasingly prosperous place until the 1970s. Between 1978 and 1982 alone, Flint's auto industry lost 16,500 jobs. The industry that had employed more than 80,000 workers in the 1960s was reduced to 8,000 workers by 2007. Michael Moore's film *Roger and Me* (1989) documents the downsizing of Flint.

the need for particular skills and experience. Labour segmentation is distinguished by particular employment relations.

The Employment Relationship

As the basic neoclassical model recognizes, labour markets are shaped by price; for example, increases in labour supply put downward pressure on wages (prices), while increases in demand cause wages to rise (Figure 5.1). If labour markets followed the perfect competition model and were perfectly fair, the lowest wages would be paid for the work of least value and the highest for the work of the highest value. In practice, however, employers and employees often bargain over the terms of employment, with outcomes depending on the relative strengths of labour and management, and the influence of relevant governing institutions, especially governments and unions (Table 5.1).

The **employment relationship** (Figure 5.2) is an agreement between the employer and employee about the organization of the work in question (working conditions,

employment relationship The agreement between employers and employees in a specific labour market regarding wages, benefits, and the organization of work.

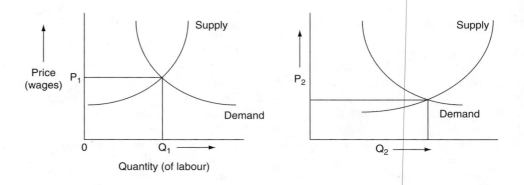

FIGURE 5.1   Labour-Market Interaction between Supply and Demand
Note: In Region A the supply of labour is higher than in Region B while demand remains the same.
Consequently, wages are lower in Region B (P2) than in Region A (P1).

control of production processes, etc.), the price that will be paid for it in the form of wages, salaries, or fees, entry (hiring) and exit (lay-off, retirement) rules, and so on. In general, employment relationships are at once mutually dependent (employers need labour; employees need jobs and income) and mutually antagonistic as each side seeks to set the terms of employment (wages, benefits, hours, etc.) and to control the way the work in question is done (tasks, speed, organization, etc.)

Employment relationships take many different forms. Some agreements are informal, consisting of little more than the understanding that employees will do what they are told to do. Others are highly structured, legally binding contracts negotiated by unions or associations on behalf of their members. Unions represent a wide variety of workers, job types, and income levels—not just blue-collar manual workers, but

TABLE 5.1 Labour Governance: The Principal Institutions	
Type	**Comments**
National governments	National laws and standards on wages and working conditions, hiring protocols, pensions, benefits, and unemployment insurance. Departments of Labour provide a "labour perspective" on national policies.
Regional governments	Regional laws on wages, working conditions, and hiring protocols including "right to work" legislation. Departments of Labour provide a "labour perspective" on regional policies.
Local governments	Special measures in response to local issues (e.g., Sunday closing).
Unions and collective bargaining	Direct negotiation of the "employment relationship" by unions organized on an international, national, or regional basis, and by industry, craft (type of job), and sometimes by firm. Union "locals" represent specific work sites or groups. Some employees' associations act like unions.
Non-structured labour bargains	Contracts and agreements in non-union firms. Contracts may or may not resemble unionized collective bargains.
International Labour Organization	Founded in 1919 under the Treaty of Versailles; became an agency of the United Nations in 1946. Promotes social justice and labour rights, especially in poor countries.
Supranational governments	European Union and its social rights legislation.
Other institutions	Religious organizations, NGOs, and households that defend and promote workers' rights.

also white-collar office workers and professionals of many kinds, including teachers, nurses, and (extraordinarily well-paid, but short career) athletes. Yet the roots of unionism lie in the effort to establish and defend the rights of low-paid, under-privileged workers by bargaining collectively on their behalf.

Labour Unions

Labour unions exist to collectively represent and empower workers in their workplaces with respect to production processes (tasks, speed, organization of work) and conditions of employment (wages, hiring, retirement, lay-off criteria, hours worked, grievance procedures, and benefits such as pensions, holidays, health care, and maternity or paternity leave) (Figure 5.2). They also seek to indirectly enhance workers' rights by supporting sympathetic political parties. Typically, members are organized in local branches or "locals" that are part of the union's governance structure; each local elects representatives for union-wide negotiations, and different unions co-operate with one another under the umbrella of a national federation.

The national federation in Canada is called the Canadian Labour Congress (CLC), representing some 3.3 million workers. In 2013, the largest union within the CLC, the Canadian Union of Public Employees (CUPE), had 627,000 members (Table 5.2). CUPE represents the changing face of Canadian unionism in two ways that are general across

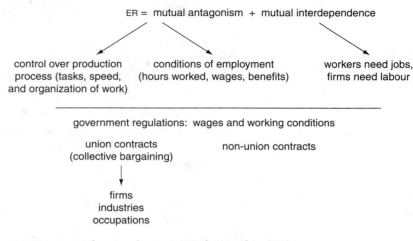

FIGURE 5.2 The Employment Relationship (ER)

Source: Based on Clark, G. 1981. "The employment relation and spatial division of labor: A hypothesis." *Annals of the Association of American Geographers* 71: 412–24.

labour unions A legal organization of workers formed to bargain with employers on its members' behalf regarding matters such as wages, benefits, and working conditions.

TABLE 5.2 Ten Major Unions of the Canadian Labour Congress

Union 2005	Membership 2013
Canadian Union of Public Employees (CUPE)	627,000 members in 2013 in a wide variety of service industries, especially in the public sector.
National Union of Public and General Employees (NUPGE)	11 component unions with 34,000 members, mainly in public-sector occupations.
Canadian Auto Workers (CAW)	Merged with CEP in 2013 to form Unifor with over 300,000 members, mainly in several private-sector industries.
Public Service Alliance of Canada (PSAC)	180,000 members, many working for the federal government.
Confédération des syndicats nationaux	300,000 workers in Quebec. Strong proponent of Quebec sovereignty.
Canadian Teachers' Federation (CTF)	An umbrella organization representing 200,000 teachers.
Canadian Union of Postal Workers (CUPW)	54,000 members, mainly but not exclusively postal workers.
Communications, Energy and Paperworkers Union of Canada (CEP)	Merged with CAW in 2013 to form Unifor with over 300,000 members, mainly in several private sector industries.
United Steelworkers of America–Canada	190,000 members, including 65,000 in steel and mining.

Source: Adapted from Canadian Social Research Links (2005); http://www.canadiansocialresearch.net/unionbkmrk.htm.

advanced market economies. First, although unions were traditionally organized around individual industries, trades, or "crafts," CUPE is primarily a service-sector union, as are seven of the leading nine unions in the CLC. Second, CUPE represents workers in a wide variety of fields, including health care, education, municipalities, libraries, universities, social services, public utilities, transportation, emergency services, and airlines. Similarly, the Canadian Auto Workers (CAW) that had previously represented workers in the transportation manufacturing sector, by 2005 represented workers in 15 different industries, including several in services, such as retail and hospital work. In 2013, CAW joined with the Communication, Energy and Paperworkers to form Unifor, illustrating a trend towards the consolidation of unions in the private sector.

The extent of unionization varies considerably among countries and regions, sectors, and industries. The rate of unionization in Canada, for instance, is approximately 2.5 times greater than in the US (Table 5.3). Rates of unionization have fallen substantially in most OECD countries: in the US, more than 30 per cent of workers were unionized in 1960, but that rate dropped to around 11 per cent in 2012, while in the UK unionization has declined from over half the workforce in the mid-1980s to 25.8 per cent in 2012. Even the Scandinavian countries, with their very high levels of unionization, have experienced notable declines. In Canada, overall unionization levels have remained more or less stable, but public-sector unions have grown while those in the private sector have declined. Declining membership inevitably means declining union power—a trend most evident in the rich market economies of the US and UK. The nature of labour unions varies among nations. The Canadian example of workers organizing to raise wages and benefits across a trade or sector (that is, achieve similar standards irrespective of companies) predominates in North America and Europe. In Japan, however, enterprise unions are organized within particular companies, align with corporate competitive interests, and are only loosely federated into common causes.

TABLE 5.3 Rates of Unionization for Selected Countries (Percentage of Wage and Salary Earners), c. 1995 and 2012

| Country | Unionization (%) | | Country | Unionization (%) | | Country | Unionization (%) | |
	1995	2011/12		1995	2011/12		1995	2011/12
Canada	32.9	26.8	Germany	28.9	18.0	Spain	18.6	15.6
Denmark	80.1	68.5	Italy	44.1	35.6	Sweden	91.1	67.5
Finland	79.3	69.0	Mexico	42.8	13.6	UK	32.9	25.8
France	9.1	7.9	South Korea	12.7	9.9	US	14.2	11.1
Japan	22.2*	18.0	Netherlands	21.7*	20.5	Poland	20.5*	14.6

* Figures are for 1999.

Sources: The 1995 data are based on International Labour Organization, *ILO Highlights Global Challenge to Trade Unions*. Press release issued 4 Nov. 1997. Copyright © International Labour Organization, 1997.
The 1999 and 2011/12 data are from OECD Trade Union Density, OECD: Stat Extracts http://stats.oecd.org/Index.aspx?DataSetCode=UN_DEN (accessed March 2014).
Note: The OECD estimates (provided for every year since 1999 are slightly less than the figures provided by national agencies when the latter are published.

Unions represent workers through **collective bargaining** where they negotiate with employers to define wages, non-wage benefits, and working conditions for their members. Once legally in place, the terms of a collective agreement cannot be arbitrarily changed. A strike (when labourers refuse to work) or lock-out (when management refuses to let labourers work) may be called if contract negotiations break down or if one of the parties to a contract believes that the terms have been violated. The legal ability to strike is itself a right that has to be negotiated.

Collective agreements vary in the degree to which they structure employment relations. In highly structured internal labour markets, **seniority** is a major factor: new entrants are restricted to the lowest grades; higher-level vacancies and new positions must be filled internally by the next most senior person. Lay-offs are also structured by seniority, where the worker with least seniority is laid off first and re-hired last. In addition, highly structured labour markets emphasize job demarcation, in which every job is defined in minute detail, resulting in numerous job categories. Less-structured collective bargains put less emphasis on seniority and job demarcation, producing fewer job categories that define worker tasks more broadly. Over time, successive collective bargains can result in extremely detailed records of work rules and rewards, especially in structured labour markets.

Government Policies

Ultimately, the rights of all workers, unionized or not, depend on government legislation. Indeed, workers have the right to unionize and bargain collectively only at the discretion of the relevant government, which simultaneously seeks to uphold the rights of employers and represent society's attitudes towards workers' rights. Thus labour policies vary substantially even among rich market economies. In Scandinavia, government and society at large have strongly supported unionization, and the government relies on unions to promote and defend workers' rights throughout the economy. In the US, by contrast, government policy tends to privilege market forces over unions; the collective bargaining process is seen as a violation of the pure market forces that favour independent, arm's-length relations. Unions are seen as constraining the freedom of business to seek competitive advantages.

The neoliberal view is that markets work best when workers and corporations represent themselves individually in the bargaining process. It tends to dismiss the reality that large corporations enjoy bargaining advantages rooted in market size, access to information, control over the production process, and options with respect to investment and location.

It is also argued that too many protections for workers reduce mobility in the labour market, particularly the willingness of people to move from region to region. Concerns over labour power are strongly felt in the US, where the federal government prohibits any union from making membership mandatory; almost half of all state governments in the US have passed "right to work" legislation prohibiting the imposition of mandatory union membership and dues. In Canada, by contrast, mandatory union membership and dues collection are supported by legislation. The relatively high overall rate of unionization in Canada reflects the fact that more than 75 per cent of public-sector employees are unionized (the rate in the private sector is less

collective bargaining The process where contracts are negotiated with employers by a labour union on behalf of a group of workers.

seniority A key principle in structured labour markets, requiring that new employees begin in the lowest level positions, and that promotion be based on the number of years an employee has spent in the job; in times of recession, those with the least seniority are normally the first to be laid off, and those with the most seniority are normally the first to be re-hired.

than 20 per cent). In the case of the health sector, employees in Canada's publicly controlled system are mainly unionized, while employees in the privately dominated US system are not unionized.

The distinction between federal and unitary systems of government (see Chapter 6) is also relevant in understanding labour policies. In unitary systems, such as the UK, labour legislation passed by the national government applies to the entire country. By contrast, in federations such as the US and Canada, the responsibility for labour policy is shared between national and regional (state or provincial) governments. As a result, it is extremely difficult to reach an agreement on any legislation involving the rights of all workers, and workers' rights almost always vary from region to region within the country. **Minimum wage** legislation—setting an hourly (or weekly or monthly) minimum wage that workers are legally entitled to—provides an illustration.

minimum wage The legally mandated minimum hourly wage that employers are required to pay in a given jurisdiction.

The UK's minimum wage policy has four levels, all of which apply across the country: a minimum wage of £6.31 per hour (in 2013) for workers over 22 years of age; a "development" rate of £5.03 per hour for workers between 18 and 21 years; a rate of £3.72 for younger workers who are not attending school; and an apprentice rate of £2.68. In Canada, each province (and territory) sets its own minimum wage, although in 2014 the variation was not great, from $9.95 in Alberta to $11.00 in Nunavut with most provinces at $10.00. The US sets a federal minimum wage standard ($7.25 per hour since 2009), but allows states to set higher minimums. Twenty-one states have introduced higher rates, with Washington State the highest in 2013 ($9.75 per hour). Some states (Minnesota, Colorado) maintain a lower rate, while several southern states (Alabama, Mississippi, and Tennessee) have no minimum wage legislation at all, but the federal law supersedes state laws (or lack of them). All the minimum wage laws provide exemptions, often for small firms.

Is government legislation as effective at protecting workers' rights as a contract negotiated through collective bargaining by unions? Although government legislation ensures that minimum standards apply to all workers, the support it provides is "distant" from actual workplaces and may not be available when workers' rights are violated. For example, workers who have been fired without cause may have to spend time and money to defend the rights that have been violated—and they may lose in court. Unions provide workers with ongoing, immediate protection of their rights through formal grievance procedures that are typically handled by their locals. Even in rich market economies with well-established legal systems, resolving labour disputes in court is a costly, uncertain, and often inequitable process, especially for low-income groups.

Workers' rights are less effectively supported in developing countries. In China, for example, trade unions are effectively controlled by the Communist party and are not generally seen as representing the interests of workers, while the court system is often regarded as capricious and corrupt. Although the International Labour Office (ILO) of the United Nations, created in 1917 and re-mandated in 1946, has become a watchdog and activist for the rights of workers around the world, its ability to intervene in such situations is limited.

Labour Market Segmentation

Theories of **labour market segmentation** describe how labour markets are structured by informal and formal institutions. In this view, the agreements, rules, and conventions that govern the buying and selling of labour are vital to understanding how labour markets actually work. In general, ideas about segmentation emphasize the particular demands for labour and experiences. Segmentation can also be anticipated by youth career expectations and choices.

> **labour market segmentation** The classification of labour markets according to different rules and conditions with respect to terms of entry, promotion, working practices, and remuneration.

Career Expectations for Young People

Young people begin to formulate their career expectations in high schools which help to prepare them for entry into the job market. Students' thinking about school and future job prospects can be crudely classified in two distinct categories: immediate and distant. Those who take the immediate, short-term view focus on getting out of school and entering the job market as quickly as possible; for many of these students, relatively short-term, on-the-job training (OJT) and learning by doing is all that is required to begin relatively routine "blue-collar" work, such as assembly-line or basic processing jobs. A significant component of immediate career choices, however, is relatively lengthy apprenticeships that can involve vocational courses along with a practical, learning-by-doing orientation. Trades jobs, such as plumbers, welders, electricians, and millwrights, are examples, and such jobs and training can provide workers with a degree of industrial mobility. Students who take the more distant, long-term view are more enthusiastic about school and choose academic courses in preparation for further studies prior to entering white-collar professional occupations, such as teaching, law, and medicine.

Class, family, and ethnic or cultural background can play an important part in students' career expectations. In the past, young people, especially those from working-class families, often took it for granted that they would move directly from school to an "immediate" working-class job, while those from middle- and upper-class families were able to anticipate professional careers requiring a longer-term "distant" logic. But within classes, family and cultural attitudes, including peer groups, can vary and play a significant part in shaping young people's career expectations. The nature of the labour market(s) in the places where they grow up can have an impact as well: expectations are likely to differ considerably between someone from a resource (e.g., coal mining or forestry) or manufacturing town and someone from an economically diverse urban setting. Ultimately, though, career choices are made by individual people with distinctive abilities and personalities. Some may conform to established patterns; others are strongly inclined to resist them. Education plays a liberating role in this regard.

Education and Income

People invest in education, training, and certification in the hope of improving their earnings. In general, such investment is rewarded. In the US, for example, median weekly incomes for workers who had not graduated from high school were substantially below average in 2013, while certification, diplomas, and degrees progressively improved income prospects (Figure 5.3). As a general rule, as the US data indicates,

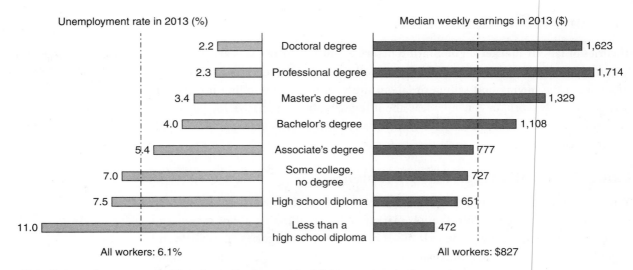

FIGURE 5.3   Earnings and Unemployment Rates for Workers Aged 25 and Over by Educational Attainment, US, 2013

Source: Current Population Survey, US Bureau of Labor Statistics, US Department of Labor.

education and training also tend to reduce the risk of unemployment and increase the length of tenure in a particular job. These patterns may change in the future, if the rising cost of education leads more people to question the value of remaining in school. Nevertheless, the link between income and education is well established and the single most important policy that governments can pursue to improve the job prospects of young people is to provide an accessible education and training system designed to reflect both the diverse needs of the labour market and the diverse interests of students. In this regard, note that the incomes received as a result of apprenticeship training for jobs such as electricians, welders, millwrights, and some computer programming, which do not require higher degrees are not identified in the data in Figure 5.3.

Job Experience and Path Dependency

Once careers are chosen, labour market segmentation is reinforced by path-dependent job experience. Most jobs require particular skills and approaches, which are reinforced over time. Familiarity with work routines increases efficiency and problem-solving abilities on the job. At the same time, individuals become acculturated by their jobs, which often have a defining influence on daily routines, networks of friends, ideas, forms of expression, and values.

Wages and benefits typically increase with tenure. The prospect of losing those accumulated rewards, together with the time and effort required to learn a new job, tends to discourage voluntary job changes, especially among older workers. In Canada, for example, from 1976–2011 the average tenure in a full-time job for workers over 55 years old was from 16 to 18 years, while the average job tenure for workers from 25 to 44 years old was around 6 years. The average job tenure for all workers during this period has hovered around 9 years, with no sign of change in this trend. Many workers remain much longer, and some spend their entire working lives in the same job. Frequent voluntary changes in "permanent" jobs are most common among young people and in relatively unskilled jobs where experience and training are not important. However, voluntary job change does not necessarily mean occupational change. Past experience and qualifications are often vital when workers change jobs.

On the other hand, when workplaces such as factories or mines close down, even highly skilled workers can have difficulty finding similar work with similar wages. Although governments and other institutions try to retrain former miners, steelworkers, and so on, success has been limited, especially with older workers. Part of the problem is that training schemes themselves do not create jobs and may not anticipate future job demands or include opportunities to make contact with potential employers. As well, well-paying jobs are especially difficult to find. Another basic problem is that older workers can experience significant difficulties in beginning new job-specific, path-dependent experiences.

Jobs vary considerably in the credentials they require. While some jobs need only short probationary periods of a few months, weeks, or even days, others demand lengthy apprenticeships, education, and training. Professions such as medicine, law, and engineering require many years of formal education, plus some "on-the-job" training, while trades such as electrical work, plumbing, and welding demand lengthy apprenticeships with some "off-the-job" training. In that regard, a benefit of our predominately service-oriented economy is the flexibility of managerial and marketing skills in changing, not only jobs, but also industries. Job changes can occur without changing occupation or discarding accumulated expertise. Still, job specialization has increased in our economy, as has the tendency to require formal certification (degrees, diplomas, etc.) indicating that workers have the appropriate skills.

Employment Segmentation in Place and Space

Patterns and practices of labour segmentation evolve in particular times and places. During the nineteenth century, the emergence of a large-scale workforce in mining and manufacturing was typically closely controlled, often in capricious ways by owners and managers. It was only modestly modified by initial employment legislation that sought to limit the exploitation of workers, including children, for example, by naming a maximum number of working hours per day and week. In the US, the development of mass production and the assembly line was reinforced by theories of **scientific management (Taylorism)** that advocated the strict control of labour, measured by time and motion studies and division of production work into a multitude of simple repetitive tasks, each performed by a different, relatively low-skilled worker. This evolution was partly shaped by reliance on large supplies of unskilled, often immigrant workers, and further influenced by the development of unionism. Among industrial countries, different types of unions had developed in the eighteenth and nineteenth centuries and gradually gained recognition. In the US, the National Labor Relations Act of 1935 was a major milestone that recognized the legality of collective bargaining in the private sector and, not without controversy (Case Study 5.1), helped stimulate **industrial unions**, such as the UAW.

scientific management (Taylorism) The practice of dividing a given job into multiple narrowly defined tasks and employing specialized labour to perform these tasks.

industrial unions A labour union that represents workers in a particular industry.

The collective bargains achieved by the UAW, and related large industrial unions, provided the archetypical employment relations of Fordism by achieving relatively high incomes and highly structured employment conditions, based on seniority, job demarcation, and grievance procedures. Indeed, collective bargains which encompassed large numbers of relatively unskilled blue-collar workers could be seen as part of a larger social contract. Workers were given a reasonable share of income (commensurate with their productivity) in return for production stability that helped create a more equitable income distribution.

Structured labour markets remain important today (including in the public sector). However, in recent years, employers have demanded greater flexibility with respect to tasks and wages, as well as hiring and firing. The arguments for and against flexibility are intimately associated with the pros and cons of globalization. Around the world, workers have seen employers' desire for flexibility as a threat to their rights and have gone on strike to protest against moves in that direction. Protests have been especially bitter at factories where management has sought to replace long-established structured collective bargains with more flexible arrangements. This flexibility will be discussed in more detail in the following sections.

The Dual Labour Market

independent labour market
The white-collar portion of the primary labour market, consisting of employees in management and R&D occupations; these workers are typically not unionized.

dependent labour market
The blue-collar portion of the primary labour market, consisting of unionized employees in production-line and related work.

part-time Employment that amounts to fewer hours than a standard full-time job (usually defined as an 8-hour day and a 40-hour week).

Under Fordism, as we noted earlier, the labour market is divided into two major segments. The primary labour market segment is characterized by relatively favourable employment relationships based on high wages, good working conditions, employment stability, structured prospects for advancement, well-established work rules, and due process in their administration. Within the primary segment, however, another important distinction can be made between an **independent labour market** consisting of people such as managers and R&D workers and a subordinate or **dependent labour market** consisting of production or office workers and trades people. The former enjoy higher levels of remuneration, greater employment stability, and better non-wage benefits than the latter. Contractual relations, the nature of negotiations, and the guidelines governing the conditions of work also vary between the independent and dependent segments of the primary labour market.

The secondary labour market is characterized by low-skilled workers in manufacturing (processing and assembly) or distribution and retail. Some employers are SMEs; others are larger firms with business strategies that depend on minimizing costs by hiring and firing as competitive conditions dictate. As a consequence, many secondary workers are employed on a temporary, **part-time**, or otherwise insecure basis. Few secondary jobs are unionized; a large proportion of them offer low wages, few (if any) benefits, and limited opportunities for advancement. Working conditions may be poor or even unsafe and employees are often subject to strict supervision. In the absence of the rewards for long tenure that workers in the primary sector enjoy, the secondary segment is characterized by high turnover rates. Disproportionate numbers of its workers are women or members of minority groups.

On the other hand, the secondary labour market is not a monolith. Some secondary workers are highly skilled, well-paid, and represented by unions, and some secondary employers treat their workers very well (even adopting union agreements as models).

There are also some secondary workers who do not wish to work in a unionized firm, and some who prefer the culture of SMEs to that of large corporations.

The relationship between big unions and big business that developed under Fordism was not coincidental. Unions gave big business a structured, stable, and productive workforce, while big business offered workers increasingly well-paid, secure jobs. In addition, the concentrations of workers in big firms made the task of union organizing relatively easy, while the collective bargaining process was not only more efficient than individual deal-making, but also more transparent in its communication of outcomes to all workers.

These mutual benefits were cemented in union contracts that improved working conditions and remuneration. Further, because they were based mainly on the principles of seniority and job demarcation, they also served to reduce competition among workers while promoting order, stability, and dignity in the workplace by giving workers clearly defined job roles and expectations. The same principles benefited management by encouraging stability in the workforce, as the rewards that came with seniority helped to "lock-in" experienced workers.

From the 1950s to the 1970s, **structured labour markets** worked well for management and labour alike. When, from time to time, primary dependent workers in mass production industries had to be laid off, the lay-offs were usually temporary and were organized according to seniority. By the early 1980s, however, factory closures and massive job downsizing in industries such as clothing, electronics, and auto manufacturing signalled the beginning of a serious decline in union power, including the power to enforce highly structured employment relationships. Employers wanted more flexible workforces.

structured labour markets
Labour markets that are formally organized by legally binding agreements between employers and employees; in Fordist labour markets the two most important structural principles are seniority and job demarcation.

Towards Flexible Labour Markets

Today, in the age of the information and communication techno-economic paradigm, firms argue that they need more flexibility to cope with a variety of circumstances: recessionary conditions, technological change that has made old job categories redundant, consumers who demand a wider range of products, and increased competition from global rivals. For North America's giant Fordist firms, this competition originated in Japan and Germany where employment was organized differently, with a greater focus on flexibility, training, and problem-solving.

In Japan after World War Two, Toyota and other companies were producing similar products with similar production technologies as US firms, but developed with different employment relations. Whereas the Fordist model recognized hundreds of different job categories, Toyota's production system developed around two basic job categories and the necessity for job rotation and constant learning. Rather than seniority, the basis for priority in lay-off periods was devotion to the company and participation in its evolution in the expectation of a greater pay-off as one got older. Instead of relying on a multi-layered management hierarchy to oversee workers, Toyota emphasized self-supervision of groups of workers. It also chose to subcontract and rely on external ("peripheral") workforces to a much greater degree than its North American counterparts did. Finally, while North American auto-makers confronted powerful industry-wide unions, Toyota enjoyed much less adversarial relations with its own firm-based unions (established after a series of confrontations in the early postwar period). In short, Toyota—having studied the Fordist employment relationship model in hopes of developing a more efficient alternative—anticipated the shift towards flexibility by several decades.

The demand for flexibility altered some of the most basic features of the old Fordist workforce. There are still two segments, but what is expected of the primary or "core" segment in particular has changed significantly. In this model, core workers must be "functionally flexible": multi-skilled, capable of performing different functions as part of a team, prepared to take responsibility for decision-making, committed to learning new skills as needs arise, and willing to adopt the best practices of competitors when necessary. In a flexible firm, job categories are far fewer than they were under Fordism. Advancement no longer depends on seniority but on performance and ability. Core workers are recognized as capable of self-supervision, and those who initiate productivity improvements are rewarded with high wages, generous benefits, and employment stability.

"Peripheral" workers, for their part, are expected to be numerically and/or financially flexible. They are hired ("subcontracted") as needed, often on a part-time or temporary basis, and even "permanent" employees' hours can be varied by adjustments in shifts, overtime work, and lay-offs. Hiring from groups willing to accept lower wages for reasons related either to excess supply or to personal circumstances provides financial flexibility. Mothers, for instance, will often trade lower wages for flexible hours. Financial flexibility may also mean subcontracting high-value work to highly skilled specialists when necessary. In such cases the purchasing firm saves the expense of developing in-house capabilities that it would not use on a regular basis, while subcontractors may actually make more money than they would as permanent staff.

In summary, the flexible firm seeks to respond quickly, easily, and cheaply to changes in their markets, production processes, or industry. The differences between the Fordist and flexible models are most significant in the case of the core segment, where emphasis on features such as teamwork, multiple skills, self-supervision, and promotion on merit stands in direct opposition to Fordist principles and has the effect of "flattening" management hierarchies. In contrast, the flexibility model has had little impact on the stability of work in the low-waged, low-skilled peripheral segment, where jobs remain vulnerable to market forces, although it has meant a significant increase in the numbers of such jobs around the world. It has, however, given rise to a new peripheral segment. Often referred to as producer services, this segment consists of well-paid specialists, such as designers and consulting engineers, whose services are often project- and contract-based and customized to meet the specific needs of clients.

New Economic Spaces and Flexibility

Labour market segmentation has important implications for local development. For example, structured Fordist labour markets contributed to stable, high-income communities but limited job mobility, and became vulnerable to change. Flexible labour markets have profound location consequences. Relocation of certain activities to **new economic spaces**—especially greenfield sites with no previous experience of such activities—can facilitate the shift towards flexibility.

Geographic separation of, say, white-collar R&D specialists and blue-collar production workers enables the firm to treat the two groups differently. As a result, it can

new economic spaces Places that either are new to any sort of economic activity and may be only sparsely populated or places that are attracting new kinds of activity for the first time, and/or areas that develop exports for the first time

offer the R&D workers in one location higher wages, more generous benefits, greater freedom for self-supervision, and more decision-making opportunities without sparking demands for similar privileges from the blue-collar workers located elsewhere. Locating in new economic spaces also allows firms to escape the expectations regarding employment relationships that have developed at existing sites. This is especially advantageous to firms when labour's expectations are rooted in union contracts that impose legally binding forms of work organization, for example, regarding seniority and job demarcation, as well as high wages and non-wage benefits such as pensions, health care, and vacation time.

In the auto industry, the opening of new manufacturing facilities in states such as Tennessee, Alabama, Kentucky, and Texas—far from the industry's traditional base in the Midwest—by MNCs such as Toyota, Honda, Nissan, Mercedes Benz, and BMW reflects the recognition by firms that it is easier to develop more flexible labour bargains with an entirely new, often non-unionized labour force in new economic spaces. Inevitably, the opening of such plants erodes union power at other locations, and threats of relocation remain a frequent part of employers' bargaining strategy in labour negotiations. Even so, the search for flexibility in new locations can fail. Thus GM chose to manufacture a new car (the Saturn) in the 1980s at a brand new location, not in Michigan but in Tennessee, in order to establish new, more flexible production and labour relations systems. However, the plant employed a union workforce, and featured a cumbersome management structure and its experiment with more flexible work organization failed; the Saturn plant was closed in 2010.

The Different Faces of Flexibility

In the second decade of the twenty-first century, the idea of flexibility evokes different responses in different places. Electronic Arts (EA) is an entertainment software company with offices in various places around the world, including Burnaby, British Columbia. EA offers a trendy, ergonomic work environment and arranges many recreational (theatre, fitness room, full-sized soccer field, tennis courts) and business services for its 1300 (2013) young (mainly male) designers, engineers, and computer geeks. Self-supervision, willingness to work long hours, and commitment to teamwork combine with enthusiasm and creativity to produce a stream of innovative games known throughout the world, such as Need for Speed, FIFA soccer, NBA, and NHL. This facility also operates EA's worldwide motion capture studio. EA is part of a creative cluster of games firms in Vancouver that rely on continuous skill development and innovation to produce important competitive advantages, which in turn provide high wages.

At about the same time, at the Computime factory in Shenzhen, China, 4,000 employees were reported to work long hours for much less than the "official" minimum wage with overtime pay virtually non-existent. Workers could be fired at will and were closely supervised; even the time spent going to the toilet was being monitored. No medical insurance was provided, though injuries were common, and factory managers had been accused of stealing wages. (One estimate suggests that in 2004, some 70 per cent of the more than 200 million migrants who left their villages for newly industrializing cities in China were owed money by their employers; the total amount of unpaid wages in that year was estimated at $156 billion.)

With such disparate working conditions it makes sense that EA workers reported they enjoy their jobs, while Computime employees told the media they work because they are desperate to support their families. Yet there are a few similarities between EA and Computime. Neither workforce is unionized. Both companies demand long hours. And both put considerable stress on their workers; EA workers occasionally quit because of it. The creative work of EA and the "sweat" work of Computime are also connected; coding for the games designed at EA is contracted out to suppliers in China.

Contracting Out

outsourcing Engaging an external supplier; also referred to as "contracting out."

Contracting out (also known as "subcontracting" or "**outsourcing**") is the practice in which a firm, instead of producing some necessary component of its product, buys it from another firm. (The contracting out of activities formerly performed internally is sometimes referred to as "vertical disintegration.") The benefit of contracting out a product or service is that it allows a firm to take advantage of external economies, whether of scale or of scope. Contracting out is often seen as representing the vicious side of labour market flexibility. Giant firms in particular are criticized for contracting out work to suppliers who cut costs by using sweated labour: overworked, underpaid, and/or exposed to unsafe labour conditions. The criticism is especially sharp when giant firms outsource to low-wage countries, work that they used to have done internally by well-paid union labour.

Among the well-known firms that have been accused of exploiting sweated labour is Nike, which contracts out all the work involved in the manufacturing of its shoes. In 2004 its contracted suppliers employed 650,000 workers in 700 factories in countries around the world, including China (124 factories), Thailand (73), South Korea (35), and Vietnam (34). Stung by criticism of its work practices, Nike responded. It documented labour abuses on its website, including physical and verbal abuse in more than a quarter of its Asian plants, restricted access to water and toilets in many plants, failure to reach minimum wage levels in a quarter of the factories, 60-hour work weeks, and punishment of those workers who refused to work overtime. It established a division to oversee the activities of its suppliers. Nike also joined the Fair Labour Association, an organization that includes footwear and clothing manufacturers, NGOs, and universities, and conducts audits to improve working conditions and employment relationships. Most MNC brand manufacturers of apparel and electronics (e.g., Apple and Levi's) have been compelled into such actions by NGO activities, as have large distributors (e.g., Walmart and Home Depot).

Manufacturing is not the only sector in which outsourcing is increasingly common. The relocation of call centres from North America to lower-wage countries such as India is part of the same trend. That highly profitable companies will spend billions on advertising while relying on sweated labour to produce the products they sell is legitimately decried. On the other hand, a ban on outsourcing to poor countries could be disastrous for workers there who have no alternative source of income. Labour market problems are not easy to resolve.

Labour Market Failures and Challenges

How labour markets work, or fail, defines development to a significant degree. The supply, quality, and stability of jobs are basic concerns facing all market societies. In this regard, governments have long been concerned with levels of **unemployment**, when the supply of workers exceeds demand, and **underemployment**, when workers can only find part-time jobs when they would rather work full-time. But there are many other labour market issues, such as income levels and gaps, fairness, discrimination, and exploitation. Workforce performance indicators provide context for appreciating labour market failures and challenges.

unemployment Occurs when workers are out of a job but are seeking regular paid employment.

underemployment Occurs when workers are employed but whose jobs do not provide the hours and income they need to support themselves and families.

Selected Workforce Indicators

The basic comparative workforce indicators annually recorded by the US Bureau of Labor on the working-age population, the participation rate, and unemployment rate for males and females in 16 countries are summarized for 2012 (Table 5.4). The working-age population is the population between the ages of 15 and 64. The unemployment rate is the percentage of the working-age population (excluding a handful of relatively small groups such as military personnel) that is actively looking for paid work. The labour force **participation rate** is the percentage of the working-age population that

participation rate In Canada, the percentage of the working-age (15–64) population (excluding residents of Aboriginal reserves, full-time members of the military, and the institutionalized population) that is in paid employment.

TABLE 5.4   Comparative Labour Force Statistics among 16 Countries 2012

Country	Working age population (000s)	Participation Rate (Female)	Unemployment Rate (Female)
USA	243, 284	60.7 (57.7)	8.1 (7.9)
Australia	18,332	66.2 (59.9)	5.2 (5.3)
Canada	27,922	66.7 (62.1)	6.3 (5.8)
France	50,782	55.9 (51.2)	10.0 (10.0)
Germany	71,274	59.2 (53.2)	5.5 (5.9)
Italy	51,729	49.0 (39.7)	10.8 (11.9)
Japan	110,752	58.4 (47.7)	3.9 (4.2)
S. Korea	41,582	61.3 (49.9)	3.2 (3.0)
Mexico	85,923	58.4 (42.0)	5.1 (5.1)
Netherlands	13,629	64.8 (58.9)	5.3 (5.2)
New Zealand	3,492	68.2 (62.6)	6.9 (7.3)
S. Africa	32,959	54.8 (48.3)	25.1 (27.8)
Spain	38,334	53.2 (59.8)	25.2 (25.5)
Sweden	7,732	65.2 (61.3)	7.9 (7.6)
Turkey	54,724	63.4 (57.2)	8.3 (10.0)
UK	50,473	63.2 (57.0)	8.0 (7.4)

Source: US Bureau of Labor, International Comparisons of Labor Force Statistics 1970–2012, 7 June 2013 http://bls.gov/ilc/home.htm
Note: The OECD estimates (provided for every year since 1999) are slightly less than the figures provided by national agencies when the latter are published.

is formally employed, and the unemployment rate is the percentage of unemployed people who are known to be actively seeking work as a share of total employment.

Among these countries the US and Japan have the biggest workforces and both have substantially increased in size since 1960. In the US, for example, the working-age population (defined as the population above 16 years) increased from 117 million to 243 million between 1960 and 2012, while Japan's increased from 65 to 110 million. Most other countries listed have recorded increases, including Canada whose workforce multiplied by more than 2.5 times in this period. Overall participation rates varied from 49 per cent (Italy) to 68.2 per cent (New Zealand) with a median rate of around 60 per cent; female rates tended to be lower. Women are more likely to have part-time jobs. However, the gap between male and female participation rates has narrowed over the decades. Note that China has the world's largest working-age population (802 million in 2010) and a (high) participation rate of around 74 per cent, making actual employment around 594 million.

As an indicator of labour market failure, unemployment rates among the 16 countries are high, with only two countries, Japan and South Korea, recording rates of less than 4 per cent in 2012. (China's unemployment rate is around 4 per cent.) The unemployment rates were even higher in 2010 or 2011. Although they fluctuate, unemployment rates since 1980 have been higher and more volatile than in the 1950s and 1960s. Generally speaking, in the 1960s rates of more than 3 percent in Europe and 5 per cent in North America were considered high, posing social and political problems. Presently, 5 per cent unemployment would be widely regarded as desirable. The Spanish, South African, Italian, and French labour markets are clearly experiencing deep problems in this regard.

Within national economies, local variations in unemployment and participation rates are significant. There are large variations in unemployment rates within regions, cities, and rural areas which are often tied to age and cultural differences. To an important degree, unemployment is a regional and local problem. However, national unemployment rates do not vary significantly by gender.

Unemployment

One major type of unemployment occurs when workers are involuntary dismissed or "laid off" from their jobs. There are three basic types of dismissal unemployment: seasonal, cyclical, and structural. Seasonal unemployment occurs on a regular basis in resource industries such as farming and logging, and in manufacturing and service activities that cater to seasonal markets. Gifts for the Christmas market, for instance, are produced in the spring and summer, sold to distributors in the late summer and fall and retailed in the fall and winter, and at each stage a workforce may be hired to deal with the bump in production. Cyclical unemployment follows business cycles; typically, unneeded workers will be temporarily laid-off during downturns. Workers who are laid off during periods of seasonal and cyclical unemployment expect to be re-hired with the next upswing in activity. However, a severe, economy-wide cyclical downturn is likely to reduce production levels and workforces for an extended period of time.

By contrast, structural unemployment occurs when workers are permanently laid off, usually as a result of the closure or downsizing of operations or the introduction of automated labour-saving machinery. A shift towards more flexible work arrangements, under which fewer workers perform a broader range of jobs, also reduces employment. Structural unemployment is typically a problem in areas formerly dominated by traditional industries that have lost jobs because of automation, the development of substitute products, and the rise of lower-cost competition using more advanced technology and cheaper labour in new economic spaces. As we have noted, finding new employment for the former specialized workers of traditional industries has been difficult, especially when they are concentrated in places where there are few alternative job opportunities. For many of these older workers, "retirement" has become the de facto option.

Youth unemployment and underemployment is a widespread, deep-seated concern that reflects on the insufficiency of the quantity and quality of jobs for school-leavers. In the UK, for instance, the government classifies a significant number of young people as "Not in Employment, Education, or Training" (NEETs). Most of these young people are isolated from mainstream society and concentrated in "tough" neighbourhoods. In 2009, between 15 and 17 per cent of 16–24 year olds were classified by the UK government as NEETs, and a government report indicated that this "disengaged population" starts to form in primary school. Many observers feel that NEETs are a product of the rapid, unplanned decline of the mining and manufacturing sectors in the UK, which traditionally provided employment for early school-leavers with few qualifications. Some other countries, such as Denmark, anticipated this problem: as structural lay-offs occurred in manufacturing, they introduced education and training programs to equip school-leavers and laid-off workers with new skills as soon as possible. However, recent recessionary conditions have exacerbated youth unemployment; across Europe in 2013, NEETs were estimated to number 14 million, while the unemployment rate among people under 25 years old was 23 per cent. Canada has less of a problem in this regard, but 2012 unemployment rates of more than 18 percent in the 15–19 year old age category and 10 per cent in the 20–24 year old age category indicates labour market failure.

Labour Market Policies in Rich Economies

The challenges facing labour markets vary enormously within and among countries and regions. Workers face very different prospects depending on whether they live in rich or poor countries, metropolitan or rural areas, large or small towns, fast-growing or declining regions. In market economies, the options change fundamentally from one generation—and techno-economic paradigm—to the next. During the boom years of Fordism, the situation of labour generally improved. Structural unemployment was typically portrayed as a regional problem resulting from the decline of old industrial and mining regions. Various policy options were debated, including whether new activities should be encouraged to locate in areas of high unemployment ("work to the workers") or laid-off workers encouraged to migrate ("workers to the work"). In recent decades, market economies have moved beyond their traditional concern for unemployment—principally involving males and workers' rights (including the right to unionize)—to consider issues such as underemployment, gender and racial bias, work-based stress and lack of worker control, access to daycare, low pay, income inequalities, and the rights of immigrant workers (Table 5.5).

TABLE 5.5 Labour Market Issues and Policies in Advanced Market Economies

Issue	Policy Emphasis	Policy Impact Concerns
Structural unemployment, de-industrialization	Retraining of laid-off workers; diversification by attracting new activities; migration subsidies.	Retraining not matched to jobs available; new activities pay low wages and fail to hire laid-off workers; migration subsidies require that migrants first find jobs elsewhere.
Seasonal unemployment	Diversification (e.g., by adding value to resource activities; extended unemployment insurance schemes).	Non-seasonal work hard to find in remote places; extended unemployment benefits reduce incentives to find work.
Under-employment	In poor countries especially, economic growth, education, and training.	Hard to detect and target groups.
Youth unemployment	Education, training	Some youth beyond reach; need to provide permanent jobs.
Gender and visible minority discrimination in hiring, occupational type, and remuneration.	Equal pay for work of equal value legislation; affirmative action; removal of glass ceilings; pressure on public-sector organizations to respond or forfeit funding.	Politicizes merit principle; reverse discrimination; targeted groups resent suggestions that their job is race- or gender-dependent.
Low or no pay	Minimum wage legislation; family wage; guaranteed annual income schemes.	Employers become hesitant to hire; guaranteed wages and incomes reduce incentives to find paid work.
Income inequality	Income redistribution through income and inheritance taxes; salary caps and control over salary ranges.	Rich people move to lower-tax or tax-free havens.
Unsafe and unhealthy working conditions	Health and safety regulations.	More difficult for small firms; bureaucracy increases.
Access to daycare	State provision or subsidies.	High cost; stay-at-home parents penalized.
Immigrant workers	Ease immigrants' requirements.	Creates competition for local worker; immigrant qualifications not recognized.

The items included on this (incomplete) list of labour market concerns are not mutually exclusive. In fact, many of them tend to appear in self-reinforcing clusters: underemployment, low pay, underrepresentation of women and visible minorities, lack of worker control, poverty among immigrant workers, and lack of access to daycare, for instance.

Governments typically seek to minimize these problems through training schemes, support for migration, affirmative action policies, minimum-wage legislation, guaranteed-income schemes, and subsidies for activities that locate in regions of high unemployment. Such measures are often controversial. Critics variously argue that they are ineffective, represent unwarranted intervention in a free market, or have unintended consequences. Affirmative action may imply that merit was not involved and be criticized for reverse discrimination. Minimum-income guarantees may lead to dependency attitudes and be deemed unfair by poorly paid workers who are excluded. Jobs allocated to immigrants may be opposed by the locally unemployed. Daycare is expensive, and often only provided to particular segments of society, etc.

The Global Employment Challenge

The ILO (2012) estimated that out of a total labour force of 3.3 billion, global unemployment levels reached nearly 202 million in 2011, an increase of 32 million over 2007 levels, and that 400 million new jobs were required simply to keep unemployment at present levels. Youth have been especially hard hit, with unemployment reaching 74.5 million in the 15–24 year old age category, an increase of 4 million since 2007 (prior to the global recessionary crash of 2008–10). Even when youth are employed, it is often part-time. Underemployment in 2011 was estimated at 700 million people.

Globally, unemployment and underemployment are greatest in poor and developing countries. While China and other East Asian countries have expanded labour forces and incomes, other developing countries, especially in Africa, the Middle East, South Asia, and Latin America have experienced high levels of unemployment and contain substantial populations of what the ILO labels the "working poor" and "vulnerable populations." Vulnerable employment basically consists of owner-operated family businesses and uses unpaid family labour; according to the ILO it amounted to 1.5 billion people in 2011. The working poor, defined as receiving an income of less than US$2.00 a day, amounted to 900 million people in 2011, even though the numbers in extreme poverty (less than $1.25 a day) have declined considerably since 2000 (primarily in East Asia). Child labour—defined as children under the age of 15 who perform work that is physically or mentally harmful and interferes with schooling—has declined but still amounted to 215 million globally, with almost 60 per cent in agriculture. The employment of people, including children, for very low wages, for long hours in poor, often hazardous conditions is often referred to as sweated labour and the workplaces as sweatshops. Policy responses to sweated and child labour are not straightforward; simply stopping labour exploitation may worsen income levels for families in the short run. Progressive labour laws and enhanced worker rights, possibly by unionization, are needed to improve working conditions while, as noted, pressure by NGOs and the ILO can help to encourage MNCs such as Nike to set standards within their supply chains. Improvements in labour productivity, as evident in China, is just as vital.

Closely related to issues of unemployment and underemployment are low pay and income inequality. We have already noted that, since the 1820s, average incomes in the world's rich countries have been rising while those in poor countries have been falling (Chapter 2). This trend has been arrested in several Asian countries, and some elsewhere, but the international income gaps between the rich and the poorest countries remain large. Moreover, income inequality has become an increasing problem in several rich market economies—most notably the US—since the 1970s, but especially since the 1980s. As the economist Robert Reich notes, drawing on census data, the richest 1 per cent of American families increased their share of the nation's total income from 9 per cent in the late 1970s to 23.5 per cent in 2007, and in 2012 the gap between the richest 1 percent and the rest was the greatest since the 1920s. The causes of this trend have been related to numerous factors, for example, declining labour power, loss of manufacturing jobs providing good incomes, financial deregulation, inappropriate tax policy especially regarding the rich, and inadequate investment in education and training. Whatever the causes, Reich and other leading economists, such as Paul Krugman and Joseph Stiglitz, believe income inequality is the central economic problem facing the US.

As these economists point out, the rich spend a smaller proportion of their income than the rest of us do and are more likely to move their money to wherever the highest returns can be found. As a result, demand for goods and services declines, as does the economy's ability to create jobs. Income inequality is also cumulative, creating social and political problems. In the past, to increase family incomes, many women have entered the workforce, many people have worked longer hours or taken on additional jobs, and many people have gone into debt. But these options are no longer as widely available as they once were. Reducing income inequality will require new policy measures that will likely require more progressive taxation (see Chapter 6) and investment in education.

Geographic Mobility and Immobility of Labour

Typically, workers move from regions of high unemployment (low demand, high supply, and downward pressure on wages) to regions with "tight" labour markets (high demand, low supply, and upward pressure on wages). Such migration, whether permanent, medium- to short-term, or seasonal, helps to lubricate the economy and has the benefit of increasing productivity by resolving spatial mismatches between labour demand and supply. However, workers (and their families) vary in their ability and desire to relocate. The reasoning behind the decision to migrate is complex, as economic considerations are usually intertwined with social and sometimes political considerations. Moreover, migration imposes costs on the regions of origin and destination alike.

The ILO (2006) estimated that in 2005 some 86 million workers (and many more people in total) around the world were living outside their country of origin or citizenship for the purposes of employment. It suggested that 20.5 million of these migrants were in North America and 28.5 million in Europe (including Russia), for a combined total of 49 million (55 per cent of the total). The male–female split was more or less equal (51–49 per cent). Since then the international migration of workers has remained significant. For example, in 2011, 12 rich countries within the OECD received 3,827,000 permanent migrants, defined as long-term migrants who may be granted residence (Table 5.6). About one-third of these migrants came from OECD countries themselves but the remaining two-thirds came from developing and poor countries, with China alone providing 529,000 migrant workers to OECD countries. Several OECD countries, such as Canada, the US, and France, have also developed temporary migration policies, mainly to attract particular types of low-waged labour. In Canada, for example, two programs (2014) that provide temporary work permits are a Seasonal Agricultural Work Program for farm workers (from Mexico and the Caribbean) and the Live-in Caregiver Program.

In addition to flows from poor to rich countries, other important geographic patterns of migration exist. There have been substantial flows of migrants from India and Pakistan and other poor countries to rich states in the Middle East, for example, to help build the massive expansion of Dubai.

Economically motivated international migration has changed the world. The migration from Europe to the New World or settler colonies of the US, Canada, Australia, New Zealand, South Africa, and other countries during the nineteenth century are

TABLE 5.6 Permanent Migration to Selected OECD Countries and the Country of Origin of Migrants, 2011

OECD Country of Destination	In-migrants (000s)	Leading Countries of Origin	Out-migrants (000s)
USA	1,052	China	529
Canada	237	Romania	310
France	211	Poland	274
Germany	233	India	240
Italy	559	Mexico	161
Netherlands	81	Philippines	159
Norway	44	USA	135
Spain	692	Germany	144
Sweden	74	Morocco	110
UK	343	UK	107
Australia	192	Pakistan	105
Japan	109	France	96

Source: OECD (2013). International Migration Outlook 2013. http://dx.doi.org/10.1787/migr_outlook-2013-en (accessed March 2014).

examples. Economic migration is also significant within countries, occurring between slow and fast growing regions and from rural to urban areas. In Canada, for example, the growth of the Athabasca oil sands in north-eastern Alberta has attracted temporary and permanent workers from every province until the oil price crash of 2015. Indeed, the economic growth of Alberta as a whole had made the province the most important destination for migration within Canada over the past decade, especially the rapidly expanding metropolitan areas of Edmonton and Calgary. However the oil price crash has arrested this growth, and unemployment has increased.

Meanwhile, in China interregional and rural–urban migration has occurred on an unprecedented scale since the beginning of economic reform and the loosening of residence restrictions (the Hukou system) that forced the majority of Chinese to remain in the countryside. It is estimated that some 400 million Chinese changed their registration from rural areas to towns and cities between 1980 and 2005, and that an additional "floating" population of perhaps 140 million (*circa* 2003) had moved without changing their registration. By 2010, the number had climbed to 230 million and it is expected that by 2020, another 300 to 500 million workers will have migrated within China.

There are limits to the geographic mobility of labour. Family ties and social networks are strongly localized. People with children in school and aging parents to care for may well be reluctant to move. Migration is costly and the rewards are uncertain: the jobs available in destination regions may not match the skills, qualifications, and preferences of potential migrants. Moreover, governments typically restrict in-migration, whether through quota systems, stringent evaluation criteria, or outright bans. The ILO (2006) has found that migrants are often subjected to workplace exploitation (for example, job and income discrimination) in rich countries, while migrants in low-paying jobs throughout the world have relatively few rights.

Finally, worker migration is a selective process that tends to be led by younger, better educated, more highly skilled, and more energetic individuals. The loss of these people has a cumulative impact on the region of origin, depriving it of the characteristics it needs most, while the region of destination benefits accordingly. For some poor countries, 30–50 per cent of their educated professionals may live elsewhere. Perhaps the most egregious example of the ill effects of this migration is the proportionally huge investments many developing countries make in educating doctors and nurses, only to see them lured away to higher-paying jobs in North America and Europe. The concept of the **brain drain** reflects these cumulative impacts (Case Study 5.2).

brain drain The loss of highly educated and qualified people from one region or country to another that offers better opportunities.

Decent Work

The most basic labour market challenge is the supply of what the ILO calls "decent jobs," especially decent jobs for youth. For labour, globalization implies both more opportunities and more competition for jobs, and therefore it places considerable downward pressure on wages. The ILO's **decent work** agenda has four strategic objectives—(1) job creation, (2) the right to work, (3) extended social protection, and (4) the promotion of dialogue and conflict resolution—and emphasizes gender and racial equality in each

decent work Full and productive employment in socially acceptable working conditions.

Case Study 5.2
BRAIN DRAIN

"Brain drain" is a shorthand term for the migration of better-educated, more highly skilled workers (e.g., engineers, scientists, professionals of all kinds) from regions with limited job opportunities to regions with expanding job opportunities. While the brain drain is usually discussed in the context of poor countries that lose their university graduates to the job markets of rich countries, it also reflects the overall nature of the migration process.

In general, regions of destination benefit from in-migration: migrants' incomes increase both the tax base and the demand for goods and services, including housing. In-migration, in other words, contributes to local multiplier processes and virtuous cycles of cumulative causation as the local region diversifies its skill-mix. For regions of origin, by contrast, out-migration contributes to a vicious cycle in which the loss of key skilled labour and potential for innovation is accompanied by a declining demand for goods and services and reductions in the local tax base. The result is a downward spiral. In some cases, out-migration may even cause unemployment rates to rise. For example, out-migration has been an enduring feature of Canada's Atlantic provinces (Newfoundland and Labrador, Prince Edward Island, New Brunswick, and Nova Scotia); yet unemployment in the region remains consistently above the Canadian average.

When migrants find work, many send part of their income "back home." In fact, the ILO estimates that migrants transfer an astonishing $250 billion a year to their home regions—a sum that amounts to more than the combined total of all official development assistance and foreign direct investment. One-third of global remittances go to China, India, Mexico, and France. A *reverse brain drain* has also occurred, as illustrated by the return flow of Indian and Chinese scientists from Silicon Valley to promote high-tech development in places like Bangalore and Beijing.

case. Although it is specifically focused on poor and developing countries, this agenda resonates with the long-standing union advocacy of the right to a job, wages sufficient to support a family, and wages that reflect labour's contribution to profits. Around the world, access to "decent" or "good" jobs remains the essential first step towards reducing poverty and inequality.

In the meantime, governments in a number of developing countries without adequate social safety nets are experimenting with programs designed to provide a minimum income to poor people so long as certain conditions are met—for instance, so long as their children attend school. For a relatively small investment of money, these conditional cash transfer (CCT) schemes not only reduce hunger but also provide an incentive to raise educational standards and thereby improve prospects in "decent" job markets. Conditional cash transfers seem to be working especially well in the rural areas of countries such as Brazil and the Philippines; unfortunately, they have been less successful in urban areas where substance abuse is widespread. A similar scheme has been tried in New York.

Conclusion

A decent job provides more than an income. It also provides the dignity that comes with work: self-respect, a sense of purpose, and a certain status and role in the community. Many of the social problems associated with poverty may have as much to do with the lack of work as they do with the lack of an adequate income. In developed countries, many of the most socially deprived and policy-challenged areas are in inner cities where people have lost the habit of working for a living.

The rights of workers have advanced enormously since the beginning of the capitalist era, often in association with social protest (Case Study 5.1). But workers' rights that are socially constructed can also be deconstructed or revised. While workers' rights are advancing in some places, they are threatened or non-existent in others.

Globally, the ILO's decent work agenda emphasizes the connections between work, poverty, and sustainable development. For poor people, decent paid work in socially acceptable and protected working conditions is vital to reducing poverty and ensuring that development can be sustained. Today, when workforces around the world are increasingly linked in both competitive and complementary relationships, international rules, regulations, and policies are essential, but they must be implemented locally. Investment in education and training is also vital to the long-term productivity and adaptability of the global workforce.

Practice Questions

1. Classify workers at a local fast food restaurant, chain-store retail outlet, or ski hill in terms of flexibility principles.
2. Discuss whether labour unions are needed in contemporary "more flexible" times.
3. How and why have your career expectations evolved since high school?

4. Assess the various dimensions of youth unemployment in rich and poor countries. Are special policy measures needed?

5. Is job discrimination an issue in your community or region? Explain. Discuss the extent of affirmative action in your community. What are its advantages and disadvantages?

6. Ask an acquaintance who has moved to your area from another country if he or she plans to go back, and why or why not.

Key Terms

labour mobility   149
Fordist (or dual) labour market   149
primary labour market   149
secondary labour market   149
flexibile labour market model   149
labour catchment area   150
employment relationship   151
labour unions   153
collective bargaining   155

seniority   155
minimum wage   156
labour market segmentation   157
scientific management
   (Taylorism)   159
industrial unions   159
independent labour market   160
dependent labour market   160
part-time   160

structured labour markets   161
new economic spaces   162
outsourcing   164
unemployment   165
underemployment   165
participation rate   165
brain drain   172
decent work   172

Recommended Resources

Atkinson, J. 1987. "Flexibility or fragmentation? The United Kingdom labour market in the 1980s." *Labour and Society* 12: 87–105.

The article that introduced the flexibility-based model discussed in this chapter.

Hayter, R. and Barnes, T. 1992. "Labour market segmentation, flexibility and recession: A British Columbian case study." *Environment and Planning D* 10: 333–53.

Elaborates on the Fordist and flexible models of labour market segmentation and assesses the extent to which labour market changes in BC are moving towards the latter model.

Herod, A. 2000. "Labour unions and economic geography" in *A Companion to Economic Geography*. Pp. 341–58. ed. Shepherd, E. and Barnes, T. Oxford: Blackwell.

A review of the literature assessing the geographical impacts of unions and the behaviour of workers, recognizing the proactive role that labour plays in shaping the economic landscape.

Peck, J. 2001. *Workfare States*. New York: Guilford.

Recognizes that labour markets develop differently according to social and political context, examines the diffusion of workfare practices beginning in the US in the 1980s, and notes the general deterioration in workers' rights.

Reich, R.B. 2010. *The Next Economy and America's Future*. New York: Knopf.

Argues that the major underlying (i.e., structural) problem with the US economy is growing income inequality, and that this trend must be reversed.

References

Clark, G. 1981. "The employment relation and spatial division of labor: A hypothesis." *Annals of the Association of American Geographers* 71: 412–24.

International Labour Office (ILO). 2006. *Facts on Labour Migration*. ILO: Geneva.

———. 2010. *Global Employment Trends*. ILO: Geneva.

Krugman, P. 2006. "The Great Wealth Transfer," *Rolling Stone*. 14 December.

Polanyi, K. 1944. *The Great Transformation*. Boston: Beacon Press.

Reich, R.B. 2010. *The Next Economy and America's Future*. New York: Knopf.

CHAPTER 6

Government

Hobbes's argument is that the alternative to government is a situation no one could seriously wish for.

Williams 2006: 5

• • •

Nothing is more important for social and economic stability and gain than effective government.

Galbraith 2000, in an interview with Barb Clapham

The overall goal of this chapter is to examine the various ways in which governments interact with markets to shape economic activities inside and outside their territorial boundaries. More specifically, this chapter's objectives are

- To explain the principles underlying government intervention in market behaviour;
- To explore the nature of national economic strategies;
- To examine the responsibilities of government at various geographic scales;
- To explain how and why governments attempt to shape international, national, and regional development; and
- To note the role that local governments play in determining local land use and the spatial routines of daily life.

The chapter has three main parts: the first focuses on the broad principles of government intervention in the economy; the second on government budgets and national macroeconomic and regional and local development policies; and the third on international forms of governance, with special attention to the construction of free trade arrangements and the problems of environmental governance.

The Role of Government

Government institutions interact at various levels—local, regional, national, international—to shape virtually every aspect of human life, including economic activity. The very idea of a society implies a degree of cohesion and stability that cannot be achieved without some form of government to maintain civil relations among the society's members, whose natural state—according to Hobbes (1651)—is one of fear and conflict over scarce resources. But governments do much more than protect their people from one another. They also manage their society's relations with other societies and play an active role in its economic life. Indeed, for Galbraith, effective government is vital to facilitate stable and equitable growth, while ineffective government would impede or prevent these goals. As globalization continues to increase the movement of information, money, goods, services, and people across political boundaries, the international and domestic responsibilities of governments are becoming increasingly interrelated and blurred.

The story of the Berlin Wall (1961–89) provides a dramatic illustration of the power of governments in changing the institutional structure of economies (Case Study 6.1). The destruction of the Wall symbolized the collapse of the single most significant political barrier to the movement of goods, services, investment, ideas, and people around the world at the time. Its fall signified the transformation of

Case Study 6.1
THE BERLIN WALL

© Shutterstock.com/gary718

At the end of the Second World War, the city of Berlin was divided into four sectors, each under the control of one of the four Allied powers: Britain, France, the US, and the Soviet Union. But relations between the Soviets and their Western allies deteriorated, and by 1949 all of Europe was divided by a heavily guarded fence, thousands of kilometres in length, known as the "iron curtain." Although Berlin was well inside the Communist eastern zone, its western portion was treated as if it were part of West Germany, and East German citizens were permitted to travel there more or less freely. As a consequence, West Berlin came to serve as a gateway for East Germans seeking to escape to the West—approximately 2.5 million in just 12 years, among them many of the regime's best-educated citizens. It was to stop this "brain drain" that a fortified fence was constructed, virtually overnight, in August 1961. It was soon replaced by a wall that surrounded West Berlin to seal it off from the surrounding territory.

For the next 28 years, the movement of people, goods, and ideas between the Marxist-inspired centrally planned economies of Eastern Europe (and China) and the market economies of the West were extremely limited. But by late 1989 the Communist régimes across Eastern Europe were collapsing amid growing public protests. On 9 November, an East German authority announced that travel restrictions were to be eased, and within hours a number of crossing points had been opened. The dismantling of the Berlin Wall heralded the removal of all barriers to trade, investment, and labour, and paved the way for German reunification in 1990. Adapting to markets and private ownership has been a massive challenge for the former East Germany, where incomes, standards of living, and job prospects were all lower than in the West, and restructuring its economy has been a priority for the German government.

Communist states, with their centrally planned economies, into more open, dynamic market economies.

Market economies celebrate the virtues of private markets and "free enterprise," and proponents of capitalism frequently express opposition to government interference. Yet governments exercise tremendous economic influence: through the huge sums they collect in tax revenues and direct back into the economy through **budget** spending; the large numbers of jobs they directly provide; the regulations they devise and enforce; and the strategic direction they give to economic development, from the local to the international level. In short, governments shape economic activity both in place and across space. As place-based institutions, national, regional, and local governments are territorially defined and mandated. In democracies, they are elected to represent and serve the interests of constituents, perhaps most importantly with respect to the functioning of the economy. At the same time, as space-based institutions, governments influence the movement of goods, people, investment, services, and information within their jurisdictions and beyond. Governments constantly seek to improve the competitiveness of their territories in the global economy and co-operate in international trade policies and organizations that may be beneficial to them.

budget An estimate of revenues and expenditures over a defined period of time.

The Nature of Government–Economy Relations

The economic roles of governments are complex and vary widely from one country to another. Even in a single country there are no hard and fast rules for determining which activities should be handled by government and which by the private sector. Activities can be either "privatized" (shifted from the public to the private sector) or "nationalized" (shifted from the private to the public sector) while some activities can be both privately and publically controlled.

Governments are institutions that have the authority to establish and uphold laws within mandated territories. Typically, government authority within countries is hierarchically and spatially structured, although the details of the structure vary among and even within national territories.

There is an important distinction between unitary and federal systems. In **unitary systems of government** (such as Great Britain's) the bulk of the power is vested in a single central government, whereas in **federal systems of government** (such as Canada's) power is divided between the central government and a number of regional governments (provincially and territorially in Canada) under which local governments operate. In general, national governments—federal or unitary—are the most influential political institutions: they oversee regional and local authorities, are responsible for relations with other states (including international trade), and represent their countries in supranational bodies, such as the World Trade Organization (WTO). Lower levels of government have their own functions, responsibilities, and powers and can be directly responsible for a large share of government expenditures. Governments at all levels also own or control enterprises that serve public purposes and provide products such as health care, education, transportation, and water supply, but which have substantial autonomy.

unitary systems of government
Political power is centralized within national governments, including with respect to the delegation of powers.

federal systems of government
Political power is constitutionally shared among national and regional levels of government.

Case Study 6.2
MONEY AND BANKS

iStockphoto.com/dynasoar

Money and market exchanges developed with civilization, first supplementing and eventually replacing earlier bartering systems in which people directly exchanged goods by mutual agreement. In China, cowry shells were being used as money by roughly 1800 BCE; coins were introduced by 400 BCE and paper money by 1100 CE. Money has a number of functions, but the principal one is to serve as a standardized currency to facilitate market exchanges of "real" goods and services. People must have confidence that the money they receive as the sellers in one exchange will not have lost value by the time they want to use it as buyers in another exchange.

In democratic market economies, governments are led by elected representatives who are supposed to pay attention to their constituents' interests. However, even the best-intended public policy is unlikely to satisfy all the diverse or "pluralistic" institutional interests that make up modern societies. Politicians can prioritize policies for political gain and can be influenced by special interest lobby groups. High-level bureaucrats also play a large role in designing policy and interpreting how policies and regulations will be implemented. Nevertheless, because most citizens recognize the necessity of government, most tolerate a fairly wide range of policies with which they do not agree. Moreover, governments themselves are diverse and complexly intertwined with other institutions in both complementary and adversarial ways. State power is real, albeit within limits, and its exercise requires scrutiny.

Principles of Government Intervention in the Economy

Perhaps the most pervasive and profound economic responsibility of government is to manage the nation's money supply and banking system (Case Study 6.2). As a standard, trustworthy unit of value, money provides a common basis for exchange and

To this end, governments and their central financial agencies attempt to influence and regulate the money supply. Broadly speaking, national financial policies seek a balance between the overall money supply and the overall demand for goods and services: "too much" money in relation to the goods and services provided leads to inflation, while "too little" money leads to stagnation.

Recognizably modern banking operations were under way in Italy as early as 1200 CE, but most of the system's growth has taken place over the last 250 years, with the rise of market economies. The Bank of England became particularly influential as a model of central banking that other countries have adopted and modified. Founded in 1694 to serve as the banker for the English government, within a century the B of E governed both the nation's banking system and its money supply. It was nationalized in 1946 and became independent in 1998, although it remains under charter to the government.

In the US, the Federal Reserve System ("the Fed") was established as the central banking system in 1913, after several public panics sparked when customers trying to withdraw their deposits found that the banks were unable to pay because the ratio between their lending and money held in deposits and securities had become too extreme. The Fed is partly public because its board of governors is appointed by the president of the US and partly private because it is owned by a consortium of banks. The Bank of Canada was created in 1934 as a private institution but was made a Crown corporation in 1938.

National banks influence the banking system via the bank rate—the interest they charge on their loans to commercial banks—and the regulations they impose on lending (for example, restrictions on the extent of loans in relation to deposits and the range of activities permitted). Although the Fed and the B of E are both national institutions focused on national policy, they anchor the agglomerations of financial organizations, in New York and London respectively, which constitute the apex of the global financial system.

is essential for markets to operate. The vast flow of money that integrates the global system over space and time is organized by financial systems comprising an array of financial institutions that are interconnected by many different kinds of markets and services. At the heart of financial systems are banks whose purpose is to mediate between societies' savers and borrowers, and to smooth out consumption and expenditure patterns by individuals and organizations over varying periods of time.

Because of money's vital importance to the functioning of economies, society looks to the government to impose authority and stability on the money system. Hence, private banks and related financial institutions are regulated by national governments in close association with central banking institutions such as the Bank of Canada or the US Federal Reserve System. In addition to controlling the supply of money and interest rates, governments also set the conditions for borrowing and lending. The poor decisions made by the UK and US governments to deregulate the banking sector, which led to the 2008 financial crisis, reflected their neoliberal convictions to empower markets (see Introduction). This deregulation, not duplicated in Canada, also demonstrates variations in government approaches to intervening in the economy.

The banking system epitomizes why governments intervene in market economies: to ensure market efficiency, accomplish social priorities, and deliver public goods (Figure 6.1).

Market Efficiency

An essential role of governments in market economies is to ensure that people and companies can freely and fairly engage in exchange. The ability of markets to set prices and allocate resources efficiently is facilitated by voluntary and informed participation, but undermined by intimidation, the withholding of behaviour, and fraudulent behaviour. This "free engagement" principle provokes complex governance and has led governments to introduce laws and regulations on many issues, for example, to prevent monopolies, prevent hiring discrimination, promote safety by restricting use of dangerous ingredients or products, and encourage the supply of information via labelling and other means. Further, to establish confidence and stability in market exchanges, governments underpin or create laws and institutions that guarantee property rights, for example, via bodies of contract laws, sophisticated land survey and property registration systems, and patents, copyrights, and other means to protect intellectual property. Governments also support legal institutions and other institutions by which complaints over the violation of rights and obligations in market exchanges can be filed.

FIGURE 6.1   **Principles of Government Intervention in Economic Activities**

Geographically, national governments typically seek to standardize rules and facilitate the free flow of goods, information, and labour within their territories, a task that is often complicated in countries with a federal system of government where regional governments have significant powers. Indeed, inter-provincial trade barriers, affecting the flows of goods, services, and labour, remain significant in Canada. Globally, the free trade agenda represents a similar attempt to standardize and facilitate market efficiencies across nations.

However, government attempts to promote market efficiencies are problematic, in part because governments are swayed by political considerations, opposition from vested interests, and constraints imposed by established arrangements and routines. As well, governments need to ensure market behaviour meets social and political goals and to provide alternatives when markets fail, most obviously with respect to public goods.

Social Priorities

Within nations, government oversight of the economy goes beyond ensuring market efficiency. Perhaps the greatest social concern is employment. Governments tend to prioritize expanding employment levels to meet the social and material aspirations and needs of people, to increase the productive capacity of the economy (because employment is a large source of taxes), and to offset the political and social problems that arise from unemployment. Government approaches to job stimulation, however, vary considerably, ranging from indirect approaches that emphasize financial stability and tax regimes favourable to business investment to the explicit identification of priorities and targeted incentives.

People also expect governments to encourage the economy to operate fairly in terms of equal opportunity, social mobility, and distribution of income. These expectations vary greatly amongst nations, especially in regard to income taxes and the redistribution of money from the wealthy to the less well-off. This redistribution has a strong geographical component, for example, redistribution from wealthy to poor regions. Such policies not only run up against ideological differences, but support for poor regions and population segments within regions runs the risk of encouraging dependency by creating a disincentive to take responsibility.

As markets develop they tend to focus only on the value that is exchanged between buyer and seller, and ignore the externalities or the social costs and benefits incurred by the various related processes of production and consumption. Responding to public needs and concerns, governments have passed laws regarding matters such as fairness in employment practices, fair competition, health, safety, and environmental protection. For example, contemporary labour market regulations have roots in a sequence of Factory Acts from the beginning of the nineteenth century in the UK that sought to protect workers and children from highly dangerous working conditions. Indeed, it may be argued that market regulations have saved capitalism from its own excesses. Governments are constantly seeking to balance the effectiveness of these regulations with the efficiencies of the market.

Public Goods

People and organizations buy many private goods, such as food, clothing, cars, household possessions, machinery, and chemicals. According to economists, private goods are exclusive—that is purchasers have exclusive control over the good—and rivalrous—

that in this context means that once consumed it is no longer available to others. The buyer of a cup of coffee, for example, has the right to drink it (or give it away or sell it) and once drunk, it is gone! In this same abstract terminology, a *pure* public good is non-exclusive and non-rivalrous—that is all members of the population can access the good or service, and consumption by one member does not deplete consumption by others (Table 6.1). While public goods may be provided by nature (usually called a commons), many also require collective effort to produce. In the absence of private supply, governments have taken it on themselves to provide many public or quasi-public goods in the form of highway networks, defence and police, fire protection, education and health systems, clean air and water, and basic research.

Whether pure public goods exist is debateable. In a spatially distributed population, for example, accessibility to public goods varies, imposing additional costs on more distant users. Further, market options are often available to provide many services otherwise controlled by the government, such as private schools and hospitals. Nevertheless, there are powerful rationales for governments in market economies to be dominant suppliers of many quasi-public goods that are largely non-exclusive and non-rivalrous in nature. In general, these goods are vital to the health, skill, training and education levels, and security of the population as a whole. Committed to serving the entire population, government-supplied ("quasi-monopoly") public goods can take advantage of economies of scale while providing stable, standardized, and socially acceptable services. In contrast, market options would provide goods on the basis of price, including profit calculations, excluding people who could not afford them. Consequently, society as a whole would be less healthy, educated, mobile, and secure, implying lower productivity, less demand for private goods and service, and more political and social problems. Simply stated, if societies do not have broad education, health services, or transport systems that get people to work, then the economy will not work to its optimum.

Note that in addition to public and private goods, "club goods" and common pool resources can be recognized according to the exclusive/rivalrous criteria (Table 6.1). Thus club goods are provided collectively to those people willing to pay a price for them, while common pool goods or resources restrict use to members in order to restrict over-exploitation and ensure sustainability (Chapter 1). This task is not easy, and can be disrupted, for example, by technological change, changes in the environment, and free-riders. Indeed quasi-public goods can also face problems of demand exceeding capacity, as evidenced by traffic congestion, water-supply cutbacks, and hospital line-ups. In such

TABLE 6.1   Public Goods in Relation to Other Goods

	Rivalrous	Non-rivalrous
Excludable	Pure Private Goods e.g., car, house, food, etc.	Club/Toll Goods e.g., toll roads, theatre, golf club
Non-excludable	Common Pool Resources e.g., inshore fishery or forests	Pure Public Goods/Commons e.g., road network, public education, atmosphere, ocean

cases governments or the organizations in control of the club goods and common pool resources face tricky problems of restricting access, building expensive new capacity, or allowing the market a greater role.

Finally, governments past and present have often chosen to establish state monopolies or state-owned companies (SOCs) in a wide range of activities from broadcasting to transportation to resource industries (see Chapter 9) that are deemed to be of strategic importance to the economy and to have significant public value (Case Study 6.3).

Case Study 6.3
STATE-OWNED COMPANIES (SOCS)

Left: Logo of Canadian Broadcasting Corporation (CBC) established in 1936

Right: Logo of the Chinese Central Television established in 1958

Sources: (left) CBC Licensing; (right) The State Administration of Press, Publication, Radio, Film and Television of the People's Republic of China

In North America, especially the US, the dominance of private corporations is widely assumed, but in much of the world government corporations dominate national economies. In China, the banks, the oil companies, and many other key sectors are oligopolies and completely dominated by what are referred to as state-owned companies (SOCs). China is not alone: key sectors are controlled by SOCs in Indonesia, Malaysia, Brazil, Russia, Saudi Arabia, and many other countries. These are not only big companies nationally, but huge internationally—the Japanese Postal Bank remains the largest bank in the world by assets, and as we shall see in chapter 8 state-controlled oil companies have become the world's largest. In Canada, SOCs are called Crown corporations, including today the Canada Mortgage and Housing Corporation (CMHC), Canadian Broadcasting Corporation (CBC) and liquor control boards. Governments may own all or a significant part of the company and usually control the board of directors.

The rationales for or origins of this public ownership preference are diverse, but founded in particular path dependencies, for example, countries that were late developers used government investment to initiate industry; provide control over natural resources that had been exploited by foreigners with little payback to the country; reflect ideological beliefs in state ownership of the entire economy or of key sectors deemed vital to national interest; support a national champion; maintain control over national cultural institutions; and so on. These origins are reflected in the sectors that SOCs predominate: energy and other resource extraction, processing, and

(Continued)

distribution; health and education; connective infrastructure such as postal systems, highways, airports and ports, water and sewage, and railways; media such as television, radio, and newspapers; investment corporations (sovereign wealth funds); and social "bads" such as gambling, alcohol, and tobacco. In many countries, SOCs serve not only strategic purposes, but are also important sources of income, to which some governments have become addicted, especially in the case of the "bads."

In recent decades, governments have privatized SOCs in several rich market economies, a process driven by the shift to neoliberal ideology and globalization. In Canada, the change in status of Air Canada and Canadian National Railways (Chapter 12) and Petro-Canada (Chapter 9) are examples. Privatization has been controversial, however, especially regarding supposed efficiency gains and the cheap "sell-off" of national assets.

National Economic Strategies

Governments, especially national governments, are vitally concerned with developing internationally competitive economies and increasing per capita incomes and associated standards of living of their populations. In recent years, pleas for an environmentally sustainable economy have become part of development mandates, albeit with varying degrees of conviction. In some market economies, such as the Scandinavian countries, egalitarian goals are important, but in others, such as the US and UK, these goals are more conflicted. Indeed, national economic strategies vary in several key respects.

A distinction is often made between domestic ("internal") and international ("external") economic strategies. The domestic/internal strategy focuses on regulation of markets, macroeconomic policies, and development of social and economic infrastructure such as education and training, industry support programs, transport and communication systems, regional development, resource exploitation, and spatial connectivity. The international/external strategy concerns trade and foreign investment, but with increasing globalization domestic and international strategies are becoming more integrated. Moreover, domestic activities are no longer only subject to national regulation, but to international trade agreements, dispute-resolution mechanisms, market-driven currency valuations, and environmental standards and protocols.

The governments of most market economies share important economic objectives. All engage in macroeconomic planning to stabilize growth, reduce the risk of economic crises, and minimize disruption when it does occur. Nevertheless, government long-term strategies towards directing their economies and intervening in market forces vary over time and space. Bearing in mind that government policies are typically influenced by pragmatic considerations, and changes in control by political parties, these variations can be appreciated as "ideal types" of national economic strategies.

Among market economies, nations where the state exerts a strong influence over the economy are often labelled *dirigiste*, a term specifically derived from French post–Second World War experience. In contrast, neoliberal economic ideologies favour minimizing government involvement. Leaving aside centrally planned economies (where the state controls most economic activities), six basic models have been used in the context of (market) state–economy relations: (1) statist, (2) corporatist,

(3) liberal, (4) neoliberal, (5) development state, and (6) welfare state. In the **statist** model, the prime exemplar of *dirigisme*, national governments play a defining role in the identification, development, and (sometimes) control of key economic sectors. In the **corporatist** model, governments co-operate with business and labour to determine strategic priorities. In liberal states, governments favour free trade and prefer as much as possible to leave economic decisions to the private sector while allowing for state intervention in markets to meet social goals. Neoliberal states narrow liberal thinking to emphasize the pre-eminence of markets for meeting economic and social goals. By contrast, the development state model supports government intervention in the economy to protect domestic "infant" industries and workers by restricting foreign imports and investment and/or targeting support for industries deemed strategically important. Finally, the **welfare state** model emphasizes the government's responsibility for delivering social justice and providing "cradle to grave" care for all citizens through publicly supported education, health care, old age pensions, social security, etc., and ensuring everyone benefits from economic development. Strong commitments to welfare state principles are often associated with "social-market" democracies, such as the Scandinavian countries. The distinctions among the six models are blurred in practice; for example, neoliberalism evolved from liberal traditions, the development state model can either imply statist or corporatist approaches, governments can change their approaches, and approaches can vary among different sectors of the economy.

Regional government strategies differ along these lines as well. In Canada, for example, the Quebec government is in the *dirigiste* camp, as symbolized by its willingness to support and subsidize key domestically controlled organizations in the public and private sector such as Caisse de Dépôt, Hydro Québec, and Bombardier, while Alberta's government is more neoliberal and open to foreign investment and not in favour of high resource taxes. In the US, in contrast to many states that are cautious and even antagonistic towards funding public goods, Massachusetts with its strong government support for public health care and education indicates a greater embrace of welfare state considerations. The innovation systems discussed in Chapter 3 reflect these national and regional differences in strategy. In the US, national funding programs such as NIH (National Institute of Health), DARPA (Defense Advanced Research Projects Agency), and incentives for academic entrepreneurship provide a broad foundation for markets to generate innovations and for clusters such as Silicon Valley to arise. Elsewhere, governments in the developmental states of Taiwan and Singapore or the pragmatic welfare states in Germany and Scandinavia provide more coordination among potential collaborators.

The Neoliberal State

The conventional liberal model, while emphasizing markets and individualism in economic behaviour, recognizes the legitimacy of government intervention in the economy in order to provide regulation, macroeconomic policies, and support for public goods such as education and health care. Neoliberalism, however, is highly critical of government participation in the economy, including support for activities that have significant public good characteristics. Deregulation, privatization, and competition are key neoliberal mantras. Thus neoliberals advocate the relaxation or outright elimination of rules and laws that constrain market activity and business behaviour, both

statism A governance model where national governments take a proactive role in directing strategic growth.

corporatism A governance model where three institutions—government, business, and labour—work together to develop a strategic plan for the economy.

welfare state A model where the government seeks to ensure that all citizens have access to the essentials (food, shelter, education, etc.) required for participation in economic and social life.

domestic and global, and the extensive privatization of activities under public control. Neoliberals are also critical of fiscal policies that seek to stimulate growth by government deficit spending and high taxes. In general, they equate government with initiative-stifling bureaucracy that constrains entrepreneurship and competitiveness in the economy. Neoliberals are especially antagonistic towards the welfare state and argue that public-sector services, such as education and health care, should be privatized.

From a development perspective, the neoliberal model argues that market forces should be the main arbitrator of competitive advantages and not government subsidization, protection, or prioritization. Internationally, neoliberalism supports free trade principles in all respects (goods, services, investment, labour, money, information) with the minimum of restrictions and believes government interference would misallocate resources.

Neoliberalism gained momentum in the 1970s and (especially) 1980s under the influence of President Ronald Reagan in the US and Prime Minister Margaret Thatcher in the UK. Their support for neoliberal principles resonated in many advanced countries, including Canada and New Zealand, as well as a few developing ones, notably Chile. Various right-wing think-tanks continue to aggressively promote neoliberal ideas. However, critics argue that neoliberal policies are pro-business and anti-labour, fail to appreciate the significance of public goods, services, and related public institutions, and lead to widening income disparities. The global financial crisis of 2008, stemming from excessive market deregulation and creating much unemployment and social distress, was a big blow to neoliberal credentials. Moreover, governments are typically pragmatic and in the industrialization of Asian powerhouses, such as South Korea and China, governments are playing vital development roles.

The Development State

In this model, governments actively stimulate industrialization by protecting infant industries; creating the necessary infrastructure, both economic (e.g., transportation and communications systems) and social (health care, education); and providing targeted support for activities that are deemed to be strategically important for long-term development. Advocates of the developmental approach argue that small-scale new industries need the protection of tariff barriers to prevent cheap imports from bigger, more efficient foreign rivals who would drive them out of business. Without such protection, infant industries would never have the chance to grow to the point where they could begin to reduce average costs of production and take advantage of economies of scale.

The role of foreign direct investment in the development state model is not so clear-cut. Some argue that FDI helps in the spread of "best practices" around the world. Others insist that FDI should be restricted because it has the effect of truncating the development of host economies (Case Study 4.4) and reducing the potential for domestic champions to emerge (Chapter 3).

Historically, all industrialized nations, including the US, Germany, and Japan, initially protected their infant industries and encouraged the rise of domestic champions to stimulate industrialization. In the late nineteenth century, under the National Policy introduced by the post-Confederation government of Sir John A. Macdonald,

Canada used tariffs to restrict imports of manufactured goods and built a transcontinental railway as part of the effort to promote western settlement, but did not restrict FDI. On the contrary, Canadian policy encouraged the establishment of branch plants that supported neither the export of manufactured goods or local R&D. Other countries such as Japan and Sweden, which restricted FDI, were considerably more successful in encouraging the emergence of domestic champions.

Japan's approach to development was comprehensive. As a "late developer," opening to the West only in the second half of the nineteenth century, Japan faced a substantial technology gap. To close the gap, the national government encouraged the study of Western institutions in many areas, from banking and education to taxation and politics. Both government and private firms, for example, sent Japanese nationals to the West to learn about current best practices and hired Western experts to fill positions in Japan. In addition, Japanese governments in the late nineteenth century identified strategic priorities, targeting industries such as shipbuilding (then on the leading edge of technological progress) for special support. Following the devastation of its industrial base in the Second World War, the Japanese government took a similar strategic approach, focusing on steel, chemicals, autos, and electronics. Japan began to outpace other advanced countries; however, the increasing diversification of industry and the shift to services have reduced the government's capacity to guide industrial development, although it still attempts to do so.

Other East Asian countries have pursued a development state model with an even more statist approach. Much of South Korea's development has taken place since the 1970s under an authoritarian government that privileged family-owned industrial conglomerates ("chaebols") and protected them against foreign competitors. Singapore's government was less brusquely authoritarian but nevertheless established state-owned companies and invited investment from foreign multinationals to initiate development. As it advanced, Singapore began to invest in education and indigenous R&D. In China and India, with communist and socialist histories, state-owned companies continue to play huge roles, even as the governments look to privatization to invigorate their economies. In China the national government has invested massively in R&D in recent years in efforts to stimulate innovation.

Government Policy Interventions and Institutions

Depending on the economic strategy, governments can wield a vast array of policies, implemented through long-standing and purpose-built administrative organizations, to achieve their objectives. We will review several of the more important policy institutions with emphasis on their different national expressions and how they generate different economic landscapes.

The Geography of Taxing and Spending

Ensuring market efficiency might be considered the essential government function, but providing social priorities and public goods overwhelmingly preoccupies a government's activities. Unfortunately they have to be paid for by tax revenues.

The geography of taxing and spending concerns the distribution of government income sources and the distribution of government spending. In other words, where the money comes from and where it goes are different and with consequences for the economy. Taxes can scare away investors and households, or taxes invested in public goods can attract investors and households. The locations of sources and destinations, while not matching, are distributed throughout the country but are only concentrated in a few nodes or regions. While the sources of income overwhelmingly derive from households and corporations, the outlets for expenditure include households, corporations, non-profits, and government entities. How a government constructs its tax regime can have implications for domestic company competitiveness in international markets and for its attractiveness to multinational company investment. How regional and local governments construct their tax regimes can have implications for the movement of people and companies within a nation. Governments also vary in how they tax and spend.

National Propensities in Revenue Generation

The proportions of GDP accounted for by taxes provide a comparative measure of the size of government within different economies. In selected OECD countries, taxes accounted for between 19.6 and 47.7 per cent of GDP in 2011 (Table 6.2). The US and Japan, the world's two largest developed economies, are near the low end of the range, while European economies are at the top end. For the OECD as a whole, the percentage of GDP accounted for by taxes was higher in 2011 than in 1975. Although the increase was modest in some cases (e.g., the UK), this trend clearly does not support the idea that government roles are waning as a result of globalization. Rather, in both relative and absolute terms, government spending has increased. Indeed, these data suggest that the role of governments generally increases with economic development. Thus in countries such as South Korea, Turkey, and Spain, which were relatively poor in 1975 but have developed rapidly since then, the share of tax revenues of GDP grew significantly by the beginning of the twenty-first century. Mexico's low level of tax revenue to GDP is consistent with its position as the poorest OECD country.

Most developed countries consistently extract 35–45 per cent of the GDP as taxes, while developing economies range from about 10 to 30 per cent. Mirroring national differences in the weight of taxation in national economies, government spending is a large share if not the majority of spending in economies. In much of Europe, for example, government spending makes up over 45 per cent of GDP. The differences among developed countries reflect two seemingly contradictory phenomena. First, all developed countries are committed to providing their citizens with a strong foundation in health, education, and security services. That is where most of the tax money goes. Second, the 10 per cent or so difference among countries is still substantial, expressed not only in significantly different taxation burdens and social service provision, but also in the national strategies, ideologies, and value systems. Universal health care and long-term parental leave are considered solid reasons for taxation in the pragmatically socialist Sweden, while the US, with its greater neoliberal influences, is conflicted over extending such benefits by imposing a greater tax burden.

TABLE 6.2 Total Tax Revenue as Percentage of GDP

Country	1975	2000	2011
Australia	26.5	31.8	26.5
Austria	37.4	43.4	42.3
Belgium	40.6	45.7	44.1
Canada	31.9	35.6	30.4
Czech Republic	n/a	39.0	34.9
Denmark	40.0	49.6	47.7
Finland	36.8	48.0	43.7
France	35.9	45.2	44.1
Germany	35.3	37.8	36.9
Greece	21.8	38.2	32.2
Hungary	n/a	39.0	37.1
Iceland	29.7	39.4	36.0
Ireland	29.1	32.2	27.9
Italy	26.1	43.2	43.0
Japan	20.8	27.1	28.6
Mexico	n/a	18.5	19.6
Netherlands	41.3	41.2	38.6
New Zealand	28.5	33.4	31.5
Poland	n/a	32.5	32.3
South Korea	14.5	23.6	25.9
Spain	18.8	35.2	32.9
Sweden	42.0	53.8	44.2
Turkey	16.0	32.3	27.8
United Kingdom	35.3	37.4	35.7
United States	25.6	29.9	24.0
OECD Total	30.3	37.2	34.1

Source: OECD. 2013. *Revenue Statistics 1965–2012*, OECD Publishing, http://www.oecd-ilibrary.org/ taxation/revenue-statistics-2013_rev_stats-2004-en-fr (Table A, Accessed March 2014)

The domestic allocation of taxing and spending power among national, regional, and local governments further reflects similarities and differences among nations. Thus, governments vary on where and how they obtain their revenues and also where and how they distribute money. The money that governments spend comes primarily from taxes. There are many types of **tax**, but there are three main sources of government revenue: (1) taxes on income (personal and corporate income taxes, capital gains tax); (2) taxes on goods and services (often referred to as sales or value-added taxes; tariffs on imports fall into the same category); and (3) taxes on property (in some places including not just real estate but also wealth and inheritances). Some taxes are levied at a fixed rate per quantity or volume; others, sometimes referred to as *ad valorem* taxes, are based on the value of the item in question. Canada's goods and services tax (GST) is an *ad valorem* tax; when introduced by the federal government in 1991 it was set at 7 per cent but since

tax A levy imposed by government on goods, services, and incomes; taxes are the source of all the money that governments spend.

2008 it has been reduced to 5 per cent. The harmonized sales tax (HST), introduced in 1997 in Newfoundland and Labrador, New Brunswick, and Nova Scotia and extended to Ontario and British Columbia in 2010 (but then terminated by the latter in 2013) and Prince Edward Island in 2013, combines the GST and provincial sales taxes and is another example of an *ad valorem* tax. Whereas provincial sales taxes are charged on all purchases made throughout the value chain, the HST applies only to the final products sold to consumers.

Note taxes are "progressive" if the *ad valorem* rate increases with ability to pay; progressive tax systems are designed to enhance social equity by requiring richer people to pay a higher rate than poorer people. A flat tax, by contrast, imposes the same percentage tax rate regardless of income, and is "regressive" as the proportion of taxes paid in relation to income declines as income increases.

National governments vary considerably in the types of taxes they rely on as sources of revenue. Personal income taxes are particularly important in North America, accounting for around 60 per cent of federal revenue in Canada in 2003–4 and in 2012–13 (Table 6.3). Taxes on goods and services are more important sources of revenue in countries like France and the UK.

Regional governments have significant budgetary powers in federal states. Indeed, in Canada in 2011 provincial governments accounted for almost 40 per cent of tax revenues collected, and in the US, state governments accounted for almost 20 per cent (Table 6.4). Mexico's regional governments, however, are poorly (directly) funded. The relative role of local governments varies considerably in terms of taxes they collect; they are especially important in Denmark, Finland, Japan, and Sweden, of little importance in Mexico, and of growing importance in several countries such as South Korea. In Norway and the UK, relative local government expenditures have both declined considerably between 1975 and 2011, in Norway's case largely replaced by nationally controlled oil wealth. Local governments generally rely on property taxes for much of their income, thus unlike national taxes, local revenues are tied to the wealth or poverty of specific places. Table 6.4 also shows that social security contributions, which are collected like a tax but allocated to a specific purpose such as health care or retirement, are especially

TABLE 6.3   Canadian Federal Government Revenue Sources, 2003–4 and 2012–13 ($ billions)

Income Source	2003–4 (%) $b	2012–13 (%) $b
1. Personal Income Tax	84.9 (45.6)	126.2 (49.0)
2. GST	28.3 (15.2)	28.8 (11.2)
3. Corporate Income Tax	27.4 (14.7)	35.0 (13.6)
4. Other Taxes/Tariffs	16.2 (8.7)	19.8 (7.7)
5. Employment Insurance Premiums (EIP)	17.5 (8.7)	20.4 (8.0)
6. Other revenues	11.8 (6.3)	26.9 (10.5)
Total:	$186.1	$256.6

Source: Department of Finance, Government of Canada. For 2003–4 data see http://www.fin.gc.ca/afr-rfa/2004/afr04_1-eng.asp#Revenues. For 2013–14 data see http://www.budget.gc.ca/2013/doc/plan/chap4-2-eng.html#a5-Outlook-for-Budgetary-Revenues

| TABLE 6.4 | Classification of Percentage Tax Revenues by Level of Government and Social Security among Selected OECD Countries 2011 (1975) | | | |

Country	Central Government	Regional Government	Local Government	Social Security
Federal				
Australia	81.3 (80.1)	15.3 (15.7)	3.4 (4.2)	0.0 (0.0)
Canada	41.5 (47.6)	39.7 (32.5)	9.7 (9.9)	9.1 (10.0)
Germany	31.7 (38.5)	21.3 (22.3)	8.0 (9.0)	38.5 (34.0)
Mexico	80.1 (n/a)	2.5 (n/a)	1.1 (n/a)	14.5 (n/a)
USA	40.6 (45.4)	20.7 (19.5)	15.9 (14.7)	22.8 (20.5)
Unitary				
Czech Rep.	54.1 (n/a)		6.6 (n/a)	44.1 (n/a)
Denmark	70.8 (68.1)		26.7 (30.4)	2.1 (0.5)
Finland	47.7 (56.0)		23.3 (23.5)	28.8 (20.4)
France	32.6 (51.2)		13.2 (7.6)	54.0 (40.6)
Greece	64.2 (67.1)		3.7 (3.4)	31.9 (29.5)
Italy	52.5 (53.2)		15.9 (0.9)	31.2 (45.9)
Japan	33.3 (45.4)		25.2 (25.6)	41.4 (29.0)
Poland	51.9 (n/a)		12.5 (n/a)	35.4 (n/a)
S. Korea	60.1 (89.0)		16.3 (10.1)	23.5 (0.9)
Norway	87.7 (50.6)		12.3 (22.4)	0.0 (27.0)
Sweden	51.3 (51.3)		35.7 (29.2)	12.6 (19.50
Turkey	63.3 (n/a)		8.8 (n/a)	27.0 (n/a)
UK	75.9 (70.5)		4.8 (11.1)	18.7 (17.5)

important in France and Germany but much less so in Canada. For many national and regional governments, personal and corporate income taxes dominate revenue sources.

Governments, just like companies, can also run their operations on money borrowed from banks and other financial institutions or from individuals in the form of government bonds. Because governments have secure future incomes in the form of taxes, financial institutions and individuals have faith in repayment and government bonds are known as the safest, if not the most lucrative, investments. The government may simply use borrowing as a means to overcome shortfalls in other forms of income (taxes), and it may use borrowing as a means to manage the economy. Governments vary borrowing levels as a means to control the money supply and interest rates in order to cool down or heat up the economy, for example, pumping money to increase demand in an underperforming economy. Of course this money has to be repaid with interest and this can become a large drain on the government and taxpayers, as servicing debt diminishes the amount of money that can be spent on services and investments. Regional and local governments can also sell bonds and incur debt, occasionally falling into bankruptcy, because, to a much greater extent than is the case with national taxation, taxpayers can remove themselves from a high debt tax situation.

Spending

Government spending can be categorized into the three areas of (1) current spending on goods and services the government uses to perform its activities (government final consumption expenditure or GFCE), (2) transfer payments to other governments and households, and (3) investments in public goods such as infrastructure, education, and health. Government spending and jobs are major contributors to national, regional, and local economies. In Canada, public sector employment accounted for 3,586,300 employees in 2014—20 per cent per cent of national employment. Budget spending is also impressive at CDN $254.1 billion in 2013 (Table 6.5). In the US, the various levels of government employed almost 21,850,000 people in 2014, or 14 per cent of the total workforce, with over 14 million in local government and 2.7 million employed by the federal government (not including 1.3 million military personnel). In 2012–13, the US federal government's budget called for $2.77 trillion in revenues and $3.45 trillion in expenditure.

Government spending is organized in typically complicated and distinctive bureaucratic structures. In Canada, government expenditures are distributed among the various levels of government, the First Nations, and the Crown corporations with large proportions of federal funds making their way back to individuals and regional and local governments (Table 6.5). About half (50.7 per cent) of the money in Canada's federal budget for 2012–13 was destined for "transfer payments," principally to individuals in the form of pensions and employment insurance (EI), and to the provinces in the form of funding for education and health care. Transfer payments are also made to support Aboriginal Peoples, farmers, R&D, foreign aid, and equalization payments. The latter are transfers from the richer to the poorer provinces to ensure that the latter are able to provide similar levels of government services. These amounted to 5 per cent of the 2003–4 budget, and 6.3 per cent in 2012–13. Just under half (47 per cent) of the government budget was allocated to federal departments, agencies, and Crown corporations in 2003–4, and slightly less (46 per cent) in 2012–13. This funding pays

TABLE 6.5 Canadian Federal Revenue Distribution, 2003–4 and 2012–13 ($ billions)

Budget Category	2003–4 $b (%)	2012–13 $b (%)
1. Public debt payments	35.0 (19.0)	29.2 (10.6)
2. Transfer payments	94.3 (51.0)	163.6 (59.3)
Transfer to people	42.0	70.3
Funding for provinces	30.0	58.4
(Equalization payments)	*9.3*	*15.4*
Other transfers	23.0	34.9
3. Other expenses	47.0 (25.0)	82.9 (30.1)
4. Budget surplus	9.1 (5.0)	–25.9
Total	$186.0	$275.7

Source: Department of Finance, Government of Canada. For 2004–5 data see: http://www.fin. gc.ca/tax-impot/2013/html-eng.asp. For 2012–13 data see: Canada Finance http://www.fin.gc.ca/ tax-impot/2013/html-eng.asp. Note that equalization payments are part of transfers to provinces.

for the federal government bureaucracy plus national defence (the military), public security (primarily the RCMP), and Crown corporations.

Central governments typically have various departments or ministries that directly support and regulate specific sectors (e.g., agriculture, forestry, and industry), while others are more regulatory-oriented (e.g., environmental, consumer departments). In federations such as the US and Canada, regional (state and provincial) governments also have substantial budgets and they too have departments that support and regulate economic activity. In British Columbia, with approximately four million people, the provincial government had budget revenues in 2012–13 of $42.5 billion (and expenses of $43.2 billion). Meanwhile local governments exist to provide various local services, ranging from education to community planning, and many have established economic development offices to stimulate growth. Money goes from the federal government to the provinces which make sure hospitals and universities are funded and also provide oversight. The provinces in turn provide funding to local governments to ensure equal levels of primary and secondary education. The US takes a different pattern, relying much more on privatized health care and local funding for schools. The differences in approaches have economic implications not only for the quality and stability of employment and funding for regional and local areas, but also because health and education levels are important foundations for regional economies.

Taxing and spending also occurs at the inter-state or international level. For example, the European Parliament and Council of Ministers has a mandate to extract revenues from its member states (based on their ability to pay) and the rest from customs duties, agricultural levies, and its share of the value-added taxes collected by members. Most of the EU's budget is spent on internal matters such as reducing disparities, promoting mobility, and supporting environmental programs within the Union. However, an important portion is directed to other parts of the world in the form of development aid, peacekeeping missions, and so on. While treaties among the EU members commit themselves to these transfers, international transfers depend more on willingness to pay, although developed countries are expected to fund the UN, World Bank, and IMF at much higher levels than emerging or less-developed countries.

Macroeconomic Policies

Macroeconomic policy addresses the aggregate performance of national economies and comprises monetary and fiscal policies. The former orchestrates and stabilizes the monetary system, and the latter directs government expenditures. Monetary policies shape money supply, interest rates (demand for money), and rules of bank lending, while the latter refers to government taxation and spending. **Macroeconomics** is the branch of economics that advises macroeconomic policy on the basis of analyses of aggregate (national) performance in terms of savings, investment, consumption, employment, and productivity. Although governments had long emphasized monetary issues, the roots of contemporary macroeconomic policies lie in response to social priorities and the realization of the importance of fiscal policies.

Responding to the Great Depression of the 1930s in the US, the Roosevelt administration undertook unprecedented control of the nation's finances and stimulated supply and demand in the economy through indirect financial support and employment

macroeconomics The branch of economics concerned with the aggregate performance of national economies in terms of savings, investment, consumption, employment, and productivity. Macroeconomic planning seeks to achieve growth and stability through adjustments to fiscal and monetary policies.

projects. Around the same time, John Maynard Keynes formalized such practices in macroeconomic theories. He proposed that governments have a responsibility to ensure stable growth, especially in off-setting high levels of unemployment and business failure during recessions or "busts."

Broadly speaking, Keynes suggested that in recessionary periods governments should use their fiscal powers to invest in "public works"—building infrastructure such as roads, dams, schools, and hospitals—even if budgetary deficits were incurred. Such public spending would create jobs, income, and multiplier effects in consumer spending (Case Study 2.2) throughout the economy, and would further stimulate private sector investments to meet new demands for goods and services, especially if interest rates— the cost of borrowing money—were lowered. Conversely, in periods of rapid growth governments should restrain inflation by raising interest rates and income taxes while reducing their own spending and pay off the debt loads acquired during recessions.

By the early 1980s, Keynesian or demand-side, fiscally-driven macroeconomic policies were falling out of favour, especially among conservative governments. Even with government deficits both unemployment and inflation had become problems. In response, supply-side theories, closely associated with the rise of neoliberalism, argued that government spending incurred debt and threatened the private sector by imposing higher taxes and higher costs by absorbing labour and other resources, and paying high wages. Rather, the supply-siders argue for lower income taxes, less government spending, and for an emphasis on monetary policy to ensure appropriately priced capital is available to the private sector. In this regard, "quantitative easing" refers to the practice of national banks increasing the money supply in an economy to facilitate business investment and consumer spending. The US, Europe, and Japan took this dramatic step, at different times, in the aftermath of the global fiscal crisis of 2008.

However, "austerity programs" that reduce government spending and services during recessions can worsen problems of unemployment and business failure. Indeed, in recent years, supply side or monetarist policies have been criticized and since the 2008 global economic crisis in particular there has been renewed interest in Keynesian ideas. There seems to be no clear consensus on appropriate macroeconomic policy, and choices among governments reflect varying mixes of ideology and pragmatism. The macroeconomic dilemma facing many market economies is indicated partly by the failure of a sustained period of very low interest rates to resolve unemployment problems. Further, government spending inevitably has limits, and increasing taxes is as controversial as lowering them, while with deepening globalization, national economic policies cannot be divorced from development elsewhere. Macroeconomic policies are also shaped by the broader roles of governments in society, and policy-makers think about how to pursue development strategies over long time horizons.

Regional and Local Development Policies

National governments began experimenting with regional policies in the 1930s, as part of the new fiscal policy to stimulate growth during the Great Depression, but with different emphases. Thus the Tennessee Valley Authority was created in 1933 to supply hydro-electric power in one of the poorest rural regions in the US, while the Prairie

Farm Rehabilitation Act of 1934 was designed to help drought-stricken farmers in western Canada. In the UK, the Special Areas Act 1934 targeted depressed coal mining and industrialized regions. By the 1950s and 1960s policies to stimulate economic growth in economically depressed or marginal regions in market economies were pervasive. They were generally "top-down" in nature: financed, designed, and organized by central government agencies. By the 1970s, "bottom-up" approaches organized by regional and local agencies were mooted and since then integrative policies involving different levels of government have become the norm. **Public–private partnerships (PPPs)** illustrate this trend, especially in relation to central city developments. At the same time, regional and local development policies are now important throughout national spaces, albeit in many different forms.

public–private partnership (PPP) An arrangement between public and private organizations in support of economic development projects.

Thus, there are policies that address problems created by forces of agglomeration, for example, related to controlling or allocating growth, preserving agricultural land, reducing traffic congestion and pollution, developing sustainable building codes, modernizing and maintaining infrastructure, coping with low income families in high cost, dynamic areas, and providing appropriate forms of governance for ever-expanding urban areas. Then there are policies that address issues of decline that result from rural depopulation, resource exhaustion and industrial restructuring, with their attendant problems of unemployment, under-employment, low incomes, and out-migration. The natures of these problems are highly varied, change over time and need to be placed in specific contexts. In providing policy approaches to these problems, different levels of government play different roles.

National Government Roles

National governments are important for providing finance, helping to establish priorities, supplying expertise, ensuring projects do not violate national or international considerations, and, where appropriate, imposing nation-wide policy frameworks. In the latter case, traditional top-down **regional development policies** that sought to stimulate growth in economically depressed regions were often justified in terms of national objectives.

regional development policy National government policies in support of regional development.

In Canada, for example, the Department of Regional Economic Expansion (DREE) was established in 1968 with a mandate to reduce regional disparities by providing infrastructure and subsidies for manufacturers to locate in "designated" (poor) regions within Atlantic Canada and Quebec, rather than in core regions elsewhere. DREE expressed the then political idea and slogan of a "Just Society" that emphasized that Canadians in all regions should enjoy job opportunities and similar incomes and levels of service. Further, it was argued that stimulating growth and reducing unemployment in the designated regions would increase national output, dampen inflationary tendencies (and associated problems of congestion, scarce public services) in growth regions, and help thwart the danger that economic problems would become sources of regionalized social and political grievance. Such policies were portrayed both as moral obligations and to the benefit of the economy as a whole.

By the 1980s such top-down policies were being rejected. In many countries, industrial incentives seemed to attract truncated branch plants with few spin-off effects, or were awarded to firms that did not need them or created over-capacity.

Many countries as a whole were experiencing significant economic problems and the sense of designated regions appeared questionable. While program emphasis was on stimulating manufacturing activities, service activities now dominated job opportunities in rich economies. The rise of neoliberal thinking opposed industrial subsidies. Further, top-down policies were deemed to be insufficiently attuned to local situations and priorities, leading to calls for "bottom-up" engagement by regional and local governments.

Nevertheless national governments remain important in a wide variety of regional, urban, and local development programs. The devolution of powers to regional governments and/or the establishment of nationally controlled regional-based bureaucracies have been important trends in many countries, and national frameworks to address regional disparities remain in operation. In Canada, federal agencies for regional development are located across the country, for example, the Western Economic Diversification Canada (headquartered in Edmonton), FedDev Ontario (headquartered in Kitchener), Canada Economic Development for Quebec Regions (headquartered in Montreal), and the Atlantic Canada Opportunities Agency (headquartered in Moncton). With subsidiary offices located throughout their regions, a key role of these agencies is to partner with provincial and local agencies and other federal government departments in setting priorities, providing funding, and implementing a wide range of projects that reflect regional imperatives. Since 1968 the federal government's equalization program has directed funds from so-called "have" to "have-not" provinces, the latter lacking the fiscal resources of the former, to equalize service provision in education and health. BC, and especially Alberta, have been the major "have" provinces, while Quebec typically receives about half the funds, the total of which was $15.4 billion in 2013 (Table 6.5). While recent studies have shown that the "have-not" provinces have higher levels of public service provision than the "have" provinces, changing the program would be politically difficult.

Regional and Local Government Roles

The rationale for villages, towns, and cities, and for provinces or states, as distinct political units is often founded in providing closer economic governance to residents. Municipal governments are especially strongly connected to local populations. While regional governments were often created to impose territorial political or administrative control, they organize common standards of governance, infrastructural networks, and other systems to support coherent economies. Although provinces or states to some degree provide the economic functions of national governments, they are more selective in tailoring governance and services to regional needs.

In recent decades, various kinds of "bottom-up" approaches have been mooted around the world where local and regional governments and related agencies take on proactive roles in initiating and implementing development. In developing countries, these approaches are linked to appropriate and immediate technologies (Chapter 3). In advanced countries, they are reflected in a proliferation of economic development offices (EDOs) in cities and towns and by the stronger engagement of state and provincial governments in spatial planning. Bottom-up approaches emphasize the importance of local knowledge in developing policies for local conditions and priorities. Theories of

local economic development (LED) and **community economic development** (CED) exemplify bottom-up approaches. LED schemes focus on increasing per capita income via economic development by promoting the interests of business. They do this by providing local entrepreneurs with basic infrastructure, low cost office and building space, advisory services, marketing help, and export promotion schemes, and other programs. A long standing effort has been to attract FDI or joint ventures by offering specialized infrastructure, serviced-land subsidies, tax breaks, marketing and information programs, and stimulation of favourable attitudes towards business. By contrast, CED features citizen participation and schemes that explicitly combine social goals with economic goals, promoting local initiative and co-operation among local actors in both public and private sectors. Community sustainability models give priority to local initiatives that promote environmental values.

The creation of local institutional capacity to identify and implement projects is a key feature of LED and CED. EDOs are often jointly funded by different levels of government, provide liaison functions, help identify opportunities, and organize marketing efforts for LED schemes. Meanwhile CED emphasizes social networks, democratic decision-making among all community stakeholders, and social activities, from organizing daycare centres or support for single mothers to creating community newspapers or festivals and gaining citizen approval for development ideas. Senior levels of government, as well as local authorities, help to fund and organize such projects.

Both LED and CED have been criticized, the former for favouring business and ignoring social issues, the latter because its participatory approach can be fractured and fractious. Efforts to build local institutional capacity have often failed to produce practical outcomes, and as local initiatives, LED and CED typically rely on senior levels of government for funding that comes with strings attached that may or may not reflect local goals. LED has also encouraged "subsidy wars" among local and regional governments over new investments. In recent decades, for example, in the US, state governments have competed vigorously for auto plant investments with each offering different terms of tax breaks, cheap energy, labour training support, infrastructure provision, and right to work laws. In Canada, the BC, Ontario, and Quebec governments compete for film industry investments, especially via tax reduction schemes. National governments also remain active in subsidy games, as revealed by the post-2008 support by the US and Canadian governments to save General Motors from failure, and the interventions by the US and UK governments to support (some) major banks.

In practice, different levels of government frequently interact with the private sector in development schemes. PPPs (P3s) are a case in point.

Public–Private Partnerships (P3s)

P3s have primarily involved co-operation between governments and private businesses in projects designed to rejuvenate or modernize city centres, many of which have been threatened by suburbanization and creation of new retail spaces, and deindustrialization that has left derelict land in central areas. P3s typically offer public benefits that justify government involvement and private benefits that justify business involvement. P3s take various forms, but essentially represent a pooling of financial, design, marketing, and development resources from public and private sources to share the risks and returns

local economic development (LED) The stimulation of local economic development by entrepreneurial initiative and leadership.

community economic development (CED) An approach to the stimulation of local economic development based mainly on citizen participation.

from large-scale investments. In practice, governments typically provide funding and private business operating control. In Canada, predecessor organizations can be traced to the 1970s although the term P3 seems to be more recent in origin. One recent report suggests that since the early 1990s, 206 projects have been built or are under construction, providing almost 300,000 direct jobs and $51.2 billion in direct economic output.

The emergence of Vancouver as one of the world's most livable cities began with the regeneration of its industrial core, an inner-city area known as False Creek. Led by local city council members and supported by provincial and federal governments, gentrification plans for the False Creek area centred on a large-scale, high-density, high-amenity residential project and the adjacent Granville Island Market; it opened in 1979 with the support of a $25-million grant from the federal government and demonstrated an early form of P3. This highly successful market, popular among residents and visitors alike, stands on land owned by the federal government and is occupied by tenants in diverse fields—retail, entertainment, the arts, even some industrial activities—all of whom rent space from a federal agency. The rejuvenation of the False Creek area was given a further boost by the 1986 Vancouver Expo, which attracted more government-sponsored projects to the vicinity (e.g., Science World).

The Canada Line—an extension of the city's transit system that provides rapid transit between downtown Vancouver and the international airport, opened in 2009— underscores how intimate a P3's connections can be when urban restructuring is planned in conjunction with the preparations for a world event such as the 2010 Winter Olympic Games. Without the games as a spur, the Canada Line probably would not have been built. All three levels of government, and various organizational spin-offs, played major roles in this megaproject, while a private joint venture company, InTransitBC, controlled by SNC-Lavalin, contributed funding in exchange for the right to operate the system under a long-term lease from the government. The Canada Line was controversial in part because it meant that local priorities for rapid transit were set aside and in part because initial plans to build the line through a tunnel were unexpectedly changed by the private contractor. Construction of a ground-level system severely disrupted local businesses for more than three years. According to media reports, over 50 small businesses were forced to relocate or close.

Another example of joint local and senior government involvement in urban restructuring comes from the UK, where deindustrialization in the early 1980s undermined the employment base of many cities, rendering much of their industrial infrastructure obsolete. Since then, various P3s, orchestrated by the national government and then taken over by new regional agencies, have transformed these cities in a relatively short period of time. Sheffield is an example, if not altogether successful (Case Study 6.4).

The first phase of restructuring, in the late 1980s and 1990s, transformed the Lower Don Valley from steel production and cheap housing into a commercial, retail, and entertainment area leaving only vestiges of heavy industry. Early in this phase, the national and local governments came into conflict over priorities, especially with respect to new retailing and a mega-mall (the Meadowhall Shopping Centre; see the map in Case Study 6.4) that the local council saw as a threat to the viability of the city centre. The national government nevertheless established the Sheffield Development Corporation (SDC) to plan the area's rejuvenation based on support for private-sector

Case Study 6.4
REJUVENATION OF THE LOWER DON VALLEY, SHEFFIELD

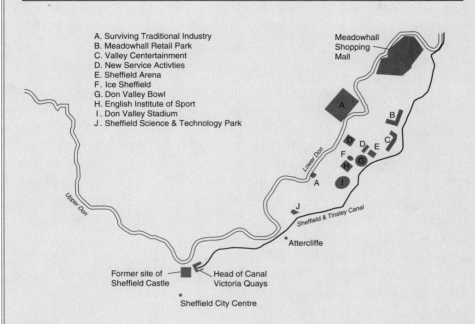

A. Surviving Traditional Industry
B. Meadowhall Retail Park
C. Valley Centertainment
D. New Service Activties
E. Sheffield Arena
F. Ice Sheffield
G. Don Valley Bowl
H. English Institute of Sport
I. Don Valley Stadium
J. Sheffield Science & Technology Park

The Lower Don Valley was at the heart of Sheffield's industrial district. Dominated by heavy industry—iron and steel manufacturing, casting, forging, and wire-making—in 1978 the area was still a dense mass of factories and poor-quality, working-class housing. By 1988, however, 40 per cent of the land was vacant and virtually all large factories had been closed and demolished. In Sheffield's case, the national and local governments adopted different, somewhat adversarial strategies. Unlike other northern cities, the Sheffield Council (dominated by the Labour party) refused to adopt the national (Conservative) government's plan for "enterprise zones" designed to attract private investment through cheap land and tax breaks.

Nevertheless, the national government established the Sheffield Development Corporation (SDC) in 1988 and spent £60 million between 1988 and 1997 to redevelop the transportation infrastructure and land quality of the Lower Don, even though the Sheffield Council feared that commercial developments there would threaten the viability of the city centre. The SDC sponsored four PPPs including the Meadowhall mega-mall, built on the site of Sheffield's largest steel works, and the flagship development of these efforts, its viability facilitated by excellent local and regional rail, bus, and tram connections. For its part, the council decided to host the World Student Games as a basis for investing in various large-scale sports and entertainment facilities, also in the Lower Don, but without the wishes and support of the national government.

In recent years, Sheffield's continuing regeneration has become a more co-operative endeavour, now focused on the city centre and formalized in an organization called Sheffield First—one of the first Local Strategic Partnerships in the UK. These efforts have featured the Victoria Quays commercial regeneration at the head of Sheffield

(Continued)

> Canal, a winter garden, related industrial museum features, and the relocation and modernization of Sheffield's market. But these initiatives, along with various proposals for distinctive "quarters" (cultural, cathedral, etc.), have been sporadic and fragmented, while plans for substantial commercial revitalization of the city centre have been downsized and remain elusive. Meanwhile, in the Don Valley, the sports stadium has been closed and knocked down. Sheffield seems to be a city in search of an identity.

investment. Substantial government financing (£101 million), supplemented by £7.5 million from the EU, was directed to upgrading roads and the land base left by collapsing industry, and to attracting new private-sector investment worth £680 million. In the past decade, the redevelopment focus has shifted to the city centre and relations between local and national government have improved. Projects in the city centre, however, have been piecemeal in development. As of 2014, some hoped-for improvements (such as a "cultural quarter") had not materialized, and major private-sector investments in retail activities remained uncertain (and downscaled). Competition with the Lower Don was also a concern.

urban regeneration The comprehensive replacement of a city's old economic structures with new ones.

The use of P3s as a basis for **urban regeneration** has drawn three main criticisms. First, critics argue that public agencies (hence taxpayers) take on most of the risks, while the private sector reaps most of the rewards. In practice, private-sector interests are not necessarily immune to the risks of P3s. Following the above-ground construction of the Canada Line, one small business (unsuccessfully) sued for damages, and as of 2014 a class-action lawsuit is pending against TransLink and the private contractor. In Sheffield, shops in the city centre suffered from the Lower Don redevelopment, and the quality of retailing in the centre has declined.

Second, critics say that P3s often impose hardships on poor people who are displaced from their communities by the construction of new facilities. Thousands lost their homes in the Lower Don Redevelopment and had to apply for housing owned by the local authority elsewhere in the city. Similarly, the gentrification of False Creek in Vancouver meant the loss of cheap (hotel) rental accommodation, displacing many low-income residents. (More recent urban renewal schemes in Sheffield and Vancouver, as elsewhere, have been required to include some low-cost housing.)

A third criticism applied to P3s is that the source of financing has a powerful influence on planning priorities. In the case of the Canada Line, the federal government's subsidy was a "take it or leave it" offer that required matching local funds; these funds were diverted from an existing regional priority, with the result that the priorities of established transportation plans were modified. And in Sheffield, the SDC's priorities were imposed on the local council. In fact, the distinctions between top-down and bottom-up planning are always somewhat blurred, since the former requires local approval as well as local agencies to ensure implementation, while the latter requires financing and strategic direction.

Local Planning and Regulation

Even within countries and regions there are wide variations in the nature of local governments, the populations they represent, and the challenges they face. The attitudes and capabilities of local governments to adapt to economic development can also vary, while relations with other levels of government can be co-operative, antagonistic, or passive. Local planning, however, is the most pervasive way local governments control development and profoundly shape people's daily routines. Local plans are most elaborate in urban areas. Typically, a **community plan** is created by the local planning department and then legally approved, rejected, or amended by locally elected officials. Sometimes advisory input is provided by a committee that may include community representatives (Figure 6.2). The plan allocates land to specific types of use or activity, such as commercial, industrial, residential, institutional, or transportation. Each of these categories is then normally subdivided. For example, residential land can be classified as single-family, duplex, high- and low-rise apartments, or mixed. Implemented through various by-laws (zoning, building, density), specifications (lot size and use), and traffic controls, this plan institutionalizes spatial patterns of land use and related behavioural patterns, such as journeys to work, school, shopping, and play.

community plan The document outlining a local authority's official development plan, including land use and zoning. Different communities call their plans by different names: "Official Plan" (Toronto); "Master Plan" (Montreal); and "Municipal Development Plan" (Calgary).

Consistent with the framework outlined in Figure 6.1, urban planning has three basic functions: (1) to maximize efficiency in markets, service provision, and transportation; (2) to minimize the "negative externalities" of incompatible land uses while maximizing the "positive externalities" that come with the concentration of compatible activities in specific areas; and (3) to provide for public goods and services such as parks, schools, hospitals, and churches. Local plans provide detailed regulations indicating which activities are and are not permitted in particular areas and how movement within the city is to be organized. Such plans are strategic and represent a long-term vision, albeit one that is subject to change.

In general, different activities have different location requirements, and combining them may cause problems. Heavy industry, for instance, is aesthetically displeasing and can pose safety and health hazards to people living nearby, while the presence of residents can be inconvenient for industry. Planning that concentrates different types of activity in designated areas facilitates positive externalities, increasing the likelihood that the occupants of the area will have similar values, interests, and needs, and reducing the costs of providing services to them.

Another focus in local planning is the provision of public goods that would be under-supplied or absent if development were left to unregulated market forces. In cities, for instance, green space contributes significantly to the quality of life. Yet in the absence of zoning regulations designed to preserve green space in the form of parks, such space would almost certainly be lost over time as individual landowners either sold it off or converted it to other uses. Dedicated zoning for schools and hospitals is essential for similar reasons.

FIGURE 6.2 **The Regulation of Urban Space: An Introductory Planning Perspective**

The principles and practices of urban planning vary not only from country to country but often between municipalities within a single metropolitan area. Variations in the rules governing matters such as setbacks from property lines or maximum building size often reflect historic differences in land use. But land use changes over time. Thus many cities in advanced market economies have revised their plans in recent years, drastically reducing the areas designated for industrial use and rezoning the land for commercial and residential purposes as part of their efforts to make local regeneration and economic development sustainable both environmentally and socially.

International Economic Policies: The Free Trade Agenda

National governments are primarily responsible for deciding how their economies will participate internationally with respect to trade, investment, labour, intellectual property, and related matters. These issues are generally discussed with respect to the free trade agenda.

Free trade agreements are generally about reducing tariff (tax) or non-tariff barriers (such as quotas, excessive bureaucracy, and regulations) on the movement of goods and other economic flows across international borders. Governments traditionally imposed or allowed such barriers to exist to protect domestic industries from foreign competitors. Since the Second World War, expanding free trade among market economies has been an international priority resulting in the first General Agreement on Tariffs and Trade (GATT) in 1947. Since 1995, the World Trade Organization (WTO) has been mandated to oversee international trade agreements and promote further liberalizing of trade among its 148 member nations. Every few years, the WTO organizes conferences or "ministerials" where national representatives discuss possibilities for moving along the free trade agenda, most recently held at Seattle in 1999, Doha in 2001, Cancún in 2003, Hong Kong in 2005, Geneva in 2009 and 2011, and Bali in 2013.

The guiding principle underlying free trade agreements is most favoured nation (MFN) status that emphasizes all trading partners should be treated equally. In practice, continuing reductions in tariffs suggest that the global economy has shifted decisively towards freer trade. In 1947, tariffs of 20 per cent or more were common; led by the rich OECD countries, average tariffs on goods among these countries had been reduced to 6.3 per cent by 1995, 3.8 per cent in 2002, and 2.8 per cent in 2010. Globally, average tariffs dropped from 15.1 to 8.1 per cent between 1995 and 2010. For the poorest 34 countries, the decline was from 17.6 (1995) to 11.2 per cent (2010). Barriers to foreign investment have also been reduced significantly. Under the WTO, free trade negotiations have increasingly focused on services, intellectual property, and reducing tariff barriers in developing and poor countries.

In practice, free trade is a political construction, embedded in rules and regulations. International trade comprises markets that organize the production and exchange of products and services that cross national borders. As this text emphasizes, markets within advanced capitalist countries are deeply regulated by various levels of governance, related to a host of issues involving not only labour, health, safety, and product labelling but newer concerns over environment and security. Since 9/11, for example, security issues have added layers of regulations on trade, adding to costs and delays at borders. Canadian exporters to the US have often been aggrieved by security-based restrictions on trade that, intended or not, have become important non-tariff barriers.

WTO agreements are lengthy, complex, and legalistic documents that impose an impressive array of general guidelines and specific rules and conventions on international transactions. The creation of the WTO itself marked a shift from a tiny, informal organization, largely "personally" controlled by its Director Generals, to a large, formal bureaucracy, headquartered in Geneva. A key initiative within the WTO was the establishment of a legally binding trade dispute mechanism (TDM), to provide assurances of the fairness of free trade. The TDM is five-step legal process that can take well over a year before a decision is made, and only national governments can participate. While WTO decisions are binding, the WTO itself has no enforcement powers. The WTO can legitimize retaliatory measures against a country that has violated a free trade agreement, but retaliation is left to aggrieved countries to initiate. Disputes typically involve counter claims that exports are subsidized and imports are restricted for protectionist reasons. In major clashes where trade flows are substantial, the WTO's TDM has often failed to provide a solution. The US–Canada softwood lumber dispute, for example, has been raging since the early 1980s, involving the Dispute Settlement Mechanism (DSM) of the WTO and NAFTA, but has only been (temporarily) resolved by three political agreements between the two federal governments. The agreement of 2006 (reinforced in 2011) imposes taxes and quotas on Canadian lumber exports to the US. The long-standing dispute between the EU and the US regarding rival claims over subsidies in the manufacture of airplanes is similarly unlikely to be resolved by WTO decision.

The process of deepening multilateral free trade within the WTO has been difficult and controversial for several reasons. First, success under GATT focused on reducing tariff barriers that were transparent, and where benefits to trading partners could be readily identified. The WTO has become more concerned with non-tariff barriers that are much more difficult than tariff barriers to identify and address, often involving matters of local preferences and the discretionary powers of customs officers. Even as tariffs are reduced, governments continue to control product flows using means such as quotas, detailed labelling requirements, health and safety regulations, or local content rules. Second, while discussions traditionally focused on trade in resources and manufactured goods, along with FDI, more recently attention has turned to services—a sector that in the past was neither a major component of trade nor protected by conventional tariff barriers. Third, since the collapse of the centrally planned economies, the group of countries seeking free trade has expanded significantly, including China and India, and increasing the number and range of interests that have to be accommodated in reaching agreement. Fourth, WTO meetings and objectives are criticized as supporting a free trade agenda favouring MNCs over local communities and for widening income gaps globally and nationally.

The Doha negotiations illustrate the problems and political nature of extending free trade principles. While the Doha Ministerial began in 2001, it was not until late 2013 that a limited agreement on "trade facilitation" was reached on measures to reduce red tape in custom procedures (a non-tariff barrier). Broader agreement on agricultural trade, especially the removal of subsidies, remains challenging. The farm-support programs in the US, EU, and Japan are substantial and their subsidized food exports threaten domestic food production in poor countries. The G33, a coalition of poor countries, has resisted lowering their trade barriers to food imports. Rather they have encouraged the WTO to treat their barriers to imports as relatively minor subsidies ("green box" subsidies in WTO terminology) that can be left in place as they don't create the trade distorting effects associated with bigger ("amber box") subsidies. Indeed, the G33 group wishes to add to existing green box policies which support research, pest and disease control, and various advisory services supplemented by settlement, land reform, and rural livelihood security programs.

In general, the pros and cons of free trade policies (Chapter 2) cast big shadows over the WTO negotiations, especially in the context of fairness for poor countries. The proponents of free trade may argue that poor countries have much to gain from trade; they also have the most to lose. For many poor countries any access to the markets of rich countries they might gain would likely be limited to a few specialized crops and resources; be vulnerable to technological change, replacement by other products, and competition from elsewhere; and be lacking in potential for diversification. Meanwhile, the shift to specialized crops for export would divert land and resources from production to meet local needs, and may be dominated by MNCs.

Regional Trading Agreements and Unions

Recent years have seen a proliferation of supranational regional trading agreements (RTAs) and unions: from 26 in 1985 to 189 by 2003 and an astonishing 377 by 2014 with more than another 200 under negotiation. The WTO's list of RTAs includes economic unions and trading agreements (see http://www.wto.org/). The European Union (EU) and the North American Free Trade Agreement (NAFTA) are the best known. RTAs are formed to promote trade and investment between and among member states by offering them terms more favourable than those offered to non-members. The push for RTAs represents both an impatience with the WTO's stalled global agreements and also a kind of co-operative protectionism.

In the EU, for instance, integration among its members is promoted ("protected") by local content rules that require the sourcing of inputs within the EU, and complex regulations specify the features (ingredients, size, place of origin, etc.) required for a product to qualify as, say, a Bavarian sausage. In practice, RTAs have indeed strengthened regional connections. Although the US was already Canada's most important trading partner by far—reflecting a degree of trade dependence not seen in any other important economy—this relationship has become even more important since the NAFTA agreement came into effect in 1994. Regional integration may even replace global links. The UK's decision to join the EU in the early 1970s required that it impose barriers on imports from fellow Commonwealth countries. These changes in turn led to others: New Zealand's trade, for instance, as it declined with the UK, became increasingly integrated with nearby Asia, especially Japan.

The Incomplete Territorialisation of Externalities Governance

Profit-oriented companies are not inherently concerned about the pollution they generate or whether their salaries provide for the health and education needs of their employees' children. Similarly, when people buy phones or soccer balls, they are concerned about cost and quality, and less about the labour and environmental conditions generated in production and consumption. Most production and consumption activities generate social costs that (economizing) corporations and individuals tend to ignore ("free ride"). Governments are expected to deal with these externalities. Can a government impose its authority on the economic agents that are causing the externalities within its own territory?

In developed countries, the effectiveness of government control on or compensation for externalities is demonstrated by safe and healthy workplaces, low levels of urban pollution, and social safety nets such as worker compensation and universal health care. These accomplishments and the institutions that ensure they are sustained have taken decades to gain acceptance and develop, and did so within particular national, regional, and local contexts. In federal countries, national governments introduce policies and administrative structures that either govern the country directly or impose responsibilities on other levels of government. Regional governments can develop their own mechanisms of externality governance, for example related to minimum wage rates, health care, and education systems, as long as they do not contravene national laws.

Increasingly territorial control is weakening. Pollution is becoming globalized, most notably through greenhouse gases and ozone-depleting substances that affect the atmospheric commons irrespective of origins. The production of other pollutants, once regionalized within countries or among a small set of countries, cross borders in increasingly large quantities and form concentrations in the global commons (e.g., plastics in the ocean gyres). More generally, territorial control has been loosened by the combination of global value chains and the competition for the location of industry. If a country imposes standards on pollution, worker safety, or health care provision, companies can go elsewhere. In both developed and developing countries, localities, regions, and nations have a record of compromising on taxes, labour rights, and environmental protection to attract companies. Countries with strong laws may benefit less ambitious countries by reducing their competition.

To address the incomplete territorialisation of externality governance, a core group of institutions and plethora of sub-institutions have been established to protect the global commons and establish common norms. They are based on agreements among nations. Although they employ a vast range of governance mechanisms they are essentially limited by the lack of territorial authority. The UN, ILO, or the WTO cannot force any country that does not sign an agreement or doesn't fulfill an agreement to change its behaviour; rather, they rely on countries to impose their own sanctions on misbehavers. The WTO allows countries to establish their own environmental and labour standards, but they cannot prevent bias toward national industries or break with the principles of comparative advantage.

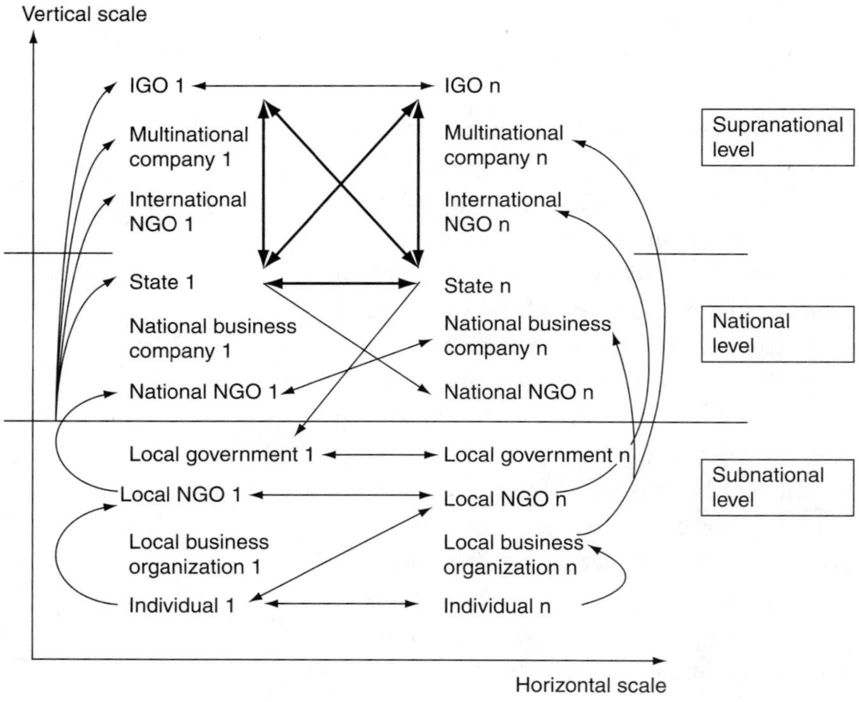

FIGURE 6.3 International Environmental Governance: Vertical and Horizontal Relations

Source: Andonova, L.B. and Mitchell, R.B. 2010. "The rescaling of global environmental politics" *Annual Review of Environment and Resources* 35: 255–82.

Many of the UN's environmental agreements, such as protection of wetlands, endangered species, or the ozone layer, have become effective over time, while others such as attempts to find a climate change agreement, have faltered. These agreements must be negotiated between the UN and nations, but within a country, the process involves a host of stakeholders at regional and local levels, across industrial sectors, and among different social agents and interests (Figure 6.3). Frustration with the UN's climate change negotiations in particular has wrought changes to what had been a vertical process to horizontal interactions among cities, companies, and NGOs. If such arrangements can achieve direct results, and may possibly influence UN–nation agreements, they imply less governance by government and a greater role for NGOs and non-profits (Chapter 7).

Conclusion

Governments at various levels exercise pervasive, multi-faceted, and powerful roles in shaping economies, domestically and internationally, while reconciling objectives of efficiency, social responsibility, and public goods. With the growth of market economies these roles have become more, not less, important, even as they vary within and among countries. All society's economic problems, from income inequality, poverty, and environmental destruction to providing local services and getting people to and from work, translate into challenges for government policy. Notwithstanding pleas for deregulation and the relentless adoption of a free trade agenda, governments continue to orchestrate, manipulate, and intercede in the economy with unintended, as well as intended, consequences. The dilemmas facing government policy are considerable, and, government agendas are inevitably shaped by political considerations and strong desires for re-election or status by leaders. In one way or another policy and economy are embedded in one another.

Practice Questions

1. Gunnar Myrdal, the famous Swedish institutional economist, claimed Canada as a model for most effective government. What country would be your nominee and why?

2. Examine the official plan for the community that you consider home. Summarize its main features. Does it have an agency dedicated to stimulating local economic development?

3. In recent years, in the US and Canada, governments have massively subsidized big auto companies and banks. Document these subsidies, using published information, for example, from newspapers. Can these subsidies be justified in terms of public good theory?

4. Select a level of government and compare and contrast its budget in 1980 and now.

5. Explain the basic concept of neoliberalism. Has this idea influenced public policy in your area?

Key Terms

budget   177
unitary systems of government   178
federal systems of government   178
statism   185
corporatism   185
welfare state   185

tax   189
macroeconomics   194
public–private partnership (PPP)   195
regional development policy   196
local economic development
(LED)   197

community economic development
(CED)   197
urban regeneration   201
community plan   201

Recommended References

Information on the Canadian government's role in the economy is available at http://www.canadianeconomy.gc.ca/english/economy/gov.html. For the US, information on the structure of the government system and the role of the federal government in the economy is available at http://www.firstgov.gov/. Many non–English speaking countries provide information on government structure, programs, and budgets in English. For Japan see http://www.mizuho-sc.com/english/ebond/government.html. For Germany see http://www.bmwi.de/English/Navigation/root.html. For Sweden see http://www.sweden.gov.se/sb/d/576.

Harvey, D. 1989. "From managerialism to entrepreneurialism: The transformation in urban governance in late capitalism." *Geografisca Annaler* 71(B): 3–17.
An early discussion of the increasingly proactive role taken by city governments in economic development.

Mazzucato, M. 2015 The Innovative State, *Foreign Affairs* January-February Issue
Argues for pro-active roles by governments in stimulating innovative activities.

Polese, M., Shearmur, R., 2006. Why some regions will decline: A Canadian case study with thoughts on local development strategies. *Papers in Regional Science* 85(1), 23–46.
Discusses community and regional responses to problems of decline.

Ray, M.D., Lamarche, R.H. and MacLachlan, I.R. 2013. Restoring the "regional" to regional policy: A regional typology of Western Canada, *Canadian Public Policy* 39: 411–29.
Stresses importance of adapting regional policies to regional circumstances.

References

Galbraith, J.K. 2000. "World of ideas: John Kenneth Galbraith" interview with Barb Clapham, printed in Canadian Investment Review, http://www.investmentreview.com/print-archives/winter-2000/world-of-ideas-john-kenneth-galbraith-729/.

Williams, G. 2006. Thomas Hobbes: Moral and Political Philosophy. *Internet Encyclopedia of Philosophy.* Accessed June 2010. http://www.iep.utm.edu/hobmoral/.

World Trade Organization. 2009. *Regional Trade Agreements Gateway.* Accessed July 2009. http://www.wto.org/english/tratop_e/region_region_htm.

CHAPTER 7

Non-profit Organizations

The individualism of current economic theory is manifest in the purely self-interested behavior it generally assumes. It has no real place for fairness, malevolence, and benevolence, nor for the preservation of human life or any other moral concern.

Daly 1994: 159

The overall goal of this chapter is to examine the economic geography of the non-profit sector and its interaction with business, government, and labour. More specifically, this chapter's objectives are

- To explore why non-profits exist in market economies;
- To examine the roles that non-profits play in both local and global development;
- To examine how advocacy groups work to change market behaviour and government policy;
- To explain how non-profits serve the non-material goals of market economies; and
- To explain the relationship between social capital and economic development.

The chapter has four main parts: the first part outlines the characteristics of non-profit organizations; the second classifies them in terms of service, expressive, and advocacy functions; the third discusses the implications for markets of environmental and human rights advocacy in particular; and the fourth explores the relationships between non-profits and economic development. As both generators and expressions of social capital, non-profits contribute to a more holistic understanding of the local–global space economy.

The Role of Non-profit Organizations

The genius of the market is that it assigns a standardized monetary value—a price—to an extraordinary variety of goods and services. But there are also many things of great value to individuals and communities that are difficult, perhaps impossible, for any market to assess in monetary terms: family, friendship, happiness, religion, security, fairness, artistic expression, altruism, fresh air, pure water, love. A remarkable development over the last few decades has been the proliferation and growth of non-profit organizations (NPOs) and non-governmental organizations (NGOs) that seek to represent and defend these "un-marketed" **social values** within economies. Often launched by social entrepreneurs, some of these organizations take direct action themselves to represent those values, while others work to persuade other institutions, especially governments, to do so. NGOs and

social values Values such as environmental health, human connection, and emotional well-being, which are facilitated by non-government and non-profit organizations.

NPOs range from tiny, informal local groups that run soup kitchens for the homeless to powerful international organizations that aggressively lobby governments and corporations on behalf of environmental or humanitarian values. The impetus in all cases is that the market often externalizes or ignores values that are social or public in nature. Sea Shepherd, for example, is a particularly aggressive organization that uses its own flotilla of ships to confront threats to marine wildlife on the high seas (Case Study 7.1).

Case Study 7.1
SEA SHEPHERD

The Sea Shepherd Conservation Society is a non-profit organization founded in 1977 by Paul Watson, who had earlier helped to create Greenpeace. Sea Shepherd is headquartered at Friday Harbour in Washington State, but its activities are global, focused on the protection of ocean wildlife—such as whales, dolphins, seals, and sharks—and their habitat. Its skull-and-crossbones flag reflects its aggressive approach, which at times has led it to scuttle whaling ships at anchor, ram and board them on the high seas, and damage propellers and drift nets. It takes its mandate from the World Charter for Nature, adopted by the UN in 1982, which calls on "states . . . public authorities, international organizations, individuals, groups and corporations" to act on behalf of international conservation law."

© William Caram/Alamy

Sea Shepherd operates (2013) four extremely well-equipped ships, the *Steve Irwin*, *Sam Simon*, *Brigitte Bardot*, and *Bob Barker*, as well as various smaller boats. A fifth, the *Farley Mowat*—named after the celebrated Canadian author and environmentalist—was seized by the Canadian government in 2008 and sold at auction in 2010 after its officers were convicted of interfering with the seal hunt in the Cabot Strait. The activities of Sea Shepherd are highly controversial even within the environmental movement. Is it eco-terrorist or eco-policeman?

Source: http://www.seashepherd.org (accessed March 2014).

NPOs and NGOs are fascinating because they serve social needs that are not met by the traditionally dominant institutions: markets, business firms, and governments. Most non-profits are strongly place-based, focused on specific problems in specific local areas. But many connect their members and carry out their activities across the globe. Some (such as Sea Shepherd) are just as electronically sophisticated and geographically mobile as their corporate opponents in coordinating their activities and communicating their messages and perhaps even less concerned about political boundaries. At the same time, non-profits raise important questions about the meaning of development and the extent to which social values can be monetized.

The **non-profit sector** also goes by such names as the non-governmental, civil society, charity, voluntary, third-sector, or social enterprise sector and the organizations that comprise the sector are remarkably varied in purpose and structure. Despite the non-profit image, however, this sector is a significant and growing part of the economy. It employs a lot of people and is responsible for a substantial proportion of the GDP in all market economies. As importantly, it makes powerful contributions to economic life by building social and human capital and shaping regulatory governance. Because the strength of non-profits varies by place and over space, they deserve to be a major topic of study in economic geography courses.

In this chapter we will examine three types of NPOs: (1) those that provide **social services**, usually on behalf of government; (2) those that perform what are sometimes described as expressive functions, encouraging self-expression, creativity, and participation; and (3) those that seek to influence the behaviour of markets and governments, often working directly to disrupt, modify, or even dismantle value chains. Among the latter are organizations such as the Sea Shepherd society (often referred to as **advocacy organizations**), but there are also organizations working to change the trading system for commodities from the developing world (e.g., coffee and chocolate) so that small producers receive a fairer price.

non-profit/non-governmental sector The terms non-profit and non-governmental are often used interchangeably because both provide a diversity of services from which they don't make a profit. The chief difference is that NGOs tend more strongly to be advocacy organizations.

social services Basic services provided by government and non-profits to people in need.

advocacy organizations Organizations that seek to change the behaviour and policies of governments, corporations, etc.

The Diversity of Non-profits

In market economies, needs of all kinds are met by four institutions: markets (and businesses), governments, households, and non-profit organizations (Figure 7.1). Markets deliver goods to those who are willing to pay the prices charged by the sellers. Governments direct and regulate markets and redistribute social wealth through taxation and the provision of public goods. The household is arguably the core social institution because it not only supplies the workers, consumers, and taxpayers that markets and governments need, but lays the groundwork for the development of both **human and social capital**. Human capital is the personal abilities and characteristics that make an individual economically productive, while social capital is the aggregate of shared values and expectations, social networks, and qualities (such as trust and willingness to co-operate) that makes economic development possible. Finally, non-profit organizations express social values and contribute to their further development by responding to the many needs that are not met by the other three institutions.

human capital The abilities and personal characteristics, formally or informally acquired, that make an individual economically productive.

social capital The aggregate of shared values and expectations, social networks, and qualities such as trust and willingness to co-operate that makes economic development possible.

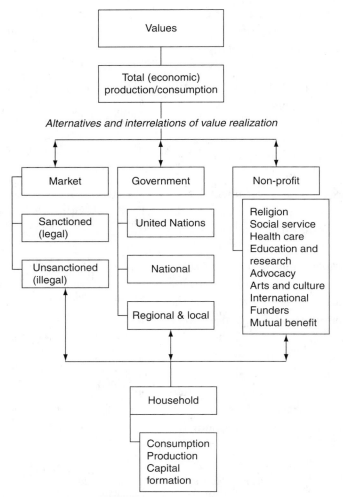

FIGURE 7.1   The Role of Non-profits in the Realization of Non-market Values

Why Do Non-profits Exist?

The development of the non-profit sector can be explained from several different perspectives. Anthropologists may see non-profits as meeting human needs for mutual support, friendship, social inclusion (and exclusion), entertainment and recreation, conflict resolution, or ritual. Sociologists may see them as helping to integrate individuals into society in general, or into particular groups and classes. Political scientists tend to see non-profits—because of their diversity, capacity for experimentation, and lack of bureaucratic constraints—as providing services that governments cannot. Economists, for their part, see the existence of the non-profit sector as a response to market failure. Economic geography integrates these varied perspectives to assess how the local–global dynamics of non-profits contribute to local development and spatial variations in the quality of life.

Although the non-profit sector is distinct from its business, government, and household counterparts, it shares characteristics with each of them. Non-profits are similar in organizational form to governments and businesses. They can grow to sizes that need large-scale funding; they employ people who want career security and advancement; and they may be led by directors who see them as vehicles for fulfilling their personal ambitions. They are also vulnerable to private agendas, factionalism, lack of accountability, and separation between leaders and grassroots members (especially volunteers). Finally, non-profits receive funds from governments, individuals, and corporations, all of which have their own ideas about the way those funds should be spent.

While non-profits are hard to define precisely, there is general agreement on a few of their central characteristics:

- Non-profit organizations are not part of formal government bureaucracies, and they have neither the coercive power, legislative authority, nor the powers of taxation of governments. Non-profits exist largely because people want to voluntarily organize them to meet social needs. In the core non-profit sector, user participation is also voluntary but is not mediated by price or cost.
- They do not generate profits for shareholders or owners. Any income left over once expenses have been paid—very largely to management and staff—is reinvested in the organization for the support of future activities. The absence of any profit motive is a key to the trust that many people place in non-profits. Indeed, voluntary labour is a significant feature of this sector.

- They are self-organizing and self-governing. Non-profits are established and run by like-minded people, and unlike governments and businesses (which are accountable to voters and shareholders respectively), depending on funding, non-profits may not require formal oversight or supervision. They exist as long as their members and supporters want them to. Nevertheless, most non-profits do have a coherent organization structure and regular meetings.
- Whatever their area of activity, they are legitimized by a highly focused sense of social purpose.

In short, non-profits resemble companies or governments in some ways and differ from them in others. Most are highly localized, although many have grown large enough in membership and funding to achieve economies of scale and scope in the functional and spatial scope of their activities.

The Size and Composition of the Non-profit Sector

The non-profit sector appears in many guises, from community hockey clubs and local "meals on wheels" groups to international aid organizations; from churches, mosques, and synagogues to industry associations and chambers of commerce; from local theatre groups to environmental advocacy organizations. Often a distinction is made between non-profits that incorporate educational and health organizations that are funded, required, and ultimately controlled by governments and "core" non-profits that operate with more autonomy of purpose and income.

However precisely defined, in developed economies, the non-profit sector accounts for an impressive amount of economic activity. Their role became evident in a pioneering international study of 35 nations (not including Canada) that showed that NPOs around the world spent approximately US$1.3 trillion in 1998—an amount equal to 5.1 per cent of global GDP (Salamon et al., 1999). If it were a country, the non-profit sector would have had the seventh largest economy in the world, employing the equivalent of 39.5 million full-time jobs (22.7 million paid and 16.8 voluntary). There are national variations in the importance of non-profits—in the Netherlands 14.4 per cent of paid and voluntary workers were engaged in the sector, while only 0.4 per cent of Mexico's labour force worked or volunteered at a non-profit. A more recent study confirms the importance and national variations in the role of non-profits, most obviously, albeit with exceptions, between rich and developing countries. In 2007, for example, the contribution of non-profits (including volunteers) to GDP was 8.1 per cent in Canada and 6.6 per cent in the US, while in Mexico and Thailand the contributions were 2.2 and 1.6 per cent (Salamon et al., 2013). All developed countries have vibrant non-profit sectors employing between 4 and 10 per cent of the workforce.

In Canada a remarkable 161,227 NPOs were recorded in 2003. This number remained stable in 2007, and probably approximates the number in 2015. Culture and recreation, including many community sports programs, was the leading area of activity by number of non-profits in 2003 (29.4 per cent) with religious and social service non-profits second and third in importance (Table 7.1). In terms of income, however, the health and education categories were most important, together accounting for 57.3 per cent of revenues received. The core NPOs received over $100 billion in revenue in 2007 and accounted for 2.5 per cent of Canadian GDP (and 7.0 per cent if hospitals,

TABLE 7.1 Activity Areas, Income, and Number of Canadian Non-profits, 2003 and 2007

Activity Area	Income $millions		Number of Non-profits
	2003	2007	2003 (%)
Culture and recreation	8,815	11,391	47,417 (29.4)
Education and research	25,767	34,439	8,284 (5.4)
Health	47,280	62,714	6,103 (3.8)
Social services	10,160	12,607	19,099 (11.8)
Environment	757	983	4,424 (2.7)
Development and housing	8,803	12,857	12,255 (7.6)
Law, advocacy, and politics	913	1,355	3,628 (2.3)
Philanthropy	3,123	5,756	15,935 (9.9)
International	1,436	1,880	1,022 (0.6)
Religion	6,885	8,062	30,679 (19.0)
Business associations, unions	6,757	8,492	8,483 (5.3)
Other	5,478	8,333	3,393 (2.1)
Total	126,174	168,869	160,722 (100)

Source: Statistics Canada, Cornerstones of Community: Highlights from the National Survey of Nonprofit and Voluntary Organizations 61-533-XIE 2004001 2003. Revised, Released 30 June, 2005.

universities, and schools are included). The core NPOs spend less of their revenue on wages and salaries, although these costs are important, and they rely on significant voluntary labour.

In the US, in 2010 there were 1,560,000 registered non-profits, just under half classified as reporting income over $5,000, including 366,086 public charities that receive donations (Table 7.2). As in the Canadian case, health and education are the most important categories in terms of revenues received but the other public charities still had over a trillion dollars revenue. In 2011 the US non-profit sector had $1.59 trillion in revenues, 72 per cent from service revenues and 22 per cent from gifts and government grants, and accounted for 5.5 per cent of GDP, about 9.2 per cent of the wages paid in the US. Almost 27 per cent of the population, or over 64 million people, provided over 15 billion hours of participation, volunteering at least once.

NPOs operate at different geographic scales, responding to a variety of social needs. In Canada, NPOs in religion, community development and housing, and sports and recreation operate mainly at the local level, while the region is the main area of operations for organizations operating in the fields of health, environment, and social services. However, some important local services are organized by national organizations. Although business and professional organizations are most important at the local level, their regional, provincial, and national activities are also significant. They have tangible economic impacts through the jobs they provide, and the money they spend, and lifestyle impacts through the values they express and help to develop.

TABLE 7.2   US Non-profits: Activity Areas, Number, and Income 2010

Activity Area	Number	Revenues ($)
All registered non-profits	1,560,000	
Reporting non-profits	618,062	2.06 trillion
Reporting public charities	366,086	1.51 trillion
Arts, culture	39,536	29.3 billion
Education	66,769	248.0 billion
Environment and animals	16,383	13.7 billion
Health	44,128	907.7 billion
Human services	124,360	196.4 billion
International	7,533	31.4 billion
Public and social benefit	43,875	74.4 billion
Religion-related	23,502	13.0 billion

Source: A.S. Blackwood, K.L. Roeger, and S.I. Pettijohn. 2012. *The Nonprofit Sector in Brief*, Urban Institute Press.

Local and regional organizations are by far the most common. Indeed, because of their strong ties to place, non-profits are often described as cornerstones of the community. Businesses and even residents may move from one community to another, but place-based institutions such as business associations, sports and arts clubs, and hospitals perpetuate local identities and constitute the fabric that holds a particular community together. Many NPOs redistribute expenditures geographically as well as socially, most obviously but by no means restricted to publically funded health, education, and social services that receive most of their funds from government sources (in Canada anywhere from 56 to 80 per cent). High levels of government funding imply limits on autonomy, modified by ability to obtain funds from other sources. Even for the core non-profits economic and social impacts are impressive, if not always obvious.

Economic Impacts

Of the three basic types of core non-profit institutions, social service operations have the most direct economic impact. Most of their funding, much from government, goes to pay employees or support local people and is therefore distributed throughout the society. In contrast, expressive organizations devoted to culture and recreation are run mainly by volunteers; their impact tends to be indirect. Finally, advocacy organizations, by seeking to change government policy in regards to human rights and the environment, often reshape value chains—their locations, who they employ, costs, and efficiencies.

Social Services

Traditionally, the basics of what we now consider social services were provided by families, churches, local charities, and local governments. With urbanization and the decline of the extended family, however, many if not all of these services had to be augmented and new providers found. Early non-profit organizations such as the Salvation Army and the YMCA stepped in to fill some of the gaps. But the range of social service needs continued to expand. Today, "social services" is a very broad category, covering everything from daycare, refugee assistance, and family counselling to local disaster relief, assistance for the homeless, support for the elderly, and services for the disabled (Table 7.3).

These diverse services have many economic ramifications. Both complex and labour-intensive, they create jobs for large numbers of highly trained, well-paid employees as well as volunteers. They redistribute large sums of money, both within local economies and from place to place around the world. They also contribute to overall well-being by helping people who otherwise might not be able to find or keep jobs participate in the economic life of society. Non-profit social services rely to a significant degree on government funding. In Canada, social services receive between 60 and 70 per cent of their funding from government sources (mostly provincial), roughly half of which come from payments for services, and half from grants and contributions. Fees for goods and services, gifts, and donations from the public are other sources of funds. If the work of unpaid volunteers is translated into monetary terms, however, total income rises significantly. Social services, such as support for the elderly and disabled, rely more on volunteers than do other NPOs.

Geographically, most social service organizations (more than 60 per cent) operate at the local level, a reflection of the personal and particular nature of the needs they

---

TABLE 7.3 Non-profit Social Services

Youth services and youth welfare Delinquency prevention services, teen pregnancy prevention, dropout prevention, youth centres and clubs, and job programs for youth. Includes YMCA, YWCA, Boy Scouts, Girl Scouts, and Big Brothers/Sisters.

Family services Includes family life education, parent education, single-parent agencies and services, and family violence shelters and services.

Services for the handicapped Includes homes, other than nursing homes, transport facilities, recreation, and other specialized services.

Services for the elderly Geriatric care, including in-home services, homemaker services, transport facilities, recreation, meal programs, and other services geared towards senior citizens, but excluding residential nursing homes.

Self-help and other personal social services Programs and services for self-help and personal development. Include support groups, personal counselling, and credit counselling/money management services.

Disaster/emergency prevention and control Preventing, predicting, controlling and alleviating disasters, educating or otherwise preparing people to cope with the effects of disasters, or providing relief to disaster victims. Includes volunteer fire departments and lifeboat services.

Temporary shelters Providing temporary shelter for the homeless. Includes traveller's aid and temporary housing.

Refugee assistance Providing food, clothing, shelter, and services to refugees and immigrants.

Income support and maintenance Providing cash assistance and other forms of direct services to persons unable to maintain a livelihood.

Material assistance Providing food, clothing, transport, and other forms of assistance. Includes food banks and clothing distribution centres.

Source: "International Classification of Non-profit Organizations: Social Services", adapted from Statistics Canada publication, *Satellite Account of Nonprofit Institutions and Volunteering*, 1997 to 2001, Catalogue 13-015-XIE2005000, http://www.statcan.gc.ca/pub/13-015-x/13-015-x2005000-eng.pdf (Accessed September 2010).

serve. However, large NPOs with the most funding, paid employment, and volunteers—the Red Cross, the YMCA and YWCA, Goodwill Industries, Big Brothers and Sisters, the Salvation Army, St John Ambulance, Meals on Wheels—are international in scope. The Salvation Army, for example, was established in 1865 and provides various forms of relief, services, and spiritual support to millions of poor people in 126 countries, with a budget in excess of $4.0 billion (2013). Its organization is decentralized along territorial lines, funded mainly by donations (45 per cent) and investment income (24 per cent), sales (15 per cent), and only 9 per cent from government sources; its paid employees are supplemented by over 4 million volunteers. Solely local NPOs can also be relatively large. For example, The Portland Hotel Society was created in 1993 as an offshoot of the Downtown Eastside Residents Association, created in 1991 to provide housing, services, and advocacy for the homeless in East Vancouver and advocacy for the supply of "clean needles" for drug users; indeed, it provided North America's first (2003) legally sanctioned injection site. In 2014, its budget was around 28 million, mainly from government sources, and it employed 453 people.

In general, social service providers respond to needs that are expressed differently in different places. Retired people, drug addicts, the poor, the unemployed, new immigrants, and children, for instance, are distributed throughout society, yet concentrated in different places. Society must therefore offer both a standard overall level of care and more focused care in particular places. Governments support NPOs as the primary providers of this care because they are (a) more flexible than government agencies encumbered by bureaucratic inertia and (b) closer to the people requiring services. In response to the AIDS crisis in the 1980s, for instance, local support groups were often up and running long before any government program.

In the 1990s, the idea that the local level was best situated to respond to social needs was widely used as a pretext for handing over the responsibility for many social services from national and regional governments to local governments and NPOs. But most non-profits do not receive adequate funding from private donations alone and have continued to need government support to varying degrees. Because donors are more likely to give to larger institutions with greater money-raising abilities, local non-profits may well lack both the money and the volunteers to respond adequately to local needs. Reliance on government funding can create problems for small non-profits because of loss of autonomy and government demands for accountability and rising standards of practice. They also become vulnerable to changing government priorities and policies. Larger non-profits may be more attractive because they offer a broader range of services and more professional staff and procedures. Centralized services also offer more anonymity for people who would prefer to discuss their health concerns (for instance) with someone from outside their community. These benefits must be weighed against increasing bureaucratization and loss of the spirit of voluntarism.

The transferal of government responsibility for the unemployed and disadvantaged to the non-profit sector also has the potential to undermine the latter's independence. According to Jennifer Wolch (1990), non-profits that deliver state-funded programs are effectively compelled to act as a "**shadow state**," supporting activities that they might not approve of and stifling, rather than expressing, dissent. The development of this idea coincided with a trend in the US, UK, and Canada towards the devolution

"shadow state" argument The idea that government funding has the effect of co-opting non-profits, discouraging dissent, and effectively forcing them to defend the status quo.

of responsibilities for various programs from the national to lower levels of government. The latter were forced to reduce their expenditures and look for alternative means of addressing social problems such as poverty, crime, homelessness, drug addiction, and unemployment. Governments often rewarded NPOs that offered creative solutions to such problems with increased funding, while imposing their own oversight and policy requirements. Some national governments may have reduced their direct provision of social services in recent decades but the demand for services still exist, probably increasingly so with rising income inequality and homelessness.

The Expressive Functions of Non-profits

Volunteer organizations—arts and culture groups and amateur sports clubs, for instance—serve what are sometimes referred to as expressive functions. Organizations of this kind focus on local communities and have very high ratios of volunteers to paid staff. Recent data is not readily available but in 2003, for instance, sports and recreation in Canada had 5,283,965 volunteers and 130,913 paid staff; 73.5 per cent of all the organizations had no paid staff at all. Clearly, society depends on "us," our friends, and neighbours to coach kids' soccer, baseball, and hockey teams and much more.

Participation in expressive functions creates bonds among community members, encourages social inclusion, establishes friendships, teaches skills, helps to define cooperation and competition, and develops lifelong interests and commitments, even career directions. (Most professional athletes, coaches, referees, sports administrators, and commentators got their start in community sports organizations.) Such participation is a vital basis of social capital.

Volunteer activity in sports and recreation also generates huge multiplier effects. These effects begin at the local level where all the volunteers, coaches, managers, parents, and players generate demand for sports equipment and sport stores, sports venues, food sales, transportation, hotels, and so on. The multiplier effect of volunteerism is amplified when we recognize that community leagues provide the indispensable components—the players and their skills—required by a whole hierarchy of commercial sport leagues. As the hard-working, talented, and outstanding few emerge from the mass of amateur players, more and more profits are extracted from them by quasi-professional university teams, professional sports leagues, and other international athletic events. The few hundred thousand dollars generated by community league teams and events escalates into billions spent on stadiums, television broadcasts, transportation and parking, hotels, food, and all the paraphernalia that is sold to fans. The scale of this impact is such that cities and regions will bet their economic vitality on sports-related investments: attracting a professional team, building a stadium, or hosting the Olympics. None of these investments would be possible without the input of volunteers at the beginning of the athletic supply chain.

Advocacy Organizations

Non-profits whose primary concern is to advocate particular policies and/or ideological positions are often referred to as non-governmental organizations (NGOs). The term recognizes that while such organizations stand apart from governments, they seek to change government policy and usually the behaviour of companies.

There are thousands of NGOs around the world. A web search for "human rights organizations" will generate well over 10,000 hits. NGOs have proliferated since the late 1960s, in response to problems that are global in scale, such as global warming, loss of biodiversity, exploitation of labour, and the threat of nuclear holocaust. Most advocacy organizations specialize in one of four areas: human rights, environmental issues, peace and security, and economic policy (Table 7.4). As noted with respect to the Portland Hotel Society, advocacy can also be part of other types of NPOs.

Advocacy NGOs can be controversial since they desire change and typically oppose vested interests. Environmental groups often have to oppose business, government, and labour in their attempts to reduce environmental damage, whether that be preserving the habitat of a local species or urging action on climate change. In the area of economic policy research, a number of think-tanks regularly take direct aim at one another in the battle for public opinion. Think-tanks specialize in formulating ideas that will have the "appropriate" electoral appeal and can be effectively communicated to voters. In the US, conservative think-tanks such as the Heritage Foundation, the American Enterprise Institute, and the Cato Institute are especially strong. In the UK, the socialist Fabian Society has helped to shape the positions of the Labour Party since the beginning of the twentieth century. In Canada, the conservative Fraser Institute and left-of-centre Canadian Centre for Policy Alternatives (CCPA) routinely publish opposing views on matters such as privatization of education, minimum wage laws, and environmental regulation.

Advocacy organizations play significant roles in policy communities and are often invited to provide expertise, ideas, and sounding boards in forums at every

TABLE 7.4   Advocacy NGOs

Category	Examples
Human rights (and relief)	Amnesty International, Oxfam, Doctors without Borders, World Vision. Amnesty International was founded in the UK in 1961 to conduct research and take action to oppose "grave abuses" involving torture (mental and physical) and discrimination, especially with respect to refugees, women, prisoners, and racial, ethnic, and religious minorities. Headquartered in London, it employs roughly 500 professional staff members and has thousands of volunteers organized around the world. AI is funded entirely by fees and donations, and does not accept any support from governments.
Environmental issues	Greenpeace, Sea Shepherd, Rainforest Action Network, Earth First! Greenpeace was founded in Vancouver in 1971 to oppose nuclear testing in Alaska and poor environmental practices generally. It remains committed to enhancing environmental values and promoting peace. It opposes deforestation around the world, whaling and sealing, and opposes nuclear power and genetic engineering. Headquartered in Amsterdam, Greenpeace has offices in 42 countries and has millions of supporters who fund it through their donations and often volunteer as activists.
Economic policy	CD Howe Institute, Fraser Institute, Canadian Centre for Policy Alternatives, (UK) Fabian Society, (US) Heritage Foundation, Hoover Institute. The Hoover Institute was founded as a conservative (libertarian) think-tank at Stanford University, San Francisco. Much of its funding comes from charitable foundations established by large corporations. The Hoover Institute promotes representative government, private enterprise, peace, and personal freedom.
Peace and security	Campaign for Nuclear Disarmament (CND), Brookings Institute, Carnegie Center for International Peace, (UN) Committee on Disarmament, Peace and Security. Famous for its "Ban the Bomb" logo, the CND was formed in the UK in 1958 to oppose nuclear, chemical, and biological weapons. In recent years, it has extended its anti-nuclear stance to include the use of nuclear power for energy generation. In 2003, it campaigned against the Iraq war.

level from local to international. Although some prefer to remain outside formal decision-making systems so as not to become compromised, others play dual roles as "virtual" members of the government bureaucracy and a "shadow" opposition. Most receive the majority of their funding from private sources: individual donors, businesses, and (increasingly) research and interest group foundations that are themselves non-profit organizations. Labour unions and chambers of commerce are other NPOs that frequently strongly oppose one another, each seeking to shape government policy.

Global–Local Dynamics of NGO Activism

Although, as we have seen, non-profits that serve local or regional populations are more numerous, non-profits serving national and international spaces are better known and more influential. In Canada, this category includes Nature Canada, the Elizabeth Fry Society, Volunteer Canada, the Royal Canadian Legion, the Canadian Union of Public Employees, the Canadian Chamber of Commerce, the National Citizens Coalition, the National Action Committee on the Status of Women, and the Canadian Civil Liberties Association. Many national organizations are federated, representing local, regional, or provincial branches of the same organization or related organizations. In addition to previously mentioned examples (Salvation Army and Sea Shepherd), international non-profits include organizations such as the Red Cross/Red Crescent, Médecins Sans Frontières (MSF; Doctors Without Borders), Greenpeace, the International Chamber of Commerce, and Oxfam.

The services provided by these organizations are diverse. The doctors of Médecins Sans Frontières provide medical care at no charge to people who are not only in desperate situations, but also in most cases could never afford to pay them. Oxfam works with local people and appropriate technologies to establish bases for economic development. Amnesty International speaks for political prisoners, and the World Wide Fund for Nature works for the protection of species that cannot defend themselves. Organizations such as these provide a much-needed counterbalance to a global economic system that is driven by market principles that too often ignore the impact of its activities on the very resources—natural and social—it depends on.

These national and international NPOs offer economies of scale in research, publications, and other services that provide valuable support for local operations, and give local voices national expression. Indeed, large NPOs are highly mobile, and can marshal resources quickly in support of engagements in remote places, for example, to provide disaster relief or to oppose habitat destruction. The economic impact of these organizations is hard to measure. In dollar terms, for example, the aid going from Canada into international NGOs is meagre. Roughly half of Canada's contributions to international non-profits come from the federal government, which through the Canadian International Development Agency (CIDA), uses non-profits to pursue overseas development goals, in much the same way that provincial and municipal governments use non-profits to deliver social services. Donations, primarily from individuals, account for a similar amount, and earned income, mainly in the form of fees for goods and services, for most of the rest. US official government aid consistently amounts to only about 0.2 per cent of GNP—far below the internationally agreed standard of 0.7 per cent—but it is estimated that private and institutional donations are more than double that amount. Both official and private donations, however, are often "directed"; as a consequence, overseas NPOs are often restricted in the uses to which they can put the money they receive.

But assessment of NPOs needs to be judged in terms of their mandates in relation to outcomes, rather than simply the size of their financial resources. A number of NPOs are primarily interested in changing the behaviour of governments, businesses, and consumers, and providing marginalized stakeholders a voice in the governance of the international economy. These NPOs are in the business of persuasion. Countries or stakeholder groups that are unable to voice their opposition to the WTO (for example, by appearing before a UN committee) can turn to international NGOs for help. Indeed it would be hard to conceptualize national or international responses to social and environmental needs without the activities of Greenpeace, the Red Cross, or Amnesty International pushing these concerns on to the agenda. Yet, the impacts of these organizations are often controversial, especially when they are seen as interfering in domestic or local affairs and imposing values from "outside" on societies that do not share them. Ethical trade and remapping resources are two themes that illustrate how NGOs are seeking to restructure value chains according to their mandates.

Ethical Trade

Recent decades have seen major shifts in production from developed to less-developed countries with fewer regulations concerning working conditions and environmental impacts. In many cases the consequences for the latter countries have been reminiscent of the Industrial Revolution in Europe, including exploitation of workers and environmental degradation. Some of these consequences can be traced directly to the transnational corporations themselves, but most have been attributed to local subcontractors or contract manufacturers. International NGOs, some well-established and some newly created for the purpose, have worked to raise awareness of these abuses among consumers in the developed world, in an effort to force the corporations to adopt more ethical practices.

The industries involved are usually labour-intensive: textiles, clothing, electronics assembly, shoes, and sporting goods. On the labour front, NGOs address issues that range from excessive working hours and inadequate health or safety precautions to sexual harassment and use of child labour. On the environmental front, NGOs argue that pollution, land degradation, and loss of biodiversity are not "externalities" but real costs that must be taken into account in corporate decision-making. Their use of the media to publicly "blame and shame" the corporations at fault has led to a number of changes in practice.

Supplier audits, for instance, have become standard since Nike and Walmart were compelled to investigate their overseas suppliers in the late 1990s. Now almost all international corporations have developed labour and environmental guidelines that their suppliers must conform to. The effectiveness of these audits, the extent of their use by individual companies, and the strictness with which companies adhere to the guidelines are open questions, but overall conditions have improved.

fair trade movement The movement to ensure that small producers in the developing world receive a fair price for their products.

The **fair trade movement** takes a different approach. It focuses on helping small farmers and artisans and their communities that produce commodities, such as coffee and textiles obtain higher and stable prices for their efforts and helps them produce in an environmentally responsible manner (Case Study 7.2). Introduced in the 1950s by international charities such as Oxfam, the movement began to expand in the late 1980s after the Dutch developed the first "fair trade" label for coffee. By 2010, more than 700 producer organizations had been created, supporting five million workers and their dependents. A certification system run by the International Fairtrade Organization ensures that fair trade principles are observed throughout the value chain. To qualify for certification, producers are required to make minimal use of agricultural chemicals (organic methods are encouraged). In return, they receive both a decent price for their crops and an additional "fair trade premium" for investment in community projects and environmentally sustainable agriculture.

The outlets for fair trade products are typically small, independent-minded retailers, but in recent years, stimulated by pressure from NPOs, growing numbers of companies such as Starbucks have begun to incorporate fair trade into their corporate culture—for instance, by including some fair trade goods in their product-mix. But unintended side effects are not uncommon. For example, supermarkets that compel suppliers to absorb the costs of compliance with fair trade codes can force them out of the market and their employees out of jobs, and an "ethical" label attached to a handful of products can serve as a public relations umbrella to shelter less progressive activities. Moreover, although fair trade increases producers' income, the increases are not enough to move them out of poverty; value chain dynamics inevitably mean prices and profits increase through the value chain, and small highly competitive suppliers have little bargaining power (see Chapter 10).

The Remapping of Resource Peripheries

Modern environmentalism traces its roots to the conservation movement that led to the establishment of the world's first national park at Yellowstone, Wyoming, in 1872. (Canada's first was Banff, founded in 1885.) However, environmental NGOs (ENGOs) have grown and proliferated especially since the late 1960s. Greenpeace's birth in 1971 is a key marker of society's response to the emergence of environmental problems during Fordism's North American apogee. Since then, environmental concerns and NGOs have gone global, driven by massive and interdependent resource exploitation, most evident in global warming, loss of biodiversity, and borderless pollution. In general ENGOs emphasize the "non-industrial" (ecological, spiritual, recreational, aesthetic) values of resources, rather than simply their commodity values, and the importance of sustainable development. ENGOs have been powerful voices for increased regulations and enforcement, for example, controls over air, water, and soil pollutants; environmental certification schemes that require sustainable resources such as forestry to be managed in sustainable ways; environmental impact analyses that over the past two decades or so have become a requirement for resource projects in many parts of the world; and the expansion of **conservation areas** that prohibit or restrict industrial exploitation.

conservation area An area in which natural processes are allowed to operate with minimal human interference.

With respect to conservation, nationally designated protected terrestrial areas increased from 2 million km^2 in 1965 to 14 million km^2 in 2006, while protected marine areas grew

Case Study 7.2

FAIR TRADE IN COFFEE

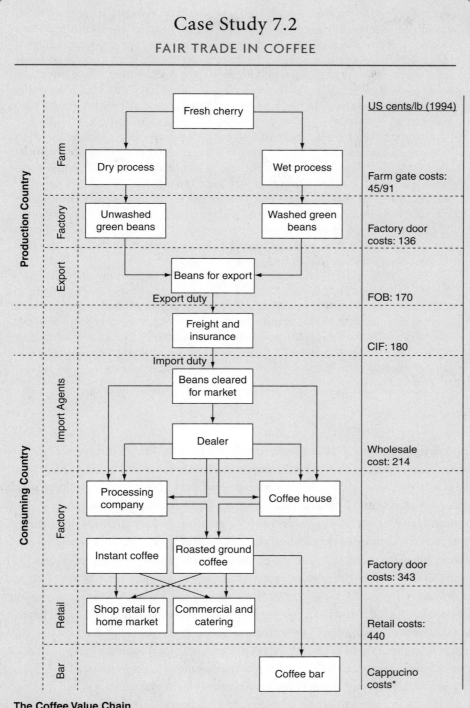

		US cents/lb (1994)
Production Country — Farm	Fresh cherry → Dry process / Wet process	Farm gate costs: 45/91
Factory	Unwashed green beans / Washed green beans	Factory door costs: 136
Export	Beans for export (Export duty)	FOB: 170
	Freight and insurance	CIF: 180
Consuming Country — Import Agents	Import duty → Beans cleared for market → Dealer	Wholesale cost: 214
Factory	Processing company / Coffee house → Instant coffee / Roasted ground coffee	Factory door costs: 343
Retail	Shop retail for home market / Commercial and catering	Retail costs: 440
Bar	Coffee bar	Cappucino costs*

The Coffee Value Chain

* Costs variable but very high. Include: overheads, advertising, other products (i.e., milk), and the "experience" of the coffee bar. (See breakdown of the price of a cup of coffee.)

Adapted from Kaplinsky, R. and Fritter, R. 2001. "Who gains from production rents as the coffee market becomes more differentiated? A value chain analysis." IDS Bulletin Paper, July 2001: 69–86.

(Continued)

Most of the world's coffee is produced by impoverished farmers in the developing world who receive only a small fraction of the high prices charged for their beans in Europe and North America. Beans grown by the farmers of the Chochu region of Ethiopia, for example, are sold either to local merchants or to a coffee co-operative that transports the coffee 400 kilometres to Addis Ababa. There prices are set at auction by "local" buyers representing the interests of MNCs such as Nestlé, Kraft, Sara Lee, and Schmuckers, and the beans are sorted, mainly by women paid less than 50 cents a day.

The objective of the fair trade movement is to ensure that growers and workers at the production end of the value chain receive a greater share of the profits from their work. The movement has roots in several organizations in the US and Europe, but it was a Dutch organization that first began to use the name in 1973. In 1997 that name was adopted by a non-profit organization that developed a certification and labelling framework for producers and marketers of produce from impoverished regions in the developing world. Development of a fair trade value chain is not straightforward, however, and needs to involve the farmers themselves, often located in very remote places, such as near Monte Picchu, Peru.

The commodities sold through fair trade today include not only coffee but also tea, cocoa, cotton, fruit, vegetables, rice, beer, wine, and honey, among others. In addition to guaranteeing a minimum price to producers, fair trade arrangements provide for investment in their communities. They also include guidelines for environmental practices that reflect the stakeholders' input; both producers and distributors are audited for compliance. By 2009, 827 producer organizations were selling 27,000 fair trade products in 70 countries, with total sales of approximately CDN$4 billion.

from virtually nothing in 1973 to just over 1 per cent of the world's oceans by 2010. Forests where biological diversity is now the designated primary-use amounted to 12 per cent of the global forest area in 2010. In the resource peripheries of rich market economies such as Canada, Australia, and New Zealand, the remapping of resources to give greater priority to environmental values has been especially evident, while in poor countries, the UN emphasizes how environmental degradation exacerbates poverty. It seeks to promote sustainable development, for example, by linking ecotourism with the conservation of environmental values in the form of clean water, protection from erosion, and long-term availability of resources. A frequent collaborator is the International Union for the Conservation of Nature (IUCN). The world's largest environmental knowledge network, with more than 1,000 member organizations (government and non-government) and almost 11,000 volunteer scientists around the world, the IUCN works with local communities as well as companies, governments, and NGOs to find practical solutions to environmental challenges.

Yet conservation is strongly resisted in many places, rich and poor, both by corporate interests and by local people who depend on access to certain resources not only for food, shelter, or employment, but for spiritual or cultural purposes. Around the world, environmental interests regularly clash with long-standing and new corporate interests, small businesses of various kinds, and traditional cultures that variously support corporate activities or claim traditional rights to resource use. The global–local dynamics are complex, especially when the NGOs involved are multinational.

In some regions, such as the Philippines, the Congo, and Indonesia, conflicts over timber, minerals, and other resources can be deadly. Even in rich peripheries, where property rights and the rule of law are well established, conflicts are often deep-seated and civil disobedience is not uncommon. Tasmania's "war in the woods" is a case in point (Case Study 7.3).

Case Study 7.3
FOREST CONFLICTS IN TASMANIA

Courtesy of Julia Affolderbach

Forest blockade ("Camp Weld") in the Weld Valley, southwest Tasmania (February 2006).

The last few decades havve witnessed intense conflicts over natural resources in Tasmania (the island state that lies off the southeastern coast of Australia). Opposition to large-scale industrial resource exploitation started to emerge in the late 1960s in response to a massive hydro generation project that called for the flooding of Lake Pedder and vast areas of adjacent forestland.

Although the effort to stop the project did not succeed, it led directly to the formation of the world's first "Green" party. Established in 1972 by activists who had worked on the anti-dam campaign, the United Tasmania Group contested a state election that same year and eventually became the Tasmanian arm of the Australian Greens. The state's environmental movement has grown considerably in numbers and strength since that time. Focusing on the protection of Tasmania's native forests, environmental NGOs criticize not only the large scale of forestry operations and their "slash and burn" practices but their focus on exports rather than value-added wood processing. It is estimated that 90 per cent of Tasmania's old-growth forests are turned into wood chips for export to Japan.

(Continued)

Many local forest campaigns have used direct action (road blockades, tree-sitting, chaining oneself to logging equipment, even sabotage) to stop forest operations temporarily. But confrontation is not likely to achieve substantive change. Thus longer-term strategies to raise public support centred on media stunts, public rallies, and lobbying of celebrities as well as politicians at the state, national, and even international level. These strategies have been very successful. In 1968 only 4.2 per cent of Tasmania was protected land; as of 2006 that figure had increased to 39.8 per cent.

Source: Based on information kindly provided by Julia Affolderbach, October 2009.

In general, ENGOs are socially sanctioned by their focused commitment to environmental values that have much public support, and for their information and exposure of environmental abuses. Their actions may be seen in the tradition of progressive social movements. Sea Shepherd's aggressive tactics have been costly and embarrassing to Japan but the NGO can claim vindication. It harassed the Japanese because they circumvented an international ban imposed in 1986 by claiming their continued hunting is research to demonstrate whaling is sustainable. However, spurred by Sea Shepherd and others, in 2014 the International Whaling commission in the Hague voted to stop Japanese whaling ships in the Antarctic. Yet, while demand for whale meat in Japan has declined drastically and employment in the whaling industry is negligible, the whaling industry is heavily subsidized and stockpiles have grown. Other NGOs have used less aggressive and dangerous tactics to oppose the hunt.

Social Capital and Development

If some NGOs routinely engage in conflict, it is the overall goal of non-profits to promote social or collective values: after all, their existence depends on their ability to attract public support. The centrality of social values to NPOs suggests that development has to be measured in terms of something more than per capita income. Although the concept of "quality of life" (QOL) is subjective, it is increasingly used as basis for understanding well-being as a package of values, rather than reducible to monetary terms such as per capita GNP. For example, an acceptable quality of life is defined not only by material wealth but by criteria related to health, education, and well-being, a sense of citizenship, of being part of a larger whole. Some of those additional criteria are related to income, of course, though not necessarily in a straightforward or linear way. Other NPOs have more to do with social needs. Perhaps the most important product of the non-profit sector is its contribution to the social capital without which no society could function, let alone develop economically.

Social Capital: Embracing Co-operative Values

The term *social capital* is relatively new, but the qualities it refers to—trust, co-operation, reciprocity, a sense of shared understanding—have long been recognized as essential to any market exchange and hence to all economic activity. Social capital often goes unrecognized because it comes into play informally, through personal interaction that has no apparent market value. Many contexts in which it is formed have nothing to do with markets, and it is not easily described in monetary terms. Rather, social capital is formed in the course of daily life, through interaction with families and friends, with co-workers and communities. The term was coined by analogy to "capital" in the conventional economic sense: investment in equipment (capital) that produces goods or services that provide returns or money (financial capital) that can be invested in stocks, bonds, etc. to bring further returns. In a related way, investments in household and education can produce skilled, educated workers representing human capital, while maintaining or enhancing the ecosystem is investing in the natural capital (or simply nature) and the ecosystem services it provides.

Like financial and productive capital, social capital becomes depleted and needs replenishment from time to time. But it differs in one important way. Whereas financial capital becomes depleted with use, social capital becomes depleted when it is *not* used; in fact, it must be used if it is to be preserved.

Perhaps the most important characteristic of social capital is its ability to produce positive externalities. Investing in social capital—spending the time and effort to maintain and expand contacts—not only produces individual benefits but helps to establish norms of behaviour and expectations of reciprocity. In this way social capital is transmitted to others both in the investor's immediate community and in the broader society. Wide diffusion of these norms creates a culture of trust that encourages efficient interaction. Reinforced by individuals' need to build and maintain their own reputations, the culture of reciprocity reduces the risk of problems such as opportunism.

On the other hand, social capital can take two forms, and one of them is not necessarily positive. Bonding within the family, sports team, or religious community offers benefits to members through reciprocity, learning and so on, but such bonding is by nature exclusive. When exclusive, bonding does not encourage the formation of social capital outside the group and may even prevent external interaction. Other types of social capital are inclusive, developed through activities that bridge differences and bring together people from different backgrounds. These ties, created through two forms of social capital, exclusive and inclusive, are described as "strong" and "weak" respectively because the former binds people tightly together in one particular form of interaction, while the latter allows and even encourages openness to new contacts and ways of thinking.

Some Local Implications

Individuals, communities, and societies that successfully invest in both weak and strong social capital stand to gain many benefits. Investment in social capital

- facilitates resolution of the "collective action" problems;
- builds habits of trust and confidence that facilitate efficient exchange;
- encourages awareness of the many ways in which our individual fates are linked and helps to foster good behaviour;
- facilitates exchange of information and development of contacts; and
- promotes a sense of connectedness that is both psychologically and physiologically beneficial.

All these benefits (or their absence) can be seen in economic interactions. Social capital is the basis of the untraded interdependencies that allow local economies to develop efficiencies through personal communication. It is also closely related to path dependence in both its positive and negative forms. Regional comparisons show that the impact of social capital is extraordinarily long-lasting. In the US, for example, social capital is strongest in the northeast, northwest, and states with strong Scandinavian backgrounds, which have long traditions of individual discipline and reciprocity in both market and non-market exchange. It is weakest in Mississippi and other areas of the south where the slave system either destroyed social capital mechanisms or prevented their formation. Whether a place succeeds in reinventing its economy, finding ways to take advantage of globalization, or chooses instead to insulate itself from it, depends not only on the quality of its resource endowments or the size of the capital infusions it receives, but also on the nature of its past social interactions.

Different regions benefit from different forms of social capital. California's "Silicon Valley" is characterized by a free flow of people from company to company, co-operation among rival companies, and respect for risk-taking and failure. The German automobile region of Baden-Württemberg has succeeded through close co-operation among carmakers and parts suppliers. In Italy, small firms such as Carlo Barbera produce world-class fashions using a production system based on a highly fragmented and specialized social division of labour coordinated by close local and family relationships, awareness of mutual need, and common labour standards. The social capital characteristic of these regions develops over time, through ongoing personal interaction, formal organizations such as industry associations and specialized schools or training programs, and government support.

The connections between social capital and economic development are complicated and change over time. Certainly development can enhance social capital by creating bridges between individuals or communities. But increasing wealth can also undermine social capital by disrupting the processes through which it is developed, especially in the household sector. Traditionally, the family was the place where values, norms, and patterns of interaction were first established. As we noted earlier in this chapter, today many of the family's traditional functions are performed by NPOs. One important reason is that, with women approaching male rates of participation in the workforce, more and more children are spending most of their days outside the household in daycare—many in centres run by non-profits. Further, commuting to work has typically increased with economic development, a trend that disrupts the family and its role as the fundamental source of social capital, and steals time away from engagement in community institutions such as schools, sports and service clubs, and political organizations. How social networks and other connections offered by the Internet will affect social capital formation has yet to be determined.

Development and "Gross National Happiness"

What are the goals of economic development? Answers to this question are powerfully place-contingent. Even within a country like Canada, however, different groups will answer differently. We have already noted that single measures such as per capita income are not really adequate today. Alternative indexes take into account various "quality of life" criteria, even attempting to measure "gross national happiness."

The idea of measuring happiness originated in one of the nominally poorest countries in the world: the Buddhist kingdom of Bhutan in the Himalayas. After centuries of isolation, the King of Bhutan decided in the 1970s to comprehensively modernize the country and permit connections with the outside. He also established a constitutional monarchy and parliamentary government in 1998. In response to growing concerns among Bhutan's population that development imposed from the outside would destroy its traditions and cherished social values, the king proposed a "happiness" index based on four criteria: (1) sustainable and equitable socio-economic development; (2) conservation of the environment; (3) preservation and promotion of culture; and (4) good governance. As might be expected, Bhutan's progress towards these goals has been slow. Nevertheless, its experiment is a powerful reminder that quality of life involves a great deal more than material wealth. It also underlines the point that development is not merely a matter of external influences, but a process in which local institutions must play a leading role. Several other indices now try to compare countries beyond the simple GDP or per capita GDP metrics. The most widely used is the UN's human development index which addresses the equality of health care and education opportunities alongside economic levels. The Happy Planet Index weighs social and economic returns within a country to the burden that it places on the global environment.

Conclusion

The non-profit, non-governmental sector performs functions that markets, businesses, and governments cannot, will not, or believe are not part of their mandate. NGOs and NPOs have a direct economic impact on employment and income from services, donations, and government subsidies. In addition, volunteers provide services that are invaluable but unremunerated. Economic governance is a second impact. NGOs such as Greenpeace and Oxfam step in when they perceive inadequacy or malfeasance on the part of either corporations or governments, and contribute to economic governance by setting moral standards. Perhaps the most important impact of civil society, however, comes in the form of social capital—the networks, values, and norms of reciprocity that determine both a society's quality of life and its economic performance. In subsequent chapters, we will see that NPOs and NGOs have become integral components of activity throughout the global economy.

Practice Questions

1. A map on p. 133 of the *Canadian Oxford World Atlas* shows the distribution of selected endangered species around the world. Choose one of these species and describe the geography of the institutions deciding its future. What institutions are hurting and helping the species and how? From what geographical bases?

2. How many NPOs are there in your community or region that advocate for the homeless and disadvantaged? What do these organizations do? How are they supported?

3. Select a high-profile international NGO and identify the geographic scope of its operations. What are its main objectives, strategies, and sources of income? How influential is it?

4. Why are ENGOs involved in so many resource conflicts? What are the sources of such organizations' legitimacy? On what grounds can they be criticized?

5. Is happiness a legitimate goal of economic development?

6. Choose a volunteer-run organization for young people in your community (choir, orchestra, sports club, etc.) and find out how it is supported. How does this organization support your community?

Key Terms

social values   209

non-profit/non-governmental sector   211

social services   211

advocacy organizations   211

human capital   211

social capital   211

"shadow state" argument   217

fair trade movement   222

conservation   222

Recommended Resources

Affolderbach, J. 2012. "Environmental bargains: Power struggles and decision-making over British Columbia's and Tasmania's old growth forests." *Economic Geography*.

Clapp, A.R. 2004. "Wilderness ethics and political ecology: Remapping the Great Bear Rainforest." *Political Geography* 23: 839–62.

An analysis of how the Great Bear Rainforest was remapped as a wilderness area, largely as a result of pressure from ENGOs.

Hayter, R. 2003. "The War in the Woods: Globalization, Post-Fordist Restructuring and the Contested Remapping of British Columbia's Forest Economy." *Annals of the Association of American Geographers* 96: 706–29.

How environmental NGOs became a countervailing power in BC.

Helliwell, J. 2007. "Well-being, social capital and public policy." *Social Indicators Research* 81: 455–96.

Discusses ways to measure well-being and social capital on a global scale.

Milligan, C., and Conradson, D., eds. 2006. *New Spaces of Health, Welfare and Governance*. Bristol: Policy Press.

Collection of geographic interpretations of the voluntary sector.

Mohan, G., and Mohan, J. 2002. "Placing social capital." *Progress in Human Geography* 26: 191–210.

Examines how social capital affects and is affected by place and local economic performance.

Pearce, J. 2009. "Social economy: Engaging as a third system." In A. Amin, A. Cameron, and R. Hudson (eds.) The Social Economy: International Perspectives on Economic Solidarity. London: Zed Books.

Discusses the role of NGOs as a third system in relation to the private and public sector systems.

Putnam, R.D. 2000. *Bowling alone: The collapse and revival of American community*. New York: Simon & Schuster.

The foremost theorist on the voluntary sector and other community organizations in creating social capital.

Skinner, M.W. 2008. "Voluntarism and long-term care in the countryside: The paradox of a threadbare sector." *Canadian Geographer* 52: 188–203.

Examines how voluntary sector providers in Canadian rural communities are responding to the local downloading of responsibilities for health and social care by senior levels of government.

Teather, E.K. 1997. "Voluntary organizations as agents in the becoming of place." *Canadian Geographer* 41: 226–305.

An early study of three voluntary organizations for rural women in Australia, Canada, and New Zealand, revealing how such organizations develop loyalty to places.

References

Daly, H.E. 1994. *For the Common Good: Redirecting the Economy Toward Community, the Environment and a Sustainable Future*. Boston: Beacon Press.

Salamon, L.M., Anheier, H.K., List, R., Toepler, S., Sokolowski, S.W., and Associates. 1999. *Global Civil Society: Dimensions of the Nonprofit Sector*. Baltimore: Johns Hopkins Center for Civil Society Studies.

Statistics Canada. 2004. *Cornerstones of Community: Highlights of the National Survey of Nonprofit and Voluntary Organizations*. Ottawa.

Wolch, J. 1990. *The Shadow State: Government and Voluntary Sector in Transition*. New York: The Foundation Centre.

PART III

Location Dynamics of Value Chains

CHAPTER 8

Cities

Cities exist for trade, they are places where people come to divide their labor, to specialize and exchange.

Ridley 2010: 158

• • •

Great cities are inherently creative.

Friedman 1970: 474

Building on the themes discussed in Parts 1 and 2, this chapter's goal is to examine cities as the integrators of markets and value chains that underpin the global economy. Cities perform these roles through the agglomerations of market activities and the institutions that make them work, and as centres of demand and production that stimulate the formation of value chains. Subsequent chapters examine the location dynamics of value chains from specific sectoral perspectives. The chapter's specific objectives are

- To explain the growth of cities in terms of their local (endogenous) and global (exogenous) dynamics;
- To outline the characteristics of cites as places, and the factors that shape their diversification;
- To explain the differences between the industrial city and the creative city;
- To introduce urban systems, and the role of hierarchical and non-hierarchical factors; and
- To explore how the internal structures and patterns of inequality within cities are affected by market forces, planning, urban politics, social relations, and environmental imperatives.

The chapter is organized in three main parts. The first part reviews the concept of agglomeration economies and discusses the nature of urban diversification and creativity. The second explores the specialization and functional interdependencies of cities within regional, national, and global urban systems. Finally, we conclude with a look at the internal structure of cities and the ways in which markets, planners, and politics shape land values, land uses, and urban structure. This chapter builds on earlier chapters with discussions of agglomeration or external (dis)economies of scale, classified as localization and urbanization economies and diseconomies (Chapter 2), and hypotheses shaping innovative behaviour (Chapter 3).

The Role of Cities

Cities are organic, living entities in which governments, businesses, labour, NGOs, and consumers actively engage with one another to exchange goods, services, and information, often co-operatively but not always. In order to function, therefore, cities depend on the interplay of institutions of all kinds, formal and informal. All the legally recognized economic activities that take place within cities are formally governed by laws and regulations that impose penalties for failure to comply, and these formal controls are reinforced by a host of informal expectations and conventions. (Illegal activity is also subject to legal sanction, of course, as well as to informal conventions.) Institutional governance makes everyday life in the city a stable, routine affair for most people. At the same time, every part of the city is subject to change from one day to the next, and such change is also part of everyday life. Familiar businesses can disappear and be replaced overnight by ones offering new services; a run-down neighbourhood that attracts a couple of new residents from a different demographic group can suddenly move upscale; a demolition site can stand empty for just a few days before new construction begins. This capacity for change points to the ongoing transformation that is an essential feature of capitalism. In market economies, institutions both create routines and facilitate change.

The reasons behind the original locations of cities are varied, often long forgotten, but it is generally agreed that their emergence was both a response and stimulus to rising agricultural productivity. The growth (and decline) of cities is intimately connected with the advance (and decline) of civilization; cities are vital organisms shaping economic development in place and space. As Jane Jacobs (1984) noted, Adam Smith's *Wealth of Nations* is literally grounded in the wealth of cities. "Cities exist for trade" and are the places where the division of labour and exchange is elaborated. What distinguishes cities from towns, villages, and hamlets is not simply their size but their diversity, creativity, and power over space—attributes that are rooted in institutions of all kinds: public and private, formal and informal, economic and non-economic.

The city is pre-eminently a place of exchange of all types: social, cultural, intellectual, and political, as well as economic. In the economic context, the primary places of exchange are the diverse markets shaped by agglomeration and their attendant economies and diseconomies of scale. The evolution of a city's markets is unique, created by the interplay of local and global dynamics. Hong Kong's renowned central business district, with its skyscrapers hugging the shoreline of one the world's busiest ports, evokes the agglomerative power of cities where the seductions of wealth, entertainment, and power are concentrated in a small area. Yet Hong Kong's development is distinctive, shaped locally by deep-water protected harbours, the once cheap virtually unoccupied land, and with a global raison d'etre defined by past roles as imperial outpost and China's main point of contact with the world (Case Study 8.1).

Cities are such a powerful force in shaping the global economy because, as you will recall (Chapter 2), agglomeration reduces transaction, transportation, and production costs. Cities make it easier to find buyers or sellers, compare prices and quality, and exchange information of all types. The distance to buyers or sellers is reduced and internal and external economies of scale are easier to integrate and coordinate. Moreover,

Case Study 8.1

A CATHEDRAL IN THE HEART OF MAMMON

An anomaly nestled in the heart of Hong Kong's aptly named "Central District," St John's Cathedral is a non-profit institution operating amidst some of the most expensive real estate in the world. But that isn't so strange. In every world city, there are buildings and spaces serving spiritual, political, or recreational purposes that have held fast against the determination of market forces to derive the highest possible value from land use. St John's is unique because it is the only piece of private property in Hong Kong. All other land is held on 50-year leases from the government of the territory (since 1997 a Special Administrative Region within China), which depends on the income from lease sales for a major portion of its revenue. (This is one of the main reasons why income taxes in Hong Kong are so low.) These departures from North American patterns illustrate the inherently institutional nature of economic geography.

Hong Kong's land market demonstrates the importance of constructing institutions to work in harmony with the market system. St John's is surrounded by the headquarters of banks, insurance companies, and real estate developers (some control the largest fortunes in Asia), and the regional offices of many MNCs. The Hong Kong Stock Exchange is a few hundred metres away. These companies and institutions organize the transactions between the Pearl River Delta region and the world. Their congregation in the Central District raises the value of land, producing profits that are turned into investments in the mainland and elsewhere. Without the lease-based property rights system, none of this would be possible. As an institution, this system has been one of Hong Kong's greatest assets (along with rule of law, respect for individual rights, free speech, and other institutions), differentiating the city from its motherland to the north and integrating it with compatible property systems

those essential attributes are facilitated and promoted by the institutions that evolve to govern the diversity of markets; untraded interdependencies within and between markets; the generation of research, design, and innovation activities; and a plethora of amenities that draw people into the agglomeration. Cities both organize dense localized (intra-urban) interactions of economic activities and land-use patterns and orchestrate the global economy within (inter-urban) systems.

Cities are not immune to economic decline, but they have innate bases of renewal available in their creative diversity, their human resources and social capital, and their infrastructure of buildings and communication networks. Over the last decades of the ICT and globalization, the industrial city as it existed throughout North America and Europe—with its close connections to a regional hinterland and exports of manufactured goods—has been replaced by the evolving postmodern service or creative city. Advanced cities of the twenty-first century export knowledge-based services and are part of **spatial innovation systems** (SIS) linking agents involved in the research, development, and transfer of new technologies around the globe. Meanwhile, the industrial city has re-agglomerated elsewhere. Rich countries are highly urbanized

spatial innovation system (SIS)
An international system linking agents around the world involved in the research, development, and transfer of new technologies.

within the global-trading system. These advantages will be tested as China develops its own property rights regime.

(Chapter 2) with over half of the world's population (2008) living in urban areas. It is through urbanization that the less developed countries strive to become rich.

In general terms, inter-urban systems that connect specialization and functional interdependencies in hierarchical and non-hierarchical relationships are becoming increasingly globalized, often in association with sharp intra-urban differences in income and standards of living. We shall see how the forces of agglomeration impose a bid-rent curve that shapes much urban form, intra-urban markets, and other economic and social activity patterns, but also how planning and government regulations attempt to balance agglomeration's power with social definitions of efficiency, equity, and environmental sustainability.

Cities Are Global and Local

Spatial agglomeration allows firms in specific industries to realize localization economies of scale, and enables firms in all industries to realize urbanization economies of scale (Chapter 2). Markets and agglomeration are interdependent, intertwined in the diverse markets that help define a city. Economic exchange of all kinds—goods, services, or labour—requires a venue, as do political, religious, and many other activities. Cities provide the infrastructure and institutions required for a myriad of complex and interdependent exchanges to occur: buildings, services, regulations, and culture. Accumulating over time, both physical infrastructure and established institutions represent powerful attractions for investment. Global cities offer the greatest variety of types of exchange, enjoy the most potential for self-sustaining (endogenous) growth, and house the greatest number of institutions with global reach.

Cities may be defined geographically by their administrative boundaries or by functional criteria based on routine behaviour, for example, commuting distances or transit networks. However defined, the economic functions, size, and complexity of a city result from the interaction of its markets and institutions with agencies beyond its boundaries in the rest of the world—or simply "globally." In this context, global refers to exogenous agencies: connections and events that are non-local. Conventionally, a city's non-local or global connections are considered in relation to a regional hinterland, an adjacent territory that often shares a common administrative framework and regional identity; the national context; and spaces beyond (Figure 8.1). Moreover, at all scales, the global–local interactions of cities are dynamic, that is the characteristics of cities as places and their connections across space change over time, typically in tandem. In addition, cities generate their own economies, driven by the demand of large populations and provided for by a host of public, private, and volunteer suppliers of goods and services.

Economic activities within cities are often classified as either basic or non-basic; **basic activities** generate export incomes and non-basic activities serve local consumption needs. In the "urban (or export)-base" model, basic or export activities are "autonomous" urban growth generators that in turn stimulate "dependent" **non-basic activities**. In this view, basic activities determine a city's (economic) identity and rationale, influence, and prosperity and their identification has been important in urban planning. Note that in this model, export income may refer to sales to foreign consumers or more generally to sales beyond a city's boundaries or regional context. The location quotient is a simple way to statistically identify basic and non-basic sectors (Case Study 8.1).

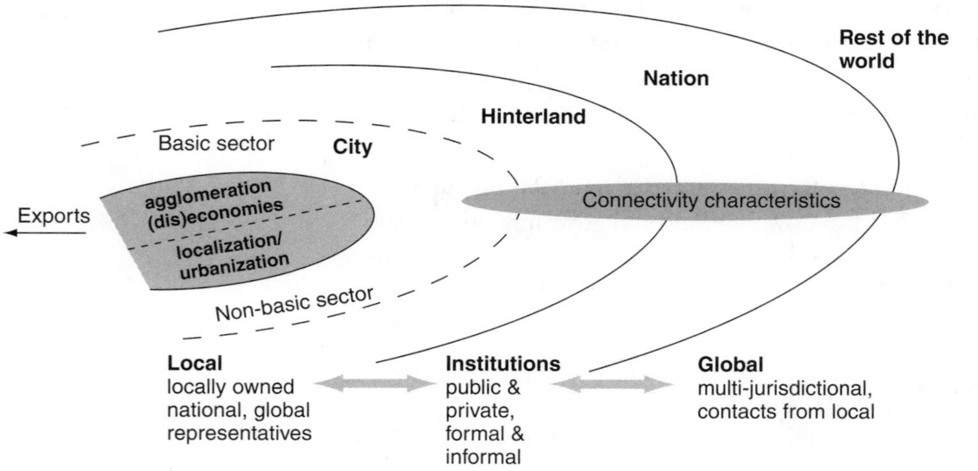

basic activities Economic activities that generate income from outside the defined territory.

non-basic activities Economic activities that serve local consumers and therefore do not bring in "export" income.

Places among Global Markets

Traditionally associated with specialized primary and manufacturing activities, basic activities in contemporary cities in established market economies increasingly comprise a wide array of services for businesses

FIGURE 8.1 The City's Global–Local Economy

(producer services) and consumers (see Chapter 12). Like goods, services earn export income for a city by their sale to consumers in foreign countries, for example, when local engineering or planning consulting firms sell their expertise in other cities or when corporate head offices charge their dispersed subsidiaries for their legal, technical, and marketing support. Services also earn export income when visitors from outside the city spend locally, for example, to attend theatres, sporting events, and museums, stay at hotels, or purchase goods in shopping malls.

Exporting is not the only cause of urban growth. The evolution of cities can be driven by "endogenous" innovative behaviour, flows of in-migrants, and large market size, which attracts investment. Innovation can in turn lead to search for economies of scale, multiplier effects, and new export strengths.

Whatever their sources of growth, cities are places of specialization and exchange and their evolution—in growth or decline—needs to be understood in terms of their local characteristics in relation to their accessibility and abilities to trade globally. In this regard, the wholesalers, trading companies, and distributors that store, move, and exchange a galaxy of goods within and among cities are the modern institutional equivalents of the merchants, such as Marco Polo, of medieval times. While the Internet facilitates the flow of information at little or no direct cost, the exchange of know-how, ideas, expertise, and confidential information remains heavily reliant on personal interaction for which the accessibilities and amenities of cities play vital roles.

Cities access markets and resources by investments in transportation and communication infrastructures that connect with adjacent towns and more distant places. Such investments reflect choices shaped by various geographical, economic, technological, and political considerations at the time they are made. As the UK began to industrialize in the late eighteenth century, landlords and local entrepreneurs built canals to allow goods (inputs and outputs) to be moved in much larger quantities and far more cheaply than by horse and cart. Subsequent investments in steam-driven railways then further greatly extended city connections. For example, the potential of Huntsman's innovation of crucible steel for Sheffield's growth was facilitated by canal and railway investments (Case Study 3.2).

In the new world territory and economic space of Canada in the nineteenth century, governments needed to take decisive roles in providing infrastructure in the absence of private money and interests. Thus Vancouver's ability to exploit the forest and mineral resources of its hinterland was stimulated by the arrival of the Canadian Pacific Railway in 1886 to provide access to national and other continental markets, as well as to imports and settlers. The CPR was part of the federal government's National Policy to promote east–west ("nation-building") linkages rather than north–south linkages with the US which implied threats of annexation. Arguably an early PPP (Chapter 6), governments basically funded the CPR through the land-grant system by which the private company received land—including 6000 acres of central Vancouver—from the provincial government as part of a payment to build the railroad. Indeed, to increase the value of its land, CPR moved the original western terminus of the railway a few miles from New Westminster to Vancouver, ensuring the latter became the core city of BC. Another important early kick-start to Vancouver's growth was the opening of the Panama Canal in 1913. In this case, Vancouver benefitted from the US government's desire to link the

Atlantic and Pacific Oceans and to reduce the time, cost, and hazard of sea transport between the west coast and the eastern US, as well as Europe. While sovereignty of the canal was transferred to Panama in 1977, the imperative to reduce transportation costs remains, as shown by a Chinese company's plans to dig a large canal in Nicaragua linking the two oceans.

During the twentieth century, the connectivity of cities has been greatly enhanced by investments in highway, airline, and information networks. In general, city economies are powerfully shaped by their national contexts: nations vary widely in terms of economic size and market potential; national governments establish rules and regulations that strongly affect the way cities participate in the global economy; and national governments are important sources of funds for large-scale infrastructure investments and for providing strategic direction. The relative location of cities in terms of accessibility to other parts of the world is also crucial to their development. In Canada, for example, Vancouver's growing links to the Pacific Rim are especially strong and reflect in its travel time and cost advantages in shipping and airline connections. That said, the advantages of Vancouver's more northerly location are offset by US coastal cities that are part of a much bigger national economy.

The relative location of cities is influenced by technological change such as larger ship sizes and longer distances flown by airplanes as well as by changes in tariff barriers, and the opening of new routes such as the Panama Canal or the Suez Canal in 1869, which affected trade between Asia to Europe. Profound changes in national context and systems of governance also affect the relative location of cities. For example, when the UK entered the EU (then the European Common Market) in 1974, British cities lost their established, often privileged connections with Commonwealth countries, while the potential for integration within Europe posed significant challenges in terms of new regulatory regimes and business and cultural practices. The traditional metal working industries of Sheffield, for example, lost many Commonwealth customers at this time but could not readily replace them with European markets. Meanwhile for Australian and New Zealand cities, access to British markets became more difficult through the establishment of tariff and non-tariff barriers and they increasingly began to look towards developing Asian trade links, strongly supported by national government policies.

Since the 1980s, the marketization and opening of the centrally planned economies of central Europe and China has had profound consequences for cities, especially within those territories but also elsewhere. Shanghai, for example, has grown from a population of 11.9 million in 1982 to 23.9 million in 2014. It is the lead city in the Shanghai River Delta region, which includes Jiangsu and Zhejiang; the region is home to just over 100 million people, and accounts for 20 per cent of China's economy. Although internationally isolated in 1980, Shanghai's connections are globally extensive and it competes with and complements Hong Kong and the Pearl River Delta region, which had an earlier start. The massive growth of Shanghai and its shift to a privatized economy—itself resulting from changes in national policy—underlines that the "local" nature of cities as places changes over time. Meanwhile Hong Kong has also continued to flourish: its Chinese community has close personal ties with the Pearl River Delta; it has a respected, stable legal system, a sophisticated population, and considerable business strengths, domestic and foreign owned.

Local Inputs into the City Economy

Cities have large populations and therefore markets. Most of the economic activity that occurs in the city is done for and by the people of the city. Houses are built, sold, and bought locally. Parents pay property taxes for teacher salaries, local roads, and police forces. Patients visit nearby doctors and hospitals. Residents want local parks, pools, gyms, nightlife, and other amenities. Imports of goods and services are significant inputs into such activities, but much value added is local. Moreover, local labour wages, rent, and other services add cost margins to most activities that generally exceed the value of imported goods and services that they incorporate. That said, local and global economies of the city are necessarily integrated. The efficiency of the city's infrastructure, the impact of governance on transaction costs, and the amenities and attractiveness of a city are all important elements to enhancing external connections. On the other hand, global forces stamp their influence on land-use planning, infrastructure, and even the politics of the city. Many cities have seen their centre of gravity shift. Firms have located to the suburbs not only for cheap land, but also to ensure reasonable housing for valued employees. Malls, airports, and other foci have had similar influences on the economic functioning and layout of the city.

The economic geography of cities as places can be explained in relation to their changing form (or morphology or internal structure); diversification of economic functions; and spatial income inequalities.

Urban Form

As Allen Scott (2008) notes, the general trend towards urbanization around the globe has been led by the growth and diffusion of large cities and mega-city regions. Between 1950 and 2005, cities with populations over a million increased from 83 (mainly in the rich countries) to 454 (many in developing countries). This growth, and associated interactions with the global economy, has required changes in, and sometimes been constrained by, the internal form of cities. With their increasing size, cities have expanded geographically by the suburbanization of housing and dispersal of economic functions, often absorbing nearby centres in the process. Increasingly large metropolitan areas and regions feature "multiple nuclei" within the city and a poly-nucleated structure of cities within the region. The multiple nuclei city evolved from established centres or new, planned suburban centres, in many instances threatening the viability of the established downtown cores or central business districts (CBDs). Physical features (mountains, lakes, seas, rivers) and political boundaries can influence the availability of accessible land for urban development, sometimes powerfully. Within such constraints, the provision of transportation networks, and related infrastructure investments, profoundly shapes the timing and spatial pattern of urban growth. In this regard, the development of subways (e.g., London since the 1860s), electrically powered tram lines and cars (since the 1890s), and the urban highway network (especially since 1950) to facilitate car, truck, and bus transit of goods and people have been influential (Chapter 13). Such connectivity is vital to facilitate the myriad of intra-urban movements of goods and people that occur for a variety of reasons *and* to connect with the city's global transportation and communication systems.

The transportation networks that define the basic morphology of cities are themselves shaped by political choices and regulations that are in turn affected by the choices made in the past from the inherited built landscape. In terms of elemental components, this inheritance comprises the street (or routeway) pattern, buildings, and their functions. Generally speaking, functions can be changed more easily than buildings, and buildings more easily than routeways. Routeways are especially expensive and contentious to change. For cities that developed prior to the automobile, accommodating its needs (and those of highway transportation in general) has proven difficult; once built, urban highway systems are difficult to further modify. Moreover, the historical and cultural values of inherited built landscapes, including street patterns (rather than simply heritage buildings), can be the basis for resistance to changes that seek new functions, buildings, development, and global links.

In these regards, cities make different policy and regulatory choices. Urban visitors in Europe and North America typically remark on an obvious difference in urban form; the CBDs of North American cities are dominated by high-rise buildings and in Europe they are not, forbidden by regulations (less so in the suburbs). At a regional scale, visitors to the Pacific Northwest will notice the proliferation of urban motorways in Seattle and their relative lack in nearby Vancouver, a difference partially reflected in the timing of development. Urban freeways were constructed in Seattle in the 1950s and early 1960s before political opposition became effective in stopping several further proposals. Vancouver drew on Seattle's experience and the freeways that had been proposed for downtown were effectively stopped in the late 1960s. Although walking, cycling, and various forms of rapid transit offer alternatives to the dominance of the car, the morphology of the city (including for the efficient movement of goods within and between cities) is still shaped by roads. Cities also need to accommodate railways, pipelines, and utility systems, and associated collection and distribution spaces.

Urban Diversification

City growth is deeply cumulative or path dependent, broadly driven by the realization of internal and external economies of scale in all aspects of the economy (Chapter 2). One significant trajectory of growth occurs through the diversification of basic activities into closely related activities through investments in related supplier or value adding activities. Companies do this by differentiating their products and responding to their customers' requests. In this trajectory, localization economies of scale are especially important as evinced through common labour pools, information sharing, and strong multiplier effects through locally linked activities (Case Study 2.3). In association, *institutional thickening* takes the form of specialized infrastructure; common marketing; and training programs, research associations, university research centres, and government departments. Quality multipliers come into play when increasing demand for information and research-intensive products and processes compels a firm to undertake R&D, additional training, or equipment purchases, or to exchange information, patents, personnel, and equipment with other firms. Localization economies build up over time, providing not only cost efficiencies and upgrading but intangible benefits such as reputation, competition, monitoring capacities, and production flexibility. Such economies, especially among small firms, are often referred to as **Marshallian externalities** in honour of Alfred Marshall, the economist who first formally described them in the late nineteenth century. Then, these economies were highlighted with respect to the development of manufacturing concentrations, such as steel and metal working activities in Sheffield, Pittsburgh, and Hamilton or textiles in New York and Montreal. Localization economies have continued to be important whether that be in the development of manufacturing agglomerations or clusters (Chapter 11), high-tech activities (Chapters 3, 11) and service activities (Chapter 12).

Indeed, city growth need not be restricted to any particular sector of the economy and its cumulative fortunes. Cities can diversify into other specialisms. This ability is strongly related to urbanization economies of scale: the benefits all firms enjoy when they locate in larger cities offering less expensive, more reliable and diverse infrastructure services (transportation, sewage, telecommunications, etc.), and—perhaps most important—access to a large and diverse labour pool, from which they can draw specialists and generalists at competitive prices. Urbanization economies also incorporate education and training programs in universities, colleges, schools, and institutes that are not limited to preparing students to work in established industries but encourage more broad-based thinking. More informally, social interactions within large cities can spark new ideas, as can migrants to the city. Urban amenities such as theatres, restaurants, parks, health care, vitality, and diversity of cultures and environments are increasingly important urbanization economies because they attract creative people. These broad-based, inter-industry influences and the formation of new firms are known as "**Jacobs externalities**," in honour of the pioneering urbanist Jane Jacobs.

Conventionally, diversification is considered favourable to urban growth and stability, providing cities with both competiveness and resilience in coping with change. Yet urban economic diversification is an ambiguous concept. For statistical purposes diversification is typically defined by the distribution of employment in an urban area compared to some benchmark distribution (Case Study 8.2). Such a definition, however, hides the internal heterogeneity of individual categories—that may or may not contain a wide range of activities—and says nothing about the extent of local linkages (and associated multiplier effects), or the nature and size distribution of business organizations or markets served. Conceptually, diversification may occur as the growth of several different activities or the replacement of existing specialisms with new ones. Diversification can also occur if an existing specialism declines while other activities become relatively more important by staying the same size. In practice, links between city diversification and growth and stability are not easy to demonstrate, or whether Marshallian or Jacobs effects are dominant.

Spatial Inequalities

Nineteenth-century industrialization created much urban squalor. The "slums" and densely packed tenements in which people lived were typically privately owned, offset by occasional "model" communities built by paternalistic mill owners to provide better living conditions for their workers and families. For various reasons, related to "welfare

Marshallian externalities (or hypothesis) Argues that innovation and growth is stimulated by specialization and intra-industry knowledge transfers, especially within agglomerations.

Jacobs externalities (or hypothesis) Argues that innovation and growth is driven by knowledge spill-overs that cross industries; and in particular that diversified cities are more innovative than specialized cities; named for Jane Jacobs.

Case Study 8.2

LOCATION QUOTIENTS AND A DIVERSIFICATION INDEX

The most sophisticated approach to measuring the economic base of a city and the multiplier effects it generates is through detailed field surveys of the purchasing and sales behaviour of firms and individuals. But field work is too expensive and time-consuming to be undertaken frequently. A simpler, albeit cruder, approach is to calculate location quotients (LQs). LQs are typically derived by comparing shares of local (urban) employment in one sector against shares of national employment in the same sector:

$$LQ = LEi/LEi \ / \ NEi/NEt$$

where *LE* = local employment;

NE = national employment;

i = industry activity; and

t = total economic activity.

The assumption is that LQs > 1 define the specialties and export-oriented (basic) activities of a city or region, while LQs < 1 define local or non-basic activities. In the hypothetical table below the shares of urban and national employment are

state" policies that improved public heath, employment, and housing conditions, the worst slums were widely eradicated across rich market economies (through the supply of public or social housing). Yet problems of spatial income inequality within cities and metropolitan areas within these economies remain, and in recent decades have widened. Thus globalization has meant increased inter-urban competition and levels of immigration, often from economically weaker countries. Wage levels and unions have been pressured with the influx of low-wage labour. Technological change has especially affected routine work in all sectors while in many cities neoliberal policies have undercut public services, as a relatively few super-rich people gain more of the economic pie. Further, as cities have grown and diversified, they have become increasingly more expensive to live in. Cities in developing countries are experiencing similar if not deeper problems, a sort of hybridization of nineteenth– and twenty-first–century issues.

Cities and Innovation

Cities are creative places, as John Friedmann recognized long ago. Yet the nature of creativity and the learning, innovation, and agglomeration processes associated with cities have changed over time (Chapter 3). A broad comparison of the industrial city of the nineteenth century and creative city of the twenty-first century points to some key differences and similarities (Figure 8.1). Innovation and creativity played significant roles in the old industrial city; as we have seen, Huntsman's innovation in crucible steel-making was a significant factor in the growth of Sheffield (Case Study 3.2). By 1900, the

provided for five sectors of the economy from which location quotients can be readily calculated.

	Sector 1 (%)	Sector 2 (%)	Sector 3 (%)	Sector 4 (%)	Sector 5 (%)
Nation	20	20	30	10	20
City	5	80	5	5	5
LQ	0.25	4.0	0.17	0.5	0.25

Once shares of employment by sector within a city and for the nation as a whole are derived, a simple index of specialization or diversification (D) can be calculated. For example, D is given by the ratio of the positive or negative sums of the differences in these shares by 100 per cent. In the example above D is 0.6. In general the higher the value of D, the less diversified the city.

Great caution has to be taken in interpreting the results of such simple statistics. Nevertheless, this approach is useful for exploring the structure of an urban economy in relation to a national benchmark, comparing urban structures over time and/or space, and raising questions about an urban structure's distinctive features. The basic and non-basic distinction has long been used in urban planning. Also, LQs provide summary indicators of patterns of specialization, or of representation and underrepresentation, that are useful for comparisons over time and space.

efforts of individual innovators were supplemented by the emergence of professionalized R&D around specific industries, which gave rise to new learning-based localization economies. Innovation and creativity, however, were focused on the technology required for manufacturing—the principal basic (export) activities of the industrial city. By contrast, in the contemporary creative city, the focus is on services (producer, consumer, and public). The knowledge content of these exports is typically high. In tandem, the workforce has become more white-collar than blue-collar, involving a different range of skills and occupations.

Other differences can be seen in the ties between cities and regional hinterlands. In the industrial city, those ties were typically strong: resource inputs and labour flowed into cities while the cities provided strategic decision-making functions, R&D, secondary manufacturing activities, transportation and distribution services for resource exports, and high-threshold goods for hinterland communities. Cities and hinterlands were strongly connected and governed around a shared-export base centred on resources and manufactured goods. But as traditional industries declined, so did traditional city–hinterland relationships and their institutional underpinnings (corporate head offices, marketing associations, R&D, trade union headquarters, etc.). The creative city is less dependent on resources from regional hinterlands; its connections are global, especially where information flows are concerned. Thus new agglomerations of creative activities in areas ranging from biotechnology to computer graphics and textile design are connected internationally and locally by flows of labour, spin-offs, and subcontracted services. In general, the spatial innovation systems linking agglomerations

creative cities Cities that are seen as centres of innovative thinking; creativity in this context is thought to be the product of factors such as education levels, economic and social diversity, networking opportunities, social tolerance, and various kinds of location amenity.

of **creative cities** are much less functionally connected to regional hinterlands than their industrial city predecessors were. In effect, the Internet and improved travel connectivity have extended localization economies to the global scale. The firms that comprise Vancouver's computer graphics cluster, for example, connect not only with one another but with Los Angeles and other film-producing cities around the world (Case Study 8.3).

The creative city no longer relies on the hinterland for industrial resources—except, perhaps, for ambitious young workers and locally grown food (Chapter 10). The hinterland can offer opportunities for recreation, cottages and second homes, and, for some people, is an alternative to the hectic, congested urban lifestyle (mainly retirees and workers who are able to rely on Internet connections). Moreover, the shift to a creative city may involve sharp changes in attitudes towards resource values, away from industrial uses to their non-industrial benefits, for example in providing scenery, ecological sanctuaries, and various outdoor experiences.

Case Study 8.3
LUX VISUAL EFFECTS: CREATING THE CREATIVE CITY

Visual Effects (Lux VFX) of Vancouver provides fascinating insights into the origins and implementation of creative ventures. Lux VFX is part of Vancouver's emerging computer software design industry, creating visual effects for the film industry, television, and ads in the form of 3D animation, CGI, and computer graphics. Lux VFX's founders, Kevin Little, Heather Paul, Harley Paul, Michael Paul, Joe Ngo, and James Halverson, met while working for Images Engine, also supplying computer graphics for the film industry, and formed their own "spin-off" company in 2005. They started by subcontracting, specifically for Zoic BC, and soon won their own contract to supply 3D animation to a Hollywood-based B-movie (horror) producer, Cinetel. Their first job was to animate the Loch Ness Monster, and Lux VFX is now working on their thirteenth contract for Cinetel. Presently, Lux VFX has a major contract with a Hollywood producer that is producing the Helix series for Sy-Fy, an episodic horror/sci-fi program in which viral experiments go wrong. Lux VFX employs 20 people, 8 permanent and 12 on contract, the latter a reflection of the business' volatility.

Although the founders considered an office in the suburbs where rents are cheaper, Lux VFX located in a tight cluster of post-production firms in inner Vancouver (near False Creek and the Granville Island market). This location facilitates co-operation with other firms, access to key external economies such as sound and editing houses and firms providing colouring, and as the airport is only 20 minutes away, easy access for visiting Hollywood executives. Vancouver has other advantages. Tax incentives (and a usually lower Canadian dollar) first attracted film producers, while Montreal and Toronto, with their own tax-supported film industries clusters, are international leaders in software design that provide important expertise to Vancouver firms. In 2013, when Ontario started offering a competing tax incentive, many of the LA-based productions quickly left Vancouver for the perceived savings in Ontario. It quickly became apparent that, although the tax incentive was better, Ontario lacked the facilities and skilled labour required, while

In the industrial city, localization economies developed through interdependent firm–supplier relations and firm rivalries. Those economies were supported by industry-specific innovations and R&D that were region-based and articulated within national innovation systems shaped by national and regional policies towards education and R&D, and industrial associations.

International transfers of knowledge were important in nineteenth-century industrialization, but they were not routine. In the creative city, agglomerations of innovative export-oriented services benefit from both immobile localization economies and mobile ones that are not constrained by national borders. The concept of the spatial innovation system reflects the growing importance of international transfers of innovation. Vancouver's emerging biotechnology industry (Figure 3.3) illustrates how agglomerations of information-intensive activities develop localization economies that are both mobile and immobile.

the flight time and change in time zone from LA was more of a factor than they had considered. Vancouver's amenities are highly regarded and attractive to talented people, while the city was also a producer of classical animated graphics in the pre-computer age.

And how did Lux VFX's founders develop their talent? In James Halverson's case, after high school, he attended a local community college (taking general arts courses). But as a "computer nerd" (in his words), in 2000 James enrolled in the newly created Vancouver Film School that offered a specialty in 3D animation, another important localization economy, without finishing his college program. He then gained local work experience in video games, before joining Images Engine. Before forming Lux VFX, James had to decline a job offer from a New York firm because without a degree he did not meet US immigration requirements. Talent is not easily formally measured.

Many thanks to James Halverson for providing and updating the information for this Case Study.

The origins of the new, globally connected creative agglomerations of services within cities can be related to two already mentioned general hypotheses. The Marshallian hypothesis suggests that the sources of innovation and creativity are essentially spin-offs from current or former areas of specialization. On the other hand, the Jacobs externalities or hypothesis suggests that innovation is usually sparked by initiatives outside those areas or by connections between industries that are only loosely related and generated by the interactions among people that are facilitated by cities. Whatever their source, according to Richard Florida (2012) the creative class is increasingly dominating the economic base of cities, especially in rich market economies. Although not easy to define, Florida distinguishes between a "super-creative core" (scientists, engineers, professors, actors, architects, researchers, planners, etc.) and "creative professionals" (lawyers, doctors, financial experts, managers, etc.). In general, the creative class thesis emphasizes the role of R&D, innovation, design, and problem-solving abilities in promoting urban growth. Related ideas refer to the enhancement of "human capital" (Chapter 7) as the basis of urban growth, while the product cycle model (Chapter 11) similarly predicts that innovative skill-intensive, problem-solving, product differentiating functions that dominate the early stages of product evolution are primarily located in metropolitan areas in rich countries.

Thus cities attract desirable, qualified, and talented employees by providing a wide range of amenities and urbanization economies, not least in education, while their diverse markets are important stimuli for research, design, and innovation. As meeting places for people and businesses, cities encourage co-operative brainstorming and incidental learning, as well as intense rivalry, all of which spur the development of new products and processes. Moreover, creative class jobs are typically high income and "clean."

The nature of the new creative industries is highly varied. Producer-based business services are the legal, accounting, engineering, exploration, and marketing services that developed around traditional industries or served established corporate head-office complexes. Some have continued to develop as autonomous export activities after the loss of the city's traditional economic base. High-tech activities such as biotechnology, telecommunications, CAD-CAM, and remote sensing are creative activities that may or may not be linked to a city's early areas of specialization. Finally, there are the many, varied "cultural industries" (computer graphics, electronic games, music production, micro-breweries), that help define the creative or—as Scott and Power (2004) call it—resurgent city. The nature and extent of creative, knowledge-based, or innovative activities varies among cities. In a North American context, Florida has sought to understand this variation in terms of the 3Ts: technology, talent, and tolerance. Thus the most creative cities have attracted the most talented people, have the strongest ability to innovate, and have the highest levels of tolerance for different lifestyles and new ideas.

Systems of Cities

Cities are located within systems of cities, towns, villages, and hamlets that exist at regional, national, and global scales; at the largest scale, they are part of a system of systems. By the 1960s, geographers talked about the functional merging of metropolitan areas and cities across broad regions within countries. Jean Gottman, for example, referred to the cities of the northeastern seaboard of the US, from Boston to Baltimore, as Megalopolis, while Peter Hall observed similar "world city" regions, centred on London, Randstadt, Holland, Paris, and Tokyo. In Allen Scott's (2008) terminology, there is now a "global mosaic of city-regions," the very largest comprising "urban super clusters" of specialist activities that service global markets. In practice, the definition of urbanized areas varies, for example, by administrative boundaries, census metropolitan areas, and population centres that may embrace mega-city regions.

Statistically, the size distribution of cities at any particular time and national or regional scale forms a "population pyramid" comprising many small towns and with few big cities. The shape of this pyramid varies, however. A relatively even distribution implies that cities change regularly in size and rank. That distribution is said to be governed by the **rank size rule**, and it seems closely applicable in some cases and time periods. On the other hand, **primate cities**, such as London and Paris, have populations that are considerably larger than the next largest cities and overshadow the entire system. Regardless of size distribution, the idea of **urban systems** recognizes that cities are linked by transportation (air, rail, sea, road) and communications (mail, telephone, Internet, television, radio) systems through which materials, information, money, and people flow. The varying sizes of cities reflect their functions and fields of specialization; in turn, the relative fortunes of cities are largely dependent on their relationships within the wider system. Indeed, national economic performance is closely tied to how well national urban systems are performing, globally as well as domestically.

Hierarchical and Non-hierarchical Relations

Urban systems are shaped by both *hierarchical* and *non-hierarchical* (or *heterarchical*) imperatives with distinctive global–local dynamics at several scales. As Christaller's central place theory argues (Chapters 1, 12), specifically with respect to the marketing of consumer services, settlement patterns are organized along hierarchical lines. Thus the larger the city, the greater the number and variety of the higher-order functions it makes available: such cities offer services that are needed infrequently, tend to be expensive, and require specialized or highly skilled people to perform them. These cities are home to specialized medical and educational services, professional sports teams, philharmonic orchestras, parliaments, high-end retail services, comprehensive museums, and specialized art galleries. They are also the central places for warehousing and logistics services. These functions require large markets to justify their existence and scale of operations, and together they create a higher-order city. Lower-order cities provide products and services that are frequently required, less expensive, and less specialized: groceries, gas, family restaurants, clothes, and so on.

But cities also include many functions whose location is not primarily or even significantly influenced by the need to provide access for consumers. The locations of resource-based activities, manufacturing, corporate headquarters, and R&D are determined by other factors: the location of the particular resources or labour skills they need to exploit, or the corporate interests they need to serve. City systems are shaped not only by hierarchical forces but by the areas of specialization of the cities within them, each of which is the product of distinct location dynamics that are heterarchical,

rank size rule A rule asserting that city sizes in a given region reveal a direct linear relationship (more precisely, a linear relationship when the logarithms of the rank of city sizes are used). In practice, there are many exceptions to this rule.

primate cities Cities that are several times larger than the next largest, and that exercise extraordinary political and economic power over national spaces; London, Paris, and Tokyo are considered to be primate cities.

urban system A set of cities (towns, etc.) that are connected to one another in such a way that changes in one will affect others; such systems can exist at various scales from local to global.

or non-hierarchical, in nature. Cities develop export specializations in specific industries in which they enjoy comparative advantages. These specializations extend a city's linkages beyond its regional or national context. In North America, leaders include New York and Toronto (financial services, communications, theatre), San Francisco (high tech and finance), Los Angeles (entertainment and aerospace), Houston and Calgary (petrochemicals), Detroit (automobiles), and Seattle (software, airplanes, wood products).

Many non-market activities that are key institutions in the evolution of cities are not arranged hierarchically. The national governments of Australia, Brazil, and Canada, for instance, are located outside the great metropolitan centres. There are many more examples of government bureaucracies that are not located in the main cities of their jurisdiction. Universities are widespread, but their distribution does not conform to strict hierarchical principles with respect to either size or excellence.

The Canadian Urban System

Urban systems have typically evolved within specific national and regional contexts underpinned by shared laws, cultures, identities, and policies. The cumulative power of agglomeration means that the leadership of urban systems is inherently stable. New York, Paris, London, and Tokyo have resided at the apex of their national urban systems for centuries, and in most cases, their economic importance has continued to grow. The next rank of cities is also relatively stable, but there is more volatility among the rankings of smaller cities within the system, particularly as regions and their industries grow or decline with changing economic, demographic, and institutional conditions.

The Canadian example is insightful because its dynamism reveals both distinctive and more general processes. Prior to the creation of the Canadian federation in 1867, urban settlement patterns in Canada largely evolved to facilitate the export of staples (resources) to British and French colonial powers. In 1871, Montreal, Quebec City, and Toronto were the three largest urban places, and other urban centres of note were in central and eastern Canada. The onset of large-scale industrialization then created a national "heartland" around the Toronto and Montreal axis where Canada's most significant manufacturing specialties developed, with the rest of country forming a hinterland in which cities grew rapidly (or not) according to the riches of their regional hinterlands. The previously mentioned National Policy of the 1870s, that featured a tariff on manufactured goods, the building of transcontinental railroads, and the agricultural settlement of the Prairies, reinforced this heartland-hinterland pattern. Thus the heartland's manufacturing specialties were protected from foreign competition and their domestic market potentials extended, while the hinterland's resource development was facilitated by immigration and improved accessibility.

With economic growth during the 1950s and 1960s, the heartland–hinterland urban settlement pattern was reinforced and modified, notably by massive shifts in staple production and exports from western Canada, with a concomitant rise in the rankings of western cities. Thus in 1971 Montreal and Toronto were the two top-ranked cities (Quebec City had dropped to seventh rank) and there were four western cities (Vancouver, Winnipeg, Calgary, and Edmonton) in the top ten, and no cities from east of Quebec (Table 8.1). Montreal and Toronto were metropoles of national and global stature with highly diversified control, service, and manufacturing functions, while nearby heartland cities had distinctive specialisms, such as steel in Hamilton and cars in Windsor. In the west, Vancouver's manufacturing and control functions were focused on staples (especially forest products and mining), while Calgary (control, finance) and Edmonton (equipment, transportation) specialized around the oil sector.

Since 1971, with the rapid rise of the service sector and declines in manufacturing, Canada's urban system has changed further. In terms of top-ten rankings, Toronto has overtaken Montreal as Canada's pre-eminent city, while the three leading western Canadian metropoles have grown rapidly. Montreal's historic primacy was based on its financial and logistic control of Canada's staple economy. Toronto rivalled Montreal in size because of its political and economic role in Canada's largest provincial economy. Toronto's manufacturing firms served all of Canada. In the 1950s and 1960s,

TABLE 8.1 Canada's Urban Population Centres by Rank 1971–2011 (Top 25)

	1971	1981	1991	2001	2011
1.	Montreal	Toronto	Toronto	Toronto	Toronto
2.	Toronto	Montreal	Montreal	Montreal	Montreal
3.	Vancouver	Vancouver	Vancouver	Vancouver	Vancouver
4.	Ottawa	Ottawa	Ottawa	Ottawa	Calgary
5	Winnipeg	Edmonton	Edmonton	Calgary	Edmonton
6.	Hamilton	Calgary	Calgary	Edmonton	Ottawa
7.	Quebec City	Winnipeg	Winnipeg	Quebec City	Quebec City
8.	Edmonton	Quebec City	Quebec City	Winnipeg	Winnipeg
9.	Calgary	Hamilton	Hamilton	Hamilton	Hamilton
10.	London	London	London	London	Kitchener
11.	St Catharines	St Catharines	St Catharines	Kitchener	London
12.	Halifax	Kitchener	Kitchener	St Catharines	Victoria
13.	Windsor	Halifax	Halifax	Halifax	St Catharines
14.	Kitchener	Windsor	Victoria	Victoria	Halifax
15.	Victoria	Victoria	Windsor	Windsor	Oshawa
16.	Oshawa	Oshawa	Oshawa	Oshawa	Windsor
17.	Sudbury	Saskatoon	Saskatoon	Saskatoon	Saskatoon
18.	Saskatoon	Regina	Regina	Regina	Regina
19.	Chicoutimi	Chicoutimi	St John's	St John's	Barrie
20.	Regina	Sudbury	Chicoutimi	Sudbury	St John's
21.	St John's	St John's	Sudbury	Chicoutimi	Abbotsford
22.	Cape Breton	Trois-Rivières	Sherbrooke	Sherbrooke	Kelowna
23.	Thunder Bay	Sherbrooke	Kingston	Barrie	Sherbrooke
24.	Kingston	Cape Breton	Trois-Rivières	Kelowna	Trois-Rivières
25.	Saint John	Thunder Bay	Saint John	Abbotsford	Guelph

Source: From Simmons, J. and Bourne, L.S. 2003. *The Canadian Urban System, 1971–2001: Responses to a Changing World*. Toronto: Centre for Urban and Community Studies, University of Toronto Research Paper 200. See also ibid. 2013. *The Canadian Urban System in 2011: Looking Back and Projecting Forward*. Idem. Research Paper 228. Reprinted with permisson of the authors and the Cites Centre at the University of Toronto.

manufacturing and services replaced staples as the foundations of the country's economy, and changes in national policies opened up borders to increased immigration and trade. Toronto and Ontario in particular benefited from these changes. The Auto Pact (1966) allowed car and auto parts to be traded duty-free between the US and Canada and expanded economies of scale that were once limited to the Canadian market. Immigrants from Asia chose to settle in Toronto and its environs. The FLQ (Front de Libération du Québec) crisis and subsequent movement for Quebec independence also weakened Montreal by pushing many of its higher-order financial services and professionals to Toronto. Montreal has renewed itself in recent decades, but the power of Toronto's agglomeration continues to diversify with the evolution of regional, national, and global economic conditions. Indeed, in 2011 what Bourne and Simmons (2013) label as "megaregions," the Toronto region with a population of 8,342,399 is about twice the size of the Montreal region (4,197,800).

Further, since 1971 the heartland-hinterland distinction has become less clear. De-industrialization and free-trade policies (notably NAFTA) have led to increased competition from the US and Mexico and caused the decline of several heartland cities, such as Hamilton and Windsor. Meanwhile high-tech activities have not only developed in the heartland—especially in Toronto, Montreal, Kitchener, and Ottawa—they have also been part of the growth of western Canadian cities. Increased links with Asia Pacific countries have also favoured the latter. Vancouver has lost much of its forest-product manufacturing base, but it has diversified into a wide range of activities. Thus Canadian industry development is not only Toronto- and Montreal-centred, but is evidenced in the expansion and integration of the Calgary and Edmonton metropoles (2011 population of 2.6 million) and in a Georgia Straight urban region linking Vancouver with the Victoria and Abbotsford metropolitan areas and beyond (2011 population of 3.1 million). Like their heartland counterparts, employment is service dominated in these regions, albeit with distinct interests in creative specialisms. Moreover, income polarisation within these megaregions is a problem, greater than contrasts between cities and the rural periphery.

In 2011, the top four megaregions alone accounted for 54.5 per cent of Canada's population (of 33.5 million). These cities also take in the lion's share of immigration to Canada, the major source of population growth. Cumulative, path dependent growth around the elaboration of agglomeration economies would indicate further polarization in megaregions in the near future. Urban trends elsewhere have similar contours.

Continental and World Cities

As the Canadian urban system has evolved, it has become increasingly integrated into continental and global urban and economic systems. In this context, the continental system is especially vital to Canada because most of its trade is with the US. The loosening of trade restrictions, the enhancement of road and rail connections, and the growth of the western cities has resulted in increasing amounts of trade flowing directly north–south. The hierarchical nature of the Canadian urban system has therefore been counterbalanced but not undermined. This pattern is reflected in airline connections.

Direct intercity connections with US cities are numerous and passenger volumes large, but these flows would be much greater if Canada were part of the US. Flows of people and trade are complemented by flows of financing and FDI. Ontario's car industry has been dominated by Detroit's decision-making for a century; in the film industry, Vancouver and Toronto are primarily lower-cost production centres for Hollywood; and substantial financing for the Alberta oil patch flows from Houston. On the other hand, Canadian banks, real estate developers, and technology firms have all made in-roads throughout the US. In absolute terms, Canadians own as much of the US as Americans own of Canada, although in relative terms American ownership of the Canadian economy is far more significant.

Within the global urban system, Toronto is the only Canadian city ranked at the alpha level, "world city," while Montreal achieves beta+ status and Vancouver at beta shows some evidence of moving up in the world (Table 8.2). Important keys to understanding the global influence of cities and variations in their world city rankings are the geographical scope and mandate of "control functions" in corporate headquarters, R&D operations, supply of specialized business services, and the provision of a wide range of services and amenities that attract consumers from around the world. World cities are very large urban areas that undertake and control a large proportion of the world's economic activity. Their influence is based on their well-developed connections with regional or national hinterlands and frequent and voluminous interactions with other cities.

TABLE 8.2   World Cities Ranking According to GaWC 2012

Alpha World Cities

Alpha ++: London, New York

Alpha +: Paris, Tokyo, Hong Kong, Singapore, Beijing, Shanghai

Alpha: Chicago, Mumbai, Milan, Moscow, Sao Paulo, Frankfurt, Toronto, Los Angeles, Madrid, Mexico City, Amsterdam, Kuala Lumpur, Brussels, Taipei

Alpha −: Seoul, San Francisco, Zürich, Buenos Aires, Johannesburg, Vienna, Istanbul, Jakarta, Warsaw, Washington, Bangkok, Miami, Melbourne, Barcelona, New Delhi, Boston, Dublin, München, Stockholm, Prague, Atlanta

Beta World Cities

B +: Berlin, Budapest, Copenhagen, Dallas, Santiago, Düsseldorf, Hamburg, Manila, Luxembourg, Montreal, Lisbon, Rome, Tel Aviv, Bangalore, Athens, Philadelphia, Houston, Beirut, Kiev, Guangzhou

B: Caracas, Geneva, Brisbane, Helsinki, Doha, Casablanca, Vancouver, Rio de Janeiro, Auckland, Stuttgart, Ho Chi Minh City, Manchester, Montevideo, Oslo, Brisbane, Bogota, Karachi, Chennai

B −: Lyon, Seattle, Birmingham, Bratislava, Detroit, Seattle, Rotterdam, Calgary, Edinburgh, Antwerp, Abu Dhabi, Perth, Calcutta, Denver, San Diego, Hanoi, Sofia, Riga, Amman, Manama, Nicosia, Quito, Rotterdam, Belgrade, Monterrey, Almaty, Shenzhen, Kuwait City, Hyderabad, Port Louis, Minneapolis, San Jose, Tunis, Lagos, Guatemala City, Panama City

Gamma World Cities (First example cited)

G +: Zagreb, etc.

G: Glasgow, etc.

G −: Nantes, etc.

Source: Reprinted from *Cities* 16, Beaverstock, J.V., P.J. Taylor, and R.G. Smith. 1999. "A roster of world cities," pp. 445–58. Copyright 1999, with permission from Elsevier.

These cities epitomize the accumulated benefits of agglomeration because many of their functions are interdependent. The GaWC is rooted in the economic geography of business functions. Thus it is based on the size and extent of the global reach of the command and control functions associated with the main and subsidiary headquarters of businesses, especially MNCS, and the localization of "higher order" business or producer services (see Chapter 12), especially financial (stock exchange, banking, insurance), advertising, law, and accountancy services. R&D, design, engineering, consulting, and business computing services would similarly fit into such an approach.

The "alpha" world cities (Table 8.2) have the greatest global reach and connect the global economy because their command and control functions are the highest-order functions in the market system. They are home to the world's largest stock markets and the clusters of business services dedicated to them, as well as the headquarters of many MNC giants. Perhaps the most important requirement for such cities is efficient and transparent governance that is capable of evolving with market demands. New York's continued pre-eminence among world financial markets has depended on the capacity of federal, state, city, and stock exchange regulatory bodies to adapt to new financial products with regulations that ensure fair access while allowing for market freedom. The resurgence of London as a global financial rival to New York occurred only after its "big bang" deregulation in 1986, when (among other changes) oversight of borrowing and lending practices and the privileges allowed to incumbent firms were significantly reduced. Hong Kong's future as the financial capital of China will depend on its ability to maintain the balance between fair market regulation and freedom. The widespread use of English as the language of business also facilitates the global power of New York and London (and is a relative advantage for Hong Kong in an Asian context).

The specialist beta and gamma cities remain key players in the global urban system, particularly in the coordination of global value chains. Much of the world's film production and distribution is controlled by the Hollywood oligopoly based in Los Angeles. Houston oil companies are among the largest global corporations because they globally integrate the exploration, pumping, and refining operations of MNCs based in beta and gamma cities. Headquarter functions have immense multiplier impacts, require a broad range of business services to support them, and draw in regional offices of the giant accounting, banking, legal, and advertising MNCs. Maintaining its cultural, environmental, infrastructure, and governance capacities is crucial for a city that hopes to retain its MNC head offices. Yet cities whose MNCs give them global influence need not be megacities. Toyota City (Japan), Portland (Oregon), Benton (Arkansas), and Cincinnati (Ohio) are not world cities, but their urban structure, environment, and governance contribute to the success and global influence of Toyota, Nike, Walmart, and Proctor and Gamble.

However, city rankings based primarily on political, cultural, and environmental characteristics may differ from the economically-driven rankings in Table 8.2. The consulting firm, Mercer, for example, ranks cities according to "quality of living" based on safety, education, health, culture, environment, and recreation considerations. In 2014, Mercer's top five were Vienna, Zürich, Auckland, München, and Vancouver. These rankings can influence the locational preferences of businesses seeking to attract key employees.

Intra-urban Land-Use Patterns

The predominant driver of a city's morphology is economic efficiency. The attractiveness of centralized access largely determines what activities are carried out where, in what type of buildings, and with what type of infrastructure. That said, cities are diverse, dynamic places that serve the myriad functions and needs of organizations and people which change over time. Thus, while the economic function of cities is served by enabling market forces to function, planning is used to provide the public goods, amenities, and welfare services not provided by the market.

Land-use patterns define how cities spatially organize themselves. Businesses compete for space and accessibility according to market criteria as they serve local or more distant customers and link with local or more distant suppliers. People compete for housing that meets work, personal, and family requirements and desires that embrace non-market functions and land uses related to recreation, religion, or green space. All land uses need to be connected by roads, paths, and bikeways that typically consume a large amount of land. Further, decisions about land-use patterns in the past cannot be easily adjusted to meet new social, economic, and technological conditions. Land can be left vacant and financially backed by speculators even in the central business districts (CBDs) of the most dynamic cities, while many institutional uses—universities, churches (Case Study 8.1), government buildings, and sometimes individual houses—need not be responsive to market forces. Urban planning, along with various levels of metropolitan and municipal government, seeks to balance economic and social goals for a highly diverse range of land uses (with associated interactions). In recent decades, these social goals have increasingly incorporated environmental considerations.

As a theoretical starting point and the predominant force in urban morphology, it is useful to ask how perfectly competitive markets shape urban land-use patterns. Then the influences of technological change—especially in relation to transportation systems, and planning and politics, including with respect to social and environmental imperatives—can be considered (Figure 8.2).

The Bid-Rent Gradient

Bid-rent theory explains how land is allocated to various uses through competitive bidding where land values are driven primarily by considerations of accessibility around the central business district (CBD). The starting model is based on neoclassical assumptions: perfectly informed and rational decision-makers operating on a homogeneous (isotropic) plane compete for land, and the only variable affecting accessibility is distance. The model's assumption that the CBD has the highest land value—because as the transportation and

bid-rent theory Explains how the allocation of land to various uses is determined by competitive bidding.

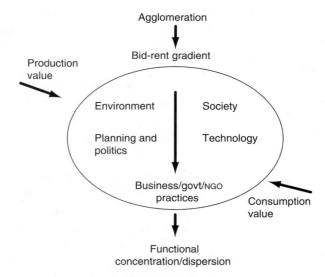

FIGURE 8.2   **The Land Market and Other Influences on the Location of Intra-urban Economic Activity**

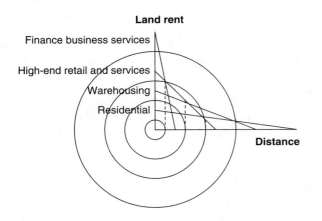

FIGURE 8.3   The Bid-Rent Allocation of Land Use

communications hub it is the most accessible—is widely confirmed in practice.

Assuming that all activities want the most central location because it offers the greatest convenience and the lowest travel costs, they distribute themselves over the idealized plane according to the amount they are able to bid (Figure 8.3). Land uses include financial and business services, as well as retail, and other commercial services, including warehousing, industrial, government, institutional, and residential uses. A competitive bidding process ensures that each parcel of land earns the highest possible rent. The bid-rent curve shows the range of prices that a particular use is willing to pay at different distances from the CBD as determined by the trade-off between the income that could be generated at the site and what it would cost to rent. At the highest point of the financial bid-rent curve, only those with very high incomes can afford to bid. With increasing distance from the CBD, declining accessibility reduces the potential for revenue generation. Since the trade-offs between accessibility, land cost, and revenue vary, depending on the use in question, as distance increases from the CBD, the bid-rent curves intersect sequentially, showing the points where cost eclipses the revenue potential of one land use, allowing another use to take over.

Core Accessibility

Accessibility is the key function of the core, provided by the twin forces of convergence and proximity. The convergence of railways, streetcars, subways, roads, and telecommunications in the downtown area makes it accessible to the greatest possible number of employees and customers. Proximity is demanded because people are reluctant to venture more than a few hundred metres from where they emerge from the subway or park their cars. Counter-intuitively, in the Internet era, the importance of access is increasing because of the need for face-to-face meetings for negotiations and collaboration on a frequent and time-efficient basis. In addition, the high accessibility of the core reduces production and transaction costs, permitting both internal and external economies of scale. For example, banks and insurance companies achieve internal economies by concentrating their employees in purpose-built buildings where transactions with customers and buyers can be conducted most efficiently.

Firms locating in city cores also benefit from urbanization and localization economies. The transportation infrastructure that provides superior accessibility is an essential urbanization economy, allowing downtown firms to employ a diverse labour force drawn from all over the city. Other infrastructure (sewage, garbage, electricity, telecommunications, etc.) and ancillary services (restaurants, designer shops, and theatres favoured by high earners in the financial sector) also locate in the core. These urbanization economies build up over time, forming a strong base in fixed overhead and externalized capital that firms exploit. Localization allows firms in the same industry but at different stages of the value chain to reduce transportation costs, increase depth of specialization, and ease interaction. When competing firms locate together, suppliers increase the scale and scope of production, reducing costs for all concerned. Thus stock markets and merchant banks draw in brokerage, legal, accounting, and insurance firms.

In many North American cities, the core evolved as the nexus of transportation beginning in the latter part of the nineteenth century; industrial, commercial, and government activities depended initially on rivers or railways for intercity transport. Land-use activities arranged themselves in concentric rings around the CBD, with some distinct enclaves created by functional interdependence: for example, the more affluent often remained close to the core, while lower-income and minority groups tended to occupy the cheaper interstices between industrial and more favoured residential lands. Manufacturing occupied land near the centre, not only because of its transportation needs but also because of its relatively high value-added component. The concentration of industrial, commercial, and government activities in the core was eventually disrupted by several factors, including

- the development of distance-defeating transportation and communication technologies, led by intra- and interurban highway networks and, more recently, the Internet;
- the disaggregation of the various stages of the product cycle (which had been functionally and spatially united in the industrial city);
- building technologies and highway and mortgage policies that encouraged suburbanization; and
- population growth that could not be accommodated inside the core. The development of residential suburbs stimulated the growth of new centres that were more accessible to suburban dwellers than the old cores.

The ability of an economic activity to compete for land on the bid-rent gradient depends on the value it generates. At the same time, the degree to which an economic activity needs the CBD depends on how that activity creates value. For example, stock market transactions, investment banking, and corporate law all require the external economies of the CBD (and the value of the land they occupy is amplified by the high values they generate). Even within a CBD, there are large differences between the financial, publishing, government, and entertainment districts. Cities with strong CBDs are generally anchored by stock markets and financial services (Toronto, New York, San Francisco, London). In cities such as Atlanta, Los Angeles, and Detroit, automobile- and telecommunications-based flexibility allows for the dispersal of economic functions away from the CBD. In Detroit, for example, GM is the only one of the big three US automakers to have its headquarters downtown (in the optimistically named Renaissance Center).

In fact, many cities have now developed multiple nuclei where CBD functions are performed. These nuclei, generated by the accessibility offered by arterial roads or highways, contain mall and business centre developments; corporate, distribution, or industrial complexes; and universities or hospitals. They create secondary or tertiary peaks in population density that give rise to bid-rent gradients in keeping with their own ability to generate demand for access and external economies. Corporate and commercial complexes are most likely to generate such demand because they produce higher value-added services and attract ancillary businesses. The spatial disaggregation of the financing, design, manufacture, and distribution of goods is reflected in the land rents supported by each activity. The value added of distribution and manufacturing activities has decreased, however, and today such nuclei are likely to attract only those companies interested in the local or urbanization economies offered by an industrial park.

The value of residential land uses reflects consumers' willingness and ability to pay for the various combinations of access and amenity that locations within the bid-rent gradients offer. Because of the high land values associated with locations near cores and nuclei, residents typically must accept increasing density in return for convenient access to work, shopping, and urban entertainment. Residents of low-density areas usually face longer distances to work, shopping, and recreation. Between these two extremes there are usually several favoured and less-favoured areas where various groups cluster for reasons related to factors such as income, environmental amenities, life cycle stage, and ethnicity.

Significance of Transportation

The importance of transportation in shaping city development needs emphasizing. When streetcars were first introduced to the industrial city, transit lines radiating from the centre made it possible for homes to be built farther away, but the fact that industry and commerce remained downtown reinforced convergence on the CBD. Nevertheless, the streetcar lines created sectors of increased accessibility that simultaneously elevated land values. Cars, trucks, and road construction wrought more radical transformations, freeing industry and labour from the core. New nuclei developed as residential, commerce, warehousing, and manufacturing centres.

The 400-series highways in central Canada and the interstate highways in the US attracted warehousing and manufacturing to the peripheries of cities, especially those ringed by perimeter highways. On the 401 north of Toronto, the warehouses begin in Halton Hills and continue past Oshawa, broken only where they are out-competed by malls and residences. Airports also brought more development to city peripheries, attracting supporting services, intermodal services, hotels, and various industries requiring rapid transport connections (Chapter 11). In Chicago, even the railway hub that stimulated the city's growth has relocated from the downtown core to the suburbs, where it operates with truck and airway services as part of a multimodal centre.

Shifting transport-based services out of the core has allowed redevelopment for commerce, residential, and recreational purposes. On the other hand, advances in communications technologies have reinforced the tendency of financial and high-end business services to cluster in the core while allowing more routine services or back-office functions to be moved to lower-priced suburban land or outsourced to other regions. However, the location parameters of infrastructure such as water supply and sewage have changed little in many decades.

Society

Demographic and social change is vital to understanding the evolving land-use patterns of urban economies. Indeed, the clustering (or segregation) by social class (income) and ethnicity has long been recognized as modifying bid-rent gradients and having economic ramifications. The co-location of wealthy households drives up property values in three ways: (1) through residents' willingness to invest in large-scale, expensive buildings (and/or expansive estates), (2) the creation of externalities that carry a certain social cachet, and (3) the diverse higher-end services that wealthy neighbourhoods attract. The value of land used for residential purposes is often greater than that of land used for producing goods or services, especially in high-density CBDs. Proposals for the construction of residential high-rises can usually out-compete proposals for the construction of office buildings or sports arenas, especially when the local government takes into account the higher tax rates it can charge residential property-owners. Higher-income groups are also typically influential and can exert the political power that will protect their enclaves from infrastructure onslaughts while enjoying superior service provision.

By contrast, ethnic enclaves have typically occupied marginal land—the only land that marginalized groups could afford—with inferior infrastructure and service provision. Some areas have seen successive waves of immigration move in, become integrated into the society, and move on, making room for a new group to move in. Land value in these areas is not necessarily low, however, since density and various business activities generate high values.

In recent decades, the numbers of single-person households in market economies, including Canada, have increased dramatically, while average household size has decreased as a result of lower fertility rates, divorce, same-sex partnerships, and the extension of the unattached or childless period of people's lives. These shifts represent both challenges and opportunities for residential demand and service delivery. For example, the segment of the population over the age of 65 spans more than 20 years of widely different activity levels and demands. A small proportion of the growing elderly population requires nursing homes or other assisted-living arrangements. Many more elderly people continue to live in their family homes, but eventually downsize to more manageable quarters or move to retirement communities. Demand for smaller homes and retirement communities is relatively new, but will grow as the relatively healthy and affluent baby-boomers retire. But the need of the elderly for access to services and transportation, and to be part of a community is important. Despite high costs, and the implied reductions in living space, downtown locations have become attractive to young people and families to provide access to jobs, amenities, and transportation.

In general, growing cities are in a state of flux, constantly adjusting to the residential, job, recreational, and transportation needs of growing populations characterized by changing democratic structures, incomes, and values. Declining cities, in contrast, can struggle to even maintain basic services, and declining tax bases imply decreased ability to rejuvenate. In both cases, the homeless, unemployed, and socially marginalized seek out niches that are often close to downtowns and offer low rental living with some accessibility.

Urban Planning

As outlined in Chapter 6 (Figure 6.2), urban planning is the process whereby local governments, led by planning departments and with help from other institutions, manage development and seek to reconcile market and social or public objectives. That is, urban planning is about both ensuring cities are efficient—in terms of how land markets and movement patterns work—and livable—in terms of meeting social values and the needs of citizens. Planning began because large-scale industrialization of cities in the nineteenth century, as in many developing countries now, occurred with little regard for how most people lived. Squalid, unhealthy living conditions were the norm, located for business convenience, usually immediately adjacent to factories. Planning emerged to organize cities to meet social as well as economic goals and developed out of varying experiments in the public and private sectors to improve working and living conditions.

"Community plans" are a typical planning approach. They outline how a neighbourhood, subdivision, municipality, or region should be designed (or modified) and serviced. They typically include zoning for different land uses; layout of roads, sewers, and other infrastructure; hospital, school, and recreational facilities; and provisions for commercial and other services. Planning requirements are detailed, restrictive, and supported by policy, tax and subsidy arrangements, and regulations, as well as specific rights and obligations that ultimately determine how any space can be used. Most plans are projected to cover a period of 20 years, subject to periodic review.

Urban planning is based on technical or "rational" analysis including population projections, multiplier effects, road loads, environmental impacts, and so on. Simultaneously, the process is inherently political and judgmental, not least because community plans require the approval of elected representatives and are often controversial. In recent decades planning processes have begun to incorporate more citizen participation and pay more attention to social, cultural, and environmental concerns. It is now legally required in most jurisdictions that governments, businesses, and NGOs inform local residents of plans, take steps to gather their opinions, and act on their concerns. Development proposals ideally should conform to existing plans, but modifications or variances are often requested in practice. These modifications may be minor, involving small adjustments to the building code or more important, for example, imposing higher densities, more restricted sitelines, or more congestion. Urban rejuvenation schemes, which in many western cities have involved the conversion of industrial uses to other uses, involve new community plans for that area. Thus major privately owned development schemes, whether featuring new housing, retailing, or other commercial use, require public infrastructure investments and planning approval. Moreover, financial resources are typically scarce so prioritizing objectives is an important feature of urban planning processes.

Urban planning both reinforces and creates the processes driving bid-rent gradients. Thus in general terms, urban plans emphasize high-density use in CBDs and low-density use in suburbs, conforming to bid-rent land value predictions, while road network extensions, the building of rapid transit systems and nodes, suburbanization plans and service centres modify existing and generate new bid-rent gradients. Planning also seeks to ensure social goals are met by zoning public spaces and parks that are protected from market forces, and by limiting negative externalities among land uses. Thus zoning schemes typically segregate large-scale industrial and wholesaling uses from residential uses to reduce hazards for residents and allow industry to function normally.

Positive externalities may be promoted by encouraging the clustering of related activities, for example, in retail centres or residential areas with common lot and building requirements that respectively allow businesses and residents with shared interests to collaborate. But zoning can also pose questions about what is socially best. For example, in CBDs should high-density residential schemes mix high income and social housing? In dynamic urban environments, residential expansion may invade spaces containing polluting industry. Should the newer residential function be prioritized, closing down the industry, and with a loss in jobs and income?

Cars as Planning Imperatives

In general, urban planning is vital to the livability of cities by ensuring the provision of public goods such as green space. Urban parks, for example, are vital for providing citizens and visitors with aesthetic, recreational, and health needs while providing important ecological values. Central Park in New York, Regents Park in London, Stanley Park in Vancouver, or Cologne's Grüngürtel ("green belt"), along with countless other parks, only survive because development is prohibited. If left to market forces these areas and their collective social values would be lost to private development. Similarly, schools, hospitals, churches, and sidewalks require similar land provisions exempting them from market forces. Indeed, for the most part the urban road network is a public good (paid for by taxes), the routes chosen and paved by governments.

Arguably, the planning issue dominating the last half a century has been to provide access for cars while minimizing their adverse impacts. Cars and highways made urban sprawl possible. Malls, big-box stores, and power centres took advantage of this and encouraged the ongoing expansion of suburban residential developments. This continual expansion is the dominant engine of the North American economy. Cars and homes fuel demand not only for stuff to stuff the house, but also for infrastructure such as roads, schools, and hospitals; and these demands in turn generate capital investment and jobs. Cars and suburbs have brought undoubted benefits in terms of making spacious, good-quality housing available to people who could not otherwise afford it. But they have also brought undoubted problems: destruction of farmland and natural environments; increased consumption of resources for infrastructure, housing, and commuting; obesity; the demise of urban commercial and manufacturing areas; loss of jobs in urban areas; and effective discrimination against those without auto-mobility. Planners attempt to address these interdependent problems by designing new places and modifying old ones to reduce dependency on the car and prioritize other values. Plans are being made to rejuvenate derelict industrial and waterfront areas; environmental regulations are being tightened and developments require cleaning up degraded and toxic sites; densification and additional transit facilities are being required as preconditions for development. These shifts in emphasis are challenging conventional business location strategies.

Car use continues to be a major planning and social concern, but it is becoming part of a growing concern over environment and livability. These tensions are particularly apparent in the transformation from industrial to creative cities. Vancouver metro, for example, developed as a resource-based industrial city and commodity exporter with strong ties to the BC hinterland and with resource production elsewhere. In recent decades, Vancouver has become "post-modern," a service town with very little large-scale resource manufacturing left, and has increasingly focused on a wide variety of creative activities. In tandem, local planners and politicians have increasingly sought to promote Vancouver as a green city, replacing car lanes with bicycle lanes and opposing resource exports (coal, oil, natural gas) that pass through its boundaries. Yet Vancouver metro's economy still has substantial jobs and income dependence, not to mention consumption needs, on resource-related activities; this debate addresses the nature of Vancouver's role in the national and global urban system. Local planners and politicians will be only one part of the institutional mix resolving these dilemmas.

From Ecological Footprint to Green Growth

The economic relationship of the city with its hinterland and the global trading system is perhaps most evocatively illustrated by its environmental dependence. Cities are not self-sustaining in their energy, water, food, materials, or even their oxygen needs. Nor are they capable of dealing with all the pollution, waste disposal, and carbon dioxide produced in the global production of everything that is consumed within a city's boundaries. The concentration of people in a relatively small place necessitates that these needs are largely met elsewhere, whether they be vegetables grown several kilometres out of town, oil pumped from the other side of the world, or greenhouse gases emitted to the atmosphere. The overall demand in terms of biologically productive space required to feed a city with energy and material needs and to absorb its wastes is called a city's ecological footprint. It is measured in hectares on a per capita basis. The term was introduced by Wackernagel and Rees and used to analyze Vancouver's global inputs and its waste and other outputs. Figure 8.4 gives per capita data for several Canadian cities. The importance of cities' ecological footprints is underscored by the

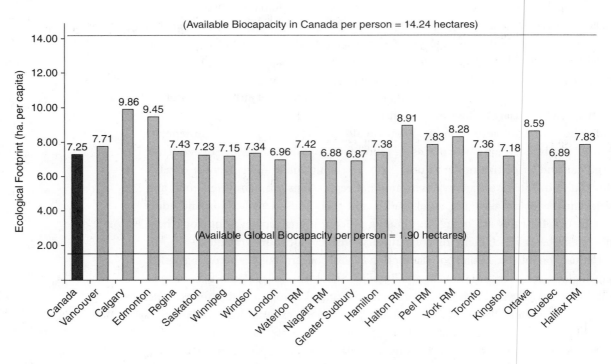

FIGURE 8.4 Canadian Municipal Ecological Footprints from Vancouver to Halifax

Source: Wilson and Anielski 2005. Anielski Management Inc.

fact that over half the world's population is now urbanized and whose wealthier citizens disproportionately consume the world's resources.

While their ecological footprints demonstrate that cities are the primary driver of global environmental impacts, the transformation of local economies is also recognized as an important means to reduce impacts. Changing land use; transportation, water, and energy infrastructures; buildings; waste disposal systems; and so on can dramatically reduce the collective environmental impacts of the city. These changes can be brought about in the design and planning stages and also in operations and retrofitting. To achieve both environmental and economic efficiency, these transformations need to be systemic in order to ensure that interdependent factors work together and that free riders don't reduce everybody else's willingness to participate. A waste recycling system illustrates these interdependencies; a city needs to coordinate activities across its space collection depots and treatment of recycled materials, and most importantly provide people and businesses with incentives to participate. The system would likely operate as some mix of public and private operations, a mix of public and private payments, and a mix of penalties and incentives for participation. A city is also likely to require governance support from the provincial (state) and national levels to ensure that, for example, products are recyclable or other cities will not shirk their responsibilities.

The most overarching means to change a city's ecological footprint is to increase city density in order to reduce material and energy needs in building and operation. This action requires a culture shift from decades of increasing suburbanization and would have dramatic consequences for land values throughout the city. The suburbs, while demanding more infrastructure and energy usage, can also be transformed through distributed energy systems (e.g., solar panels), electric vehicles or hydrogen vehicle fuel stations, and recycling systems. Alternatively, mixed land-use zoning can reduce commuting, enhance the integration of nature into urban areas, and reduce the heat island effect. Other strategies to reduce municipal environmental impacts include green building regulations for new builds and retrofits; district heating and cooling systems; support for public transport, introduction of systems, such as bus rapid transit or bicycle lanes and lending systems, and disincentives to driving (e.g., congestion charges); combined water and wastewater management; use of landfills for gas extraction and solar energy panel covers; and so on.

Municipal planners usually initiate such systems, but they face enormous practical challenges: for example, achieving compatibility with traditional objectives such as providing housing and increasing employment. Planners also work to overcome the limitations of the still dominant zoning approach to planning, such as zoning economic functions to separate them due to perceived incompatibilities rather than to integrate them for environmental solutions. Planners typically count on businesses, NGOs, and government to undertake the activities necessary to bring their designs to fruition and carry them forward. To coordinate these activities, motivations of stakeholders must be understood and appropriate incentives and disincentives created and implemented. Stakeholders need to participate in design, implementation, and monitoring of solutions. Financing is usually a key to realizing changes, and creative solutions have been sought; for example, to transform many of its old and environmentally inefficient residential towers, Toronto created a fund collateralized by future tax payments, and Tokyo created the world's first municipal greenhouse gas emissions trading scheme. China asked its most advanced cities to pilot an emissions trading scheme before introducing one to the country as a whole. Hundreds of cities around the world share information about best practices through forums such as the C40, International Council for Local Environmental Initiatives (ICLEI) and Cool Cities.

Cities have also realized that they can improve their economy through green growth. The OECD has identified four main ways this can be done: (1) Increase jobs through activities such as the retrofitting of buildings and provision of public transport, (2) Increase the attractiveness of a city by improving air and water quality and by enhancing nature's integration, (3) Increase local production of green goods and services both through exports and by displacing imports, and (4) Increase the value of urban land by making the city a healthier, more enjoyable place to live.

Politics, Real Estate, and Social Housing

Bid-rent theory, with its assumptions of perfect competition, provides useful insights on market allocations of land and is complemented by planning processes that have strong democratic features. Yet, business and political power and leadership have long been and remain extremely influential in shaping urban land-use patterns. In North America, for example, westward settlement was facilitated by the land-grant system where big railway companies organized the land and accessibility to define CBDs, as already noted in Vancouver's case. In Europe, the landed aristocracy were important in shaping how "their" settlements developed, and in many places they remain owners of large segments of land, including urban spaces.

In contemporary times, real estate developers play crucial roles in the urban economy. These roles vary, as illustrated by US and Canadian experience. In the US, suburbanization primarily meant segregation along income and ethnic lines and the relocation of relatively high-income, white families. As a result, many US downtowns lost almost all vitality. Income and ethnic variations, as well as poverty, exist in Canada, but social segregation has been less common in Canadian cities than in their US counterparts, and residential neighbourhoods in downtown areas have never been abandoned. In Canada, multi-level political agencies traditionally have been stronger advocates of vibrant CBDs than in the US. As well, Canadian real estate developers have seen downtown areas as a stable source of profit and therefore have been willing to invest in enhancing them.

Until the mid-1990s, most of the high-quality buildings in Canada's largest CBDs were controlled either by developers such as Olympia and York, Cadillac-Fairview, Oxford, and Trizec, or by major banks. Although a number of those buildings have since been taken over by pension funds looking for stable investments to reinforce their real estate portfolios, the developers have contributed to the enhancement of CBDs in several ways: by preserving historic buildings or developing amenities in return for government permission to exceed density limits; by moving tenants from older to newer buildings; by maintaining high standards in their buildings; by encouraging synergies resulting from mixed uses; and by generally raising the reputation of individual CBDs by associating their buildings with well-known and respected brands.

In the suburbs, another type of developer predominates in the provision of detached or semi-detached houses. Residential developers are praised for bringing spacious but affordable housing to the masses and reviled for imposing tracts of cookie-cutter houses over what was once farmland or natural habitat. Developers achieve economy of scale by adapting mass production techniques to on-site house construction and coordinating the activities of all the players involved in the housing industry: planners, real estate agents, contractors, banks, insurers, and so on. These tasks pose challenges; coordination with planners alone may involve road building, access to highways, sewage collection and treatment, and provision of schools and hospitals. Similarly complex coordination is required at every stage, from land assembly (preparation) and construction (production) to marketing and sales (distribution), maintenance, and financing. But the benefits are significant: market stabilization, cost efficiencies, and ability to manage externalities. Moreover, the development process is embedded in a set of institutions that reach back decades, relating to mortgage provision, rising union wages, and highway construction, and is tied into the development of commercial and industrial centres. The provision of these homes and primary family assets has become a pillar of the North American economy and foundation of consumption patterns (Chapter 14).

There are many debates and conflicts about how cities should be developed in modern-market economies, from proposals regarding specific sites, to broader development plans, rezoning, transportation corridors, the role of public transportation, recreation facilities, and social housing. These debates occur in open forums in city council meetings, in forums for citizen participation, in the circulation of planning proposals, and in more private meetings involving city politicians, bureaucracies, and real estate and related business representatives. Yet decisions are also made with little or no public consultation. Sometimes a distinction can be made between development-minded and more public-minded mayors and councils. While the former ally themselves with market forces and powerful business interests, the latter envisage cities as places to live for people of all ages and income categories, and accordingly give greater emphasis to the provision of public goods such as parks, transit systems, and social housing.

Regardless of ideological persuasion, the provision of decent housing for low-income families has long been a challenge for cities. The market and bid-rent curves allocate residential land to those who can afford to pay and excludes those who cannot. The marginalized poor are reduced to living in low-rent or even rent-free dwellings in tenements, slums, shanty towns, boarding rooms, and cheap hotels that are over-crowded, lack facilities, and may be unhygienic, anti-social, and even dangerous. Ironically, because some of these quarters pack in a lot of people on a square foot basis they can be quite profitable for landlords. Other options include temporary shelters on the street. In response many city, regional, and national governments have introduced social housing—subsidized accommodation—for low-income people. Some minimum level of living conditions is a right of citizenship. In addition, social housing can be regarded as a public good that benefits the economy as a whole by encouraging a healthier, more contented, and productive workforce while removing reasons for social and political protest. Early experiments in social housing in the nineteenth century were led by a few enlightened factory owners who built "model" communities for their workers not simply for altruistic reasons but to develop a stable, efficient and dedicated workforce. More substantial provision of social housing, however, has depended on government subsidization and control, such as illustrated by the garden city movement, and, on a much bigger scale, council housing in England (Case Study 8.4).

Indeed, the need to provide low-cost housing remains a big problem in all urban environments, not least dynamic, rich cities where rising land values, gentrification, and the attraction of "spectacles," such as major sporting events, can make situations worse for low-income people by the removal of the cheap rental accommodation. Solutions to these problems are not easy, but some form of local government intervention would seem essential; for example, by specific zoning regulations, requiring developers to incorporate social housing as part of their broader plans, changing rules to make rentals easier in family homes, or allowing "spaces" between existing properties to be filled. Another possibility might be to use vacant land or land that contains disused buildings for social-housing purposes. While such land is a widespread feature of many cities (and small towns) it is often owned by corporations and rich individuals, as a form of land banking, who are waiting for the appropriate time to sell and realize the value of their assets as defined by bid-rent curves.

Case Study 8.4
COUNCIL HOUSING IN ENGLAND

The provision of council housing estates, built and controlled by local city councils, was a major social experiment introduced throughout the UK in the 1920s and 1930s. These estates were large, often containing 20,000–30,000 people, and radically improved the living conditions of working people living in cramped conditions, with families often renting single rooms, close to industrial operations. Council estates, such as Shiregreen, provided suburban living and relatively spacious homes with gardens with access to schools. The council allocated houses on the basis of considerations such as family size and income, collected rent, organized requests for relocation, and implemented modernization (e.g., the introduction of small fridges) on a collective basis. Council housing may be seen as part of the welfare state (Chapter 6), while the implementation of privatization in the 1980s reflects neoliberal thinking of the importance of private ownership (e.g., as an incentive to improve homes and accumulate wealth). Privatization nevertheless faced problems related to the ingrained attitudes towards renting, especially among older people, and significant unemployment. Privatization has meant the purchase of houses or a shift in rental collection to a private agency. House prices vary, but remain relatively cheap. Nineteen Homestead Road has been bought, evidenced by the new gate, front door frame, and car space.

Conclusion

It is often said that the growing urbanization of the world's population, and the uneven concentration of much of the world's economic activities in relatively small areas, testifies to the importance of geography. The fact that organizations and people compete for land use within cities, as well as many associated conflicts, reinforces this point. Further, cities are the organizing nodes of global and local value chains that not only connect cities and city systems internally and with each other, but with resource-based, low-wage peripheries and communities. The rest of this book addresses these value chains—within and among cities, and between cities and peripheries—along sectoral lines (energy, agricultural, manufacturing, services, and transportation) and ends by returning to consumption, also largely city based.

The value chain connects cities and influences their internal structures, most notably through the location of economic activities and their impact on land value. The fact that this introduction to the economic geography of cities has drawn concepts and themes from virtually all the preceding chapters reflects the vital roles that cities, towns, and villages play in coordinating the global economic system. As the places where local and global institutions intersect, the largest cities orchestrate the global economy.

19 Homestead Road, the birthplace of the elder author, is located in the Shiregreen estate, Sheffield, and part of a row of four houses ("Four block"). June 2011.

Practice Questions

1. What is the economic base of your city? To what extent has it diversified?
2. Identify a recent, well-publicized land-use conflict in your city or town. Explain the conflict in terms of its location and the interests of the principal actors involved. How was the conflict resolved?
3. Neither Berlin nor Shanghai is usually considered an alpha city, yet they are both extraordinarily dynamic places. In what category would you place these cities in 1980 and now? Explain.
4. Where was the original site of your town or city (or nearest town or city)? Summarize (a) regional, national, and global connections and (b) local, institutional, and population developments. Look at a small scale aerial photograph of your city that shows its shape from a distance. What economic influences have given it that shape?
5. What initiatives has your city taken to become more sustainable in its use of resources—a "green economy"?
6. Where do low-income families live in your city? What, if any, provisions have been made for social housing?

Key Terms

spatial innovation system   236
basic activities   238
non-basic activities   238
Marshallian externalities
   (or hypothesis)   243

Jacobs externalities or
   hypothesis   243
creative cities   246
   rank size rule   249
   primate cities   249

urban system   249
bid-rent theory   255

Recommended Resources

Beaverstock, J., Taylor, P., and Smith, R. 1999. "A roster of world cities." *Cities* 16: 445–58.
   A classification scheme for the world's cities.

Filion, P. 2010. "Growth and decline in the Canadian urban system: The impact of emerging economic, policy and demographic trends." *Geojournal* 75: 517–38.
   Insightful overview with lots of supporting data.

Hutton, T. 2009. "The inner city as site of cultural production sui generis: A review essay." *Geography Compass*: 600–29.

A review of the role that cultural industries play in redefining inner cities, with particular reference to London, San Francisco, Singapore, and New York.

Jacobs, J. 1984. *Cities and the Wealth of Nations*. New York: Random House.
   Argues that cities rather than nation-states are the chief organizers of the global economy.

References

Bourne, L. 2006. "Understanding change in cities: A personal research path." *The Canadian Geographer* 51: 121–38.

Florida, R. 2012. *The Rise of the Creative Class Revisited*. New York: Basic Books.

Friedmann, J. 1970. "Review symposium: The economy of cities. Jane Jacobs (New York: 1969)." *Urban Affairs Review* 5: 474–80.

OECD. 2013. *Green Growth in Cities, OECD Green Growth Studies*, OECD Publishing.

Ridley, M. 2010. *The Rational Optimist*. New York: Harp Collins.

Scott, A.J. 2008. *Social Economy of the Metropolis*. Oxford: Oxford University Press.

Scott, A.J. and Power, D. 2004. *Cultural Industries and the Production of Culture*. Taylor & Francis.

Simmons. J. and Bourne, L. 2003. *The Canadian Urban System 1871–2001: Responses to a Changing World*. Research Paper no. 200. Toronto: University of Toronto Centre for Urban and Community Studies.

———. 2013. *The Canadian Urban System in 2011: Looking Back and Projecting Forward*. Research Paper no. 228. Toronto: University of Toronto Centre for Urban and Community Studies.

Wackernagel, M. and Rees, W. 1996. *Our Ecological Footprint: Reducing Human Impact on the Earth*. Gabriola Island, BC: New Society Publishers.

CHAPTER 9

Energy and Mineral Resources

Resources are not, they become.

Zimmerman 1939

The overall goal of this chapter is to examine the place of resources in the value cycle and its implications for local development, especially in the context of non-renewable mineral resources. More specifically, this chapter's objectives are

- To investigate how resource industrialization affects the non-market values of nature;
- To illustrate the importance of energy resources in global trade;
- To explore domestic and international policy influences on the search for renewable energy sources;
- To examine the economic, technological, political, social, and environmental factors that drive the resource cycle;
- To explore the reasons why resource-based value chains are dominated by large-scale operations;
- To examine the reasons why differing approaches to integration of investor- and state-owned companies are important; and
- To explore the implications for local development of resource exploration.

This chapter focuses on the **non-renewable resources** of energy and minerals because coal, oil, and natural gas in particular are the lifeblood of the economy. Yet **renewable resources**, such as tuna and Tasmanian forests that we previously discussed (Chapter 7), need to be kept in mind for two reasons. First, the pattern of exploitation is similar. Minerals and fossil fuels are known as non-renewable resources because they cannot be replaced once exploited and that fact has huge implications on companies and communities. The consequences may be similar for renewable resources such as forests and fisheries if they are not managed sustainably. Second, the exploitation of non-renewable resources is often in conflict with the exploitation, appreciation, or creation of renewable resources. Mining, drilling, refining, and consuming ores and fossil fuels damages the ecological basis of renewable resources and because of that, society is attempting to develop alternatives, whether they be renewable energy or new materials. Keeping the comparability between non-renewable and renewable resource exploitation in mind enables you to see the broader utility of the concepts introduced in the chapter. Keeping the tension between the two in mind enables a better understanding of how resource exploitation evolves in general.

non-renewable resources Finite resources (such as fossil fuels) that are depleted with use. For the most part, non-renewable resources are minerals, non-organic, and organic matter transformed by geological processes over a long period of time and concentrated in particular places. Ores such as bauxite, iron, or gold are minerals, as are oil and natural gas even though their origins are biotic.

renewable resources Resources (such as forests or fish) that are not necessarily depleted with use—as long as the rate of use does not exceed the rate of replacement. Many renewable resources, such as forests or fisheries, are biotic and they require limitations on exploitation and management to ensure sustained use, while other renewables, such as wind and solar energy, require technological development.

The chapter is in three main parts. The first part distinguishes the market and non-market values of resources, and examines the role of resources in international trade, resource governance issues, and the search for sustainable energy development. The second part explores the resource cycle model and the organization of resource-based value chains. Finally, the third part looks at the relationships between resources and development.

Contradictions of Resource Exploitation

Our sophisticated and overwhelmingly service-based economy remains dependent on what we generally call natural resources. Resources are essential to economic development at every scale—national, regional, and local—and will be so for a long time. Natural resources, however, as E.W. Zimmerman pointed out, are culturally created. A naturally occurring material such as coal or oil does not become a "resource" until a particular culture finds a use for it, which in turn depends on the level of technological and economic development that culture has reached. Yet natural resources are limited both in the amount available and in the locations where they are found, and in the great majority of cases these locations are distant from the various production and consumption centres to which they are destined. Mineral resources are not evenly distributed around the world and major economically viable deposits are relatively few. A significant part of world trade is driven by the need to gain access to resources from a small number of fixed locations. Thus resources can deservedly be called the geographic foundation of the value cycle and integrating them into global production and distribution systems raises formidable economic and organizational challenges.

Though essential as industrial inputs, resources may have more important economic considerations. Environmental resources, marketed and un-marketed (with those two highly interrelated), sustain economic activities such as agriculture, the circular use of materials and energy, and tourism, while enhancing residential and workplace quality of life. Concerns for the non-market values of resources impose policy controls over exploitation. While this chapter focuses primarily on conventional resource use, the unconventional role of resources will be discussed more in later chapters on services and consumption. To start, let's observe the contradictions of resource exploitation.

Natural resources may seem to be "provided freely" by nature, but as Polanyi said of labour, they are "fictitious commodities" that have not evolved to serve markets (Chapter 5). Their conversion to serve markets is an inherently social process that has four profound consequences for the characteristics and impacts of their exploitation. First, is ownership. Although nature provides natural resources freely, society has to establish some form of ownership to ensure orderly and economically secure exploitation. However, the natural endowment characteristic of resources results in a general sense of ownership on the part of the public. Hence government policy-makers often seek to ensure that their resource policies return benefits to society as a whole, not just to private exploitation. A singular outcome of this tendency is for state-owned companies (SOCs) to be particularly powerful in the resource sector.

Second, the potential for resource exploitation is interdependent with the evolution of techno-economic regimes. Thus, after the invention and diffusion of the internal combustion engine, oil exploration and refining expanded, complex corporate organizations evolved to find and deliver it, and world politics changed dramatically. The oil era will end only when technologies to convert wind and sunlight into electricity, fusion reactors, or biofuels through artificial photosynthesis have been developed at a reasonable cost. Conversely, there is enough carbon in the ground to keep humanity fuelled by fossil fuels forever if the technology and economics are right. Natural resources are not only socially created, but socially destroyed and are substitutable.

Third, just as exploitation has to be socially acceptable, the non-market values of resources—ecological, recreational, spiritual, and aesthetic—have to be considered alongside or before exploitation. ENGOs have been leading advocates of the importance of nature's non-market values, urging policy-makers to give them higher priority (Chapter 7). To that end, ecological economists have created the concepts of "natural capital" and "ecosystem services" in the hope they will be given more consideration in an economic system that has long denied them value. Public support for nature's non-industrial values has also grown, from a quality of life perspective in regard to the present generation to a sustainability focus for future generations. In practice, people tend to feel differently about natural resources than they do about other commodities. Most resource-development policy proposals, therefore, promise non-industrial benefits as well as industrial ones, measures to protect the environment, and sometimes limits on corporate ownership and control.

Fourth, most resource exploitation embodies contradictions that threaten its continuation. While resource-endowed regions seek to expand exploitation, resource-deficient regions have significant incentive to reduce their resource dependency by increasing efficiency in resource use. Moreover, since the extraction, refining, transport, and use of fossil fuels and mineral resources all harm the environment, society has a strong incentive to minimize their exploitation and use. On the one hand, resource exploitation can lead to virtuous cycles of rapid growth, high **resource rents** (payments to government for the right to use resources on public land), relatively high incomes for workers, and diversification of economic activity. On the other hand, many resource regions find themselves trapped or locked-in to particular resource cycles, unable to translate resource wealth into sustainable, high-income development once the particular resource they depend on runs out, is made technologically redundant, or becomes an environmental pariah. The contradictions of resource use, for regional economies and for the global economy as a whole, are illustrated in the conflicted development of the Athabasca Oil/Tar Sands in northern Alberta (Case Study 9.1).

The oil/tar sands deposits are massive, but the pattern of their development is not unusual. Typically, the exploitation of resources begins with high-quality deposits in locations that are relatively accessible. Then, as those deposits are depleted, the focus shifts to less accessible locations. This shift is reflected in the **resource cycle**: an evolutionary pattern that moves from discovery through an initial period of slow growth to one of rapid growth, followed by a plateau and eventual decline. Exactly how this cycle plays out will depend on many factors, including market **boom–bust cycles**, technological changes, social and environmental attitudes, and broader shifts in geopolitics, all of which present their own challenges for the companies and communities involved.

resource rents Payments made to government for the right to exploit publicly owned resources; also known as "resource royalties."

resource cycle The long-term pattern associated with resource exploitation, in which an initial period of rapid growth is followed by a plateau and eventual decline.

boom–bust cycles Alternating periods of rapid growth and recession.

Case Study 9.1

THE ATHABASCA OIL SANDS

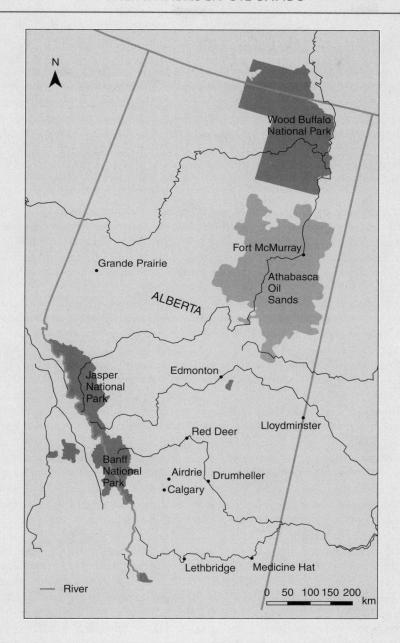

N

Wood Buffalo
National Park

Grande Prairie

Fort McMurray

Athabasca
Oil
Sands

ALBERTA

Jasper
National
Park

Edmonton

Lloydminster

Red Deer

Banff
National
Park

Airdrie   Drumheller

Calgary

Lethbridge   Medicine Hat

— River

0   50 100 150 200
                    km

The Athabasca oil sands were a resource to the Cree who used the tarry bitumen deposits they found there to waterproof their canoes. Their potential as a source of fossil fuel was not recognized until the early twentieth century. R&D to develop

(Continued)

that potential began in the 1920s, but four decades of improvements in processing technology were needed before the sands' petrochemical potential could be exploited. Even then, the oil sands did not become profitable until the quantities of conventional oil available from cheaper, more accessible sources began to decline, oil prices increased accordingly, and—crucially—extraction methods were developed that permitted production on a massive scale while substantially externalizing the environmental costs. The Athabasca project is the largest of three oil sands projects in Alberta's northern forests. Today the project has become a lightning rod for environmental opposition to the resource extraction industry.

The environmental costs are substantial. Muskeg bogs and boreal forests—habitat for many birds and mammals, including endangered woodland caribou—are destroyed. For every barrel of oil produced, 2009 estimates claim that two to four barrels of contaminated water are dumped into tailing ponds and 80 kg of greenhouse gases are released into the atmosphere. However, the industry is working to reduce these figures (for example, by sequestering carbon in underground mines).

Today it is estimated that the Athabasca oil sands contain at least 35 billion barrels of surface mineable bitumen and 98 billion barrels of bitumen recoverable by other methods. Canada's oil reserves are now believed to be third behind those of Saudi Arabia and Venezuela. In 2013, production from all Alberta's oil sands areas was 1.98 million barrels per day and expected to climb to 3.7 million barrels by 2020 and 5.2 million barrels by 2030. The oil sands' economic impact in Alberta includes 146,000 jobs and for every oil sands job, another job through indirect association and 1.5 jobs through induced association. By 2013, Alberta had collected $302 billion in oil sands–related taxes and in addition was collecting royalties of $4.4 billion per year. It expected those royalties to climb to $18.2 billion by 2023 and to exceed $600 billion by 2040 (2013 Canadian dollars). Fort McMurray—the service centre for the region—was a remote outpost of a few hundred people in the 1950s; by 2010 it had a population of 76,000. In Canada, the economic impacts include 514,000 jobs (2014) from direct, indirect, and induced effects of the construction of new projects and the operation of new and existing projects, and total taxes paid to the federal government of $574 billion (2013 dollars).

Economic implications of the Athabasca oil sands for Canada also include the value of the Canadian dollar, which is tied to oil prices. With oil seeming relatively scarce and expensive oil sands exploitation possible, the value of the Canadian dollar increased, rendering other Canadian exports more expensive (and imports cheaper). The reality of the oil-dollar linkage was exposed in 2014 when oil flooded international markets and the Canadian dollar sank, with new supplies coming from the US where industry introduced "fracking" technology, and then from established cheap Middle East sources that sought to defend their market share. For the Athabasca oil sands, however, as prices have fallen, environmental opposition has increased, in particular to the building of the Keystone pipeline that would facilitate oil exports to the US and the Northern Gateway pipeline that would cross British Columbia to serve Asian markets. Aboriginal land claims have further conflicted with the latter proposal. Contraction rather than expansion may be the oil sands future.

Source: Pembina Institute (2006); Boreal Songbird Initiative (2007); Canadian Association of Petroleum Producers (2010) http://www.energy.alberta.ca/oilsands/791.asp (2015).

Nature's Values

The market values of natural resources are readily expressed through price and cost calculations. If a resource can be delivered to consumers for a price that covers the value attached to the resource plus production and transportation costs, and provides a profit margin that takes into account financing and the risks facing suppliers, then exploitation can proceed as far as the market is concerned. The many (economic, technological, organizational, and political) factors that affect such market relations can be elaborated in relation to resource cycle dynamics. Nature's non-market values, however, are harder to assess. Indeed, the onset of large-scale industrialization and urbanization 250 years ago largely ignored the impacts of resource exploitation on natural capital. Nature seemed to be infinitely abundant with clean air, water, and biotic and mineral resources. With modest government control most resources were in effect regarded as a type of commons known as open access resources (Table 6.1). In association, there were widespread beliefs in nature's self-healing properties. In many regions, for example, sustained yield in forestry largely depended on self-generation, while many cities and industrial operations assumed the oceans would "naturally" break down effluent discharges. Indeed, the city of Victoria in British Columbia still makes this assumption.

In recent decades, however, there has been growing alarms for the depletion of natural capital around the globe. The concept of ecosystem services provides a measuring stick for the implications of these losses. According to the Millennium Ecosystem Assessment (htt://www.unep.org/maweb/en/Framework.aspx), ecosystems services are benefits people derive from the interaction of living and non-living things in terms of nutrient cycles and energy flows. The word *benefits* understates the contribution as ecosystem services are the foundation upon which much human activity ultimately depends. Ecosystem services include marketable products such as fish and other food, wood, water, minerals, and energy; services that produce those products such as nutrient dispersal, seed dispersal, soil generation, and pollination; and regulating services such as carbon sequestration, flood and water control, cleansing of air, soil, and water, and pest control (Table 9.1).

Estimating the value of these services is not straightforward and depends on aggregating the results of detailed scientific studies in different ecosystems in monetary terms. These values can be partly illustrated by assessing the implications of cost to society arising from the loss of these services. For example, wetlands and coral reefs provide protection from storms that would otherwise damage property, remove soil, and threaten lives. Forests similarly protect water sheds and water tables, absorb carbon, and protect soils. The removal of wetlands, the decline of coral reefs, and deforestation all impose costs on society. A study by Constanza et al. (2014) has estimated that globally ecosystem services amount to an astounding $142.7 trillion dollars (2014), much bigger than global GDP. The fact that many of these services are consistently downgraded or destroyed means that the losses to global GDP are also huge and were estimated at $4.3–20.2 trillion per year between 1997 and 2011. Although there are criticisms, this study has stimulated similar estimates of ecosystem service values at regional levels. The David Suzuki Foundation provides estimates for Canadian regions (http://www.davidsuzuki.org).

TABLE 9.1	Classification of Ecosystem Services	
Section	**Division**	**Group**
Provisioning	Nutrition	Biomass Water
	Materials	Biomass, Fibre Water
	Energy	Biomass-based energy sources Mechanical energy
Regulation & Maintenance	Mediation of waste, toxics, and other nuisances	Mediation by biota Mediation by ecosystems
	Mediation of flows	Mass flows Liquid flows Gaseous/air flows
	Maintenance of physical, chemical, biological conditions	Lifecycle maintenance, habitat and gene pool protection Pets and disease control Soil formation and composition Water conditions Atmospheric composition and climate regulation
Cultural	Physical and intellectual interactions with ecosystems and land-/seascapes [environmental settings]	Physical and experiential interactions Intellectual and representational interactions
	Spiritual, symbolic, and other interactions with ecosystems and land-/seascapes [environmental settings]	Spiritual and/or emblematic Other cultural outputs

Source: Haines-Young, R. and M. Potschin (2013). *Common International Classification of Ecosystem Services* (*CICES*), Version 4.3. *Report to the European Environment Agency*. Available at: www.cices.eu

The Lifeblood of the Global Economy

Global trade in "visible" goods (raw materials and manufactured goods), amounted to exports of $175.98 trillion in 2012—four times as much as "non-visible" (service) trade ($43.48 trillion). Together, fuel and mining products accounted for $43.5 trillion of the visible trade total (agricultural products accounted for another $1.43 trillion). Some of these materials are processed prior to export as resource commodities (e.g., iron ore may be pelletized). There are also many manufactured goods (e.g., forest products, chemicals, iron, and steel) that include major resource inputs.

Among resource exports, oil stands out (Table 9.2). In 2012, it accounted for over a third of the global trade in resources. Export values for crude oil in 2012, not counting oil products, were over four times higher than those for natural gas and ten times higher than those for other commodities, such as copper and aluminum—the next most important resource categories in international trade. (The value of crude oil exports was 18–20 times greater than that of the largest agricultural category, fruit and nuts, and almost 40 times larger than that of wheat exports.) Clearly, the international trade in fossil-based energy is massive. The Middle East is easily the dominant oil exporter, serving all major global consuming regions in 2012 (Figure 9.1a). Other important export flows from Russia and its former satellites, Canada, Latin America (especially Venezuela), North Africa (especially Algeria), and West Africa (especially Nigeria) are to adjacent continental markets. In 2012,

TABLE 9.2   Selected Commodities, Value of Global Exports and Imports, 2012

Commodity	Top three exporters	Imports ($b)	Exports ($b)	Top 5 nations Imports (%)	Top 5 nations Exports (%)
Wheat	US, Australia, Canada	48.2	48.7	24.5	63.0
Rice	India, Thailand, Vietnam	23.8	24.4	29.2	75.1
Sugar	Brazil, Thailand, India	45.3	44.5	24.3	51.9
Tea	Sri Lanka, Kenya, China	7.8	8.2	24.7	60.6
Coffee	Brazil, Germany, Vietnam	40.0	39.2	43.9	47.1
Cocoa	Nigeria, Cote d'Ivoire, Netherlands	17.5	19.5	49.6	66.4
Fruits/nuts	US, Spain, Chile	92.7	88.4	37.4	38.1
Rubber	Nigeria, Thailand, Indonesia	28.5	36.4	58.5	88.4
Wool	Australia, China, New Zealand	6.8	6.6	71.0	72.8
Cotton	US, India, Australia	24.3	22.6	72.8	74.9
Oil seeds	US, Brazil, Canada	79.7	75.0	65.9	74.6
Wood, rough	US, Russia, New Zealand	17.3	12.6	67.8	48.8
Wood, simply worked	Canada, Russia, Sweden	38.8	36.3	44.0	49.0
Fish	Norway, China, US	61.0	57.8	41.0	41.3
Copper	Chile, Peru, Australia	53.7	51.9	78.5	69.5
Aluminum	Australia, Brazil, US	16.6	14.0	55.3	70.4
Oil, crude	Saudi Arabia, Russia, Nigeria	1,722.6	1,562.3	55.3	46.8
Oil, non-crude	Russia, US, Singapore	904.4	1,016.3	32.1	39.9
Natural gas	Russia, Qatar, Norway	370.2	327.5	55.3	61.9
Iron ore	Australia, Brazil, South Africa	158.6	126.7	83.5	81.1
Coal	Australia, Indonesia, US	145.2	127.1	64.3	80.4

Source: Adapted from United Nations, International Merchandise Trade Statistics http://comtrade.un.org/pb/CommodityPagesNew.aspx?y=2012.

the leading oil product exporters were the US, Russia, and the Middle East. The main importing regions for crude oil in 2012 were the US (22.0 per cent), Europe (24.6 per cent), and Pacific Asia (around 33.5 per cent) (Figure 9.1b). These flows were supplemented by important intra-regional flows within North America (mainly from Canada and Venezuela to the US), Eurasia (mainly from Russia to western Europe), and Pacific Asia (Indonesia to China, Australia, and Japan), and by exports from Nigeria in West Africa to both North America and Asia. The importance of oil to the international economy is further revealed by a casual glance at the list of the world's largest corporations (Table 9.3). In 2013, six of the top ten were energy companies (five were oil companies); another two were car manufacturers; and another was Walmart, which depends on cars to get customers to and from its stores. There are more oil and vehicle companies in the next 10 largest corporations.

Again using data from 2012, trade in natural gas was mainly intra-regional, with Russia and its former satellites the dominant supplier, principally to European markets. Europe and Eurasia also obtained gas imports from North Africa and the Middle East.

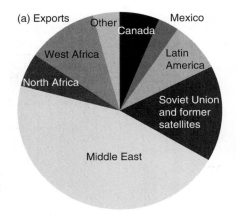

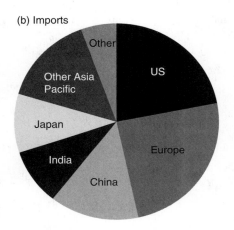

FIGURE 9.1 Crude Oil Exports and Imports by Selected World Region 2012

Source: Adapted from *Canadian Oxford School Atlas*, edited by Quentin H. Stanford and Linda Masci Linton (9e, OUP, 2008), copyright © Oxford University Press, reprinted by permission of Oxford University Press.

All parts of the world import some gas from the Middle East (Figure 9.2a and b). The international trade in coal is smaller; Indonesia emerged as the largest exporter by 2012 with Australia second in importance (Figure 9.3a). Asian countries are the biggest importers (Figure 9.3b). The largest exports move from Australia (mainly to Japan and Europe) and from South Africa to Europe; both Canada and the US also serve important export markets (Figure 9.3). However, the great bulk (84 per cent) of the coal produced around the world is consumed domestically; China, the US, and India are particularly big consumers of their own coal.

The great majority of resource exports go to rich countries to fuel industrial needs for everything from construction steel to rare metals for electronics. Most economically advanced countries enjoyed access to significant domestic resources in their early industrialization, and expansion often required supplementation by imports. Even though the US was blessed with a sufficiency of most of the minerals it needed for industrial development it had to eventually import, either for lack of supply or because of lower costs. Japan had enough coal and iron for initial industrialization, but rapidly became dependent on imports of those resources and most everything else, especially oil. Concurrent with its rise as the world's workshop, China has become a major consumer of resources. Although China has a substantial resource base of its own, rapid industrialization has drained a large proportion of the accessible reserves and its imports have increased accordingly. Since 2010, it has become the world's largest consumer of energy; in 2013, China accounted for one-third of the world's increase in oil consumption; and in 2014, it surpassed the US in terms of oil consumption. Other resource imports also increased. South Korea has also become a notable resource importer in recent decades as its industrial power has grown, while India is an awakening import giant.

Fossil fuels are the lifeblood of the global economy. Oil, coal, and gas (Table 9.4) accounted for almost 90 per cent of the primary energy (energy source as found in nature, before refinement or conversion) consumed around the world in 1980 and remained at 86.91 per cent in 2012. Although oil has become slightly less important than it once was, it is still the leading primary energy source. However, the trend in global consumption of conventional energy—towards larger shares for coal and natural gas—has been relatively stable since 1983 and is likely to continue in the immediate future (Table 9.4). Coal and natural gas reserves are more substantial and accessible than oil reserves. The primary use of both coal and oil is to produce electricity; coal continues to fuel most of the electricity generation in the US, China, India, and Germany. Alternative energy sources are increasing but still only provide 13.1 per cent of global consumption needs. Most of the remaining energy is generated by hydroelectric and nuclear power; the latter is now roughly equal to the former in importance.

In terms of global production patterns, the Middle East was easily the top oil-producing region in 2012, while Europe and Eurasia led the way in natural gas and the Asia

Rank	Company	Revenues ($ billions)	Employment
	TABLE 9.3 Oil's Dominance among the Fortune 500 World's Largest Firms, 2013		
1	Royal Dutch Shell*	481.7	94,364
2	Walmart Stores	469.2	2,200,000
3	Exxon Mobil*	449.9	88,000
4	China Petroleum Corp.*	428.2	1,015,039
5	China National Petroleum*	408.6	1,656,465
6	BP*	338.3	85,700
7	State Grid Corp. of China#	298.4	849,594
8	Toyota Motor Corp.	256.7	333,498
9	Volkswagen	246.6	549,763
10	Total S*	234.3	97,176
11	Chevron*	233.9	62,000
12	Glencore Xstrata#	214.4	61,000
13	Japan Post Holdings	190.9	209,000
14	Samsung Electronics	178.6	236,000
15	E.ON AG	169.8	72,083
16	Phillips 66*	169.6	13,500
17	ENI SpA*	167.9	77,838
18	Berkshire Hathaway Inc.	162.4	288,500
19	Apple	156.5	76,100
20	AXA	154.6	94,364

Source: Adapted from FORTUNE Magazine, www.fortune.com/global500/2013/ (accessed September 2015), © 2013 Time Inc. Used under license.
* Oil companies; # Other resource and utility companies

Pacific region dominated coal production. North America, (western) Europe, Eurasia, and to a slightly lesser extent the Asia Pacific, were important producers of all three commodities. The same regions also dominated consumption, and a significant amount of energy trade took place within each of them.

The development of hydraulic fracking and horizontal drilling technologies is allowing the exploitation of shale-based oil and natural gas reserves and raising questions about radical changes in energy production and trade. Some forecasts indicate that the US could become self-sufficient in gas and oil and even become an important exporter. However, rapid expansion of these sources has been criticized on economic and environmental grounds. Indeed, fossil fuel exploitation and transportation in general are subject to widespread environmental opposition because of their contributions to global warming and threats to local ecosystems, with the Athabasca oil/tar sands being a particularly favoured target (Case Study 9.1). But the environmental issues of fossil fuel use are much more general in scope. What is needed is more comprehensive and long-run thinking that would allow the global consumption patterns, evident in Table 9.4, to be restructured towards more sustainable sources before global temperature rises become inevitable and dramatically change ecosystems. Will targeting the Athabasca oil/tar sands, and the various pipeline proposals emanating from it, lead

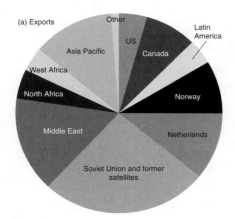

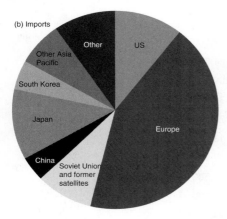

FIGURE 9.2   Gas Exports and Imports by Selected World Region 2012

Source: Adapted from *Canadian Oxford School Atlas*, edited by Quentin H. Stanford and Linda Masci Linton (9e, OUP, 2008), copyright © Oxford University Press, reprinted by permission of Oxford University Press.

to more exploitation and tanker traffic in the polar regions? And how will the massive economic benefits of this project be replaced?

The most important reason behind these trading patterns is the fact that energy resources are fixed in their locations. Investment has to occur where resources are located in order to extract them. Those locations are often deficient in the transportation and urban infrastructure needed to access and export resources. Extraction becomes economically viable only when the value of the resource is high enough to compensate for the investment required. A second important reason is the fact that resources are difficult to transport and need processing into a more useable form, sometimes near sources, sometimes near markets, sometimes between and accessible to both.

Political Governance and Resource Rights

Resource exploitation takes place within a governance regime that is both domestic and global.

Resource Property Rights

"Open access" resources have no property rights attached to them. Historically, especially since large-scale industrialization, many resources, including air and water, have been exploited and degraded without constraint or regulation. The North American bison, hunted to virtual extinction in the nineteenth century, is one example of an open access resource; the fish stocks decimated by unregulated deep-sea fishing are another. Property rights, however, regulate resource use and may be assigned to resources that are owned either privately (by individuals or corporations) or in common (by specified groups of people or by governments). Property rights, which confer the benefits arising from resource ownership in accordance with stipulated obligations, are based on three principles (Chapter 1). First, they are exclusive: the owner of any property or resource has the sole right to its use. Second, security of the right to use resources is provided by law and cannot be arbitrarily removed. Third, property rights are transferable: owners are entitled to sell or trade their property rights. We associate these characteristics with private property, but the right to use land and resources comes in other forms as well. **Common property**, for example, means that access is limited to a specified group, and that use within that group is regulated. Thus it does *not* mean that anyone can have "open access," as is sometimes supposed. An unusual feature of the laws governing mineral resources in most countries is that the owners (private or public) of the lands where they occur can sell the underground rights to exploit them while retaining control of the surface. In some countries, the government can sell underground rights without

common property (or common pool) resources Resources with property rights assigned to a specified group.

regard for the wishes or interests of the private owners of the surface. The contemporary fracking frenzy in North Dakota, for example, is tapping underground shale oil and gas supplies whose rights were sold by farmers decades ago, often for very little.

In the case of oil and gas, property-rights allocation is complicated by the mobility of those substances under the earth's surface and the possibility of extraction from different sites. The early days of the American oil industry saw rampant drilling as owners of different surface sites competed to extract as much oil as possible from the ground below the surface, which fell under a common property regime that lacked the necessary regulatory power. In general, resources governed by inadequate or inappropriately regulated private or common property rights regimes are just as vulnerable as open access resources. The emergence of market economies precipitated the privatization of property rights, which in geographic terms was dramatically expressed by the commercialization of agriculture beginning in the late eighteenth century. Traditional common property rights were alienated and de facto ownership through occupation was legally formalized throughout Europe and the New World. Mineral rights have been part of this privatization trend. However, government ownership and control of land remains strong in most countries. In Canada, only 11 per cent of the land is privately owned: the rest is designated Crown land (publicly owned): 41 per cent of the total belongs to the federal Crown and 48 per cent to the provincial Crown. In the US, 28.8 per cent of all land is federally owned; in addition, each state owns a significant portion of its territory, although the proportions vary widely. In the former communist and socialist countries (China, India, Russia, etc.) and authoritarian states (Saudi Arabia, Libya, Zimbabwe, etc.) government control over resources is formally more extensive, even if the ability to enforce control is lacking in some cases.

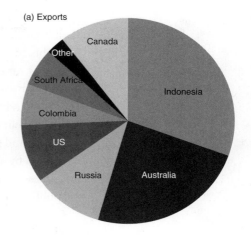

(a) Exports

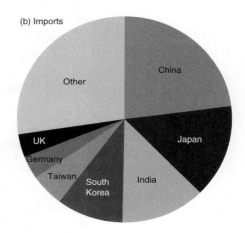

(b) Imports

FIGURE 9.3   Coal Exports and Imports by Selected World Region 2012

Source: Adapted from *Canadian Oxford School Atlas*, edited by Quentin H. Stanford and Linda Masci Linton (9e, OUP, 2008), copyright © Oxford University Press, reprinted by permission of Oxford University Press.

TABLE 9.4   World Consumption of Primary Energy Resources, 1980, 2006, and 2012

Fuel type	Million tons of oil equivalent (MTOE)		
	1980 (%)	2006 (%)	2012 (%)
Oil	2,981.3 (44.9)	3,958.9 (35.2)	4,130.5 (33.1)
Gas	1,295.8 (19.5)	2,557.9 (22.8)	2,987.1 (23.9)
Coal	1,804.2 (27.2)	3,278.0 (29.2)	3,730.1 (29.9)
Hydroelectric	384.6 (5.8)	689.0 (6.1)	831.1 (6.7)
Nuclear	161.0 (2.4)	653.2 (5.8)	560.4 (4.5)
Other	6.8 (0.1)	95.0 (0.8)	237.4 (1.9)
Total	6,633.7	11,233.7	12,476.6

Source: BP Statistical Review of World Energy 2013. http://www.bp.com/en/global/corporate/about-bp/energy-economics/statistical-review-of-world-energy-2013.html (accessed June 2014).

Each nation defines property rights in its own way and with different classifications, but all limit the owners' discretion in use, either directly (through government's control over rights to resource exploitation) or indirectly (through zoning and environmental regulations, for example). Even governments that favour privatization of resources often impose use-based restrictions. Resource contracts typically specify time horizons, renewal possibilities, rent (tax) payments, and minimum or maximum levels of production, and provide guidelines designed to reduce environmental impacts. Moreover, in all countries, both public and private lands are subject to myriad demands for competing land uses and surrounded by a maze of zoning and other regulations. Underlying government's policy approaches to both public and private rights is an admittedly varying sense that resources are a public inheritance that requires wise stewardship in the public interest.

Governments generally maintain that resources are public goods that should be used for national and regional economic development. Indeed, for peripheral regions and poor countries, natural resources are often the only area in which they have a comparative advantage. The desire for development encourages every region to exploit all available resources, often as fast as possible. To facilitate resource projects, governments are frequently prepared either to build or to subsidize both economic infrastructure, such as transportation facilities, and social infrastructure such as housing, hospitals, and schools. MNCs are offered lucrative resource contracts to ensure investment, and state-owned companies may be created to meet development goals when necessary (see Part 3).

Geopolitics

In the name of security and national interest, governments have frequently sought to achieve stable and controlled security of access to essential resources. Indeed, a major motive for European colonization in the nineteenth century was to secure control over resources, and relationships established in the colonial era are still evident. Thus oil companies based in the UK were influential in Saudi Arabia, Kuwait, and Iraq; US-based MNCs had become established in the region by the 1920s; and French oil companies developed strong ties with Algeria. With ancient roots in what used to be called Mesopotamia, the present state of Iraq was the product of an agreement between two colonial powers, France and the UK, and was controlled by the latter from 1920 until it was granted independence in 1932. Part of the rationale for the American and British invasions of Iraq in 1991 and 2003 was the lure of Iraq's vast oil wealth, in which then-president Saddam Hussein had encouraged rival foreign interests to invest. Since the 2003 invasion, oil corporations in both the US and the UK have been assisted in their efforts to gain access to Iraqi oil reserves by their own national governments, which have lobbied the friendly government they installed in Iraq for the necessary legislation. The same collaborative arrangements have been extended to the French oil giant Total (in co-operation with Chevron).

Another type of geopolitical influence can be seen in Canada's role as a stable and secure supplier of resources to the US. In the name of "continentalism," US-based MNCs were provided ready access to Canadian mineral and forest resources that have in large part been developed specifically to serve the US market. Even so, the recent expansion of shale oil and gas fracking in the US has been partly justified on national security grounds, and to reduce dependence on Canadian (and Mexican) imports. At the same time, energy self-sufficiency can generate jobs and reduce balance of payments deficits.

On the other hand, colonial legacies and the unfavourable conditions imposed by MNCs have led many countries to promote the use of their natural resources for local development and limit the truncating negative impacts of foreign ownership, such as loss of profits, **value-added**, and local R&D. Among the approaches that have been taken are the following:

- nationalization of the resource sector, and/or creation of domestic champions;
- formation of resource cartels whose members agree to fix production levels to maintain or increase prices; for example, the **Organization of Petroleum Exporting Countries (OPEC)** established in 1960 by Iran, Iraq, Kuwait, Saudi Arabia, and Venezuela (seven more countries have joined since);
- introduction of export restrictions or bans on the export of raw resources in order to encourage more local "value-added" processing;
- increases in the rents paid to government for resource exploitation; and
- creation of safeguards to protect local entrepreneurs' access to resources.

These policies are controversial, not least because they are seen as contravening the principles of free trade, and in most cases they have proven difficult to implement. OPEC's relative success reflects its members' dominant position with respect to global oil reserves.

Even more controversial is the use of resources by governments in the service of political objectives. Washington's support of Israel in the Yom Kippur War of 1973 led OPEC to impose an embargo on oil exports to the US. Both Venezuela and Russia repeatedly used oil as a political lever in the first decade of the twenty-first century. In response, the largest market countries (particularly the US) and their MNCs maintain reserves of oil and other essential commodities; seek to diversify their suppliers; and are always on the alert for social and political events that could disrupt supply. The presence of American land, sea, and air forces in the Middle East is a reflection of these geopolitical concerns. Oil is not the only strategic mineral. China holds 30 percent of the reserves of several rare earth metals (gadolinium, cerium, europium, lanthanum) that are vital components of electronic products such as mobile phones, cameras, and hybrid cars. However, as a low cost (but environmentally damaging) producer, by 2009, China was producing over 90 per cent of global supplies. In that same year it cut exports drastically, claiming it needed to halt environmental destruction and consolidate small producers. Other countries feared China was trying to manipulate the market, and the Japanese thought China was reacting to their territorial dispute. Soon Australia, Canada, the US, Malaysia, and Russia, which had virtually been priced out of production (a surprising stop from a geopolitical perspective), were investing in new sites, recycling, and developing alternative technologies. Their complaint to the WTO was accepted in 2015 and China agreed to remove export controls on rare earths.

As with oil, governments are concerned with the security of supply of a wide range of resources. Japan is particularly vulnerable to disruptions in supply, and its giant trading companies or *sogo shosha* (Japanese companies that trade in a wide range of products and materials), often with explicit or implicit government support,

value-added The value added to basic materials or services through the application of labour and capital.

Organization of Petroleum Exporting Countries (OPEC) A cartel set up to control production levels among members to ensure price maintenance.

have sought to obtain resources from multiple sources around the world. For example, the sogo shosha have been instrumental in shipping coal to Japan from sources in Australia, Canada, South Africa, and the US.

Environmental Impacts

Directly or indirectly, the mineral resource industry has been responsible for many of humanity's greatest offences against the environment. The chief offence is the production of greenhouse gases. Together, the extraction, transport, refining, and use of petroleum and coal account for the majority of the greenhouse gas emissions attributable to human activity. There is little alternative but to reduce such emissions while working as quickly as possible to replace fossil fuels with other energy sources. The second greatest offence is the destruction of habitat and consequent loss of biodiversity caused by mine development and operation. Underground and particularly surface mining destroys large areas through excavation, dumping of extracted materials, and leaching of hazardous waste into water tables and river systems. Both these offences are direct consequences of the externalization of environmental considerations, and both can be seen in the Athabasca oil/tar sands project (Case Study 9.1).

Most oil spills are slightly different from the above externalities in that they are accidental. But their destructive impact—not only on the environment but on economic activity and every aspect of human life that depends on it—is no less devastating. The blow-out of the BP oil well in the Gulf of Mexico in 2010 was a recent large event in a series of disasters that includes the wreck of the *Torrey Canyon* supertanker off the English coast in 1967, the blow-out of the Ixtoc oil rig in Mexico's Bay of Campeche in 1979, and the running aground of the *Exxon Valdez* tanker in 1989 off the Alaska coast. The BP incident, in which an estimated 5 million barrels of oil were released into the waters of the Gulf, is believed to be the worst accidental oil spill in history—although it is possible that others have gone unreported.

Energy Transitions: Sources and Systems

Many renewable energy (RE) candidates are vying to replace fossil fuels. The most well known are: hydro-power; photovoltaic solar (converting sunshine into electricity); solar thermal (heating up a liquid to drive a turbine or for another purpose); wind power; tidal power; biofuels; and nuclear. It is also possible to use carbon capture and storage (or sequestration; CCS) to make coal or other fossil fuels greener. Most of these alternatives continue civilization's electrification. Two other important elements complementing RE electricity generation are storage and coordination of supply across smart grids (especially to balance out intermittent RE supply). Hydrogen, high volume batteries, pumped water, heat, and ice are examples of storage. Smart grids must be able to accept, compensate for, and redistribute energy derived from intermittent sources; require market and computerized control of the grid, transmission lines capable of dealing with large fluctuations in volume, geographic scale, and diversity of supply (including imports and exports); and demand responses by users (e.g., heat pumps, fuel cells, and electric vehicles that can be charged and charge the grid according to fluctuations). From an economic/resource geography perspective, the most interesting outcome of this shift to electrification will be the increased regionalization, even localization of energy supply. The long-standing resource periphery nature of fossil fuel supply will be significantly altered.

Energy sources are pivot points for change that embrace systems of infrastructure, business, policy, and consumption cultures that are built on energy types. *Energy transitions* describes a systematic transition towards RE which, in essence, is akin to a shift in techno-economic paradigm in specific sectoral and geographic contexts, but occurring through a multi-scalar process. Five features of energy transitions can be noted:

1. *Site*: the basis of the energy transition for many places will be the potentials for RE production and storage available within the region. Wind, hydro, solar, tidal, bio-fuel, or underground thermal resources all vary greatly by region and each region is likely to advance its transition by designing systems to utilize the best mix of resources available. Obvious examples include Denmark's capture of its abundant wind power and the escalating use of solar in the US southwest.

2. *Situation*: no region will be completely dependent on internal resources, but will import electricity from elsewhere. The viability of a smart grid within a region will depend on interconnectivity with other grids that allow flexible responses of imports and exports according to fluctuations in solar and wind energy outputs. Some regions may seem absolutely favoured in terms of situation (e.g., New England benefits from Quebec hydro-power more readily than Los Angeles does from British Columbia hydro-power). However, situation is relative to technology and economics. Thus improvements in high voltage direct current transmission or Los Angeles wealth may allow imports from British Columbia or wind power from the mid-west.

3. *Socio-economic capacities*: technological, financial, and managerial capacities within a region are crucial to developing or adapting RE technologies within a region. These include capacities within firms, held by the workforce, supported by educational institutions and within government regulatory and policy entities. Equally important are the social-cultural attitudes to RE, the awareness of the problem, the willingness act, and roles played by government, NGOs, and business association forms of governance. Most pointedly, policies, regulations, and standards have to be introduced and require buy-in from or compulsion of all stakeholders.

4. *Embeddedness and path dependency*: RE will be replacing infrastructures, policies, modes of consumption, and vested interests that are deeply embedded within the region and are likely to resist change. The speed and efficiency of a transition will depend on the socio-economic will for change gaining the upper hand. That said, the new forms of RE are likely to bear the legacy of past modes of consumption. For example, a city built around cars and private transport is more likely to favour conversion to electric cars rather than to rely solely on electric buses or subways.

5. *Multi-scalar influences*: whatever happens within a region will be affected by national and international influences. The UN's framework convention on climate change as interpreted and turned into policies by national governments is the pre-eminent driver of change (although it can be said that many cities are developing strategies because of the ineffectiveness of global and national governance). International ENGOs also help to shape national and regional policies. At the more practical level, the science and technologies that will enable energy transitions are

global R&D efforts, particularly through academic exchange, but also in the labs of multi-national enterprises that have the scale and long-term capacity to develop them. Although most energy transitions will arise from regional contexts, most of the technologies utilized, whether electric vehicles, green buildings, or refrigerators, will be designed and built by MNCs. In many places the energy transition has been promoted by promising the benefits of green development, but as in most industries, the eventual winner in RE provision turns out to be MNEs (see Case Study 9.2 for more on green development).

Case Study 9.2
GREEN DEVELOPMENT AND ITS DILEMMAS

David Cooper/GetStock.com

Green development serves two important purposes: achieving environmental goals and fostering environmental industries. Governments around the world, particularly in developed countries, have begun looking to green development as a way to sustain or revive their industrial bases. Supplying their populations' own needs was not the only reason these and other governments invested in green development strategies (while imposing costs on their own consumers). They also hoped to capture international markets. Competition among them has helped to drive the development of green technology. But not every competitor can win.

In the 1990s, Japan built the world's largest photovoltaic (PV) industry by offering subsidies for the purchase of solar panels from Japanese manufacturers. Between 1994 and 2004, solar energy capacity increased from 31 to 1000 megawatts, and Japan produced 60 per cent of the world's PV cells. However, even by 2005 the industry hadn't become self-sustaining and a more free-market oriented government eliminated subsidies. Through the first decade of the twenty-first century, Japan was complacent in its superior energy conservation technologies relative to much of the world and relied on nuclear power. The Fukushima disaster reawakened Japan's search for RE—a "feed-in tariff" (FIT) program was introduced, the electricity grid was opened to competition, investments in solar farms raised Japan to the world's second largest market, and it started to develop floating wind power platforms. At the same time, however, the "nuclear village" of vested interests in business, regulatory institutions, and politicians are not letting the nuclear industry fade away, but are working to reopen nuclear reactors and to build and export more.

Around the same time Japan was boosting the PV industry, Denmark took the lead in wind power as a result of a grassroots movement and a government commitment to reduce the country's reliance on fossil fuels and nuclear power. The Danish initiative depends on localized power generation by more than 2000 co-operatives and utilities. Collaboration between industry and government on R&D led to the development of what became the world's largest wind turbine industry. With strong government, industry, and social support, by 2014 Denmark was obtaining 25 per cent of its total energy consumption from renewable sources and 10 per cent of its exports were from the renewable sector.

Germany, although not noted for its sunshine, decided to invest in solar power in 2000. The motive was not only to gain an additional source of electricity, but to develop a PV manufacturing industry. The government established a FIT system whereby electric utilities would pay above-market prices for electricity generated by people who installed PV panels (and other renewable energy systems). From the introduction of the law until 2006, the yearly production value of the industry grew from 450 million to 4.9 billion euros; companies invested 500 million euros annually in the construction, expansion, and modernization of factories; and employment rose by 50,000 jobs. Exports include planning, construction, equipment, systems engineering, operation, monitoring, finance packages and training, and cross-sectorial consulting services, such as technical consultation, feasibility studies, environmental impact studies, audits, and instruments for measurement. For a while, Germany's combination of policies made it the world's PV production and consumption centre, before being undermined by Chinese competition. Even so, Germany's energy transition is on course to produce 40 percent of electricity by RE by 2020. This "Energiewende" is propelled by the overwhelming support of Germany's consumers and small businesses to purchase and produce RE.

Established only in 2005, China's PV industry was the world's largest by 2010, driven by demand in Europe and North America, while PV usage in China remained limited. Chinese demand did drive the expansion of wind turbine production, however; and the government required 70 per cent local content for any wind development. China's emergence as the powerhouse of green energy equipment has been supported in part by national policies, but also by large regional external economies of scale, as well as local governments and investors who have provided funding and space with a readiness not seen in many developed countries. China's rise eliminated dozens if not hundreds of small solar companies in Europe and the US, but China faced its own shakeout after the global financial crisis—over-production and subsequent price destruction—and from trade disputes initiated by the Europeans, Japanese, and Americans. The government then revived China's and the world's PV industries by changing its policies to favour PV as one solution to its air pollution and greenhouse gas emission problems.

(Continued)

Even in the US, regional and national green development policies encountered great difficulties—with spectacular failures of champion firms such as Solyndra. On the other hand, California has been leading the alternative energy transition in North America for decades. Many of its low-carbon actions were born of efforts to reduce air pollution from power plants and internal combustion vehicles. One institution, Southern California's Air Resources Board can be credited with imposing standards that transformed the world automobile industry. California has found that reducing pollution is economically beneficial, drawing in tens of billions of dollars in investment and providing 500,000 jobs across 40,000 companies. While several companies such as Solyndra have disappeared, giants have arisen from the turmoil. The firm SolarCity has made it possible for homeowners and businesses to reduce upfront costs of installation and reduce energy costs while the installation pays for itself over time. Industry has become local or regional, with more emphasis on assembly and installation than on making the panels. This model is becoming so successful, that another contradiction is arising—most homes or businesses remain hooked up to the grid, but do not pay much for electricity nor the maintenance of the grid. Thus costs become higher for the local generator and their customers.

Other North American regional efforts have not been as successful. In 2006, Ontario adopted the most aggressive FIT system in North America with the aim of creating 50,000 renewable energy jobs in three years. Although the FIT program was successful in attracting applicants and producing electricity, a review was undertaken in 2009 because the local multiplier impacts were weaker than had been expected. The program was rethought to subsidize local firms, remove caps on project size, and stipulate "made in Ontario" content. Thus, although the Ontario government contracted with the Korean firm Samsung to develop 2.5 gigawatts of solar and wind power, the establishment of manufacturing capacity in the province by both Samsung and its component suppliers was a condition of the deal. Other countries opposed Ontario's attempt to foster and protect the industry, and won their appeal to the WTO. The program also faced strong social and political objections for everything from the cost of the FIT program to local opposition to the siting of wind farms. That said, Ontario is one of the few jurisdictions in the world that has weaned itself from coal power generation.

For an early study promoting green development, see Hajer (1995); see also the discussion of the "Green paradigm" in Chapter 3.

But what happens to resource industries in an era of regional energy transitions? First, hydro-electricity remains the world's greatest and most dependable source of RE (even though it has many damaging environmental impacts). Many hydro-electric sites are within economic range of consumption centres (e.g., Hydro-Quebec) and industry is attracted to sites of cheap electricity (e.g., aluminum refining in Iceland and Microsoft's data centres in Washington State). Other large-scale electricity production locations also offer locations close to consumers and industry: the huge wind resource that extends from Texas to Alberta, and the solar power of the Sahara, a short distance from Europe. Fossil fuels remain another possibility through the use of carbon capture and storage (CCS) technologies, although this technology remains very expensive, energy inefficient, and hazardous, with leakage from geological sequestration formations. CCS, like hydro, has other environmental threats (e.g., habitat and landscape destruction, pollution).

Other important implications and changes from this use of the resource periphery include a greater distribution of benefits to landholders on which wind turbines or solar installations are situated, continued income flowing into the periphery from construction and maintenance, and skill transferrals. Several of these benefits accrue from the fact that solar and wind energy make use of surface rights that belong to landowners, whereas mineral extraction often does not.

The Resource Cycle Meets the Value Chain

Resource industries differ from other kinds of economic activity in how they are embedded in nature: resource exploitation directly changes nature and nature has a direct impact on resource activities. Moreover, the expected value of the production from any site has to be considered in the context of a complex value chain that includes exploration, production, transportation, distribution, and economic development. Finally, the resource cycle itself poses a unique set of challenges.

The Resource Cycle

The resource cycle model proposes three main stages in the long-term exploitation of a resource: (1) discovery, (2) production, and (3) abandonment. The discovery phase includes exploration, development, and assessment of the quantity available at a particular site. Rapid expansion in the production stage is followed by a levelling-off and then rapid decline and, when the supply is completely depleted (or no longer economic), abandonment. An idealized pattern of oil field exploitation can be imagined as an overall "peaked" boom and bust cycle for the region as a whole, with overlapping cycles in sub-regions and individual wells, each of which reaches a plateau as a prelude to decline. As peak production is reached in established regions, their vulnerability to decline is typically reinforced by the rise of new supply areas. US conventional (i.e., non-fracking) production as a whole began to decline in 1970 and imports have exceeded domestic output since 1990; the giant Cantarell Field in Mexico began to decline in 2006; North Sea oil has been in decline since 1999; and Alberta's conventional oil production has been decreasing for some time. While the resource cycle model applies most clearly to non-renewable resources such as oil, it is also relevant to renewable resources such as fish and forests, when the rate of depletion is higher than the rate of replacement.

When the most accessible deposits in an area have been exhausted, they must be replaced with less accessible ones that cost more to extract. Average costs of production increase accordingly, even though the quality may be lower. In the sedimentary oil basin of central Alberta, production started in 1947 and peaked in 1973. Since then production from this basin has declined and become more expensive. The costs of extracting natural gas from this field have also increased: in 2006, almost twice as many wells had to be drilled to equal the 1996 production level. Meanwhile, production in northern Alberta's Athabasca oil/tar sands region (Case Study 9.1) increased. But costs are high: in 2009, the operating cost of a new oil sands mine was between $9 and $14 a barrel, depending on the technology used. Oil from Iraq and Saudi Arabia cost just $1 a barrel to produce and the cost in central Alberta and the US was around $6 a barrel.

For expansion of the Athabasca oil/tar sands to be economically feasible, a price of at least $50 a barrel is required.

While there is an inherent tendency for costs to increase over the course of the resource cycle, real prices are unlikely to increase in the same way. Thus resource producers often experience a "**cost-price squeeze**" as new, perhaps lower-cost, sources are developed or consuming industries turn to different resources as less expensive substitutes. In the case of oil, it seemed like demand from emerging markets in the 2000's reflected a paradigmatic turning point when the demand for oil continually pushed higher as global supplies were expected to be depleted or at least become more expensive. However, fracking returned the real cost of oil back to a declining trajectory—at least for a while. Among the challenges facing the developers of the Athabasca oil sands are the sustainability of high oil prices, the potential for cheaper substitutes to emerge, growing awareness of the need to make more efficient use of energy, and radical price swings in this cyclical industry.

Resource cycles vary in duration; some last only a few years, others many decades, even centuries, and in some cases mines that were once abandoned can be reopened as technological and economic conditions change. Shorter-term variations also occur in response to business cycles: sharp fluctuations in the level of economic activity over relatively short periods of time (several months to a year). Alternating periods of growth (strong demand) and recession (weak demand), each lasting anywhere from a few months to several years, may be reflected in shifts within the overall pattern of the resource cycle. During these business cycles, individual operations move in and out of profitability and production (see Figure 1.4). In general, the duration of the resource cycles for any particular place depends on how costs and prices are affected by a wide variety of factors, including the scale (size) and extent of resources; the rate of depletion; access to markets; competition from lower-cost supply regions; the availability of substitutes; technological changes that may permit the opening of new supply regions; and the further exploitation of existing supplies.

Ultimately, resource cycles are limited not by particular sources, but by the finite nature of reserves, and how the latter are defined. In the case of oil, established regions, for example within Texas, Alberta, and Mexico, have experienced declining production since 2009 as conventional oil supplies, loosely defined as pools of oil that flow naturally or can be pumped directly to the surface, have decreased. However, technological developments have allowed the exploitation of unconventional sources of oil, such as from oil sands, shales, sandstones, deep water, ultra-deep water, and (potentially) the Arctic. From this perspective, supply is not a problem for the foreseeable future. However, there is a growing public frustration with increased oil production because of its impact on global warming, and for other harmful impacts on the environment. Oil (and coal and gas) may yet peak because of declining demand.

Industrial Organization of Resource Exploitation

The resource cycle is a powerful influence on the organization of production. Resource operations are typically large in scale and remote in location. These factors make it all the more important that firms try to minimize their risks and capture the potential in the other segments of the value chain, such as processing of the raw resource. Both

cost-price squeeze The situation in which producers face increasing average costs of production and declining prices for their products.

private (investor-owned) and public (state-owned) resource firms are more likely than firms in other sectors to pursue internal integration. Investor-owned companies (IOCs) seek to secure diverse sources of supply and develop core competencies that are not tied to any particular mineral source. State-owned companies, by contrast, tend to focus on achieving benefits for the nation or region. Although these companies vary widely in motivation and effectiveness, they are mandated to maximize investments and opportunities that will provide sustained economic advantages.

Small firms still play a role in some resource industries, especially agriculture, fishing, and forestry. As chapter 10 also shows, corporate integration has been an important trend in these industries as well. However, small firms tend to be more strongly linked to local economies than MNCs, while state-owned companies seek economies of scale with strong local connections (Figure 9.4).

Why Integration Is Imperative

Three factors above all make internal integration—vertical and horizontal—imperative for any mineral resource company, whether it is owned by investors or by the state.

First, mining operations are extremely expensive to establish and run. Before mining itself can begin, extensive exploration and test drilling must be carried out. Then complex, capital-intensive technology is required to reduce the per-unit costs of processing (refining, crushing, smelting, pelletizing, etc.). Pipelines, tanker ships, and

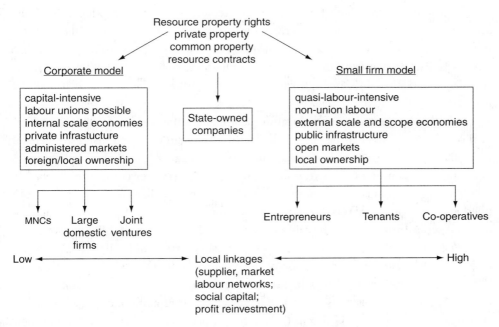

FIGURE 9.4 **The Organization of Industrialized Resource Exploitation: Basic Institutional Types**

For a pioneering discussion see Baldwin (1956) who developed this model in the context of agriculture based on the distinction between the plantation (corporate) and entrepreneurial (small firm) models.

Source: Reprinted from Hayter, R. 2009. "Resource Industries." *International Encyclopedia of Human Geography*, Volume 9, Kitchin, R., Thrift, N. (eds.), 381–9, with permission from Elsevier.

special rail cars must also be put in place, and distribution and marketing networks established (there are 160,000 gas stations in the US alone). To reach the scale of production necessary to reduce per-unit costs, massive investment is required.

Second, resource activities typically generate secondary products, by-products, or hazardous waste streams that must be transformed into other products. Coal and oil, for instance, are usually found in conjunction with natural gas. Therefore companies specializing in either of those commodities must be prepared to deal with gas as well.

Third, as we have noted, every commodity declines in value over time. Mineral resources are no exception to that rule. Therefore it is important to capture whatever value-added may be available downstream in the value chain as a result of further processing and distribution activities. These realities apply to both investor- and state-owned companies. At present, the two typically differ not only in motivation but also in the degree to which their value chains are internally integrated. IOCs control every stage from production and refining to distribution and sales, whereas state-owned companies focus on production, national refining, and sales. These differences largely reflect the fact that many of the Western oil companies were global in scope before 1950; the growth of the state-owned companies has in most cases been more recent.

Multinational corporations (investor-owned companies)

Perhaps the most familiar examples of vertical integration are the oil giants—Shell, Exxon-Mobil, Chevron, BP—that extract, refine, and market oil products. But they are not the only ones. Vertical integration is standard among the giants in the aluminum sector as well: companies such as Alcoa (Case Study 9.3), Rio Tinto Alcan, and Reynolds Packaging. The largest mining companies, such as BHP Billiton, Anglo-American, and Rio Tinto, are highly diversified, operating across all sub-sectors of the mineral resources field.

In highly integrated oil and aluminum companies, the degree of internal integration of resource flows can be very high. In 2012, the firm obtained just 2 per cent of its bauxite supplies from independent mines (Case Study 9.3) and obtained about 25 per cent of its alumina from external sources in 2006. In other companies, there is a much stronger reliance on resource supplies from external sources, involving both open and relational markets. Steel companies, for instance, often purchase coal, iron ore, and scrap from sources they do not own, or own only in part. Highly diversified trading companies (*sogo shosha*) such as Mitsubishi and Mitsui have played powerful roles in the geographic diversification of the resource supplies needed by the Japanese industry. More generally, if it is in the interest of traders, resource consumers, and consuming nations to establish diverse sources of supply and maintain supply stability at reasonable cost, then it is also in the interest of producing regions to develop their resources. Unfortunately, such diversification can lead to over-supply, as the BC coal industry discovered in the 1980s when Japanese investments in Tumbler Ridge were forsaken for cheaper coal elsewhere.

The key advantage that resource MNCs enjoy is their ability to leverage economies of scale across hundreds of production sites, multiple transportation networks, and

Case Study 9.3

ALCOA'S GLOBAL VALUE CHAIN

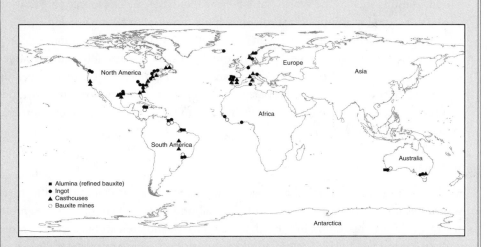

Alcoa is the world's third largest aluminium company, with 31,000 employees in 350 locations in almost 40 countries and revenues of US$23 billion in 2013. While Alcoa has downsized since 2006, and sold its wrapping paper business, it remains highly integrated. In addition to controlling the mines—in Australia, Brazil, Guinea, Jamaica, and Suriname—that supply its raw material (bauxite), it has refineries, smelters, and casthouses (where the aluminum is formed into ingots, rolls, and coils) on almost every continent. As the map shows, these facilities are all located either on or near coastlines. Easy access to tide water gives the company considerable "location flexibility": that is, flexibility with regard to both supply sources and markets around the world. Since smelting is an extremely energy-intensive process, smelters are typically located close to cheap, large supplies of energy; Alcoa's largest smelter is in Baie Comeau, Quebec, which uses hydroelectric power. In 2012, Alcoa used 55.2 million metric tonnes of bauxite, 45 million from its own mines, 7.1 million tonnes from related mines, and just 1.1 million tonnes from independent sources.

Alcoa has more than 200 other locations, mainly in the US and western Europe, where it manufactures end-products that include wheels, fastening systems, and high precision castings for the transportation, aerospace, and oil industries. Its head office (since 2006) is in New York but its longer established and much bigger "operational" head office remains in Pittsburgh. Alcoa's R&D expenses were US$192 million in 2013 (0.9 per cent of sales).

thousands of distribution outlets. The massive scale and integration of MNCs generate other advantages as well, among them the following:

- Ability to invest in exploration and R&D: In addition to finding and developing new resource locations, resource giants have the capacity to develop new technologies;
- Ability to charge for goods and services provided to affiliated operations around the world: MNCs, for example, require their subsidiaries to pay for head-office and R&D services, and there is little or no public scrutiny of these charges;

- Ability to manage site risks: Access to alternative sources means that MNCs do not have to halt production in the event of labour disruption or political instability; they can simply move their operations elsewhere;
- International support: Governments support free trade and FDI, and multilateral organizations such as the World Bank believe that large-scale resource exploitation by MNCs can play an important role in helping developing countries; and
- Investment incentives: Governments in many countries, rich and poor, welcome multinational resource projects that promise to create many jobs, attract large-scale investment, and produce substantial export income. Governments seeking to attract FDI offer generous property rights regimes with low taxes and royalties in return for large-scale development by MNCs with expertise in production and access to international markets.

Joint Ventures

Joint venture organizations involving two or more companies are widespread features of several resource sectors, such as oil and gas, coal, and forestry. Syncrude, for example, is one of the largest operators in the Athabasca oil/tar sand area and is a joint venture among multiple Canadian, US, and Chinese interests, producing over 100 million barrels of crude oil in 2011 while directly employing 5500 people. The rationales for joint venture arrangements in major resource development vary and include the sharing of financial risks, marketing responsibilities, and complementary expertise. In the latter context, many joint ventures involve local partners with knowledge of local social, political, economic, and even environmental conditions and foreign partners with production and marketing expertise plus financial resources. From this perspective, local partners can reduce the spatial entry barriers facing foreign investors, while the latter can reduce the fixed costs and risks facing local companies in developing new technology and building marketing connections (see Chapter 4). Local partners may also reduce criticisms of foreign ownership of resources. Joint ventures typically do not have the mandate to grow further, unless all partners agree, and partners may be bought out. Especially in the oil sector the "local partners" may be state-owned, which adds to the legitimacy of foreign investment.

State-Owned Companies (SOCs)

Although investor-owned resource MNCs have long been the principal players in market economies, state-owned companies are also important. Canada, for instance, has used them extensively. Petro-Canada was a Crown corporation, created in 1975 to serve as a domestic champion. When privatized in 1991, the federal government remained an owner, foreign investment was limited to 25 per cent, and no single private corporation was permitted to own more than a 10 per cent share. In 2004, however, Ottawa sold its interest and in 2009 Petro-Canada was acquired by the US-based Suncor. Similarly, the Potash Corporation of Saskatchewan is a provincial Crown corporation established in 1975 to manage Saskatchewan's potash mines. It was privatized in 1989 and now controls almost a quarter of the world's potash. Canada's commitment to free trade and its dependence on the US help to explain its reluctance to take a more nationalistic stance regarding its resources. There was also tension between provincial control of the resources and federal control over these SOCs.

Globally, the most significant trend towards resource nationalization has been in the oil sector and within developing countries, especially since 1970. Until then the global oil industry was dominated by the "Seven Sisters," a group of US- and UK-based highly integrated MNCs, several of which remain active (Exxon, BP, Shell, Chevron). SOCs, many of them established specifically to counterbalance the power of MNCs and to ensure that more of the benefits of oil exploitation accrue to the state, now control the majority of reserves and are the biggest producers (Figure 9.5). Several noteworthy SOCs are omitted from this figure including Statoil of Norway, China's Sinopec, and Malaysia's Petronas. Although oil nationalization was once regarded as highly threatening to western economies, the reality has become accepted, and joint ventures between IOCs and SOCs are more frequent. Even so, oil nationalization by OPEC and others has encouraged the MNCs to explore for oil and gas elsewhere including the polar regions and to exploit domestic fracking.

State control of mineral resources can be used to advance a variety of national objectives:

- Reducing energy costs for consumers;
- Job creation: Employment per unit of production in SOCs is well above industry benchmarks and averages;
- Economic development (both directly through production and other activities and indirectly through infrastructure development and transfer of organizational and business know-how);
- Earning export income;
- Foreign policy: State control of production has often been used to exert pressure on other governments, particularly in the case of oil;
- Ensuring security of supply: China, for example, cannot rely on its domestic oil reserves for all its needs. Therefore the state-owned China National Petroleum Company has exploration and production operations in at least 10 other countries (including Canada) to supply industrial and individual consumers at home. Direct control of overseas sources ensures that their production goes only to China;
- Capturing value added: SOCs are more likely than investor-owned MNCs to invest in downstream activities in producing regions. In the oil sector, for example, MNCs prefer to locate their refineries close to their markets in rich economies, while SOCs tend to locate them in the producing regions.
- Sovereign wealth funds can be readily created by state owned companies and used for investment domestically and elsewhere as national priorities dictate.

Small Firms in the Resource Cycle

Although giant oil companies are the backbone of the global resource industry, independent firms continue to operate at various levels. A multitude of smaller firms specialize in oil and mineral exploration. In the 1990s, independent refiners in the US increased their market share from 8 to 23 per cent (the majors still controlled 60 per cent, and imports

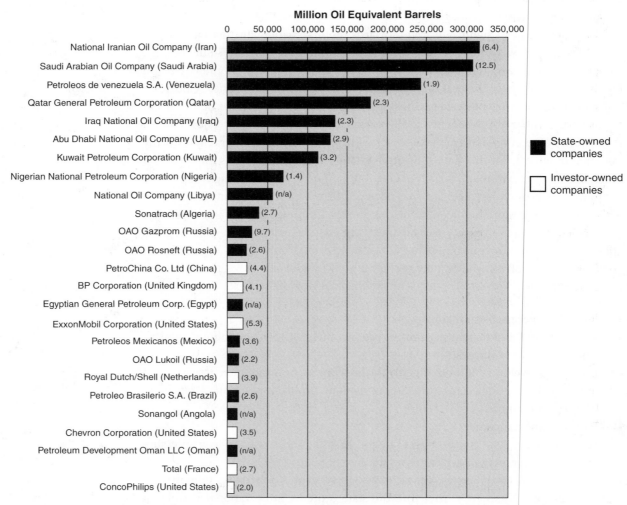

FIGURE 9.5   World's Largest Oil and Gas Companies by Size of Reserves (and Daily Production*) 2010
*Daily production is in million oil equivalent barrels.

Source: Petrostrategies http://www.petrostrategies.org/Links/worlds_largest_oil_and_gas_companies.htm (accessed June 2014).

accounted for another 10 per cent). And even though distribution is mainly controlled by refiners (majors and independents), independent retailers had 17 per cent of the market. The fact that more than 10,000 US and global companies support the major oil and gas companies' operations with equipment, services, supplies, and design and engineering illustrates the competitive and co-operative dynamic between giant and smaller firms.

In India—the world's third largest producer of coal—88 per cent of total production in the early 2000s came from mines operated by the state-owned Coal India Ltd. There is only one other significant company, although countless illegal entrepreneurs eke out a precarious living by selling "waste" coal and tailings. In China, production was fragmented among 20,000 mines that once employed six million people, but thousands of these have been closed down and millions of miners have had to find other work as the government has consolidated and cleaned the industry. Globally there are about 100 million people active in small-scale or artisanal mining, usually extracting rare metals for industrial purposes. The quantities produced in this way are not great, but the income is important. Unfortunately, these (often illegal) activities can have profound consequences both for the environment and for human health and safety.

Commodity Markets

Commodity markets are exchanges that set prices for raw materials or primary products such as oil, wheat, and soy beans. The practices and regulations of modern commodity markets are rooted in traditions established at the agricultural markets of Chicago in the late nineteenth century. A key requirement of most commodity markets, especially those dealing in commodities that cannot be directly inspected, is that products are qualitatively standardized. Thus oil is benchmarked according to a standard called West Texas Intermediate—a high-quality light oil with particular characteristics (e.g., sulphur content) that is mined and refined in the US. Its price is an indicator for other benchmark prices that are normally lower (e.g., North Sea Bent Crude, OPEC reference basket).

There are two main types of commodity transaction. **Spot transactions** are completed (more or less) on the spot, for delivery within two days. Transactions conducted at a particular point in time at an agreed-on price but involving delivery at some later date are called **futures**. Futures markets allow both sellers and buyers flexibility in meeting their obligations: for example, the buyer of a futures contract has the option of selling this obligation to a third party. However, futures are open to speculation. The most important commodities exchanges are the London Metal Exchange, the New York Mercantile Exchange (for metals and energy), and the Chicago Board of Trade (for agricultural commodities).

Commodity markets used to be physical places with trading floors and "pits" for traders, but have been replaced by electronic exchanges that allow transactions to occur anywhere across cyberspace. In some sectors, such as agriculture, commodity markets are declining in favour of long-term contracts, but in coal and oil they are increasing, stimulated by the deregulation of electric utilities. Carbon emission credit exchanges, such as the European emissions trading system, were established to create a market for carbon and to spur innovation, but have suffered from an oversupply of credits.

spot transactions Transactions completed (more or less) on the spot, for delivery within two days.

futures Market contracts at an agreed-on price that involve delivery at some later date.

Resource-Based Development

Many places around the world have counted on their natural resources to bring economic growth and development. There are high-income resource peripheries, and some countries, regions, and cities that once relied on resources for development have successfully transformed their economies towards manufacturing and services. Yet, resource-based development is often hard to sustain, and is especially problematical in poor countries where export dependence on resources is high.

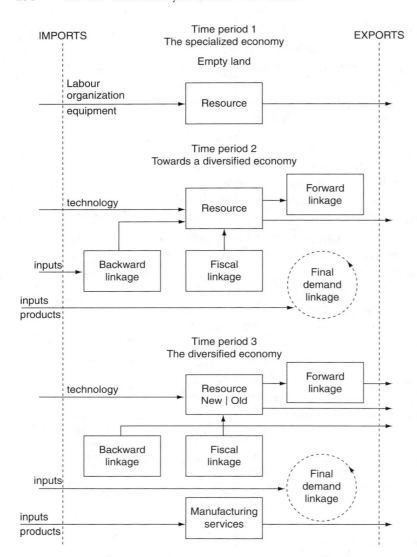

FIGURE 9.6   A Virtuous Model of Resource Development via Diversification

Notes: For definition of linkages, see text and Case Study 2.2

backward linkages Investments in activities that supply inputs to the resource (export) sector.

forward linkages Investments in activities that buy inputs from the (export) resource sector.

final demand linkages Investments in activities that provide goods and services for local consumption.

Resources: Advantage or Curse?

In principle, resource exploitation should spur development not only through the income it generates but through various multiplier effects and other spin-offs that come with population growth and infrastructure creation. In an ideal scenario, an "empty" land with resources that are in demand elsewhere will attract investment, entrepreneurship, and corporate organization, along with a labour supply and government support for the establishment of the necessary social and economic infrastructure (e.g., a transportation system to provide access to markets). Resource exports will generate income for labour and capital, as well as tax revenue for government. Then the circulation of this money will generate multiplier effects throughout the economy. Over time, resource production will stimulate investment and diversification in other activities through the creation of several types of linkages (Figure 9.6). **Backward linkages** are formed by companies that supply inputs to the resource industry; **forward linkages** are formed by companies that work with the outputs of the resource activity. In the case of the oil industry, the manufacture of equipment such as oil derricks and the provision of certain specialized services, such as oil firefighting or legal services, would represent backward linkages, while a refinery to process the oil would represent a forward linkage.

In this idealized scenario (Figure 9.6), population growth promotes both final demand and fiscal linkages. **Final demand linkages** are created by demand for products for local consumption; as population and income increase, increasingly high-order "central place" goods and services become viable (Chapter 1). Fiscal linkages—the taxes paid to governments by individuals and businesses—provide jobs, infrastructure, and public goods such as health care and education, which in time will add to the region's innovation potential. Linked activities and multiplier effects are likely to be concentrated in cities, where agglomeration economies will increase the potential for new forms of growth (outside the resource sector). Meanwhile, the resource sector itself may benefit from technological innovations that, for example, reduce the costs of existing operations or lead to new uses for the local resources. In time, the range of activity will diversify to the point that the resource industry is just one among many pillars of the regional economy. In North America, agglomerations of resource-linked activities are still evident in cities such as Houston and Calgary with their concentrations of oil-sector head offices and R&D operations, as well as service and equipment suppliers. Many other cities, including Montreal, Toronto, and Vancouver, have diversified well beyond their resource roots.

Unfortunately, such diversification is elusive, and places that are rich in resources can often find themselves trapped in dependency on resources that eventually decline. In Canada, the pioneering scholar, historian, and economic geographer Harold Innis proposed that a wealth of natural resources might not be developed as fully as possible in the national interest, and may even hinder development. Innis recognized how Canada's economy, patterns of settlement, and transportation networks had been defined by dependence on the exploitation and export of a handful of "staple" commodities: fur, fish, timber, wheat, and minerals. Exports to "metropolitan powers"—initially the UK, but increasingly and overwhelmingly the US—fuelled Canada's economic development. The "**staple trap**," as Innis saw it, lay in the fact that development based on staples was uneven, as every export boom was eventually followed by a bust, and diversification based on them (including the development of forward, backward, and fiscal linkages) was limited, when it occurred at all. The life-cycle model of Canadian resource towns anticipates stalled development and long-term decline (Figure 9.7).

In this model, towns established for the purpose of exploiting a particular resource experience rapid initial growth. The moment when the population stabilizes and resource exports are in full swing marks the beginning of a period of maturity that lasts until the resource activity begins to decline—a period that can vary considerably in duration. At that point the town faces two broad possibilities: it can try to adapt to the new reality and diversify, or, as laid-off residents leave to seek new opportunities elsewhere, it can begin "winding down" the local services it provides. Eventually, the town may be entirely abandoned. Diversification, for resource towns, usually means finding a way to generate service jobs outside the resource sector. Towns close enough to metropolitan areas may be able to reinvent themselves as bedroom or retirement communities. More often, hopes for diversification focus on tourism, sometimes in connection with the town's resource history.

For isolated towns with little or no tradition of small-firm entrepreneurship, however, the prospects for success are not great. Traditionally, resource towns in Canada developed under the control of a single company. Those established since 1950 have been formally planned to provide living conditions on a

> **staple trap** The trap experienced by economies that depend on resources for development and find their opportunities for diversification to be limited.

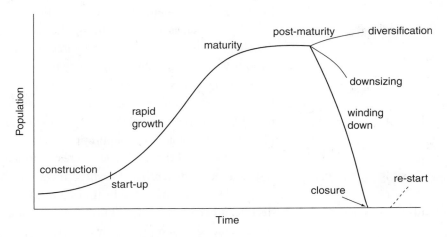

FIGURE 9.7 Life Cycle of Canadian Resource Towns

Source: Based on Lucas (1971); Bradbury and St Martin (1983).

par with metropolitan standards. But no amount of planning can make resource towns invulnerable. Decline is not automatic, of course. Some former resource communities have survived because they were absorbed into expanding urban areas nearby; others, because they have managed to develop other activities such as tourism, sports and recreation, or health care and education; still others simply eke out a living as best they can with a much smaller, poorer population.

Diversification becomes increasingly difficult as the resource cycle enters its final phase, when the resource is becoming depleted and increasingly costly to extract. Isolation from markets, lack of urbanization economies, and non-local industry ownership compound the difficulty, as does the fact that specialized workers are less likely to seek retraining than they are to look elsewhere for opportunities in the same resource sector. Similarly on the global scale, many poor nations with significant resource wealth have tended not to perform as well as other economies, apparently suffering from what is sometimes called a "**resource curse**" (a concept similar to Innis's "staple trap"). Thus resource economies may become trapped or cursed into their resource specialisms related to the **Dutch disease**, the booming and busting nature of development that respectively offset interest and abilities to diversify, by high levels of foreign ownership that emphasizes raw material exports; by highly specialized expertise; and by dedicated transportation and related infrastructure projects limited to the purposes of resource extraction.

Why do resource economies not take advantage of boom times to develop the various linkages anticipated in the virtuous model? In the case of backward linkages, opportunities will be scarce if the multinational giant in charge prefers to get what it needs—services, equipment, R&D—from established suppliers in its own home country; the host government may choose to accommodate such preferences to facilitate exploitation of the resource and support free-trade principles. Opportunities for forward linkages may be similarly thwarted if the firms prefer to invest closer to major markets, and transportation costs are likely to discourage firms from locating value-added activities in resource peripheries, remote from markets. Especially in mining, resource projects are often capital-intensive, employing relatively little labour. Even if wages are high, the fact that overall spending power is not great in such circumstances limits the development of final demand linkages.

Fiscal linkages are also damaged by resource booms. According to the Dutch disease hypothesis, a boom in the resource sector is likely to undermine other sectors in various ways: by causing the currency to increase in value (which will make manufacturing and service exports uncompetitive); and by encouraging governments to focus their support on the resource industry, to the detriment of other sectors. The problem of failing to diversify while times are good is reinforced if government spends its surplus rents on frivolous or inappropriate projects, or if corruption is widespread.

Boom and bust cycles also create difficulties for diversification, entrenching resource dependence. There is little incentive for governments, labour, or communities to diversify during boom times. But as the resource becomes depleted and rents decline, it becomes harder to find the financing necessary for diversification. In addition, resource economies often experience declining terms of trade in international markets: that is, falling prices for their exports and rising prices for the manufactured goods and services they have to import. Competition among resource suppliers and

resource curse The idea that a rich resource base may actually limit the ability of a poor country to achieve sustainable development and diversification; similar to Innis's "staple trap" thesis.

Dutch disease The tendency of resource booms to undercut other activities (such as manufacturing) by causing the national currency to increase in value, rendering their exports more expensive and imports cheaper.

the development of substitute products further contribute to declining terms of trade. Meanwhile, people living in resource communities typically experience so many business cycles that it can be hard for them to distinguish a "winding down" from a temporary downturn that will soon be followed by another upswing. Resource dependence thus poses deep-seated challenges for long-term development.

Differing Interests of Resource Peripheries and Metropoles

Yet decline is not inevitable, and even remote resource towns may be able to revive their failing economies. In this regard, the resource strategies of the resource suppliers (national and regional) and the resource consumers (national and regional) both overlap and differ. The latter seek stable and low-cost supplies to support domestic industry, high standards of living, and national security needs, while the former seek stable markets that generate the highest possible revenues that can be used to promote their own industrial, consumer, and security needs. In practice, the interest of resource peripheries in developing local resource-based value chains conflicts with the interest of consuming core regions over three main issues: resource rents, value-added, and rate of exploitation.

The governments receiving resource rents have often chosen to keep the rates low at first in order to encourage exploitation of the resource. In time, however, they typically choose to increase their rates, especially when supplies begin to run out. Such increases naturally increase the cost of those resources to consuming regions, potentially posing threats of inflation and recession.

At the same time, there is often a tug-of-war between supply and consuming regions for value-added activity. Resource regions typically want to add value to the resources they produce before export, in order to broaden their economic base. But since consuming regions typically want to import those resources precisely to support their own existing value-added activities, they may impose tariffs on value-added imports but not on raw materials. They may also enjoy considerable bargaining power if they have access to alternative resource supplies.

Resource supply and consuming regions often have strongly differing views on how quickly a resource should be exploited. Typically, consuming regions prefer faster rates of exploitation, because increasing supply means lower prices. In contrast, resource regions often prefer slower rates of growth in order to reduce inflationary effects, extend the life of the resource, provide time horizons long enough that resource rents can be used to build social capital and fund diversification opportunities, and in general to ensure that high profits are not dissipated. In the 1970s and 1980s, for example, there were major disagreements between the Newfoundland provincial government and the Canadian federal government over how quickly the Hibernia offshore oil fields should be exploited. And more recently, the OPEC cartel has not been willing to significantly increase oil supply simply to satisfy the desire of market economies to reduce oil prices.

Conclusion

The tension between humans and nature is especially apparent in the use of non-renewable resources for material and energy needs. Thus global environmental change, indexed simply by CO_2 emissions, is requiring new efforts to define "the wise use of resources." More realistic pricing—for instance, pricing that includes some form of tax on carbon emissions in order to internalize the environmental cost—is crucial to encourage more efficient use of resources. Perhaps most fundamentally, R&D and innovation need to be even more strongly focused on resource use and the development of alternative fuels with fewer, less harmful consequences for the environment.

Practice Questions

1. Figure 9.1 (from the *Canadian Oxford World Atlas*) shows that the OPEC countries, especially in the Middle East, are the main sources of oil exports. Has OPEC's policy of limiting production been successful in promoting development? Would it be beneficial for global development if those market restrictions were removed? Discuss the implications for (a) the oil industry; (b) the Canadian economy; or (c) global warming if the Athabasca oil/tar sands were closed down.

2. Is there a case for public ownership of resources? Do you think nature should be a public trust? What are the general patterns of resource ownership and control in your region?

3. When you turn on the water tap in your kitchen or press the light switch in your living room, where do the water and electricity come from? How do they reach you? Have the nature and organization of these supplies changed in recent years? How? Are you aware of any changes to come in the future?

4. What sorts of programs has your region introduced to reduce energy use and promote the use of more sustainable sources? How successful have they been?

5. Should your region go nuclear? Should the globe? Is energy self-sufficiency a good idea?

6. This chapter (and text) has not had the space to contemplate water as a resource. Can you sketch out a chapter that would address the global economic geography of fresh water? How is water supply sourced and regulated in your community?

Key Terms

non-renewable resources   270
renewable resources   270
resource rents   272
resource cycle   272
boom–bust cycles   272
common property (or common pool) resources   280

value-added   283
Organization of Petroleum Exporting Countries (OPEC)   283
cost-price squeeze   290
spot transactions   297
futures   297
backward linkages   298

forward linkages   298
final demand linkages   298
staple trap   299
resource curse   300
Dutch disease   300

Recommended Resources

Auty, R.M. 2001. *Resource Abundance and Economic Development*. Oxford: Oxford University Press.
    Discusses the resource curse and explores the evidence for and against the idea that resource riches lead to economic development in the context of developing countries.

Bakker, K. and Bridge G. 2006. "Material worlds? Resource geographies and the 'matter of nature.'" *Progress in Human Geography* 30: 1: 1–23.
    Advanced discussion of current thinking about the relationship between the material world of nature and human behaviour.

Bradshaw, M.J. 2007. "The 'greening' of global project financing: The case of the Sakhalin-II off-shore oil and gas project." *The Canadian Geographer* 51: 255–79.

A case study of how environmental NGOs are holding oil MNCs to account.

Bridge, G. 2001. "Resources Geography." *International Encyclopedia of the Social and Behavioral Sciences* 132: 66–9. Elsevier Science Limited.

A concise introduction to resource geography, including the meaning of "resources."

Bridge, G., Bouzarovski, S., Bradshaw, M., and Eyre, N. 2013. "Geographies of energy transition: Space, place and the low-carbon economy," *Energy Policy* 53: 331–40.

An excellent overview of the geography of energy transitions.

Campbell, C.J. and Laherrère, J.H. 1998. "The end of cheap oil." *Scientific American* March: 78–83.

A lucid warning regarding the decline in global production of oil, paralleling Duncan and Youngquist (1998).

Clapp, R.A. 1998. "The resource cycle in forestry and fishing." *The Canadian Geographer* 42: 129–44.

Explains the resource cycle model and notes its application to two renewable resource industries.

Duncan, R.C. and Youngquist, W. 1998. *The World Petroleum Life-Cycle*. http://dieoff.com/page133.htm (accessed 4 Aug. 2009).

A forecast of global oil production, written by industry insiders, that so far has proven remarkably accurate.

Freudenburg, W.R. 1992. "Addictive economies: Extractive industries and vulnerable localities in a changing world economy." *Rural Sociology* 57: 305–32.

Excellent review of the economic and non-economic factors that limit the diversification of resource towns throughout North America.

Hanink, D. 2000. "Resources." Pp. 227–41 in Sheppard, E. and Barnes, T.J., eds. *A Companion to Economic Geography*. Oxford: Blackwell.

A useful guide to key concepts and issues in resource development.

Hardin, G. 1968. "The tragedy of the commons." *Scientific American* 162: 1243–8.

A seminal article on the human tendency to over-exploit what Hardin calls "the commons" but are actually open access resources.

Hayter, R. and Barnes, T. 1990. "Innis' staple theory, exports, and recession: British Columbia 1981–86." *Economic Geography* 66: 156–73.

A review of Innis's staple theory as a distinctive Canadian model of development and its relevance to the effects of the severe recession of the early 1980s on British Columbia's economy.

Lahiri-Dutt, K. 2003. "Informal coal mining in Eastern India: Evidence from the Raniganj Coalbelt." *Natural Resources Forum* 27: 68–77.

Small-scale coal mining is important to the lives of thousands in India, even though the practice is discouraged both by the state and by private companies.

Le Billon, P. 2001. "The political ecology of war: Natural resources and armed conflict." *Political Geography* 20: 561–84.

Examines the extent to which natural resource exploitation is at the centre of "real" wars in poor countries.

Parker, P. 1997. "Canada-Japan coal trade: An alternative form of the staple production model." *The Canadian Geographer* 41: 248–66.

A still-relevant analysis of the way Japanese trading companies and coal users have developed coal supplies, with specific reference to Canada.

Patchell, J. and Hayter, R. 2013. "Environmental and evolutionary economic geography: Time for EEG^2?" *Geografiska Annaler*: Series B 95: 111–30.

The authors' multi-scalar perspective on how to combine evolutionary economic geography with environmental geography.

Stern, P. and Hall, P. 2010. "Historical limits: Narrowing possibilities in 'Ontario's most historic town'." *The Canadian Geographer* 54: 209–27.

Assesses the efforts made by the mining town of Cobalt to transform and brand itself as a tourist destination and concludes that this strategy has precluded other options.

Watkins, M.H. 1963. "A staple theory of economic growth." *The Canadian Journal of Economic and Political Science* 29: 141–8.

A classic study summarizing Innis's model of Canadian economic development.

Yergi, D. 1992. *The Prize: The Epic Quest for Oil, Money and Power*. New York: Free Press.

A detailed account of the interweaving of corporate and national interests in the historical development of the oil industry in the Middle East.

Zimmerman, E.W. 1956. *World Resources and Industries*. New York: Harper Row.

A classic study, first published in 1939, and widely used as a textbook in economic geography until the 1960s.

References

Constanza, R. et al. 2014. "Changes in the global value of ecosystem services." *Global Environmental Change* 26: 152–8.

Zimmerman, E.W. 1956. *World Resources and Industries*. New York: Harper Row.

CHAPTER 10

Agriculture

[I]t would seem infinitely better, if a very large number of these wage-earners in cities (so absolutely helpless and dependent . . . on what capital is prepared to allow them in wages, [who are] forced to pay . . . at the contractor's store whatever price he chooses to charge for life's necessaries) were living in the country where they could produce at least the necessaries of life for themselves . . .

Bryce 1914

• • •

But the tide will never turn. Back to the land is an idle dream. We can no more restore the pastoral age than we can go back to the spindle and the loom.

Howe 1915

The overall goal of this chapter is to examine the economic geography of the agricultural sector and its role in the global economy. More specifically, this chapter's objectives are

- To explore the diversity of agricultural activity around the world;
- To introduce von Thünen's land-rent model and its applicability at various scales from local to global;
- To explore the different meanings of food security in developing and developed countries;
- To examine the role government subsidies play in agriculture;
- To look at the organization of agricultural markets;
- To investigate some of the environmental and health concerns associated with the food industry; and
- To explore current alternatives to conventional, industrialized agriculture.

The chapter is organized in three main parts. The first part is a conceptual discussion of market, political, and cultural forces that influence agriculture at the local and global scales. The second emphasizes agricultural organization including the links between domestic and global markets, the "price-taking" role of independent farmers and the pressures for integration and concentration in food production, processing, and distribution. The third is a look at some of the regional and community dynamics that shape agriculture.

Although forestry and fisheries differ from agriculture in some important ways, they also share several characteristics with agriculture; we have included them in our discussion where appropriate.

Geographic Differences in Agriculture

Have you ever stopped at a roadside stand to buy fresh corn from a family farm? There are similar stands in Japan where passers-by simply deposit payment into a box for the produce they take. In Tunisia, children will wave you down, selling bread made from their family's wheat. On a train in Vietnam, a brief stop at any station is an opportunity for dozens of local fruit growers to board in hopes of selling their produce. In France, winegrowers will invite you into their cellars for a glass of *vin du terroir*. All these examples point to the farmers' dilemma—their markets are far away and it is costly to deliver produce to them. In the few instances where customers come to them, they can save transportation costs, cut out the middlemen, and charge close to retail prices. These examples typify the geographical tensions between places of production, spaces of consumption, and the distribution systems that link them (Figure 10.1).

Around the world, agricultural production is regionally specialized, reflecting the influence of multiple factors that range from **environmental conditions** and local cultural traditions to economic considerations of cost and market access, regional economic development, and technological advancement, to national governance structures and international trade patterns. The economic geography of agriculture involves much more than different crops and livestock, however. The relative contribution of agriculture to GDP in different countries varies greatly. In advanced countries such as Canada, that contribution is fairly minor, even though both output and export numbers are huge. In less-developed countries, much more of the population is involved in agriculture, but their productivity is much lower. In no industry are the consequences of different market–institution–innovation configurations as important as they are in agriculture.

environmental conditions The state of nature in terms of soil, landforms, vegetation, waterways, and climate in particular places.

That different crops are grown in different places reflects opportunities for competitive advantage. The markets for agricultural produce, however, are often far from the producer. Farmers must calculate the cost of transporting their crops to market and then decide whether it is worth it to grow them at all. These costs help determine what crops farmers grow and where.

Because individual farmers are distributed over space and produce relatively small volumes, it is difficult for them to find economies of scale. Assembly of land into bigger, fewer farms is one response. Another way of improving efficiency is for farmers to collectively form **co-operatives** to handle downstream segments of the value chain, such as transportation, storage, and marketing. Increasingly, however, these tasks are being taken over by a few large, powerful firms. For consumers, this organization has resulted in an ever-greater proliferation of inexpensive and standardized agro-industrial products. For farmers, as value-added is increasingly captured by other segments of the value chain, returns have diminished and the communities that used to flourish around clusters of related farming activities have withered.

co-operative An organization or business jointly owned and operated by its members.

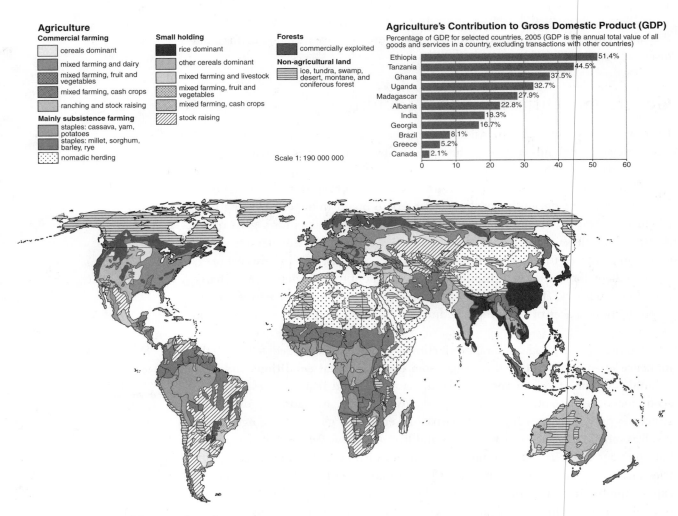

Agriculture

Commercial farming
- cereals dominant
- mixed farming and dairy
- mixed farming, fruit and vegetables
- mixed farming, cash crops
- ranching and stock raising

Mainly subsistence farming
- staples: cassava, yam, potatoes
- staples: millet, sorghum, barley, rye
- nomadic herding

Small holding
- rice dominant
- other cereals dominant
- mixed farming and livestock
- mixed farming, fruit and vegetables
- mixed farming, cash crops
- stock raising

Forests
- commercially exploited

Non-agricultural land
- ice, tundra, swamp, desert, montane, and coniferous forest

Scale 1: 190 000 000

Agriculture's Contribution to Gross Domestic Product (GDP)

Percentage of GDP for selected countries, 2005 (GDP is the annual total value of all goods and services in a country, excluding transactions with other countries)

Country	%
Ethiopia	51.4%
Tanzania	44.5%
Ghana	37.5%
Uganda	32.7%
Madagascar	27.9%
Albania	22.8%
India	18.3%
Georgia	16.7%
Brazil	8.1%
Greece	5.2%
Canada	2.1%

FIGURE 10.1   World Agricultural Map

Source: Adapted from *Canadian Oxford School Atlas*, edited by Quentin H. Stanford and Linda Masci Linton (9e, OUP, 2008), copyright © Oxford University Press, reprinted by permission of Oxford University Press.

Local and Global Markets: Disparities and Dilemmas

Markets have become the predominant determinant of where and how agriculture is practised. For thousands of years, farm produce was destined almost exclusively for local markets. Over the past two centuries, however, national and international markets have become increasingly important.

Market versus Nature

It is tempting to relate agricultural specializations to natural variations (in soils, climate, topography). Indeed, such variations are important through their impacts on cost and revenue structures. Even if nature's bounties were uniformly distributed, geographic variations in agricultural activity will occur with distance from markets. Explanation of this phenomenon began in 1826 when German economist and land-owner **Johann Heinrich von Thünen** suggested that agricultural land-use patterns

Johann Heinrich von Thünen, The German landowner and economist whose book *The Isolated State* (1826) introduced agricultural land-use theory (and by extension urban land-use theory).

would vary systematically around an urban market. Von Thünen's original model assumes that the land surrounding a town is uniform in every respect, from soil fertility and relief to transportation access, and that all farmers (each one an agrarian *Homo economicus*) seek to maximize their **location rent** to obtain the highest possible income from their land. With distance from markets, location rent for each activity is reduced by transportation costs. (At the margin of cultivation, revenues are offset by costs, which are assumed to include **normal profits**: the minimum necessary to make cultivation worthwhile.) Under the pressure of competition to maximize location rent, each farmer seeks to increase yields (income) per hectare by investing in labour and capital on land closest to the town, where high production costs are offset by low transportation costs. With distance from the market, farmers seek to reduce transportation and costs of production. They choose crops with lower yields per hectare and with lower input requirements in terms of labour, capital, and transportation costs per hectare.

In von Thünen's idealized landscape, if revenues per hectare and transportation cost gradients vary by type of activity, then these choices create a pattern of concentric circles of different crops as distance from town increases. Farming types characterized by high inputs, incomes, and yields per hectare are labelled **intensive agriculture**, illustrated by fruit and vegetable production, including that in greenhouses. Farming types characterized by low inputs, incomes, and yields per hectare are labelled **extensive agriculture**, illustrated by sheep and cattle ranching. In some cases, the same crop, for example, wheat, can be either intensively or extensively grown, the choice being dependent on location rent. That is, this model predicts intensive farming near markets and extensive farming on the margin, where transportation costs are significant.

Homogeneous plains, of course, rarely exist in reality. Some areas around a given market will be more productive than others. Just north of Toronto, for instance, the Holland Marsh area is renowned for its vegetable crops; yet to the west, the Niagara Escarpment makes certain areas unsuitable for farming of any kind. Areas with rivers, railways, or roads enjoy much easier access to markets than those without, and therefore attract more productive activities. Competition from neighbouring towns also tends to disrupt von Thünen's perfectly concentric circles. Nevertheless, even today the broad patterns of agricultural land use around many urban centres in western countries, such as in southern Ontario, resemble his model to some degree.

As commercial agriculture expanded rapidly during the nineteenth century, along with new transportation systems (canals, railways, steamships), particularly to engulf the "New World," patterns of intensive farming near markets and extensive farming on the margins developed on national and global scales. Thus prairie and mid-western wheat and sheep ranching in Australia and New Zealand and cattle ranching in Argentina developed (extensively) to serve growing demands in Europe and eastern cities in North America. As von Thünen's theory recognizes, location rent is influenced by differences in natural endowments and human expertise in their exploitation; the spatial patterns of production have distinct characteristics in different regions.

Furthermore, patterns of regional specialization are not fixed. Even when environmental conditions remain the same, many factors, such as changes in market preferences, transportation costs, government policy or regulations (e.g., regarding the use of pesticides and herbicides, subsidies), competitive pressures, or farming technology, can

location rent The surplus income earned on a unit of land over and above that earned on the marginal unit of land where revenues are offset by costs.

normal profit The minimum profit necessary to make productive activity worthwhile, theoretically counted as a cost.

intensive agriculture Farming characterized by relatively high inputs (labour, capital, fertilizer, and seed per unit of land) and relatively high incomes per hectare.

extensive agriculture Farming characterized by relatively low inputs (labour, capital, fertilizer, and seed per unit of land) and relatively low incomes per hectare.

dramatically alter established patterns. In general, as transportation costs were reduced in the twentieth century, their impact on location rent and agricultural specialization has been reduced, and intensive farming in distant peripheries has become viable. Technology, of course, is hugely influential. Refrigeration has meant that the market gardens that once surrounded cities are now thousands of miles away in California, Mexico, Spain, or year round warm climates. Biotechnology has become one of the most powerful forces for change in the geography of agriculture. Meanwhile, factors such as government subsidies, variations in fertility, and the costs of labour, land, equipment, and other inputs (seed, fertilizer, and pesticides) have become more influential in shaping agricultural land uses. In the US, for example, corn acreage has been increasingly shifted to soybeans and, in 2014, a record 84.8 million acres of soybeans were planted, almost the same area as "king" corn; soybeans have experienced higher prices, strong export markets in China, and have lower fertilizer costs. Corn's acreage, however, has been maintained in part because it is the main feedstock for biodiesel and ethanol fuels that the US has sought to develop, by R&D programs and subsidies, for energy self-sufficiency and environmental reasons.

In general, regional specialization has developed and changed in response to market demands. Thus South India and the Malukus (Spice Islands) came to specialize in spice production specifically in order to supply markets in Europe, which lacked the environmental conditions to grow them. Among Canada's regional specialties are grain and cattle on the prairies, cattle in the cordillera and the foothills, fruit and vegetables in the Okanagan Valley and Niagara Peninsula, and dairy farming in the St Lawrence Lowlands. In the US, the Midwest specializes in corn and soy; California in fruit, vegetables, nuts, cotton, rice, and wine; Florida in citrus fruit, vegetables, and sugar cane; and so on. With a few notable exceptions—including potatoes, tomatoes, and

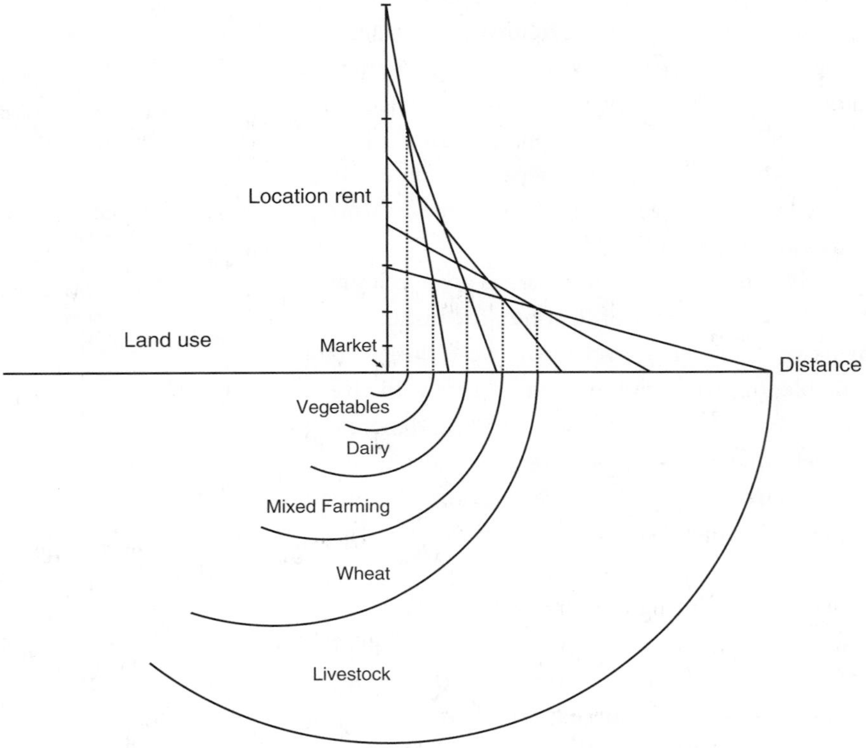

FIGURE 10.2   **Defining Land Use by Location Rent**

On an isotropic plain around a market centre different agricultural activities will locate according to the highest location rent that can be generated on a unit of land. That sum is defined by the total revenues minus total costs.

LR = Y[(p – a)] – Yfk

LR = location rent per hectare
Y = yield or quantity produced per hectare
P = price per unit of output
a = costs of production
k = shipping distance from farm to market
f = shipping cost per unit of output

corn—almost all these crops originated elsewhere (primarily in Europe) and were imposed on the landscape.

Specialization has intensified with the increase in global trade. Relatively low value commodities, such as soybeans, cake of soybeans, wheat, and maize are exported in large quantities with low unit value returns (Table 10.1). Processing adds value to exports. Rich market economies continue to be important commodity exporters and (a) potentially an important source of foodstuffs to poor countries and (b) a threat to less efficient farmers in poor countries (Table 10.2). Supported by rich subsidies, France was a bigger wheat exporter in 2011 than either Australia or Canada. As expected, many of the largest import- ers of agricultural goods in 2011 were rich market economies. However, China was the largest importer of four of the top six commodities (Table 10.3).

Food Security and Subsidies

From 1996 to 2006, agriculture's share of world employment dropped from 41.9 to 36.1 per cent, and to about 33.5 per cent in 2012 and it still represents over a billion people. The vast majority of the world's agricultural workers live in South and East Asia and in sub-Saharan Africa, (where two-thirds of the labour force is employed in agriculture). In these regions, much agricultural production is not destined for any market, however

TABLE 10.1 World's Most Traded Agricultural Commodities, 2011

rk	Commodity	Quantity (tonnes)	Value (1000 $)	Unit value ($/tonne)
1	Soybeans	90,813,977	51,403,325	566
2	Wheat	147,205,956	51,184,264	348
3	Food (prepared, not otherwise specified)	13,416,474	49,892,030	3,719
4	Palm oil	36,589,672	42,034,273	1,149
5	Maize	108,067,148	36,342,489	336
6	Rubber natural dry	7,179,256	33,765,962	4,703
7	Wine	10,004,329	33,041,355	3,303
8	Coffee, green	6,445,688	28,303,554	4,391
9	Bever. dist. alcohol	4,074,220	27,945,091	6,859
10	Cake of soybeans	63,593,084	27,458,049	432
11	Meat-cattle boneless (beef & veal)	4,931,836	26,246,728	5,322
12	Cigarettes	1,003,748	25,381,577	25,287
13	Cheese of whole cow milk	4,764,853	24,670,883	5,178
14	Cotton lint	7,856,760	23,177,384	2,950
15	Sugar raw centrifugal	33,838,303	22,649,899	669
16	Pastry	6,958,052	22,542,345	3,240
17	Chocolate (processed, not otherwise specified)	4,717,528	22,429,658	4,755
18	Chicken meat	11,391,477	21,792,056	1,913
19	Pork	5,260,397	18,300,081	3,479
20	Sugar refined	21,921,611	16,694,636	762

Source: Food and Agriculture Organization on the UN trade statistics database (FAOSTAT). Top Imports—World—2011. http://faostat. fao.org/site/342/default.aspx. Reproduced with permission.

TABLE 10.2	Major Commodity Exporters, 2011		
Rank	Area	Commodity	Value (1000 $)
1	USA	Soybeans	17,563,868
2	Malaysia	Palm oil	17,452,177
3	Indonesia	Palm oil	17,261,248
4	Brazil	Soybeans	16,327,287
5	USA	Maize	13,982,404
6	Indonesia	Natural rubber	11,735,105
7	Brazil	Sugar raw centrifugal	11,548,786
8	USA	Wheat	11,134,659
9	Thailand	Natural rubber	10,634,724
10	France	Wine	9,941,495
11	Argentina	Cake of soybeans	9,906,725
12	USA	Cotton lint	8,425,179
13	UK	Bev. dist. alcohol	8,330,057
14	Brazil	Coffee, green	8,000,416
15	Brazil	Chicken meat	7,063,214
16	France	Wheat	6,738,299
17	Italy	Wine	6,075,404
18	Canada	Wheat	5,742,111
19	Australia	Wheat	5,709,036
20	Brazil	Cake of soybeans	5,697,860

Adapted from FAOSTAT trade statistics database. Major commodity exporters—2007 http://faostat. fao.org/site/342/default.aspx

subsistence agriculture Farming to feed the family, with little or no surplus for sale.

local. Rather, the norm is **subsistence agriculture**, intended primarily for household consumption. Subsistence farming is small in scale, varied (since more than one crop is required to sustain a family through the year), and labour-intensive. In developed countries, by contrast, only 1 to 3 per cent of the workforce is employed in agriculture, and their direct contribution to GDP is similarly small. Agriculture in these countries is large in scale, capital-intensive, highly specialized, and destined for large domestic and international urban markets.

Food insecurity remains one of the human population's most pressing problems. In 2013, the UN's Food and Agriculture Organization (FAO) (http://www.fao.org) estimated that approximately 842 million people, especially in South and East Asia, and Africa, suffer from chronic undernourishment, about three-quarters in rural areas. As many more suffer from the hidden hunger of inadequate micronutrients.

The reasons for hunger are complex, including poor soils, inadequate rainfall, overpopulation, war, disease, lack of education (for women especially), and government corruption, to name a few. Inadequate access to markets is also an important factor. Even subsistence farmers will occasionally have a small surplus that they might barter or sell—if they could transport it to the appropriate market. In many places,

Rank	Area	Commodity	Value (1000 $)
TABLE 10.3	Major Commodity Importers, 2011		
1	China	Soybeans	29,726,067
2	China	Cotton lint	9,466,067
3	China	Natural rubber	8,572,961
4	USA	Coffee, green	7,081,860
5	India	Palm oil	6,765,572
6	China	Palm oil	6,634,042
7	USA	Bever. distil. alcohol	6,399,268
8	Japan	Cigarettes	5,758,924
9	Japan	Maize	5,347,247
10	Japan	Pork	5,205,930
11	USA	Wine	5,046,034
12	Germany	Coffee, green	4,902,386
13	USA	Natural rubber	4,837,811
14	UK	Wine	4,781,924
15	Germany	Cheese, whole cow milk	3,997,915
16	USA	Beer of barley	3,795,971
17	Japan	Natural rubber	3,783,464
18	Germany	Wine	3,252,589
19	Egypt	Wheat	3,199,207
20	UK	Food Prep Nes	3,111,616

however, underinvestment in infrastructure (transportation, communications, storage, etc.) leaves small farmers unable to sell their produce and earn income to invest in productivity improvements. Alleviating these problems is one goal of the UN's millennium development program which calls on both developed and developing countries to improve nutrition, agricultural productivity, natural resource conservation, social safety nets, and market access at the community level.

Food security is also a problem for developed countries, in at least two ways. First, most rely to some extent on imports of many foods that they either cannot grow at all or cannot grow as economically as other places; Japan and the UK depend on imports for approximately half their populations' caloric needs. Second, it is essential to ensure the safety of the food that people consume. Thus even though agriculture is a major industry, subject to all the forces of competitive efficiency, it is also subject to a high degree of government regulation for health reasons. That regulation has to be national in scope because food from any one source, domestic or imported, can be distributed across the country. Central authorities are required to trace the location of food-borne illnesses.

The political economy of agriculture involves more than food security. In Europe and North America, independent farmers are a significant voting constituency, with

food security Assurance of access to food of sufficient caloric and nutritional value in order to sustain a healthy life.

organizations that represent a powerful lobby at every level of government from local to national. A second powerful lobby, sometimes in concert with farmers' organizations and sometimes in conflict with them, is the agri-food industry. As a result of the pressure exerted by these lobbies, agriculture in developed countries is highly subsidized—to the point that payments from governments make up a major share of farm income. Subsidies may also be offered for reasons related to self-sufficiency and maintenance of rural employment and levels. An index of these subsidies prepared by the FAO estimated Canadian governments' support for all commodities at 1 per cent in 2007 of their total value. The US was estimated at 10 per cent and the EU at 26 per cent. In 2007–8, Canadian federal and provincial subsidies amounted to about 40 per cent of the agricultural sector's GDP. While many countries subsidize agriculture, the biggest support continues to be provided by the EU, US, and Japan. Direct or indirect payments to farmers account for the largest share of subsidies, while spending on research and inspection is also important.

Agricultural subsidies are expensive, but there is rarely a huge outcry against their cost, partly out of respect for farmers and the cultural traditions they represent and partly because of support for local production, and perhaps for a degree of self-sufficiency. However, developing countries and their supporters criticize subsidies in developed countries because subsidies keep prices artificially low. Subsidies not only make it difficult for less-developed countries to export their own produce, but encourage those countries to import from the developed world, undercutting the prices that local producers can charge.

Much of the developed world faces the inverse problem of the developing world: it produces and consumes too much food. But quantity is not the only problem. With increasing prosperity has come a major shift in the types of foods consumed—away from vegetables, fruits, and grains and towards meat, dairy, and processed foods that are typically high in fat, salt, and sugar and low in nutritional value. As a consequence, the number of overweight people in the world (estimated at roughly 1.3 billion) surpassed the number of the undernourished (800 million) around the year 2000 (Popkin, 2007). Worse, although the majority of overweight people live in developed countries, their lifestyle is increasingly being exported to the developing world, with similar consequences. A poor processed-food–based diet, combined with decreasing exercise, is now making more and more people in the developing world vulnerable to the same chronic diseases (diabetes, heart disease, etc.) that afflict so many in the West.

The nutritionist Marion Nestle (2007) suggests that pressure to increase shareholder value has led food-processing and grocery companies to (a) produce and sell more food of progressively lower quality and (b) conduct sophisticated research into what ingredients and marketing techniques are most likely to encourage compulsive eating. This problem has been exacerbated by government support for R&D in food processing and agro-chemistry. There has been much less support for R&D in the area of fruit and vegetables.

Agriculture and the Environment

Agriculture, forestry, and fishing all expropriate natural ecosystems for human use. Cropland today takes up 10 per cent of the world's surface, and grazing and pastureland

30 to 40 per cent. Bryant (1997) estimates that the pre-agricultural world contained some 62,203 million square kilometres of forests; of that total, half has been eliminated and only 22 per cent is intact old growth. Virtually all of the earth's marine and aquatic systems are exploited by fishing or aquaculture. The consequences include habitat loss, declining wildlife populations (to the point of extinction in some cases), and ongoing loss of biodiversity. The latter problem is exacerbated by a widespread preference for **monocultures** not only in agriculture but in forestry and fisheries. Unbroken expanses of a single food crop (wheat, rice, soybean) or commercially valuable tree (palm, rubber, pine) leave few niches for wildlife. As for aquaculture, commercial fish farms take over habitat from wild species; become breeding grounds for diseases and parasites that are fatal to them; and, in many cases, rely on wild fish to feed the farmed stock.

monoculture Cultivation of a single type of crop.

Monocultures require enormous inputs of pesticides (insecticides, fungicides, herbicides) and fertilizers. Among the consequences are contamination of both soil and water; soil breakdown and erosion; sedimentation and eutrophy of lakes and rivers; hypoxia of coastal areas; and poisoning of consumers by chemical residues. Other destructive practices associated with industrial agriculture include the intensive use of antibiotics to maintain the health of animals kept in overcrowded spaces, and irrigation, which over time causes salts to accumulate in the soil to the point that it becomes unusable. The use of chemical fertilizers and pesticides has made it possible to feed a human population of over seven billion—some two billion more than the earth could have sustained otherwise. But that intervention—the so-called Green Revolution—has had serious environmental consequences (Case Study 10.1).

Agriculture has the second largest **carbon footprint** of any economic sector, after electricity production. The burning of forests for agricultural purposes releases carbon into the atmosphere, and the crops or range land that replace them have much smaller capacities for storing carbon dioxide and reflect radiation. In addition to the carbon dioxide and nitrogen oxide released through the use of fossil fuels in agriculture, methane emissions from livestock and rice cultivation are a significant factor in global warming. Climate change, incurred by agriculture and other activities, of course has a kickback effect. Shifts in temperature and precipitation regimes have significant implications for agricultural yields and production costs around the world. In California, for example, the rapid expansion of almond production and other nuts, their consumption encouraged by health benefits, is now being arrested by massive water requirements, the availability and cost of which has become prohibitive for many farmers. Elsewhere in the US, cattle herds have been reduced in size because of declining grasslands, and rising costs of feed. The impacts of climate change are predicted to be most severe in tropical and sub-tropical developing countries.

carbon footprint A measure of the environmental impact of a given activity in terms of the amount of carbon it causes to be released into the atmosphere.

The economic ramifications of environmental degradation are profound at every level. First, environmental degradation directly reduces the productivity of the land, which in turn reduces the already limited food available to the poorest subsistence-level farmers. The consequences may not be quite so devastating for those selling their produce at markets, but even if they have enough food to eat, their ability to invest in improving the productivity of their operations will be reduced. Environmental degradation is a primary contributor to the vicious circle that keeps one billion of the world's people in hunger.

Second, environmental degradation imposes higher costs on consumers. The additional inputs required to make degraded land productive, for instance, mean higher costs; prices inevitably rise as supplies are reduced; and certain foods may simply disappear from the market (as is happening now with growing numbers of wild fish species).

Third, environmental degradation imposes externalities not only on the current population in the form of health costs, loss of amenities, and so on, but also on future generations.

Fourth, environmental degradation impairs the ability of the ecosystem to perform countless functions essential to life on earth, including water cycling, photosynthesis, carbon storage, pollination, and seed dispersal. Although a particular instance of pollution may not be noticed outside the local area, the cumulative impact of all the environmental degradation that takes place in a day, let alone a year, is global. Many potential disasters have no doubt been prevented by laws and regulations put in place by local, regional, and national governments. At the international level, however, environmental governance is very weak. The failure to establish governance is largely responsible for the collapse of fish stocks around the world; one group of leading scientists (Worm et al. 2006) has suggested that the world's oceans will be essentially dead by 2050.

As with all industrial sectors, agriculture has to be made sustainable, but there are many visions and contradictions expressed by the term *sustainable agriculture*. To many it is represented by organic agriculture—renouncing the use of chemical pesticides, fertilizers, antibiotics, and genetically modified organisms (GMOs)—and adopting alternative practices such as integrated pest management, use of natural fertilizers, crop rotation, polyculture, and so on. Advocates of this version of sustainable agriculture believe the food is healthier, provides more support for family farms and communities, and protects animal welfare with practices such as free range and pasture grazing. The parameters of organic farming also make it more sensitive to local environments and markets than the monocultures imposed by chemical and energy inputs.

In a somewhat antagonistic view of sustainable agriculture, others claim that GMOs, advanced pesticides and fertilizers, and other high-tech approaches are the only way to sustainability because the organic route cannot feed the globe's growing population. The claim is that advanced techniques will allow us to produce food more intensively and return more biosphere to nature. Arguments and evidence for both sides are continually produced, and both approaches are dynamic and intermesh to a great degree. For example no-tillage agriculture and the development of perennial cropping require significant research, yet the outcomes would be acceptable to both perspectives. The same could be said for urban farming, whether of the garden lot or hydroponic varieties. Most importantly, both views of sustainable agriculture have strong impacts on the reform of the conventional, or industrial, version of agriculture that predominates in developed countries and the subsistence and low productivity agriculture that predominates in developing countries. We will see these interactions throughout the chapter.

Organization: From Old MacDonald to McDonald's

The supply side of agricultural markets, in developing and developed countries alike, is the quintessential expression of market competition. Tens of thousands of farmers may be competing to sell the same commodity at any time. On the demand side, the number of potential consumers is much higher. However, the number of firms that provide the transportation, wholesaling, and retailing links between farmers and consumers is constantly shrinking. The few giants that remain are increasing their control over inputs into the value chain as well as sales.

In market terms, the relationship between farmers and distributors (processors, etc.) is one of price takers and price makers. Because there are so many competing farmers, they have no power to set market prices: they have to take whatever price they can get. Moreover, because of the competition, small changes in demand will cause large fluctuations in price. By contrast, because the large distributors control significant portions of the market, they have all the power they need to set prices and to impose the cost of fluctuating demand on farmers.

The tension between open markets (farmers) and administered markets (distributors, processors, equipment suppliers) is the single most important influence on the configuration of the agricultural supply chain. And that tension is increased by the simple fact that farmers are distributed over space, which means that their production must be gathered and transported to urban markets. In this section we will look at these different types of markets and the variations of their organization from global to local scales.

Interdependence between Domestic and Global Markets

Most countries still produce most of the food they consume, and agricultural markets in almost all countries continue to be dominated by domestic conditions. How a commodity such as wheat or milk is priced is determined through the complex interplay of domestic production, demand, distribution, and regulation. Food prices at any particular time depend not only on the cost of inputs such as land, fertilizer, and livestock feed, but on recent weather conditions; current prices for agricultural products destined for uses other than human consumption, such as animal feed or biofuel; and overall levels of technology (capital intensity). Maintaining reserve stocks of commodities helps to moderate price fluctuations resulting from weather, for instance. Domestic demand for food for human consumption is shaped by population, income, and taste, as well as the prices of competing products imported from other countries.

Although domestic markets largely determine the demand and prices within most countries, international markets have a strong influence as well (Figure 10.3). The prices of export commodities depend on international markets. If international prices sink, then

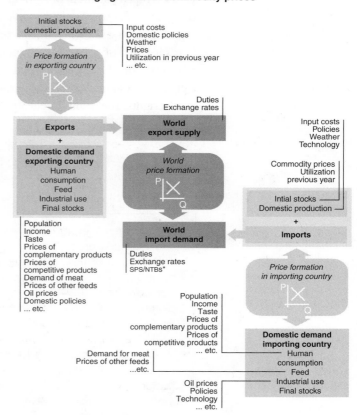

Factors affecting agricultural commodity prices

*SPS/NTBs: Agreement on the Application of Sanitary and Phytosanitary Measures/Non-tariff Barriers.

FIGURE 10.3   Domestic and International Agricultural Market Interdependence

Source: Food and Agriculture Organization of the United Nations, 2009, produced by the Electronic Publishing Policy and Support Branch, Knowledge and Communication Department, FAO, *The State of Agricultural Commodity Markets*, 2009. http://www.fao.org/3/a-i0854e.pdf. Reproduced with permission.

domestic supply will increase, while prices and incomes fall. Importing countries benefit when international prices decrease and suffer when international prices rise. International prices depend in large part on commodities traders, who in transacting sales for a significant proportion of global production, set prices for the majority of production (see Chapter 9). It is to counter the power of these international markets and the relative handful of companies that dominate them that institutions such as the International Fairtrade Organization exist (see Chapter 7). Canada, although it is only a mid-ranking producer of food, is a major exporter and highly dependent on international markets.

Despite the occasional spike (such as the one caused by the sudden demand for biofuels in 2008), global real prices (adjusted for inflation) for food have been declining for several decades as a result of improvements in seeds, fertilizers, pesticides, and equipment; diffusion of agricultural knowledge; and expansion of the land area under cultivation. Most important, however, were the research and development programs that led to the Green Revolution (Case Study 10.1), and more recently to genetically modified (GM) food. GM food is created by merging genes from different species. It was first approved by the US Federal Department of Agriculture in 1994 for the Flavr Savr tomato. (The possibility of DNA transfer was first discovered in 1946.) GM foods have since expanded considerably, especially in the US where most corn, soybeans, and cotton is GM. The rationale for GM foods is based on their impacts on increasing yields and reducing the need for pesticides and herbicides. However, GM seed production is overwhelmingly controlled by Monsanto, a giant MNC, so seed costs are high, and Monsanto is extremely pro-active in protecting its patent rights. The health implications of GM foods continue to be debated by NGOs.

Case Study 10.1

THE GREEN REVOLUTION

The term *green revolution* was coined in the late 1960s to refer to the rapid increases in agricultural production achieved in parts of Asia and Latin America through a combination of new, high-yield grain varieties (mainly rice, wheat, and corn) and mechanization. The research program that developed the new cultivars was established in the mid-1940s and carried out in Mexico and the Philippines under the direction of the American agronomist Norman Borlaug with funding first from the Rockefeller Foundation and later from a number of international agencies.

The green revolution was particularly successful in India, which had suffered a series of disastrous food shortages. But it has been widely criticized on both environmental and socio-economic grounds. Environmentalists argue that mechanization depends on the use of non-renewable fossil fuels and that the technological inputs required for intensive cultivation of a single crop—irrigation, chemical fertilizers, heavy doses of pesticides—have had serious consequences for soils, water supplies, and biodiversity, as well as human health. Other critics have focused on the social and economic implications of the green revolution. For instance, it has been suggested that mechanization has removed an important source of employment for farm workers; that the costs of those high-tech inputs have widened the gaps between rich and poor; and that the need for them has increased developing countries' dependence on the developed world.

Vertical and Horizontal Integration of Production

Agricultural production is a segment in the value chain of both human and animal food, and various industrial products. In the case of fresh produce or meat, the agricultural component *is* the product. For many processed foods and industrial goods, however, agriculture provides only the raw materials, which will be transformed at later stages of the production process. Similar agricultural goods may be grown for all these markets concurrently (e.g., potatoes are grown for direct consumption, for use in processed foods, and to produce starch for the textile, paper, glues, pharmaceuticals, and bioplastics industries). Although primary production is crucial to the agro-industry value chain, it accounts for a surprisingly small percentage of its overall value. In Canada, the agri-industry system accounts for roughly 8 per cent of GDP and employs 2.1 million people (2013). Within that system, however, primary agriculture represents only 1.3 per cent of GDP and 1.8 per cent of national employment. Upstream input and service suppliers (e.g., equipment or fertilizer suppliers) represent 0.7 per cent of GDP and 0.4 per cent of employment, while downstream food and beverage processing accounts for 2 and 1.7 per cent; food retail and wholesale represents 2.6 and 3.7 per cent; and food service 1.5 and 5 per cent.

As we noted in the opening of this section, the agriculture value chain is characterized by a vastly greater number of competitors in the production (farming) segment than in either the upstream (supply of inputs) and downstream (processing, transportation, distribution, etc.) segments. The oligopolies that dominate those segments integrate the value chain. A handful of upstream–downstream MNC conglomerates exploit the synergies between these agricultural activities by combining and coordinating functions from gene production to processing (Figure 10.4). Firms such as

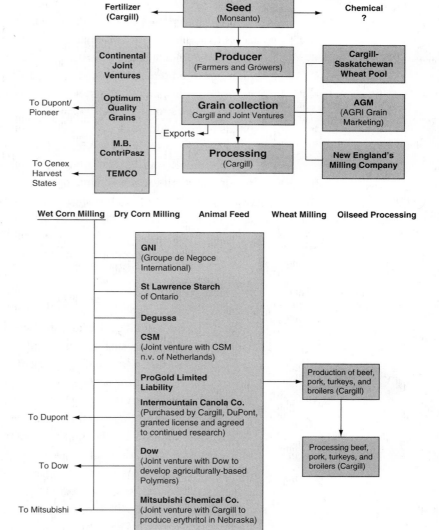

FIGURE 10.4 The Monsanto/Cargill Cluster

Source: Adapted from Food Circles Networking Project at http://www.foodcircles.missouri.edu/cargill.pdf

Cargill—which in 2013 had operations in 66 countries, employed 140,000, and earned revenues of $136.7 billion—integrate storage, transport, and processing of finished products (especially generic and "house brands" for retailers), while providing raw and semi-finished materials to other food processors. Cargill is the largest privately held corporation in the US.

The food processors make many familiar brands for grocery and convenience stores, restaurants, and vending machines. However, they rarely extend their operations into grocery retailing or food services. Food retailing is dominated by another set of firms, which concentrate mainly on domestic markets. The grocers integrate upstream directly by manufacturing certain products (Loblaw is Canada's largest baker) and by contracting with farmers for vegetable and fruit production. Canada's four largest grocery chains now control 75 per cent of sales, though there are roughly 4000 independent grocery stores. By providing information on matters such as food safety and supply chain efficiency, the Canadian Federation of Independent Grocers gives smaller firms access to some of the economies of scale enjoyed by the big chains. Increasingly, grocers must compete with department stores, drug stores, gas stations, and other non-traditional food retailers, which together have captured about 15 per cent of the market.

Food service companies seek a greater degree of integration. Full- and quick-service restaurants together receive 36 per cent of the money that the average Canadian spends on food, although only 8 per cent of meals are eaten there. Although competition is greater in the food service industry than in most segments of the value chain, there has been significant horizontal (e.g., by Cara Operations) and vertical (e.g., by Tim Hortons) integration. The degree of monopoly is much higher in specific sub-sectors; McDonald's, for instance, has approximately 90 per cent of the US hamburger market. Many big food service companies are integrated backwards into food preparation and processing. Some also exercise considerable control over transportation, storage, and even agricultural production, through relational contracting and quality standards.

Recycling

Recycling is not considered part of the agricultural value chain, even though the agro-industry produces significant amounts of waste—not only packaging but organic materials and sewage sludge. Geographically and ecologically, the outstanding characteristic of the latter two waste flows is that the nutrients they contain are not returned to their various sources. Instead, most become landfill in locations far from their point of origin. Yet most organic waste can be recycled, while sewage sludge, after processing, can be used for fertilizer or to produce methane for energy. Organic kitchen waste (vegetable and meat) currently makes up about 40 per cent of municipal domestic waste and is usually disposed of in landfill, although it can be composted for fertilizer or converted into either methane or syngas. Cans, glass containers, plastics, and so on can go into different waste streams for recycling; such materials are more likely than organics to enter international material markets. While there is little downstream integration of distributors or upstream producers into recycling of organic waste, firms are increasingly held responsible for their packaging, especially in jurisdictions (e.g., Europe) with "take-back" legislation.

Waste collection and disposal are the responsibility of municipal and regional governments. Over the last few decades, most municipalities have outsourced these services to private companies. In turn, in order to achieve economies of scale, the industry has consolidated in the hands of a few large firms. Waste Management, Browning-Ferris, and Allied Waste now control large shares of the North American waste collection and disposal market.

Advantages of Integration

In farming as in any other industrial sector, the driving force behind integration is the desire for economies of scale. For example:

- *Facilities and equipment*: Needs range from grain-handling facilities and concentrations of feedlots to storage transportation (trucks, barges, and railway cars with specialized design, refrigeration, and other equipment) and processing facilities, to retail outlets for sales to consumers.
- *R&D*: Different agricultural sub-sectors may require biotechnology, chemical and mechanical engineering, and other capacities for breeding and genetic modification, or for the development of pesticides, fungicides, and fertilizers. Furthermore, since the profit margins on commodities tend to be low, finding ways to add value requires constant innovation: for instance, developing new food products or applications (pharmaceuticals, nutraceuticals, cosmetics, textiles, etc.).
- *Marketing*: Marketing is required simply because agricultural production takes place far from urban markets. The majority of agricultural produce is branded either as packaged goods or as part of another good or service (grocery, restaurant, etc.).
- *Management*: Coordination requirements increase with the scale and diffusion of operations. (Cocoa, for instance, is produced and purchased in several countries on three continents; then it must be processed into chocolate in several countries on two other continents.) Greater managerial capacities give corporations significant information advantages over farmers, particularly with respect to pricing and bargaining with competing suppliers.
- *Finance*: Agriculture and forestry face distinctive risks related to weather, disease, environmental change, and politics. Managing such risks requires financial resources. One of the most outstanding characteristics of large agribusiness firms is their power to buy and sell in quantity, which gives them leverage to extract lower prices within the value chain. Financial power also allows them to take advantage of merger and acquisition opportunities.
- *Supply chain accountability*: Health or environmental problems in the food-supply chain often lead to increased regulation. Although many small farmers have been put out of business because they couldn't meet stringent new regulations, various health alarms (*E. coli*, bovine spongiform encephalopathy, etc.) in recent years have made it imperative that the origins of every food product can be traced. That task is easier if there are only a few trusted suppliers. Ironically, organic farmers have to pay for certification, whereas conventional farmers are free to externalize their environmental impacts and do not even have to pay the additional labelling costs that organic farmers do.
- *Standardization*: Mass-produced food products require inputs that are qualitatively standardized, and food-safety regulations amplify this tendency. Such standardization is far easier for large, integrated firms. The same emphasis on standardization or—as farmers increasingly refer to it—"commoditization" drives the need to differentiate through R&D and product innovation.

The forces compelling integration and consolidation are at work in and among all segments of the agriculture value chain. Almost all chicken, beef, and pork production in North America is now controlled by a handful of companies. Although this is not the case with produce and grains, the pressure to eliminate seasonal fluctuations, to have smooth production and distribution levels, to standardize and brand particular products, and to industrialize agriculture is pervasive. McDonald's has led this trend for several decades (Case Study 10.2).

The Primary Producer Segment

Old MacDonald's farm is no more or at least not much more. Small mixed farms, raising chickens, pigs, and cows, with a garden and assorted crops, have been

---

Case Study 10.2
INTEGRATING MCDONALD'S

McDonald's describes its franchise system as a three-legged stool consisting of the corporation, its suppliers, and its owner/operators. The corporation develops the products, runs the production, distribution, and training systems, and arranges the purchases of all inputs to achieve economies of scale for 36,000 restaurants, 69 million customers per day, and annual sales of US$27 billion (2014).

McDonald's only deals with suppliers that can deliver on a massive scale. While some of these firms are multinational, many more are regionally based; this not only reduces costs but makes it possible to accommodate cultural differences in taste. In Canada, only 100 firms supply 1,440 restaurants with a division of Cargill supplying all beef and chicken. When McDonald's expanded into India, it worked with a limited number of suppliers to bring them up to its standards. McDonald's maintains strict control over all suppliers of the food it uses in order to meet its own quality standards and comply with the health and environmental regulations imposed by government. It is also subject to NGO governance; for example, it has banned the use of beef from cattle reared on land cleared from rainforests and of chicken from birds whose beaks have been cut off.

Third-party logistics firms supply both transportation and distribution centres (DCs). There are about 180 such centres around the world, each controlling about 400 stock-keeping units and delivering to approximately 200 restaurants. This system is constantly being upgraded to lighten the stock- and record-keeping burden of the restaurants, while ensuring that their supply needs are met automatically.

McDonald's makes its money from its franchisees who return 4 per cent of monthly sales and rent to the franchisor. To qualify, franchisees must undertake a rigorous training program and put up 30 per cent of capital to purchase the store. McDonald's sets the standards required in food preparation, food service, and employee training. Increasingly, to meet environmental demands, McDonald's is also taking responsibility for the recycling of its packaging and (where possible) composting food waste. A significant part of McDonald's success is the matching of technological food preparation with easily trained labour and with self-service by the customers. Thus, it integrates operations from field to rubbish bin to reduce costs and build profits. Fast-food workers have become increasingly restive about their returns from this system, but without a union's countervailing power they are in much the same price- (wage-) taking position as competing farmers.

replaced by specialized operations focusing on a single product and making intensive use of costly equipment, chemicals, and facilities. Many small farmers in Canada are hobby farmers with alternative sources of income. The number of farmers in Canada, as elsewhere, has dropped dramatically over the last century. Yet the primary sector of the agricultural value chain—that is, actual production of food—remains highly populated. Farmers face formidable challenges related to competition, changing markets, technology and policy, natural hazards, and growing concern over environmental impacts. Beef production in southern Alberta, and the shift in grain markets from export to domestic sources of feed, illustrate these challenges (Case Study 10.3).

Case Study 10.3
FEEDLOTS IN LETHBRIDGE COUNTY, ALBERTA

Feedlots are specialized farms that feed and prepare beef cattle for slaughter. They procure feeder cattle from ranches and dairy farms and after 90–150 days "on feed" these feeder cattle are shipped to meat-packing plants where they are slaughtered and processed into beef. Some feedlots are owned by ranchers (forward integration), others by meat-packing enterprises (backward integration), but many are independent, making them vulnerable to daily fluctuations in grain and cattle markets. To avoid this risk, many feedlots either operate on a "custom" basis for a daily fee or have a "forward contract" with a packing plant to supply a specified number of cattle with particular weight and quality attributes.

This feedlot pen is being cleaned out by a specialized front-end loader and manure spreader. The feedlot operator has contracted this service from a custom corral-cleaning firm whose services include scraping excavation and loading

(Continued)

of manure into trucks, as well as hauling and spreading or stockpiling the manure subject to provincial environmental regulations. The feedlot operator thus avoids owning expensive and specialized equipment that is only used occasionally, reduces feedlot labour force requirements, and relies on specialized contractors to apply manure in the precise concentrations needed to maximize barley yields, consistent with regulations.

Lethbridge County boasts the largest concentration of beef cattle in any Canadian municipality. Large feedlots holding 10,000 to 30,000 head of feeder cattle first appeared in Alberta in the 1980s, stimulated by agricultural policy. With the end of freight-rate subsidies for grain exports (known as the Crow Rate), farmers began to grow feed grain for domestic consumption by cattle. The Alberta government compensated farmers for loss of the subsidy, especially in areas where irrigated water was available. Lethbridge County feedlots are also dependent on imported corn from the American Midwest and various by-products such as dried distillers' grains from methanol and biodiesel plants, and by-products from milling sugar beets. The environmental burden posed by feedlot agriculture is locally contentious, related to foul odours, the heavy application of manure on limited crop land, the potential for groundwater contamination, and the possibility that overland runoff from manure stock piles and manure lagoons may be gradually degrading rivers and streams. In response, Alberta's Natural Resources Conservation Board has established a provincial code of practice for manure handling.

This information kindly provided by Ian MacLachlan, October 2014

Engel's Law A theory proposing that the proportion of income that people spend on food declines as their income increases.

The Challenges That Farmers Face

The Great Transformation shifted the emphasis in agriculture from subsistence to commodity production. Farmers were caught up in capitalism's drive for constant economic growth and continually increasing productivity. At the same time, the relative value of agricultural production began to decline. In 1857, a German statistician named Ernst Engel found that, as people's income rises, the proportion of it that they are willing to spend on food declines; even if they spend more than they did in the past, the increase does not keep pace with the increase in income. Today Canadians on average spend less than 10 per cent of their incomes on groceries (Figure 10.5). Farmers have been drawn into this virtuous (or vicious, depending on your viewpoint) cycle by several interrelated realities.

Engel's Law is just one of several interrelated realities that make life difficult for agricultural producers. Others include:

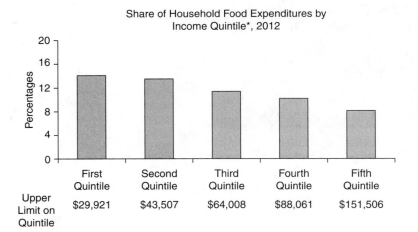

Share of Household Food Expenditures by Income Quintile*, 2012

	First Quintile	Second Quintile	Third Quintile	Fourth Quintile	Fifth Quintile
Upper Limit on Quintile	$29,921	$43,507	$64,008	$88,061	$151,506

FIGURE 10.5   Engel's Law in Canada: Proportion of Household Income Spent on Food, 2012

*Quintile: Households are ranked in ascending order by total household income and are then divided into five groups, each comprising 20 per cent of households. The 1st quintile is the lowest 20 per cent of households and the 5th quintile is the highest 20 per cent of households.

Source: Based on Statistics Canada and Agriculture and Agri-Food Canada (AAFC) calculations.

- *Demand for increasing returns*: Every time farmers borrow money to invest in their operations, they are charged interest. To be justified, therefore, any capital investment must generate sufficient returns not only to pay back the loan, but to cover all the costs associated with it. In this way the pressure for increased productivity becomes constant.
- *Capital intensity*: Farmers have sought productivity increases primarily through investment in machinery, automation, expansion, and inputs such as improved seed and breeding stock, fertilizers, and pesticides, and, perhaps most important, increases in acreage. Output figures have soared with these inputs, but so have farmers' costs.
- *Lack of control over prices and risks*: Since the industrialization of agriculture, the number of firms in the accumulating, storage, and processing segments of the value chain has decreased even more dramatically than the number of farmers. With less competition among buyers of their produce, the market power of farmers has declined. Furthermore, demand is both price- and income-inelastic—people are not willing either to buy greater volumes or to pay more per unit—while supply varies from year to year, depending on weather, pests, and so on. Yet farmers have no control over the risks they are exposed to or the prices they receive for their produce. In bad years, many can't produce enough to benefit from the higher prices that come with scarcity; while in good years, prices are forced down by the increase in supply.
- *Lack of control over input costs*: While most farmers either serve open markets or sell to much bigger powerful organizations, inputs of seed, fertilizer, and equipment are concentrated in the hands of a few sellers who can control prices. Monsanto, for example, owns over 80 per cent of GM seeds grown locally.

Contracting, Co-operatives, Marketing Boards

The transactions between farmers and their downstream buyers are rarely handled through open markets. Instead, they are mediated by relational contracting, co-operatives, and marketing boards.

Most grocery and fast-food chains and food processors will contract directly with large farms. This *relational contracting* reduces risk by giving buyers greater control over quantity, price, quality, delivery timeliness, and variety than they would have if they relied on open markets. At the same time, it assures farmers a market for their produce, and if the profit margin is sufficient, to plan investments in inputs. Relational contracting involves a high level of information exchange and operational coordination between the parties to the transaction. But there are risks, because buyers commit themselves to large purchases from a source that is vulnerable to any number of variables, including weather conditions. Farmers, for their part, have to make dedicated investments that may be of little use if the relationship ends, so the terms of the contract need to be carefully specified. In a type of relational contract known as a *production contract*, the buyer actually owns the commodity and possibly other inputs, while the farmer supplies the production services along with other inputs such as facilities, equipment, and possibly feed. Both transaction partners are compensated according to their inputs, but generally production contracts give the buyer greater control. In another type of relational contract known as a *market contract*, the buyer still stipulates quality, price, and so on, but the farmer owns and manages the commodity and inputs.

Relational contracts are most popular among agro-processing companies and large commercial farms, particularly in the US. Contracts grew from an 11 per cent share of the market in 1969 to 41 per cent in 2005, with 40 per cent of the contracting concentrated in the hog and poultry industries. For example, Tyson, the world's second largest processor and distributor of chicken, pork, and beef, had almost 6,800 poultry farms under contract in 2013, and generated revenues of over \$34 billion. Other brand manufacturers such as Kraft, Kellogg's, and Heinz also contract extensively. The practice is vital to permit standardization in restaurant chains. Kentucky Fried Chicken has to ensure that all its drumsticks are a specific size and shape. Farmers may also sign contracts upstream with suppliers of inputs such as seed or pesticides. As the world's principal supplier of genetically modified seeds and the pesticides they require, Monsanto sells to farmers under contracts with strict stipulations. Hundreds of farmers have been taken to court for violating Monsanto's prohibitions on storing, using, or selling seeds produced by the previous year's crop. To further protect its intellectual property rights, the company has even sued farmers who never bought seeds from it, let alone signed any contract—simply for having Monsanto seeds growing on their property, regardless of whether they were being intentionally cultivated.

Since the nineteenth century, the most common vehicle for farmers seeking to improve their position in the value cycle has been the *co-operative*. Three types of co-operatives provide economies of scale and upstream and downstream integration that give farmers a way out of their price-taker predicament in buying and selling situations. *Marketing co-operatives* bring together their members' production for branding, distribution, and retail sale. The Alberta Wheat Pool, Manitoba Pool Elevators, Saskatchewan Wheat Pool, and United Grain Growers collected their members' grain crops, stored them in co-operative–owned elevators, and sold them on world markets. Beginning in the 1990s, these co-ops were amalgamated and privatized as the Viterra Corporation. In 2009, Viterra purchased an Australian company, largely in order to gain the scale necessary to compete with Cargill and Archer-Daniels-Midland. Co-ops control 59 per cent of the dairy market in Canada, processing and distributing under names such as Gay-Lea in Ontario and Scotsburn in the Maritimes. In Quebec, Co-op Fédérée handles 65 per cent of poultry slaughtering and half of retail poultry sales. Co-ops also operate in the vegetable and fruit regions of the Maritimes, Ontario, and British Columbia. In 2007, sales by Canada's 1309 agricultural co-operatives amounted to more than \$8.9 billion at home and abroad. Regionally focused supply co-ops such as Co-op Atlantic, Federated Co-operatives Ltd, and GROWMARK Inc. reduce the costs of fertilizers and chemicals, feed, and seed.

Co-operatives are a foundation of EU agriculture, and in the US there are close to 4000 co-ops with a total membership of more than three million. Political activities are a core function of co-ops, which inform members regarding policies that will affect them and present a united voice in the policy-making process: strengthening the farmers' position in the value chain; gaining financial, R&D, export, and other support; and ensuring sufficient welfare services. Co-ops not only get involved in politics, but also form political parties. In Canada before the Second World War, several provincial governments were formed by co-op–led parties, notably the United Farmers of Alberta (UFA). UFA members were among the founders of the Co-operative Commonwealth

Federation (CCF; the forerunner of the NDP), which developed a number of policies that Canadians and other nationalities now take for granted, including old-age pensions and universal health care.

Co-operatives use their political influence to attract government backing for their efforts to improve the terms of trade for farmers. Marketing boards are organizations authorized by national and provincial governments to manage supply and demand by controlling production levels and prices. The objective is to stabilize price fluctuations related to production cycles, weather variations, and so on. Quotas are often set to control production as well as entry into the industry. Today there are marketing boards for more than 80 sub-sectors in Canada, the most famous being the wheat, dairy, and egg marketing boards. With the free market–oriented (neoliberal) philosophy of the Conservative government, the Canadian Wheat Marketing board lost its monopsony to buy grain in 2012 and then was majority privatized with foreign ownership in 2015. The US does not have marketing boards; however, a group of producers can vote to have the local or state government bring in certain measures or regulations that will be binding on all members of the industry in their region. Government collection and publication of price information also helps to level the field between farmers and downstream firms with better knowledge of current prices.

Regional Renewal and Relocalization

Rural–urban migration has accompanied industrialization and post-industrialization. It has been a spatially segmented and chronologically phased process, taking place over the course of two centuries in Europe and North America and continuing today dramatically in China, with India and Africa soon to follow. Rural depopulation has been pushed by increasing capital intensity (which has eliminated many jobs in agriculture, forestry, and fishing) and pulled by the agglomeration power of cities offering diverse opportunities for employment in the manufacturing and service industries. Canada is a good example. Until 1921, it was a predominantly rural nation. Since then, the urban population has climbed to 78 per cent (2010). In the wake of this migration, rural communities, along with the towns and villages that used to provide them with services, have lost their critical mass and in some cases disappeared. Combatting this decline in population requires that primary producers discover ways of adding value to their products and avoiding the commodification trap. To that end, they have sought to sell their products both on short value chains connecting to local consumers and on long value chains connecting them to distant urban markets.

Current Trends

In advanced countries, the contributions of the agricultural value chains to overall GDP remain important. In Canada in 2012, for example, together, the food service, retail and wholesale, processing, input and service suppliers, and primary agriculture segments of the agricultural-based value chain represented roughly 6.7 per cent of GDP and around 12 per cent of employment. The contributions of primary agriculture to the GDP, however, were only about 1.3 and 1.6 per cent respectively (Figure 10.6). The primary

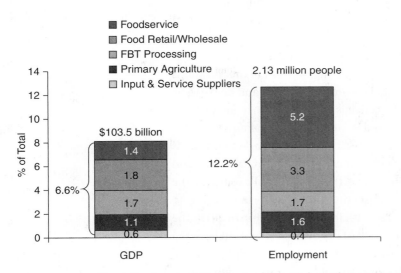

FIGURE 10.6   Contributions of the Agriculture Value Chain to GDP and Employment, Canada, 2012

Source: Based on Statistics Canada and AAFC Calculations

goal has been to produce more using more land and fewer people. Farms are becoming larger even as operating margins (per unit profits) decline.

The agricultural sectors in the US and EU have followed similar, though not identical, trajectories. Farm numbers in the US declined from four million in 1960 to about two million in 1990, and have remained stable since then. However, concentration of production is much higher. Only 2 per cent of farms produce more than 50 per cent of all agricultural commodities. Incorporated agribusiness is more prevalent, as are "family farms" with dozens of full-time and seasonal workers (some of whom may not be legally employed).

For decades, the EU's Common Agricultural Policy promoted a more "rational" approach to agriculture based on large, capital-intensive farms. However, tradition helped to maintain smaller operations, and policy was eventually changed to accommodate them. In 2010, there were 12.2 million farms in the EU, of which about half had less than 2 hectares, while a small number of large farms (over 100 hectares)—2.7 per cent of holdings—controlled half the farmland. Farm sizes and employment numbers vary from country to country, depending on earlier farm policies and date of entry into the EU. In the UK, rationalization policies adopted by the government of Margaret Thatcher helped to expand average farm size, but Italian policies support much smaller farms. France is a mix: although its rationalization policies and large land area have given it the highest proportion in the EU of farms over 50 hectares (37 per cent), other policies support small and traditional farming. Many new entrants to the EU have seen their farm numbers drop dramatically because the EU market demands greater scale and efficiency.

In the developing world, agriculture is still dominated by small landholders producing for both local and global markets. But sales in local markets, as we noted earlier, are influenced by the prices on global markets. Developing countries often complain to the WTO that the subsidies provided to farmers in the EU, North America, and Australia have the effect of reducing prices in developing countries and impoverishing small landholders there. Meanwhile, MNCs continue to move into developing countries both as buyers and as organizers of production. Whether the governments of these countries encourage or discourage trade and FDI depends on their governance capacities, historic influences, and economic approach.

Demise and Diversification of Rural Communities

Ironically, efficiency gains on farms have negative multiplier effects in small communities (see Case Study 2.2). Localization economies suffer because fewer farmers mean less demand for local suppliers of goods and services, from feed and machinery stores to banks. Larger farms often go to larger towns for such services. Urbanization and rural infrastructure economies go into decline as reductions in demand lead to decreases in retail sales and student numbers, and fewer people walking into local health clinics. Government maintenance of services and infrastructure becomes cost inefficient when demand drops below a minimum level. Schools and clinics have to be shut down. Often these trends are reinforced when regional superstores attract business away from main streets or governments centralize services in urban centres. Residents who can no longer get the services they need move away, exacerbating the local brain drain and the loss of community vitality.

The long established trend of rural population decline in most advanced countries has to varying degrees been arrested and even reversed, but not for farmers per se. The increase is the result of the extension of commuting from fast-growing metropolitan areas, by retirees looking for alternative lifestyles and where tourism and recreation opportunities have expanded. Over the past two decades the idea of post-productivism, a term that broadly indicates a decline in the role of production and related social values in rural areas, has also been debated.

From Regional Commodity Exports to Regions of Origin

The ability of farmers to increase their profits and income, without becoming caught up in the trap of constant scale increases, depends on their ability to differentiate their products. Differentiation can be either vertical (different grades of a particular quality such as sweetness or redness in strawberries) or horizontal (different varieties of strawberry or the same variety grown in different locations, each of which produces a unique taste). In essence, primary producers need to build reputations for both quality and distinctiveness.

Marketing co-operatives exemplify farmers working together to enhance quality. The differentiation impact is especially strong when the co-op distributes its products under its own brand name (e.g., the Gay Lea dairy co-operative in Ontario, the Blue Diamond almond co-operative in California, or the Sunkist co-operative representing citrus growers in California and Arizona). Most co-ops, especially in Canada, are deeply rooted in specific regions and remain committed to them. In general, marketing co-operatives focus on vertical differentiation and compete with other commodity providers on that basis. Although it was not unusual in the past for co-ops to expand their production to other locations, more and more producers are focusing on geographical origin as the feature that distinguishes their product from all rivals.

Many famous products take their names from their places of origin, including cheeses like Brie and Parmigiano-Reggiano, wines such as Champagne and Chianti, and specialty meats such as Parma ham. These products are so widely known that their names came to be the generic terms for all products of their type, however inferior. Yet for a place name to effectively capture value-added, producers must be able to limit its use. National and international regulations restricting the use of such names to the authentic regional products have been under development over the past century. Many such names are now recognized by the WTO as having a status equivalent to trademarks. France led the way with its "appellation of origin" (AOC, *Appellation d'origine contrôlée*) system. Originally developed by winegrowers, it was adopted for use by cheesemakers, vegetable and fruit growers, and meat producers. Now products of a similar type that are produced elsewhere are forbidden to use the names of the authentic products. Many other countries have developed similar systems. Formal use of **geographical indications** in North America is limited to wine—in Canada, the system is called Vintners Quality Assurance (VQA)—but PEI potatoes, Nova Scotia lobster, and Florida oranges could lay informal claim to a similar status.

Acquiring a geographical indication motivates producers in a particular territory to organize as a community working together for collective goals. The primary objectives are to build the reputation of the territory's products and to ensure that use of the appellation is restricted to local producers and authentic products of the territory. To protect and enhance the value of the appellation, producers typically develop quality control, quantity controls, and the capacity to defend the trademark in court, in addition to organizing marketing and R&D.

While territories in many agricultural sectors have focused on building their collective reputation, wine territories have also developed the capacity to enhance the individual reputation of each winegrower. It is an interdependent process, as in places such as Bordeaux, the reputations of individual winegrowers have been critical to building the reputation of the territory. Co-operatives have been integral to the development of collective reputations, often representing 25–50 per cent of a region's production. However, where independent winegrowers, cheesemakers, and others strive to strongly differentiate their products, co-operatives have typically blended or standardized their members' products, though they increasingly are working to allow greater differentiation, whether by labelling a members' produce or by some other distinction.

Territorial organizations provide economies of scale in collective reputation and economies of scope in R&D, distribution, and marketing that let individual producers enjoy the advantages of integration (including branding) without increasing the scale of their operations. Of course, to have the authority to enforce the rules decided by the majority of their members, such organizations need a mandate from the national government (Case Study 10.4).

In addition to differentiating products for sale in urban markets, a distinct territorial identity can bring urban markets to the producers and augment agriculture with other activities. Farmers can increase their direct sales significantly if, in addition to offering visitors the opportunity to taste and buy their produce, they can offer other types of activity (recreation, entertainment, etc.). Agritourism helps to revive both local and urban/rural infrastructure economies as tourism generates business for restaurants, historical attractions, and so on; it may also attract more producers, who will in turn increase demand for various goods and services. There is an increasing awareness that if producers are subsidized to grow crops or leave them fallow, they can preserve the ecological and scenic functions of the countryside. The EU in particular promotes multifunctional agriculture, but other types of rural business, such as forestry and fishing operations, can also combine their main activity with tourism.

Rise of the Greenbelt

Urban development has typically meant the expropriation of agricultural land for residential, industrial, and commercial purposes. True to von Thünen's theory, the rents

---

geographical indication A kind of legal trademark given to growers in a specific region; in some cases a geographical indication comes with regulations on cultivation and processing methods.

Case Study 10.4
ITALY'S CHIANTI CLASSICO TERRITORY

Chianti is the name of the wine-growing region in the hills between the cities of Florence and Siena in Tuscany. For more than two centuries, the name referred to a small portion of that area that gave birth to a distinctive and eponymous wine. When wineries outside the area started to use the Chianti name, winegrowers within the area began an 80-year battle to win recognition for a territorial trademark (a "geographical indication") and governance over the winegrowers who want to use the trademark. The Italian government never rescinded its 1929 decision to allow other areas and producers to use the Chianti name, compelling the original Chianti district to create a new identity and territory under the name of Chianti Classico. To that end, they created borders that delimited who could use the trademark, common winegrower practices, marketing efforts, and other collective activities. When estate winegrowing began to flourish in the 1970s, this territorial identity added significant value to the Classico wines, facilitating investment in improving facilities and practices. The area attracted entrepreneurs, and a local workforce that had been largely lost to urban migration was replaced.

While the men of Chianti Classico concentrated on the techniques of the cellars, the women realized tourism depended on a warm reception on the estates and throughout the territory and set up an organization for that purpose. By the 1980s, many producers were renovating their properties to accommodate tourists. Soon, the Classico district became a prime destination for the "slow food" movement as well as tourists aspiring to learn Tuscan cuisine. The winegrowers' association created an organization to preserve the territory's cultural heritage of churches, farmhouses, and castles. Meanwhile, forests were protected against the expansion of vineyards and towns. In this way agriculture, tourism, and environmental preservation reinforced one another.

The winegrowing territory encompasses parts of the two provinces of Firenze and Siena, five complete communes, and segments of another four. Although the objectives of the winegrowers usually coincide with those of other citizens and politicians, from time to time they come into conflict over some industrial or urban development issues. However, the territory, supported by the regional, national, and EU governments, has become a model of how multifunctional agriculture can work.

© iStockphoto.com/clodio

from these land uses—higher than any agricultural rents—have driven farming ever farther from the urban core. Land of the highest agricultural quality has been swallowed up for more profitable uses. Adding to the power of the market in such cases are municipal planners anxious to annex more land to increase the city's tax base. Up-to-date statistics are hard to find, but losses of typically the best agricultural land to urbanization is reported as an important problem in many countries, not least in North America and the EU. Meanwhile, farmers seeking to expand their cropland to meet growing market demand are forced into increasingly marginal areas.

Ironically, demand for locally grown vegetables, fruit, meat, and dairy products has never been higher. **Locavore** demand is increasingly met not only by farm shops, roadside stands, and pick-your-own operations, but also by urban farmers' markets where face-to-face encounters between farmer and consumer add both economic and social value to the transaction. Economic and social value-added also coincide in **community supported agriculture**, in which a number of consumers buy shares in an individual farmer's production for the year and receive regular deliveries of produce throughout the growing season. In this way the farmer is guaranteed an income that may even allow her to invest in improving her operation. The resurgence of local production is reinforced by growing demand for organic food, support for environmental values, and resistance to industrialized agriculture. The planting of an organic garden in the White House reflects the increasing respectability of urban farming.

Non-governmental organizations, farmers' groups, and, ultimately, governments have worked to preserve **greenbelts**, where land around cities is zoned for agricultural, natural, or recreational uses. In Britain, a greenbelt around London was first proposed in the first half of the nineteenth century and was formally established in 1938. It is the largest of 14 greenbelts that now protect approximately 13 per cent of England's land area. Other notable greenbelts surround Copenhagen, the cities of Portland and San Francisco in the US, and the lands of the former border between East and West Germany. Canada has three greenbelts: in the city of Ottawa; in southern Ontario's "golden horseshoe"; and in BC (the Agricultural Land Reserve is not a continuous "belt," but a series of protected areas in different parts of the province). Common threats in greenbelt areas include urban sprawl, resource extraction (especially for gravel), and road-building. Deft coordination of municipal and regional authorities is required to ensure that short-term financial need is not allowed to prevail over the long-term need for green space. Support is also needed for local agriculture and marketing, recognition of the multifunctional (especially ecological) benefits of agriculture and nature, education of producers and consumers, and accessibility for urban residents.

locavore Someone who chooses (as far as possible) to consume locally grown food, in order to support local farmers and minimize the environmental costs of transporting food from farm to market.

community supported agriculture A system in which consumers (usually local) contract with farmers to buy their produce.

greenbelt Land around cities that is zoned for agricultural, natural, or recreational uses.

Conclusion

The essential challenge for agriculture is to deliver perishable produce to distant markets when production is diffused over vast spaces. This creates an organizational problem that markets and institutions have sorted out in a variety ways. Von Thünen's classic explanation of land use around an urban market as the product of the relationship between distance and value is still relevant. However, as urban markets have expanded both in size and in complexity of demands, distance from the market and economies of scale have increased. Within a vastly expanded spatial differentiation of agricultural production, there are several ways of integrating the value chain from the farmer's field through collection, storage, and processing to distribution and marketing. But in recent decades a few giant firms have come to dominate their markets. This evolution has reduced food costs while increasing dramatically the range of options available in both the grocery and restaurant sectors. Unfortunately, it has also been accompanied by environmental damage, nutritional misdirection, and increasing pressure on the economic viability of the independent farm in both developed and developing countries. Farmers have recognized the weakness inherent in their location on the value chain and have collectively organized themselves to obtain the same economy of scale and integration benefits of large corporations. They have also found means to decommoditize their production through collective and individual branding.

Practice Questions

1. Map the global geography of sugar production and trade. If sugar offers little if any nutrient value, then why is consumption of sugar so big? Which companies control the production and distribution of sugar?
2. Map the global geography of tobacco production. Explain variations in the per capita consumption of cigarettes.
3. Compare the export geography of wheat for Canada and France.
4. How does your community dispose of household and garden waste?
5. Should the countryside be protected from urban expansion? Explain the pros and cons of greenbelts and agricultural reserves.

Key Terms

environmental conditions   305
co-operative   305
von Thünen, Johann Heinrich   306
location rent   307
normal profit   307
intensive agriculture   307

extensive agriculture   307
subsistence agriculture   310
food security   311
monoculture   312
carbon footprint   312
Engel's Law   322

geographical indication   328
locavore   330
community supported
   agriculture   330
greenbelt   330

Recommended Resources

Almstedt, A. 2013. Post-productivism in rural areas: A contested concept. In Lundmark, L. and Sandström, C. eds. *Natural Resources and Regional Development Theory*. Umeå: Umeå University.
Reviews the debate over post-productivism.

Carter-Whitney, M. 2008. *Ontario's Greenbelt in an International Context: Comparing Ontario's Greenbelt to Its Counterparts in Europe and North America*. Toronto: Canadian Institute for Environmental Law and Policy.
A comparison of greenbelts in Ontario and elsewhere.

Kedron, P. and Bagchi-Sen, S. 2011. "A study of the renewable energy sector within Iowa." *Annals of the Association of American Geographers* 101: 1–15.

Explains the evolution of Iowa's ethanol industry especially in relation to government policies, R&D and innovation, the role of farmers, and agri-processors.

Maye, D., Holloway, L., and Kneafsey, M., eds. 2007. *Alternative Food Geographies. Representation and Practice.* Amsterdam: Elsevier.

Examples of several alternative food value chains.

Morgan, K., Marsden, T., and Murdoch, J. 2006. *Worlds of Food. Place, Power, and Provenance in the Food Chain.* Oxford: Oxford University Press.

An overview of food value chains and their governance.

Patchell, J. 2008. "The region as organization: Differentiation and collectivity in Bordeaux, Napa, and Chianti Classico." In Stringer, C. and Le Heron, R. eds. *Agri-food Commodity Chains and Globalising Networks.* Aldershot, England, and Burlington, VT: Ashgate.

Three wine-growing territories that have inverted market power within the food value chain.

Wrigley, N. 2001. "The consolidation wave in US food retailing: A European perspective." *Agribusiness* 17: 489.

An overview and critique of the consolidation and backward integration of the food sector, from grocery stores back to the field.

References

Bryant, D., Nielsen, D, and Tangley, L. 1997. *The Last Frontier Forests.* Washington, DC: World Resources Institute; http://pdf.wri.org/lastfrontierforests.pdf.

Bryce, P.H. 1914. *Effects Upon Public Health and Natural Prosperity from Rural Depopulation and Abnormal Increase of Cities.* Read before the General Sessions, American Public Health Association, Jacksonville, Fla., 1 Dec. 1914.

Howe, F.C. 1915. *The Modern City and Its Problems.* Charles Scribner's Sons, p. 6.

MacLachlan, I. 2005. "Feedlot Growth in Southern Alberta: A Neo-Fordist Interpretation." In Gilg, A., Yarwood, R., Essex, S., Smithers, J., and Wilson, R. eds. *Rural Change and Sustainability: Agriculture, the Environment and Communities.* CABI Publishing, pp. 28–47.

Nestle, M. 2007. "Eating made simple." *Scientific American.* September: 60–9.

UN Millennium Project. 2005. *Halving Hunger: It Can Be Done.* Summary version of the report of the Task Force on Hunger. New York: The Earth Institute at Columbia University.

Worm, B., et al. 2006. "Impacts of biodiversity loss on ocean ecosystem services." *Science* 314 (5800): 787–90.

CHAPTER 11

Manufacturing

No country can pay its way without a strong industrial
[manufacturing] sector.

Fingleton 2006

The overall goal of this chapter is to analyze the contemporary location dynamics of manufacturing activities. More specifically, this chapter's objectives are

- To examine the tendencies towards concentration and dispersal in manufacturing locations;
- To explore how location affects the profitability of manufacturing;
- To explain the changing location dynamics and organization of manufacturing-based value chains;
- To explore the shift from value chains to value cycles in goods producing industries; and
- To explain the implications of value chain dynamics for location and development.

The chapter is organized in three main parts. The first part focuses on overall patterns of manufacturing and the forces that shape the location of individual operations; the second examines how location impacts on product cycles; and the third looks at clustering and congregations in agglomerations.

Geographic Differences in Manufacturing

In 2009, Toyota ended GM's 77-year reign as the world's largest automobile manufacturer in terms of both sales and production. Toyota still held that position at the end of 2014, but both GM and Volkswagen were almost as large. They and a few other groups, such as Hyundai, Ford, Fiat-Chrysler, and Renault-Nissan, dominate the global automobile market. In this competitive oligopolistic market, each of these giant MNCs have located subsidiaries, bought out local companies, and sought market share in all the world's major markets.

On its way to the top, Toyota revolutionized the way manufacturing is done. Other companies sought to emulate its innovative production systems (including GM), often closing down their existing plants and opening new ones, often in new economic spaces (see Chapter 5) with no experience in the auto industry. While Toyota itself has rarely closed a big factory, its geographical expansion around the world and the location of its manufacturing operations in new places has been vital to its success (Case Study 11.1).

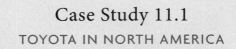

Case Study 11.1
TOYOTA IN NORTH AMERICA

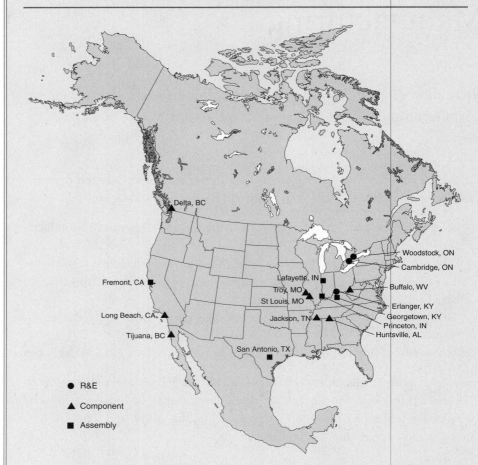

The geography of the Japanese auto industry is broadly similar to its Western counterpart in that it is characterized by agglomeration: multiple suppliers clustered around a few core firms. Toyota manufactured its first cars in the 1930s and began exporting in the 1950s. However, with the exception of a facility built in Brazil in 1959,

Its expansion indicates the diverse range of location factors underlying why particular nations, regions, and sites are chosen.

Manufacturing is the fabrication, processing, or assembly of materials and components to create a material good. Although understood as making things, manufacturing embraces value-adding processes that not only assemble or in some way transform material parts and components, but also incorporate services such as logistics, marketing, design, and R&D. Normally the manufactured product has a value greater than the sum of its constituent elements and contributing processes. Manufacturing activities are highly diverse, but there is a basic distinction between primary and secondary manufacturing.

its production continued to be concentrated at home until the 1980s, even though American firms had been establishing foreign branch plants for decades. Indeed, for most of Toyota's history, its production has been concentrated in the city of Toyota (prior to 1959 called Koromo). Toyota's international expansion began in earnest in 1984 when it entered a joint venture with General Motors to establish an assembly plant in Fremont, California. Additional assembly facilities were soon established in the US and Canada, and then in Europe, partly in order to avoid the political backlash attracted by exports from Japan. Since then, Toyota has expanded not only production and distribution but also R&D and design on a global scale, most recently in China and other parts of Asia. Although the Fremont plant was closed down in 2010 (with a loss of 4700 jobs), it has been re-opened by the Tesla Motor company for the production of electric vehicles. Between 2003 and 2013, Toyota's annual auto assembly total in North America alone increased from more than 1.28 to 1.86 million vehicles, the latter being a record level of production and showing how the firm has recovered from its recall problems of 2010–11.

In 2008, Toyota opened its eighth North American plant in Woodstock, Ontario. The plant, which includes an R&D centre and cost approximately $1 billion to build, was designed to produce the RAV-4 SUV, which until then had been produced only in Japan. The decision to locate a plant in Ontario was based in part on Canada's health-care system, which was expected to save the company between $4 and $5 per worker per day. In addition, Toyota Canada's president indicated that training costs were lower in Canada than in the US. On the other hand, higher training costs in the US are offset by larger government subsidies, lower-wage rates, and use of non-union labour. Following the sale of its Fremont plant, Toyota opened its eighth North American assembly plant in Blue Springs, Mississippi, to manufacture Corollas (previously produced at Fremont).

In general, Toyota has sought to balance its investments, both between Canada and the US and between North America and the rest of the world, to ensure market access and to increase its market share. It now has production facilities in a number of developing countries. Having established its first Chinese factory in 2000, the company began ramping up production to take advantage of China's vast market potential. By 2010, it was building its seventh assembly plant there. In 2009–11, however, Toyota suffered a recall crisis, which it interpreted as a loss of quality control due to too rapid expansion. Hence it put a moratorium on factory building. That moratorium was lifted in 2015 after Toyota had developed a new manufacturing platform (Toyota New Global Architecture) that would reduce the cost of building both factories and cars.

Primary manufacturing converts raw materials or commodities into more processed commodities: for instance, refining crude oil, slaughtering poultry, milling grain, or sawing logs. Since activities of this kind have already been discussed in preceding chapters, this chapter mainly focuses on **secondary manufacturing**, in which manufactured components are transformed into an array of products such as computers, airplanes, automobiles, buses, ships, industrial machinery, pipelines, chemicals, tools, furniture, clothing, and all the things we need for leisure and recreation (skis, hiking boots, snowboards, surfing boards, sailboats, kayaks, mountaineering equipment, hockey sticks, soccer balls, and specialized clothing).

primary manufacturing Activities based mainly on processing of raw materials.

secondary manufacturing Activities that add value to already processed materials.

Manufacturing has long been regarded as an engine of economic growth. The Industrial Revolution that began in the late eighteenth century was defined by the vast, rapid increases in manufacturing activities and jobs, and powered the rise to global domination of the rich market economies. Indeed, "developed," "advanced," or "industrialized" countries became synonymous labels. This association has been reinforced in recent decades by the key role that manufacturing has played in the emergence of newly industrialized countries (NICs) such as South Korea and Taiwan. Manufacturing remains the foundation of China's and India's plans to transform themselves into modern, urban societies. Meanwhile, manufacturing continues to matter to rich economies as a source of the innovation, productivity, jobs, skills, and exports necessary for any country to survive in the global economy. Among the innovation priorities for manufacturing today is the quest for ways to minimize harmful impacts on the environment.

Manufacturing activities are organized within value chains that are typically dominated by giant corporations, but incorporate the activities of thousands of large and small suppliers. Manufacturing-based value chains have become more elaborate in recent decades. Traditionally, manufacturing was concentrated in rich market economies with relatively simple links to the rest of the world, either for resources or small-scale market outlets. Then in the 1960s and '70s, some large companies began to relocate the manufacturing operations of some products that had reached the "mature" phase of their life cycle to poorer countries. In so doing a new international division of labour was created. More recently, as globalization has intensified, value chains have become more geographically nuanced with location decisions increasingly based on constantly shifting combinations of labour cost, skill, and creativity. Moreover, once-dominant western (and Japanese) companies are now rivaled by MNCs based in recently industrialized countries. Among auto companies in 2014, South Korea's Hyundai and China's SAIC Motors, the latter being an SOC (Chapter 6), were ranked eighth and tenth largest.

The Shifting Geography of Manufacturing

The location dynamics of manufacturing vary by spatial scale and time period. Globally, in recent decades, as an important feature of what is meant by globalization, the most striking trend has been the growth of competitive manufacturing activities in the new (since the 1960s) economic spaces of Asia Pacific. Since the dismantling of the Berlin Wall (Case Study 6.1), industrial expansion and modernization among eastern European countries is also noteworthy and has been strongly focused on re-integration within the expanded EU.

Global Perspective

On the eve of the Industrial Revolution, in the mid-eighteenth century, the distribution of manufacturing activities around the world roughly corresponded to the distribution of population. The biggest countries, India and China, were the biggest manufacturers but manufacturing was small-scale, conducted in homes and workshops, and often rural-based. Then came western innovations in textiles, steel-making, steam engines, and railways and manufacturing exploded in the West. The UK led the way in both innovation and production, accounting for 10 per cent of global production by 1830 and 23 per cent by 1880. The US, Germany, France, and other European countries quickly expanded their manufacturing sectors. By the mid-1800s, the distribution of wealth among the world's nations closely reflected the distribution of manufacturing, and by 1900 rich market western economies (plus Japan) accounted for more than 92 per cent of global manufacturing. In the twentieth century, the US became the world's pre-eminent manufacturer, and the organization, production system, and management–labour relations that characterized Fordism became the leading template for mass production (Chapter 3).

However, the Fordist template began to be undermined in the 1970s when Japan, the single non-Western industrial power, and Germany sparked a dramatic reorientation of manufacturing around the world towards more flexible forms of mass production (Figure 11.1; Table 3.1). In general terms, Japanese and German MNCs gave greater emphasis to *functionally flexible* core workers (Chapter 6) and relations with suppliers that were hierarchical in nature (Figure 4.3), i.e., that involved highly stable, direct transactions with a few leading, innovative suppliers who in turn contracted out lower value activities to extensive populations of SMEs (see Case Study 11.4 for details on the Toyota example). These systems of flexible mass production more readily responded to market fluctuations, technological change, and competition. In some

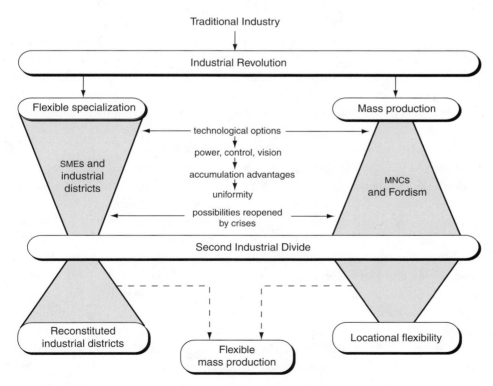

MNCs = Multinational corporations
SMEs = Small and medium enterprises

FIGURE 11.1   Mass Production versus Flexible Specialization

regions, *flexible specialization* evolved around localized concentrations of SMEs that formed networks of specialized and diverse production to provide another response to market and technological dynamism. In contrast, production flexibility has also meant the search for low-wage production sites, and the employment of *peripheral labour forces*, hired and fired as needed (Chapter 5). Geographically the search for flexible manufacturing and employment has resulted in seemingly contradictory patterns of both dispersal and concentration.

When they began increasing their exports in the 1960s, Japanese manufacturers used the expertise they had developed in an intensely competitive national market to threaten domestic manufacturers in North America and Europe. American and, later, European and Japanese companies responded by searching out less costly production sites in places such as Mexico (the **maquiladora**) and the four Asian NICs or "Tigers" (South Korea, Taiwan, Hong Kong, and Singapore). With the help of various government programs designed to both stimulate the development of domestic firms and attract foreign investment, the Tigers, along with a few other NICs, rapidly became global exporters and then innovators of secondary manufacturing goods.

Spurred on by the success of the Tigers, the world's most populous countries then fell into the embrace of market economics. China opened its doors to foreign investment in the late 1970s, and less than 30 years later (in 2001) became a member of the WTO. India, while never as staunchly Marxist as China, governed its economy with a mix of socialism and nationalism that has only more recently been reformed. To provide jobs and development to over 2.5 billion people, the two countries, especially China, have looked to manufacturing for global markets. As a consequence, hundreds of millions of workers have entered manufacturing and more than doubled the sector's total labour force. Now much of the manufacturing for corporations based in established market economies, particularly in industries such as electronics, shoes, textiles, and furniture manufacturing, has relocated to China and India. Those countries have become leaders in manufacturing themselves, and are now serving their own populations as well as competing internationally.

The shift from developed to developing countries, and the increasing competitiveness of the latter, is indicated by their growing share of exports since the 1980s (Table 11.1). Although the old market economies, led by Germany, the US, and Japan, are still big exporters, the share of manufactured exports from Asia Pacific countries (outside Japan) has risen spectacularly. In 1990, China's share of manufactured exports was less than Canada's modest contributions, but by 2012, it had become the world's largest export manufacturer, accounting for 16.8 per cent of global exports by value. In 2012, Greater China (China plus Hong Kong and Chinese Taipei) accounted for 22.7 per cent of the total, while South Korea's exports add another 4 per cent of the global total; Singapore and other Asia Pacific countries (e.g., Malaysia, Thailand) are also important exporters. Elsewhere, only Mexico has become a new export manufacturer of note in recent decades, bearing in mind EU data "hides" national contributions and that some other countries have highly specialized export-based manufacturing, such as textiles in Bangladesh.

The evolution of auto production illustrates broad shifts in global manufacturing geography (Table 11.2). In 1900, the global auto industry produced a total of roughly 10,000 vehicles, mostly in the US and Germany. By the 1920s, the US dominated global

maquiladora A foreign-owned branch plant in one of the export processing zones, often along the Mexico–US border but also located in interior locations, that draws on low-waged labour and tariff-free imports of inputs for goods designed for export.

TABLE 11.1   Export Value of Manufacturing by Country, 1980–2012

Selected Countries and Regions	Value ($B) 1980	Value ($B) 1990	Value ($B) 2000	Value ($B) 2012
United States	142.2	290.5	648.9	1102
Japan	122.7	275.1	449.7	710
China	8.7	44.3	219.9	1925
EU 15	553.7	1203.3	1901.8	
EU 27			2012.0	4385
Germany[1]	162.1	375.7	483.2	1052 est
France	81.1	161.3	272.8	595 est
UK	78.7	146.7	233.0	508 est
Canada	30.0	73.3	175.6	211
South Korea	15.7	60.6	154.9	463
Brazil	7.5	16.1	31.7	82
Russia			24.8	101
Mexico	4.4	25.3	138.6	269
Hong Kong[2]	18.0	75.6	192.5	423
Singapore[2]	8.3	37.5	117.7	283
Poland	11.7	8.5	25.4	In EU 27
India				180
Chinese Taipei				262
World Total	1092.4	2391.2	4696.2	11490

[1] Data for 1980 cover only the Federal Republic of Germany; data for 1990 cover both Federal Republic and the German Democratic Republic.
[2] Includes re-exports. The estimates for Germany, France, and the UK in 2012 are based on extrapolating their share of EU 27 trade to 2012.
Source: Adapted from World Trade Organization Statistics Database

production, followed (albeit at a great distance) by Canada and the UK. After the war, most of the major industrialized nations had at least two important auto manufacturers (Ford, GM, Chrysler; Volkswagen, BMW, Mercedes; Renault, Citroen, Peugeot; Saab, Volvo; Fiat, Alfa Romeo; Rover, Leyland; Toyota, Nissan, Honda, Mitsubishi, Mazda), and provided tariff protection for their operations. This competition, together with national protectionism and the common corporate strategy of establishing branch plants in foreign countries, had the effect of limiting exports of autos. (Industries such as steel and appliances were similar in both structure and location.)

Japan's exports of small vehicles took off after the oil shock of 1973. Japan became the world's second largest producer in 1975 and overtook the US in 1980. But its export success was not completely welcomed in the West. When, in response to political pressure, the US introduced voluntary export quotas and the EU erected a number of tariff and non-tariff barriers, Japanese manufacturers were forced to establish production facilities in those jurisdictions in the 1980s to gain market access. Recently, Germany and South Korea have challenged Japan's dominance in terms of both exports and branch plants, while China and India are emerging as powerful producers serving their

TABLE 11.2 Auto Production (000s) in Selected Countries, 1900–2012

	1900	1925	1950	1975	1990	2000	2012
North America							
US	4	4,266	8,006	8,965	9,780	12,800	10,781
Mexico			22	361	804	1,936	3,002
Canada				1,442	1,922	2,962	2,464
Western Europe							
Germany	2	63	306	3,186	4,661	5,527	5,649
France				2,861	3,295	3,348	1,968
Italy		49	128	1,459	1,875	1,738	672
UK		167	784	1,648	1,296	1,814	1,577
Sweden			18	367	336	301	163
Spain				814	1,679	3,033	1,979
Asia-Pacific							
Japan				6,941	13,487	10,141	9,943
China					709[1]	2,069	19,272
South Korea				36	1,322	3,115	4,562
India						801	4,145
Brazil				930	914	1,682	3,343
Russia				1,964	2,000	1,206	2,232
World Total	10	4,901	10,577	32,998	44,165	58,374	84,141

[1] 1991

Source: Industry Canada: Statistical Review of the Canadian Automotive Industry: 1998 Edition, International Organization of Motor Vehicle Manufacturers (OICA), http://oica.net, China Automotive Industry Yearbook 2003, Morales (1994: 16).

own fast-growing domestic markets and increasing their exports. South Korea, India, and China had virtually no presence as vehicle producers in 1975; yet by 2012, they accounted for one-third of the (much bigger) global total.

Location Trends: An Overview

There are some industries that will establish themselves virtually everywhere as soon as the local market reaches a certain size. Every community, for instance, needs gravel for road construction; fortunately (since stone is expensive to transport), a source of supply is almost always available nearby, in the form of a riverbed or of rock that can be crushed. At a larger geographical scale, production of drinks made from concentrate is dispersed internationally to minimize the cost of transporting an input that can be found everywhere (water). Ubiquitous or dispersed manufacturing was common in food processing until economies of scale, improvements in transportation and refrigeration, and distribution-integration encouraged corporate consolidation and geographic concentration. In 1961, for example, there were 1,710 dairies in Canada; by 2013, there were only 279, 14 per cent of them owned by Saputo, Agropur, and Parmalat. Together, the three giants process 75 per cent of the country's milk; the remaining 25 per cent is produced by the many smaller dairies. The many local newspapers of earlier times are in decline because of the Internet, although most local markets still support at least one paper. On the whole, manufacturing based in industrial regions has displaced ubiquitous manufacturing.

The economic historian Sydney Pollard (1981) described European industrialization as a "regional phenomenon" of geographically uneven growth in which neighbouring cities and towns coalesced to form functionally related industrial **conurbations** (aggregations of urban areas). Industrialization in North America and Japan was similarly regional. In North America, conurbations with distinctive fields of specialization formed a gigantic manufacturing belt that extended from the northeastern Atlantic coast (Boston, New York) down to Philadelphia, through Pittsburgh to Chicago and Detroit in the Midwest, and north to Quebec and southern Ontario. In Japan, Tokyo and Osaka anchored an east–west axis of industrialization. The unparalleled scale of China's industrialization since the beginning of economic liberalization in 1979 is epitomized by the conurbations that developed in regions such as Beijing–Tianjin, Shanghai, and the Pearl River Delta in the first decades of its market transformation (Case Study 11.2). Today the interior of the country, particularly along the Yangtze River, is developing into another industrial powerhouse.

Around the world, different manufacturing regions have different areas of specialization. North America's manufacturing belt has deindustrialized but still has concentrations of glass and ceramics, automobile assembly and parts manufacturing, transportation, heavy machinery, machine tools, steel, petrochemicals, and food processing. Other industry concentrations include carpets in Georgia; aerospace assembly and engineering in Los Angeles, Seattle, Houston, and Montreal; and electronics in Silicon Valley. Elsewhere there are regional specialties in motorcycles and musical instruments (Hamamatsu, Japan), ships (Incheon, South Korea), sport cars (southern Germany and northern Italy), airplanes (southwest France), and memory chips (Taiwan). Every manufacturing region is a complex of many firms engaged in one or more industries and usually related through localization and urbanization economies.

Infrastructure and the spatial structure of activities vary from region to region and over time. In the early days of industrialization, location was often dictated by access to raw materials, inputs such as coal, and waterways for transportation. Then agglomeration economies led to geographic concentration in selected cities. In time, changes in products, process technologies, and transportation modes (both for production and for labour) encouraged various industries to locate in suburban areas, where the availability of land made it possible to begin designing factories as single-floor operations. **Suburbanization**, which began around 1900 for large cities such as New York, took off after the Second World War, as more people acquired cars, and the suburbs provided a supply of female labour. Patterns of regional agglomeration changed again when manufacturing began to spread outwards from metropolitan areas to more distant rural places, which typically offered cheaper land, less costly and/or non-union labour, and good transportation access; this **non-metropolitan industrialization** became a significant trend in the US in the 1950s. The shift of manufacturing from the rustbelt to the sunbelt was a related trend in the US in the latter decades of the twentieth century; high-tech entrepreneurs were particularly attracted to places with strong recreational

conurbation A large, functionally connected region formed through the coalescence of two or more neighbouring towns; if the towns are industrial, so is the conurbation.

suburbanization The shift in economic activities (or housing) outwards from central cities but still within the built-up area or commuting zone.

non-metropolitan industrialization The dispersal of economic activities away from metropolitan areas to more distant rural places.

Case Study 11.2
THE PEARL RIVER DELTA

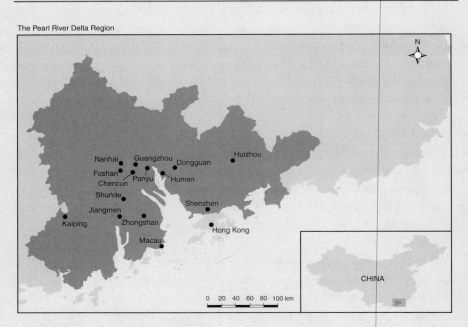

The Pearl River Delta Region

In 1979, the Chinese government decided that fundamental economic reform was necessary if China was to catch up with the rest of the world after three decades of Maoist central planning and extremely limited contacts with market economies. Among the first to take advantage of this change were some light manufacturing (cutlery, toys, electronics) businesses that had driven the growth of Hong Kong in the 1950s. These manufacturers quickly expanded their operations into nearby cit-

and cultural amenities. Similar shifts away from established industrial centres occurred in other market economies as well, especially with the internationalization of manufacturing through FDI. In Japan, some operations were shifted to rural areas, but the trend was nowhere near as strong as in the US; most manufacturing continued to be concentrated in the core cities and because of upgrading among SMEs, overall production became increasingly concentrated.

In the global dispersal of manufacturing, export processing zones (EPZs) have been a widespread policy tool used by developing countries to attract FDI. EPZs are typically located in new economic spaces, sometimes far from established centres, provide social and economic infrastructure in the form of railways, roads, and port facilities, allow MNCs to import inputs duty free, usually so long as the product is exported, and attract low-wage labour from rural areas. In some cases, such as in the Pearl River Delta and Mexico's maquiladoras, EPZs have become the centres of new agglomerations. Note, however, that while some countries, such as Mexico, have emphasized EPZs as the central plank of their industrialization strategy, countries such as South Korea have

ies such as Shenzhen, which in 1984 was designated the first of what became five special economic zones expected to lead China's modern industrialization. In those zones inputs are imported without tariffs so long as they are used to make goods for exports, FDI has freedom of entry, and taxes are low. Large MNCs have been attracted to the region as have their major suppliers. Since the 1990s, the Pearl River Delta has grown enormously. Although the region (including Hong Kong and Macau) makes up just 0.2 per cent of China's land area, by 2002, it accounted for 9 per cent of the country's GDP and more than 33 per cent of its exports. The registered largely rural population of over 16 million in 1980 has been estimated to be close to 70 million in 2015 and living in new industrialized cities.

The repatriation of Hong Kong in 1997 formally reinforced its position as the decision-making centre and entrepôt for the Delta region (and beyond). On the mainland, individual towns have become large manufacturing cities with distinctive areas of specialization: electronics, computers, and peripherals for Dongguan; autos, parts, and other transportation equipment for Guangzhou; laser diodes, digital electronics, CD-ROMs, telephones, batteries, and circuit boards for Huizhou; textiles, aluminum products for Nanhai; sporting goods, textiles, garments, and motorcycles for Panyu; industrial ceramics for Foshan. Shenzhen has become the leading city in the region, not only for its electronics export performance, but increasingly for its generation of MNCs and its innovative capacities. The dynamism of the region is such that its renown as the world's workshop is likely soon to be supplanted by recognition of its ability to generate new products and industrial processes.

The PRD's export success has not been without problems, however. Rapid urban expansion has been fuelled by millions of poor in-migrants, typically from rural areas, who work for low pay in poor conditions and are denied the same benefits (health care, education, and even sanitation) as that of holders of municipal residence rights (*hukou*). Urbanization has also led to extreme congestion, while inadequate regulation and enforcement have permitted serious industrial pollution of the region's air and water, and degradation of farmland. At the root of a large proportion of these negative externalities is North American demand for cheap imports.

For additional detail see Lin (1997) and Enright et al. (2007).

favoured domestic companies. Others like China, India, and Taiwan, have sought more balanced approaches.

Finally, **deindustrialization**—the *absolute*, permanent loss of a number of (secondary) manufacturing jobs in a particular region over a lengthy period of time—is a significant location trend. The closure of resource-based manufacturing (and mining) operations has often undermined isolated towns and regions. The term deindustrialization, however, is primarily used in relation to the decline of once powerful industrial regions, including the industrial conurbations of northern England, Germany's Ruhr region, and the North American manufacturing belt. Nationally, the UK deindustrialized earlier and deeper than other countries, losing half its manufacturing jobs between 1966 and the early 1980s, a trend that has since continued. In the US and Germany, deindustrialization of their manufacturing belts has been to some extent compensated by growth in other regions of their economies. In deindustrialized cities, where successful, rejuvenation has generally involved the service sector, fundamentally changing their identity. Some cities such as Detroit are still desperately seeking to revive. The reasons for deindustrialization are

deindustrialization Substantial net loss of manufacturing jobs in a particular region over a lengthy period of time.

complex, but it has occurred primarily in regions that have been locked into obsolescent technologies, institutional arrangements, and attitudes, and that have found themselves unable to compete with new economic spaces.

Locating Manufacturing Facilities

It might be supposed that choosing a new manufacturing location is simply based on minimizing costs. However, this task is complicated because there are many different location conditions, some of which cannot be quantified, and the search for lower costs is subject to the ability of manufacturers to generate values (profits) among different places and access markets. How do firms trade off, for example, a low wage and relatively low skilled labour force against a highly skilled and highly productive labour force? Further, such contemplations can be affected by many other factors, such as spatial variations in transportation costs and services, urbanization and localization economies, energy costs, taxation levels, subsidies, political stability, and amenities. Since it is unlikely that any one location will be the "best" in terms of all relevant location criteria, decision-makers need to trade off the various pros and cons. Given that many **location factors** are hard to quantify and are inherently subjective, these trade-offs are judgmental in practice. As the Toyota example reveals, individual firms can locate similar facilities in different places based on different trade-offs (Case Study 11.1). Moreover, big firms may have the power to alter location conditions and base their location choice after negotiations with host governments (and other agencies) for lower taxes and wages, and weaker regulations.

Generally speaking, new site location choices vary according to forms of business segmentation (Chapter 4), techno-economic possibilities (Chapter 3), industry type, and spatial scale. In the latter context, for example, the mix and relative importance of location factors will depend on whether decision-makers are considering choice of continent, nation, region, community, or sites within a community. Especially at broader regional scales, the focus of this chapter, the traditional cost-minimizing approach as developed by Alfred Weber in 1929, remains a useful point of departure for understanding manufacturing location patterns.

Cost-Minimizing Approach to Manufacturing Location

The least-cost model, in its simplest form, assumes a perfectly competitive market in which each firm's decisions are made by a perfectly informed and rational *Homo economicus*. This perfect decision-maker determines which locations are best suited to minimize costs and maximize profits. In the long run, competition enforces an economically rational location pattern since factories placed in non-viable locations by less than perfect decision-makers will ultimately fail. Analytically, the model first identifies the location that offers the lowest possible transportation cost and then assesses the influence of labour, agglomeration, and other location factors as appropriate.

Effects of Transportation Costs

Procurement costs are the costs of transporting inputs (both materials and services) to the factory; distribution costs are the costs of transporting outputs to con-

location factors The advantages and disadvantages of different locations to firms.

procurement costs The logistical costs of transporting inputs (both material supplies and services) to the factory.

sumers. Typically, total transportation costs increase with distance, while average costs taper off because of fixed-cost effects (Figure 11.2). While this is not always the case, the effects of transportation costs on industrial location generally depend on where the input and output markets are located and how the inputs are physically changed by processing. Weber classified those inputs available virtually everywhere as "ubiquitous"; those undergoing no change in physical characteristics as a result of processing as "pure"; and those undergoing some change in physical characteristics as a result of processing as "impure." For example, the optimum (cost-minimizing) location for a factory that serves one market and uses **ubiquitous inputs** would be at the market itself, where distribution costs would be nil (the model assumes that procurement costs are zero everywhere). If a factory serving a single market uses one **pure input**, and transportation costs vary linearly (in direct proportion) with distance, the optimal location can be anywhere between these two locations. By contrast, for a factory that uses ubiquitous inputs and serves multiple spatially distributed buyers, the optimal location is the midpoint or **median** between the markets, where total distribution costs are minimized. As a result of fixed costs in facilities and related charges, the average cost of transportation will gradually taper off over distance. Such tapering effects are especially evident in rail, ocean, and pipeline networks, which involve large fixed costs in the form of equipment and terminal facilities (Chapter 13).

In the case of **impure inputs**, the changes in physical characteristics caused by processing will determine the optimal location. Thus for activities based on inputs that lose weight or become less perishable, such as smelting or fish processing, procurement costs are higher than distribution costs and are "input-oriented." For activities in which the inputs gain weight or become more perishable, the distribution costs are higher and are "output-oriented"; examples include paper box manufacturing and soft drink bottling. If a factory uses more than one impure input from more than one source, then the minimum transportation cost location will be intermediate between input sources and output markets. The "pull" of the various location factors is measured by **isotims**: lines of equal transportation cost around each location factor (procurement cost around input sources, distribution cost around each market) to all (or selected) locations across the cost

ubiquitous inputs In Weberian location theory, inputs that are available everywhere.

pure inputs In Weberian location theory, inputs that do not change physically during processing.

median The middle point of a distribution in which half the observations are on either side.

impure inputs In Weberian location theory, inputs that change physically during processing.

isotims Lines of equal transportation costs around each location condition.

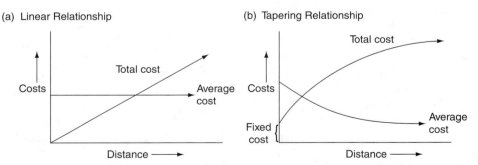

(a) Linear Relationship

(b) Tapering Relationship

Average cost per tonne km = total cost of transportation per tonne / distance travelled. In 11.2a, total costs are defined only by variable costs (e.g., labour, fuel) which increase with distance, while average costs remain the same. In 11.2b, total costs are defined by fixed costs (e.g., terminal facilities), as well as variable costs, so that total costs increase but taper off with distance, and average costs per tonne km decline.

FIGURE 11.2   **Relationships between Transportation Costs and Distance**

surface (Figure 11.3a). Isotims increase with distance around each location factor. If the transportation cost isotims relevant to each location factor are summed at each location on the diagram a total transportation cost surface can be derived and the minimum transportation cost location ("P") identified. If procurement costs are higher than distribution costs, which is the case in the hypothetical example in Figure 11.3, then P will be pulled closer to the input sources, away from the market. **Isodapanes** ("cost contours") link locations with the same total transportation costs and reveal the transportation cost

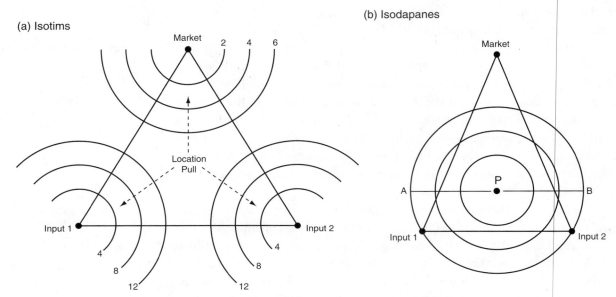

(a) Isotims

(b) Isodapanes

Isotims are lines of equal transportation costs around each location factor. The $4, $8, and $12 per tonne kilometre isotims are shown for Input 1 and 2 and the $2, $4, and $6 per tonne kilometre isotim for the market.

By adding isotims, a total transportation cost surface is derived. P is the lowest transportation cost point. Isodapanes are lines of (increasing) equal transportation cost around P.

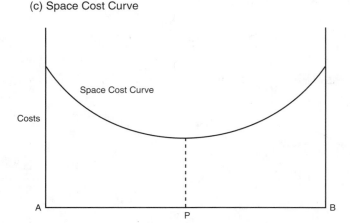

(c) Space Cost Curve

This figure is a transect from A to B (Fig 11.3b).

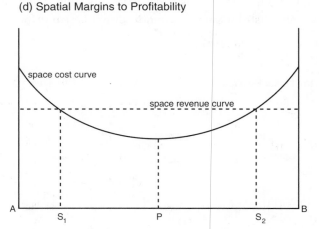

(d) Spatial Margins to Profitability

S_1–S_2 define the spatial limits to profitability (where revenues exceed costs).

FIGURE 11.3   The Weberian Location Triangle

penalty of not locating at P (Figure 11.3b). A transect across this cost surface, for example from A to B on Figure 11.3b, is a space–cost curve (Figure 11.3c).

Labour, Agglomeration, and Other Factors

For Weber, labour and agglomeration economies exerted powerful effects on industrial location. The minimum transportation cost location (P) may not be the best choice because it may not minimize labour costs (L) or maximize external economies (E) for any particular activity. For example, assume that there is a single location in this landscape that offers low labour costs—say, because its labour supply is particularly large and competitive (Figure 11.4a). Whether the factory should locate at P or L (or E) depends on whether the labour cost saving would offset the transportation cost penalty of not being at P. Analytically, the answer to this question depends on the **critical isodapane**, the transportation cost penalty that equals the labour-cost savings. If L is inside the critical isodapane, the factory should relocate there, since the saving in labour costs will more than offset the increase in transportation costs. If L is outside the critical isodapane, the factory should be located at P. Similarly, if the factory owner is faced with a choice between P and a location (E) that maximizes agglomeration economies of scale, if the cost savings at the latter more than offset the transportation-cost penalty, then the optimal location is at E. The least-cost location suggested by agglomeration economies may differ from both P and L (Figure 11.4b).

Other location factors in the form of energy costs, environmental costs, building or utility costs, taxation levels, or government subsidies can be similarly mapped as

critical isodapane The isodapane that equals the costs of another location condition, such as labour.

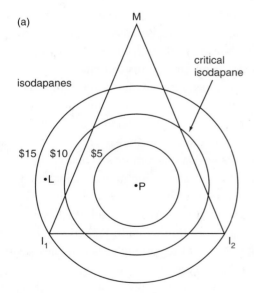

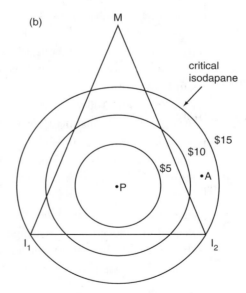

(a)

P = Minimum transportation point. Cheap labour cost location reduces average cost by $10. Should the factory choose P or L?

(b)

Agglomeration economies available reduce average costs by $15. Should the factory choose P or A?

FIGURE 11.4 Cheap Labour Sites and Agglomeration Economies in Relation to the Minimum Transport Cost Location

cost surfaces and aggregated to create an increasingly complex total cost surface. In practice, cost variations can be expressed at different scales—regionally, nationally, or internationally—and cost calculations can be developed for individual nodes. For a factory to earn a profit, its revenues must be greater than its costs. Revenues might be the same in every location, but are more likely to vary. As long as revenues exceed costs, any location is within the **spatial margin of profitability** (Figure 11.3d).

spatial margin of profitability The range of alternative locations in which a given activity can be viable.

Substitution and Its Location Implications

Whatever the scale and range of location factors included, an important implication of the least-cost model is that location choice can be understood as an exercise in substitution. Firms that locate near markets substitute lower distribution costs for higher procurement costs, and firms that locate at L or E trade off higher transportation costs for either lower labour costs or greater agglomeration economies. Any location decision depends on context and the trade-offs may come in various configurations. For example, firms may trade off the benefits of locations that offer lower labour costs against those offering higher labour productivity; the disadvantages of a community with high taxes or few agglomeration economies against benefits in the form of services or government subsidies; or higher transportation costs against lower energy costs. Long-term survival requires that the costs inherent in the firm's trade-off decision remain competitive.

substitution principle In location theory, the ability to trade off one location condition for another.

The **substitution principle** recognizes the flexibility of market relations across space and is what makes alternative locations viable. Factories producing similar products can survive in different locations by drawing on different combinations of location advantages and market (and non-market) relations. The fact that different locations offer different substitution possibilities allows multiple manufacturers of similar products to survive. In the case of Toyota (Case Study 11.1), both Woodstock, Ontario, and the southern US offer labour advantages, though of very different kinds.

location conditions Variations in costs and revenues and other attributes over space.

Furthermore, the importance of various **location conditions** can change significantly over time. For example, transportation costs are generally less important today than they were 100 years ago. At that time, manufacturing was dominated by "heavy" industries that consumed large quantities of raw materials, generated considerable waste, and produced finished goods that were often difficult to transport for distribution. Under those conditions, transportation costs were a much more serious consideration than they are today since the development of transportation networks, including highways, and increased efficiencies in the processing of materials. The rise of "light" manufacturing characterized by low weight and high value per unit volume has also helped to reduce the relative significance of transportation costs. In Weberian terms, the relative decline in transportation costs implies that the spacing of isodapanes has widened and the penalties for not locating at P have decreased.

Broadly speaking, the enhancement of transportation networks is a key condition underlying globalization. Yet access to transportation facilities remains essential, and increasing concern for environmental sustainability is likely to draw new attention to transportation costs in terms of pollution (see Chapter 13). Since transportation has become relatively less important as a location factor, labour and agglomeration economies have become more important. The advances in transportation that have made globalization possible have revealed sharp variations in labour characteristics.

Location Choice

Weberian theory suggests that location decisions are entirely rational, made on the basis of given cost conditions. In practice, however, "real-world" decision-makers make their choices in a context of limited information, bounded rationality, past experience, and preferences that are shaped by existing corporate structures and strategies. New location choices are strongly influenced by many market-related and non-market–related factors and by intangible, qualitative features that cannot be quantified in the precise way that cost structures can.

Decision-makers translate location conditions into location factors of varying importance. They often choose locations that offer access to cultural, educational, recreational, or environmental amenities, for example, to help attract and maintain a highly skilled and educated labour force (Table 11.3). If labour costs are important in choosing

TABLE 11.3   Institutional Typology of Location Conditions

Location conditions: as market relations (access to)	Comments on tangible and intangible features
Consumers	Distribution costs. Personal contact, interaction with consumers, variations in tastes and requirements, threats from rivals.
Labour	Wages and non-wage benefits, hiring and training costs. Skill, availability, unionization, attitudes, and values.
Suppliers: material inputs	Delivered costs. Personal contact with suppliers, quality, reliability, just-in-time possibilities.
Suppliers: equipment	Delivered (fixed) costs. Maintenance and problem-solving services. Quality, reliability.
Suppliers: services	Costs. Personal contact with suppliers, quality, reliability.
Transportation	Freight rates. Reliability, frequency, diversity, damage, availability.
Land	Costs (fixed). Serviced/unserviced, size and shape, level, soil and ground conditions.
Buildings	Costs (fixed or rental) for new or existing buildings, property taxes. Availability, size and shape, accessibility.
Housing	Market prices in existing housing, costs of supplying housing. Quality and availability.
Energy and utilities	Costs including taxes for power, water, sewage, etc. Diversity, reliability, availability.
General government services	Corporate and personal income taxes to governments for providing public goods (security, health, education). Reliability, quality.
Agglomeration: localization economies	The "cost" advantages derived from the agglomeration of related activities (e.g., access to common, skilled labour pools). Cf. diseconomies and disadvantages.
Agglomeration: urbanization economies	The "cost" advantages derived from location in large cities (e.g., access to diverse labour pools). Cf. diseconomies and disadvantages.
As non-market relations	
Untraded interdependencies	Advantages (disadvantages) of co-operative (unco-operative) behaviour and information exchange. Trust and malfeasance.
Industrial associations	Associations financed by members (often supplemented by government funds) to provide collective services, e.g., marketing.
Amenities: social	Entertainment, culture, and sports functions and facilities.
Amenities: environmental	Aesthetic and recreational considerations; air quality.
Domestic government policies and attitudes	Investment subsidies, tax relief, training subsidies. Zoning restrictions, building permit requirements. Proactive, passive, development attitudes.
International government policies	Trade policy (tariffs and quotas), foreign investment policies, political stability, reliability of legal systems.
Environmental and social policies	Air, water, and land use regulations. Labour laws, health care, worker compensation, corporate liability. ENGO activities.

location, so too are the skills, attitudes, and problem-solving abilities of the labour force. In boom times especially, labour availability may be a vital consideration, regardless of cost. Unionization must also be taken into account with firms taking both positive and negative perspectives on union-dominated labour forces. In some countries and regions, political instability or government corruption can be a severe disincentive to investment. While agglomeration economies are considered to be important, they are not easy to quantify in monetary terms. In general, recognition of the importance of intangible or "soft" location conditions leads to new trade-offs. Firms may have to decide whether they would prefer lower wages or a more skilled and productive labour force; lower taxes or the amenities and services that only higher taxes can buy.

Location decision-making also differs fundamentally between SMEs and MNCs; the former typically locate new facilities in their home region with little or no formal analysis of options, while the latter are likely to be more rigorous, contemplate options and employ location specialists (Chapter 4). Moreover, MNCs have the resources, know-how, and experience to negotiate deals with local, regional, and national governments regarding taxes, infrastructure provision, support for labour training, energy cost reductions, cheap land prices, or tariff reductions on inputs that are imported. Thus MNCs can play one community against another in seeking subsidies of various kinds. But communities themselves often seek to out-bid each other for investment. Over the last 30 years in the US, for example, states and towns have been extremely proactive in competing for the new auto assembly plants of German and Japanese MNCs. MNCs have also increasingly sought to locate headquarter locations in order to minimize taxes (Chapter 4). The implication of these kinds of behaviours—not limited to manufacturing—is that location conditions are not "given" but negotiated and manipulated.

Product cycle and value chain dynamics further complicate location decisions.

Location, Product Cycle Dynamics, and Value Chains

The value chains within which manufacturers operate are constantly being reconfigured, principally because of product cycle dynamics. Many different types of products and industries depend on innovation that improves efficiency, enhances value, or both. Innovation processes are highly variable among manufacturers. Innovation can be stimulated by the market or problem solving, thus depending largely on the ingenuity, experience, and awareness of individual managers and engineers, and without connection to formal R&D. In other instances, R&D can be engaged on a temporary basis, utilizing the firm's existing human resources, to meet a particular contingency. Many firms rely on the process innovations of machinery and equipment manufacturers.

In some industries, however, innovation is part of formal and institutionalized R&D. Product innovation is followed by a predictable series of stages: rapid growth, maturity, declining growth, various degrees of obsolescence, and possibly renewed growth through product differentiation (Figure 3.1). During this evolution, technology shifts from laboratory equipment and pilot plants (R&D stage) to relatively small-scale operations (innovation stage) to large-scale plants (mature stage) using standardized machinery designed for mass production. In practice, the innovation process is not necessarily linear. Instead of beginning with basic R&D and ending in commercialization, innovation often works "backwards" from marketing awareness to R&D. Innovation can also be stimulated by sharing information back and forth between stages. While process innovations can increase the efficiency and quality of existing products, they can also open up wide-ranging, unanticipated opportunities for product development. The recent development of 3-D printing is a fascinating case in point. In general, product cycle dynamics, innovation, and location are intimately related.

The product cycle perspective on value chains provides insight into industries as diverse as apparel, detergents, electronics, automobiles, and airplanes. In these industries, innovation and maturation drive spatial organization and restructuring, albeit in different ways. In most cases, a small number of core companies (usually MNCs) integrate the value chain through key activities such as branding, finance, design, assembly, quality control, and control of distribution. Recently, though, factors such as the global spread of R&D as well as manufacturing activity, growing policy interest in innovation as a basis for economic growth, and increasing levels of technological sophistication have made value-chain development more complex. Geographically, product cycle dynamics may encourage either dispersal or concentration, and more generally are contributing towards more complex spatial divisions of labour, modifying the traditional relationship between cores and peripheries.

Product Cycle Dynamics and Location Dispersal

The location implications of the product cycle were first explored in the 1960s by scholars investigating why US-based MNCs in research-intensive industries, such as electronics, were investing in poor, low-wage countries. The product cycle model then became a generalized template for explaining why companies were dispersing branch plants in poor countries and in rural areas within rich countries as part of the trend towards non-metropolitan industrialization. In the original model (Figure 3.1), the relative importance of various inputs changes through the life of the product, and these changes encourage the dispersal of production. Thus in the early stages (R&D; initial manufacturing of the innovative product), the business must be located in a rich country, particularly in a metropolitan agglomeration, so that it can have access to highly trained professionals (scientists, engineers, technicians) and skilled labour. Advanced countries offer large, high-income markets of consumers willing to try new products, while agglomerations facilitate access to external (urbanization and localization) economies—universities, producer services (engineering consultants, accountants, lawyers, etc.), specialized engineering firms, machinery manufacturers, suppliers of venture (risk) capital, and various amenities helpful to attract key workers. Moreover, R&D and design are key "control functions" through which MNCs coordinate their activities globally, and they are typically centralized (partly in order to maintain secrecy and partly to realize economies of scale).

In the later stages of the product cycle, when the product has become standardized, is being mass-produced, and the markets for it are better known, the firm looks for different input and location conditions. In general, as their products mature, firms need to reduce the *average* costs of producing them. One important way of reducing costs is to replace skilled labour with relatively unskilled labour and rely on proven technology to exploit internal economies of scale in large factories. In this model, as manufacturing becomes more capital-intensive and standardized or mature, geographic differences in labour costs become decisive. In location analysis terms, the spatial margins of profitability broadens as the need for specialized labour and external economies of scale (provided by location in a major agglomeration) is replaced by the need for less-skilled labour (available virtually anywhere, but cheaper in less-developed countries) and internal economies of scale at the level both of the factory and of the firm. Firm-level internal economies of scale are mobile geographically, since MNCs can easily provide head-office services (marketing, accounting, strategic direction) to affiliated operations dispersed in locations that offer basic economic overhead capital, such as buildings and land, as well as transportation and communication facilities.

The basic product cycle model essentially depicts a "filtering down" of investment to cheap labour locations. In rich countries, such branch plants are often found in specially created industrial parks, often in less-developed areas within the country; while in poor countries, branch plants have typically been located in special economic zones, such as Shenzhen in China. Moreover, low-cost locations only make sense given access to markets, as Toyota's global expansions reveal (see Case Study 11.1). In a modified version of this model of dispersal, firms seek out new economic spaces to establish new forms of often non-unionized labour relations to replace existing sites where labour relations have become difficult or obsolete and high in cost.

Branch Plants and Local Development

From the perspective of local development, the tendency of branch plants to access cheap labour has been criticized for two sets of reasons. First, because branch plants rely on inputs from affiliated plants and suppliers based in their parent companies' home countries (Case Study 4.4), they do not generate significant multiplier effects. Rather they contribute towards "truncated economic structures" that lack high-level corporate functions including R&D and global marketing. Second, host countries that rely on supplying cheap labour to branch plants leave themselves vulnerable to competition from new economic spaces offering even cheaper labour or bigger subsidies. The prospect of a "race to the bottom," with branch plants continually relocating to lower-wage locations, is a concern even in rich market economies, where lagging regions, such as Atlantic Canada and northeast England that have looked to branch plants for economic rejuvenation, have been disappointed. Mexico has seen hundreds of maquiladoras close since the late 1990s, in part because their wages have risen; other places (especially in Asia) have become more competitive.

On the other hand, relocation is not inevitable: many maquiladoras have survived wage increases. The reality is that location decisions are rarely based on the cost of labour alone. Other labour characteristics that can make a given location more or less attractive include productivity and unionization. Moreover, once a branch plant has been established, it typically becomes a significant fixed and sunk cost for its parent firm. Furthermore, as we have seen, access to markets—whether at global or local scales—is an imperative. One reason so many branch plants have been located in Mexico's border region is the proximity of US markets, to which they gained access through NAFTA. Today, more and more European and North American companies are locating in China to gain access to its vast market. Some companies that took the plunge earlier, such as Volkswagen, have reaped a large market share.

In-situ Location Adjustments

Once established, existing manufacturing sites can exist for long periods of time, occasionally expanded, modernized, or rationalized, or even changed to entirely new uses. Since the early 1950s, assembly of Boeing's commercial airplanes has been concentrated in Seattle, Washington. While its subcontracting patterns have long reached out across North America—and increasingly to Japan—and the company has built branch plants elsewhere and moved its headquarters to Chicago (2001), aircraft assembly is still strongly concentrated in the Seattle area. The original production sites of Ford and GM in Michigan, Volkswagen in Wolfsburg, and Toyota in Toyota Town have remained important despite international expansion.

In fact, more capital investment occurs at existing (brownfield) sites than at new ones to expand production of the existing product-mix using similar or more efficient technologies; to create more value added by replacing or extending existing product-lines; to engage new lines of production; or to rationalize production by reducing plant capacity, the number of product-lines operated, and the number of workers. Multi-plant and multinational firms sometimes re-allocate activities among their various locations. Existing locations attract capital investment for several reasons. Existing fixed and sunk capital (buildings, equipment, infrastructure) will have been depreciated (paid off), while new locations demand entirely new fixed investments. It is hard, often impossible to move capital investments. Existing locations have skilled workforces that would take time and money to develop elsewhere; they may also offer a variety of additional advantages that could be hard to duplicate elsewhere, such as access to resources or transportation cost advantages.

An important strategy for manufacturers to accommodate rising costs is by adding value to their products so that they can command higher prices. In the 1950s in Taiwan, for example, electronic firms began by producing extremely cheap electronic products, first for small domestic markets and then for export markets. Eventually, however, rising wages in Taiwan pushed low-cost electronics manufacturing to relocate to China, while Taiwan came to focus on higher-quality value-added products for brand-name companies such as the domestic Acer. Most of the Japanese and American branch plants that were established in Taiwan in the 1960s and 1970s as low-cost operations are still present, but are now producing higher-value products requiring a higher skilled, better-paid labour force. Singapore's textile industry has experienced a similar evolution, moving from the manufacture of cheap (no-brand) products to the design and production of high-quality clothing destined for sale either under the manufacturer's own name or to major chains.

Companies frequently seek to capture value through the patented branding of products. Companies that design, engineer, and manufacture a significant part of the products sold under brands they own are referred to as brand manufacturers or original equipment manufacturers (OEMs). In some industries, however, OEMs confusingly refer to contract manufacturers that are supplied with design specifications by core firms. If they improve their value chain capacities by conducting design, they are referred to as original design manufacturers (ODMs). If they start making and selling their own brands, then they are referred to as original brand manufacturers (OBMs). It might also be noted that retailers such as Walmart and Sears contract with the original manufacturers of various brand-name products (clothes, appliances, food, etc.)

to produce them for sale under the Walmart or Sears name. (In such cases the original manufacturers lose the value-added of the brand.)

Supplier Coordination

producer-driven value chains
Firm–supplier relations that are organized by manufacturing companies.

buyer-driven value chains
Firm–supplier relations that are organized by service companies such as Walmart.

Conventionally, a distinction is made between **producer-** and **buyer-driven value chains** with respect to the coordination of suppliers by powerful core firms. In producer-driven chains, the core firms are typically producers (Boeing, General Motors, Toyota), while in buyer-driven chains the core firms are retailers, branded manufacturers, and marketers (Walmart, the Gap, Nike). In an increasing number of cases, production is organized in a buyer-driven value chain. Electronics manufacturers (Apple, Nokia) and shoe manufacturers (Nike, Adidas) generally undertake R&D, design, and prototype manufacturing themselves, although they may leave the last to the contract manufacturer. Commercial firms such as Walmart and JC Penney usually provide only specifications and leave design to the manufacturer.

One way of reducing costs and/or finding value-added opportunities is for powerful MNCs to coordinate the efforts of suppliers in relational market arrangements through contracts that specify price, quantity, quality, and delivery schedules. Suppliers may be extremely small "jobbers" that provide customized products or batch manufacturers that supply products in small quantities. Some suppliers are MNCs themselves, especially in producer-driven chains. The specific division of labour through a value chain depends on a variety of factors, including how efficiently each process in the chain may be separated from other processes, the volume of production, the degree of standardization desired, and the capacity of the core firm to coordinate tasks internally and externally. In the business-to-business transactions that make up the manufacturing value chain, suppliers of products that are standardized or adaptable for a variety of clients are often in the best position to achieve economies of scale, which will also benefit core firms that can bargain to take advantage of their efficiency gains. Coordination serves five general objectives:

1. Controlling volumes of production;
2. Appropriate scheduling and timely delivery;
3. Cost control;
4. Quality control; and
5. Environmental sustainability and corporate social responsibility.

Improvements in database, computer, and communication technologies have made all these tasks easier.

The "Make or Buy" Decision

In organizing its value chain, a core firm must first decide whether to make what it needs internally—in which case it would create an administered market—or to buy supplies from other firms. Although they are still important, administered markets are less common than they once were, in part because external suppliers are increasingly accustomed to integration with others and in part because they have become increasingly efficient.

If the firm decides to find a supplier on the open market, it will incur transaction costs. But these costs can be reduced if the firm is co-located in an agglomeration with several competing suppliers. If the firm wants to work with distant suppliers, especially to discover new providers, a traditional but still effective and economical way to make contact is to attend trade fairs and exhibitions such as CeBIT, the information and communication technology fair held yearly in Hanover, Germany, or the twice-yearly Canton Fair in Guangzhou, China (Case Study 11.3). There are thousands of these events held every year, in virtually all industries. Supported by industry associations and governments (municipal to national), they provide a venue for buyers and suppliers not only

Case Study 11.3
THE CANTON FAIR

Hanhanpeggy/Dreamstime.com

Have you ever wondered where all the merchandise at your local discount store comes from? Chances are it came from the Canton Fair. Held twice yearly, in spring and fall, the Canton Fair was established in 1957 and survived the Cultural Revolution (1966–8) as China's only venue for trade with the rest of the world. Since the liberalization of China's economy, beginning in 1979, it has become the world's largest trade fair. In spring 2009—during the depths of the global recession—it had over 20,000 exhibitors, 165,000 foreign buyers, and conducted $26 billion in business over two weeks. In 2013, there were over 200,000 foreign buyers while over 25,000 enterprises set up almost 60,000 booths.

The fair is particularly important for new traders who want to know what China has to offer and can't visit every part of the nation. Representing small or medium businesses, these traders are usually looking for a few attractive items to sell back home. Its primary purpose, however, is to promote the formation of new business relationships. Typically, big retailers meet potential suppliers at the fair, then visit their factories, and deal directly with them.

to meet one another but to see the latest trends in their industries, compare competing technologies and prices, develop new deals, and assess their own competitive positions.

Advances in information and communication technologies have also improved the functioning of open markets. Companies develop databases in which suppliers are graded on criteria such as cost, quality, delivery, and sustainability, and the databases linked into supply chain management systems. Similar databases are created by industry associations and businesses such as the Alibaba Group. Some firms also use the web for combinatorial package bidding (which allows firms to bid on a package of items rather than just individual items) and reverse auctions (in which sellers compete by lowering bids to provide a product). The most sophisticated firms have taken such auctions a step further, allowing buyers and suppliers to engage in electronic trading.

On the other hand, the firm might turn to a few key suppliers with whom it has already developed close relationships. Relational markets are important in manufacturing because they facilitate close co-operation. The Japanese automobile industry in particular demonstrated the success of interfirm co-operation with suppliers on R&D, investments, pricing schedules, and quality control, encouraging rapid diffusion of the practice to other countries and industries. The main contexts for relational markets are still regional or sub-national agglomeration, but the global expansion of value chains also relies on efficient relational markets.

Apple's success in design and system development is in part attributable to the fact that its staff worked closely (and in secret) with several trusted collaborators (original design manufacturers, contract manufacturers, component and material specialists) from around the globe. Similarly, Walmart works closely with its main suppliers. Most MNCs maintain substantive resources to work with their suppliers and periodically reward the best performers with recognition and increased sales.

Environmental Responsibilities: From Value Chains to Value Cycles

The need to reduce the environmental impacts of economic activities is a recurrent theme throughout this book. Among economic activities, however, manufacturing carries a heavier responsibility: most environmental problems are strongly related to the production, use, and disposal/recycling of goods. Whether we are concerned about waste and recycling, carbon emissions, or hazardous substances, solutions to these environmental problems will depend on changing the design, material inputs, and production processes of goods.

Most products have multiple environmental issues at every stage of the value chain and very often dealing with an issue will require a coordinated response throughout the value chain. Communication and computer electronics, for example, use substantial resources and energy in their manufacture. They also use and release many hazardous substances in production that expose labour and communities to health risks and require facilities to deal with storage and disposal. When electronics leave the factory, they incorporate hazardous substances that enter homes and businesses. There are packaging and transportation impacts as parts and components and final products are shipped around the world. Since companies and consumers use energy to run the devices and networks, **server farms** are the fastest growing demand for electricity.

In addition, electronics are notorious for e-waste, an end-of-life issue, posing significant policy challenges for all levels of government. The US and China respectively generated 10 million tonnes and 11.1 million tonnes of e-waste in 2013. One estimate suggested that the US only recycled about two-thirds of electronic products discarded, while fewer than 10 per cent of mobile phones are dismantled and reused. The local disposal or recycling is further challenged by shortening product life cycles. Indeed, e-waste streams from rich to poor countries, where local regulations are weak, remains a problem, often flouting international conventions regarding the movement of hazardous substances (Figure 11.5). Without proper regulation, handling e-waste that contains toxic substances such as lead, cadmium, and mercury poses serious health problems to the people, including children, who go through it looking for aluminum, copper, and steel to strip and sell for reuse. E-waste also poses environmental problems related, for example, to the release of dioxins, carcinogens, and heavy metals; use of non-renewable resources; and loss of material.

FIGURE 11.5   E-waste Trade and Recycling in China

Peter Essick/Aurora Photos

server farms A collection of hundreds and even thousands of computers to serve large-scale demands on a 24-7 basis. They require large amounts of power to run and keep cool.

Value-Cycle Approach

Environmental problems have increasingly been addressed by changes to design and manufacturing processes to create global value cycles (Figure 11.6). The conceptual basis of this approach is to transform value chains into a set of connected, cyclical processes that eliminate or recycle waste at each stage of production, distribution, and consumption, and to "close the loop" on the entire production process. The value-cycle approach (also known as "cradle-to-cradle" production or industrial ecology) requires both the transformation of existing products and the development of new ones. Like the segments of value chains, the stages of the value cycle generate value. In the value cycle, however, value is generated by processes and products that have gone unrecognized in the past—by-products (formerly waste), resource savings, pollution prevention, insurance savings, brand building on the basis of green reputations, and improvements in employee health, retention, and productivity, among others.

The transformation of value chain into value cycles requires four interrelated elements. First, lifecycle analysis (LCA) is needed to assess all the environmental impacts of the product and the processes involved in its manufacture, upstream and downstream (extraction, processing, parts and component manufacture and assembly, distribution, use by final consumer, and disposal or recycling), as well as transportation, packaging, and related infrastructure. Second, design for environment (DfE) uses LCA to design products that incorporate technologies and practices that minimize or eliminate environmental impacts through the life cycle. DfE may require reform of existing technologies, replacement of some materials, or both; it also requires innovation. Third, environmental governance is required because, even though many of the technologies needed to conserve energy and reduce pollution exist, most firms will continue to externalize environmental costs unless they are explicitly forbidden to do so. International, national, and regional government policies must support value creation throughout the cycle and prevent externalities from escaping the system. Finally, a firm needs an overall corporate strategy that seeks to integrate its operations with other activities in the value cycle, both upstream and downstream.

One firm cannot control the entire value cycle. Therefore a high degree of inter-firm and inter-institutional interaction and collaboration is required. The producers responsible for both direct and indirect impacts have the information necessary for LCA and DfE and must share it. The geographic challenges to imposing governance on global value cycles are formidable, as "free-rider" countries can become pollution havens for non-compliant companies. Europe and Japan rely on extended product responsibility laws and regulations regarding waste in electric and electronic equipment to address this problem. North America's approach has been looser, but environmental legislation has increased exponentially over the last few decades. Moreover, the developing countries have recognized that in addition to manufacturing for the developed world they have been absorbing its pollution and waste. NGOs have helped to bring this situation to their attention and to raise the awareness of developed-country consumers and governments. For these reasons—and in expectation of stricter regulation to come, particularly on greenhouse gases—virtually all manufacturing MNCs now design their products and processes with the value cycle in mind.

Geographic Trends

In a global economy tied together by value chains, the environmental impacts occurring within the different stages of extraction, processing, production, distribution, consumption, and recycling are necessarily occurring in different locations. Transformation of these activities to reduce their impacts both modifies existing and creates new value chain configurations. Indeed, the recycling of materials for industrial re-use—a long-established trend that preceded the formal development of the value-cycle approach—has varying locational consequences. For example, the use of scrap metal has allowed the survival of some steel firms in established locations where access to iron ore or ingots is no longer economical, while the use of recycled newspapers in paper production has increased the viability of new large urban-based locations away from conventional forest-based locations.

The varying locational implications arising from the design of entirely new processes and products within value cycles is revealed in the previously discussed case of BMW's development of the all-electric i3 series (Chapter 2). In this case, environmental concerns have stimulated the search for suppliers and processing sites. For example, carbon fibre is sourced from suppliers in Japan and the fibre made into parts in Washington State to access cheap hydro-power. Similarly, Tesla and Panasonic chose Nevada as the site for a "Gigafactory" for lithium batteries for Tesla's planned mass produced affordable car because the state provided access to abundant supplies of renewable energy and $1.25 billion in subsidies. After all, the environmental vision of EVs doesn't make much sense without use of renewable energy in manufacturing. The two companies were in discussions with several southwest US-state governments on the matter. As a related trend, Google, Microsoft, and Apple have made commitments to be 100 per cent supplied by renewable energy and are siting new server farms near renewable energy sites to counter the fact that they are the dominant source of increased electricity consumption.

Companies also need production facilities and supplier facilities that reduce environmental impacts, as well as securing the talent needed to develop environmental technologies and design for the environment. In Toyota's development of the hybrid EV, it generated talent and technology internally while working with new R&D firm liaisons, government laboratories, and existing tier 1 suppliers. In so doing, it helped to cement the automotive cluster and Nagoya's greater industrial agglomeration, and now it is establishing branch plants and hybrid technology R&D overseas. Still Toyota sees fuel cells as the future, and in this regard, in addition to its Japanese government and industrial partners, it is working closely with universities, companies, and government in California because of the state's commitment to the hydrogen option. One vexing issue is how to encourage consumers and governments to recycle in ways that make sense in monetary terms. In some jurisdictions with take-back laws, companies and consumers are compelled to share the cost and incomes from recycling and thus a viable market is established. In other jurisdictions, companies wanting to follow through on a global commitment to recycling are forced to establish their own recycling systems; hence Dell, Apple, and Samsung offer free end-of-life return policies to consumers.

Brief reference to two "success" stories underscores the challenges facing the development of products that meet value-cycle expectations. First, the 1987 Montreal Protocol demanded that chemical companies stop producing CFCs that science had shown were seriously damaging the earth's ozone layer. Dupont, the world's biggest chemical (US-based) MNC, was already working on alternatives and speeded up its development program that, in turn, stimulated oligopolistic rivals such as (UK-based) ICI to do likewise. CFCs are used in aerosols and for refrigeration and cooling purposes and were first replaced by hydrochlorofluorocarbons (HCFCs), and more successfully by fluorocarbons (FCs) and hydrofluorocarbons (HFCs). The problem of ozone depletion has been resolved—NASA has estimated that the ozone layer will be back to normal by 2050—and HCFCs are to be phased out by 2030. Yet, FCs and HFCs still contribute to global warming, and their use needs to be reduced or eliminated by the development of alternative resources and/or processes.

Second, as already noted, the development of EVs by BMW, Toyota, and Tesla, along with related initiatives by other auto companies, has been greatly stimulated by laudable desires to reduce energy consumption. Yet, even as vehicles are becoming more fuel-efficient, more are being produced—global production of autos almost doubled from 1990 to 2012 (Table 11.2).

Clustering and Congregation

A premise of this chapter is that manufacturing moves to the places that suit it best, whether in relation to one dominant or a range of location factors. But what does that mobility of manufacturing capital mean to places that manufacture? Must they always dwell under the threat of livelihoods and communities disrupted? The short answer is yes. A more complex response recognizes that manufacturing can capture deeply rooted benefits of external economies through the evolution of industrial clusters.

Although manufacturing is mobile, the geography of manufacturing continues to be anchored by regionally and locally defined external economies. As we saw in Chapter 8, at the larger spatial scale of large cities or regions of cities, these external economies are called agglomerations. As in the case of Detroit, those economies can unravel or, as in the case of the Pearl River Delta, they can evolve at a rapid pace. But the key to building successful agglomerations is usually found at smaller spatial scales and among distinct sectors called industrial districts or clusters.

Calculating external economies of clusters is not easy. In addition to the challenges of assessing and aggregating the benefits of fixed assets, access to experienced labour pools, and related firms, clusters generate untraded interdependencies, such as interpersonal networks, co-operative behaviour, information-sharing, and the positive externalities generated by competition, including innovation. Nor are all agglomerations and clusters equally successful. If the components of a cluster are only weakly connected, or operate more or less independently, the advantages they can enjoy will be limited. Moreover, clusters are not islands, but must be integrated into global value chains.

Given the inherent diversity among and within the different industrial sectors that define clusters, there have been many attempts to define them. Thus, there are "hub and spoke" clusters dominated by a single firm, and flexible specialized clusters with many SMEs, and variations in between. In policy terms, Porter's "diamond interpretation" of clusters has been influential (Case Study 11.5). The next sections use two global apex agglomerations to illustrate two core types of clusters and their propensity to evolve new institutional mixes. The stakes are high, because like companies, clusters compete with one another, but whereas companies can move for competitive advantage, regions must invigorate clusters or suffer decline. Finally, we discuss whether it is possible for governments to stimulate the growth of manufacturing clusters.

Silicon Valley and Its Progeny

California's Santa Clara Valley, near San Francisco, used to produce very good wine. Today it is better known as Silicon Valley, the model for high technology "parks," "clusters," or "complexes" around the globe. Silicon Valley, however, has a distinct history in which pioneering individuals, Stanford University, and the defense policies of the US government all played prominent roles.

Stanford was founded in 1891 in Palo Alto, where the Federal Telegraph Co. employed the inventor Lee de Forest to develop vacuum tubes; and military contracts funded research and production in the region in the years before the First World War. Several decades later, in the heyday of Fordism, Silicon Valley had moved into the forefront of the movement that became ICT, producing innovations that ranged from radar and silicon transistors, microprocessors, and personal computers, to Internet infrastructure, browsers, and search engines. Stanford itself played a seminal role, as did Fred Terman, its Dean of Engineering in the 1940s and 1950s, who championed the efforts of his students when they set out to establish their own companies: Hewlett Packard (HP), Shockley Transistor (which pioneered the use of silicon in transistors), Fairchild Semiconductor, and Intel. Throughout this evolution, start-ups, large established firms, military research and procurement, and university research have intertwined. Specialized venture capital investors, lawyers, and real estate developers have also played their part.

The cluster of activities centred at Stanford became a powerful centripetal force, attracting high-tech firms as well as engineering, investment, and management talent from around the world. A culture of information diffusion, collaboration, and competition encourages people to transfer from one company to another; start up new companies that either fail or get sold; then recombine, sharing information within and between companies, and mediating their relationships both informally and through legal contracts. In 2006, the San Francisco Bay area as a whole, including Silicon Valley and San Jose, was home to almost 400,000 highly paid high-tech jobs—the largest concentration in the US, and one with a remarkable record of innovation. In addition to Intel and HP, major players in the region include Apple, Google, Facebook, Cisco, eBay, Adobe, Oracle, and Yahoo.

Even in Silicon Valley, however, the functional structure of the agglomeration has changed since the 1960s (Figure 11.6). In general, manufacturing has declined (the processing and assembly of silicon chips has been relocated elsewhere, for instance), while

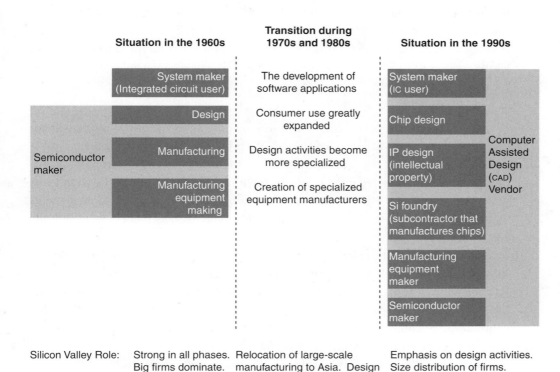

FIGURE 11.6   Changes in the Organization of the Semiconductor Industry

Source: Based on McCann, P and T. Arita. 2002. Edward Elgar Publishing, http://www.e-elgar.com. Reprinted with permission.

R&D (especially design of new computer software) has increased. Assembly activities, which were once significant, have declined because of increasing land and labour costs, high taxes, and environmental constraints, among other factors. In the 1950s and 1960s the bigger firms such as Hewlett Packard established production elsewhere, including a small-scale operation in Taiwan to serve small local markets. In the 1970s and 1980s, mass production of Silicon Valley's innovations was increasingly shifted to lower-wage locations in Asia for the purpose of export to destinations that included the US. In the 1990s, wafer processing and innovative manufacturing moved to Singapore and Europe in search of more highly skilled labour. And in the new millennium, Valley firms have been relocating to China to take advantage of its technological upgrading as well as its markets. In contrast, Japanese firms have kept their wafer processing operations in Japan, relocating only their assembly work to lower-wage locations.

Inspired by Silicon Valley, similar clusters of research-intensive, high-tech activity developed rapidly in places such as Taiwan (Hsinchu Science Park) and Bangalore, India, under the direction of Asian-born scientists and engineers with Silicon Valley experience. Hsinchu and Bangalore have also benefited from government policies designed to assist domestic firms, for example by establishing research institutes and providing building and transportation infrastructure. The fact that former "poor" countries now play leading roles in R&D and innovation as well as manufacturing indicates that successful Asian clusters have not only taken on important value-added roles but also have "reversed" the dynamics of the product cycle model. Taiwan, for instance, began its career in the electronics industry as a branch–plant production site for MNCs that concentrated their R&D work in the US, Japan, and the EU. Now, electronics R&D is conducted in Taiwan and Taiwanese firms send their routine production work to China.

Toyota City

Toyota developed out of a textile machinery manufacturer in the cotton and textile producing area around the city of Nagoya. With the rise in car manufacturing in the 1930s, Toyota chose a new economic space, in the nearby town of Koromo, for production and eventually renamed it Toyota Town. The Toyota production system (TPS) has since evolved as a tight interrelation between internal production improvements within Toyota and its suppliers organized by relational markets that are hierarchically structured (Case Study 11.4; Case Study 11.1). The most important and the majority of suppliers developed or located in proximity to the core firm. The two core ideas of the TPS are human control in the context of automated production and **just-in-time (JIT)** delivery of products and parts, which are produced only when needed. Among the features central to the TPS were innovations in equipment setup, teamwork, quality control, worker control over continuous flow, limited pay differentials between workers and managers, a bonus system, and enterprise unions.

In and around Toyota Town, Toyota is supplied by tier 1 contractors, which in turn are supplied by tier 2 suppliers, and so on, with each layer of transactions organized on a just-in-time basis. Inter-firm relations are very stable, encouraging suppliers to invest in R&D and skill development, but also competitive, as Toyota typically has two or three potential suppliers of any particular component. The system is highly efficient.

just-in-time (JIT) A system in which parts and components are delivered to the factory just prior to their use.

Case Study 11.4
TOYOTA'S INTER-FIRM RELATIONS AND LOCATIONS

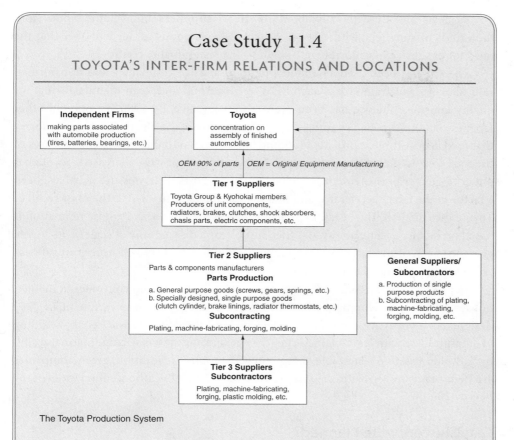

The Toyota Production System

The core of the Toyota production system is the parent firm together with 13 other firms, some of which were spun off from Toyota and some of which it invested in. Each has many subsidiary firms and plants, but the company's headquarters, R&D, and main production facilities are in Toyota Town. Next in proximity are the 217 suppliers of components, parts, and materials that make up the Kyohokai group and the 125 equipment and distribution suppliers that make up the Eihokai. Of the members of the Kyohokai, 121 are located in the Tokai area around Toyota Town, 65 in the Kanto area around Tokyo, and 31 in the Kansai area around Osaka. Each plant is supplied by a smaller number of firms. The components or modules made by each of the Kyohokai members are assembled from the parts and components made by dozens of suppliers who in turn rely on other suppliers. The second- and third-tier suppliers are usually close to the buying firm. This hierarchy is important because it reduces complexity at each level, ensuring that Toyota plants deal with only a limited number of components, and facilitates JIT delivery.

Toyota's relational markets are both co-operative and competitive. The transaction begins with collaboration on the design and engineering of components or material and continues throughout the life of a car model. Some companies are trusted to design the products they supply to Toyota themselves; others are provided with Toyota's own designs. Contracts take into consideration the investments that each firm has to make in development and stipulate that costs must decrease over time. Toyota will often contract with two or more firms to ensure competition not only in the bidding process, but throughout the product life cycle. Suppliers are graded on performance and either rewarded or penalized in the next contract.

tier 1 suppliers The most important suppliers who have direct contact and business relations with core firms.

Toyota deals directly with only about 200 **tier 1 suppliers**, but obtains inputs from thousands of suppliers, while the JIT system delivers goods as needed, obviating the need for expensive inventories and allowing for greater quality control.

The TPS has been used as a model by thousands of companies and transformed into management frameworks such as "re-engineering" and "lean manufacturing." Its quality control influence has been even more pervasive. By 1980, Toyota and other Japanese automakers had so undermined the viability of the North American industry that the US government provided funding for research into the organizational bases of their success. As global competitiveness in the auto industry has increased, Toyota has put even greater pressure on its suppliers to reduce costs. This pressure may have been a factor in the faulty accelerator and braking mechanisms that led to the mass recall of Toyota cars in 2010. This recall is especially damaging for a brand whose reputation is based on quality and reliability. Are these problems a diseconomy of scale? Did Toyota give too much or too little decision-making autonomy to its North American subsidiary? Does the TPS need to be re-conceived?

In practice, the TPS has evolved considerably since 1950 and continues to involve, not least by the development of EVs. In tandem, the Japanese government in 2009 designated Toyota City as an "Eco-model City" and since then subsidized the building of smart houses with state-of-the-art energy generating systems, including (naturally) hybrid plug-ins, and renewable energy power stations. In addition, green companies are being encouraged with the goal of reducing the 1990 carbon emission levels by half by 2050.

Can Clustering Be Planned?

Governments have been trying to promote manufacturing agglomerations for some time. In rich market economies, industrial parks were a common feature of regional policies designed to attract branch plants; proponents argued that a concentration of enterprises would both justify the high fixed costs of buildings and transportation facilities, and generate local multiplier effects. Similar thinking inspired the creation of special economic zones in poor countries. In the 1970s, governments hoping to create their own Silicon Valleys began to establish high-tech parks as homes for technologically sophisticated industries led by electronics and biotechnology. These efforts have intensified in recent years, stimulated and formalized by the cluster framework proposed by Michael Porter to analyze the competitiveness of a region or locality in the global context (Case Study 11.5).

Clusters are seen as generating a dynamism among firms that leads to consistent development of internationally successful products and sustained regional development. Maintaining vibrant manufacturing clusters is not easy, however. In Canada's case, attempts by governments to create competitive clusters from scratch have been largely disappointing. Biopharmaceuticals are an example of this. Outstanding biomedical research is being conducted at several Canadian universities, as well as public and private institutes; both Toronto and Montreal have production facilities; and Canadians consume lots of pharmaceuticals. Yet the great majority of those drugs are sold by MNCs. Although Canada produces significant quantities of generic drugs, it is not converting its research into branded products, and biomedical clusters remain small in scale. Nor can

Case Study 11.5
PORTER'S DIAMOND

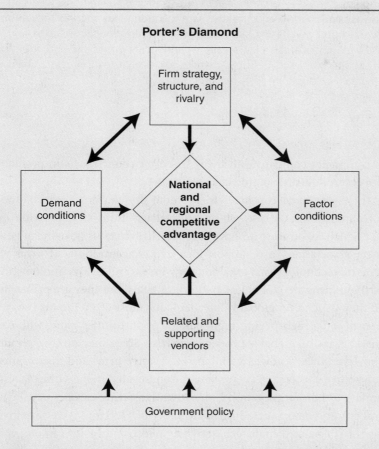

Porter's Diamond

According to Michael Porter (1990), the industrial specializations that drive national and regional development are shaped by factors that can be classified into four broad categories:

1. *Factor conditions*: given factors such as resources and population and their development by investments in infrastructure, education and knowledge, capital equipment, and finance;
2. *Demand conditions*: size and sophistication of home markets, which compel companies to develop new technologies and scale to gain an international advantage;
3. *Related and supporting industries*: the quality of and competition among suppliers in the industry or related industries; and
4. *Firm strategy, structure, and rivalry*: the appropriateness of firm goals and organization in the context of whatever domestic rivalry may exist.

He argued that this "diamond" model was a better basis for understanding national and regional development than the economists' traditional model of comparative advantage (Case Study 2.4). While the latter emphasized "given" resource endowments as the basis for comparative advantage, Porter's diamond model highlights the competitive advantages that can be created over time, for example, through

(Continued)

investment in education (a factor condition) and entrepreneurial initiative (a firm strategy variable). Porter's diamond also recognizes that government can play a role in developing a nation's competitive advantage, for example, by encouraging the formation of clusters of economic activity at national and regional levels. Also, political decisions (positive and negative) are made in areas such as regulation (from finance to safety), purchasing, subsidies, infrastructure, education, taxes, and leadership. As well, chance can play a role in influencing economic activities: external events such as wars and political changes, disruptive technological changes, and shifts in financial or commodity markets.
Source: Based on Porter 2001.

Canadian companies compare with their American counterparts in profitability since they do not control the distribution chain.

Nevertheless, Canada continues to promote the cluster approach. The National Research Council supports research and university–industry interaction in 18 locations, 11 of which it considers to have reached cluster status. Provincial governments and industry associations also support localized concentrations of firms working in areas such as biotechnology (in Saskatoon) or wireless and GPS technology (in Calgary). Several of these firms are global leaders in their fields. But they are primarily research firms, and they depend on global value chains for most of the inputs into their activities. Such firms offer high-paying, high-skill job opportunities, along with connections both to markets and to global sources of expertise. Since they rarely interact with one another, however, these knowledge-intensive, creative firms and institutions have *not* formed innovative clusters: they have simply congregated in the same places.

A similar case of congregation rather than clustering can be seen in Vancouver, where local entrepreneurs, including the founder of Electronic Arts, are involved in a wide range of specialized high-tech activities (electronics, biotechnology, aerospace, submarines, printing, fuel cells, imaging and software equipment, and GIS). Most of their firms are small, however; the largest, MacDonald Dettwiler (GIS), employs about 1500 people. Talented people congregate in Vancouver, attracted by its universities, recreational opportunities, multicultural mix, amenities, and so on. But their firms have limited engagement either with one another or with manufacturers (only prototype or small-lot manufacturing is done in Vancouver).

Conclusion

This chapter has revealed the complexity of the forces shaping the location dynamics of manufacturing activities. Understanding these dynamics is important because manufacturing is vital to economic development, a major factor in the transfer of technology, and an important component of international trade. Location and profitability are closely intertwined. At the same time, manufacturing activities have options when it comes to location: the same activity can be carried out in many different places, each of which has its own advantages and disadvantages.

Practice Questions

1. Where are the main manufacturing areas in your community? How did they originate?

2. Several rich countries, or large regions of rich countries, have deindustrialized in recent decades. Is this a source of concern?

3. Should your community seek to protect and/or develop its manufacturing base?

4. Explain the main clusters of manufacturing activity in your community.

5. How would you define green manufacturing? Can you provide examples from your community?

Key Terms

primary manufacturing   335
secondary manufacturing   335
maquiladora   338
conurbation   341
suburbanization   341
non-metropolitan
    industrialization   341
deindustrialization   343
location factors   344

procurement costs   344
ubiquitous inputs   345
pure inputs   345
median   345
impure inputs   345
isotims   345
isodapane   346
critical isodapane   347
spatial margin of profitability   348

substitution principle   348
location conditions   348
producer-driven value chains   354
buyer-driven value chains   354
server farms   356
just-in-time (JIT)   362
tier 1 suppliers   362

Recommended Resources

Edgington, D. and Hayter, R. 2013. "The in situ upgrading of Japanese electronics firms in Malaysian Industrial clusters." *Economic Geography* 89: 227–59.

    Reviews literature on clusters, and stresses importance of in-situ location adjustments.

Gertler, M. 2004. *Manufacturing Culture: The Institutional Geography of Industrial Practice*. Oxford: Oxford University Press.

    A comparative investigation of industrial practices in Germany and Canada, and how German manufacturing practices have been modified or resisted in Ontario.

Hayter, R. 1999. *Industrial Location Dynamics: The Firm, the Factory and Production System*.

    Out of print textbook on manufacturing location, can be accessed at: http://www.sfu.ca/geography/wp-content/uploads/2010/08/dynamics_hayter.jpg

Lepawsky, J. and McNabb, C. 2010. "Mapping international flows of electronic waste." *The Canadian Geographer* 54: 177–95.

    Argues that international flows of electronic waste are only partly explained by the "pollution haven" hypothesis (in which rich countries use poor countries as dumps for their waste).

Lyons, D., Rice, M., and Wachal, R. 2009. "Circuits of scrap: Closed loop industrial ecosystems and the geography of US international recyclable material flows 1995–2005." *The Geographic Journal* 175: 286–300.

    Argues that "closed loop" recycling systems need to be considered at global as well as local, regional, and national scales.

Patchell, J. 1997. "Creating the Japanese electric vehicle industry: The challenges of uncertainty and cooperation." *Environment and Planning* A 31: 997–1016.

    Discusses co-operation among firms and other agencies in developing new auto technology.

Potter, A. and Watts, H.D. 2010. "Evolutionary agglomeration theory: Increasing returns, diminishing returns, and the industry life cycle." *Journal of Economic Geography*: 1–39.

    An analysis of agglomeration economies and diseconomies over time in Sheffield, an early centre of the Industrial Revolution.

Wainwright, O. 2014. "Work begins on the world's first 3D printed house." *The Guardian*, March 28.

References

Aritsa, T. and McCann, P. 2002. "The relationship between the spatial and hierarchical organization of multiplant firms: Observations from the global semiconductor industry." Pp. 319–63 in P. McCann, ed. *Industrial Location Dynamics.* Cheltenham: Edward Elgar.

Enright, M.J., Scott, E.E., and Chang, K. 2007. *Regional Powerhouse: The Greater Pearl River Delta and the Rise of China.* Singapore: John Wiley.

Fingleton, E. 2006. "Manufacturing matters." *Fortune* http://money.cnn.com/magazines/fortune/fortune_archive/2006/03/06/8370712/index.htm (accessed 5 Oct. 2009).

Lin, G.C.S. 1997. *Red Capitalism in South China: Growth and Development of the Pearl River Delta.* Vancouver: University of British Columbia Press.

Pollard, S. 1981. *Peaceful Conquest: The Industrialization of Europe 1760–1970.* Oxford: Oxford University Press.

Porter, M. 1990. *The Competitive Advantage of Nations.* New York: Free Press.

Weber, A. 1929. *Alfred Weber's Theory of the Location of Industries*, translated by C.J. Friedrich. Chicago: University of Chicago Press.

CHAPTER 12

Services

The first and simplest characteristic of a post-industrial society is that the majority of the labor force is no longer engaged in agriculture or manufacturing but in services.

Bell 1973: 15

The overall goal of this chapter is to investigate the location dynamics of services and their implications for local and regional development. More specifically, this chapter's objectives are

- To examine the role of market accessibility in the location patterns of service activities with particular reference to spatial distribution and clustering;
- To emphasize the diverse composition of service activities and their role as job and growth generators;
- To explore the organizational and location dynamics of service activities within cities, especially with respect to retail and consumer services;
- To examine the rationale for public services, especially with respect to education and health; and
- To explore the role of producer services in local and international value chains.

The chapter is divided into four main parts. The first part defines services and discusses their organization and location in reference to accessibility and integration. The following three parts discuss the service sector's basic sub-sectors: consumer, producer, and public services.

The Development of the Service Industry

The service or "tertiary" sector encompasses a vast range of occupations, from consumer-oriented fields such as retail, finance, and tourism to the many support services—research, engineering, advertising—required by the primary (raw materials) and secondary (manufacturing) sectors. The fact that these activities are termed "post-industrial," however, does not mean that they are disconnected from the resource or manufacturing industries. In general, the service activities central to Bell's definition of **post-industrialism** are those that respond to the diversification and amplification of society's needs by increasing the information content of economic activity. Bell emphasized education and health but there are an astounding number of others. A huge diversity can be found in areas we categorize broadly as business or personal

post-industrialism The phase in the development of a market economy when services replace manufacturing as the dominant driver of the economy and the source of jobs.

Case Study 12.1
FEEDING THE KNOWLEDGE ECONOMY

Aerial photo of Simon Fraser University which is located on top of Burnaby Mountain, 370 metres above sea-level.

services or even in more narrow categories such as environmental services or tourism. The response to social needs generates new services in several ways. Sometimes there is a collective recognition of the need to develop new services, as in the case of educational institutions devoted to research and the formation of a flexible, innovative workforce (Case Study 12.1). More often, service generation occurs in less obvious forms; for instance, a salesperson or hairdresser with up-to-date information on products and styles may be able to offer consumers creative ideas. Although some services have little or no information content, it is the need for specialized information that has driven the massive growth of the **service sector**.

Traditionally, only primary and secondary activities were considered basic: that is, as sources of external (export) income for national and local economies (Chapter 8). Tertiary (service) activities were conventionally classified as non-basic because they focused mainly on local markets. Today, however, services are seen as important contributors to exports as well. Thus regional and local public policies are designed to generate employment and develop competitive advantages in the service industries.

Defining Services

The burgeoning tertiary sector is both a consequence and a prerequisite of a maturing, complex economy that is increasingly knowledge-based or information-intensive. The

service sector The service sector is often defined negatively as not producing material goods such as manufactured goods. Rather, services engage distribution and selling, information providing, creative, maintenance, and decision making activities that facilitate and control goods production and meet personal and social needs.

In Canada, as elsewhere, the 1960s saw massive growth in higher education. For the first time since the turn of the century, established universities were expanded and many new institutions were created in an effort to meet the increasing demand for better-educated workforces in rapidly growing economies. Simon Fraser University (SFU), opened in 1965, employs more than 3,300 faculty and staff to serve over 35,000 students on three campuses. The University of Victoria, opened around the same time, is similar in size, while the University of British Columbia, founded in 1908, at the end of the first phase of modern university growth, has more than 10,000 faculty and staff to serve a student body of almost 50,000 on two campuses. Community colleges were also established throughout the province, five of which gained university status after 1988. The University of Northern British Columbia, established in 1990, is a recent addition to the province's higher education system.

Universities are big employers in the local context, and in terms of research, teaching, and training they are vital to the nation's long-term economic health. In recent years, governments have encouraged them to take on a more direct role in innovation by establishing "technology-oriented complexes" on university campuses. In British Columbia, for example, the provincial government created four "Discovery Parks" in the 1980s, including one at SFU (on the photo located just off the ring road on the bottom, right-centre). If this particular Discovery Park had limited success attracting high tech companies, Simon Fraser (along with most universities) has since developed active entrepreneurial incubator programs to stimulate such developments.

service sector defines a maturing economy because it is the home of all the specialized activities required to meet the increasingly specialized needs of individuals and business. If the service sector offers many of society's most tedious, routinized, low-wage jobs—cleaning, washing dishes, flipping burgers—it also offers many of the most interesting and well-paid. Fields such as education, medicine, financial management, computer technology, design, engineering, and journalism require not only extensive training and education but also ongoing, lifelong learning.

In the US and Canada, service-sector employment more than doubled between 1970 and 2012–14 (Table 12.1) and accounted for 79.9 and 79.0 per cent of total employment respectively in 2012–14. The so-called goods-producing sector—the primary, manufacturing, and construction (and sometimes utilities) sectors—comprise the remainder. However, many (producer) services in business, transportation, finance, trade, self-employment, and even government (e.g., departments related to agriculture, mining, forestry, industry, and business) supply inputs to the value chains of goods; a broad definition of goods production is likely to account for over half of the employment levels in modern economies.

The traditional mainstays (retail and wholesale trade) grew, but not as fast as the service sector as a whole. Most growth occurred in areas such as health care, education, legal and technical professions, insurance, and real estate (Table 12.2). Services are also booming in emerging market economies. They already account for more than half of

TABLE 12.1   Employment by Broad Sector Groupings, US and Canada 1970 and 2012

Category	United States		Canada	
	1970	**2012**	**1970**	**2014**
Primary	4,140	2,913	674	664
Manufacturing	17,850	11,918	1,751	1,696
Construction	3,650	5,641	389	1,368
Sub-total	25,640	20,472	2,814	3,728
SERVICES TOTAL	46,105	116,067	5,065	14,044
Other (self-employed)	4,933	8,816		
TOTAL EMPLOYMENT	76,678	145,356	7,879	17,772

Source: US Bureau of Labor Statistics and Statistics Canada http://www40.statcan.gc.ca/l01/cst01/econ40-eng

Notes: The 1970 figures for the primary sector in Canada have been slightly modified to add to the total listed (since the official figures for each sector do not add to the total indicated). Also, utilities are classified as part of services.

India's GDP and are approaching that level in China. Indeed, India has made services its lead export industry. As China reaches the limits of the growth it can expect from low-end manufacturing, its hopes are increasingly focused on service occupations.

Differences and Commonalities

Despite their economic importance, the default definition of services is that they are activities belonging to neither the primary nor the secondary sector. It has been suggested that tertiary services should be disaggregated to create two additional sectors: quaternary (activities involving knowledge creation and information processing) and quinary (top-level decision-making activities) with the tertiary label referring to retail and wholesale. But there is no broad agreement on these terms. In practice, industrial classification systems divide services into sub-sectors such as retail, wholesale, education, health, and finance.

At the same time, the relationships between "non-material" services and the primary and secondary sectors are important. Service activities such as engineering, design, and regulation are integral to "material" activities such as mining and manufacturing through various creative, regulatory, and routine inputs. Service industries such as health care, recreation, and hospitality rely heavily on manufactured and agricultural goods. The tertiary sector in general is just as dependent on equipment, materials, and energy as the resource and manufacturing sectors are. In other words, services are part of value chains, in some cases controlling them and in others contributing to them. Whether a particular economic activity is predominantly material or non-material in nature is often hard to determine. The unifying factor is that all of them add value. The key reason for distinguishing services from other types of activity is that the value they add is non-material. In the past, service-sector jobs were often distinguished from other types by the fact that the labour they involved was not primarily physical. However, since the 1970s, computerization has eliminated much of the manual labour required in the

TABLE 12.2 Employment in Services in the US (2012) and Canada (May 2014) (000s)

Category	United States (%)	Canada (%)
Utilities	554 (0.5)	156 (1.1)
Wholesale trade	5,673 (4.9)	601 (4.2)
Retail trade	14,875 (12.8)	2,134 (15.0)
Transportation/warehousing	4,415 (3.8)	906 (6.4)
Information	2,678 (2.3)	In Information, culture, etc.
Financial activities	7,786 (6.7)	1,082 (7.6)
Professional, scientific, technical	In prof. and bus.	1,366 (9.6)
Business, building, and other support services (US)/Professional and business services (Canada)	17,930 (15.4)	751 (5.3)
Education services	3,347 (2.9)	1,346 (9.4)
Health care and social assistance	16,972 (14.6)	2,234 (15.7)
Leisure and hospitality (US)/Information, culture and recreation (Canada)	13,746 (11.8)	798 (5.6)
Accommodation	See Leisure	1,160 (8.1)
Public administration	21,917 (18.9)	955 (6.7)
Other services	6,175 (5.3)	772 (5.4)
Total Services	116,067	14,261

Source: Compiled from US Bureau of Labor Statistics and Statistics Canada. The US data are year (2012) averages, and incorporate author's estimates of "self-employed." The Canada data are a monthly (May) estimate for 2014. Note in official Canadian data utilities are placed in the goods producing sector.

goods-producing sectors as well. There are also many service activities that do involve hard physical work—for example, trades such as plumbing or the movement of goods.

Traditionally, the geography of service activities has focused on retailing, which in most cases requires proximity to consumers, even today. Service providers either locate themselves where interaction is needed or they provide whatever technological means are necessary to permit long-distance interaction. These principles hold for all three types of service: consumer (traditional retailing, e-tailing, personal care, travel arrangements), producer (legal, accounting, and consulting services for business), and public (health care, education).

Accessibility, Integration, and Innovation

The pre-eminent insight of Christaller's central place theory (Chapter 1) is the need to strike a balance between accessibility and efficiency in the provision of services. Low-order services have small ranges (consumers will travel only short distances to buy them) and low-threshold requirements (a relatively small population will provide sufficient demand for the service to be viable). Grocery stores and personal care businesses such as barbershops are classic examples: small-scale operations that do not require large investments are widely distributed and serve relatively few customers. Higher-order services, by contrast, have higher demand thresholds, larger ranges,

goods-producing sector
Statistically, the primary, manufacturing, and construction sectors. In theory, activities that exploit resources and add value to produce tangible goods, including many services.

and higher investment requirements. Therefore, they are concentrated in a relatively small number of more accessible central places. As different services require progressively higher thresholds and/or levels of investment, a hierarchy of services develops. According to Christaller's **marketing principle**, the result is a **nested hierarchy** of centres (Case Study 12.2; Table 1.1).

Christaller developed his model in the context of agricultural regions. Since his time, the massive population shift from rural to urban regions has reduced threshold populations in the former and increased the concentration of services in the latter. This trend has been strongly reinforced by increasing economies of scale in services themselves: larger, more efficient stores offer lower prices and so increase both the range of

marketing principle In Christaller's central place theory, the principle that services should be located so as to maximize access for consumers.

nested hierarchy The highest level of the hierarchy serves the entire population with the highest-order goods, as well as all lower-order goods to nearby consumers, while lower levels in the hierarchy serve progressively lower-order goods in progressively smaller catchment areas.

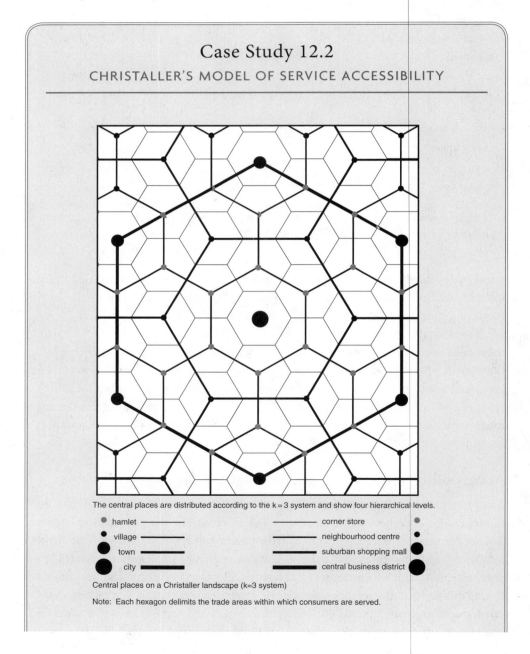

Case Study 12.2
CHRISTALLER'S MODEL OF SERVICE ACCESSIBILITY

The central places are distributed according to the k = 3 system and show four hierarchical levels.

- hamlet ——————
- village ——————
- town ——————
- city ——————

- corner store ——————
- neighbourhood centre ——————
- suburban shopping mall ——————
- central business district ——————

Central places on a Christaller landscape (k=3 system)

Note: Each hexagon delimits the trade areas within which consumers are served.

In this neoclassical model, the largest, most accessible central places offer the full range of services; somewhat less accessible medium-sized central places offer medium- and low-order goods, and the least accessible, lowest-order central places offer only the lowest-order services with the lowest ranges and threshold populations. The model assumes that consumers will travel to the nearest centre that supplies the good or service they want. The hexagonal shape of market areas eliminates overlap and allows consumers in all parts of the region to be serviced. The marketing principle is also labelled the k = 3 system, as each centre serves the equivalent of three market areas associated with the next lower-order centre. For example, the A order centre supplies A order goods to its exclusive B order market area plus one third of six B order market areas. For Christaller, this spatial hierarchy is complicated by the transportation principle, which recognizes that consumer accessibility is modified by transportation networks that channel traffic in particular directions on particular routes, and the administrative principle, which recognizes that political boundaries (local and international) further modify spatial hierarchies, for example, by imposing barriers to trade and different regulations.

goods and the amounts of time and money that consumers are willing to spend on shopping trips. The Christaller model also reflects the distribution of services within a city, everything from lower-order, high accessibility convenience stores and gas stations to higher-order, centrally located finance, medical, and entertainment services. In both cases, the tendency towards concentration is heightened by localization economies, since consumers will be willing to travel greater distances if providers of the same, similar, and related higher-order services cluster together.

Service concentrations reduce transaction costs for consumers by increasing opportunities for comparison and permitting greater certainty of choice. The advantages of concentration are reinforced when providers of similar services adopt differentiation strategies, increasing the need for comparison, offer shared (or free-riding) advantages in advertising and security, while allowing rival firms to keep an eye on one another and adjust their own strategies as necessary. At the same time, concentration reduces costs for suppliers, who will ultimately be compelled by competition to pass those savings on to retailers, who may pass them on to consumers. Examples of localization economies among retail activities are easy to find in specialized restaurant, entertainment, and shopping areas. In Canadian cities, for instance, Mountain Equipment Co-op often provides an anchor for a variety of competitive-but-interdependent outdoor stores. Tokyo's Akihabara district is the world's largest and most diverse concentration of electronic goods.

The distribution of services over space is the result of a competition-driven quest for efficiency of access and reductions in transaction costs. That efficiency is augmented by internal and external economies of scale and scope. The distribution of similar services over space provides ample opportunities for efficiencies in purchasing, logistics, marketing, and management. In fact, the drive for integration efficiencies is strongly evident in the service sector, yet it can be subtle and taken for granted. For example, the concentration of many competitive and complementary service firms at the neighbourhood or street level is made possible by the governance and infrastructure provided by public authorities and firms servicing a much broader area. Services may also be concentrated under one roof and management, as is the case in malls. This horizontal and vertical integration is critical because services are subject to constant innovation and hence to the realities of the product cycle. Innovation in the service sector is perhaps most evident in retailing, illustrated by new types of stores and the rise of e-commerce, but innovation has also created entirely new service activities and skills, such as software design, while changing others. Innovation in turn implies new challenges for integration within value chains.

Consumer Services

central business district (CBD)
The downtown and usually the most accessible area within a city, defined by commercial and retail activities.

In North America, service distribution and land-use patterns were initially based on the concentration of lower-order services (corner stores, street-front shopping strips) at the local or neighbourhood level and higher-order services (department stores, specialty stores and services such as musical instruments, professional sports and entertainment, medical specialists, accountants, etc.) in the **central business district (CBD)** of towns and cities. Since the 1950s, suburbanization has greatly modified those parameters. The increasing use of cars by consumers and truck-based logistics by service providers has resulted in greater dispersal of higher-order services within continually expanding metropolitan regions. In the US, the suburban shopping mall had become a significant feature of the retailing landscape by 1960, and this trend soon spread to Canada. In recent decades, big-box stores (Walmart is the classic example) and power retail centres (defined as retail areas covering a minimum of 23,000 square metres and containing at least three big-box stores as well as other retailers, served by common parking) have further diversified the location dynamics of retailing. At the same time, many downtown areas have been rejuvenated with the construction of high-density, often high-income housing that increases demand for services. In general, services are increasingly concentrated in metropolitan areas and declining in rural areas. Along the way, concentration has taken different forms at different scales (Figure 12.1). Important services such as automobile sales and service (which together account for 20 per cent of retail activity) show a parallel tendency to cluster together.

Consumer Service Formats

consumer services Services that supply individual (household) consumers.

The concentration of **consumer services** can be seen in retail strips, shopping centres, and power retail operations. Retail strips develop organically along main roads, often in response to pedestrian and transit-based passenger transport, though suburban strips invariably offer more parking than downtown strips do. They typically begin with lower-order services, but often acquire some specialty services, attracting larger markets, over time; this is especially common in the case of ethnic-oriented strips such as Chinatowns. They always offer banking, postal, and other services that integrate the locality with the broader economic space. In recent years, they have attracted chain operations (drugstores, convenience, hardware, and fast-food outlets) as well. A shopping mall or centre is an enclosed space, sometimes larger than 350,000 square metres in size that is built, owned, and managed by a single organization. It generally provides a variety of retail and other services, encouraging both competition within the

The Dawn of the Planned Shopping Centre
local automotive-oriented shopping centres developed, small-scale unenclosed community centres, independently developed

The Catalytic Shopping Centre
development of super-regional centres, greenfield sites, shopping centres as the catalyst for future suburban growth in major markets

The Era of the Power Centre
large-format retailers (big-box stores clustered in power centres, US-style retailing, economies of scale, functional shopping, auto-dependent

pre-1940s | 1950s | 1960s | 1970s | 1980s | 1990s | 2000s+

The Age of the Department Store
low consumer mobility, high-order goods purchased downtown, dominance of department stores, local convenience shopping

Shopping Centres and "Planned" Communities
housing stock and enclosed shopping provision devloped simultaneously, planned development of shopping centre industry

The Entertainment Shopping Centre
mixing of retail and entertainment, multi-screen theatres, new development and redevelopment of existing centres

Power Retail Integration
increased presence of big-box stores within existing shopping centres (often into former department store space), co-location of power centres and shopping centres

FIGURE 12.1   Evolution of the Canadian Retail System

Source: Buliung, R. and Hernandez, T. 2009. *Places to Shop and Places to Grow: Power, Retail, Consumer Travel Behaviour, and Urban Growth Management in the Greater Toronto Area*, Neptis Foundation (neptis.org).

same category of services and multi-purpose shopping; the larger malls also function as entertainment centres.

A power centre is anchored by three or more big-box stores offering much of the variety found in the many stores of a shopping centre or retail strip. They are labelled category-killers because of their threat to specialty stores. Giants such as Walmart aim to offer all the retail and many of the personal services available from smaller independent operations, while Canadian Tire, Home Depot, and Ikea compete against specialists in a limited number of categories such as hardware, lumber, and automotive, or home furnishings and decoration. Power retailers typically offer lower prices, more own-brands, and fewer high-end brands than their smaller counterparts. Initially, big boxes were independently located on arterial roads, but with increasing popularity they have come to congregate together in competition and complementarity. An area with several power centres within a square kilometre of one another is termed a power node.

Which service format develops in a given area depends on a complex set of factors including demographic structure (household types), social class and income, ethnicity and lifestyles, age of neighbourhood/development, technology, corporate strategy, infrastructure development, and attitudes towards zoning and community development. That said, and despite the speed with which the service landscape changes, established infrastructure does not disappear overnight. In the Greater Toronto Area (GTA) for example, between 1996 and 2006 there was only a small decline in the total area occupied by small, unenclosed malls, strips, and free-standing stores. Shopping centres in the GTA have declined significantly, however.

The big-box format developed simultaneously in North America and Europe, beginning in the mid-1980s; Canada's first big-box store was Walmart (Case Study 4.1). Big boxes are now a common feature of Canadian cities in a variety of locations: as stand-alone facilities on suburban highways, as part of retail strips, or in rejuvenated central city areas. They have succeeded in increasing their share of consumer retail

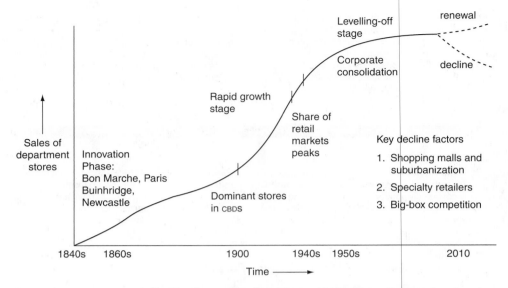

FIGURE 12.2   Generalized Life Cycle of the Downtown Department Store

markets because of innovations in store design (layout, lighting, etc.) and the selection of sites that provide easy access and free parking. In some instances, they have displaced malls and they have crushed department stores (Figure 12.2; Case Study 12.3). Although the big-box model assumes that consumers will be travelling to them by car, some stores are now being established in inner-city areas.

The shift towards power retail illustrates a shortening of the retail life cycle. Department stores took roughly 80 years (1860–1940) to reach their peak market share; the corresponding period for supermarkets was 35 years (1930–65) and for "do-it-yourself" (DIY) stores between 15 and 20 years (1960s–1980s). In Canada, big boxes have taken less than 15 years to reach a dominant market share. Retail locations also wax and wane. Suburban shopping malls initially took away market share from CBDs; in the US they had obtained the same market share by 1965. Recently, though, suburban malls themselves have suffered. In the two years that followed the mortgage crisis of 2007, 400 of America's 2000 largest malls closed down. While the recession may have been the immediate cause, overbuilding and competition from other retailing locations were contributing factors.

E-commerce: New Story, Old Theme

The Internet has displaced the catalogue and is increasingly competing with strictly bricks-and-mortar companies. Jeff Bezos launched Amazon.com in 1995—one year after the US government decided to let private companies develop the Net (Figure 12.2). By 1998 Amazon had begun to diversify beyond books and develop a platform where individuals and companies could advertise, sell, and collect payments. By 2008, sales were $19.16 billion and $75 billion in 2014; however, its net profits have remained in the tens of millions. About 60 per cent of its employees work in areas such as management, software development, and editorial work; the rest are employed in its distribution centres (10 in the US, 1 in Canada, and another 14 around the world in 2014).

Case Study 12.3
LIFE CYCLES OF TWO CANADIAN RETAIL ICONS

Feature	
	Eaton's
Origins	Founded in 1869 in Toronto (population 70,000) by Timothy Eaton, an Irish immigrant. The new store featured "fixed" prices (no haggling) and money-back guarantees. Eaton had previously been part-owner of a general store in a nearby community. By 1907, Eaton's comprised two major stores plus a rapidly growing mail-order business established in 1884; its catalogue, known as "The Homesteader's Bible," became a real threat to local shopkeepers in remote places.
Growth	By the 1950s, Eaton's had more than 90 retail stores across Canada, enjoyed a 50 per cent share of department store sales, and employed almost 25,000 people. It was a paternalistic employer, providing workers with housing, hospitals, and summer camps but paying low wages and resisting unionization.
Decline	In 1976, the mail catalogue business was closed to reduce costs as catalogue demand declined. Despite the successful opening of a new flagship store in Toronto in 1977, Eaton's was losing market share to both established competitors and the big-box stores. In 1997, it had just 12.4 per cent of department store sales and, in 1999, it declared bankruptcy. The creditors were owed $419 million.
Current status	Sears Canada tried unsuccessfully to revive some stores in the best locations. The last store closed in 2002.
	The Bay
Origins	The Hudson's Bay Company (HBC), the first commercial corporation in North America, was formed in 1670 with a charter from the king of England to organize the fur trade within the huge Hudson's Bay watershed. In 1881, the HBC opened its first department store (in Winnipeg) as well as a catalogue business (never as big as Eaton's). Renamed "The Bay" in the 1960s.
Growth	In 1970, the head office was relocated from London, England, to Winnipeg and in 1977 to Toronto. It became national in scope in the 1960s with the acquisition of Morgan's. In the 1970s, The Bay acquired several major rivals, notably Simpson's, Zeller's, and Fields, and in the 1990s, Woodwards of Vancouver and K-mart Canada. The Bay has also launched specialty stores, such as Home Outfitters and operates almost 100 stores across Canada.
Decline	Although it is now the dominant department store in Canada, The Bay (like Eaton's) experienced profitability problems especially after 1997.
Current status	In 2006, The Bay was acquired by Jerry Zucker who became the first US citizen to head the company.

Note: Canada's first department store was opened in 1866 in Montreal.

The dot-com bust of 2000 buried many e-tailers. However, since then, e-commerce has grown rapidly, dominated by companies such as Amazon and eBay. They are followed by Apple and rapidly rising Asian giants like Alibaba, Tencent, and Rakuten. That said, Internet sales are a boon for small companies that are able to reach a global market—often using the services of the previously mentioned giants. Online sales have grown to close to $300 billion in the US by 2013 and over $900 billion globally, and the rate of growth is astonishing. In 2007, Amazon was only the twenty-fifth largest retailer in the US with 14 billion in sales compared to number one Walmart's $380 billion, but in 2013, it was eleventh with $67 billion in sales, while Walmart was at $468 billion. Yet, the top ten retailers are still brick and mortar purveyors of groceries, dry goods, and drugs—Kroger, Target, Costco, Home Depot, Walgreen, CVS Caremark, Lowe's, Safeway, and, of course, McDonald's. Clearly, consumers still prefer conventional local retail outlets where they can check the ripeness of the mangos or ask the pharmacist for an opinion; however, it is still early days for e-commerce. The competition between these two forms of distribution recalls an earlier battle. In the late nineteenth century, catalogue companies, led by Sears, first cut into the revenues of local stores, then developed department stores, product brands, and many other services. It was in response to the threat posed by Sears that Canadian department stores like Eaton's and Simpson's developed their own catalogues. Many brick-and-mortar companies are converting to multi-channel retailers, selling via the Internet and catalogues as well. (Source of e-commerce statistics: https://nrf.com.)

Centrism

Suburbanization played a critical role in shaping service provision. As the suburbs were developed, centrist-minded planners (who aim to maximize consumer access to service centres) sought to recreate the spatial hierarchy of services associated with CBDs on a multi-nucleated model. At the heart of many developments was the local shopping mall, which offered consumers free parking, shelter from the weather, and a broad selection of goods and services while providing retailers with security, economies of scale, and easy access to potential customers. Planning for new stores, shopping centres, clusters, or retail strips focused on accessibility and localization economies in the context of projections for residential development. In contrast to CBDs, which had developed informally, shopping malls were formally negotiated between local government, planners, developers, and major tenants. (Power retailers are less dependent on planning, but have taken advantage of existing high-access locations.)

centrism An approach to planning that seeks to maximize consumer access to service centres.

Centrist planning often takes into account social as well as development objectives: thus most malls are designed not only to stimulate economic activity but to create jobs and provide places for socializing. Yet **centrism** is problematic because, in reality, the thresholds and ranges of all services do not necessarily fit neatly into a few fixed hierarchical levels, as the Christaller model assumes. Thresholds and ranges are also influenced by changes in factors such as local population numbers, income levels, and transportation networks. The decisions formalized in development plans and zoning ordinances have long-term impacts; but to predict the appropriate balance of high-density housing, office space, sports facilities, hotels, and retail clusters for a time years in the future is not easy. Moreover, the low wages and limited multiplier effects of retail operations mean that planning centred on shopping malls is not likely to provide significant economic benefits.

The rise to power of retail giants revealed the weaknesses of centrist planning. Giant retailers, such as Walmart and Tesco, "leak" income from localities through payments for imported goods and head-office services, as well as profit-taking. This leakage reduces the multiplier effects available to the local community, while damaging the unique social and economic basis of CBDs and, perhaps, other retail strips. For example, high-order functions and small and locally owned businesses in CBDs have often been threatened by large-scale retail developments in other parts of the city (a widespread problem in Canada, the US, and the UK). In Canada, medium-sized cities (with populations between 70,000 and 700,000), such as Brantford, Hamilton, Kitchener, and Peterborough in Ontario, have sought to revitalize CBDs with shopping mall developments, but have had limited success. In many cases, the support given to malls or big-box stores amounts to discrimination against local businesses and reinforces their dominance in consumer markets (Case Study 12.4). In addition, "new" jobs in retailing, if they are truly non-basic, displace and reduce existing ones, not necessarily in the same location. Increasingly, the expansion of power retailing is interfering with planners' efforts to achieve new objectives, such as reducing car dependency and increasing environmental sustainability. In the case of the Greater Toronto Area (GTA), power retailing is expanding in conflict with attempts to establish a greenbelt in southern Ontario. On the other hand, new shopping centres can be serviced by rapid transit, as is the case with Sheffield's Meadow Hall complex, that replaced a major steel works, and which is served by tram, bus, and rail lines (Case Study 6.4).

Business Organization

The vast majority of consumer services are small and independent (see Table 4.3). Establishing a gift shop, fitness club, restaurant, convenience store, or financial consultancy, often requires a significant financial investment, but experience, expertise, and willingness to work are usually the most important inputs into service businesses. Although the vast majority of service businesses are independent, they are integrated into value chains, and because their buying and selling is strongly localized, they extend value creation (via multiplier effects) into local communities. SME wholesalers and retailers play vital roles not only in the distribution of goods, but in the discovery and promotion of new local products. Their success depends on their ability to coordinate the supply chain, assess demand, and offer supporting services (e.g., store presentations in retail). Other service businesses, such as travel agencies, funeral homes, and automobile repair shops, are integrated into a broad range of value chains as well. Because many of the transactions that small firms make are open-market, all parties must be able to trust each other and the sustainability of their businesses. But small service firms can also develop long-term relationships (i.e., relational markets) with particular suppliers whose products they use and promote. For example, bed and breakfast (B&B) operations in the UK that are located on popular hiking routes have developed stable relationships with trip organizers such as Contour. The organizer deals directly with the hikers, thus reducing the transaction costs both for the latter and for the B&B. Along routes with large-scale flows of hikers, specialist companies such as Packhorse and Sherpa in the UK (and elsewhere) can be subcontracted to move the hikers' bags and backpacks from one B&B to the next. Governments and small businesses also collaborate in relational markets: for example, governments contract with local stores to

Case Study 12.4

TESCO: THE POWER OF A RETAIL GIANT

Picture kindly provided by Elizabeth Garnham, April 2010

Built in 2004, this Tesco Extra near Nottingham, England, has a floor space of roughly 10,000 square metres. While the site had already been zoned for a smaller store owned by a rival, Tesco obtained permission for its store relatively easily, conditional on providing underground parking (for those not shopping by bus).

Tesco is the largest grocery chain in the UK, with a market share of 31 per cent in 2010, although this has declined to 28.7 per cent in 2014 with the arrival of German discounters, Aldi and Lidl. A parliamentary competition commission concluded in 2008 that independent convenience stores and grocers were "unlikely to survive" past 2015. In the UK, independent stores have declined significantly, and the New

serve as the retail agents for government-controlled services such as the post office, lotteries, and public transit.

The majority of service firms remain small, but a dynamic small firm can become national in scope (Case Study 12.5). A few even extend corporate integration globally.

Corporate Integration

Economies of scale and scope are essential in the service sector, and large firms grab the lion's share of revenue and profits. Big-box stores are able to offer a vast variety of products, stock them in prodigious amounts, and sell them at lower prices than smaller stores can. What enables them to undercut smaller retailers is their ability to reduce

Economics Foundation (2006) has estimated that 50 small specialist shops closed every week between 1997 and 2002. This decline has been arrested since 2009 especially in CBDS, although turnover of small stores remains high. There are also growing concerns that giants such as Tesco abuse their power and enjoy a number of unfair advantages over their smaller rivals. For example:

- *Predatory pricing practices*: Big chains are able to set prices below cost for selected goods that smaller rivals cannot match.
- *Predatory supplier practices*: Chains demand prices from suppliers that are too low to support adequate wages for workers; purchase stock from places with minimal employee rights and low wage conditions; and can readily discard small suppliers that do not comply with their demands for price reductions. Small suppliers are often reluctant to complain about corporate intimidation, because they fear losing the giants' business. Meanwhile, the public is unlikely to be outraged by consumer prices that are too low.
- *Aggressive expansion tactics*: By creating excess capacity, giants can force smaller rivals to close, while they themselves are able to write off temporary losses against taxes.
- *Local planning practices that give priority to large-scale developments*: Malls and power centres dominated by the retail chains enjoy the advantages of consumer access via publicly funded highways and do not pay for the public costs (negative externalities) associated with congestion and pollution.
- *Increasing regulation* (e.g., regarding product labelling, employment relations, accounting, and recording practices): Giant retailers are more capable of absorbing the rising costs of compliance than their smaller rivals.
- *Regulatory bias in favour of the giant chains*: Despite Tesco's already massive market share, the UK's anti-monopoly commission allowed the company to acquire a major stake in the 24-hour convenience-store market on the grounds that such stores belong to a different category from other grocery operations.
- *Land banking* by big companies (Chapter 8) keeps some sites empty, effectively pre-empting other uses, private or social, for example, in housing that may create consumer demands. A recent report, for example, indicated that Tesco controlled 310 separate sites without a store in the UK.

Even so, as Tesco's decline in market share reveals, competition among the giants is fierce.

costs by buying in great quantities; use advanced logistics systems to control product flow and stocking; and spread the costs of advertising over larger revenues. Large size may also allow taking unfair advantage of small firms (see Case Study 12.6). As integration is achieved at various scales (regional, national, international), each expansion offers the possibility of increased returns and also increased complexity in management. A few fast-food chains have been internationalized to extraordinary degrees (Table 12.3). Large service firms build both product and store/location portfolios, each of which is subject to product cycles. For example, clothes retailers and hair-styling chains have to change the styles they offer on a seasonal basis, and may have to modify the nature of their stores in different locations, depending on their target markets.

Stockbyte/Thinkstock

Case Study 12.5
GROWING THE GOODLIFE

GoodLife Fitness started out in 1979 as a single 185 square-metre club called #1 Nautilus Fitness in London, Ontario. Today, with over 300 clubs and more than 12,000 employees, it is Canada's largest fitness chain. The company has won Canada's 50 Best

The efficiencies of integration in Canadian retail sales had by 2003 resulted in 78 per cent of sales dominated by 94 corporations controlling 330 chains (Gomez-Insausti, 2006). The top three companies control 3,200 locations and 25 per cent of the total market, while the largest 30 firms control 14,000 locations and 68 per cent of the market. The degree to which a particular sector is dominated by its top three or four firms varies: in general merchandise, the top chains had 86 per cent of market share (2011); the proportions were somewhat smaller in food (62 per cent), home furnishings (35 per cent), and apparel (19 per cent). Big chains are constantly changing locations, closing some stores and opening others. In the early 2000s, the trend towards power retailing led the big chains to abandon many smaller locations, which were taken up by smaller chains such as The Brick. Although most large companies operate more than one chain in different price categories, this tendency is especially pronounced in apparel. Integration of supply chains has allowed chains such as the Gap, H&M, and Zara to penetrate markets around the world. By 2010, these international retailers controlled more than 40 per cent of the Canadian domestic market.

Many service companies are vertically integrated (e.g., Weston). Others pursue a more hybrid type of coordination. Franchising, for instance, creates a sort of relational

Managed Companies award for seven consecutive years and has been recognized with Canada's 10 Most Admired Corporate Cultures award. Although its initial niche was equipment-based training, from the outset, the core of the business has been its service orientation. In 2006, GoodLife received an international Entrepreneur of the Year citation for its customer-driven focus. The founder and owner, David Patchell-Evans, graduated in physical education and business, and insisted that all instructors have a phys-ed background as well. Instilling that culture in instructors through training and the provision of career opportunities was an industry innovation that propelled growth across the country.

GoodLife is constantly revising its portfolio: removing squash courts to make room for aerobics; replacing stair masters with exercise bikes; combining nutrition and personal training. A wide mix of products, appealing to a diverse range of clients, is continually being developed. The store and location portfolio is also refreshed, with small and big clubs, clubs for women only, clubs focused on cardio fitness, and locations that include malls and grocery stores, downtown suburban retail strips, and even a train station and Toronto's airport.

The location of the original club was a matter of chance, the result of taking over from someone who wanted out of the business. Early expansion took two forms. In some cases, new clubs were established in locations determined by basic demographic analysis; in others, poorly run clubs in good locations were taken over. Later expansion followed the agglomeration model, through association with complementary service businesses such as Loblaw (Canada's largest food distributor). Close to GoodLife's thirtieth anniversary, Patchell-Evans finally decided to explore a more scientific approach to location. Then he bought 100 existing clubs.

Entrepreneurial flexibility and a measure of audacity have been the keys to GoodLife's success. Patchell-Evans would not have been able to deliver that success, however, if his mom didn't mind the books.

market. There are two basic types of franchise system. In product-and-trade-name systems, the franchisee simply acquires the right to use the trade name, trademark, and/or product of the franchisor. The franchisor could be a manufacturer, wholesaler, or retailer whose objective is to increase the distribution of the product in question. Franchise operations of this kind are common in the auto and truck, soft drink, tire, gasoline, and beer industries.

Business-format franchising is considerably more complicated. It is common in the restaurant, food-service, hotel/motel, printing, retailing, and real estate sectors and is gaining popularity among banks, doctors, lawyers, dentists, accountants, and opticians. In addition to obtaining an established service or product, the franchisee gains access to technical and managerial assistance, as well as quality control standards; has less need to pay for capital and operating expenses; and enjoys increased opportunities for growth. The franchisor also benefits in several ways, gaining not only royalties, but rapid expansion, highly motivated owner-operators, and the capacity to develop economies of scale in purchasing, logistics, and so on; and ease of entry into new communities. Disadvantages for both sides include the risks of overdependence, loss of freedom, shirking of responsibilities, and free-riding. Territorial

TABLE 12.3   Evolution of Fast Food: McDonald's, Kentucky Fried Chicken, and Starbucks

Institution	Origins	Growth	Size 2013/2014
McDonald's	The MacDonald brothers opened a restaurant in 1940 in San Bernardino, California, and in 1948 introduced the "speedy-service system," the basic innovation of the fast-food industry. Ray Kroc, the corporate founder, opened a restaurant in Des Plaines, Illinois.	The first McDonald's in Canada opened in 1967 in Richmond, BC; the first offshore franchise opened in the Netherlands in 1971. McDonald's went public in 1965. Eighty per cent of outlets are franchisees.	Total franchised sales of $70 billion from 35,000 restaurants in 119 countries (2013). Employs 1.8 million people worldwide.
Kentucky Fried Chicken	Harland Sanders opened his first restaurant in 1936, in Corbin, Kentucky, and introduced his chicken recipe in 1940. In 1952, the first KFC franchise opened in Salt Lake City, Utah.	By 1960, there were 400 KFC restaurants in the US and Canada. By 1971, there were 3,500 restaurants worldwide and 6,000 by 1979.	Approximately $23 billion, 19,000 restaurants in 118 countries (2013). A subsidiary of Yumi Brands (which also controls A&W and Pizza Hut), the world's largest restaurant system.
Starbucks	The first Starbucks store opened in Seattle, Washington, in 1971. The original stores sold only coffee beans. Starbucks did not become a café chain until 1987, when it was bought by a former director of the company named Howard Schultz.	In 1987, there were 17 Starbucks locations, including the first in Canada in Vancouver. By 1991, there were 116 locations (US and Canada). First overseas location was in Japan (1996) and 4,709 locations worldwide by 2001.	Revenues of $14.9 billion (2013) and 20,000 stores by 2014 in 64 countries, operated by over 200,000 "partners" (employees).

The oldest fast-food chain in the US is said to be A&W. The first outlet (selling root beer) opened in 1922 in Lodi, California.

management, balancing competition, and necessary thresholds for each franchise are important.

The hospitality industry has developed a variety of relational forms. For example, Hilton is an international hotel chain with several brands (Hilton, Hampton, Conrad, Embassy Suites) and ownership of key locations and many franchises; Best Western, by contrast, is a voluntary chain consisting of some 4000 independent hotels gathered under one brand name; and there are dozens of local B&B consortia or associations in Canada that help their members with marketing and management. In general, chains usually begin as local operations that offer consumers some new combination of convenience, function, layout, atmosphere, or taste. They then extend this entry advantage to new locations, while continuing to develop new products, ramp up economies of scale and scope, and accumulate experience in location planning.

Shopping Clones

Together, the internationalization of retailing and other services and the diffusion of planned shopping malls have had the effect of standardizing the shopping experience for consumers around the world. For many consumers the arrival of retail chains is welcome, however, this standardization has also generated hostility among people who resent the loss of local identity and control. In response to these losses, some local governments have sought to limit the numbers of outlets of individual chains (especially "formula restaurants") in their jurisdictions. In Arcata, California, for example, in 2002 the city council passed an ordinance that modified the municipal code to define formula restaurants and permit no more than nine of them in the town. Several other towns in California, Washington State, Maine, Rhode Island, Wisconsin, and Utah have passed similar ordinances in the hope of preserving local charm and diversity. The small town of Springfield, Utah (population 500), which passed its ordinance in 2006, was (unsuccessfully) sued by the Subway franchisee. The standard counterclaims—that chains create jobs and respond to consumer demand for lower prices and greater convenience—are not necessarily true: prices are not always lower in supermarkets; some consumers are inconvenienced by centralization tendencies; and new large stores often eliminate as many jobs as they create. Moreover, as Case Study 12.4 suggests, there are reasons to believe that competition between retail chains and local stores is unfair, not least because of local and national regulations that favour the former.

The Non-basic to Basic Shift: The Tourism Example

As noted, basic activities serve export (or external) markets, while non-basic activities meet local consumption needs (Chapter 8). Basic activities are considered especially important because they bring income into the local community from outside, stimulating local development through their multiplier effects. The standard examples are primary- and secondary-sector activities (resources, manufacturing) whose products are sold to external markets. By contrast, service (tertiary) activities have been considered non-basic because they are locally focused and do not tap into external sources of income. In practice, however, the distinction between basic and non-basic is blurred, and all kinds of services have become significant basic activities that serve export markets, contributing towards the rapid growth of "**invisible trade**" (see Table 12.4 and the next section on producer services). In fact, the financial sectors of London and New York have long exported their services, while tourism is an increasingly important generator of export income in many places around the world (Figure 12.3).

invisible trade International trade in services.

In the case of tourism, many (if not most) of the services that visitors spend money on—restaurants, transportation, attractions such as museums—exist primarily to serve local needs. They reflect local tastes and interests and local taxpayers usually fund them. Yet business travellers and tourists use them as well: think of the Louvre in Paris, the British Museum in London, or the Smithsonian Institute in Washington. The extent to which these services are formally recognized as basic activities and supported as essential generators of income and employment depends on the service providers' ability to coordinate their efforts to gain community acceptance and persuade government of their importance. Consistent with the post-industrial thesis, some of the fastest-growing segments of the tourist industry have a strong educational component, most

TABLE 12.4   Value of Global Merchandise and Service Exports, 1980–2012 ($b)

	1980	1990	2000	2006	2012
Merchandise exports ($m)	2,034	3,449	6,545	12,083	17,930
Service exports ($m)	365	780	1,493	2,755	4,150
Ratio of services to merchandise/total trade (%)	17.9/15.2	22.6/18.2	22.8/18.5	22.8/18.6	23.1/18.8

Values are in current dollars, unadjusted for inflation.
Source: Data obtained via the WTO International Trade Statistics Database at http://www.wto.org/english/res_e/statis_e/statis_e.htm

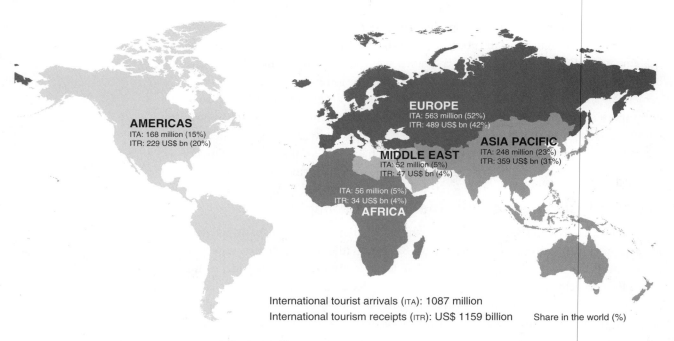

AMERICAS
ITA: 168 million (15%)
ITR: 229 US$ bn (20%)

EUROPE
ITA: 563 million (52%)
ITR: 489 US$ bn (42%)

ASIA PACIFIC
ITA: 248 million (23%)
ITR: 359 US$ bn (31%)

MIDDLE EAST
ITA: 52 million (5%)
ITR: 47 US$ bn (4%)

ITA: 56 million (5%)
ITR: 34 US$ bn (4%)
AFRICA

International tourist arrivals (ITA): 1087 million
International tourism receipts (ITR): US$ 1159 billion   Share in the world (%)

FIGURE 12.3   Tourist Destinations by Arrivals and Receipts

Source: United Nations World Tourism Organization (UNWTO)

Case Study 12.6
THE CULTURAL HERITAGE INDUSTRY

Rievaulx Abbey is a thirteenth-century Cistercian monastery in North Yorkshire operated by English Heritage.

notably the cultural heritage industry (museums, historic buildings, war memorials, etc.) and ecotourism (wildlife reserves, guided walks, wilderness experiences). These developments have involved both private- and public-sector initiatives, as well as NGOs (Case Study 12.6).

In most places, tourist services develop organically to serve the domestic market; domestic tourist expenditures in Canada, for instance, outweigh foreign tourist expenditures four-to-one. But other countries such as Greece, Thailand, and the Bahamas are highly dependent on tourist dollars. In 2012, France, the US, Spain, and China were the leading destinations by arrivals, while the US, Spain, France, and China were the leading destinations by expenditures. The US earned $139.6 billion from tourists in 2013, while Canada earned $17.6 billion. Overall, global patterns in tourism are much more strongly regionalized. The overwhelming majority of visitors to France and Spain, for instance, come from other European countries; to the US from other North American countries; and to China from other Asian countries. Perhaps France offers the best illustration of the tourism's dual nature as both a non-basic and basic service. The features that attract tourists to France were clearly designed for the local population, and although the French are willing to enhance their cultural landscape for tourists, they are not willing to turn it into Disneyland to do so (with one exception that proves the rule).

One of the main reasons cited by foreign tourists for visiting the UK is cultural heritage, with London being the main centre visited. However, the National Trust and English Heritage have developed an extensive network of historic sites in rural areas that feature churches, monasteries, castles, country houses, monuments, museums, buildings, and relic landscape features from former times (iron age forts, burial grounds, and stonework) that attract domestic and foreign visitors. Cities and towns retaining the layout, architecture, and engineering principles of the past can be enjoyed by wandering around, observation, and reading public information sources. Both National Trust and English Heritage have a strong education commitment and provide many information-based services (books, brochures, guided tours, recorders, displays, and operations of old equipment). More than just manors and monuments, cultural heritage provides windows into how people lived, worked, and died in past times. It has also become part of the curriculum of primary schools.

The economics of cultural heritage are complex. It is provided to tourists by governments, private businesses, and non-profits, usually at a price, but generally it is subsidized and often offered free. These payments, government subsidies, and donor funds have the direct economic impact of additional jobs and income to rural areas, even though usually on a seasonal basis. However, most are dependent on volunteers to supply labour, guide services, and management. Most of these people do the job out of altruism and interest. Yet, there are significant positive externalities from the heritage industry which make it worth supporting. Local hotels, pubs, and restaurants benefit from the tourist influx; even residential land values bask in the aura supplied by heritage.

Producer Services

Producer services are non-material inputs purchased by business. Examples include design, engineering, accounting, advertising, and customer service. Many of these activities are knowledge-intensive and require face-to-face collaboration between suppliers and customers, although call centres and back-office operations are routinized and can be dispersed. Producer services have been the fastest growing sector in the developed countries for several decades now and are a growing source of export income.

Economies of Specialization and Outsourcing

Firms buy the services of external suppliers—from engineers to temporary office workers—in order to make short-term use of their expertise without spending the time and money that would be required to develop the same expertise in-house or incur the many costs that come with taking on full-time employees. Some firms will hire third-party specialists to provide independent auditor's statements, legal documentation, or environmental accreditation; others will rely on temporary office workers or call-centre employees to perform routinized tasks. Purchasing these services only as they are needed gives organizations of all types—firms, government agencies, NGOs—flexibility with few if any sunk costs.

Producer services serve specialized markets and realize economies of scale and scope at global scales. For example, the four accountancy firms that are global in scope—Deloitte and Touche, Ernst & Young, KPMG, and PricewaterhouseCoopers—provide a wide variety of managerial support services (auditing, tax preparation, mergers and acquisitions, due diligence) and together perform the audits of 78 per cent of US public companies controlling 99 per cent of the annual sales of publicly listed companies. The majority of producer services firms, however, are small, independent operations run by entrepreneurs working either alone or in partnerships.

The market for producer services is necessarily relational, requiring a high level of collaboration. Face-to-face interaction is the norm, and the reputation of the service provider depends on the success of each relationship. The need for face-to-face interaction encourages clustering, especially when several independent specialists are working on the same project in different capacities, for example, in relation to advertising, engineering, software and design. On the other hand, service providers who are particularly distinguished in their field may arrange meetings at their own convenience, rather than the client's; this is often the case with well-known lawyers and designers. High-level personal interaction is still an important driver of international travel, although improved communications technologies have made it possible to meet in virtual space. Many companies use both external and internal producer services, complementing external expertise with institutional knowledge.

Overall, invisible (services) trade grew from 15.2 ($365 billion) to 23.1 per cent ($4,150 billion) of total global trade between 1980 and 2012; services now account for over a quarter of total exports (Table 12.4). Service exports are concentrated in the rich market economies, especially the US and UK (Table 12.5); in 2012 these two countries accounted for 20.7 per cent of global commercial service exports. Moreover, in 2012 the US and UK enjoyed massive trade surpluses (exports minus imports) in

TABLE 12.5   Leading National Exporters and Importers of Commercial Services, 2012

Exporters	Value ($b)	Share (%)	Importers	Value ($b)	Share (%)
United States	621	14.3	United States	411	9.9
UK	280	6.4	Germany	293	7.1
Germany	257	5.9	China	280	6.7
France	211	4.8	Japan	175	4.2
China	190	4.4	UK	174	4.2
Japan	142	3.3	France	172	4.1
India	141	3.2	India	127	3.1
Spain	136	3.1	Netherlands	119	2.9
Netherlands	131	3.0	Singapore	118	2.8
Hong Kong	123	2.8	Ireland	112	2.7
Ireland	116	2.7	South Korea	107	2.6
Singapore	112	2.6	Canada	105	2.5
South Korea	110	2.5	Italy	105	2.5
Italy	103	2.4	Russia	104	2.5
Belgium	95	2.2	Belgium	92	2.2
Switzerland	90	2.1	Spain	89	2.1
Canada	78	1.8	Brazil	78	1.9
Sweden	76	1.7	Australia	63	1.5
Luxembourg	70	1.6	UAE	63	1.5
Denmark	65	1.5	Denmark	57	1.4
Austria	61	1.4	Hong Kong	57	1.4
Global services total	4,350	100.0		4,150	100.0
(Total trade)	(22,080)			(22,338)	

Source: WTO. 2012. *International Trade Statistics 2013*, Table 1.9, page 26. http://www.wto.org/english/res_e/statis_e/its2013_e/its2013_e.pdf

services, respectively $210 and $106 billion. Although the top four service exporters in that year were all established advanced countries, China had risen to fifth place ahead of Japan, while South Korea, India, Singapore, and Hong Kong were among the top 20. Hong Kong (with its close ties to south China) and Singapore (with its close ties to the Association of Southeast Asian Nations [ASEAN] countries) complement one another while competing (along with Shanghai) to provide business services throughout Asia. Hong Kong's surplus on service trade amounted to an impressive $24.2 billion. France and Spain are two other countries with large service trade surpluses. Clearly, while service exporters in the US and UK are aided by the fact that English is the global language of business, those in France and Spain also benefit from their countries' histories as colonial powers whose languages are still widely spoken in their former colonies.

Finally, it is worth noting that the service sector attracts a significant amount of foreign investment. More than one-third of the FDI flowing into and out of the US in 1997, 2007, and 2008 was in the service sector.

Local Development

Producer services have greater potential than consumer services to move into the "basic activity" category. Producer services tend to be knowledge-intensive activities that agglomerate in order to facilitate collaboration, gain proximity to universities, share infrastructure, and reap the rewards of readily accessible pools of skilled labour. Each set of producer services supports urban or regional intra-industry clusters, contributes to inter-industry clusters (activities linking different industries), and helps to build positive cumulative causation. In Canada, the tendency of producer services to agglomerate in Ontario is clear in terms of both revenue and employment (Table 12.6). Toronto's ICT cluster—including software development, semiconductors, communications equipment, wireline services, wireless services, and creative and content development—is the third largest in North America and includes both intra- and inter-industry elements. Local, national, and international companies do business with a variety of Toronto-based clusters in other fields such as finance, business services, design, film, and television. A similar, though smaller, concentration of producer services exists in Alberta, where a cluster of engineering and surveying/mapping firms is the largest in Canada and serves the province's oil patch.

The evolution of producer services is closely linked to the growth of the so-called creative industries that have tended to cluster in cities, sometimes in highly localized concentrations. Deriving localization economies from shared information and access to a highly educated, well-trained, and innovative labour pool, these clusters give high-cost regions the competitive advantages they need to survive because, in addition to providing

TABLE 12.6 Producer Services in Canada, 2012, Operating Revenues $million (%)

Services	Atlantic Canada	Quebec	Ontario	Manitoba	Saskatchewan	Alberta	British Columbia
Accounting	625.7 (4.2)	2,883.9 (19.3)	6,260.8 (41.9)	448.3 (3.0)	395.7 (2.6)	2,283.1 (15.3)	2,035.8 (13.1)
Advertising and related	116.4 (1.6)	1,641.9 (23.0)	4,112.9 (57.7)	97.1 (1.3)	75.5 (1.1)	471.5 (6.6)	606.1 (8.5)
Architectural	110.5 (3.4)	694.3 (21.4)	1,271.7 (39.2)	92.1 (2.8)	77.1 (2.4)	503.7 (15.5)	469.7 (14.5)
Software development and computers	573.8 (1.6)	8,428.0 (23.4)	18,150.5 (50.4)	519.3 (1.4)	431.2 (1.2)	4,101.1 (11.4)	3,417.6 (9.5)
Consulting	191.6 (2.0)	1,834.4 (19.3)	4,273.7 (45.1)	161.7 (1.7)	118.1 (1.2)	1,724.6 (18.2)	1,146.9 (12.1)
Specialized design	13.3 (1.2)	180.1 (16.2)	587.4 (52.7)	16.6 (1.5)	n/a	139.6 (12.5)	168.6 (15.1)
Employment	297.9 (2.6)	1,534.8 (13.4)	5,731.4 (50.0)	70.8 (0.2)	83.0 (0.5)	3,063.4 (26.7)	675.3 (5.9)
Engineering	1,164.9 (4.1)	5,689.1 (20.3)	6,831.0 (24.4)	329.3 (1.2)	812.3 (2.9)	9,900.2 (34.8)	3,603.8 (12.7)
Surveying and mapping	45.6 (1.5)	339.8 (11.4)	394.7 (13.3)	35.2 (1.2)	143.9 (4.8)	1,603.9 (54.0)	295.9 (10.0)

Source: Based on data from Statistics Canada

products and services that are valuable, they support high incomes. Knowledge-based clusters are varied and include long-established professional groups in law, financing, and accounting, as well as ground-breaking efforts in the arts, computer graphics, textile design, fashion, electronic games, energy, biotechnology, nanotechnology, and so on. The common denominator is the ability to innovate. The common attractions are like-minded people and a high-amenity environment. As in Silicon Valley, governments and industry associations try to encourage the development of such clusters by ensuring that the appropriate conditions are in place: infrastructure, amenities, taxation policy, and so on. But success is not guaranteed.

Governments of many cities, towns, and regions now offer location incentives to attract firms specializing in back-office and call-centre services. Although many of these businesses are now based in less expensive parts of the world (especially India), locations in rich economies are still preferred for operations that require delivery of physical products or materials. To attract a Staples call-centre and 500 jobs to the Halifax suburb of Lower Sackville, the province of Nova Scotia invested $1.5 million for recruiting potential employees through the Nova Scotia Community College, and provided a 10 per cent payroll rebate for five years, up to a maximum of $7 million. Peterborough, Ontario, offers similar inducements and has the advantage of proximity to Toronto. Nevertheless, Toronto itself remains the largest call-centre location in North America; ranking first in overall investment, it has some 2,900 centres employing more than 11,000 staff.

Offshore outsourcing of producer services reduces costs significantly. Although distance poses its challenges, India has overcome many of them, largely because it offers a highly educated workforce that is willing to undercut the wage levels of developed countries. Initially, offshore operations were limited to routinized and relatively simple consumer services, but improvements in technological capacity have enabled companies to take advantage of higher-end services at a discount. Ireland similarly used a highly educated workforce to attract knowledge-intensive services. Governments in both India and Ireland supported the development of producer services with policy, money, and infrastructure, but the greatest factor in their success may well have been the labour force's proficiency in English. Language has been an important consideration in Japan as well: many Japanese companies have outsourced producer services to Dalian in northeastern China—an area that, having been occupied by Japan from 1904 to 1945, has retained a proficiency in Japanese.

It is debatable whether the dispersal of export-oriented services creates a spatial division of labour similar to the one associated with the manufacturing product cycle model, specifically with respect to the dispersal of routinized operations to low-cost locations. First, market access remains a location imperative that constrains possibilities for cost minimization to a greater degree than in manufacturing. Second, maturing services do not tend to move towards standardized products and routine processes in the way that manufactured goods do: in fact, specialization and market niches often become more important over time. Third, many service products are custom-made and therefore relatively insensitive to price competition. Fourth, technological change in data processing often reduces labour requirements, offsetting the need to find cheap labour. In fact, service providers tend to be more immediately affected by changes in communications technology than manufacturers are by changes in capital equipment. Because building and equipment needs are simpler and less expensive, back offices tend to be easier to establish than manufacturing branch plants—and easier to close down.

Public Services

In advanced market economies, virtually all government functions fall into the "service" category. Government-owned manufacturing and resource companies are now rare and most of the ones that existed in the past, such as Petro-Canada and the Potash Corporation of Saskatchewan, have been privatized (nationally owned resource and manufacturing companies are still important in many countries, however; see Chapters 6 and 9). The extent to which services should be provided by governments is also debated; moreover, the boundaries between private and public service provision are often blurred. Indeed, the same service is often available from both private and public providers and governments usually fund private and NPOs to provide services. In every society there are various services—national defence, law creation and enforcement, local planning, roads, street lighting—that are virtually always provided by the public sector. In addition, advanced economies provide education and health care as public services while also allowing private suppliers. From Japan to the UK and North America, debates continue to rage over which sector, public or private, should control those services, as well as national rail services, local bus services, social housing, and water and energy supply.

Arguments for public provision of services, most impassioned with respect to education and health, centre on the principle that they are public goods and, as such, must be available to all, regardless of ability to pay (Table 6.1). The reasoning behind this principle may be in part moral, but it is also practical. Education and health care produce positive externalities; they benefit not only the individual but the society as a whole. Conversely, if education and health care (and other services with "public good" characteristics) were privatized, they would be provided only to those who could afford them. The positive externalities of universal coverage would be lost, and the society as a whole would suffer accordingly. Objections to state control of schools and health care are rooted partly in an ideological preference for markets to function with as little government involvement as possible and partly in concerns regarding inefficiency and bureaucracy.

Education and Health

Public services are by definition accessible to all (relevant) members of the public. By contrast, private businesses make their services available only in exchange for payment. Governments subsidize education and health care because all members of the society do not have the same ability to pay.

Advocates of public funding for schools and medical services emphasize the socioeconomic benefits of good health and education. Some of those benefits relate directly to economic development, since a healthier, better-educated workforce will tend to be more productive, innovative, and flexible. Similar to network effects, if every individual is healthier and better educated, that improves the economy for everyone. Other benefits may be less explicitly economic in nature but are no less important to quality of life and the effective functioning of a society. For example, there is a negative correlation between education and crime rates. At the same time, in a society that expects rewards to be earned on the basis of merit, equality of access to education and health care must be considered a fundamental right. Access to education and health care is essential if people from lower-income families are to have the opportunity to compete and catch up with the rest.

In evolutionary perspective, education and health care are both causes and effects of economic development. Economic development increases a society's ability to invest in education and health, while increases in intellectual capital enhance a society's ability to develop competitive advantages. Academic research and university–industry collaboration are obvious examples. But these would not be possible without the foundation of human capital developed through preschool, primary, and secondary education.

Typically, the costs of supplying public services are borne by the public through a progressive taxation system in which the rates charged depend on the individual's ability to pay: thus the rich pay higher taxes than the poor, effectively subsidizing the delivery of public services to those who could not afford to buy them from private suppliers. Typically, in countries with publicly funded education and health care, participation is compulsory. Thus in Canada school attendance is compulsory up to a certain age, although families can choose private rather than public schools (if they can afford the fees). Similarly, all employees and business owners are required to pay into the public health care if they are earning a taxable income. Opting out of the system is difficult for doctors, although they can relocate elsewhere, and not possible for individuals, although they can pay for private options in Canada and especially in the US.

National Variations

Market economies laid the foundations of universal education in the nineteenth century and developed their health systems incrementally throughout the twentieth. All developed countries now spend at least 5 per cent of their GDP on education when both public and private funds are taken into account. (Canada and Scandinavia spend roughly 7 per cent, and consistently rank among the top ten in international comparisons of student abilities). Health expenditures are also consistently high, and infant mortality has been reduced to between 3 and 6 deaths per 1000 births.

However, most developing countries have relatively rudimentary education systems. Illiteracy rates typically run above 30 per cent—as high as 60 per cent in some cases. Similarly in health care, the ratios of physicians and hospital beds to population are far lower in poor than in rich countries, and infant mortality rates can be extremely high. These unfortunate realities reflect both limited urban and institutional evolution. Large segments of the developing world's population still live in rural areas and lack access to the educational and health resources that cities develop over time. Universal education and health care are both public goods that require not only sophisticated and complex institutions to deliver, but a high level of social consensus and cost sharing to initiate and maintain.

Even in the developed world there are significant differences both in how these institutions are constructed. The fulcrum is a balance between public and private funding: it is a question of whether education should be conceived of as a purely public good or should the person who benefits the most from the education pay for it. Of course the debate is nuanced by access to education, equity in opportunity, disproportionate burdens of upfront costs and loans, etc. How countries attempt to resolve the conundrum drives many differences in organization. In education, all developed countries have established universal access to primary and secondary schools and broad meritocratic entrance to university. This potential progression is the basis of the economic system, offers equity of opportunity and mobility, and helps legitimize political economic systems.

In many countries, schools originated from pre-existing institutions with literacy traditions such as churches or temples. Local communities, trades, philanthropists, and private interests also initiated schools and universities, but today primary and secondary education is overwhelmingly publicly supported and administered in developed societies. As in Canada, the state controls most primary and secondary schools through direct funding and administration. That said, the funding and governance of schools mixes public, private, and non-profit interests. In the US, Japan, Australia, and the UK private education systems educate from 10 to 20 per cent of the student population, with more students opting for private secondary schools. Most of the private schools in the US are actually religious institutions, but their student numbers and private school enrolments in general have been declining. Some countries (or regions) forbid public funds to be spent on private schools, while others have no problem with that approach. Different economic philosophies produce different outcomes. The term private school is a misleading one because most have charity and non-profit status, yet are club goods—people donate (pay) for services that will be shared among a self-selected group. Other economic structures are also used. In the US, impatience with public school performance stimulated the development of charter schools that allow students to take their public funding to the school of their choice. In Switzerland and Netherlands, some secondary or non-tertiary schools receive funding from companies as part of apprenticeship or co-operative-type programs.

At the university level, all advanced countries provide subsidies and in some countries tuition is free, but students are generally required to pay part of the cost themselves. That burden has been increasing in the last two decades, and the UK especially has seen a shift from public to tuition-based funding. European countries favour the free tuition model, while the Anglo-American countries favour the paid tuition model, but with supporting student loans and grants. Another distinction is between private and public universities. The model in Canada and Europe is for public-supported universities with some private donations and some level of tuition. In the US, the majority of students attend public universities in one of the state systems, but another 20 per cent attend private universities, including most of the elite research universities and liberal arts colleges. The elite universities are non-profits, funded by high tuitions and donations that through investment have been converted into endowment funds. In Japan, the elite universities are publicly funded, but the majority of students go to expensive private universities. In Canada, the total revenue for universities was $37.4 billion in 2009, and of that tuition only accounted for $7.6 billion. Provincial and Federal support amounted to $17.2 billion and $3.3 billion respectively and the majority of the rest of the income came from sales of other goods and services, investment returns, private donations, and a small amount from local governments. One of the arguments for not providing higher education as a public service is based on the expectation that university graduates will be compensated for the fees they have paid when they enter the job market (Figure 5.3). As fees continue to rise, more and more students today support their studies with part- or even full-time jobs—a trend that affects the nature of their educational experience most obviously with respect to the time available for their studies.

Health-care provision is a somewhat different public good. It is a form of collective insurance that must be paid for on an ongoing basis to be there when we need it. It presents a free-rider problem insofar that people don't want to pay for insurance now when they don't see the immediate return (for example, healthy young people paying for a system that disproportionately supports older people). It also presents a perverse moral hazard dilemma insofar that insurers will wish to exclude people with health problems. To overcome these problems Canada and most developed nations have universal health care operated by the state. The free-rider problem is removed by extracting payment through taxes and the moral hazard problem is removed by insuring everyone regardless of their health problems and even regardless of behaviours (such as smoking) that incur health problems. In Canada, the federal government sets national health standards and transfers some funding to the provinces and the provinces supply the rest of the funding and administer the program. Hospitals, doctors, and governments engage in the financial administration of the system, leaving individuals or households largely removed these concerns.

The US was for many years the only advanced economy in the world that did not provide universal health insurance. It depended on insurance premiums paid (by individuals and corporations on behalf of their employees) to medical insurance companies to fund private health care and the federal government provided subsidized Medicare to the aged and Medicaid to some categories of the poor. A large proportion of Americans—15.4 per cent of the population in 2009—had no medical insurance at all; more than half of those people didn't have access to family doctors; and before the mortgage crisis, half the personal bankruptcies in the US could be traced to the catastrophic costs of medical care. After two decades of debate the Affordable Care Act was passed in 2010, reflecting the path dependence of the American reliance on private insurance companies and private hospitals. To remove the free-rider problem, it requires all individuals to become insured and provides subsidies to those who previously could not afford insurance. However, individuals must choose among several public and private options for insurance. In other words, a market for insurance has been created instead of relying on a single payer (the state). To remove the moral hazard problem, insurers are compelled to take on all applicants irrespective of health history and according to local or regional standards.

Domestic Variations

An important distinction is between federal and unitary systems of government (Chapter 6). In unitary countries such as England, Italy, and France, the primary responsibility for the structure and funding of the school and health systems lies with the national government. By contrast, in federal systems, education and health are typically controlled by regional (state or provincial) governments with allowances for variations between states or provinces in matters of health and education. There may also be some variation between and within municipalities or regions within a state or province. In the US, state governments vary widely in the funding they allocate to education (Figure 12.4). These variations reflect cultural attitudes, political ideologies, and economic conditions.

(a) Per-Pupil Spending on Primary and Secondary Education, 2010–11 (current dollars)

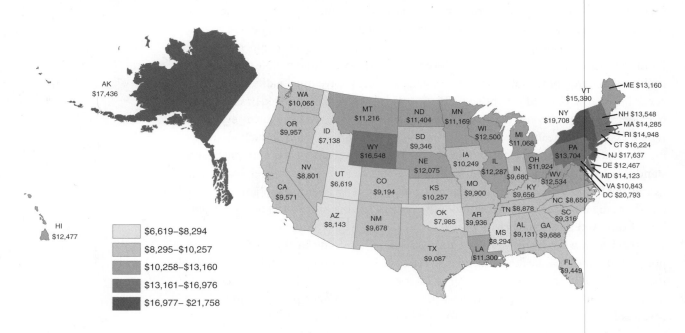

(b) Distribution of Support to Schools by National, Regional, and Local Levels, 2010–11

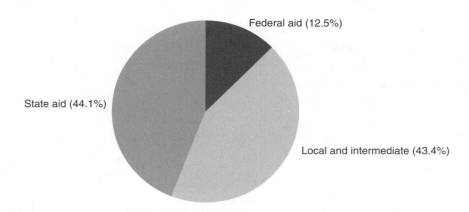

FIGURE 12.4   Expenditures on Schools in the US

Source: Digest of Education Statistics, Institute for Education Sciences, US Department of Education, http://nces.ed.gov/programs/digest/d09/tables/dt09_172.asp

There is much less variation in Canada, where the federal government uses its provincial equalization program to direct additional funds to the poorer provinces in order to ensure more-or-less equal levels of service across the country. Federal systems allow for variations between states or provinces in matters of health and education. There may also be some variation between and within municipalities or regions within a state or province. Canadian provinces provide a high degree of equity in education because they fund each school board on an equal per student basis—indeed, one study has found that students in the poorer provinces are now better supported than their counterparts in the richer ones (Eisen et al., 2010). Although the provinces have a significant oversight role with respect to curriculum and teaching certification, each school board has a degree of autonomy and funding to respond to its local needs. However, there is growing concern that fundraising by schools can increase disparities between wealthier and poorer areas, especially in terms of school equipment and facilities.

In the US, local differences are especially noticeable between affluent suburbs and poor inner-city districts in the same metropolitan area because a large part of the money used to fund schools comes from the local tax base. The federal government provides some additional monies to inner-city schools to make up the difference. But it is not enough in light of the additional challenges that inner-city schools face: more diverse student populations, with higher proportions of minority children and children "at risk" of dropping out because of delinquency, crime, or pregnancy; larger class sizes; and limited resources such as computers. Not surprisingly, a recent study of trends across the US found that performance indicators such as graduation rates consistently favour the suburbs (Swanson 2009).

Health care even in a unitary state such as Britain with its National Health Service (NHS) can vary because of institutional factors in hospitals, clinics, and so on. In order to instill some market discipline into the NHS, patients are allowed to see ratings on doctors and hospitals and have some choice. A federal system allows health-care services to be constructed at a regional level. Thus in Canada, the provinces control health services and each province has reorganized its services a number of times in the few decades that universal care has been provided. The political- and social-capital capacities to build such institutions should not be underestimated. The reason why the US federal government has stepped into health care so markedly is that only a few states have established their own systems.

Accessibility and Efficiency

Irrespective of whether health and education services are delivered from a unitary or federal platform, this level of organization cannot solve all the geographical dilemmas of public goods provision. Accessibility, and its ramifications for efficiency, matter. Areas with limited access to medical and education services also have low incomes, poor living conditions and lifestyles, high unemployment, and low levels of educational achievement. The conjunction of these realities reflects the vicious power of circular and cumulative causation. Poor health and inadequate education limit access to decent employment, which in turn limits access to better health, education, and living conditions. Providing health care and education as public services improves poor people's chances of obtaining better jobs. In turn, increasing productive employment increases the tax revenue available to support health and education as public goods, producing a virtuous cycle of circular causation.

The distribution of public service facilities such as schools and hospitals is much the same as the distribution of retail and other service facilities insofar as it reflects a balance between the distances that people are willing to travel to access them and the threshold populations required to support their services. Thus primary schools and general practitioners are more widely and evenly distributed across the landscape than research hospitals and universities. The analogy is not perfect, however, because the distribution of these services is not determined entirely by market considerations. Rather, the society determines the minimal level of service that must be offered in each area. In Canada, governments use taxation and other policies to uphold the principle of universal and equal access by ensuring that all regions have the services they need, and that market forces do not threaten the integrity of the public health and education systems. Per capita spending on primary and secondary schools is about $10,000 among all the provinces, but in the northern territories it climbs to $15,000–20,000. Nevertheless, these systems face resource limitation, and their organization, including the balance between private and public supply, continues to be a matter of debate.

> **public services** Services supplied by the government itself or through NGOs, usually based on a public good or externality rationale (i.e., either the private sector can't provide the service or provision of the service will have beneficial effects on society as a whole).

In practice, access to **public services** could never be truly equal, if only because populations are distributed across space. Someone who lives an hour away from the local hospital will always be less well served than someone who lives only five minutes away; the cost of attending university will always be higher for students from rural areas—who incur both travel and accommodation expenses—than for those who live near their schools. The fact that populations are unevenly distributed compounds the problem since to give sparsely and densely populated regions equal access to hospitals or universities would be prohibitively expensive. In Canada, another concern is the limited access to health and education services among Aboriginal Peoples living on reserves in remote areas, and their poorer health in comparison to urban counterparts. Since the resources that taxpayers are willing to spend on public services are limited, governments must constantly adjust the levels of service they provide, in response to changes in social expectations as well as demographic and economic conditions.

Conclusion

The service sector can be divided into three basic sub-sectors. Consumer services provide individual (household) consumers with groceries, haircuts, bartending, and entertainment, among other things. Producer and related creative services—corporate legal services, business accounting, software design—are primarily about information. Services must be accessible to their clients, but convenience has to be weighed against costs. Costs can be reduced by clustering related services and improving integration over space. The demands made by consumers reverberate through value chains, creating greater demand for both consumer and producer services. Producer services in particular tend to agglomerate in a complex and geographically distinct manner. Finally, public services exist to provide services with public good characteristics that the private sector is either not capable of providing at all or cannot provide at a level sufficient to ensure the realization of positive externalities that benefit society as a whole. Education and health care are pre-eminent among public services because of their influence, direct and indirect, on the economy.

Practice Questions

1. Map the distribution of big box stores in your municipality. What are the implications for local, small businesses?

2. Classify retail activities by function in a planned shopping mall in your municipality. How important are locally owned businesses? Classify retail activities on a commercial street or highway in your municipality. How important are locally owned businesses?

3. In your community would you be in favour of restricting "formula" restaurants or big-box retailers?

4. Map all the coffee shops in your municipality. Which firms have used what location strategies and why? Where in your community would you choose to locate your own independent coffee shop?

5. Identify and map a key export-oriented producer service in your community. Explain its origins and location pattern.

7. How do public and private schools in your region differ in funding (and tax allowances, etc.) and distribution?

8. Should health services be freely available to all, regardless of income?

Key Terms

post-industrialism   369
service sector   370
goods-producing sector   373
marketing principle   374

nested hierarchy   374
central business district (CBD)   376
consumer services   376
centrism   380

invisible trade   388
public services   400

Recommended Resources

Bain, A.L. 2006. "Resisting the creation of forgotten places: Artistic production in Toronto neighbourhoods." *The Canadian Geographer* 50 (4): 417–31.
An analysis of clustering among artistic communities that are part of the creative city idea, but not necessarily well-paid.

Beyers, W.B. 2005. "Producer services." Pp. 382–3 in Warf, B., ed. *Encyclopedia of Human Geography*. Thousand Oaks, CA: Sage.
A concise introduction to producer services.

Birkin, M., Clarke, G., and Clarke, M.N. 2002. *Retail Geography and Intelligent Network Planning*. Chichester: Wiley.
The most recent update on the geography and geographic tools of service outlet distribution.

Bryson, J.R. and Daniels, P.W. 2007. *The Handbook of Service Industries*. Cheltenham: Edward Elgar.
After reading Beyer's overview of producer services, come here for details on the sub-sectors and issues.

Buliung, R., and Hernandez, T. 2009. *Places to Shop and Places to Grow*. Toronto: Neptis Foundation.

Power retail explained in the context of sustainability ambitions.

Filon, P. and Hammond K. 2006. "The failure of shopping malls as a tool of downtown revitalization in mid-size urban areas." *Plan Canada* (Winter): 49–52.
Assesses the status of shopping malls built to rejuvenate CBDs in numerous mid-size towns in Ontario; most have faced declining fortunes.

Hernandez, T. and Simmons, J. (2006). "Evolving retail landscapes: Power retail in Canada." *The Canadian Geographer / Le Géographe canadien* 50: 465–86. doi: 10.1111/j.1541-0064.2006.00158.x
Discusses how the Canadian retail landscape has been transformed by the growth and clustering of big box retailers into a range of "power retail" developments.

Illeris, S. 1996. *The Service Economy: A Geographical Approach*. New York: John Wiley.

An examination of consumer, producer, and public services and how they relate to one another.

Leslie, D., and Reimer, S. 2006. "Situating design in the Canadian household furniture industry." *The Canadian Geographer* 50 (3): 319–41.

A case study that explores the "creative cluster" hypothesis in the context of a traditional industry.

Schultz, C. 2002. "Environmental service providers, knowledge transfer, and the greening of industry." Pp. 165–86 in Hayter, R. and Le Heron, R.B., eds. *Knowledge, Industry and Environment*. London: Ashgate.

A clear introduction to the nature and scope of environmental services, especially as provided by small, specialist firms to manufacturing industry.

United Nations. Manual on Statistics of International Trade in Services. http://unstats.un.org/unsd/tradeserv/TFSITS/MSITS/m86_english.pdf. Not statistics themselves but definitions, etc.

Wilson, K. and Cardwell, N. 2012. "Urban Aboriginal health: Examining inequalities between Aboriginal and non-Aboriginal populations in Canada." *The Canadian Geographer* 56 (1) 98–116.

World Trade Organization. http://stat.wto.org/StatisticalProgram/WSDBStatProgram Home.aspx?Language=E. Detailed data on services and goods.

References

Bell, D. 1973. *The Coming of Post-industrial Society*. New York: Basic Books.

Eisen, B., Milke, M., and Slywchuk, G. 2010. *The Real Have-Nots in Confederation: Ontario, Alberta and British Columbia*. Winnipeg: Frontier Centre for Public Policy.

Gomez-Insausti, R. 2006. "Canada's leading retailers: Latest trends and strategies for their major chains." *Canadian Journal of Regional Science* 3: 359–74.

New Economics Foundation. 2006. *Submission to the Competition Commission Inquiry into the Groceries Market*.

Swanson, C.B. 2009. *Closing the Graduation Gap: Educational and Economic Conditions in America's Largest Cities*. Betheseda: Editorial Projects in Education.

United Nations. *Manual on Statistics of International Trade in Services*. http://unstats.un.org/unsd/tradeserv/TFSITS/MSITS/m86_english.pdf.

World Trade Organization. http://stat.wto.org/StatisticalProgram/WSDBStatProgram Home.aspx?Language=E.

CHAPTER 13

Transportation and Communication Networks

Geography has been effective in determining the grooves of economic life through its effect on transportation and communication.

Innis 1946: 87

The overall goal of the chapter is to explain transportation and communications networks as the infrastructure of economic organization. More specifically, this chapter's objectives are

- To examine the importance of fixed costs and economies of scale in network operations;
- To explore the role of government regulation, and sometimes control, in creating and operating networks as public goods;
- To explain the principles of spatial interaction—that is, the three conditions required for the flow of goods and services across space to occur;
- To examine the spatial measures of network efficiency in terms of connectivity and related indexes of network design;
- To explore the issues of connectivity at different geographic scales; and
- To examine the environmental costs of networks and some of the ideas that have been proposed to reduce these costs.

The chapter is organized in three main sections. The first explores network economies, especially in terms of spatial interaction, transport modes, and logistics. The second focuses on network costs, connectivity, pricing, regulation, and design. The third examines networks at interregional and intra-urban scales, and also considers some environmental implications related to transportation and communication.

The Role of Transportation and Communication Networks

Transportation and communication networks are the lifelines of trade, regional development, and globalization—the "grooves of economic life." They are the infrastructure that underlies the spatial division of labour and the wealth (and poverty) of nations. Road, rail, highway, canal, sea, air, pipeline, and information networks connect all parts of the globe. They disperse economic activity to remote places and reinforce the wealth, power, and prestige of the largest, most accessible centres. Perhaps more than any other sector, people's lives have been changed dramatically by technologies that have reduced the time and cost of long-distance transportation and communications. From the railways and canal systems of the Industrial Revolution to the fibre optics and Internet of the ICT, global communications and transportation have moved goods, people, and information faster and cheaper than before—usually by orders of magnitude. Heavy, durable goods now circumnavigate the globe in days rather than years. Fruits and vegetables are flown from one corner of the world to another in a matter of hours at a cost low enough to make importing them economically viable. Information and money are transmitted instantaneously via the Internet. Although these sectors have evolved greatly, they are capital intensive and their sunk costs have long-lasting implications for both economic development (negative as well as positive) and the environment (largely negative). In addition to technological innovation, transportation and communication efficiency is generated increasing use of network economies.

network externalities or effects
The changes in benefits (gains or losses) to a user of a network when the number of other users changes. Also referred to as demand-side economies of scale.

Economic efficiency in most sectors derives from merging scale and scope economies. These two benefits manifest through communication and transportation as **network externalities** or **network effects**. Simply stated, as a network grows—increasing the

Case Study 13.1
OUT OF THE BOX

Advances in technology have been the keys to the improvements in transportation and communications that in turn have propelled successive techno-economic paradigms. The locomotive, the car, and the Internet were all cutting-edge technologies in their day. But more mundane innovations can also be revolutionary. The container—a standardized steel box—transformed international trade by making it easier, and much less costly, to transfer goods from one mode of transportation to another. If it is smaller than a car, chances are good that any manufactured item you use has spent some time inside one of these boxes that enables it to travel on an intermodal network of ships, trucks, trains, and airplanes.

Containerization was not invented by any one person. The practice developed as different rail and shipping companies across Europe and North America came up with their own ways of saving on terminal costs. The White Pass and Yukon Route—a small, independent railway running between Skagway, Alaska, and Whitehorse, Yukon—was a North American pioneer. In 1956, it developed a standardized box that allowed for efficient transfer from ships to trains, and vice-versa. Efforts to establish global standards for containers began with industry association discussions in the 1960s, and the first ISO standards were created in the 1970s. What they were trying to create was a network, a system with multiple nodes, and a common means of interaction where a diversity of companies could participate. However, individual governments' regulation of the various modes in isolation delayed the arrival of truly global standardization until the age of deregulation in the 1980s and 1990s. Today, a handful of container types and sizes are used to enable efficient transfer of cargo from one mode of transport to another. Many different vehicles such as stack trains or container ships specialize in the use of containers.

numbers of users and connectivity among them—so does the network's capacity to generate value and reduce costs. These benefits are highlighted by the burgeoning of the Internet and the generation of social media, but they are important in utilities and transportation as well, the latter especially with the advent of inter-modality (Case Study 13.1). Through transport and communications, other sectors are metamorphizing into networks that allow firms to flexibly connect thousands of suppliers, distributors or researchers, or conversely allow consumers to reach out to global markets.

Networks

Moving goods, people, and information is an enormous area of economic activity that employs millions of people, requires massive investments, and consumes enormous amounts of resources (including land and energy). The industrial sectors directly contributing to networks—that is, those responsible for moving materials and energy, people, products, and information—are the utilities, transportation, and information sectors of the economy. In Canada, the three sectors account for shares of 2.5, 4.1, and 6.2 per cent respectively of total GDP, for a total of 12.8 per cent. In the US,

The containerization network has made it possible to integrate both different modes of transport and different standards within a single mode (e.g., different railway track gauges). Of course, as simple as containers themselves may be, container shipping today relies on many more sophisticated technologies—GPS, RFID (radio frequency identification), bar codes, and so on—for tracking and delivery.

© Dennis Chang—Singapore/Alamy

Container terminal, Singapore.

the figures are 1.3, 3.4, and 4.8, for a total of 9.5 per cent. These figures don't take into account the manufacturing of vehicles or computers and communication equipment or the construction of roads or the laying of Internet cables.

Transportation and communication are a direct result of exchange—people, goods, and information have to move for markets to function. Movement may occur by people congregating in a place to exchange or by the good or service being shipped or transmitted by sellers in one place to buyers in another. In essence, this movement or interaction is compelled by the complementarity of interests between buyer and seller. As such, it has generally being conceived of as a linear process—things moving from one place to another along a path of least cost (if not the shortest). To buyers and sellers, minimizing transportation costs and time may be the goal, but the providers of transportation and communications services seek profits, revenues, and growth.

Transfers of various kinds are rarely simple or direct, and usually depend on a common infrastructure that serves huge volumes of exchanges going any number of directions. Any package of goods or information may have to travel through multiple nodes or hubs via several different modes of transport or transmission, and can only do so because network infrastructures support a plethora of other goods and information. To ensure coordination of activities, various transportation and communication systems must overlap and need to function not simply as linear transportation and communication paths between two points, but as networks that allow economic exchange to occur in a myriad of competing configurations. Containerization was a vitally important innovation, for example, that facilitated the movement of goods across ship, rail, and trucking networks (Case Study 13.1).

Principles of Spatial Interaction

Economic interaction over space requires three conditions: complementarity, transferability, and lack of intervening opportunity. **Complementarity** exists when the production of a specific good or service in one place is matched by demand for it in another place. Complementarity stimulates the flow over space not only of goods and services, knowledge and investment, but of people—everyone from economic migrants to tourists.

The principle of **transferability** refers to the cost of transferring a particular good or service between two **nodes** or points: if it cannot be transferred at a reasonable cost, the exchange will not take place. As a general rule, costs increase with distance from the point of production or demand; therefore transactions tend to decrease in volume and value as distance increases between places. This tendency, known as **distance decay**, occurs along the length of any route between two nodes, and it also occurs with distance from a route (e.g., away from a main road). Distance decay can be seen even in telecommunications: the great majority of email and phone traffic takes place over relatively short distances.

In the case of goods, the total cost to transport increases with distance, even if average costs decline (see tapering effect), until transfer ceases to be worthwhile. Transferability may also be limited by physical, technical, political, or cultural barriers. The degree to which these factors limit trade depends on the mode of transport in question: coal freight, for instance, will be affected by the nature of the terrain, whereas

complementarity The condition of spatial interaction whereby the demand for a product in one place is matched by supply of a product in another, thus stimulating exchange.

transferability A condition of spatial interaction; exchange can occur between two places when revenues cover the costs of both production and movement.

nodes The points connected by network links (roads, wires, wireless connections, etc.).

distance decay The tendency for the spatial movement of goods, people, and information to decline as distance increases.

electronic money transfers require a certain level of technology. (One important difference between electronic communications networks and traditional transport modes is that once the initial set-up costs are covered, electronics communication networks can accommodate vast numbers of potential transaction partners at little extra cost.)

Finally, the likelihood of interaction between any two places will be reduced if **intervening opportunities**—alternative sources of supply or demand—exist at any point between them. Increasing distance makes intervening opportunities more likely, a phenomenon that holds true for telephone and Internet communications along with trade in goods and services. Telecommunications in general remain strongly localized, despite the advent of the Internet and low long-distance phone rates.

intervening opportunity (lack of) The condition of spatial interaction whereby exchange between two places will not occur if there is an alternative source of supply or demand between them.

Network Economies and Externalities

Transportation and communication infrastructures typically involve massive fixed costs that imply that conventionally defined economies of scale (reductions in average cost with increased scale of activity) are (potentially) significant (Chapter 2; later this chapter). Economies of scope are also typically strong if, for example, a particular route can be readily used for different purposes.

However, few nodes or places are connected only as pairs, but are connected directly and indirectly with many other places. These multiple connections define networks formed whether in relation to places or firms connected by transport, utilities, and communications systems or people using the same systems for social networks. Network externalities or impacts arise when increased usage occurs at little or no additional or marginal cost, but benefits or values are provided to all users and/or the network operators. The telephone system is an example: if two people have telephones, then just one link is possible; if five people have telephones, then ten links are possible, etc. The growth of e-commerce that allows all users to trade and interact with one another is a further obvious example of network externalities. Similarly, in transportation and communications, network externalities are promoted when network components are compatible. Railway track gauge (width) and signalling systems must be the same throughout the network, or else goods will have to be unloaded and loaded again. Early railway networks in the UK, for example, featured different gauges that limited efficiency until they were standardized. Similarly, Internet transactions are possible only when computers speak the same digital language. Thus the potential for network economies depends on the compatibility of technical standards whether they are container box sizes (Case Study 13.1) or file formats (e.g., PDF).

As a combination of both hardware and software components, which must connect a diversity of nodes to be effective, networks require large economies of scale. Network economies of scale differ from conventional internal economies of scale, however, because increasing use (additions to scale) is often delivered by independent users or agents joining the network. Note that while economies of scale are conventionally classified in terms of average cost (Chapter 2) network externalities are demand driven and are sometimes labelled as such.

In general, network externalities or effects are positive when increases in the number of people or firms using the network increases its value for users and its connectivity. Increasing connectivity enhances value to the user, while more users increases the returns to the organization(s) who set up the network by extracting payment from them or charging others (e.g., advertisers) who want to access them. The network effect is negative when increasing use and connectivity leads to congestion on some or all of the network. Negative effects can also occur if increasing use leads to environmental problems and/or safety and health hazards. In Canada, for example, with pipelines at full capacity, oil shipments have been diverted onto the rail network, a switch that became apparent with the explosion of 72 railcars in Lac-Mégantic, Quebec, in July 2013, causing 63 deaths and massive community damage. The railcars had been left parked and unattended for the night, but then rolled down the hill into the town. In part, this accident occurred because regulations had not fully recognized the implications of the sudden use of the rail network for oil shipments.

Networks can be built privately if a firm (or consortium of firms) can provide the necessary mix of capital, expertise, and resources. In most cases, however, some public input is required. Government involvement—both financial and regulatory—tends to be greater in transportation and communications than in other industries for two reasons: (1) the costs of establishing and operating such networks are high, and (2) the public goods provided by them are substantial. As the Lac-Mégantic case reveals, the negative externalities of transportation can also be substantial. While the companies that use networks are driven by market demand, government is driven by the need to ensure equitable access and limit negative externalities.

In some cases, these standards are made available for common use by competitors and customers (either for free or at cost) and in others they are held exclusively by a single company. A company that is able to capture the use of the network in some way will benefit accordingly. For example, everyone who uses the standard Microsoft operating system benefits with each new user who expands the pool of people with whom they can communicate. Although Microsoft doesn't benefit from the increased flow through the system, it captures income from sales of the operating system itself. Microsoft's operating system became the de facto standard because any computer manufacturer could use it, and despite their diverse origins, the computers were able to communicate with one another. By contrast, Apple chose to bundle its computer hardware and software together and forbid other manufacturers from using the latter. If Apple was unable to capture the benefits of expansion for its software in network externalities, it benefited when it created a common platform for downloading and paying for music. Companies such as Linux and Ubuntu take the opposite approach, offering open standard software, created voluntarily by communities of developers, that is non-proprietary and free for anyone to use.

The Internet has made the understanding of network economies and externalities essential. For example, many Internet companies, such as Google, Facebook, or Twitter, knew that initially they had to give away their product to establish the network and then try to find some way for people to pay them (and monetize the network). Fortunately, for those Silicon Valley companies, the real hardware and software network—the Internet—already existed and they were able to build a user network within it. Although the Internet receives much attention, network economies of scale and effects are as important to the more conventional transportation, utilities, and communication modes. Compatibility of technical and managerial standards are what matters. Compatibility allows firms of all sizes to take part in networks, both as providers of network services and as users.

For example, thousands of small- and medium-sized trucking, freight forwarding, storage, and delivery companies operate within logistics systems and thousands of software companies and hardware companies produce the applications and components used in computer or mobile phone-based communication systems.

All the same, for efficient operation, networks often require large organizations to coordinate operations. In intercity networks, large firms and governments can generate further economies of scale, known as corridor efficiencies, by providing the main road, rail lines, and pipelines, in adjacent networks. In providing regional feeder or distribution systems, they also provide spatial economies of scope by ensuring access to a broader hinterland. An advantage of MNCs is how they can use these networks to connect to their own networks of facilities, suppliers, distributors, and customers. Both scale and scope economies are sought by customers who want to capture the benefits of corridor efficiencies while retaining access to diverse destinations. Although a network allows for coordination among firms through open market and relational transactions, such exchanges require frequent transfer of ownership, legal responsibility, information exchange, and monitoring. The coordination demands are such that large sections of networks tend to be vertically and horizontally integrated within firms.

Network Configurations

The configuration of any network depends in part on the number of places or spaces that it needs to connect. The most fundamental unit of **connectivity** (and the basis for much network analysis) is the route between two nodes (Figure 13.1a). A node may be anything from a single firm or factory to an urban agglomeration. In addition to connecting major nodes, however, a network has to spread over space to provide access for those who are not concentrated in the cities but dispersed over the landscape. Outside cities, demand for infrastructure such as roads, railways, telephones, or Internet connections is relatively diffuse, and the per-person costs of construction and operation are higher than in high-density urban centres. As connectivity increases, so does access to goods and services. The substantial costs of increasing connectivity are usually subsidized to some extent by government or by companies (private, SOC, or co-operative). In this context, subsidy refers to the use of income from serving less costly customers to supporting more costly ones. Whether a network receives subsidization depends on the situation, influences including sector, context, stage of development, public good, and nationalistic concerns. For example, some countries support the rapid rollout of high-speed Internet networks in order to strengthen national competitiveness, while others leave Internet networks and resolving the digital divide to the market.

Spatial networks can be configured to balance various needs and goals. In the absence of any need for access to or from other places between two cities, a single route connecting them is efficient (Figure 13.1a). If there are more settlements, dispersed farms, or residences between the two towns that have to be connected then network costs escalate. Network planners are also challenged to minimize costs both for the investors (private and public) who pay for the infrastructure and for the eventual users, whether they are individuals, businesses, or government agencies, such as public transit operations or the post office (Figure 13.1b). In practice, most territorial spaces are centred on nodes surrounded by a ranked order of smaller locations and served by hierarchical networks that

connectivity The degree to which places in a region are linked or accessible to one another; may be measured in terms of cost, time, and/or capacity.

draw flows of people, information, and goods into the nodes for access to functions and transferral to other nodes (Figure 13.1c). Hierarchical systems, however, often mean that connections between peripheral nodes are limited. Minimizing distance-based costs is usually important, although environmental and political considerations may mean that longer-distance routes are chosen. Intervening opportunities also bring significant permutations to any simple distance-based hierarchic ordering.

Historically, the configuration of continental spatial networks was determined by the existing river systems because they provided the most efficient form of transport; later, railways and roads often followed those systems because they were relatively free of physical barriers. In the North American west, however, spatial networks both within cities and across the plains were developed in a grid pattern that was favoured to save on survey costs and reduce the transaction costs of distributing land. In Canada, for example, most country roads from Ontario going west follow the grid of concession roads and are only occasionally dissected by highways built to draw the flows of traffic into larger cities more efficiently (Figure 13.1d). In contrast to the organically evolved cities of Europe and Asia, North American cities are remarkable for the planned regularity not only of their roads but also their sewers, transmission lines, and telecommunication cables.

Terrestrial (conventional) radio and television communications do not follow a fixed network pattern. Instead, stations broadcast their signals uniformly over space, providing equal access to everyone with the appropriate receiver except those living in the shadow of some barrier to transmission (Figure 13.1e). Internet and cellular phone networks combine the grid and broadcast systems. The Internet is a network of networks in which millions of individual computers are connected by modems, local area networks (LANs), or wi-fi (wireless) to Internet service providers (ISPs), which in turn are connected with higher-order network access providers (Figure 13.1f). Remarkably,

(a) Single route network

City A   City B

(b) Route cost minimization between two centres and among smaller centres on a linear space

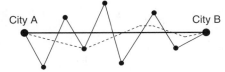

City A   City B

(c) Hierarchic networks providing access over space

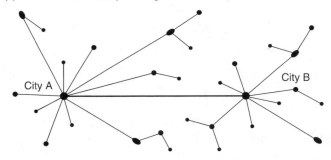

City A   City B

(d) Survey grid dominated network

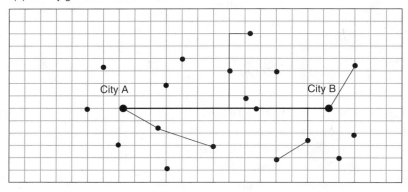

FIGURE 13.1 **Networks**   *(Continued)*

the critical function of the ISP is to ensure low-cost access to a network that is otherwise used for free. A crucial requirement of the Internet's operation over cyberspace is the designation of each computer's Internet Protocol (IP) address, which identifies each computer by place. This system of assigning addresses was initially controlled by the US government, but was assigned to the non-profit Internet Corporation for Assigned Names and Numbers (ICANN) in 1998.

Cellular phone companies provide access to cells of space within a region and manage transfers of signals between cells. The merging of fixed and mobile telecommunications technologies and the capacity to combine linear and areal access marks another tremendous advance in communications technology. Satellite networks provide the greatest coverage, allowing access from anywhere on the earth's surface and permitting constant monitoring of any place, vehicle, or person.

(e) Broadcasting over space

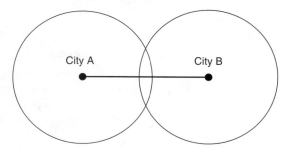

(f) Internet hierarchy of networks

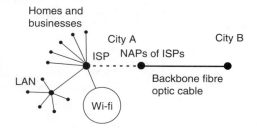

FIGURE 13.1   **Networks** *(Continued)*

Network Analysis and Design

The kind of analysis required to design a network from scratch or to improve an existing one varies. Government planners setting out to design a road network focus on collective needs, but railway, airline, bus, logistics, and telecommunication companies will balance the potential profits of serving a broad variety of customers with the greater surety of developing higher volume connections between larger nodes. In either case, the goals relate to minimizing overall distance, time, and costs while ensuring adequate service both for passengers and freight in terms of accessibility, reliability, comfort, safety, prevention of loss or damage, flexibility, and frequency. Planners employed by a large company, by contrast, focus on optimizing the company's own use by devising supply and distribution networks within the public and private networks available to them. Individuals or households have to devise only a few routes—to work, the gym, or the cottage—a task now made easier by map applications that help them find the most direct, least costly, or least crowded route.

The efficiency of network structure is quantitatively assessed using graph theory. Graph theory translates a *topographical* map of an actual network into a matrix of origins and destinations in the form of a **topological map** that summarizes the network in terms of nodes, distances, and linkages (Figure 13.1). Important measures that summarize topological properties include connectivity, accessibility, and **circuity** (Case Study 13.2).

Less abstractly, the flows of goods, people, and information over a network are measured in terms of tonnes, passengers, bites, trips per day, people per car, volumes per year, and so on. This information can be translated to matrices and topographical maps to provide total network loads, average loads, and for comparison, among nodes and routes. The analysis of flows is important to determining the capacity of the network or of routes between nodes, scheduling, identifying congestion tipping points, etc. In

topological map A network described by nodes and edges or links between nodes.

circuity A measure of the degree to which the paths that make up a network deviate from the straight path.

Case Study 13.2

SIMPLE MEASURES OF NETWORK FORM

Actual networks, such as road, rail, or pipeline systems, can be reduced to "graphs" of various types made up of vertices and edges (Figure 13.1), the former representing places (centres or nodes) and the latter the routes or links between places. Reducing a network to a graph makes some simple measures of connectivity, accessibility, and circuity possible. These measures are useful for tracing the evolution of a network over time or for comparing networks between two regions.

"Connectivity" refers to the linkages among the different nodes in a network. The simplest measure of a network's connectivity is the number of connections it contains. This **cyclomatic number** (μ), is determined as follows:

$$\mu = e - v + p$$

where e is the number of edges;
v is the number of vertices; and
p is the number of connected components

A value of 0 occurs when p is one or more, and there are a minimum number of edges between vertices. As the number of edges increases, so does μ, indicating increasing connectivity. In Figure 13.1a, for example, there are two vertices, one edge and just one connected component. Thus μ is zero. If more edges are added to connect A and B in Figure 13.1a then μ would increase.

The **beta index** or ratio (β) is the average number of edges per vertex:

$$\beta = e/v$$

This index ranges from zero to three, higher numbers indicating increasing connectivity. In Figure 13.1a β is 0.5; if A and B were disconnected, β would be zero.

cyclomatic number A measure of the connectivity of a network.

beta index A measure of the connectivity of a network.

combination with forecasts for demand, flow analysis is used to create models for the development of networks and the optimization of routes through networks. The balance between serving dominant nodes and providing connectivity over space is reflected in the network density of an area: the total distance (in kilometres) of linkages or routes divided by the area (square kilometres). The network density of an area usually depends on a combination of physical geography, level and type of development, and political will.

Many different analytical models are used to devise optimal networks and optimal routes, as well as to manage traffic on both. For example, the gravity model is used to analyze the pull of different nodes, such as cities or zones within cities and the effect of their attraction on network traffic. Many models and techniques are used to predict flows through networks in order to facilitate practical planning from both public and private perspectives (one theory uses ant behaviour to determine routes for trucking or air travel). Despite the many means available to optimize networks, however, uncertainties remain. Demand generation and fluctuation are still difficult to determine accurately. And the unusual degree of government involvement in transportation

There are also simple indexes that assess the status of individual vertices in a network. The associated number, for example, indicates the number of edges to a vertex from the vertex that is most remote to it. Centres with lower associated numbers are more accessible than those with higher associated numbers. In Figure 13.1b, for example, A and B both have associated numbers of three, while all other vertices have higher scores. The accessibility of an individual node in terms of actual distances is measured by the Shimbel index:

$$Aj = \sum_{i=1}^{v} d_{ij}$$

where A is the accessibility of a node i;
dij is the shortest distance between i and j; and all the vertices connecting to j are summed.

The index for network accessibility is:

$$D(V) = \sum_{i=1}^{V} \sum_{j=1}^{v} d_{ij}$$

$D(V)$ is the sum of all the shortest paths in the network.

Finally, circuity is a measure of the degree to which the paths that make up a network deviate from the straight path ("Euclidean distance"). The basic measure of circuity is the ratio of actual distance between two vertices and the straight line distance. A network is efficient when it reduces circuity. A node with high circuity will be penalized because more roundabout paths mean higher transport costs, which are a disincentive to trade.

and communications means that political decision-making plays a greater role in these industries than in most others. Consequently, the parameters for network design and use range from concerns for equitable access to partisan interests. Ultimately, those parameters reflect particular political-economic frameworks.

Modes, Media, Intermodality, and Logistics

Distinctions between different transportation *modes* and communications *media* are reflected in different technologies and managerial parameters. It is not difficult to identify the differences in markets and operations between rail freight and commercial television broadcasting. The first depends on strictly fixed lines, is comparatively efficient at transporting bulk goods long distances, and is paid for by users according to volume and distance carried. The latter disseminates information indiscriminately over space and is paid for by charging sponsors to advertise their products to viewers or by charging for TV ownership (as in the UK and Japan). The different transportation modes or communication media play significant roles in the economy and are often designated as distinct industrial sectors. In the reality of operations, however, many modes and media operate in parallel or together. For example, continental railways were developed in tandem with telegraph communications because the latter enhanced managerial control over scheduling, loading, and so on. Telecommunications technologies have improved this control immensely. Airplanes today rarely fly with empty seats or too much food because passenger scheduling systems allow accurate matching of plane sizes and schedules to passenger demand. In the same way, global positioning systems allow trucking companies to monitor their fleets 24/7 and ensure that their equipment and labour are in constant use.

modes of transportation Alternative ways of moving goods or people between two places.

Intermodality is the combination of different transportation modes. Most goods today are transferred through several **modes of transportation** as they travel from production to consumption, often via containers (Case Study 13.1). Toys made in Guangdong are first packed into containers that are loaded on trucks which are driven to Hong Kong's harbour; there the containers are loaded on ships that carry them to ports elsewhere; then they are loaded either onto trucks for delivery to nearby destinations or onto rail cars for longer trips, which may end in a warehouse where the consignment is divided for shipment to other warehouses or retailers. There are countless possible combinations of transport modes. When multiple shipments are combined for transport (an important economy of scale), telecommunication technologies such as bar coding and radio-frequency identification (RFID) facilitate identification and real-time control—essential when the individual parts of the shipment are separated for delivery.

The ultimate example of intermodality is the *logistics* system. Logistics is defined as the process of planning, implementing, and controlling the efficient, cost-effective flow of raw materials, in-process inventory, finished goods, and related information from the point of origin to the point of consumption. It includes, not only transportation and communication, but also storage, handling, organizing, packaging, and, increasingly, design, manufacturing, and other means of adding value. The central idea is that greater coordination throughout the product cycle improves efficiency. Conventionally, logistics focused on the forward transfer of products from producers through distributors to end customers. Today, in response to environmental concerns, **reverse logistics** may be required to handle collection of the product (and/or its packaging) for reuse, recycling, or disposal. Logistics services are provided by specialist third- or fourth-party firms, some of which began as transportation specialists and some of which moved into logistics as part of their information management services. Manufacturers and distributors also handle logistics in-house as a complement to their other economies of scale. In practice, most firms use both in-house and third-party logistics services, although the outsourcing of logistics has been one of the strongest business trends of the early twenty-first century.

reverse logistics Organization of the collection of products (and their packaging) for reuse, recycling, or disposal.

Costs, Pricing, and Location

Fixed and Variable Costs

Transportation and communications are among the most capital-intensive of industries, but they offer substantial economies of scale based on fixed costs. Typically, economies of scale in networks are captured by the cube-square law, massing of reserves, and bulk transactions (see Case Study 2.1). Tandem trailer trucks and the Super Jumbo Airbus reduce overall labour requirements, while wider pipelines and bigger tankers improve energy efficiency. Fixed costs, especially investments in capital-intensive industries, require above-average payback times. The costs of handling at the end of a route are known as terminal costs and can be especially high for modal transfers. Smaller loading bay requirements and the fact that roads are constructed and maintained at public expense mean that trucking has lower-fixed costs than railway transport, which requires more elaborate, larger loading areas and facilities, as well as track maintenance over hundreds or thousands of kilometres.

Running costs are variable costs that change in proportion to the level of activity. In transportation, they are easily identified as the line-haul costs of fuel, labour, and so on, which increase with distance travelled. The ratio of fixed to variable costs for broadcasters is particularly high because the level of service they provide is constant over their entire broadcast area. Among the value-added services that can be offered and charged to customers within a well-coordinated logistics system are extra-fast delivery, measures to prevent damage, tracking, storage, and refrigeration. Especially in the movement of goods, high fixed and/or terminal costs imply that average costs per tonne-kilometre will decline with distance. Thus the average (not total) transportation cost of moving commodities by rail, ship, or barge typically declines with distance, a tendency often referred to as the "**tapering effect**." In road haulage, in which fixed costs are less important, the tapering effect is less significant. The prices paid for by consumers for transportation services, however, may or may not directly reflect the actual costs incurred.

Tracking the variable costs and imposing prices based on them is a difficult managerial task for firms. To reduce these transaction costs, service providers (firms and governments) may charge uniform rates, regardless of location (although this means that consumers closer to the service subsidize those farther away). Another strategy is to set prices according to zones. Telephone, courier services, and post offices use this strategy. Two standard business-to-business strategies with differing approaches to achieving transport efficiencies are **c.i.f. (cost, insurance, freight)** and **f.o.b. (free on board)**. In c.i.f. pricing, the exporter pays the costs of delivery to a ship, as well as the insurance costs and freight charges for shipping, but can achieve economies of scale by combining the shipping costs of many customers. Under the f.o.b. system, the purchasing firm pays the on-board freight charges in order to have more control over transport costs. (While customers of transportation services, such as road haulage or railroad companies, want to minimize costs, private suppliers are profit-oriented and often charge "whatever the market will bear.")

A uniform delivered pricing (u.d.p.) system, however, charges the same price to all consumers regardless of distance travelled. A national price for postage stamps reflects such a system. Zonal pricing is a stepped form of uniform delivered pricing, often practised by passenger bus companies. In contrast, basing-point pricing systems mean that consumers pay for the price of the good plus cost of transportation from the basing point regardless of the actual origin of the good. In these systems, the transportation charges may not reflect the transportation charges incurred and in this sense are considered "discriminatory." In a u.d.p. sytem, for example, short distance exchanges pay more than the transportation costs incurred, while more distant consumers pay less. Generally speaking, these kinds of pricing systems are administratively easier to operate, and in some

tapering effect When the average cost of transportation declines with distance.

c.i.f. (cost, insurance, and freight) A pricing system in which the price charged to buyers incorporates the costs of the good, insurance, and freight charges up to the destination port.

f.o.b. (free on board) A pricing system that means buyers pay the on-board costs of transportation from the point of shipping.

cases may involve a public good rationale, for example, in charging all consumers the same price for mailing letters no matter where they live. In this regard, a difficult policy problem is how to charge for passenger services on buses or ferries, or even whether these services should be provided for remotely located people.

A Location Imperative

In general, the massive fixed costs of transportation and communication networks respond to and reinforce geographic concentrations in the most accessible places. This tendency is reinforced by capital inertia: once in place, such networks are difficult if not impossible to relocate. For a network with high fixed costs to be viable, demand for it must be high enough to reduce average costs and price (or taxation) levels. And (in a chicken-and-egg relationship), consumer demand depends on accessibility. In general, inertia effects are reinforced by intermodality and competition. The need for inter-modal transfers establishes dominant nodes around ports and airports, as competitors within and among modes are attracted to the same high-demand routes.

By contrast, remote, sparsely populated areas often have difficulty attracting investment in transportation and communications simply because demand levels do not justify the expense. It is no coincidence that East Africa is the last major region in the world to be connected to the fibre optic cable network. Yet access to roads, power, water, and information may be considered a right of citizenship, and it is a necessary (though not sufficient) condition for development: for example, a new highway in a remote area may or may not stimulate economic activity, but is certainly a prerequisite for it. Together, huge fixed costs, important implications for development, and concern for equity, as well as environmental externalities, have required governments to be closely involved in the creation and maintenance of transportation and communication networks.

Network Externalities and Public Goods

The rationale for government participation in economic activity is strong when that activity has public good characteristics and involves significant externalities with long-term implications for development, efficiency, and equity. Governments build and operate road networks because they create positive externalities in the form of equal "free" (that is, taxpayer-supported) access and promote easy interaction for economic and social purposes. The highway system could be managed by private firms, but markets exclude those who cannot afford to pay, and society as a whole would suffer if the network were allowed to decline or if access to it were restricted. Indeed, market economies continue to debate the extent to which various networks—from railways to water supply systems—should be owned or regulated by government. Internet services, for example, are supplied mainly by private companies, but to ensure equity, governments provide free access in public facilities such as libraries.

The most ambiguous aspect of transportation and communication pricing derives from the uncertainties of funding public goods and services such as roads, sewers and water supply, or national broadcasting networks. The relevant government may have a reasonable idea of the costs associated with building and operating these networks, but those costs are not made evident to users who pay for those services only indirectly, through the tax system. Which mode or medium receives support varies from country to country, as does the level of support they receive. The US federal government supported the construction of railways with land grants, encouraged the creation of the telephone company AT&T, and built the interstate highway system. But it never supported public radio or television, or operated railways and airlines, to the degree that Britain, France, Japan, and Canada did. On the other hand, continental Europe and Japan charge drivers directly to use high-speed freeways.

Instead of running public services themselves, many governments have chosen to grant a monopoly to a company and rely on regulation rather than market competition for governance. Since the 1980s, however, belief in market governance and neoliberalization has increased and many state corporations have been privatized and monopolies opened to competition. Still, government control of sectors such as roads and highways remains monopolistic. Intermodal and logistics companies must work within the framework created by government, including its physical configuration, operational efficiency, and fee structure. Furthermore, most modes and media continue to be highly regulated because governments recognize the need to improve access to and control the use of public goods, especially in cases where freedom of access encourages free-riding.

Government intervention is needed to address the negative externalities of network supply and operation. One obvious example associated with transportation is air pollution. Combustion of gasoline, diesel, and coal in various transport modes produces numerous pollutants. Although emission levels have been reduced by bans on additives such as lead, and the introduction of catalytic converters, cleaner fuels, and improved engines, greater reductions are needed, especially on greenhouse gas emissions. Carbon taxes and carbon credit trading are two market-based regulations encouraging private companies to reduce their emissions and stimulate them to internalize the costs of these externalities. Other transportation externalities include habitat destruction, congestion, noise, water pollution, and the introduction of alien species. Among the externalities associated with communications are the potential threats to health posed by electromagnetic fields and the visual degradation of the landscape.

Transportation, Communications, and Scale

Networks exist at different geographic scales. A basic distinction is between international and domestic systems. Some international networks cover vast distances, requiring the use of ships, airplanes, undersea cables, and satellites. Others involve smaller distances and fewer physical barriers, but may be equally subject to trade barriers in the form of tariffs and regulatory requirements.

Within national borders, the demand and supply conditions for transport and communications often vary dramatically from one place to another. Many national governments have invested heavily in transportation and communications infrastructure as a prerequisite for economic development. Building ports and canals, roads, railways, and pipelines is essential for any country that wants to encourage economic activity and participate in international trade.

Interurban Networks

The connections between international and national networks are important because they make trade possible between city-regions that propel the global economy. These connections force national networks to comply with international treaties and other countries' regulations; for example, Ontario was compelled to allow bigger trucks on its roads after bigger trucks came into use in major American manufacturing states. That said, each country has its own history of transportation and communications development, and its own distinctive mix of intercity transportation modes.

In North America, rail is the dominant mode of transport for bulky, low-value goods, although inland waterways, coastal shipping, and pipelines are also important. In the US, trucks move almost as much freight as pipelines and shipping combined, and the value of the goods they carry is vastly greater. For passenger travel, most Americans and Canadians alike rely on cars for travel to most places within a day's drive; otherwise, they would rather fly than take the train. Intercity trains are significant only on the American east coast, and they account for only about 1.5 per cent of intercity travel in Canada.

In Europe and Japan, proximity to the sea means that bulk freight often travels by ship. Otherwise, both economies rely primarily on trucking, although European canal systems allow some inland shipping as well. (Japan has no navigable waterways.) Both Japan and Europe rely on high-speed trains for a significant proportion of intercity passenger travel. Today, however, the trains face tough competition. The Europeans have built high-speed highway networks and reduced intercity airfares below the cost of the taxi ride to the airport. The Japanese have tunneled and bridged their mountainous archipelago to provide equitable access to a high-speed road network. With only a few major population centres, they also use aircraft, although ticket prices are much higher than in Europe or North America. China, which built its civilization on river and canal transportation, has built the world's largest high-speed railway and highway networks in order to develop its twenty-first century economy. Still, transport on the Yangtze and Yellow rivers and the rivers feeding into the Pearl River Delta remains important.

Government Ownership and Regulation

Unlike most industries, transportation and communications have historically been closely tied to government, although the nature of government's involvement has changed over time. Indeed in the network and public good requirements of many sectors, for a long time, resulted in them being considered as natural monopolies that required government ownership or regulation. The transcontinental railways of North America could not have been built without government funding and land grants. And the Internet began when the US Defense Department's Advanced Research Project Agency (ARPA) funded theoretical research and connecting hardware (Case Study 3.1). Most other countries, wanting to play catch-up in development to the UK and USA, developed national champions and often monopolies in railways, telecommunications, television, airlines, and so on. In the fifties and sixties, for example, most countries (145, except the US) believed it had to have a national carrier airline. In the 1980s, economic recession and the diseconomies of scale inherent in massive organizations precipitated a dramatic disinvestment in transportation and communications by governments around the world. The liberalization of Canada's transportation system began with the National Transportation Act of 1987, which sought to reduce regulatory restrictions (e.g., regarding airline ownership), encourage competition, and generally move Ottawa's policy more into line with Washington's. In addition, communication monopolies such as the Bell telephone systems in both Canada and the US had their government-permitted monopolies broken up. This process was hastened by rapidly evolving cable, digital, Internet, and mobile communications.

Even if governments give up or never had direct ownership and administration of networks, they can still exert significant control. In regard to safety, security, and environment, the government's regulatory role is as important today as ever. On the railways, for example, signals, load limitations, flammability, toxicity, and so on are subject to government regulation and oversight. Airlines are also subject to regulations governing everything from air traffic control protocols and safety inspections to prohibition of smoking on board. In some cases, national control achieves extra-territoriality, as when US federal air regulations control not only its US based airlines in foreign airspace, but also foreign airlines in US airspace, and by acting as a de facto world regulatory standard. In the case of the Internet, the US government has ceded control over the critical issue of domain names to an international governing body. In regard to some global networks, the UN has stepped in to provide governance, for example, the IMO (International Marine Organization) coordinates safety, technical, security, and environmental issues for the world's many shipping companies. That said, companies can also establish networks through collective agreement as has been done on many technical standards for mobile phones. Other standards have been set by competition, as Qualcomm did with its Code-Division Multiple Access (CDMA) standards for mobile phone semiconductors.

Intermodality

Intercity networks operate primarily on a modal basis. Most railway companies are focused on their rail operations, trucking companies on trucking, airlines on moving people and freight through the air, and shippers on bulk and containers. However, intermodality and coordination among modes are being integrated into their operations. For example, North America's largest railway, Union and Pacific, specializes most of its rolling stock in commodities, but also gains about 20 per cent of its revenue from intermodal activities. Similarly, while the majority of international shipping remains committed to commodities such as oil, gas, chemicals, coal, wheat, and other commodities, the greatest value added is derived from the intermodal services of carrying, loading, and unloading containers. The Danish company Maersk leads a handful of companies that dominate intercontinental intermodality. In the case of Federal Express (FEDEX) and United Parcel Service (UPS), intermodality is a primary business orientation.

Standardized containers are the basis of intermodality because they allow the greatest possible efficiency in the transfer of goods between transport modes. Different standards are used depending on whether the transfer is from ship to truck or from plane to truck and the standards are defined by the relevant industry associations. Intermodality is also important for passengers. Air passengers, for instance, need various connections to get from airport to town, and the entire North American air system depends on the availability of car-based ground transport. All North American airports are designed not only to accommodate vast parking facilities and drop-off locations, but also to accommodate the car rental industry (Case Study 13.3). European and Asian airports put more emphasis on passenger rail connections.

Whether a transport network uses one mode or several, coordination of activities within it depends on telecommunications primarily Internet-based. Some of the most important companies in the transportation business today began in software. The need to integrate transportation, communication, and the flow of goods, people, and information related to any product suggests that the future of transportation and communication lies in the logistics industry.

Logistics specialists coordinate the supply chain, minimizing the costs of transportation and warehousing, ensuring timely parts supply and assembly, and reducing environmental impacts. The firms involved are diverse, including transport, telecommunications, third-party logistics firms, and producers and distributors that are large enough to support their own logistics systems. They all integrate their activities within common networks that are constructed from private, public, and associational elements. Technology of course plays an important role in the evolution of the logistics industry, the sophistication of Internet-based and GIS telecommunications, and the simplicity of standardizing boxes that ease transferability.

Industry Concentration

The belief in natural monopolies and government ownership may have weakened, but the economies available in the many activities of the transportation, communication, and utilities sectors have stimulated a high degree of industry concentration. Concentration has been led in the North American trucking industry by logistics firms such as FEDEX and UPS that have sophisticated systems for tracking packages and larger freight throughout their intermodal network. There is however a large number of relatively similar sized competitors in the trucking industry who, while compelled to develop logistics capacities, are still provided with their core infrastructure network, the publicly subsidized roads and highways. In the airline industry, it is very difficult to link all pairs of cities with direct flights, while at the same time customers want more convenience in flight scheduling, minimum baggage difficulties, and so on. Sophisticated control systems are required to match plane capacities with passenger numbers, baggage handling, and scheduling. Airlines have tried to overcome some of these coordination problems by establishing regional hubs and airport networks, but in spite of that there has been a long-time trend to co-operation and mergers among airlines. The once highly populated US airline industry is now dominated by five companies (Delta 16.3 per cent; Southwest 15.8; United 15.6; American 12.6; and US Airways 8.4) (http://www.transtats.bts.gov accessed 20 June 2014). The issue is greater at a global scale, but national regulations and biases prevent mergers to a great degree. Airlines responded by forming alliances, and three of these—Star Alliance, SkyTeam, and Oneworld—control about 80 per cent of the air traffic over the Atlantic and Pacific oceans. The trend towards oligopoly in the network industry is also demonstrated by the car rental industry. Whether the network imperative is a satisfactory explanation for increasing concentration or whether corporate consolidation of networks is good for consumers are, of course, other questions.

Intra-urban Networks

The size and concentration of the markets for transportation and communications inside a major city allow for economies of scale that only the highest-volume intercity businesses can attain. Cities have multiple cable, cellular phone, and ISP providers from which customers can choose. Companies can run their own delivery services or outsource delivery to a number of firms. Several commuting modes may be available. Urban spaces are filled with networks with relatively high levels of economic efficiency; indeed achieving the benefits of urban economies requires the development of these networks. That said, there are important considerations in the private and public development of overlapping and efficient networks. Most notable are the social and economic needs for equity in accessibility and mobility; the relationship of transportation to urban form; and the impact of externalities such as congestion and pollution.

Accessibility

Mobility—the ability to travel to work, shops, or other activity sites—would seem to go hand in hand with accessibility. But this is not always the case. Increasing mobility in the form of car travel, for instance, makes it possible for workplaces and stores to be located at greater distances and in more concentrated form. As a result, most people who in the past would have walked to local shops or places of work must now travel long distances to work, shop, and socialize. (No wonder many people now socialize on cyberspace.) This separation of activities can raise equity issues. Lack of a car or convenient, affordable public transportation seriously inhibits people's ability to participate in the economy, forces them to incur travel costs on necessities such as shopping (or pay a premium for local purchases), and reduces both social and recreational opportunities. Not simply a question of equity, lack of participation inhibits overall economic development.

A basic measurement of accessibility is

$$Aj = \sum_j O_j d_{ij}^{-b}$$

where Ai is the accessibility of person I, O is the number of opportunities at distance j from person i's home and dij is a measure of separation between i and j for travel time, costs, or distance. The same equation can be used to measure the accessibility of one place to another by defining Ai as the place and Oj as the number of opportunities available in the place; and b is the rate at which accessibility declines with distance.

The tendency, in modern cities, for the population to be distributed across the urban space is premised on the benefits associated with dispersed residential and commercial location (e.g., amenities, customer access, lower rents). Opposing this centrifugal force are centripetal forces such as the convenience and internal economies of scale that become possible when people and economic activities are brought together in one central place. Most cities have one dominant node and a network of secondary nodes connected by high-volume corridors. Such corridors tend to attract some economic activity themselves, but they may repel businesses if negative externalities such as air and noise pollution outweigh convenience. Thus in terms of both nodes and corridors, economic land use may mitigate diffused access by concentrating activities, but those activities can create their own problems.

Traffic congestion, for instance, is a negative externality arising from the provision of roads as public goods, and the resulting low cost of use for car drivers (Case Study 13.3). Of course, congestion can also occur on transit lines leading into urban centres. In cities such as Tokyo, Osaka, and Mumbai, it is not unusual for urban rail lines to operate at more than 300 per cent of capacity. Commuters suffer the consequences, from delays

Case Study 13.3
URBAN CONGESTION COSTS

Urban congestion affects the quality of life of commuters, imposes environmental costs (as gridlocked traffic wastes energy, produces greenhouse gases, imposes economic costs through "lost time" by commuters), and discourages investments in new business operations. It is not easy to estimate these costs. The first systematic attempt in Canada was conducted by Transport Canada for nine cities in 2002, and a recent study provides data for 2006.

Annual Total Costs of Congestion by City (2006, $millions)

Urban area	At 50 per cent threshold	At 70 per cent threshold
Quebec City	$63	$108
Montreal	$697	$910
Ottawa-Gatineau (all)	$220	$380
Toronto	$1,298	$2,014
Hamilton (all)	$13	$37
Winnipeg	$73	$125
Calgary	$149	$180
Edmonton	$85	$120
Vancouver	$518	$755
Total all urban areas	$3,116	$4,629

Source: Transport Canada Study: The Cost of Urban Traffic Congestion in Canada. Ottawa: Transport Canada 2012. See http:// http://www.comt.ca (accessed July 2014).

Globally, according to this study, Canadian cities have not performed well either. Thus in 2011, Vancouver lost its rank as the world's most liveable city largely because

in service to increased likelihood of disease transmission. A central problem with public goods that invite overuse, such as roads and transit systems, is that increasing their capacity only encourages additional use. The air pollution caused by the various forms of road travel and the generation of power for electrified transit is exacerbated by traffic congestion. Transportation and communication facilities take up land that would otherwise be available for parks and gardens while disrupting neighbourhoods.

Land-Use Planning

Land-use planning has the intention of alleviating some of the problems outlined above, but has also recreated similar problems in new locales. Early industrial cities developed in a radial pattern because economic activity was concentrated in or near the downtown and because streetcars—the best transportation technology available at the time—could be most efficiently deployed with one central interchange in the

of rising traffic congestion. Toronto also fares poorly in terms of traffic congestion among comparable global cities.

In this study, Transport Canada primarily emphasized an "engineering" approach that focused on the direct features of congestion especially the duration of peak congestion periods, trip purpose, unit fuel prices, values of time, and unit green gas mitigation costs. Congestion costs were estimated according to 50, 60 and 70 per cent of free flow speeds to be expected on particular highways. On major highways, for example, where 100 km per hour is permissible, speeds of 70 km per hour or less were deemed to reflect congestion. The study found that more than 90 per cent of urban congestion costs during these periods stemmed from the value of the time lost to auto travellers, while the remaining costs were the value of the fuel consumed (around 7 or 8 per cent) and the greenhouse gases (GHG) emitted (around 2 or 3 per cent). While these estimates of costs, totalling between $3.1 and $4.6 billion, may seem high, they are conservative because they do not include the costs of non-recurrent congestion (i.e., congestion caused by random events, such as bad weather, accidents, and stalled vehicles), freight traffic, congestion during non-peak hours, and other congestion-related costs such as noise and stress.

Beijing has become a particular centre of concern for traffic congestion, as private car ownership has increased from virtually zero in 1985 to over five million by 2014. Car ownership is increasing rapidly, as is truck transportation on a road network that has reached the bursting point. In September 2010, the media reported that some commuters were stuck in traffic jams lasting several days. Beijing has introduced a lottery to get car licences and plans to limit car numbers to six million by 2016.

Reducing urban congestion and its impacts is a challenge even in cities with substantial rapid transit systems. In 2003, London, England, introduced a traffic congestion tax that is imposed on all vehicles entering the central city, using an automatic licence-plate recognition system, with fines for non-payment. Dealing with these issues is a priority for most cities and as cities experiment with solutions they exchange ideas with one another. For example, Curitiba in Brazil developed a bus rapid transit system that incorporates the boarding, ticketing, and dedicated route efficiencies of subways, but on normal roads and at a fraction of the cost. Variations of system have been adopted in cities such as Guangzhou, Jakarta, and Chicago.

same downtown area. Later technological and economic developments, particularly the diffusion of automobile ownership, made it possible for cities to break free from the radial pattern and develop multiple nodes. Close attention was paid to the planning of these nodes and, above all, the access routes to them. Planners in North American cities strongly favoured the separation of land uses and connecting them through automobile-based transport. Thus large residential areas developed, largely devoid of pedestrian-accessible shopping but within a short drive of malls. Sometimes residential development took precedence and subsequently attracted commercial and industrial centres. Firms located in those centres for reasonable rents and for reasonable housing for employees. In other cases, commercial or industrial developments were the focus, and residential communities were developed in conjunction. Agglomeration economies helped to strengthen these centres, while the consequent bid-rent curves kept residential functions at a low level. In this way, the sprawl out from central cities and the creation of the suburban, multi-nucleated, and edge-city structure was, for the most part, planned.

The planned or suburbanized North American city has been successful in providing the cultural norm of an affordable detached home for families, generating family-based capital formation with housing as a vehicle, low rents for commercial and industrial enterprises linked with affordable labour, and arguably low-cost shopping, most recently from big box stores. However, there are many direct, opportunity, and externalized costs arising from this structure. The direct costs arise from supplying the roads, utilities (sewers, electricity, cable, Internet, etc.), schools, and other infrastructure and services to widespread residential, commercial, and industrial functions. Supplying such infrastructure and services incurs much greater resource and energy use, and does so in both the construction and operation phases. The dispersed structure and car reliance is also blamed for various ills, ranging from the inhibition of community formation to obesity. Residents in areas that are segregated by poverty or other factors have difficulty finding public or private transport to jobs. Nor has the development of these new centres solved the congestion problem. Since the cost of the roads that provided access to those centres was not visible or directly imposed on users, demand for additional roads was constant. And as the numbers of roads have increased, so have the numbers of cars and drivers.

The major alternative approach to suburban planning is called "smart growth" in North America and compact cities in Europe. The two key features of the approach are greater encouragement of mixed uses and transit-centred development. Other features include making the urban system more pedestrian and bicycle friendly, using distributed energy and waste recycling, increasing green space and conserving resources, and improving the mix of residential types (i.e., not segregated by income, ethnicity, etc.). Increasing social capital and citizen involvement in the development and operation of what should be a community is a key factor in realizing these ambitions. Undoubtedly, the smart growth and compact city ideas are influential, and not only by making cities more dense or transit oriented. Low-density residential areas, new and old, are benefiting—after all, it is not that difficult to allow commercial operations within a residential area or establish bike lanes on wide roads. However, there are difficulties, especially, in pre-existing suburbs. For one, it is not easy to persuade drivers to take transit, especially when they compare the costs in time and convenience. Most importantly, although it is easy to advocate mixed uses and thereby reduce commuting times, building community intra-actions and so on, the external economies and bid-rent curves that concentrate jobs in one place and relegate residential functions to others are difficult to overcome. This is not only a suburban problem. An irony of the increasing densification and amelioration of urban cores is their increasing attractiveness, hence predominance of wealthier residents and the requirement that others working in the city are forced into longer commutes. The balance between jobs and residences is not easy to obtain or plan for. Vancouver exemplifies the difficulties in trying to balance smart planning and density objectives within a urban system that remains car-based.

Transportation Networks and the Environment

The negative externalities produced by transportation networks include air, soil, and water pollution, habitat loss, and more than a million traffic deaths around the world every year. Perhaps most problematically, transportation is a leading cause of greenhouse gas emissions, and its impacts are rising at a higher rate than those of other activities such as electricity generation.

Petroleum accounts for 95 per cent of the fuel used for transportation, and although its use has been made more efficient, further advances and substitutions are urgently needed if global warming is to be slowed down and the social and economic benefits of transportation maintained. The greatest reductions will have to be achieved in the developed countries, which are still the dominant users of transportation fuel. Yet global development means that expansion of freight and passenger transportation will be dramatic. China is now the world's largest producer and market for cars, and its air travel industry is growing at double-digit rates. In general, highway driving uses the most energy and contributes the most to carbon emissions (Figure 13.2).

Many technologies are being developed to reduce the environmental impacts of transportation. For road vehicles, these include biofuels and flexfuel cars, hydrogen production and fuel cell vehicles, battery improvements, and electric vehicles. Ship hulls and engines are being redesigned, and airplanes are being designed to match more efficient engines and composite materials. Innovative use of telecommunications

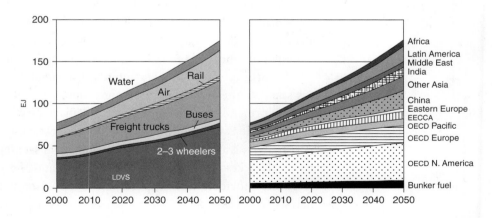

FIGURE 13.2   Projection of Transport Energy Consumption by Mode and Region

Notes: EJ = energy, joules; EECCA = Eastern Europe, Caucasus, and Central Asia; LDVs = Light duty vehicles

Source: Adapted from Metz, B., O.R. Davidson, P.R. Bosch, R. Dave, and L.A. Meyer (eds), *Contribution of Working Group III to the Fourth Assessment Report of the Intergovernmental Panel on Climate Change, 2007*. Cambridge University Press, Cambridge, United Kingdom and New York, NY, USA. Figure 5.3 at: http://www.ipcc.ch/publications_and_data/ar4/wg3/en/ch5s5-2-2.html

will also be crucial: for example, intelligent transportation systems can improve traffic flow, reducing congestion and pollution; and increasing reliance on telecommuting and teleconferencing can reduce the actual numbers of vehicles on the roads. Decreasing GHGs will also require changes in land use to increase residential density, provide more transit and non-motorized options, and reduce commuting distances and times. Achieving these changes will require that governments find the political will to implement some policies that are likely to be unpopular—for example, incorporating the cost of transport externalities in prices.

A crucial consideration in the redesign of transport and communication systems will be network operations. An efficient network requires connectivity between nodes, accessibility throughout the system, and complementarity of facilities. Cars and trucks must be able to move freely through the road system, just as information needs to flow freely around the Internet. Each individual transportation or communications network is also a part of an intermodal network in which goods, people, and information move from one medium to another. Integrating a new technology or practice into this intermodal system will be a matter of competition and co-operation among governments, businesses, industry associations, NGOs, and consumers. The mass introduction of electric vehicles, for example, would require not only a network infrastructure of standardized recharging facilities or battery exchanges throughout the transportation system, but also a concerted effort by the handful of companies that make the majority of the world's cars. The world's vehicle manufacturers are well on their way to developing low-GHG vehicles, but rapid deployment probably requires a formal international agreement in which individual countries agree to impose carbon taxes or other measure to eliminate free riding on the environment.

Conclusion

Transportation and communication networks are the biggest infrastructure investments made by society, and their operations account for a large proportion of any economy's output. Yet the cost of these systems is outweighed by the benefits of reducing the friction of distance so as to facilitate economic activity in general.

The principles of complementarity, transferability, and intervening opportunities stipulate the basic conditions for economic activity to occur over space. Based on these principles, the connectivity and efficiency of any network can be analyzed, from a simple two-node relationship to the most complex hierarchy of interactions. Such analysis is useful for government planners seeking to provide more equitable, less costly access for its citizens, businesses, and the government's own activities. Corporate providers of transportation and communication services also use such analysis to develop their own proprietary networks or to devise the most efficient means to combine public and private modes. Mapping applications use similar software to get people to desired locations.

Because economic interaction depends on efficient transportation and communications, governments tend to play a more active regulatory role in these industries than they do in others, working to ensure equitable access and minimize negative externalities. The extent to which government should influence networks that are for the most part privately operated and used remains contentious. The interplay between public and private governance occurs both between cities and within them. Between cities, the emphasis is on efficient connection; within them, on providing equitable access and environmental sustainability.

Practice Questions

1. Make a topological map of the road network and an origin–destination graph of 10 cities, towns, or villages in your region. Calculate the cyclomatic number and beta index.

2. How many students in your class use public transit to travel to and from school? How many car pool, ride bikes, walk, or drive a car? Does this pattern vary by location of origin? Are you aware of changing trends?

3. Should (virtually) all citizens be served by some form of public transportation regardless of where they live?

4. Analyze your email and text messages in terms of geographic distance (local, regional, national, international). Is distance-decay a factor or has the tyranny of distance been removed?

5. Should urban bus services be privately or publicly owned?

6. How has your community sought to reduce the environmental impacts of auto and truck transportation?

Key Terms

network externalities or effects   404
complementarity   406
transferability   406
nodes   406
distance decay   406
intervening opportunity   407

connectivity   409
topological map   411
circuity   411
cyclomatic number   412
beta index   412
modes of transportation   414

reverse logistics   414
tapering effect   415
c.i.f. (cost, insurance, and freight)   415
f.o.b. (free on board)   415

Recommended Resources

Behan, K., Maoh, H., and Kanaroglou, P. 2008. "Smart growth strategies, transportation and urban sprawl: Simulated futures for Hamilton, Ontario." *The Canadian Geographer* 52: 291–308. Discusses the impacts of road networks on accessibility within cities and outlines future scenarios.

Black, W.R. 2003. *Transportation: A Geographical Analysis*. New York: Guilford Press.
An excellent, accessible introduction to the field of transportation geography.

Bunting, T., and Filion, P. 2006. *Canadian Cities in Transition: Local Through Global Perspectives*. Don Mills, Ont.: Oxford University Press.

A reader on Canadian cities with chapters discussing intercity and intracity network issues.

Hanson, S., and Giuliano, G. 2004. *The Geography of Urban Transportation*. New York: Guilford Press.
The leading reader on the geography of urban transportation, emphasizing accessibility and sustainability.

Hensher, D.A. 2004. *Handbook of Transport Geography and Spatial Systems*. Kidlington, Oxford, UK: Elsevier.
Concise answers to all the questions you may have.

Kellerman, A. 2002. *The Internet on Earth: A Geography of Information*. Hoboken, NJ: J. Wiley.

A pioneering discussion of how the Internet changes geography.

Knowles, R.D. 2004. "Impacts of privatizing Britain's rail passenger services—franchising, refranchising, and ten year transport plan targets." *Environment and Planning* A 36: 2065–87.

Examines the varying geographical and other impacts of a major privatization scheme.

Kobayashi, K., Lakshmanan, T.R., and Anderson, W.P. 2006. *Structural Change in Transportation and Communications in the Knowledge Society*. Cheltenham: Edward Elgar.

Several (fairly technical) articles dealing with networks in a variety of modes and situations.

McCalla, R.J., Slack, B.J., and Comtois, C. 2004. "Dealing with globalization at the regional and local level: The case of contemporary containerization." *The Canadian Geographer* 48: 473–87.

Discusses the challenges facing the companies that move containers of goods around the world, noting that co-operation has gone further in shipping across oceans than over land.

Rodrigue, J.P., Comtois, C., and Slack, B. 2013. *The Geography of Transportation Systems*. London: Routledge.

An excellent introduction to the field; especially strong on logistics, intermodality, and environmental issues.

References

Innis, H.A. 1946. *Political Economy in the Modern State*. Toronto: Ryerson.

CHAPTER 14

Consumption

The unidirectional flow of instruction from consumer to market to producer may be denoted the Accepted Sequence. We have seen that this sequence does not hold. And we have now isolated a formidable apparatus . . . causing its reversal. The mature corporation . . . has means for managing what the consumer buys at the prices which it controls.

Galbraith 1967: 211

The overall goal of this chapter is to integrate consumption into the economic geography of value chains (and cycles). We do this by showing how the institutions of the value chain shape consumption and are shaped by consumption. More specifically, this chapter's objectives are

- To examine consumption as a social activity and as both a cause and an effect of production;
- To explore alternative theories of consumer choice;
- To explain how consumption is shaped across space, including the influences of advertising and the Internet; and
- To explain the household/home as a place of consumption.

The chapter is in three main parts. The first part reviews various theories regarding the nature of the consumer and consumer choice, particularly in the context of households. The second examines how "**space pervaders**" and "**netvigators**" influence demand over space, and the third examines consumption in place—private and public—along with some of the critical social issues that surround consumption.

Geographic Differences in Consumption

The end point of every value chain is **consumption**, underscoring its position as the last chapter in this book. Yet, to the degree that consumer demand determines what is produced, consumption is also the beginning. Indeed, growing commitments to recycling and reuse means the beginnings and ends of value chains are increasingly linked as value cycles (Chapter 11). Moreover, consumption plays a pivotal role in the economic system. As economists since Keynes have emphasized, aggregate demand is closely related to aggregate investment, levels of production, and the nature of business cycles. These relationships are expressed as inflation when consumers spend too

space pervaders Advertisers who attempt to persuade consumers by diffusing their messages throughout space.

netvigators Consumers who use the Internet to search for goods and services.

consumption The purchase of goods and services for personal or household use.

much and push up prices, and as recessions when consumer spending fails to maintain production and employment levels (Chapter 6). Over time, changes in demand lead to changes in production. In addition to signalling what the market wants in the way of products, consumer feedback today often includes environmental and social expectations of company operations. But consumers as individuals are not "sovereign," and their choices are not purely independent governed only by personal taste. Rather, consumption is a socialized, experiential activity in which consumers' choices affect one another and are shaped by the corporations that produce goods and services and by government policies that seek to stimulate consumption in recessions or dampen it during inflationary booms. Even an apparently simple recreational activity, such as bass fishing, can suggest the complexity of the interaction between consumption and production (Case Study 14.1).

Geography is critical to understanding consumption because it explains the spatial relationship between producer and consumer. Production necessarily develops some range of economies of scale and scope in place (usually a series of places) that subsequently requires distribution over space to locations of consumption. The household and other venues of consumption, whether night clubs or fitness clubs, are much more

Case Study 14.1
BASS FISHING

© Jim West/Alamy

numerous than the locations of production. Partly because of their sunk costs and partly because of their desire for growth, producers are constantly trying to tell consumers to accept what they offer, as well as listening to what consumers tell them. For this reason, the geography of services (Chapter 12) cannot be complete without two additional geographies: one focusing on the means that producers use to shape consumption over space, and the other, on the way consumer demand is shaped in *particular places* of consumption. Fortunately for producers, consumers and their demands are concentrated in cities.

Christaller's central place model (Case Study 12.2), like all models that assume perfect competition, takes **consumer sovereignty** for granted. In this conventional view—the one that Galbraith called the "**Accepted Sequence**"—consumers are entirely autonomous agents who drive producers to respond to their needs or (as economists often refer to them) "wants." It was Veblen who first scathingly criticized the principle of consumer sovereignty—that in fact, consumer demand was not independent but interdependent and manipulated. Since his time, the shaping of consumer demand by society and production imperatives has been a central theme of institutional economics. Above all, perhaps, the Accepted Sequence is undermined by the pervasive influence that giant corporations exercise over consumers.

consumer sovereignty The principle that consumers make independent choices and are the autonomous drivers of production.

Accepted Sequence Galbraith's term for the neoclassical idea (consistent with consumer sovereignty) that influence flows in a single direction from consumers to markets to producers.

Bass fishing is both a popular sport and a big business. In the US alone, some 30 million bass enthusiasts (20 per cent of them women) generate approximately $20 billion dollars in economic activity every year. Fishing rods and tackle, fish- and depth-finders are minor expenses compared to the big-ticket items. The special "bass boat" in the photo above would have cost roughly $40,000, depending on horsepower. Then there are the costs of transportation and lodging for fishing trips and competitions. An amateur can spend more than a thousand dollars on each competition, a professional more than two thousand. What accounts for this behaviour?

Obviously, bass fishing provides some sense of purpose, satisfaction, and identity. But the actual fishing is only part of the experience. The paraphernalia are fun, and the sleek, powerful boat guarantees a rush even on the days when the fish aren't biting. In the off-season, there are books to read and videos to watch. Above all, though, bass fishing is a social activity, enjoyed with friends and family.

Of course, no one would be fishing if the fish weren't there. Bass (there are several species) are native to North America, but the sport has become so popular that they have been introduced in more than 60 countries—and with them all the environmental problems that come with the introduction of alien species. Bass fishing illustrates the complexity of both consumer behaviour and of its ramifications within a global economic space.

Demand: Decision-Making by Consumers

Until relatively recently, economic geography was dominated by a focus on production. But whether considered as an autonomous, independent force driving production or the result of manipulation by powerful producers, consumer behaviour requires investigation. Consumer choice theory is an appropriate place to start.

Theories of Consumer Choice

The "Accepted Sequence," in which consumers let markets know what they want and markets pass this information back to producers, is one of the features that distinguish traditional neoclassical models of consumer choice from the marketing and institutional models (Table 14.1). In Christaller's central place landscape, for example, the sovereign consumer possesses perfect information about competing products and possible substitutes and, after weighing all the relevant costs and benefits, makes the most rational choice. In spatial terms, rationality means choosing the good that costs the least, available in the nearest centre; in economic terms, it means buying any particular good in the largest quantity possible—until the cost of the good outweighs its utility or the satisfaction that the consumer can expect from consuming it. The consumer's buying power is so great, according to this neoclassical view, that it compels companies to provide choice in quality and cost of products—in effect, giving consumers the upper hand over producers. That said, modern neoclassical thinkers have adapted the concept of consumer rationality in various ways to take into account factors such as the quality of the information available and the capacity of individuals to use it, or the influence of experience and social capital on decision-making. These adaptations are bringing the neoclassical and institutional perspectives closer together.

More than a century ago, Veblen proposed that purchase decisions were determined not by rational utilitarian logic, but by social conditions and expectations. In

TABLE 14.1 Consumer Behaviour Theories			
	Neoclassical	**Marketing**	**Institutional**
Theoretical premises	Consumers are represented by *Homo economicus*, who is perfectly rational and has all the information required to select the best options.	Households are made up of consumers with money to spend and different preferences. Spending patterns reflect the influence of factors such as demographics.	Consumers' behaviour is shaped by various psychological, social, environmental, and cultural parameters.
Implications for spatial activity	Consumers are attracted to least-distance (least-cost) locations.	Products and places of consumption depend on life-cycle stage, identity, and group situation. Market niches and segmentation create varied consuming patterns.	Consumers' choices of places and products are shaped by group inclusion and exclusion. Market niches and segmentation create varied consuming patterns.
Explanatory models in economic geography	Central place theory.	Geodemographics, mental maps.	Actor network theory, and other explanations of social relations.
Links to producers	Arm's-length relations; consumers are completely independent.	Producers respond to and influence the consumers.	Producers influence consumer behaviour through advertising, etc.

particular, he argued that the "**conspicuous consumption**" of luxury goods and services by members of the emerging "**leisure class**" was motivated not by any functional need, but merely by the desire to flaunt their wealth. In addition, he argued that the upper classes exercised a powerful influence on consumption through what he called "**status emulation**," in which consumers of lower social strata sought to increase their own status and self-worth by following the trends set by the strata above them. He also identified a number of patterns characteristic of different types of consumers: a willingness to pay higher prices for a commodity to enhance status ("Veblen effects"); a tendency to copy the behaviour of peers ("bandwagon effects"); a preference for goods that can only be appreciated by cultural elites ("snob effects"); and a preference for simpler, more austere lifestyles ("countersnobbery effects"). The idiom "keeping up with the Joneses" is a variant of bandwagon effects that stresses the desire of people to consume goods similar to those of neighbours or friends. Much advertising is based on shaping these desires.

In summary, consumers' choices and values are shaped by the collective views and habits of the culture that surrounds them. In the past, institutional explanations of consumer choice generally focused on these cultural influences, often described as filters that exist between production and the market, but the emphasis in recent years has shifted to the filters that exist between the consumer and the market. Different schools of thought emphasize different influences: identity, family, ethnicity, workplace, gender relations, class, dominant culture, or subcultures. The manifestations of these influences may be as trivial as the choice between chopsticks and cutlery or as weighty as the choice facing a parent when they have to decide whether to send their kids to a public, private, or religious school, or home-school them.

The institutional approach fully recognizes the power of producers to shape consumer tastes. The purpose of the massive expenditures on advertising by large corporations is to channel consumer choice in the direction of their products. Given the similarly large fixed costs of R&D, production, and marketing networks, effective advertising is imperative. The institutional approach still leaves room for individual decision-making, however; in fact, it is driven by the idea that individuals desire authenticity and use consumption as a way of developing their identity.

The neoclassical and institutional decision-making approaches are complementary with respect to making choices among competing products and substitutes. The marketing perspective borrows liberally from neoclassical and institutional perspectives in understanding consumer behaviour. Thus marketers know how useful it is to understand, for example, a specific ethnic market, a tourist's desire for authenticity, or a teenager's need to belong, and they appreciate research that reveals how important "coolness" and diversity are to urban development.

There is a countervailing force, however. Although all of us are consumers, consumption has negative connotations for many people who believe that their choices have social implications or meaning. For example, some people may refuse to buy goods produced by sweated labour or products associated with negative environmental externalities. In this regard, consumer choice involves questions of morality and ethics, and some institutional perspectives draw out these concerns.

Thus while consumer choice in the central place model is highly structured and orderly, the marketing and institutional perspectives take a more ambiguous view.

conspicuous consumption The ostentatious purchase of goods by the leisure class to underline their importance and power.

leisure class Veblen's term for the rich, powerful elite who control society.

status emulation The tendency to mimic the consumption behaviour of higher-status or higher-income groups.

They argue that consumer choice is shaped by the interplay of three forces. The first is the human desire for meaning and self-expression. We may not see every purchase as an opportunity to express (or shape) our identity, but not many people will buy even a pair of jeans without giving some thought to what their choice will say about them. Second are the social, economic, and political forces that shape our behaviour both informally—through the influence of family, ethnicity, class, and so on—and formally through institutions such as the law (e.g., the reality of the formal limitations on consumer choice is clear in the legal restrictions that surround private health care in Canada). Finally, consumption is shaped in a variety of ways by natural forces. Notwithstanding technological change, there are limits to the quantities of resources available for humans to consume. These limits become clear every time a local or global resource supply is exhausted, which unfortunately has happened to many of the world's fish stocks (Chapter 9). Even with resources that seem to be abundant, such as coal, the environmental impacts of consumption are cause for concern. If the ways in which humans consume resources (and dispose of the resulting waste) have not been seriously limited until now, they soon will be.

Space and place provide the contexts and connections through which these forces operate, presenting consumers with opportunities for conformity, adaptation, and rebellion. Space is the medium through which business, government, and NPOs distribute their wares and messages—goods, services, advertising, and information. How consumers perceive those wares depends on the lenses they look through, which are shaped by all the factors outlined above. At the same time, the character of particular places and spaces, as they themselves have been shaped by nature and social forces, also imposes its own influence on the consumer.

Consumers and Households: Who, What, and Where

Although neoclassical economics emphasizes the individual nature of consumer demand, in reality many purchases are made on behalf of other people—above all, other members of the consumer's **household**. The fact that governments collect data on household rather than individual spending points to the inherently social nature of consumption.

The nature of households in the western world has changed dramatically over the last century. A mere three generations ago, families were much larger: not only were there more children, but most families included grandparents or other relatives as well. The middle half of the twentieth century saw the decline of the extended family household and the rise of the nuclear family. As norms changed, however, divorce rates increased, women entered the workforce in higher numbers, and the nuclear family began to give way to single-parent families, blended families, and single households. Today, dramatic shifts are occurring in two of the dominant economic variables: population growth and number (and nature) of households. Where once it was assumed that populations would continue to grow, many countries (including Canada) now have below-replacement birth rates and depend on immigration for growth. Meanwhile, the numbers of households are increasing, partly because of increases in family breakups, people living longer, and people staying single longer.

household The occupant(s) of a particular home or dwelling unit. Sometimes households are interpreted as an individual consumer and sometimes the household members are disaggregated.

The family remains the dominant social unit, but the number of non-family–based households has increased rapidly. Consumption is also socialized outside the family in the various group settings where we spend much of our time: at work, in clubs, on package tours, and so on. Perhaps the best example of the way consumption is socialized is the intense, inescapable social and corporate pressure to buy gifts for weddings, birthdays, and holidays such as Christmas. A good deal of consumption also takes place outside the home, in stores, theatres, restaurants, cafés, bars, stadiums, amusement parks, and so on. In many, perhaps all, of these public contexts, consumption involves human interaction and has an experiential component that goes far beyond the simple acts of choosing something and paying for it. It may even help to create the personal and social identity of the consumer. Virtual places, as well, are emerging as important parts of the consumption landscape. Sites offering entertainment of some kind are leading the way, but information and advice are also available on virtually any topic. Of course, much of the "free" information on offer is provided via the same advertising-sponsored model pioneered by newspapers and television; the web has simply replaced one-minute commercials with pop-ups or banner ads (and collecting your personal information).

What consumers buy often depends on the context in which they consume. The power of social relations is reflected in the tendency for households of similar composition, with similar incomes, tastes, and consuming patterns, to cluster together in certain neighbourhoods. To reveal these similarities, a marketing science known as geodemographics uses information from sources such as the census and sales reports to segment populations by job type, income, family structure, consumption habits, interests, and activities. The Canadian geodemographic company Environics Analytics is a leader in this field as it seeks to provide data on consumers while respecting privacy (Case Study 14.2). Environics Analytics has used census data to identify more than 68 distinct types of household and classify every postal code across the country according to its dominant type. Profiles of this kind include everything from demographics, income, and education to lifestyle preferences and social values. Such information is invaluable for marketers, allowing them to pinpoint their targets—young singles, families with kids, dinks (double income, no kids), retirees—with considerable precision. Although geodemographics is used primarily by business for marketing purposes, governments, NPOs, and political groups use the same information to target their marketing efforts to the appropriate audiences: for instance, to help health officials get important information to people who need it or help an environmental NGO reach the specific people whose behaviour they want to change or whose support they want to enlist.

Consumption over Space

The relationship between producers and consumers is dynamic. Producers constrained by their need to achieve economies of scale and scope act from a spatial perspective. They need to persuade consumers who are widely dispersed to buy products that must be homogeneous or differentiated within a limited range. Advertising is the most obvious means of persuasion, penetrating all places of acquisition and consumption.

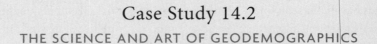

Case Study 14.2
THE SCIENCE AND ART OF GEODEMOGRAPHICS

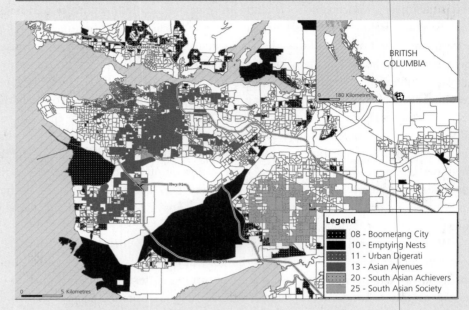

Ethnic Diversity in Vancouver

Source: Environics Analytics 2015, PRIZM5.

Certain maps show where specific groups of people are located within a city or country and can have amusing names to depict groups. In this map, for example, the "urban digerati" neighbourhoods are filled with recent high-rise apartments and condos with close access to fitness clubs, boutiques, coffee bars, micro-brew pubs, etc. Typically, residents have not started families, are socially active, and are attracted by the latest fashions. While interested in acquiring goods, they are sensitive to ecological concerns. As another example, South Asian Achievers have emerged as a fast-growing segment of family-filled households in new suburban subdivisions. These middle-aged, relatively recent immigrants—two-thirds are foreign-born—are characterized by mixed educations, upper-middle-incomes, and child-centred lifestyles. Tony Lea and Michael Weiss introduce the science and art of producing geodemographics.

The Science of the Geodemographics Industry—Tony Lea

In this conflicted age of high demand for data and deep concern about privacy, many marketers have turned to a popular technique that permits useful inferences about a person's educational level, ethnicity, household income, or monthly coffee expenditure. To determine such demographic and socio-economic variables, many take the average value for a small neighbourhood as a good approximation for a typical resident. Given the high cost of collecting these data about customers and prospects, this approach is very helpful. It is also the elevator version of the value of geodemography—the study of people at small geographic levels. In fact, a 40-year-old industry has developed in most Western countries that packages census and similar data to sell to analysts and marketers. If the data are unavailable at a small scale, they can be modelled to smaller levels of geography using increasingly sophisticated methods. The theory behind geodemography is based on the use of small-area averages. With geodemography, one shouldn't use the mean income for a large city, postal code, or FSA area in Canada to represent each household inside these large areas; the number would be meaningless. But data from small areas are required for this approach.

Geodemography produces a range of information that's sought after by many marketers. It can answer such questions as: Why do higher income households buy a certain wine more frequently than less affluent consumers? How come South Asian customers disproportionately buy Honda and Acura cars? And are people with university degrees the best prospects for new high-rise condos in certain kinds of neighbourhoods? Marketers can answer these questions using inferences from survey research of a sample of households and by asking questions about their demographics and marketplace preferences. This approach is relatively inexpensive when the goods are already included in a commercial syndicated survey. If the data have to be collected anew, then the process can be much more costly. Marketers can use the census and large official surveys—like the National Household Survey (NHS) in Canada—to create elaborate demographic profiles of people and households using the smallest geographic areas for which full demographic data are available: US census block groups and Canadian census dissemination areas (DAs). Of course, data on actual customers can be used to establish consumption and preferences. The data can be collected using residential addresses or postal codes tied to small census areas. The demographic and consumption data do not have to be collected for the same areas or at the same time to yield very valuable information. Geography allows these data sets to be joined, matched, or correlated. The use of small geographical areas, at which data are aggregated, serves as the "lingua franca" in this realm—an important though often misunderstood role.

The process of geodemographic analysis is the same as looking at two choropleth (colour-coded) maps of small areas that show almost the same pattern. If a map showing high-school educational achievement is similar to another depicting the frequency of attending casinos, it is usually reasonable to assume that there is a relationship. Then we can use high-school education as the basis for targeting additional neighbourhoods in marketing casinos. This connection is called an "ecological inference." Note that we are not examining people or households, which would make the correlation more legitimate, but using small areas. In fact, we are correlating average small-area demographics and average purchase behaviour. There are some situations where this process can lead to an incorrect inference—called an ecological fallacy—where one concludes that people and households will exhibit the same trends and relationships as small areas. (One project observed that small areas that were predominantly black were highly correlated with reading a sports magazine. A more detailed follow-up examination showed that the heaviest readers were white people living in black areas. But practice over the last several decades indicates that this process works well.)

In many countries, companies have developed user-friendly geodemographic segmentation or cluster systems that capture the key variables for use in predicting preferences and consumption. Each small census-based neighbourhood (or DA) in Canada has been statistically assigned to one cluster. In Canada, Environics Analytics' PRIZM (Potential Rating Index of Zip Markets) cluster system consists of 68 clusters or lifestyle types. These clusters help classify and define neighbourhoods

(Continued)

by their demographics, social values, and behavioural attributes, and have been given unique and descriptive nicknames like "Pets and PCs," "Cosmopolitan Elite," and "Les Chics." Marketers can then assign cluster segments to customer data that includes an address or geocode, and profile the records against any benchmark to create a cluster profile to identify the discriminating factors of the buyers. This process is helpful in its own right for many reasons. For example, one can find the best clusters for titanium tennis rackets and then target market the best prospect areas in the whole country or a particular region.

These days, Environics Analytics can access a growing number of large syndicated surveys that annually ask people or households a wide range of questions regarding product preferences, activities, and product consumption rates. For each product type or activity (using ketchup, driving pick-up trucks, carrying cell phones, booking travel online, or attending basketball games), responders can be assigned to these geodemographic clusters because of their collected addresses. In addition, each small neighbourhood has been assigned to a PRIZM cluster, which allows the data to be aggregated by cluster. We can then easily score each cluster by the average buy rate or percentage of a community saying "yes" to a survey question. So even if there is a small survey with only 5,000 or 10,000 respondents nationwide, we can still get some strong evidence about people or households living in clusters who buy the product or participate in the activity. Through the PRIZM segment, we can link product usage and lifestyle activities to variables in other surveys and subject areas. EA has now PRIZM-profiled almost 20,000 variables from such surveys. Each PRIZM profile is displayed in a report showing a list of clusters, in some sensible order, which shows the number of customers, their buy rates and their index relative to a benchmark—typically the percentage of households in this cluster. This report provides a powerful and cost-effective toolset for marketers of all kinds. Not only can they learn which clusters will buy special ski bindings, tofu, or ipods based on past behaviour, they can also use the information to design better marketing campaigns. With PRIZM permitting consumption of X to be related to so many other variables (the "lingua franca"), clients who have data on their customers' addresses can also know a good deal of reasonably inferred information about what else their customers prefer and buy, or dislike and avoid.

From Hockeyville to Stilettos & Sneakers: The Art of Naming Lifestyle Segments—Michael J. Weiss

Call it Lifestyle Nomenclature—the art (and occasional science) of naming consumer segments and target groups. Any business involved in classifying people into lifestyle types and target groups needs to come up with descriptive nicknames for them. Great segment names—monikers that are evocative, memorable, succinct, and clever—are one reason why marketing analytics has captured the imagination of marketers and sociologists alike.

Developing a winning segment name starts with the basics: the demographic, psychographic, or behavioural characteristics that is important to the client. If the client sells consumer electronics, for example, segment names could reflect household size and tech acceptance, because they are two major drivers of electronics purchases. Then for every target group, analysts compile a fat Excel workbook with survey results, market research, and census data. Throw in a dash of imagination and a few pop culture references, and you get names like You & I Tunes (singles and couples outfitted with the latest mobile devices), Plugged-In Families (large families with all manner of audio-video gadgets) and Dial-Up Duos (mature couples content with their slow Internet connections).

For a financial services client, lifestage and affluence are key factors, so segment names like Fiscal Rookies, Prosperous Parents, and Retiree Chic could be appropriate. Some years ago, I worked on a project for a credit card company that wanted nicknames reflecting only customer behaviour in the marketplace. Among the winners: Markdown Mavens, Mall-Landia and—for a particular breed of Manhattan fashionista who walked to work in Nikes carrying her Manolo Blahniks in her Coach bag—Stilettos & Sneakers.

Over the past twenty years, I've had the opportunity to help name the lifestyle types of more than a dozen segmentation systems and target groups for scores of client projects at Environics Analytics. But I don't develop these names alone. In addition to reviewing empirical data, I devour everything related to contemporary culture, from cable TV programs to supermarket tabloids. If you find all the magazines in the pocket of your airline seat ripped to shreds, it's a good bet I was there, collecting ads and articles for inspiration. McMansion Elite, Twilight Years and Yuptowns all came from what I call mile-high research.

My file of potential lifestyle names has grown over the years and currently numbers over 3,950. Many are simply variations on a theme (Kids & Careers, Kids & Commuters), and some have been roundly rejected, including a personal favourite, Movers & Shakespeares, which I imagine to be a segment of young liberal arts graduates who move frequently—or perhaps heavy-hitter CEOs with advanced degrees in English literature. One day, I figure, a company will call needing a name for a segment of club-store shoppers (Families in Bulk), café habitués (Berets & Baguettes) or obsessive video uploaders (YouTubers). And on that day, I will be ready.

We are grateful to Tony Lea, senior vice president and chief methodologist, Environics Analytics; and Michall J. Weiss, vice president of marketing, Environics Analytics. The firm's web-site is: http://www.environicsanalytics.ca/about-us/about-the-team

Producers also use places such as chain and department stores, giant outlets or factories or amusement parks to attract people to their products. In this sense, producers invert the spatial relationship, attracting consumers to them. In a twisted attempt to be cute, we label companies and other organizations that have to deal with distance in their attempts to change consumption patterns as "space pervaders."

Consumers have their greatest influence on what is produced by exercising their power of selection, but that power is limited by information asymmetries and transaction costs that are exacerbated by distance. Since advertising tends to be more persuasive than informational, consumers who wish to make an informed choice must search out the relevant information themselves, comparing similar products and, for more complex goods, perhaps even reading some literature on the subject. The Internet has reduced the costs of this kind of research dramatically. But there are inherent technical complexities in many products that make it difficult for consumers to determine their quality, safety, or environmental implications with any certainty. To counter that limitation, consumers can turn to a host of rating agencies, consumer protection laws, standards organizations, and NGO critiques. It is, of course, in the best interest of producers to develop feedback mechanisms that allow them to monitor consumers' choices, minimize their transaction costs, and modify products as necessary in a timely manner. In honour of the consumer's digital empowerment, we label them "netvigators."

Space Pervaders

The most obvious way of pervading space is through advertising. Corporate spending on advertising is huge. In 2013, the consumer products company Proctor and Gamble was the largest advertiser in the US at $4,830 million, with the next nine at $2–3,000 million and the total expenditure of the top 25 an impressive $51,047 million (Advertising Age 2014). Proctor and Gamble took the top spot internationally at $10,615 million, followed by its consumer goods competitor Unilever at $7,413 million, and the world's top 25 advertisers spending a total of $72,712 million. A key goal of advertising is to promote a company's specific **brands** and develop consumer loyalty to the point that the brand becomes the product. Advertising can be done through a variety of media: television, newspaper, magazines, billboards, radio, direct mail, and the Internet. For many decades, television has been the most important medium for selling a single product to a nationwide market as it achieves economies of scale by reducing the per capita costs of reaching consumers. To reach specific demographic or geographical segments, however, direct mail or Google ads may be more effective. In general, the Internet is the fastest-growing medium for advertising with a growth rate of 18.2 per cent in the US in 2013. The Internet now has a 21.7 per cent share of the market compared to television's 39 per cent (Adage 2014).

brands Labels that link a product to a particular corporation; branding is especially effective when consumers conflate brand and product.

Indeed, advertising has become increasingly pervasive. On conventional TV programming in North America, for example, almost one-third of viewing time (2014) is given over to adverts, more than twice as much as in the late 1960s. Thus for US adults who watch TV on average for about 4.5 hours (2010–2013), about 90 minutes are spent being bombarded by ads. During the Super Bowl, advertising time is longer than game-time and may be more anticipated than the game; indeed there is a lot of media and public attention drummed up as to what type of ads will be created for the event.

Producers have two basic ways of stimulating demand. One way is through informational advertising, which allows customers to make decisions based on facts about objective characteristics. The other is through transformational advertising, which uses images, styles, or projected meanings to influence the attitudes of consumers. Because the first option empowers consumers to evaluate a prospective purchase, it is often seen in a more positive light than the second, which is criticized for creating new wants and encouraging excessive, even harmful consumption. The extent to which an advertisement resonates with a consumer depends not only on the information or images it uses, but also on the consumer's sense of identity and social references. For that reason, perhaps the most important function of advertising is to address and shape consumer identity and its association with particular places and contexts.

Advertising that creates place or context usually accomplishes two goals simultaneously: (1) it establishes a shared public understanding that allows us to imagine placing ourselves in a situation similar to the one we see on the page or the screen, and (2) at the same time, in seeming to take us there, it sets us apart from others or with people we'd like to be with. It puts us in a group that is clearly differentiated from other groups. To create these places, contexts, and meanings, advertisers assemble symbols from the realms of nature and social relations. Nature can be decontextualized by advertising, used in a generic manner to enhance the value of whatever product is associated with it or it can use the cachet of particular places to enhance the product through association. Social relations are defined through style—of clothes, cars, manners—and positioned against supporting and contrasting backdrops. Indirectly, the language of symbols leads consumers to both a meaning they can grasp and a product they can purchase. Businesses are not the only institutions that use advertising to create meaning and overcome space: so do governments and NGOs (Figure 14.1).

The conquest of space is assisted by the fundamental way producers have reoriented and restructured products as brands. Branding is not new—artisans have been signing their products for centuries—but the advent of long-distance rail transport made it imperative. Brands evolved to meet the geographical challenge of communicating quality, distinctiveness, and identity when production became separated from consumption. Kellogg's and Proctor and Gamble were among the first North American brands, beginning with advertisements focused on specific products. Now a brand creates an association not only with a particular product, but with an image and the organization and reputation behind it. Brand equity is the value of the image and the loyalty that consumers feel for a product or company, and corporations work hard to develop it. Once it has established itself, a company will develop an umbrella brand covering a range of products, relying on brand equity and the loyalty associated with it to make the introduction of new products easier. Specific products from Apple, Proctor and Gamble, Yum Brands, and Toyota come and go, but consumers around the world remain loyal to the brands themselves.

The rise of corporate branding and its global impact is paralleled by the consolidation of advertising companies. Before 1980, small private advertising companies served the needs of regional or national markets. Now an oligopoly of global holding companies—WPP, Omnicom, Publicis, the Interpublic Group, and Dentsu—control dozens of individually named subsidiaries. Massive consolidation has left just a few firms, each of which integrates a broad range of operations: broadcasting (television and radio), cable programming, Internet, film, video sales and rentals, music, publishing, even theme parks. These firms marry the sale of content and advertising: that is, consuming the product requires consuming the advertising.

FIGURE 14.1   NGOs Build the Message and the Brand across the Globe

A different spatial strategy is to bring the production system as well as the product to consumers. This is what chain stores, restaurants, and service outlets such as Tim Hortons, The Keg, Sylvan Learning Centre, and Jiffylube do. Whether these operations are franchised or directly owned, most of them follow strictly standardized production processes, in order to ensure that customers will find exactly the same products at every outlet and to maintain the chain's reputation. The exceptions are chains that specialize in customized services (e.g., financial, tax, or travel advice). In every case, chain operations rely on large-scale advertising. McDonald's spends one-quarter of its revenues on advertising.

The products themselves are not the only standardized elements in chain operations. Their customers also become part of a standardized co-production. In a fast-food chain, for instance, consumers line up, make their choices from an overhead signboard; order their food and wait for it by the counter; carry their trays to a table; eat (often without cutlery); then clean up and put the garbage in the trash can. They are given cues to draw out these behaviours (signboards instead of menus, appropriately placed garbage cans and signs). These cues are reinforced in part by the absence of alternatives and in part by the store's layout and use of furnishings to channel them into the desired patterns of movement and behaviour, including the (short) length of time they stay. Customers in turn routinize these behaviours.

Disneyland

A more complete inversion of the spatial production–consumption process occurs when consumers are attracted to destinations that are themselves part of the production process. Places like theme parks and casinos use a strategy known as experiential marketing. Walt Disney World, for instance, is a complex of attractions—live entertainment, rides, events, restaurants, hotels, golf courses—linked by symbols and images reflecting technological optimism, nostalgia, myth, and patriotism. These symbols and images are part of the business of exporting Mickey and friends. They drive sales of movies, TV shows, and videos through various media channels, as well as toys and other merchandise available through direct outlets and other distributors. Exports of Disney products from sites of production both reinforce and are reinforced by imports of consumers to sites of production.

Disney has become a model for other types of experiential place-making. Experiential marketing is the creation of permanent, travelling, or large event attractions that draw consumers into an environment designed explicitly to immerse them in the corporation's brand values. The idea is to overcome the limitations of advertising by giving a company a venue to build the mythology of its brand. Examples include not only Niketown or Mattel's American Girl Places, but also factory tours at Ford's now environmentally friendly River Rouge plant and family finger-painting at Crayola factories. Many of these venues even charge consumers a fee for the opportunity to expose themselves to intensive advertising.

Interaction with customers does not have to be so elaborate, however. A priority of marketers is to survey customers and conduct focus groups to identify any problems with a product and invite suggestions for improvements or even new products. Companies also use direct sales outlets to give customers the opportunity to work with their products and provide valuable insights into actual use.

Netvigators

The ability to overcome space is not limited to producers. The advent of the Internet has dramatically changed the spatial relationships between producers and consumers. Advertising through print and broadcast media can reach consumers across vast spaces, but in order to achieve the necessary economies of scale, such advertising usually features only a small sample of a company's products. If the products advertised are not available nearby, potential consumers may face large costs for travel to higher-order service nodes. And consumers who wish to buy products that are not advertised are likely to face large transaction costs.

The Internet both reduces the transaction costs of the consumer search and makes it possible for consumers to order products from anywhere in the world. Initially, search engines were somewhat haphazard, but advances in technology at Yahoo and Google have made them much more efficient. As consumers are empowered to search product details and corporate sites for comparison, companies respond by increasing the sophistication of their sites and offering more informational advertising. The Internet allows consumers to compare products at very little cost in terms of either money or time, and without having to consult a salesperson. It also makes it possible to read product reviews and search for free advice. To a large extent the most successful e-commerce companies are those that have done the most to enhance the choices and information available to customers and respond to them rapidly, so-called customer-centric. For example, most large e-commerce companies promote reviews of their products by customers, and to retain credibility, regulate how the reviews are done. Corporate websites are important, and so are the various social media.

One of the Internet's greatest advantages is the technological means to meet consumers' demand for variety. Traditional bricks-and-mortar retailers can stock only a fraction of the thousands (in some cases, millions) of items offered by their online competition. In the past, the top 20 per cent of products accounted for roughly 80 per cent of sales (a phenomenon known as Pareto's 80/20 rule), but this is no longer the case: now a small quantity of niche products not found in conventional stores account for 30 to 40 per cent of sales (Figure 14.2). This "long tail

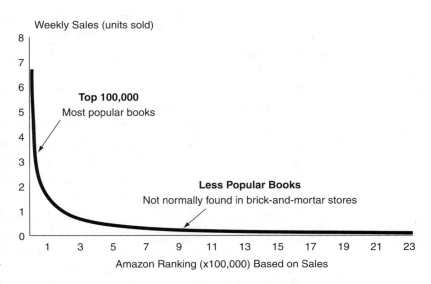

FIGURE 14.2   The Long Tail Phenomenon

While a typical large brick-and-mortar store may offer between 40,000 and 100,000 books, the online retailer offers 3 million.

Source: Based on Brynjolfsson, B., J. Hu, and D.M. Smith. 2006 "From niches to riches: Anatomy of the long tail," MIT Sloan Management Review 47 (4).

phenomenon" shows promise for niche products, and may stimulate the development of new products, strategies, and industrial structures. New industrial structures are especially evident in publishing and music, where digitalization enables authors, musicians, and software designers to publish and distribute their work online with or without the assistance of intermediaries. Unfortunately for these producers, the Internet provides copious means to find digital content and access it without payment.

More conventional firms have responded to their e-tailing competitors through collaboration and by developing their own online presence and sales systems. Increasingly, though, consumers are gaining power as they move from retail place to e-tail space. In addition to comparison shopping in person, consumers can choose to search on the web but buy in store; search in store but buy on the web; or search on the web and buy on the web. This trend is driven by consumers' increasing awareness of how much time and effort it costs them to physically shop around and how those costs compare with the costs of having products sent to them. E-commerce is growing at an exponential rate—approximately 25 per cent a year in many countries, including Canada. E-commerce is also blurring the boundaries of distribution and consumption as Internet enabled devices become both the tool for buying and consuming books, apps, games, music, movies etc. The role of venues such as stores and homes for buying and consuming is decreasing.

Consumer opportunities open up when an industry discovers an efficient new technology or approach to using the web. But the potential for consumers to pursue those opportunities is constrained by institutional factors. Canada lags behind the US in uptake of Internet shopping. So far, it has failed to produce any significant e-tailers, and most Canadians still tend to do their e-shopping south of the border. The reasons may include conservatism on the part of Canadian retailers, government bureaucracy, the fee structure of telecom companies, lack of competition in telecommunication, and failure of the private and public sectors to support the diffusion of high-bandwidth infrastructure.

Finally, in the context of consumer–producer interactions over space, the changing use of advertising should be noted. The Internet has given a second life to advertising's role in media sponsorship. Just as TV networks are supported by commercials and newspapers are supported by print ads, so the Internet is supported by pop-ups, banners, and Google keyword placements. The main differences between the old and new advertising media lie in the space they cover, the numbers of actively searching consumers they are able to reach, and the feedback they provide to producers. Website visits, "cookies," and online purchases all contribute to the flow of information that producers receive, and data collection agencies offer additional information about individual consumer and segment buying practices. Individual companies can track the preferences of individual consumers and offer suggestions either by email or when they visit a site. At the same time, the website itself eclipses the broadcast commercial or print-ad as a vehicle for persuasion, since consumers navigating through it are compelled to follow its structure and can be subjected to any number of messages projected in animation, film, print, or sound.

Consumption of Places and Space

Most of what we consider consumption has to do with goods, such as phones, cars, and tennis rackets, and services, such as haircuts, financial advice, and physiotherapy. Of course, we need places and spaces to consume these products. However, we also consume places and spaces individually, as households, and collectively. These locations are in fact some of our largest private and public investments, venues to dispose of income, and shapers of complementary consumption. Topping these locations is the home; purchasing and renting of homes remains the bellwether of economic activity. We discuss home choice first as it anchors much of our other consumption. Outside our homes we need another venue to socialize, which is paid for by taxes, admission ticket, or the price of a coffee, beer, or some other indirect funding. In particular we focus on the consumption that occurs in private spaces of shops and shopping malls, and in the public spaces of shopping streets and neighbourhoods. Recreational consumption can also occur in public spaces (e.g., parks, trails) or private spaces (e.g., Disneyland).

Private Places

The home, especially the house, is the dominant driver of the modern space-economy. For most people, a house, a condominium, or some other structure is the largest purchase they will make. Housing is actually one of the most variegated of products, for high and low income and asset levels, from standardized to completely unique pieces of architecture. Those who rent are also creating a demand for housing that is similarly produced for many levels of income and in a diverse array of choices. Moreover, housing generates a multiplicity of complementary demand: furniture, appliances, insurance, and taxes, not to mention maintenance and remodelling. It also generates vast expenses for infrastructure and maintenance. The housing industry, from project planning and land purchase, until the last nails are driven in, is so fundamental to economic life that house sales are a standard indicator of an economy's health. Funding for infrastructure—malls, hospitals, schools—depends on the demand for houses, whether the source of the funding is public or private. An outstanding feature of many developed countries, especially in North America, is the way a common preference for a detached house in a suburban location has shaped urban geography. This trend has continued throughout the twentieth century into the twenty-first despite the fracturing of families and the decreasing size of the average household. Indeed, houses have consistently become bigger and more important both as living places and as symbols of identity. They seem to confirm the neoclassical idea that there is no limit to human wants. In the 1950s, the average new single-detached home in the US was well under 1000 square feet, but in 2014 they broke 2600 square feet. Despite the increase in the average size of newly built houses, since the financial crisis of 2007, there has been a trend to smaller homes and movement back to city centres. This trend has several distinct causes: baby boomers downsizing after kids have left and entering retirement (and need to pay for retirement), the cost and inconvenience of commuting, the attractiveness of more condensed centres, and the decreasing ability of young people to buy their first place in the suburbs. Whether people from cultures accustomed to more modest dwellings are limited by appetite or other parameters is an interesting question.

The choice of home is influenced by many variables, including income, family size, age, local amenities, quality of services such as health care and schools, closeness to work, type of community, class, security, and access to golf courses or shopping centres. Parameters such as government housing policies, zoning, or ethnic relations may also play important roles. For most people, house choice is a localized activity, shaped by vacancies available within a certain price range; search costs and the location preferences or biases of the purchaser; and anchor points, such as workplaces or previous residences in the same area. On the other hand, many households migrate to other parts of a country or the globe specifically to find a particular type of home or setting. In other cases, the decision to relocate is sparked by the need to find work, a job transfer, or a chance to sell high in one place and buy lower in another.

Two of the most important influences on the home purchase/rental decision are the location of schools and the ethnicity of the neighbourhood. A US study found that 90 per cent of people consider school boundaries important in their house-buying decision and that over 50 per cent of people would pay from 1 to 20 per cent more than their housing budget in order to find a home in a good school area (realtor.com). Moreover, to secure the right school area, people would give up other neighbourhood amenities such as pools, shopping, parks, and private amenities such as garages, backyards, and bedrooms. Busing or boarding children at distant schools is another way of making the school location choice. Unfortunately, many, if not most, people cannot afford to make such a choice. Their schools also suffer relatively when wealthier people congregate in particular school areas and take their resources with them or send their children and resources to other school areas.

Historically, ethnic and economic influences combine in home purchases, as captured in the names given to many areas in the geodemographic segmentation. It occurs both as a process of segregation and congregation. In the past when immigrants moved to North America and Europe, they were both relatively poorer and lacked knowledge of the regional markets and institutions, moreover legal and non-legal means were used to prevent them from moving into some urban areas. They often had to move to areas of lower rent, or rather with rents they could afford; size and congestion lowered the rents. Ironically, returns to slum landowners could be higher than seemingly more expensive areas because of the density of occupation. Despite these conditions, these areas were often the seedbeds for economic success because people of similar ethnic groups shared social capital and often extended financial capital. All major North American cities have ethnic enclaves with the greatest diversity and largest enclaves in the portal cities such as New York, Toronto, Los Angeles, San Francisco, and Vancouver. Some of the older enclaves, such as Toronto's Chinatown, Little Italy, or the Danforth (Greek) remain enclaves primarily in terms of food and other businesses, while the bulk of the original immigrants have moved up and out. The majority of second and third generation tend to diffuse within the urban landscape, with house choices determined more by income level and amenity preference than ethnic background; however, some secondary enclaves also evolve in the suburbs and exurbs. New enclaves are still established by more recent and relatively less well-off immigrants, such as Toronto's

Jamaicans and Bengalis. On the other hand, a more novel phenomenon is the wealthy immigrant house hunter. Spurred on by emigrant-country government programs that provide wealthy investors with a shortcut to residency and citizenship, many Anglo-American cities have been targeted by the emerging wealthy in emerging countries. These people are looking for cleaner environments, better education and health care, stable political–economic institutions and a safe place to invest their money. Toronto, Vancouver, and Sydney have attracted newly wealthy Chinese, while London has been a magnet for the super-rich from Russia, India, the Middle East, and elsewhere.

Location of course is only one aspect of a home's value and meaning. Although a housing purchase is often assumed to reflect the buyer's values and identity, in fact, this is often not the case. In North America, most new housing projects are driven by economies of scale at every stage from planning, financing, and infrastructure to design, materials purchasing, components manufacturing, and on-site assembly. The result is a collection of mass-produced houses with a limited number of minor variations in size and design. It is important to note that such developments are generally undertaken on a speculative basis, in the expectation that the investment will pay off as market values appreciate over time. As the housing bubble of the early twenty-first century showed, however, finding enough financially qualified consumers to absorb the supply may be difficult.

In North America, personalization is a gradual process that doesn't begin until after the house has been bought. Thus a subdivision that begins as cookie cutter houses, twenty years later has taken on much more diversity. In Japan, on the other hand, differentiation begins at the outset based on interaction between consumer and builder in the design of a house. Most houses are built by local contractors using a post-and-beam construction method that allows them to tailor a house to suit the purchaser. National construction firms have made significant inroads into this market, mainly by developing flexible production systems that enabled them to develop features that customers demand. On the other hand, local contractors adopt innovations that the national builders have been able to develop because of their R&D, scale, and interaction with a greater number of customers.

In addition to offering opportunities for self-expression through interior design, furnishings, and decoration, houses also increasingly provide meaning to their occupants as venues for entertainment, whether passive (home theatres) or active (fully equipped kitchens for adventurous cooks). One of the reasons for houses growing so large is that many activities in the home have become atomized, that is, family members want their own space to do their own thing. In these ways and more, the household drives the purchase of many products. Moreover, many of the individual consumer's tastes and ideas about what is appropriate for consumption are formed within the household setting. The multifaceted social reproduction that takes place in the home has a direct impact on consumption, whether the young person who is formed in the home accepts or rejects the patterns established there. In any event, consumption patterns change as the dependent child becomes a semi-independent teenager, then a single person, cohabiting partner, parent, single parent, and so on. This dynamism of the household setting—how its demand changes over time—is something marketers are keen to understand and anticipate. A young single may be happy with a small apartment in a trendy area of town. Double income no kid couples (dinks) may prefer a gentrified townhouse near upscale shops. A family with kids is likely to want a house with a yard, while a retired couple may choose a cottage with no yard that is located next to a golf course or ski area. In every case, the baggage of the past, the conditions of the present, and expectations for the future shape consumption.

Finally, house location sets parameters for the household's spatial activity patterns with a strong influence on spending. Consumers' awareness of specific services—retail, recreational, health, and so on—depends in part on the location of those services in relation to the home. More generally, individuals' mental maps—their tacit knowledge of where activities and landscape features are located and how they are connected—is strongly localized around the places where they live and, to a lesser degree, work. The household is thus the key mediator between the spaces of production and advertising and the social places where goods and services are offered.

Private and Public Social Spaces

Most products and services consumed in the household are purchased through commercial establishments, agencies, or contractors. A store is a private place that invites members of the public to enter in the hope that they will spend some money there. But a store is also a place within a larger place: a residential street, a shopping street, mall, a CBD, a village, a fast-food strip. In the case of a mall, the surrounding place may be private, with no purpose other than separating customers from their money. On the other hand, a shopping street or a CBD serves numerous functions other than commerce: transportation, socializing, skateboarding. Consumers may go to such places with the objective of making a purchase, but they may also use the place for any number of other reasons or purposes. Stores and malls, as enclosed places designed for profit-seeking, are able to shape consumer behaviour, restrict deviant activities, and keep people focused on spending money. Public spaces have less capacity to control behaviour (although they can be policed part of the time) and as a result they are much more open to combining commercial and social functions.

At the same time, a store (or the mall or street that contains it) is a place within a space. The size of that space depends in part on the individual store's power to attract customers and in part on the drawing power of the businesses around it. Most malls are designed to hold a fixed number of stores and serve a fixed-sized market. Many are planned in conjunction with residential developments and are therefore strongly predisposed to accommodate specific patterns of spatial activity. In a typical mall created to be accessed by car, entry and exit routes are deliberately designed to "capture" consumers. In North America, the infrastructure that supports these spatial activity patterns is provided in part by the government that builds the roads and in part by the private developers of the mall. In other countries, however, the majority of metropolitan commuters rely on transit. In Japan, for example, companies such as Tokyu in Tokyo and Hankyu in Osaka operate both commuter railway lines and the department stores located at their most significant terminals and neighbourhoods. In Hong Kong, the government funded the subway system (Mass Transit Railway) through the development of very dense mixed residential and commercial projects—that is, the sale and rental of apartments, malls, offices, and stores help to fund the subway network.

Strip malls and shopping streets fed by public transit also rely on public infrastructure. The infrastructure for an arterial access road with gas stations, fast-food joints, and office buildings will be different from the infrastructure for a residential street two or three blocks away. Access to these open commercial areas is not restricted or controlled in the way that access to a shopping mall is. Such areas are also more open to change as the surrounding environment changes since the numbers, sizes, and types of businesses are not fixed and there is usually no need to seek official approval from a mall owner or planning department for changes to a particular business. Informal commercial areas are increasingly recognized as a city's main source of diversity and vitality. In the 1960s, Jane Jacobs decried the homogenization imposed by rationalized urban planning. She argued that vibrant communities required dense, mixed, and organically generated urban areas. Her views have gained wide acceptance. The influence of diversity and the creation of subcultures are increasingly linked not only with the prosperity of cities and regions, but also with advances in high technology because creative people of all sorts tend to congregate together.

A strong indicator of a city's potential for the development of creative clusters is the presence of a "bohemian" culture. But a bohemia is rarely if ever created in a mall: they form organically in localized areas that are mixing bowls of ethnicities, incomes, ages, rents, services, buildings. As a consequence, there is little that planners can do to create a bohemia.

At the opposite end of the spectrum from bohemian flea markets and charity shops are the types of stores found in malls. Whatever their specialties, the great majority of these stores are chain operations. Chains control the layout of the store, the inventory, and how the store is managed. Increasingly chains are global operations that use real-time information systems to ensure that consumer demands are quickly converted into responses at the production end of the supply chain. One reason for Walmart's dominance is its reliance on brand-name producers to do their own advertising and OEM suppliers to create products that it can sell as its own or in-store brands. Brand products need to sell on their own merits, although chain stores do collaborate with producers on promotions. Stores and malls occasionally use promotional gimmicks to attract shoppers. They may also rent space to various entertainment, food, or other services, to take advantage of the public's interest in them.

Above all, however, there are two features of both conventional retail stores and malls that highlight the use of place to shape consumer behaviour: layout and imagery. In general, the layout of a store or mall is designed to channel customers past as much merchandise as possible, in the hope of spurring additional purchases. Entry and exit points are in separate locations, and goods in high demand are placed at the back of the store. Malls are designed for equal access from several entrances and encourage flow-through by situating a department store as an anchor at one end and another anchor such as a grocery store at the other. Both stores and malls take care to place complementary and competing products or stores close to each other. Both also emphasize security and seek to control the pace of traffic (Figure 14.3), but the increasing use of surveillance technologies and behaviour analysis allows for more effective layout and imagery.

The image that a store or mall seeks to present depends on the population segment it targets. Architecture, lighting, air conditioning, colours, artwork, and sound are only some

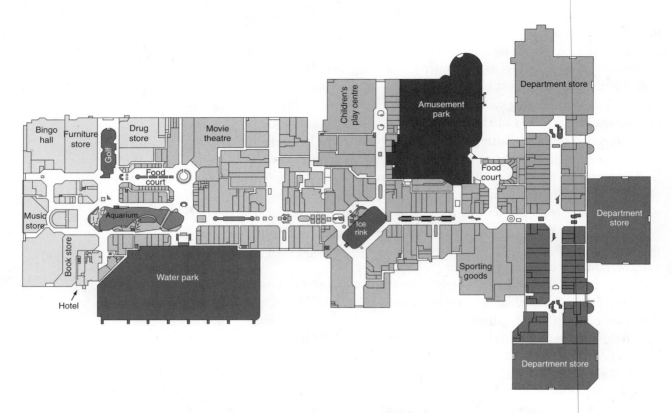

FIGURE 14.3   **The West Edmonton Mall: Shopping, Entertainment, Fantasy**

of the elements that can be used to create an atmosphere conducive to consumption. For a store, that atmosphere may be anything from the no-frills look of a discount store to the high style of a shop specializing in designer fashions. The range is equally broad for malls. One of the largest and most diverse, the West Edmonton Mall presents itself as a fantasy land, combining the explicit entertainment of rides, swimming pools, and restaurants with the implicit entertainment that many people find in shopping. Geographical analysis of malls often focuses on the way they use imagery to foster consumerism or exclude undesirable groups such as street people. On the other hand, some studies have found that social groups can use such places of consumption as sources for the development of identity or venues for social interaction and performance.

Understanding the design and operations of malls in particular is important because they have become models for what were once less formally planned public spaces. The significance of mall design is magnified when we consider that these places of consumption are also places of social reproduction—of segregation or mixing, for instance, prepackaged fantasy or imaginative inspiration.

Separation of Production and Consumption: Society and Environment

Most research on consumer behaviour is done from a marketing perspective: that is, it focuses on finding out how people can be persuaded to consume more. But there are other perspectives. Some researchers explore how individual consumers' decisions are influenced by psychological, cultural, and class factors, among others. Some important research has found that—contrary to neoclassical theory—demand for certain goods will actually increase as the price rises because rarity accords additional social status. Other research has suggested that within a capitalist society consumption can become a substitute for community life. Moreover, rather than enjoying the benefits of rising production by enjoying greater leisure, people have become trapped in a work-and-spend cycle, where the benefits of higher production accrue in greater income rather than greater free time. Economic geography adds to these critical perspectives by focusing on the spatial separation of production and consumption and emphasizing the importance of looking at consumption not as the end point of a value chain but as an inflection point in a value cycle. Understanding this spatial relationship is particularly important to understand the consumer's responsibility for environmental impacts throughout the value cycle.

Upstream in the Value Cycle

Consumers usually see only a final product and have little, if any, idea of the social or environmental costs of its production. One of the primary functions of NGOs (Chapter 7) is to reveal the impacts of corporate practices upstream in the value chain to consumers downstream. Organizations seeking to raise awareness of the environmental impact of consumption use a variety of tools and media to get their message across. For example, the "ecological footprint" is used to estimate the total demand—for food, shelter, energy— placed on the environment by an individual, company, or nation. More pointedly, labels such as Energy Star, Fair Trade, Forest Stewardship Council, Marine Stewardship Council, and Bird Friendly Coffee tell consumers what they can and can't consume with peace of mind. Changing consumer choices sends a message to producers, and consumers themselves may communicate directly with producers to tell them to change their activities. Governments also impose governance upstream in the value chain, through consumer protection agencies and regulatory activities (e.g., drug, environmental protection).

Informing consumers about the upstream impacts of their purchases is a way of addressing an important information asymmetry (a basic market failure). Changing practices that are socially or environmentally unacceptable imposes new production and transaction costs (e.g., for cleaner production methods, alternative materials, traceability, quality controls, higher wages, better working conditions). Perhaps the greatest challenge, however, is to persuade consumers to pay higher prices in the hope that producers will internalize the externalities produced in earlier segments of the value chain. Although consumers may drive many of these changes, implementing them requires complex interaction on the part of firms, labour forces, and local governance institutions. Understanding how demands for environmental sustainability are constructed in place and over space is increasingly critical to understanding the value cycle.

Downstream in the Value Cycle

Consumers also play a critical role in determining downstream environmental impacts through the products they choose to buy, how they use them, and the degree to which they participate in reuse or recycling. In other words, consumers are critical to closing the loop on material and energy flows.

At the consumption stage, consumers are increasingly aware of the need to choose products that are as environmentally friendly as possible and to make efficient use of them. Whether the product is a car, a computer, or a household cleanser, it is likely to have its greatest environmental impact at the consumption stage. Perhaps the most obvious role for consumers, however, is the one they can play in reverse logistics (Chapter 13). Recovery of products for reuse or recycling depends on the participation of consumers, who are increasingly requested, offered incentives, or compelled by government regulations to recycle waste materials. Recycling requires some education, and not only in the types of material or grades of plastic that the local recycling program will accept: consumer participation often requires instruction in the reasons behind the inclusion of recycling fees in the prices of some products or the exclusion of certain materials from landfill or incinerators, as well as the general implications of their purchases on the product cycle.

Consumption: Too Much and Too Little?

But there is deepening concern, especially in the already rich market economies but by no means limited to them, that consumption is becoming excessive and threatening environmental and social values. For environmentalists, excessive consumption is threatening the planet because production and consumption is exceeding sustainability thresholds. For other critics, excessive consumption or hyper-materialism is linked to the breakdown of spiritual values, more ethical development goals, and community-spirited behaviour. These criticisms are often further associated with the rise of neoliberalism since the 1980s and policies that seek to replace public goods with markets (Chapters 6, 12). At the same time, low-income levels don't allow for adequate consumption, that is, people cannot meet basic human needs, allow for a modicum of equality of opportunity for people or otherwise fail to meet the general expectations of the society. Most importantly, inadequate consumption is a problem for lesser-developed countries, but it is also a problem within rich countries.

The policy challenges of market economies, as expressed by over- and inadequate consumption, are not straightforward. At an urban scale, restructuring and gentrification has transformed many former industrial cities (places of production) into places of consumption that are green and anti-productivist, but not anti-materialist. Thus many rich cities around the world, well-illustrated by Vancouver, Canada, desire to be "green" and have variously sought to combat pollution and congestion, reduce GHGs, maintain green areas, develop bicycle lanes, and seek clean producer service, design, and R&D activities while removing or at least constraining traditional "heavy" industry. Yet if these policies are applauded by residents, restraints on consumption are harder to discern. Erstwhile, green residents still typically wish to ski, kayak, canoe, sail, hike, mountain climb, and run, all of which require specialized clothes and equipment; invest large sums in their main homes with their multiple carports; build second and third homes for occasional use; and travel extensively. On the other hand, such activities and propensities stimulate the economy and provide jobs for an ever-increasing global population. Since the onset of market economies, circa 1800, the global population has increased from about one billion to seven billion (2014).

For an increasing number of people, increasingly living in cities, increased consumption has been a goal and reality.

The market system and its evolving division of labour, driven by and reinforcing technological and institutional innovations, has permitted this global, albeit unequal, intensification of consumption with its attendant threats to environmental and social values. Innovation-inspired divisions of labour also provide a basis for optimism for addressing the challenges of over-consumption, inadequate consumption, environmental sustainability, and social integrity. In this regard, we can interpret nations, regions, and cities (and other localities) as (complex) institutional arrangements and experiments that in varying ways are seeking to address these challenges. Economic geography has an important role in understanding these variations.

Conclusion

The role of consumers in value chains and cycles is ambiguous, but demonstrably multi-faceted and of influence in all places of the value cycle. Consumer demands influence production, making some products wildly successful and causing others to flop. But producers also influence consumers, most obviously through advertising and branding. In terms of economic development, increasing consumption is widely seen as both necessary, in that it creates jobs, and desirable, in that it is a sign of progress towards a higher standard of living for all. In rich countries, consumer spending continues to be used as an important indicator of economic health, and during recessions consumers are typically encouraged to spend more to stimulate production. In developing countries, it is hoped that the low wages that typically provide an initial competitive advantage will eventually be raised in order to increase consumption and by direct implication the standard of living.

Practice Questions

The following questions ask you to explore your own behaviour as a consumer.

1. With increasing income, people often purchase second (and more homes) for recreational, renting (income generating), and risk-sharing purposes. Should such purchases that add to unused housing stock be more strongly regulated—that is, restricted?
2. With respect to one of your outdoor recreational activities, what goods do you need and purchase? What factors influence your choice of brands?
3. Assess the importance of status emulation in the purchase of goods among your peer group.
4. Choose one area of habitual activity—e.g., the food you normally eat at dinner, your journey to school, or your weekend recreational activities—and construct an ecological footprint for it.
5. Identify options for ethical purchasing in your community? How important is ethical purchasing to you?

Key Terms

space pervaders   429
netvigators   429
consumption   429
consumer sovereignty   431

Accepted Sequence   431
conspicuous consumption   433
leisure class   433
status emulation   433

household   434
brands   440

Recommended Resources

Advertising Age. 2010. "Database of 100 leading national advertisers." http://adage.com/marketertrees/2010. (Accessed Oct. 2010).

A good source of data on advertising expenditures by corporations in the US.

Carlin, G. "George Carlin talks about 'stuff'." http://www.youtube.com/watch?v=MvgN5g CuLac. (Accessed 5 Nov. 2009).

A brilliant skit offering a highly original, insightful take on the consumption of goods ("stuff") for household use.

Crewe, L. 2000. "Geographies of retailing and consumption." *Progress in Human Geography* 24: 275–90.

A review paper that emphasizes the power of retailing and consumption in the changing economic geographies of market economies.

Goss, J. 1993. "The magic of the mall: An analysis of form, function, and meaning in the contemporary retail built environment." *Annals of the Association of American Geographers* 83: 18–47.

Discusses malls as meeting places, experiential places, and controlled spaces, as well as markets seeking profits.

Jacobs, J. 1961. *The Death and Life of American Cities*. New York: Random House.

A critique of urban planning, especially with respect to land-use zoning and urban renewal schemes.

Jones, K., and Simons, J. 1993. *Location, Location, Location: Analyzing the Retail Environment*. Scarborough, ON: Nelson Canada.

Still the pre-eminent guide to understanding retail location in Canada.

Mansvelt, J. 2005. *Geographies of Consumption*. London: Sage.

Compares and synthesizes theories of consumption and space.

Zukin, S. 2010. *Naked City: Death and Life of Authentic Urban Places*. Oxford: Oxford University Press.

Latest book that reveals the power of consumption, among other considerations, in shaping cities.

References

Brynjolfsson, E., Hu, Y., and Smith, M. 2006. "From niches to riches: Anatomy of the long tail." *MIT Sloan Management Review* 47: 4.

Galbraith, J. 1967. *The New Industrial State*. Boston: Houghton Mifflin.

Sack, R. 1992. *Place, Modernity, and the Consumer's World: A Relational Framework for Geographical Analysis*. Baltimore: Johns Hopkins University Press.

Glossary

Accepted Sequence Galbraith's term for the neoclassical idea (consistent with consumer sovereignty) that influence flows in a single direction from consumers to markets to producers.

administered market Exchange of goods or services that is internal to a particular (usually large) organization; also referred to as hierarchical, integrated, or internal.

advocacy organizations Organizations that seek to change the behaviour and policies of governments, corporations, etc.

agglomeration The spatial concentration of related manufacturing and service activities; also increasingly used to refer to an industrial district. Also known as **clustering**.

anti-monopoly laws Laws that prevent excessive corporate concentration in an industry and/or unfair business practices.

areal differentiation An approach to human geography that refers to regions as unique places shaped by distinct evolutionary interactions among economic, political, cultural, and environmental forces.

backward linkages Investments in activities that supply inputs to the resource (export) sector.

basic activities Economic activities that generate income from outside the defined territory.

behavioural theory of the firm A theory that sees firms as using their limited information and bounded rationality to reach decisions that are acceptable rather than optimal. See also **satisficers**.

beta index A measure of the connectivity of a network.

bid-rent theory Explains how the allocation of land to various uses is determined by competitive bidding.

boom–bust cycles Alternating periods of rapid growth and recession.

brain drain The loss of highly educated and qualified people from one region or country to another that offers better opportunities.

branch plants A plant or factory that is owned and controlled by a company whose head office is elsewhere.

brands Labels that link a product to a particular corporation; branding is especially effective when consumers conflate brand and product.

budget An estimate of revenues and expenditures over a defined period of time.

business (capitalist) firm A legally recognized private-sector organization that employs at least two people to provide a good or service for sale at a profit.

business segmentation The grouping of businesses by type in terms of routines, structures, and strategies.

buyer-driven value chains Firm-supplier relations that are organized by service companies such as Walmart.

carbon footprint A measure of the environmental impact of a given activity in terms of the amount of carbon it causes to be released into the atmosphere.

central business district (CBD) The downtown and usually the most accessible area within a city, and defined by commercial and retail activities.

central place theory A theory of the location of suppliers in response to the distribution of consumers whereby minimizing distances to markets is the predominant imperative.

centrism An approach to planning that seeks to maximize consumer access to service centres.

ceteris paribus Latin for "all other things being equal"; a standard phrase in abstract neoclassical theorizing, used to rule out extraneous factors that could affect the outcome of a prediction.

c.i.f. (cost, insurance, and freight) A pricing system in which the price charged to buyers incorporates the costs of the good, insurance, and freight charges up to the destination port.

circuity A measure of the degree to which the paths that make up a network deviate from the straight path.

circular and cumulative causation Refers to social processes that are interdependent and self-reinforcing in particular specialized directions for lengthy periods of time. Originally developed by Myrdal in the 1940s and 1950s to explain engrained patterns of poverty and racism in the US, and then with reference to rich and poor nations and regions on a global stage.

clusters Another term for "agglomeration": the spatial concentration of related manufacturing and service activities; also increasingly used to refer to an industrial district.

codified knowledge Knowledge that can be communicated precisely through language or formulas (e.g., in textbooks or patents).

collective bargaining The process where contracts are negotiated with employers by a labour union on behalf of a group of workers.

common property (or common pool) resources Resources with property rights assigned to a specified group.

community economic development (CED) An approach to the stimulation of local economic development based mainly on citizen participation.

community plan The document outlining a local authority's official development plan, including land use and zoning. Different communities call their plans by different names: "Official Plan" (Toronto); "Master Plan" (Montreal); "Municipal Development Plan" (Calgary).

community supported agriculture A system in which consumers (usually local) contract with farmers to buy their produce.

comparative advantage model Proposes that a region or nation (or firm or individual) should specialize in the goods and services that can be produced more efficiently than others, specifically, with lower opportunity costs.

competitive advantage Proposes that businesses and governments should actively seek out and shape opportunities for specialization and trade.

complementarity The condition of spatial interaction whereby the demand for a product in one place is matched by supply of a product in another, thus stimulating exchange.

connectivity The degree to which places in a region are linked or accessible to one another; may be measured in terms of cost, time, and/or capacity.

conservation area   An area in which natural processes are allowed to operate with minimal human interference.

conspicuous consumption   The ostentatious purchasing of goods by the leisure class to underline their importance and power.

consumer services   Services that supply individual (household) consumers.

consumer sovereignty   The principle that states consumers make independent choices and are the autonomous driver of production.

consumption   The purchase of goods and services for personal or household use.

conurbation   A large, functionally connected region formed through the coalescence of two or more neighbouring towns; if the towns are industrial, so is the conurbation.

co-operative   An organization or business jointly owned and operated by its members.

coordination costs   The expenses of administering market places and the search costs of market participants.

corporate strategy   Allocation of investment to meet long-term corporate goals (growth, profit, market share, downsizing, restructuring).

corporatism   A governance model in which three institutions—government, business, and labour—work together to develop a strategic plan for the economy.

cost-price squeeze   The situation in which producers face increasing average costs of production and declining prices for their products.

countervailing power   Refers to major economic and non-economic institutions that constrain the power of "big business."

creative cities   Cities that are seen as centres of innovative thinking; creativity in this context is thought to be the product of factors such as economic and social diversity, networking opportunities, social tolerance, and various kinds of location amenity.

critical isodapane   The isodapane that equals the costs of another location condition such as labour.

Crown corporation   The Canadian term for an organization owned by the government but operated on an arm's-length basis, more or less like a business.

cyclomatic number   A measure of the connectivity of a network.

decent work   The goal of the United Nations' current campaign for full and productive employment in socially acceptable working conditions.

deindustrialization   Substantial net loss of manufacturing jobs in a particular region over a certain time.

demand curve (schedule)   A graph showing the quantity of products that consumers are able and willing to buy at different prices.

dependent labour market   The blue-collar portion of the primary labour market, consisting of unionized employees in production-line and related work.

diseconomies of scale   Costs and disadvantages associated with agglomerations of economic activity.

distance decay   The tendency for the spatial movement of goods, people, and information to decline as distance increases.

division of labour   The allocation of specific jobs to different people, and by extension to equipment and places; also referred to as "specialization" of labour.

domestic champions   A large, locally (as opposed to foreign-) owned firm that helps to advance strategic industries in terms of innovation, employment, local supply networks, and exports. Some latecomer firms are domestic champions.

Dutch disease   The tendency of resource booms to undercut other activities (such as manufacturing) by causing the currency to increase in value.

economies of scale   A reduction in average costs of production achieved by increasing the scale of a particular activity.

economies of scope   A reduction in average costs of production achieved by using existing resources to perform tasks over a range of product categories.

embedded   A term emphasizing the interrelatedness of economic and non-economic factors or processes.

employment relationship   The agreement between employers and employees in a specific labour market regarding wages, benefits, and the organization of work.

endogenous forces   In the context of local development, initiatives driven by local, internal, or domestic actors. The birth of new locally based businesses and the local generation of innovations are illustrations.

Engel's Law   A theory proposing that the proportion of income that people spend on food declines as their income increases.

entry advantages   Advantages, based on accumulated expertise that allow firms to establish operations in new regions.

environmental conditions   The state of nature in terms of soil, landforms, vegetation, waterways, and climate in particular places.

exogenous forces   Forces that drive local development from non-local, external, or global forces, no matter how the local is defined. Export demands, foreign competition, and foreign-based MNCs are illustrations.

extensive agriculture   Farming characterized by relatively low inputs (labour, capital, fertilizer, and seed per unit of land) and relatively low incomes per hectare.

external (corporate) growth   Growth through acquisition of existing plants and facilities.

external economies (diseconomies) of scale   Reductions (increases) in average costs of production resulting from location in urban–industrial centres of economic activity rather than small places.

fair trade movement   The movement to ensure that small producers in the developing world receive a fair price for their products.

federal systems of government   Political power is constitutionally shared among national and regional levels of government.

final demand linkages   Investments in activities that provide goods and services for local consumption.

firm-level internal economies of scale   Reductions in average costs of production resulting from increasing levels of output.

fixed costs   Costs that remain the same regardless of scale of activity.

flexible model of labour markets   A workplace model where labour is organized according to flexibility principles that are variously defined in terms of multi-tasking, multi-skilling, team work, financial discretion, job intensification, and ease of hiring and firing.

f.o.b. (free on board)   A pricing system that means buyers pay the on-board costs of transportation from the point of shipping.

food security   Assurance of access to food of sufficient caloric and nutritional value in order to sustain a healthy life.

Fordism   The TEP that dominated the half-century from the 1920s to the 1970s, characterized by mass production based on internal economies of scale organized by giant firms typically employing unionized labour.

Fordist (or dual) labour market model   A model developed to explain labour markets that evolved in the US (and elsewhere) after the 1950s, based on the differences between primary and secondary labour markets.

foreign direct investment (FDI)   Investment by MNCs in foreign countries in operations that they own and control.

formal institutions   Structured constraints on human behaviour; for example, the laws at the societal level and corporate regulations at the organizational level.

forward linkages   Investments in activities that buy inputs from the (export) resource sector.

franchise   A business owned and operated by an individual or partnership but controlled according to the system of rules and conventions established by a core firm.

free trade   Market exchanges without political interference in the form of tariff and non-tariff barriers. The market exchanges refer to goods, services, and investment.

futures   Market contracts at an agreed-on price that involve delivery at some later date.

general purpose technologies (GPTs)   Major innovations (such as the steam engine) that have a transformative effect on economic life.

geographic multipliers   The ratio of total economic impacts to the initial economic impact of a new development within a region.

geographical indication   A kind of legal trademark given to growers in a specific region; in some cases a geographical indication comes with regulations on cultivation and processing methods.

globalization   A term with multiple meanings, in economic geography for example, often used as shorthand to refer to the current stage of capitalist development, the intensification of international networks of all kinds; qualitative changes in the relationships between markets and governments; and the increasing role of exogenous forces in national and regional development. But it also implies the expansion of environmental problems to a truly global scale.

goods-producing sector   Statistically, the primary, manufacturing, and construction sectors. In theory, activities that exploit resources and add value to produce tangible goods.

gravity model   A model used to summarize interactions between places, based on the assumption that bigger and nearer places generate more activity.

greenbelt   Land around cities that is zoned for agricultural, natural, or recreational uses.

Greenfield investment   Investment in new locations where entirely new facilities can be constructed.

green TEP   A TEP in which economic development would meet the challenge of environmental sustainability as well as economic goals related to jobs and income.

gross domestic product (GDP)   The total value of market-related activity produced within a nation over a period of time, usually a year; Gross National Product (GNP) consists of GDP plus whatever income was earned by citizens working abroad.

hierarchical subcontracting   Contracting out to a limited number of important suppliers of key components, who in turn engage other suppliers to provide the good or service in question.

home economies   The countries or regions where MNCs that control FDI in host economies have their head offices.

Homo economicus   "Economic man," the personification of rationality in neoclassical economics; the ideal economic actor, in possession of all the information necessary for decision-making, who seeks to maximize profits and minimize costs.

horizontal integration   Expansion in the same or related lines of business. Expansion in related lines of business is sometimes called horizontal diversification.

household   The occupant(s) of a particular home or dwelling unit. Sometimes households are interpreted as an individual consumer and sometimes household members are disaggregated.

human capital   The abilities and personal characteristics, formally or informally acquired, that make an individual economically productive.

immobile economies   External economies that are available only when the firms concerned are in close proximity to one another.

impure inputs   In Weberian location theory, inputs that change physically during processing.

incremental innovation   Organizational and technological innovation with relatively small, almost imperceptible impacts on productivity.

independent labour markets   The white-collar portion of the primary labour market, consisting of employees in management and R&D occupations; these workers are typically not unionized.

industrial districts   An agglomeration of related economic activities within a particular place; traditionally used in the context of manufacturing activities.

industrial unions   A labour union that represents workers in a particular industry.

informal institutions   Tacit, culturally rooted constraints on human behaviour.

information and communication techno-economic paradigm (ICT)   The TEP that has developed since the 1970s, characterized by flexibility in mass production, specialization, and labour, exploiting economies of scope as well as scale.

innovation diffusion   The spread of an innovation over time (temporal diffusion) and space (spatial diffusion).

innovation system   The interlocking network of institutions that creates and markets innovations, applying scientific and technological knowledge to enhance competitive advantage. The core of innovation systems is a "triple helix" of corporate, government, and university R&D.

institutional theory of the firm   Focuses on how and why firms grow with emphasis on the structures and strategies of large and giant firms run by technostructures with the power to influence the behaviour of other organizations.

institutional thickness   The range of external economies provided to local firms by institutions (private and public) such as industrial associations or government planning departments.

institutions   The rules, conventions, habits and routines that organize human behaviour, formally and informally.

intensive agriculture Farming characterized by relatively high inputs (labour, capital, fertilizer, and seed per unit of land) and relatively high incomes per hectare.

intermediate or appropriate technology Techniques and organizational practices that are based on local resources and markets and understood and controlled by local people. The term *intermediate* suggests the potential for innovation in traditional societies.

internal (corporate) growth Investment in new plants and facilities.

intervening opportunity The condition of spatial interaction whereby exchange between two places will not occur if there is an alternative source of supply or demand between them.

investor-owned companies (IOCs) Private-sector companies that are financed by individual and group-based shareholders representing individuals.

invisible hand Adam Smith's description of the way market self-regulation operates under conditions of (preferably perfect) competition.

invisible trade International trade in services.

isodopane Lines of equal transportation costs around the minimum transportation cost point.

isotims Lines of equal transportation costs around each location condition.

Jacobs externalities Argues that innovation and growth is driven by knowledge spill-overs that cross industries; and in particular that diversified cities are more innovative than specialized cities; named for Jane Jacobs.

just-in-time (JIT) A system in which parts and components are delivered to the factory just prior to their use.

key factor industries Industries that are inexpensive and abundant, with multiple applications throughout the economy, such as oil during Fordism.

Kondratieff cycles Long waves of economic growth and decline lasting about 50 years.

labour catchment areas The geographic area within where local labour markets operate, centred on a spatial concentration of workplaces within the same commuting zone.

labour market A (metaphorical) market in which the suppliers (workers) offer to provide their labour to the demanders (employers) in exchange for payment of some kind—usually wages, salaries, or fees, and in some cases non-wage benefits as well. The price of labour varies with supply and demand.

labour market segmentation The classification of labour market according to different rules and conditions with respect to terms of entry, promotion, working practices, and remuneration.

labour mobility Voluntary movement of workers from one region and job to another, whether drawn by opportunity or pushed by lack of jobs.

labour unions A legal organization of workers formed to bargain with employers on its members' behalf regarding matters such as wages, benefits, and working conditions.

large firms (LFs) Firms with roughly 500 to 10,000 employees, typically focused on a limited range of products and serving global markets.

latecomer firm A locally owned firm that has emerged in growing industries in latecomer countries, that is, countries that have industrialized successfully relatively recently, including by exports. Large latecomer firms are sometimes labelled domestic champions.

lateral subcontracting Contracting out directly to all suppliers.

leisure class Veblen's term for the rich, powerful elite who control society.

local economic development (LED) The stimulation of local economic development by entrepreneurial initiative and leadership.

localization economies (diseconomies) of scale Reductions in average costs of production (or the gain of economic advantages) resulting from location in agglomerations of related economic activities.

location conditions Variations in costs and revenues and other attributes over space.

location factors The advantages and disadvantages of different locations.

location rent The surplus income earned on a unit of land over and above that earned on the marginal unit of land where revenues are offset by costs.

locavore Someone who chooses (as far as possible) to consume locally grown food, in order to support local farmers and minimize the environmental costs of transporting food from farm to market.

logistics Management of the flow of goods and services from points of origin to points of destination.

macroeconomics The branch of economics concerned with the aggregate performance of national economies in terms of savings, investment, consumption, employment, and productivity. Macroeconomic planning seeks to achieve growth and stability through adjustments to fiscal and monetary policies.

managerial structures How decision-making functions within a firm are coordinated across space and in places.

maquiladora A foreign-owned branch plant in one of the export processing zones, often along the Mexico–US border but also located in interior locations, that draws on low-waged labour and tariff-free imports of inputs for goods designed for export.

market economies Societies that rely on market institutions to organize the production of goods and services for consumption.

market failure Refers to situations where markets do not work properly (e.g., when competition does not exist or when market behaviour results in unacceptable environmental or social consequences).

marketing principle In Christaller's central place theory, the principle that services should be located so as to maximize access for consumers.

markets Institutions for the price-based exchange of goods and services.

Marshallian hypothesis Argues that innovation and growth is stimulated by specialization and intra-industry knowledge transfers, especially within agglomerations.

median The middle point of a distribution in which half the observations are on either side.

minimum efficient (optimal) scale (MES/MOS) The lowest level of output at which a plant or firm minimizes average costs that are low enough to make production viable.

minimum wage The legally mandated minimum hourly wage that employers are required to pay in a given jurisdiction.

modes of transportation Alternative ways of moving goods or people between two places.

monoculture Cultivation of a single type of crop.

monopoly A market controlled by a single supplier.

moral hazard   A form of opportunism where people take risks they would otherwise avoid because they are insured against losses; such changes in behaviour are usually detrimental to the interests of the insurer.

multinational corporations (MNCs)   Firms with operations (branch plants, mines, offices, etc.) in at least two countries; typically produce a variety of products, have decentralized decision-making structures, and enjoy substantial internal economies of scale both in production and functional areas, such as marketing; often employ tens of thousands of people and have annual sales in the multi-billions.

multiplier   How an initial spurt of growth in one activity generates growth in other activities.

negative externalities   Costs that fall on third parties and the environment that were not taken into account in the decisions made by market actors.

neoclassical economics   Refers to mainstream economics that developed in the nineteenth century as a formal, abstract interpretation of economic behaviour that is rational, self-interested, and perfectly informed.

neoclassical theory of the firm   A theory based on the *Homo economicus* model, in which firms are assumed to make optimal decisions based on complete information and perfect rationality.

neoliberalism   The school of thought that emphasizes the power of markets to meet society's needs and argues for minimal government interference in the economy; neoliberals often advocate deregulation and privatization of the services provided by the public sector.

nested hierarchy   The highest level of the hierarchy serves the entire population with the highest-order goods, as well as all lower-order goods to nearby consumers, while lower levels in the hierarchy serve progressively lower-order goods in progressively smaller catchment areas.

netvigators   Consumers who use the Internet to search for goods and services.

network economies   The efficiency gains that become possible when all the components of a network are properly integrated.

network externalities   The changes in benefits (gains or losses) to a user of a network when the number of other users changes. Also referred to as demand-side economies of scale.

new economic spaces   Places that either are new to any sort of economic activity or have some experience with other economic activities but are beginning to attract entirely new ones.

New International Division of Labour (NIDL)   A term initially used in the 1970s to refer to the establishment of export-based secondary manufacturing in developing countries. The NIDL has evolved considerably since then.

nodes   The points connected by network links (roads, wires, wireless connections, etc.).

non-basic activities   Economic activities that serve local consumers and therefore do not bring in "export" income.

non-market responses   Government policies designed to solve market problems by means that do not target market actors (e.g., providing social support for third parties negatively affected by market activity, or creating conservation areas from which market activity is excluded).

non-metropolitan industrialization   The dispersal of economic activities away from metropolitan areas to more distant rural places.

non-profit/non-government sector   The terms non-profit and non-government are often used interchangeably because both provide a diversity of services from which they don't take a profit. The chief difference between the two is that non-governmental organizations tend more strongly to be advocacy organizations.

non-renewable resources   Finite resources (such as fossil fuels) that are depleted with use.

non-tariff barriers   Restrictions on imports using means other than taxation (e.g., quotas, punitive inspection requirements).

normal profit   The minimum profit necessary to make productive activity worthwhile, theoretically considered as a cost.

oligopoly   A market dominated by relatively few suppliers; a highly concentrated oligopoly has very few suppliers, while a weak one has many suppliers).

open market   A market made up of numerous small, independent buyers and sellers engaged in voluntary interactions.

operating-level economies of scale   Reductions in average costs of production or operation resulting from increasing the size of individual factories, stores or other types of operation.

opportunity costs   The potential benefits that are lost when one alternative is chosen and the other is rejected.

optimal decisions   Decisions and behaviour that maximize or minimize some specified objective (e.g., maximizing profit or minimizing cost); associated with neoclassical theory.

Organization of Petroleum Exporting Countries (OPEC)   A cartel set up to control production levels among members to ensure price maintenance.

organizational innovation   An innovation that changes the way economic activity is governed or organized.

outsourcing   Engaging an external supplier; also referred to as "contracting out."

participation rate   In Canada, the percentage of the working-age (15–65) population (excluding residents of Aboriginal reserves, full-time members of the military, and the institutionalized population) that is in paid employment.

part-time   Employment that amounts to fewer hours than a standard full-time job (usually defined as an eight-day and a 40-hour week).

path dependency   The tendency for initial decisions and processes to continue influencing subsequent decisions and processes.

perfect competition   The ideal market model in neoclassical economic theory in which fair competition between multiple buyers and sellers would ensure self-regulation.

place   A particular territory, locality, region, or neighbourhood where people live and work.

plant-level diseconomies of scale   Increases in average costs of production resulting from increasing levels of output within a plant or factory.

post-industrialism   The phase in the development of a market economy when services replace manufacturing as the dominant driver of the economy and the source of jobs (a definition that unfortunately equates industry with manufacturing).

primary labour market   The privileged segment of Fordist labour markets comprising salaried and production workers in large firms who receive higher wages, job security, enhanced health care, pensions, and other benefits and whose work conditions are highly structured.

primary manufacturing   Activities based mainly on processing of raw materials.

primate cities Cities that are several times larger than the next largest, and that exercise extraordinary political and economic power over national spaces; London, Paris, and Tokyo are considered to be primate cities.

procurement costs The logistical costs of transporting inputs (both material supplies and services) to the factory.

producer-driven value chains Firm–supplier relations that are organized by manufacturing companies.

product differentiation Modification of a product to create technologically or stylistically similar but different products to serve distinct market niches, usually resulting in economies of scope.

product life cycle A metaphorical linking of the evolutionary trajectories of products with life-cycle stages, such as birth, growth, maturity, old age, and death.

product maturity The stage in the product life cycle when mass production technology has been developed, market potentials are predictable, and low-cost, relatively unskilled labour is the key location condition.

property rights The rights attached to ownership of any asset, physical or otherwise, and the associated income or benefit.

protectionism Protecting domestic industries by imposing penalties and restrictions on market exchanges between countries.

public goods Goods and services that are available to everyone in a society either at no direct charge or at prices below the actual cost of supplying them.

public–private partnership (PPP) An arrangement between public and private organizations in support of economic development projects.

public services Services supplied by the government itself or through NGOs, usually based on a public good or externality rationale (i.e., either the private sector can't provide the service or provision of the service will have beneficial effects on society as a whole).

pure inputs In Weberian location theory, inputs that do not change physically during processing.

radical innovation An organizational or technological innovation with large-scale impact on productivity throughout an industrial sector or social group.

range (of a good or service) The distance that consumers are willing to travel to buy a particular good or service.

range of tolerance The location options available to firms for particular types of adjustments or the adjustments available to firms at particular locations (see also **spatial margin of profitability**).

rank size rule A rule asserting that city sizes in a given region reveal a direct linear relationship (more precisely, a linear relationship when the logarithms of city sizes are used). In practice there are many exceptions to this rule.

regional development policy National government policies in support of regional development.

region (geographic) A territory or area with certain characteristics—physical, social, political, functional, etc.—that distinguish it from contiguous territories.

relational market A type of market characterized by repeated transactions between firms or other partners that are mediated by formal contracts and informal behavioural expectations. Information exchange and collaboration are important features in such relationships, for example, between core firms and suppliers.

renewable resources Resources (such as forests or fish) that are not necessarily depleted with use—as long as the rate of use does not exceed the rate of replacement.

research and development (R&D) Systematic, learning-based creative work that increases society's stock of knowledge and levels of technological capability.

resource curse The idea that a rich resource base may actually limit the ability of a poor country to achieve sustainable development and diversification; similar to Innis's "staple trap" thesis.

resource cycle The long-term pattern associated with resource exploitation, in which an initial period of rapid growth is followed by a plateau and eventual decline.

resource rents Payments made to governments for the right to exploit publicly owned resources; also known as "resource royalties."

reverse logistics Organization of the collection of products (and their packaging) for reuse, recycling, or disposal.

satisficers Decision-makers who make choices intended to meet aspiration levels. The choices are based on the availability of, and ability to interpret, information.

scientific management (Taylorism) The practice of dividing a given job into multiple narrowly defined tasks and employing specialized labour to perform these tasks.

secondary labour market The less-privileged segment of the Fordism labour market, where jobs are typically non-unionized, insecure, and poorly paid, and where working conditions are often poor or hazardous. Most employers in this segment are SMEs; disproportionate numbers of employees are women, minorities, and students.

secondary manufacturing Activities that process already manufactured inputs.

seedbed hypothesis Argues that new firms are typically located close to their founders' home place, typically where they have lived for a long time. Many new firms begin in the homes of founders and move to nearby locations when they need more space.

seniority A key principle in structured labour markets, requiring that new employees begin in the lowest level positions, and that promotion be based on the number of years an employee has spent in the job; in times of recession, those with the least seniority are normally the first to be laid off, and those with the most seniority are normally the first to be re-hired.

service sector The service sector is primarily defined by what its participants do not do: their main focus is not on producing goods (such as agriculture or manufacturing). It provides functions that support or use such goods, or may be almost disconnected from them, but the physical and intellectual labour of services is not actually engaged in making goods.

"shadow state" argument The idea that government funding has the effect of co-opting non-profits, discouraging dissent, and effectively forcing them to defend the status quo.

small and medium enterprises (SMEs) Relatively small firms, owned and controlled by individual entrepreneurs and employing no more than 500 people (usually fewer). SMEs typically serve local markets only and many have just one plant.

social capital The aggregate of shared values and expectations, social networks, and qualities such as trust and willingness to co-operate that makes economic development possible.

social division of labour The distribution of tasks among firms in an economy. In practice, this term is often used when small firms play important roles.

social services Basic services provided by government to people in need.

social values Values such as environmental health, human connection, and emotional well-being, which are for the most part provided by government and non-profit organizations.

space The area over which people and activities are linked; regions, nations, continents, or even the globe itself could all be defined as spaces, depending on the context.

space pervaders Advertisers who attempt to persuade consumers by diffusing their messages throughout space.

space–time and space–cost distanciation The cost and time of moving goods, information, services, and people across space.

spatial division of labour The allocation of specialized tasks among and within places and across space.

spatial entry barriers The costs and uncertainties that face firms contemplating investment in a new region or country.

spatial innovation system (SIS) An international system linking agents around the world involved in the research, development and transfer of new technologies.

spatial margin of profitability The range of alternative locations in which a given activity can be viable (see also **range of tolerance**).

spot transactions Transactions completed (more or less) on the spot, for delivery within two days.

standard of living A measure of people's ability to live a healthy and productive life in accordance with acceptable social norms; often approximated by per capita GDP in international comparisons.

staple trap The trap experienced by economies that depend on resources for development and find their opportunities for diversification to be limited.

statism A governance model in which national governments take a proactive role in directing strategic growth.

state-owned companies (SOC) Public-sector owned and controlled companies. Typically control is exercised through a quasi-independent organization that in many ways resembles a private company but operates within a government-defined mandate and reports to a minister. In Canada, these organizations are called "Crown corporations."

status emulation The tendency to mimic the consumption behaviour of higher-status or higher-income groups.

structured labour markets Labour markets that are formally organized by legally binding agreements between employers and employers; in Fordist labour markets the two most important structural principles are seniority and job demarcation.

subsistence agriculture Family farming to feed the family, with little or no surplus for sale.

substitution principle In location theory, the ability to trade off one location condition for another.

suburbanization The shift in economic activities (or housing) outwards from central cities but still within the built-up area or commuting zone.

sunk costs Costs that cannot be recovered in the sense that they represent past investments, for example, by a business in R&D, or a consumer's purchase of a theatre ticket, that cannot be reversed. But for most businesses sunk costs are recovered by pricing strategies, cost accounting, and/or by increasing volume.

supply curve (schedule) A graph showing the quantity of products that suppliers are able and willing to supply at different prices.

sustainable development Economic growth that can be sustained over long periods of time without harming the environment.

tacit knowledge Knowledge that is acquired through observation, imitation, and hands-on experience and is difficult to communicate verbally.

tariffs Taxes imposed on the goods and services imported into a country.

tax A levy imposed by government on goods, services, and incomes; taxes are the source of all the money that governments spend.

techno-economic paradigm (TEP) Strategic, long-term (roughly 50-year) shifts or waves in the economy driven by paradigmatic innovations, and associated with particular key factor industries, productivity principles, leading industries, labour relations, R&D models, and organizational structures.

technological capability The ability (within an economy or firm) to generate commercially profitable innovations.

technological innovation The first commercial introduction of new methods of production ("processes") or new products that typically involves manufacturing, processing, communication, or transportation activities.

technostructure The top-level management and supporting staff that control large firms and MNCs; a concept associated with institutional theories.

threshold population The minimum population required to ensure that demand for a good is sufficient to make its production economically viable.

tier 1 suppliers The most important suppliers who have direct contact and business relations with core firms.

topological map A network described by nodes and edges or links between nodes.

transaction costs Generally refer to the costs of establishing and maintaining systems of rules and rights, for example, with respect to markets.

transferability A condition of spatial interaction; exchange can occur between two places when revenues cover the costs of both production and movement.

transnational corporations (TNCs) A term often used as a synonym for MNC. However, TNCs were originally defined as corporations with no loyalty or commitment to any particular home region.

ubiquitous inputs In Weberian location theory, inputs that are available everywhere.

underemployed Workers who are employed but whose jobs do not provide the hours and income they need to support themselves and their families.

unemployed Workers who are out of a job but are seeking regular paid employment.

unitary systems of government Political power is centralized within national governments, including with respect to the delegation of powers.

urbanization economies of scale Reductions in average costs of production (or economic advantages) resulting from location in urban agglomerations that provide diverse labour pools, amenities, economic and social infrastructures, market exchanges and meeting places, and regional and international connections.

urban regeneration The comprehensive replacement of a city's old economic structures with new ones.

urban system   A set of cities (towns, etc.) that are connected to one another in such a way that changes in one will affect others; such systems can exist at various scales from local to global.

value-added   The value added to basic materials or services through the application of labour and capital.

value chain   The complex set of interrelated activities required to produce a product (a good or a service) and distribute it to consumers.

variable costs   Costs that vary with scale of activity.

vertical integration   Expansion either upstream, towards sources of supply (backward vertical integration), or downstream, towards markets (forward vertical integration).

vicious cycle   A cycle of self-reinforcing processes that reinforce unfavourable socio-economic changes (in poor regions, for example, outmigration and disinvestment further reduce the prospects for economic development).

virtuous cycle   A cycle of processes that reinforce favourable socio-economic changes (for example, the ability of core regions to attract more people). Investment reinforces their attractiveness for further economic development.

von Thünen, Heinrich   The German landowner and economist whose book *The Isolated State* (1826) introduced agricultural land-use theory (and by extension urban land-use theory).

welfare state   The system (more comprehensive in some countries than others) where the government seeks to ensure that all citizens have access to the essentials (food, shelter, education, etc.) required for participation in economic and social life.

Index

Italicized page numbers refer to definitions.

Selected Abbreviations

CAW	Canadian Auto Workers
CBD	central business district
CCT	conditional cash transfer
CED	community economic development
CLC	Canadian Labour Congress
CUPE	Canadian Union of Public Employees
DfE	design for environment
DREE	Department of Regional Economic Expansion (Canada)
EDO	economic development offices
ENGO	Environmental Non-government organization
EOC	economic overhead capital
EPZ	export processing zones
FDI	foreign direct investment
GATT	General Agreement on Tariffs and Trade
GDP	gross domestic product
GERD	Gross Expenditures on R&D
GNP	gross national product
ICLEI	International Council for Local Environmental Initiatives
ICT	information and communication technologies
ILO	International Labour Organization
JIT	just in time
LCA	life-cycle analysis
LED	local economic development
MES	minimum efficient scale
MNC	multi-national corporations
NGO	non-government organizations
NIC	newly industrialized countries
NIDL	new international division of labour
OBM	original brand manufacturers
ODM	original design manufacturers
OECD	Organization for Economic Co-operation and Development
OEM	original equipment manufacturers
OJT	on-the job training
PPP or P3s	public–private partnerships
PPP	purchasing power parity
RTA	regional trading agreements
SIS	spatial innovation systems
SME	small-medium enterprises
SOC	social overhead capital
SOE	state-owned enterprises
TDM	trade dispute mechanism
TEP	techno-economic paradigm
TPS	Toyota production system
WHO	World Health Organization
WLO	World Labour Organization

FIFTH EDITION

FUNDAMENTALS OF
PHYSICS

FIFTH EDITION

FUNDAMENTALS OF
PHYSICS

PART 2

DAVID HALLIDAY
University of Pittsburgh

ROBERT RESNICK
Rensselaer Polytechnic Institute

JEARL WALKER
Cleveland State University

JOHN WILEY & SONS, INC.

New York • Chichester • Brisbane • Toronto • Singapore

ACQUISITIONS EDITOR Stuart Johnson
DEVELOPMENTAL EDITOR Rachel Nelson
SENIOR PRODUCTION SUPERVISOR Cathy Ronda
PRODUCTION ASSISTANT Raymond Alvarez
MARKETING MANAGER Catherine Faduska
ASSISTANT MARKETING MANAGER Ethan Goodman
DESIGNER Dawn L. Stanley
MANUFACTURING MANAGER Mark Cirillo
PHOTO EDITOR Hilary Newman
COVER PHOTO William Warren/Westlight
ILLUSTRATION EDITOR Edward Starr
ILLUSTRATION Radiant/Precision Graphics

This book was set in Times Roman by Progressive Information Technologies, and printed and bound by Von Hofmann Press. The cover was printed by Phoenix Color Corp.

ISBN 0-471-14854-7

Printed in the United States of America

10 9 8 7 6 5 4

Hello There!

You are about to begin your first college level physics course. You may have heard from friends and fellow students that physics is a difficult course, especially if you don't plan to go on to a career in the hard sciences. But that doesn't mean it has to be difficult for you. The key to success in this course is to have a good understanding of each chapter before moving on to the next. When learned a little bit at a time, physics is straightforward and simple. Here are some ideas that can make this text and this class work for you:

- Read through the **Sample Problems** and solutions carefully. These problems are similar to many of the end-of-chapter exercises, so reading them will help you solve homework problems. In addition, they offer a look at how an experienced physicist would approach solving the problem.

- Try to answer the **Checkpoint** questions as you read through the chapter. Most of these can be answered by thinking through the problem, but it helps if you have some scratch paper and a pencil nearby to work out some of the harder questions. Hold the answer page at the back of the book with your thumb (or a tab or bookmark) so you can refer to it easily. The end-of-chapter **Questions** are very similar— you can use them to quiz yourself after you've read the whole chapter.

- Use the **Review & Summary** sections at the end of each chapter as the first place to look for formulas you might neeed to solve homework problems. These sections are also helpful in making study sheets for exams.

- The biggest tip I can give you is pretty obvious. Do your homework! Understanding the homework problems is the best way to master the material and do well in the course. Doing a lot of homework problems is also the best way to review for exams. And by all means, consult your classmates whenever you are stuck. Working in groups will make your studying more effective.

The study of basic physics is required for degrees in Engineering, Physics, Biology, Chemistry, Medicine, and many other sciences because the fundamentals of physics are the framework on which every other science is built. Therefore, a solid grasp of basic physical principles will help you understand upper-level science courses and make your study of these courses easier.

I took introductory physics because it was a prerequisite for my B.S. in Mechanical Engineering. Even though engineering and pure physics are worlds apart, I find myself using this book as a reference almost every day. I urge you to keep it after you have finished your course. The knowledge you will gain from this book and your introductory physics course is the foundation for all other sciences. This is the primary reason to take this class seriously and be successful in it.

Best of luck!

Josh Kane

Josh Kane

Preface

For four editions, *Fundamentals of Physics* has been successful in preparing physics students for careers in science and engineering. The first three editions were coauthored by the highly regarded team of David Halliday and Robert Resnick, who developed a groundbreaking text replete with conceptual structure and applications. In the fourth edition, the insights provided by new coauthor, Jearl Walker, took the text into the 1990s and met the challenge of guiding students through a time of tremendous advances and a ferment of activity in the science of physics. Now, in the fifth edition, we have expanded on the conventional strengths of the earlier editions and enhanced the applications that help students forge a bridge between concepts and reasoning. We not only *tell* students how physics works, we *show* them, and we give them the opportunity to show us what they have learned by testing their understanding of the concepts and applying them to real-world scenarios. Concept checkpoints, problem solving tactics, sample problems, electronic computations, exercises and problems — all of these skill-building signposts have been developed to help students establish a connection between conceptual theories and application. The students reading this text today are the scientists and engineers of tomorrow. It is our hope that the fifth edition of *Fundamentals of Physics* will help prepare these students for future endeavors by contributing to the enhancement of physics education.

CHANGES IN THE FIFTH EDITION

Although we have retained the basic framework of the fourth edition of *Fundamentals of Physics,* we have made extensive changes in portions of the book. Each chapter and element has been scrutinized to ensure clarity, currency, and accuracy, reflecting the needs of today's science and engineering students.

Content Changes

Mindful that textbooks have grown large and that they tend to increase in length from edition to edition, we have reduced the length of the fifth edition by combining several chapters and pruning their contents. In doing so, six chapters have been rewritten completely, while the remaining chapters have been carefully edited and revised, often extensively, to enhance their clarity, incorporating ideas and suggestions from dozens of reviewers.

- *Chapters 7 and 8 on energy* (and sections of later chapters dealing with energy) have been rewritten to provide a more careful treatment of energy, work, and the work–kinetic energy theorem. As the same time, the text material and problems at the end of each chapter still allow the instructor to present the more traditional treatment of these subjects.

- *Temperature, heat, and the first law of thermodynamics* have been condensed from two chapters to one chapter (*Chapter 19*).

- *Chapter 21 on entropy* now includes a statistical mechanical presentation of entropy that is tied to the traditional thermodynamical presentation.

- *Chapters on Faraday's law and inductance* have been combined into one new chapter (*Chapter 31*).

- *Treatment of Maxwell's equations* has been streamlined and moved up earlier into the chapter on magnetism and matter (*Chapter 32*).

- *Coverage of electromagnetic oscillations and alternating currents* has been combined into one chapter (*Chapter 33*).

- *Chapters 39, 40, and 41 on quantum mechanics* have been rewritten to modernize the subject. They now include experimental and theoretical results of the last few years. In addition, quantum physics and special relativity are introduced in some of the early chapters in short sections that can be covered quickly. These early sections lay some of the groundwork for the ''modern physics'' topics that appear later in the extended version of the text and add an element of suspense about the subject.

New Pedagogy

In the interest of addressing the needs of science and engineering students, we have added a number of new pedagogical features intended to help students forge a bridge between concepts and reasoning and to marry theory with practice. These new features are designed to help students test their understanding of the material. They were also developed to help students prepare to apply the information to exam questions and real-world scenarios.

- To provide opportunities for students to check their understanding of the physics concepts they have just read, we have placed **Checkpoint** questions within the chapter presentations. Nearly 300 Checkpoints have been added to help guide the student away from common errors and misconceptions. All of the Checkpoints require decision making and reasoning on the part of the student (rather than computations requiring calculators) and focus on the key points of the physics that students need to understand in order to tackle the exercises and problems at the end of each chapter. Answers to all of the Checkpoints are found in the back of the book, sometimes with extra guidance to the student.

- Continuing our focus on the key points of the physics, we have included additional **Checkpoint-type questions** in the Questions section at the end of each chapter. These new questions require decision making and reasoning on the part of the student; they ask the student to organize the physics concepts rather than just plug numbers into equations. Answers to the odd-numbered questions are now provided in the back of the book.

- To encourage the use of computer math packages and graphing calculators, we have added an **Electronic Computation problem section** to the Exercises and Problems sections of many of the chapters.

These new features are just a few of the pedagogical elements available to enhance the student's study of physics. A number of tried-and-true features of the previous edition have been retained and refined in the fifth edition, as described below.

CHAPTER FEATURES

The pedagogical elements that have been retained from previous editions have been carefully planned and crafted to motivate students and guide their reasoning process.

- *Puzzlers* Each chapter opens with an intriguing photograph and a "puzzler" that is designed to motivate the student to read the chapter. The answer to each puzzler is provided within the chapter, but it is not identified as such to ensure that the student reads the entire chapter.

- *Sample Problems* Throughout each chapter, sample problems provide a bridge from the concepts of the chapter to the exercises and problems at the end of the chapter. Many of the nearly 400 sample problems featured in the text have been replaced with new ones that more sharply focus on the common difficulties students experience in solving the exercises and problems. We have been especially mindful of the mathematical difficulties students face. The sample problems also provide

an opportunity for the student to see how a physicist thinks through a problem.

- *Problem Solving Tactics* To help further bridge concepts and applications and to add focus to the key physics concepts, we have refined and expanded the number of problem solving tactics that are placed within the chapters, particularly in the earlier chapters. These tactics provide guidance to the students about how to organize the physics concepts, how to tackle mathematical requirements in the exercises and problems, and how to prepare for exams.

- *Illustrations* Because the illustrations in a physics textbook are so important to an understanding of the concepts, we have altered nearly 30 percent of the illustrations to improve their clarity. We have also removed some of the less effective illustrations and added many new ones.

- *Review & Summary* A review and summary section is found at the end of each chapter, providing a quick review of the key definitions and physics concepts *without* being a replacement for reading the chapter.

- *Questions* Approximately 700 thought-provoking questions emphasizing the conceptual aspects of physics appear at the ends of the chapters. Many of these questions relate back to the checkpoints found throughout the chapters, requiring decision making and reasoning on the part of the student. Answers to the odd-numbered questions are provided in the back of the book.

- *Exercises & Problems* There are approximately 3400 end-of-chapter exercises and problems in the text, arranged in order of difficulty, starting with the exercises (labeled "E"), followed by the problems (labeled "P"). Particularly challenging problems are identified with an asterisk (*). Those exercises and problems that have been retained from previous editions have been edited for greater clarity; many have been replaced. Answers to the odd-numbered exercises and problems are provided in the back of the book.

VERSIONS OF THE TEXT

The fifth edition of *Fundamentals of Physics* is available in a number of different versions, to accommodate the individual needs of instructors and students alike. The Regular Edition consists of Chapters 1 through 38 (ISBN 0-471-10558-9). The Extended Edition contains seven additional chapters on quantum physics and cosmology (Chapters 1–45) (ISBN 0-471-10559-7). Both editions are available as single, hardcover books, or in the alternative versions listed on page ix:

- Volume 1—Chapters 1–21 (Mechanics/Thermodynamics), cloth, 0-471-15662-0
- Volume 2—Chapters 22–45 (E&M and Modern Physics), cloth, 0-471-15663-9
- Part 1—Chapters 1–12, paperback, 0-471-14561-0
- Part 2—Chapters 13–21, paperback, 0-471-14854-7
- Part 3—Chapters 22–33, paperback, 0-471-14855-5
- Part 4—Chapters 34–38, paperback, 0-471-14856-3
- Part 5—Chapters 39–45, paperback, 0-471-15719-8

The Extended edition of the text is also available on CD ROM.

SUPPLEMENTS

The fifth edition of *Fundamentals of Physics* is supplemented by a comprehensive ancillary package carefully developed to help teachers teach and students learn.

Instructor's Supplements

- *Instructor's Manual* by J. RICHARD CHRISTMAN, U.S. Coast Guard Academy. This manual contains lecture notes outlining the most important topics of each chapter, as well as demonstration experiments, and laboratory and computer exercises; film and video sources are also included. Separate sections contain articles that have appeared recently in the *American Journal of Physics* and *The Physics Teacher*.
- *Instructor's Solutions Manual* by JERRY J. SHI, Pasadena City College. This manual provides worked-out solutions for all the exercises and problems found at the end of each chapter within the text. *This supplement is available only to instructors.*
- *Solutions Disk.* An electronic version of the Instructor's Solutions Manual, for instructors only, available in TeX for Macintosh and Windows™.
- *Test Bank* by J. RICHARD CHRISTMAN, U.S. Coast Guard Academy. More than 2200 multiple-choice questions are included in the Test Bank for *Fundamentals of Physics*.

- *Computerized Test Bank.* IBM and Macintosh versions of the entire Test Bank are available with full editing features to help you customize tests.
- *Animated Illustrations.* Approximately 85 text illustrations are animated for enhanced lecture demonstrations.
- *Transparencies.* More than 200 four-color illustrations from the text are provided in a form suitable for projection in the classroom.

Student's Supplements

- *A Student's Companion* by J. RICHARD CHRISTMAN, U.S. Coast Guard Academy. Much more than a traditional study guide, this student manual is designed to be used in close conjunction with the text. The Student's Companion is divided into four parts, each of which corresponds to a major section of the text, beginning with an overview "chapter." These overviews are designed to help students understand how the important topics are integrated and how the text is organized. For each chapter of the text, the corresponding Companion chapter offers: Basic Concepts, Problem Solving, Notes, Mathematical Skills, and Computer Projects and Notes.
- *Solutions Manual* by J. RICHARD CHRISTMAN, U.S. Coast Guard Academy and EDWARD DERRINGH, Wentworth Institute. This manual provides students with complete worked-out solutions to 30 percent of the exercises and problems found at the end of each chapter within the text.
- **Interactive Learningware** by JAMES TANNER, Georgia Institute of Technology, with the assistance of GARY LEWIS, Kennesaw State College. This software contains 200 problems from the end-of-chapter exercises and problems, presented in an interactive format, providing detailed feedback for the student. Problems from Chapter 1 to 21 are included in Part 1, from Chapters 22 to 38 in Part 2. The accompanying workbooks allow the student to keep a record of the worked-out problems. The Learningware is available in IBM 3.5″ and Macintosh formats.
- **CD Physics.** The entire Extended Version of the text (Chapters 1–45) is available on CD ROM, along with the student solutions manual, study guide, animated illustrations, and Interactive Learningware.

Acknowledgments

A textbook contains far more contributions to the elucidation of a subject than those made by the authors alone. J. Richard Christman, of the U.S. Coast Guard Academy, has once again created many fine supplements for us; his knowledge of our book and his recommendations to students and faculty are invaluable. James Tanner, of Georgia Institute of Technology, and Gary Lewis, of Kennesaw State College, have provided us with innovative software, closely tied to the text's exercises and problems. J. Richard Christman, of the U.S. Coast Guard Academy, and Glen Terrell, of the University of Texas at Arlington, contributed problems to the Electronic Computation sections of the text. Jerry Shi, of Pasadena City College, performed the Herculean task of working out solutions for every one of the Exercises and Problems in the text. We thank John Merrill, of Brigham Young University, and Edward Derringh, of the Wentworth Institute of Technology for their many contributions in the past. We also thank George W. Hukle of Oxnard, California, for his check of the answers at the back of the book.

At John Wiley, publishers, we have been fortunate to receive strong coordination and support from our former editor, Cliff Mills. Cliff guided our efforts and encouraged us along the way. When Cliff moved on to other responsibilities at Wiley, we were ably guided to completion by his successor, Stuart Johnson. Rachel Nelson has coordinated the developmental editing and multilayered preproduction process. Catherine Faduska, our senior marketing manager, and Ethan Goodman, assistant marketing manager, have been tireless in their efforts on behalf of this edition. Jennifer Bruer has built a fine supporting package of ancillary materials. Monica Stipanov and Julia Salsbury managed the review and administrative duties admirably.

We thank Lucille Buonocore, our able production manager, and Cathy Ronda, our production editor, for pulling all the pieces together and guiding us through the complex production process. We also thank Dawn Stanley, for her design; Brenda Griffing, for her copy editing; Edward Starr, for managing the line art program; Lilian Brady, for her proofreading; and all other members of the production team.

Stella Kupferburg and her team of photo researchers, particularly Hilary Newman and Pat Cadley, were inspired in their search for unusual and interesting photographs that communicate physics principles beautifully. We thank Boris Starosta and Irene Nunes for their careful development of a full-color line art program, for which they scrutinized and suggested revisions of every piece. We also owe a debt of gratitude for the line art to the late John Balbalis, whose careful hand and understanding of physics can still be seen in every diagram.

We especially thank Edward Millman for his developmental work on the manuscript. With us, he has read every word, asking many questions from the point of view of a student. Many of his questions and suggested changes have added to the clarity of this volume. Irene Nunes added a final, valuable developmental check in the last stages of the book.

We owe a particular debt of gratitude to the numerous students who used the fourth edition of *Fundamentals of Physics* and took the time to fill out the response cards and return them to us. As the ultimate consumers of this text, students are extremely important to us. By sharing their opinions with us, your students help us ensure that we are providing the best possible product and the most value for their textbook dollars. We encourage the users of this book to contact us with their thoughts and concerns so that we can continue to improve this text in the years to come. In particular, we owe a special debt of gratitude to the students who participated in a final focus group at Union College in Schenecdaty, New York: Matthew Glogowski, Josh Kane, Lauren Papa, Phil Tavernier, Suzanne Weldon, and Rebecca Willis.

Finally, our external reviewers have been outstanding and we acknowledge here our debt to each member of that team:

MARIS A. ABOLINS
Michigan State University

BARBARA ANDERECK
Ohio Wesleyan University

ALBERT BARTLETT
University of Colorado

MICHAEL E. BROWNE
University of Idaho

TIMOTHY J. BURNS
Leeward Community College

JOSEPH BUSCHI
Manhattan College

PHILIP A. CASABELLA
Rensselaer Polytechnic Institute

RANDALL CATON
Christopher Newport College

J. RICHARD CHRISTMAN
U.S. Coast Guard Academy

ROGER CLAPP
University of South Florida

W. R. CONKIE
Queen's University

PETER CROOKER
University of Hawaii at Manoa

WILLIAM P. CRUMMETT
*Montana College of Mineral
Science and Technology*

EUGENE DUNNAM
University of Florida

ROBERT ENDORF
University of Cincinnati

F. PAUL ESPOSITO
University of Cincinnati

JERRY FINKELSTEIN
San Jose State University

ALEXANDER FIRESTONE
Iowa State University

ALEXANDER GARDNER
Howard University

ANDREW L. GARDNER
Brigham Young University

JOHN GIENIEC
*Central Missouri State
University*

JOHN B. GRUBER
San Jose State University

ANN HANKS
American River College

SAMUEL HARRIS
Purdue University

EMILY HAUGHT
Georgia Institute of Technology

LAURENT HODGES
Iowa State University

JOHN HUBISZ
North Carolina State University

JOEY HUSTON
Michigan State University

DARRELL HUWE
Ohio University

CLAUDE KACSER
University of Maryland

LEONARD KLEINMAN
University of Texas at Austin

EARL KOLLER
Stevens Institute of Technology

ARTHUR Z. KOVACS
Rochester Institute of Technology

KENNETH KRANE
Oregon State University

SOL KRASNER
University of Illinois at Chicago

PETER LOLY
University of Manitoba

ROBERT R. MARCHINI
Memphis State University

DAVID MARKOWITZ
University of Connecticut

HOWARD C. MCALLISTER
University of Hawaii at Manoa

W. SCOTT MCCULLOUGH
Oklahoma State University

JAMES H. MCGUIRE
Tulane University

DAVID M. MCKINSTRY
Eastern Washington University

JOE P. MEYER
Georgia Institute of Technology

ROY MIDDLETON
University of Pennsylvania

IRVIN A. MILLER
Drexel University

EUGENE MOSCA
United States Naval Academy

MICHAEL O'SHEA
Kansas State University

PATRICK PAPIN
San Diego State University

GEORGE PARKER
North Carolina State University

ROBERT PELCOVITS
Brown University

OREN P. QUIST
South Dakota State University

JONATHAN REICHART
SUNY—Buffalo

MANUEL SCHWARTZ
University of Louisville

DARRELL SEELEY
Milwaukee School of Engineering

BRUCE ARNE SHERWOOD
Carnegie Mellon University

JOHN SPANGLER
St. Norbert College

ROSS L. SPENCER
Brigham Young University

HAROLD STOKES
Brigham Young University

JAY D. STRIEB
Villanova University

DAVID TOOT
Alfred University

J. S. TURNER
University of Texas at Austin

T. S. VENKATARAMAN
Drexel University

GIANFRANCO VIDALI
Syracuse University

FRED WANG
Prairie View A&M

ROBERT C. WEBB
Texas A&M University

GEORGE WILLIAMS
University of Utah

DAVID WOLFE
University of New Mexico

We hope that our words here reveal at least some of the wonder of physics, the fundamental clockwork of the universe. And, hopefully, those words might also reveal some of our awe of that clockwork.

DAVID HALLIDAY
6563 NE Windermere Road
Seattle, WA 98105

ROBERT RESNICK
Rensselaer Polytechnic Institute
Troy, NY 12181

JEARL WALKER
Cleveland State University
Cleveland, OH 44115

HOW TO USE THIS BOOK:

You are about to begin what could be one the most exciting course that you will undertake in college. It offers you the opportunity to learn what makes our world "tick" and to gain insight into the role physics plays in our everyday lives. This knowledge will not come without some effort, however, and this book has been carefully designed and written with an awareness of the kinds of difficulties and challenges you may face. Therefore, before you begin, we have provided a visual overview of some of the key features of the book that will aid in your studies.

Chapter Opening Puzzlers

Each chapter opens with an intriguing example of physics in action. By presenting high-interest applications of each chapters concepts, the puzzlers are intended to peak your interest and motivate you to read the chapter.

In 1977, Kitty O'Neil set a dragster record by reaching 392.54 mi/h in a sizzling time of 3.72 s. In 1958, Eli Beeding Jr. rode a rocket sled from a standstill to a speed of 72.5 mi/h in an elapsed time of 0.04 s (less than an eye blink). How can we compare these two rides to see which was more exciting (or more frightening)—by final speeds, by elapsed times, or by some other quantity?

SAMPLE PROBLEM 2-6

(a) When Kitty O'Neil set the dragster records for the greatest speed and least elapsed time, she reached 392.54 mi/h in 3.72 s. What was her average acceleration?

SOLUTION: From Eq. 2-7, O'Neil's average acceleration was

$$\bar{a} = \frac{\Delta v}{\Delta t} = \frac{392.54 \text{ mi/h} - 0}{3.72 \text{ s} - 0}$$

$$= +106 \frac{\text{mi}}{\text{h} \cdot \text{s}}, \qquad \text{(Answer)}$$

where the motion is taken to be in the positive *x* direction. In

Answers to Puzzlers

All chapter-opening puzzlers are answered later in the chapter, either in text discussion or in a sample problem.

If the car
r, the bob
oninertial

w wish to
accelera-
the stan-
e use) the
been as-

ictionless
at by trial
accelera-
definition,
ody has a

body by

CHECKPOINT **1**: In the figure, two perpendicular forces $\mathbf{F}_1$ and $\mathbf{F}_2$ are combined in six different ways. Which ways may be used to correctly determine the net force $\Sigma\mathbf{F}$?

(a)   (b)   (c)

(d)   (e)

Checkpoints

Checkpoints appear throughout the text, focusing on the key points of physics you will need to tackle the exercises and problems found at the end of each chapter. These checkpoints help guide you away from common errors and misconceptions.

Checkpoint Questions

Checkpoint-type questions at the end of each chapter ask you to organize the physics concepts rather than plug numbers into equations. Answers to the odd-numbered questions are provided in the back of the book.

ey puck in

$\mathbf{v} = -2t\mathbf{i}$

ponents of
ion vector
nd and t is
-2 and 3?
cal projec-
t identical
n the same
nal speeds

(c)

peed (a) a

$4.9\mathbf{j}$ (x is
). Has the

9. Figure 4-25 shows three paths for a kicked football. Ignoring the effects of air on the flight, rank the paths according to (a) time of flight, (b) initial vertical velocity component, (c) initial horizontal velocity component, and (d) initial speed. Place the greatest first in each part.

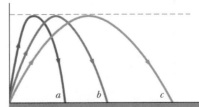

FIGURE 4-25 Question 9.

10. Figure 4-26 shows the velocity and acceleration of a particle at a particular instant in three situations. In which situation, and at that instant, is (a) the speed increasing, (b) the speed decreasing, (c) the speed not changing, (d) $\mathbf{v}\cdot\mathbf{a}$ positive, (e) $\mathbf{v}\cdot\mathbf{a}$ negative, and (f) $\mathbf{v}\cdot\mathbf{a} = 0$?

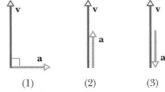

FIGURE 4-26 Question 10.

Sample Problems

The sample problems offer you the opportunity to work through the physics concepts just presented. Often built around real-world applications, they are closely coordinated with the end-of-chapter Questions, Exercises, and Problems.

d the
(This
ed in
e two

ation
bject
, Fig.

ılong
ction.
e +**j**

SAMPLE PROBLEM 4-1

The position vector for a particle is initially

$$\mathbf{r}_1 = -3\mathbf{i} + 2\mathbf{j} + 5\mathbf{k}$$

and then later is

$$\mathbf{r}_2 = 9\mathbf{i} + 2\mathbf{j} + 8\mathbf{k}$$

(see Fig. 4-2). What is the displacement from $\mathbf{r}_1$ to $\mathbf{r}_2$?

SOLUTION: Recall from Chapter 3 that we add (or subtract) two vectors in unit-vector notation by combining the components, axis by axis. So Eq. 4-2 becomes

$$\Delta \mathbf{r} = (9\mathbf{i} + 2\mathbf{j} + 8\mathbf{k}) - (-3\mathbf{i} + 2\mathbf{j} + 5\mathbf{k})$$
$$= 12\mathbf{i} + 3\mathbf{k}. \qquad \text{(Answer)}$$

The displacement vector is parallel to the xz plane, because it lacks any y component, a fact that is easier to pick out in the numerical result than in Fig. 4-2.

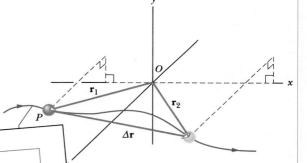

Sample Problem 4-1. The displacement $\Delta \mathbf{r} =$
d of $\mathbf{r}_1$ to the head of $\mathbf{r}_2$.

PROBLEM SOLVING TACTICS

TACTIC 1: *Reading Force Problems*
Read the problem statement several times until you have a good mental picture of what the situation is, what data are given, and what is requested. In Sample Problem 5-1, for example, you should tell yourself: "Someone is pushing a sled. Its speed changes, so acceleration is involved. The motion is along a straight line. A force is given in one part and asked for in the other, and so the situation looks like Newton's second law applied to one-dimensional motion."

If you know what the problem is about but don't know what to do next, put the problem aside and reread the text. If you are hazy about Newton's second law, reread that section. Study the sample problems. The one-dimensional-motion parts of Sample Problem 5-1 and the constant acceleration should send you back to Chapter 2 and especially to Table 2-1, which displays all the equations you are likely to need.

TACTIC 2: *Draw Two Types of Figures*
You may need two figures. One is a rough sketch of the actual real-world situation. When you draw the forces on it, place the tail of each force vector either on the boundary of or within the body feeling that force. The other figure is a free-body diagram in which the forces on a *single* body are drawn, with the body represented with a dot or a sketch. Place the tail of each force vector on the dot or sketch.

TACTIC 3: *What I em?*

Problem Solving Tactics

Careful attention has been paid to helping you develop your problem-solving skills. Problem-solving tactics are closely related to the sample problems and can be found throughout the text, though most fall within the first half. The tactics are designed to help you work through assigned homework problems and prepare for exams. Collectively, they represent the stock in trade of experienced problem solvers and practicing scientists and engineers.

Review and Summary

Review & Summary sections at the end of each chapter review the most important concepts and equations.

REVIEW & SUMMARY

Conservative Forces

A force is a **conservative force** if the net work it does on a particle moving along a closed path from an initial point and then back to that point is zero. Or, equivalently, it is conservative if its work on a particle moving between two points does not depend on the path taken by the particle. The gravitational force (weight) and the spring force are conservative forces; the kinetic frictional force is a **nonconservative force.**

Potential Energy

A **potential energy** is energy that is associated with the configuration of a system in which a conservative force acts. When the conservative force does work W on a particle within the system, the change ΔU in the potential energy of the system is

$$\Delta U = -W. \qquad (8\text{-}1)$$

If the particle moves from point x_i to point x_f, the change in the potential energy of the system is

$$\Delta U = -\int_{x_i}^{x_f} F(x)\, dx. \qquad (8\text{-}6)$$

Gravitational Potential Energy

The potential energy associated with a system consisting of the Earth and a nearby particle is the **gravitational potential energy.** If the particle moves from height y_i to height y_f, the change in the gravitational potential energy of the particle–Earth system is

$$\Delta U = mg(y_f - y_i) = mg\,\Delta y. \qquad (8\text{-}7)$$

If the **reference position** of the particle is set as $y_i = 0$ and the corresponding gravitational potential energy of the system is set as $U_i = 0$, then the gravitational potential energy U when the particle is at any position y is

$$U = mgy. \qquad (8\text{-}9)$$

in which the subscripts refer to different instants during an transfer process. This conservation can also be written as

$$\Delta E = \Delta K + \Delta U = 0.$$

Potential Energy Curves

If we know the **potential energy function** $U(x)$ for a sy which a force F acts on a particle, we can find the force

$$F(x) = -\frac{dU(x)}{dx}.$$

If $U(x)$ is given on a graph, then at any value of x, the fo the negative of the slope of the curve there and the kinetic of the particle is given by

$$K(x) = E - U(x),$$

where E is the mechanical energy of the system. A **turnin** is a point x where the particle reverses its motion (there, The particle is in **equilibrium** at points where the slope $U(x)$ curve is zero (there, $F(x) = 0$).

Work by Nonconservative Forces

If a nonconservative applied force F does work on particl part of a system having a potential energy, then the wor done on the system by F is equal to the change ΔE in the m ical energy of the system:

$$W_{app} = \Delta K + \Delta U = \Delta E. \qquad (8\text{-}2)$$

If a kinetic frictional force f_k does work on an obj change ΔE in the total mechanical energy of the object system containing it is given by

$$\Delta E = -f_k d,$$

in which d is the displacement of the object during the wo

Exercises and Problems

A hallmark of this text, nearly 3400 end-of-chapter exercises and problems are arranged in order of difficulty, starting with the exercises (labeled "E"), followed by the problems (labeled "P"). Particularly difficult problems are identified with an asterisk (*). Answers to all the odd-numbered exercises and problems are provided in the back of the book. New electronic computation problems, which require the use of math packages and graphing calculators, have been added to many of the chapters.

FIGURE 10-44 Problem 56.

57P. Two 22.7 kg ice sleds are placed a short distance apart, one directly behind the other, as shown in Fig. 10-45. A 3.63 kg cat, standing on one sled, jumps across to the other and immediately back to the first. Both jumps are made at a speed of 3.05 m/s relative to the ice. Find the final speeds of the two sleds.

FIGURE 10-45 Problem 57.

58P. The bumper of a 1200 kg car is designed so that it can just absorb all the energy when the car runs head-on into a solid wall at 5.00 km/h. The car is involved in a collision in which it runs at 70.0 km/h into the rear of a 900 kg car moving at 60.0 km/h in the same direction. The 900 kg car is accelerated to 70.0 km/h as a result of the collision. (a) What is the speed of the 1200 kg car immediately after impact? (b) What is the ratio of the kinetic energy absorbed in the collision to that which can be absorbed by the bumper of the 1200 kg car?

59P. A railroad freight car weighing 32 tons and traveling at 5.0

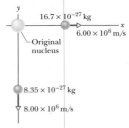

FIGURE 10-46 Exercise 62.

63E. In a game of pool, the cue ball strikes anothe at rest. After the collision, the cue ball moves at 3.5 line making an angle of 22.0° with its original dir tion, and the second ball has a speed of 2.00 m/s angle between the direction of motion of the secon original direction of motion of the cue ball and (b speed of the cue ball. (c) Is kinetic energy conserv

64E. Two vehicles A and B are traveling west and tively, toward the same intersection, where they co together. Before the collision, A (total weight 2700 with a speed of 40 mi/h and B (total weight 3600 lt of 60 mi/h. Find the magnitude and direction of the v (interlocked) vehicles immediately after the collisi

65E. In a game of billiards, the cue ball is given a V and strikes the pack of 15 stationary balls. All engage in nume____ ____ and ball–cushion col time lat____ ____ some accident) all

Brief Contents

Contents

CHAPTER 18

WAVES—II   *425*

How does a bat detect a moth in total darkness?

CHAPTER 19

TEMPERATURE, HEAT, AND THE FIRST LAW OF THERMODYNAMICS   *453*

Why are black robes worn in extremely hot climates?

CHAPTER 20

THE KINETIC THEORY OF GASES   *484*

When a room's air is heated, does the internal energy of the air increase?

CHAPTER 21

ENTROPY AND THE SECOND LAW OF THERMODYNAMICS *509*

What in the world gives direction to time?

APPENDICES *A1*

ANSWERS TO CHECKPOINTS AND ODD-NUMBERED QUESTIONS, EXERCISES AND PROBLEMS *AN1*

INDEX *I1*

FIFTH EDITION

FUNDAMENTALS OF
PHYSICS

13
Equilibrium and Elasticity

Rock climbing may be the ultimate physics exam. Failure can mean death, and even "partial credit" can mean severe injury. For example, in a long chimney climb, in which your shoulders are pressed against one wall of a wide, vertical fissure and your feet are pressed against the opposite wall, you need to rest occasionally or you will fall due to exhaustion. Here the exam consists of a single question: What can you do to relax your push on the walls in order to rest? If you relax without considering the physics, the walls will not hold you up. So, what is the answer to this life-and-death, one-question exam?

13-1 EQUILIBRIUM

Consider these objects: (1) a book resting on a table, (2) a hockey puck sliding across a frictionless surface with constant velocity, (3) the rotating blades of a ceiling fan, and (4) the wheel of a bicycle that is traveling along a straight path at constant speed. For each of these four objects:

1. The linear momentum **P** of its center of mass is constant.

2. Its angular momentum **L** about its center of mass, or about any other point, is also constant.

We say that such objects are in **equilibrium.** The two requirements for equilibrium are then

$$\mathbf{P} = \text{a constant} \quad \text{and} \quad \mathbf{L} = \text{a constant}. \quad (13\text{-}1)$$

Our concern in this chapter is with situations in which the constants in Eq. 13-1 are in fact zero. That is, we are concerned largely with objects that are not moving in any way—either in translation or in rotation—in the reference frame from which we observe them. Such objects are in **static equilibrium.** Of the four objects mentioned at the beginning of this section, only one—the book resting on the table—is in static equilibrium.

The balancing rock of Fig. 13-1 is another example of an object that, for the present at least, is in static equilib-

rium. It shares this property with countless other structures, such as cathedrals, houses, filing cabinets, and taco stands, that remain stationary over time.

As we discussed in Section 8-5, if a body returns to a state of static equilibrium after having been displaced from it by a force, the body is said to be in *stable* static equilibrium. A marble placed at the bottom of a hemispherical bowl is an example. If, however, a small force can displace the body and end the equilibrium, the body is in *unstable* static equilibrium.

For example, suppose we balance a domino as in Fig. 13-2*a*, with the domino's center of mass vertically above a supporting edge. The torque about the supporting edge due to the domino's weight **W** is zero, because the line of action of **W** is through that edge. Since the weight then cannot cause the domino to rotate about the edge, the domino is in equilibrium. Of course, even a slight force on it due to some chance disturbance ends the equilibrium: as the weight's line of action moves to one side of the supporting edge (as in Fig. 13-2*b*), the torque due to the weight increases the domino's rotation. Thus the domino in Fig. 13-2*a* is in unstable static equilibrium.

The domino in Fig. 13-2*c* is not quite as unstable. To topple this domino, a force would have to rotate it through and beyond the balance position of Fig. 13-2*a*, in which the center of mass is above a supporting edge. So, a slight force will not topple this domino, but a vigorous flick of the finger against the domino certainly will. (If we arrange a chain of such upright dominos, a finger flick against the first can cause the whole chain to fall.)

FIGURE 13-1 A balancing rock near Petrified Forest National Park in Arizona. Although its perch seems precarious, the rock is in static equilibrium.

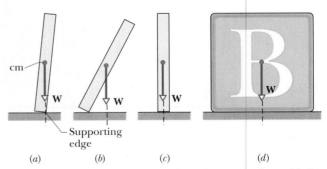

FIGURE 13-2 (*a*) A domino balanced on one edge, with its center of mass vertically above that edge. The domino's weight **W** is directed through the supporting edge. (*b*) If the domino is rotated even slightly from the balanced orientation, its weight **W** causes a torque that increases the rotation. (*c*) A domino upright on a narrow side is somewhat more stable than the domino in (*a*). (*d*) A square block is even more stable.

FIGURE 13-3 A construction worker balanced above New York City is in static equilibrium but is more stable parallel to the beam than perpendicular to it.

The child's square block in Fig. 13-2d is even more stable because its center of mass would have to be moved even farther to get it to pass above a supporting edge. A flick of the finger may not topple the block. (This is why you never see a chain of toppling square blocks.) The worker in Fig. 13-3 is like both the domino and the square block: parallel to the beam, his stance is wide and he is stable; perpendicular to the beam, his stance is narrow and he is unstable (and at the mercy of a chance gust of wind).

The analysis of static equilibrium is very important in engineering practice. The design engineer must isolate and identify all the external forces and torques that may act on a structure and, by good design and wise choice of materials, make sure that the structure will tolerate these loads. Such analysis is necessary to make sure, for example, that bridges do not collapse under their traffic and wind loads, and that the landing gear of aircraft will survive the shock of rough landings.

13-2 THE REQUIREMENTS OF EQUILIBRIUM

The translational motion of a body is governed by Newton's second law in its linear momentum form, given by Eq. 9-28 as

$$\sum \mathbf{F}_{ext} = \frac{d\mathbf{P}}{dt}. \qquad (13-2)$$

If the body is in translational equilibrium, that is, if **P** is a constant, then $d\mathbf{P}/dt = 0$ and we must have

$$\sum \mathbf{F}_{ext} = 0 \qquad \text{(balance of forces).} \qquad (13-3)$$

The rotational motion of a body is governed by Newton's second law in its angular momentum form, given by Eq. 12-37 as

$$\sum \boldsymbol{\tau}_{ext} = \frac{d\mathbf{L}}{dt}. \qquad (13-4)$$

If the body is in rotational equilibrium, that is, if **L** is a constant, then $d\mathbf{L}/dt = 0$ and we must have

$$\sum \boldsymbol{\tau}_{ext} = 0 \qquad \text{(balance of torques).} \qquad (13-5)$$

Thus the two requirements for a body to be in equilibrium are as follows:

1. The vector sum of all the external forces that act on the body must be zero.

2. The vector sum of all the external torques that act on the body, measured about *any* possible point, must also be zero.

These requirements obviously hold for *static* equilibrium as well as the more general equilibrium in which **P** and **L** are constant but not zero.

Equations 13-3 and 13-5, as vector equations, are each equivalent to three independent scalar equations, one for each direction of the coordinate axes:

Balance of forces	Balance of torques	
$\sum F_x = 0$	$\sum \tau_x = 0$	
$\sum F_y = 0$	$\sum \tau_y = 0$	(13-6)
$\sum F_z = 0$	$\sum \tau_z = 0$	

For convenience, we have dropped the subscript ext.

We shall simplify matters by considering only situations in which the forces that act on the body lie in the *xy* plane. This means that the only torques that can act on the body must tend to cause rotation around an axis parallel to the *z* axis. With this assumption, we eliminate one force equation and two torque equations from Eq. 13-6, leaving

$$\sum F_x = 0 \qquad \text{(balance of forces)}, \qquad (13\text{-}7)$$

$$\sum F_y = 0 \qquad \text{(balance of forces)}, \qquad (13\text{-}8)$$

$$\sum \tau_z = 0 \qquad \text{(balance of torques)}. \qquad (13\text{-}9)$$

Here, F_x and F_y are, respectively, the x and the y components of the external forces that act on the body, and τ_z represents the torques that these forces exert either about the z axis or about *any* axis parallel to it.

A hockey puck sliding at constant velocity over ice satisfies Eqs. 13-7, 13-8, and 13-9 and is thus in equilibrium *but not in static equilibrium*. For static equilibrium, the linear momentum **P** of the puck must be not only constant but also zero; the puck must be resting on the ice. Thus there is another requirement for static equilibrium:

---

3. The linear momentum **P** of the body must be zero.

CHECKPOINT 1: The figure gives six overhead views of a uniform rod on which two or more forces act perpendicularly to the rod. If the magnitudes of the forces are adjusted properly (but kept nonzero), in which situations can the rod be in static equilibrium?

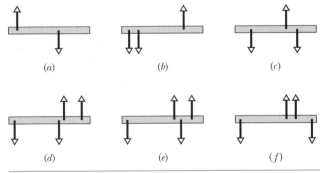

(a)   (b)   (c)

(d)   (e)   (f)

13-3 THE CENTER OF GRAVITY

The weight **W** of an extended body is the vector sum of the gravitational forces acting on the individual elements (the atoms) of the body. Instead of considering all those individual elements, we can say that a single force **W** effectively acts at a single point called the **center of gravity** (cg) of the body. By *effectively* we mean that if the forces on the individual elements were somehow turned off and force **W** at the center of gravity turned on, the net force and the net torque (about any point) acting on the body would not change.

Michel Menin walks a taut wire 10,335 ft above French farmland, adjusting the position of a heavy pole to keep the center of mass of the Menin–pole system over the wire in spite of wind gusts.

Up until now, we have assumed that the weight **W** acts at the center of mass (cm) of the body, which is equivalent to assuming that the center of gravity is at the center of mass. Here we show that this assumption is valid *provided* the acceleration **g** due to the gravitational force is constant over the body.

Figure 13-4a shows an extended body, of mass M, and one of its elements, of mass m_i. Each such element has weight $m_i\mathbf{g}_i$, where $\mathbf{g}_i$ is the acceleration due to the gravitational force at the location of the element. In Fig. 13-4a, each weight $m_i\mathbf{g}_i$ produces a torque τ_i on the element about the origin O. Using Eq. 11-32 ($\tau = r_\perp F$), we can write torque τ_i as

$$\tau_i = x_i m_i g_i, \qquad (13\text{-}10)$$

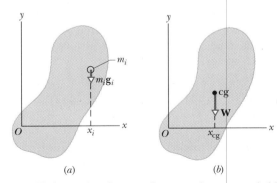

FIGURE 13-4 (a) An element of mass m_i in an extended body has weight $m_i\mathbf{g}_i$, with moment arm x_i about the origin O of a coordinate system. (b) The weight **W** of a body is said to act at the center of gravity (cg) of the body. The moment arm of **W** about O is equal to x_{cg}.

where x_i is the moment arm $r_\perp$ of force $m_i\mathbf{g}_i$. The net torque on all the elements of the body is then

$$\tau_{\text{net}} = \sum \tau_i = \sum x_i m_i g_i. \qquad (13\text{-}11)$$

Figure 13-4b shows the body's weight $\mathbf{W}$ acting at the body's center of gravity. From Eq. 11-32, the torque about O due to $\mathbf{W}$ is

$$\tau = x_{\text{cg}}W, \qquad (13\text{-}12)$$

where x_{cg} is the moment arm of $\mathbf{W}$. The body's weight $\mathbf{W}$ is equal to the sum of the weights $m_i\mathbf{g}_i$ of its elements. So we can substitute $\sum m_i g_i$ for W in Eq. 13-12 to write

$$\tau = x_{\text{cg}} \sum m_i g_i. \qquad (13\text{-}13)$$

By definition, the torque due to weight $\mathbf{W}$ acting at the center of gravity is equal to the net torque due to the weights of all the individual elements of the body. So, τ in Eq. 13-13 is equal to τ_{net} in Eq. 13-11, and we can write

$$x_{\text{cg}} \sum m_i g_i = \sum x_i m_i g_i \qquad (13\text{-}14)$$

If the accelerations g_i are all equal, we can cancel them from the two sides of Eq. 13-14. Substituting the body's mass M for $\sum m_i$ on the left side of Eq. 13-14 and then rearranging yield

$$x_{\text{cg}} = \frac{1}{M} \sum x_i m_i. \qquad (13\text{-}15)$$

With Eq. 9-4 we see that the right side of Eq. 13-15 gives the coordinate x_{cm} for the center of mass. We can now write

$$x_{\text{cg}} = x_{\text{cm}}. \qquad (13\text{-}16)$$

Thus, the body's center of gravity and its center of mass have the same x coordinate.

We can generalize this result for three dimensions by using vector notation. The generalized result is: If the gravitational acceleration is constant over a body, the body's center of gravity is at its center of mass. For the rest of this book, we shall assume that these points coincide.

From Eq. 13-12, we see that the torque due to the weight of a body is zero only if the moment arm x_{cg} is zero. This means that if the body is suspended from some arbitrary point S about which it can rotate, it rotates (due to the torque $\tau = x_{\text{cg}}W$ about S) until x_{cg} is zero. Its center of gravity is then vertically below the suspension point, as in Figs. 13-5a and b, and the body is in equilibrium. If the body is suspended at its center of gravity, as in Fig. 13-5c, then for any orientation of the body, x_{cg} is zero and the body is in equilibrium.

13-4 SOME EXAMPLES OF STATIC EQUILIBRIUM

In this section we examine six sample problems involving static equilibrium. In each, we select a system of one or more objects to which we apply the equations of equilibrium (Eqs. 13-7, 13-8, and 13-9). The forces involved in the equilibrium are all in the xy plane, which means that the torques involved are parallel to the z axis. Thus, in applying Eq. 13-9, the balance of torques, we select an axis parallel to the z axis about which to calculate the torques. Although Eq. 13-9 is satisfied for *any* such choice of axis, we shall see that certain choices simplify the application of Eq. 13-9 by eliminating one or more unknown force terms.

SAMPLE PROBLEM 13-1

A uniform beam of length L whose mass m is 1.8 kg rests with its ends on two digital scales, as in Fig. 13-6a. A uniform block whose mass M is 2.7 kg rests on the beam, its center a distance $L/4$ from the beam's left end. What do the scales read?

We choose as our system the beam and the block, taken together. Figure 13-6b is a free-body diagram for this system, showing all the forces that act on it. The scales push upward at the ends of the beam with forces $\mathbf{F}_l$ and $\mathbf{F}_r$. The magnitudes of these two forces are the scale readings that we seek. The weight of the beam, $m\mathbf{g}$, acts downward at the beam's center of mass. Similarly, $M\mathbf{g}$, the weight of the block, acts downward at *its* center of mass. In Fig. 13-6b, the block is represented by a dot within the boundary of the beam, and the vector $M\mathbf{g}$ is drawn with its tail on that dot. (In drawing Fig. 13-6b from 13-6a, the vector $M\mathbf{g}$ is shifted in the direction it points, that is, along its line of action. The shift does not alter $M\mathbf{g}$, or a torque due to $M\mathbf{g}$ about any axis perpendicular to the figure.)

Our system is in static equilibrium so that the balance of forces equations (Eqs. 13-7 and 13-8) and the balance of torques equation (Eq. 13-9) apply. We solve this problem in two equivalent ways.

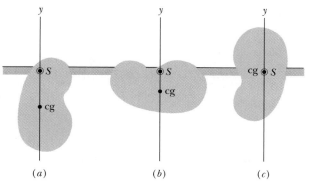

FIGURE 13-5 A body free to rotate about a suspension point S will rotate until its center of gravity is vertically below S as in (a) and (b), unless the center of gravity is at S as in (c).

FIRST SOLUTION: The forces have no x components, so Eq. 13-7, which is $\Sigma F_x = 0$, provides no information. Equation 13-8 gives, for the magnitudes of the y components,

$$\sum F_y = F_l + F_r - Mg - mg = 0. \qquad (13\text{-}17)$$

We have two unknown forces (F_l and F_r), but we cannot find them separately because we have only one equation. Fortunately, we have another equation at hand, namely, Eq. 13-9, the balance of torques equation.

We can apply Eq. 13-9 to *any* axis perpendicular to the plane of Fig. 13-6. Let us choose an axis through the left end of the beam. We take as positive those torques that—acting alone—would produce a counterclockwise rotation of the beam about our axis. We then have, from Eq. 13-9,

$$\sum \tau_z = (F_l)(0) - (Mg)(L/4) - (mg)(L/2) + (F_r)(L)$$
$$= 0,$$

or
$$F_r = (g/4)(2m + M)$$
$$= (\tfrac{1}{4})(9.8 \text{ m/s}^2)(2 \times 1.8 \text{ kg} + 2.7 \text{ kg})$$
$$= 15 \text{ N}. \qquad \text{(Answer)} \quad (13\text{-}18)$$

Note how choosing an axis that passes through the application point of one of the unknown forces, F_l, eliminates that force

from Eq. 13-9 and allows us to solve directly for the other force. *Such a choice can help simplify a problem.*

If we solve Eq. 13-17 for F_l and substitute known quantities, we find

$$F_l = (M + m)g - F_r$$
$$= (2.7 \text{ kg} + 1.8 \text{ kg})(9.8 \text{ m/s}^2) - 15 \text{ N}$$
$$= 29 \text{ N}. \qquad \text{(Answer)}$$

SECOND SOLUTION: As a check, let us solve this problem in a different way, by applying the balance of torques equation about two different axes. Choosing first an axis through the left end of the beam, as we did above, we find Eq. 13-18 and the solution $F_r = 15$ N.

For an axis passing through the right end of the beam, Eq. 13-9 yields

$$\sum \tau_z = - (F_l)(L) + (Mg)(3L/4) + (mg)(L/2) + (F_r)(0)$$
$$= 0.$$

Solving for F_l, we find

$$F_l = (g/4)(2m + 3M)$$
$$= (\tfrac{1}{4})(9.8 \text{ m/s}^2)(2 \times 1.8 \text{ kg} + 3 \times 2.7 \text{ kg})$$
$$= 29 \text{ N}, \qquad \text{(Answer)}$$

in agreement with our earlier result. Note that the length of the beam enters this problem not explicitly but only as it affects the mass of the beam.

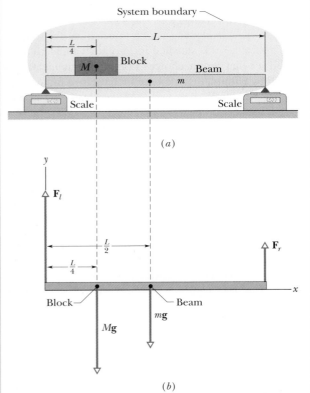

FIGURE 13-6 Sample Problem 13-1. (*a*) A beam of mass m supports a block of mass M. The system boundary is marked. (*b*) A free-body diagram, showing the forces that act on the system *beam + block*.

C**HECKPOINT 2:** The figure gives an overhead view of a uniform rod in static equilibrium. (a) Can you find the magnitudes of unknown forces $\mathbf{F}_1$ and $\mathbf{F}_2$ by balancing the forces? (b) If you wish to find the magnitude of force $\mathbf{F}_2$ by using a single equation, where should you place a rotational axis? (c) The magnitude of $\mathbf{F}_2$ turns out to be 65 N. What then is the magnitude of $\mathbf{F}_1$?

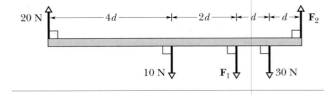

SAMPLE PROBLEM 13-2

A bowler holds a bowling ball whose mass M is 7.2 kg in the palm of his hand. As Fig. 13-7*a* shows, his upper arm is vertical and his lower arm is horizontal. What forces must the biceps muscle and the bony structure of the upper arm exert on the lower arm? The forearm and hand together have a mass m

of 1.8 kg; the needed dimensions are as shown in Fig. 13-7a.

SOLUTION: Our system is the lower arm and the bowling ball, taken together. Figure 13-7b shows a free-body diagram of the system. (The ball is represented by a dot within the boundary of the lower arm; the ball's weight vector $M\mathbf{g}$ has its tail on that dot. In drawing Fig. 13-7b from 13-7a, the vector $M\mathbf{g}$ is shifted along its line of action; the shift does not alter $M\mathbf{g}$, or a torque due to $M\mathbf{g}$ about any axis perpendicular to the figure.) The unknown forces are $\mathbf{T}$, the force exerted by the biceps muscle, and $\mathbf{F}$, the force exerted by the upper arm on the lower arm. The forces are all vertical.

From Eq. 13-8, which is $\Sigma F_y = 0$, we find

$$\sum F_y = T - F - mg - Mg = 0. \qquad (13\text{-}19)$$

Applying Eq. 13-9 about an axis through O at the elbow, and taking torques that would cause counterclockwise rotations as positive, we obtain

$$\sum \tau_z = (F)(0) + (T)(d) - (mg)(D) - (Mg)(L)$$
$$= 0. \qquad (13\text{-}20)$$

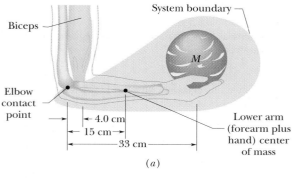

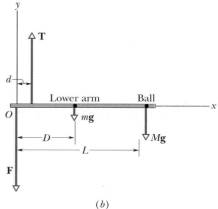

(a)

(b)

FIGURE 13-7 Sample Problem 13-2. (a) A hand holds a bowling ball. The system boundary is marked. (b) A free-body diagram of the *lower arm + ball* system, showing the forces that act. The vectors are not to scale; the powerful forces exerted by the biceps muscle and the elbow joint are many times larger than either weight.

By choosing our axis to pass through point O, we have eliminated the variable F from Eq. 13-20. Solved for T, it yields

$$T = g\,\frac{mD + ML}{d}$$
$$= (9.8 \text{ m/s}^2)\,\frac{(1.8 \text{ kg})(15 \text{ cm}) + (7.2 \text{ kg})(33 \text{ cm})}{4.0 \text{ cm}}$$
$$= 648 \text{ N} \approx 650 \text{ N}. \qquad \text{(Answer)}$$

Thus the biceps muscle must pull up on the forearm with a force that is about nine times larger than the weight of the bowling ball—holding the ball as in Fig. 13-7a is difficult.

If we now solve Eq. 13-19 for F and substitute known quantities into it, we find

$$F = T - g(M + m)$$
$$= 648 \text{ N} - (9.8 \text{ m/s}^2)(7.2 \text{ kg} + 1.8 \text{ kg})$$
$$= 560 \text{ N}, \qquad \text{(Answer)}$$

which is about eight times the weight of the bowling ball.

SAMPLE PROBLEM 13-3

A ladder whose length L is 12 m and whose mass m is 45 kg rests against a wall. Its upper end is a distance h of 9.3 m above the ground, as in Fig. 13-8a. The center of mass of the ladder is one-third of the way up the ladder. A firefighter whose mass M is 72 kg climbs the ladder until her center of mass is halfway up. Assume that the wall, but not the ground, is frictionless. What forces are exerted on the ladder by the wall and by the ground?

SOLUTION: Figure 13-8b shows a free-body diagram of the *firefighter + ladder* system. (The firefighter is represented by a dot within the boundary of the ladder; her weight vector $M\mathbf{g}$ has its tail on that dot. In drawing Fig. 13-8b from 13-8a, the vector $M\mathbf{g}$ is shifted along its line of action; the shift does not alter $M\mathbf{g}$, or a torque due to $M\mathbf{g}$ about any axis perpendicular to the figure.) The wall exerts a horizontal force $\mathbf{F}_w$ on the ladder; it can exert no vertical force because the wall–ladder contact is assumed to be frictionless. The ground exerts a force $\mathbf{F}_g$ on the ladder, with horizontal component F_{gx} (due to friction) and vertical component F_{gy} (the usual normal force). We choose coordinate axes as shown, with the origin O at the point where the ladder meets the ground. The distance a from the wall to the foot of the ladder is readily found from

$$a = \sqrt{L^2 - h^2} = \sqrt{(12 \text{ m})^2 - (9.3 \text{ m})^2} = 7.58 \text{ m}.$$

From Eqs. 13-7 and 13-8, the balance of forces equations, we have for the system, respectively,

$$\sum F_x = F_w - F_{gx} = 0 \qquad (13\text{-}21)$$

and

$$\sum F_y = F_{gy} - Mg - mg = 0. \qquad (13\text{-}22)$$

Equation 13-22 yields

$$F_{gy} = g(M + m) = (9.8 \text{ m/s}^2)(72 \text{ kg} + 45 \text{ kg})$$
$$= 1146.6 \text{ N} \approx 1100 \text{ N}. \qquad \text{(Answer)}$$

We next balance the torques acting on the system, choosing an axis through O, perpendicular to the plane of the figure. The moment arms about O for $\mathbf{F}_w$, $M\mathbf{g}$, $m\mathbf{g}$, $\mathbf{F}_{gx}$ and $\mathbf{F}_{gy}$ are h, $a/2$, $a/3$, zero, and zero, respectively. The zero moment arms for $\mathbf{F}_{gx}$ and $\mathbf{F}_{gy}$ mean that these forces produce zero torque about O. From Eq. 13-9, the balance of torques equation, we then have

$$\sum \tau_z = -(F_w)(h) + (Mg)(a/2) + (mg)(a/3)$$
$$= 0. \qquad \text{(13-23)}$$

Solving Eq. 13-23 for F_w, we find that

$$F_w = \frac{ga(M/2 + m/3)}{h}$$
$$= \frac{(9.8 \text{ m/s}^2)(7.58 \text{ m})(72/2 \text{ kg} + 45/3 \text{ kg})}{9.3 \text{ m}}$$
$$= 407 \text{ N} \approx 410 \text{ N}. \qquad \text{(Answer)}$$

From Eq. 13-21 we then have

$$F_{gx} = F_w = 410 \text{ N}. \qquad \text{(Answer)}$$

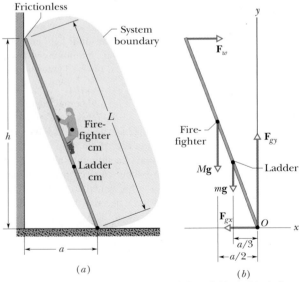

(a)

(b)

FIGURE 13-8 Sample Problems 13-3 and 13-4. (*a*) A firefighter climbs halfway up a ladder that is leaning against a frictionless wall. The ground is not frictionless. (*b*) A free-body diagram, showing the forces that act on the firefighter–ladder system. The origin O of a coordinate system is chosen at the point of application of the unknown force $\mathbf{F}_g$ (whose vector components $\mathbf{F}_{gx}$ and $\mathbf{F}_{gy}$ are shown). This choice simplifies the solution for the other unknown force, $\mathbf{F}_w$.

SAMPLE PROBLEM 13-4

In Sample Problem 13-3, let the coefficient of static friction μ_s between the ladder and the ground be 0.53. How far up the ladder can the firefighter go before the ladder starts to slip?

SOLUTION: The forces that act have the same labels as in Fig. 13-8. Let q be the fraction of the ladder's length the firefighter climbs before slipping occurs (she is then a horizontal distance qa from O). At the onset of slipping, we have

$$F_{gx} = \mu_s F_{gy}, \qquad \text{(13-24)}$$

in which F_{gx} is the static frictional force (usually symbolized as f_s) and F_{gy} is the normal force (usually symbolized as N).

If we apply Eq. 13-9, the balance of torques equation, about an axis through O we have, also at the onset of slipping,

$$\sum \tau_z = -(F_w)(h) + (Mg)(qa) + (mg)(a/3) = 0,$$

or

$$F_w = \frac{ga}{h} (\tfrac{1}{3}m + Mq). \qquad \text{(13-25)}$$

This equation shows us that as the firefighter climbs the ladder, that is, as q increases, the force F_w exerted by the wall must increase if equilibrium is to be maintained. To find q at the onset of slipping, we must first find F_w.

Equation 13-7, the balance of forces equation for the x direction, gives

$$\sum F_x = F_w - F_{gx} = 0.$$

If we combine this equation with Eq. 13-24 we find, for the ladder at the onset of slipping,

$$F_w = F_{gx} = \mu_s F_{gy}. \qquad \text{(13-26)}$$

From Eq. 13-8, the balance of forces equation for the y direction, we have

$$\sum F_y = F_{gy} - Mg - mg,$$

or

$$F_{gy} = (M + m)g. \qquad \text{(13-27)}$$

Combining Eqs. 13-26 and 13-27, we find

$$F_w = \mu_s g(M + m). \qquad \text{(13-28)}$$

If, finally, we combine Eqs. 13-25 and 13-28 and solve for q, we have

$$q = \frac{\mu_s h}{a} \frac{(M + m)}{M} - \frac{m}{3M} \qquad \text{(13-29)}$$
$$= \frac{(0.53)(9.3 \text{ m})}{7.6 \text{ m}} \frac{(72 \text{ kg} + 45 \text{ kg})}{72 \text{ kg}} - \frac{45 \text{ kg}}{(3)(72 \text{ kg})}$$
$$= 0.85. \qquad \text{(Answer)}$$

The firefighter can climb 85% of the way up the ladder before it starts to slip.

You can show from Eq. 13-29 that the firefighter can climb all the way up the ladder (which corresponds to $q = 1$) without slipping if $\mu_s > 0.61$. On the other hand, the ladder will slip when weight is put on the first rung (which corresponds to $q \approx 0$) if $\mu_s < 0.11$.

CHECKPOINT **3:** In the figure, a stationary 5 kg rod AC is held against a wall by a rope and friction between rod and wall. The uniform rod is 1 m long, and angle $\theta = 30°$. (a) If you are to find the magnitude of the force **T** on the rod from the rope with a single equation, at what labeled point should a rotational axis be placed? With that choice of axis and counterclockwise torques positive, what is the sign of (b) the torque τ_w due to the rod's weight and (c) the torque τ_r due to the pull on the rod by the rope? (d) Is τ_r greater than, less than, or equal to τ_w?

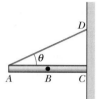

SAMPLE PROBLEM 13-5

Figure 13-9a shows a safe, whose mass M is 430 kg, hanging by a rope from a boom whose dimensions a and b are 1.9 m and 2.5 m, respectively. The boom's uniform beam has a mass m of 85 kg; the mass of the horizontal cable is negligible.

(a) Find the tension T in the cable.

SOLUTION: Figure 13-9b is a free-body diagram of the beam, which we take as our system. The forces acting on it are the tension **T** exerted by the cable, the weight $M\mathbf{g}$ of the safe, the weight $m\mathbf{g}$ of the beam itself (acting at the center of the beam), and the force components F_h and F_v exerted on the beam by the hinge that fastens the beam to the wall.

Let us apply Eq. 13-9, the balance of torques equation, to an axis through the hinge perpendicular to the plane of the figure. Taking torques that would cause counterclockwise rotations as positive, we have

$$\sum \tau_z = (T)(a) - (Mg)(b) - (mg)(\tfrac{1}{2}b) = 0.$$

By our wise choice of an axis, we have eliminated the unknown forces F_h and F_v from this equation (since they create no torque about that axis), leaving only the unknown force T. Solving for T yields

$$T = \frac{gb(M + \tfrac{1}{2}m)}{a}$$

$$= \frac{(9.8 \text{ m/s}^2)(2.5 \text{ m})(430 \text{ kg} + 85/2 \text{ kg})}{1.9 \text{ m}}$$

$$= 6090 \text{ N} \approx 6100 \text{ N.} \qquad \text{(Answer)}$$

(b) Find the force components F_h and F_v exerted on the beam by the hinge.

SOLUTION: We now apply the balance of forces equations. From Eq. 13-7 we have

$$\sum F_x = F_h - T = 0,$$

and so $\qquad F_h = T = 6090 \text{ N} \approx 6100 \text{ N.} \qquad$ (Answer)

From Eq. 13-8 we have

$$\sum F_y = F_v - mg - Mg = 0,$$

and so

$$F_v = g(m + M) = (9.8 \text{ m/s}^2)(85 \text{ kg} + 430 \text{ kg})$$

$$= 5047 \text{ N} \approx 5000 \text{ N.} \qquad \text{(Answer)}$$

(c) What is the magnitude F of the net force exerted by the hinge on the beam?

SOLUTION: From the figure we see that

$$F = \sqrt{F_h^2 + F_v^2}$$

$$= \sqrt{(6090 \text{ N})^2 + (5047 \text{ N})^2} \approx 7900 \text{ N.} \quad \text{(Answer)}$$

Note that F is substantially greater than either the combined weights of the safe and the beam, 5000 N, or the tension in the horizontal wire, 6100 N.

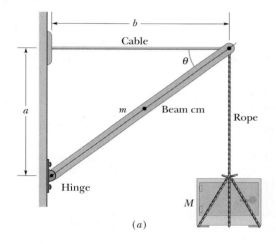

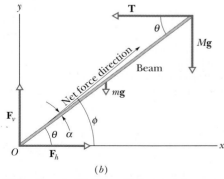

FIGURE 13-9 Sample Problem 13-5. (a) A heavy safe is hung from a boom consisting of a horizontal steel cable and a uniform beam. (b) A free-body diagram for the beam. Note that the net force acting on the beam due to $\mathbf{F}_v$ and $\mathbf{F}_h$ does not point directly along the beam axis.

(d) What is the angle α between the direction of the net force **F** exerted on the beam by the hinge and the center line of the beam?

SOLUTION: From the figure we see that

$$\theta = \tan^{-1}\frac{a}{b} = \tan^{-1}\frac{1.9 \text{ m}}{2.5 \text{ m}} = 37.2°,$$

and

$$\phi = \tan^{-1}\frac{F_v}{F_h} = \tan^{-1}\frac{5047 \text{ N}}{6090 \text{ N}} = 39.6°.$$

So

$$\alpha = \phi - \theta = 39.6° - 37.2° = 2.4°. \quad \text{(Answer)}$$

If the weight of the beam were small enough to neglect, you would find $\alpha = 0$; that is, the hinge force would point directly along the beam axis.

SAMPLE PROBLEM 13-6

In Fig. 13-10, a rock climber with mass $m = 55$ kg rests during a "chimney climb," pressing only with her shoulders and feet against the walls of a fissure of width $w = 1.0$ m. Her center of mass is a horizontal distance $d = 0.20$ m from the wall against which her shoulders are pressed. The coefficient of static friction between her shoes and the wall is $\mu_1 = 1.1$, and between her shoulders and the wall it is $\mu_2 = 0.70$.

(a) What minimum horizontal push must the climber exert on the walls to keep from falling?

SOLUTION: Her horizontal push against the wall is the same at her shoulders and at her feet. That push must be equal in magnitude to the normal force **N** on her due to the wall at each of those places. Thus there is no net horizontal force on her and Eq. 13-7, $\Sigma F_x = 0$, is satisfied.

As her weight $m\mathbf{g}$ acts to slide her down the walls, static frictional forces $\mathbf{f}_1$ and $\mathbf{f}_2$ automatically act upward at feet and shoulders, respectively, to counter the tendency to slide. As long as she does not fall, Eq. 13-8 ($\Sigma F_y = 0$) is satisfied and we have

$$f_1 + f_2 = mg. \quad (13\text{-}30)$$

Let us assume that the climber initially pushes hard against the walls, and then slowly relaxes her push. As she decreases her push, the magnitude N of the normal force decreases, and so do the products $\mu_1 N$ and $\mu_2 N$ that limit the static friction at her feet and shoulders, respectively (see Eq. 6-1).

Suppose N is reduced until the limit $\mu_1 N$ equals the magnitude f_1 of the frictional force at her feet and the limit $\mu_2 N$ equals the magnitude f_2 of the frictional force at her shoulders. She is then on the verge of slipping at both places. If she were to relax her push any more, the resulting decrease in the limits $\mu_1 N$ and $\mu_2 N$ would leave $f_1 + f_2$ less than mg, and she would fall. Thus when she is on the verge of slipping, Eq. 13-30 becomes

$$\mu_1 N + \mu_2 N = mg, \quad (13\text{-}31)$$

which yields

$$N = \frac{mg}{\mu_1 + \mu_2} = \frac{(55 \text{ kg})(9.8 \text{ m/s}^2)}{1.1 + 0.70} = 299 \text{ N} \approx 300 \text{ N}.$$

Thus her minimum horizontal push must be about 300 N.

(b) For that push, what must be the vertical distance h between her feet and shoulders if she is to be stable?

SOLUTION: To satisfy Eq. 13-9, $\Sigma\tau = 0$, the forces acting on her must not create a net torque about *any* rotation axis. Let us consider a rotational axis perpendicular to the page at her shoulders. The net torque about that axis is given by

$$\sum\tau = -f_1 w + Nh + mgd = 0. \quad (13\text{-}32)$$

Solving this for h, setting $f_1 = \mu_1 N$, and substituting $N = 299$ N and other known values, we find

$$h = \frac{f_1 w - mgd}{N} = \frac{\mu_1 Nw - mgd}{N} = \mu_1 w - \frac{mgd}{N}$$

$$= (1.1)(1.0 \text{ m}) - \frac{(55 \text{ kg})(9.8 \text{ m/s}^2)(0.20 \text{ m})}{299 \text{ N}}$$

$$= 0.739 \text{ m} \approx 0.74 \text{ m}. \quad \text{(Answer)}$$

We would find the same required value of h if we chose any other rotation axis perpendicular to the page, such as one at her feet.

(c) What are the magnitudes of the frictional forces supporting the climber?

SOLUTION: For $N = 299$ N, we have

$$f_1 = \mu_1 N = (1.1)(299 \text{ N})$$
$$= 328.9 \text{ N} \approx 330 \text{ N}, \quad \text{(Answer)}$$

and from Eq. 13-30 we then have

$$f_2 = mg - f_1 = (55 \text{ kg})(9.8 \text{ m/s}^2) - 328.9 \text{ N}$$
$$= 210.1 \text{ N} \approx 210 \text{ N}. \quad \text{(Answer)}$$

(d) Is she stable if she exerts the same force (299 N) on the walls when her feet are higher, with $h = 0.37$ m?

SOLUTION: Again $N = 299$ N, and the net torque on the climber about any rotation axis must be zero. For the axis at her shoulders, we obtain f_1 from Eq. 13-32:

$$f_1 = \frac{Nh + mgd}{w}$$

$$= \frac{(299 \text{ N})(0.37 \text{ m}) + (55 \text{ kg})(9.8 \text{ m/s}^2)(0.20 \text{ m})}{1.0 \text{ m}}$$

$$= 218 \text{ N}.$$

This is less than the limit $\mu_1 N$ ($= 329$ N) and thus is possible to attain.

Next we use Eq. 13-30 to find the value of f_2 that will ensure that $\Sigma F_y = 0$:

$$f_2 = mg - f_1 = (55 \text{ kg})(9.8 \text{ m/s}^2) - 218 \text{ N} = 321 \text{ N}.$$

This exceeds the limit $\mu_2 N$ ($= 209$ N) and thus is impossible to attain with a push of 299 N. The only way the climber can avoid slipping when $h = 0.37$ m (or for any value of h less than 0.74 m) is by pushing harder than 299 N on the walls so as to increase the limit $\mu_2 N$.

Similarly, if $h > 0.74$ m, she must also exert a force greater than 299 N on the walls to be stable. Here, then, is the advantage of knowing the physics before you climb a chimney. When you need to rest, you will avoid the (dire) error of novice climbers who place their feet too high or too low. Instead, you will know that there is a "best" distance between shoulders and feet, requiring the least push, and giving you a good chance to rest.

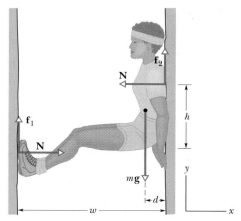

FIGURE 13-10 Sample Problem 13-6. The forces on a climber resting in a rock chimney. The push of the climber on the chimney walls gives rise to the normal forces **N** (which are equal in magnitude) and the frictional forces f_1 and f_2.

13-5 INDETERMINATE STRUCTURES

For the problems of this chapter, we have only three independent equations at our disposal, usually two balance of forces equations and one balance of torques equation about a given axis. Thus if a problem has more than three unknowns, we cannot solve it.

It is easy to find such problems. In Sample Problems 13-3 and 13-4, for example, we could have assumed that there is friction between the wall and the top of the ladder. Then there would have been a vertical frictional force acting where the ladder touches the wall, making a total of four unknown forces. With only three equations, we could not have solved this problem.

Consider also an unsymmetrically loaded car. What are the forces—all different—on the four tires? Again, we cannot find them because we have only three independent equations with which to work. Similarly, we can solve an equilibrium problem for a table with three legs but not one with four legs. Problems like these, in which there are more unknowns than equations, are called **indeterminate.**

And yet, solutions to indeterminate problems exist in the real world. If you rest the tires of the car on four platform scales, each scale will register a definite reading, the sum of the readings being the weight of the car. What is eluding us in our efforts to solve equations for the individual scale readings?

The problem is that we have assumed—without making a great point of it—that the bodies to which we apply the equations of static equilibrium are perfectly rigid. By this we mean that they do not deform when forces are applied to them. Strictly, there are no such bodies. The tires of the car, for example, deform easily under load until the car settles into a position of static equilibrium.

We have all had experience with a wobbly restaurant table, which we usually level by putting folded paper under one of the legs. If a big enough elephant sat on such a table, however, you may be sure that if the table did not collapse, it would deform just like the tires of a car. Its legs would all touch the floor, the forces acting upward on the table legs would all assume definite (and different) values as in Fig. 13-11, and the table would no longer wobble. But how do we find the values of those forces acting on the legs?

To solve such indeterminate equilibrium problems, we must supplement equilibrium equations with some knowledge of *elasticity,* the branch of physics and engineering that describes how real bodies deform when forces are applied to them. The next section provides an introduction to this subject.

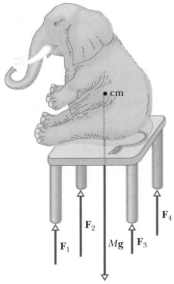

FIGURE 13-11 The table is an indeterminate structure. The four forces on the table legs are different in magnitude and cannot be found from the laws of static equilibrium alone.

CHECKPOINT **4:** A horizontal uniform bar of weight 10 N is to hang from a ceiling by two wires that exert upward forces F_1 and F_2. The figure shows four arrangements for the wires. Which arrangements, if any, are indeterminate (so that we cannot solve for numerical values of F_1 and F_2)?

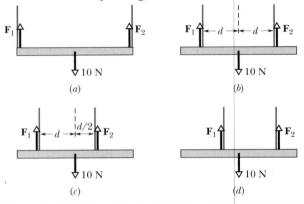

13-6 ELASTICITY

When a large number of atoms come together to form a metallic solid, such as an iron nail, they settle into equilibrium positions in a three-dimensional *lattice,* a repetitive arrangement in which each atom has a well-defined equilibrium distance from its nearest neighbors. The atoms are held together by interatomic forces that are represented by springs in Fig. 13-12.* The lattice is remarkably rigid, which is another way of saying that the "interatomic springs" are extremely stiff. It is for this reason that we perceive many ordinary objects such as metal ladders, tables, and spoons as perfectly rigid. Of course, some ordinary objects, such as garden hoses or rubber gloves, do not strike us as rigid at all. The atoms that make up these objects *do not* form a rigid lattice like that of Fig. 13-12 but are aligned in long flexible molecular chains, each chain being only loosely bound to its neighbors.

All real "rigid" bodies are to some extent **elastic,** which means that we can change their dimensions slightly by pulling, pushing, twisting, or compressing them. To get a feeling for the orders of magnitude involved, consider a vertical steel rod 1 m long and 1 cm in diameter. If you hang a subcompact car from the end of such a rod, the rod

---

*Ordinary metal objects, such as an iron nail, are made up of grains of iron, each grain formed as a more or less perfect lattice, such as that of Fig. 13-12. The forces between the grains are much weaker than the forces that hold the lattice together, so that rupture usually occurs along grain boundaries.

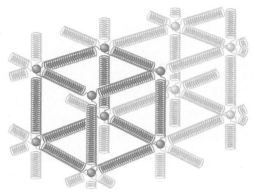

FIGURE 13-12 The atoms of a metallic solid are distributed on a repetitive three-dimensional lattice. The springs represent interatomic forces.

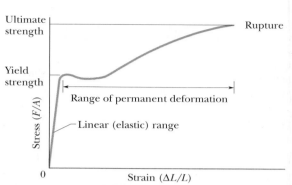

FIGURE 13-14 A stress-strain curve for a steel test specimen such as that of Fig. 13-15. The specimen deforms permanently when the stress is equal to the *yield strength* of the material. It ruptures when the stress is equal to the *ultimate strength* of the material.

will stretch, but only by about 0.5 mm, or 0.05%. Furthermore, the rod will return to its original length when the car is removed.

If you hang two cars from the rod, the rod will be permanently stretched and will not recover its original length when you remove the load. If you hang three cars from the rod, the rod will break. Just before rupture, the elongation of the rod will be less than 0.2%. Although deformations of this size seem small, they are important in engineering practice. (Whether a wing under load will stay on an airplane is obviously important.)

Figure 13-13 shows three ways that a solid might change its dimensions when forces act on it. In Fig. 13-13a, a cylinder is stretched. In Fig. 13-13b, a cylinder is deformed by a force perpendicular to its axis, much as one might deform a pack of cards or a book. In Fig. 13-13c, a

solid object, placed in a fluid under high pressure, is compressed uniformly on all sides. What the three deformation types have in common is that a **stress,** or deforming force per unit area, produces a **strain,** or unit deformation. In Fig. 13-13, *tensile stress* (associated with stretching) is in (*a*), *shearing stress* is in (*b*), and *hydraulic stress* is in (*c*).

The stresses and the strains take different forms in the three cases of Fig. 13-13, but—over the range of engineering usefulness—stress and strain are proportional to each other. The constant of proportionality is called a **modulus of elasticity,** so that

$$\text{stress} = \text{modulus} \times \text{strain.} \qquad (13\text{-}33)$$

Figure 13-14 shows the relation between stress and strain for a steel test cylinder such as that of Fig. 13-15. In a

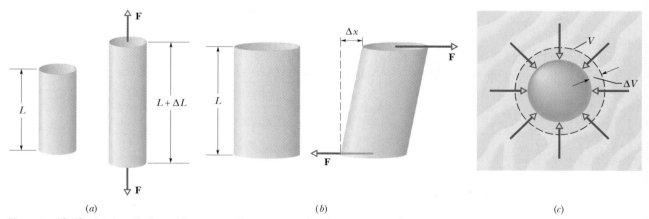

| (a) | (b) | (c) |

FIGURE 13-13 (*a*) A cylinder subject to *tensile stress* stretches by an amount ΔL. (*b*) A cylinder subject to *shearing stress* deforms by an amount Δx, somewhat like a pack of playing cards would. (*c*) A solid sphere subject to uniform *hydraulic stress* from a fluid shrinks in volume by an amount ΔV. All the deformations shown are greatly exaggerated.

FIGURE 13-15 A test specimen, used to determine a stress–strain curve such as that of Fig. 13-14. The change that occurs in a certain length L is measured in a tensile stress–strain test.

standard test, the tensile stress on a test cylinder is slowly increased from zero to the point where the cylinder fractures, and the strain is carefully measured and graphed. For a substantial range of applied stresses, the stress–strain relation is linear, and the specimen recovers its original dimensions when the stress is removed; it is here that Eq. 13-33 applies. If the stress is increased beyond the **yield strength** S_y of the specimen, the specimen becomes permanently deformed. If the stress continues to increase, the specimen eventually ruptures, at a stress called the **ultimate strength** S_u.

Tension and Compression

For simple tension or compression, the stress is defined as F/A, the force divided by the area over which it acts (the force is perpendicular to the area, as you can see in Fig. 13-13a). The strain, or unit deformation, is then the dimensionless quantity $\Delta L/L$, the fractional (or sometimes percentage) change in the length of the specimen. If the specimen is a long rod and the stress does not exceed the yield strength, then not only the entire rod but also every section of it experiences the same strain when a given stress is applied. Because the strain is dimensionless, the modulus in Eq. 13-33 has the same dimensions as the stress, namely, force per unit area.

The modulus for tensile and compressive stresses is called the **Young's modulus** and is represented in engi-

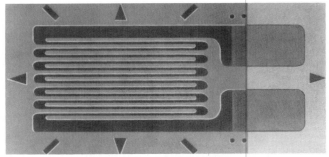

FIGURE 13-16 A strain gauge of overall dimensions 9.8 mm by 4.6 mm. The gauge is fastened with adhesive to the object whose strain is to be measured; it experiences the same strain as the object. The electrical resistance of the gauge varies with the strain, permitting strains up to about 3% to be measured.

neering practice by the symbol E. Equation 13-33 becomes

$$\frac{F}{A} = E\frac{\Delta L}{L}. \qquad (13\text{-}34)$$

The strain $\Delta L/L$ in a specimen can often be measured conveniently with a *strain gauge* (Fig. 13-16). These simple and useful devices, which can be attached directly to operating machinery with adhesives, are based on the principle that the electrical properties of the gauge are dependent on the strain it undergoes.

Although the Young's modulus for an object may be almost the same for tension and compression, the object's ultimate strength may well be different for the two types of stress. Concrete, for example, is very strong in compression but is so weak in tension that it is almost never used in that manner. Table 13-1 shows the Young's modulus and other elastic properties for some materials of engineering interest.

TABLE 13-1
SOME ELASTIC PROPERTIES OF SELECTED MATERIALS OF ENGINEERING INTEREST

MATERIAL	DENSITY ρ (kg/m³)	YOUNG'S MODULUS E (10^9 N/m²)	ULTIMATE STRENGTH S_u (10^6 N/m²)	YIELD STRENGTH S_y (10^6 N/m²)
Steel[a]	7860	200	400	250
Aluminum	2710	70	110	95
Glass	2190	65	50[b]	—
Concrete[c]	2320	30	40[b]	—
Wood[d]	525	13	50[b]	—
Bone	1900	9[b]	170[b]	—
Polystyrene	1050	3	48	—

[a] Structural steel (ASTM-A36).   [b] In compression.   [c] High strength.   [d] Douglas fir.

Shearing

In the case of shearing, the stress is also a force per unit area, but the force vector lies in the plane of the area rather than at right angles to it. The strain is the dimensionless ratio $\Delta x/L$, with the quantities defined as shown in Fig. 13-13b. The corresponding modulus, which is given the symbol G in engineering practice, is called the **shear modulus.** For shearing, Eq. 13-33 is written as

$$\frac{F}{A} = G\frac{\Delta x}{L}. \qquad (13\text{-}35)$$

Shearing stresses play a critical role in the buckling of shafts that rotate under load and in bone fractures caused by bending.

Hydraulic Stress

In Fig. 13-13c, the stress is the fluid pressure p on the object, which, as you will see in Chapter 15, is a force per unit area. The strain is $\Delta V/V$, where V is the original volume of the specimen and ΔV is the absolute value of the change in volume. The corresponding modulus, with symbol B, is called the **bulk modulus** of the material. The object is said to be under *hydraulic compression,* and the pressure can be called the *hydraulic stress.* For this situation, we write Eq. 13-33 as

$$p = B\frac{\Delta V}{V}. \qquad (13\text{-}36)$$

The bulk modulus of water is 2.2×10^9 N/m², and that of steel is 16×10^{10} N/m². The pressure at the bottom of the Pacific Ocean, at its average depth of about 4000 m, is 4.0×10^7 N/m². The fractional compression $\Delta V/V$ of a volume of water due to this pressure is 1.8%; that for a steel object is only about 0.025%. In general, solids—with their rigid atomic lattices—are less compressible than liquids, in which the atoms or molecules are less tightly coupled to their neighbors.

SAMPLE PROBLEM 13-7

A structural steel rod has a radius R of 9.5 mm and a length L of 81 cm. A force F of 6.2×10^4 N (about 7 tons) stretches it along its length.

(a) What is the stress in the rod?

SOLUTION: From its definition, the stress is

$$\text{stress} = \frac{F}{A} = \frac{F}{\pi R^2} = \frac{6.2 \times 10^4\ \text{N}}{(\pi)(9.5 \times 10^{-3}\ \text{m})^2}$$

$$= 2.2 \times 10^8\ \text{N/m}^2. \qquad \text{(Answer)}$$

The yield strength for structural steel is 2.5×10^8 N/m², so that this rod is dangerously close to its yield strength.

(b) What is the elongation of the rod under this load? What is the strain?

SOLUTION: From Eq. 13-34, using the result we have just calculated and the value of E for steel (Table 13-1), we obtain

$$\Delta L = \frac{(F/A)L}{E} = \frac{(2.2 \times 10^8\ \text{N/m}^2)(0.81\ \text{m})}{2.0 \times 10^{11}\ \text{N/m}^2}$$

$$= 8.9 \times 10^{-4}\ \text{m} = 0.89\ \text{mm}. \qquad \text{(Answer)}$$

Thus the strain is

$$\frac{\Delta L}{L} = \frac{8.9 \times 10^{-4}\ \text{m}}{0.81\ \text{m}}$$

$$= 1.1 \times 10^{-3} = 0.11\%. \qquad \text{(Answer)}$$

SAMPLE PROBLEM 13-8

The femur, which is the principal bone of the thigh, has a minimum diameter in an adult male of about 2.8 cm, corresponding to a cross-sectional area A of 6×10^{-4} m². At what compressive load would it break?

SOLUTION: From Table 13-1 we see that the ultimate strength S_u for bone in compression is 170×10^6 N/m². The compressive force at fracture is the force F that produces the stress S_u; so

$$F = S_u A = (170 \times 10^6\ \text{N/m}^2)(6 \times 10^{-4}\ \text{m}^2)$$

$$= 1.0 \times 10^5\ \text{N}. \qquad \text{(Answer)}$$

This is 23,000 lb or 11 tons. Although this is a large force, it can be encountered during, for example, an unskillful parachute landing on hard ground. The force need not be sustained to break the bone; a few milliseconds will do it.

SAMPLE PROBLEM 13-9

A table has three legs that are 1.00 m in length and a fourth leg that is longer by a distance $d = 0.50$ mm, so that the table wobbles slightly. A heavy steel cylinder whose mass M is 290 kg is placed upright on the table (whose mass is much less than M) so that all four legs compress and the table no longer wobbles. The legs are wooden cylinders whose cross-sectional area A is 1.0 cm². The Young's modulus E for the wood is 1.3×10^{10} N/m². Assume that the tabletop remains level and that the legs do not buckle. With what force does the floor push upward on each leg?

SOLUTION: We take the table plus steel cylinder as our system. The situation is like Fig. 13-11, except we now have a steel cylinder on the table. If the tabletop remains level, each of the three short legs must be compressed by the same amount (call it ΔL_3) and thus by the same force F_3. The single long leg must be compressed by a larger amount ΔL_4, by a larger force F_4, and we must have

$$\Delta L_4 = \Delta L_3 + d. \qquad (13\text{-}37)$$

We can rewrite Eq. 13-34 as $\Delta L = FL/EA$. We then use this relationship to substitute for L_4 and L_3 in Eq. 13-37, letting L represent, in turn, the initial length of the three short legs (1.0 m) and the approximate initial length of the long leg (also 1.0 m). Equation 13-37 becomes

$$F_4L = F_3L + dAE. \qquad (13\text{-}38)$$

From Eq. 13-8, the balance of forces in the vertical direction, we have for our system

$$\sum F_y = 3F_3 + F_4 - Mg = 0. \qquad (13\text{-}39)$$

If we solve Eqs. 13-38 and 13-39 for the unknown force F_3, we find

$$F_3 = \frac{Mg}{4} - \frac{dAE}{4L}$$

$$= \frac{(290 \text{ kg})(9.8 \text{ m/s}^2)}{4}$$

$$- \frac{(5.0 \times 10^{-4} \text{ m})(10^{-4} \text{ m}^2)(1.3 \times 10^{10} \text{ N/m}^2)}{(4)(1.00 \text{ m})}$$

$$= 711 \text{ N} - 163 \text{ N} = 548 \text{ N} \approx 550 \text{ N}. \qquad \text{(Answer)}$$

From Eq. 13-39 we then find

$$F_4 = Mg - 3F_3 = (290 \text{ kg})(9.8 \text{ m/s}^2) - 3(548 \text{ N})$$

$$\approx 1200 \text{ N}. \qquad \text{(Answer)}$$

You can show that to reach their equilibrium configuration, the three short legs were each compressed by 0.42 mm and the single long leg by 0.92 mm, the difference being 0.50 mm.

CHECKPOINT **5:** The figure shows a horizontal block that is suspended by two wires, A and B, which are identical except for their original lengths. The center of mass of the block is closer to wire B than to wire A. (a) Measuring torques about the block's center of mass, state whether the torque due to wire A is greater than, less than, or equal to the torque due to wire B. (b) Which wire exerts more force on the block? (c) If the wires are now equal in length, which one was originally shorter?

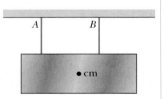

REVIEW & SUMMARY

Static Equilibrium

A rigid body at rest is said to be in **static equilibrium.** For such a body, the vector sum of the external forces acting on it is zero:

$$\sum \mathbf{F}_{\text{ext}} = 0 \qquad \text{(balance of forces).} \qquad (13\text{-}3)$$

If all the forces lie in the xy plane, this vector equation is equivalent to two scalar component equations:

$$\sum F_x = 0 \quad \text{and} \quad \sum F_y = 0 \qquad \begin{array}{c}\text{(balance of}\\ \text{forces).}\end{array} \qquad (13\text{-}7, \ 13\text{-}8)$$

Static equilibrium also implies that the vector sum of the external torques acting on the body about *any* point is zero, or

$$\sum \tau_{\text{ext}} = 0 \qquad \text{(balance of torques).} \qquad (13\text{-}5)$$

If the forces lie in the xy plane, all torque vectors are parallel to the z axis, and Eq. 13-5 is equivalent to the single scalar component equation

$$\sum \tau_z = 0 \qquad \text{(balance of torques).} \qquad (13\text{-}9)$$

Center of Gravity

The gravitational force acts individually on all the individual elements of a body. The net effect of all the individual actions may be found by imagining an equivalent total gravitational force Mg

acting at a specific point called the **center of gravity.** If the gravitational acceleration $\mathbf{g}$ is the same for all the particles in a body, the center of gravity is at the center of mass.

Elastic Moduli

Three **elastic moduli** are used to describe the elastic behavior (deformations) of objects as they respond to the forces that act on them. The **strain** (fractional change in length) is linearly related to the applied **stress** (force per unit area) by the modulus in each case. The general relation is

$$\text{stress} = \text{modulus} \times \text{strain}. \qquad (13\text{-}33)$$

Tension and Compression

When an object is under tension or compression, Eq. 13-33 is written as

$$\frac{F}{A} = E \frac{\Delta L}{L}, \qquad (13\text{-}34)$$

where $\Delta L/L$ is the strain of the object, F is the magnitude of the applied force $\mathbf{F}$ causing the strain, A is the cross-sectional area over which $\mathbf{F}$ is applied (perpendicular to A, as in Fig. 13-13a), and E is the **Young's modulus** for the object. The stress is F/A.

Shearing

When an object is under a shearing stress, Eq. 13-33 is written as

$$\frac{F}{A} = G \frac{\Delta x}{L},$$ (13-35)

where $\Delta x/L$ is the strain of the object, Δx is the displacement of one end of the object in the direction of the applied force **F** (as in Fig. 13-13b), and G is the **shear modulus** of the object. The stress is F/A.

Hydraulic Stress

When an object undergoes *hydraulic compression* due to a stress exerted by a surrounding fluid, Eq. 13-33 is written as

$$p = B \frac{\Delta V}{V},$$ (13-36)

where p is the pressure (*hydraulic stress*) on the object due to the fluid, $\Delta V/V$ (the strain) is the absolute value of the fractional change in the object's volume due to that pressure, and B is the **bulk modulus** of the object.

QUESTIONS

1. Figure 13-17 shows an overhead view of a uniform stick on which four forces act. Suppose we choose a rotational axis through point O, calculate the torques about that axis due to the forces, and find that these torques balance. Will the torques balance if, instead, the rotational axis is chosen to be at (a) point A, (b) point B, or (c) point C? (d) Suppose, instead, that we find that the torques about point O do not balance. Is there another point about which the torques will balance?

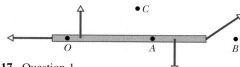

FIGURE 13-17 Question 1.

2. Figure 13-18 shows four overhead views of rotating uniform disks that are sliding across a frictionless floor. Three forces, of magnitude F, $2F$, or $3F$, act on each disk; each force is applied either at the rim, at the center, or halfway between rim and center. The force vectors rotate along with the disks, and, in the "snapshots" of Fig. 13-18, point left or right. Which disks are in equilibrium?

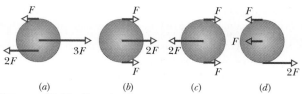

FIGURE 13-18 Question 2.

3. Figure 13-19 shows overhead views of two structures on which three forces act. The directions of the forces are as indicated. If the magnitudes of the forces are adjusted properly (but kept nonzero), which structure can be in static equilibrium?

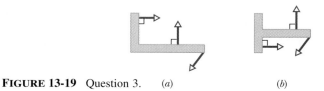

FIGURE 13-19 Question 3.   (a)   (b)

4. Figure 13-20 shows a mobile of toy penguins hanging from a ceiling. Each crossbar is horizontal, has negligible mass, and extends three times as far to the right of the wire supporting it as to the left. Penguin 1 has mass $m_1 = 48$ kg. What are the masses of the other penguins?

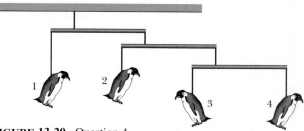

FIGURE 13-20 Question 4.

5. Figure 13-21 shows an overhead view of a metal square lying on a frictionless floor. Three forces, which are drawn to scale, act at the corners of the square. (a) Is the first requirement of equilibrium (in Eq. 13-1) satisfied? (b) Is the second requirement of equilibrium satisfied? (c) If the answer to either (a) or (b) is no, could a fourth force acting on the square then satisfy both requirements of equilibrium?

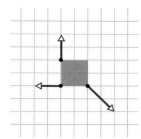

FIGURE 13-21 Question 5.

6. (a) How many different free-standing towers can you build with three of the small (four-projection) Lego blocks? (Two such blocks are stackable one directly above the other, or one displaced half a block length left or right of the other. Mirror-image arrangements count as one.) How many such towers are in (b) stable equilibrium and (c) unstable equilibrium (center of mass almost over an edge)? (d) Which arrangement is most stable (hardest to topple)? Why?

7. Figure 13-22 shows four arrangements in which a painting is suspended from a wall with two identical lengths of wire. The wires in Fig. 13-22b and c all make the same angle with the horizontal. Rank the four arrangements according to the tension in the wires, largest first.

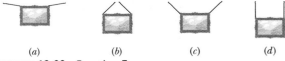

(a)   (b)   (c)   (d)

FIGURE 13-22   Question 7.

8. A ladder leans against a frictionless wall but is prevented from falling because of friction between it and the ground. Suppose you shift the base of the ladder toward the wall. Tell whether the following become larger, smaller, or stay the same: (a) the normal force on the ladder from the ground, (b) the force on the ladder from the wall, (c) the static frictional force on the ladder from the ground, and (d) the maximum value $f_{s,max}$ of the static frictional force.

9. A physical therapist gone wild has constructed the (stationary) assembly of massless pulleys and cords seen in Fig. 13-23. One long cord wraps around all the pulleys and shorter cords suspend pulleys from the ceiling or weights from the pulleys. Except for one, the weights (in newtons) are indicated. (a) What is that last weight? (*Hint:* When a cord loops halfway around a pulley as here, it pulls on the pulley with a net force that is twice the tension in the cord.) (b) What is the tension in the short cord labeled with T?

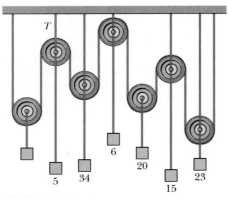

FIGURE 13-23   Question 9.

10. Three piñatas hang from the (stationary) assembly of massless pulleys and cords seen in Fig. 13-24. One long cord runs from the ceiling at the right to the lower pulley at the left. Several shorter cords suspend pulleys from the ceiling or piñatas from the pulleys. The weights (in newtons) of two piñatas are given.

(a) What is the weight of the third piñata? (*Hint:* When a cord loops around a pulley, it pulls on the pulley with a net force that is twice the tension in the cord.) (b) What is the tension in the short cord labeled with T?

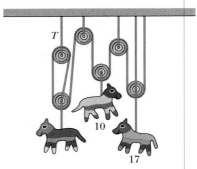

FIGURE 13-24   Question 10.

11. (a) In Checkpoint 3, to express τ_r in terms of T, should you use $\sin \theta$ or $\cos \theta$? (b) If angle θ is decreased (by shortening the rope but still keeping the rod horizontal), does the torque τ_r required for equilibrium become larger, smaller, or stay the same? (c) Does the corresponding force T become larger, smaller, or stay the same?

12. The table gives the areas of three surfaces and the magnitude of a force that is applied perpendicular to the surface and uniformly across it. Rank the surfaces according to the stress on them, greatest first.

	AREA	FORCE
Surface A	$0.5A_0$	$2F_0$
Surface B	$2A_0$	$4F_0$
Surface C	$3A_0$	$6F_0$

13. The table gives the initial lengths of three rods and the changes in their length when forces are applied to their ends to put them under strain. Rank the rods according to their strain, greatest first.

	INITIAL LENGTH	CHANGE IN LENGTH
Rod A	$2L_0$	ΔL_0
Rod B	$4L_0$	$2\Delta L_0$
Rod C	$10L_0$	$4\Delta L_0$

EXERCISES & PROBLEMS

SECTION 13-4 Some Examples of Static Equilibrium

1E. An eight-member family, whose weights in pounds are indicated in Fig. 13-25, is balanced on a seesaw. What is the number of the person who causes the largest torque, about the rotation axis at *fulcrum f*, directed (a) out of the page and (b) into the page?

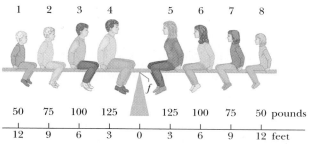

| 1 | 2 | 3 | 4 | | 5 | 6 | 7 | 8 |

| 50 | 75 | 100 | 125 | | 125 | 100 | 75 | 50 pounds |

| 12 | 9 | 6 | 3 | 0 | 3 | 6 | 9 | 12 feet |

FIGURE 13-25 Exercise 1.

2E. A certain nut is known to require forces of 40 N exerted on its shell from both sides to crack it. What force components $F_\perp$, perpendicular to the handles, will be required when the nut is placed in the nutcracker shown in Fig. 13-26?

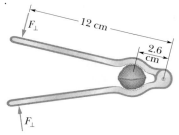

$F_\perp$ — 12 cm —
2.6 cm
$F_\perp$

FIGURE 13-26 Exercise 2.

3E. The leaning Tower of Pisa (see Fig. 13-27) is 55 m high and 7.0 m in diameter. The top of the tower is displaced 4.5 m from the vertical. Treat the tower as a uniform, circular cylinder. (a)

FIGURE 13-27 Exercise 3. The leaning Tower of Pisa (the photograph is not distorted).

What additional displacement, measured at the top, will bring the tower to the verge of toppling? (b) What angle with the vertical will the tower make at that moment?

4E. A particle is acted on by forces given, in newtons, by $F_1 = 10i - 4j$ and $F_2 = 17i + 2j$. (a) What force F_3 balances these forces? (b) What direction does F_3 have relative to the x axis?

5E. A bow is drawn until the tension in the string is equal to the force exerted by the archer. What is the angle between the two parts of the string?

6E. In Fig. 13-28, a man is trying to get his car out of mud on the shoulder of a road. He ties one end of a rope tightly around the front bumper and the other end tightly around a utility pole 60 ft away. He then pushes sideways on the rope at its midpoint with a force of 125 lb, displacing the center of the rope 1.0 ft from its previous position, and the car barely moves. What force does the rope exert on the car? (The rope stretches somewhat.)

FIGURE 13-28 Exercise 6.

7E. A rope, assumed massless, is stretched horizontally between two supports that are 3.44 m apart. When an object of weight 3160 N is hung at the center of the rope, the rope is observed to sag by 35.0 cm. What is the tension in the rope?

8E. The system in Fig. 13-29 is in equilibrium, but it begins to slip if any additional mass is added to the 5.0 kg object. What is the coefficient of static friction between the 10 kg block and the plane on which it rests?

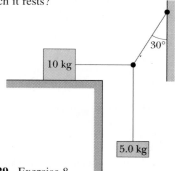

FIGURE 13-29 Exercise 8.

9E. A scaffold of mass 60 kg and length 5.0 m is supported in a horizontal position by one vertical cable at each end. A window washer of mass 80 kg stands at a point 1.5 m from one end. What is the tension in the cable (a) closest to the window washer and (b) farthest away from the window washer?

10E. A beam is carried by three men, one man at one end and the other two supporting the beam between them on a crosspiece placed so that the load is equally divided among the three men.

Where is the crosspiece placed? (Neglect the mass of the crosspiece.)

11E. A uniform cubical crate is 0.750 m on each side and weighs 500 N. It rests on the floor with one edge against a very small, fixed obstruction. At what height above the floor must a horizontal force of 350 N be applied to the crate to just tip it?

12E. A uniform sphere of weight W and radius r is held in place by a rope attached to a frictionless wall a distance L above the center of the sphere, as in Fig. 13-30. Find (a) the tension in the rope and (b) the force exerted on the sphere by the wall.

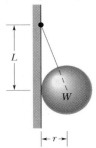

FIGURE 13-30 Exercise 12.

13E. An automobile with a mass of 1360 kg has 3.05 m between the front and rear axles. Its center of gravity is located 1.78 m behind the front axle. Determine (a) the force exerted on each of the front wheels (assumed the same) and (b) the force exerted on each of the back wheels (assumed the same) by the level ground.

14E. A 160 lb man is walking across a level bridge and stops one-fourth of the way from one end. The bridge is uniform and weighs 600 lb. What are the vertical forces exerted on the bridge by its supports at (a) the far end and (b) the near end?

15E. A diver of weight 580 N stands at the end of a 4.5 m diving board of negligible weight. The board is attached to two pedestals 1.5 m apart, as shown in Fig. 13-31. What are the magnitude and direction of the force on the board from (a) the left pedestal and (b) the right pedestal? (c) Which pedestal is being stretched, and which compressed?

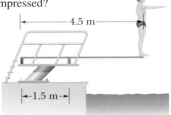

FIGURE 13-31 Exercise 15.

16E. A meter stick balances horizontally on a knife-edge at the 50.0 cm mark. With two nickels stacked over the 12.0 cm mark, the stick is found to balance at the 45.5 cm mark. A nickel has a mass of 5.0 g. What is the mass of the meter stick?

17E. A 75 kg window cleaner uses a 10 kg ladder that is 5.0 m long. He places one end on the ground 2.5 m from a wall, rests the upper end against a cracked window, and climbs the ladder. He climbs 3.0 m up the ladder when the window breaks. Neglecting friction between the ladder and window and assuming that the base of the ladder did not slip, find (a) the force exerted on the

window by the ladder just before the window breaks and (b) the magnitude and direction of the force exerted on the ladder by the ground just before the window breaks.

18E. Figure 13-32 shows the anatomical structures in the lower leg and foot that are involved in standing tiptoe with the heel raised off the floor so that the foot effectively contacts the floor at only one point, shown as P in the figure. Calculate, in terms of a person's weight W, the forces that must be exerted on the foot by (a) the calf muscle (at A) and (b) the lower-leg bones (at B) when the person stands tiptoe on one foot. Assume that $a = 5.0$ cm and $b = 15$ cm.

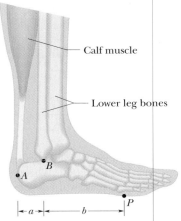

FIGURE 13-32 Exercise 18.

19P. By means of a turnbuckle G, bar AB of the square frame $ABCD$ in Fig. 13-33 is put in tension, as if its ends A and B were subject to the horizontal, outward forces **T** shown. Determine the forces on the other bars; identify those bars that are in tension and those that are being compressed. The diagonals AC and BD pass each other freely at E. Symmetry considerations can lead to considerable simplification in this and similar problems.

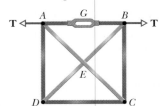

FIGURE 13-33 Problem 19.

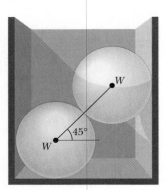

FIGURE 13-34 Problem 20.

20P. Two identical, uniform, frictionless spheres, each of weight W, rest in a rigid rectangular container as shown in Fig. 13-34. Find, in terms of W, the forces acting on the spheres due to (a) the container surfaces and (b) one another, if the line of centers of the spheres makes an angle of 45° with the horizontal.

21P. An 1800 lb construction bucket is suspended by a cable A that is attached at O to two other cables B and C, making angles of 51° and 66° with the horizontal (Fig. 13-35). Find the tension in (a) cable A, (b) cable B, and (c) cable C. (*Hint:* To avoid solving two equations in two unknowns, position the axes as shown in the figure.)

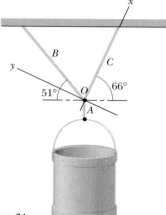

FIGURE 13-35 Problem 21.

22P. The force **F** in Fig. 13-36 is just sufficient to hold the 14 lb block and weightless pulleys in equilibrium. There is no appreciable friction. Calculate the tension T in the upper cable.

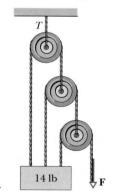

FIGURE 13-36 Problem 22.

23P. The system in Fig. 13-37 is in equilibrium with the string in the center exactly horizontal. Find (a) tension T_1, (b) tension T_2, (c) tension T_3, and (d) angle θ.

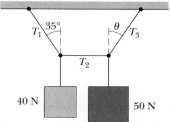

FIGURE 13-37 Problem 23.

24P. A balance is made up of a rigid, massless rod supported at and free to rotate about a point not at the center of the rod. It is balanced by unequal weights placed in the pans at each end of the rod. When an unknown mass m is placed in the left-hand pan, it is balanced by a mass m_1 placed in the right-hand pan; and when the mass m is placed in the right-hand pan, it is balanced by a mass m_2 in the left-hand pan. Show that $m = \sqrt{m_1 m_2}$.

25P. A 15 kg weight is being lifted by the pulley system shown in Fig. 13-38. The upper arm is vertical, whereas the forearm makes an angle of 30° with the horizontal. What forces are being exerted on the forearm by (a) the triceps muscle and (b) the upper-arm bone (the humerus)? The forearm and hand together have a mass of 2.0 kg with a center of mass 15 cm (measured along the arm) from the point where the forearm and upper-arm bones are in contact. The triceps muscle pulls vertically upward at a point 2.5 cm behind the contact point.

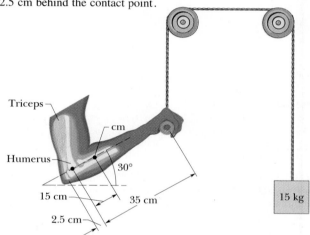

FIGURE 13-38 Problem 25.

26P. A 50.0 kg uniform square sign, 2.00 m on a side, is hung from a 3.00 m rod of negligible mass. A cable is attached to the end of the rod and to a point on the wall 4.00 m above the point where the rod is fixed to the wall (Fig. 13-39). (a) What is the tension in the cable? What are the (b) horizontal and (c) vertical components of the force exerted by the wall on the rod?

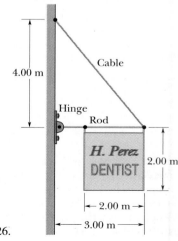

FIGURE 13-39 Problem 26.

27P. In Fig. 13-40, a 55 kg rock climber is in a lie-back climb along a fissure, with hands pulling on one side of the fissure and feet pressed against the opposite side. The fissure has width $w = 0.20$ m, and the center of mass of the climber is a horizontal distance $d = 0.40$ m from the fissure. The coefficient of static friction between hands and rock is $\mu_1 = 0.40$, and between boots and rock it is $\mu_2 = 1.2$. (a) What is the least horizontal pull by the hands and push by the feet that will keep him stable? (b) For the horizontal pull of (a), what must be the vertical distance h between hands and feet? (c) If the climber encounters wet rock, so that μ_1 and μ_2 are reduced, what happens to the answers to (a) and (b), respectively?

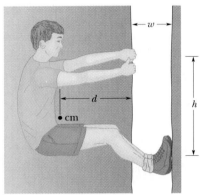

FIGURE 13-40   Problem 27.

28P. Forces F_1, F_2, and F_3 act on the structure of Fig. 13-41, shown in an overhead view. We wish to put the structure in equilibrium by applying a force, at a point such as P, whose vector components are F_h and F_v. We are given that $a = 2.0$ m, $b = 3.0$ m, $c = 1.0$ m, $F_1 = 20$ N, $F_2 = 10$ N, and $F_3 = 5.0$ N. Find (a) F_h, (b) F_v, and (c) d.

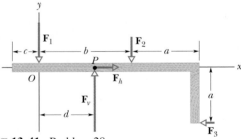

FIGURE 13-41   Problem 28.

29P. In Fig. 13-42, what magnitude of force **F** applied horizontally at the axle of the wheel is necessary to raise the wheel over an obstacle of height h? Take r as the radius of the wheel and W as its weight.

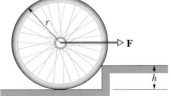

FIGURE 13-42   Problem 29.

30P. A trap door in a ceiling is 0.91 m square, has a mass of 11 kg, and is hinged along one side with a catch at the opposite side. If the center of gravity of the door is 10 cm toward the hinged side from the door's center, what forces must (a) the catch and (b) the hinge sustain?

31P. Four identical uniform bricks, each of length L, are put on top of one another (Fig. 13-43) in such a way that part of each extends beyond the one beneath. Find, in terms of L, the maximum values of (a) a_1, (b) a_2, (c) a_3, (d) a_4, and (e) h, such that the stack is in equilibrium.

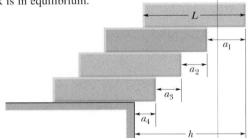

FIGURE 13-43   Problem 31.

32P. One end of a uniform beam that weighs 50.0 lb and is 3.00 ft long is attached to a wall with a hinge. The other end is supported by a wire (see Fig. 13-44). (a) Find the tension in the wire. What are the (b) horizontal and (c) vertical components of the force of the hinge on the beam?

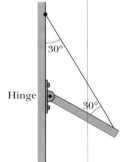

FIGURE 13-44   Problem 32.

33P. The system in Fig. 13-45 is in equilibrium. A mass of 225 kg hangs from the end of the uniform strut whose mass is 45.0 kg. Find (a) the tension T in the cable and the (b) horizontal and (c) vertical force components exerted on the strut by the hinge.

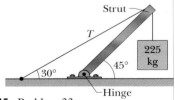

FIGURE 13-45   Problem 33.

34P. A door 2.1 m high and 0.91 m wide has a mass of 27 kg. A hinge 0.30 m from the top and another 0.30 m from the bottom each support half the door's weight. Assume that the center of gravity is at the geometrical center of the door and determine the (a) vertical and (b) horizontal force components exerted by each hinge on the door.

35P. A nonuniform bar of weight W is suspended at rest in a horizontal position by two massless cords as shown in Fig. 13-46. One cord makes the angle $\theta = 36.9°$ with the vertical; the other makes the angle $\phi = 53.1°$ with the vertical. If the length L of the bar is 6.10 m, compute the distance x from the left-hand end of the bar to its center of gravity.

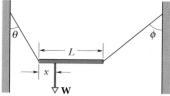

FIGURE 13-46   Problem 35.

36P. In Fig. 13-47, a thin horizontal bar AB of negligible weight and length L is pinned to a vertical wall at A and supported at B by a thin wire BC that makes an angle θ with the horizontal. A weight W can be moved anywhere along the bar; its position is defined by the distance x from the wall to its center of mass. As a function of x, find (a) the tension in the wire, and the (b) horizontal and (c) vertical components of the force exerted on the bar by the pin at A.

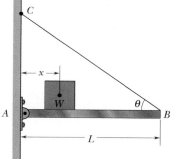

FIGURE 13-47   Problems 36 and 37.

37P. In Fig. 13-47, suppose the length L of the uniform bar is 3.0 m and its weight is 200 N. Also, let $W = 300$ N and $\theta = 30°$. The wire can withstand a maximum tension of 500 N. (a) What is the maximum possible distance x before the wire breaks? With W placed at this maximum x, what are the (b) horizontal and (c) vertical components of the force exerted on the bar by the pin at A?

38P. Two uniform beams, A and B, are attached to a wall with hinges and then loosely bolted together as in Fig. 13-48. Find the

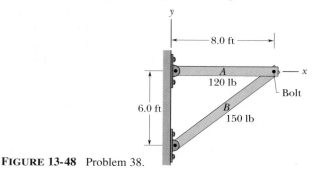

FIGURE 13-48   Problem 38.

horizontal and vertical components of the force on (a) beam A due to its hinge, (b) beam A due to the bolt, (c) beam B due to its hinge, and (d) beam B due to the bolt.

39P. Four identical, uniform bricks of length L are stacked on a table in two ways, as shown in Fig. 13-49 (compare with Problem 31). We seek to maximize the overhang distance h in both arrangements. Find the optimum distances a_1, a_2, b_1, and b_2, and calculate h for the two arrangements. (See "The Amateur Scientist," *Scientific American*, June 1985, for a discussion and an even better version of arrangement (b).)

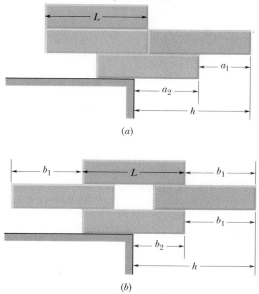

FIGURE 13-49   Problem 39.

40P. A uniform plank, with length $L = 20$ ft and weight $W = 100$ lb, rests on the ground and against a frictionless roller at the top of a wall of height $h = 10$ ft (see Fig. 13-50). The plank remains in equilibrium for any value of $\theta \geq 70°$ but slips if $\theta < 70°$. Find the coefficient of static friction between the plank and the ground.

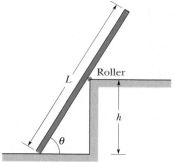

FIGURE 13-50   Problem 40.

41P. For the stepladder shown in Fig. 13-51, sides AC and CE are each 8.0 ft long and hinged at C. Bar BD is a tie-rod 2.5 ft long, halfway up. A man weighing 192 lb climbs 6.0 ft along the ladder. Assuming that the floor is frictionless and neglecting the weight of the ladder, find (a) the tension in the tie-rod and the forces exerted on the ladder by the floor at (b) A and (c) E. (*Hint:* It will help to isolate parts of the ladder in applying the equilibrium conditions.)

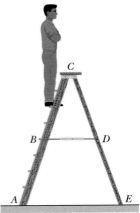

FIGURE 13-51   Problem 41.

42P. A uniform cube of side length L rests on a horizontal floor. The coefficient of static friction between cube and floor is μ. A horizontal pull P is applied perpendicular to one of the vertical faces of the cube, at a distance h above the floor on the vertical midline of the cube face. As P is slowly increased, the cube will either (a) begin to slide or (b) begin to tip. What is the condition on μ for (a) to occur? For (b)? (*Hint:* At the onset of tipping, where is the normal force located?)

43P. A cubical box is filled with sand and weighs 890 N. We wish to "roll" the box by pushing horizontally on one of the upper edges. (a) What minimum force is required? (b) What minimum coefficient of static friction between box and floor is required? (c) Is there a more efficient way to roll the box? If so, find the smallest possible force that would have to be applied directly to the box to roll it. (*Hint:* See the hint for Problem 42.)

44P. A crate, in the form of a cube with edge lengths of 4.0 ft, contains a piece of machinery whose design is such that the center of gravity of the crate and its contents is located 1.0 ft above its geometrical center. The crate rests on a ramp that makes an angle θ with the horizontal. As θ is increased from zero, an angle will be reached at which the crate will either start to slide down the ramp or tip over. Which event will occur (a) when the coefficient of static friction between ramp and crate is 0.60 and (b) when it is 0.70? In each case, give the angle at which the event occurs. (*Hint:* See the hint for Problem 42.)

45P*. A car on a horizontal road makes an emergency stop by applying the brakes so that all four wheels lock and skid along the road. The coefficient of kinetic friction between tires and road is 0.40. The separation between the front and rear axles is 4.2 m, and the center of mass of the car is located 1.8 m behind the front axle and 0.75 m above the road; see Fig. 13-52. The car weighs 11 kN. Calculate (a) the braking deceleration of the car, (b) the normal force on each wheel, and (c) the braking force on each wheel. (*Hint:* Although the car is not in translational equilibrium, it *is* in rotational equilibrium.)

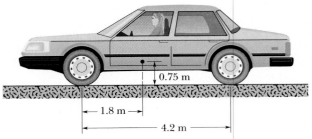

FIGURE 13-52   Problem 45.

SECTION 13-6   Elasticity

46E. Figure 13-53 shows the stress–strain curve for quartzite. What are (a) the Young's modulus and (b) the approximate yield strength for this material?

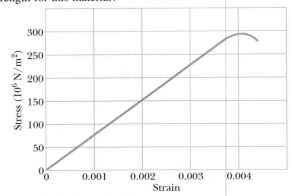

FIGURE 13-53   Exercise 46.

47E. After a fall, a 95 kg rock climber finds himself dangling from the end of a rope that had been 15 m long and 9.6 mm in diameter but which has stretched by 2.8 cm. For the rope, calculate (a) the strain, (b) the stress, and (c) the Young's modulus.

48E. A mine elevator is supported by a single steel cable 2.5 cm in diameter. The total mass of the elevator cage plus occupants is 670 kg. By how much does the cable stretch when the elevator is (a) at the surface, 12 m below the elevator motor, and (b) at the bottom of the shaft, which is 350 m deep? (Neglect the mass of the cable.)

49E. Suppose the (square) beam in Fig. 13-9a is of Douglas fir. What must be its thickness to keep the compressive stress on it to $\frac{1}{6}$ of its ultimate strength? (See Sample Problem 13-5.)

50E. A horizontal aluminum rod 4.8 cm in diameter projects 5.3 cm from a wall. A 1200 kg object is suspended from the end of the rod. The shear modulus of aluminum is 3.0×10^{10} N/m². Neglecting the rod's weight, find (a) the shear stress on the rod and (b) the vertical deflection of the end of the rod.

51E. A solid copper cube has an edge length of 85.5 cm. How much pressure must be applied to the cube to reduce the edge length to 85.0 cm? The bulk modulus of copper is 1.4×10^{11} N/m².

52P. A tunnel 150 m long, 7.2 m high, and 5.8 m wide (with a flat roof) is to be constructed 60 m beneath the ground. (See Fig. 13-54.) The tunnel roof is to be supported entirely by square steel columns, each with a cross-sectional area of 960 cm². The density of the ground material is 2.8 g/cm³. (a) What is the total weight that the columns must support? (b) How many columns are needed to keep the compressive stress on each column at one-half its ultimate strength?

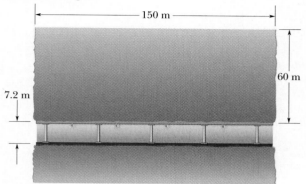

FIGURE 13-54 Problem 52.

53P. A rectangular slab of slate rests on a 26° incline; see Fig. 13-55. The slab has dimensions 43 m long, 2.5 m thick, and 12 m wide. Its density is 3.2 g/cm³. The coefficient of static friction between the slab and the underlying rock is 0.39. (a) Calculate the component of the slab's weight acting parallel to the incline. (b) Calculate the static frictional force on the slab. By comparing (a) and (b), you can see that the slab is in danger of sliding and is prevented from doing so only by chance protrusions between the slab and the underlying rock. (c) To stabilize the slab, bolts are driven perpendicular to the incline. If each bolt has a cross-sectional area of 6.4 cm² and will snap under a shearing stress of 3.6×10^8 N/m², what is the minimum number of bolts needed? Assume that the bolts do not affect the normal force.

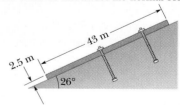

FIGURE 13-55 Problem 53.

54P. In Fig. 13-56, a lead brick rests horizontally on cylinders A and B. The areas of the top faces of the cylinders are related by $A_A = 2A_B$; the Young's moduli of the cylinders are related by $E_A = 2E_B$. The cylinders had identical lengths before the brick was placed on them. What fraction of the brick's weight is supported (a) by cylinder A and (b) by cylinder B? The horizontal distances between the center of mass of the brick and the centerlines of the cylinders are d_A for cylinder A and d_B for cylinder B. (c) What is the ratio d_A/d_B?

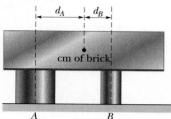

FIGURE 13-56 Problem 54.

55P. In Fig. 13-57, a 103 kg uniform log hangs by two steel wires, A and B, both of radius 1.20 mm. Initially, wire A was 2.50 m long and 2.00 mm shorter than wire B. The log is now horizontal. What forces are exerted on it by (a) wire A and (b) wire B? (c) What is the ratio d_A/d_B?

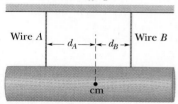

FIGURE 13-57 Problem 55.

56P. Figure 13-58 is an overhead view of a rigid rod that turns about a vertical axle until the identical rubber stoppers A and B are forced against rigid walls at distances r_A and r_B from the axle. Initially the stoppers touch the walls, without being compressed. Then force F is applied perpendicular to the rod at a distance R from the axle. Find expressions for the forces compressing (a) stopper A and (b) stopper B?

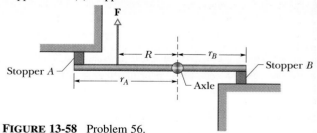

FIGURE 13-58 Problem 56.

14
Gravitation

Hydra

Centaurus

Sagittarius

Andromeda
Galaxy

Large Magellanic Cloud

The Milky Way galaxy is a disk-shaped collection of dust, planets, and billions of stars, including our Sun and solar system. The force that binds it or any other galaxy together is the same force that holds Earth's moon in orbit and you on Earth—the gravitational force. That force is also responsible for one of nature's strangest objects, the black hole, a star that has completely collapsed onto itself. The gravitational force near a black hole is so strong that not even light can escape it. But if that is the case, how can a black hole be detected?

14-1 THE WORLD AND THE GRAVITATIONAL FORCE

The drawing that opens this chapter shows our view of the Milky Way galaxy: we are near the edge of the disk of the galaxy, about 26,000 light-years (2.5×10^{20} m) from its center, which in the drawing lies in the star collection known as Sagittarius. Our galaxy is a member of the Local Group of galaxies, which includes the Andromeda galaxy (Fig. 14-1) at a distance of 2.3×10^6 light-years, and several closer dwarf galaxies, such as the Large Magellanic Cloud shown in the opening drawing.

The Local Group is part of the Local Supercluster of galaxies. Measurements taken during and since the 1980s suggest that the Local Supercluster and the supercluster consisting of the clusters Hydra and Centaurus are all moving toward an exceptionally massive region called the Great Attractor. This region appears to be about 150 million light-years away, on the opposite side of the Milky Way from us, past the clusters Hydra and Centaurus.

The force that binds together these progressively larger structures, from star to galaxy to supercluster, and may be drawing them all toward the Great Attractor, is the gravitational force. That force not only holds you on Earth but also reaches out across intergalactic space.

FIGURE 14-1 The Andromeda galaxy. Located 2.3×10^6 light-years from us, and faintly visible to the naked eye, it is very similar to our home galaxy, the Milky Way.

14-2 NEWTON'S LAW OF GRAVITATION

Physicists like to examine seemingly unrelated phenomena to show that a relationship can be found if they are examined closely enough. This search for unification has been going on for centuries. In 1665, the 23-year-old Isaac Newton made a basic contribution to physics when he showed that the force that holds the Moon in its orbit is the same force that makes an apple fall. We take this so much for granted now that it is not easy for us to comprehend the ancient belief that the motions of earthbound bodies and heavenly bodies were different in kind and were governed by different laws.

Newton concluded that not only does Earth attract an apple and the Moon but every body in the universe attracts every other body; this tendency of bodies to move toward each other is called **gravitation.** Newton's conclusion takes a little getting used to, because the familiar attraction of Earth for earthbound bodies is so great that it overwhelms the attraction that earthbound bodies have for each other. For example, Earth attracts an apple with a force of several ounces. You also attract a nearby apple (and it attracts you), but the force of attraction is less than the weight of a speck of dust.

Quantitatively, Newton proposed a *force law* that we call **Newton's law of gravitation:** every particle attracts any other particle with a **gravitational force** whose magnitude is given by

$$F = G \frac{m_1 m_2}{r^2}$$

(Newton's law of gravitation).   (14-1)

Here m_1 and m_2 are the masses of the particles, r is the distance between them, and G is the **gravitational constant,** whose value is now known to be

$$G = 6.67 \times 10^{-11} \ \mathrm{N \cdot m^2/kg^2}$$
$$= 6.67 \times 10^{-11} \ \mathrm{m^3/kg \cdot s^2}. \quad (14\text{-}2)$$

As Fig. 14-2 shows, a particle m_2 attracts a particle m_1 with a gravitational force **F** that is directed toward particle m_2.

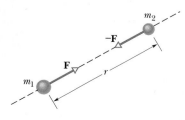

FIGURE 14-2 Two particles, of masses m_1 and m_2 and with a separation of r, attract each other according to Newton's law of gravitation, Eq. 14-1. The forces of attraction, **F** and $-$**F**, are equal in magnitude and are in opposite directions.

And particle m_1 attracts particle m_2 with a gravitational force $-\mathbf{F}$ that is directed toward m_1. The forces $\mathbf{F}$ and $-\mathbf{F}$ form an action–reaction pair and are opposite in direction but equal in magnitude. They depend on the separation of the two particles, but not on their location: the particles could be in a deep cave or in deep space. And $\mathbf{F}$ and $-\mathbf{F}$ are not altered by the presence of other bodies, even if those bodies lie between the two particles we are considering.

The strength of the gravitational force, that is, how strongly two given masses at a given separation attract each other, depends on the value of the gravitational constant G. If G—by some miracle—were suddenly multiplied by a factor of 10, you would be crushed to the floor by Earth's attraction. If it were divided by this factor, Earth's attraction would be weak enough for you to jump over a tall building.

Although Newton's law of gravitation applies strictly to particles, we can also apply it to real objects as long as the sizes of the objects are small compared to the distance between them. The Moon and Earth are far enough apart so that, to a good approximation, we can treat them both as particles. But what about an apple and Earth? From the point of view of the apple, the broad and level Earth, stretching out to the horizon beneath the apple, certainly does not look like a particle.

Newton solved the apple–Earth problem by proving an important theorem called the *shell theorem:*

> A uniform spherical shell of matter attracts a particle that is outside the shell as if all the shell's mass were concentrated at its center.

Earth can be thought of as a nest of such shells, one within another, and each attracting a particle outside Earth's surface as if the mass of that shell were at the center of the shell. Thus, from the apple's point of view, Earth *does* behave like a particle, located at the center of Earth and having a mass equal to that of the planet.

Suppose, as in Fig. 14-3, that Earth pulls down on an apple with a force of 0.80 N. The apple must then pull up on Earth with a force of 0.80 N, which we take to act at the

center of Earth. Although the forces are matched in magnitude, they produce different accelerations when the apple is released. For the apple, the acceleration is about 9.8 m/s², the familiar acceleration of a falling body near Earth's surface. For Earth, the acceleration measured in a reference frame attached to the center of mass of the apple–Earth system is only about 1×10^{-25} m/s².

CHECKPOINT 1: A particle is to be placed, in turn, outside four objects, each of mass m: (1) a large uniform solid sphere, (2) a large uniform spherical shell, (3) a small uniform solid sphere, and (4) a small uniform shell. In each situation, the distance between the particle and the center of the object is d. Rank the objects according to the magnitude of the gravitational force they exert on the particle, greatest first.

14-3 GRAVITATION AND THE PRINCIPLE OF SUPERPOSITION

Given a group of particles, we find the net (or resultant) gravitational force exerted on any one of them by using the **principle of superposition.** This is a general principle that says a net effect is the sum of the individual effects. Here, the principle means that we first compute the gravitational force that acts on our selected particle due to each of the other particles, in turn. We then find the net force by adding these forces vectorially, as usual.

For n interacting particles, we can write the principle of superposition for gravitational forces as

$$\mathbf{F}_1 = \mathbf{F}_{12} + \mathbf{F}_{13} + \mathbf{F}_{14} + \mathbf{F}_{15} + \cdots + \mathbf{F}_{1n}. \quad (14\text{-}3)$$

Here $\mathbf{F}_1$ is the net force on particle 1 and, for example, $\mathbf{F}_{13}$ is the force exerted on particle 1 by particle 3. We can express this equation more compactly as a vector sum:

$$\mathbf{F}_1 = \sum_{i=2}^{n} \mathbf{F}_{1i}. \quad (14\text{-}4)$$

What about the gravitational force exerted on a particle by a real extended object? The force is found by dividing the object into units small enough to treat as particles and then using Eq. 14-4 to find the vector sum of the forces exerted on the particle by all the units. In the limiting case, we can divide the extended object into differential units of mass dm, each of which exerts only a differential force $d\mathbf{F}$ on the particle. In this limit, the sum of Eq. 14-4 becomes an integral and we have

$$\mathbf{F}_1 = \int d\mathbf{F}, \quad (14\text{-}5)$$

FIGURE 14-3 The apple pulls up on Earth just as hard as Earth pulls down on the apple.

in which the integral is taken over the entire extended object. If the object is a uniform sphere or a spherical shell, we can avoid the integration of Eq. 14-5 by assuming that the object's mass is concentrated at the object's center and using Eq. 14-1.

SAMPLE PROBLEM 14-1

Figure 14-4a shows an arrangement of five particles, with masses $m_1 = 8.0$ kg, $m_2 = m_3 = m_4 = m_5 = 2.0$ kg, and with $a = 2.0$ cm and $\theta = 30°$. What is the net gravitational force $\mathbf{F}_1$ acting on m_1 due to the other particles?

SOLUTION: From Eq. 14-4 we know that $\mathbf{F}_1$ is the vector sum of forces $\mathbf{F}_{12}$, $\mathbf{F}_{13}$, $\mathbf{F}_{14}$, and $\mathbf{F}_{15}$, which are the gravitational forces acting on m_1 due to the other particles. Because m_2 and m_4 are equal and because both are a distance $r = 2a$ from m_1, we have from Eq. 14-1

$$F_{12} = F_{14} = \frac{Gm_1 m_2}{(2a)^2}. \qquad (14\text{-}6)$$

Similarly, since m_3 and m_5 are equal and are both a distance $r = a$ from m_1, we have

$$F_{13} = F_{15} = \frac{Gm_1 m_3}{a^2}. \qquad (14\text{-}7)$$

Figure 14-4b is a free-body diagram for m_1. It and Eq. 14-6 show that $\mathbf{F}_{12}$ and $\mathbf{F}_{14}$ are equal in magnitude but opposite in direction; thus those forces cancel. Inspection of Fig. 14-4b and Eq. 14-7 reveals that the x components of $\mathbf{F}_{13}$ and $\mathbf{F}_{15}$ also cancel, and that their y components are identical in magnitude and both point toward increasing y. Thus $\mathbf{F}_1$ points toward increasing y, and its magnitude is twice the y component of $\mathbf{F}_{13}$:

$$
\begin{aligned}
F_1 &= 2F_{13} \cos\theta = 2\frac{Gm_1 m_3}{a^2}\cos\theta \\
&= 2\frac{(6.67 \times 10^{-11}\ \text{m}^3/\text{kg}\cdot\text{s}^2)(8.0\ \text{kg})(2.0\ \text{kg})}{(0.020\ \text{m})^2}\cos 30° \\
&= 4.6 \times 10^{-6}\ \text{N}. \qquad\qquad\qquad\text{(Answer)}
\end{aligned}
$$

Note that the presence of m_5 along the line between m_1 and m_4 does not alter the gravitational force exerted by m_4 on m_1.

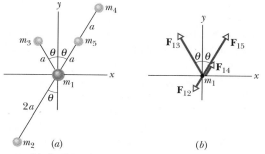

FIGURE 14-4 Sample Problem 14-1. (a) An arrangement of five particles. (b) The forces acting on the particle of mass m_1 due to the other four particles.

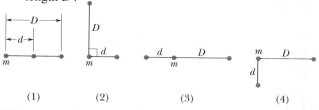

SAMPLE PROBLEM 14-2

In Fig. 14-5, a particle of mass $m_1 = 0.67$ kg is a distance $d = 23$ cm from one end of a uniform rod with length $L = 3.0$ m and mass $M = 5.0$ kg. What is the magnitude of the gravitational force $\mathbf{F}_1$ on the particle due to the rod?

SOLUTION: We consider a differential mass dm of the rod, located a distance r from m_1 and occupying a length dr along the rod. From Eq. 14-1, we may write the differential gravitational force dF_1 on m_1 due to dm as

$$dF_1 = \frac{Gm_1}{r^2}\, dm. \qquad (14\text{-}8)$$

The direction of this force is to the right in Fig. 14-5. In fact, because m_1 is located on a line through the length of the rod, each differential force dF_1 due to each differential mass dm in the rod is directed to the right. Thus, we can find the magnitude F_1 of the net force on m_1 simply by summing the magnitudes of these differential forces. We do so by integrating Eq. 14-8 along the length of the rod. (If m_1 were not on the central axis of the rod, the differential force vectors would point in different directions and the net force would have to be found by vector summation.)

The right-hand side of Eq. 14-8 contains two variables, r and dm. Before we can integrate, we must eliminate dm from that equation; we do so by using the fact that the rod is uniform in density to write

$$\frac{dm}{dr} = \frac{M}{L}. \qquad (14\text{-}9)$$

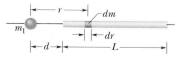

FIGURE 14-5 Sample Problem 14-2. A particle of mass m_1 is a distance d from one end of a rod of length L. A differential mass dm of the rod is a distance r from m_1.

This allows us to substitute $dm = (M/L)\,dr$ in Eq. 14-8. Then we integrate as indicated in Eq. 14-5 to get

$$F_1 = \int dF_1 = \int_d^{L+d} \frac{Gm_1}{r^2} \frac{M}{L}\,dr = \frac{Gm_1 M}{L} \int_d^{L+d} \frac{dr}{r^2}$$

$$= -\frac{Gm_1 M}{L}\left[\frac{1}{r}\right]_d^{L+d} = -\frac{Gm_1 M}{L}\left[\frac{1}{L+d} - \frac{1}{d}\right]$$

$$= \frac{Gm_1 M}{d(L+d)}$$

$$= \frac{(6.67 \times 10^{-11}\ \text{m}^3/\text{kg}\cdot\text{s}^2)(0.67\ \text{kg})(5.0\ \text{kg})}{(0.23\ \text{m})(3.0\ \text{m} + 0.23\ \text{m})}$$

$$= 3.0 \times 10^{-10}\ \text{N}. \hspace{2cm} \text{(Answer)}$$

PROBLEM SOLVING TACTICS

TACTIC 1: *Drawing Gravitational Force Vectors*

When you are given a diagram of particles, such as Fig. 14-4a, and asked to find the net gravitational force on one of them, you should usuall● draw a free-body diagram showing only the particle of concern and the forces *it alone* experiences, as in Fig. 14-4b. If, instead, you choose to superimpose the vectors on the given diagram, be sure to draw the vectors with either their tails (preferably) or their heads on the particle experiencing those forces. If you draw the vectors elsewhere, you invite confusion. And confusion is guaranteed if you draw the vectors on the particles *causing* the forces on the particle of concern.

TACTIC 2: *Simplifying a Sum of Forces with Symmetry*

In Sample Problem 14-1 we used the symmetry of the situation to reduce the time and amount of calculation involved in the solution. By realizing that m_2 and m_4 are positioned symmetrically about m_1, and thus that $\mathbf{F}_{12}$ and $\mathbf{F}_{14}$ cancel, we avoided calculating either force. And by realizing that the x components of $\mathbf{F}_{13}$ and $\mathbf{F}_{15}$ cancel and that their y components are identical and add, we saved even more effort.

CHECKPOINT 3: In the figure, what is the direction of the net gravitational force on the particle of mass m_1 due to the other particles, each of mass m, that are arranged symmetrically relative to the y axis?

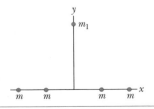

14-4 GRAVITATION NEAR EARTH'S SURFACE

Let us ignore the rotation of Earth for a moment and assume the planet to be a nonrotating uniform sphere of mass M. The magnitude of the gravitational force acting on a particle of mass m, located outside Earth a distance r from Earth's center, is then given by Eq. 14-1 as

$$F = G\frac{Mm}{r^2}. \hspace{2cm} (14\text{-}10)$$

If the particle is released, it will fall toward the center of Earth, as a result of the gravitational force F, with an acceleration we shall call the **gravitational acceleration** a_g. Newton's second law tells us that F and a_g are related by

$$F = ma_g. \hspace{2cm} (14\text{-}11)$$

Now, substituting F from Eq. 14-10 into Eq. 14-11 and solving for a_g, we find

$$a_g = \frac{GM}{r^2}. \hspace{2cm} (14\text{-}12)$$

Table 14-1 shows values of a_g computed for various altitudes above Earth's surface.

The gravitational acceleration a_g computed with Eq. 14-12 is not the same as the free-fall acceleration g that we would measure for the falling particle (and that we have approximated as 9.80 m/s² near Earth's surface). The two accelerations differ for three reasons: (1) Earth is not uniform, (2) it is not a perfect sphere, and (3) it rotates. Moreover, because g differs from a_g, the weight mg of the particle differs from the gravitational force on the particle as given by Eq. 14-10 for the same three reasons. Let us now examine those reasons.

1. *Earth is not uniform.* The density of Earth varies radially as shown in Fig. 14-6, and the density of the crust (or outer section) of Earth varies from region to region over Earth's surface. Thus g varies from region to region.

2. *Earth is not a sphere.* Earth is approximately an ellipsoid, flattened at the poles and bulging at the equator. Its equatorial radius is greater than its polar radius by 21 km.

TABLE 14-1 VARIATION OF a_g WITH ALTITUDE

ALTITUDE (km)	a_g (m/s²)	ALTITUDE EXAMPLE
0	9.83	Mean Earth surface
8.8	9.80	Mt. Everest
36.6	9.71	Highest manned balloon
400	8.70	Space shuttle orbit
35,700	0.225	Communications satellite

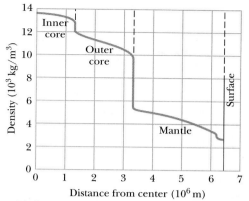

FIGURE 14-6 The density of Earth as a function of distance from the center. The limits of the solid inner core, the largely liquid outer core, and the solid mantle are shown, but the crust of Earth is too thin to show clearly on this plot.

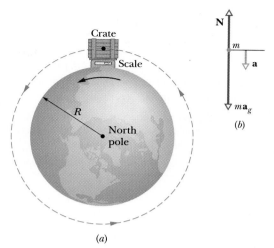

FIGURE 14-7 (*a*) A crate resting on a platform scale at Earth's equator as Earth rotates. The view is along Earth's rotational axis, looking down on the north pole. (*b*) A free-body diagram for the crate. The crate is in uniform circular motion and is thus accelerated toward the center of Earth. The gravitational force on it has magnitude ma_g. The normal force **N** from the scale has magnitude mg, where g is the free-fall acceleration.

Thus a point at the poles is closer to the dense core of Earth than is a point on the equator. This is one reason the free-fall acceleration g increases as one proceeds, at sea level, from the equator toward the poles.

3. *Earth is rotating.* The rotation axis runs through the north and south poles of Earth. An object located on Earth's surface anywhere except at those poles must rotate in a circle about the rotation axis and thus must have a centripetal acceleration that points toward the center of the circle. This centripetal acceleration requires a centripetal force that also points toward that center.

To see how Earth's rotation causes the free-fall acceleration g to differ from the gravitational acceleration a_g, and causes weight to differ from the gravitational force given by Eq. 14-10, we analyze a simple situation in which a crate of mass m rests on a scale at the equator. Figure 14-7*a* shows this situation as viewed from a point in space above the north pole.

Figure 14-7*b* is a free-body diagram for the crate. The centripetal acceleration **a** of the crate points toward the center of the circle it travels, which is coincident with the center of Earth (assumed spherical). Earth exerts a gravitational force on the crate with magnitude ma_g (Eq. 14-11). The scale exerts an upward normal force **N** on the crate. Applying Newton's second law to the crate, with the positive direction being radially inward toward Earth's center, we find

$$\sum F = ma_g - N = ma. \qquad (14\text{-}13)$$

The magnitude of **N** is the reading on the scale and thus is equal to the weight mg of the crate. Substituting mg for N in Eq. 14-13 yields

$$ma_g - mg = ma, \qquad (14\text{-}14)$$

which shows that the weight mg of the crate differs from the magnitude ma_g of the gravitational force acting on the crate. Dividing Eq. 14-14 by m shows also that g is different from a_g by an amount equal to the centripetal acceleration a.

For the centripetal acceleration a we can substitute $\omega^2 R$, where ω is the angular speed of Earth's rotation and R is the radius of the circular path taken by the crate. (R is approximately the radius of Earth.) For ω we can substitute $2\pi/T$, where $T = 24$ h is approximately the time for one rotation of Earth. When we make these substitutions and divide by m, Eq. 14-14 gives us

$$a_g - g = \omega^2 R = \left(\frac{2\pi}{T}\right)^2 R \qquad (14\text{-}15)$$

$$= 0.034 \text{ m/s}^2.$$

Thus the free-fall acceleration g (9.78 m/s² ≈ 9.8 m/s²) measured on the equator of the real, rotating planet is slightly less than the gravitational acceleration a_g due strictly to the gravitational force. The difference between g and a_g becomes progressively smaller as the crate is moved to progressively greater latitudes, because the crate then travels in a smaller circle, making R in Eq. 14-15 smaller. So, often we can approximate the free-fall acceleration g with the gravitational acceleration a_g. And we can approximate the weight mg of an object with the gravitational force (given by Eq. 14-10).

SAMPLE PROBLEM 14-3

Consider a pulsar, a collapsed star of extremely high density, with a mass M equal to that of the Sun (1.98×10^{30} kg), a radius R of only 12 km, and a rotational period T of 0.041 s. At its equator, by what percentage does the free-fall acceleration g differ from the gravitational acceleration a_g?

SOLUTION: To find a_g on the surface of the pulsar, we use Eq. 14-12, with R replacing r and with M being the mass of the pulsar. Substituting the known values gives

$$a_g = \frac{GM}{R^2} = \frac{(6.67 \times 10^{-11} \text{ m}^3/\text{kg} \cdot \text{s}^2)(1.98 \times 10^{30} \text{ kg})}{(12,000 \text{ m})^2}$$

$$= 9.2 \times 10^{11} \text{ m/s}^2.$$

Dividing Eq. 14-15 (which applies to any rotating body) by a_g and substituting the known data, we now find

$$\frac{a_g - g}{a_g} = \left(\frac{2\pi}{T}\right)^2 \frac{R}{a_g} = \left(\frac{2\pi}{0.041 \text{ s}}\right)^2 \frac{12,000 \text{ m}}{9.2 \times 10^{11} \text{ m/s}^2}$$

$$= 3.1 \times 10^{-4} = 0.031\%. \qquad \text{(Answer)}$$

Even though a pulsar rotates extremely rapidly, its rotation reduces the free-fall acceleration from the gravitational acceleration only slightly, because its radius is so small.

SAMPLE PROBLEM 14-4

(a) An astronaut whose height h is 1.70 m floats feet "down" in an orbiting space shuttle at a distance $r = 6.77 \times 10^6$ m from the center of Earth. What is the difference in the gravitational acceleration at her feet and at her head?

SOLUTION: Equation 14-12 tells us the gravitational acceleration at any distance r from the center of Earth:

$$a_g = \frac{GM_E}{r^2}, \qquad (14\text{-}16)$$

where M_E is Earth's mass. We cannot simply apply Eq. 14-16 twice, first with, say, $r = 6.77 \times 10^6$ m for the feet and then with $r = 6.77 \times 10^6$ m + 1.70 m for the head. If we do, a calculator will give us the same value for a_g twice, and thus a difference of zero, because h is so small compared to r. Instead, we differentiate Eq. 14-16 with respect to r, obtaining

$$da_g = -2\frac{GM_E}{r^3} dr, \qquad (14\text{-}17)$$

where da_g is the differential change in the gravitational acceleration due to a differential change dr in r. For the astronaut, $dr = h$ and $r = 6.77 \times 10^6$ m. Substituting data into Eq. 14-17, we find

$$da_g = -2\frac{(6.67 \times 10^{-11} \text{ m}^3/\text{kg} \cdot \text{s}^2)(5.98 \times 10^{24} \text{ kg})}{(6.77 \times 10^6 \text{ m})^3} (1.70 \text{ m})$$

$$= -4.37 \times 10^{-6} \text{ m/s}^2. \qquad \text{(Answer)}$$

This result means that the gravitational acceleration of the astronaut's feet toward Earth is slightly greater than the gravitational acceleration of her head toward Earth. This difference in acceleration tends to stretch her body, but the difference is so small that the stretching is unnoticeable.

(b) If the astronaut is now feet "down" at the same orbital radius r of 6.77×10^6 m about a black hole of mass $M_h = 1.99 \times 10^{31}$ kg (which is 10 times our Sun's mass), what is the difference in the gravitational acceleration at her feet and head? The black hole has a surface (called the *horizon* of the black hole) of radius $R_h = 2.95 \times 10^4$ m. Nothing, not even light, can escape from the surface or anywhere inside it. Note that the astronaut is (wisely) well outside the surface (at $r = 229R_h$).

SOLUTION: We can again use Eq. 14-17 if we substitute $M_h = 1.99 \times 10^{31}$ kg for M_E. We find

$$da_g = -2\frac{(6.67 \times 10^{-11} \text{ m}^3/\text{kg} \cdot \text{s}^2)(1.99 \times 10^{31} \text{ kg})}{(6.77 \times 10^6 \text{ m})^3} (1.70 \text{ m})$$

$$= -14.5 \text{ m/s}^2. \qquad \text{(Answer)}$$

This means that the gravitational acceleration of the astronaut's feet toward the black hole is noticeably larger than that of her head. The resulting tendency to stretch her body would be bearable but quite painful. If she drifted closer to the black hole, the stretching tendency would increase drastically.

14-5 GRAVITATION INSIDE EARTH

Newton's shell theorem can also be applied to a situation in which a particle is located *inside* a uniform shell, to show the following:

A uniform shell of matter exerts no *net* gravitational force on a particle located inside it.

If the density of Earth were uniform, the gravitational force acting on a particle would be a maximum at Earth's surface. The force would, as we expect, decrease as the particle moved outward. If the particle were to move inward, perhaps down a deep mine shaft, the gravitational force would change for two reasons. (1) It would tend to increase because the particle would be moving closer to the center of Earth. (2) It would tend to decrease because the shell of material lying outside the particle's radial position would not exert any net force on the particle.

For a uniform Earth, the second influence prevails and the force on the particle steadily decreases to zero as the particle approaches the center of Earth. However, for the real (nonuniform) Earth, the force on the particle actually increases as the particle begins to descend. The force reaches a maximum at a certain depth; only then does it begin to decrease as the particle descends further.

SAMPLE PROBLEM 14-5

Suppose a tunnel runs through Earth from pole to pole, as in Fig. 14-8. Assume that Earth is a nonrotating, uniform sphere. Find the gravitational force on a particle of mass m dropped into the tunnel when it reaches a distance r from Earth's center.

SOLUTION: The force that acts on the particle is associated only with the mass M' of Earth that lies within a sphere of radius r. The portion of Earth that lies outside this sphere does not exert any net force on the particle.

Mass M' is given by

$$M' = \rho V' = \rho \frac{4\pi r^3}{3}, \qquad (14\text{-}18)$$

in which V' is the volume occupied by M' (which lies within the dashed line in Fig. 14-8), and ρ is the assumed uniform density of Earth.

The force acting on the particle is then, using Eqs. 14-1 and 14-18,

$$F = -\frac{GmM'}{r^2} = -\frac{Gm\rho 4\pi r^3}{3r^2} = -\left(\frac{4\pi mG\rho}{3}\right) r$$
$$= -Kr, \qquad \text{(Answer)} \quad (14\text{-}19)$$

in which K, a constant, is equal to $4\pi mG\rho/3$. We have inserted a minus sign to indicate that the force $\mathbf{F}$ and the displacement $\mathbf{r}$ are in opposite directions, the former being toward the center of Earth and the latter away from that point. Thus Eq. 14-19 tells us that the force acting on the particle is proportional to the displacement of the particle but oppositely directed, precisely the situation for Hooke's law.

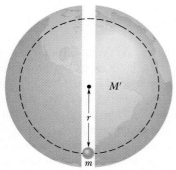

FIGURE 14-8 Sample Problem 14-5. A particle is dropped into a tunnel drilled through Earth.

14-6 GRAVITATIONAL POTENTIAL ENERGY

In Section 8-3, we discussed the gravitational potential energy of a particle–Earth system. We were careful to keep the particle near Earth's surface, so that we could regard the gravitational force as constant. We then chose some reference configuration of the system as having a gravitational potential energy of zero. Often, in this configuration the particle was on Earth's surface. For particles not on Earth's surface, the gravitational potential energy decreased when the separation between the particle and Earth decreased.

Here, we broaden our view and consider the gravitational potential energy U of two particles, of masses m and M, separated by a distance r. We again choose a reference configuration with U equal to zero. However, to simplify the equations, the separation distance r in the reference configuration is now large enough to be approximated as *infinite*. As before, the gravitational potential energy decreases when the separation decreases. Since $U = 0$ for $r = \infty$, the potential energy is negative for any finite separation and becomes progressively more negative as the particles move closer together.

With these facts in mind and as we shall justify below, we take the gravitational potential energy of the two-particle system to be

$$U = -\frac{GMm}{r} \qquad \text{(gravitational potential energy).} \quad (14\text{-}20)$$

Note that $U(r)$ approaches zero as r approaches infinity and that for any finite value of r, the value of $U(r)$ is negative.

The potential energy given by Eq. 14-20 is a property of the system of two particles rather than of either particle alone. There is no way to divide this energy and say that so much belongs to one particle and so much to the other. Nevertheless, if $M \gg m$, as is true for Earth and a baseball, we often speak of "the potential energy of the baseball." We can get away with this because, when a baseball moves in the vicinity of Earth, changes in the potential energy of the baseball–Earth system appear almost entirely as changes in the kinetic energy of the baseball, since changes in the kinetic energy of Earth are too small to be measured. Similarly, in Section 14-8 we shall speak of "the potential energy of an artificial satellite" orbiting Earth, because the satellite's mass is so much smaller than Earth's mass. When we speak of the potential energy of bodies of comparable mass, however, we have to be careful to treat them as a system.

If our system contains more than two particles, we consider each pair of particles in turn, calculate the gravi-

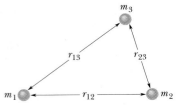

FIGURE 14-9 Three particles exert gravitational forces on one another. The gravitational potential energy of the system is the sum of the gravitational potential energies of each of the three possible pairs of the particles.

tational potential energy of that pair with Eq. 14-20 as if the other particles were not there, and then algebraically sum the results. Applying Eq. 14-20 to each of the three pairs of Fig. 14-9, for example, gives the potential energy of the system as

$$U = -\left(\frac{Gm_1m_2}{r_{12}} + \frac{Gm_1m_3}{r_{13}} + \frac{Gm_2m_3}{r_{23}} \right). \quad (14\text{-}21)$$

A *globular cluster* (Fig. 14-10) in the constellation Sagittarius is a good example of a naturally occurring system of particles. It contains about 70,000 stars, which can be paired in 2.5×10^9 different ways. Contemplation of this structure suggests the enormous amount of gravitational potential energy stored in the universe.

Proof of Eq. 14-20

Let a baseball, starting from rest at a great (infinite) distance from Earth, fall toward point P, as in Fig. 14-11. The potential energy of the baseball–Earth system is initially

FIGURE 14-10 A globular cluster, like this one in the constellation Sagittarius, contains tens of thousands of stars in an almost spherical collection. There are many such clusters in our Milky Way galaxy, some of which are visible with a small telescope.

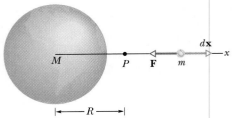

FIGURE 14-11 A baseball of mass m falls toward Earth from infinity, along a radial line (an x axis) passing through point P at a distance R from the center of Earth.

zero. When the baseball reaches P, the potential energy is the negative of the work W done by the gravitational force as the baseball moves to P from its distant position. Thus, from Eqs. 8-1 and 8-6 (generalized to vector form), we have

$$U = -W = -\int_{\infty}^{R} \mathbf{F}(x) \cdot d\mathbf{x}. \quad (14\text{-}22)$$

The limits on the integral are the initial distance of the baseball, which we have taken to be infinitely great, and its final distance R.

The vector $\mathbf{F}(x)$ in Eq. 14-22 points radially inward toward the center of Earth in Fig. 14-11, and the vector $d\mathbf{x}$ points radially outward, the angle ϕ between them being 180°. Thus

$$\mathbf{F}(x) \cdot d\mathbf{x} = F(x)(\cos 180°)(dx)$$
$$= -F(x)\, dx. \quad (14\text{-}23)$$

For $F(x)$ in Eq. 14-23, we now substitute from Newton's law of gravitation (Eq. 14-1) written as a function of x rather than r, obtaining

$$\mathbf{F}(x) \cdot d\mathbf{x} = -\frac{GMm}{x^2}\, dx.$$

Putting that result into Eq. 14-22, we obtain

$$U = \int_{\infty}^{R} \left(\frac{GMm}{x^2} \right) dx = -\left[\frac{GMm}{x} \right]_{\infty}^{R} = -\frac{GMm}{R},$$

which, with a change in symbol from R to r, is Eq. 14-20.

Equation 14-20 holds no matter what path the baseball takes in moving in toward Earth. Consider a path made up of small steps, as in Fig. 14-12. No work is done along circular steps like AB, because along them the (radial) gravitational force is perpendicular to the displacement. The total work done along all the radial steps, such as BC, is the same as the work done in moving along a single radial line, as in Fig. 14-11. Thus the work done on a particle by the gravitational force when the particle moves between any two points is independent of the path that the

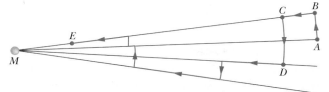

FIGURE 14-12 The work done by the gravitational force as a baseball moves from A to E is independent of the path followed.

particle follows and depends only on the particle's initial and final positions. The work may be computed simply as the negative of the difference in potential energy between the two points:

$$W = -\Delta U = -(U_f - U_i), \quad (14\text{-}24)$$

where U_f and U_i are the potential energies associated with the final and initial positions. This is what we mean when we say, as we did in Chapter 8, that the gravitational force is *conservative*. And, as discussed there, if the work *did* depend on the path, as it does for frictional forces, there would be no potential energy associated with the gravitational force.

Potential Energy and Force

In the proof of Eq. 14-20, we derived the potential energy function U from the force function F. We should be able to go the other way, that is, to start from the potential energy function and derive the force function. Guided by Eq. 8-19, we can write

$$F = -\frac{dU}{dr} = -\frac{d}{dr}\left(-\frac{GMm}{r}\right)$$
$$= -\frac{GMm}{r^2}. \quad (14\text{-}25)$$

This is just Newton's law of gravitation (Eq. 14-1). The minus sign indicates that the force on mass m points radially inward, toward mass M.

Escape Speed

If you fire a projectile upward, usually it will slow, stop momentarily, and return to Earth. There is, however, a certain minimum initial speed that will cause it to move upward forever, theoretically coming to rest only at infinity. This initial speed is called the **escape speed.**

Consider a projectile of mass m, leaving the surface of a planet (or some other astronomical body or system) with escape speed v. It has a kinetic energy K given by $\frac{1}{2}mv^2$ and a potential energy U given by Eq. 14-20:

$$U = -\frac{GMm}{R},$$

in which M is the mass of the planet and R its radius.

When the projectile reaches infinity, it stops and thus has no kinetic energy. It also has no potential energy because this is our zero-potential-energy configuration. Its total energy at infinity is therefore zero. From the principle of conservation of energy, its total energy at the planet's surface must also have been zero, so

$$K + U = \tfrac{1}{2}mv^2 + \left(-\frac{GMm}{R}\right) = 0.$$

This yields

$$v = \sqrt{\frac{2GM}{R}}. \quad (14\text{-}26)$$

The escape speed does not depend on the direction in which a projectile is fired. However, attaining that speed is easier if the projectile is fired in the direction the launch site is moving as Earth rotates about its axis. For example, rockets are launched eastward at Cape Canaveral to take advantage of the Cape's eastward speed of 1500 km/h.

Equation 14-26 can be applied to find the escape speed of a projectile from any astronomical body, provided we substitute the mass of the body for M and the radius of the body for R. Table 14-2 shows escape speeds from some astronomical bodies.

CHECKPOINT **4:** You move a ball of mass m away from a sphere of mass M. (a) Does the gravitational potential energy of the ball–sphere system increase or decrease? (b) Is positive or negative work done by the gravitational force between the ball and the sphere?

TABLE 14-2 SOME ESCAPE SPEEDS

BODY	MASS (kg)	RADIUS (m)	ESCAPE SPEED (km/s)
Ceres[a]	1.17×10^{21}	3.8×10^5	0.64
Earth's moon	7.36×10^{22}	1.74×10^6	2.38
Earth	5.98×10^{24}	6.37×10^6	11.2
Jupiter	1.90×10^{27}	7.15×10^7	59.5
Sun	1.99×10^{30}	6.96×10^8	618
Sirius B[b]	2×10^{30}	1×10^7	5200
Neutron star[c]	2×10^{30}	1×10^4	2×10^5

[a] The most massive of the asteroids.

[b] A *white dwarf* (a star in a final stage of evolution) that is a companion of the bright star Sirius.

[c] The collapsed core of a star that remains after that star has exploded in a *supernova* event.

SAMPLE PROBLEM 14-6

An asteroid, headed directly toward Earth, has a speed of 12 km/s relative to the planet when it is at a distance of 10 Earth radii from Earth's center. Ignoring the effects of the terrestrial atmosphere on the asteroid, find the asteroid's speed when it reaches Earth's surface.

SOLUTION: Because the mass of an asteroid is much less than that of Earth, we can assign the gravitational potential energy of the asteroid–Earth system to the asteroid alone, and we can neglect any change in the speed of Earth relative to the asteroid during the asteroid's fall. Because we are to ignore the effects of the atmosphere on the asteroid, the mechanical energy of the asteroid is conserved during its fall; that is,

$$K_f + U_f = K_i + U_i,$$

where K and U are the kinetic energy and gravitational potential energy of the asteroid, and the subscripts i and f refer to initial values (at a distance of 10 Earth radii) and final values (at a distance of one Earth radius).

Let m represent the mass of the asteroid, M the mass of Earth ($= 5.98 \times 10^{24}$ kg), and R the radius of Earth ($= 6.37 \times 10^6$ m). Using Eq. 14-20 for the gravitational potential energy and $\frac{1}{2}mv^2$ for K, we have

$$\tfrac{1}{2}mv_f^2 - \frac{GMm}{R} = \tfrac{1}{2}mv_i^2 - \frac{GMm}{10R}.$$

Rearranging and substituting known values, we find

$$v_f^2 = v_i^2 + \frac{2GM}{R}\left(1 - \frac{1}{10}\right)$$

$$= (12 \times 10^3 \text{ m/s})^2$$

$$+ \frac{2(6.67 \times 10^{-11} \text{ m}^3/\text{kg} \cdot \text{s}^2)(5.98 \times 10^{24} \text{ kg})}{6.37 \times 10^6 \text{ m}}\, 0.9$$

$$= 2.567 \times 10^8 \text{ m}^2/\text{s}^2,$$

and thus

$$v_f = 1.60 \times 10^4 \text{ m/s} = 16 \text{ km/s}. \qquad \text{(Answer)}$$

At this speed, the asteroid would not have to be particularly large to do considerable damage. As an example, if it were only 5 m across, the impact could release about as much energy as the nuclear explosion at Hiroshima. Alarmingly, about 500 million asteroids of this size are near Earth's orbit, and in 1994 one of them apparently penetrated Earth's atmosphere and exploded at an altitude of 20 km near a remote South Pacific island (setting off nuclear-explosion warnings on six military satellites). An asteroid 500 m across (there may be a million of them near Earth's orbit) could end modern civilization and almost eliminate humans worldwide.

14-7 PLANETS AND SATELLITES: KEPLER'S LAWS

The motions of the planets, as they seemingly wander against the background of the stars, have been a puzzle since the dawn of history. The "loop-the-loop" motion of Mars, shown in Fig. 14-13, was particularly baffling. Johannes Kepler (1571–1630), after a lifetime of study, worked out the empirical laws that govern these motions. Tycho Brahe (1546–1601), the last of the great astronomers to make observations without the help of a telescope, compiled the extensive data from which Kepler was able to derive the three laws of planetary motion that now bear his name. Later, Newton (1642–1727) showed that his law of gravitation leads to Kepler's empirical laws.

We discuss each of Kepler's laws in turn. Although we apply the laws here to planets orbiting the Sun, they hold equally well for satellites, either natural or artificial, orbiting Earth or any other massive central body.

1. THE LAW OF ORBITS: All planets move in elliptical orbits, with the Sun at one focus.

Figure 14-14 shows a planet, of mass m, moving in such an orbit around the Sun, whose mass is M. We assume that $M \gg m$, so that the center of mass of the planet–Sun system is virtually at the center of the Sun.

The orbit in Fig. 14-14 is described by giving its **semimajor axis** a and its **eccentricity** e, the latter defined

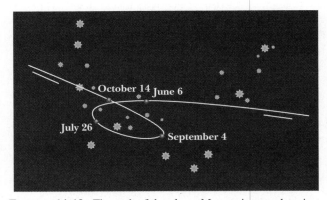

FIGURE 14-13 The path of the planet Mars as it moved against a background of the constellation Capricorn during 1971. Its position on four selected days is marked. Both Mars and Earth are moving in orbits around the Sun so that we see the position of Mars relative to us; this sometimes results in an apparent loop in the path of Mars.

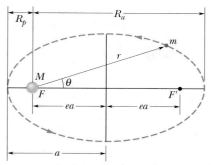

FIGURE 14-14 A planet of mass m moving in an elliptical orbit around the Sun. The Sun, of mass M, is at one focus F of the ellipse. The other, or "empty," focus is F'. Each focus is a distance ea from the center, e being the eccentricity of the ellipse. The semimajor axis a of the ellipse, the perihelion (nearest the Sun) distance R_p, and the aphelion (farthest from the Sun) distance R_a are also shown.

so that ea is the distance from the center of the ellipse to either focus F or F'. *An eccentricity of zero corresponds to a circle,* in which the two foci merge to a single central point. The eccentricities of the planetary orbits are not large, so—sketched on paper—the orbits look circular. The eccentricity of the ellipse of Fig. 14-14, which has been exaggerated for clarity, is 0.74. The eccentricity of Earth's orbit is only 0.0167.

2. THE LAW OF AREAS: A line that connects a planet to the Sun sweeps out equal areas in equal times.

Qualitatively, this second law tells us that the planet will move most slowly when it is farthest from the Sun and most rapidly when it is nearest to the Sun. As it turns out, Kepler's second law is totally equivalent to the law of conservation of angular momentum. Let us prove it.

The area of the shaded wedge in Fig. 14-15a closely approximates the area swept out in time Δt by a line connecting the Sun and the planet, which are separated by a distance r. The area ΔA of the wedge is approximately the area of a triangle with base $r\,\Delta\theta$ and height r. Thus $\Delta A \approx \frac{1}{2}r^2\,\Delta\theta$. This expression for ΔA becomes more exact as Δt (hence $\Delta\theta$) approaches zero. The instantaneous rate at which area is being swept out is then

$$\frac{dA}{dt} = \frac{r^2}{2}\frac{d\theta}{dt} = \frac{r^2\omega}{2}, \qquad (14\text{-}27)$$

in which ω is the angular speed of the rotating line connecting Sun and planet.

Figure 14-15b shows the linear momentum $\mathbf{p}$ of the planet, along with its components. From Eq. 12-27, the magnitude of the angular momentum $\mathbf{L}$ of the planet about the Sun is given by the product of r and the component of $\mathbf{p}$ perpendicular to r, or

$$L = rp_\perp = (r)(mv_\perp) = (r)(m\omega r)$$
$$= mr^2\omega, \qquad (14\text{-}28)$$

where we have replaced $v_\perp$ with its equivalent ωr (Eq. 11-16). Eliminating $r^2\omega$ between Eqs. 14-27 and 14-28 leads to

$$\frac{dA}{dt} = \frac{L}{2m}. \qquad (14\text{-}29)$$

If dA/dt is constant, as Kepler said it is, then Eq. 14-29 means that L must also be constant—angular momentum is conserved. So Kepler's second law is indeed equivalent to the law of conservation of angular momentum.

3. THE LAW OF PERIODS: The square of the period of any planet is proportional to the cube of the semimajor axis of its orbit.

To see this, consider a circular orbit with radius r (the radius is equivalent to the semimajor axis of an ellipse).

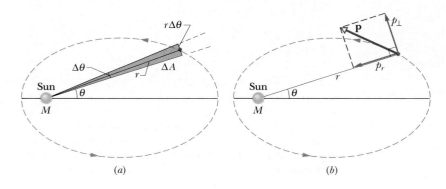

FIGURE 14-15 (a) In time Δt, the line r connecting the planet to the Sun (of mass M) sweeps through an angle $\Delta\theta$, sweeping out an area ΔA. (b) The linear momentum $\mathbf{p}$ of the planet and its components.

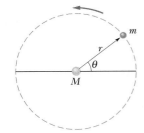

FIGURE 14-16 A planet of mass m moving around the Sun in a circular orbit of radius r.

Applying Newton's second law, $F = ma$, to the orbiting planet in Fig. 14-16 yields

$$\frac{GMm}{r^2} = (m)(\omega^2 r). \qquad (14\text{-}30)$$

Here we have substituted from Eq. 14-1 for the force F and used Eq. 11-21 to substitute $\omega^2 r$ for the centripetal acceleration. If we replace ω with $2\pi/T$ (Eq. 11-18), where T is the period of the motion, we obtain Kepler's third law:

$$T^2 = \left(\frac{4\pi^2}{GM}\right) r^3 \qquad \text{(law of periods).} \quad (14\text{-}31)$$

The quantity in parentheses is a constant, its value depending only on the mass of the central body.

Equation 14-31 holds also for elliptical orbits, provided we replace r with a, the semimajor axis of the ellipse.

On February 7, 1984, at a height of 102 km above Hawaii and with a speed of about 29,000 km/h, Bruce McCandless stepped (untethered) into space from a space shuttle and became the first human satellite.

TABLE 14-3
KEPLER'S LAW OF PERIODS FOR THE SOLAR SYSTEM

PLANET	SEMIMAJOR AXIS a (10^{10} m)	PERIOD T (y)	T^2/a^3 (10^{-34} y^2/m^3)
Mercury	5.79	0.241	2.99
Venus	10.8	0.615	3.00
Earth	15.0	1.00	2.96
Mars	22.8	1.88	2.98
Jupiter	77.8	11.9	3.01
Saturn	143	29.5	2.98
Uranus	287	84.0	2.98
Neptune	450	165	2.99
Pluto	590	248	2.99

This law predicts that the ratio T^2/a^3 has essentially the same value for every planetary orbit around a given massive body. Table 14-3 shows how well it holds for the orbits of the planets of the solar system.

CHECKPOINT 5: Satellite 1 is in a certain circular orbit about a planet, while satellite 2 is in a larger circular orbit. Which satellite has (a) the longer period and (b) the greater speed?

SAMPLE PROBLEM 14-7

A satellite in circular orbit at an altitude h of 230 km above Earth's surface has a period T of 89 min. What mass of Earth follows from these data?

SOLUTION: We apply Kepler's law of periods to the satellite–Earth system. Solving Eq. 14-31 for M yields

$$M = \frac{4\pi^2 r^3}{GT^2}. \qquad (14\text{-}32)$$

The radius r of the satellite orbit is

$$r = R + h = 6.37 \times 10^6 \text{ m} + 230 \times 10^3 \text{ m}$$
$$= 6.60 \times 10^6 \text{ m},$$

in which R is the radius of Earth. Substituting this value and the period T into Eq. 14-32 yields

$$M = \frac{(4\pi^2)(6.60 \times 10^6 \text{ m})^3}{(6.67 \times 10^{-11} \text{ m}^3/\text{kg} \cdot \text{s}^2)(89 \times 60 \text{ s})^2}$$
$$= 6.0 \times 10^{24} \text{ kg.} \qquad \text{(Answer)}$$

In the same way, we could find the mass of our Sun from the period and radius of Earth's orbit (assumed circular), and the mass of Jupiter from the period and orbital radius of any one of its moons (without knowing the mass of that moon).

SAMPLE PROBLEM 14-8

Comet Halley orbits about the Sun with a period of 76 years and, in 1986, had a distance of closest approach to the Sun, its *perihelion distance* R_p, of 8.9×10^{10} m. Table 14-3 shows that this is between the orbits of Mercury and Venus.

(a) What is the comet's farthest distance from the Sun, its *aphelion distance* R_a?

SOLUTION: We can find the semimajor axis of the orbit of comet Halley from Eq. 14-31. Substituting a for r in that equation and solving for a yield

$$a = \left(\frac{GMT^2}{4\pi^2} \right)^{1/3}. \tag{14-33}$$

If we substitute the mass M of the Sun, 1.99×10^{30} kg, and the period T of the comet, 76 years or 2.4×10^9 s, into Eq. 14-33 we find that $a = 2.7 \times 10^{12}$ m.

Study of Fig. 14-14 shows that $R_a + R_p = 2a$, or

$$R_a = 2a - R_p$$
$$= (2)(2.7 \times 10^{12} \text{ m}) - 8.9 \times 10^{10} \text{ m}$$
$$= 5.3 \times 10^{12} \text{ m}. \qquad \text{(Answer)}$$

Table 14-3 shows that this is just a little less than the semimajor axis of the orbit of Pluto.

(b) What is the eccentricity of the orbit of comet Halley?

SOLUTION: Figure 14-14 shows that $ea = a - R_p$, or

$$e = \frac{a - R_p}{a} = 1 - \frac{R_p}{a}$$
$$= 1 - \frac{8.9 \times 10^{10} \text{ m}}{2.7 \times 10^{12} \text{ m}} = 0.97. \qquad \text{(Answer)}$$

This cometary orbit, with an eccentricity approaching unity, is a long thin ellipse.

SAMPLE PROBLEM 14-9

Observations of the light from a certain star indicate that it is part of a binary (two-star) system. This visible star has orbital speed $v = 270$ km/s, orbital period $T = 1.70$ days, and approximate mass $m_1 = 6M_s$, where M_s is the Sun's mass, 1.99×10^{30} kg. Assuming that the visible star and its companion object, which is dark and unseen, are in circular orbits (see Fig. 14-17), determine the approximate mass m_2 of the dark object.

SOLUTION: As in the two particle systems of Section 9-2, the center of mass of this two-star system lies along a line connecting their centers. As with the freely rotating bodies and systems of Chapter 12, this two-star system rotates about its center of mass. In Fig. 14-17, the center of mass is labeled point O; the visible star and dark object orbit O with orbital

radii r_1 and r_2, respectively, and have a separation of $r = r_1 + r_2$. From Eq. 14-1, the gravitational force on the visible star from the dark object is

$$F = \frac{Gm_1 m_2}{r^2}.$$

Applying Newton's second law, $F = ma$, to the visible star then yields

$$\frac{Gm_1 m_2}{r^2} = m_1 a = (m_1)(\omega^2 r_1), \tag{14-34}$$

where ω is the angular speed of the visible star and $\omega^2 r_1$ is its centripetal acceleration toward O.

We can obtain another equation in these same variables by locating the center of mass O. Because O is a distance r_1 from the visible star, we can use Eq. 9-1 to write

$$r_1 = \frac{m_2 r}{m_1 + m_2}.$$

This yields

$$r = r_1 \frac{m_1 + m_2}{m_2}. \tag{14-35}$$

Now substituting Eq. 14-35 for r and $2\pi/T$ for ω into Eq. 14-34 gives us, after some algebraic manipulation,

$$\frac{m_2^3}{(m_1 + m_2)^2} = \frac{4\pi^2}{GT^2} r_1^3. \tag{14-36}$$

We still have two unknowns, m_2 and r_1. We can, however, find r_1 from the orbital motion of the visible star: the orbital period T is equal to the circumference of the orbit $(2\pi r_1)$ divided by the speed v of the planet. This gives us

$$T = \frac{2\pi r_1}{v},$$

or

$$r_1 = \frac{vT}{2\pi}. \tag{14-37}$$

If we substitute $m_1 = 6M_s$ and Eq. 14-37 for r_1, then Eq. 14-36 becomes

$$\frac{m_2^3}{(6M_s + m_2)^2} = \frac{v^3 T}{2\pi G}$$

$$= \frac{(2.7 \times 10^5 \text{ m/s})^3 (1.70 \text{ days})(86,400 \text{ s/day})}{(2\pi)(6.67 \times 10^{-11} \text{ N} \cdot \text{m}^2/\text{kg}^2)}$$

$$= 6.90 \times 10^{30} \text{ kg},$$

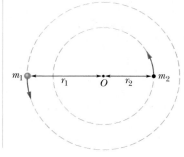

FIGURE 14-17 Sample Problem 14-9. A visible star with mass m_1 and a dark, unseen object with mass m_2 orbit around the center of mass of the two-star system at O.

or $$\frac{m_2^3}{(6M_s + m_2)^2} = 3.47 M_s. \qquad (14\text{-}38)$$

We can solve this cubic equation for m_2, or, since we are working with approximate masses anyway, we can substitute integer multiples of M_s for m_2 until we find

$$m_2 \approx 9M_s. \qquad \text{(Answer)}$$

The data here approximate those for the binary system LMC X-3 in the Large Magellanic Cloud (shown in the figure that begins this chapter). From other data, the dark object is known to be especially compact: it may be a star that collapsed under its own gravitational pull to become a neutron star or a black hole. Since a neutron star cannot have a mass larger than about $2M_s$, the result $m_2 \approx 9M_s$ strongly suggests that the dark object is a black hole.

Thus, one can detect the presence of a black hole if it is part of a binary system with a visible star whose mass, orbital speed, and orbital period can be measured.

14-8 SATELLITES: ORBITS AND ENERGY (OPTIONAL)

As a satellite orbits Earth on its elliptical path, both its speed, which fixes its kinetic energy K, and its distance from the center of Earth, which fixes its gravitational potential energy U, fluctuate with fixed periods. However, the mechanical energy E of the satellite remains constant. (Since the satellite's mass is so much smaller than Earth's mass, we assign U and E of the Earth–satellite system to the satellite alone.)

The potential energy is given by Eq. 14-20 and is

$$U = -\frac{GMm}{r}$$

(with $U = 0$ for infinite separation). Here r is the radius of the orbit, assumed for the time being to be circular.

To find the kinetic energy of a satellite in a circular orbit, we write Newton's second law, $F = ma$, as

$$\frac{GMm}{r^2} = m\frac{v^2}{r}, \qquad (14\text{-}39)$$

where v^2/r is the centripetal acceleration of the satellite. Then from Eq. 14-39, the kinetic energy is

$$K = \tfrac{1}{2}mv^2 = \frac{GMm}{2r}, \qquad (14\text{-}40)$$

which shows us that for a satellite in a circular orbit,

$$K = -\frac{U}{2} \quad \text{(circular orbit)}. \qquad (14\text{-}41)$$

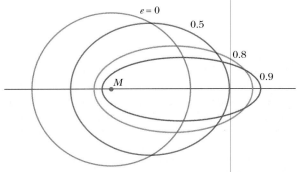

FIGURE 14-18   Four orbits about an object of mass M. All four orbits have the same semimajor axis a and thus correspond to the same total mechanical energy E. Their eccentricities e are marked.

The total mechanical energy of the orbiting satellite is

$$E = K + U = \frac{GMm}{2r} - \frac{GMm}{r}$$

or $$E = -\frac{GMm}{2r} \quad \text{(circular orbit)}. \qquad (14\text{-}42)$$

This tells us that for a satellite in a circular orbit, the total energy E is the negative of the kinetic energy K:

$$E = -K \quad \text{(circular orbit)}. \qquad (14\text{-}43)$$

For a satellite in an elliptical orbit of semimajor axis a, we can substitute a for r in Eq. 14-42 to find the mechanical energy as

$$E = -\frac{GMm}{2a} \quad \text{(elliptical orbit)}. \qquad (14\text{-}44)$$

Equation 14-44 tells us that the total energy of an orbiting satellite depends only on the semimajor axis of its orbit and not on its eccentricity e. For example, four orbits with the same semimajor axis are shown in Fig. 14-18; the same satellite would have the same total mechanical energy E in all four orbits. Figure 14-19 shows the variation of K, U, and E with r for a satellite moving in a circular orbit about a massive central body.

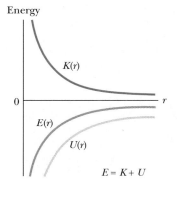

FIGURE 14-19   The variation of kinetic energy K, potential energy U, and total energy E with radius r for a satellite in a circular orbit. For any value of r, the values of U and E are negative, the value of K is positive, and $E = -K$. As r approaches infinity, all three energy curves approach a value of zero.

CHECKPOINT **6:** In the figure, a space shuttle is initially in a circular orbit of radius r about Earth. At point P, the pilot briefly fires a forward-pointing thruster to decrease the shuttle's kinetic energy K and mechanical energy E. (a) Which of the dashed elliptical orbits shown in the figure will the shuttle then take? (b) Is the orbital period T of the shuttle (the time to return to P) then greater than, less than, or the same as in the circular orbit?

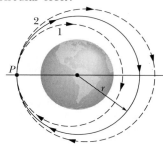

SAMPLE PROBLEM 14-10

A playful astronaut releases a bowling ball, of mass $m = 7.20$ kg, into a circular orbit about Earth at an altitude h of 350 km.

(a) What is the mechanical energy E of the ball in its orbit?

SOLUTION: The orbital radius r is given by

$$r = R + h = 6370 \text{ km} + 350 \text{ km} = 6.72 \times 10^6 \text{ m},$$

in which R is the radius of Earth. From Eq. 14-42, the mechanical energy is

$$E = -\frac{GMm}{2r}$$

$$= -\frac{(6.67 \times 10^{-11} \text{ N} \cdot \text{m}^2/\text{kg}^2)(5.98 \times 10^{24} \text{ kg})(7.20 \text{ kg})}{(2)(6.72 \times 10^6 \text{ m})}$$

$$= -2.14 \times 10^8 \text{ J} = -214 \text{ MJ}. \qquad \text{(Answer)}$$

(b) What was the mechanical energy E_0 of the ball on the launchpad at Cape Canaveral? From there to the orbit, what was the change ΔE in the ball's mechanical energy?

SOLUTION: On the launchpad, the ball has some kinetic energy because of the rotation of Earth, but we can show that this is small enough to neglect. Thus the total energy E_0 is equal to the potential energy U_0, which is given by Eq. 14-20:

$$E_0 = U_0 = -\frac{GMm}{R}$$

$$= -\frac{(6.67 \times 10^{-11} \text{ N} \cdot \text{m}^2/\text{kg}^2)(5.98 \times 10^{24} \text{ kg})(7.20 \text{ kg})}{6.37 \times 10^6 \text{ m}}$$

$$= -4.51 \times 10^8 \text{ J} = -451 \text{ MJ}. \qquad \text{(Answer)}$$

You might be tempted to say that the potential energy of the ball on Earth's surface is zero. Recall, however, that our zero-potential-energy configuration is one in which the ball is removed to a great distance from Earth. You also might be tempted to use Eq. 14-42 to find E_0, but recall that Eq. 14-42 applies to a satellite *in orbit*.

The *increase* in the mechanical energy of the ball from launchpad to orbit was

$$\Delta E = E - E_0 = (-214 \text{ MJ}) - (-451 \text{ MJ})$$

$$= 237 \text{ MJ}. \qquad \text{(Answer)}$$

You can buy this amount of energy from your utility company for a few dollars.

14-9 EINSTEIN AND GRAVITATION

Principle of Equivalence

Einstein once said: "I was . . . in the patent office at Bern when all of a sudden a thought occurred to me: 'If a person falls freely, he will not feel his own weight.' I was startled. This simple thought made a deep impression on me. It impelled me toward a theory of gravitation."

Thus Einstein tells us how he began to form his **general theory of relativity.** The fundamental postulate of this theory about gravitation (the gravitating of objects toward each other) is called the **principle of equivalence,** which says that gravitation and acceleration are equivalent. If a physicist were locked up in a small box as in Fig. 14-20, he would not be able to tell whether the box was at

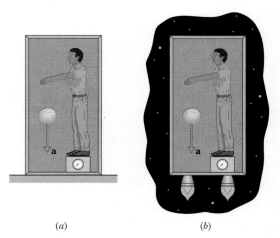

(a)   (b)

FIGURE 14-20 (a) A physicist in a box resting on Earth sees a cantaloupe falling with acceleration $a = 9.8$ m/s². (b) If he and the box accelerate in deep space at 9.8 m/s², the cantaloupe has the same acceleration relative to him. It is not possible, by doing experiments within the box, for the physicist to tell which situation he is in. For example, the platform scale on which he stands reads the same weight in both situations.

rest on Earth (and subject only to Earth's gravitational force), as in Fig. 14-20a, or accelerating through interstellar space at 9.8 m/s² (and subject only to the force producing that acceleration), as in Fig. 14-20b. In both situations he would feel the same and would read the same value for his weight on a scale. Moreover, if he watched an object fall past him, the object would have the same acceleration relative to him in both situations.

Curvature of Space

We have thus far explained gravitation as due to a force between masses. Einstein showed that, instead, gravitation is due to a curvature (or shape) of space that is caused by the masses. (As is discussed later in this book, space and time are entangled; so, the curvature of which Einstein spoke is really a curvature of *spacetime,* the combined four dimensions of our universe.)

Picturing how space (such as vacuum) can have curvature is difficult. An analogy might help: Suppose that

from orbit we watch a race in which two boats begin on the equator with a separation of 20 km and head due south (Fig. 14-21a). To the sailors, the boats travel along flat, parallel paths. However, with time the boats draw together until, nearer the south pole, they touch. The sailors in the boats can interpret this drawing together in terms of a force acting on the boats. However, we can see that the boats

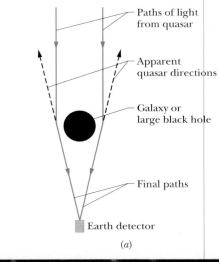

(a)

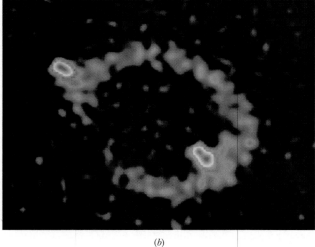

(b)

FIGURE 14-22 (a) Light from a distant quasar follows curved paths around a galaxy or a large black hole because the mass of the galaxy or black hole has curved the adjacent space. If the light is detected, it appears to have originated along the backward extensions of the final paths (dashed lines). (b) The Einstein ring known as MG1131+0456 on the computer screen of a telescope. The source of the light (actually, radio waves, which is a form of invisible light) is far behind the large, unseen galaxy that produces the ring; a portion of the source appears as the two bright spots seen along the ring.

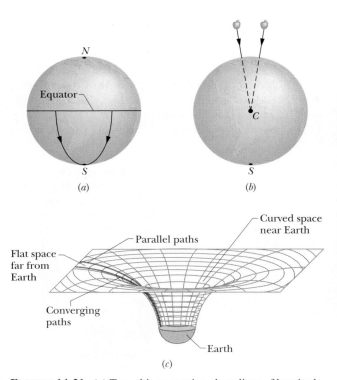

FIGURE 14-21 (a) Two objects moving along lines of longitude toward the south pole converge because of the curvature of Earth's surface. (b) Two objects falling freely near Earth move along lines that converge toward the center of Earth because of the curvature of space near Earth. (c) Far from Earth (and other masses), space is flat and parallel paths remain parallel. Close to Earth, the parallel paths begin to converge because space is curved by Earth's mass.

draw together simply because of the curvature of Earth's surface. We can see this because we are viewing the race from ''outside'' that surface.

Figure 14-21*b* shows a similar race: two horizontally separated apples are dropped from the same height above Earth. Although the apples may appear to travel along parallel paths, they actually move toward each other because they both fall toward the Earth's center. We can interpret the motion of the apples in terms of the gravitational force on the apples by Earth. But we can also interpret the motion in terms of a curvature of the space near Earth, due to the presence of Earth's mass. This time we cannot see the curvature because we cannot get ''outside'' the curved space, as we got ''outside'' the curved Earth in the boat example. But we can depict the curvature with a drawing like Fig. 14–21*c*; there the apples would move along a surface that curves toward Earth because of Earth's mass.

When light passes near Earth, its path bends slightly because of the curvature of space there, an effect called *gravitational lensing*. When it passes a more massive structure, like a galaxy or a black hole having large mass, its path can be bent more. If such a massive structure is between us and a quasar (an extremely bright, extremely distant source of light), the light from the quasar can bend around the massive structure and toward us (Fig. 14-22*a*). Then, because the light seems to be coming from slightly different directions in the sky, we see the same quasar in those different directions. In some situations, the quasars we see blend together to form a giant luminous arc, which is called an *Einstein ring* (Fig. 14-22*b*).

Is gravitation best attributed to the curvature of space-time due to the presence of masses or to a force between masses? Or should it be attributed to the actions of a type of fundamental particle called a *graviton,* as conjectured in some modern physics theories? We do not know.

REVIEW & SUMMARY

The Law of Gravitation

Any particle in the universe attracts any other particle with a **gravitational force** whose magnitude is

$$F = G\frac{m_1 m_2}{r^2} \qquad \text{(Newton's law of gravitation),} \qquad (14\text{-}1)$$

where m_1 and m_2 are the masses of the particles and r is their separation. The **gravitational constant** G is a universal constant whose value is $6.67 \times 10^{-11} \text{ N} \cdot \text{m}^2/\text{kg}^2$.

The Gravitational Behavior of Uniform Spherical Shells

Equation 14-1 holds only for particles. The gravitational force between extended bodies must generally be found by adding (integrating) the individual forces on individual particles within the bodies. However, if either of the bodies is a uniform spherical shell or a spherically symmetric solid, the net gravitational force it exerts on an *external* object may be computed as if all the mass of the shell or body were located at its center.

Superposition

Gravitational forces obey the **principle of superposition;** that is, the total force $\mathbf{F}_1$ on a particle labeled as particle 1 is the sum of the forces exerted on it by all other particles taken one at a time:

$$\mathbf{F}_1 = \sum_{i=2}^{n} \mathbf{F}_{1i}, \qquad (14\text{-}4)$$

in which the sum is a vector sum of the forces $\mathbf{F}_{1i}$ exerted on particle 1 by particles 2, 3, $\cdots$, n. The gravitational force $\mathbf{F}_1$ exerted on a particle by an extended body is found by dividing the body into units of differential mass dm, each of which exerts a differential force $d\mathbf{F}$ on the particle, and then integrating to find the sum of those forces:

$$\mathbf{F}_1 = \int d\mathbf{F}. \qquad (14\text{-}5)$$

Gravitational Acceleration

The *gravitational acceleration* a_g of a particle (of mass m) is due solely to the gravitational force acting on it. When the particle is at distance r from the center of a uniform, spherical body of mass M, the magnitude of the gravitational force on the particle is given by Eq. 14-1. In addition, by Newton's second law,

$$F = ma_g, \qquad (14\text{-}11)$$

and thus a_g is given by

$$a_g = \frac{GM}{r^2}. \qquad (14\text{-}12)$$

Free-Fall Acceleration and Weight

The actual free-fall acceleration $\mathbf{g}$ of a particle near Earth differs slightly from the gravitational acceleration $\mathbf{a}_g$, and the weight $m\mathbf{g}$ of the particle differs from the gravitational force (Eq. 14-1) acting on the particle, because Earth is not uniform or spherical and because Earth rotates.

Gravitation Within a Spherical Shell

A uniform shell of matter exerts no gravitational force on a particle located inside it. This means that if a particle is located inside a uniform solid sphere of matter at distance r from its center, the gravitational force exerted on the particle is due only to the mass

M' that lies within a sphere of radius r. This mass M' is given by

$$M' = \rho \frac{4\pi r^3}{3}, \qquad (14\text{-}18)$$

where ρ is the density of the sphere.

Gravitational Potential Energy

The gravitational potential energy $U(r)$ of a system of two particles, with masses M and m and separated by a distance r, is the negative of the work that would be done by the gravitational force of either particle acting on the other as the separation between the particles changes from infinite (very large) to r. This energy is

$$U = -\frac{GMm}{r} \qquad \begin{array}{l}\text{(gravitational} \\ \text{potential energy).}\end{array} \qquad (14\text{-}20)$$

Potential Energy of a System

If a system contains more than two particles, the total gravitational potential energy is the sum of terms representing the potential energies of all the pairs; as an example, we have for three particles, of masses m_1, m_2, and m_3,

$$U = -\left(\frac{Gm_1 m_2}{r_{12}} + \frac{Gm_1 m_3}{r_{13}} + \frac{Gm_2 m_3}{r_{23}} \right). \qquad (14\text{-}21)$$

Escape Speed

An object will escape the gravitational pull of an astronomical body of mass M and radius R if the object's speed near the body's surface is at least equal to the **escape speed,** given by

$$v = \sqrt{\frac{2GM}{R}}. \qquad (14\text{-}26)$$

Kepler's Laws

Gravitational attraction holds the solar system together and makes possible orbiting Earth satellites, both natural and artificial. Such motions are governed by Kepler's three laws of planetary motion, all of which are direct consequences of Newton's laws of motion and gravitation:

1. *The law of orbits.* All planets move in elliptical orbits with the Sun at one focus.

2. *The law of areas.* A line joining any planet to the Sun sweeps out equal areas in equal times (a statement equivalent to conservation of angular momentum).

3. *The law of periods.* The square of the period T of any planet about the Sun is proportional to the cube of the semimajor axis a of the orbit. For circular orbits with radius r, the semimajor axis a is replaced by r and the law is written as

$$T^2 = \left(\frac{4\pi^2}{GM} \right) r^3 \qquad \text{(law of periods)}, \qquad (14\text{-}31)$$

where M is the mass of the attracting body—the Sun in the case of the solar system. This result is generally valid for elliptical planetary orbits, when the semimajor axis a is inserted in place of the circular radius r.

Energy in Planetary Motion

When a planet or satellite with mass m moves in a circular orbit with radius r, its potential energy U and kinetic energy K are given by

$$U = -\frac{GMm}{r} \quad \text{and} \quad K = \frac{GMm}{2r}. \qquad (14\text{-}20,\ 14\text{-}40)$$

The mechanical energy $E = K + U$ is

$$E = -\frac{GMm}{2r}. \qquad (14\text{-}42)$$

For an elliptical orbit of semimajor axis a,

$$E = -\frac{GMm}{2a}. \qquad (14\text{-}44)$$

Einstein's View of Gravitation

Einstein pointed out that gravitation and acceleration are equivalent. This **principle of equivalence** led him to a theory of gravitation (the **general theory of relativity**) that explains gravitational effects in terms of a curvature of space.

QUESTIONS

1. In Fig. 14-23, two particles, of masses m and $2m$, are fixed in place on an axis. (a) Where on the axis can a third particle of mass $3m$ be placed (other than at infinity) so that the net gravitational force on it from the first two particles is zero: to the left of the first two particles, to their right, between them but closer to the more massive particle, or between them but closer to the less massive particle? (b) Does the answer change if the third particle has, instead, a mass of $16m$? (c) Is there a point off the axis at which the net force on the third particle would be zero?

2. In Fig. 14-24, a central particle is surrounded by two circular rings of particles, at radii r and R, with $R > r$. All the particles have mass m. What are the magnitude and direction of the net gravitational force on the central particle due to the particles in the rings?

FIGURE 14-23
Question 1.

FIGURE 14-24
Question 2.

3. In Fig. 14-25, a central particle of mass M is surrounded by a square array of other particles, separated by either distance d or $d/2$ along the perimeter of the square. What are the magnitude and direction of the net gravitational force on the central particle due to the other particles?

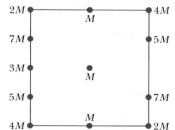

FIGURE 14-25
Question 3.

4. Figure 14-26 shows a particle of mass m that is moved from an infinite distance to the center of a ring of mass M, along the central axis of the ring. For the trip, how does the magnitude of the gravitational force on the particle due to the ring change?

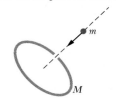

FIGURE 14-26
Question 4.

5. Figure 14-27 shows four arrangements of a particle of mass m and one or more uniform rods of mass M and length L, each a distance d from the particle. Rank the arrangements according to the magnitude of the net gravitational force on the particle from the rods, greatest first.

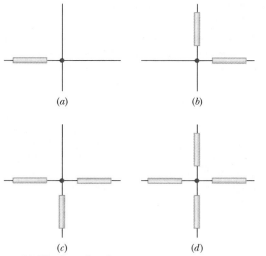

FIGURE 14-27   Question 5.

6. Rank the four systems of equal-mass particles in Checkpoint 2 according to the absolute value of the gravitational potential energy of the system, greatest first.

7. In Fig. 14-28, a particle of mass m (not shown) is to be moved from an infinite distance to one of the three possible locations a, b, and c. Two other particles, of masses m and $2m$, are fixed in place. Rank the three possible locations according to the work done by the net gravitational force on the moving particle due to the fixed particles, greatest first.

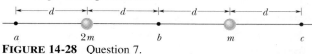

FIGURE 14-28   Question 7.

8. In Fig. 14-29, a particle of mass m is initially at point A, at distance d from the center of one uniform sphere and distance $4d$ from the center of another uniform sphere, both of mass $M \gg m$. State whether, if you moved the particle to point D, the following would be positive, negative, or zero: (a) the change in the gravitational potential energy of the particle, (b) the work done by the net gravitational force on the particle, (c) the work done by your force. (d) What are the answers if, instead, the move were from point B to point C?

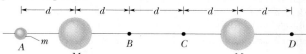

FIGURE 14-29   Question 8.

9. Reconsider the situation of Question 8. Would the work done by you be positive, negative, or zero if you moved the particle (a) from A to B, (b) from A to C, (c) from B to D? (d) Rank those moves according to the absolute value of the work done by your force, greatest first.

10. The following are the masses and radii of three planets: planet A, $2M$ and R; planet B, $3M$ and $2R$; planet C, $4M$ and $2R$. Rank the planets according to the escape speeds from their surfaces, greatest first.

11. Figure 14-30 shows six paths by which a rocket orbiting a moon might move from point a to point b. Rank the paths according to (a) the corresponding change in the gravitational potential energy of the rocket–moon system and (b) the net work done on the rocket by the gravitational force from the moon, greatest first.

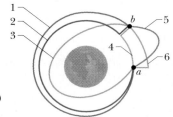

FIGURE 14-30
Question 11.

12. Which of the orbital paths (for, say, a spy satellite) in Fig. 14-31 do not require continuous adjustments by booster rockets? Orbit 1 is at latitude 60°; orbit 2 is in an equatorial plane; orbit 3 is about the center of Earth, between latitudes 60° N and 60° S.

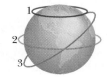

FIGURE 14-31
Question 12.

13. A satellite, with speed v_1 and mass m, is in a circular orbit about a planet of mass M_1. Another satellite, with speed v_2 and mass $2m$, is in a circular orbit of the same radius about a planet of mass M_2. Is M_2 greater than, less than, or equal to M_1 (a) if the satellites have the same period and (b) if $v_2 > v_1$?

EXERCISES & PROBLEMS

SECTION 14-2 Newton's Law of Gravitation

1E. What must the separation be between a 5.2 kg particle and a 2.4 kg particle in order for their gravitational attraction to be 2.3×10^{-12} N?

2E. Some believe that the positions of the planets at the time of birth influence the newborn. Others deride this belief and claim that the gravitational force exerted on a baby by the obstetrician is greater than that exerted by the planets. To check this claim, calculate and compare the gravitational force exerted on a 3 kg baby (a) by a 70 kg obstetrician who is 1 m away and roughly approximated as a point mass, (b) by the massive planet Jupiter ($m = 2 \times 10^{27}$ kg) at its closest approach to Earth ($= 6 \times 10^{11}$ m), and (c) by Jupiter at its greatest distance from Earth ($= 9 \times 10^{11}$ m). (d) Is the claim correct?

3E. The Sun and Earth each exert a gravitational force on the Moon. What is the ratio $F_{\text{Sun}}/F_{\text{Earth}}$ of these two forces? (The average Sun–Moon distance is equal to the Sun–Earth distance.)

4E. One of the *Echo* satellites consisted of an inflated spherical aluminum balloon 30 m in diameter and of mass 20 kg. Suppose a meteor having a mass of 7.0 kg passes within 3.0 m of the surface of the satellite. What is the gravitational force on the meteor from the satellite at the closest approach?

5P. A mass M is split into two parts, m and $M - m$, which are then separated by a certain distance. What ratio m/M maximizes the gravitational force between the parts?

SECTION 14-3 Gravitation and the Principle of Superposition

6E. How far from Earth must a space probe be along a line toward the Sun so that the Sun's gravitational pull on the probe balances Earth's pull?

7E. A spaceship is on a straight-line path between Earth and its moon. At what distance from Earth is the net gravitational force on the spaceship zero?

8P. What is the percent change in the acceleration of Earth toward the Sun when the alignment of Earth, Sun, and Moon changes from an eclipse of the Sun (with the Moon between Earth and Sun) to an eclipse of the Moon (Earth between Moon and Sun)?

9P. Four spheres, with masses $m_1 = 400$ kg, $m_2 = 350$ kg, $m_3 = 2000$ kg, and $m_4 = 500$ kg, have (x, y) coordinates of $(0, 50$ cm$)$, $(0, 0)$, $(-80$ cm$, 0)$, and $(40$ cm$, 0)$, respectively. What is the net gravitational force $\mathbf{F}_2$ on m_2 due to the other masses?

10P. In Fig. 14-32a, four spheres form the corners of a square whose side is 2.0 cm long. What are the magnitude and direction of the net gravitational force from them on a central sphere with mass $m_5 = 250$ kg?

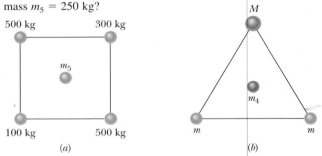

(a)   (b)

FIGURE 14-32 Problems 10 and 11.

11P. In Fig. 14-32b, two spheres of mass m and a third sphere of mass M form an equilateral triangle, and a fourth sphere of mass m_4 is at the center of the triangle. The net gravitational force on m_4 from the three other spheres is zero; what is M in terms of m?

12P. Two spheres with masses $m_1 = 800$ kg and $m_2 = 600$ kg are separated by 0.25 m. What is the net gravitational force (in both magnitude and direction) from them on a 2.0 kg sphere located 0.20 m from m_1 and 0.15 m from m_2?

13P. The masses and coordinates of three spheres are as follows: 20 kg, $x = 0.50$ m, $y = 1.0$ m; 40 kg, $x = -1.0$ m, $y = -1.0$ m; 60 kg, $x = 0$ m, $y = -0.50$ m. What is the magnitude of the gravitational force on a 20 kg sphere located at the origin due to the other spheres?

14P. Figure 14-33 shows two uniform rods of the same length L and mass M, lying on an axis with separation d. Using the result of Sample Problem 14-2, set up an integral (with integration limits and ready for integration) to find the gravitational force of one rod on the other.

$\vdash\!\!-\!\!-\!\!- L \!\!-\!\!-\!\!-\!\!\dashv\!\!\vdash\!\!-\!\!- d \!\!-\!\!-\!\!\dashv\!\!\vdash\!\!-\!\!- L \!\!-\!\!-\!\!\dashv$

FIGURE 14-33 Problem 14.

15P. Figure 14-34 shows a spherical hollow inside a lead sphere of radius R; the surface of the hollow passes through the center of the sphere and "touches" the right side of the sphere. The mass of the sphere before hollowing was M. With what gravitational force does the hollowed-out lead sphere attract a small sphere of mass m that lies at a distance d from the center of the lead sphere, on the straight line connecting the centers of the spheres and of the hollow?

FIGURE 14-34 Problem 15.

SECTION 14-4 Gravitation Near Earth's Surface

16E. Calculate the gravitational acceleration on the surface of the Moon from values of the mass and radius of the Moon found in Appendix C.

17E. At what altitude above Earth's surface would the gravitational acceleration be 4.9 m/s²?

18E. You weigh 120 lb at sidewalk level outside the World Trade Center in New York City. Suppose that you ride from this level to the top of one of its 1350 ft towers. Ignoring Earth's rotation, how much less would you weigh there (because you are slightly farther from the center of Earth)?

19E. A typical neutron star may have a mass equal to that of the Sun but a radius of only 10 km. (a) What is the gravitational acceleration at the surface of such a star? (b) How fast would an object be moving if it fell from rest through a distance of 1.0 m on such a star? (Assume the star does not rotate.)

20E. An object lying on Earth's equator is accelerated (a) toward the center of Earth because Earth rotates, (b) toward the Sun because Earth revolves around the Sun in an almost circular orbit, and (c) toward the center of our galaxy because the Sun moves about the galactic center. For the latter, the period is 2.5×10^8 y and the radius is 2.2×10^{20} m. Calculate these three accelerations as multiples of $g = 9.8$ m/s².

21E. (a) What will an object weigh on the Moon's surface if it weighs 100 N on Earth's surface? (b) How many Earth radii must this same object be from the center of Earth if it is to weigh the same as it does on the Moon?

22P. If g is to be determined by dropping an object through a distance of exactly 10 m, what percent error in the measurement of the time of fall results in a 0.1% error in the value of g?

23P. The fastest possible rate of rotation of a planet is that for which the gravitational force on material at the equator just barely provides the centripetal force needed for the rotation. (Why?) (a) Show that the corresponding shortest period of rotation is

$$T = \sqrt{\frac{3\pi}{G\rho}},$$

where ρ is the uniform density of the spherical planet. (b) Calculate the rotation period assuming a density of 3.0 g/cm³, typical of many planets, satellites, and asteroids. No astronomical object has ever been found to be spinning with a period shorter than that determined by this analysis.

24P. In Fig. 14-35, identical masses m hang from strings on a balance at the surface of Earth. The strings have negligible mass

and differ in length by h. Assume that Earth is spherical, with density $\rho = 5.5$ g/cm³. (a) Show that the difference ΔW in the weights, due to one mass being closer to Earth than the other, is $8\pi G\rho mh/3$. (b) Find the difference in length that will give a ratio $\Delta W/W = 1 \times 10^{-6}$, where W is either weight.

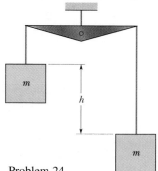

FIGURE 14-35 Problem 24.

25P. A body is suspended from a spring scale in a ship sailing along the equator with a speed v. (a) Show that the scale reading will be very close to $W_0(1 \pm 2\omega v/g)$, where ω is the angular speed of Earth and W_0 is the scale reading when the ship is at rest. (b) Explain the $\pm$ sign.

26P. Certain neutron stars (extremely dense stars) are believed to be rotating at about 1 rev/s. If such a star has a radius of 20 km, what must be its minimum mass so that material on its surface remains in place during the rapid rotation?

27P. The radius R_h and mass M_h of a black hole are related by $R_h = 2GM_h/c^2$, where c is the speed of light. Assume that the gravitational acceleration a_g of an object at a distance $r_o = 1.001R_h$ from the center of a black hole is given by Eq. 14-12 (it is, for large black holes). (a) Find an expression for a_g at r_o in terms of M_h. (b) Does a_g at r_o increase or decrease with an increase of M_h? (c) What is a_g at r_o for a very large black hole whose mass is 1.55×10^{12} times the solar mass of 1.99×10^{30} kg? (d) If the astronaut of Sample Problem 14-4 is at r_o with her feet toward this black hole, what is the difference in gravitational acceleration at head and feet? (e) Is the tendency to stretch the astronaut severe?

SECTION 14-5 Gravitation Inside Earth

28E. Two concentric shells of uniform density having masses M_1 and M_2 are situated as shown in Fig. 14-36. Find the force on a particle of mass m when the particle is located at (a) $r = a$, (b) $r = b$, and (c) $r = c$. The distance r is measured from the center of the shells.

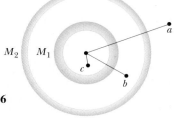

FIGURE 14-36
Exercise 28.

29E. With what speed would mail pass through the center of Earth if it were dropped down the tunnel of Sample Problem 14-5?

30E. Show that, at the bottom of a vertical mine shaft dug to depth D, the measured value of a_g will be

$$a_g = a_{gs} \left(1 - \frac{D}{R}\right),$$

a_{gs} being the surface value. Assume that Earth is a uniform sphere of radius R.

31P. A solid sphere of uniform density has a mass of 1.0×10^4 kg and a radius of 1.0 m. What is the gravitational force due to the sphere on a particle of mass m located at a distance of (a) 1.5 m and (b) 0.50 m from the center of the sphere? (c) Write a general expression for the gravitational force on m at a distance $r \leq 1.0$ m from the center of the sphere.

32P. A uniform solid sphere of radius R produces a gravitational acceleration of a_g on its surface. At what two distances from the center of the sphere is the gravitational acceleration $a_g/3$? (*Hint:* Consider distances both inside and outside the sphere.)

33P. Figure 14-37 shows, not to scale, a cross section through the interior of Earth. Rather than being uniform throughout, Earth is divided into three zones: an outer *crust,* a *mantle,* and an inner *core.* The dimensions of these zones and the masses contained within them are shown on the figure. Earth has a total mass of 5.98×10^{24} kg and a radius of 6370 km. Ignore rotation and assume that Earth is spherical. (a) Calculate a_g at the surface. (b) Suppose that a bore hole (the *Mohole*) is driven to the crust–mantle interface at a depth of 25 km; what would be the value of a_g at the bottom of the hole? (c) Suppose that Earth were a uniform sphere with the same total mass and size. What would be the value of a_g at a depth of 25 km? (See Exercise 30.) (Precise measurements of a_g are sensitive probes of the interior structure of Earth, although results can be clouded by local density variations.)

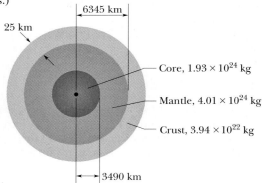

FIGURE 14-37 Problem 33. Not to scale.

SECTION 14-6 Gravitational Potential Energy

34E. (a) What is the gravitational potential energy of the two-particle system in Exercise 1? If you triple the separation between the particles, how much work is done (b) by the gravitational force between the particles and (c) by you?

35E. (a) In Problem 9, remove m_1 and calculate the gravitational potential energy of the remaining three-particle system. (b) If m_1 is then put back in place, is the potential energy of the four-particle system more or less than that of the system in (a)? (c) In (a), is the work done by you to remove m_1 positive or negative? (d) In (b), is the work done by you to replace m_1 positive or negative?

36E. In Problem 5, what ratio m/M gives the least gravitational potential energy for the system?

37E. The mean diameters of Mars and Earth are 6.9×10^3 km and 1.3×10^4 km, respectively. The mass of Mars is 0.11 times Earth's mass. (a) What is the ratio of the mean density of Mars to that of Earth? (b) What is the value of g on Mars? (c) What is the escape speed on Mars?

38E. A spaceship is idling at the fringes of our galaxy, 80,000 light-years from the galactic center. What is the ship's escape speed from the galaxy? The mass of the galaxy is 1.4×10^{11} times that of our sun. Assume, for simplicity, that the matter forming the galaxy is distributed in a uniform sphere.

39E. Calculate the amount of energy required to escape from (a) Earth's moon and (b) Jupiter relative to that required to escape from Earth.

40E. Show that the escape speed from the Sun at Earth's distance from the Sun is $\sqrt{2}$ times the speed of the Earth in its orbit, assumed to be a circle. (This is a specific case of a general result for circular orbits: $v_{esc} = \sqrt{2}v_{orb}$.)

41E. A particle of comet dust with mass m is a distance R from Earth's center and a distance r from the Moon's center. If Earth's mass is M_E and the Moon's mass is M_m, what is the total gravitational potential energy of the particle–Earth system and the particle–Moon system?

42E. Upon "burning out," a large star can collapse under its own gravitational force to become a black hole. The star's surface then has a radius R_s such that the removal of a mass m from the surface to infinity would require work equal to the mass energy mc^2 of that mass. If M_s is the star's mass, show with Newton's law of gravitation that $R_s = GM_s/c^2$. (The correct value for R_s is actually twice this. To get it, we would have to apply Einstein's general theory of relativity rather than Newton's law.)

43P. The three spheres in Fig. 14-38, with masses $m_1 = 800$ g, $m_2 = 100$ g, and $m_3 = 200$ g, have their centers on a common line, with $L = 12$ cm and $d = 4.0$ cm. You move the middle sphere until its center-to-center separation from m_3 is $d = 4.0$ cm. How much work is done on m_2 (a) by you and (b) by the net gravitational force on m_2 due to m_1 and m_3?

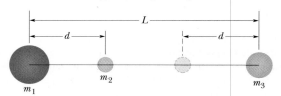

FIGURE 14-38 Problem 43.

44P. A rocket is accelerated to a speed of $v = 2\sqrt{gR_e}$ near Earth's surface (where Earth's radius is R_e) and then coasts upward. (a) Show that it will escape from Earth. (b) Show that very far from Earth its speed is $v = \sqrt{2gR_e}$.

45P. (a) What is the escape speed on a spherical asteroid whose radius is 500 km and whose gravitational acceleration at the surface is 3.0 m/s²? (b) How far from the surface will a particle go if it leaves the asteroid's surface with a radial speed of 1000 m/s? (c) With what speed will an object hit the asteroid if it is dropped from 1000 km above the surface?

46P. Zero, a hypothetical planet, has a mass of 5.0×10^{23} kg, a radius of 3.0×10^6 m, and no atmosphere. A 10 kg space probe is to be launched vertically from its surface. (a) If the probe is launched with an initial energy of 5.0×10^7 J, what will be its kinetic energy when it is 4.0×10^6 m from the center of Zero? (b) If the probe is to achieve a maximum distance of 8.0×10^6 m from the center of Zero, with what initial kinetic energy must it be launched from the surface of Zero?

47P. In a double-star system, two stars of mass 3.0×10^{30} kg each rotate about the system's center of mass at a radius of 1.0×10^{11} m. (a) What is their common angular speed? (b) If a meteoroid passes through this center of mass perpendicular to the orbital plane of the stars, what value must its speed exceed at that point if it is to escape to "infinity" from the star system?

48P. Two neutron stars are separated by a distance of 10^{10} m. They each have a mass of 10^{30} kg and a radius of 10^5 m. They are initially at rest with respect to each other. (a) How fast are they moving when their separation has decreased to one-half its initial value? (b) How fast are they moving just before they collide?

49P. A projectile is fired vertically from Earth's surface with an initial speed of 10 km/s. Neglecting air drag, how far above the surface of Earth will it go?

50P. A sphere of matter, of mass M and radius a, has a concentric cavity of radius b, as shown in cross section in Fig. 14-39. (a) Sketch a curve of the gravitational force F exerted by the sphere on a particle of mass m, located a distance r from the center of the sphere, as a function of r in the range $0 \leq r \leq \infty$. Consider $r = 0$, b, a, and ∞ in particular. (b) Sketch the corresponding curve for the potential energy $U(r)$ of the system.

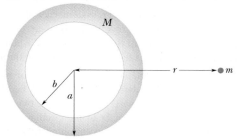

FIGURE 14-39   Problem 50.

51P. A 20 kg mass is located at the origin, and a 10 kg mass is located on the x axis at $x = 0.80$ m. The 10 kg mass is released from rest while the 20 kg mass is held in place at the origin. (a) What is the gravitational potential energy of the two-mass system immediately after the 10 kg mass is released? (b) What is the

kinetic energy of the 10 kg mass after it has moved 0.20 m toward the 20 kg mass?

52P*. Several planets (Jupiter, Saturn, Uranus) possess nearly circular surrounding rings, perhaps composed of material that failed to form a satellite. In addition, many galaxies contain ring-like structures. Consider a homogeneous ring of mass M and radius R. (a) What gravitational attraction does it exert on a particle of mass m located a distance x from the center of the ring along its axis? See Fig. 14-40. (b) Suppose the particle falls from rest as a result of the attraction of the ring of matter. Find an expression for the speed with which it passes through the center of the ring.

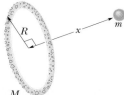

FIGURE 14-40   Problem 52.

53P*. The gravitational force between two particles with masses m and M, initially at rest at great separation, pulls them together. Show that at any instant the speed of either particle relative to the other is $\sqrt{2G(M + m)/d}$, where d is their separation at that instant. (*Hint:* Use the laws of conservation of energy and conservation of linear momentum.)

SECTION 14-7   Planets and Satellites: Kepler's Laws

54E. The mean distance of Mars from the Sun is 1.52 times that of Earth from the Sun. From Kepler's law of periods, calculate the number of years required for Mars to make one revolution about the Sun; compare your answer with the value given in Appendix C.

55E. The planet Mars has a satellite, Phobos, which travels in an orbit of radius 9.4×10^6 m with a period of 7 h 39 min. Calculate the mass of Mars from this information.

56E. Determine the mass of Earth from the period T and the radius r of the Moon's orbit about Earth: $T = 27.3$ days and $r = 3.82 \times 10^5$ km. Assume the Moon orbits the center of Earth rather than the center of mass of the Earth–Moon system.

57E. Our Sun, with mass 2.0×10^{30} kg, revolves about the center of the Milky Way galaxy, which is 2.2×10^{20} m away, once every 2.5×10^8 years. Assuming that each of the stars in the galaxy has a mass equal to that of our Sun, that the stars are distributed uniformly in a sphere about the galactic center, and that our Sun is essentially at the edge of that sphere, estimate roughly the number of stars in the galaxy.

58E. A satellite is placed in a circular orbit with a radius equal to one-half the radius of the Moon's orbit. What is its period of revolution in lunar months? (A lunar month is the period of revolution of the Moon.)

59E. (a) What linear speed must an Earth satellite have to be in a circular orbit at an altitude of 160 km? (b) What is the period of revolution?

60E. Most asteroids revolve around the Sun between Mars and Jupiter. However, several "Apollo asteroids" with diameters of about 30 km move in orbits that cross the orbit of Earth. The orbit of one of these asteroids is shown to scale in Fig. 14-41. By taking measurements directly from the figure, calculate the asteroid's period of revolution in years.

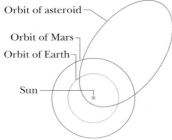

FIGURE 14-41 Exercise 60.

61E. A satellite, moving in an elliptical orbit, is 360 km above Earth's surface at its farthest point and 180 km above at its closest point. Calculate (a) the semimajor axis and (b) the eccentricity of the orbit. (*Hint:* See Sample Problem 14-8.)

62E. The Sun's center is at one focus of Earth's orbit. How far from this focus is the other focus? Express your answer in terms of the solar radius, 6.96×10^8 m. The eccentricity of Earth's orbit is 0.0167, and the semimajor axis may be taken to be 1.50×10^{11} m. See Fig. 14-14.

63E. (a) Using Kepler's third law (Eq. 14-31), express the gravitational constant G in terms of the astronomical unit as a length unit, the solar mass as a mass unit, and the year as a time unit. (One astronomical unit = 1 AU = 1.496×10^{11} m. One solar mass = $1M_s = 1.99 \times 10^{30}$ kg. One year = 1 y = 3.156×10^7 s.) (b) What form does Kepler's third law take in these units?

64E. A satellite hovers over a certain spot on the equator of (rotating) Earth. What is the altitude of its orbit (called a *geosynchronous orbit*)?

65E. A comet that was seen in April 574 by Chinese astronomers on a day known by them as the Woo Woo day was spotted again in May 1994. Assume the time between observations is the period of the Woo Woo day comet and take its eccentricity as 0.11. What are (a) the semimajor axis of the comet's orbit and (b) its greatest distance from the Sun in terms of the mean orbital radius R_P of Pluto.

66E. In 1993 the spacecraft *Galileo* sent home an image (Fig. 14-42) of asteroid 243 Ida and an orbiting tiny moon (now known as Dactyl), the first confirmed example of an asteroid–moon system. In the image, the moon, which is 1.5 km wide, is 100 km from the center of the asteroid, which is 55 km long. The shape of the moon's orbit is not well known; assume it is circular with a period of 27 h. (a) What is the mass of the asteroid? (b) The volume of the asteroid, measured from the *Galileo* images, is 14,100 km³. What is the density of the asteroid?

67P. Assume that the satellite of Exercise 64 is in orbit at the longitude of Chicago. You are in Chicago (latitude 47.5°) and

FIGURE 14-42 Exercise 66. A picture from the spacecraft *Galileo* shows a tiny moon orbiting asteroid 243 Ida.

want to pick up the signals the satellite broadcasts. In what direction should you point your antenna?

68P. In 1610, Galileo used his telescope to discover four prominent moons around Jupiter. Their mean orbital radii a and periods T are

NAME	a (10^8 m)	T (days)
Io	4.22	1.77
Europa	6.71	3.55
Ganymede	10.7	7.16
Callisto	18.8	16.7

(a) Plot $\log a$ (y axis) against $\log T$ (x axis) and show that you get a straight line. (b) Measure the slope of the line and compare it with the value that you expect from Kepler's third law. (c) Find the mass of Jupiter from the intercept of this line with the y axis.

69P. Show how, guided by Kepler's third law (Eq. 14-31), Newton could deduce that the force holding the Moon in its orbit, assumed circular, depends on the inverse square of the distance from the center of Earth.

70P. In a certain binary-star system, each star has the same mass as our Sun, and they revolve about their center of mass. The distance between them is the same as the distance between Earth and the Sun. What is their period of revolution in years?

71P. A certain triple-star system consists of two stars, each of mass m, revolving about a central star of mass M in the same circular orbit of radius r (Fig. 14-43). The two stars are always at opposite ends of a diameter of the circular orbit. Derive an expression for the period of revolution of the stars.

72P. (a) What is the escape speed from the Sun for an object in Earth's orbit (of orbital radius R) but far from Earth? (b) If an object already has a speed equal to Earth's orbital speed, what additional speed must it be given to escape as in (a)? (c) Suppose an object is launched from Earth in the direction of Earth's orbital

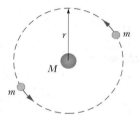

FIGURE 14-43 Problem 71.

motion. What speed must it be given during the launch so that when it is far from Earth, but still at a distance of about R from the Sun, it can escape from the Sun? (This is the speed required for an Earth-launched object to escape from the Sun.)

73P*. Three identical stars of mass M are located at the vertices of an equilateral triangle with side L. At what speed must they move if they all revolve under the influence of one another's gravitational force in a circular orbit circumscribing the triangle while still preserving the equilateral triangle?

74P*. A satellite in a circular orbit had been intended to hover over a certain spot on Earth's surface. Through error, however, the satellite's orbital radius was made 1.0 km too large for this to happen. At what rate and in what direction does the point directly below the satellite move across Earth's surface?

SECTION 14-8 Satellites: Orbits and Energy

75E. An asteroid, whose mass is 2.0×10^{-4} times the mass of Earth, revolves in a circular orbit around the Sun at a distance that is twice Earth's distance from the Sun. (a) Calculate the period of revolution of the asteroid in years. (b) What is the ratio of the kinetic energy of the asteroid to that of Earth?

76E. Consider two satellites, A and B, of equal mass m, moving in the same circular orbit of radius r around Earth, of mass M_E, but in opposite senses of rotation and therefore on a collision course (see Fig. 14-44). (a) In terms of G, M_E, m, and r, find the total mechanical energy $E_A + E_B$ of the two-satellite-plus-Earth system before collision. (b) If the collision is completely inelastic so that the wreckage remains as one piece of tangled material (mass $= 2m$), find the total mechanical energy immediately after collision. (c) Describe the subsequent motion of the wreckage.

FIGURE 14-44 Exercise 76.

77P. Two Earth satellites, A and B, each of mass m, are to be launched into circular orbits about Earth's center. Satellite A is to orbit at an altitude of 4000 mi. Satellite B is to orbit at an altitude of 12,000 mi. The radius of Earth R_E is 4000 mi. (a) What is the

ratio of the potential energy of satellite B to that of satellite A, in orbit? (b) What is the ratio of the kinetic energy of satellite B to that of satellite A, in orbit? (c) Which satellite has the greater total energy if each has a mass of 14.6 kg? By how much?

78P. Use the conservation of mechanical energy and Eq. 14-44 to show that if an object is in an elliptical orbit about a planet, then its distance r from the planet and speed v are related by

$$v^2 = GM \left(\frac{2}{r} - \frac{1}{a} \right).$$

79P. Use the result of Problem 78 and data contained in Sample Problem 14-8 to calculate (a) the speed v_p of comet Halley at perihelion and (b) its speed v_a at aphelion. (c) Using conservation of angular momentum relative to the Sun, find the ratio of the comet's perihelion distance R_p to its aphelion distance R_a in terms of v_p and v_a.

80P. (a) Does it take more energy to get a satellite up to 1000 mi above Earth than to put it in circular orbit once it is there? (Take Earth's radius to be 4000 mi.) (b) What about 2000 mi? (c) What about 3000 mi?

81P. One way to attack a satellite in Earth orbit is to launch a swarm of pellets in the same orbit as the satellite but in the opposite direction. Suppose a satellite in a circular orbit 500 km above Earth's surface collides with a pellet having mass 4.0 g. (a) What is the kinetic energy of the pellet in the reference frame of the satellite? (b) What is the ratio of this kinetic energy to the kinetic energy of a 4.0 g bullet from a modern army rifle with a muzzle velocity of 950 m/s?

82P. Consider a satellite in a circular orbit about Earth. State how the following properties of the satellite depend on the radius r of its orbit: (a) period, (b) kinetic energy, (c) angular momentum, and (d) speed.

83P. What are (a) the speed and (b) the period of a 220 kg satellite in an approximately circular orbit 640 km above the surface of Earth? Suppose the satellite loses mechanical energy at the average rate of 1.4×10^5 J per orbital revolution. Adopting the reasonable approximation that the trajectory is a "circle of slowly diminishing radius," determine the satellite's (c) altitude, (d) speed, and (e) period at the end of its 1500th revolution. (f) What is the magnitude of the average retarding force? (g) Is angular momentum around Earth's center conserved for the satellite or the satellite–Earth system?

84P. The orbit of Earth about the Sun is *almost* circular: the closest and farthest distances are 1.47×10^8 km and 1.52×10^8 km, respectively. Determine the corresponding variations in (a) total energy, (b) potential energy, (c) kinetic energy, and (d) orbital speed. (*Hint:* Use the laws of conservation of energy and conservation of angular momentum.)

85P. In a shuttle craft of mass $m = 2000$ kg, Captain Janeway orbits a planet of mass $M = 5.98 \times 10^{24}$ kg, in a circular orbit of radius $r = 6.80 \times 10^6$ m. What are (a) the period of the orbit and (b) the speed of the shuttle craft? Janeway briefly fires a forward-pointing thruster, reducing her speed by 1.00%. Just then, what are (c) the speed, (d) the kinetic energy, (e) the gravitational potential energy, and (f) the mechanical energy of the shuttle craft? (g) What is the semimajor axis of the elliptical orbit now taken by the craft? (h) What is the difference between the period of the original circular orbit and that of the new elliptical orbit, and which orbit has the smaller period?

SECTION 14-9 Einstein and Gravitation

86E. In Fig. 14-20*b*, the scale on which the 60 kg physicist stands reads 220 N. How long will the cantaloupe take to reach the floor if the physicist drops it from rest (relative to himself), 2.1 m from the floor?

87P. Figure 14-45 shows the walls of a duct in a ship in deep space; the ship has an acceleration $\mathbf{a} = (2.5 \text{ m/s}^2)\mathbf{i}$. An electron is sent across the duct's width of 3.0 cm, from the origin shown and with an initial velocity of $\mathbf{v}_0 = (0.40 \text{ m/s})\mathbf{j}$. In unit vector notation and relative to the ship, what are (a) the displacement of the electron at the end of its flight and (b) its velocity just before impact with the far wall?

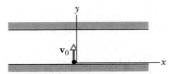

FIGURE 14-45 Problem 87.

Electronic Computation

88. A 6000 kg planetary probe is in a circular solar orbit with a radius of 108×10^6 km (the orbital radius of Venus). The space agency wants to put the probe into the same orbit as Earth, a circular orbit with a radius of 150×10^6 km. The first step is to increase the probe's speed so that the probe is in an elliptical orbit with a perihelion distance equal to the radius of its initial orbit and an aphelion distance equal to the radius of its final orbit. The mass of the Sun is 1.99×10^{30} kg. (a) Calculate the increases in speed and energy that are required. Plot the elliptical orbit. (In polar coordinates, with the origin at the center, the equation for an ellipse is

$$r^2 = \frac{r_p r_a (r_p + r_a)^2}{(r_p + r_a)^2 \sin^2 \theta + 4 r_p r_a \cos^2 \theta},$$

where r_p is the perihelion distance and r_a is the aphelion distance.) (b) When the aphelion position is reached, the speed of the probe is again changed to put it into the desired circular orbit. What changes in the speed and energy are required?

89. In your calculator, make a list of the periods T and a separate list for the semimajor axes a for the planets in Table 14-3. Multiply the list for T by an appropriate factor so that the unit of T is seconds. (a) Store values of T^2 and a^3 in two new lists. Have the calculator do a linear regression fit of T^2 versus a^3. From the parameters of the fit and using the known value of G, determine the mass of the Sun. (b) Store values of log T in a list and log a in another list. Have the calculator plot log T versus log a and then do a linear regression fit of the plot. From the parameters of this fit and using the known value of G, again determine the mass of the Sun.

15
Fluids

The force exerted by water on the body of a descending diver increases noticeably, even for a relatively shallow descent to the bottom of a swimming pool. However, in 1975, using scuba gear with a special gas mixture for breathing, William Rhodes emerged from a chamber that had been lowered 1000 ft into the Gulf of Mexico, and he then swam to a record depth of 1148 ft. Strangely, a novice scuba diver practicing in a swimming pool might be in more danger from the force exerted by the water than was Rhodes. And occasionally, novice scuba divers die because they have neglected that danger. What is this potentially lethal risk?

15-1 FLUIDS AND THE WORLD AROUND US

Fluids—which include both liquids and gases—play a central role in our daily lives. We breathe and drink them, and a rather vital fluid circulates in the human cardiovascular system. There is the fluid ocean and the fluid atmosphere.

In a car, there are fluids in the tires, the gas tank, the radiator, the combustion chambers of the engine, the exhaust manifold, the battery, the air conditioning system, the windshield wiper reservoir, the lubrication system, and the hydraulic system. (*Hydraulic* means operated via a liquid.) The next time you see a large piece of earthmoving machinery, count the hydraulic cylinders that permit the machine to do its work. Large jet planes have scores of them.

We use the kinetic energy of a moving fluid in windmills, and the potential energy of another fluid in hydroelectric power plants. Given time, fluids carve the landscape. We often travel great distances just to watch fluids move. Perhaps it is time to see what physics can tell us about fluids.

15-2 WHAT IS A FLUID?

A **fluid,** in contrast to a solid, is a substance that can flow. Fluids conform to the boundaries of any container in which we put them. They do so because a fluid cannot sustain a force that is tangential to its surface. (In the more formal language of Section 13-6, a fluid is a substance that flows because it cannot withstand a shearing stress. It can, however, exert a force in the direction perpendicular to its surface.) Some materials, such as pitch, take a long time to conform to the boundaries of a container, but they do so eventually; thus we classify them as fluids.

You may wonder why we lump liquids and gases together and call them fluids. After all (you may say), liquid water is as different from steam as it is from ice. Actually, it is not. Ice, like other crystalline solids, has its constituent atoms organized in a fairly rigid three-dimensional array called a crystalline lattice. In neither steam nor liquid water, however, is there any such orderly long-range arrangement.

15-3 DENSITY AND PRESSURE

When we discuss rigid bodies, we are concerned with particular lumps of matter, such as wooden blocks, baseballs, or metal rods. Physical quantities that we find useful, and in whose terms we express Newton's laws, are *mass* and *force*. We might speak, for example, of a 3.6 kg block acted on by a 25 N force.

With fluids, we are more interested in properties that vary from point to point in the extended substance than with properties of specific lumps of that substance. It is more useful to speak of **density** and **pressure** than of mass and force.

Density

To find the density ρ (rho) of a fluid at any point, we isolate a small volume element ΔV around that point and measure the mass Δm of the fluid contained within that element. The **density** is then

$$\rho = \frac{\Delta m}{\Delta V}. \qquad (15\text{-}1)$$

In theory, the density at any point in a fluid is the limit of this ratio as the volume element ΔV at that point is made smaller and smaller. In practice, we assume that a fluid sample is large compared to atomic dimensions and thus "smooth" (with uniform density), rather than "lumpy" with atoms. This assumption allows us to write Eq. 15-1 in the form $\rho = m/V$, where m and V are the mass and volume of the sample.

Density is a scalar property; its SI unit is the kilogram per cubic meter. Table 15-1 shows the densities of some

TABLE 15-1 SOME DENSITIES

MATERIAL OR OBJECT		DENSITY (kg/m^3)
Interstellar space		10^{-20}
Best laboratory vacuum		10^{-17}
Air:	20°C and 1 atm	1.21
	20°C and 50 atm	60.5
Styrofoam		1×10^2
Water:	20°C and 1 atm	0.998×10^3
	20°C and 50 atm	1.000×10^3
Seawater: 20°C and 1 atm		1.024×10^3
Whole blood		1.060×10^3
Ice		0.917×10^3
Iron		7.9×10^3
Mercury		13.6×10^3
Earth:	average	5.5×10^3
	core	9.5×10^3
	crust	2.8×10^3
Sun:	average	1.4×10^3
	core	1.6×10^5
White dwarf star (core)		10^{10}
Uranium nucleus		3×10^{17}
Neutron star (core)		10^{18}
Black hole (1 solar mass)		10^{19}

substances and the average densities of some objects. Note that the density of a gas (see Air in the table) varies considerably with pressure, but the density of a liquid (see Water) does not. That is, gases are readily *compressible* but liquids are not.

Pressure

Let a small pressure-sensing device be suspended inside a fluid-filled vessel, as in Fig. 15-1a. The sensor (Fig. 15-1b) consists of a piston of area ΔA riding in a close-fitting cylinder and resting against a spring. A readout arrangement allows us to record the amount by which the (calibrated) spring is compressed by the surrounding fluid, thus indicating the magnitude ΔF of the force that acts on the piston. We define the **pressure** exerted by the fluid on the piston as

$$p = \frac{\Delta F}{\Delta A}. \qquad (15\text{-}2)$$

In theory, the pressure at any point in the fluid is the limit of this ratio as the area ΔA of the piston, centered on that point, is made smaller and smaller. However, if the force is uniform over a flat area A, we can write Eq. 15-2 as $p = F/A$. (When we say a force is uniform over an area, we mean that the same force would be measured at every point of the area.)

We find by experiment that at a given point in a fluid at rest, the pressure p defined by Eq. 15-2 has the same value no matter how the pressure sensor is oriented. Pressure is a scalar, having no directional properties. It is true that the force acting on the piston of our pressure sensor is a vector, but Eq. 15-2 involves only the *magnitude* of that force, a scalar quantity.

The SI unit of pressure is the newton per square meter, which is given a special name, the **pascal** (Pa). In metric countries, tire pressure gauges are calibrated in kilopascals. The pascal is related to some other common (non-SI) pressure units as follows:

$$1 \text{ atm} = 1.01 \times 10^5 \text{ Pa} = 760 \text{ torr} = 14.7 \text{ lb/in.}^2.$$

The *atmosphere* (atm) is, as the name suggests, the approximate average pressure of the atmosphere at sea level. The *torr* (named for Evangelista Torricelli, who invented the mercury barometer in 1674) was formerly called the *millimeter of mercury* (mm Hg). The pound per square inch is often abbreviated psi. Table 15-2 shows some pressures.

TABLE 15-2 SOME PRESSURES

	PRESSURE (Pa)
Center of the Sun	2×10^{16}
Center of Earth	4×10^{11}
Highest sustained laboratory pressure	1.5×10^{10}
Deepest ocean trench (bottom)	1.1×10^8
Spike heels on a dance floor	1×10^6
Automobile tire[a]	2×10^5
Atmosphere at sea level	1.0×10^5
Normal blood pressure[a,b]	1.6×10^4
Best laboratory vacuum	10^{-12}

[a] Pressure in excess of atmospheric pressure.

[b] The systolic pressure, corresponding to 120 torr on the physician's pressure gauge.

SAMPLE PROBLEM 15-1

A living room has floor dimensions of 3.5 m and 4.2 m and a height of 2.4 m.

(a) What does the air in the room weigh?

SOLUTION: If V is the volume of the room and ρ is the density of air at 1 atm (see Table 15-1), then the weight W of the air is

$$W = mg = \rho V g$$
$$= (1.21 \text{ kg/m}^3)(3.5 \text{ m} \times 4.2 \text{ m} \times 2.4 \text{ m})(9.8 \text{ m/s}^2)$$
$$= 418 \text{ N} \approx 420 \text{ N.} \qquad \text{(Answer)}$$

This is about 94 lb.

(b) What force does the atmosphere exert on the floor of the room?

SOLUTION: The force is

$$F = pA = (1.0 \text{ atm})\left(\frac{1.01 \times 10^5 \text{ N/m}^2}{1 \text{ atm}}\right)(3.5 \text{ m})(4.2 \text{ m})$$
$$= 1.5 \times 10^6 \text{ N.} \qquad \text{(Answer)}$$

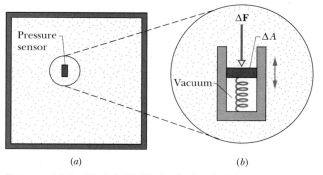

FIGURE 15-1 (a) A fluid-filled vessel containing a small pressure sensor, shown in (b). The pressure is measured by the relative position of the movable piston in the sensor.

This force (≈ 170 tons) is the weight of a column of air covering the floor and extending all the way to the top of the atmosphere. It is equal to the force that would be exerted on the floor if (in the absence of the atmosphere) the room were filled with mercury to a depth of 30 in. Why doesn't this enormous force break the floor?

15-4 FLUIDS AT REST

Figure 15-2*a* shows a tank of water—or other liquid—open to the atmosphere. As every diver knows, the pressure *increases* with depth below the air–water interface. The diver's depth gauge, in fact, is a pressure sensor much like that of Fig. 15-1*b*. As every mountaineer knows, the pressure *decreases* with altitude as one ascends into the atmosphere. The pressures encountered by the diver and the mountaineer are usually called *hydrostatic pressures,* because they are due to fluids that are static (at rest).

Let us look first at the increase in pressure with depth below the water's surface. We set up a vertical y axis, its origin being at the air–water interface and the direction of increasing y being up. Consider a water sample contained in an imaginary right circular cylinder of horizontal base area A, and let y_1 and y_2 (both of which are *negative* numbers) be the depths below the surface of the upper and the lower cylinder faces, respectively.

Figure 15-2*b* shows a free-body diagram for the water in the cylinder. The water sample is in equilibrium, its weight (downward) being exactly balanced by the difference between the force of magnitude $F_2 = p_2A$ acting upward on its lower face and the force of magnitude $F_1 = p_1A$ acting downward on its upper face. Thus

$$F_2 = F_1 + W. \qquad (15\text{-}3)$$

The volume V of the cylinder is $A(y_1 - y_2)$. Thus the mass m of the water in the cylinder is $\rho A(y_1 - y_2)$, in which ρ is the uniform density of water. The weight W is then $mg = \rho Ag(y_1 - y_2)$. Substituting this for W in Eq. 15-3 and substituting for F_1 and F_2 yield

$$p_2A = p_1A + \rho Ag(y_1 - y_2),$$

or

$$p_2 = p_1 + \rho g(y_1 - y_2). \qquad (15\text{-}4)$$

This equation can be used to find pressures both in a liquid and in the atmosphere. For the former, suppose we seek the pressure p at a depth h below the liquid surface. Then we choose level 1 to be the surface, level 2 to be a distance h below it (as in Fig. 15-3), and p_0 to represent the atmospheric pressure on the surface. We then substitute

$$y_1 = 0, \quad p_1 = p_0 \quad \text{and} \quad y_2 = -h, \quad p_2 = p$$

into Eq. 15-4, which becomes

$$p = p_0 + \rho gh \qquad \text{(pressure at depth } h). \quad (15\text{-}5)$$

Note that the pressure at a given depth depends on that depth but not on any horizontal dimension. Thus Eq. 15-5 holds no matter what the shape of the containing vessel. If the bottom surface of the vessel is at depth h, then Eq. 15-5 gives the pressure p on that surface.

In Eq. 15-5, p is said to be the total pressure or **absolute pressure** at level 2. To see why, note in Fig. 15-3 that the pressure p at level 2 consists of two contributions: (1) p_0, the pressure due to the atmosphere, which bears down on the liquid, and (2) ρgh, the pressure due to the liquid above level 2, which bears down on level 2. In general, the difference between an absolute pressure and an atmospheric pressure is called the **gauge pressure.** So, for the situation of Fig. 15-3, the gauge pressure is ρgh.

Equation 15-4 also holds above the liquid surface: it gives the atmospheric pressure at a given distance above level 1 in terms of the atmospheric pressure p_1 at level 1

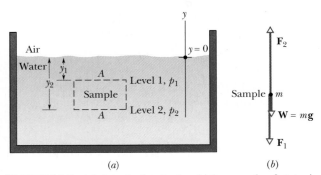

(*a*)

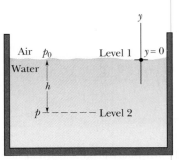

FIGURE 15-2 (*a*) A tank of water in which a sample of water is contained in an imaginary cylinder of horizontal base area A. (*b*) A free-body diagram of the water sample. The water in the sample is in static equilibrium, its weight being balanced by the net upward buoyant force that acts on it.

FIGURE 15-3 The pressure p increases with depth h below the water surface according to Eq. 15-5.

(*assuming* that the atmospheric density is uniform over that distance). For example, to find the atmospheric pressure at a distance d above level 1 in Fig. 15-3, we substitute

$$y_1 = 0, \quad p_1 = p_0 \quad \text{and} \quad y_2 = d, \quad p_2 = p.$$

Then with $\rho = \rho_{air}$, we obtain

$$p = p_0 - \rho_{air}gd.$$

CHECKPOINT **1:** The figure shows four containers of olive oil. Rank them according to the pressure at depth h, greatest first.

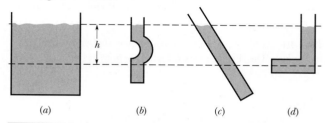

(a)   (b)   (c)   (d)

SAMPLE PROBLEM 15-2

(a) An enterprising diver reasons that if a typical snorkel tube, which is 20 cm long, works, a tube 6.0 m long should also work. If he foolishly uses such a tube (Fig. 15-4), what is the pressure difference Δp between the external pressure on him and the air pressure in his lungs? Why is he in danger?

SOLUTION: First consider the diver at depth $L = 6.0$ m without the snorkel tube. The external pressure on him is given by Eq. 15-5 as

$$p = p_0 + \rho gL.$$

His body adjusts to that pressure by contracting slightly until the internal pressures are in equilibrium with the external pressure. In particular, the average blood pressure increases, and the average air pressure in his lungs increases until it equals the external pressure p.

If he then foolishly uses the 6.0 m tube to breathe, the pressurized air in his lungs will be expelled upward through the tube to the atmosphere, and the air pressure in his lungs will rapidly drop to atmospheric pressure p_0. Assuming he is in fresh water of density 1000 kg/m³, the pressure difference Δp between the lower pressure within his lungs and the higher pressure on his chest will then be

$$\Delta p = p - p_0 = \rho gL$$
$$= (1000 \text{ kg/m}^3)(9.8 \text{ m/s}^2)(6.0 \text{ m})$$
$$= 5.9 \times 10^4 \text{ Pa.} \qquad \text{(Answer)}$$

This pressure difference, about 0.6 atm, is sufficient to collapse the lungs and force the still-pressurized blood into them, a process known as lung squeeze.

(b) A novice scuba diver practicing in a swimming pool takes enough air from his tank to fully expand his lungs before abandoning the tank at depth L and swimming to the surface. He ignores instructions and fails to exhale during his ascent. When he reaches the surface, the pressure difference between the external pressure on him and the air in his lungs is 70 torr. From what depth did he start? What potentially lethal danger does he face?

SOLUTION: When he fills his lungs at depth L, the external pressure on him (and the air pressure within his lungs) is again given by Eq. 15-5 as

$$p = p_0 + \rho gL.$$

As he ascends, the external pressure on him decreases, until it is atmospheric pressure p_0 at the surface. His blood pressure also decreases, until it is normal. But because he does not exhale, the air pressure in his lungs remains at the value it had at depth L. So, at the surface, the pressure difference between the higher pressure in his lungs and the lower pressure on his chest is given by

$$\Delta p = p - p_0 = \rho gL,$$

from which we find

$$L = \frac{\Delta p}{\rho g}$$
$$= \frac{70 \text{ torr}}{(1000 \text{ kg/m}^3)(9.8 \text{ m/s}^2)} \left(\frac{1.01 \times 10^5 \text{ Pa}}{760 \text{ torr}} \right)$$
$$= 0.95 \text{ m.} \qquad \text{(Answer)}$$

The pressure difference of 70 torr (about 9% of atmospheric pressure) is sufficient to rupture the diver's lungs and force air from them into the depressurized blood, which then carries the air to the heart, killing the diver. If the diver follows instructions and gradually exhales as he ascends, he allows the pressure in his lungs to equalize with the external pressure, and then there is no danger.

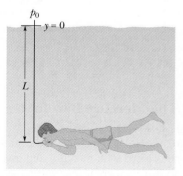

FIGURE 15-4 Sample Problem 15-2. DON'T TRY THIS with a tube that is longer than a standard snorkel tube or the attempt to breathe through the tube could kill you. The reason is that, because the external (water) pressure on your chest can be so much greater than the internal (air) pressure, you might not be able to expand your lungs to inhale.

SAMPLE PROBLEM 15-3

The U-tube in Fig. 15-5 contains two liquids in static equilibrium: water of density ρ_w is in the right arm, and oil of unknown density ρ_x is in the left. Measurement gives $l = 135$ mm and $d = 12.3$ mm. What is the density of the oil?

SOLUTION: If the pressure at the oil–water interface in the left arm is p_{int}, then the pressure in the right arm at the level of that interface must also be p_{int}, because the left and right arms are connected by water below the level of the interface. In the right arm, the interface is a distance l below the free surface of the *water* and we have, from Eq. 15-5,

$$p_{int} = p_0 + \rho_w gl \quad \text{(right arm)}.$$

In the left arm, the interface is a distance $l + d$ below the free surface of the *oil* and we have, again from Eq. 15-5,

$$p_{int} = p_0 + \rho_x g(l + d) \quad \text{(left arm)}.$$

Equating these two expressions and solving for the unknown density yield

$$\rho_x = \rho_w \frac{l}{l + d} = (1000 \text{ kg/m}^3) \frac{135 \text{ mm}}{135 \text{ mm} + 12.3 \text{ mm}}$$

$$= 916 \text{ kg/m}^3. \qquad \text{(Answer)}$$

Note that the answer does not depend on the atmospheric pressure p_0 or the free-fall acceleration g.

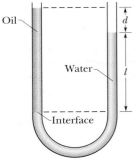

FIGURE 15-5 Sample Problem 15-3. The oil in the left arm stands higher than the water in the right arm because the oil is less dense than the water. Both fluid columns produce the same pressure P_{int} at the level of the interface.

15-5 MEASURING PRESSURE

The Mercury Barometer

Figure 15-6a shows a very basic *mercury barometer,* a device used to measure the pressure of the atmosphere. The long glass tube is filled with mercury and inverted with its open end in a dish of mercury, as the figure shows. The space above the mercury column contains only mercury vapor, whose pressure is so small at ordinary temperatures that it can be neglected.

We can use Eq. 15-4 to find the atmospheric pressure p_0 in terms of the height h of the mercury column. We choose level 1 of Fig. 15-2 to be that of the air–mercury

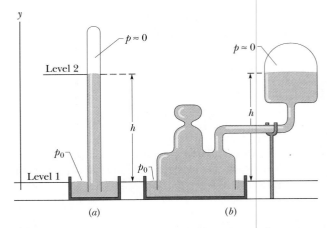

FIGURE 15-6 (a) A mercury barometer. (b) Another mercury barometer. The distance h is the same in both cases.

interface and level 2 to be that of the top of the mercury column, as labeled in Fig. 15-6a. We then substitute

$$y_1 = 0, \quad p_1 = p_0 \quad \text{and} \quad y_2 = h, \quad p_2 = 0$$

into Eq. 15-4, finding that

$$p_0 = \rho gh, \qquad (15\text{-}6)$$

where ρ is the density of the mercury.

For a given pressure, the height h of the mercury column does not depend on the cross-sectional area of the vertical tube. The fanciful mercury barometer of Fig. 15-6b gives the same reading as that of Fig. 15-6a; all that counts is the vertical distance h between the mercury levels.

Equation 15-6 shows that, for a given pressure, the height of the column of mercury depends on the value of g at the location of the barometer and on the density of mercury, which varies with temperature. The column height (in millimeters) is numerically equal to the pressure (in torr) *only* if the barometer is at a place where g has its accepted standard value of 9.80665 m/s² *and* the temperature of the mercury is 0°C. If these conditions do not prevail (and they rarely do), small corrections must be made before the height of the mercury column can be transformed into a pressure.

The Open-Tube Manometer

An *open-tube manometer* (Fig. 15-7) measures the gauge pressure of a gas. It consists of a U-tube containing a liquid, with one end of the tube connected to the vessel whose gauge pressure we wish to measure and the other end open to the atmosphere. We can use Eq. 15-4 to find the gauge pressure in terms of the height h shown in Fig. 15-7. Let us choose levels 1 and 2 as shown in Fig. 15-7. We then

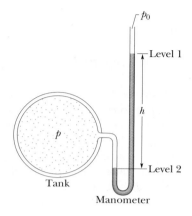

FIGURE 15-7 An open-tube manometer, connected so that the gauge pressure of the gas in the tank on the left is read. The right arm of the U-tube is open to the atmosphere.

substitute

$$y_1 = 0, \quad p_1 = p_0 \quad \text{and} \quad y_2 = -h, \quad p_2 = p$$

into Eq. 15-4, finding that

$$p_g = p - p_0 = \rho g h. \tag{15-7}$$

The gauge pressure p_g is directly proportional to h.

The gauge pressure can be positive or negative, depending on whether $p > p_0$ or $p < p_0$. In inflated tires or the human circulatory system, the (absolute) pressure is greater than atmospheric pressure, so the gauge pressure is a positive quantity, sometimes called the *overpressure*. If you suck on a straw to pull fluid up the straw, the (absolute) pressure in your lungs is actually less than atmospheric pressure. The gauge pressure in your lungs is then a negative quantity.

SAMPLE PROBLEM 15-4

The column in a mercury barometer has a measured height h of 740.35 mm. The temperature is $-5.0°C$, at which temperature the density of mercury is 1.3608×10^4 kg/m³. The free-fall acceleration g at the site of the barometer is 9.7835 m/s². What is the atmospheric pressure in pascals and in torr?

SOLUTION: From Eq. 15-6 we have

$$p_0 = \rho g h$$
$$= (1.3608 \times 10^4 \text{ kg/m}^3)(9.7835 \text{ m/s}^2)(0.74035 \text{ m})$$
$$= 9.8566 \times 10^4 \text{ Pa}. \quad \text{(Answer)}$$

Barometer readings are usually expressed in torr, where 1 torr is the pressure exerted by a column of mercury 1 mm high at a place where g has an accepted standard value of 9.80665 m/s² and at a temperature (0.0°C) at which mercury has a density of 1.35955×10^4 kg/m³. Thus from Eq. 15-6,

$$1 \text{ torr} = (1.35955 \times 10^4 \text{ kg/m}^3)(9.80665 \text{ m/s}^2)$$
$$\times (1 \times 10^{-3} \text{ m})$$
$$= 133.326 \text{ Pa}.$$

Applying this conversion factor yields, for the atmospheric pressure recorded on the barometer,

$$p_0 = 9.8566 \times 10^4 \text{ Pa} = 739.29 \text{ torr}. \quad \text{(Answer)}$$

Note that the pressure in torr (739.29 torr) is numerically close to—but otherwise differs significantly from—the height h of the mercury column expressed in mm (740.35 mm).

15-6 PASCAL'S PRINCIPLE

When you squeeze one end of a tube of toothpaste, you are watching **Pascal's principle** in action. This principle is also the basis for the Heimlich maneuver, in which a sharp pressure increase properly applied to the abdomen is transmitted to the throat, forcefully ejecting food lodged there. The principle was first stated clearly in 1652 by Blaise Pascal (for whom the unit of pressure is named):

> A change in the pressure applied to an enclosed incompressible fluid is transmitted undiminished to every portion of the fluid and to the walls of the containing vessel.

Demonstrating Pascal's Principle

Consider the case in which the fluid is an incompressible liquid contained in a tall cylinder, as in Fig. 15-8. The cylinder is fitted with a piston on which a container of lead shot rests. The atmosphere, container, and shot exert pressure p_{ext} on the piston and thus on the liquid. The pressure p at any point P in the liquid is then

$$p = p_{ext} + \rho g h. \tag{15-8}$$

Let us add a little more lead shot to the container to increase p_{ext} by an amount Δp_{ext}. The quantities ρ, g, and h in Eq. 15-8 are unchanged, so the pressure change at P is

$$\Delta p = \Delta p_{ext}. \tag{15-9}$$

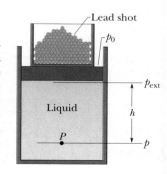

FIGURE 15-8 Weights loaded onto the piston create a pressure p_{ext} at the top of the enclosed (incompressible) liquid. If p_{ext} is increased, by adding more weights, the pressure increases by the same amount at all points within the liquid.

This pressure change is independent of h, so it must hold for all points within the liquid, as Pascal's principle states.

Pascal's Principle and the Hydraulic Lever

Figure 15-9 shows how Pascal's principle can be made the basis of a hydraulic lever. In operation, let an external force of magnitude F_i be exerted downward on the left-hand (or input) piston, whose area is A_i. An incompressible liquid in the device then exerts an upward force of magnitude F_o on the righthand (or output) piston, whose area is A_o. To keep the system in equilibrium, an external load (not shown) must exert a downward force of magnitude F_o on the output piston. The force F_i applied on the left and the downward force F_o exerted by the load on the right produce a change Δp in the pressure of the liquid that is given by

$$\Delta p = \frac{F_i}{A_i} = \frac{F_o}{A_o},$$

so
$$F_o = F_i \frac{A_o}{A_i}. \qquad (15\text{-}10)$$

Equation 15-10 shows that the output force F_o exerted on the load must be larger than the input force F_i if $A_o > A_i$, as is the case in Fig. 15-9.

If we move the input piston downward a distance d_i, the output piston moves upward a distance d_o, such that the same volume V of the incompressible liquid is displaced at both pistons. Then

$$V = A_i d_i = A_o d_o,$$

which we can write as

$$d_o = d_i \frac{A_i}{A_o}. \qquad (15\text{-}11)$$

This shows that, if $A_o > A_i$ (as in Fig. 15-9), the output piston moves a smaller distance than the input piston moves.

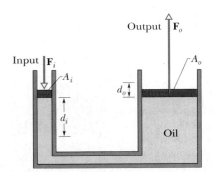

FIGURE 15-9   A hydraulic arrangement, used to magnify a force $\mathbf{F}_i$. The work done by $\mathbf{F}_i$, however, is not magnified and is the same for both the input and output forces.

From Eqs. 15-10 and 15-11 we can write the output work as

$$W = F_o d_o = \left(F_i \frac{A_o}{A_i}\right)\left(d_i \frac{A_i}{A_o}\right) = F_i d_i, \qquad (15\text{-}12)$$

which shows that the work W done *on* the input piston by the applied force is equal to the work W done *by* the output piston in lifting the load placed on it.

We see here that a given force exerted over a given distance can be transformed to a larger force exerted over a smaller distance. The product of force and distance remains unchanged so that the same work is done. However, there is often tremendous advantage in being able to exert the larger force. Most of us, for example, cannot lift an automobile and thus welcome the availability of a hydraulic jack, even though we have to pump the handle farther than the automobile rises. In this device, the displacement d_i is accomplished not in a single stroke but over a series of small strokes.

15-7 ARCHIMEDES' PRINCIPLE

Figure 15-10 shows a student in a swimming pool, manipulating a very thin plastic sack filled with water. She will find that it is in static equilibrium, tending neither to rise nor to sink. Yet the water in the sack has weight and—for that reason—should sink. The weight must be balanced by an upward force whose magnitude is equal to the weight of the water in the sack.

This upward **buoyant force** $\mathbf{F}_b$ is exerted on the water in the sack by the water that surrounds the sack. This buoyant force exists because—as you have already seen—the pressure in the water increases with depth below the surface, so the pressure near the bottom of the sack is greater than the pressure near the top.

Let us now remove the sack of water. Figure 15-11a shows the forces acting at the hole in the water formerly occupied by the sack. The buoyant force $\mathbf{F}_b$, which points up, is the vector sum of all these forces.

FIGURE 15-10   A thin-walled plastic sack of water is in static equilibrium in the pool. Its weight must be balanced by a net upward force exerted on the sack by the water surrounding it.

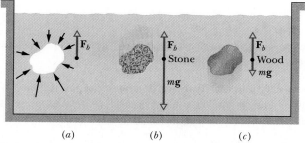

FIGURE 15-11 (*a*) The water surrounding the hole in the water exerts a resultant upward buoyant force on whatever fills the hole. (*b*) For a stone of the same volume as the hole, the weight exceeds the buoyant force. (*c*) For a lump of wood of the same volume, the weight is less than the buoyant force.

Let us fill the hole in Fig. 15-11*a* with a stone of exactly the same dimensions, as in Fig. 15-11*b*. *The same upward buoyant force that acted on the water-filled sack will act on the stone.* However, this force is too small to balance the weight of the stone, so the stone will sink. Even though the stone sinks, the water's buoyant force reduces its apparent weight, making the stone easier to lift as long as it is under water.

If we now fill the hole of Fig. 15-11*a* with a block of wood of the same dimensions, as in Fig. 15-11*c*, the same upward buoyant force acts on the wood. This time, however, the upward force is *greater* than the weight of the wood, and the wood will rise toward the surface. We summarize these facts with **Archimedes' principle**:

> A body fully or partially immersed in a fluid is buoyed up by a force equal to the weight of the fluid that the body displaces.

It is this principle that explains floating. When, say, the piece of wood in Fig. 15-11*c* rises enough to break through the water's surface, it displaces less water than when it was submerged. According to Archimedes' principle, its buoyant force decreases. The wood continues to rise out of the water until the buoyant force acting on it has decreased to exactly the weight of the wood. The wood is then in static equilibrium; it is floating.

Recall that we say the weight of an object acts effectively at the object's center of mass. Similarly, we can say the buoyant force on a fully or partially immersed object acts effectively at a point called the **center of buoyancy.** This point is located where the center of mass of the displaced fluid would have been, had it not been displaced. If a uniformly dense object is fully submerged, then the center of buoyancy and the object's center of mass coincide. If the object is only partially submerged, those points are separated; then the buoyant force can produce a torque on the object about its center of mass.

CHECKPOINT 2: A penguin floats first in a fluid of density ρ_0, then in a fluid of density $0.95\rho_0$, and then in a fluid of density $1.1\rho_0$. (a) Rank the densities according to the buoyant force on the penguin, greatest first. (b) Rank the densities according to the amount of fluid displaced by the penguin, greatest first.

SAMPLE PROBLEM 15-5

The "tip of the iceberg" in popular speech has come to mean a small visible fraction of something that is mostly hidden. For real icebergs, what is this fraction?

In the late evening of August 21, 1986, something (possibly a volcanic tremor) disturbed Cameroon's Lake Nyos, which has a high concentration of dissolved carbon dioxide. The disturbance caused that gas to form bubbles. Being lighter than the surrounding fluid (the water), those bubbles were buoyed to the surface, where they released the carbon dioxide. The gas, being heavier than the surrounding fluid (now the air), rushed down the mountainside like a river, asphyxiating 1700 persons and the scores of animals seen here.

SOLUTION: The weight of an iceberg of total volume V_i is

$$W_i = \rho_i V_i g,$$

where $\rho_i = 917$ kg/m³ is the density of ice.

The weight of the displaced seawater, which is the buoyancy force F_b, is

$$W_w = F_b = \rho_w V_w g,$$

where $\rho_w = 1024$ kg/m³ is the density of seawater and V_w is the volume of the displaced water, that is, the submerged volume of the iceberg. For the floating iceberg, these two forces are equal, or

$$\rho_i V_i g = \rho_w V_w g.$$

From this equation, we find that the fraction we seek is

$$\text{frac} = \frac{V_i - V_w}{V_i} = 1 - \frac{V_w}{V_i} = 1 - \frac{\rho_i}{\rho_w}$$

$$= 1 - \frac{917 \text{ kg/m}^3}{1024 \text{ kg/m}^3}$$

$$= 0.10 \text{ or } 10\%. \qquad \text{(Answer)}$$

SAMPLE PROBLEM 15-6

A spherical, helium-filled balloon has a radius R of 12.0 m. The balloon, support cables, and basket have a mass m of 196 kg. What maximum load M can the balloon carry? Take $\rho_{He} = 0.160$ kg/m³ and $\rho_{air} = 1.25$ kg/m³; the volume of air displaced by the load is negligible.

SOLUTION: The weight of the displaced air, which is the buoyant force, and the weight of the helium in the balloon are

$$W_{air} = \rho_{air} V g \quad \text{and} \quad W_{He} = \rho_{He} V g,$$

in which $V \ (= 4\pi R^3 / 3)$ is the volume of the balloon.

At equilibrium, from Archimedes' principle,

$$W_{air} = W_{He} + mg + Mg$$

or

$$M = (\tfrac{4}{3}\pi)(R^3)(\rho_{air} - \rho_{He}) - m$$

$$= (\tfrac{4}{3}\pi)(12.0 \text{ m})^3(1.25 \text{ kg/m}^3 - 0.160 \text{ kg/m}^3)$$

$$- 196 \text{ kg}$$

$$= 7690 \text{ kg}. \qquad \text{(Answer)}$$

An object with this mass would weigh 17,000 lb at sea level.

15-8 IDEAL FLUIDS IN MOTION

The motion of *real fluids* is very complicated and not yet fully understood. Instead, we shall discuss the motion of an **ideal fluid,** which is simpler to handle mathematically and

yet provides useful results. Here are four assumptions that we make about our ideal fluid:

1. *Steady flow* In *steady* (or *laminar*) *flow* the velocity of the moving fluid at any fixed point does not change with time, either in magnitude or in direction. The gentle flow of water near the center of a quiet stream is steady; that in a chain of rapids is not. Figure 15-12 shows a transition from steady flow to *nonsteady* (or *turbulent*) flow for a rising stream of smoke. The speed of the smoke particles increases as they rise and, at a certain critical speed, the flow changes from steady to nonsteady.

2. *Incompressible flow* We assume, as we have already done for fluids at rest, that our ideal fluid is incompressible. That is, its density has a constant, uniform value.

3. *Nonviscous flow* Roughly speaking, the viscosity of a fluid is a measure of how resistive the fluid is to flow. For example, thick honey is more resistive to flow than water, and so honey is said to be more viscous than water. Viscosity is the fluid analog of friction between solids; both are mechanisms by which the kinetic energy of moving objects can be transferred to thermal energy. In the absence of friction, a block could glide at constant speed along a horizontal surface. In the same way, an object moving through a nonviscous fluid would experience no *viscous drag force,* that is, no resistive force due to viscosity. The British scientist Lord Rayleigh noted that in an ideal fluid a ship's propeller would not work but, on the other hand, a ship (once set into motion) would not need a propeller!

FIGURE 15-12 At a certain point, the rising flow of smoke and heated gas changes from steady to turbulent.

4. *Irrotational flow* Although it need not concern us further, we also assume that the flow is *irrotational*. To test for this property, let a tiny grain of dust move with the fluid. Although this test body may (or may not) move in a circular path, in irrotational flow the test body will not rotate about an axis through its own center of mass. For a loose analogy, the motion of a Ferris wheel is rotational; that of its passengers is irrotational.

15-9 STREAMLINES AND THE EQUATION OF CONTINUITY

Figure 15-13 shows streamlines traced out by dye injected into a moving fluid; Fig. 15-14 shows similar streamlines revealed by smoke. A **streamline** is the path traced out by a tiny fluid element, which we may call a fluid "particle." As the fluid particle moves, its velocity may change, both in magnitude and in direction. However, its velocity vector at any point is always tangent to the streamline at that point (Fig. 15-15). Streamlines never cross because, if they did, a fluid particle arriving at the intersection would have two velocities simultaneously, an impossibility.

In flows like that of Figs. 15-13 and 15-14, we can isolate a *tube of flow* whose boundary is made up of streamlines. Such a tube acts like a pipe because any fluid particle that enters it cannot escape through its walls; if a particle did escape, we would have a case of streamlines crossing each other.

Figure 15-16 shows two cross sections, of areas A_1 and A_2, along a thin tube of flow. Let us station ourselves at B and monitor the fluid, moving with speed v_1, for a short time interval Δt. During this interval, a fluid particle moves

FIGURE 15-14 Smoke reveals streamlines in airflow past a car in a wind-tunnel test.

a small distance $v_1 \, \Delta t$, and a volume ΔV of fluid, given by

$$\Delta V = A_1 v_1 \, \Delta t,$$

passes through area A_1.

Assume that the fluid is incompressible and cannot be created or destroyed. Thus in this same time interval, the same volume of fluid must pass point C, farther down the tube of flow. If the speed there is v_2, this means that

$$\Delta V = A_1 v_1 \, \Delta t = A_2 v_2 \, \Delta t,$$

or

$$A_1 v_1 = A_2 v_2.$$

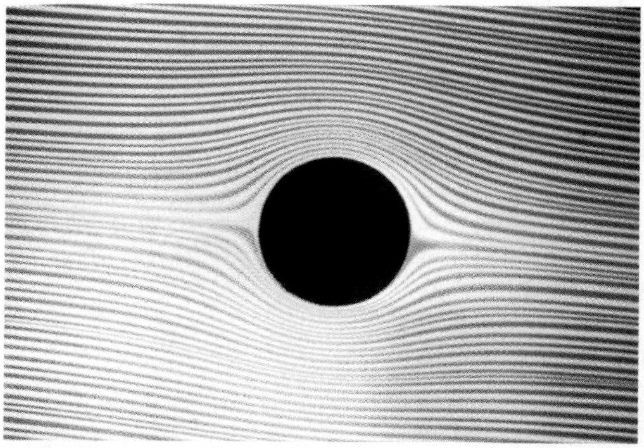

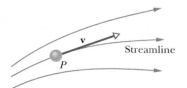

FIGURE 15-15 A fluid particle P traces out a streamline as it moves. The velocity vector of the particle is tangent to the streamline at every point.

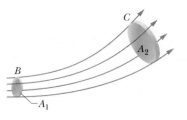

FIGURE 15-16 A tube of flow is defined by the streamlines that form its boundary. The flow rate of fluid must be the same for all cross sections of the tube of flow.

FIGURE 15-13 The steady flow of a fluid around a cylinder, as revealed by a dye tracer.

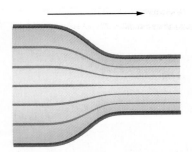

FIGURE 15-17   When a channel, such as a pipe, constricts, the streamlines draw closer together, signaling an increase in the fluid velocity. The arrow shows the direction of flow.

Thus, along the tube of flow,

$$R = Av = \text{a constant}, \qquad (15\text{-}13)$$

in which R, whose SI unit is the cubic meter per second, is the **volume flow rate** of the fluid. Equation 15-13 is called the **equation of continuity** for fluid flow. It tells us that the flow is faster in the narrower parts of a tube of flow, where the streamlines are closer together, as in Fig. 15-17.

Equation 15-13 is actually an expression of the law of conservation of mass in a form useful in fluid mechanics. In fact, if we multiply R by the (assumed uniform) density of the fluid, we get the quantity $Av\rho$, which is the **mass flow rate,** whose SI unit is the kilogram per second. Then Eq. 15-13 effectively tells us that the mass that flows through point B in Fig. 15-16 each second must be equal to the mass that flows through point C each second.

CHECKPOINT 3: The figure shows a pipe and gives the volume flow rate (in cm³/s) and the direction of flow for all but one section. What are the volume flow rate and the direction of flow for that section?

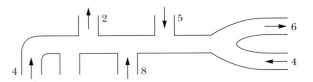

SAMPLE PROBLEM 15-7

The cross-sectional area A_0 of the aorta (the major blood vessel emerging from the heart) of a normal resting person is 3 cm², and the speed v_0 of the blood is 30 cm/s. A typical capillary (diameter $\approx$ 6 μm) has a cross-sectional area A of 3×10^{-7} cm² and a flow speed v of 0.05 cm/s. How many capillaries does such a person have?

SOLUTION: All the blood that passes through the capillaries must have passed through the aorta so that, from Eq. 15-13,

$$A_0v_0 = nAv,$$

where n is the number of capillaries. Solving for n yields

$$n = \frac{A_0v_0}{Av} = \frac{(3 \text{ cm}^2)(30 \text{ cm/s})}{(3 \times 10^{-7} \text{ cm}^2)(0.05 \text{ cm/s})}$$

$$= 6 \times 10^9 \text{ or 6 billion.} \qquad \text{(Answer)}$$

You can easily show that the combined cross-sectional area of the capillaries is about 600 times the area of the aorta.

SAMPLE PROBLEM 15-8

Figure 15-18 shows how the stream of water emerging from a faucet "necks down" as it falls. The cross-sectional area A_0 is 1.2 cm², and A is 0.35 cm². The two levels are separated by a vertical distance $h = 45$ mm. At what rate does water flow from the tap?

SOLUTION: From the equation of continuity (Eq. 15-13) we have

$$A_0v_0 = Av, \qquad (15\text{-}14)$$

where v_0 and v are the water velocities at the corresponding levels. From Eq. 2-23 we can also write, because the water is falling freely with acceleration g,

$$v^2 = v_0^2 + 2gh. \qquad (15\text{-}15)$$

Eliminating v between Eqs. 15-14 and 15-15 and solving for v_0, we obtain

$$v_0 = \sqrt{\frac{2ghA^2}{A_0^2 - A^2}}$$

$$= \sqrt{\frac{(2)(9.8 \text{ m/s}^2)(0.045 \text{ m})(0.35 \text{ cm}^2)^2}{(1.2 \text{ cm}^2)^2 - (0.35 \text{ cm}^2)^2}}$$

$$= 0.286 \text{ m/s} = 28.6 \text{ cm/s}.$$

The volume flow rate R is then

$$R = A_0v_0 = (1.2 \text{ cm}^2)(28.6 \text{ cm/s})$$

$$= 34 \text{ cm}^3/\text{s}. \qquad \text{(Answer)}$$

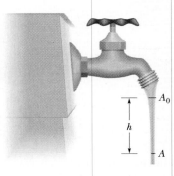

FIGURE 15-18   Sample Problem 15-8. As water falls from a tap, its speed increases. Because the flow rate must be the same at all cross sections, the stream must "neck down."

15-10 BERNOULLI'S EQUATION

Figure 15-19 represents a tube of flow (or an actual pipe, for that matter) through which an ideal fluid is flowing at a steady rate. In a time interval Δt, suppose that a volume of fluid ΔV, colored purple in Fig. 15-19a, enters the tube at the left (or input) end and an identical volume, colored green in Fig. 15-19b, emerges at the right (or output) end. The emerging volume must be the same as the entering volume because the fluid is incompressible, with an assumed constant density ρ.

Let y_1, v_1, and p_1 be the elevation, speed, and pressure of the fluid entering at the left, and y_2, v_2, and p_2 be the corresponding quantities for the fluid emerging at the right. By applying the principle of conservation of energy to the fluid, we shall show that these quantities are related by

$$p_1 + \tfrac{1}{2}\rho v_1^2 + \rho g y_1 = p_2 + \tfrac{1}{2}\rho v_2^2 + \rho g y_2. \quad (15\text{-}16)$$

We can also write this equation as

$$p + \tfrac{1}{2}\rho v^2 + \rho g y = \text{a constant}. \quad (15\text{-}17)$$

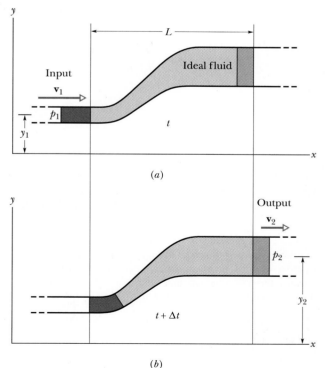

(a)

(b)

FIGURE 15-19 Fluid flows at a steady rate through a length L of a tube, from the input end at the left to the output end at the right. From time t in (a) to time $t + \Delta t$ in (b), the amount of fluid shown in purple enters the input end and the equal amount shown in green emerges from the output end.

Equations 15-16 and 15-17 are equivalent forms of **Bernoulli's equation,** after Daniel Bernoulli, who studied fluid flow in the 1700s.* Like the equation of continuity (Eq. 15-13), Bernoulli's equation is not a new principle but simply the reformulation of a familiar principle in a form more suitable to fluid mechanics. As a check, let us apply Bernoulli's equation to fluids at rest, by putting $v_1 = v_2 = 0$ in Eq. 15-16. The result is

$$p_2 = p_1 + \rho g(y_1 - y_2),$$

which is Eq. 15-4.

A major prediction of Bernoulli's equation emerges if we take y to be a constant ($y = 0$, say) so that the fluid does not change elevation as it flows. Equation 15-16 then becomes

$$p_1 + \tfrac{1}{2}\rho v_1^2 = p_2 + \tfrac{1}{2}\rho v_2^2, \quad (15\text{-}18)$$

which tells us that:

> If the speed of a fluid particle increases as it travels along a horizontal streamline, the pressure of the fluid must decrease, and conversely.

Put another way, where the streamlines are relatively close together (that is, where the velocity is relatively great), the pressure is relatively low, and conversely.

The link between a change in speed and a change in pressure makes sense if you consider a fluid particle. When the particle nears a narrow region, the higher pressure behind it accelerates it so that it then has a greater speed in the narrow region. And when it nears a wide region, the higher pressure ahead of it decelerates it so that it then has a lesser speed in the wide region.

Bernoulli's equation is strictly valid only to the extent that the fluid is ideal. If viscous forces are present, thermal energy will be involved. We take no account of this in the derivation that follows.

Proof of Bernoulli's Equation

Let us take as our system the entire volume of the (ideal) fluid shown in Fig. 15-19. We shall apply the principle of conservation of energy to this system as it moves from its initial state (Fig. 15-19a) to its final state (Fig. 15-19b). The fluid lying between the two vertical planes separated by a distance L in Fig. 15-19 does not change its properties during this process; we need be concerned only with changes that take place at the input and the output ends.

*For irrotational flow (which we assume), the constant in Eq. 15-17 has the same value for all points within the tube of flow; the points do not have to lie along the same streamline. Similarly, the points 1 and 2 in Eq. 15-16 can lie anywhere within the tube of flow.

We apply energy conservation in the form of the work–kinetic energy theorem,

$$W = \Delta K, \tag{15-19}$$

which tells us that the change in the kinetic energy of our system must equal the net work done on the system. The change in kinetic energy results from the change in speed between the ends of the pipe and is

$$\Delta K = \tfrac{1}{2}\Delta m\, v_2^2 - \tfrac{1}{2}\Delta m\, v_1^2$$
$$= \tfrac{1}{2}\rho\, \Delta V(v_2^2 - v_1^2), \tag{15-20}$$

in which $\Delta m\, (= \rho\, \Delta V)$ is the mass of the fluid that enters at the input end and leaves at the output end during a small time interval Δt.

The work done on the system arises from two sources. The work W_g done by the weight ($\Delta m\, \mathbf{g}$) of mass Δm during the vertical lift of the mass from the input level to the output level is

$$W_g = -\Delta m\, g(y_2 - y_1)$$
$$= -\rho g\, \Delta V(y_2 - y_1). \tag{15-21}$$

This work is negative because the upward displacement and the downward weight point in opposite directions.

Work W_p must also be done *on* the system (at the input end) to push the entering fluid into the tube and *by* the system (at the output end) to push forward the fluid ahead of the emerging fluid. In general, the work done by a force of magnitude F, acting on a fluid sample contained in a tube of area A to move the fluid through a distance Δx, is

$$F\, \Delta x = (pA)(\Delta x) = p(A\, \Delta x) = p\, \Delta V.$$

The work done on the system is then $p_1\, \Delta V$, and the work done by the system is $-p_2\, \Delta V$. Their sum W_p is

$$W_p = -p_2\, \Delta V + p_1\, \Delta V$$
$$= -(p_2 - p_1)\Delta V. \tag{15-22}$$

The work–kinetic energy theorem of Eq. 15-19 now becomes

$$W = W_g + W_p = \Delta K.$$

Substituting from Eqs. 15-20, 15-21, and 15-22 yields

$$-\rho g\, \Delta V(y_2 - y_1) - \Delta V(p_2 - p_1) = \tfrac{1}{2}\rho\, \Delta V(v_2^2 - v_1^2).$$

This, after a slight rearrangement, matches Eq. 15-16, which we set out to prove.

SAMPLE PROBLEM 15-9

Ethanol of density $\rho = 791$ kg/m³ flows smoothly through a horizontal pipe that tapers (as in Fig. 15-17) in cross-sectional area from $A_1 = 1.20 \times 10^{-3}$ m² to $A_2 = A_1/2$. The pressure difference Δp between the wide and narrow sections of pipe is 4120 Pa. What is the volume flow rate R of the ethanol?

SOLUTION: Rearranging Eq. 15-18 (Bernoulli's equation for level flow) yields

$$p_1 - p_2 = \tfrac{1}{2}\rho v_2^2 - \tfrac{1}{2}\rho v_1^2 = \tfrac{1}{2}\rho(v_2^2 - v_1^2), \tag{15-23}$$

where subscripts 1 and 2 refer to the wide and narrow sections of pipe, respectively. Equation 15-13 (the continuity equation) tells us that the flow is faster in the narrower section. Here that means $v_2 > v_1$. Equation 15-23 then tells us that $p_1 > p_2$.

Equation 15-13 also tells us the volume flow rate R is the same in the wide and narrow sections. So

$$R = v_1 A_1 = v_2 A_2.$$

These equations, with $A_2 = A_1/2$, give us

$$v_1 = \frac{R}{A_1} \quad\text{and}\quad v_2 = \frac{R}{A_2} = \frac{2R}{A_1}.$$

Substituting these expressions into Eq. 15-23, setting $p_1 - p_2 = \Delta p$, and rearranging give us

$$\Delta p = \tfrac{1}{2}\rho\left(\frac{4R^2}{A_1^2} - \frac{R^2}{A_1^2}\right) = \frac{3\rho R^2}{2A_1^2}.$$

Solving for R, we find

$$R = A_1\sqrt{\frac{2\Delta p}{3\rho}}$$
$$= 1.20 \times 10^{-3}\text{ m}^2\sqrt{\frac{(2)(4120\text{ Pa})}{(3)(791\text{ kg/m}^3)}}$$
$$= 2.24 \times 10^{-3}\text{ m}^3\text{/s.} \qquad\text{(Answer)}$$

SAMPLE PROBLEM 15-10

In the old West, a desperado fires a bullet into an open water tank (Fig. 15-20), creating a hole a distance h below the water surface. What is the speed v of the water emerging from the hole?

SOLUTION: This situation is essentially that of water moving (downward) with speed V through a wide pipe (the tank) of cross-sectional area A and then moving (horizontally) with speed v through a narrow pipe (the hole) of cross-sectional area a. From Eq. 15-13, we know that

$$R = av = AV$$

and thus that

$$V = \frac{a}{A}\, v.$$

Because $a \ll A$, we see that $V \ll v$.

We can also relate v to V (and to h) through the Bernoulli equation (Eq. 15-16). We take the level of the hole as our reference level for measuring elevations (and thus gravitational potential energy). Noting that the pressure at the top of the tank and at the bullet hole is the atmospheric pressure p_0 (because both places are exposed to the atmosphere), we write Eq. 15-16 as

$$p_0 + \tfrac{1}{2}\rho V^2 + \rho g h = p_0 + \tfrac{1}{2}\rho v^2 + \rho g(0). \quad (15\text{-}24)$$

(Here the top of the tank is represented by the left side of the equation, and the hole by the right side. The zero on the right

indicates that the hole is at our reference level.) Before we solve Eq. 15-24 for v, we can use our result that $V \ll v$ to simplify it: We assume that V^2, and thus the term $\tfrac{1}{2}\rho V^2$ in Eq. 15-24, is negligible compared to the other terms, and we drop it. Solving the remaining equation for v then yields

$$v = \sqrt{2gh}. \qquad \text{(Answer)}$$

This is the same speed that an object would have when falling a height h from rest.

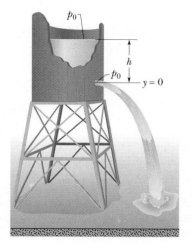

FIGURE 15-20 Sample Problem 15-10. Water pours through a hole in a water tank, at a distance h below the water surface. The water pressure at the surface and at the hole is atmospheric pressure p_0.

CHECKPOINT 4: Water flows smoothly through the pipe shown in the figure, descending in the process. Rank the four numbered sections of pipe according to (a) the volume flow rate R through them, (b) the flow speed v through them, and (c) the water pressure p within them, greatest first.

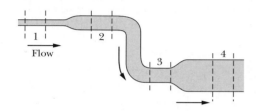

REVIEW & SUMMARY

Density

The **density** ρ of any material is defined as its mass per unit volume:

$$\rho = \frac{\Delta m}{\Delta V}. \qquad (15\text{-}1)$$

Usually, where a material sample is large compared with atomic dimensions, we can write Eq. 15-1 as $\rho = m/V$.

Fluid Pressure

A **fluid** is a substance that can flow; it conforms to the boundaries of its container because it cannot withstand shearing stress. It can, however, exert a force perpendicular to its surface. That force is described in terms of **pressure** p:

$$p = \frac{\Delta F}{\Delta A}, \qquad (15\text{-}2)$$

in which ΔF is the force acting on a surface element of area ΔA. If the force is uniform over a flat area, Eq. 15-2 can be written as $p = F/A$. The force resulting from fluid pressure at a particular point in a fluid has the same magnitude in all directions. *Gauge pressure* is the difference between the actual pressure (or *absolute pressure*) at a point and the atmospheric pressure.

Pressure Variation with Height and Depth

Pressure in a fluid at rest varies with vertical position y. For y measured positive upward,

$$p_2 = p_1 + \rho g(y_1 - y_2). \qquad (15\text{-}4)$$

The pressure is the same for all points at the same level. If h is the *depth* of a fluid sample below some reference level at which the pressure is p_0, Eq. 15-4 becomes

$$p = p_0 + \rho g h. \qquad (15\text{-}5)$$

Pascal's Principle

Pascal's principle, which can be derived from Eq. 15-4, states that a change in the pressure applied to an enclosed fluid is transmitted undiminished to every portion of the fluid and to the walls of the containing vessel.

Archimedes' Principle

The surface of an immersed object is acted on by forces associated with the fluid pressure. The vector sum of those forces (called the **buoyant force**) acts vertically upward, through the center of mass of the displaced fluid (the **center of buoyancy**). Archimedes' principle states that the magnitude of the buoyant force is equal to the weight of the fluid displaced by the object. When an object floats, its weight is equal to the buoyant force acting on it.

Flow of Ideal Fluids

An *ideal fluid* is incompressible and lacks viscosity, and its flow is steady and irrotational. A **streamline** is the path followed by an individual fluid particle. A *tube of flow* is a bundle of streamlines. The principle of conservation of mass shows that the flow within any tube of flow obeys the **equation of continuity:**

$$R = Av = \text{a constant}, \tag{15-13}$$

in which R is the **volume flow rate,** A the cross-sectional area of the tube of flow at any point, and v the speed of the fluid at that point, assumed to be constant across A. The **mass flow rate** $Av\rho$ is also constant.

Bernoulli's Equation

Applying the principle of conservation of mechanical energy to the flow of an ideal fluid leads to **Bernoulli's equation:**

$$p + \tfrac{1}{2}\rho v^2 + \rho g y = \text{a constant} \tag{15-17}$$

along any tube of flow.

QUESTIONS

1. Figure 15-21 shows a tank filled with water. Five horizontal floors and ceilings are indicated; all have the same area and are located at distances L, $2L$, or $3L$ below the top of the tank. Rank the floors and ceilings according to the force on them due to the water, greatest first.

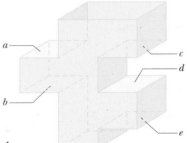

FIGURE 15-21   Question 1.

2. Figure 15-22 shows four situations in which a red liquid and a gray liquid are in a U-tube. In one situation the liquids cannot be in static equilibrium. (a) Which situation is that? (b) For the other three situations, assume static equilibrium. For each, is the density of the red liquid greater than, less than, or equal to the density of the gray liquid?

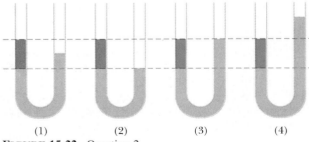

FIGURE 15-22   Question 2.

3. The containers of Fig. 15-23 have the same level of water and same base area, and are made of the same material. (a) Rank the containers (and contents) according to their weights on a scale, greatest first. (b) Rank the containers according to the pressure of the water on the base of the container. (c) Does Eq. 15-2 indicate that the answers to (a) and (b) are inconsistent? This apparent inconsistency is known as the *hydrostatic paradox.*

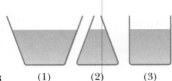

FIGURE 15-23   Question 3.

4. We fully submerge an irregular 3 kg lump of material in a certain fluid. The fluid that would have been in the space now occupied by the lump has a mass of 2 kg. (a) When we release the lump, does it move upward, move downward, or remain in place? (b) If we next fully submerge the lump in a less dense fluid and again release it, what does it do?

5. Figure 15-24 shows four solid objects floating in corn syrup. Rank the objects according to their density, greatest first.

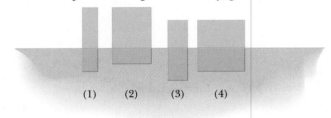

FIGURE 15-24   Question 5.

6. Figure 15-25 shows three identical open-top containers filled to the brim with water; toy ducks float in two of them. Rank the containers and contents according to their weight, greatest first.

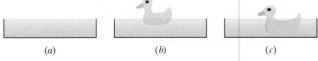

FIGURE 15-25   Question 6.

7. A boat with an anchor on board floats in a swimming pool that is somewhat wider than the boat. Does the water level in the pool move upward, move downward, or remain the same if the anchor is (a) dropped into the water or (b) thrown onto the surrounding ground? (c) Does the water level in the pool move upward, move downward, or remain the same if, instead, a cork is dropped from the boat into the water, where it floats?

8. Three gas bubbles of identical sizes are submerged in water: bubble 1 is filled with hydrogen, bubble 2 is filled with helium, and bubble 3 is filled with carbon dioxide. Rank the bubbles according to the magnitude of the buoyant force acting on them, greatest first.

9. A block of wood floats at a certain level in a pail of water in a stationary elevator. Does the block float higher, lower, or at the same level when the elevator cab (a) moves upward at constant speed, (b) moves downward at constant speed, (c) accelerates upward, and (d) accelerates downward with an acceleration magnitude less than 9.8 m/s²?

10. A container of water is suspended from a spring balance. Does the weight reading increase, decrease, or remain the same

when we, without spilling any water, (a) lower a heavy metal object via string into the water and (b) float a cork in the water?

11. Figure 15-26 shows two rectangular, uniformly dense blocks that are floating in water. Both blocks have been tipped by hand and then released. For each block, (a) does buoyancy cause a clockwise torque or a counterclockwise torque about the block's center of mass and (b) will the block right itself or tip over more?

FIGURE 15-26 Question 11.

EXERCISES & PROBLEMS

SECTION 15-3 Density and Pressure

1E. Convert a density of 1.0 g/cm³ to kg/m³.

2E. Three liquids that will not mix are poured into a cylindrical container. The volumes and densities of the liquids are 0.50 L, 2.6 g/cm³; 0.25 L, 1.0 g/cm³; and 0.40 L, 0.80 g/cm³. What is the force acting on the bottom of the container due to these liquids? One liter = 1 L = 1000 cm³. (Ignore the contribution due to the atmosphere.)

3E. Find the pressure increase in the fluid in a syringe when a nurse applies a force of 42 N to the syringe's circular piston, which has a radius of 1.1 cm.

4E. You inflate the front tires on your car to 28 psi. Later, you measure your blood pressure, obtaining a reading of 120/80, the readings being in mm Hg. In metric countries (which is to say, most of the rest of the world), these pressures are customarily reported in kilopascals (kPa). In kilopascals, what are (a) your tire pressure and (b) your blood pressure?

5E. An office window has dimensions 3.4 m by 2.1 m. As a result of the passage of a storm, the outside air pressure drops to 0.96 atm, but inside the pressure is held at 1.0 atm. What net force pushes out on the window?

6E. A fish maintains its depth in fresh water by adjusting the air content of porous bone or air sacs to make its average density the same as that of the water. Suppose that with its air sacs collapsed, a fish has a density of 1.08 g/cm³. To what fraction of its expanded body volume must the fish inflate the air sacs to reduce its density to that of water?

7P. An airtight box having a lid with an area of 12 in.² is partially evacuated. If a force of 108 lb is required to pull the lid off the box and the outside atmospheric pressure is 15 lb/in.², what is the air pressure in the box?

8P. In 1654 Otto von Guericke, inventor of the air pump, gave a demonstration before the noblemen of the Holy Roman Empire in which two teams of eight horses could not pull apart two evacu-

ated brass hemispheres. (a) Assuming that the hemispheres have thin walls, so that R in Fig. 15-27 may be considered both the inside and outside radius, show that the force F required to pull apart the hemispheres is $F = \pi R^2 \, \Delta p$, where Δp is the difference between the pressures outside and inside the sphere. (b) Taking R equal to 1.0 ft and the inside pressure as 0.10 atm, find the force the teams of horses would have had to exert to pull apart the hemispheres. (c) Why are two teams of horses used? Would not one team have proved the point just as well if the hemispheres were attached to a sturdy wall?

FIGURE 15-27 Problem 8.

SECTION 15-4 Fluids at Rest

9E. Calculate the hydrostatic difference in blood pressure between the brain and the foot in a person of height 1.83 m. The density of blood is 1.06×10^3 kg/m³.

10E. Find the pressure, in pascals, 150 m below the surface of the ocean. The density of seawater is 1.03 g/cm³, and the atmospheric pressure at sea level is 1.01×10^5 Pa.

11E. The sewer outlets of a house constructed on a slope are 8.2 m below street level. If the sewer is 2.1 m below street level, find the minimum pressure differential that must be created by the sewage pump to transfer waste of average density 900 kg/m³.

12E. Figure 15-28 displays the *phase diagram* of carbon, showing the ranges of temperature and pressure in which carbon will crystallize either as diamond or graphite. What is the minimum depth at which diamonds can form if the temperature at that depth is 1000°C and the subsurface rocks have density 3.1 g/cm³? Assume that, as in a fluid, the pressure is due to the weight of material lying above.

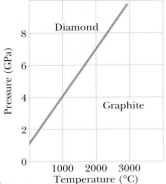

FIGURE 15-28 Exercise 12.

13E. The human lungs can operate against a pressure differential of up to about $\frac{1}{20}$ of an atmosphere. If a diver uses a snorkel for breathing, about how far below water level can she or he swim?

14E. A swimming pool has the dimensions 80 ft × 30 ft × 8.0 ft. (a) When it is filled with water, what is the force (resulting from the water alone) on the bottom, on the short sides, and on the long sides? (b) If you are concerned with the possibility that the concrete walls and floor will collapse, is it appropriate to take the atmospheric pressure into account? Why?

15E. (a) Find the total weight of water on top of a nuclear submarine at a depth of 200 m, assuming that its (horizontal cross-sectional) hull area is 3000 m². (b) What water pressure would a diver experience at this depth? Express your answer in atmospheres. Do you think that occupants of a damaged submarine at this depth could escape without special equipment? Assume the density of seawater is 1.03 g/cm³.

16E. Crew members attempt to escape from a damaged submarine 100 m below the surface. What force must they apply to a pop-out hatch, which is 1.2 m by 0.60 m, to push it out? Assume that the density of the ocean water is 1025 kg/m³.

17E. A simple open U-tube contains mercury. When 11.2 cm of water is poured into the right arm of the tube, how high above its initial level does the mercury rise in the left arm?

FIGURE 15-29 Exercise 18.

18E. A cylindrical barrel has a narrow tube fixed to the top, as shown with dimensions in Fig. 15-29. The vessel is filled with water to the top of the tube. Calculate the ratio of the hydrostatic force exerted on the bottom of the barrel to the weight of the water contained inside the barrel. Why is the ratio not equal to one? (Ignore the presence of the atmosphere.)

19P. Two identical cylindrical vessels with their bases at the same level each contain a liquid of density ρ. The area of either base is A, but in one vessel the liquid height is h_1, and in the other h_2. Find the work done by the gravitational force in equalizing the levels when the two vessels are connected.

20P. (a) Consider a container of fluid subject to a *vertical upward* acceleration of magnitude a. Show that the pressure variation with depth in the fluid is given by

$$p = \rho h(g + a),$$

where h is the depth and ρ is the density. (b) Show also that if the fluid as a whole undergoes a *vertical downward* acceleration of magnitude a, the pressure at a depth h is given by

$$p = \rho h(g - a).$$

(c) What is the pressure if the fluid is in free fall?

21P. In analyzing certain geological features, it is often appropriate to assume that the pressure at some horizontal *level of compensation,* deep inside Earth, is the same over a large region and is equal to the pressure exerted by the weight of the overlying material. That is, the pressure on the level of compensation is given by the fluid pressure formula. This model requires, for example, that mountains have *roots* (Fig. 15-30). Consider a mountain 6.0 km high. The continental rocks have a density of 2.9 g/cm³, and beneath the continent is the mantle, with a density of 3.3 g/cm³. Calculate the depth D of the root. (*Hint:* Set the pressure at points a and b equal; the depth y of the level of compensation will cancel out.)

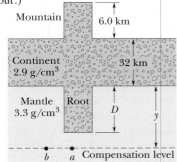

FIGURE 15-30
Problem 21.

22P. In Fig. 15-31, the ocean is about to overrun the continent. Find the depth h of the ocean using the level-of-compensation method shown in Problem 21.

23P. Water stands at a depth D behind the vertical upstream face of a dam, as shown in Fig. 15-32. Let W be the width of the dam. (a) Find the resultant horizontal force exerted on the dam by the gauge pressure of the water and (b) the net torque owing to that gauge pressure about a line through O parallel to the width of the dam. (c) Find the moment arm of the resultant horizontal force about the line through O.

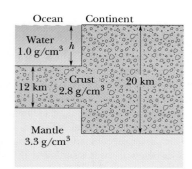

FIGURE 15-31
Problem 22.

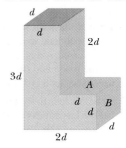

FIGURE 15-32
Problem 23.

24P. The L-shaped tank shown in Fig. 15-33 is filled with water and is open at the top. If $d = 5.0$ m, what are (a) the force on face A and (b) the force on face B due to the water?

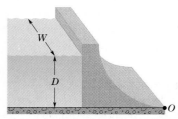

FIGURE 15-33
Problem 24.

25P. Figure 15-34 shows a dam and part of the freshwater reservoir backed up behind it. The dam is made of concrete of density 3.2 g/cm³ and has the dimensions shown on the figure. (a) The force exerted by the water pushes horizontally on the dam face, and this is resisted by the force of static friction between the dam and the bedrock foundation on which it rests. The coefficient of friction is 0.47. Calculate the factor of safety against sliding, that is, the ratio of the maximum possible friction force to the force exerted by the water. (b) The water also tries to rotate the dam about a line running along the base of the dam through point A; see Problem 23. The torque resulting from the weight of the dam acts in the opposite sense. Calculate the factor of safety against rotation, that is, the ratio of the torque owing to the weight of the dam to the torque exerted by the water.

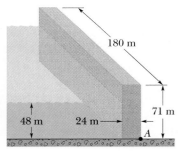

FIGURE 15-34
Problem 25.

SECTION 15-5 Measuring Pressure

26E. Calculate the height of a column of water that gives a pressure of 1 atm at the bottom. Assume that $g = 9.80$ m/s².

27E. To suck lemonade of density 1000 kg/m³ up a straw to a maximum height of 4.0 cm, what minimum gauge pressure (in atmospheres) must you produce in your lungs?

28P. What would be the height of the atmosphere if the air density (a) were constant and (b) decreased linearly to zero with height? Assume a sea-level density of 1.3 kg/m³.

SECTION 15-6 Pascal's Principle

29E. A piston of small cross-sectional area a is used in a hydraulic press to exert a small force $\mathbf{f}$ on the enclosed liquid. A connecting pipe leads to a larger piston of cross-sectional area A (Fig. 15-35). (a) What force $\mathbf{F}$ will the larger piston sustain? (b) If the small piston has a diameter of 1.5 in. and the large piston one of 21 in., what weight on the small piston will support 2.0 tons on the large piston?

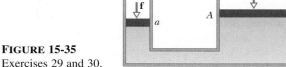

FIGURE 15-35
Exercises 29 and 30.

30E. In the hydraulic press of Exercise 29, through what distance must the large piston be moved to raise the small piston a distance of 3.5 ft?

SECTION 15-7 Archimedes' Principle

31E. A tin can has a total volume of 1200 cm³ and a mass of 130 g. How many grams of lead shot could it carry without sinking in water? The density of lead is 11.4 g/cm³.

32E. A boat floating in fresh water displaces 8000 lb of water. (a) How many pounds of water would this boat displace if it were floating in salt water of density 68.6 lb/ft³? (b) Would the volume of water displaced change? If so, by how much?

33E. About one-third of the body of a physicist swimming in the Dead Sea will be above the water line. Assuming that the human body density is 0.98 g/cm³, find the density of the water in the Dead Sea. (Why is it so much greater than 1.0 g/cm³?)

34E. An iron anchor appears 200 N lighter in water than in air. (a) What is the volume of the anchor? (b) How much does it weigh in air? The density of iron is 7870 kg/m³.

35E. An object hangs from a spring balance. The balance registers 30 N in air, 20 N when this object is immersed in water, and 24 N when the object is immersed in another liquid of unknown density. What is the density of that other liquid?

36E. A cubical object of dimensions $L = 2.00$ ft on a side and weight $W = 1000$ lb in a vacuum is suspended by a rope in an open tank of liquid of density $\rho = 2.00$ slugs/ft³ as in Fig. 15-36. (a) Find the total downward force exerted by the liquid and the atmosphere on the top of the object. (b) Find the total upward force on the bottom of the object. (c) Find the tension in the rope. (d) Calculate the buoyant force on the object using Archimedes' principle. What relation exists among all these quantities?

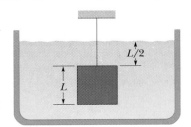

FIGURE 15-36
Exercise 36.

37E. A block of wood floats in water with two-thirds of its volume submerged. In oil the block floats with 0.90 of its volume submerged. Find the density of (a) the wood and (b) the oil.

38E. It has been proposed to move natural gas from the North Sea gas fields in huge dirigibles, using the gas itself to provide lift. Calculate the force required to tether such an airship to the ground for off-loading when it is fully loaded with 1.0×10^6 m³ of gas at a density of 0.80 kg/m³. (The weight of the airship is negligible by comparison.)

39E. A blimp is cruising slowly at low altitude, filled as usual with helium gas. Its maximum useful payload, including crew and cargo, is 1280 kg. How much more payload could the blimp carry if you replaced the helium with hydrogen? (Why not do it?) The volume of the helium-filled interior space is 5000 m³. The density of helium gas is 0.16 kg/m³, and the density of hydrogen is 0.081 kg/m³.

40E. A helium balloon is used to lift a 40 kg payload to an altitude of 27 km, where the air density is 0.035 kg/m³. The balloon has a mass of 15 kg, and the density of the gas in the balloon is 0.0051 kg/m³. What is the volume of the balloon? Neglect the volume of the payload.

41P. A hollow sphere of inner radius 8.0 cm and outer radius 9.0 cm floats half-submerged in a liquid of density 800 kg/m³. (a) What is the mass of the sphere? (b) Calculate the density of the material of which the sphere is made.

42P. A hollow spherical iron shell floats almost completely submerged in water. The outer diameter is 60.0 cm, and the density of iron is 7.87 g/cm³. Find the inner diameter.

43P. An iron casting containing a number of cavities weighs 6000 N in air and 4000 N in water. What is the volume of the cavities in the casting? The density of iron is 7.87 g/cm³.

44P. (a) What is the minimum area of the top surface of a slab of ice 0.30 m thick floating on fresh water that will hold up an automobile of mass 1100 kg? (b) Does it matter where the car is placed on the block of ice?

45P. Three children, each of weight 80 lb, make a log raft by lashing together logs of diameter 1.0 ft and length 6.0 ft. How many logs will be needed to keep them afloat? Take the density of wood to be 50 lb/ft³.

46P. Assume the density of brass weights to be 8.0 g/cm³ and that of air to be 0.0012 g/cm³. What percent error arises from

neglecting the buoyancy of air in weighing an object of mass m and density ρ on a beam balance?

47P. A car has a total mass of 1800 kg. The volume of air space in the passenger compartment is 5.00 m³. The volume of the motor and front wheels is 0.750 m³, and the volume of the rear wheels, gas tank, and trunk is 0.800 m³; water cannot enter these areas. The car is parked on a hill; the handbrake cable snaps and the car rolls down the hill into a lake (Fig. 15-37). (a) At first, no water enters the passenger compartment. How much of the car, in cubic meters, is below the water surface with the car floating as shown? (b) As water slowly enters, the car sinks. How many cubic meters of water are in the car as it disappears below the water surface? (The car remains horizontal, owing to a heavy load in the trunk.)

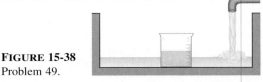

FIGURE 15-37 Problem 47.

48P. A block of wood has a mass of 3.67 kg and a density of 600 kg/m³. It is to be loaded with lead so that it will float in water with 0.90 of its volume immersed. What mass of lead is needed (a) if the lead is on top of the wood and (b) if the lead is attached below the wood? The density of lead is 1.13×10^4 kg/m³.

49P. You place a glass beaker, partially filled with water, in a sink (Fig. 15-38). The beaker has a mass of 390 g and an interior volume of 500 cm³. You now start to fill the sink with water and you find, by experiment, that if the beaker is less than half full, it will float; but if it is more than half full, it remains on the bottom of the sink as the water rises to its rim. What is the density of the material of which the beaker is made?

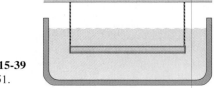

FIGURE 15-38
Problem 49.

50P. What is the acceleration of a rising hot-air balloon if the ratio of the air density outside the balloon to that inside is 1.39? Neglect the mass of the balloon fabric and the basket.

51P. A metal rod of length 80 cm and mass 1.6 kg has a uniform cross-sectional area of 6.0 cm². Due to a nonuniform density, the center of mass of the rod is 20 cm from one end of the rod. The rod is suspended in a horizontal position in water by ropes attached to both ends as shown in Fig. 15-39. (a) What is the ten-

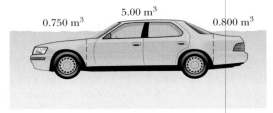

FIGURE 15-39
Problem 51.

sion in the rope closer to the center of mass? (b) What is the tension in the rope farther from the center of mass?

52P*. The tension in a string holding a solid block below the surface of a liquid (of density greater than the solid) is T_0 when the containing vessel (Fig. 15-40) is at rest. Show that when the vessel has an upward vertical acceleration of magnitude a, the tension T is equal to $T_0(1 + a/g)$.

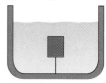

FIGURE 15-40   Problem 52.

SECTION 15-9 Streamlines and the Equation of Continuity

53E. Figure 15-41 shows the merging of two streams to form a river. One stream has a width of 8.2 m, depth of 3.4 m, and current speed of 2.3 m/s. The other stream is 6.8 m wide and 3.2 m deep, and flows at 2.6 m/s. The width of the river is 10.5 m, and the current speed is 2.9 m/s. What is its depth?

FIGURE 15-41   Exercise 53.

54E. The water flowing through a 0.75 in. (inside diameter) pipe flows out through three 0.50 in. pipes. (a) If the flow rates in the three smaller pipes are 7.0, 5.0, and 3.0 gal/min, what is the flow rate in the 0.75 in. pipe? (b) What is the ratio of the speed of water in the 0.75 in. pipe to that in the pipe carrying 7.0 gal/min?

55E. A garden hose having an internal diameter of 0.75 in. is connected to a (stationary) lawn sprinkler that consists merely of an enclosure with 24 holes, each 0.050 in. in diameter. If the water in the hose has a speed of 3.0 ft/s, at what speed does it leave the sprinkler holes?

56P. Water is pumped steadily out of a flooded basement at a speed of 5.0 m/s through a uniform hose of radius 1.0 cm. The hose passes out through a window 3.0 m above the waterline. What is the power of the pump?

57P. A river 20 m wide and 4.0 m deep drains a 3000 km² land area in which the average precipitation is 48 cm/y. One-fourth of this rainfall returns to the atmosphere by evaporation, but the remainder ultimately drains into the river. What is the average speed of the river current?

SECTION 15-10 Bernoulli's Equation

58E. Water is moving with a speed of 5.0 m/s through a pipe with a cross-sectional area of 4.0 cm². The water gradually descends 10 m as the pipe increases in area to 8.0 cm². (a) What is the speed at the lower level? (b) If the pressure at the upper level is 1.5×10^5 Pa, what is the pressure at the lower level?

59E. Models of torpedoes are sometimes tested in a horizontal pipe of flowing water, much as a wind tunnel is used to test model airplanes. Consider a circular pipe of internal diameter 25.0 cm and a torpedo model, aligned along the axis of the pipe, with a diameter of 5.00 cm. The torpedo is to be tested with water flowing past it at 2.50 m/s. (a) With what speed must the water flow in the part of the pipe that is unconstricted by the model? (b) What will the pressure difference be between the constricted and unconstricted parts of the pipe?

60E.   A water intake at a pump storage reservoir (Fig. 15-42) has a cross-sectional area of 8.00 ft². The water flows in at a speed of 1.33 ft/s. At the generator building 600 ft below the intake point, the cross-sectional area is smaller than at the intake and the water flows out at 31.0 ft/s. What is the difference in pressure, in pounds per square inch, between inlet and outlet?

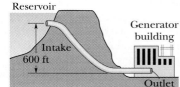

FIGURE 15-42
Exercise 60.

61E. A water pipe having a 1.0 in. inside diameter carries water into the basement of a house at a speed of 3.0 ft/s and a pressure of 25 lb/in.². If the pipe tapers to 0.50 in. and rises to the second floor 25 ft above the input point, what are (a) the speed and (b) the water pressure at the second floor?

62E. How much work is done by pressure in forcing 1.4 m³ of water through a pipe having an internal diameter of 13 mm if the difference in pressure at the two ends of the pipe is 1.0 atm?

63E. In a horizontal oil pipeline that has a constant cross-sectional area, the pressure decrease between two points 1000 ft apart is 5.0 lb/in.². What is the energy loss per cubic foot of oil per foot of pipe?

64E. A tank of large area is filled with water to a depth $D = 1.0$ ft. A hole of cross section $A = 1.0$ in.² in the bottom of the tank allows water to drain out. (a) What is the rate at which water flows out in cubic feet per second? (b) At what distance below the bottom of the tank is the cross-sectional area of the stream equal to one-half the area of the hole?

65E. Suppose that two tanks, 1 and 2, each with a large opening at the top, contain different liquids. A small hole is made in the side of each tank at the same depth h below the liquid surface, but the hole in tank 1 has half the cross-sectional area of the hole in tank 2. (a) What is the ratio ρ_1/ρ_2 of the densities of the liquids if the mass flow rate is the same for the two holes? (b) What is the ratio of the volume flow rates from the two tanks? (c) To what height above the hole in the second tank should liquid be added or drained to equalize the volume flow rates?

66E. Air flows over the top of an airplane wing of area A with speed v_t and past the underside of the wing (also of area A) with speed v_u. Show that in this simplified situation Bernoulli's equation predicts that the magnitude L of the upward lift force on the wing will be

$$L = \tfrac{1}{2}\rho A(v_t^2 - v_u^2),$$

where ρ is the density of the air.

67E. If the speed of flow past the lower surface of an airplane wing is 110 m/s, what speed of flow over the upper surface will give a pressure difference of 900 Pa between upper and lower surfaces? Take the density of air to be 1.30×10^{-3} g/cm³, and see Exercise 66.

68E. An airplane has a wing area (each wing) of 10.0 m². At a certain airspeed, air flows over the upper wing surface at 48.0 m/s and over the lower wing surface at 40.0 m/s. What is the mass of the plane? Assume that the plane travels at constant velocity, that the air density is 1.20 kg/m², and that lift effects associated with the fuselage and tail assembly are small. Discuss the lift if the airplane, flying at the same airspeed, is (a) in level flight, (b) climbing at 15°, and (c) descending at 15°. (See Exercise 66.)

69P. Water flows through a horizontal pipe and is delivered into the atmosphere at a speed of 15 m/s as shown in Fig. 15-43. The diameters of the left and right sections of the pipe are 5.0 cm and 3.0 cm, respectively. (a) What volume of water is delivered into the atmosphere during a 10 min period? (b) What is the flow speed of the water in the left section of the pipe? (c) What is the gauge pressure in the left section of the pipe?

FIGURE 15-43 Problem 69.

70P. An opening of area 0.25 cm² in an otherwise closed beverage keg is 50 cm below the level of the liquid (of density 1.0 g/cm³) in the keg. What is the speed of the liquid flowing through the opening if the gauge pressure in the air space above the liquid is (a) zero and (b) 0.40 atm?

71P. In a hurricane, the air (density 1.2 kg/m³) is blowing over the roof of a house at a speed of 110 km/h. (a) What is the pressure difference between inside and outside that tends to lift the roof? (b) What would be the lifting force on a roof of area 90 m²?

72P. The windows in an office building are of dimensions 4.00 m by 5.00 m. On a stormy day, air is blowing at 30.0 m/s past a window on the top floor. Calculate the net force on the window. The air density is 1.23 kg/m³ there.

73P. A sniper fires a rifle bullet into a gasoline tank, making a hole 50 m below the surface of the gasoline. The tank was sealed and is under 3.0 atm absolute pressure, as shown in Fig. 15-44. The stored gasoline has a density of 660 kg/m³. At what speed v does the gasoline begin to shoot out of the hole?

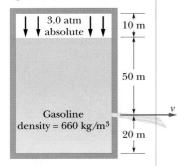

FIGURE 15-44
Problem 73.

74P. The fresh water behind a reservoir dam is 15 m deep. A horizontal pipe 4.0 cm in diameter passes through the dam 6.0 m below the water surface, as shown in Fig. 15-45. A plug secures the pipe opening. (a) Find the friction force between plug and pipe wall. (b) The plug is removed. What volume of water flows out of the pipe in 3.0 h?

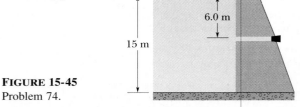

FIGURE 15-45
Problem 74.

75P. A tank is filled with water to a height H. A hole is punched in one of the walls at a depth h below the water surface (Fig. 15-46). (a) Show that the distance x from the base of the tank to the point at which the resulting stream strikes the floor is given by $x = 2\sqrt{h(H - h)}$. (b) Could a hole be punched at another depth to produce a second stream that would have the same range? If so, at what depth? (c) At what depth should the hole be placed to make the emerging stream strike the ground at the maximum distance from the base of the tank?

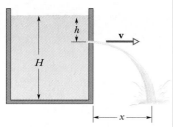

FIGURE 15-46 Problem 75.

76P. A *siphon* is a device for removing liquid from a container. It operates as shown in Fig. 15-47. Tube ABC must initially be filled, but once this has been done, liquid will flow through the tube until the liquid surface in the container is level with the tube opening at A. The liquid has density ρ and negligible viscosity. (a) With what speed does the liquid emerge from the tube at C? (b) What is the pressure in the liquid at the topmost point B? (c) Theoretically, what is the greatest possible height h_1 that a siphon can lift water?

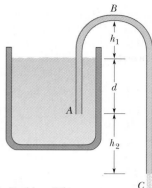

FIGURE 15-47 Problem 76.

77P. A *Venturi meter* is used to measure the flow speed of a fluid in a pipe. The meter is connected between two sections of the pipe (Fig. 15-48); the cross-sectional area A of the entrance and exit of the meter matches the pipe's cross-sectional area. Between the entrance and exit, the fluid flows from the pipe with speed v and then through a narrow "throat" of cross-sectional area a with speed V. A manometer connects the wider portion of the meter to the narrower portion. The change in the fluid's speed is accompanied by a change Δp in the fluid's pressure, which causes a height difference h of the liquid in the two arms of the manometer. (a) By applying Bernoulli's equation and the equation of continuity to points 1 and 2 in Fig. 15-48, show that

$$v = \sqrt{\frac{2a^2 \, \Delta p}{\rho(A^2 - a^2)}},$$

where ρ is the density of the fluid. (b) Suppose that the fluid is water; that the cross-sectional diameters are 10 in.² in the pipe, 5.0 in.² in the throat; that the pressure is 8.0 lb/in.² in the pipe and 6.0 lb/in.² in the throat. What is the rate of water flow in cubic feet per second?

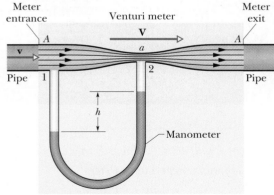

FIGURE 15-48 Problems 77 and 78.

78P. Consider the Venturi tube of Problem 77 and Fig. 15-48 without the manometer. Let A equal $5a$. Suppose that the pressure p_1 at A is 2.0 atm. (a) Compute the values of v at A and V at a that would make the pressure p_2 at a equal to zero. (b) Compute the corresponding volume flow rate if the diameter at A is 5.0 cm. The phenomenon that occurs at a when p_2 falls to nearly zero is known as cavitation. The water vaporizes into small bubbles.

79P. A pitot tube (Fig. 15-49) is used to determine the airspeed of an airplane. It consists of an outer tube with a number of small holes B (four are shown); the tube is connected to one arm of a U-tube. The other arm of the U-tube is connected to a hole, A, at the front end of the device, which points in the direction the plane is headed. At A the air becomes stagnant so that $v_A = 0$. At B, however, the speed of the air presumably equals the airspeed of the aircraft. (a) Use Bernoulli's equation to show that

$$v = \sqrt{\frac{2\rho g h}{\rho_{\text{air}}}},$$

where v is the airspeed of the plane and ρ is the density of the liquid in the U-tube. (b) The tube contains alcohol and indicates a level difference h of 26.0 cm. What is the plane's speed relative to the air? The density of the air is 1.03 kg/m³ and that of alcohol is 810 kg/m³.

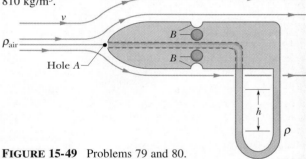

FIGURE 15-49 Problems 79 and 80.

80P. A pitot tube (see Problem 79) on a high-altitude aircraft measures a differential pressure of 180 Pa. What is the airspeed if the density of the air is 0.031 kg/m³?

16
Oscillations

Just as the third game of the 1989 World Series was about to begin near San Francisco, seismic waves from a magnitude 7.1 earthquake near Loma Prieta, 100 km distant, hit the area, causing extensive damage and killing 67 people. The photograph shows part of a 1.4 km stretch of the Nimitz Freeway, where dozens died when an upper deck collapsed onto a lower deck, trapping motorists. Obviously, the collapse was due to violent shaking by the seismic waves. But why was that particular stretch so severely damaged when the rest of the freeway, almost identical in construction, escaped collapse?

16-1 OSCILLATIONS

We are surrounded by oscillations—motions that repeat themselves. There are swinging chandeliers, boats bobbing at anchor, and the surging pistons in the engines of cars. There are oscillating guitar strings, drums, bells, diaphragms in telephones and speaker systems, and quartz crystals in wristwatches. Less evident are the oscillations of the air molecules that transmit the sensation of sound, the oscillations of the atoms in a solid that convey the sensation of temperature, and the oscillations of the electrons in the antennas of radio and TV transmitters.

Oscillations are not confined to material objects such as violin strings and electrons. Light, radio waves, x rays, and gamma rays are also oscillatory phenomena. You will study such oscillations in later chapters and will be helped greatly there by analogy with the mechanical oscillations that you are about to study here.

Oscillations in the real world are usually *damped;* that is, the motion dies out gradually, transferring mechanical energy to thermal energy by the action of frictional forces. Although we cannot totally eliminate such loss of mechanical energy, we can replenish the energy from some source. The children in Fig. 16-1, for example, know that by swinging their legs or torsos they can "pump" the swing and maintain or enhance the oscillations. In doing this, they transfer biochemical energy to mechanical energy of the oscillating systems.

FIGURE 16-1 A child soon learns how to maintain the oscillations of a swing by transferring energy into the swing's motion.

16-2 SIMPLE HARMONIC MOTION

Figure 16-2 shows a sequence of "snapshots" of a simple oscillating system, a particle moving repeatedly back and forth about the origin of the x axis. In this section we simply describe the motion. Later, we shall discuss how to attain such motion.

One important property of oscillatory motion is its **frequency,** or number of oscillations that are completed each second. The symbol for frequency is f, and its SI unit is the **hertz** (abbreviated Hz), where

$$1 \text{ hertz} = 1 \text{ Hz} = 1 \text{ oscillation per second}$$
$$= 1 \text{ s}^{-1}. \qquad (16\text{-}1)$$

Related to the frequency is the **period** T of the motion, which is the time for one complete oscillation (or **cycle**). That is,

$$T = \frac{1}{f}. \qquad (16\text{-}2)$$

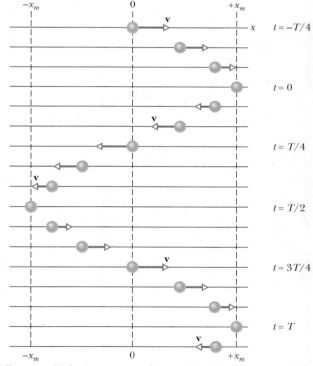

FIGURE 16-2 A sequence of "snapshots" (taken at equal time intervals) showing the position of a particle as it oscillates back and forth about the origin along the x axis, between the limits $+ x_m$ and $- x_m$. The vector arrows are scaled to indicate the speed of the particle. The speed is maximum when the particle is at the origin and zero when it is at $\pm x_m$. If the time t is chosen to be zero when the particle is at $+ x_m$, then the particle returns to $+ x_m$ at $t = T$, where T is the period of the motion. The motion is then repeated.

Any motion that repeats itself at regular intervals is called **periodic motion** or **harmonic motion.** We are interested here in motion that repeats itself in a particular way, namely, like that in Fig. 16-2. It turns out that for such motion the displacement x of the particle from the origin is given as a function of time by

$$x(t) = x_m \cos(\omega t + \phi) \quad \text{(displacement),} \quad (16\text{-}3)$$

in which x_m, ω, and ϕ are constants. This motion is called **simple harmonic motion** (SHM), a term that means that the periodic motion is a sinusoidal function of time.

The quantity x_m in Eq. 16-3, a positive constant whose value depends on how the motion was started, is called the **amplitude** of the motion; the subscript m stands for *maximum* because the amplitude is the magnitude of the maximum displacement of the particle in either direction. The cosine function in Eq. 16-3 varies between the limits ± 1, so the displacement $x(t)$ varies between the limits $\pm x_m$, as Fig. 16-2 shows.

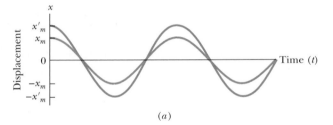

(a)

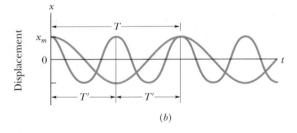

(b)

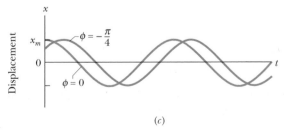

(c)

FIGURE 16-3 In all three cases, the blue curve is obtained from Eq. 16-3 with $\phi = 0$. (a) The red curve differs from the blue curve *only* in that its amplitude x'_m is greater. (b) The red curve differs from the blue curve *only* in that its period is $T' = T/2$. (c) The red curve differs from the blue curve *only* in that $\phi = -\pi/4$ rad rather than zero.

The time-varying quantity $(\omega t + \phi)$ in Eq. 16-3 is called the **phase** of the motion, and the constant ϕ is called the **phase constant** (or **phase angle**). The value of ϕ depends on the displacement and velocity of the particle at $t = 0$. For the $x(t)$ plots of Fig. 16-3a, the phase constant ϕ is zero (compare the plots and Eq. 16-3 for $t = 0$).

It remains to interpret the constant ω. The displacement $x(t)$ must return to its initial value after one period T of the motion. That is, $x(t)$ must equal $x(t + T)$ for all t. To simplify our analysis, let us put $\phi = 0$ in Eq. 16-3. From that equation we then have

$$x_m \cos \omega t = x_m \cos [\omega(t + T)].$$

The cosine function first repeats itself when its argument (the phase) has increased by 2π rad, so that we must have, in the equation above,

$$\omega(t + T) = \omega t + 2\pi$$

or,

$$\omega T = 2\pi.$$

Thus from Eq. 16-2,

$$\omega = \frac{2\pi}{T} = 2\pi f. \quad (16\text{-}4)$$

The quantity ω is called the **angular frequency** of the motion; its SI unit is the radian per second. (To be consistent, then, ϕ must be in radians.) Figure 16-3 compares $x(t)$ for two simple harmonic motions that differ either in amplitude, in period (and thus in frequency and angular frequency), or in phase constant.

CHECKPOINT 1: A particle undergoing simple harmonic oscillation of period T (like that in Fig. 16-2) is at $-x_m$ at time $t = 0$. Is it at $-x_m$, at $+x_m$, at 0, between $-x_m$ and 0, or between 0 and $+x_m$ when (a) $t = 2.00T$, (b) $t = 3.50T$, and (c) $t = 5.25T$?

The Velocity of SHM

By differentiating Eq. 16-3, we can find an expression for the velocity of a particle moving with simple harmonic motion. That is,

$$v(t) = \frac{dx}{dt} = \frac{d}{dt}\,[x_m \cos(\omega t + \phi)]$$

or

$$v(t) = -\omega x_m \sin(\omega t + \phi) \quad \text{(velocity).} \quad (16\text{-}5)$$

Figure 16-4a is a plot of Eq. 16-3 with $\phi = 0$. Figure 16-4b shows Eq. 16-5, also with $\phi = 0$. Analogous to the amplitude x_m in Eq. 16-3, the positive quantity ωx_m in Eq.

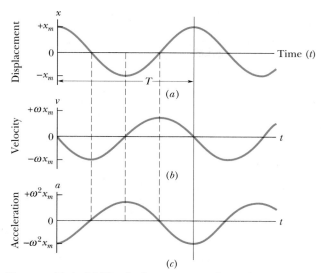

FIGURE 16-4 (a) The displacement $x(t)$ of a particle oscillating in SHM with phase angle ϕ equal to zero. The period T marks one complete oscillation. (b) The velocity $v(t)$ of the particle. (c) The acceleration $a(t)$ of the particle.

16-5 is called the **velocity amplitude** v_m. As you can see in Fig. 16-4b, the velocity of the oscillating particle varies between the limits $\pm v_m = \pm \omega x_m$. Note also in that figure that the curve of $v(t)$ is *shifted* (to the left) from the curve of $x(t)$ by one-quarter period: when the magnitude of the displacement is greatest (that is, $x(t) = x_m$), the magnitude of the velocity is least (that is, $v(t) = 0$). And when the magnitude of the displacement is least (that is, zero), the magnitude of the velocity is greatest (that is, $v_m = \omega x_m$).

The Acceleration of SHM

Knowing the velocity $v(t)$ for simple harmonic motion, we can find an expression for the acceleration of the oscillating particle by differentiating once more. Thus we have, from Eq. 16-5,

$$a(t) = \frac{dv}{dt} = \frac{d}{dt}[-\omega x_m \sin(\omega t + \phi)]$$

or

$$a(t) = -\omega^2 x_m \cos(\omega t + \phi) \quad \text{(acceleration).} \quad (16\text{-}6)$$

Figure 16-4c is a plot of Eq. 16-6 for the case $\phi = 0$. The positive quantity $\omega^2 x_m$ in Eq. 16-6 is called the **acceleration amplitude** a_m. That is, the acceleration of the particle varies between the limits $\pm a_m = \pm \omega^2 x_m$, as Fig. 16-4c shows. Note also that the curve of $a(t)$ is shifted (to the left) by one-quarter period relative to the curve of $v(t)$.

We can combine Eqs. 16-3 and 16-6 to yield

$$a(t) = -\omega^2 x(t), \quad (16\text{-}7)$$

which is the hallmark of simple harmonic motion: the ac-

celeration is proportional to the displacement but opposite in sign, and the two quantities are related by the square of the angular frequency. Thus, as Fig. 16-4 shows, when the displacement has its greatest positive value, the acceleration has its greatest negative value, and conversely. When the displacement is zero, the acceleration is also zero.

PROBLEM SOLVING TACTICS

TACTIC 1: *Phase Angles*

Note the effect of the phase angle ϕ on a plot of $x(t)$. When $\phi = 0$, $x(t)$ has a graph like that in Fig. 16-4a, a typical cosine curve. A *negative* value for ϕ shifts the curve *rightward* along the t axis (as in Fig. 16-3c), while a *positive* value shifts it *leftward*.

Two plots of SHM with different phase angles are said to have a *phase difference;* or, each is said to be *phase-shifted* from the other, or *out of phase* with the other. The curves in Fig. 16-3c, for example, have a phase difference of $\pi/4$ rad.

Because SHM repeats after each period T and the cosine function repeats after each 2π rad, one period T represents a phase difference of 2π rad. In Fig. 16-4, $x(t)$ is phase-shifted to the right from $v(t)$ by one-quarter period, or $-\pi/2$ rad; it is shifted to the right from $a(t)$ by one-half period, or $-\pi$ rad. A phase shift of 2π rad causes a curve of SHM to coincide with itself; that is, it looks unchanged.

16-3 THE FORCE LAW FOR SIMPLE HARMONIC MOTION

Once we know how the acceleration of a particle varies with time, we can use Newton's second law to learn what force must act on the particle to give it that acceleration. If we combine Newton's second law and Eq. 16-7, we find, for simple harmonic motion,

$$F = ma = -(m\omega^2)x. \quad (16\text{-}8)$$

This result—a force proportional to the displacement but opposite in sign—is familiar. It is Hooke's law,

$$F = -kx, \quad (16\text{-}9)$$

for a spring, the spring constant here being

$$k = m\omega^2. \quad (16\text{-}10)$$

We can in fact take Eq. 16-9 as an alternative definition of simple harmonic motion. It says:

Simple harmonic motion is the motion executed by a particle of mass m subject to a force that is proportional to the displacement of the particle but opposite in sign.

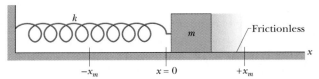

FIGURE 16-5 A linear simple harmonic oscillator. Like the particle of Fig. 16-2, the block moves in simple harmonic motion once it has been pulled to the side and released. Its displacement is then given by Eq. 16-3.

The block–spring system of Fig. 16-5 forms a **linear simple harmonic oscillator** (linear oscillator, for short), where "linear" indicates that F is proportional to x rather than to some other power of x. The angular frequency ω of the simple harmonic motion of the block is related to the spring constant k and the mass m of the block by Eq. 16-10, which yields

$$\omega = \sqrt{\frac{k}{m}} \qquad \text{(angular frequency).} \qquad (16\text{-}11)$$

By combining Eqs. 16-4 and 16-11, we can write, for the **period** of the linear oscillator of Fig. 16-5,

$$T = 2\pi \sqrt{\frac{m}{k}} \qquad \text{(period).} \qquad (16\text{-}12)$$

Equations 16-11 and 16-12 tell us that a large angular frequency (and thus a small period) goes with a stiff spring (large k) and a light block (small m).

Every oscillating system, be it the linear oscillator of Fig. 16-5, a diving board, or a violin string, has some element of "springiness" and some element of "inertia," or mass, and thus resembles a linear oscillator. In the linear oscillator of Fig. 16-5, these elements are located in separate parts of the system, the springiness being entirely in the spring, which we assume to be massless, and the inertia being entirely in the block, which we assume to be rigid. In a violin string, however, the two elements are both within the string itself, as you will see in Chapter 17.

CHECKPOINT **2:** Which of the following relationships between the force F on a particle and the particle's position x implies simple harmonic oscillation: (a) $F = -5x$, (b) $F = -400x^2$, (c) $F = 10x$, (d) $F = 3x^2$?

SAMPLE PROBLEM 16-1

A block whose mass m is 680 g is fastened to a spring whose spring constant k is 65 N/m. The block is pulled a distance $x = 11$ cm from its equilibrium position at $x = 0$ on a frictionless surface and released from rest at $t = 0$.

(a) What force does the spring exert on the block just before the block is released?

SOLUTION: From Hooke's law

$$F = -kx = -(65 \text{ N/m})(0.11 \text{ m})$$
$$= -7.2 \text{ N.} \qquad \text{(Answer)}$$

The minus sign reminds us that the spring force acting on the block, which points back toward the origin, is opposite the displacement of the block, which points away from the origin.

(b) What are the angular frequency, the frequency, and the period of the resulting oscillation?

SOLUTION: From Eq. 16-11 we have

$$\omega = \sqrt{\frac{k}{m}} = \sqrt{\frac{65 \text{ N/m}}{0.68 \text{ kg}}} = 9.78 \text{ rad/s}$$
$$\approx 9.8 \text{ rad/s.} \qquad \text{(Answer)}$$

The frequency follows from Eq. 16-4, which yields

$$f = \frac{\omega}{2\pi} = \frac{9.78 \text{ rad/s}}{2\pi} = 1.56 \text{ Hz} \approx 1.6 \text{ Hz.} \qquad \text{(Answer)}$$

The period follows from Eq. 16-2, which yields

$$T = \frac{1}{f} = \frac{1}{1.56 \text{ Hz}} = 0.64 \text{ s} = 640 \text{ ms.} \qquad \text{(Answer)}$$

(c) What is the amplitude of the oscillation?

SOLUTION: As we discussed in Section 8-4, the mechanical energy of a spring–block system like that in Fig. 16-5 is conserved because friction is not involved. Since the block is released from rest 11 cm from its equilibrium point, it has kinetic energy of zero whenever it is again 11 cm from that point. Thus its maximum displacement is 11 cm; that is,

$$x_m = 11 \text{ cm.} \qquad \text{(Answer)}$$

(d) What is the maximum speed of the oscillating block?

SOLUTION: From Eq. 16-5 we see that the velocity amplitude is

$$v_m = \omega x_m = (9.78 \text{ rad/s})(0.11 \text{ m})$$
$$= 1.1 \text{ m/s.} \qquad \text{(Answer)}$$

This maximum speed occurs when the oscillating block is rushing through the origin; compare Figs. 16-4a and 16-4b, where you can see that the speed is a maximum whenever $x = 0$.

(e) What is the magnitude of the maximum acceleration of the block?

SOLUTION: From Eq. 16-6 we see that the acceleration amplitude is

$$a_m = \omega^2 x_m = (9.78 \text{ rad/s})^2 (0.11 \text{ m})$$
$$= 11 \text{ m/s}^2. \qquad \text{(Answer)}$$

This maximum acceleration occurs when the block is at the ends of its path. At those points, the force acting on the block has its maximum magnitude; compare Figs. 16-4a and 16-4c, where you can see that the magnitudes of the displacement and acceleration are maximum at the same times.

(f) What is the phase constant ϕ for the motion?

SOLUTION: At $t = 0$, the moment of release, the displacement of the block has its maximum value x_m and the velocity of the block is zero. If we put these *initial conditions,* as they are called, into Eqs. 16-3 and 16-5, we find

$$1 = \cos \phi \quad \text{and} \quad 0 = \sin \phi,$$

respectively. The smallest angle that satisfies both these requirements is

$$\phi = 0. \qquad \text{(Answer)}$$

(Any angle that is an integer multiple of 2π rad also satisfies these requirements.)

This equation has two solutions:

$$\phi = -25° \quad \text{and} \quad \phi = 155°.$$

You will see in part (c) how to choose between them.

(c) What is the amplitude x_m of the motion?

SOLUTION: From Eq. 16-13 we have, provisionally putting $\phi = 155°$,

$$x_m = \frac{x(0)}{\cos \phi} = \frac{-0.0850 \text{ m}}{\cos 155°} = 0.094 \text{ m}$$
$$= 9.4 \text{ cm}. \qquad \text{(Answer)}$$

If we had chosen $\phi = -25°$, we would have found $x_m = -9.4$ cm. However, the amplitude of the motion must always be a *positive* constant, so $-25°$ cannot be the correct phase constant. We must therefore have, in part (b),

$$\phi = 155°. \qquad \text{(Answer)}$$

SAMPLE PROBLEM 16-2

At $t = 0$, the displacement $x(0)$ of the block in a linear oscillator like that of Fig. 16-5 is -8.50 cm. Its velocity $v(0)$ then is -0.920 m/s, and its acceleration $a(0)$ is $+47.0$ m/s².

(a) What are the angular frequency ω and the frequency f of this system?

SOLUTION: If we put $t = 0$ in Eqs. 16-3, 16-5, and 16-6, we find

$$x(0) = x_m \cos \phi, \qquad (16\text{-}13)$$

$$v(0) = -\omega x_m \sin \phi, \qquad (16\text{-}14)$$

and

$$a(0) = -\omega^2 x_m \cos \phi. \qquad (16\text{-}15)$$

These three equations contain three unknowns, namely, x_m, ϕ, and ω. We should be able to find all three, but here we need only ω.

If we divide Eq. 16-15 by Eq. 16-13, the result is

$$\omega = \sqrt{-\frac{a(0)}{x(0)}} = \sqrt{-\frac{47.0 \text{ m/s}^2}{-0.0850 \text{ m}}}$$
$$= 23.5 \text{ rad/s}. \qquad \text{(Answer)}$$

The frequency f follows from Eq. 16-4 and is

$$f = \frac{\omega}{2\pi} = \frac{23.5 \text{ rad/s}}{2\pi} = 3.74 \text{ Hz}. \qquad \text{(Answer)}$$

(b) What is the phase constant ϕ?

SOLUTION: If we divide Eq. 16-14 by Eq. 16-13, we find

$$\frac{v(0)}{x(0)} = \frac{-\omega x_m \sin \phi}{x_m \cos \phi} = -\omega \tan \phi.$$

Solving for $\tan \phi$, we find

$$\tan \phi = -\frac{v(0)}{\omega x(0)} = -\frac{-0.920 \text{ m/s}}{(23.5 \text{ rad/s})(-0.0850 \text{ m})}$$
$$= -0.461.$$

SAMPLE PROBLEM 16-3

In Fig. 16-6a, a uniform bar with mass m lies symmetrically across two rapidly rotating, fixed rollers, A and B, with distance $L = 2.0$ cm between the bar's center of mass and each roller. The rollers, whose directions of rotation are shown in the figure, slip against the bar with coefficient of kinetic friction $\mu_k = 0.40$. Suppose the bar is displaced horizontally by a distance x, as in Fig. 16-6b, and then released. What is the angular frequency ω of the resulting horizontal simple harmonic (back and forth) motion of the bar?

SOLUTION: To find ω, we find an expression for the horizontal acceleration a of the bar as a function of x and compare it with Eq. 16-7. We do so by applying Newton's second law vertically and horizontally; then we use the law in angular form about the contact point between the bar and roller A.

The vertical forces acting on the bar are its weight $m\mathbf{g}$ and supporting forces $\mathbf{F}_A$ due to roller A and $\mathbf{F}_B$ due to roller B. Since there is no net vertical force acting on the bar, Newton's second law gives us

$$\sum F_y = F_A + F_B - mg = 0. \qquad (16\text{-}16)$$

The horizontal forces acting on the bar are the kinetic frictional forces $f_{kA} = \mu_k F_A$ (toward the right) due to roller A

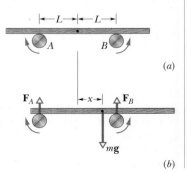

FIGURE 16-6 Sample Problem 16-3. (*a*) A bar is in equilibrium on two rotating rollers, *A* and *B*, that slip beneath it. (*b*) The bar is displaced from equilibrium by a distance *x* and then released.

and $f_{kB} = -\mu_k F_B$ (toward the left) due to roller B. Horizontally, Newton's second law gives us

$$\mu_k F_A - \mu_k F_B = ma,$$

or

$$a = \frac{\mu_k F_A - \mu_k F_B}{m}. \qquad (16\text{-}17)$$

The bar experiences no net torque about an axis perpendicular to the plane of Fig. 16-6 through the contact point between the bar and roller A. So Newton's second law for torques about that axis gives us

$$\sum \tau = F_A(0) + F_B 2L - mg(L + x) + f_{kA}(0) + f_{kB}(0)$$
$$= 0, \qquad (16\text{-}18)$$

where the forces $\mathbf{F}_A$, $\mathbf{F}_B$, mg, $\mathbf{f}_{kA}$, and $\mathbf{f}_{kB}$ have moment arms about that axis of 0, $2L$, $L + x$, 0, and 0, respectively.

Solving Eqs. 16-16 and 16-18 for F_A and F_B, we find

$$F_A = \frac{mg(L - x)}{2L} \quad \text{and} \quad F_B = \frac{mg(L + x)}{2L}.$$

Substituting these results into Eq. 16-17 yields

$$a = -\frac{\mu_k g}{L} x. \qquad (16\text{-}19)$$

Comparison of Eq. 16-19 with Eq. 16-7 reveals that the bar must be undergoing simple harmonic motion with angular frequency ω given by

$$\omega^2 = \frac{\mu_k g}{L}.$$

Thus

$$\omega = \sqrt{\frac{\mu_k g}{L}} = \sqrt{\frac{(0.40)(9.8 \text{ m/s}^2)}{0.020 \text{ m}}}$$
$$= 14 \text{ rad/s.} \qquad \text{(Answer)}$$

PROBLEM SOLVING TACTICS

TACTIC 2: *Identifying SHM*

In Sample Problem 16-3 we were able to identify the motion as linear simple harmonic motion once you reached Eq. 16-19. In linear SHM the acceleration a and displacement x are related by an equation of the form

$$a = -(\text{a positive constant})x,$$

which says that the acceleration is proportional to the displacement from the equilibrium position but is in the opposite direction. Once you find such an expression, you can immediately compare it to Eq. 16-7, identify the positive constant as being equal to ω^2, and so quickly get an expression for the angular frequency of the motion. With Eq. 16-4 you then can find the period T and the frequency f.

As you will see in Sample Problem 16-8, the same technique can be used to identify angular SHM. In such motion, the angular acceleration α and angular displacement θ are re-

lated by an equation of the form

$$\alpha = -(\text{a positive constant})\theta,$$

which says that the angular acceleration is proportional to the angular displacement from the equilibrium position but is in the opposite direction. As for linear SHM, you can identify the positive constant with ω^2 to find ω, T, and f.

In some problems you might derive an expression for the force F as a function of displacement x. If the motion is linear SHM, the force and displacement are related by

$$F = -(\text{a positive constant})x,$$

which says that the force is proportional to the displacement but is in the opposite direction. Once you have found such an expression, you can immediately compare it to Eq. 16-9 and identify the positive constant as being k. If you know the mass that is involved, you can then use Eqs. 16-11, 16-12, and 16-4 to find the angular frequency ω, period T, and frequency f.

You can similarly identify angular SHM. In such motion the torque τ and angular displacement θ are related by

$$\tau = -(\text{a positive constant})\theta,$$

which says that the torque is proportional to the angular displacement from the equilibrium position but is in the opposite direction.

16-4 ENERGY IN SIMPLE HARMONIC MOTION

In Chapter 8 we saw that the energy of a linear oscillator shuttles back and forth between kinetic and potential forms, while their sum—the mechanical energy E of the oscillator—remains constant. We now consider this situation quantitatively.

The potential energy of a linear oscillator like that of Fig. 16-5 is associated entirely with the spring. Its value depends on how much the spring is stretched or compressed, that is, on $x(t)$. We can use Eqs. 8-11 and 16-3 to find

$$U(t) = \tfrac{1}{2}kx^2 = \tfrac{1}{2}kx_m^2 \cos^2(\omega t + \phi). \qquad (16\text{-}20)$$

Note carefully that a function in the form $\cos^2 A$ (as here) is the same as $(\cos A)^2$ but is *not* the same as $\cos A^2$, which means $\cos (A^2)$.

The kinetic energy of the system is associated entirely with the block. Its value depends on how fast the block is moving, that is, on $v(t)$. We can use Eq. 16-5 to find

$$K(t) = \tfrac{1}{2}mv^2 = \tfrac{1}{2}m\omega^2 x_m^2 \sin^2(\omega t + \phi). \qquad (16\text{-}21)$$

If we use Eq. 16-11 to substitute k/m for ω^2, we can write

Eq. 16-21 as

$$K(t) = \tfrac{1}{2}mv^2 = \tfrac{1}{2}kx_m^2 \sin^2(\omega t + \phi). \quad (16\text{-}22)$$

The mechanical energy follows from Eqs. 16-20 and 16-22 and is

$$E = U + K$$
$$= \tfrac{1}{2}kx_m^2 \cos^2(\omega t + \phi) + \tfrac{1}{2}kx_m^2 \sin^2(\omega t + \phi)$$
$$= \tfrac{1}{2}kx_m^2 [\cos^2(\omega t + \phi) + \sin^2(\omega t + \phi)].$$

For any angle α,

$$\cos^2 \alpha + \sin^2 \alpha = 1.$$

Thus the quantity in the square brackets above is unity and we have

$$E = U + K = \tfrac{1}{2}kx_m^2. \quad (16\text{-}23)$$

The mechanical energy of a linear oscillator is indeed a constant, independent of time. The potential energy and kinetic energy of the linear oscillator are shown as functions of time in Fig. 16-7a, and as functions of displacement in Fig. 16-7b.

You might now understand why an oscillating system normally contains an element of springiness and an ele-

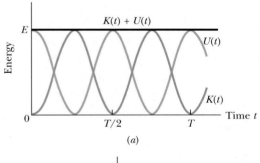

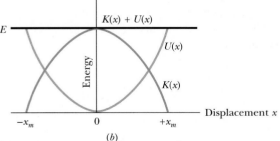

FIGURE 16-7 (a) The potential energy $U(t)$, kinetic energy $K(t)$, and mechanical energy E as functions of time, for a linear harmonic oscillator. Note that all energies are positive and that the potential energy and the kinetic energy peak twice during every period. (b) The potential energy $U(x)$, kinetic energy $K(x)$, and mechanical energy E as functions of position, for a linear harmonic oscillator with amplitude x_m. For $x = 0$ the energy is all kinetic, and for $x = \pm x_m$ it is all potential.

ment of inertia: it uses the former to store its potential energy and the latter to store its kinetic energy. The system of Sample Problem 16-3 is an exception, because no potential energy is involved. Instead, the energy transferred to the kinetic energy of the bar by the rollers actually varies over time.

C HECKPOINT 3: In Fig. 16-5, the block has a kinetic energy of 3 J and the spring has an elastic potential energy of 2 J when the block is at $x = +2.0$ cm. (a) What is the kinetic energy at $x = 0$? What are the elastic potential energies at (b) $x = -2.0$ cm and (c) $x = -x_m$?

SAMPLE PROBLEM 16-4

(a) What is the mechanical energy of the linear oscillator of Sample Problem 16-1?

SOLUTION: We find, substituting data from Sample Problem 16-1 into Eq. 16-23,

$$E = \tfrac{1}{2}kx_m^2 = (\tfrac{1}{2})(65 \text{ N/m})(0.11 \text{ m})^2$$
$$= 0.393 \text{ J} \approx 0.39 \text{ J.} \qquad \text{(Answer)}$$

This value remains constant throughout the motion.

(b) What is the potential energy of this oscillator when the block is halfway to its end point, that is, when $x = \pm\tfrac{1}{2}x_m$?

SOLUTION: For any displacement, the potential energy of a spring–block system is given by $U = \tfrac{1}{2}kx^2$. Here,

$$U = \tfrac{1}{2}kx^2 = \tfrac{1}{2}k(\tfrac{1}{2}x_m)^2 = \tfrac{1}{4}(\tfrac{1}{2}kx_m^2)$$
$$= \tfrac{1}{4}E = (\tfrac{1}{4})(0.393 \text{ J}) = 0.098 \text{ J.} \quad \text{(Answer)}$$

(c) What is the kinetic energy of the oscillator when $x = \tfrac{1}{2}x_m$?

SOLUTION: We find this from

$$K = E - U$$
$$= 0.393 \text{ J} - 0.098 \text{ J} \approx 0.30 \text{ J.} \quad \text{(Answer)}$$

Thus, at this point during the oscillation, 25% of the energy is in potential form and 75% in kinetic form.

16-5 AN ANGULAR SIMPLE HARMONIC OSCILLATOR

Figure 16-8 shows an angular version of a simple harmonic oscillator; the element of springiness or elasticity is associated with the twisting of a suspension wire rather than the extension and compression of a spring as we previously had. The device is called a **torsion pendulum,** with *torsion* referring to the twisting.

FIGURE 16-8 An angular simple harmonic oscillator, or torsion pendulum, is an angular version of the linear simple harmonic oscillator of Fig. 16-5. The disk oscillates in a horizontal plane; the reference line oscillates with angular amplitude θ_m. The twist in the suspension wire stores potential energy as a spring does and provides the restoring torque.

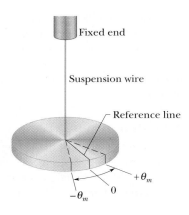

Fixed end

Suspension wire

Reference line

$+\theta_m$

$-\theta_m$

0

If we rotate the disk in Fig. 16-8 from its rest position (where the reference line is at 0) and release it, it will oscillate about that position in **angular simple harmonic motion.** Rotating the disk through an angle θ in either direction introduces a restoring torque given by

$$\tau = -\kappa\theta. \qquad (16\text{-}24)$$

Here κ (Greek *kappa*) is a constant, called the **torsion constant,** that depends on the length, diameter, and material of the suspension wire.

Comparison of Eq. 16-24 with Eq. 16-9 leads us to suspect that Eq. 16-24 is the angular form of Hooke's law, and that we can transform Eq. 16-12, which gives the period of linear SHM, into an equation for the period of angular SHM: we replace the spring constant k in Eq. 16-12 with its equivalent, the constant κ of Eq. 16-24, and we replace the mass m in Eq. 16-12 with *its* equivalent, the rotational inertia I of the oscillating disk. These replacements lead to

$$T = 2\pi\sqrt{\frac{I}{\kappa}} \qquad \text{(torsion pendulum)} \quad (16\text{-}25)$$

for the period of the angular simple harmonic oscillator, or torsion pendulum.

SAMPLE PROBLEM 16-5

As Fig. 16-9a shows, a thin rod whose length L is 12.4 cm and whose mass m is 135 g is suspended at its midpoint from a long wire. Its period T_a of angular SHM is measured to be 2.53 s. An irregularly shaped object, which we call object X, is then hung from the same wire, as in Fig. 16-9b, and its period T_b is found to be 4.76 s.

(a) What is the rotational inertia of object X about its suspension axis?

SOLUTION: In Table 11-2(*e*), the rotational inertia of a thin rod about a perpendicular axis through its midpoint is given as $\frac{1}{12}mL^2$. Thus we have

$$I_a = \tfrac{1}{12}mL^2 = (\tfrac{1}{12})(0.135 \text{ kg})(0.124 \text{ m})^2$$
$$= 1.73 \times 10^{-4} \text{ kg} \cdot \text{m}^2.$$

Now let us write Eq. 16-25 twice, once for the rod and once for object X:

$$T_a = 2\pi\sqrt{\frac{I_a}{\kappa}} \quad \text{and} \quad T_b = 2\pi\sqrt{\frac{I_b}{\kappa}}.$$

Here the subscripts refer to Figs. 16-9*a* and 16-9*b*. The constant κ, which is a property of the wire, is the same for both figures; only the periods and the rotational inertias differ.

Let us square each of these equations, divide the second by the first, and solve the resulting equation for I_b. The result is

$$I_b = I_a\frac{T_b^2}{T_a^2} = (1.73 \times 10^{-4} \text{ kg} \cdot \text{m}^2)\frac{(4.76 \text{ s})^2}{(2.53 \text{ s})^2}$$
$$= 6.12 \times 10^{-4} \text{ kg} \cdot \text{m}^2. \qquad \text{(Answer)}$$

(b) What would be the period of oscillation if both objects were fastened together and hung from the wire, as in Fig. 16-9c?

SOLUTION: We can again write Eq. 16-25 twice, this time as

$$T_a = 2\pi\sqrt{\frac{I_a}{\kappa}} \quad \text{and} \quad T_c = 2\pi\sqrt{\frac{I_c}{\kappa}}.$$

After dividing again and putting $I_c = I_a + I_b$, we find

$$T_c = T_a\sqrt{\frac{I_c}{I_a}} = T_a\sqrt{\frac{I_a + I_b}{I_a}} = T_a\sqrt{1 + \frac{I_b}{I_a}}$$
$$= (2.53)\sqrt{1 + \frac{6.12 \times 10^{-4} \text{ kg} \cdot \text{m}^2}{1.73 \times 10^{-4} \text{ kg} \cdot \text{m}^2}}$$
$$= 5.39 \text{ s}. \qquad \text{(Answer)}$$

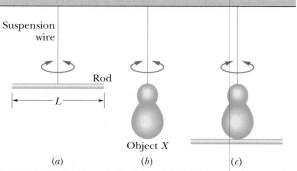

Suspension wire

Rod

L

Object X

(*a*)   (*b*)   (*c*)

FIGURE 16-9 Sample Problem 16-5. Three torsion pendulums consisting of a wire and (*a*) a rod, (*b*) an irregularly shaped object, and (*c*) the rod and the object rigidly connected.

16-6 PENDULUMS

We turn now to a class of simple harmonic oscillators in which the springiness is associated with the gravitational force rather than with the elastic properties of a twisted wire or a compressed or stretched spring.

The Simple Pendulum

If you hang an apple at the end of a long thread fixed at its upper end, and then set the apple swinging back and forth a small distance, you easily see that the apple's motion is periodic. Is it, in fact, simple harmonic motion? To idealize this situation, we consider a **simple pendulum,** which consists of a particle of mass m (called the *bob* of the pendulum) suspended from an unstretchable, massless string of length L, as in Fig. 16-10a. The bob is free to swing back and forth in the plane of the page, to the left and right of a vertical line through the point at which the upper end of the string is fixed.

The element of inertia in this pendulum is the mass m of the particle, and the element of springiness is in the gravitational attraction between the particle and Earth. Potential energy can be associated with the varying vertical distance between the swinging particle and Earth; we may view that varying distance as the varying length of a "gravitational spring."

The forces acting on the particle, shown in Fig. 16-10b, are its weight $m\mathbf{g}$ and the tension $\mathbf{T}$ in the string. We resolve $m\mathbf{g}$ into a radial component $mg \cos \theta$ and a component $mg \sin \theta$ that is tangent to the path taken by the particle. This tangential component is a restoring force, because it always acts opposite the displacement of the particle so as to bring the particle back toward its central location, the

equilibrium position ($\theta = 0$), where it would be at rest were it not swinging. We write the restoring force as

$$F = -mg \sin \theta, \qquad (16\text{-}26)$$

where the minus sign indicates that F acts opposite the displacement.

If we assume that the angle θ in Fig. 16-10 is small, then $\sin \theta$ is very nearly equal to θ in radians. [For example, when $\theta = 5.00°$ ($= 0.0873$ rad), $\sin \theta = 0.0872$, a difference of only about 0.1%.] Also, the displacement s of the particle measured along its arc is equal to $L\theta$. Thus, for small θ, Eq. 16-26 becomes

$$F \approx -mg\theta = -mg \, \frac{s}{L} = -\left(\frac{mg}{L} \right) s. \quad (16\text{-}27)$$

A glance back at Eq. 16-9 shows that we again have Hooke's law, with the displacement now being arc length s instead of x. Thus *if a simple pendulum swings through a small angle,* it is a linear oscillator like the block–spring oscillator of Fig. 16-5; that is, it undergoes simple harmonic motion. Now the amplitude of the motion is measured as the **angular amplitude** θ_m, the maximum angle of swing. And the spring constant k is mg/L, the effective spring constant of the pendulum's gravitational spring.

By substituting mg/L for k in Eq. 16-12, we find, for the period of a simple pendulum,

$$T = 2\pi \sqrt{\frac{m}{k}} = 2\pi \sqrt{\frac{m}{mg/L}} \qquad (16\text{-}28)$$

or

$$T = 2\pi \sqrt{\frac{L}{g}} \qquad \text{(simple pendulum).} \quad (16\text{-}29)$$

Equation 16-29 holds only if the angular amplitude θ_m is small (which is what we assume in the exercises and problems for this chapter unless otherwise stated).

The element of inertia seems to be missing in Eq. 16-29 because the period is independent of the mass of the particle. This comes about because the element of springiness, which is the gravitational spring constant mg/L, is itself proportional to the mass of the particle, and the two masses cancel in Eq. 16-28. Figure 8-7 shows how energy shuttles back and forth between potential and kinetic forms during every oscillation of a simple pendulum.

The Physical Pendulum

Most pendulums in the real world are not even approximately "simple." Figure 16-11 shows a generalized **physical pendulum,** as we shall call realistic pendulums, with its weight $m\mathbf{g}$ acting at its center of mass C.

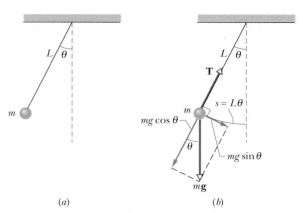

(a)   (b)

FIGURE 16-10 (a) A simple pendulum. (b) The forces acting on the bob are its weight $m\mathbf{g}$ and the tension $\mathbf{T}$ in the string. The tangential component $mg \sin \theta$ of the weight is a restoring force that brings the pendulum back to the central position.

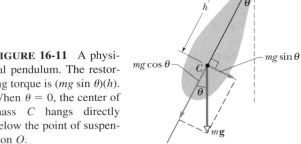

FIGURE 16-11 A physical pendulum. The restoring torque is $(mg \sin \theta)(h)$. When $\theta = 0$, the center of mass C hangs directly below the point of suspension O.

When the pendulum of Fig. 16-11 is displaced through an angle θ in either direction from its equilibrium position, a restoring torque appears. This torque acts about an axis through the suspension point O in Fig. 16-11 and has the magnitude

$$\tau = -(mg \sin \theta)(h). \qquad (16\text{-}30)$$

Here $mg \sin \theta$ is the tangential component of the weight $m\mathbf{g}$, and h (which is equal to OC) is the moment arm of this force component. The minus sign indicates that the torque is a restoring torque. That is, the torque always acts to reduce the angle θ to zero.

We once more decide to limit our interest to small amplitudes, so that $\sin \theta \approx \theta$. Then Eq. 16-30 becomes

$$\tau \approx -(mgh)\theta. \qquad (16\text{-}31)$$

Comparison with Eq. 16-24 shows that we again have Hooke's law in angular form and that the physical pendulum is an angular harmonic oscillator. Thus a physical pendulum undergoes simple harmonic motion *if* the angular amplitude θ_m of its motion is small. The term mgh of Eq. 16-31 is analogous to the torsion constant κ of Eq. 16-24. Substituting mgh for κ in Eq. 16-25, we find

$$T = 2\pi \sqrt{\frac{I}{mgh}} \quad \text{(physical pendulum)} \qquad (16\text{-}32)$$

for the period of a physical pendulum when θ_m is small. Here I is the rotational inertia of the pendulum (about an axis through its point of support and perpendicular to its plane of swing), and h is the distance between the point of support and the center of mass of the swinging pendulum.

A physical pendulum will not swing if we hang it by its center of mass. Formally, this corresponds to putting $h = 0$ in Eq. 16-32. That equation then predicts $T \to \infty$, which implies that such a pendulum will never complete one swing.

Corresponding to any physical pendulum that oscillates about a given suspension point O with period T is a simple pendulum of length L_0 with the same period T. We can find L_0 with Eq. 16-29. The point along the physical pendulum at distance L_0 from point O is called the *center of oscillation* of the physical pendulum for the given suspension point.

The physical pendulum of Fig. 16-11 includes the simple pendulum as a special case. For the latter, h is the length L of the string and I is mL^2. Making these substitutions in Eq. 16-32 leads to

$$T = 2\pi \sqrt{\frac{I}{mgh}} = 2\pi \sqrt{\frac{mL^2}{mgL}} = 2\pi \sqrt{\frac{L}{g}},$$

which is exactly Eq. 16-29, the expression for the period of a simple pendulum.

Measuring g

We can use a physical pendulum to measure the free-fall acceleration g. (Countless thousands of such measurements have been made during geophysical prospecting.)

To analyze a simple case, take the pendulum to be a uniform rod of length L, suspended from one end. For such a pendulum, h in Eq. 16-32, the distance between the suspension point and the center of mass, is $\frac{1}{2}L$. Table 11-2(f) tells us that the rotational inertia of this pendulum about a perpendicular axis through one end is $\frac{1}{3}mL^2$. If we put $h = \frac{1}{2}L$ and $I = \frac{1}{3}mL^2$ in Eq. 16-32 and solve for g, we find

$$g = \frac{8\pi^2 L}{3T^2}. \qquad (16\text{-}33)$$

Thus by measuring L and the period T, we can find the value of g. (If precise measurements are to be made, a number of refinements are needed, such as swinging the pendulum in an evacuated chamber.)

CHECKPOINT 4: Three physical pendulums, of masses m_0, $2m_0$, and $3m_0$, have the same shape and size and are suspended at the same point. Rank the masses according to the periods of the pendulums, greatest period first.

SAMPLE PROBLEM 16-6

A meter stick, suspended from one end, swings as a physical pendulum, as in Fig. 16-12a.

(a) What is its period of oscillation?

SOLUTION: From Table 11-2(f) we see that the rotational inertia of a rod or stick of length L about a perpendicular axis through one end is $\frac{1}{3}mL^2$. The distance h from the point of suspension to the center of mass, which is point C in Fig.

16-12a, is $\frac{1}{2}L$. If we substitute these two quantities into Eq. 16-32, we find

$$T = 2\pi\sqrt{\frac{I}{mgh}} = 2\pi\sqrt{\frac{\frac{1}{3}mL^2}{mg(\frac{1}{2}L)}} = 2\pi\sqrt{\frac{2L}{3g}} \quad (16\text{-}34)$$

$$= 2\pi\sqrt{\frac{(2)(1.00\text{ m})}{(3)(9.8\text{ m/s}^2)}} = 1.64\text{ s}. \quad \text{(Answer)}$$

(b) For the stick of Fig. 16-12a, what is the distance L_0 between the suspension point O of the stick and the center of oscillation of the stick?

SOLUTION: To locate the center of oscillation of the meter stick, we find the length L_0 of a simple pendulum (Fig. 16-12b) having the same period T as the meter stick. Setting Eqs. 16-29 and 16-34 equal yields

$$T = 2\pi\sqrt{\frac{L_0}{g}} = 2\pi\sqrt{\frac{2L}{3g}}.$$

You can see by inspection that

$$L_0 = \frac{2}{3}L = (\frac{2}{3})(100\text{ cm}) = 66.7\text{ cm}. \quad \text{(Answer)}$$

In Fig. 16-12a, point P marks this distance from suspension point O. Thus point P is the stick's center of oscillation for the given suspension point.

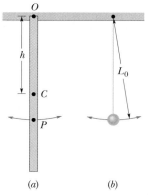

(a)   (b)

FIGURE 16-12 Sample Problem 16-6. (a) A meter stick suspended from one end as a physical pendulum. (b) A simple pendulum whose length L is chosen so that the periods of the two pendulums are equal. Point P on the pendulum of (a) marks the center of oscillation, a distance L_0 from the suspension point.

SAMPLE PROBLEM 16-7

A disk whose radius R is 12.5 cm is suspended, as a physical pendulum, from a point at distance h from its center C (Fig. 16-13). Its period T is 0.871 s when $h = R/2$. What is the free-fall acceleration g at the location of the pendulum?

SOLUTION: The rotational inertia I_{cm} of a disk about its central axis is $\frac{1}{2}mR^2$. From the parallel-axis theorem, the rotational inertia about an axis parallel to the central axis and through the point of suspension O, as in Fig. 16-13, is

$$I = I_{cm} + mh^2 = \frac{1}{2}mR^2 + m(\frac{1}{2}R)^2 = \frac{3}{4}mR^2.$$

If we put $I = \frac{3}{4}mR^2$ and $h = \frac{1}{2}R$ in Eq. 16-32, we find

$$T = 2\pi\sqrt{\frac{I}{mgh}} = 2\pi\sqrt{\frac{\frac{3}{4}mR^2}{mg(\frac{1}{2}R)}} = 2\pi\sqrt{\frac{3R}{2g}}.$$

Solving for g then gives us

$$g = \frac{6\pi^2R}{T^2} = \frac{(6\pi^2)(0.125\text{ m})}{(0.871\text{ s})^2} = 9.76\text{ m/s}^2. \quad \text{(Answer)}$$

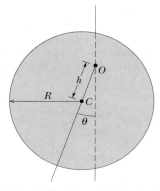

FIGURE 16-13 Sample Problem 16-7. A physical pendulum consisting of a uniform disk suspended from a point (O) that is halfway from the center of the disk to the rim.

SAMPLE PROBLEM 16-8

In Fig. 16-14, a penguin (obviously skilled in aquatic sports) dives from a uniform board that is hinged at the left and attached to a spring at the right. The board has length $L = 2.0$ m and mass $m = 12$ kg; the spring constant k is 1300 N/m. When the penguin dives, it leaves the board and spring oscillating with a small amplitude. Assume that the board is stiff enough not to bend, and find the period T of the oscillations.

SOLUTION: The spring applies a time-varying torque τ (about the hinge) to the board, which undergoes a time-varying angular acceleration α about the hinge. Since a spring is involved, we might guess that simple harmonic motion is also involved, but we will not draw that conclusion yet. Instead, we use Eqs. 11-30 and 11-35 to write

$$\tau = LF\sin 90° = I\alpha. \quad (16\text{-}35)$$

Here I is the rotational inertia of the board about the hinge, F is the force on the right end of the board from the spring, and 90° is the angle between the length of the board and the direction of that force.

We may treat the board as being a thin rod pivoted about one end, so that from Table 11-2(f), $I = mL^2/3$. The force F from the spring is $F = -kx$, where x is the vertical displacement of the right end of the board.

Substituting these expressions for I and F into the rightmost equality of Eq. 16-35 and evaluating sin 90°, we find

$$-Lkx = \frac{mL^2\alpha}{3}. \qquad (16\text{-}36)$$

We now have a mixture of linear displacement x (vertically) and rotational acceleration α (about the hinge). We can rewrite Eq. 16-36 in terms of just the rotational motion of the right end of the board about the hinge by using Eq. 11-15,

$$s = \theta r.$$

Here θ is the angle of the board's rotation about the hinge, $r = L$ is the radius of the rotation, and s is the arc length, which is approximated by the vertical displacement x (for small θ). We can rewrite this as $x = \theta L$ and then substitute for x in Eq. 16-36 to find

$$-Lk\theta L = \frac{mL^2\alpha}{3},$$

which becomes

$$\alpha = -\frac{3k}{m}\theta. \qquad (16\text{-}37)$$

Equation 16-37 is an angular version of Eq. 16-7. It tells us that the board does indeed undergo simple harmonic motion, with angular acceleration α and angular displacement θ. Comparing Eqs. 16-37 and 16-7 reveals that

$$\omega^2 = \frac{3k}{m}$$

and thus that $\omega = \sqrt{3k/m}$. Using Eq. 16-4, $\omega = 2\pi/T$, we then have

$$T = 2\pi\sqrt{\frac{m}{3k}} = 2\pi\sqrt{\frac{12\text{ kg}}{3(1300\text{ N/m})}}$$

$$= 0.35\text{ s}. \qquad \text{(Answer)}$$

Perhaps surprisingly, the period is independent of the board's length L.

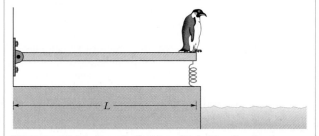

FIGURE 16-14 Sample Problem 16-8. The dive by the penguin causes the board and spring to oscillate; the board pivots about the hinge at the left.

16-7 SIMPLE HARMONIC MOTION AND UNIFORM CIRCULAR MOTION

In 1610, Galileo, using his newly constructed telescope, discovered the four principal moons of Jupiter. Over weeks of observation, each moon seemed to him to be moving back and forth relative to the planet in what today we would call simple harmonic motion; the disk of the planet was the midpoint of the motion. The record of Galileo's observations, written in his own hand, is still available. A. P. French of MIT used Galileo's data to work out the position of the moon Callisto relative to Jupiter. In the results shown in Fig. 16-15, the circles are based on Galileo's observations and the curve is a best fit to the data. The curve strongly suggests Eq. 16-3, the displacement function for SHM. A period of about 16.8 days can be measured from the plot.

Actually, Callisto moves with essentially constant speed in an essentially circular orbit around Jupiter. Its true motion—far from being simple harmonic —is uniform circular motion. What Galileo saw—and what you can see with a good pair of binoculars and a little patience—is the projection of this uniform circular motion on a line in the plane of the motion. We are led by Galileo's remarkable observations to the conclusion that simple harmonic motion is uniform circular motion viewed edge-on. In more formal language:

> Simple harmonic motion is the projection of uniform circular motion on a diameter of the circle in which the latter motion occurs.

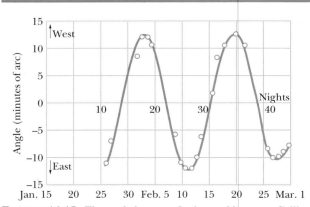

FIGURE 16-15 The angle between Jupiter and its moon Callisto as seen from Earth. The circles are based on Galileo's 1610 measurements. The curve is a best fit, strongly suggesting simple harmonic motion. At Jupiter's mean distance, 10 minutes of arc corresponds to about 2×10^6 km. (Adapted from A. P. French, *Newtonian Mechanics*, W. W. Norton & Company, New York, 1971, p. 288.)

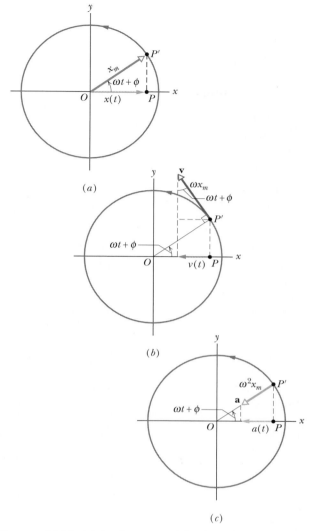

(a)

(b)

(c)

FIGURE 16-16 (a) A reference particle P' moving with uniform circular motion in a reference circle of radius x_m. Its projection P on the x axis executes simple harmonic motion. (b) The projection of the velocity **v** of the reference particle is the velocity of SHM. (c) The projection of the acceleration **a** of the reference particle is the acceleration of SHM.

Figure 16-16a gives an example. It shows a *reference particle P'* moving in uniform circular motion with (constant) angular speed ω in a *reference circle*. The radius x_m of the circle is the magnitude of the particle's position vector. At any time t, the angular position of the particle is $\omega t + \phi$, where ϕ is the angular position at $t = 0$.

The projection of particle P' onto the x axis is a point P, which we take to be a second particle. The projection of the position vector of particle P' onto the x axis gives the location $x(t)$ of P. Thus we find

$$x(t) = x_m \cos(\omega t + \phi),$$

which is precisely Eq. 16-3. Our conclusion is correct. If reference particle P' moves in uniform circular motion, its projection particle P moves in simple harmonic motion.

This relation throws new light on the angular frequency ω of simple harmonic motion, and you can see more clearly how the "angular" originates. The quantity ω is simply the constant angular speed with which the reference particle P' moves along its reference circle. The phase constant ϕ has a value determined by the position of reference particle P' along its reference circle at $t = 0$.

Figure 16-16b shows the velocity of the reference particle. The magnitude of the velocity vector is ωx_m, and its projection on the x axis is

$$v(t) = -\omega x_m \sin(\omega t + \phi),$$

which is exactly Eq. 16-5. The minus sign appears because the velocity component of P in Fig. 16-16b points to the left, in the direction of decreasing x.

Figure 16-16c shows the acceleration of the reference particle. The magnitude of the acceleration vector is $\omega^2 x_m$ and its projection on the x axis is

$$a(t) = -\omega^2 x_m \cos(\omega t + \phi),$$

which is exactly Eq. 16-6. Thus whether we look at the displacement, the velocity, or the acceleration, the projection of uniform circular motion is indeed simple harmonic motion.

16-8 DAMPED SIMPLE HARMONIC MOTION (OPTIONAL)

A pendulum will swing hardly at all under water, because the water exerts a drag force on the pendulum that quickly eliminates the motion. A pendulum swinging in air does better, but still the motion dies out because the air exerts a drag force on the pendulum (and friction acts at its support), transferring energy from the pendulum's motion.

When the motion of an oscillator is reduced by an external force, the oscillator and its motion are said to be **damped.** An idealized example of a damped oscillator is shown in Fig. 16-17: a block with mass m oscillates on a spring with spring constant k. From the mass, a rod extends to a vane (both assumed massless) that is submerged in a liquid. As the vane moves up and down, the liquid exerts an inhibiting drag force on it and thus on the entire oscillating system. With time, the mechanical energy of the block–spring system decreases, as energy is transferred to thermal energy of the liquid and vane.

Let us assume that the liquid exerts a **damping force** $\mathbf{F}_d$ that is proportional in magnitude to the velocity **v** of the

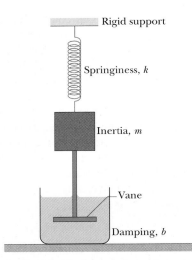

FIGURE 16-17 An idealized damped simple harmonic oscillator. A vane immersed in a liquid exerts a damping force on the oscillating block.

vane and block (an assumption that is accurate if the vane moves slowly). Then

$$F_d = -bv, \qquad (16\text{-}38)$$

where b is a **damping constant** that depends on the characteristics of both the vane and the liquid and has the SI unit of kilogram per second. The minus sign indicates that $\mathbf{F}_d$ opposes the motion. The total force acting on the block is then

$$\sum F = -kx - bv,$$

or, if we set $v = dx/dt$,

$$\sum F = -kx - b\frac{dx}{dt}. \qquad (16\text{-}39)$$

Substituting this total force into Newton's second law yields the differential equation

$$m\frac{d^2x}{dt^2} + b\frac{dx}{dt} + kx = 0,$$

whose solution is

$$x(t) = x_m e^{-bt/2m} \cos(\omega' t + \phi), \qquad (16\text{-}40)$$

where ω', the angular frequency of the damped oscillator, is given by

$$\omega' = \sqrt{\frac{k}{m} - \frac{b^2}{4m^2}}. \qquad (16\text{-}41)$$

If $b = 0$ (there is no damping), then Eq. 16-41 reduces to Eq. 16-11 ($\omega = \sqrt{k/m}$) for the angular frequency of an undamped oscillator, and Eq. 16-40 reduces to Eq. 16-3 for

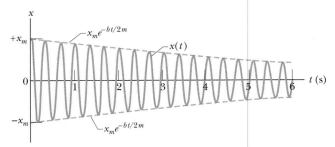

FIGURE 16-18 The displacement function $x(t)$ for the damped oscillator of Fig. 16-17. See Sample Problem 16-9. The amplitude, which is $x_m\, e^{-bt/2m}$, decreases exponentially with time.

the displacement of an undamped oscillator. If the damping constant is not zero but small (so that $b \ll \sqrt{km}$), then $\omega' \approx \omega$.

We can regard Eq. 16-40 as a cosine function whose amplitude, which is $x_m\, e^{-bt/2m}$, gradually decreases with time, as Fig. 16-18 suggests. For an undamped oscillator, the mechanical energy is constant and is given by Eq. 16-23: $E = \frac{1}{2}kx_m^2$. If the oscillator is damped, the mechanical energy is not constant but decreases with time. If the damping is small, we can find $E(t)$ by replacing x_m in Eq. 16-23 with $x_m\, e^{-bt/2m}$, the amplitude of the damped oscillations. Doing so, we find

$$E(t) \approx \tfrac{1}{2}kx_m^2\, e^{-bt/m}, \qquad (16\text{-}42)$$

which tells us that the mechanical energy decreases exponentially with time.

SAMPLE PROBLEM 16-9

For the damped oscillator of Fig. 16-17, $m = 250$ g, $k = 85$ N/m, and $b = 70$ g/s.

(a) What is the period of the motion?

SOLUTION: Because $b \ll \sqrt{km} = 4.6$ kg/s, the period is approximately that of the undamped oscillator. From Eq. 16-12, we then have

$$T = 2\pi\sqrt{\frac{m}{k}} = 2\pi\sqrt{\frac{0.25\text{ kg}}{85\text{ N/m}}} = 0.34\text{ s.} \quad \text{(Answer)}$$

(b) How long does it take for the amplitude of the damped oscillations to drop to half its initial value?

SOLUTION: The amplitude at time t is displayed in Eq. 16-40 as $x_m\, e^{-bt/2m}$. It has the value x_m at $t = 0$. Thus we must find the value of t for which

$$x_m\, e^{-bt/2m} = \tfrac{1}{2}x_m.$$

Let us divide both sides by x_m and take the natural logarithm of the equation that remains. We find

$$\ln \tfrac{1}{2} = \ln(e^{-bt/2m}) = -bt/2m,$$

or

$$t = \frac{-2m \ln \tfrac{1}{2}}{b} = \frac{-(2)(0.25 \text{ kg})(\ln \tfrac{1}{2})}{0.070 \text{ kg/s}}$$

$$= 5.0 \text{ s}. \qquad \text{(Answer)}$$

Note, from the answer we found in (a), that this is about 15 periods of oscillation.

(c) How long does it take for the mechanical energy to drop to one-half its initial value?

SOLUTION: From Eq. 16-42 we see that the mechanical energy at time t is $\tfrac{1}{2}kx_m^2 \, e^{-bt/m}$. It has the value $\tfrac{1}{2}kx_m^2$ at $t = 0$. Thus we must find the value of t for which

$$\tfrac{1}{2}kx_m^2 \, e^{-bt/m} = \tfrac{1}{2}(\tfrac{1}{2}kx_m^2).$$

If we divide by $\tfrac{1}{2}kx_m^2$ and solve for t as we did above, we find

$$t = \frac{-m \ln \tfrac{1}{2}}{b} = \frac{-(0.25 \text{ kg})(\ln \tfrac{1}{2})}{0.070 \text{ kg/s}} = 2.5 \text{ s}. \quad \text{(Answer)}$$

This is exactly half the time calculated in (b), or about 7.5 periods of oscillation. Figure 16-18 was drawn to illustrate this sample problem.

CHECKPOINT **5:** Here are three sets of values for the spring constant, damping constant, and mass for the damped oscillator of Fig. 16-17. Rank the sets according to the time for the mechanical energy to decrease to one-fourth of its initial value, greatest first.

Set 1	$2k_0$	b_0	m_0
Set 2	k_0	$6b_0$	$4m_0$
Set 3	$3k_0$	$3b_0$	m_0

16-9 FORCED OSCILLATIONS AND RESONANCE

A person swinging passively in a swing is an example of *free oscillation*. If a kind friend pulls or pushes the swing periodically, as in Fig. 16-19, we have *forced, or driven, oscillations*. There are now *two* angular frequencies with which to deal: (1) the *natural* angular frequency ω of the system, which is the angular frequency at which it would oscillate if it were suddenly disturbed and then left to oscillate freely, and (2) the angular frequency ω_d of the external driving force.

We can use Fig. 16-17 to represent an idealized forced simple harmonic oscillator if we allow the structure marked ''rigid support'' to move up and down at a variable

FIGURE 16-19 Two frequencies are suggested in this painting by Nicholas Lancret: (1) the natural frequency at which the lady —left to herself—would swing, and (2) the frequency at which her friend tugs on the rope. If these two frequencies match, there is resonance.

angular frequency ω_d. A forced oscillator oscillates at the angular frequency ω_d of the driving force, and its displacement $x(t)$ is given by

$$x(t) = x_m \cos(\omega_d t + \phi), \qquad (16\text{-}43)$$

where x_m is the amplitude of the oscillations.

How large the displacement amplitude x_m is depends on a complicated function of ω_d and ω. The velocity amplitude v_m of the oscillations is easier to describe: it is greatest when

$$\omega_d = \omega \qquad \text{(resonance)}, \qquad (16\text{-}44)$$

a condition called **resonance.** Equation 16-44 is also *approximately* the condition at which the displacement amplitude x_m of the oscillations is greatest. Thus if you push a swing at its natural frequency, the displacement and velocity amplitudes will increase to large values, a fact that children learn quickly by trial and error. If you push at other frequencies, either higher or lower, the displacement and velocity amplitudes will be smaller.

Figure 16-20 shows how the displacement amplitude of an oscillator depends on the angular frequency ω_d of the driving force, for three values of the damping coefficient b. Note that for all three the amplitude is approximately greatest when $\omega_d/\omega = 1$, that is, when the resonance condition of Eq. 16-44 is satisfied. The curves of Fig. 16-20 show that the smaller the damping, the taller and narrower is the *resonance peak*.

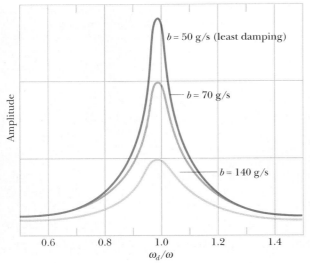

FIGURE 16-20 The displacement amplitude x_m of a forced oscillator varies as the angular frequency ω_d of the driving force is varied. The amplitude is greatest approximately at $\omega_d/\omega = 1$, the resonance condition. The curves here correspond to three values of the damping constant b.

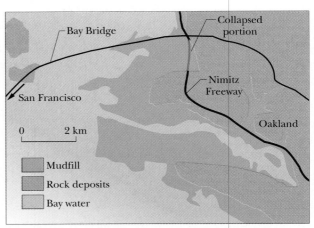

FIGURE 16-21 The geological structure in the Oakland portion of the San Francisco Bay area. The collapsed stretch of the Nimitz Freeway is indicated. (Adapted from ''Sediment-Induced Amplification and the Collapse of the Nimitz Freeway,'' by S. E. Hough et al., *Nature,* April 26, 1990.)

All mechanical structures have one or more natural frequencies, and if a structure is subjected to a strong external driving force that matches one of these frequencies, the resulting oscillations of the structure may rupture it. Thus, for example, aircraft designers must make sure that none of the natural frequencies at which a wing can vibrate matches the angular frequency of the engines at cruising speed. It obviously would be dangerous for a wing to flap violently at certain engine speeds.

The collapse of a 1.4 km stretch of the Nimitz Freeway, shown in this chapter's opening photograph, is an example of the destruction that resonance can produce. Oscillations of the ground due to the earthquake's seismic waves had their greatest velocity amplitude at an angular frequency of about 9 rad/s, which almost exactly matched a natural angular frequency of the individual horizontal sections of the freeway. The collapse was confined to a particular 1.4 km stretch because that section of the freeway was built on a loosely structured mudfill whose velocity amplitude during the quake was *at least five times larger* than that of the rock deposits under the rest of the freeway (Fig. 16-21).

REVIEW & SUMMARY

Frequency

Every oscillatory, or periodic, motion has a *frequency* f (the number of oscillations per second) measured, in the SI system, in hertz:

$$1 \text{ hertz} = 1 \text{ Hz} = 1 \text{ oscillation per second}$$
$$= 1 \text{ s}^{-1}. \quad (16\text{-}1)$$

Period

The *period* T is the time required for one complete oscillation, or **cycle.** It is related to the frequency by

$$T = \frac{1}{f}. \quad (16\text{-}2)$$

Simple Harmonic Motion

In *simple harmonic motion* (SHM), the displacement $x(t)$ of a particle from its equilibrium position is described by the equation

$$x = x_m \cos(\omega t + \phi) \quad \text{(displacement)}, \quad (16\text{-}3)$$

in which x_m is the **amplitude** of the displacement, the quantity $(\omega t + \phi)$ is the **phase** of the motion, and ϕ is the **phase constant.** The **angular frequency** ω is related to the period and frequency of the motion by

$$\omega = \frac{2\pi}{T} = 2\pi f \quad \text{(angular frequency)}. \quad (16\text{-}4)$$

Differentiating Eq. 16-3 leads to equations for the particle's velocity and acceleration during SHM as functions of time:

$$v = -\omega x_m \sin(\omega t + \phi) \quad \text{(velocity)} \quad (16\text{-}5)$$

and

$$a = -\omega^2 x_m \cos(\omega t + \phi) \quad \text{(acceleration)}. \quad (16\text{-}6)$$

In Eq. 16-5, the positive quantity ωx_m is the **velocity amplitude** v_m of the motion. In Eq. 16-6, the positive quantity $\omega^2 x_m$ is the **acceleration amplitude** a_m of the motion.

The Linear Oscillator

A particle with mass m that moves under the influence of a Hooke's law restoring force given by $F = -kx$ exhibits simple harmonic motion with

$$\omega = \sqrt{\frac{k}{m}} \quad \text{(angular frequency)} \quad (16\text{-}11)$$

and

$$T = 2\pi \sqrt{\frac{m}{k}} \quad \text{(period)}. \quad (16\text{-}12)$$

Such a system is called a **linear simple harmonic oscillator.**

Energy

A particle in simple harmonic motion has, at any time, kinetic energy $K = \frac{1}{2}mv^2$ and potential energy $U = \frac{1}{2}kx^2$. If no friction is present, the mechanical energy $E = K + U$ remains constant even though K and U change.

Pendulums

Examples of devices that undergo simple harmonic motion are the **torsion pendulum** of Fig. 16-8, the **simple pendulum** of Fig. 16-10, and the **physical pendulum** of Fig. 16-11. Their periods of oscillation for small oscillations are, respectively,

$$T = 2\pi \sqrt{I/\kappa}, \quad (16\text{-}25)$$

$$T = 2\pi \sqrt{L/g}, \quad (16\text{-}29)$$

and

$$T = 2\pi \sqrt{I/mgh}. \quad (16\text{-}32)$$

In each case, the period involves an inertial term divided by a "springiness" term that measures the strength of the oscillator's restoring force.

Simple Harmonic Motion and Uniform Circular Motion

Simple harmonic motion is the projection of uniform circular motion onto the diameter of the circle in which the latter motion occurs. Figure 16-16 shows that all parameters of circular motion (position, velocity, and acceleration) project to the corresponding values for simple harmonic motion.

Damped Harmonic Motion

The mechanical energy E in a real oscillating system decreases during the oscillations because external forces, such as a drag force, inhibit the oscillations and transfer mechanical energy to thermal energy. The real oscillator and its motion are then said to be **damped.** If the **damping force** is given by $F_d = -bv$, where v is the velocity of the oscillator and b is a **damping constant,** then the displacement of the oscillator is given by

$$x(t) = x_m e^{-bt/2m} \cos(\omega' t + \phi), \quad (16\text{-}40)$$

where ω', the angular frequency of the damped oscillator, is given by

$$\omega' = \sqrt{\frac{k}{m} - \frac{b^2}{4m^2}}. \quad (16\text{-}41)$$

If the damping constant is small ($b \ll \sqrt{km}$), then $\omega' \approx \omega$, where ω is the angular frequency of the undamped oscillator. For small b, the mechanical energy E of the oscillator is given by

$$E(t) \approx \frac{1}{2}kx_m^2 e^{-bt/m}. \quad (16\text{-}42)$$

Forced Oscillations and Resonance

If an external driving force with angular frequency ω_d acts on an oscillating system with *natural* angular frequency ω, the system oscillates with angular frequency ω_d. The velocity amplitude v_m of the system is greatest when

$$\omega_d = \omega, \quad (16\text{-}44)$$

a condition called **resonance.** The amplitude x_m of the system is (approximately) greatest under the same condition.

QUESTIONS

1. Which of the following relationships between the acceleration a and the displacement x of a particle involve SHM: (a) $a = 0.5x$, (b) $a = 400x^2$, (c) $a = -20x$, (d) $a = -3x^2$?

2. The acceleration $a(t)$ of a particle undergoing SHM is graphed in Fig. 16-22. (a) Which of the labeled points corresponds to the particle being at $-x_m$? (b) At point 4, is the velocity of the particle positive, negative, or zero? (c) At point 5, is the particle at $-x_m$, at $+x_m$, at 0, between $-x_m$ and 0, or between 0 and $+x_m$?

3. A particle oscillates according to

$$x = x_m \cos(\omega t + \phi).$$

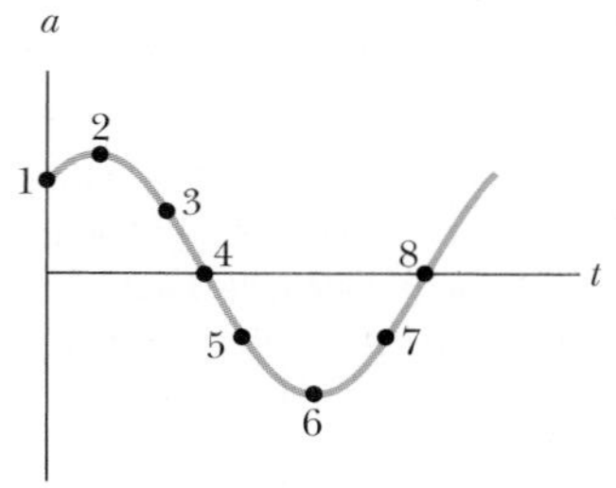

FIGURE 16-22   Question 2.

At $t = 0$, is the particle at $-x_m$, at $+x_m$, at 0, between $-x_m$ and 0, or between 0 and $+x_m$ if ϕ is (a) $\pi/2$, (b) $-\pi/3$, (c) $-3\pi/4$, and (d) $3\pi/4$?

4. Which of the following describe ϕ for the SHM of Fig. 16-23a: (a) $-\pi < \phi < -\pi/2$, (b) $\pi < \phi < 3\pi/2$, (c) $-3\pi/2 < \phi < -\pi$?

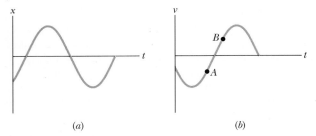

(a)   (b)

FIGURE 16-23 Questions 4 and 5.

5. The velocity $v(t)$ of a particle undergoing SHM is graphed in Fig. 16-23b. Is the particle momentarily stationary, headed toward $-x_m$, or headed toward $+x_m$ at (a) point A on the graph and (b) point B? Is the particle at $-x_m$, at $+x_m$, at 0, between $-x_m$ and 0, or between 0 and $+x_m$ when its velocity is represented by (c) point A and (d) point B? Is the speed of the particle increasing or decreasing at (e) point A and (f) point B?

6. What is the phase difference of the two linear oscillators, with identical masses and spring constants, in (a) Fig. 16-24a and (b) Fig. 16-24b? (c) What is the phase difference between the red oscillator in Fig. 16-24a and the green oscillator in Fig. 16-24b?

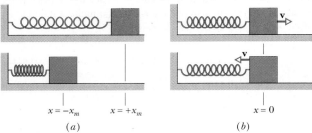

$x = -x_m$ $x = +x_m$ $x = 0$

(a)   (b)

FIGURE 16-24 Question 6.

7. A spring and a block are arranged to oscillate (1) on a frictionless horizontal surface as in Fig. 16-5, (2) on a frictionless slope of 45° (with the block at the lower end of the spring), and (3) vertically while hanging from a ceiling. Rank the arrangements according to (a) the rest length of the spring and (b) the frequency of oscillation, greatest first.

8. Three springs each hang from a ceiling with a stationary block attached to the lower end. The blocks (with $m_1 > m_2 > m_3$) stretch the springs by equal distances. Each block–spring system is then set into vertical SHM. Rank the masses according to the period of oscillation, greatest first.

9. Figure 16-25 shows three arrangements of a block attached to identical springs that are in the relaxed state when the block is centered. Rank the arrangements according to the frequency of oscillation of the block, greatest first.

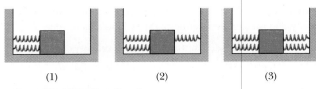

(1)   (2)   (3)

FIGURE 16-25 Question 9.

10. A mass m is suspended from a spring with spring constant k. The spring is then cut in half and the same mass is suspended from one of the halves. Which, the original spring or the half-spring, gives a greater frequency of oscillation when the mass is put into vertical SHM?

11. When mass m_1 is hung from spring A and a smaller mass m_2 is hung from spring B, the springs are stretched by the same distance. If the two spring–mass systems are then put into vertical SHM with the same amplitude, which system will have more energy?

12. Do the following increase, decrease, or remain the same if the amplitude of a simple harmonic oscillator is doubled: (a) the period, (b) the spring constant, (c) the total energy, (d) the maximum speed, (e) the maximum acceleration?

13. Figure 16-26 shows three physical pendulums consisting of identical uniform spheres of the same mass that are rigidly connected by identical rods of negligible mass. Each pendulum is vertical and can pivot about suspension point O. Rank the pendulums according to period of oscillation, greatest first.

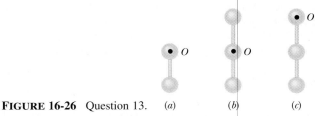

FIGURE 16-26 Question 13.   (a)   (b)   (c)

14. A pendulum suspended from the ceiling of an elevator cab has period T when the cab is stationary. Is the period larger, smaller, or the same when the cab moves (a) upward with constant speed, (b) downward with constant speed, (c) downward with constant upward acceleration, (d) upward with constant upward acceleration, (e) upward with constant downward acceleration $a = g$, and (f) downward with constant downward acceleration $a = g$?

15. A pendulum mounted in a cart has period T when the cart is stationary and on a horizontal plane. Is the period larger, smaller, or the same if the cart is on a plane inclined at angle θ (Fig. 16-27)?

FIGURE 16-27
Question 15.

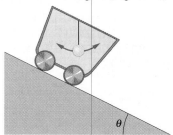

and (a) stationary, (b) moving down the plane with constant speed, (c) moving up the plane with constant speed, (d) moving up the plane with constant acceleration directed up the plane, (e) moving down the plane with constant acceleration directed up the plane, (f) moving down the plane with acceleration $a = g \sin \theta$ directed down the plane, and (g) moving up the plane with acceleration $a = g \sin \theta$ directed down the plane?

16. You are to build the oscillation transfer device shown in Fig. 16-28. It consists of two spring–block systems hanging from a flexible rod. When the spring of system 1 is stretched and then released, the resulting SHM of system 1 at frequency f_1 oscillates the rod. The rod then provides a driving force on system 2, at the same frequency f_1. For components you have to choose from four springs with spring constants k of 1600, 1500, 1400, and 1200

N/m, and four blocks with masses m of 800, 500, 400, and 200 kg. Without written calculation, determine which spring should go with which block in each of the two systems to maximize the amplitude of oscillations in system 2.

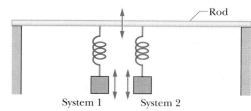

FIGURE 16-28   Question 16.

EXERCISES & PROBLEMS

SECTION 16-3   The Force Law for Simple Harmonic Motion

1E. An object undergoing simple harmonic motion takes 0.25 s to travel from one point of zero velocity to the next such point. The distance between those points is 36 cm. Calculate (a) the period, (b) the frequency, and (c) the amplitude of the motion.

2E. An oscillating mass–spring system takes 0.75 s to begin repeating its motion. Find (a) the period, (b) the frequency in hertz, and (c) the angular frequency in radians per second.

3E. A 4.00 kg block hangs from a spring, extending it 16.0 cm from its unstretched position. (a) What is the spring constant? (b) The block is removed and a 0.500 kg body is hung from the same spring. If the spring is then stretched and released, what is its period of oscillation?

4E. An oscillator consists of a block of mass 0.500 kg connected to a spring. When set into oscillation with amplitude 35.0 cm, it is observed to repeat its motion every 0.500 s. Find (a) the period, (b) the frequency, (c) the angular frequency, (d) the spring constant, (e) the maximum speed, and (f) the maximum force exerted on the block.

5E. The vibration frequencies of atoms in solids at normal temperatures are of the order of 10^{13} Hz. Imagine the atoms to be connected to one another by springs. Suppose that a single silver atom in a solid vibrates with this frequency and that all the other atoms are at rest. Compute the effective spring constant. One mole of silver (6.02×10^{23} atoms) has a mass of 108 g.

6E. What is the maximum acceleration of a platform that vibrates with an amplitude of 2.20 cm at a frequency of 6.60 Hz?

7E. A loudspeaker produces a musical sound by means of the oscillation of a diaphragm. If the amplitude of oscillation is limited to 1.0×10^{-3} mm, what frequencies will result in the acceleration of the diaphragm exceeding g?

8E. The scale of a spring balance which reads from 0 to 32.0 lb is 4.00 in. long. A package suspended from the balance is found to oscillate vertically with a frequency of 2.00 Hz. (a) What is the spring constant? (b) How much does the package weigh?

9E. A 20 N weight is hung from the bottom of a vertical spring, causing the spring to stretch 20 cm. (a) What is the spring constant? (b) This spring is now placed horizontally on a frictionless table. One end of it is held fixed, and the other end is attached to a 5.0 N weight. The weight is then moved (stretching the spring) and released from rest. What is the period of oscillation?

10E. A 50.0 g mass is attached to the bottom of a vertical spring and set vibrating. If the maximum speed of the mass is 15.0 cm/s and the period is 0.500 s, find (a) the spring constant of the spring, (b) the amplitude of the motion, and (c) the frequency of oscillation.

11E. A particle with a mass of 1.00×10^{-20} kg is vibrating with simple harmonic motion with a period of 1.00×10^{-5} s and a maximum speed of 1.00×10^{3} m/s. Calculate (a) the angular frequency and (b) the maximum displacement of the particle.

12E. A small body of mass 0.12 kg is undergoing simple harmonic motion of amplitude 8.5 cm and period 0.20 s. (a) What is the maximum value of the force acting on it? (b) If the oscillations are produced by a spring, what is the spring constant?

13E. In an electric shaver, the blade moves back and forth over a distance of 2.0 mm. The motion is simple harmonic, with frequency 120 Hz. Find (a) the amplitude, (b) the maximum blade speed, and (c) the maximum blade acceleration.

14E. A loudspeaker diaphragm is vibrating in simple harmonic motion with a frequency of 440 Hz and a maximum displacement of 0.75 mm. What are (a) the angular frequency, (b) the maximum speed, and (c) the maximum acceleration?

15E. An automobile can be considered to be mounted on four identical springs as far as vertical oscillations are concerned. The springs of a certain car are adjusted so that the vibrations have a frequency of 3.00 Hz. (a) What is the spring constant of each spring if the mass of the car is 1450 kg and the weight is evenly distributed over the springs? (b) What will be the vibration frequency if five passengers, averaging 73.0 kg each, ride in the car? (Again, consider an even distribution of weight.)

16E. A body oscillates with simple harmonic motion according to the equation

$$x = (6.0 \text{ m}) \cos[(3\pi \text{ rad/s})t + \pi/3 \text{ rad}].$$

At $t = 2.0$ s, what are (a) the displacement, (b) the velocity, (c) the acceleration, and (d) the phase of the motion? Also, what are (e) the frequency and (f) the period of the motion?

17E. A particle executes linear SHM with frequency 0.25 Hz about the point $x = 0$. At $t = 0$, it has displacement $x = 0.37$ cm and zero velocity. For the motion, determine (a) the period, (b) the angular frequency, (c) the amplitude, (d) the displacement at time t, (e) the velocity at time t, (f) the maximum speed, (g) the maximum acceleration, (h) the displacement at $t = 3.0$ s, and (i) the speed at $t = 3.0$ s.

18E. The piston in the cylinder head of a locomotive has a stroke (twice the amplitude) of 0.76 m. If the piston moves with simple harmonic motion with an angular frequency of 180 rev/min, what is its maximum speed?

19P. Figure 16-29 shows an astronaut on a body-mass measuring device (BMMD). Designed for use on orbiting space vehicles, its purpose is to allow astronauts to measure their mass in the "weightless" conditions in Earth orbit. The BMMD is a spring-mounted chair; an astronaut measures his or her period of oscillation in the chair; the mass follows from the formula for the period of an oscillating block–spring system. (a) If M is the mass of the astronaut and m the effective mass of that part of the BMMD that also oscillates, show that

$$M = (k/4\pi^2)T^2 - m,$$

where T is the period of oscillation and k is the spring constant. (b) The spring constant was $k = 605.6$ N/m for the BMMD on Skylab Mission Two; the period of oscillation of the empty chair was 0.90149 s. Calculate the effective mass of the chair. (c) With an astronaut in the chair, the period of oscillation became 2.08832 s. Calculate the mass of the astronaut.

20P. A 2.00 kg block hangs from a spring. A 300 g body hung below the block stretches the spring 2.00 cm farther. (a) What is the spring constant? (b) If the 300 g body is removed and the block is set into oscillation, find the period of the motion.

21P. The end point of a spring vibrates with a period of 2.0 s when a mass m is attached to it. When this mass is increased by 2.0 kg, the period is found to be 3.0 s. Find the value of m.

22P. The end of one of the prongs of a tuning fork that executes simple harmonic motion of frequency 1000 Hz has an amplitude of 0.40 mm. Find (a) the maximum acceleration and (b) the maximum speed of the end of the prong. Find (c) the acceleration and (d) the speed of the end of the prong when the end has a displacement of 0.20 mm.

23P. A 0.10 kg block oscillates back and forth along a straight line on a frictionless horizontal surface. Its displacement from the origin is given by

$$x = (10 \text{ cm}) \cos[(10 \text{ rad/s})t + \pi/2 \text{ rad}].$$

(a) What is the oscillation frequency? (b) What is the maximum speed acquired by the block? At what value of x does this occur? (c) What is the maximum acceleration of the block? At what value of x does this occur? (d) What force, applied to the block, results in the given oscillation?

24P. At a certain harbor, the tides cause the ocean surface to rise and fall a distance d in simple harmonic motion, with a period of 12.5 h. How long does it take for the water to fall a distance $d/4$ from its maximum height?

25P. Two blocks ($m = 1.0$ kg and $M = 10$ kg) and a spring ($k = 200$ N/m) are arranged on a horizontal, frictionless surface as shown in Fig. 16-30. The coefficient of static friction between the two blocks is 0.40. What is the maximum possible amplitude of simple harmonic motion of the spring–blocks system if no slippage is to occur between the blocks?

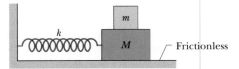

FIGURE 16-30 Problem 25.

26P. A block is on a horizontal surface (a shake table) that is moving horizontally with simple harmonic motion of frequency 2.0 Hz. The coefficient of static friction between block and surface is 0.50. How great can the amplitude of the SHM be if the block is not to slip along the surface?

27P. A block is on a piston that is moving vertically with simple harmonic motion. (a) If the SHM has period 1.0 s, at what amplitude of motion will the block and piston separate? (b) If the piston has an amplitude of 5.0 cm, what is the maximum frequency for which the block and piston will be in contact continuously?

28P. An oscillator consists of a block attached to a spring ($k = 400$ N/m). At some time t, the position (measured from the system's equilibrium location), velocity, and acceleration of the

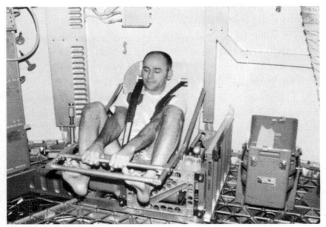

FIGURE 16-29 Problem 19.

block are $x = 0.100$ m, $v = -13.6$ m/s, and $a = -123$ m/s². Calculate (a) the frequency of oscillation, (b) the mass of the block, and (c) the amplitude of the motion.

29P. A simple harmonic oscillator consists of a block of mass 2.00 kg attached to a spring of spring constant 100 N/m. When $t = 1.00$ s, the position and velocity of the block are $x = 0.129$ m and $v = 3.415$ m/s. (a) What is the amplitude of the oscillations? What were the (b) position and (c) velocity of the mass at $t = 0$ s?

30P. A massless spring hangs from the ceiling with a small object attached to its lower end. The object is initially held at rest in a position y_i such that the spring is at its rest length. The object is then released from y_i and oscillates up and down, with its lowest position being 10 cm below y_i. (a) What is the frequency of the oscillation? (b) What is the speed of the object when it is 8.0 cm below the initial position? (c) An object of mass 300 g is attached to the first object, after which the system oscillates with half the original frequency. What is the mass of the first object? (d) Relative to y_i, where is the new equilibrium (rest) position with both objects attached to the spring?

31P. Two particles oscillate in simple harmonic motion along a common straight line segment of length A. Each particle has a period of 1.5 s, but they differ in phase by $\pi/6$ rad. (a) How far apart are they (in terms of A) 0.50 s after the lagging particle leaves one end of the path? (b) Are they then moving in the same direction, toward each other, or away from each other?

32P. Two particles execute simple harmonic motion of the same amplitude and frequency along the same straight line. They pass each other moving in opposite directions each time their displacement is half their amplitude. What is the phase difference between them?

33P. Two identical springs are attached to a block of mass m and to fixed supports as shown in Fig. 16-31. Show that the frequency of oscillation on the frictionless surface is

$$f = \frac{1}{2\pi}\sqrt{\frac{2k}{m}}.$$

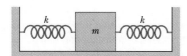

FIGURE 16-31 Problems 33 and 34.

34P. Suppose that the two springs in Fig. 16-31 have different spring constants k_1 and k_2. Show that the frequency f of oscillation of the block is then given by

$$f = \sqrt{f_1^2 + f_2^2},$$

where f_1 and f_2 are the frequencies at which the block would oscillate if connected only to spring 1 or only to spring 2.

35P. Two springs are joined and connected to a mass m as shown in Fig. 16-32. The surface is frictionless. If the springs both have force constant k, show that the frequency of oscillation of m is

$$f = \frac{1}{2\pi}\sqrt{\frac{k}{2m}}.$$

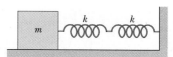

FIGURE 16-32 Problem 35.

36P. A block weighing 14.0 N, which slides without friction on a 40.0° incline, is connected to the top of the incline by a massless spring of unstretched length 0.450 m and spring constant 120 N/m, as shown in Fig. 16-33. (a) How far from the top of the incline does the block stop? (b) If the block is pulled slightly down the incline and released, what is the period of the ensuing oscillations?

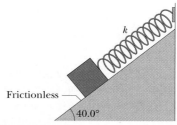

Frictionless

40.0°

FIGURE 16-33 Problem 36.

37P. A uniform spring whose unstretched length is L has a spring constant k. The spring is cut into two pieces of unstretched lengths L_1 and L_2, with $L_1 = nL_2$. (a) What are the corresponding spring constants k_1 and k_2 in terms of n and k? (b) If a block is attached to the original spring, as in Fig. 16-5, it oscillates with frequency f. If the spring is replaced with the piece L_1 or L_2, the corresponding frequency is f_1 or f_2. Find f_1 and f_2 in terms of f.

38P. Three 10,000 kg ore cars are held at rest on a 30° incline on a mine railway using a cable that is parallel to the incline (Fig. 16-34). The cable is observed to stretch 15 cm just before the coupling between the lower cars breaks, detaching the lowest car. Assuming that the cable obeys Hooke's law, find (a) the frequency and (b) the amplitude of the resulting oscillations of the remaining two cars.

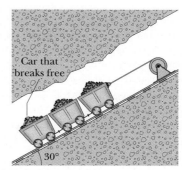

Car that breaks free

30°

FIGURE 16-34 Problem 38.

39P. To alleviate the traffic congestion between two cities such as Boston and Washington, DC, engineers have proposed building a rail tunnel along a chord line connecting the cities (Fig. 16-35). A train, unpropelled by any engine and starting from rest, would fall through the first half of the tunnel and then move up

FIGURE 16-35 Problem 39.

the second half. Assuming Earth to be a uniform sphere and ignoring air drag and friction, (a) show that the travel between cities is half a cycle of simple harmonic motion and (b) find the travel time between cities.

SECTION 16-4 Energy in Simple Harmonic Motion

40E. Find the mechanical energy of a block–spring system having a spring constant of 1.3 N/cm and an amplitude of 2.4 cm.

41E. An oscillating block–spring system has a mechanical energy of 1.00 J, an amplitude of 10.0 cm, and a maximum speed of 1.20 m/s. Find (a) the spring constant, (b) the mass of the block and (c) the frequency of oscillation.

42E. A 5.00 kg object on a horizontal frictionless surface is attached to a spring with spring constant 1000 N/m. The object is displaced from equilibrium 50.0 cm horizontally and given an initial velocity of 10.0 m/s back toward the equilibrium position. (a) What is the frequency of the motion? What are (b) the initial potential energy of the block–spring system, (c) the initial kinetic energy, and (d) the amplitude of the oscillation?

43E. A vertical spring stretches 9.6 cm when a 1.3 kg block is hung from its end. (a) Calculate the spring constant. This block is then displaced an additional 5.0 cm downward and released from rest. Find (b) the period, (c) the frequency, (d) the amplitude, and (e) the maximum speed of the resulting SHM.

44E. A (hypothetical) large slingshot is stretched 1.50 m to launch a 130 g projectile with speed sufficient to escape from Earth (11.2 km/s). Assume the elastic slingshot obeys Hooke's law. (a) What is the spring constant of the device, if all the elastic potential energy is converted to kinetic energy? (b) Assume that an average person can exert a force of 220 N. How many people would be required to stretch the slingshot?

45E. When the displacement in SHM is one-half the amplitude x_m, what fraction of the total energy is (a) kinetic energy and (b) potential energy? (c) At what displacement, in terms of the amplitude, is the energy of the system half kinetic energy and half potential energy?

46E. A block of mass M, at rest on a horizontal frictionless table, is attached to a rigid support by a spring of constant k. A bullet of mass m and velocity v strikes the block as shown in Fig. 16-36. The bullet remains embedded in the block. Determine (a) the velocity of the block immediately after the collision and (b) the amplitude of the resulting simple harmonic motion.

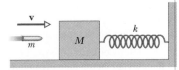

FIGURE 16-36 Exercise 46.

47P. A 3.0 kg particle is in simple harmonic motion in one dimension and moves according to the equation

$$x = (5.0 \text{ m})\cos[(\pi/3 \text{ rad/s})t - \pi/4 \text{ rad}].$$

(a) At what value of x is the potential energy of the particle equal to half the total energy? (b) How long does it take the particle to move to this position x from the equilibrium position?

48P. A 10 g particle is undergoing simple harmonic motion with an amplitude of 2.0×10^{-3} m. The maximum acceleration experienced by the particle is 8.0×10^3 m/s²; the phase constant is $-\pi/3$ rad. (a) Write an equation for the force on the particle as a function of time. (b) What is the period of the motion? (c) What is the maximum speed of the particle? (d) What is the total mechanical energy of this simple harmonic oscillator?

49P. A massless spring with spring constant 19 N/m hangs vertically. A body of mass 0.20 kg is attached to its free end and then released. Assume that the spring was unstretched before the body was released. Find (a) how far below the initial position the body descends, and (b) the frequency and (c) the amplitude of the resulting motion, assumed to be simple harmonic.

50P. A 4.0 kg block is suspended from a spring with a spring constant of 500 N/m. A 50 g bullet is fired into the block from directly below with a speed of 150 m/s and is embedded in the block. (a) Find the amplitude of the resulting simple harmonic motion. (b) What fraction of the original kinetic energy of the bullet appears as mechanical energy in the harmonic oscillator?

51P*. A solid cylinder is attached to a horizontal massless spring so that it can roll without slipping along a horizontal surface (Fig. 16-37). The spring constant k is 3.0 N/m. If the system is released from rest at a position in which the spring is stretched by 0.25 m, find (a) the translational kinetic energy and (b) the rotational kinetic energy of the cylinder as it passes through the equilibrium position. (c) Show that under these conditions the center of mass of the cylinder executes simple harmonic motion with period

$$T = 2\pi\sqrt{\frac{3M}{2k}},$$

where M is the mass of the cylinder. (*Hint:* Find the time derivative of the total mechanical energy.)

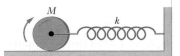

FIGURE 16-37 Problem 51.

SECTION 16-5 An Angular Simple Harmonic Oscillator

52E. A flat uniform circular disk has a mass of 3.00 kg and a radius of 70.0 cm. It is suspended in a horizontal plane by a vertical wire attached to its center. If the disk is rotated 2.50 rad about the wire, a torque of 0.0600 N · m is required to maintain the disk in this position. Calculate (a) the rotational inertia of the disk about the wire, (b) the torsion constant, and (c) the angular frequency of this torsion pendulum when it is set oscillating.

53P. A 95 kg solid sphere with a 15 cm radius is suspended by a vertical wire attached to the ceiling of a room. A torque of 0.20 N · m is required to twist the sphere through an angle of 0.85 rad. What is the period of oscillation when the sphere is released from this position?

54P. An engineer has an odd-shaped object of mass 10 kg and needs to find its rotational inertia about an axis through its center of mass. The object is supported with a wire along the desired axis. The wire has a torsion constant $\kappa = 0.50$ N · m. If this torsion pendulum oscillates through 20 complete cycles in 50 s, what is the rotational inertia of the odd-shaped object?

55P. The balance wheel of a watch oscillates with an angular amplitude of π rad and a period of 0.500 s. Find (a) the maximum angular speed of the wheel, (b) the angular speed of the wheel when its displacement is $\pi/2$ rad, and (c) the angular acceleration of the wheel when its displacement is $\pi/4$ rad.

SECTION 16-6 Pendulums

56E. What is the length of a simple pendulum whose period is 1.00 s at a point where $g = 32.2$ ft/s^2?

57E. A 2500 kg demolition ball swings from the end of a crane, as shown in Fig. 16-38. The length of the swinging segment of cable is 17 m. (a) Find the period of swing, assuming that the system can be treated as a simple pendulum. (b) Does the period depend on the ball's mass?

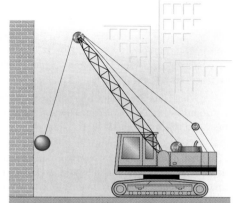

FIGURE 16-38
Exercise 57.

58E. What is the length of a simple pendulum that marks seconds by completing a full swing from left to right and then back again every 2.0 s?

59E. If a simple pendulum with length 1.50 m makes 72.0 oscillations in 180 s, what is the acceleration of gravity at its location?

60E. Two oscillating systems that you have studied are the block–spring and the simple pendulum. You can prove an interesting relation between them. Suppose that you hang a weight on the end of a spring, and when the weight is at rest, the spring is stretched a distance h as in Fig. 16-39. Show that the frequency of this block–spring system is the same as that of a simple pendulum whose length is h.

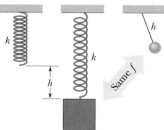

FIGURE 16-39 Exercise 60.

61E. A performer, seated on a trapeze, is swinging back and forth with a period of 8.85 s. If she stands up, thus raising the center of mass of the system *trapeze + performer* by 35.0 cm, what will be the new period of the trapeze? Treat *trapeze + performer* as a simple pendulum.

62E. A simple pendulum with length L is swinging freely with small angular amplitude. As the pendulum passes its central (or equilibrium) position, its cord is suddenly and rigidly clamped at its midpoint. In terms of the original period T of the pendulum, what will the new period be?

63E. A physical pendulum consists of a meter stick that is pivoted at a small hole drilled through the stick a distance x from the 50 cm mark. The period of oscillation is observed to be 2.5 s. Find the distance x.

64E. A pendulum is formed by pivoting a long thin rod of length L and mass m about a point on the rod that is a distance d above the center of the rod. (a) Find the period of this pendulum in terms of d, L, m, and g, assuming that it swings with small amplitude. What happens to the period if (b) d is decreased, (c) L is increased, or (d) m is increased?

65E. A physical pendulum consists of a uniform solid disk (of mass M and radius R) supported in a vertical plane by a pivot located a distance d from the center of the disk (Fig. 16-40). The disk is displaced by a small angle and released. Find an expression for the period of the resulting simple harmonic motion.

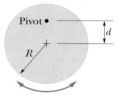

FIGURE 16-40 Exercise 65.

66E. A uniform circular disk whose radius R is 12.5 cm is suspended, as a physical pendulum, from a point on its rim. (a) What is its period of oscillation? (b) At what radial distance $r < R$ is there a point of suspension that gives the same period?

67E. A pendulum consists of a uniform disk with radius 10.0 cm and mass 500 g attached to a uniform rod with length 500 mm and mass 270 g (Fig. 16-41). (a) Calculate the rotational inertia of the pendulum about the pivot. (b) What is the distance between the pivot and the center of mass of the pendulum? (c) Calculate the period of oscillation.

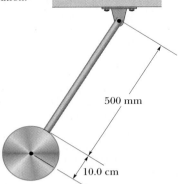

FIGURE 16-41
Exercise 67.

68E. (a) If the physical pendulum of Sample Problem 16-6 is inverted and suspended at point P, what is its period of oscillation? (b) Is the period now greater than, less than, or equal to its previous value?

69E. In Sample Problem 16-6, we saw that a physical pendulum has a center of oscillation at distance $2L/3$ from its point of suspension. Show that the distance between the point of suspension and the center of oscillation for a physical pendulum of any form is I/mh, where I and h have the meanings assigned to them in Eq. 16-32, and m is the mass of the pendulum.

70E. A meter stick swinging from one end oscillates with a frequency f_0. What would be the frequency, in terms of f_0, if the bottom half of the stick were cut off?

71P. A stick with length L oscillates as a physical pendulum, pivoted about point O in Fig. 16-42. (a) Derive an expression for the period of the pendulum in terms of L and x, the distance from the point of support to the center of mass of the pendulum. (b) For what value of x/L is the period a minimum? (c) Show that if $L = 1.00$ m and $g = 9.80$ m/s², this minimum is 1.53 s.

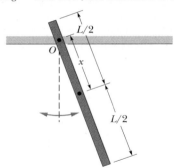

FIGURE 16-42 Problem 71.

72P. The center of oscillation of a physical pendulum has this interesting property: if an impulsive force (assumed horizontal and in the plane of oscillation) acts at the center of oscillation, no reaction is felt at the point of support. Baseball players (and players of many other sports) know that unless the ball hits the bat at this point (called the "sweet spot" by athletes), the reaction due to the impact will sting their hands. To prove this property, let the stick in Fig. 16-12a simulate a baseball bat. Suppose that a horizontal force $\mathbf{F}$ (due to impact with the ball) acts toward the right at P, the center of oscillation. The batter is assumed to hold the bat at O, the point of support of the stick. (a) What acceleration does point O undergo as a result of $\mathbf{F}$? (b) What angular acceleration is produced by $\mathbf{F}$ about the center of mass of the stick? (c) As a result of the angular acceleration in (b), what linear acceleration does point O undergo? (d) Considering the magnitudes and directions of the accelerations in (a) and (c), convince yourself that P is indeed the "sweet spot."

73P. The fact that g varies from place to place over Earth's surface drew attention when Jean Richer in 1672 took a pendulum clock from Paris to Cayenne, French Guiana, and found that it lost 2.5 min/day. If $g = 9.81$ m/s² in Paris, what is g in Cayenne?

74P. A scientist is making a precise measurement of g at a certain point in the Indian Ocean (on the equator) by timing the swings of a pendulum of accurately known construction. To provide a stable base, the measurements are conducted in a submerged submarine. It is observed that a slightly different result for g is obtained when the submarine is moving eastward than when it is moving westward, the speed in each case being 16 km/h. Account for this difference and calculate the fractional error $\Delta g/g$ in g for either travel direction.

75P. A long uniform rod of length L and mass m is free to rotate in a horizontal plane about a vertical axis through its center. A spring with force constant k is connected horizontally between one end of the rod and a fixed wall, as shown in the overhead view of Fig. 16-43. When the rod is in equilibrium, it is parallel to the wall. What is the period of the small oscillations that result when the rod is rotated slightly and released?

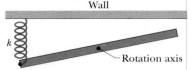

FIGURE 16-43 Problem 75.

76P. What is the frequency of a simple pendulum 2.0 m long (a) in a room, (b) in an elevator accelerating upward at a rate of 2.0 m/s², and (c) in free fall?

77P. A simple pendulum of length L and mass m is suspended in a car that is traveling with constant speed v around a circle of radius R. If the pendulum undergoes small oscillations in a radial direction about its equilibrium position, what will its frequency of oscillation be?

78P. For a simple pendulum, find the angular amplitude θ_m at which the restoring torque required for simple harmonic motion deviates from the actual restoring torque by 1.0%. (See "Trigonometric Expansions" in Appendix E.)

79P. The bob on a simple pendulum of length R moves in an arc of a circle. (a) By considering that the acceleration of the bob as it moves through its equilibrium position is that for uniform circular motion (mv^2/R), show that the tension in the string at that position is $mg(1 + \theta_m^2)$ if the angular amplitude θ_m is small. (See "Trigonometric Expansions" in Appendix E.) (b) Is the tension at other positions of the bob larger, smaller, or the same?

80P. A wheel is free to rotate about its fixed axle. A spring is attached to one of its spokes a distance r from the axle, as shown in Fig. 16-44. (a) Assuming that the wheel is a hoop of mass m and radius R, obtain the angular frequency of small oscillations of this system in terms of m, R, r, and the spring constant k. How does the result change if (b) $r = R$ and (c) $r = 0$?

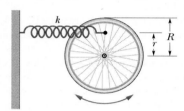

FIGURE 16-44 Problem 80.

81P. A 2.5 kg disk, 42 cm in diameter, is supported by a massless rod, 76 cm long, which is pivoted at its end, as in Fig. 16-45. (a) The massless torsion spring is initially not connected. What then is the period of oscillation? (b) The torsion spring is now connected so that, in equilibrium, the rod hangs vertically. What should be the torsional constant of the spring so that the new period of oscillation is 0.50 s shorter than before?

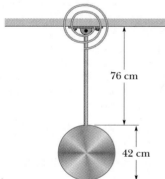

76 cm

42 cm

FIGURE 16-45 Problem 81.

82P. A physical pendulum has two possible pivot points A and B; point A has a fixed position, and B is adjustable along the length of the pendulum, as shown in Fig. 16-46. The period of the pendulum when suspended from A is found to be T. The pendulum is then reversed and suspended from B, which is moved until the pendulum again has period T. Show that the free-fall acceleration g is given by

$$g = \frac{4\pi^2 L}{T^2},$$

in which L is the distance between A and B for equal periods T. (Note that g can be measured in this way without knowing the rotational inertia of the pendulum or any of its dimensions except L.)

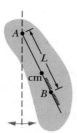

FIGURE 16-46 Problem 82.

83P*. A uniform rod with length L swings from a pivot as a physical pendulum. How far, in terms of L, should the pivot be from the center of mass to minimize the period?

SECTION 16-8 Damped Simple Harmonic Motion

84E. The amplitude of a lightly damped oscillator decreases by 3.0% during each cycle. What fraction of the energy of the oscillator is lost in each full oscillation?

85E. In Sample Problem 16-9, what is the ratio of the amplitude of the damped oscillations to the initial amplitude when 20 full oscillations have elapsed?

86E. For the system shown in Fig. 16-17, the block has a mass of 1.50 kg and the spring constant is 8.00 N/m. The damping force is given by $-b(dx/dt)$, where $b = 230$ g/s. Suppose that the block is initially pulled down a distance 12.0 cm and released. (a) Calculate the time required for the amplitude to fall to one-third of its initial value. (b) How many oscillations are made by the block in this time?

87P. A damped harmonic oscillator consists of a block ($m = 2.00$ kg), a spring ($k = 10.0$ N/m), and a damping force $F = -bv$. Initially, it oscillates with an amplitude of 25.0 cm; because of the damping, the amplitude falls to three-fourths of this initial value at the completion of four oscillations. (a) What is the value of b? (b) How much energy has been "lost" during these four oscillations?

88P. (a) In Eq. 16-39, find the ratio of the maximum damping force ($-b \, dx/dt$) to the maximum spring force ($-kx$) during the first oscillation for the data of Sample Problem 16-9. (b) Does this ratio change appreciably during later oscillations?

89P. Assume that you are examining the characteristics of the suspension system of a 2000 kg automobile. The suspension "sags" 10 cm when the weight of the entire automobile is placed on it. In addition, the amplitude of oscillation decreases by 50% during one complete oscillation. Estimate the values of k and b for the spring and shock absorber system of one wheel, assuming each wheel supports 500 kg.

SECTION 16-9 Forced Oscillations and Resonance

90E. For Eq. 16-43, suppose the amplitude x_m is given by

$$x_m = \frac{F_m}{[m^2(\omega_d^2 - \omega^2)^2 + b^2\omega_d^2]^{1/2}},$$

where F_m is the (constant) amplitude of the external oscillating force exerted on the spring by the rigid support in Fig. 16-17. At resonance, what are (a) the amplitude and (b) the velocity amplitude of the oscillating object?

91P. A 2200 lb car carrying four 180 lb people travels over a rough ''washboard'' dirt road with corrugations 13 ft apart which causes the car to bounce on its spring suspension.. The car bounces with maximum amplitude when its speed is 10 mi/h. The car now stops, and the four people get out. By how much does the car body rise on its suspension owing to this decrease in weight?

Electronic Computation

92. You see two carts connected by a spring and oscillating on a horizontal air track. You happen to know that the spring constant of the spring is $k = 50.0$ N/m. You use Newton's second law to determine that the period of oscillation is the same for each cart and is related to the masses of the carts by

$$T = 2\pi \sqrt{\frac{m_1 m_2}{k(m_1 + m_2)}}.$$

Using a position detector, you determine that the positions of the carts as functions of time can be expressed as $x_1(t) = 2.70[1 - \cos(18.0t)]$ and $x_2(t) = 10.70 + 1.29 \cos(18.0t)$, where the coordinates are in centimeters and the time is in seconds. (a) Use conservation of momentum to find the masses of the carts. (b) Generate a table of the values of x_1 and x_2 for every 0.01 s from $t = 0$ to $t = 35$ s. For each of these times, also calculate the position of the center of mass, the total momentum of the two-cart system, and the force of the spring on each cart. Verify that the center of mass does not move, that the total momentum is conserved, and that the forces on the carts have the same magnitude but opposite directions. (c) Use the table to find the equilibrium length of the spring.

93. A 2.0 kg block is attachd to the end of a spring with a spring constant of 350 N/m and forced to oscillate by an applied force $F = (15$ N$) \sin(\omega t)$, where $\omega = 35$ rad/s. The damping constant is $b = 15$ kg/s. At $t = 0$, the block is at rest with the spring at its rest length. (a) Use numerical integration to plot the displacement of the block for the first 1.0 s of its motion. Use the motion near the end of the 1.0 s interval to estimate the amplitude, period, and angular frequency. Repeat the calculation for (b) $\omega = \sqrt{k/m}$ and (c) $\omega = 20$ rad/s.

When a beetle moves along the sand within a few tens of centimeters of this sand scorpion, the scorpion immediately turns toward the beetle and dashes to it (for lunch). The scorpion can do this without seeing (it is nocturnal) or hearing the beetle. How can it so precisely locate its prey?

17-1 WAVES AND PARTICLES

Two ways to get in touch with a friend in a distant city are to write a letter and to use the telephone.

The first choice (the letter) involves the concept of "particle": a material object moves from one point to another, carrying with it information and energy. Most of the preceding chapters deal with particles or with systems of particles.

The second choice (the telephone) involves the concept of "wave," the subject of this chapter and the next. In a wave, information and energy move from one point to another but no material object makes that journey. In your telephone call, a sound wave carries your message from your vocal cords to the telephone. There, an electromagnetic wave takes over, passing along a copper wire or an optical fiber or through the atmosphere, possibly by way of a communications satellite. At the receiving end there is another sound wave, from a telephone to your friend's ear. Although the message is passed, nothing that you have touched reaches your friend. Leonardo da Vinci understood about waves when he wrote of water waves: "it often happens that the wave flees the place of its creation, while the water does not; like the waves made in a field of grain by the wind, where we see the waves running across the field while the grain remains in place."

Particle and *wave* are the two great concepts in classical physics, in the sense that we seem able to associate almost every branch of the subject with one or the other. The two concepts are quite different. The word *particle* suggests a tiny concentration of matter capable of transmitting energy. The word *wave* suggests just the opposite, namely, a broad distribution of energy, filling the space through which it passes. The job at hand is to put aside particles for a while and to learn something about waves.

17-2 TYPES OF WAVES

Waves are of three main types:

1. *Mechanical waves.* These waves are most familiar because we encounter them almost constantly; common examples include water waves, sound waves, and seismic waves. All these waves have certain central features: they are governed by Newton's laws, and they can exist only within a material medium, such as water, air, and rock.

2. *Electromagnetic waves.* These waves are less familiar, but you use them constantly; common examples include visible and ultraviolet light, radio and television waves, microwaves, x rays, and radar waves. These waves require no material medium to exist. Light waves from stars, for example, travel through the vacuum of space to reach us.

All electromagnetic waves travel through a vacuum at the same speed c, given by

$$c = 299,792,458 \text{ m/s} \quad \text{(speed of light).} \quad (17\text{-}1)$$

3. *Matter waves.* Although these waves are commonly used in modern technology, their type is probably very unfamiliar to you. Electrons, protons, and other fundamental particles, and even atoms and molecules, travel as waves. Because we commonly think of these things as constituting matter, these waves are called matter waves.

Much of what we discuss in this chapter applies to waves of all kinds. However, for specific examples we shall refer to mechanical waves.

17-3 TRANSVERSE AND LONGITUDINAL WAVES

A wave sent along a stretched, taut string is the simplest mechanical wave. If you give one end of a stretched string a single up-and-down jerk, a wave in the form of a single *pulse* travels along the string, as in Fig. 17-1a. This pulse and its motion can occur because the string is under ten-

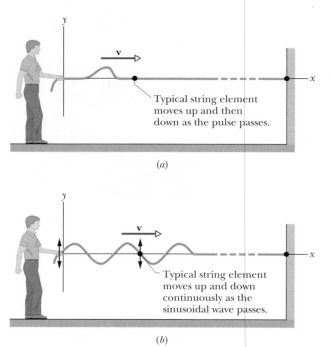

(a)

(b)

FIGURE 17-1   (a) Sending a single pulse down a long stretched string. (b) Sending a continuous sinusoidal wave down the string. Because the oscillations of any string element (represented here by a dot) are perpendicular to the direction in which the wave moves, the wave is a *transverse wave*.

sion. When you pull your end of the string upward, it begins to pull upward on the adjacent section of the string via tension between the two sections. As the adjacent section moves upward, it begins to pull the next section upward, and so on. Meanwhile, you have pulled down on your end of the string. So, as each section moves upward in turn, it begins to be pulled back downward by neighboring sections that are already on the way down. The net result is that a distortion in the string's shape (the pulse) moves along the string at some velocity **v**.

If you move your hand up and down in continuous simple harmonic motion, a continuous wave travels along the string at velocity **v**. Because the motion of your hand is a sinusoidal function of time, the wave has a sinusoidal shape at any given instant, as in Fig. 17-1*b*. That is, the wave has the shape of a sine curve or a cosine curve.

We consider here only an ''ideal'' string, in which no frictionlike forces within the string cause the wave to die out as it travels along the string. In addition, we assume that the string is so long that we need not consider a wave rebounding from the far end.

One way to study the waves of Fig. 17-1 is to monitor the **wave form** (shape of the wave) as it moves to the right. Alternatively, we can monitor the motion of an element of the string as the element oscillates up and down while the wave passes through it. We would find that the displacement of every such oscillating string element is *perpendicular* to the direction of travel of the wave, as indicated in Fig. 17-1. This motion is said to be **transverse,** and the wave is said to be a **transverse wave.**

Figure 17-2 shows how a sound wave can be produced by a piston in a long, air-filled pipe. If you suddenly move the piston rightward and then leftward, you can send a pulse of sound along the pipe. The rightward motion of the piston moves the elements of air next to it rightward, changing the air pressure there. The increased air pressure then pushes rightward on the elements of air somewhat

farther along the pipe. Once they have moved rightward, the elements move back leftward. Thus the motion of the air and the change in air pressure travel rightward along the pipe as a pulse.

If you push and pull on the piston in simple harmonic motion, as is being done in Fig. 17-2, a sinusoidal wave travels along the pipe. Because the motion of the elements of air is parallel to the direction of the wave's travel, the motion is said to be **longitudinal,** and the wave is said to be a **longitudinal wave.** In this chapter we concentrate on transverse waves, and string waves in particular; in Chapter 18 we shall concentrate on longitudinal waves, and sound waves in particular.

Both a transverse wave and a longitudinal wave are said to be **traveling waves** because the wave travels from one point to another, as from one end of the string to the other end in Fig. 17-1 or from one end of the pipe to the other end in Fig. 17-2. Note that it is the wave that moves between the two points and not the material (string or air) through which the wave moves.

The sand scorpion shown in the photograph opening this chapter uses waves of both transverse and longitudinal motion to locate its prey. When a beetle even slightly disturbs the sand, it sends pulses along the sand's surface (Fig. 17-3). One set of pulses is longitudinal, traveling with speed $v_l = 150$ m/s. A second set is transverse, traveling with speed $v_t = 50$ m/s.

The scorpion, with its eight legs spread roughly in a circle about 5 cm in diameter, intercepts the faster longitudinal pulses first and learns the direction of the beetle: it is in the direction of whichever leg is disturbed earliest by the pulses. The scorpion then senses the time interval Δt between that first interception and the interception of the slower transverse waves and uses it to determine the distance d to the beetle. This distance is given by

$$\Delta t = \frac{d}{v_t} - \frac{d}{v_l},$$

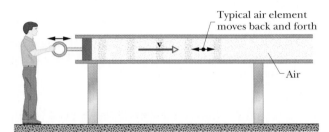

FIGURE 17-2 A sound wave is set up in an air-filled pipe by moving a piston back and forth. Because the oscillations of an element of the air (represented by the black dot) are parallel to the direction in which the wave travels, the wave is a *longitudinal wave.*

FIGURE 17-3 A beetle's motion sends fast longitudinal pulses and slower transverse pulses along the sand's surface. The sand scorpion first intercepts the longitudinal pulses; here, it is the rear-most right leg that senses the pulses earliest.

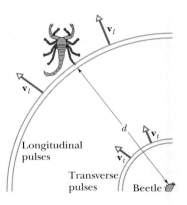

and it turns out to be

$$d = (75 \text{ m/s}) \, \Delta t.$$

For example, if $\Delta t = 4.0$ ms, then $d = 30$ cm, which gives the scorpion a perfect fix on the beetle.

17-4 WAVELENGTH AND FREQUENCY

To completely describe a wave on a string (and the motion of any element along its length), we need a function that gives the shape of the wave. This means that we need a relation in the form $y = h(x, t)$, in which y is the transverse displacement of any string element as a function h of the time t and the position x of the element along the string. In general, a sinusoidal shape like the wave in Fig. 17-1b can be described with h being either a sine function or a cosine function; both give the same general shape for the wave. In this chapter we use the sine function.

For a sinusoidal wave like that in Fig. 17-1b, traveling toward increasing values of x, the transverse displacement y of a string element at position x at time t is given by

$$y(x, t) = y_m \sin(kx - \omega t), \qquad (17\text{-}2)$$

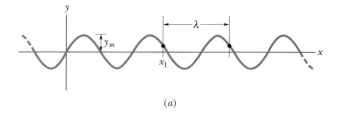

(a)

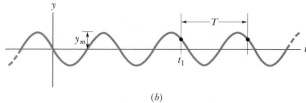

(b)

FIGURE 17-4   (a) A snapshot of a string at $t = 0$ as the sinusoidal wave of Eq. 17-2 travels along it. A typical wavelength λ is shown: it is the horizontal distance between two successive points on the string (represented by dots) with the same displacement. The amplitude y_m is also shown: it is the maximum displacement of the string. (b) A graph showing the displacement of a string element at $x = 0$ as a function of time as the sinusoidal wave passes through the element. A typical period T is shown: it is the time between identical displacements of the string element (represented by dots).

Here y_m is the **amplitude** of the wave; the subscript m stands for maximum, because the amplitude is the magnitude of the maximum displacement of the string element in either direction parallel to the y axis. (Hence, y_m is always a positive quantity.) The quantities k and ω are constants whose meanings we are about to discuss. The quantity $kx - \omega t$ is called the **phase** of the wave.

You may wonder why, of the infinite variety of wave forms that are available, we pick the sinusoidal wave of Eq. 17-2 for detailed study. This is in fact a wise choice because, as you will see in Section 17-8, *all* wave forms—including the pulse of Fig. 17-1a—can be constructed by adding sinusoidal waves of carefully selected amplitudes and values of k. Understanding sinusoidal waves is thus the key to understanding waves of any shape.

Because Eq. 17-2 has two independent variables (x and t), we cannot represent the dependent variable y on a single two-dimensional graph. We would need a videotape to show it fully and in real time. We can, however, learn a lot from plots like those of Fig. 17-4.

Wavelength and Angular Wave Number

Figure 17-4a shows how the transverse displacement y of Eq. 17-2 varies with position x at an instant, arbitrarily called $t = 0$. That is, the figure is a "snapshot" of the wave at that instant. With $t = 0$, Eq. 17-2 becomes

$$y(x, 0) = y_m \sin kx \qquad (t = 0). \qquad (17\text{-}3)$$

Figure 17-4a is a plot of this equation; it shows the shape of the actual wave at time $t = 0$.

The **wavelength** λ of a wave is the distance (along the direction of the wave) between repetitions of the wave shape. A typical wavelength is marked in Fig. 17-4a. By definition, the displacement y is the same at both ends of this wavelength, that is, at $x = x_1$ and $x = x_1 + \lambda$. Thus, by Eq. 17-3,

$$y_m \sin kx_1 = y_m \sin k(x_1 + \lambda)$$
$$= y_m \sin(kx_1 + k\lambda). \qquad (17\text{-}4)$$

A sine function begins to repeat itself when its angle (or argument) is increased by 2π rad; so Eq. 17-4 will be accurate only if $k\lambda = 2\pi$, or if

$$k = \frac{2\pi}{\lambda} \qquad \text{(angular wave number)}. \qquad (17\text{-}5)$$

We call k the **angular wave number** of the wave; its SI unit is the radian per meter. (Note that the symbol k here does *not* represent a spring constant as previously.)

Period, Angular Frequency, and Frequency

Figure 17-4b shows how the displacement y of Eq. 17-2 varies with time t at a fixed position, taken to be $x = 0$. If you were to monitor the string, you would see that the single element of the string at that position moves up and down in simple harmonic motion given by Eq. 17-2 with $x = 0$:

$$y(0, t) = y_m \sin(-\omega t)$$
$$= -y_m \sin \omega t \qquad (x = 0). \qquad (17\text{-}6)$$

Here we have made use of the fact that $\sin(-\alpha) = -\sin\alpha$, where α is any angle. Figure 17-4b is a plot of this equation; it *does not* show the shape of the wave.

We define the **period** of oscillation T of a wave to be the time interval between repetitions of the motion of an oscillating string element (at any position x). A typical period is marked in Fig. 17-4b. Applying Eq. 17-6 to both ends of this time interval and equating the results yield

$$-y_m \sin \omega t_1 = -y_m \sin \omega(t_1 + T)$$
$$= -y_m \sin(\omega t_1 + \omega T). \qquad (17\text{-}7)$$

This can be true only if $\omega T = 2\pi$, or if

$$\omega = \frac{2\pi}{T} \qquad \text{(angular frequency).} \qquad (17\text{-}8)$$

We call ω the **angular frequency** of the wave; its SI unit is the radian per second.

The **frequency** f of the wave is defined as $1/T$ and is related to the angular frequency ω by

$$f = \frac{1}{T} = \frac{\omega}{2\pi} \qquad \text{(frequency).} \qquad (17\text{-}9)$$

Like the frequency of simple harmonic motion in Chapter 16, this frequency f is a number of oscillations per unit time — here, made by a string element as the wave moves through it. As in Chapter 16, f is usually measured in hertz or its multiples.

CHECKPOINT 1: The figure shows a snapshot of three waves traveling along a string. The phases for the waves are given by (a) $2x - 4t$, (b) $4x - 8t$, and (c) $8x - 16t$. Which phase corresponds to which wave in the figure?

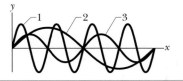

17-5 THE SPEED OF A TRAVELING WAVE

Figure 17-5 shows two snapshots of the wave of Eq. 17-2, taken a small time interval Δt apart. The wave is traveling in the direction of increasing x (to the right in Fig. 17-5), the entire wave pattern moving a distance Δx in that direction during the interval Δt. The ratio $\Delta x/\Delta t$ (or, in the differential limit, dx/dt) is the **wave speed** v. How can we find its value?

As the wave in Fig. 17-5 moves, each point of the moving wave form (such as point A) retains its displacement y. (Points on the string do not retain their displacement, but points on the wave *form* do.) For each such point, Eq. 17-2 tells us that the argument of the sine function must be a constant:

$$kx - \omega t = \text{a constant.} \qquad (17\text{-}10)$$

Note that although this argument is constant, both x and t are changing. In fact, as t increases, x must also, to keep the argument constant. This confirms that the wave pattern is moving toward increasing x.

To find the wave speed v, we take the derivative of Eq. 17-10, getting

$$k\frac{dx}{dt} - \omega = 0$$

or
$$\frac{dx}{dt} = v = \frac{\omega}{k}. \qquad (17\text{-}11)$$

Using Eq. 17-5 ($k = 2\pi/\lambda$) and Eq. 17-8 ($\omega = 2\pi/T$), we can rewrite the wave speed as

$$v = \frac{\omega}{k} = \frac{\lambda}{T} = \lambda f \qquad \text{(wave speed).} \qquad (17\text{-}12)$$

The equation $v = \lambda/T$ tells us that the wave speed is one wavelength per period: the wave moves a distance of one wavelength in one period of oscillation.

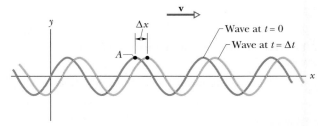

FIGURE 17-5 A snapshot of the traveling wave of Eq. 17-2 at $t = 0$ and at a later time $t = \Delta t$. During the time interval Δt, the entire curve shifts a distance Δx to the right.

Equation 17-2 describes a wave moving in the direction of increasing x. We can find the equation of a wave traveling in the opposite direction by replacing t in Eq. 17-2 with $-t$. This corresponds to the condition

$$kx + \omega t = \text{a constant}, \quad (17\text{-}13)$$

which (compare Eq. 17-10) requires that x *decrease* with time. Thus, a wave traveling toward decreasing x is described by the equation

$$y(x, t) = y_m \sin(kx + \omega t). \quad (17\text{-}14)$$

If you analyze the wave of Eq. 17-14 as we have just done for the wave of Eq. 17-2, you will find for its velocity

$$\frac{dx}{dt} = -\frac{\omega}{k}. \quad (17\text{-}15)$$

The minus sign (compare Eq. 17-11) verifies that the wave is indeed moving in the direction of decreasing x and justifies our switching the sign of the time variable.

Consider now a wave of generalized shape, given by

$$y(x, t) = h(kx \pm \omega t), \quad (17\text{-}16)$$

where h represents *any* function, the sine function being one possibility. Our analysis above shows that all waves in which the variables x and t enter in the combination $kx \pm \omega t$ are traveling waves. Furthermore, all traveling waves *must* be of the form of Eq. 17-16. Thus $y(x, t) = \sqrt{ax + bt}$ represents a possible (though perhaps physically a little bizarre) traveling wave. The function $y(x, t) = \sin(ax^2 - bt)$, on the other hand, does *not* represent a traveling wave.

SAMPLE PROBLEM 17-1

A sinusoidal wave traveling along a string is described by

$$y(x, t) = 0.00327 \sin(72.1x - 2.72t), \quad (17\text{-}17)$$

in which the numerical constants are in SI units (0.00327 m, 72.1 rad/m, and 2.72 rad/s).

(a) What is the amplitude of this wave?

SOLUTION: By comparison with Eq. 17-2, we see that

$$y_m = 0.00327 \text{ m} = 3.27 \text{ mm}. \quad \text{(Answer)}$$

(b) What are the wavelength, period, and frequency of this wave?

SOLUTION: By inspection of Eq. 17-17, we see that

$$k = 72.1 \text{ rad/m} \quad \text{and} \quad \omega = 2.72 \text{ rad/s}.$$

From Eq. 17-5 we have

$$\lambda = \frac{2\pi}{k} = \frac{2\pi \text{ rad}}{72.1 \text{ rad/m}} = 0.0871 \text{ m}$$

$$= 8.71 \text{ cm}. \quad \text{(Answer)}$$

From Eq. 17-9 we have

$$T = \frac{2\pi}{\omega} = \frac{2\pi \text{ rad}}{2.72 \text{ rad/s}} = 2.31 \text{ s}. \quad \text{(Answer)}$$

From Eq. 17-9 we have

$$f = \frac{1}{T} = \frac{1}{2.31 \text{ s}} = 0.433 \text{ Hz}. \quad \text{(Answer)}$$

(c) What is the speed of this wave?

SOLUTION: From Eq. 17-12 we have

$$v = \frac{\omega}{k} = \frac{2.72 \text{ rad/s}}{72.1 \text{ rad/m}} = 0.0377 \text{ m/s}$$

$$= 3.77 \text{ cm/s}. \quad \text{(Answer)}$$

Note that the quantities calculated in (b) and (c) are independent of the amplitude of the wave.

(d) What is the displacement y at $x = 22.5$ cm and $t = 18.9$ s?

SOLUTION: From Eq. 17-17 we obtain

$$y = 0.00327 \sin(72.1 \times 0.225 - 2.72 \times 18.9)$$

$$= (0.00327 \text{ m}) \sin(-35.1855 \text{ rad})$$

$$= (0.00327 \text{ m})(0.588)$$

$$= 0.00192 \text{ m} = 1.92 \text{ mm}. \quad \text{(Answer)}$$

Thus the displacement is positive. (Be sure to change your calculator mode to radians before evaluating the sine.)

SAMPLE PROBLEM 17-2

In Sample Problem 17-1d, we showed that the transverse displacement y of an element of the string for the wave of Eq. 17-17 at $x = 0.225$ m and $t = 18.9$ s is 1.92 mm.

(a) What is u, the transverse speed of the same element of the string, at that place and at that time? (This speed, which is associated with the transverse oscillation of an element of the string, is in the y direction. Do not confuse it with v, the constant speed at which the *wave form* travels along the x axis.)

SOLUTION: The generalized equation of the wave of Eq. 17-17 is Eq. 17-2:

$$y(x, t) = y_m \sin(kx - \omega t). \quad (17\text{-}18)$$

In this expression, let us hold x constant but allow t (for the time being) to be a variable. We then take the derivative of this equation with respect to t, obtaining*

$$u = \frac{\partial y}{\partial t} = -\omega y_m \cos(kx - \omega t). \quad (17\text{-}19)$$

---

*A derivative taken while one (or more) of the variables is treated as a constant is called a *partial derivative* and is represented by the symbol $\partial/\partial x$, rather than d/dx.

Substituting numerical values from Sample Problem 17-1, we obtain

$$u = (-2.72 \text{ rad/s})(3.27 \text{ mm}) \cos(-35.1855 \text{ rad})$$
$$= 7.20 \text{ mm/s}. \qquad \text{(Answer)}$$

Thus, at $t = 18.9$ s, the element of string at $x = 22.5$ cm is moving in the direction of increasing y, with a speed of 7.20 mm/s.

(b) What is the transverse acceleration a_y at that position and at that time?

SOLUTION: From Eq. 17-19, treating x as a constant but allowing t (for the time being) to be a variable, we find

$$a_y = \frac{\partial u}{\partial t} = -\omega^2 y_m \sin(kx - \omega t).$$

Comparison with Eq. 17-18 shows that we can write this as

$$a_y = -\omega^2 y.$$

We see that the transverse acceleration of an oscillating string element is proportional to its transverse displacement but opposite in sign, completely consistent with the action of the element: namely, it is moving transversely in simple harmonic motion. Substituting numerical values yields

$$a_y = -(2.72 \text{ rad/s})^2(1.92 \text{ mm})$$
$$= -14.2 \text{ mm/s}^2. \qquad \text{(Answer)}$$

Thus, at $t = 18.9$ s, the element of string at $x = 22.5$ cm is displaced from its equilibrium position by 1.92 mm in the increasing y direction and has an acceleration of magnitude 14.2 mm/s² in the decreasing y direction.

C HECKPOINT **2:** Here are the equations of three waves:
(1) $y(x, t) = 2 \sin(4x - 2t)$,
(2) $y(x, t) = \sin(3x - 4t)$,
(3) $y(x, t) = 2 \sin(3x - 3t)$.
Rank the waves according to their (a) wave speed and (b) maximum transverse speed, greatest first.

PROBLEM SOLVING TACTICS

TACTIC 1: *Evaluating Large Phases*
Sometimes, as in Sample Problems 17-1d and 17-2, an angle much greater than 2π rad (or 360°) crops up and you are asked to find its sine or cosine. Adding or subtracting an integral multiple of 2π rad to such an angle does not change the value of any of its trigonometric functions. In Sample Problem 17-1d, for example, the angle is -35.1855 rad. Adding $(6)(2\pi \text{ rad})$ to this angle yields

$$-35.1855 \text{ rad} + (6)(2\pi \text{ rad}) = 2.51361 \text{ rad},$$

an angle less than 2π rad that has the same trigonometric functions as -35.1855 rad (Fig. 17-6). As an example, the sine of both 2.51361 rad and -35.1855 rad is 0.588.

Your calculator will reduce such large angles for you automatically. Caution: Do not round off large angles if you intend to take their sines or cosines. In taking the sine of a very large angle, you are throwing away most of the angle and taking the sine of what is left over. If, for example, you were to round -35.1855 rad to -35 rad (a change of 0.5% and normally a reasonable step), you would be changing the sine of the angle by 27%. Also, if you change a large angle from degrees to radians, be sure to use an exact conversion factor (such as $180° = \pi$ rad) rather than an approximate one (such as $57.3° \approx 1$ rad).

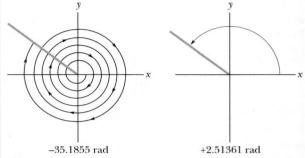

FIGURE 17-6 The two angles are different but all their trigonometric functions are identical.

17-6 WAVE SPEED ON A STRETCHED STRING

The speed of a wave is related to the wave's wavelength and frequency by Eq. 17-12, but *it is set by the medium.* If a wave is to travel through a medium such as water, air, steel, or a stretched string, it must cause the particles of that medium to oscillate as it passes. For that to happen, the medium must possess both inertia (so that kinetic energy can be stored) and elasticity (so that potential energy can be stored). These two properties determine how fast the wave can travel in the medium. And conversely, it should be possible to calculate the speed of the wave through the medium in terms of these properties. We do so now for a stretched string, in two ways.

Dimensional Analysis

In dimensional analysis we examine carefully the dimensions of all the physical quantities that enter into a given situation. In this case, we seek a speed v, which has the dimension of length divided by time, or LT^{-1}.

The inertia characteristic of a stretched string is the mass of a string element, which is represented by the mass m of the string divided by the length l of the string. We call this ratio the *linear density* μ of the string. Thus $\mu = m/l$, its dimension being mass divided by length, ML^{-1}.

You cannot send a wave along a stretched string without further stretching the string. The tension in the string does this stretching and must therefore represent the elastic characteristic of the string. The tension τ is a force and has the dimension (think about $F = ma$) MLT^{-2}. (We would normally use the symbol T for tension but we want to avoid confusion with our use of this symbol for the period of oscillation.)

The problem here is to combine μ (dimension ML^{-1}) and τ (dimension MLT^{-2}) in such a way as to generate v (dimension LT^{-1}). A little juggling of various combinations suggests

$$v = C\sqrt{\frac{\tau}{\mu}}, \qquad (17\text{-}20)$$

in which C is a dimensionless constant. The drawback of dimensional analysis is that it cannot give the value of such a dimensionless constant. In our second approach to determine the speed, you will see that Eq. 17-20 is indeed correct and that $C = 1$.

Derivation from Newton's Second Law

Instead of the sinusoidal wave of Fig. 17-1b, let us consider a single symmetrical pulse such as that of Fig. 17-7. For convenience, we choose a reference frame in which the pulse remains stationary. That is, we run along with the pulse, keeping it constantly in view. In this frame, the string appears to move past us, from right to left in Fig. 17-7, with speed v.

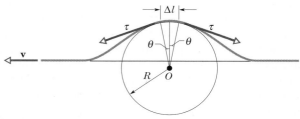

FIGURE 17-7 A symmetrical pulse, viewed from a reference frame in which the pulse is stationary and the string appears to move right to left with speed v. We find speed v by applying Newton's second law to a string element of length Δl, located at the top of the pulse.

Consider a small segment of the pulse, of length Δl, forming an arc of a circle of radius R. A force equal in magnitude to the string tension τ pulls tangentially on this segment at each end. The horizontal components of these forces cancel, but the vertical components add to form a restoring force **F**. In magnitude,

$$F = 2\tau \sin \theta \approx \tau(2\theta) = \tau\frac{\Delta l}{R} \qquad \text{(force).} \quad (17\text{-}21)$$

We have used here the approximation $\sin \theta \approx \theta$ for the small angles θ in Fig. 17-7; from that figure, we have also used $2\theta = \Delta l/R$.

The mass of the segment is given by

$$\Delta m = \mu\,\Delta l \qquad \text{(mass).} \quad (17\text{-}22)$$

At the moment shown in Fig. 17-7, the string element Δl is moving in an arc of a circle. Thus it has a centripetal acceleration toward the center of that circle, given by

$$a = \frac{v^2}{R} \qquad \text{(acceleration).} \quad (17\text{-}23)$$

Equations 17-21, 17-22, and 17-23 contain the elements of Newton's second law. Combining them in the form

$$\text{force} = \text{mass} \times \text{acceleration}$$

gives

$$\frac{\tau\,\Delta l}{R} = (\mu\,\Delta l)\frac{v^2}{R}.$$

Solving this equation for the speed v yields

$$v = \sqrt{\frac{\tau}{\mu}} \qquad \text{(speed),} \quad (17\text{-}24)$$

in exact agreement with Eq. 17-20 if the constant C in that equation is given the value unity. Equation 17-24 gives the speed of the pulse in Fig. 17-7 and the speed of any other wave on the same string under the same tension.

Equation 17-24 tells us that the speed of a wave along a stretched ideal string depends only on the characteristics of the string and not on the frequency of the wave. The *frequency* of the wave is fixed entirely by whatever generates the wave (for example, the person in Fig. 17-1b). The *wavelength* of the wave is then fixed by Eq. 17-12 in the form $\lambda = v/f$.

CHECKPOINT 3: The figure shows two situations in which the same string is put under tension by a suspended mass of 5 kg. In which situation will the speed of waves sent along the string be greater?

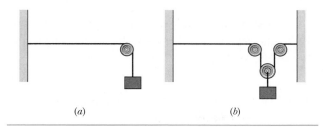

(a)   (b)

SAMPLE PROBLEM 17-3

In Fig. 17-8, a stranded climber has hooked himself onto a makeshift rope lowered by a rescuer. The rope between the climber and rescuer consists of two sections: section 1 with length l_1 and linear density μ_1, and section 2 with length $l_2 = 2l_1$ and linear density $\mu_2 = 4\mu_1$. The climber happens to pluck the bottom end of the rope (as a "ready" signal) at the same time the rescuer plucks the top end.

(a) What is the speed v_1 of the resulting pulses in section 1, in terms of their speed v_2 in section 2?

SOLUTION: We assume that the mass of the rope sections is negligible relative to the mass of the climber. Thus the tension τ in the rope is equal to the weight of the climber and is the same in the two sections. From Eq. 17-24, speeds v_1 and v_2 are given by

$$v_1 = \sqrt{\frac{\tau}{\mu_1}} \quad \text{and} \quad v_2 = \sqrt{\frac{\tau}{\mu_2}}. \qquad (17\text{-}25)$$

Dividing the first expression by the second, and substituting $\mu_2 = 4\mu_1$, we find

$$\frac{v_1}{v_2} = \sqrt{\frac{\tau}{\mu_1}}\sqrt{\frac{\mu_2}{\tau}} = \sqrt{\frac{\mu_2}{\mu_1}} = \sqrt{\frac{4\mu_1}{\mu_1}} = 2,$$

or $\qquad\qquad v_1 = 2v_2.$ (Answer) (17-26)

(b) In terms of l_2, at what distance below the rescuer do the two pulses pass through each other?

SOLUTION: We can simplify our calculations by first deciding whether the pulses pass each other above or below the knot joining the two sections. Let t be the time the pulses take to reach each other. From Eq. 17-26, we know that the climber's pulse moves through section 1 twice as fast as the rescuer's pulse moves through section 2. Because $l_2 = 2l_1$, we also know that the climber's pulse must travel half as far as the rescuer's pulse to reach the knot. So the climber's pulse must reach the knot first, and the point of passing must be above the knot. Let us assume that the pulses pass at a distance d below the rescuer at time t.

To reach the point of passing, the rescuer's pulse must travel downward through a distance d at speed v_2 in time t. So

$$t = \frac{d}{v_2}. \qquad (17\text{-}27)$$

To reach the point of passing, the climber's pulse must travel upward through distance l_1 at speed v_1 and then through distance $l_2 - d$ at speed v_2, all in time t. So

$$t = \frac{l_1}{v_1} + \frac{l_2 - d}{v_2}. \qquad (17\text{-}28)$$

Substituting t from Eq. 17-27 into Eq. 17-28, and setting $l_1 = l_2/2$ and $v_1 = 2v_2$, we find

$$\frac{d}{v_2} = \frac{l_1}{v_1} + \frac{l_2 - d}{v_2} = \frac{l_2/2}{2v_2} + \frac{l_2 - d}{v_2}.$$

Multiplying through by v_2 and rearranging yield

$$d = \tfrac{5}{8}l_2. \qquad \text{(Answer)}$$

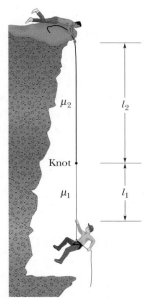

FIGURE 17-8 Sample Problem 17-3. A stranded climber hangs from a rope consisting of two sections. His rescurer has secured the rope at the top.

17-7 ENERGY AND POWER OF A TRAVELING STRING WAVE

When we set up a wave on a stretched string, we provide energy for the motion of the string. As the wave moves away from us, it transports that energy as both kinetic energy and elastic potential energy. Let us consider each form in turn.

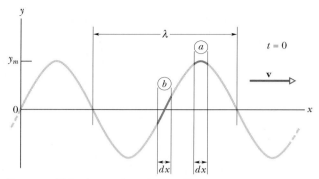

FIGURE 17-9 A snapshot of a traveling wave on a string at time $t = 0$. String element a is at displacement $y = y_m$, and string element b is at displacement $y = 0$. The kinetic energy of the string element at each position depends on the transverse velocity of the element. The potential energy depends on the amount by which the string element is stretched as the wave passes through it.

Kinetic Energy

An element of the string of mass dm, oscillating transversely in simple harmonic motion as the wave passes through it, has kinetic energy associated with its transverse velocity **u**. When the element is rushing through its $y = 0$ position (Fig. 17-9), its transverse velocity—and thus its kinetic energy—is a maximum. When the element is at its extreme position $y = y_m$, its transverse velocity—and thus again its kinetic energy—is zero.

Elastic Potential Energy

To send a sinusoidal wave along a previously straight string, the wave must necessarily stretch the string. As a string element of length dx oscillates transversely, its length must increase and decrease in a periodic way if the string element is to fit the sinusoidal wave form. Elastic potential energy is associated with these length changes, just as for a spring.

When the string element is at its $y = y_m$ position (Fig. 17-9), its length has its normal undisturbed value dx, so its elastic potential energy is zero. However, when the element is rushing through its $y = 0$ position, it is stretched to its maximum extent, and its elastic potential energy then is a maximum.

The oscillating string element thus has both its maximum kinetic energy and its maximum elastic potential energy at $y = 0$. In the snapshot of Fig. 17-9, the regions of the string at maximum displacement have no energy, and the regions at zero displacement have maximum energy. As the wave travels along the string, the tension in the string continuously does work to transfer energy from regions with energy to regions with no energy.

The Rate of Energy Transmission

The kinetic energy dK associated with a string element of mass dm is given by

$$dK = \tfrac{1}{2}\, dm\, u^2, \tag{17-29}$$

where u is the transverse speed of the oscillating string element, given by Eq. 17-19 as

$$u = \frac{\partial y}{\partial t} = -\omega y_m \cos(kx - \omega t). \tag{17-30}$$

Using this relation and putting $dm = \mu\, dx$, we rewrite Eq. 17-29 as

$$dK = \tfrac{1}{2}(\mu\, dx)(-\omega y_m)^2 \cos^2(kx - \omega t). \tag{17-31}$$

Dividing Eq. 17-31 by dt gives the rate at which the kinetic energy of a string element changes, and thus the rate at which kinetic energy is carried along by the wave. The ratio dx/dt that then appears on the right of Eq. 17-31 is the wave speed v, so we obtain

$$\frac{dK}{dt} = \tfrac{1}{2}\mu v \omega^2 y_m^2 \cos^2(kx - \omega t). \tag{17-32}$$

The *average* rate at which kinetic energy is transported is

$$\overline{\left(\frac{dK}{dt}\right)} = \tfrac{1}{2}\mu v \omega^2 y_m^2 \,\overline{\cos^2(kx - \omega t)}$$
$$= \tfrac{1}{4}\mu v \omega^2 y_m^2, \tag{17-33}$$

where an overhead bar means an average value of the quantity. In Eq. 17-33 we have taken the average over an integer number of wavelengths and have used the fact that the average value of the square of a cosine function over an integer number of wavelengths is $\tfrac{1}{2}$.

Elastic potential energy is also carried along with the wave, and at the same average rate given by Eq. 17-33. Although we shall not examine the proof, you should recall that, in an oscillating system such as a pendulum or a spring–block system, the average kinetic energy and the average potential energy are indeed equal.

The **average power,** which is the average rate at which energy of both kinds is transmitted by the wave, is then

$$\overline{P} = 2\,\overline{\left(\frac{dK}{dt}\right)} \tag{17-34}$$

or, from Eq. 17-33,

$$\overline{P} = \tfrac{1}{2}\mu v \omega^2 y_m^2 \qquad \text{(average power).} \tag{17-35}$$

The factors μ and v in this equation depend on the material and tension of the string. The factors ω and y_m depend on the process that generates the wave. The dependence of the average power of a wave on the square of its amplitude and also on the square of its angular frequency is a general result, true for waves of all types.

SAMPLE PROBLEM 17-4

A string has a linear density μ of 525 g/m and is stretched with a tension τ of 45 N. A wave whose frequency f and amplitude y_m are 120 Hz and 8.5 mm, respectively, is traveling along the string. At what average rate is the wave transporting energy along the string?

SOLUTION: Before finding $\overline{P}$ from Eq. 17-35, we must calculate the angular frequency ω and the wave speed v. From Eq. 17-9,

$$\omega = 2\pi f = (2\pi)(120 \text{ Hz}) = 754 \text{ rad/s}.$$

From Eq. 17-24 we have

$$v = \sqrt{\frac{\tau}{\mu}} = \sqrt{\frac{45 \text{ N}}{0.525 \text{ kg/m}}} = 9.26 \text{ m/s}.$$

Equation 17-35 then yields

$$\overline{P} = \tfrac{1}{2}\mu v \omega^2 y_m^2$$
$$= (\tfrac{1}{2})(0.525 \text{ kg/m})(9.26 \text{ m/s})(754 \text{ rad/s})^2(0.0085 \text{ m})^2$$
$$= 100 \text{ W}. \qquad \text{(Answer)}$$

17-8 THE PRINCIPLE OF SUPERPOSITION FOR WAVES

It often happens that two or more waves pass simultaneously through the same region. When we listen to a concert, for example, sounds from many instruments fall simultaneously on our eardrums. The electrons in the antennas of our radio and TV receivers are set in motion by a whole array of signals from different broadcasting centers. The water of a lake or harbor may be churned up by the wakes of many boats.

Suppose that two waves travel simultaneously along the same stretched string. Let $y_1(x, t)$ and $y_2(x, t)$ be the displacements that the string would experience if each wave acted alone. The displacement of the string when both waves act is then

$$y'(x, t) = y_1(x, t) + y_2(x, t), \qquad (17\text{-}36)$$

the sum being an algebraic sum. This summation of displacements along the string means

---

Overlapping waves algebraically add to produce a **resultant wave.**

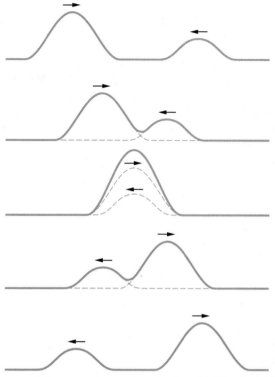

FIGURE 17-10 A series of snapshots that show two pulses traveling in opposite directions along a stretched string. The superposition principle applies as the pulses move through each other.

This is another example of the **principle of superposition,** which says that when several effects occur simultaneously, their net effect is the sum of the individual effects.

Figure 17-10 shows a sequence of snapshots of two pulses traveling in opposite directions on the same stretched string. When the pulses overlap, the resultant pulse is their sum. Moreover, each pulse moves through the other, as if the other were not present:

Overlapping waves do not in any way alter the travel of each other.

Fourier Analysis

French mathematician Jean Baptiste Fourier (1786–1830) explained how the principle of superposition can be used to analyze nonsinusoidal wave forms. He showed that any wave form can be represented as the sum of a large number of sinusoidal waves, of carefully chosen frequencies and amplitudes. English physicist Sir James Jeans expressed it well:

[Fourier's] theorem tells us that every curve, no matter what its nature may be, or in what way it was originally obtained, can be exactly reproduced by superposing a sufficient number of simple harmonic [sinusoidal] curves—in brief, every curve can be built up by piling up waves.

Figure 17-11 shows an example of a *Fourier series,* as such sums are called. The sawtooth curve in Fig. 17-11*a* shows the variation with time (at position $x = 0$) of the wave we wish to represent. The Fourier series that represents it can be shown to be

$$y(t) = -\frac{1}{\pi} \sin \omega t - \frac{1}{2\pi} \sin 2\omega t$$
$$-\frac{1}{3\pi} \sin 3\omega t \cdots, \qquad (17\text{-}37)$$

in which $\omega = 2\pi/T$, where T is the period of the sawtooth curve. The green curve of Fig. 17-11*a*, which represents the sum of the first six terms of Eq. 17-37, matches the sawtooth curve rather well. Figure 17-11*b* shows these six terms separately. By adding more terms, we can approximate the sawtooth curve as closely as we wish.

You can see now why we spent so much time analyzing the behavior of a sinusoidal wave. When we understand that, Fourier's theorem will open the door to all other wave shapes.

17-9 INTERFERENCE OF WAVES

Suppose we send two sinusoidal waves of the same wavelength and amplitude in the same direction along a stretched string. The superposition principle applies. What resultant wave does it predict for the string?

The resultant wave depends on the extent to which the waves are *in phase* (in step) with respect to each other, that is, how much one wave form is shifted from the other wave form. If the waves are exactly in phase (so that the peaks and valleys of one are exactly aligned with those of the other without any shift), they combine to double the displacement of either wave acting alone. If they are exactly out of phase (the peaks of one are exactly aligned with the valleys of the other), they cancel everywhere and the string remains straight. We call this phenomenon of combining and canceling of waves **interference,** and the waves are said to **interfere.** (These terms refer only to the displacements of the waves; the travel of the waves is unaffected.)

Let one wave traveling along a stretched string be given by

$$y_1(x, t) = y_m \sin(kx - \omega t) \qquad (17\text{-}38)$$

and another, shifted from the first, by

$$y_2(x, t) = y_m \sin(kx - \omega t + \phi). \qquad (17\text{-}39)$$

These waves have the same angular frequency ω (and thus the same frequency f), the same angular wave number k (and thus the same wavelength λ), and the same amplitude y_m. They travel in the same direction, that of increasing x, with the same speed, given by Eq. 17-24. They differ only by a constant angle ϕ, which we call the **phase constant.** These waves are said to be *out of phase* by ϕ or to have a *phase difference* of ϕ, or one wave is said to be *phase-shifted* from the other by ϕ.

From the principle of superposition (Eq. 17-36), the combined wave has displacement

$$y'(x, t) = y_1(x, t) + y_2(x, t)$$
$$= y_m \sin(kx - \omega t) + y_m \sin(kx - \omega t + \phi). \quad (17\text{-}40)$$

In Appendix E we see that we can write the sum of the sines of two angles α and β as

$$\sin \alpha + \sin \beta = 2 \sin \tfrac{1}{2}(\alpha + \beta) \cos \tfrac{1}{2}(\alpha - \beta). \quad (17\text{-}41)$$

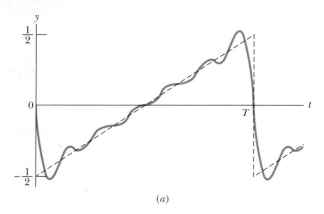

(a)

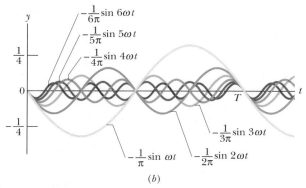

(b)

FIGURE 17-11 (*a*) The dashed sawtooth curve is approximated by the green curve, which is the result of evaluating the first six terms in the sum of Eq. 17-37. (Evaluating more terms would give closer approximations.) (*b*) Separate displays of the first six terms of that equation, each a sinusoidal function.

Applying this relation to Eq. 17-40 yields

$$y'(x, t) = [2y_m \cos \tfrac{1}{2}\phi] \sin(kx - \omega t + \tfrac{1}{2}\phi). \quad (17\text{-}42)$$

The resultant wave is thus also a sinusoidal wave traveling in the direction of increasing x. It is the only wave you would actually see on the string (you would *not* see the two combining waves of Eqs. 17-38 and 17-39).

> If two sinusoidal waves of the same amplitude and wavelength travel in the *same* direction along a stretched string, they interfere to produce a resultant sinusoidal wave traveling in that direction.

The resultant wave differs from the original waves in two respects: (1) its phase constant is $\tfrac{1}{2}\phi$, and (2) its amplitude y'_m is the quantity in the brackets in Eq. 17-42:

$$y'_m = 2y_m \cos \tfrac{1}{2}\phi. \quad (17\text{-}43)$$

If $\phi = 0$ rad (or 0°), the two combining waves are exactly in phase (as in Fig. 17-12a). Then Eq. 17-42 reduces to

$$y'(x, t) = 2y_m \sin(kx - \omega t) \quad (\phi = 0). \quad (17\text{-}44)$$

Note that the amplitude of the resultant wave is twice the amplitude of either combining wave. This is the greatest amplitude the resultant wave can have, because the cosine term in Eq. 17-42 and Eq. 17-43 has its greatest value (unity) when $\phi = 0$. Interference that produces the greatest possible amplitude is called *fully constructive interference*.

If $\phi = \pi$ rad (or 180°), the combining waves are exactly out of phase (as in Fig. 17-12b). Then $\cos \tfrac{1}{2}\phi$ becomes $\cos \pi/2 = 0$, and the amplitude of the resultant wave (Eq. 17-43) is zero. We then have for all values of x and t,

$$y'(x, t) = 0 \quad (\phi = \pi \text{ rad}). \quad (17\text{-}45)$$

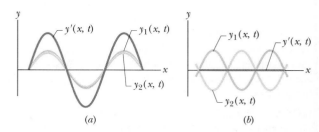

FIGURE 17-12 Two identical waves, $y_1(x, t)$ and $y_2(x, t)$, traveling in the same direction along a string, interfere to give a resultant wave $y'(x, t)$. (*a*) If the waves are exactly in phase, they undergo fully constructive interference and produce a resultant wave of twice their own amplitude. (*b*) If they are exactly out of phase, they undergo fully destructive interference and the string remains straight.

Now, although we sent two waves along the string, we see no motion of the string. This type of interference is called *fully destructive interference.*

A phase difference $\phi = 2\pi$ rad (or 360°) corresponds to a shift of one wave relative to the other wave by a distance of one wavelength. So phase differences can also be described in terms of wavelengths. For example, in Fig. 17-12b the waves may be said to be 0.50 wavelength out of phase. Table 17-1 shows some other examples of phase differences and the interference they produce. Note that when interference is neither fully constructive nor fully destructive, it is called *intermediate interference.* The amplitude of the resultant wave is then intermediate between 0 and $2y_m$.

Two waves with the same wavelength are in phase if their phase difference is zero or any integer number of wavelengths. Thus the integer part of any phase difference *expressed in wavelengths* may be discarded. For example, a phase difference of 0.40 wavelength is equivalent in every way to one of 2.40 wavelengths, and so the simpler of the two numbers can be used in computations.

TABLE 17-1 **PHASE DIFFERENCES AND RESULTING INTERFERENCE TYPES**[a]

PHASE DIFFERENCE, IN			AMPLITUDE OF RESULTANT WAVE	TYPE OF INTERFERENCE
DEGREES	RADIANS	WAVELENGTHS		
0	0	0	$2y_m$	Fully constructive
120	$2\pi/3$	0.33	y_m	Intermediate
180	π	0.50	0	Fully destructive
240	$4\pi/3$	0.67	y_m	Intermediate
360	2π	1.00	$2y_m$	Fully constructive
865	15.1	2.40	$0.60y_m$	Intermediate

[a]The phase difference is between two otherwise identical waves, with amplitude y_m, moving in the same direction.

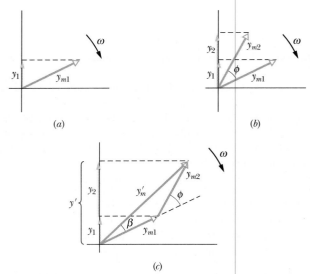

SAMPLE PROBLEM 17-5

Two identical waves, moving in the same direction along a stretched string, interfere with each other. The amplitude y_m of each wave is 9.8 mm, and the phase difference ϕ between them is 100°.

(a) What is the amplitude y_m' of the resultant wave due to the interference of these two waves, and what type of interference occurs?

SOLUTION: From Eq. 17-43 we have for the amplitude

$$y_m' = 2y_m \cos \tfrac{1}{2}\phi$$
$$= (2)(9.8 \text{ mm}) \cos(100°/2)$$
$$= 13 \text{ mm}. \qquad \text{(Answer)}$$

Because the phase difference is between 0 and 180°, the interference is intermediate.

(b) What phase difference, in radians and in wavelengths, will give the resultant wave an amplitude of 4.9 mm?

SOLUTION: From Eq. 17-43 we have

$$y_m' = 2y_m \cos \tfrac{1}{2}\phi,$$

or $\qquad\qquad 4.9 \text{ mm} = (2)(9.8 \text{ mm}) \cos \tfrac{1}{2}\phi,$

which gives us (with a calculator in the radian mode)

$$\phi = 2 \cos^{-1} \frac{4.9 \text{ mm}}{(2)(9.8 \text{ mm})}$$
$$= \pm 2.636 \text{ rad} \approx \pm 2.6 \text{ rad}. \qquad \text{(Answer)}$$

There are two solutions because we can obtain the same resultant wave by letting the first wave *lead* (travel ahead of) or *lag* (travel behind) the second wave by 2.6 rad. In wavelengths, the phase difference is

$$\frac{\phi}{2\pi \text{ rad/wavelength}} = \frac{\pm 2.636 \text{ rad}}{2\pi \text{ rad/wavelength}}$$
$$= \pm 0.42 \text{ wavelength.} \quad \text{(Answer)}$$

CHECKPOINT **4:** Here are four other possible phase differences between the two waves of Sample Problem 17-5, expressed in wavelengths: 0.20, 0.45, 0.60, and 0.80. Rank them according to the amplitude of the resultant wave, greatest first.

17-10 PHASORS

We can represent a string wave (or any other type of wave) vectorially with a **phasor.** In essence, a phasor is a vector that has a magnitude equal to the amplitude of the wave and that rotates around an origin; the angular speed of the phasor is equal to the angular frequency ω of the wave. For

FIGURE 17-13 (*a*) A phasor of magnitude y_{m1} rotating about an origin at angular speed ω represents a sinusoidal wave. The phasor's projection y_1 on the vertical axis represents the displacement of a point through which the wave passes. (*b*) A second phasor, of magnitude y_{m2} and rotating at a constant angle ϕ to the first phasor, represents a second wave, with a phase constant ϕ. (*c*) The resultant wave of the two waves is represented by the vector sum y_m' of the two phasors. The projection y' on the vertical axis represents the displacement of a point as that resultant wave passes through it.

example, the wave

$$y_1(x, t) = y_{m1} \sin(kx - \omega t) \qquad (17\text{-}46)$$

is represented by the phasor shown in Fig. 17-13*a*. The magnitude of the phasor is the amplitude y_{m1} of the wave. As the phasor rotates around the origin at angular speed ω, its projection y_1 on the vertical axis varies sinusoidally, from a maximum of y_{m1} through zero to a minimum of $-y_{m1}$. This variation corresponds to the sinusoidal variation in the displacement y_1 of any point along the string as the wave passes through it.

When two waves travel along the same string in the same direction, we can represent them and their resultant wave in a *phasor diagram*. The phasors in Fig. 17-13*b* represent the wave of Eq. 17-46 and a second wave given by

$$y_2(x, t) = y_{m2} \sin(kx - \omega t + \phi). \qquad (17\text{-}47)$$

The angle between the two phasors in Fig. 17-13*b* is equal to the phase constant ϕ in Eq. 17-47. This angle is constant because the phasors rotate at the same angular speed ω, the angular frequency of the two waves.

Because waves y_1 and y_2 have the same angular wave number k and angular frequency ω, we know that their resultant is of the form

$$y'(x, t) = y_m' \sin(kx - \omega t + \beta), \qquad (17\text{-}48)$$

where y_m' is the amplitude of the resultant wave and β is its phase constant. To find the values of y_m' and β, we would have to sum the two combining waves, as we did to obtain Eq. 17-42.

To do this on a phasor diagram, we vectorially add the two phasors at any instant during their rotation, as in Fig. 17-13c where phasor y_{m2} has been shifted. The magnitude of the vector sum equals the amplitude y_m' in Eq. 17-48. The angle between the vector sum and the phasor for y_1 equals the phase constant β in Eq. 17-48.

Note that in contrast to the method of Section 17-9, using phasors allows us to combine waves *even if their amplitudes are different.*

SAMPLE PROBLEM 17-6

Two waves $y_1(x, t)$ and $y_2(x, t)$ have the same wavelength and travel in the same direction along a string. Their amplitudes are $y_{m1} = 4.0$ mm and $y_{m2} = 3.0$ mm, and their phase constants are 0 and $\pi/3$ rad, respectively. What are the amplitude y_m' and phase constant β of the resultant wave?

SOLUTION: Because the waves travel along the same string, Eq. 17-24 tells us they must have the same wave speed v. Then, because they also have the same wavelength (and thus the same angular wave number k), Eq. 17-12 tells us that they must have the same angular frequency ω. Thus they can be represented by two phasors rotating at the same angular speed ω about an origin, as in Fig. 17-13b. The angle ϕ between the phasors is $\pi/3$ rad.

To find the vector sum of the two phasors as in Fig. 17-13c, we are free to draw them at any instant during their rotation. To simplify the vector addition, we choose to draw them as shown in Fig. 17-14a. We now add the phasors as we would any vectors (Fig. 17-14b). The horizontal component of resultant phasor y_m' is

$$y_{mh}' = y_{m1} \cos 0 + y_{m2} \cos \pi/3$$
$$= 4.0 \text{ mm} + (3.0 \text{ mm}) \cos \pi/3 = 5.50 \text{ mm}.$$

The vertical component of y_m' is

$$y_{mv}' = y_{m1} \sin 0 + y_{m2} \sin \pi/3$$
$$= 0 + (3.0 \text{ mm}) \sin \pi/3 = 2.60 \text{ mm}.$$

Thus the resultant wave has an amplitude of

$$y_m' = \sqrt{(5.50 \text{ mm})^2 + (2.60 \text{ mm})^2}$$
$$= 6.1 \text{ mm} \qquad \text{(Answer)}$$

and a phase constant of

$$\beta = \tan^{-1} \frac{2.60 \text{ mm}}{5.50 \text{ mm}} = 0.44 \text{ rad.} \qquad \text{(Answer)}$$

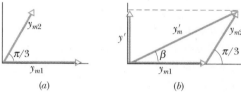

FIGURE 17-14 Sample Problem 17-6. (a) Two phasors of magnitudes y_{m1} and y_{m2} and with phase difference $\pi/3$. (b) Vector addition of these phasors at any instant during their rotation gives the magnitude y_m' of the phasor for the resultant wave.

17-11 STANDING WAVES

In the preceding two sections, we discussed two sinusoidal waves of the same wavelength and amplitude traveling *in the same direction* along a stretched string. What if they travel in opposite directions? We can again find the resultant wave by applying the superposition principle.

Figure 17-15 suggests the situation graphically. It shows the two combining waves, one traveling to the left in

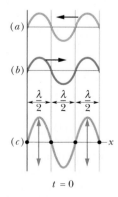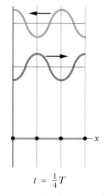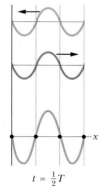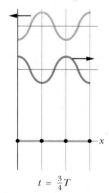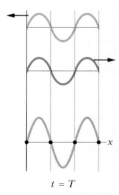

$$t = 0 \qquad t = \tfrac{1}{4}T \qquad t = \tfrac{1}{2}T \qquad t = \tfrac{3}{4}T \qquad t = T$$

FIGURE 17-15 How traveling waves produce standing waves. (a) and (b) represent snapshots of two waves of the same wavelength and amplitude, traveling in opposite directions, at five different instants during a single period of oscillation. (c) Their su-perposition at the five instants. Note the nodes and antinodes in (c), the nodes being represented by dots. There are no nodes or antinodes in the traveling waves (a) and (b).

Fig. 17-15*a*, the other to the right in Fig. 17-15*b*. Figure 17-15*c* shows their sum, obtained by applying the superposition principle graphically. The outstanding feature of the resultant wave is that there are places along the string, called **nodes**, where the string is permanently at rest. Four such nodes are marked by dots in Fig. 17-15*c*. Halfway between adjacent nodes are **antinodes,** where the amplitude of the resultant wave is a maximum. Wave patterns such as that of Fig. 17-15*c* are called **standing waves** because the wave patterns do not move left or right: the locations of the maxima and minima do not change.

> If two sinusoidal waves of the same amplitude and wavelength travel in *opposite* directions along a stretched string, their interference with each other produces a standing wave.

To analyze a standing wave, we represent the two combining waves with the equations

$$y_1(x, t) = y_m \sin(kx - \omega t) \qquad (17\text{-}49)$$

and
$$y_2(x, t) = y_m \sin(kx + \omega t). \qquad (17\text{-}50)$$

The principle of superposition gives, for the combined wave,

$$y'(x, t) = y_1(x, t) + y_2(x, t)$$
$$= y_m \sin(kx - \omega t) + y_m \sin(kx + \omega t).$$

Applying the trigonometric relation of Eq. 17-41 leads to

$$y'(x, t) = [2y_m \sin kx]\cos \omega t. \qquad (17\text{-}51)$$

This is *not* a traveling wave because it is not of the form of Eq. 17-16. Equation 17-51 describes a standing wave.

The quantity $2y_m \sin kx$ in the brackets of Eq. 17-51 can be viewed as the amplitude of oscillation of the string element that is located at position *x*. However, since an amplitude is always positive and $\sin kx$ can be negative, we take the absolute value of the quantity $2y_m \sin kx$ to be the amplitude at *x*.

In a traveling sinusoidal wave, the amplitude of the wave is the same for all string elements. That is not true for a standing wave, in which the amplitude *varies with position*. In the standing wave of Eq. 17-51, for example, the amplitude is zero for values of *kx* that give $\sin kx = 0$. Those values are

$$kx = n\pi, \qquad \text{for } n = 0, 1, 2, \cdots . \qquad (17\text{-}52)$$

Substituting $k = 2\pi/\lambda$ in this equation and rearranging, we get

$$x = n\frac{\lambda}{2}, \qquad \text{for } n = 0, 1, 2, \cdots .$$
$$\text{(nodes)}, \qquad (17\text{-}53)$$

as the positions of zero amplitude—the nodes—for the standing wave of Eq. 17-51. Note that adjacent nodes are separated by $\lambda/2$, half a wavelength.

The amplitude of the standing wave of Eq. 17-51 has a maximum value of $2y_m$, which occurs for values of *kx* that give $|\sin kx| = 1$. Those values are

$$kx = \tfrac{1}{2}\pi, \tfrac{3}{2}\pi, \tfrac{5}{2}\pi, \cdots$$
$$= (n + \tfrac{1}{2})\pi, \qquad \text{for } n = 0, 1, 2, \cdots . \qquad (17\text{-}54)$$

Substituting $k = 2\pi/\lambda$ in Eq. 17-54 and rearranging, we get

$$x = \left(n + \frac{1}{2}\right)\frac{\lambda}{2}, \qquad \text{for } n = 0, 1, 2, \cdots$$
$$\text{(antinodes)}, \qquad (17\text{-}55)$$

as the positions of maximum amplitude—the antinodes—of the standing wave of Eq. 17-51. The antinodes are one-half wavelength apart and are located halfway between pairs of nodes.

Reflections at a Boundary

We can set up a standing wave in a stretched string by allowing a traveling wave to be reflected from the far end of the string. The incident (original) wave and the reflected wave can then be described by Eqs. 17-49 and 17-50, respectively, and they can combine to form a pattern of standing waves.

In Fig. 17-16, we use a single pulse to show how such reflections take place. In Fig. 17-16*a*, the string is fixed at its left end. When the pulse arrives at that end, it exerts an upward force on the support (the wall). By Newton's third law, the support exerts an equal but opposite force on the string. This reaction force generates a pulse at the support, which travels back along the string in the direction opposite that of the incident pulse. In a ''hard'' reflection of this kind, there must be a node at the support because the string is fixed there. The reflected and incident pulses must have opposite signs, so as to cancel each other at that point.

In Fig. 17-16*b*, the left end of the string is fastened to a light ring that is free to slide without friction along a rod. When the incident pulse arrives, the ring moves up the rod. As the ring moves, it pulls on the string, stretching the string and producing a reflected pulse with the same sign and amplitude as the incident pulse. Thus in such a ''soft'' reflection, the incident and reflected pulses reinforce each other, creating an antinode at the end of the string; the

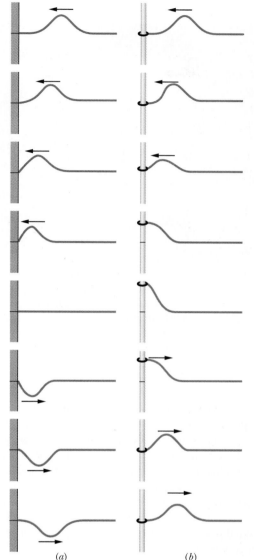

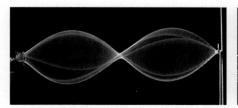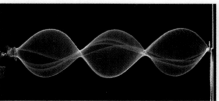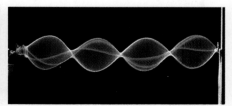

(a)   (b)

FIGURE 17-16 (a) A pulse incident from the right is reflected at the left end of the string, which is tied to a wall. Note that the sign of the reflected pulse is reversed. (b) Here the left end of the string is tied to a ring that can slide without friction up and down the rod. The pulse is reflected without a change of sign.

maximum displacement of the ring is twice the amplitude of either of these pulses.

CHECKPOINT 5: Two waves with the same amplitude and wavelength interfere in three different situations to produce resultant waves with the following equations: (1) $y'(x, t) = 4 \sin(5x - 4t)$

(2) $y'(x, t) = 4 \sin(5x) \cos(4t)$

(3) $y'(x, t) = 4 \sin(5x + 4t)$

In which situation are the two combining waves traveling (a) toward increasing x, (b) toward decreasing x, and (c) in opposite directions?

17-12 STANDING WAVES AND RESONANCE

If (say) the left end of a stretched string is oscillated sinusoidally with the other end fixed, the oscillation sends a continuous traveling wave rightward along the string. The frequency of the wave is that of the oscillation. The wave reflects at the fixed end and travels leftward back through itself. The right-going wave and the left-going wave then interfere with each other.

For certain frequencies, the interference produces a standing wave pattern (or **oscillation mode**) with nodes and large antinodes like those in Fig. 17-17. Such a standing wave is said to be produced at **resonance,** and the string is said to *resonate* at these certain frequencies, called **resonant frequencies.** If the string is oscillated at some frequency other than a resonant frequency, a standing wave is not set up. Then the interference of the right-going wave and the left-going wave results in only small (perhaps imperceptible) oscillations of the string.

Consider a similar situation in which a string, such as a guitar string, is stretched between two clamps separated by a fixed distance L, and the string is somehow made to oscillate at a resonant frequency to set up a standing wave pattern. Since each end of the string is fixed, there must be a node at each end. The simplest pattern that meets this

FIGURE 17-17 Stroboscopic photographs reveal (imperfect) standing wave patterns on a string being made to oscillate by a vibrator at the left end. The patterns occur at certain frequencies of oscillation.

requirement is that in Fig. 17-18a, which shows the string at both its extreme displacements (one solid and one dashed). Note that there is only one antinode, which is at the center of the string. Also note that there is half a wavelength in the length L, which we take to be the string's length. Thus, for this pattern, $\lambda/2 = L$. This condition tells us that if the left-going and right-going traveling waves are to set up this pattern by their interference, they must have the wavelength $\lambda = 2L$.

A second simple pattern meeting the requirement of fixed ends is shown in Fig. 17-18b. This pattern has three nodes and two antinodes. For the left-going and right-going waves to set it up, they must have a wavelength $\lambda = L$. A third pattern is shown in Fig. 17-18c. It has four nodes and three antinodes, and the wavelength is $\lambda = \frac{2}{3}L$. You could continue the progression by drawing increasingly more complicated patterns. In each step of the progression, the pattern would have one more node and one more antinode than the preceding step, and an additional $\lambda/2$ would be fitted into the distance L.

The relation between λ and L for a standing wave on a string can be summarized as

$$\lambda = \frac{2L}{n}, \qquad \text{for } n = 1, 2, 3, \cdots. \quad (17\text{-}56)$$

The resonant frequencies that correspond to these wavelengths follow from Eq. 17-12:

$$f = \frac{v}{\lambda} = n\frac{v}{2L}, \qquad \text{for } n = 1, 2, 3, \cdots. \quad (17\text{-}57)$$

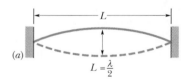

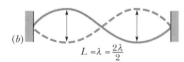

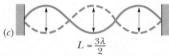

FIGURE 17-18 A string, stretched between two clamps, is made to oscillate in standing wave patterns. (a) The simplest possible pattern consists of one *loop,* which refers to the composite shape formed by the string in its extreme displacements (the solid and dashed lines). (b) The next simplest pattern has two loops. (c) The next pattern has three loops.

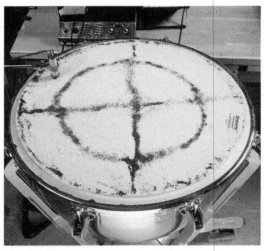

FIGURE 17-19 One of many possible standing wave patterns for a kettledrum head, made visible by dark powder sprinkled on the drumhead. As the head is set into oscillation at a single frequency by a mechanical vibrator at the upper left of the photograph, the powder collects at the nodes, which are circles and straight lines (rather than points) in this two-dimensional example.

Here v is the speed of the traveling waves on the string.

Equation 17-57 tells us that the resonant frequencies are integer multiples of the lowest resonant frequency, $f = v/2L$, which corresponds to $n = 1$. The oscillation mode with that lowest frequency is called the *fundamental mode* or the *first harmonic.* The *second harmonic* is the oscillation mode with $n = 2$; the *third harmonic* is that with $n = 3$; and so on. The frequencies associated with these modes are often labeled f_1, f_2, f_3, and so on. The collection of all possible oscillation modes is called the **harmonic series,** and n is called the **harmonic number** of the nth harmonic.

The phenomenon of resonance is common to all oscillating systems and can occur in two and three dimensions. For example, Fig. 17-19 shows a two-dimensional standing wave pattern on the oscillating head of a kettledrum.

SAMPLE PROBLEM 17-7

In Fig. 17-20, a string, tied to a sinusoidal vibrator at P and running over a support at Q, is stretched by a block of mass m. The separation L between P and Q is 1.2 m, the linear density of the string is 1.6 g/m, and the frequency f of the vibrator is fixed at 120 Hz. The amplitude of the motion at P is small enough for that point to be considered a node. A node also exists at Q.

(a) What mass m allows the vibrator to set up the fourth harmonic on the string?

SOLUTION: The resonant frequencies are given by Eq. 17-57 as

$$f = \frac{v}{2L}\, n, \qquad \text{for } n = 1, 2, 3, \cdots . \quad (17\text{-}58)$$

We need to set the tension τ in the string so that the vibrator frequency is equal to the frequency of the fourth harmonic as given by this equation.

The speed v of waves on the string is given by Eq. 17-24 as

$$v = \sqrt{\frac{\tau}{\mu}} = \sqrt{\frac{mg}{\mu}}, \quad (17\text{-}59)$$

where the tension τ in the string is equal to the weight mg of the block. Substituting v from Eq. 17-59 into Eq. 17-58, setting $n = 4$ for the fourth harmonic, and solving for m give us

$$m = \frac{4L^2 f^2 \mu}{n^2 g} \quad (17\text{-}60)$$

$$= \frac{(4)(1.2\text{ m})^2(120\text{ Hz})^2(0.0016\text{ kg/m})}{(4)^2(9.8\text{ m/s}^2)}$$

$$= 0.846\text{ kg} \approx 0.85\text{ kg}. \quad \text{(Answer)}$$

(b) What standing wave mode is set up if $m = 1.00$ kg?

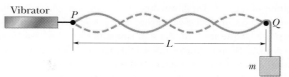

FIGURE 17-20 Sample Problem 17-7. A string under tension connected to a vibrator. For a fixed vibrator frequency, standing wave patterns will occur for certain values of the string tension.

SOLUTION: If we insert this value of m into Eq. 17-60 and solve for n, we find that $n = 3.7$, an impossibility since n must be an integer. Thus, with $m = 1.00$ kg, the vibrator cannot set up a standing wave on the string, and any oscillation of the string will be small, perhaps even imperceptible.

CHECKPOINT **6:** In the following series of resonant frequencies, one frequency (lower than 400 Hz) is missing: 150, 225, 300, 375 Hz. (a) What is the missing frequency? (b) What is the frequency of the seventh harmonic?

PROBLEM SOLVING TACTICS

TACTIC 2: *Harmonics on a String*

When you need to obtain information about a certain harmonic on a stretched string of given length L, first draw that harmonic (as in Fig. 17-18). If you are asked about, say, the fifth harmonic, you need to draw five loops between the fixed support points. That would mean that five loops, each of length $\lambda/2$, occupy the length L of the string. Thus, $5(\lambda/2) = L$, and $\lambda = 2L/5$. You can then use Eq. 17-12 ($f = v/\lambda$) to find the frequency of the harmonic.

Keep in mind that the wavelength of a harmonic is set only by the length L of the string, but the frequency depends also on the wave speed v, which is set by the tension and the linear density of the string via Eq. 17-24.

REVIEW & SUMMARY

Transverse and Longitudinal Waves

Mechanical waves can exist only in material media and are governed by Newton's laws. **Transverse** mechanical waves, like those on a stretched string, are waves in which the particles of the medium oscillate perpendicular to the wave's direction of travel. Waves in which the particles of the medium oscillate parallel to the wave's direction of travel are **longitudinal** waves.

Sinusoidal Waves

A sinusoidal wave moving in the $+x$ direction has the mathematical form

$$y(x, t) = y_m \sin(kx - \omega t), \quad (17\text{-}2)$$

where y_m is the **amplitude** of the wave, k the **angular wave**

number, ω the **angular frequency,** and $kx - \omega t$ the **phase.** The wavelength λ is related to k by

$$k = \frac{2\pi}{\lambda}. \quad (17\text{-}5)$$

The **period** T and **frequency** f of the wave are related to ω by

$$\frac{\omega}{2\pi} = f = \frac{1}{T}. \quad (17\text{-}9)$$

Finally, the **wave speed** v is related to these other parameters by

$$v = \frac{\omega}{k} = \frac{\lambda}{T} = \lambda f. \quad (17\text{-}12)$$

Equation of a Traveling Wave

In general, any function of the form

$$y(x, t) = h(kx \pm \omega t) \qquad (17\text{-}16)$$

can represent a **traveling wave** with a wave speed given by Eq. 17-12 and a wave shape given by the mathematical form of h. The plus sign is for a wave traveling in the $-x$ direction, and the minus sign for a wave in the $+x$ direction.

Wave Speed on Stretched String

The speed of a wave on a stretched string with tension τ and linear density μ is

$$v = \sqrt{\frac{\tau}{\mu}}. \qquad (17\text{-}24)$$

Power

The **average power,** or average rate at which energy is transmitted by a sinusoidal wave on a stretched string, is given by

$$\overline{P} = \tfrac{1}{2}\mu v\omega^2 y_m^2. \qquad (17\text{-}35)$$

Superposition of Waves

When two or more waves traverse the same medium, the displacement of any particle of the medium is the sum of the displacements that the individual waves would give it. This is called **superposition.**

Fourier Series

With a *Fourier series,* any wave can be constructed as the superposition of appropriate sinusoidal waves.

Interference of Waves

Two sinusoidal waves on the same string exhibit **interference,** adding or canceling according to the principle of superposition. If the two are traveling in the same direction and have the same amplitude y_m and frequency (hence the same wavelength) but differ in phase by a **phase constant** ϕ, the result is a single wave with this same frequency:

$$y'(x, t) = [2y_m \cos \tfrac{1}{2}\phi] \sin(kx - \omega t + \tfrac{1}{2}\phi). \qquad (17\text{-}42)$$

If $\phi = 0$, the waves are in phase and their interference is (fully) constructive; if $\phi = \pi$ rad, they are out of phase and their interference is destructive.

Phasors

A wave $y(x, t)$ can be represented with a *phasor.* This is a vector that has a magnitude equal to the amplitude y_m of the wave and rotates about an origin with an angular speed equal to the angular frequency ω of the wave. The projection of the rotating phasor on a vertical axis gives the displacement y of the wave.

Standing Waves

The interference of two identical sinusoidal waves moving in opposite directions produces **standing waves.** For a string with fixed ends, the standing wave is given by

$$y'(x, t) = [2y_m \sin kx]\cos \omega t. \qquad (17\text{-}51)$$

Standing waves are characterized by fixed positions of zero displacement called **nodes** and fixed positions of maximum displacement called **antinodes.**

Resonance

Standing waves on a string can be set up by reflection of traveling waves from the ends of the string. If an end is fixed, it must be the position of a node. This limits the frequencies at which standing waves will occur on a given string. Each possible frequency is a **resonant frequency,** and the corresponding standing wave pattern is an **oscillation mode.** For a stretched string of length L with fixed ends, the resonant frequencies are

$$f = \frac{v}{\lambda} = n\,\frac{v}{2L}, \qquad \text{for } n = 1, 2, 3, \cdots. \qquad (17\text{-}57)$$

The oscillation mode corresponding to $n = 1$ is called the *fundamental mode* or the *first harmonic;* the mode corresponding to $n = 2$ is the *second harmonic;* and so on.

QUESTIONS

1. What is the wavelength of the (strange) wave in Fig. 17-21, where each segment of the wave has length d?

FIGURE 17-21 Question 1.

2. A string is put under tension and then three different sinusoidal waves are separately sent along it. Figure 17-22 gives plots of the resulting displacement versus time for an element on the string.

Rank the three plots according to the wavelength of the corresponding wave, greatest first.

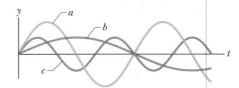

FIGURE 17-22 Question 2.

3. In Fig. 17-23, four strings are placed under tension by one or two suspended blocks, all of the same mass. Strings A, B, and C

have the same linear density; string D has a greater linear density. Rank the strings according to the speed that waves will have when sent along them, greatest first.

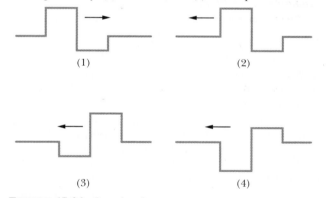

FIGURE 17-23   Question 3.

4. In Fig. 17-24, wave 1 consists of a rectangular peak of height 4 units and width d, and a rectangular valley of depth 2 units and width d. The wave travels rightward along an x axis. Waves 2, 3, and 4 are similar waves, with the same heights, depths, and widths, that will travel leftward along that axis and through wave 1. With which choice will the interference give, for an instant, (a) the deepest valley, (b) a flat line, and (c) a level peak $2d$ wide?

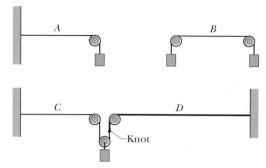

FIGURE 17-24   Question 4.

5. If you start with two waves traveling in phase on a string and then somehow phase-shift one of them by 5.4 wavelengths, what type of interference will occur on the string?

6. Here are some more phase differences between the two waves of Sample Problem 17-5, expressed in radians: $\pi/4$, $7\pi/4$, $-\pi/4$, and $-7\pi/4$. (a) Without written calculation, rank them in terms of the amplitude of the resultant wave, greatest first. (b) What type of interference occurs with each phase difference?

7. The amplitudes and phase differences for four pairs of waves of equal wavelengths are (a) 2 mm, 6 mm, and π rad; (b) 3 mm, 5 mm, and π rad; (c) 7 mm, 9 mm, and π rad; (d) 2 mm, 2 mm, and 0 rad. Each pair travels in the same direction along the same string. Without written calculation, rank the four pairs according to the amplitude of their resultant wave, greatest first. (*Hint:* Construct phasor diagrams.)

8. (a) Three waves equal in amplitude and wavelength travel along a string in the same direction. Take one to have a phase constant of $0°$. What are the phase constants of the other waves if all three together produce fully destructive interference? (*Hint:* Construct a phasor diagram.) (b) What are the phase constants if there are four such waves? Now there are two (different) answers.

9. If you set up the seventh harmonic on a string, (a) how many nodes are present, and (b) is there a node, antinode, or some intermediate state at the midpoint? If you next set up the sixth harmonic, (c) is its resonant wavelength longer or shorter than that for the seventh harmonic, and (d) is the resonant frequency higher or lower?

10. (a) In Sample Problem 17-7 and Fig. 17-20, if we gradually increase the mass of the block the (frequency remains fixed), new resonant modes appear. Do the harmonic numbers of the new resonant modes increase or decrease from one to the next? (b) Is the shift from one resonant mode to the next gradual, or does each resonant mode disappear well before the next one appears?

11. A string is stretched between two fixed supports separated by distance L. (a) For which harmonic numbers is there a node at the point located $L/3$ from one of the supports? Is there a node, antinode, or some intermediate state at a point located $2L/5$ from one support if (b) the fifth harmonic is set up and (c) the tenth harmonic is set up?

12. Strings A and B have identical lengths and linear densities, but string B is under greater tension than is string A. Figure 17-25 shows four situations, (a) through (d), in which standing wave patterns exist on the two strings. In which situations is there the possibility that strings A and B are oscillating at the same resonant frequency?

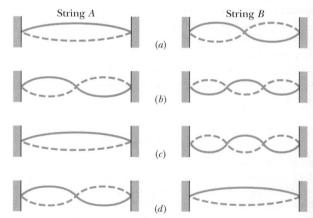

FIGURE 17-25   Question 12.

13. Two strings of equal lengths but unequal linear densities are tied together with a knot and stretched between two supports. A particular frequency happens to produce a standing wave on each length, with a node at the knot, as shown in Fig. 17-26. Which string has the greater linear density?

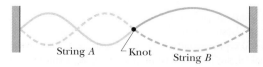

FIGURE 17-26   Question 13.

14. Do the wavelength and frequency, respectively, of the second harmonic on a string stretched between two supports increase, decrease, or remain the same if we (a) increase the distance between the supports without increasing the tension, (b) increase the tension in the string, and (c) switch to a string of greater linear density?

15. Guitar players know that prior to a show, a guitar must be played and the strings then tightened, because during the first few minutes of playing, the strings warm and loosen slightly. Does that loosening increase or decrease the resonant frequencies of the strings?

EXERCISES & PROBLEMS

SECTION 17-5 The Speed of a Traveling Wave

1E. A wave has a speed of 240 m/s and a wavelength of 3.2 m. What are the (a) frequency and (b) period of the wave?

2E. A wave has angular frequency 110 rad/s and wavelength 1.80 m. Calculate (a) the angular wave number and (b) the speed of the wave.

3E. The speed of electromagnetic waves in vacuum is 3.0×10^8 m/s. (a) Wavelengths of (visible) light waves range from about 400 nm in the violet to about 700 nm in the red. What is the range of frequencies of light waves? (b) The range of frequencies for shortwave radio (for example, FM radio and VHF television) is 1.5–300 MHz. What is the corresponding wavelength range? (c) X rays are also electromagnetic. Their wavelength range extends from about 5.0 nm to about 1.0×10^{-2} nm. What is the frequency range for x rays?

4E. A sinusoidal wave travels along a string. The time for a particular point to move from maximum displacement to zero is 0.170 s. What are the (a) period and (b) frequency? (c) The wavelength is 1.40 m; what is the wave speed?

5E. Write the equation for a wave traveling in the negative direction along the x axis and having an amplitude of 0.010 m, a frequency of 550 Hz, and a speed of 330 m/s.

6E. A traveling wave on a string is described by

$$y = 2.0 \sin \left[2\pi \left(\frac{t}{0.40} + \frac{x}{80} \right) \right],$$

where x and y are in centimeters and t is in seconds. (a) For $t = 0$, plot y as a function of x for $0 \le x \le 160$ cm. (b) Repeat (a) for $t = 0.05$ s and $t = 0.10$ s. (c) From your graphs, what is the wave speed, and in which direction ($+x$ or $-x$) is the wave traveling?

7E. Show that $y = y_m \sin(kx - \omega t)$ may be written in the alternative forms

$$y = y_m \sin k(x - vt), \qquad y = y_m \sin 2\pi \left(\frac{x}{\lambda} - ft \right),$$

$$y = y_m \sin \omega \left(\frac{x}{v} - t \right), \qquad y = y_m \sin 2\pi \left(\frac{x}{\lambda} - \frac{t}{T} \right).$$

8E. A single pulse, whose wave shape is given by the function $h(x - 5t)$, is shown in Fig. 17-27 for $t = 0$. Here x is in centimeters and t is in seconds. What are the (a) speed and (b) direction of travel of the pulse? (c) Plot $h(x - 5t)$ as a function of x for $t = 2$ s. (d) Plot $h(x - 5t)$ as a function of t for $x = 10$ cm.

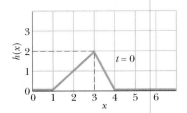

FIGURE 17-27
Exercise 8.

9E. Show (a) that the maximum transverse speed of a particle in a string owing to a traveling wave is given by $u_{max} = \omega y_m = 2\pi f y_m$ and (b) that the maximum transverse acceleration is $a_{y,max} = \omega^2 y_m = 4\pi^2 f^2 y_m$.

10E. The equation of a transverse wave traveling along a string is given by

$$y = (2.0 \text{ mm}) \sin[(20 \text{ m}^{-1})x - (600 \text{ s}^{-1})t].$$

(a) Find the amplitude, frequency, velocity, and wavelength of the wave. (b) Find the maximum transverse speed of a particle in the string.

11E. (a) Write an expression describing a sinusoidal transverse wave traveling on a cord in the $+y$ direction with an angular wave number of 60 cm^{-1}, a period of 0.20 s, and an amplitude of 3.0 mm. Take the transverse direction to be the z direction. (b) What is the maximum transverse speed of a point on the cord?

12P. The equation of a transverse wave traveling along a very long string is given by $y = 6.0 \sin(0.020\pi x + 4.0\pi t)$, where x and y are expressed in centimeters and t is in seconds. Determine (a) the amplitude, (b) the wavelength, (c) the frequency, (d) the speed, (e) the direction of propagation of the wave, and (f) the maximum transverse speed of a particle in the string. (g) What is the transverse displacement at $x = 3.5$ cm when $t = 0.26$ s?

13P. (a) Write an equation describing a sinusoidal transverse wave traveling on a cord in the $+x$ direction with a wavelength of 10 cm, a frequency of 400 Hz, and an amplitude of 2.0 cm. (b) What is the maximum speed of a point on the cord? (c) What is the speed of the wave?

14P. Prove that if a transverse sinusoidal wave is traveling along a string, then the slope at any point of the string is equal to the ratio of the particle speed to the wave speed at that point.

15P. A transverse sinusoidal wave of wavelength 20 cm is moving along a string toward increasing x. The transverse displacement of the string particle at $x = 0$ as a function of time is shown in Fig. 17-28. (a) Make a rough sketch of one wavelength of the wave (the portion between $x = 0$ and $x = 20$ cm) at time $t = 0$. (b) What is the velocity of propagation of the wave? (c) Write the equation for the wave with all the constants evaluated. (d) What is the transverse velocity of the particle at $x = 0$ at $t = 5.0$ s?

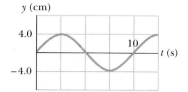

FIGURE 17-28
Problem 15.

16P. A sinusoidal wave of frequency 500 Hz has a velocity of 350 m/s. (a) How far apart are two points that differ in phase by $\pi/3$ rad? (b) What is the phase difference between two displacements at a certain point at times 1.00 ms apart?

SECTION 17-6 Wave Speed on a Stretched String

17E. What is the speed of a transverse wave in a rope of length 2.00 m and mass 60.0 g under a tension of 500 N?

18E. The heaviest and lightest strings in a certain violin have linear densities of 3.0 and 0.29 g/m. What is the ratio, heavier to lighter, of the diameters of these strings, assuming that they are made of the same material?

19E. The speed of a transverse wave on a string is 170 m/s when the string tension is 120 N. To what value must the tension be changed to raise the wave speed to 180 m/s?

20E. The tension in a wire clamped at both ends is doubled without appreciably changing its length. What is the ratio of the new to the old wave speed for transverse waves in this wire?

21E. Show that, in terms of the tensile stress S and the volume density ρ, the speed v of transverse waves in a wire is given by

$$v = \sqrt{\frac{S}{\rho}}.$$

22E. The equation of a transverse wave on a string is

$$y = (2.0 \text{ mm}) \sin[(20 \text{ m}^{-1})x - (600 \text{ s}^{-1})t].$$

The tension in the string is 15 N. (a) What is the wave speed? (b) Find the linear density of this string in grams per meter.

23E. The linear density of a string is 1.6×10^{-4} kg/m. A transverse wave is propagating on the string and is described by the following equation: $y = (0.021 \text{ m}) \sin[(2.0 \text{ m}^{-1})x + (30 \text{ s}^{-1})t]$. (a) What is the wave speed? (b) What is the tension in the string?

24E. What is the fastest transverse wave that can be sent along a steel wire? Allowing for a reasonable safety factor, the maximum tensile stress to which steel wires should be subjected is 7.0×10^8 N/m². The density of steel is 7800 kg/m³. Show that your answer does not depend on the diameter of the wire.

25P. A stretched string has a mass per unit length of 5.0 g/cm and a tension of 10 N. A sinusoidal wave on this string has an amplitude of 0.12 mm and a frequency of 100 Hz and is traveling toward decreasing x. Write an equation for this wave.

26P. For a sinusoidal transverse wave on a stretched cord, find the ratio of the maximum particle speed (the speed with which a single particle in the cord moves transverse to the wave) to the wave speed. If a wave having a certain frequency and amplitude is sent along a cord, would this speed ratio depend on the material of which the cord is made, such as wire or nylon?

27P. A sinusoidal transverse wave is traveling along a string toward decreasing x. Figure 17-29 shows a plot of the displacement as a function of position at time $t = 0$. The string tension is 3.6 N, and its linear density is 25 g/m. Find (a) the amplitude, (b) the wavelength, (c) the wave speed, and (d) the period of the wave. (e) Find the maximum speed of a particle in the string. (f) Write an equation describing the traveling wave.

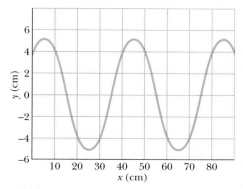

FIGURE 17-29 Problem 27.

28P. A sinusoidal wave is traveling on a string with speed 40 cm/s. The displacement of the particles of the string at $x = 10$ cm is found to vary with time according to the equation $y = (5.0 \text{ cm}) \sin[1.0 - (4.0 \text{ s}^{-1})t]$. The linear density of the string is 4.0 g/cm. What are (a) the frequency and (b) the wavelength of the wave? (c) Write the general equation giving the transverse displacement of the particles of the string as a function of position and time. (d) Calculate the tension in the string.

29P. In Fig. 17-30a, string 1 has a linear density of 3.00 g/m, and string 2 has a linear density of 5.00 g/m. They are under tension owing to the hanging block of mass $M = 500$ g. (a) Cal-

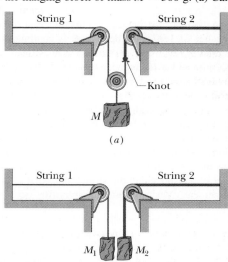

FIGURE 17-30 Problem 29.   (b)

culate the wave speed in each string. (b) The block is now divided into two blocks (with $M_1 + M_2 = M$) and the apparatus rearranged as shown in Fig. 17-30b. Find M_1 and M_2 such that the wave speeds in the two strings are equal.

30P. A wire 10.0 m long and having a mass of 100 g is stretched under a tension of 250 N. If two pulses, separated in time by 30.0 ms, are generated, one at each end of the wire, where will the pulses first meet?

31P. The type of rubber band used inside some baseballs and golf balls obeys Hooke's law over a wide range of elongation of the band. A segment of this material has an unstretched length l and a mass m. When a force F is applied, the band stretches an additional length Δl. (a) What is the speed (in terms of m, Δl, and the force constant k) of transverse waves on this stretched rubber band? (b) Using your answer to (a), show that the time required for a transverse pulse to travel the length of the rubber band is proportional to $1/\sqrt{\Delta l}$ if $\Delta l \ll l$ and is constant if $\Delta l \gg l$.

32P*. A uniform rope of mass m and length l hangs from a ceiling. (a) Show that the speed of a transverse wave on the rope is a function of y, the distance from the lower end, and is given by $v = \sqrt{gy}$. (b) Show that the time a transverse wave takes to travel the length of the rope is given by $t = 2\sqrt{l/g}$.

SECTION 17-7 Energy and Power of a Traveling String Wave

33E. Energy is transmitted at rate P_1 by a wave of frequency f_1 on a string with tension τ_1. What is the new energy transmission rate P_2 in terms of P_1 (a) if the tension of the string is increased to $\tau_2 = 4\tau_1$ and (b) if, instead, the frequency of the wave is decreased to $f_2 = f_1/2$?

34E. A string along which waves can travel is 2.70 m long and has a mass of 260 g. The tension in the string is 36.0 N. What must be the frequency of traveling waves of amplitude 7.70 mm in order that the average power be 85.0 W?

35P. A transverse sinusoidal wave is generated at one end of a long, horizontal string by a bar that moves up and down through a distance of 1.00 cm. The motion is continuous and is repeated regularly 120 times per second. The string has linear density 120 g/m and is kept under a tension of 90.0 N. Find (a) the maximum value of the transverse speed u and (b) the maximum value of the transverse component of the tension. (c) Show that the two maximum values calculated above occur at the same phase values for the wave. What is the transverse displacement y of the string at these phases? (d) What is the maximum rate of energy transfer along the string? (e) What is the transverse displacement y when this maximum transfer occurs? (f) What is the minimum rate of energy transfer along the string? (g) What is the transverse displacement y when this minimum transfer occurs?

SECTION 17-9 Interference of Waves

36E. Two identical traveling waves, moving in the same direction, are out of phase by $\pi/2$ rad. What is the amplitude of the resultant wave in terms of the common amplitude y_m of the two combining waves?

37E. What phase difference between two otherwise identical traveling waves, moving in the same direction along a stretched string, will result in the combined wave having an amplitude 1.50 times that of the common amplitude of the two combining waves? Express your answer in degrees, radians, and wavelengths.

38P. Two sinusoidal waves, identical except for phase, travel in the same direction along a string and interfere to produce a resultant wave given by $y'(x, t) = (3.0 \text{ mm}) \sin(20x - 4.0t + 0.820 \text{ rad})$, with x in meters and t in seconds. What are (a) the wavelength λ of the two waves, (b) the phase difference between them, and (c) their amplitude y_m?

SECTION 17-10 Phasors

39E. Determine the amplitude of the resultant wave when two sinusoidal string waves having the same frequency and traveling in the same direction are combined, if their amplitudes are 3.0 cm and 4.0 cm and they have phase constants of 0 and $\pi/2$ rad, respectively.

40E. The amplitudes of two sinusoidal waves traveling in the same direction along a stretched string are 3.0 and 5.0 mm; their phase constants are $0°$ and $70°$, respectively. The waves have the same wavelength. What are (a) the amplitude and (b) the phase constant of the resultant wave?

41E. Two sinusoidal waves of the same wavelength travel in the same direction along a stretched string with amplitudes of 4.0 and 7.0 mm and phase constants of 0 and 0.8π rad, respectively. What are (a) the amplitude and (b) the phase constant of the resultant wave?

42P. Two sinusoidal waves of the same period, with amplitudes of 5.0 and 7.0 mm, travel in the same direction along a stretched string; they produce a resultant wave with an amplitude of 9.0 mm. The phase constant of the 5.0 mm wave is 0. What is the phase constant of the 7.0 mm wave?

43P. Three sinusoidal waves of the same frequency travel along a string in the positive direction of an x axis. Their amplitudes are y_1, $y_1/2$, and $y_1/3$, and their phase constants are 0, $\pi/2$, and π, respectively. What are (a) the amplitude and (b) the phase constant of the resultant wave? (c) Plot the wave form of the resultant wave at $t = 0$, and discuss its behavior as t increases.

44P. Four sinusoidal waves travel in the positive x direction along the same string. Their frequencies are in the ratio $1:2:3:4$, and their amplitudes are in the ratio $1:\frac{1}{2}:\frac{1}{3}:\frac{1}{4}$, respectively. When $t = 0$, at $x = 0$, the first and third waves are $180°$ out of phase with the second and fourth. Plot the resultant wave form when $t = 0$, and discuss its behavior as t increases.

SECTION 17-12 Standing Waves and Resonance

45E. A string under tension τ_i oscillates in the third harmonic at frequency f_3, and the waves on the string have wavelength λ_3. If the tension is increased to $\tau_f = 4\tau_i$ and the string is again made to oscillate in the third harmonic, what then are (a) the frequency of oscillation in terms of f_3 and (b) the wavelength of the waves in terms of λ_3?

46E. A nylon guitar string has a linear density of 7.2 g/m and is under a tension of 150 N. The fixed supports are 90 cm apart. The string is oscillating in the standing wave pattern shown in Fig. 17-31. Calculate the (a) speed, (b) wavelength, and (c) frequency of the traveling waves whose superposition gives this standing wave.

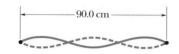

FIGURE 17-31 Exercise 46.

47E. Two sinusoidal waves with identical wavelengths and amplitudes travel in opposite directions along a string with a speed of 10 cm/s. If the time interval between instants when the string is flat is 0.50 s, what is the wavelength of the waves?

48E. When played in a certain manner, the lowest resonant frequency of a certain violin string is concert A (440 Hz). What are the frequencies of the second and third harmonics of that string?

49E. A string fixed at both ends is 8.40 m long and has a mass of 0.120 kg. It is subjected to a tension of 96.0 N and set oscillating. (a) What is the speed of the waves on the string? (b) What is the longest possible wavelength for a standing wave? (c) Give the frequency of that wave.

50E. The equation of a transverse wave traveling along a string is

$$y = 0.15 \sin(0.79x - 13t),$$

in which x and y are in meters and t is in seconds. (a) What is the displacement y at $x = 2.3$ m, $t = 0.16$ s? (b) Write the equation of a wave that, when added to the given one, would produce standing waves on the string. (c) What is the displacement of the resultant standing wave at $x = 2.3$ m, $t = 0.16$ s?

51E. A 120 cm length of string is stretched between fixed supports. What are the three longest possible wavelengths for traveling waves on the string that can produce standing waves? Sketch the corresponding standing waves.

52E. A 125 cm length of string has a mass of 2.00 g. It is stretched with a tension of 7.00 N between fixed supports. (a) What is the wave speed for this string? (b) What is the lowest resonant frequency of this string?

53E. What are the three lowest frequencies for standing waves on a wire 10.0 m long having a mass of 100 g, which is stretched under a tension of 250 N?

54E. A 1.50 m wire has a mass of 8.70 g and is held under a tension of 120 N. The wire is held rigidly at both ends and set into vibration. Calculate (a) the speed of waves on the wire, (b) the wavelengths of the waves that produce one- and two-loop standing waves on the string, and (c) the frequencies of the waves that produce one- and two-loop standing waves.

55E. String A is stretched between two clamps separated by distance l. String B, with the same linear density and under the same tension as string A, is stretched between two clamps separated by distance $4l$. Consider the first eight harmonics of string B. Which, if any, has a resonant frequency that matches a resonant frequency of string A?

56P. A string is stretched between fixed supports separated by 75.0 cm. It is observed to have resonant frequencies of 420 and 315 Hz, and no other resonant frequencies between these two. (a) What is the lowest resonant frequency for this string? (b) What is the wave speed for this string?

57P. Two waves are propagating on the same very long string. A generator at one end of the string creates a wave given by

$$y = (6.0 \text{ cm}) \cos \frac{\pi}{2} [(2.0 \text{ m}^{-1})x + (8.0 \text{ s}^{-1})t],$$

and one at the other end creates the wave

$$y = (6.0 \text{ cm}) \cos \frac{\pi}{2} [2.0 \text{ m}^{-1})x - (8.0 \text{ s}^{-1})t].$$

(a) Calculate the frequency, wavelength, and speed of each wave. (b) Find the points on the string at which there is no motion (the nodes). (c) At which points is the motion of the string a maximum (the antinodes)?

58P. A string oscillates according to the equation

$$y' = (0.50 \text{ cm}) \left[\sin \left(\frac{\pi}{3} \text{ cm}^{-1} \right) x \right] \cos[(40\pi \text{ s}^{-1})t].$$

(a) What are the amplitude and speed of the two waves (identical except for direction of travel) whose superposition gives this oscillation? (b) What is the distance between nodes? (c) What is the speed of a particle of the string at the position $x = 1.5$ cm when $t = \frac{9}{8}$ s?

59P. Two transverse sinusoidal waves travel in opposite directions along a string. Each wave has an amplitude of 0.30 cm and a wavelength of 6.0 cm. The speed of a transverse wave on the string is 1.5 m/s. Plot the shape of the string at times $t = 0$ (arbitrary), $t = 5.0$, $t = 10$, $t = 15$, and $t = 20$ ms.

60P. Two pulses travel along a string in opposite directions, as in Fig. 17-32. (a) If the wave speed v is 2.0 m/s and the pulses are 6.0 cm apart at $t = 0$, sketch the patterns when t is equal to 5.0, 10, 15, 20, and 25 ms. (b) What has happened to the energy of the pulses at $t = 15$ ms?

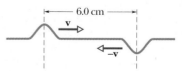

FIGURE 17-32 Problem 60.

61P. Two waves on a string are described by these equations:

$$y_1 = (0.10 \text{ m}) \sin 2\pi [(0.50 \text{ m}^{-1})x + (20 \text{ s}^{-1})t],$$
$$y_2 = (0.20 \text{ m}) \sin 2\pi [(0.50 \text{ m}^{-1})x - (20 \text{ s}^{-1})t].$$

Sketch the total response for the point on the string at $x = 3.0$ m; that is, plot y versus t for that value of x.

62P. A string 3.0 m long is oscillating as a three-loop standing wave whose amplitude is 1.0 cm. The wave speed is 100 m/s. (a) What is the frequency? (b) Write equations for two waves that, when combined, will result in this standing wave.

63P. Vibration from a 600 Hz tuning fork sets up standing waves in a string clamped at both ends. The wave speed for the string is 400 m/s. The standing wave has four loops and an amplitude of 2.0 mm. (a) What is the length of the string? (b) Write an equation for the displacement of the string as a function of position and time.

64P. In an experiment on standing waves, a string 90 cm long is attached to the prong of an electrically driven tuning fork that oscillates perpendicular to the length of the string at a frequency of 60 Hz. The mass of the string is 0.044 kg. What tension must the string be under (weights are attached to the other end) if it is to vibrate in four loops?

65P. Show that the maximum kinetic energy in each loop of a standing wave produced by two traveling waves of identical amplitudes is $2\pi^2\mu y_m^2 fv$.

66P. An aluminum wire, of length $l_1 = 60.0$ cm, cross-sectional area 1.00×10^{-2} cm², and density 2.60 g/cm³, is joined to a steel wire, of density 7.80 g/cm³ and the same cross-sectional area. The compound wire, loaded with a block of mass $m = 10.0$ kg, is arranged as in Fig. 17-33 so that the distance l_2 from the joint to the supporting pulley is 86.6 cm. Transverse waves are set up in the wire by using an external source of variable frequency; a node is located at the pulley. (a) Find the lowest frequency of excitation for which standing waves are observed such that the joint in the wire is a node. (b) How many nodes are observed at this frequency?

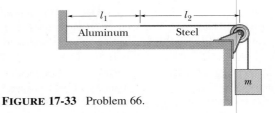

FIGURE 17-33 Problem 66.

Electronic Computation

67. Two waves,

$$y_1 = (2.50 \text{ mm})\sin[(25.1 \text{ rad/m})x - (440 \text{ rad/s})t]$$

and

$$y_2 = (1.50 \text{ mm})\sin[(25.1 \text{ rad/m})x + (440 \text{ rad/s})t],$$

travel along a stretched string. (a) Plot $y'(x, t) = y_1(x, t) + y_2(x, t)$ as a function of t for $x = 0, \lambda/8, \lambda/4, 3\lambda/8,$ and $\lambda/2$, where λ is the wavelength. The graphs should extend from $t = 0$ to a little over one period. (b) The result is the superposition of a standing wave and a traveling wave. In which direction does the traveling wave travel? How can you change the original waves so the resultant is the superposition of standing and traveling waves with the same amplitudes as before but with the traveling wave moving in the opposite direction? (c) Use your graphs to find the place at which the amplitude of vibration is the largest and the smallest. (d) How are the maximum and minimum amplitudes related to the amplitudes 2.50 mm and 1.50 mm of the original two traveling waves?

18
Waves—II

This horseshoe bat not only can locate a moth flying in total darkness but can also determine the moth's relative speed, to home in on the insect. How does the detection system work? And how can a moth "jam" the system or otherwise reduce its effectiveness?

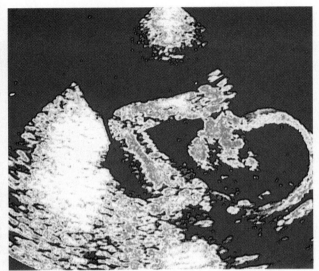

FIGURE 18-1 An ultrasound image of a fetus searching for a thumb to suck.

18-1 SOUND WAVES

As we saw in Chapter 17, mechanical waves are waves that require a material medium to exist. There are two types of mechanical waves: *transverse waves* involve oscillations perpendicular to the direction in which the wave travels; *longitudinal waves* involve oscillations parallel to the direction of wave travel.

A **sound wave** can be roughly defined as any longitudinal wave. Seismic prospecting teams use such waves to probe Earth's crust for oil. Ships carry sound ranging gear (sonar) to detect underwater obstacles. Submarines use sound waves to stalk other submarines, largely by listening for the characteristic noises produced by the propulsion system. Figure 18-1, a computer-processed image of a fetal head, shows how sound waves can be used to explore the

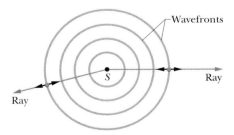

FIGURE 18-2 A sound wave travels from a point source S through a three-dimensional medium. The wavefronts form spheres centered on S; the rays are radial to S. The short, double-headed arrows indicate that elements of the medium oscillate parallel to the rays.

soft tissues of the human body. In this chapter we shall focus on sound waves that travel through the air and that are audible to people.

Figure 18-2 illustrates several ideas that we shall use in our discussions. Point S represents a tiny sound source, called a *point source*, that emits sound waves in all directions. The *wavefronts* and *rays* indicate the direction of travel and the spread of the sound waves. **Wavefronts** are surfaces over which the oscillations of the air due to the sound wave have the same value; such surfaces are represented by whole or partial circles in a two-dimensional drawing for a point source. **Rays** are directed lines perpendicular to the wavefronts that indicate the direction of travel of the wavefronts. The short double arrows superimposed on the rays of Fig. 18-2 indicate that the longitudinal oscillations of the air are parallel to the rays.

Near the point source of Fig. 18-2, the wavefronts are spherical and are spreading out in three dimensions, and the waves are said to be *spherical*. As the wavefronts move outward and their radii become larger, their curvature decreases. Far from the source, we approximate the wavefronts as straight lines and the waves are said to be *planar*.

18-2 THE SPEED OF SOUND

The speed of any mechanical wave, transverse or longitudinal, depends on both an inertial property of the medium (to store kinetic energy) and an elastic property of the medium (to store potential energy). Thus we can generalize Eq. 17-24, which gives the speed of a transverse wave along a stretched string, by writing

$$v = \sqrt{\frac{\tau}{\mu}} = \sqrt{\frac{\text{elastic property}}{\text{inertial property}}}, \qquad (18\text{-}1)$$

where (for transverse waves) τ is the tension in the string and μ is the string's linear density. If the medium is air, we can guess that the inertial property, corresponding to μ, is the volume density ρ of air. What shall we put for the elastic property?

In a stretched string, potential energy is associated with the periodic stretching of the string elements as the wave passes through them. As a sound wave passes through air, potential energy is associated with periodic compressions and expansions of small volume elements of the air. The property that determines the extent to which an element of the medium changes its volume as the pressure (force per unit area) applied to it is increased or decreased is the **bulk modulus** B, as defined (from Eq. 13-36)

$$B = -\frac{\Delta p}{\Delta V/V} \qquad \text{(definition of } B\text{).} \qquad (18\text{-}2)$$

TABLE 18-1 THE SPEED OF SOUND[a]

MEDIUM	SPEED (m/s)	MEDIUM	SPEED (m/s)
Gases		*Solids*	
Air (0°C)	331	Aluminum	6420
Air (20°C)	343	Steel	5941
Helium	965	Granite	6000
Hydrogen	1284		
Liquids			
Water (0°C)	1402		
Water (20°C)	1482		
Seawater[b]	1522		

[a]0°C and 1 atm pressure, except where noted.
[b]At 20°C and 3.5% salinity.

Here $\Delta V/V$ is the fractional change in volume produced by a change in pressure Δp. As explained in Section 15-3, the SI unit for pressure is the newton per square meter, which is given a special name, the *pascal* (Pa). From Eq. 18-2 we see that the unit for B is also the pascal. The signs of Δp and ΔV are always opposite: when we increase the pressure on a fluid element (Δp positive), its volume decreases (ΔV negative). We include a minus sign in Eq. 18-2 so that B will always be a positive quantity. Now substituting B for τ and ρ for μ in Eq. 18-1 yields

$$v = \sqrt{\frac{B}{\rho}} \qquad \text{(speed of sound)} \qquad (18\text{-}3)$$

for a medium with bulk modulus B and density ρ. Table 18-1 lists the speed of sound in various media.

The density of water is almost 1000 times greater than the density of air. If this were the only relevant factor, we would expect from Eq. 18-3 that the speed of sound in water would be considerably less than the speed of sound in air. However, Table 18-1 shows us that the reverse is true. We conclude (again from Eq. 18-3) that the bulk modulus of water must be more than 1000 times greater than that of air. This is indeed the case. Water is much more incompressible than air, which (see Eq. 18-2) is another way of saying that its bulk modulus is much greater.

Formal Derivation of Eq. 18-3

We now derive Eq. 18-3 by direct application of Newton's laws. Let a single pulse in which air is compressed travel (from right to left) with speed v through the air in a long tube, like that in Fig. 17-2. Let us run along with the pulse at that speed, so that the pulse appears to stand still in our reference frame. Figure 18-3a shows the situation as it is viewed from that frame. The pulse is standing still, and air is moving at speed v through it from left to right.

Let the pressure of the undisturbed air be p and the pressure inside the pulse be $p + \Delta p$, where Δp is positive owing to the compression. Consider a slice of air of thickness Δx and face area A, moving toward the pulse at speed v. As this element of air enters the pulse, its leading face encounters a region of higher pressure, which slows it to speed $v + \Delta v$, in which Δv is negative. This slowing is complete when the rear face reaches the pulse, which requires time interval

$$\Delta t = \frac{\Delta x}{v}. \qquad (18\text{-}4)$$

Let us apply Newton's second law to the element. During Δt, the average force on the element's trailing face is pA toward the right, and the average force on the leading

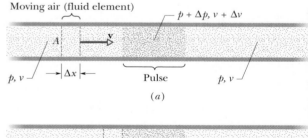

FIGURE 18-3 A pulse (compression) is sent down a long air-filled tube. The reference frame of the figure is chosen so that the pulse is at rest and the air moves from left to right. (*a*) A slice of air of width Δx moves toward the pulse with speed v. (*b*) The leading face of the slice enters the pulse. The forces acting on the leading and trailing faces (due to air pressure) are shown.

face is $(p + \Delta p)A$ toward the left (Fig. 18-3b). So the average net force on the element during Δt is

$$F = pA - (p + \Delta p)A$$
$$= -\Delta p\, A \quad \text{(net force)}. \quad (18\text{-}5)$$

The minus sign indicates that the net force on the fluid element points to the left in Fig. 18-3b. The volume of the element is $A\,\Delta x$, so with the aid of Eq. 18-4, we can write its mass as

$$\Delta m = \rho A\,\Delta x = \rho A v\,\Delta t \quad \text{(mass)}. \quad (18\text{-}6)$$

Then the average acceleration of the element during Δt is

$$a = \frac{\Delta v}{\Delta t} \quad \text{(acceleration)}. \quad (18\text{-}7)$$

From Newton's second law $(F = ma)$, we have, from Eqs. 18-5, 18-6, and 18-7,

$$-\Delta p\, A = (\rho A v\,\Delta t)\frac{\Delta v}{\Delta t},$$

which we can write as

$$\rho v^2 = -\frac{\Delta p}{\Delta v / v}. \quad (18\text{-}8)$$

The air that occupies a volume $V\,(= Av\,\Delta t)$ outside the pulse is compressed by an amount $\Delta V\,(= A\,\Delta v\,\Delta t)$ as it enters the pulse. Thus

$$\frac{\Delta V}{V} = \frac{A\,\Delta v\,\Delta t}{Av\,\Delta t} = \frac{\Delta v}{v}. \quad (18\text{-}9)$$

Substituting Eq. 18-9 and then Eq. 18-2 into Eq. 18-8 leads to

$$\rho v^2 = -\frac{\Delta p}{\Delta v / v} = -\frac{\Delta p}{\Delta V / V} = B.$$

Solving for v yields Eq. 18-3 for the speed of the air toward the right in Fig. 18-3, and thus for the actual speed of the pulse toward the left.

SAMPLE PROBLEM 18-1

One clue used by your brain to determine the direction of a source of sound is the time delay Δt between the arrival of the sound at the ear closer to the source and the arrival at the farther ear. Assume that the source is distant so that a wavefront from it is approximately straight when it reaches you, and let D represent the separation between your ears.

(a) Find an expression that gives Δt in terms of D and the angle θ between the direction of the source and the forward direction.

SOLUTION: The situation is shown in Fig. 18-4 for wavefronts approaching you from a source that is located in front of

you and to your right. The time delay Δt is due to the distance d that each wavefront must travel to reach your left ear (L) after it reaches your right ear (R). From Fig. 18-4, we find

$$\Delta t = \frac{d}{v} = \frac{D \sin \theta}{v}, \quad \text{(Answer)} \quad (18\text{-}10)$$

where v is the speed of sound in air. Based on a lifetime of experience, your brain correlates each detected value of Δt (from zero to the maximum value) with a value of θ (from zero to 90°) for the direction of the sound source.

(b) Suppose that you are submerged in water at 20°C when a wavefront arrives from directly to your right. Based on the time-delay clue, at what angle θ from the forward direction will the source seem to be?

SOLUTION: We find the time delay Δt_w in this situation with Eq. 18-10, substituting $\theta = 90°$ and using v_w, the speed of sound in water, instead of v, the speed of sound in air:

$$\Delta t_w = \frac{D \sin 90°}{v_w} = \frac{D}{v_w}. \quad (18\text{-}11)$$

Since v_w is about four times v, Δt_w is about one-fourth the maximum time delay in air. Based on experience, your brain will process the time delay as if it occurred in air. Thus the sound source will appear to be at an angle θ smaller than 90°. To find that apparent angle, we substitute the time delay D/v_w from Eq. 18-11 for Δt in Eq. 18-10, obtaining

$$\frac{D}{v_w} = \frac{D \sin \theta}{v}. \quad (18\text{-}12)$$

Substituting $v = 343$ m/s and $v_w = 1482$ m/s (from Table 18-1) into Eq. 18-12, we then find

$$\sin \theta = \frac{v}{v_w} = \frac{343 \text{ m/s}}{1482 \text{ m/s}} = 0.231$$

and thus

$$\theta = 13°. \quad \text{(Answer)}$$

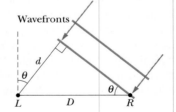

FIGURE 18-4 Sample Problem 18-1. A wavefront travels a distance d $(= D \sin \theta)$ farther to reach the left ear (L) than to reach the right ear (R).

18-3 TRAVELING SOUND WAVES

Here we examine the displacements and pressure variations associated with a sinusoidal sound wave passing through air. Figure 18-5a displays such a wave traveling rightward through a long air-filled tube. Recall that we can produce such a wave by sinusoidally moving a piston at the

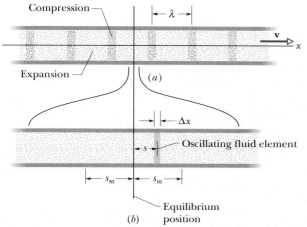

Compression

Expansion

λ

(a)

Δx

Oscillating fluid element

s

s_m s_m

Equilibrium
(b) position

FIGURE 18-5 (a) A sound wave, traveling through a long air-filled tube with speed v, consists of a moving, periodic pattern of expansions and compressions of the air. The wave is shown at an arbitrary instant. (b) A horizontally expanded view of a short piece of the tube. As the wave passes, a fluid element of thickness Δx oscillates left and right in simple harmonic motion about its equilibrium position. At the instant shown in (b), the element happens to be displaced a distance s to the right of its equilibrium position. Its maximum displacement, either right or left, is s_m.

left end of the tube (as in Fig. 17-2). The piston's rightward motion moves the element of air next to it and compresses that air; the piston's leftward motion allows the element of air to move back to the left and the pressure to decrease. As each element of air pushes on the next element in turn, the right–left motion of the air and the change in its pressure travel along the tube as a sound wave.

Consider a thin element of air of thickness Δx, located at a position x along the tube. As the wave passes through x, the element of air oscillates left and right in simple harmonic motion about its equilibrium position (Fig. 18-5b). Thus the oscillations of each air element due to the passing sound wave are like those of a string element due to a transverse wave, except that the air element oscillates *longitudinally* rather than *transversely*. To describe the displacement $s(x, t)$ of an element of air from its equilibrium position, we can use either a sine function or a cosine function. In this chapter we use a cosine function:

$$s(x, t) = s_m \cos(kx - \omega t). \quad (18\text{-}13)$$

Here s_m is the **displacement amplitude**, that is, the maximum displacement of the air element to either side of its equilibrium position (Fig. 18-5b).* The angular wave

---
*For the transverse displacement of an element of a stretched string, we used the symbol $y(x, t)$. We use the notation $s(x, t)$ here to avoid writing $x(x, t)$ for the longitudinal displacement of an element of air.

number k, angular frequency ω, frequency f, wavelength λ, speed v, and period T for a sound (longitudinal) wave are defined and interrelated exactly as for a transverse wave, except that λ is now the distance (again along the direction of travel) in which the pattern of compression and expansion due to the wave begins to repeat itself (see Fig. 18-5a). (We assume s_m is much less than λ.)

As the wave moves, the air pressure at any position x in Fig. 18-5a varies sinusoidally, as we prove below. To describe this variation we write

$$\Delta p(x, t) = \Delta p_m \sin(kx - \omega t). \quad (18\text{-}14)$$

A negative value of Δp in Eq. 18-14 corresponds to an expansion of the air, and a positive value to a compression. Here Δp_m is the **pressure amplitude**, which is the maximum increase or decrease in pressure due to the wave; Δp_m is normally very much less than the pressure p present when there is no wave. As we shall prove, the pressure amplitude Δp_m is related to the displacement amplitude s_m in Eq. 18-13 by

$$\Delta p_m = (v\rho\omega) s_m. \quad (18\text{-}15)$$

Figure 18-6 shows plots of Eqs. 18-13 and 18-14 at $t = 0$; with time, the two curves would move rightward along the horizontal axes. Note that the displacement and pressure variation are $\pi/2$ rad (or 90°) out of phase. Thus, for example, the pressure variation is zero when the displacement is a maximum.

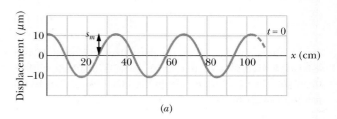

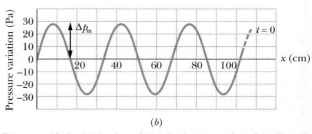

FIGURE 18-6 (a) A plot of the displacement function (Eq. 18-13) for $t = 0$. (b) A similar plot of the pressure variation function (Eq. 18-14). Both plots are for a 1000 Hz sound wave whose pressure amplitude is at the threshold of pain; see Sample Problem 18-2.

$\mathcal{C}$HECKPOINT **1:** When the oscillating fluid element in Fig. 18-5b is moving rightward through the point of zero displacement, is the pressure at its equilibrium value, just beginning to increase, or just beginning to decrease?

Derivation of Eqs. 18-14 and 18-15

Figure 18-5b shows an oscillating element of air of cross-sectional area A and thickness Δx, with its center displaced from its equilibrium position by distance s.

From Eq. 18-2 we can write, for the pressure variation in the displaced element,

$$\Delta p = -B \frac{\Delta V}{V}. \qquad (18\text{-}16)$$

The quantity V in Eq. 18-16 is the volume of the element, given by

$$V = A \, \Delta x. \qquad (18\text{-}17)$$

The quantity ΔV in Eq. 18-16 is the change in volume that occurs when the element is displaced. This volume change comes about because the displacements of the two faces of the element are not quite the same, differing by some amount Δs. Thus we can write the change in volume as

$$\Delta V = A \, \Delta s. \qquad (18\text{-}18)$$

Substituting Eqs. 18-17 and 18-18 into Eq. 18-16 and passing to the differential limit yield

$$\Delta p = -B \frac{\Delta s}{\Delta x} = -B \frac{\partial s}{\partial x}. \qquad (18\text{-}19)$$

The symbols ∂ indicate that the derivative in Eq. 18-19 is a *partial derivative* which tells us how s changes with x when the time t is fixed. From Eq. 18-13 we then have, treating t as a constant,

$$\frac{\partial s}{\partial x} = \frac{\partial}{\partial x} [s_m \cos(kx - \omega t)] = -k s_m \sin(kx - \omega t).$$

Substituting this quantity for the partial derivative in Eq. 18-19 yields

$$\Delta p = B k s_m \sin(kx - \omega t),$$

which is Eq. 18-14, which we set out to prove, with $\Delta p_m = B k s_m$.

Using Eq. 18-3, we can now write

$$\Delta p_m = (Bk) s_m = (v^2 \rho k) s_m.$$

Equation 18-15, which we also promised to prove, follows at once if we eliminate k by using $v = \omega/k$ (Eq. 17-12).

SAMPLE PROBLEM 18-2

The maximum pressure amplitude Δp_m that the human ear can tolerate in loud sounds is about 28 Pa (which is very much less than the normal air pressure of about 10^5 Pa). What is the displacement amplitude s_m for such a sound in air of density $\rho = 1.21$ kg/m³, at a frequency of 1000 Hz?

SOLUTION: From Eq. 18-15 we have

$$s_m = \frac{\Delta p_m}{v \rho \omega} = \frac{\Delta p_m}{v \rho (2\pi f)}$$

$$= \frac{28 \text{ Pa}}{(343 \text{ m/s})(1.21 \text{ kg/m}^3)(2\pi)(1000 \text{ Hz})}$$

$$= 1.1 \times 10^{-5} \text{ m} = 11 \ \mu\text{m}. \qquad \text{(Answer)}$$

The displacement amplitude for even the loudest sound that the ear can tolerate is obviously very small, being about one-seventh the thickness of this page.

The pressure amplitude Δp_m for the *faintest* detectable sound at 1000 Hz is 2.8×10^{-5} Pa. Proceeding as above leads to $s_m = 1.1 \times 10^{-11}$ m or 11 pm, which is about one-tenth the radius of a typical atom. The ear is indeed a sensitive detector of sound waves. The ear, in fact, can detect a pulse of sound whose total energy is as small as a few electron-volts, about the same as the energy required to remove an outer electron from a single atom.

18-4 INTERFERENCE

Figure 18-7 shows two point sources S_1 and S_2 that emit sound waves of wavelength λ in phase; that is, the emerging waves reach their maximum displacement values simultaneously. Suppose also that the sound waves pass through some common point P while traveling in (approximately) the same direction. If they have traveled along paths with identical lengths to reach point P, they will be in phase there. However, if they have traveled along paths with different lengths, as in Fig. 18-7, they may not be in phase. Their phase difference at point P depends on the **path length difference** ΔL between the two paths.

> When two waves travel along different paths, the path length difference ΔL can alter their phase difference.

FIGURE 18-7   Two point sources S_1 and S_2 emit spherical sound waves in phase. The rays indicate that the waves pass through a common point P.

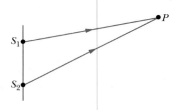

We find the phase difference ϕ between the waves by recalling (from Section 17-4) that a phase difference of 2π rad corresponds to one wavelength. Thus we can write the proportion

$$\frac{\phi}{2\pi} = \frac{\Delta L}{\lambda}, \qquad (18\text{-}20)$$

from which

$$\phi = \frac{\Delta L}{\lambda} 2\pi. \qquad (18\text{-}21)$$

Like transverse waves, sound waves undergo constructive interference and destructive interference. They undergo fully constructive interference when ϕ is zero or an integer multiple of 2π, that is, when

$$\phi = m(2\pi), \qquad m = 0, 1, 2, \cdots$$
$$\text{(fully constructive interference)}. \qquad (18\text{-}22)$$

They undergo fully destructive interference when ϕ is an odd multiple of π, that is, when

$$\phi = (m + \tfrac{1}{2})(2\pi), \qquad m = 0, 1, 2, \cdots$$
$$\text{(fully destructive interference)}. \qquad (18\text{-}23)$$

From Eq. 18-21, we see that these conditions correspond to

$$\Delta L = m\lambda, \qquad m = 0, 1, 2, \cdots$$
$$\text{(fully constructive interference)}. \qquad (18\text{-}24)$$

and

$$\Delta L = (m + \tfrac{1}{2})\lambda, \qquad m = 0, 1, 2, \cdots$$
$$\text{(fully destructive interference)}. \qquad (18\text{-}25)$$

SAMPLE PROBLEM 18-3

In Fig. 18-8a, two point sources S_1 and S_2, which are in phase and separated by a distance $D = 1.5\lambda$, emit identical sound waves of wavelength λ.

(a) What is the phase difference of the waves from S_1 and S_2 at point P_1, which lies on the perpendicular bisector of the distance D, but at a distance greater than D from the sources. What type of interference occurs at P_1?

SOLUTION: To reach P_1, waves emitted by S_1 and S_2 travel along paths that are not in the same direction. However, because P_1 is farther from the sources than the source separation D, we can approximate the directions of the waves as being the same. Furthermore, those waves travel identical distances to reach P_1. Thus the path length difference ΔL is 0, and Eq. 18-21 tells us that

$$\phi = \frac{\Delta L}{\lambda} 2\pi = 0. \qquad \text{(Answer)}$$

The value $\phi = 0$ satisfies Eq. 18-22 with $m = 0$, and so it corresponds to fully constructive interference. We get the same result with Eq. 18-24: the path length difference $\Delta L = 0$ satisfies that equation, with $m = 0$.

(b) What are the phase difference and type of interference at point P_2 in Fig. 18-8a?

SOLUTION: Point P_2 is on the line that extends through S_1 and S_2. So to reach P_2, waves emitted by S_1 must travel a distance D farther than those emitted by S_2. From Eq. 18-21 with $\Delta L = D = 1.5\lambda$, we find

$$\phi = \frac{\Delta L}{\lambda} 2\pi = \frac{1.5\lambda}{\lambda} 2\pi = 3\pi \text{ rad.} \qquad \text{(Answer)}$$

This value for ϕ satisfies Eq. 18-23 with $m = 1$, and so it corresponds to fully destructive interference. We reach the same conclusion with Eq. 18-25: because $\Delta L = 1.5\lambda$, that equation is satisfied with $m = 1$. Note that whether the phase difference is expressed in radians or wavelengths, the result is independent of the distance between P_2 and the source S_2.

(c) Figure 18-8b shows a circle with a radius much greater than D, centered on the midpoint between sources S_1 and S_2. What is the number of points N around this circle at which the interference is fully constructive?

SOLUTION: From (a), we know that the path length difference is $\Delta L = 0$ at points a and b in Fig. 18-8b, where the

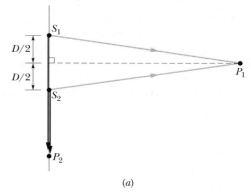

(a)

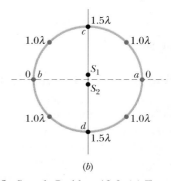

(b)

FIGURE 18-8 Sample Problem 18-3. (a) Two point sources S_1 and S_2, separated by distance D, emit spherical sound waves in phase. The waves travel equal distances to reach point P_1. Point P_2 is on the line extending through S_1 and S_2. (b) The phase difference (in terms of wavelengths) between the waves from S_1 and S_2, at eight points on a large circle around the sources.

circle intersects the perpendicular bisector of a line through the sources. From (b), we know that the path length difference is $\Delta L = 1.5\lambda$ at points c and d where the circle intersects a straight line through the sources. These results indicate that there must be intermediate points along the circle where $\Delta L = 1.0\lambda$; fully constructive interference occurs at those points. Without locating these four points exactly, we can indicate their presence as in Fig. 18-8b. We then count the points of fully constructive interference around the circle, finding

$$N = 6. \qquad \text{(Answer)}$$

CHECKPOINT 2: In Sample Problem 18-3, if the distance D between sources S_1 and S_2 were, instead, equal to 4λ, what type of interference would occur, and what value of m would be associated with it, at (a) point P_1 and (b) point P_2?

18-5 INTENSITY AND SOUND LEVEL

If you have ever tried to sleep while someone played loud music nearby, you are well aware that there is more to sound than frequency, wavelength, and speed. There is also intensity. The **intensity** I of a sound wave at a surface is the average rate per unit area at which energy is transferred by the wave through or onto the surface. We can write this as

$$I = \frac{P}{A}, \qquad (18\text{-}26)$$

where P is the rate of energy transfer (power) of the sound wave and A is the area of the surface intercepting the sound. The intensity I is related to the displacement amplitude s_m of the sound wave by

$$I = \tfrac{1}{2}\rho v\omega^2 s_m^2, \qquad (18\text{-}27)$$

which we shall derive shortly.

Variation of Intensity with Distance

How intensity varies with distance from a real sound source is often complex: some real sources (like loudspeakers) may beam sound in particular directions, and the environment usually produces echoes (reflected sound waves) that overlap the direct sound waves. In some situations, however, we can ignore echoes and assume that the sound source is a point source that emits the sound *isotropically,* that is, with equal intensity in all directions. The wavefronts spreading from such an isotropic point source S at a particular instant are shown in Fig. 18-9.

Let us assume that the mechanical energy of the sound waves is conserved as they spread from this source. Let us

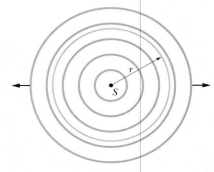

FIGURE 18-9 A point source S emits sound waves uniformly in all directions. The waves pass through an imaginary sphere of radius r that is centered on S.

also center an imaginary sphere of radius r on the source, as shown in Fig. 18-9. All the energy emitted by the source must pass through the surface of the sphere. Thus, the rate at which energy is transferred through the surface by the sound waves must equal the rate at which energy is emitted by the source (that is, the power P_s of the source). From Eq. 18-26, the intensity I at the sphere must then be

$$I = \frac{P_s}{4\pi r^2}, \qquad (18\text{-}28)$$

where $4\pi r^2$ is the area of the sphere. Equation 18-28 tells us that the intensity of sound from an isotropic point source decreases with the square of the distance r from the source.

CHECKPOINT 3: The figure indicates three small patches 1, 2, and 3 that lie on the surfaces of two imaginary spheres; the spheres are centered on an isotropic point source of sound S. The rates at which energy is transmitted through the three patches by the sound waves are equal. Rank the patches according to (a) the intensity of the sound on them and (b) their area, greatest first.

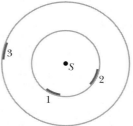

The Decibel Scale

You saw in Sample Problem 18-2 that the displacement amplitude at the human ear ranges from about 10^{-5} m for the loudest tolerable sound to about 10^{-11} m for the faintest detectable sound, a ratio of 10^6. From Eq. 18-27 we see that the intensity of a sound varies as the *square* of its amplitude, so the ratio of intensities at these two limits of the human auditory system is 10^{12}. Humans can hear over an enormous range of intensities.

Sound can cause the wall of a drinking glass to oscillate. If the sound produces a standing wave of oscillations and if the intensity of the sound is large enough, the glass will shatter.

Derivation of Eq. 18-27

Consider, in Fig. 18-5a, a thin slice of air of thickness dx, area A, and mass dm, oscillating back and forth as the sound wave of Eq. 18-13 passes through it. The kinetic energy dK of the slice of air is

$$dK = \tfrac{1}{2}dm\, v_s^2. \qquad (18\text{-}30)$$

Here v_s is not the speed of the wave but the speed of the oscillating element of air, obtained from Eq. 18-13 as

$$v_s = \frac{\partial s}{\partial t} = -\omega s_m \sin(kx - \omega t).$$

Using this relation and putting $dm = \rho A\, dx$ allow us to rewrite Eq. 18-30 as

$$dK = \tfrac{1}{2}(\rho A\, dx)(-\omega s_m)^2 \sin^2(kx - \omega t). \quad (18\text{-}31)$$

Dividing Eq. 18-31 by dt gives the rate at which kinetic energy moves along with the wave. As we saw in Chapter 17 for transverse waves, the ratio dx/dt is the wave speed, so we have

$$\frac{dK}{dt} = \tfrac{1}{2}\rho A v \omega^2 s_m^2 \sin^2(kx - \omega t). \qquad (18\text{-}32)$$

The *average* rate at which kinetic energy is transported is

$$\overline{\left(\frac{dK}{dt}\right)} = \tfrac{1}{2}\rho A v \omega^2 s_m^2 \,\overline{\sin^2(kx - \omega t)}$$
$$= \tfrac{1}{4}\rho A v \omega^2 s_m^2. \qquad (18\text{-}33)$$

To obtain this equation, we have used the fact that the average value of the square of a sine (or a cosine) function over one wavelength is $\tfrac{1}{2}$.

We assume that *potential* energy is carried along with the wave at this same average rate. The wave intensity I, which is the average rate per unit area at which energy of both kinds is transmitted by the wave, is then, from Eq. 18-33,

$$I = \frac{2\overline{(dK/dt)}}{A} = \tfrac{1}{2}\rho v \omega^2 s_m^2,$$

which is Eq. 18-27, the equation we set out to derive.

We deal with such an enormous range of values by using logarithms. Consider the relation

$$y = \log x,$$

in which x and y are variables. It is a property of this equation that if we *multiply* x by 10, then y increases by 1. To see this, we write

$$y' = \log(10x) = \log 10 + \log x = 1 + y.$$

Similarly, if we multiply x by 10^{12}, y increases by only 12.

Thus instead of speaking of the intensity I of a sound wave, it is much more convenient to speak of its **sound level** β, defined as

$$\beta = (10\text{ dB}) \log \frac{I}{I_0}. \qquad (18\text{-}29)$$

Here dB is the abbreviation for **decibel,** the unit of sound level, a name that was chosen to recognize the work of Alexander Graham Bell. I_0 in Eq. 18-29 is a standard reference intensity ($= 10^{-12}$ W/m^2), chosen because it is near the lower limit of the human range of hearing. For $I = I_0$, Eq. 18-29 gives $\beta = 10 \log 1 = 0$, so our standard reference level corresponds to zero decibels. Then β increases by 10 dB every time the sound intensity increases by an order of magnitude (a factor of 10). Thus, $\beta = 40$ corresponds to an intensity that is 10^4 times the standard reference level. Table 18-2 lists the sound levels for a variety of environments.

TABLE 18-2 **SOME SOUND LEVELS (dB)**

Hearing threshold	0	Rock concert	110
Rustle of leaves	10	Pain threshold	120
Conversation	60	Jet engine	130

SAMPLE PROBLEM 18-4

An electric spark jumps along a straight line of length $L = 10$ m, emitting a pulse of sound that travels radially outward from the spark. (The spark is said to be a *line source* of sound.) The power of the emission is $P_s = 1.6 \times 10^4$ W.

(a) What is the intensity I of the sound when it reaches a distance $r = 12$ m from the spark?

SOLUTION: Let us center an imaginary cylinder of radius $r = 12$ m and length $L = 10$ m (open at both ends) on the spark, as shown in Fig. 18-10. The rate at which energy passes through the cylinder must equal the rate P_s at which energy is emitted by the source. From Eq. 18-26, the intensity I at the cylindrical surface must then equal the power P_s divided by the area $2\pi rL$ of that surface:

$$I = \frac{P_s}{2\pi rL}. \qquad (18\text{-}34)$$

This tells us that the intensity of the sound from a line source decreases with distance r (and not with the square of distance r as for a point source). Substituting the given data, we find

$$I = \frac{1.6 \times 10^4 \text{ W}}{2\pi(12\text{ m})(10\text{ m})} = 21.2 \text{ W/m}^2 \approx 21 \text{ W/m}^2. \quad \text{(Answer)}$$

(b) At what rate P_d is sound energy intercepted by an acoustic detector of area $A_d = 2.0$ cm², aimed at the spark and located a distance $r = 12$ m from the spark?

SOLUTION: From Eq. 18-26, we know that

$$I = \frac{P_d}{A_d}.$$

Solving for P_d and substituting the intensity I from (a) and the given value for A_d, we find

$$P_d = (21.2 \text{ W/m}^2)(2.0 \times 10^{-4} \text{ m}^2) = 4.2 \text{ mW}. \quad \text{(Answer)}$$

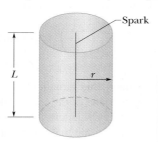

FIGURE 18-10 Sample Problem 18-4. A spark along a straight path of length L emits sound waves radially outward. The waves pass through an imaginary cylinder of radius r and length L that is centered on the spark.

SAMPLE PROBLEM 18-5

In 1976, the Who set a record for the loudest concert: the sound level 46 m in front of the speaker systems was $\beta_2 = 120$ dB. What is the ratio of the intensity I_2 of the band at that spot to the intensity I_1 of a jackhammer operating at sound level $\beta_1 = 92$ dB?

SOLUTION: Let us write the ratio of the two intensities as

$$\frac{I_2}{I_1} = \frac{I_2/I_0}{I_1/I_0}.$$

Taking logarithms on each side and multiplying through by 10 dB, we have

$$(10 \text{ dB}) \log \frac{I_2}{I_1} = (10 \text{ dB}) \log \frac{I_2}{I_0} - (10 \text{ dB}) \log \frac{I_1}{I_0}.$$

From Eq. 18-29 we then see that the terms on the right are β_2 and β_1. So

$$(10 \text{ dB}) \log \frac{I_2}{I_1} = \beta_2 - \beta_1. \qquad (18\text{-}35)$$

Note that the *ratio* of two intensities corresponds to a *difference* in their sound levels. Substituting given data yields

$$(10 \text{ dB}) \log \frac{I_2}{I_1} = 120 \text{ dB} - 92 \text{ dB} = 28 \text{ dB}$$

and

$$\log \frac{I_2}{I_1} = \frac{28 \text{ dB}}{10 \text{ dB}} = 2.8.$$

Taking the antilog of both sides gives us

$$\frac{I_2}{I_1} = 630. \qquad \text{(Answer)}$$

Thus, the Who was *very* loud.

Temporary exposure to sound intensities as great as those of a jackhammer and the 1976 Who concert results in a temporary reduction of hearing. Repeated or prolonged exposure can result in permanent reduction of hearing (Fig. 18-11). Loss of hearing is a clear risk for anyone continually listening to, say, heavy metal at high volume.

FIGURE 18-11 Sample Problem 18-5. Peter Townshend of the Who, playing in front of a speaker system. His repeated and prolonged exposure to high-intensity sound, especially when playing into a speaker to produce feedback, resulted in a permanent reduction in his hearing.

18-6 SOURCES OF MUSICAL SOUND

Musical sounds can be set up by oscillating strings (guitar, piano, violin), membranes (kettledrum, snare drum), air columns (flute, oboe, pipe organ, and the fujara of Fig. 18-12), wooden blocks or steel bars (marimba, xylophone), and many other oscillating bodies. Most instruments involve more than a single oscillating part. In the violin, for example, not only the strings but also the body of the instrument participates in producing the music.

Recall from Chapter 17 that standing waves can be set up on a stretched string that is fixed at both ends. They arise because waves traveling along the string are reflected back onto the string at each end. If the wavelength of the waves is suitably matched to the length of the string, the superposition of waves traveling in opposite directions produces a standing wave pattern (or oscillation mode). The wavelength required of the waves for such a match is one that corresponds to a *resonant frequency* of the string. The advantage of setting up standing waves is that the string then oscillates with a large, sustained amplitude, pushing back and forth against the surrounding air and thus generating a noticeable sound wave with the same frequency as the oscillations of the string. This production of sound is of obvious importance to, say, a guitarist.

We can set up standing waves of sound in an air-filled pipe in a similar way. As sound waves travel through the air in the pipe, they reflect at each end and back through the pipe. (The reflection occurs even if an end is open, but the reflection is not as complete as when the end is closed.) If the wavelength of the sound waves is suitably matched to the length of the pipe, the superposition of waves traveling in opposite directions through the pipe sets up a standing wave pattern. The wavelength required of the sound waves for such a match is one that corresponds to a resonant frequency of the pipe. The advantage of such a standing wave is that the air in the pipe oscillates with a large, sustained amplitude, emitting at any open end a sound wave that has the same frequency as the oscillations in the pipe. This emission of sound is of obvious importance to, say, an organist.

Many other aspects of standing sound wave patterns are similar to those of string waves: the closed end of a pipe is like the fixed end of a string in that there must be a node (zero displacement) there. And the open end of a pipe is like the end of a string attached to a freely moving ring, as in Fig. 17-16*b*, in that there must be an antinode there. (Actually, the antinode for the open end of a pipe is located slightly beyond the end, but we shall not dwell on that detail here.)

The simplest standing wave pattern that can be set up in a pipe with two open ends is shown in Fig. 18-13*a*. There is an antinode across each open end, as required. There is also a node across the middle of the pipe. An easier way of representing this standing longitudinal sound wave is shown in Fig. 18-13*b*—by drawing it as a standing transverse string wave.

The standing wave pattern of Fig. 18-13*a* is called the *fundamental mode* or *first harmonic*. For it to be set up, the sound waves in a pipe of length *L* must have a wavelength

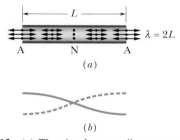

(a)

(b)

FIGURE 18-13 (*a*) The simplest standing wave pattern of displacement for (longitudinal) sound waves in a pipe with both ends open has an antinode (A) across each end and a node (N) across the middle. (The displacements represented by the double arrows are greatly exaggerated.) (*b*) The corresponding standing wave pattern for (transverse) string waves.

FIGURE 18-12 The air column within a fujara oscillates when that traditional Slovakian instrument is played.

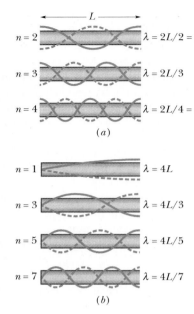

FIGURE 18-14 Standing wave patterns for string waves superimposed on pipes to represent standing sound wave patterns in the pipes. (*a*) With *both* ends of the pipe open, any harmonic can be set up in the pipe. (*b*) But with only *one* end open, only odd harmonics can be set up.

given by $L = \lambda/2$, so that $\lambda = 2L$. Several more standing sound wave patterns for a pipe with two open ends are shown in Fig. 18-14*a* using string wave representations. The *second harmonic* requires sound waves of wavelength $\lambda = L$. The *third harmonic* requires wavelength $\lambda = 2L/3$. And so on.

More generally, the resonant frequencies for a pipe of length L with two open ends correspond to the wavelengths

$$\lambda = \frac{2L}{n}, \qquad n = 1, 2, 3, \cdots, \qquad (18\text{-}36)$$

where n is called the *harmonic number*. The resonant frequencies are then given by

$$f = \frac{v}{\lambda} = \frac{nv}{2L}, \qquad n = 1, 2, 3, \cdots$$

$$\text{(pipe, two open ends),} \qquad (18\text{-}37)$$

where v is the speed of sound.

Figure 18-14*b* shows (using string wave representations) some of the standing sound wave patterns that can be set up in a pipe with only one open end. As required, across the open end there is an antinode and across the closed end there is a node. The simplest pattern requires sound waves having a wavelength given by $L = \lambda/4$, so that $\lambda = 4L$. The next simplest pattern requires a wavelength given by $L = 3\lambda/4$, so that $\lambda = 4L/3$. And so on.

More generally, the resonant frequencies for a pipe of length L with only one open end correspond to the wavelengths

$$\lambda = \frac{4L}{n}, \qquad n = 1, 3, 5, \cdots, \qquad (18\text{-}38)$$

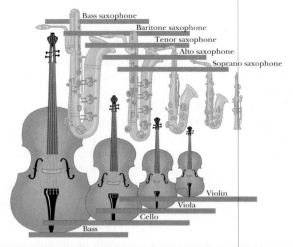

FIGURE 18-15 The saxophone and violin families, showing the relations between instrument length and frequency range. The frequency range of each instrument is indicated by a horizontal bar along a frequency scale suggested by the keyboard at the bottom; the frequency increases toward the right.

in which the harmonic number n *must be an odd number.* The resonant frequencies are then given by

$$f = \frac{v}{\lambda} = \frac{nv}{4L}, \qquad n = 1, 3, 5, \cdots$$

$$\text{(pipe, one open end).} \qquad (18\text{-}39)$$

Note again that only odd harmonics can exist in a pipe with one open end. For example, the second harmonic, with $n = 2$, cannot be set up in such a pipe. Note also that for such a pipe the adjective in a phrase such as "the third harmonic" still refers to the harmonic number n (and not to, say, the third possible harmonic).

The length of a musical instrument reflects the range of frequencies over which the instrument is designed to function: smaller length implies higher frequencies. Figure 18-15, for example, shows the saxophone and violin families, with their frequency ranges suggested by the piano keyboard. Note that, for every instrument, there is overlap with its higher- and lower-frequency neighbors.

In any oscillating system that gives rise to a musical sound, whether it be a violin string or the air in an organ pipe, the fundamental and one or more of the higher harmonics are usually generated simultaneously. The resultant sound is the superposition of these components. For different instruments, the higher harmonics have different intensities, which accounts for the different sounds produced by different instruments playing the same note. Figure 18-16 shows, for example, the resultant wave forms when the same note, with the same fundamental frequency, is sounded on three different instruments.

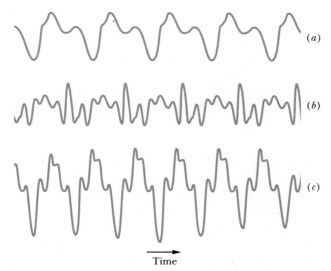

FIGURE 18-16 The wave forms produced by (*a*) a flute, (*b*) an oboe, and (*c*) a saxophone when they all play the same note, with the same first harmonic frequency.

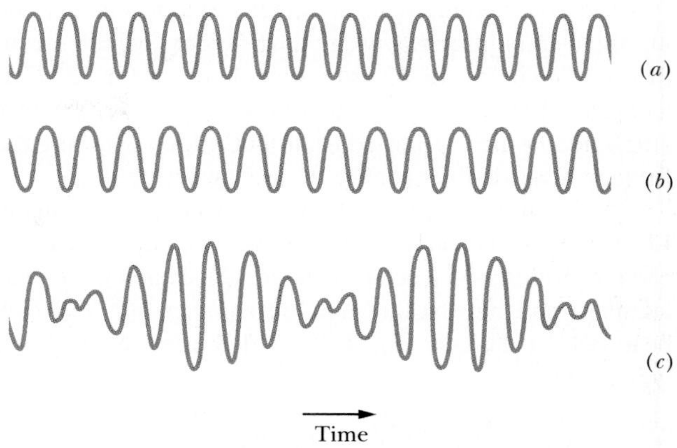

FIGURE 18-17 (*a, b*) The pressure variations Δp of two sound waves that are detected separately. The frequencies of the waves are nearly equal. (*c*) The resultant pressure variation if the two waves are detected simultaneously.

SAMPLE PROBLEM 18-6

Weak background noises from a room set up the fundamental standing wave in a cardboard tube of length $L = 67.0$ cm with two open ends. Assume that the speed of sound in the air within the tube is 343 m/s.

(a) What frequency do you hear from the tube if you jam your ear against one end?

SOLUTION: With your ear closing one end, the fundamental frequency is given by Eq. 18-39 with $n = 1$:

$$f = \frac{v}{4L} = \frac{343 \text{ m/s}}{(4)(0.670 \text{ m})} = 128 \text{ Hz}. \quad \text{(Answer)}$$

If the background noises set up any higher harmonics, such as the third harmonic, you may also hear frequencies that are *odd* multiples of 128 Hz.

(b) What frequency do you hear from the tube if you move your head away enough so that the tube has two open ends?

SOLUTION: With both ends open, the fundamental frequency is given by Eq. 18-37 with $n = 1$:

$$f = \frac{v}{2L} = \frac{343 \text{ m/s}}{(2)(0.670 \text{ m})} = 256 \text{ Hz}. \quad \text{(Answer)}$$

If the background noises set up any higher harmonics, such as the second harmonic, you may also hear frequencies that are *integer* multiples of 256 Hz. However, sound with a frequency of 128 Hz is no longer emitted by the tube.

CHECKPOINT **4:** Pipe A, with length L, and pipe B, with length $2L$, both have two open ends. Which harmonic of pipe B has the same frequency as the fundamental of pipe A?

18-7 BEATS

If we listen, a few minutes apart, to two sounds whose frequencies are, say, 552 and 564 Hz, most of us cannot tell one from the other. However, if the sounds reach our ears simultaneously, what we hear is a sound whose frequency is 558 Hz, the *average* of the two combining frequencies. We also hear a striking variation in the intensity of this sound: it increases and decreases in slow, wavering *beats* that repeat at a frequency of 12 Hz, the *difference* between the two combining frequencies. Figure 18-17 shows this **beat** phenomenon.

Let the time-dependent variations of the displacements due to two sound waves at a particular location be

$$s_1 = s_m \cos \omega_1 t \quad \text{and} \quad s_2 = s_m \cos \omega_2 t. \quad \text{(18-40)}$$

We have assumed, for simplicity, that the waves have the same amplitude. According to the superposition principle, the resultant displacement is

$$s = s_1 + s_2 = s_m(\cos \omega_1 t + \cos \omega_2 t).$$

Using the trigonometric identity (see Appendix E)

$$\cos \alpha + \cos \beta = 2 \cos \tfrac{1}{2}(\alpha - \beta) \cos \tfrac{1}{2}(\alpha + \beta)$$

allows us to write the resultant displacement as

$$s = 2s_m \cos \tfrac{1}{2}(\omega_1 - \omega_2)t \cos \tfrac{1}{2}(\omega_1 + \omega_2)t. \quad \text{(18-41)}$$

If we write

$$\omega' = \tfrac{1}{2}(\omega_1 - \omega_2) \quad \text{and} \quad \omega = \tfrac{1}{2}(\omega_1 + \omega_2), \quad \text{(18-42)}$$

we can then write Eq. 18-41 as

$$s(t) = [2s_m \cos \omega' t] \cos \omega t. \quad \text{(18-43)}$$

We now assume that the angular frequencies ω_1 and ω_2 of the combining waves are almost equal, which means that $\omega \gg \omega'$ in Eq. 18-42. We can then regard Eq. 18-43 as a cosine function whose angular frequency is ω and whose amplitude (which is not constant but varies with angular frequency ω') is the quantity in the brackets.

A maximum amplitude will occur whenever cos $\omega't$ in Eq. 18-43 has the value $+1$ or -1, which happens twice in each repetition of the cosine function. Because cos $\omega't$ has angular frequency ω', the angular frequency ω_{beat} at which beats occur is $\omega_{\text{beat}} = 2\omega'$. Then, with the aid of Eq. 18-42, we can write

$$\omega_{\text{beat}} = 2\omega' = (2)(\tfrac{1}{2})(\omega_1 - \omega_2) = \omega_1 - \omega_2.$$

Because $\omega = 2\pi f$, we can recast this as

$$f_{\text{beat}} = f_1 - f_2 \qquad \text{(beat frequency).} \qquad (18\text{-}44)$$

Musicians use the beat phenomenon in tuning their instruments. If an instrument is sounded against a standard frequency (for example, the lead oboe's reference A) and tuned until the beat disappears, then the instrument is in tune with that standard. In musical Vienna, concert A (440 Hz) is available as a telephone service for the benefit of the city's many professional and amateur musicians.

SAMPLE PROBLEM 18-7

You wish to tune the note A_3 on a piano to its proper frequency of 220 Hz. You have available a tuning fork whose frequency is 440 Hz. How should you proceed?

SOLUTION: These two frequencies are too far apart to produce beats. Recall that in our analysis of Eq. 18-43 we assumed that the two combining frequencies were reasonably close to each other. However, note that the second harmonic of A_3 is 2×220 or 440 Hz.

Let us say that the actual piano string is mistuned such that its fundamental frequency is not exactly 220 Hz. You listen for beats between the fundamental frequency of the tuning fork and the second harmonic of A_3, and you hear a beat frequency of 6 Hz. You then change the tension in the corresponding string until the beat note disappears. The string is then in tune.

C̶HECKPOINT **5:** In Sample Problem 18-7, you tighten the string and the beat frequency increases from 6 Hz. Should you continue to tighten the string or should you loosen the string to put the string in tune?

18-8 THE DOPPLER EFFECT

A police car is parked by the side of the highway, sounding its 1000 Hz siren. If you are also parked by the highway, you will hear that same frequency. But if there is relative motion between you and the police car, either toward or away from each other, you will hear a different frequency. For example, if you are driving *toward* the police car at 120 km/h (about 75 mi/h), you will hear a *higher* frequency (1096 Hz, an *increase* of 96 Hz). If you are driving *away from* the police car at that same speed, you will hear a *lower* frequency (904 Hz, a *decrease* of 96 Hz).

These motion-related frequency changes are examples of the **Doppler effect.** The effect was proposed (although not fully worked out) in 1842 by Austrian physicist Johann Christian Doppler. It was tested experimentally in 1845 by Buys Ballot in Holland, ''using a locomotive drawing an open car with several trumpeters.''

The Doppler effect holds not only for sound waves but also for electromagnetic waves, including microwaves, radio waves, and visible light. Police use the Doppler effect with microwaves to determine the speed of a car: a radar unit beams microwaves of a certain frequency f toward the oncoming car. The microwaves that reflect from the metal portions of the car back to the radar unit have a higher frequency f' owing to the motion of the car relative to the radar unit. The radar unit captures the difference between f' and f and converts it into the speed of the car, which is then displayed for the operator to see. However, the displayed speed is accurate only if the car is moving directly toward (or away from) the radar unit. (Misalignment reduces f' and also the displayed speed.)

In the analyses that follow, we restrict ourselves to sound waves, and we take as a reference frame the body of air through which these waves travel. This means that we

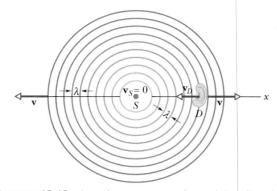

FIGURE 18-18 A stationary source of sound S emits spherical wavefronts, shown one wavelength apart, that expand outward at speed v. A sound detector D, represented by an ear, moves with velocity $\mathbf{v}_D$ toward the source. The detector senses a higher frequency because of its motion.

measure the speeds of a source S of sound waves and a detector D of those waves *relative to that body of air*. (Unless otherwise stated, the body of air is stationary relative to Earth, so the speeds can also be measured relative to Earth.) We shall assume that S and D move either directly toward or directly away from each other, at speeds less than the speed of sound.

We shall now derive equations for the Doppler effect for two particular situations: (1) for the detector moving and the source stationary and (2) for the source moving and the detector stationary. Then we shall combine the equations for these particular situations to find the *general Doppler effect equation*, which holds for both those situations and also when both detector and source are moving.

Detector Moving; Source Stationary

In Fig. 18-18, a detector D (represented by an ear) is moving at speed v_D toward a stationary source S that emits spherical wavefronts, of wavelength λ and frequency f, moving at the speed v of sound in air. The wavefronts are drawn one wavelength apart. The frequency detected by detector D is the rate at which D intercepts wavefronts (or individual wavelengths). If D were stationary, that rate would be f, but since D is moving into the wavefronts, the rate of interception is greater, and thus the detected frequency f' is greater than f.

Let us for the moment consider the situation in which D is stationary (Fig. 18-19). In time t, the wavefronts move to the right a distance vt. The number of wavelengths in that distance vt is the number of wavelengths intercepted by D in time t, and that number is vt/λ. The rate at which D intercepts wavelengths, which is the frequency f detected by D, is

$$f = \frac{vt/\lambda}{t} = \frac{v}{\lambda}. \tag{18-45}$$

In this situation, with D stationary, there is no Doppler effect: the frequency detected by D is the frequency emitted by S.

Now let us again consider the situation in which D moves opposite the wavefronts (Fig. 18-20). In time t, the wavefronts move to the right a distance vt as previously, but now D moves to the left a distance $v_D t$. Thus in this time t, the distance moved by the wavefronts relative to D is $vt + v_D t$. The number of wavelengths in this relative distance $vt + v_D t$ is the number of wavelengths intercepted by D in time t, and is $(vt + v_D t)/\lambda$. The *rate* at which D intercepts wavelengths in this situation is the frequency f', given by

$$f' = \frac{(vt + v_D t)/\lambda}{t} = \frac{v + v_D}{\lambda}. \tag{18-46}$$

From Eq. 18-45, we have $\lambda = v/f$. Then Eq. 18-46 becomes

$$f' = \frac{v + v_D}{v/f} = f\frac{v + v_D}{v}. \tag{18-47}$$

Note that in Eq. 18-47, f' must be greater than f unless $v_D = 0$ (the detector is stationary).

Similarly, we can find the frequency detected by D if D moves away from the source. In this situation, the wavefronts move a distance $vt - v_D t$ relative to D in time t, and f' is given by

$$f' = f\frac{v - v_D}{v}. \tag{18-48}$$

In Eq. 18-48, f' must be less than f unless $v_D = 0$.

We can summarize Eqs. 18-47 and 18-48 with

$$f' = f\frac{v \pm v_D}{v} \qquad \text{(detector moving; source stationary).} \tag{18-49}$$

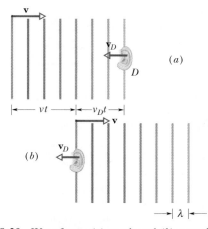

FIGURE 18-20 Wavefronts (*a*) reach and (*b*) pass detector D, which moves opposite the wavefronts. In time t, the wavefronts move a distance vt to the right and D moves a distance $v_D t$ to the left.

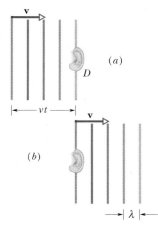

FIGURE 18-19 Wavefronts of Fig. 18-18, assumed planar, (*a*) reach and (*b*) pass a stationary detector D; they move a distance vt to the right in time t.

You can determine which sign applies in Eq. 18-49 by remembering the physical result: when the detector moves toward the source, the frequency is greater (*toward* means *greater*), which requires a plus sign in the numerator. Otherwise, a minus sign is required.

Source Moving; Detector Stationary

Let detector D be stationary with respect to the body of air, and let source S move toward D at speed v_S (Fig. 18-21). The motion of S changes the wavelength of the sound waves it emits, and thus the frequency detected by D.

To see this change, let T ($= 1/f$) be the time between the emission of any pair of successive wavefronts W_1 and W_2. During T, wavefront W_1 moves a distance vT and the source moves a distance $v_S T$. At the end of T, wavefront W_2 is emitted. In the direction in which S moves, the distance between W_1 and W_2, which is the wavelength λ' of the waves moving in that direction, is $vT - v_S T$. If D detects those waves, it detects frequency f' given by

$$f' = \frac{v}{\lambda'} = \frac{v}{vT - v_S T} = \frac{v}{v/f - v_S/f}$$

$$= f\frac{v}{v - v_S}. \qquad (18\text{-}50)$$

Note that f' must be greater than f unless $v_S = 0$.

In the direction opposite that taken by S, the wavelength λ' of the waves is $vT + v_S T$. If D detects those

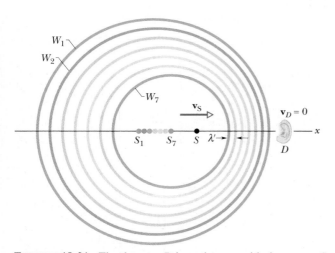

FIGURE 18-21 The detector D is stationary, with the source S moving toward it at speed v_S. Wavefront W_1 was emitted when the source was at S_1, wavefront W_7 when it was at S_7. At the moment depicted, the source is at S. The detector perceives a higher frequency because the moving source, chasing its own wavefronts, emits a reduced wavelength λ' in the direction of its motion.

waves, it detects frequency f' given by

$$f' = f\frac{v}{v + v_S}. \qquad (18\text{-}51)$$

Now f' must be less than f unless $v_S = 0$.

We can summarize Eqs. 18-50 and 18-51 with

$$f' = f\frac{v}{v \mp v_S} \qquad \begin{array}{c}\text{(source moving;}\\ \text{detector stationary).}\end{array} \qquad (18\text{-}52)$$

You can determine which sign applies in Eq. 18-52 by remembering the physical result: when the source moves toward the detector, the frequency is greater (*toward* means *greater*), which requires a minus sign in the denominator. Otherwise a plus sign should be used.

General Doppler Effect Equation

We can combine Eqs. 18-49 and 18-52 to produce the general Doppler effect equation, in which both the source and the detector can be moving with respect to the air mass. Replacing f in Eq. 18-52 (the frequency of the source) with f' of Eq. 18-49 (the frequency associated with motion of the detector) leads to

$$f' = f\frac{v \pm v_D}{v \mp v_S} \qquad \begin{array}{c}\text{(general Doppler}\\ \text{effect).}\end{array} \qquad (18\text{-}53)$$

Putting $v_S = 0$ in Eq. 18-53 reduces it to Eq. 18-49, and putting $v_D = 0$ reduces it to Eq. 18-52. The plus and minus signs are determined as in Eqs. 18-49 and 18-52 (*toward* means *greater*).

The Doppler Effect at Low Speeds

The Doppler effects for a moving detector (Eq. 18-49) and for a moving source (Eq. 18-52) are different, even though the detector and the source may be moving at the same speed. However, if the speeds are low enough (that is, if $v_D \ll v$ and $v_S \ll v$), the frequency changes produced by these two motions are essentially the same.

By using the binomial theorem (see Tactic 2 of Chapter 7), you can show that Eq. 18-53 can be written in the form

$$f' \approx f\left(1 \pm \frac{u}{v}\right) \qquad \text{(low speeds only),} \qquad (18\text{-}54)$$

in which u ($= |v_S \pm v_D|$) is the *relative* speed of the source with respect to the detector. The rule for signs remains the same. If the source and the detector are moving *toward* each other, we anticipate a *greater* frequency; this requires that we choose the plus sign in Eq. 18-54. On the other hand, if the source and the detector are moving away from each other, we anticipate a frequency decrease and choose the minus sign in Eq. 18-54.

CHECKPOINT **6:** The figure indicates the motion of a sound source and a detector for six situations in stationary air. For each situation, is the detected frequency greater than or less than the emitted frequency, or can't we tell without more information?

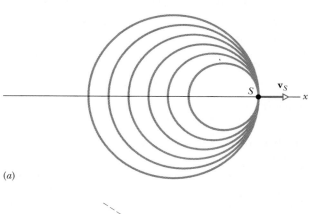

(a)

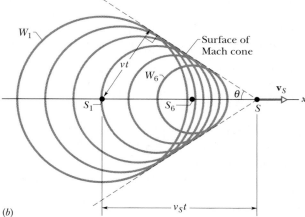

(b)

FIGURE 18-22 (*a*) A source of sound *S* moves at speed v_S equal to the speed of sound and thus as fast as the wavefronts it generates. (*b*) A source *S* moves at speed v_S faster than the speed of sound and thus faster than the wavefronts. When the source was at position S_1 it generated wavefront W_1, and at position S_6 it generated W_6. All the spherical wavefronts expand at the speed of sound v and bunch along the surface of a cone called the Mach cone, forming a shock wave. The surface of the cone has half-angle θ and is tangent to all the wavefronts.

Supersonic Speeds; Shock Waves

If a source is moving toward a stationary detector at a speed equal to the speed of sound, that is, if $v_S = v$, Eq. 18-52 predicts that the detected frequency f' will be infinitely great. This means that the source is moving so fast that it keeps pace with its own spherical wavefronts, as Fig. 18-22*a* suggests. What happens when the speed of the source *exceeds* the speed of sound?

For such *supersonic* speeds, Eq. 18-52 no longer applies. Figure 18-22*b* depicts the spherical wavefronts that originated at various positions of the source. The radius of any wavefront in this figure is vt, where v is the speed of sound and t is the time that has elapsed since the source emitted that wavefront. Note that all the wavefronts bunch along a V-shaped envelope in Fig. 18-22*b*, which in three dimensions is a cone called the *Mach cone*. A **shock wave** is said to exist along the surface of this cone, because the bunching of wavefronts causes an abrupt rise and fall of air pressure as the surface passes through any point. From Fig. 18-22*b*, we see that the half-angle θ of the cone, called the *Mach cone angle*, is given by

$$\sin \theta = \frac{vt}{v_S t} = \frac{v}{v_S} \qquad \text{(Mach cone angle).} \qquad (18\text{-}55)$$

The ratio v_S/v is called the *Mach number*. When you hear that a particular plane has flown at Mach 2.3, it means that its speed was 2.3 times the speed of sound in the air through which the plane was flying. The shock wave generated by a supersonic aircraft or projectile (Fig. 18-23) produces a burst of sound, called a *sonic boom,* in which the air pressure first suddenly increases and then suddenly decreases below normal before returning to normal.

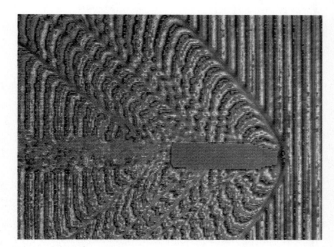

FIGURE 18-23 A false-color, high-speed image of a 20 mm bullet traveling at about Mach 1.3. Note the prominent Mach cone produced by the nose of the bullet and the secondary cones produced by irregular features along the sides.

SAMPLE PROBLEM 18-8

A toy rocket moves at a speed of 242 m/s directly toward a stationary pole (through stationary air) while emitting sound waves at frequency $f = 1250$ Hz.

(a) What frequency f' is sensed by a detector that is attached to the pole?

SOLUTION: To find f', we shall use Eq. 18-53 for the general Doppler effect. Because the detector is stationary, we set $v_D = 0$. And because the sound source (the rocket) is moving *toward* the detector, we choose the minus sign in the denominator. Then substituting $v = 343$ m/s from Table 18-1 and the given data, we find that the detected frequency is

$$f' = f\frac{v}{v - v_S} = (1250 \text{ Hz})\frac{343 \text{ m/s}}{343 \text{ m/s} - 242 \text{ m/s}}$$

$$= 4245 \text{ Hz} \approx 4250 \text{ Hz.} \qquad \text{(Answer)}$$

As a quick check on the answer, recall the physical result: when a source moves *toward* a stationary detector, the detected frequency (here, 4245 Hz) should be *greater* than the emitted frequency (here, 1250 Hz).

(b) Some of the sound reaching the pole reflects back to the rocket, which has an onboard detector. What frequency f'' does it detect?

SOLUTION: The pole now acts as the source of sound in that it reflects sound waves, producing an echo. The frequency of the waves from the pole is the same as the frequency of the waves "sensed" by the pole, namely, $f' = 4245$ Hz. Because now the source (the pole) is stationary, we set $v_S = 0$ in Eq. 18-53. And because now the detector (the onboard detector) is moving toward the new source, we choose the plus sign in the numerator. So the frequency sensed by the onboard detector is

$$f'' = f'\frac{v + v_D}{v} = (4245 \text{ Hz})\frac{343 \text{ m/s} + 242 \text{ m/s}}{343 \text{ m/s}}$$

$$= 7240 \text{ Hz.} \qquad \text{(Answer)}$$

Again, as a quick check on the answer, recall the physical result: when a detector moves *toward* a stationary source, the detected frequency (here, 7240 Hz) should be *greater* than the emitted frequency (here, 4245 Hz).

CHECKPOINT 7: In Sample Problem 18-8, if the air is moving toward the pole at speed 20 m/s, (a) what value for the source speed v_S should be used in the solution of part a, and (b) what value and sign for the detector speed v_D should be used there?

SAMPLE PROBLEM 18-9

Bats navigate and search out prey by emitting, and then detecting reflections of, ultrasonic waves, which are sound waves with frequencies greater than what can be heard by a human. Suppose a horseshoe bat flies toward a moth at speed $v_b = 9.0$ m/s, while the moth flies toward the bat with speed $v_m = 8.0$ m/s. From its nostrils, the bat emits ultrasonic waves of frequency f_{be} that reflect from the moth back to the bat with frequency f_{bd}. The bat adjusts the emitted frequency f_{be} until the returned frequency f_{bd} is 83 kHz, at which the bat's hearing is best.

(a) What is the frequency f_m of the waves heard and reflected by the moth when f_{bd} is 83 kHz?

SOLUTION: To find f_m, we use Eq. 18-53, with the moth being the source (of reflected waves with frequency f_m) and the bat being the detector (of the echo with frequency $f_{bd} = 83$ kHz). Since the detector moves toward the source (with speed v_b), we choose the plus sign in the numerator of Eq. 18-53. Since the source moves toward the detector (with speed v_m), we choose the minus sign in the denominator. We then have

$$f_{bd} = f_m\frac{v + v_b}{v - v_m},$$

or $\qquad 83 \text{ kHz} = f_m\dfrac{343 \text{ m/s} + 9.0 \text{ m/s}}{343 \text{ m/s} - 8.0 \text{ m/s}},$

from which

$$f_m = 78.99 \text{ kHz} \approx 79 \text{ kHz.} \qquad \text{(Answer)}$$

(b) What is the frequency f_{be} emitted by the bat when f_{bd} is 83 kHz?

SOLUTION: We again use Eq. 18-53, but now with the bat as source (at frequency f_{be}) and the moth as detector (of frequency f_m). Since the detector moves toward the source (with speed v_m), we choose the plus sign in the numerator of Eq. 18-53. Since the source moves toward the detector (with speed v_b), we choose the minus sign in the denominator. We then have

$$f_m = f_{be}\frac{v + v_m}{v - v_b}$$

or $\qquad 78.99 \text{ kHz} = f_{be}\dfrac{343 \text{ m/s} + 8.0 \text{ m/s}}{343 \text{ m/s} - 9.0 \text{ m/s}},$

from which

$$f_{be} = 75 \text{ kHz.} \qquad \text{(Answer)}$$

The bat determines the relative speed of the moth (17 m/s) from the 8 kHz (= 83 kHz − 75 kHz) it must lower its emitted frequency to hear an echo with a frequency of 83 kHz (at which it hears best). Some moths evade capture by flying away from the direction in which they hear ultrasonic waves. That choice of flight path reduces the frequency difference between what the bat emits and what it hears, and then the bat may not notice the echo. Some moths avoid capture by clicking to produce their own ultrasonic waves, thus "jamming" the detection system and confusing the bat.

18-9 THE DOPPLER EFFECT FOR LIGHT

It is tempting to apply to light waves the Doppler effect equation (Eq. 18-53) developed for sound waves in the preceding section, simply substituting c, the speed of light, for v, the speed of sound. We must avoid this temptation.

The reason is that sound waves require a medium through which to travel, but light waves do not. Thus, although the speed of sound is always measured with respect to the medium, the speed of light is not. Moreover, the speed of light always has the same value c, in all directions and in all inertial frames. For these reasons, Einstein's theory of special relativity shows that the Doppler effect for light depends only on the relative motion between the source of the light and a detector.

Even though the Doppler equations for light and for sound are different, at low enough speeds they reduce to the same approximate result. (In fact, *all* the predictions of special relativity theory reduce to their *classical* counterparts at low enough speeds.) Thus, Eq. 18-54, with v replaced by c, holds for light waves if $u \ll c$, where u is the relative speed of the source and the detector. That is, as a close approximation,

$$f' = f(1 \pm u/c) \qquad \text{(light waves; } u \ll c\text{).} \qquad (18\text{-}56)$$

If the source and the detector are *approaching* each other, we anticipate a frequency *increase,* and our rule for signs calls for the plus sign in Eq. 18-56.

In Doppler observations in astronomy (which involve light rather than sound), the wavelength is easier to measure than the frequency. This suggests that we replace f' and f in Eq. 18-56 with c/λ' and c/λ, respectively. Doing so, we find

$$\lambda' = \lambda(1 \pm u/c)^{-1} \approx \lambda(1 \mp u/c).$$

We can write this as

$$\frac{\lambda' - \lambda}{\lambda} = \mp \frac{u}{c}$$

or as

$$u = \frac{\Delta\lambda}{\lambda} c \qquad \text{(light waves; } u \ll c\text{),} \qquad (18\text{-}57)$$

in which $\Delta\lambda$ is the *magnitude* (no signs) of the Doppler wavelength shift.

Equation 18-57 gives us a way to determine the relative speed of a light source and a detector from a measured wavelength shift. If the wavelength decreases (called a *blue shift* because the blue portion of the visible spectrum has the shortest wavelengths), the frequency necessarily increases. That means the source and detector are approaching each other. If the wavelength increases (a *red shift*), the source and detector are moving away from each

other. Astronomers have measured the wavelength shifts of light reaching Earth from distant stars and galaxies. They have discovered that light from *all* the distant galaxies has undergone a red shift, indicating that all those galaxies are moving away from us. Moreover, the more distant a galaxy is, the faster it is moving.

SAMPLE PROBLEM 18-10

Figure 18-24a shows curves of intensity versus wavelength for light reaching us from interstellar gas on two opposite sides of galaxy M87 (Fig. 18-24b). One curve peaks at 499.8 nm; the other at 501.6 nm. The gas orbits the core of the galaxy at a radius $r = 100$ light-years, apparently moving toward us on one side of the core and moving away from us on the opposite side.

(a) Which curve corresponds to the gas moving toward us? What is the speed of the gas relative to us (and relative to the galaxy's core)?

SOLUTION: If the gas were not moving around the galaxy's core, the light from it would be detected at some wavelength λ (set by the emission process and the motion of the galaxy away from us). However, the motion of the gas changes the detected wavelength via the Doppler effect, increasing the wavelength for the gas moving away from us and decreasing it for the gas moving toward us. Thus, the curve peaking at 501.6 nm corresponds to motion away from us, and that peaking at 499.8 nm corresponds to motion toward us.

Let us assume that the increase and the decrease in wavelength due to the motion of the gas are equal in magnitude. Then the unshifted wavelength λ must be the average of the two shifted wavelengths:

$$\lambda = \frac{501.6 \text{ nm} + 499.8 \text{ nm}}{2} = 500.7 \text{ nm.}$$

The Doppler shift $\Delta\lambda$ of the light from the gas moving away from us is then

$$\Delta\lambda = 501.6 \text{ nm} - 500.7 \text{ nm} = 0.90 \text{ nm.}$$

Substituting this and $\lambda = 500.7$ nm into Eq. 18-57, we find that the speed of the gas is

$$u = \frac{\Delta\lambda}{\lambda} c = \frac{0.90 \text{ nm}}{501.6 \text{ nm}} 3.0 \times 10^8 \text{ m/s}$$

$$= 5.39 \times 10^5 \text{ m/s.} \qquad \text{(Answer)}$$

(b) The gas orbits the core of the galaxy because it experiences a gravitational force due to the mass M of the core. What is that mass in multiples of the Sun's mass M_S ($= 1.99 \times 10^{30}$ kg)?

SOLUTION: From Eq. 14-1, the gravitational force on an orbiting gas element of mass m at orbital radius r is

$$F = \frac{GMm}{r^2}.$$

Applying Newton's second law to the gas element and substituting the centripetal acceleration u^2/r for a, we have

$$\frac{GMm}{r^2} = ma = \frac{mu^2}{r}.$$

Solving this for M and substituting known data, we find

$$
\begin{aligned}
M &= \frac{u^2 r}{G} \\
&= \frac{(5.39 \times 10^5 \text{ m/s})^2(100 \text{ ly})(9.46 \times 10^{15} \text{ m/ly})}{6.67 \times 10^{-11} \text{ N} \cdot \text{m}^2/\text{kg}^2} \\
&= 4.12 \times 10^{39} \text{ kg} = (2.1 \times 10^9)M_S. \quad \text{(Answer)}
\end{aligned}
$$

This result tells us that a mass equivalent to two billion suns has been compacted into the core of the galaxy, strongly suggesting that a "supermassive" black hole occupies the core.

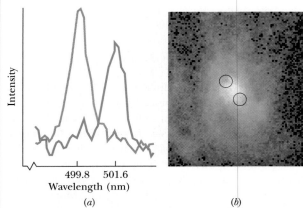

FIGURE 18-24 Sample Problem 18-10. (*a*) Plots of intensity versus wavelength for light emitted by gas on opposite sides of galaxy M87. (*b*) The central region of M87. The circles indicate the locations of the gas whose intensity is given in (*a*). The center of M87 is halfway between the circles.

REVIEW & SUMMARY

Sound Waves

Sound waves are longitudinal mechanical waves that can travel through solids, liquids, or gases. The speed v of a sound wave in a medium having **bulk modulus** B and density ρ is

$$v = \sqrt{\frac{B}{\rho}} \quad \text{(speed of sound).} \tag{18-3}$$

In air at 20°C, the speed of sound is 343 m/s.

A sound wave causes a longitudinal displacement s of a mass element in a medium as given by

$$s = s_m \cos(kx - \omega t), \tag{18-13}$$

where s_m is the **displacement amplitude** (maximum displacement) from equilibrium, $k = 2\pi/\lambda$, and $\omega = 2\pi f$, λ and f being the wavelength and frequency, respectively, of the sound wave. The sound wave also causes a pressure change Δp of the medium from the equilibrium pressure:

$$\Delta p = \Delta p_m \sin(kx - \omega t), \tag{18-14}$$

where the **pressure amplitude** is

$$\Delta p_m = (v\rho\omega)s_m. \tag{18-15}$$

Interference

The interference of two sound waves with identical wavelengths passing through a common point depends on their phase difference ϕ there. If the sound waves were emitted in phase and are traveling in approximately the same direction, ϕ is given by

$$\phi = \frac{\Delta L}{\lambda} 2\pi, \tag{18-21}$$

where ΔL is their **path length difference** (the difference in the distances traveled by the waves to reach the common point). Conditions for fully constructive and fully destructive interference of the waves are given by

$$\phi = m2\pi, \quad m = 0, 1, 2, \cdots$$
$$\text{(fully constructive interference)} \tag{18-22}$$

and

$$\phi = (m + \tfrac{1}{2})2\pi, \quad m = 0, 1, 2, \cdots$$
$$\text{(fully destructive interference).} \tag{18-23}$$

These conditions correspond to

$$\Delta L = m\lambda, \quad m = 0, 1, 2, \cdots$$
$$\text{(fully constructive interference)} \tag{18-24}$$

and

$$\Delta L = (m + \tfrac{1}{2})\lambda, \quad m = 0, 1, 2, \cdots$$
$$\text{(fully destructive interference).} \tag{18-25}$$

Sound Intensity

The **intensity** I of a sound wave at a surface is the average rate per unit area at which energy is transferred by the wave through or onto the surface:

$$I = \frac{P}{A}, \tag{18-26}$$

where P is the rate of energy transfer (power) of the sound wave and A is the area of the surface intercepting the sound. The intensity I is related to the displacement amplitude s_m of the sound wave by

$$I = \tfrac{1}{2}\rho v\omega^2 s_m^2. \tag{18-27}$$

The intensity at a distance r from a point source that emits sound waves of power P_s is

$$I = \frac{P_s}{4\pi r^2}. \tag{18-28}$$

Sound Level in Decibels

The *sound level* β in *decibels* (dB) is defined as

$$\beta = (10 \text{ dB}) \log \frac{I}{I_0}, \qquad (18\text{-}29)$$

where I_0 ($= 10^{-12}$ W/m^2) is a reference intensity level to which all intensities are compared. For every factor-of-10 increase in intensity, 10 dB is added to the sound level.

Standing Wave Patterns in Pipes

Standing sound wave patterns can be set up in pipes. A pipe open at both ends will resonate at frequencies

$$f = \frac{v}{\lambda} = \frac{nv}{2L}, \qquad n = 1, 2, 3, \cdots$$

$$\text{(pipe, two open ends),} \quad (18\text{-}37)$$

where v is the speed of sound in the air in the pipe. For a pipe closed at one end and open at the other, the resonant frequencies are

$$f = \frac{v}{\lambda} = \frac{nv}{4L}, \qquad n = 1, 3, 5, \cdots$$

$$\text{(pipe, one open end).} \quad (18\text{-}39)$$

Beats

Beats arise when two waves having slightly different frequencies, f_1 and f_2, are detected together. The beat frequency is

$$f_{\text{beat}} = f_1 - f_2. \qquad (18\text{-}44)$$

The Doppler Effect

The *Doppler effect* is a change in the observed frequency of a wave when the source or the detector moves relative to the medium. For sound the observed frequency f' is given in terms of the source frequency f by

$$f' = f \frac{v \pm v_D}{v \mp v_S}, \qquad \begin{array}{c}\text{(general Doppler}\\ \text{effect),}\end{array} \quad (18\text{-}53)$$

where v_D is the speed of the detector relative to the medium, v_S is that of the source, and v is the speed of sound in the medium. The signs are chosen such that f' tends to be *greater* for motion (of detector or source) "toward" and *less* for motion "away."

Shock Wave

If the source speed relative to the medium exceeds the speed of sound in the medium, the Doppler equation no longer applies. In such a case, shock waves result. The half angle θ (see Fig. 18-22) of the wavefront is given by

$$\sin \theta = \frac{v}{v_S} \qquad \text{(Mach cone angle).} \quad (18\text{-}55)$$

Doppler Effect for Light

When there is a relative speed u between a light source and a detector, the detected frequency f' of the light is

$$f' = f(1 \pm u/c), \qquad (18\text{-}56)$$

where f is the frequency that would be detected were there no relative motion. The relative speed u is related to the Doppler shift in wavelength $\Delta\lambda$ by

$$u = \frac{\Delta\lambda}{\lambda} c, \qquad (18\text{-}57)$$

where λ is the wavelength that would be detected were there no relative motion. If the source and detector move toward each other, the wavelength shift $\Delta\lambda$ is negative; if they move away from each other, $\Delta\lambda$ is positive.

QUESTIONS

1. Figure 18-25 shows the paths taken by two pulses of sound that begin simultaneously and then race each other through equal distances in air. The only difference between the paths is that a region of hot (low density) air lies along path 2. Which pulse wins the race?

FIGURE 18-25 Question 1.

2. A sound wave of wavelength λ and displacement amplitude s_m begins to travel down a passageway (a tube, the opening of the ear, etc.). When a small device in the passageway detects this wave, it issues a second sound wave (said to be *antisound*) that is able to cancel the first wave, so that nothing is heard at the far end of the passageway. For such cancellation, what must be (a) the direction of travel, (b) the wavelength, and (c) the displacement amplitude of the second wave? (d) What must be the phase difference between the two waves? (Such antisound devices are used to eliminate unwanted sound in a noisy environment.)

3. In Fig. 18-26, two point sources S_1 and S_2, which are in phase, emit identical sound waves of wavelength 2.0 m. In terms of wavelengths, what is the phase difference between the waves arriving at point P if (a) $L_1 = 38$ m and $L_2 = 34$ m, and (b) $L_1 = 39$ m and $L_2 = 36$ m? (c) Assuming that the source separation is much smaller than L_1 and L_2, what type of interference occurs at P in situations (a) and (b), respectively?

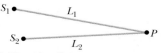

FIGURE 18-26 Question 3.

4. In Fig. 18-27, sound waves of wavelength λ are emitted by a point source S and travel to a detector D directly along path 1 and via reflection from a panel along path 2. Initially, the panel is almost along path 1 and the waves arriving at D along the two paths are almost exactly in phase. Then the panel is moved away from path 1 as shown until the waves arriving at D are exactly out of phase. What then is the path length difference Δd of the waves along the two paths?

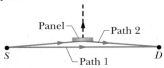

FIGURE 18-27 Question 4.

5. In Fig. 18-28, two point sources S_1 and S_2, which are in phase, emit identical sound waves of wavelength λ, and point P is at equal distances from them. Then S_2 is moved directly away from P by a distance equal to $\lambda/4$. Are the waves at P then exactly in phase, exactly out of phase, or do they have some intermediate phase relation if (a) S_1 is moved directly toward P by a distance equal to $\lambda/4$ and (b) S_1 is moved directly away from P by a distance equal to $3\lambda/4$?

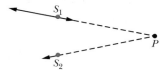

FIGURE 18-28 Question 5.

6. In Sample Problem 18-3 and Fig. 18-8a, the waves arriving at point P_1 on the perpendicular bisector are exactly in phase; that is, the waves from S_1 and S_2 always tend to move an element of air at P_1 in the same direction. Let the intersection of the perpendicular bisector and the line through S_1 and S_2 be point P_3. (a) Are the waves arriving at P_3 exactly in phase, exactly out of phase, or do they have some intermediate relation? (b) What is the answer if we increase the separation between the sources to 1.7λ?

7. Without using a calculator, determine the increase in a sound level when the intensity of the sound source becomes 10^7 times what it was previously.

8. A standing sound wave in a pipe has five nodes and five antinodes. (a) How many open ends does the pipe have? (b) What is the harmonic number n for this standing wave?

9. The sixth harmonic is set up in a pipe. (a) How many open ends does the pipe have (it has at least one)? (b) Is there a node, antinode, or some intermediate state at the midpoint?

10. (a) When an orchestra warms up, the players' warm breath increases the temperature of the air within the wind instruments (and thus decreases the density of that air). Do the resonant frequencies of those instruments increase or decrease? (b) When the slide of a slide trombone is pushed outward, do the resonant frequencies of the instrument increase or decrease?

11. Pipe A has length L and one open end. Pipe B has length $2L$ and two open ends. Which harmonics of pipe B have a frequency that matches a resonant frequency of pipe A?

12. Figure 18-29 shows a stretched string of length L and pipes a, b, c, and d of lengths L, $2L$, $L/2$, and $L/2$, respectively. The string's tension is adjusted until the speed of waves on the string equals the speed of sound waves in the air. The fundamental mode of oscillation is then set up on the string. In which pipe will the sound produced by the string cause resonance, and what oscillation mode will that sound set up?

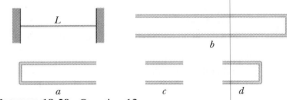

FIGURE 18-29 Question 12.

13. You are given four tuning forks. The fork with the lowest frequency oscillates at 500 Hz. By striking two tuning forks at a time, you can produce the following beat frequencies: 1, 2, 3, 5, 7, and 8 Hz. What are the possible frequencies of the other three forks? (There are two sets of answers.)

14. A friend rides, in turn, the rims of three fast merry-go-rounds while holding a sound source that emits isotropically at a certain frequency. You stand far from each merry-go-round. The frequency you hear for each of your friend's three rides varies as the merry-go-round rotates. The variations in frequency for the three rides are given by the three curves in Fig. 18-30. Rank the curves according to (a) the linear speed v of the sound source, (b) the angular speeds ω of the merry-go-rounds, and (c) the radii r of the merry-go-rounds, greatest first.

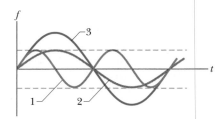

FIGURE 18-30 Question 14.

EXERCISES & PROBLEMS

Where needed in the problems, use

speed of sound in air = 343 m/s = 1125 ft/s

and

density of air = 1.21 kg/m³

unless otherwise specified.

SECTION 18-2 The Speed of Sound

1E. A rule for finding your distance in miles from a lightning flash is to count seconds from the time you see the flash until you hear the thunder and then divide the count by five. (a) Explain this rule and determine the percent error in it at 20°C, assuming that the sound travels to you along a straight path. (b) Devise a similar rule for obtaining the distance in kilometers.

2E. A column of soldiers, marching at 120 paces per minute, keep in step with the music of a band at the head of the column. It is observed that the soldiers in the rear end of the column are striding forward with the left foot when musicians are advancing with the right. What is the length of the column approximately?

3E. You are at a large outdoor concert, seated 300 m from the speaker system. The concert is also being broadcast live via satellite (at the speed of light). Consider a listener 5000 km away who receives the broadcast. Who hears the music first, you or the listener, and by what time difference?

4E. Two spectators at a soccer game in Montjuic Stadium see, and a moment later hear, the ball being kicked on the playing field. The time delay for one spectator is 0.23 s and for the other 0.12 s. Sight lines from each spectator to the player kicking the ball meet at an angle of 90°. (a) How far is each spectator from the player? (b) How far are the spectators from each other?

5E. The average density of Earth's crust 10 km beneath the continents is 2.7 g/cm³. The speed of longitudinal seismic waves at that depth, found by timing their arrival from distant earthquakes, is 5.4 km/s. Use this information to find the bulk modulus of Earth's crust at that depth. For comparison, the bulk modulus of steel is about 16×10^{10} Pa.

6E. What is the bulk modulus of oxygen at standard temperature (0°C) and pressure (1 atm) if under these conditions of temperature and pressure 1 mol (32.0 g) of oxygen occupies 22.4 L and the speed of sound in the oxygen is 317 m/s?

7P. An experimenter wishes to measure the speed of sound in an aluminum rod 10 cm long by measuring the time it takes for a sound pulse to travel the length of the rod. If results good to four significant figures are desired, how precisely must the length of the rod be known and how closely must she be able to resolve time intervals?

8P. The speed of sound in a certain metal is V. One end of a long pipe of that metal of length L is struck a hard blow. A listener at the other end hears two sounds, one from the wave that has traveled along the pipe and the other from the wave that has traveled through the air. (a) If v is the speed of sound in air, what time interval t elapses between the arrivals of the two sounds? (b) Suppose that $t = 1.00$ s and the metal is steel. Find the length L.

9P. A man strikes a long aluminum rod at one end. Another man, at the other end with his ear close to the rod, hears the sound of the blow twice (once through air and once through the rod), with a 0.120 s interval between. How long is the rod?

10P. Earthquakes generate sound waves inside Earth. Unlike a gas, Earth can experience both transverse (S) and longitudinal (P) sound waves. Typically, the speed of S waves is about 4.5 km/s and that of P waves 8.0 km/s. A seismograph records P and S waves from an earthquake. The first P waves arrive 3.0 min before the first S waves (Fig. 18-31). Assuming the waves traveled in a straight line, how far away did the earthquake occur?

11P. A stone is dropped into a well. The sound of the splash is heard 3.00 s later. What is the depth of the well?

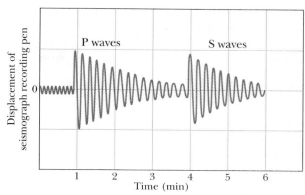

FIGURE 18-31   Problem 10.

SECTION 18-3   Traveling Sound Waves

12E. The audible frequency range for normal hearing is from about 20 Hz to 20 kHz. What are the wavelengths of sound waves at these frequencies?

13E. The shortest wavelength emitted by a bat is about 3.3 mm. What is the corresponding frequency?

14E. Diagnostic ultrasound of frequency 4.50 MHz is used to examine tumors in soft tissue. (a) What is the wavelength in air of such a sound wave? (b) If the speed of sound in tissue is 1500 m/s, what is the wavelength of this wave in tissue?

15E. (a) A conical loudspeaker has a diameter of 15.0 cm. At what frequency will the wavelength of the sound it emits in air be equal to its diameter? Be ten times its diameter? Be one-tenth its diameter? (b) Make the same calculations for a speaker of diameter 30.0 cm.

16E. Remarkably detailed images of transistors can be formed by an acoustic microscope. If the sound waves in such a microscope have a frequency of 4.2 GHz and a speed (in the liquid helium in which the specimen is immersed) of 240 m/s, what is their wavelength?

17P. (a) A continuous sinusoidal longitudinal wave is sent along a very long coiled spring from an oscillating source attached to it. The frequency of the source is 25 Hz, and the distance between successive points of maximum expansion in the spring is 24 cm. Find the wave speed. (b) If the maximum longitudinal displacement of a particle in the spring is 0.30 cm and the wave moves in the negative direction of an x axis, write the equation for the wave. Let the source be at $x = 0$ and the displacement at $x = 0$ be zero when $t = 0$.

18P. The pressure in a traveling sound wave is given by the equation

$$\Delta p = (1.5 \text{ Pa}) \sin \pi[(1.00 \text{ m}^{-1})x - (330 \text{ s}^{-1})t].$$

Find (a) the pressure amplitude, (b) the frequency, (c) the wavelength, and (d) the speed of the wave.

19P. Two sound waves, from two different sources with the same frequency, 540 Hz, travel at a speed of 330 m/s. The sources are in phase. What is the phase difference of the waves at a point that is 4.40 m from one source and 4.00 m from the other? The waves are traveling in the same direction.

20P. At a certain point in space, two waves produce pressure variations given by

$$\Delta p_1 = \Delta p_m \sin \omega t,$$

$$\Delta p_2 = \Delta p_m \sin(\omega t - \phi).$$

What is the pressure amplitude of the resultant wave at this point when $\phi = 0$, $\phi = \pi/2$, $\phi = \pi/3$, and $\phi = \pi/4$?

SECTION 18-4 Interference

21P. In Fig. 18-32, two loudspeakers, separated by a distance of 2.00 m, are in phase. Assume the amplitudes of the sound from the speakers are approximately the same at the position of a listener, who is 3.75 m directly in front of one of the speakers. (a) For what frequencies in the audible range (20–20,000 Hz) does the listener hear a minimum signal? (b) For what frequencies is the signal a maximum?

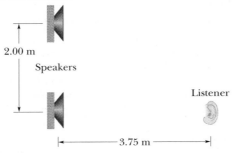

FIGURE 18-32 Problem 21.

22P. Two loudspeakers are located 11.0 ft apart on the stage of an auditorium. A listener is seated 60.0 ft from one and 64.0 ft from the other. A signal generator drives the two speakers in phase with the same amplitude and frequency. The transmitted frequency is swept through the audible range (20–20,000 Hz). (a) What are the three lowest frequencies at which the listener will hear a minimum signal because of destructive interference? (b) What are the three lowest frequencies at which the listener will hear a maximum signal?

23P. Two point sources of sound waves of identical wavelength λ and amplitude are separated by distance $D = 2.0\lambda$. The sources are in phase. (a) How many points of maximum signal (maximum constructive interference) lie along a large circle around the sources? (b) How many points of minimum signal (destructive interference)?

24P. A sound wave of 40.0 cm wavelength enters the tube shown in Fig. 18-33 at the source end. What must be the smallest radius r such that a minimum will be heard at the detector end?

FIGURE 18-33
Problem 24.

Source   Detector

25P. In Fig. 18-34, a point source S of sound waves lies near a reflecting wall AB. A sound detector D intercepts sound ray R_1 traveling directly from S. It also intercepts sound ray R_2 that reflects from the wall such that the *angle of incidence* θ_i is equal to the *angle of reflection* θ_r. Find two frequencies at which there is maximum constructive interference of R_1 and R_2 at D. (Reflection of sound by the wall does not alter the phase of the sound wave.)

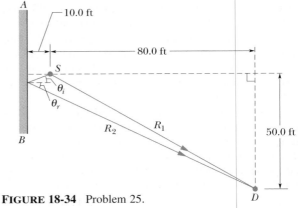

FIGURE 18-34 Problem 25.

26P*. Two point sources, separated by a distance of 5.00 m, emit sound waves at the same amplitude and frequency (300 Hz), but the sources are exactly out of phase. At what points along the line between the sources do the sound waves result in maximum oscillations of the air molecules? (*Hint:* One such point is midway between the sources.)

SECTION 18-5 Intensity and Sound Level

27E. A 1.0 W point source emits sound waves isotropically. Assuming that the energy of the waves is conserved, what is the intensity (a) 1.0 m from the source and (b) 2.5 m from the source?

28E. A source emits sound waves isotropically. The intensity of the waves 2.50 m from the source is 1.91×10^{-4} W/m². Assuming that the energy of the waves is conserved, what is the power of the source?

29E. A sound wave of frequency 300 Hz has an intensity of 1.00 μW/m². What is the amplitude of the air oscillations caused by this wave?

30E. Two sounds differ in sound level by 1.00 dB. What is the ratio of the greater intensity to the smaller intensity?

31E. A certain sound level is increased by 30 dB. By what multiple is (a) its intensity increased and (b) its pressure amplitude increased?

32E. A salesperson claimed that a stereo system had a maximum audio power of 120 W. Testing the system with several speakers set up so as to simulate a point source, the consumer noted that she could get as close as 1.2 m with the volume full on before the sound hurt her ears. Should she report the firm to the Consumer Product Safety Commission?

33E. A certain loudspeaker system emits sound isotropically with a frequency of 2000 Hz and an intensity of 0.960 mW/m² at a distance of 6.10 m. Assume that there are no reflections. (a) What is the intensity at 30.0 m? At 6.10 m, what are (b) the displacement amplitude and (c) the pressure amplitude?

34E. The source of a sound wave has a power of 1.00 μW. If it is a point source, (a) what is the intensity 3.00 m away and (b) what is the sound level in decibels at that distance?

35E. (a) If two sound waves, one in air and one in (fresh) water, are equal in intensity, what is the ratio of the pressure amplitude of the wave in water to that of the wave in air? Assume the water and the air are at 20°C. (See Table 15-1). (b) If the pressure amplitudes are equal instead, what is the ratio of the intensities of the waves?

36P. (a) Show that the intensity I of a wave is the product of the wave's energy per unit volume u and its speed v. (b) Radio waves travel at a speed of 3.00×10^8 m/s. Find u for a radio wave 480 km from a 50,000 W source, assuming the wavefronts to be spherical.

37P. Assume that a noisy freight train on a straight track emits a cylindrical, expanding sound wave. Assuming that the air absorbs no energy, how does the amplitude s_m of the wave depend on the perpendicular distance r from the source?

38P. A sound wave travels out uniformly in all directions from a point source. (a) Justify the following expression for the displacement s of the medium at any distance r from the source:

$$s = \frac{b}{r} \sin k(r - vt),$$

where b is a constant. Consider the speed, direction of propagation, periodicity, and intensity of the wave. (b) What are the dimensions of the constant b?

39P. Find the ratios of (a) the intensities, (b) the pressure amplitudes, and (c) the particle displacement amplitudes for two sounds whose sound levels differ by 37 dB.

40P. At a distance of 10 km, a 100 Hz horn, assumed to be a point source, is barely audible. At what distance would it begin to cause pain?

41P. You are standing at a distance D from a source that emits sound waves equally in all directions. You walk 50.0 m toward the source and observe that the intensity of these waves has doubled. Calculate the distance D.

42P. A point source emits 30.0 W of sound isotropically. A small microphone of effective cross-sectional area 0.750 cm^2 is located 200 m from the source. Calculate (a) the sound intensity there and (b) the power intercepted by the microphone.

43P. In a test, a subsonic jet flies overhead at an altitude of 100 m. The sound intensity on the ground as the jet passes overhead is 150 dB. At what altitude should the plane fly so that the ground noise is no greater than 120 dB, the threshold of pain? Ignore the finite time required for the sound to reach the ground.

44P. An audio engineer has designed a loudspeaker that is spherical and emits sound isotropically. The speaker emits 10 W of acoustic power into a room with completely absorbent walls, floor, and ceiling (an *anechoic chamber*). (a) What is the intensity (W/m^2) of the sound waves 3.0 m from the center of the source? (b) How does the amplitude of the waves at 4.0 m compare with that at 3.0 m from the center of the source?

45P. Figure 18-35 shows an air-filled, acoustic interferometer, used to demonstrate the interference of sound waves. S is an oscillating diaphragm; D is a sound detector, such as the ear or a microphone. Path SBD can be varied in length, but path SAD is fixed. At D, the sound wave coming along path SBD interferes with that coming along path SAD. In one demonstration, the sound intensity at D has a minimum value of 100 units at one position of B and continuously climbs to a maximum value of 900 units when B is shifted by 1.65 cm. Find (a) the frequency of the sound emitted by the source and (b) the ratio of the amplitude at D of the SAD wave to that of the SBD wave. (c) How can it happen that these waves have different amplitudes, considering that they originate at the same source?

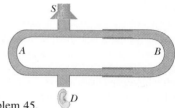

FIGURE 18-35   Problem 45.

SECTION 18-6   Sources of Musical Sound

46E. A sound wave of frequency 1000 Hz propagating through air has a pressure amplitude of 10.0 Pa. What are the (a) wavelength, (b) particle displacement amplitude, and (c) maximum particle speed? (d) An organ pipe open at both ends has this frequency as a fundamental. How long is the pipe?

47E. A sound wave in a fluid medium is reflected at a barrier so that a standing wave is formed. The distance between nodes is 3.8 cm, and the speed of propagation is 1500 m/s. Find the frequency of the sound wave.

48E. A violin string 15.0 cm long and fixed at both ends oscillates in its $n = 1$ mode. The speed of waves on the string is 250 m/s, and the speed of sound in air is 348 m/s. What are (a) the frequency and (b) the wavelength of the emitted sound wave?

49E. A violin string, oscillating in its fundamental mode, generates a sound wave with wavelength λ. By what multiple must the tension be increased if the string, still oscillating in its fundamental mode, is to generate a sound wave with wavelength $\lambda/2$?

50E. Organ pipe A, with both ends open, has a fundamental frequency of 300 Hz. The third harmonic of organ pipe B, with one end open, has the same frequency as the second harmonic of pipe A. How long are (a) pipe A and (b) pipe B?

51E. The water level in a vertical glass tube 1.00 m long can be adjusted to any position in the tube. A tuning fork vibrating at 686 Hz is held just over the open top end of the tube. At what positions of the water level will there be resonance?

52E. (a) Find the speed of waves on a violin string of mass 800 mg and length 22.0 cm if the fundamental frequency is 920 Hz. (b) What is the tension in the string? For the fundamental, what is the wavelength of (c) the waves on the string and (d) the sound waves emitted by the string?

53P. A certain violin string is 30 cm long between its fixed ends and has a mass of 2.0 g. The "open" string (no applied finger) sounds an A note (440 Hz). (a) To play a C (523 Hz), how far down the string must one place a finger? (b) What is the ratio of the wavelength of the string waves required for an A note to that required for a C note? (c) What is the ratio of the wavelength of the sound wave for an A note to that for a C note?

54P. A string on a cello has length L, for which the fundamental frequency is f. (a) By what length l must the string be shortened by fingering to change the fundamental frequency to rf? (b) What is l if $L = 0.80$ m and $r = 1.2$? (c) For $r = 1.2$, what is the ratio of the wavelength of the new sound wave emitted by the string to that of the wave emitted before fingering?

55P. In Fig. 18-36, S is a small loudspeaker driven by an audio oscillator and amplifier, adjustable in frequency from 1000 to 2000 Hz only. The tube D is a piece of cylindrical sheet-metal pipe 18.0 in. long and open at both ends. (a) If the speed of sound in air is 1130 ft/s at the existing temperature, at what frequencies will resonance occur in the pipe when the frequency emitted by the speaker is varied from 1000 Hz to 2000 Hz? (b) Sketch the standing waves (using the convention of Fig. 18-13b) for each resonant frequency.

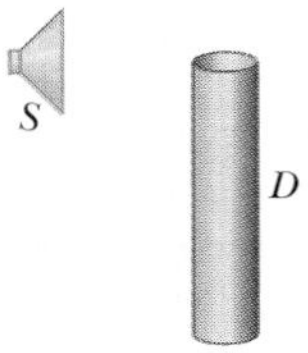

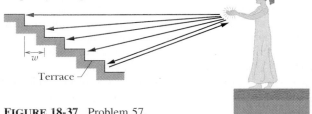

FIGURE 18-36 Problem 55.

56P. A well with vertical sides and water at the bottom resonates at 7.00 Hz and at no lower frequency. The air in the well has a density of 1.10 kg/m³ and a bulk modulus of 1.33×10^5 Pa. How deep is the well?

57P. A hand-clap on stage in an amphitheater (Fig. 18-37) sends out sound waves that scatter from terraces of width $w = 0.75$ m. The sound returns to the stage as a periodic series of pulses, one from each terrace; the parade of pulses sounds like a played note. (a) Assuming that all the rays are horizontal, find the frequency at which the pulses return (that is, the frequency of the perceived note). (b) If the width w of the terraces were smaller, would the frequency be higher or lower?

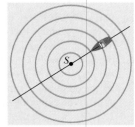

FIGURE 18-37 Problem 57.

58P. A tube 1.20 m long is closed at one end. A stretched wire is placed near the open end. The wire is 0.330 m long and has a mass of 9.60 g. It is fixed at both ends and vibrates in its fundamental mode. By resonance, it sets the air column in the tube into oscillation at that column's fundamental frequency. Find (a) that frequency and (b) the tension in the wire.

59P. The period of a pulsating variable star may be estimated by considering the star to be executing *radial* longitudinal pulsations in the fundamental standing wave mode; that is, the star's radius varies periodically with time, with a displacement antinode at the star's surface. (a) Would you expect the center of the star to be a displacement node or antinode? (b) By analogy with a pipe with one open end, show that the period of pulsation T is given by

$$T = \frac{4R}{v},$$

where R is the equilibrium radius of the star and v is the average sound speed. (c) Typical white dwarf stars are composed of material with a bulk modulus of 1.33×10^{22} Pa and a density of 10^{10} kg/m³. They have radii equal to 9.0×10^{-3} solar radius. What is the approximate pulsation period of a white dwarf?

60P. A violin string 30.0 cm long with linear density 0.650 g/m is placed near a loudspeaker that is fed by an audio oscillator of variable frequency. It is found that the string is set into oscillation only at the frequencies 880 and 1320 Hz as the frequency of the oscillator is varied over the range 500–1500 Hz. What is the tension in the string?

SECTION 18-7 Beats

61E. A tuning fork of unknown frequency makes three beats per second with a standard fork of frequency 384 Hz. The beat frequency decreases when a small piece of wax is put on a prong of the first fork. What is the frequency of this fork?

62E. The A string of a violin is a little too tightly stretched. Four beats per second are heard when the string is sounded together with a tuning fork that is oscillating accurately at concert A (440 Hz). What is the period of the violin string oscillation?

63P. Two identical piano wires have a fundamental frequency of 600 Hz when kept under the same tension. What fractional increase in the tension of one wire will lead to the occurrence of 6 beats/s when both wires oscillate simultaneously?

64P. You have five tuning forks that oscillate at different frequencies. Give (a) the maximum number and (b) the minimum number of different beat frequencies you can produce by using the forks two at a time.

SECTION 18-8 The Doppler Effect

65E. A source S generates circular waves on the surface of a lake; the pattern of wave crests is shown in Fig. 18-38. The speed

FIGURE 18-38 Exercise 65.

of the waves is 5.5 m/s, and the crest-to-crest separation is 2.3 m. You are in a small boat heading directly toward S at a constant speed of 3.3 m/s with respect to the shore. What frequency of the waves do you observe?

66E. Trooper B is chasing speeder A along a straight stretch of road. Both are moving at a speed of 100 mi/h. Trooper B, failing to catch up, sounds his siren again. Take the speed of sound in air to be 1100 ft/s and the frequency of the source to be 500 Hz. What is the Doppler shift in the frequency heard by speeder A?

67E. A whistle used to call a dog has a frequency of 30 kHz. The dog, however, ignores it. The owner of the dog, who cannot hear sounds above 20 kHz, wants to use the Doppler effect to make certain that the whistle is working. She asks a friend to blow the whistle from a moving car while the owner remains stationary and listens. (a) How fast would the car have to move and in what direction for the owner to hear the whistle at 20 kHz? (b) Repeat for a whistle frequency of 22 kHz instead of 30 kHz.

68E. The 16,000 Hz whine of the turbines in the jet engines of an aircraft moving with speed 200 m/s is heard at what frequency by the pilot of a second craft trying to overtake the first at a speed of 250 m/s?

69E. An ambulance with a siren emitting a whine at 1600 Hz overtakes and passes a cyclist pedaling a bike at 8.00 ft/s. After being passed, the cyclist hears a frequency of 1590 Hz. How fast is the ambulance moving?

70E. In 1845, Buys Ballot first tested the Doppler effect for sound. He put a trumpet player on a flatcar drawn by a locomotive and another player near the tracks. If each player blew a 440 Hz note and there were 4.0 beats/s as they approached each other, what was the speed of the flatcar?

71E. The speed of light in water is about three-fourths the speed of light in vacuum. A beam of high-speed electrons from a beta-tron emits Cerenkov radiation in water; the wavefront of this light forms a cone of angle 60°. Find the speed of the electrons in the water.

72E. A bullet is fired with a speed of 2200 ft/s. Find the angle made by the shock cone with the line of motion of the bullet.

73P. Two identical tuning forks can oscillate at 440 Hz. A person is located somewhere on the line between them. Calculate the beat frequency as measured by this individual if (a) she is standing still and the tuning forks both move to the right at 30.0 m/s, and (b) the tuning forks are stationary and the listener moves to the right at 30.0 m/s.

74P. A whistle of frequency 540 Hz moves in a circle of radius 2.00 ft at an angular speed of 15.0 rad/s. What are (a) the lowest and (b) the highest frequencies heard by a listener a long distance away at rest with respect to the center of the circle?

75P. A plane flies at 1.25 times the speed of sound. The sonic boom reaches a man on the ground exactly 1 min after the plane passed directly overhead. What is the altitude of the plane? Assume the speed of sound to be 330 m/s.

76P. A jet plane passes overhead at a height of 5000 m and a speed of Mach 1.5. (a) Find the Mach cone angle. (b) How long

after the jet has passed directly overhead will the shock wave reach the ground? Use 331 m/s for the speed of sound.

77P. Figure 18-39 shows a transmitter and receiver of waves contained in a single instrument. It is used to measure the speed u of a target object (idealized as a flat plate) that is moving directly toward the unit, by analyzing the waves reflected from the target. (a) Show that the frequency f_r of the reflected waves at the receiver is related to their source frequency f_s by

$$f_r = f_s \left(\frac{v + u}{v - u} \right),$$

where v is the speed of the waves. (b) In a great many practical situations, $u \ll v$. In this case, show that the equation above becomes

$$\frac{f_r - f_s}{f_s} \approx \frac{2u}{v}.$$

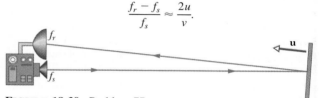

FIGURE 18-39 Problem 77.

78P. A stationary motion detector sends sound waves of 0.150 MHz toward a truck approaching at a speed of 45.0 m/s. What is the frequency of the waves reflected back to the detector?

79P. A siren emitting a sound of frequency 1000 Hz moves away from you toward the face of a cliff at a speed of 10 m/s. Take the speed of sound in air as 330 m/s. (a) What is the frequency of the sound you hear coming directly from the siren? (b) What is the frequency of the sound you hear reflected off the cliff? (c) What is the beat frequency between the two sounds? Is it perceptible (it must be less than 20 Hz to be perceived)?

80P. A person on a railroad car blows a trumpet note at 440 Hz. The car is moving toward a wall at 20.0 m/s. Calculate (a) the frequency of the sound as received at the wall and (b) the frequency of the reflected sound arriving back at the source.

81P. A French submarine and a U.S. submarine move head-on during maneuvers in motionless water in the North Atlantic (Fig. 18-40). The French sub moves at 50.0 km/h, and the U.S. sub at 70.0 km/h. The French sub sends out a sonar signal (sound wave in water) at 1000 Hz. Sonar waves travel at 5470 km/h. (a) What is the signal's frequency as detected by the U.S. sub? (b) What frequency is detected by the French sub in the signal reflected back to it by the U.S. sub?

FIGURE 18-40 Problem 81.

82P. A source of sound waves of frequency 1200 Hz moves to the right with a speed of 98.0 ft/s relative to the air. Ahead of it is a reflecting surface moving to the left with a speed of 216 ft/s relative to the air. Take the speed of sound in air to be 1080 ft/s and find (a) the wavelength of the sound emitted toward the reflector by the source, (b) the number of wavefronts per second arriving at the reflecting surface, (c) the speed of the reflected waves, (d) the wavelength of the reflected waves, and (e) the number of reflected wavefronts per second arriving at the source.

83P. In a discussion of Doppler shifts of ultrasonic waves used in medical diagnosis, the authors remark: "For every millimeter per second that a structure in the body moves, the frequency of the incident ultrasonic wave is shifted approximately 1.30 Hz per MHz." What speed of the ultrasonic waves in tissue do you deduce from this statement?

84P. An acoustic burglar alarm consists of a source emitting waves of frequency 28.0 kHz. What will be the beat frequency of waves reflected from an intruder walking at an average speed of 0.950 m/s directly away from the alarm?

85P. A bat is flitting about in a cave, navigating via ultrasonic bleeps. Assume that the sound emission frequency of the bat is 39,000 Hz. During one fast swoop directly toward a flat wall surface, the bat is moving at 0.025 times the speed of sound in air. What frequency does the bat hear reflected off the wall?

86P. A submarine, near the water surface, moves north at speed 75.0 km/h in a northbound current of speed 30.0 km/h, where both speeds are relative to the ocean floor. The sub emits a sonar signal (sound wave), of frequency $f = 1000$ Hz and speed 5470 km/h, that is detected by a destroyer north of the sub. What is the difference between the detected frequency and f if the destroyer (a) drifts with the current at 30.0 km/h and (b) is stationary relative to the ocean floor?

87P. A 2000 Hz siren and a civil defense official are both at rest with respect to the ground. What frequency does the official hear if the wind is blowing at 12 m/s (a) from source to observer and (b) from observer to source?

88P. Two trains are traveling toward each other at 100 ft/s relative to the ground. One train is blowing a whistle at 500 Hz. (a) What frequency will be heard on the other train in still air? (b) What frequency will be heard on the other train if the wind is blowing at 100 ft/s toward the whistle and away from the listener? (c) What frequency will be heard if the wind direction is reversed?

89P. A girl is sitting near the open window of a train that is moving at a velocity of 10.00 m/s to the east. The girl's uncle stands near the tracks and watches the train move away. The locomotive whistle emits sound at frequency 500.0 Hz. The air is still. (a) What frequency does the uncle hear? (b) What frequency does the girl hear? A wind begins to blow from the east at 10.00 m/s. (c) What frequency does the uncle now hear? (d) What frequency does the girl now hear?

SECTION 18-9 The Doppler Effect for Light

90E. Figure 18-41 is a graph of the intensity versus wavelength for light reaching Earth from galaxy NGC 7319, which is about 3×10^8 light-years away. The most intense light was emitted by the oxygen in NGC 7319. In a laboratory that emission is at wavelength $\lambda = 513$ nm, but in the light from NGC 7319 it has been shifted to 525 nm due to the Doppler effect (all the emissions from NGC 7319 have been shifted). (a) What is the speed of NGC 7319 relative to Earth? (b) Is the relative motion toward or away from our planet?

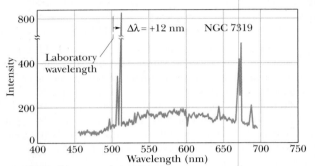

FIGURE 18-41 Exercise 90.

91E. Certain wavelengths in the light from a galaxy in the constellation Virgo are observed to be 0.4% longer than the corresponding light from terrestrial sources. What is the radial speed of this galaxy with respect to Earth? Is it approaching or receding?

92E. In the "red shift" of radiation from a distant galaxy, a certain radiation, known to have a wavelength of 434 nm when observed in the laboratory, has a wavelength of 462 nm. (a) What is the speed of the galaxy (along the line of sight) relative to Earth? (b) Is the galaxy approaching or receding?

93E. Assuming that Eq. 18-57 holds, find how fast you would have to go through a red light to have it appear green. Take 620 nm as the wavelength of red light and 540 nm as the wavelength of green light.

94P. The period of rotation of the Sun at its equator is 24.7 days; its radius is 7.00×10^5 km. What Doppler wavelength shift is expected for light with wavelength 550 nm emitted from the edge of the Sun's disk?

95P. An Earth satellite, transmitting at a frequency of 40 MHz (exactly), passes directly over a radio receiving station at an altitude of 400 km and at a speed of 3.0×10^4 km/h. Plot the change in frequency attributable to the Doppler effect as a function of time, counting $t = 0$ as the instant the satellite is over the station. (*Hint:* The speed u in the Doppler formula is not the actual speed of the satellite but its component in the direction of the station. Neglect the curvature of Earth and that of the satellite orbit.)

96P. Microwaves, which travel at the speed of light, are reflected from a distant airplane approaching the wave source. It is found that when the reflected waves are beat against the waves radiating from the source, the beat frequency is 990 Hz. If the microwaves are 0.100 m in wavelength, what is the approach speed of the airplane?

19

Temperature, Heat, and the First Law of Thermodynamics

An object with a black surface usually heats up more than one with a white surface when both are in sunlight. Such is true of the robes worn by Bedouins in the Sinai desert: black robes heat up more than white robes. Why then would a Bedouin ever wear a black robe? Wouldn't that actually decrease the chance of survival in the harsh desert environment?

19-1 THERMODYNAMICS

With this chapter, we leave the subject of mechanics and begin a new subject—*thermodynamics*. Mechanics deals with the mechanical energy of systems and is governed by Newton's laws. Thermodynamics deals with an internal energy of systems —thermal energy—and is governed by a new set of laws, which you will come to know in this and the next two chapters.

The central concept of thermodynamics is temperature. This word is so familiar that most of us—because of our built-in sense of hot and cold—tend to be overconfident in our understanding of it. Our "temperature sense" is in fact not always reliable. On a cold winter day, for example, an iron railing seems much colder to the touch than a wooden fence post, yet both are at the same temperature. This difference in our sense perception comes about because iron removes energy from our fingers more quickly than wood does. Here, we shall develop the concept of temperature from its foundations, without relying in any way on our temperature sense.

Temperature is one of the seven SI base quantities. Physicists measure temperature on the **Kelvin scale,** which is marked in units called *kelvins*. Although the temperature of a body apparently has no upper limit, it does have a lower limit; this limiting low temperature is taken as the zero of the Kelvin temperature scale. Room temperature is about 290 kelvins, or 290 K as we write it, above this

absolute zero. Figure 19-1 shows the wide range over which temperatures are determined.

When the universe began, some 10–20 billion years ago, its temperature was about 10^{39} K. As the universe expanded it cooled, and it has now reached an average temperature of about 3 K. We on Earth are a little warmer than that because we happen to live near a star. Without our Sun, we too would be at 3 K (or rather, we could not exist).

19-2 THE ZEROTH LAW OF THERMODYNAMICS

The properties of many bodies change as we alter their temperature, perhaps by moving them from a refrigerator to a warm oven. To give a few examples: as their respective temperatures increase, the volume of a liquid increases, a metal rod grows a little longer, and the electrical resistance of a wire increases, as does the pressure exerted by a confined gas. We can use any one of these properties as the basis of an instrument that will help us to pin down the concept of temperature.

Figure 19-2 shows such an instrument. Any resourceful engineer could design and construct it, using any one of the properties listed above. The instrument is fitted with a digital readout display and has the following properties: if you heat it with a Bunsen burner, the displayed number starts to increase; if you then put it into a refrigerator, the displayed number starts to decrease. The instrument is not calibrated in any way, and the numbers have (as yet) no physical meaning. The device is a *thermoscope* but not (as yet) a *thermometer.*

Suppose that, as in Fig. 19-3a, we put the thermoscope (which we shall call body T) into intimate contact with another body (body A). The entire system is confined within a thick-walled insulating box. The numbers displayed by the thermoscope roll by until, eventually, they come to rest (let us say the reading is "137.04") and no further change takes place. In fact, we suppose that every measurable property of body T and of body A has assumed a stable, unchanging value. Then we say that the two bodies are in *thermal equilibrium* with each other. And

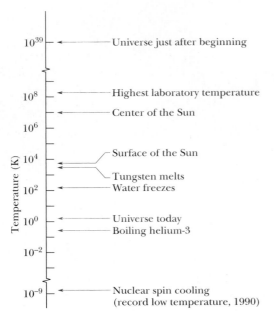

FIGURE 19-1 Some temperatures on the Kelvin scale. Temperature $T = 0$ corresponds to $10^{-\infty}$ and cannot be plotted on this logarithmic scale.

FIGURE 19-2 A thermoscope. The numbers increase when the device is heated and decrease when it is cooled. The thermally sensitive element could be —among many possibilities—a coil of wire whose electrical resistance is measured and displayed.

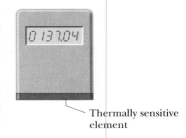

Thermally sensitive element

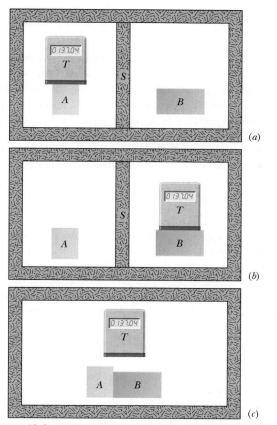

FIGURE 19-3 (*a*) Body *T* (a thermoscope) and body *A* are in thermal equilibrium. (Body *S* is a thermally insulating screen.) (*b*) Body *T* and body *B* are also in thermal equilibrium, at the same reading of the thermoscope. (*c*) If (*a*) and (*b*) are true, the zeroth law of thermodynamics states that body *A* and body *B* will also be in thermal equilibrium.

even though the displayed readings for body *T* have not been calibrated, we conclude that bodies *T* and *A* must be at the same (unknown) temperature.

Suppose that we next put body *T* in intimate contact with body *B* (Fig. 19-3*b*) and find that the two bodies come to thermal equilibrium *at the same reading of the thermoscope*. Then bodies *T* and *B* must be at the same (still unknown) temperature. If we now put bodies *A* and *B* into intimate contact (Fig. 19-3*c*), will they immediately be in thermal equilibrium with each other? Experimentally, we find that they are.

The experimental facts shown in Fig. 19-3 are summed up in the **zeroth law of thermodynamics:**

> If bodies *A* and *B* are each in thermal equilibrium with a third body *T*, then they are in thermal equilibrium with each other.

In less formal language, the message of the zeroth law is: "Every body has a property called **temperature.** When

two bodies are in thermal equilibrium, their temperatures are equal." And vice versa. We can now make our thermoscope (the third body *T*) into a thermometer, confident that its readings will have physical meaning. All we have to do is calibrate it.

We use the zeroth law constantly in the laboratory. If we want to know whether the liquids in two beakers are at the same temperature, we measure the temperature of each with a thermometer. We do not need to bring the two liquids into intimate contact and observe whether they are or are not in thermal equilibrium.

The zeroth law, which has been called a logical afterthought, came to light only in the 1930s, long after the first and the second law of thermodynamics had been discovered and numbered. Because the concept of temperature is fundamental to those two laws, the law that establishes temperature as a valid concept should have the lowest number. Hence the zero.

19-3 MEASURING TEMPERATURE

Let us see how we define and measure temperatures on the Kelvin scale. Equivalently, let us see how to calibrate a thermoscope so as to make it into a usable thermometer.

The Triple Point of Water

To set up a temperature scale, we pick some reproducible thermal phenomenon and, quite arbitrarily, assign a certain Kelvin temperature to its thermal environment. That is, we select a *standard fixed point* and give it a standard fixed-point temperature. We could, for example, select the freezing point or the boiling point of water but, for various technical reasons, we do not. Instead we select the **triple point of water.**

Liquid water, solid ice, and water vapor (gaseous water) can coexist, in thermal equilibrium, at only one set of values of pressure and temperature. Figure 19-4 shows a triple-point cell, in which this so-called triple point of

FIGURE 19-4 A triple-point cell, in which solid ice, liquid water, and water vapor coexist in thermal equilibrium. By international agreement, the temperature of this mixture has been defined to be 273.16 K. The bulb of a constant-volume gas thermometer is shown inserted into the well of the cell.

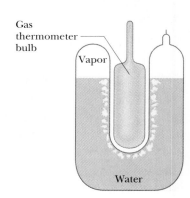

water can be achieved in the laboratory. By international agreement, the triple point of water has been assigned a value of 273.16 K as the standard fixed-point temperature for the calibration of thermometers. That is,

$$T_3 = 273.16 \text{ K} \qquad \text{(triple-point temperature)}, \qquad (19\text{-}1)$$

in which the subscript 3 reminds us of the triple point. This agreement also sets the size of the kelvin as 1/273.16 of the difference between absolute zero and the triple-point temperature of water.

Note that we do not use a degree mark in reporting Kelvin temperatures. It is 300 K (not 300°K), and it is read "300 kelvins" (not "300 degrees Kelvin"). The usual SI prefixes apply. Thus 0.0035 K is 3.5 mK. No distinction in nomenclature is made between Kelvin temperatures and temperature differences. Thus we can write, "the boiling point of sulfur is 717.8 K" and "the temperature of this water bath was raised by 8.5 K."

The Constant-Volume Gas Thermometer

So far, we have not discussed the particular temperature-sensitive physical property on which, by international agreement, we shall base our thermometer. Should it be the length of a metal rod, the electrical resistance of a wire, the pressure exerted by a confined gas, or something else? The choice is important because different choices lead to different temperatures for—to give one example—the boiling point of water. For reasons that will emerge below, the standard thermometer, against which all other thermometers are to be calibrated, is based on the pressure exerted by a gas confined to a fixed volume.

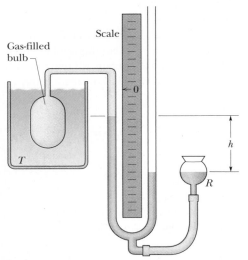

FIGURE 19-5 A constant-volume gas thermometer, its bulb immersed in a bath whose temperature T is to be measured.

Figure 19-5 shows such a **(constant-volume) gas thermometer.** It consists of a gas-filled bulb made of glass, quartz, or platinum (depending on the temperature range over which the thermometer is to be used) connected by a tube to a mercury manometer. By raising and lowering reservoir R, the mercury level on the left can always be brought to the zero of the manometer scale, thus assuring that the volume of the confined gas remains constant. The temperature of any body in thermal contact with the bulb is then defined to be

$$T = Cp, \qquad (19\text{-}2)$$

in which p is the pressure exerted by the gas and C is a constant. The pressure p is calculated from the relation

$$p = p_0 - \rho g h, \qquad (19\text{-}3)$$

in which p_0 is the atmospheric pressure, ρ is the density of the mercury in the manometer, and h is the measured level difference of mercury in the two arms of the tube.

With the bulb of the gas thermometer immersed in a triple-point cell, as in Fig. 19-4, we have

$$T_3 = Cp_3, \qquad (19\text{-}4)$$

in which p_3 is the pressure reading under this condition. Eliminating C between Eqs. 19-2 and 19-4 leads to

$$T = T_3 \left(\frac{p}{p_3} \right)$$
$$= (273.16 \text{ K}) \left(\frac{p}{p_3} \right) \qquad \text{(provisional).} \quad (19\text{-}5)$$

Equation 19-5 is not yet our final definition of a temperature measured with a gas thermometer. We have said nothing about what gas (or how much gas) we are to place in the thermometer. If our thermometer is used to measure some temperature, such as the boiling point of water, we find that different gases in the bulb give slightly different measured temperatures. However, as we use smaller and smaller amounts of gas to fill the bulb, the readings converge nicely to a single temperature, no matter what gas we use. Figure 19-6 shows this comforting convergence.*

Thus we write, as our final recipe for measuring temperature with a gas thermometer,

$$T = (273.16 \text{ K}) \left(\lim_{m \to 0} \frac{p}{p_3} \right). \qquad (19\text{-}6)$$

---

*For pressure units, we shall use units introduced in Section 15-3. The SI unit for pressure is the newton per square meter, which is called the pascal (Pa). The pascal is related to other common pressure units by

$$1 \text{ atm} = 1.01 \times 10^5 \text{ Pa} = 760 \text{ torr} = 14.7 \text{ lb/in.}^2.$$

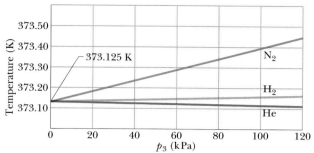

FIGURE 19-6 Temperatures calculated from Eq. 19-5 for a constant-volume gas thermometer whose bulb is immersed in boiling water. Different gases were used within the bulb, each at a variety of densities (which gave varying values for p_3). Note that all readings converge in the limit of zero density to a temperature of 373.125 K.

This instructs us to measure an unknown temperature T as follows. Fill the bulb with an arbitrary mass m of *any* gas (for example, nitrogen) and measure p_3 (using a triple-point cell) and p, the gas pressure at the temperature being measured. Calculate the ratio p/p_3. Then repeat both measurements with a smaller amount of gas in the bulb, and again calculate this ratio. Continue this way, using smaller and smaller amounts of gas, until you can extrapolate to the ratio p/p_3 that you would find if there were approximately no gas in the bulb. Calculate the temperature T by substituting that extrapolated ratio into Eq. 19-6. (The temperature measured in this way is the *ideal gas temperature*.)

If temperature is to be a truly fundamental physical quantity, one in which the laws of thermodynamics may be expressed, it is absolutely necessary that its definition be independent of the properties of specific materials. It would not do, for example, to have a quantity as basic as temperature depend on the expansivity of mercury, the electrical resistivity of platinum, or any other such "handbook" property. We chose the gas thermometer as our standard instrument precisely because no such specific properties of materials are involved in its operation. Use *any* gas—you will always get the same result.

19-4 THE CELSIUS AND FAHRENHEIT SCALES

So far, we have discussed only the Kelvin scale, used in basic scientific work. In nearly all countries of the world, the Celsius scale (formerly called the centigrade scale) is the scale of choice for popular and commercial use and much scientific use. Celsius temperatures are measured in degrees, and the Celsius degree has the same size as the kelvin. However, the zero of the Celsius scale is shifted to a more convenient value than absolute zero. If T_C represents a Celsius temperature, then

$$T_C = T - 273.15°. \qquad (19-7)$$

In expressing temperatures on the Celsius scale, the degree symbol is commonly used. Thus we write 20.00°C for a Celsius reading but 293.15 K for a Kelvin reading.

The Fahrenheit scale, used in the United States, employs a smaller degree than the Celsius scale and a different zero of temperature. You can easily verify both these differences by examining an ordinary room thermometer on which both scales are marked. The relation between the Celsius and Fahrenheit scales is

$$T_F = \tfrac{9}{5} T_C + 32°, \qquad (19-8)$$

where T_F is Fahrenheit temperature. Transferring between these two scales can be done easily by remembering a few corresponding points (such as the freezing and boiling points of water; see Table 19-1) and by making use of the fact that 9 degrees on the Fahrenheit scale equals 5 degrees on the Celsius scale. Figure 19-7 compares the Kelvin, Celsius, and Fahrenheit scales.

We use the letters C and F to distinguish measurements and degrees on the two scales. Thus,

$$0°C = 32°F$$

SAMPLE PROBLEM 19-1

The bulb of a gas thermometer is filled with nitrogen to a pressure of 120 kPa. What provisional value (see Fig. 19-6) would this thermometer yield for the boiling point of water and what is the error in this value?

SOLUTION: From Fig. 19-6, the curve for nitrogen shows that, at 120 kPa, the provisional temperature for the boiling point of water would be about 373.44 K. The actual boiling point of water (found by extrapolation on Fig. 19-6) is 373.125 K. Thus using the provisional temperature leads to an error of 0.315 K, or 315 mK.

TABLE 19-1 SOME CORRESPONDING TEMPERATURES

TEMPERATURE	°C	°F
Boiling point of water[a]	100	212
Normal body temperature	37.0	98.6
Accepted comfort level	20	68
Freezing point of water[a]	0	32
Zero of Fahrenheit scale	≈ −18	0
Scales coincide	−40	−40

[a]Strictly, the boiling point of water on the Celsius scale is 99.975°C, and the freezing point is 0.00°C. Thus there is slightly less than 100 C° between those two points.

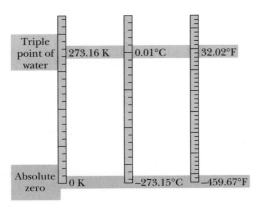

FIGURE 19-7 The Kelvin, Celsius, and Fahrenheit temperature scales compared.

means that 0° on the Celsius scale measures the same temperature as 32° on the Fahrenheit scale, whereas

$$5 \, C° = 9 \, F°$$

means that a temperature difference of five Celsius degrees (note the degree symbol is *after* the letter) is equivalent to a temperature difference of nine Fahrenheit degrees.

SAMPLE PROBLEM 19-2

Suppose you come across old scientific notes that describe a temperature scale called Z on which the boiling point of water is 65.0°Z and the freezing point is −14.0°Z.

(a) What temperature change ΔT on the Z scale would correspond to a change of 53.0 F°?

SOLUTION: To find a conversion factor between the two scales, we can use the boiling and freezing points of water. On the Z scale, the temperature difference between those two points is 65.0°Z − (−14.0°Z), or 79.0 Z°. On the Fahrenheit scale, it is 212°F − 32.0°F, or 180 F°. Thus a change of 79.0 Z° equals a change of 180 F°. For a change of 53.0 F°, we can now write

$$\Delta T = 53.0 \, F° = 53.0 \, F° \left(\frac{79.0 \, Z°}{180 \, F°} \right)$$

$$= 23.3 \, Z°. \qquad \text{(Answer)}$$

(b) To what temperature on the Fahrenheit scale would a temperature $T = -98.0°Z$ correspond?

SOLUTION: The freezing point of water is −14.0°Z, so the difference between T and the freezing point is 84.0 Z°. To convert this difference to Fahrenheit degrees, we write

$$\Delta T = 84.0 \, Z° \left(\frac{180 \, F°}{79.0 \, Z°} \right) = 191 \, F°.$$

Thus T is 191 F° below the freezing point of water and is, on the Fahrenheit scale,

$$T = 32.0°F − 191 \, F° = -159°F. \qquad \text{(Answer)}$$

CHECKPOINT 1: The figure shows three temperature scales with the freezing and boiling points of water indicated. (a) Rank the size of a degree on these scales, greatest first. (b) Rank the following temperatures, highest first: 50°X, 50°W, and 50°Y.

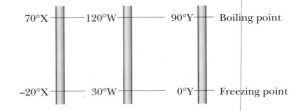

PROBLEM SOLVING TACTICS

TACTIC 1: *Temperature Changes*

Between the boiling and freezing points of water, there are (approximately) 100 kelvins and 100 Celsius degrees. Thus a kelvin is the same size as a Celsius degree. From this or from Eq. 19-7, we then know that any temperature change is the same number whether expressed in kelvins or Celsius degrees. For example, a temperature change of 10 K is equivalent to a change of 10 C°.

Between the boiling and freezing points of water, there are 180 Fahrenheit degrees. Thus 180 F° = 100 K, and a Fahrenheit degree must be 100 K/180 F°, or 5/9, the size of a kelvin or Celsius degree. From this or from Eq. 19-8, we then know that any temperature change expressed in Fahrenheit degrees must be $\frac{9}{5}$ times that same temperature change expressed in either kelvins or Celsius degrees. For example, in Fahrenheit degrees, a temperature change of 10 K is (9/5)(10 K), or 18 F°.

You should take care not to confuse a *temperature* with a temperature *change*. A temperature of 10 K is certainly not the same as one of 10°C or 18°F but, as above, a temperature *change* of 10 K is the same as one of 10 C° or 18 F°.

19-5 THERMAL EXPANSION

You can often loosen a tight metal jar lid by holding it under a stream of hot water. Both the metal of the lid and the glass of the jar expand as the hot water adds energy to their atoms. (With the added energy, the atoms can move a bit farther from each other than usual, against the spring-like interatomic forces that hold every solid together.) However, because the atoms in the metal move farther apart than those in the glass, the lid expands more than the jar and thus is loosened.

FIGURE 19-8 Railroad tracks in Asbury Park, New Jersey, distorted because of thermal expansion on a very hot July day.

Such **thermal expansion** is not always desirable, as Fig. 19-8 suggests. To preclude buckling, therefore, expansion slots are placed in bridges to accommodate roadway expansion on hot days. Dental materials used for fillings must be matched in their thermal expansion properties to those of tooth enamel (otherwise consuming hot coffee or cold ice cream would be quite painful). In aircraft manufacture, however, rivets and other fasteners are often cooled in dry ice before insertion and then allowed to expand to a tight fit.

Thermometers and thermostats may be based on the differences in expansion between the components of a *bimetal strip* (Fig. 19-9). And the familiar liquid-in-glass thermometers are based on the fact that liquids such as mercury or alcohol expand to a different (greater) extent than do their glass containers.

TABLE 19-2 SOME COEFFICIENTS OF LINEAR EXPANSION[a]

SUBSTANCE	α (10^{-6}/C°)	SUBSTANCE	α (10^{-6}/C°)
Ice (at 0°C)	51	Steel	11
Lead	29	Glass (ordinary)	9
Aluminum	23	Glass (Pyrex)	3.2
Brass	19	Diamond	1.2
Copper	17	Invar[b]	0.7
Concrete	12	Fused quartz	0.5

[a]Room temperature values except for the listing for ice.

[b]This alloy was designed to have a low coefficient of expansion. The word is a shortened form of "invariable."

Linear Expansion

If the temperature of a metal rod of length L is raised by an amount ΔT, its length is found to increase by an amount

$$\Delta L = L\alpha \, \Delta T, \qquad (19\text{-}9)$$

in which α is a constant called the **coefficient of linear expansion.** The coefficient α has the unit "per degree" or "per kelvin" and depends on the material. If we rewrite Eq. 19-9 as

$$\alpha = \frac{\Delta L/L}{\Delta T}, \qquad (19\text{-}10)$$

we see that α is the fractional increase in length per unit change in temperature. Although α varies somewhat with temperature, for most practical purposes it can be taken as constant for a particular material. Table 19-2 shows some coefficients of linear expansion.

The thermal expansion of a solid is like (three-dimensional) photographic enlargement. Figure 19-10*b*

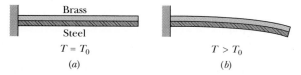

FIGURE 19-9 (*a*) A bimetal strip, consisting of a strip of brass and a strip of steel welded together, at temperature T_0. (*b*) The strip bends as shown at temperatures above this reference temperature. Below the reference temperature the strip bends the other way. Many thermostats operate on this principle, making and breaking an electrical contact as the temperature rises and falls.

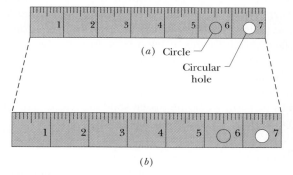

FIGURE 19-10 The same steel rule at two different temperatures. When it expands, every dimension is increased in the same proportion. The scale, the numbers, the thickness, and the diameters of the circle and circular hole are all increased by the same factor. (The expansion has been exaggerated for clarity.)

shows the (exaggerated) expansion of a steel rule after its temperature is increased from that of Fig. 19-10a. Equation 19-9 applies to every linear dimension of the rule, including its edge, thickness, diagonals, and the diameters of the circle etched on it and the circular hole cut in it. If the disk cut from that hole originally fits snugly in the hole, it will continue to fit snugly if it undergoes the same temperature increase as the rule.

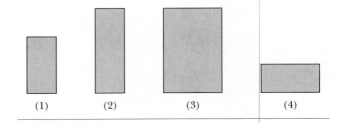

(1)   (2)   (3)   (4)

Volume Expansion

If all dimensions of a solid expand with temperature, the volume of that solid must also expand. For liquids, volume expansion is the only meaningful expansion parameter. If the temperature of a solid or liquid whose volume is V is increased by an amount ΔT, the increase in volume is found to be

$$\Delta V = V\beta\,\Delta T, \qquad (19\text{-}11)$$

where β is the **coefficient of volume expansion** of the solid or liquid. The coefficients of volume expansion and linear expansion for a solid are related by

$$\beta = 3\alpha. \qquad (19\text{-}12)$$

The most common liquid, water, does not behave like other liquids. Above about 4°C, water expands as the temperature rises, as we would expect. Between 0 and about 4°C, however, water *contracts* with increasing temperature. Thus, at about 4°C, the density of water passes through a maximum. At all other temperatures, the density of water is less than this maximum value.

This behavior of water is the reason that lakes freeze from the top down rather than from the bottom up. As the water on the surface is cooled from, say, 10°C toward the freezing point, it becomes denser ("heavier") than the lower water and sinks to the bottom. Below 4°C, however, further cooling makes the water then on the surface *less* dense ("lighter") than the lower water, so it stays on the surface until it freezes. If lakes froze from the bottom up, the ice so formed would tend not to melt completely during the summer, being insulated by the water above. After a few years, many bodies of open water in the temperate zones of Earth would be frozen solid all year round. And aquatic life as we know it could not exist.

CHECKPOINT **2:** The figure shows four rectangular metal plates, with sides of L, $2L$, or $3L$. They are all made of the same material, and their temperature is to be increased by the same amount. Rank the plates according to the expected increase in (a) their vertical heights and (b) their areas, greatest first.

SAMPLE PROBLEM 19-3

A steel wire whose length L is 130 cm and whose diameter d is 1.1 mm is heated to an average temperature of 830°C and stretched taut between two rigid supports. What tension develops in the wire as it cools to 20°C?

SOLUTION: First let us calculate the amount by which the wire would shrink if it were allowed to cool freely. From Eq. 19-9 and Table 19-2, we can write the amount of shrinkage as

$$\Delta L = L\alpha\,\Delta T = (1.3\text{ m})(11 \times 10^{-6}/\text{C}°)(830°\text{C} - 20°\text{C})$$
$$= 1.16 \times 10^{-2}\text{ m} = 1.16\text{ cm}.$$

However, the wire is not permitted to shrink. We therefore calculate what force would be required to stretch the wire by this amount. From Eq. 13-34 we have

$$F = \frac{\Delta L\, EA}{L} = \frac{\Delta L\, E(\pi/4)d^2}{L},$$

in which E is Young's modulus for steel (see Table 13-1) and A is the cross-sectional area of the wire. Substitution yields

$$F = (1.16 \times 10^{-2}\text{ m})(200 \times 10^9\text{ N/m}^2)(\pi/4)$$
$$\times \frac{(1.1 \times 10^{-3}\text{ m})^2}{1.3\text{ m}}$$
$$= 1700\text{ N.} \qquad \text{(Answer)}$$

Can you show that this answer is independent of the length of the wire?

Sometimes the bulging brick walls of old buildings are supported by running a steel stress rod from outside wall to outside wall, through the building. The rod is then heated, and nuts on the outside walls are tightened. As the rod cools and contracts, tension develops in it, helping to keep the walls from bulging outward.

SAMPLE PROBLEM 19-4

On a hot day in Las Vegas, an oil trucker loaded 9785 gal of diesel fuel. He encountered cold weather on the way to Payson, Utah, where the temperature was 41 F° lower than in Las Vegas, and where he delivered his entire load. How many gallons did he deliver? The coefficient of volume expansion for diesel fuel is $9.5 \times 10^{-4}/\text{C}°$, and the coefficient of linear expansion for his steel truck tank is $11 \times 10^{-6}/\text{C}°$.

SOLUTION: From Eq. 19-11,

$$\Delta V = V\beta \, \Delta T$$

$$= (9785 \text{ gal})(9.5 \times 10^{-4}/\text{C}°)(-41 \text{ F}°)\left(\frac{5 \text{ C}°}{9 \text{ F}°}\right)$$

$$= -212 \text{ gal.}$$

Thus the amount delivered was

$$V_{\text{del}} = V + \Delta V = 9785 \text{ gal} - 212 \text{ gal}$$

$$= 9573 \text{ gal} \approx 9600 \text{ gal.} \qquad \text{(Answer)}$$

Note that the thermal expansion of the steel tank has nothing to do with the problem. Question: Who paid for the "missing" diesel fuel?

PROBLEM SOLVING TACTICS

TACTIC 2: *Units for Temperature Changes*

The coefficient of linear expansion α is defined to be the fractional change in length (a dimensionless number) per unit change in temperature. In Table 19-2, that unit change in temperature is expressed in Celsius degrees (C°). Since any temperature *change* expressed in Celsius degrees is numerically the same when expressed in kelvins, the values of α in Table 19-2 could just as well be expressed in kelvins.

For example, the value of α for steel can be written as either $11 \times 10^{-6}/\text{C}°$ or $11 \times 10^{-6}/\text{K}$. This means that we could have substituted either expression for α in Sample Problem 19-3. We could have made a similar substitution for β in Sample Problem 19-4.

19-6 TEMPERATURE AND HEAT

If you take a can of cola from the refrigerator and leave it on the kitchen table, its temperature will rise—rapidly at first but then more slowly—until the temperature of the cola equals that of the room (the two are then in thermal equilibrium). In the same way, the temperature of a cup of hot coffee, left sitting on the table, will fall until it also reaches room temperature.

In generalizing this situation, we describe the cola or the coffee as a *system* (with temperature T_S) and the relevant parts of the kitchen as the *environment* (with temperature T_E) of that system. Our observation is that if T_S is not equal to T_E, then T_S will change (T_E may also change some) until the two temperatures are equal and thus thermal equilibrium is reached.

Such a change in temperature is due to the transfer of a form of energy between the system and its environment. This energy is *internal energy* (or *thermal energy*), which is the collective kinetic and potential energies associated

with the random motions of the atoms, molecules, and other microscopic bodies within an object. The transferred internal energy is called **heat** and is symbolized Q. Heat is *positive* when internal energy is transferred to a system from its environment (we say that heat is absorbed). Heat is *negative* when internal energy is transferred from a system to its environment (we say that heat is released or lost).

This transfer of energy is shown in Fig. 19-11. In the situation of Fig. 19-11a, in which $T_S > T_E$, internal energy is transferred from the system to the environment, so Q is negative. In Fig. 19-11b, in which $T_S = T_E$, there is no such transfer, Q is zero, and heat is neither released nor absorbed. In Fig. 19-11c, in which $T_S < T_E$, the transfer is to the system from the environment, so Q is positive.

We are led then to this definition of heat:

Heat is the energy that is transferred between a system and its environment because of a temperature difference that exists between them.

Recall that energy can also be transferred between a system and its environment by means of *work W*, which we always associate with a force acting on a system during a displacement of the system. Heat and work, unlike temperature, pressure, and volume, are not intrinsic properties

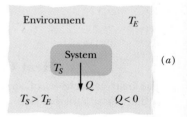

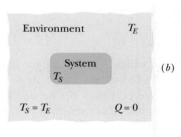

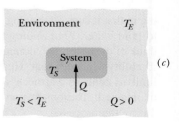

FIGURE 19-11 If the temperature of a system exceeds that of its environment as in (a), heat Q is lost by the system to the environment until thermal equilibrium (b) is established. (c) If the temperature of the system is below that of the environment, heat is absorbed by the system until thermal equilibrium is established.

of a system. They have meaning only as they describe the transfer of energy into or out of a system. Thus it is proper to say: "During the last 3 min, 15 J of heat was transferred to the system from its environment" or "During the last minute, 12 J of work was done on the system by its environment." It is meaningless to say: "This system contains 450 J of heat" or "This system contains 385 J of work."

Before scientists realized that heat is transferred energy, heat was measured in terms of its ability to raise the temperature of water. Thus the **calorie** (cal) was defined as the amount of heat that would raise the temperature of 1 g of water from 14.5°C to 15.5°C. In the British system, the corresponding unit of heat was the **British thermal unit** (Btu), defined as the amount of heat that would raise the temperature of 1 lb of water from 63°F to 64°F.

In 1948, the scientific community decided that since heat (like work) is transferred energy, the SI unit for heat should be the same as that for energy, namely, the **joule.** The calorie is now defined to be 4.1860 J (exactly) with no reference to the heating of water. (The "calorie" used in nutrition, sometimes called the Calorie (Cal), is really a kilocalorie.) The relations among the various heat units are

$$1 \text{ cal} = 3.969 \times 10^{-3} \text{ Btu} = 4.186 \text{ J}. \quad (19\text{-}13)$$

19-7 THE ABSORPTION OF HEAT BY SOLIDS AND LIQUIDS

Heat Capacity

The **heat capacity** C of an object (for example, a Pyrex coffee pot or a marble slab) is the proportionality constant between an amount of heat and the change in temperature that this heat produces in the object. Thus

$$Q = C(T_f - T_i), \quad (19\text{-}14)$$

in which T_i and T_f are the initial and final temperatures of the object. Heat capacity C has the unit of energy per degree or energy per kelvin. The heat capacity C of, say, a marble slab used in a bun warmer might be 179 cal/C°, which we can also write as 179 cal/K or as 747 J/K.

The word "capacity" in this context is really misleading in that it suggests analogy with the capacity of a bucket to hold water. *That analogy is false,* and you should not think of the object as "containing" heat or being limited in its ability to absorb heat. Heat transfer can proceed without limit as long as the necessary temperature difference is maintained. The object may, of course, melt or vaporize during the process.

Specific Heat

Two objects made of the same material, say marble, will have heat capacities proportional to their masses. It is therefore convenient to define a "heat capacity per unit mass" or **specific heat** c that refers not to an object but to a unit mass of the material of which the object is made. Equation 19-14 then becomes

$$Q = cm(T_f - T_i). \quad (19\text{-}15)$$

Through experiment we would find that although the heat capacity of the particular marble slab mentioned above is 179 cal/C° (or 747 J/K), the specific heat of marble itself (in that slab or in any other marble object) is 0.21 cal/g·C° (or 880 J/kg·K).

From the way the calorie and the British thermal unit were initially defined, the specific heat of water is

$$c = 1 \text{ cal/g·C°} = 1 \text{ Btu/lb·F°}$$
$$= 4190 \text{ J/kg·K}. \quad (19\text{-}16)$$

Table 19-3 shows the specific heats of some substances at room temperature. Note that the value for water is relatively high. The specific heat of any substance actually depends somewhat on temperature, but the values in Table 19-3 apply reasonably well in a range of temperatures near room temperature.

TABLE 19-3 SPECIFIC HEATS OF SOME SUBSTANCES AT ROOM TEMPERATURE

SUBSTANCE	SPECIFIC HEAT cal/g·K	SPECIFIC HEAT J/kg·K	MOLAR SPECIFIC HEAT J/mol·K
Elemental Solids			
Lead	0.0305	128	26.5
Tungsten	0.0321	134	24.8
Silver	0.0564	236	25.5
Copper	0.0923	386	24.5
Aluminum	0.215	900	24.4
Other Solids			
Brass	0.092	380	
Granite	0.19	790	
Glass	0.20	840	
Ice (−10°C)	0.530	2220	
Liquids			
Mercury	0.033	140	
Ethyl alcohol	0.58	2430	
Seawater	0.93	3900	
Water	1.00	4190	

CHECKPOINT **3:** A certain amount of heat Q will warm 1 g of material A by 3 C° and 1 g of material B by 4 C°. Which material has the greater specific heat?

Molar Specific Heat

In many instances the most convenient unit for specifying the amount of a substance is the mole (mol), where

$$1 \text{ mol} = 6.02 \times 10^{23} \text{ elementary units}$$

of *any* substance. Thus 1 mol of aluminum means 6.02×10^{23} atoms (the atom being the elementary unit), and 1 mol of aluminum oxide means 6.02×10^{23} molecules of the oxide (because the molecule is the elementary unit of a compound).

When quantities are expressed in moles, the specific heat must also involve moles (rather than a mass unit); it is then called a **molar specific heat.** Table 19-3 shows the values for some elemental solids (each consisting of a single element) at room temperature.

Note that the molar specific heats of all the elements listed in Table 19-3 have about the same value at room temperature, namely, 25 J/mol · K. As a matter of fact, the molar specific heats of all solids increase toward that value as the temperature increases. (Some substances, such as carbon and beryllium, do not reach this limiting value until temperatures well above room temperature. Other substances may melt or vaporize before they reach this limit.)

When we compare two substances on a molar basis, we are comparing samples that contain the same number of elementary units. The fact that, at high enough temperatures, all solid elements have the same molar specific heat tells us that atoms of all kinds—whether they be aluminum, copper, uranium, or anything else—absorb heat in the same way.

An Important Point

In determining and then using the specific heat of any substance, we need to know the conditions under which heat transfer occurs. For solids and liquids, we usually assume that the sample is under constant pressure (usually atmospheric) during the heat transfer. It is also conceivable that the sample is held at constant volume while the heat is absorbed. This means that thermal expansion of the sample is prevented by applying external pressure. For solids and liquids, this is very hard to arrange experimentally but the effect can be calculated, and it turns out that the specific heats under constant pressure and constant volume for any solid or liquid differ usually by no more than a few percent. Gases, as you will see, have quite different values for their specific heats under constant-pressure conditions and under constant-volume conditions.

Heats of Transformation

When heat is absorbed by a solid or liquid, the temperature of the sample does not necessarily rise. Instead, the sample may change from one *phase,* or *state,* (that is, solid, liquid, or gas) to another. Thus ice may melt and water may boil, absorbing heat in each case without a temperature change. In the reverse processes (water freezing, steam condensing), heat is released by the sample, again while the temperature of the sample remains constant.

The amount of heat per unit mass that must be transferred when a sample completely undergoes a phase change is called the **heat of transformation** L. So when a sample of mass m completely undergoes a phase change, the total heat transferred is

$$Q = Lm. \tag{19-17}$$

When the phase change is from liquid to gas (then the sample must absorb heat) or from gas to liquid (then the sample must release heat), the heat of transformation is called the **heat of vaporization** L_V. For water at its normal boiling or condensation temperature,

$$L_V = 539 \text{ cal/g} = 40.7 \text{ kJ/mol}$$
$$= 2256 \text{ kJ/kg}. \tag{19-18}$$

When the phase change is from solid to liquid (then the sample must absorb heat) or from liquid to solid (then the sample must release heat), the heat of transformation is called the **heat of fusion** L_F. For water at its normal freezing or melting temperature,

$$L_F = 79.5 \text{ cal/g} = 6.01 \text{ kJ/mol}$$
$$= 333 \text{ kJ/kg}. \tag{19-19}$$

Table 19-4 shows the heats of transformation for some substances.

SAMPLE PROBLEM 19-5

A candy bar has a marked nutritional value of 350 Cal. How many kilowatt-hours of energy will it deliver to the body as it is digested?

SOLUTION: The Calorie in this case is a kilocalorie so that

$$\text{energy} = (350 \times 10^3 \text{ cal})(4.19 \text{ J/cal})$$
$$= (1.466 \times 10^6 \text{ J})(1 \text{ W} \cdot \text{s/J})$$
$$\times (1 \text{ h}/3600 \text{ s})(1 \text{ kW}/1000 \text{ W})$$
$$= 0.407 \text{ kW} \cdot \text{h}. \tag{Answer}$$

This amount of energy would keep a 100 W light bulb burning

TABLE 19-4 SOME HEATS OF TRANSFORMATION

	MELTING		BOILING	
SUBSTANCE	MELTING POINT (K)	HEAT OF FUSION L_F (kJ/kg)	BOILING POINT (K)	HEAT OF VAPORIZATION L_V (kJ/kg)
Hydrogen	14.0	58.0	20.3	455
Oxygen	54.8	13.9	90.2	213
Mercury	234	11.4	630	296
Water	273	333	373	2256
Lead	601	23.2	2017	858
Silver	1235	105	2323	2336
Copper	1356	207	2868	4730

for 4.1 h. To burn up this much energy by exercise, a person would have to jog about 3 or 4 mi.

A generous human diet corresponds to about 3.5 kW·h per day, which represents the absolute maximum amount of work that a human can do in one day. In an industrialized country, this amount of energy can be purchased for perhaps 35 cents.

SAMPLE PROBLEM 19-6

(a) How much heat is needed to take ice of mass $m = 720$ g at $-10°C$ to a liquid state at $15°C$?

SOLUTION: To answer, we must consider three steps. Step 1 is to raise the temperature of the ice from $-10°C$ to the melting point at $0°C$. We use Eq. 19-15, with the specific heat c_{ice} of ice given in Table 19-3. For this step, the initial temperature T_i is $-10°C$ and the final temperature T_f is $0°C$. We find

$$Q_1 = c_{ice}m(T_f - T_i)$$
$$= (2220 \text{ J/kg·K})(0.720 \text{ kg})[0°C - (-10°C)]$$
$$= 15{,}984 \text{ J} \approx 15.98 \text{ kJ}.$$

Step 2 is to melt the ice (there can be no change in temperature until this step is completed). We now use Eqs. 19-17 and 19-19, finding

$$Q_2 = L_Fm = (333 \text{ kJ/kg})(0.720 \text{ kg}) \approx 239.8 \text{ kJ}.$$

Step 3 is to raise the temperature of the now liquid water from $0°C$ to $15°C$. We again use Eq. 19-15, but now with the specific heat c_{liq} of liquid water given in Table 19-3. In this step the initial temperature T_i is $0°C$ and the final temperature T_f is $15°C$. We find

$$Q_3 = c_{liq}m(T_f - T_i)$$
$$= (4190 \text{ J/kg·K})(0.720 \text{ kg})(15°C - 0°C)$$
$$= 45{,}252 \text{ J} \approx 45.25 \text{ kJ}.$$

The total required heat Q_{tot} is the sum of the amounts required in the three steps:

$$Q_{tot} = Q_1 + Q_2 + Q_3$$
$$= 15.98 \text{ kJ} + 239.8 \text{ kJ} + 45.25 \text{ kJ}$$
$$\approx 300 \text{ kJ}. \qquad \text{(Answer)}$$

Note that the heat required to melt the ice is much larger than the heat required to raise the temperature of either the ice or the liquid water.

(b) If we supply the ice with a total heat of only 210 kJ, what then are the final state and temperature of the water?

SOLUTION: From step 1, we know that 15.98 kJ is needed to raise the temperature of the ice to the melting point. The remaining heat Q_{rem} is then 210 kJ $-$ 15.98 kJ, or about 194 kJ. From step 2, we can see that this amount of heat is insufficient to melt all the ice. We can find the mass m of ice that is melted by the heat Q_{rem} by using Eqs. 19-17 and 19-19:

$$m = \frac{Q_{rem}}{L_F} = \frac{194 \text{ kJ}}{333 \text{ kJ/kg}} = 0.583 \text{ kg} \approx 580 \text{ g}.$$

Thus the mass of the ice that remains is 720 g $-$ 580 g, or 140 g. Since not all the ice is melted, the temperature of the ice–liquid water must be $0°C$. Hence we have

580 g water   and   140 g ice, at $0°C$.   (Answer)

SAMPLE PROBLEM 19-7

A copper slug whose mass m_c is 75 g is heated in a laboratory oven to a temperature T of $312°C$. The slug is then dropped into a glass beaker containing a mass $m_w = 220$ g of water. The heat capacity C_b of the beaker is 45 cal/K. The initial temperature T_i of the water and the beaker is $12°C$. What is the final temperature T_f of the slug, the beaker, and the water when thermal equilibrium is reached?

SOLUTION: We take as our system *water + beaker + copper slug*. No heat enters or leaves this system, so the algebraic sum of the internal heat transfers that occur must be zero. There are three such transfers:

$$\text{for the water:} \quad Q_w = m_w c_w (T_f - T_i);$$

$$\text{for the beaker:} \quad Q_b = C_b (T_f - T_i);$$

$$\text{for the copper:} \quad Q_c = m_c c_c (T_f - T).$$

The temperature difference is written — in all three cases — as the final temperature minus the initial temperature (T_i for the water and beaker, and T for the copper). We do this even though we realize that Q_w and Q_b are positive (indicating that heat is added to the initially cool water and beaker) and that Q_c is negative (indicating that heat is released by the initially hot copper slug). Doing this allows us to write

$$Q_w + Q_b + Q_c = 0. \tag{19-20}$$

Substituting the heat transfer expressions into Eq. 19-20 yields

$$m_w c_w (T_f - T_i) + C_b (T_f - T_i) \\ + m_c c_c (T_f - T) = 0. \tag{19-21}$$

Temperatures enter Eq. 19-21 only as differences. Thus, because the intervals on the Celsius and Kelvin scales are identical, we can use either of these scales in this equation. Solving Eq. 19-21 for T_f, we have

$$T_f = \frac{m_c c_c T + C_b T_i + m_w c_w T_i}{m_w c_w + C_b + m_c c_c}.$$

The numerator is, with Celsius temperatures,

$$(75 \text{ g})(0.092 \text{ cal/g} \cdot \text{K})(312°C) + (45 \text{ cal/K})(12°C) \\ + (220 \text{ g})(1.00 \text{ cal/g} \cdot \text{K})(12°C) \\ = 5332.8 \text{ cal},$$

and the denominator is

$$(220 \text{ g})(1.00 \text{ cal/g} \cdot \text{K}) + 45 \text{ cal/K} \\ + (75 \text{ g})(0.092 \text{ cal/g} \cdot \text{K}) \\ = 271.9 \text{ cal/C°}.$$

So, we then have

$$T_f = \frac{5332.8 \text{ cal}}{271.9 \text{ cal/C°}} = 19.6°C \approx 20°C. \quad \text{(Answer)}$$

From the given data you can show that

$$Q_w \approx 1670 \text{ cal}, \quad Q_b \approx 342 \text{ cal}, \quad Q_c \approx -2020 \text{ cal}.$$

Apart from rounding errors, the algebraic sum of these three heat transfers is indeed zero, as Eq. 19-20 requires.

19-8 A CLOSER LOOK AT HEAT AND WORK

Here we look in some detail at how heat and work are exchanged between a system and its environment. Let us take as our system a gas confined to a cylinder with a movable piston, as in Fig. 19-12. The upward force on the piston due to the pressure of the confined gas is equal to the weight of lead shot loaded onto the top of the piston. The walls of the cylinder are made of insulating material that does not allow any heat transfer. The bottom of the cylinder rests on a reservoir for thermal energy, a *thermal reservoir* (perhaps a hot plate) whose temperature T you can control by turning a knob.

The system (the gas) starts from an *initial state i*, described by a pressure p_i, a volume V_i, and a temperature T_i. You want to change the system to a *final state f*, described by a pressure p_f, a volume V_f, and a temperature T_f. The procedure by which you change the system from its initial state to its final state is called a *thermodynamic process*. During such a process, heat may be transferred into the system from the thermal reservoir (positive heat) or vice versa (negative heat). And work is done by the system to raise the loaded piston (positive work) or lower it (negative work). We assume that all such changes occur slowly, with the result that the system is always in (approximate) thermal equilibrium, (that is, every part of the system is always in thermal equilibrium with every other part).

Suppose that you remove a few lead shot from the piston of Fig. 19-12, allowing the gas to push the piston

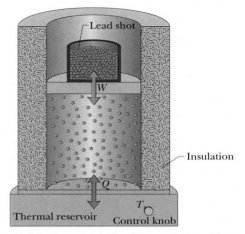

FIGURE 19-12 A gas is confined to a cylinder with a movable piston. Heat Q can be added to, or withdrawn from, the gas by regulating the temperature T of the adjustable thermal reservoir. Work W can be done by the gas by raising or lowering the piston.

and remaining shot upward through a differential displacement $d\mathbf{s}$ with an upward force $\mathbf{F}$. Since the displacement is tiny, we can assume that $\mathbf{F}$ is constant during the displacement. Then $\mathbf{F}$ has a magnitude that is equal to pA, where p is the pressure of the gas and A is the face area of the piston. The differential work dW done by the gas during the displacement is

$$dW = \mathbf{F} \cdot d\mathbf{s} = (pA)(ds) = p(A\ ds)$$
$$= p\ dV, \qquad (19\text{-}22)$$

in which dV is the differential change in the volume of the gas owing to the movement of the piston. When you have removed enough shot to allow the gas to change its volume from V_i to V_f, the total work done by the gas is

$$W = \int dW = \int_{V_i}^{V_f} p\ dV. \qquad (19\text{-}23)$$

During the change in volume, the pressure and temperature of the gas may also change. To evaluate the integral in Eq. 19-23 directly, we would need to know how pressure varies with volume for the actual process by which the system changes from state i to state f.

There are actually many ways to take the gas from state i to state f. One way is shown in Fig. 19-13a, which is a plot of the pressure of the gas versus its volume and which is called a p-V diagram. In Fig. 19-13a, the curve indicates that the pressure decreased as the volume increased. The integral of Eq. 19-23 (and thus the work W done by the gas) is represented by the shaded area under the curve between points i and f. Regardless of what exactly we did to take the gas along the curve, that work is positive, owing to the fact that the gas increased its volume by forcing the piston upward.

Another way to get from state i to state f is shown in Fig. 19-13b: there the change takes place in two steps— the first from state i to state a, and the second from state a to state f.

Step ia of this process is carried out at constant pressure, which means that you leave undisturbed the lead shot that rides on top of the piston in Fig. 19-12. You cause the volume to increase (from V_i to V_f) by slowly turning up the temperature control knob, raising the temperature of the gas to some higher value T_a. (Increasing the temperature increases the force by the gas on the piston, moving it upward.) During this process, positive work is done by the expanding gas (to lift the loaded piston) and heat is added to the system from the thermal reservoir (in response to the arbitrarily small temperature differences that you create as you turn up the temperature). This heat is positive because it is added to the system.

Step af of the process of Fig. 19-13b is carried out at

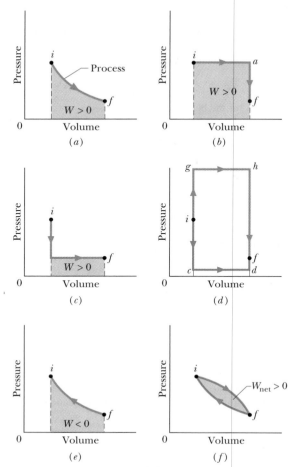

FIGURE 19-13   (a) The system of Fig. 19-12 goes from an *initial state i* to a *final state f* by means of a *thermodynamic process*. The area marked W represents the work done *by* the system during this process. The work is positive, because the process proceeds toward greater volume. (b) Another process for moving between the same two states; the work is now greater than that in (a). (c) Still another process, requiring less (positive) work. (d) The work can be made as small as you like (path $icdf$) or as large as you like (path $ighf$). (e) When the volume is reduced (by some external force), the work done *by* the system is negative. (f) The net work done by the system during a (closed) cycle is represented by the enclosed area, which is the difference in the areas beneath the two curves that make up the cycle.

constant volume, so you must wedge the piston, preventing it from moving. And then you must use the control knob to decrease the temperature so that the pressure drops from p_a to its final value p_f. During this process, heat is lost by the system to the thermal reservoir.

For the overall process iaf, the work W, which is positive and is carried out only during step ia, is represented by the shaded area under the curve. Heat is transferred during both steps ia and af, with a net heat transfer Q.

Figure 19-13c shows a process in which the two steps above are carried out in reverse order. The work W in this case is smaller than for Fig. 19-13b, as is the net heat absorbed. Figure 19-13d suggests that you can make the work done by the gas as small as you want (by following a path like *icdf*) or as large as you want (by following a path like *ighf*).

To sum up: a system can be taken from a given initial state to a given final state by an infinite number of processes. Heat may or may not be involved, and in general, the work W and the heat Q will have different values for different processes. We say that heat and work are *path-dependent* quantities.

Figure 19-13e shows an example in which negative work is done by a system as some external force compresses the system, reducing its volume. The absolute value of the work done is still equal to the area beneath the curve, but because the gas is *compressed,* the work done by the gas is negative.

Figure 19-13f shows a *thermodynamic cycle* in which the system is taken from some initial state i to some other state f and then back to i. The net work done by the system during the cycle is the sum of the positive work done during the expansion and the negative work done during the compression. In Fig. 19-13f, the net work is positive because the area under the expansion curve (i to f) is greater than the area under the compression curve (f to i).

CHECKPOINT **4:** The p-V diagram shows six curved paths (connected by vertical paths) that can be followed by a gas. Which two of them should be part of a closed cycle if the net work done by the gas is to be at its maximum positive value?

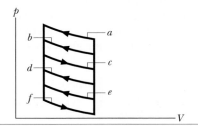

19-9 THE FIRST LAW OF THERMODYNAMICS

You have just seen that when a system changes from a given initial state to a given final state, both the work W done and the heat Q exchanged depend on the nature of the process. Experimentally, however, we find a surprising thing. *The quantity $Q - W$ is the same for all processes.* It depends only on the initial and final states and does not

depend at all on how the system gets from one to the other. All other combinations of Q and W, including Q alone, W alone, $Q + W$, and $Q - 2W$, are *path dependent;* only the quantity $Q - W$ is not.

The quantity $Q - W$ must represent a change in some intrinsic property of the system. We call this property the *internal energy E_{int}* and we write

$$\Delta E_{int} = E_{int,f} - E_{int,i}$$
$$= Q - W \quad \text{(first law).} \quad (19\text{-}24)$$

Equation 19-24 is the **first law of thermodynamics.** If the thermodynamic system undergoes only a differential change, we can write the first law as*

$$dE_{int} = dQ - dW \quad \text{(first law).} \quad (19\text{-}25)$$

The internal energy E_{int} of a system tends to increase if energy is added via heat Q and tends to decrease if energy is lost via work W done by the system.

In Chapter 8, we discussed the principle of energy conservation as it applies to isolated systems, that is, to systems in which no energy enters or leaves the system. The first law of thermodynamics is an extension of that principle to systems that are *not* isolated. In such cases, energy may be transferred into or out of the system as either work W or heat Q. In our statement of the first law of thermodynamics above, we assume that there are no changes in the kinetic energy or the potential energy of the system as a whole; that is, $\Delta K = \Delta U = 0$.

Before this chapter, the term *work* and the symbol W always meant the work done *on* a system. But starting with Eq. 19-22 and continuing through the next two chapters about thermodynamics, we focus on the work done *by* a system, such as the gas in Fig. 19-12.

The work done *on* a system is always the negative of the work done *by* the system. So if we rewrite Eq. 19-24 in terms of the work W_{on} done *on* the system, we have $\Delta E_{int} = Q + W_{on}$. This tells us the following: the internal energy of a system tends to increase if heat is absorbed by the system or if positive work is done *on* the system. Conversely, the internal energy tends to decrease if heat is lost by the system or if negative work is done *on* the system.

---

*Here dQ and dW, unlike dE_{int}, are not true differentials. That is, there are no such functions as $Q(p, V)$ and $W(p, V)$ that depend only on the state of the system. The quantities dQ and dW are called *inexact differentials* and are usually represented by the symbols $đQ$ and $đW$. For our purposes, we can treat them simply as infinitesimally small energy transfers.

19-10 SOME SPECIAL CASES OF THE FIRST LAW OF THERMODYNAMICS

Here we look at four different thermodynamic processes, in each of which a certain restriction is imposed on the system. We then see what consequences follow when we apply the first law of thermodynamics to the process.

1. *Adiabatic processes.* An adiabatic process is one that occurs so rapidly or occurs in a system that is so well insulated that *no transfer of heat* occurs between the system and its environment. Putting $Q = 0$ in the first law (Eq. 19-24) then leads to

$$\Delta E_{int} = -W \qquad \text{(adiabatic process).} \qquad (19\text{-}26)$$

This tells us that if work is done *by* the system (that is, if W is positive), the internal energy of the system decreases by the amount of work. Conversely, if work is done *on* the system (that is, if W is negative), the internal energy of the system increases by that amount.

Figure 19-14 shows an idealized adiabatic process. Heat cannot enter or leave the system because of the insulation. Thus the only way energy can be transferred between the system and its environment is by work. If we remove shot from the piston and allow the gas to expand, the work done by the system (the gas) is positive and the internal energy of the gas decreases. If, instead, we add shot and compress the gas, the work done by the system is negative and the internal energy of the gas increases.

2. *Constant-volume processes.* If the volume of a system (such as a gas) is held constant, that system can do no work. Putting $W = 0$ in the first law (Eq. 19-24) yields

$$\Delta E_{int} = Q \qquad \text{(constant-volume process).} \qquad (19\text{-}27)$$

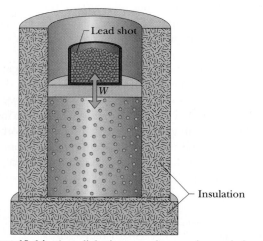

FIGURE 19-14 An adiabatic expansion can be carried out by slowly removing lead shot from the top of the piston. Adding lead shot reverses the process at any stage.

Thus if heat is added to a system (that is, if Q is positive), the internal energy of the system increases. Conversely, if heat is removed during the process (that is, if Q is negative), the internal energy of the system must decrease.

3. *Cyclical processes.* There are processes in which, after certain interchanges of heat and work, the system is restored to its initial state. In that case, no intrinsic property of the system—including its internal energy—can possibly change. Putting $\Delta E_{int} = 0$ in the first law (Eq. 19-24) yields

$$Q = W \qquad \text{(cyclical process).} \qquad (19\text{-}28)$$

Thus the net work done during the process must exactly equal the net amount of heat transferred; the store of internal energy of the system remains unchanged. Cyclical processes form a closed loop on a pressure–volume plot, as in Fig. 19-13f. We shall discuss such processes in some detail in Chapter 21.

4. *Free expansion.* These are adiabatic processes in which no work is done on or by the system. Thus $Q = W = 0$ and the first law requires that

$$\Delta E_{int} = 0 \qquad \text{(free expansion).} \qquad (19\text{-}29)$$

Figure 19-15 shows how such an expansion can be carried out. A gas, which is in thermal equilibrium within itself, is initially confined by a closed stopcock to one half of an insulated double chamber; the other half is evacuated. The stopcock is opened, and the gas expands freely to fill both halves of the chamber. No heat is transferred to or from the gas because of the insulation. No work is done by the gas because it rushes into a vacuum, its motion unopposed by any counteracting pressure.

A free expansion differs from all other processes we have considered because it cannot be done slowly and in a controlled way. As a result, at any given instant during the sudden expansion, the gas is not in thermal equilibrium and its pressure is not the same everywhere. So, although

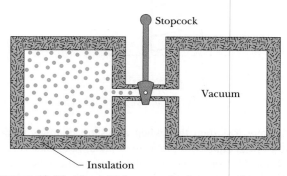

FIGURE 19-15 The initial stage of a free-expansion process. After the stopcock is opened, the gas eventually reaches an equilibrium final state, filling both chambers.

TABLE 19-5 **THE FIRST LAWS OF THERMODYNAMICS:**
FOUR SPECIAL CASES

The Law: $\Delta E_{int} = Q - W$ (Eq. 19-24)

PROCESS	RESTRICTION	CONSEQUENCE
Adiabatic	$Q = 0$	$\Delta E_{int} = -W$
Constant volume	$W = 0$	$\Delta E_{int} = Q$
Closed cycle	$\Delta E_{int} = 0$	$Q = W$
Free expansion	$Q = W = 0$	$\Delta E_{int} = 0$

we can plot the initial and final states on a *p-V* diagram, we cannot plot the expansion itself.

Table 19-5 summarizes the characteristics of the processes of this section.

$\mathbb{C}$HECKPOINT **5:** For one complete cycle as shown in the *p-V* diagram, are (a) ΔE_{int} for the gas and (b) the net heat transfer *Q* positive, negative, or zero?

SAMPLE PROBLEM 19-8

Let 1.00 kg of liquid water at 100°C be converted to steam at 100°C by boiling at standard atmospheric pressure (1.00 atm or 1.01×10^5 Pa). The volume changes from an initial value of 1.00×10^{-3} m^3 as a liquid to 1.671 m^3 as steam (Fig. 19-16).

(a) How much work is done by the system this process?

SOLUTION: The work is given by Eq. 19-23. Because the pressure is constant (at 1.01×10^5 Pa) during the boiling process, we can take *p* outside the integral, obtaining

$$W = \int_{V_i}^{V_f} p \, dV = p \int_{V_i}^{V_f} dV = p(V_f - V_i)$$

$$= (1.01 \times 10^5 \text{ Pa})(1.671 \text{ m}^3 - 1.00 \times 10^{-3} \text{ m}^3)$$

$$= 1.69 \times 10^5 \text{ J} = 169 \text{ kJ}. \qquad \text{(Answer)}$$

The result is positive, indicating that work is done *by* the system on its environment in lifting the weighted piston of Fig. 19-16.

(b) How much heat must be added to the system during the process?

SOLUTION: Since there is no temperature change, but only a phase change, we use Eqs. 19-17 and 19-18:

$$Q = L_V m = (2260 \text{ kJ/kg})(1.00 \text{ kg})$$

$$= 2260 \text{ kJ}. \qquad \text{(Answer)}$$

The result is positive, which indicates that heat is *added to* the system, as we expect.

(c) What is the change in the internal energy of the system during the boiling process?

SOLUTION: We find this from the first law (Eq. 19-24):

$$\Delta E_{int} = Q - W = 2260 \text{ kJ} - 169 \text{ kJ}$$

$$\approx 2090 \text{ kJ} = 2.09 \text{ MJ}. \qquad \text{(Answer)}$$

This quantity is positive, indicating that the internal energy of the system has increased during the boiling process. This energy goes into separating the H$_2$O molecules, which strongly attract each other in the liquid state.

We see that, when water is boiled, about 7.5% (= 169 kJ/2260 kJ) of the added heat goes into the work of pushing back the atmosphere. The rest goes into internal energy that is added to the system.

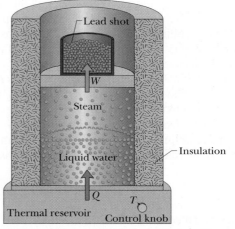

FIGURE 19-16 Sample Problem 19-8. Water boiling at constant pressure. Heat is added from the thermal reservoir until the liquid water has changed completely into steam. Work is done by the expanding gas as it lifts the loaded piston.

19-11 HEAT TRANSFER MECHANISMS

We have discussed the transfer of heat between a system and its environment, but we have not yet described how that transfer takes place. There are three transfer mechanisms: conduction, convection, and radiation.

Conduction

If you leave a metal poker in a fire for any length of time, its handle will get hot. Energy is transferred from the fire to the handle by **conduction** along the length of the poker. The vibration amplitudes of the atoms and electrons of the metal at the fire end of the poker become relatively large because of the high temperature of their environment.

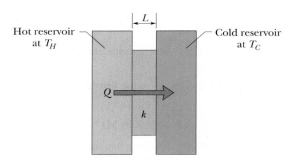

FIGURE 19-17 Thermal conduction. Heat is transferred from a reservoir at temperature T_H to a cooler reservoir at temperature T_C through a conducting slab of thickness L and thermal conductivity k.

TABLE 19-6 SOME THERMAL CONDUCTIVITIES[a]

	$k(\mathrm{W/m \cdot K})$		$k(\mathrm{W/m \cdot K})$
Metals		*Building Materials*	
Stainless steel	14	Polyurethane foam	0.024
Lead	35	Rock wool	0.043
Aluminum	235	Fiberglass	0.048
Copper	401	White pine	0.11
Silver	428	Window glass	1.0
Gases			
Air (dry)	0.026		
Helium	0.15		
Hydrogen	0.18		

[a]Conductivities change somewhat with temperature. The given values are at room temperature.

These increased vibrational amplitudes, and thus the associated energy, are passed along the poker, from atom to atom, during collisions between adjacent atoms. In this way, a region of rising temperature extends itself along the poker to the handle.

Consider a slab of face area A and thickness L, whose faces are maintained at temperatures T_H and T_C by a hot reservoir and a cold reservoir, as in Fig. 19-17. Let Q be the heat that is transferred through the slab, from its hot face to its cold face, in time t. Experiment shows that the rate of heat transfer H (the amount per unit time) is given by

$$H = \frac{Q}{t} = kA\frac{T_H - T_C}{L}, \qquad (19\text{-}30)$$

in which k, called the *thermal conductivity,* is a constant that depends on the material of which the slab is made. Large values of k define good heat conductors, and conversely. Table 19-6 gives thermal conductivities of some common metals, gases, and building materials.

Thermal Resistance to Conduction (*R*-Value)

If you are interested in insulating your house or in keeping cola cans cold on a picnic, you are more concerned with poor heat conductors than with good ones. For this reason, the concept of *thermal resistance R* has been introduced into engineering practice. The *R*-value of a slab of thickness L is defined as

$$R = \frac{L}{k}. \qquad (19\text{-}31)$$

Thus the lower the thermal conductivity of the material of which a slab is made, the higher the *R*-value of the slab. Note that R is a property attributed to a slab of a specified thickness, not to a material. The commonly used unit for R

(which, in the United States at least, is almost never stated) is the square foot–Fahrenheit degree-hour per British thermal unit ($\mathrm{ft^2 \cdot F° \cdot h/Btu}$). (Now you know why the unit is rarely stated.)

Combining Eqs. 19-30 and 19-31 leads to

$$H = A\frac{T_H - T_C}{R}, \qquad (19\text{-}32)$$

which allows one to calculate the rate of heat flow through a slab if its *R*-value, its area, and the temperature difference between its faces are known.

Conduction Through a Composite Slab

Figure 19-18 shows a composite slab, consisting of two materials having different thicknesses L_1 and L_2 and different thermal conductivities k_1 and k_2. The temperatures of the outer surfaces of the slab are T_H and T_C. Each face of

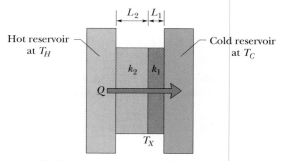

FIGURE 19-18 Heat is transferred at a steady rate through a composite slab made up of two different materials with different thicknesses and different thermal conductivities. The steady-state temperature at the interface of the two materials is T_X.

the slab has area A. Let us derive an expression for the rate of heat transfer through the slab under the assumption that the transfer is a *steady-state* process; that is, the temperatures everywhere in the slab and the rate of heat transfer do not change with time.

In the steady state, the rates of heat transfer through the two materials are equal. This is the same as saying that the heat conducted through one material in a certain time must be equal to that conducted through the other material in the same amount of time. Were this not true, temperatures in the slab would be changing and we would not have a steady-state situation. Letting T_X be the temperature of the interface between the two materials, we can now use Eq. 19-30 to write

$$H = \frac{k_2 A (T_H - T_X)}{L_2} = \frac{k_1 A (T_X - T_C)}{L_1}. \quad (19\text{-}33)$$

Solving Eq. 19-33 for T_X yields, after a little algebra,

$$T_X = \frac{k_1 L_2 T_C + k_2 L_1 T_H}{k_1 L_2 + k_2 L_1}. \quad (19\text{-}34)$$

Substituting this expression for T_X into either equality of Eq. 19-33 yields

$$H = \frac{A(T_H - T_C)}{L_1/k_1 + L_2/k_2}. \quad (19\text{-}35)$$

Equation 19-31 reminds us that $L/k = R$.

We can extend Eq. 19-35 to apply to any number n of materials making up a slab:

$$H = \frac{A(T_H - T_C)}{\Sigma\,(L/k)} = \frac{A(T_H - T_C)}{\Sigma\,R}. \quad (19\text{-}36)$$

The summation sign in the denominators tells us to add the values of L/k or R for all the materials.

CHECKPOINT 6: The figure shows the face and interface temperatures of a composite slab consisting of four materials, of identical thicknesses, through which the heat transfer is steady. Rank the materials according to their thermal conductivities, greatest first.

25°C | 15°C | 10°C | −5.0°C | −10°C

| a | b | c | d |

Convection

When you look at the flame of a candle or a match, you are watching thermal energy being transported upward by **convection.** Such heat transfer occurs when a fluid, such as

A football rally is illuminated by a fierce bonfire. Heated air and hot gases from the fire rise, and cooler air flows into the base of the fire.

air or water, is in contact with an object whose temperature is higher than that of the fluid. The temperature of the fluid that is in contact with the hot object increases, and (in most cases) the fluid expands and thus becomes less dense. Because it is now lighter than the surrounding cooler fluid, this expanded fluid rises because of buoyant forces. Some of the surrounding cooler fluid then flows so as to take the place of the rising warmer fluid and is itself warmed. The process can then continue.

Convection is part of many natural processes. Atmospheric convection plays a fundamental role in determining global climate patterns and daily weather variations. Glider pilots and birds alike seek rising thermals (currents of warm air) that keep them aloft. Huge energy transfers take place within the oceans by the same process. Finally, energy is transported to the surface of the Sun from the nuclear furnace at its core by enormous cells of convection, in which hot gas rises to the surface along the cell core and cooler gas around the core descends below the surface.

Radiation

The third method of heat transfer between an object and its environment is via electromagnetic waves (visible light is one kind of electromagnetic wave). Heat transferred in this

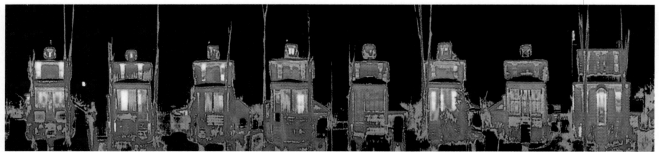

FIGURE 19-19 A false-color thermogram reveals the rate at which energy is radiated by houses along a street. The rates, from largest to smallest, are color coded as white, red, pink, blue, and black. You can tell where there is insulation in the walls, a heavy curtain over a window, and a higher air temperature at the ceiling on the second floor.

way is often called **thermal radiation** to distinguish it from electromagnetic *signals* (as in, say, television broadcasts) and from nuclear radiation (energy and particles emitted by nuclei). (To "radiate" generally means to emit.) When you stand in bright sunshine, you are warmed by absorbing thermal radiation from the Sun. No medium is required for heat transfer via radiation.

The rate P_r at which an object emits energy via electromagnetic radiation (that is, the power P_r of the radiating object) depends on its surface area A and the temperature T of that area in kelvins and is given by

$$P_r = \sigma \varepsilon A T^4. \qquad (19\text{-}37)$$

Here $\sigma = 5.6703 \times 10^{-8}$ W/m²·K⁴ is called the *Stefan–Boltzmann constant* after Josef Stefan (who discovered Eq. 19-37 experimentally in 1879) and Ludwig Boltzmann (who derived it theoretically soon after). The symbol ε represents the *emissivity* of the object's surface, which has a value between 0 and 1, depending on the composition of the surface. A surface with the maximum emissivity of 1.0 is said to be a *blackbody radiator,* but such a surface is possible only in theory. Note again that the temperature in Eq. 19-37 must be in kelvins so that a temperature of absolute zero corresponds to no radiation. Note also that every object whose temperature is above 0 K—including people—emits thermal radiation. (See Fig. 19-19.)

The rate P_a at which an object absorbs energy via thermal radiation from its environment, which we take to be at uniform temperature T_{env} (in kelvins), is

$$P_a = \sigma \varepsilon A T_{\text{env}}^4. \qquad (19\text{-}38)$$

The emissivity ε in Eq. 19-38 is the same as that in Eq. 19-37. An idealized blackbody radiator, with $\varepsilon = 1$, will absorb all the radiated energy it intercepts (rather than sending a portion back away from itself through reflection or scattering).

Because an object will radiate energy to the environment while it absorbs energy from the environment, the object's net rate P_n of energy exchange due to thermal radiation is

$$\begin{aligned} P_n &= P_a - P_r \\ &= \sigma \varepsilon A (T_{\text{env}}^4 - T^4). \qquad (19\text{-}39) \end{aligned}$$

The emissivity of black cloth is greater than that of white cloth; hence, by Eq. 19-39 a black robe will absorb more net energy from sunshine than a white robe and thus will be at a higher temperature. In fact, research has shown that in the hot desert, a black Bedouin robe can be 6 C° higher in temperature than a similar white robe. Why, then, would someone who must avoid overheating to survive in the harsh desert wear a black robe?

FIGURE 19-20 Convection up through the hotter black robe is more vigorous than that up through the cooler white robe. (After "Why Do Bedouins Wear Black Robes in Hot Deserts?" by A. Shkolnik, C. R. Taylor, V. Finch, and A. Borut, *Nature*, Vol. 283, January 24, 1980, pp. 373–374.)

The answer is that the black robe, which is itself at a higher temperature than a comparable white robe, does indeed warm the air inside the robe. That warmer air then rises faster and leaves through the porous fabric, while external air is drawn into the robe through its open bottom (Fig. 19-20). Thus the black fabric enhances air circulation under the robe, and that keeps the Bedouin from getting any hotter than a person in a white robe. In fact, the continuous breeze blowing past his body may even make him feel more comfortable.

SAMPLE PROBLEM 19-9

A composite slab (see Fig. 19-18) whose face area A is 26 ft² is made up of 2.0 in. of rock wool (1.0 in. has an R-value of 3.3) and 0.75 in. of white pine (1.0 in. has an R-value of 1.3). The temperature difference between the faces of the slab is 65 F°. What is the rate of heat transfer through the slab?

SOLUTION: The R-value for the 2.0 in. of rock wool is 3.3 × 2.0 or 6.6 ft²·°F·h/Btu. For the 0.75 in. of wood it is 1.3 × 0.75 or 0.98, in the same units. The composite slab thus has an R-value of 6.6 + 0.98 or 7.58 ft²·F°·h/Btu. Substitution into Eq. 19-36 yields

$$H = \frac{A(T_H - T_C)}{\Sigma R} = \frac{(26 \text{ ft}^2)(65 \text{ F}°)}{7.58 \text{ ft}^2 \cdot \text{F}° \cdot \text{h/Btu}}$$

$$= 223 \text{ Btu/h} \approx 220 \text{ Btu/h } (= 65 \text{ W}). \quad \text{(Answer)}$$

Thus at this temperature difference, each such insulating slab would transmit heat continuously at the rate of 65 W.

SAMPLE PROBLEM 19-10

Figure 19-21 shows the cross section of a wall made of white pine of thickness L_a and brick of thickness $L_d (= 2.0L_a)$, sandwiching two layers of unknown material with identical thicknesses and thermal conductivities. The thermal conductivity of the pine is k_a and that of the brick is $k_d (= 5.0k_a)$. The face area A of the wall is unknown. Heat conduction through the wall has reached the steady state; the only known interface temperatures are $T_1 = 25°C$, $T_2 = 20°C$, and $T_5 = -10°C$.

(a) What is interface temperature T_4?

SOLUTION: We cannot find T_4 by simply applying Eq. 19-30 layer by layer, starting with the pine and working our way rightward, because we do not know enough about the intermediate layers. However, since the heat conduction has reached the steady state, we know that the rate of conduction H_a through the pine must equal the rate of conduction H_d through the brick. From Eq. 19-30 and Fig. 19-21, we can write these rates as

$$H_a = k_a A \frac{T_1 - T_2}{L_a} \quad \text{and} \quad H_d = k_d A \frac{T_4 - T_5}{L_d}.$$

Setting $H_a = H_d$ and solving for T_4 yield

$$T_4 = \frac{k_a L_d}{k_d L_a}(T_1 - T_2) + T_5.$$

Letting $L_d = 2.0L_a$ and $k_d = 5.0k_a$, and inserting the known temperatures, we find

$$T_4 = \frac{k_a(2.0L_a)}{(5.0k_a)L_a}(25°C - 20°C) + (-10°C)$$

$$= -8.0°C. \quad \text{(Answer)}$$

(b) What is interface temperature T_3?

SOLUTION: Now that we know T_4, we can find T_3, even though we know little about the intermediate layers. (In fact, at this point you might be able to guess the answer.) Since the heat conduction process is in the steady state, the rate of conduction H_b through layer b is equal to the rate of conduction H_c through layer c. Then, from Eq. 19-30,

$$k_b A \frac{T_2 - T_3}{L_b} = k_c A \frac{T_3 - T_4}{L_c}.$$

Because the thermal conductivities k_b and k_c of the layers are equal and so are their thicknesses L_b and L_c, we have

$$T_2 - T_3 = T_3 - T_4,$$

which gives us

$$T_3 = \frac{T_2 + T_4}{2} = \frac{20°C + (-8.0°C)}{2}$$

$$= 6.0°C. \quad \text{(Answer)}$$

This shows that since the intermediate layers have identical thermal conductivities, a point midway across them has a temperature that is midway between the temperatures of their outside surfaces.

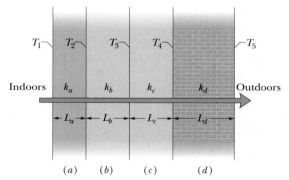

FIGURE 19-21 Sample Problem 19-10. A wall of four layers through which there is steady-state heat transfer.

SAMPLE PROBLEM 19-11

During an extended wilderness hike, you have a terrific craving for ice. Unfortunately, the air temperature drops to only 6.0°C each night—too high to freeze water. However, because a clear, moonless night sky acts like a blackbody radiator at a temperature of $T_s = -23°C$, perhaps you can make ice by letting a shallow layer of water radiate energy to such a sky. To start, you thermally insulate a container from the ground by placing a poorly conducting layer of, say, foam rubber or straw beneath it. Then you pour water into the container, forming a thin, uniform layer with mass $m = 4.5$ g, top surface area $A = 9.0$ cm^2, depth $d = 5.0$ mm, emissivity $\varepsilon = 0.90$, and initial temperature 6.0°C. Find the time required for the water to freeze via radiation. Can the freezing be accomplished during one night?

SOLUTION: If you are to make ice via thermal radiation, the radiation must first reduce the water temperature from 279 K (= 6.0°C) to the freezing point of 273 K. Using Eq. 19-15 and Table 19-3, we find the associated energy exchange to be

$$Q_1 = cm(T_f - T_i)$$
$$= (4190 \text{ J/kg·K})(4.5 \times 10^{-3} \text{ kg})(273 \text{ K} - 279 \text{ K})$$
$$= -113 \text{ J}.$$

This energy must be radiated away by the water to drop its temperature to the freezing point.

An additional amount of energy Q_2 must be radiated away to change the phase of the water. Using Eqs. 19-17 and 19-19, and inserting a minus sign to show that the energy is lost by the water, we find

$$Q_2 = -mL_F = -(4.5 \times 10^{-3} \text{ kg})(3.33 \times 10^5 \text{ J/kg})$$
$$= -1499 \text{ J}.$$

The total amount of energy Q to be radiated away by the water is thus

$$Q = Q_1 + Q_2 = -113 \text{ J} - 1499 \text{ J} = -1612 \text{ J}.$$

The water will absorb energy from the sky while it is radiating energy to the sky. The net rate of energy exchange P_n is given by Eq. 19-39. The time t required to radiate energy Q is then

$$t = \frac{Q}{P_n} = \frac{Q}{\sigma \varepsilon A(T_s^4 - T^4)}. \tag{19-40}$$

Although the temperature T of the water decreases slightly while the water is cooling, we can approximate T as being the freezing point, 273 K. With $T_s = 250$ K, the denominator of Eq. 19-40 is

$$(5.67 \times 10^{-8} \text{ W/m}^2 \cdot \text{K}^4)(0.90)(9.0 \times 10^{-4} \text{ m}^2)$$
$$\times [(250 \text{ K})^4 - (273 \text{ K})^4] = -7.57 \times 10^{-2} \text{ J/s},$$

and Eq. 19-40 gives us

$$t = \frac{-1612 \text{ J}}{-7.57 \times 10^{-2} \text{ J/s}} = 2.13 \times 10^4 \text{ s} = 5.9 \text{ h}. \quad \text{(Answer)}$$

Because t is less than a night, freezing water by having it radiate to the dark sky is feasible. In fact, in some parts of the world people used this technique long before the introduction of electric freezers.

REVIEW & SUMMARY

Temperature; Thermometers

Temperature is an SI base quantity related to our sense of hot and cold. It is measured with a thermometer, which contains a working substance with a measurable property, such as length or pressure, that changes in a regular way as the substance becomes hotter or colder.

Zeroth Law of Thermodynamics

When a thermometer and some other object are placed in contact with each other, they eventually reach thermal equilibrium. The reading of the thermometer is then taken to be the temperature of the other object. The process provides consistent and useful temperature measurements because of the **zeroth law of thermodynamics:** if bodies A and B are each in thermal equilibrium with a third body C (the thermometer), then A and B are in thermal equilibrium with each other.

The Kelvin Temperature Scale

Temperature is measured, in the SI system, on the **Kelvin scale.** The scale is established by defining the temperature of the *triple point* of water to be 273.16 K. Other temperatures are then defined by use of a *constant-volume gas thermometer,* in which the pressure of a constant-volume sample of a gas is proportional to its temperature. Since different gases give consistent results only at very low densities, we define the *temperature T* as measured with a gas thermometer to be

$$T = (273.16 \text{ K}) \left(\lim_{m \to 0} \frac{p}{p_3} \right). \tag{19-6}$$

Here T is measured in kelvins, p_3 and p are the pressures of the gas at 273.16 K and the measured temperature, respectively, and m is the mass of the gas in the thermometer.

Celsius and Fahrenheit Scales

The Celsius temperature scale is defined by

$$T_C = T - 273.15°, \tag{19-7}$$

and the Fahrenheit temperature scale is defined by

$$T_F = \tfrac{9}{5}T_C + 32°. \tag{19-8}$$

Thermal Expansion

All objects change size with changes in temperature. For a temperature change ΔT, a change ΔL in any linear dimension L is given by

$$\Delta L = L\alpha\,\Delta T, \tag{19-9}$$

in which α is the **coefficient of linear expansion.** The change ΔV in the volume V of a solid or liquid is

$$\Delta V = V\beta\,\Delta T. \tag{19-11}$$

Here $\beta = 3\alpha$ is the material's **coefficient of volume expansion.**

Heat

Heat Q is energy that is transferred between a system and its environment because of a temperature difference between them. It can be measured in **joules** (J), **calories** (cal), **kilocalories** (Cal or kcal), or **British thermal units** (Btu), with

$$1 \text{ cal} = 3.969 \times 10^{-3} \text{ Btu} = 4.186 \text{ J}. \tag{19-13}$$

Heat Capacity and Specific Heat

If heat Q is added to an object, the temperature change $T_f - T_i$ is related to Q by

$$Q = C(T_f - T_i), \tag{19-14}$$

in which C is the **heat capacity** of the object. If the object has mass m, then

$$Q = cm(T_f - T_i), \tag{19-15}$$

where c is the **specific heat** of the material making up the object. The **molar specific heat** of a material is the heat capacity per mole, or per 6.02×10^{23} elementary units of the material.

Heat of Transformation

Heat supplied to a material may change the material's physical state or phase—for example, from solid to liquid or from liquid to gas. The amount of heat required per unit mass to change the phase (but not the temperature) of a particular material is its **heat of transformation** L. Thus

$$Q = Lm. \tag{19-17}$$

The **heat of vaporization** L_V is the amount of energy per unit mass that must be added to vaporize a liquid or that must be removed to condense a gas. The **heat of fusion** L_F is the amount of energy per unit mass that must be added to melt a solid or that must be removed to freeze a liquid.

Work Associated with Volume Change

A gas may exchange energy with its surroundings through work. The amount of work W done by a gas as it expands or contracts from an initial volume V_i to a final volume V_f is given by

$$W = \int dW = \int_{V_i}^{V_f} p\,dV. \tag{19-23}$$

The integration is necessary because the pressure p may vary during the volume change.

First Law of Thermodynamics

The principle of conservation of energy for a thermodynamic process is expressed in the **first law of thermodynamics,** which may assume either of the forms

$$\Delta E_{\text{int}} = E_{\text{int},f} - E_{\text{int},i} = Q - W \quad \text{(first law)} \tag{19-24}$$

or

$$dE_{\text{int}} = dQ - dW \quad \text{(first law)}. \tag{19-25}$$

E_{int} represents the internal energy of the material, which depends only on its state (temperature, pressure, and volume). Q represents the heat exchanged by the system with its surroundings; Q is positive if the system gains heat and negative if the system loses heat. W is the work done *by* the system; W is positive if the system expands against some external force exerted by the surroundings, and negative if the system contracts because of some external force. *Both Q and W are path dependent; ΔE_{int} is path independent.*

Applications of the First Law

The first law of thermodynamics finds application in several special cases:

$$\begin{aligned}
\textit{adiabatic processes:} &\quad Q = 0, \quad \Delta E_{\text{int}} = -W \\
\textit{constant-volume processes:} &\quad W = 0, \quad \Delta E_{\text{int}} = Q \\
\textit{cyclical processes:} &\quad \Delta E_{\text{int}} = 0, \quad Q = W \\
\textit{free expansion processes:} &\quad Q = W = \Delta E_{\text{int}} = 0
\end{aligned}$$

Conduction, Convection, and Radiation

The rate H at which heat is *conducted* through a slab whose faces are maintained at temperatures T_H and T_C is

$$H = \frac{Q}{t} = kA\frac{T_H - T_C}{L}, \tag{19-30}$$

in which A and L are the face area and length of the slab, and k is the thermal conductivity of the material.

Convection occurs when temperature differences cause a heat transfer by motion within a fluid. *Radiation* is heat transfer via the emission of electromagnetic energy. The rate P_r at which an object emits energy via thermal radiation is

$$P_r = \sigma\varepsilon AT^4, \tag{19-37}$$

where σ (= 5.6703×10^{-8} W/m$^2\cdot$K^4) is the Stefan–Boltzmann constant, ε is the emissivity of the object's surface, A is its surface area, and T is its surface temperature (in kelvins). The rate P_a at which an object absorbs energy via thermal radiation from its environment, which is at the uniform temperature T_{env} (in kelvins), is

$$P_a = \sigma\varepsilon AT_{\text{env}}^4. \tag{19-38}$$

QUESTIONS

1. Figure 19-22 shows three temperature scales with the freezing and boiling points of water indicated. Rank a change of 25 R°, 25 S°, and 25 U° according to the corresponding change in temperature, greatest first.

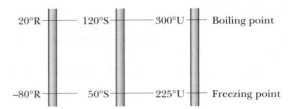

20°R — 120°S — 300°U — Boiling point

−80°R — 50°S — 225°U — Freezing point

FIGURE 19-22 Question 1.

2. A rod, initially at room temperature, is to be heated and cooled in six steps. The changes in its length, expressed in terms of some basic unit, will be +7, +5, +3, −4, −6, and −4. (a) Is the final temperature of the rod equal to room temperature, above it, or below it? (b) Is there a sequence of the steps such that the rod is at room temperature after one of the intermediate steps?

3. The table gives the initial length L, change in temperature ΔT, and change in length ΔL of four rods. Rank the rods according to their coefficients of thermal expansion, greatest first.

ROD	L (m)	ΔT (C°)	ΔL (m)
a	2	10	4×10^{-4}
b	1	20	4×10^{-4}
c	2	10	8×10^{-4}
d	4	5	4×10^{-4}

4. Rank the Celsius, Kelvin, and Fahrenheit temperature scales according to the heat required to raise the temperature of one gram of water by one degree on each scale, greatest first.

5. Materials A, B, and C are solids that are at their melting temperatures. Material A requires 200 J to melt 4 kg. Material B requires 300 J to melt 5 kg. And material C requires 300 J to melt 6 kg. Rank the materials according to their heats of fusion, greatest first.

6. Figure 19-23 shows four paths on a p-V diagram along which a gas can be taken from state i to state f. Rank the paths according to (a) the change ΔE_{int}, (b) the work W done by the gas, and (c) the magnitude of the heat transfer Q, greatest first.

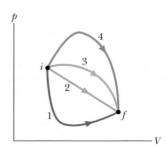

FIGURE 19-23
Question 6.

7. Figure 19-24 shows two closed cycles on p-V diagrams for a gas. The three parts of cycle 1 are of the same length and shape as those of cycle 2. For each cycle, should the cycle be traversed clockwise or counterclockwise if (a) the net work W done by the gas is to be positive and (b) the net heat transfer Q between the gas and the environment is to be positive?

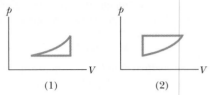

(1)   (2)

FIGURE 19-24 Questions 7 and 8.

8. For which cycle in Fig. 19-24, traversed clockwise, is (a) W greater and (b) Q greater?

9. Figure 19-25 shows a composite slab of three different materials, a, b, and c, with identical thicknesses and with thermal conductivities $k_b > k_a > k_c$. The heat transfer through them is nonzero and steady. Rank the materials according to the temperature difference ΔT across them, greatest first.

FIGURE 19-25
Question 9.

10. Figure 19-26 shows three different arrangements of materials 1, 2, and 3 to form a wall. The thermal conductivities are $k_1 > k_2 > k_3$. The left side of the wall is 20 C° higher than the right side. Rank the arrangements according to (a) the (steady state) rate of energy transfer through the wall and (b) the temperature difference across material 1, greatest first.

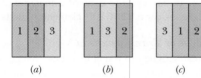

1 2 3   1 3 2   3 1 2

FIGURE 19-26
Question 10.

(a)   (b)   (c)

11. During an icicle's growth, its outer surface is covered with a thin sheath of liquid water that slowly seeps downward to form drops, one at a time, that hang at the tip (Fig. 19-27). Each drop straddles a thin tube of liquid water that extends up into the icicle, toward its *root* (its top). As the water at the top of the tube gradually freezes, energy is released. Is that energy conducted radially

FIGURE 19-27
Question 11.

outward through the ice, downward through the water to the hanging drop, or upward toward the root? (Assume that the air temperature is below 0°C.)

12. Figure 19-28 shows a horizontal cross section (top view) of a square room surrounded on four sides by thick walls. The walls are all made of the same material and all have the same face area. They have thicknesses of either L, $2L$, or $3L$ as shown, and they are well insulated along their lengths. The faces forming the room are maintained at 5°C, and the conduction of energy outward through the walls is steady. Rank the walls according to the rate of conduction through them, greatest first.

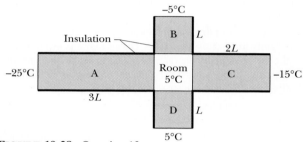

FIGURE 19-28 Question 12.

13. Suppose a block of wood and a block of metal are at the *same* temperature. When the blocks feel cold to your fingers, the metal feels colder than the wood; when the blocks feel hot to your fingers, the metal feels hotter than the wood. At what temperature will the blocks feel the same?

14. The following solid objects, made of the same material, are maintained at a temperature of 300 K in an environment whose temperature is 350 K: a cube of edge length r, a sphere of radius r, and a hemisphere of radius r. Rank the objects according to the net rate at which thermal radiation is exchanged with the environment, greatest first.

15. The following pairs give the temperatures of an object and its environment, respectively, for three situations: (1) 300 K and 350 K; (2) 350 K and 400 K; and (3) 400 K and 450 K. Without computation, rank the situations according to the object's net rate of energy exchange P_n, greatest first.

EXERCISES & PROBLEMS

SECTION 19-3 Measuring Temperature

1E. To measure temperatures, physicists and astronomers often measure how the intensity of electromagnetic radiation emitted by an object varies with wavelength. The wavelength λ_{max} at which the intensity is greatest is related to T, the temperature of the object in kelvins, by

$$\lambda_{max}T = 0.2898 \text{ cm} \cdot \text{K}.$$

In 1965, microwave radiation peaking at $\lambda_{max} = 0.107$ cm was discovered coming in all directions from space. To what temperature does this correspond? The interpretation of this *background radiation* is that it is left over from some 15 billion years ago, soon after the universe began.

2E. Suppose the temperature of a gas at the boiling point of water is 373.15 K. What then is the limiting value of the ratio of the pressure of the gas at that boiling point to its pressure at the triple point of water? (Assume the volume of the gas is the same at both temperatures.)

3P. Two constant-volume gas thermometers are assembled, one using nitrogen and the other using hydrogen. Both contain enough gas so that $p_3 = 80$ kPa. What is the difference between the pressures in the two thermometers if both are inserted into a water bath at the boiling point? Which gas is at higher pressure?

4P. A particular gas thermometer is constructed of two gas-containing bulbs, each of which is put into a water bath, as shown in Fig. 19-29. The pressure difference between the two bulbs is measured by a mercury manometer as shown. Appropriate reservoirs, not shown in the diagram, maintain constant gas volume in the two bulbs. There is no difference in pressure when both baths are at the triple point of water. The pressure difference is 120 torr when one bath is at the triple point and the other is at the boiling point of water. Finally, the pressure difference is 90.0 torr when one bath is at the triple point and the other is at an unknown temperature to be measured. What is the unknown temperature?

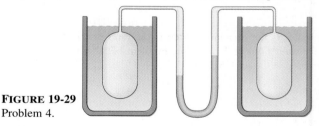

FIGURE 19-29
Problem 4.

SECTION 19-4 The Celsius and Fahrenheit Scales

5E. At what temperature is the Fahrenheit scale reading equal to (a) twice that of the Celsius and (b) half that of the Celsius?

6E. If your doctor tells you that your temperature is 310 kelvins above absolute zero, should you worry? Explain your answer.

7E. (a) In 1964, the temperature in the Siberian village of Oymyakon reached a value of −71°C. What temperature is this on the Fahrenheit scale? (b) The highest officially recorded temperature in the continental United States was 134°F in Death Valley, California. What is this temperature on the Celsius scale?

8E. (a) The temperature of the surface of the Sun is about 6000 K. Express this on the Fahrenheit scale. (b) Express normal human body temperature, 98.6°F, on the Celsius scale. (c) In the continental United States, the lowest officially recorded temperature is −70°F at Rogers Pass, Montana. Express this on the Celsius scale. (d) Express the normal boiling point of oxygen, −183°C, on the Fahrenheit scale. (e) At what Celsius temperature would you find a room to be uncomfortably warm?

9E. At what temperature do the following pairs of scales read the same: (a) Fahrenheit and Celsius (verify the listing in Table 19-1), (b) Fahrenheit and Kelvin, and (c) Celsius and Kelvin?

10P. Suppose that on a temperature scale X, water boils at −53.5°X and freezes at −170°X. What would a temperature of 340 K be on the X scale?

11P. It is an everyday observation that hot and cold objects cool down or warm up to the temperature of their surroundings. If the temperature difference ΔT between an object and its surroundings ($\Delta T = T_{obj} - T_{sur}$) is not too great, the rate of cooling or warming of the object is proportional, approximately, to this temperature difference; that is,

$$\frac{d\,\Delta T}{dt} = -A(\Delta T),$$

where A is a constant. The minus sign appears because ΔT decreases with time if ΔT is positive and increases if ΔT is negative. This is known as *Newton's law of cooling*. (a) On what factors does A depend? What are its dimensions? (b) If at some instant $t = 0$ the temperature difference is ΔT_0, show that it is

$$\Delta T = \Delta T_0 e^{-At}$$

at a later time t.

12P. The heater of a house breaks down one day when the outside temperature is 7.0°C. As a result, the inside temperature drops from 22°C to 18°C in 1.0 h. The owner fixes the heater and adds insulation to the house. Now she finds that, on a similar day, the house takes twice as long to drop from 22°C to 18°C when the heater is not operating. What is the ratio of the constant A in Newton's law of cooling (see Problem 11) after the insulation is added to the value before?

SECTION 19-5   Thermal Expansion

13E. A steel rod has a length of exactly 20 cm at 30°C. How much longer is it at 50°C?

14E. An aluminum flagpole is 33 m high. By how much does its length increase as the temperature increases by 15 C°?

15E. The Pyrex glass mirror in the telescope at the Mt. Palomar Observatory has a diameter of 200 in. The temperature ranges from −10°C to 50°C on Mt. Palomar. Determine the maximum change in the diameter of the mirror.

16E. A circular hole in an aluminum plate is 2.725 cm in diameter at 0.000°C. What is its diameter when the temperature of the plate is raised to 100.0°C?

17E. An aluminum-alloy rod has a length of 10.000 cm at 20.000°C and a length of 10.015 cm at the boiling point of water. (a) What is the length of the rod at the freezing point of water? (b) What is the temperature if the length of the rod is 10.009 cm?

18E. (a) What is the coefficient of linear expansion of aluminum per Fahrenheit degree? (b) Use your answer in (a) to calculate the change in length of a 20 ft aluminum rod if the rod is heated from 40°F to 95°F.

19E. Soon after Earth was formed, heat released by the decay of radioactive elements raised the average internal temperature from 300 to 3000 K, at about which value it remains today. Assuming an average coefficient of volume expansion of 3.0×10^{-5} K^{-1}, by how much has the radius of Earth increased since the planet was formed?

20E. The Stanford linear accelerator contains hundreds of brass disks tightly fitted into a steel tube. The system was assembled by cooling the disks in dry ice (at −57.00°C) to enable them to slide into the close-fitting tube. If the diameter of a disk is 80.00 mm at 43.00°C, what is its diameter in the dry ice?

21E. A glass window is exactly 20 cm by 30 cm at 10°C. By how much has its area increased when its temperature is 40°C?

22E. At 20°C, a brass cube has an edge length of 30 cm. What is the increase in the cube's surface area when it is heated from 20°C to 75°C?

23E. Find the change in volume of an aluminum sphere with an initial radius of 10 cm when the sphere is heated from 0.0°C to 100°C.

24E. What is the volume of a lead ball at 30°C if the ball's volume at 60°C is 50 cm³?

25E. By how much does the volume of an aluminum cube 5.00 cm on an edge increase when the cube is heated from 10.0°C to 60.0°C?

26E. An aluminum cup of 100 cm³ capacity is filled with glycerin at 22°C. How much glycerin, if any, will spill out of the cup if the temperature of the cup and glycerin is raised to 28°C? (The coefficient of volume expansion of glycerin is 5.1×10^{-4}/C°.)

27E. A steel rod at 25.0°C is bolted securely at both ends and then cooled. At what temperature will it rupture? Use Table 13-1.

28P. At 20°C, a rod is exactly 20.05 cm long on a steel ruler. Both the rod and the ruler are placed in an oven at 270°C, where the rod now measures 20.11 cm on the same ruler. What is the coefficient of thermal expansion for the material of which the rod is made?

29P. A steel rod is 3.000 cm in diameter at 25°C. A brass ring has an interior diameter of 2.992 cm at 25°C. At what common temperature will the ring just slide onto the rod?

30P. The area A of a rectangular plate is ab. Its coefficient of linear expansion is α. After a temperature rise ΔT, side a is longer by Δa and side b is longer by Δb (Fig. 19-30). Show that if we neglect the small quantity $\Delta a\,\Delta b/ab$, then $\Delta A = 2\alpha A\,\Delta T$.

31P. Density is mass divided by volume. If the volume V is temperature dependent, so is the density ρ. Show that a small change in density $\Delta\rho$ with a change in temperature ΔT is given by

$$\Delta\rho = -\beta\rho\,\Delta T,$$

where β is the coefficient of volume expansion. Explain the minus sign.

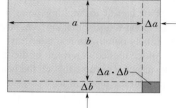

FIGURE 19-30
Problem 30.

32P. When the temperature of a metal cylinder is raised from 0.0°C to 100°C, its length increases by 0.23%. (a) Find the percent change in density. (b) What is the metal? Use Table 19-2.

33P. Show that when the temperature of a liquid in a barometer changes by ΔT, and the pressure is constant, the height h changes by $\Delta h = \beta h \Delta T$, where β is the coefficient of volume expansion. Neglect the expansion of the glass tube.

34P. When the temperature of a copper penny is raised by 100 C°, its diameter increases by 0.18%. To two significant figures, give the percent increase in (a) the area of a face, (b) the thickness, (c) the volume, and (d) the mass of the penny. (e) Calculate its coefficient of linear expansion.

35P. A pendulum clock with a pendulum made of brass is designed to keep accurate time at 20°C. What will be the error, in seconds per hour, if the clock operates at 0.0°C?

36P. As a result of a temperature rise of 32 C°, a bar with a crack at its center buckles upward (Fig. 19-31). If the fixed distance L_0 is 3.77 m and the coefficient of linear expansion of the bar is $25 \times 10^{-6}/C°$, find the rise x of the center.

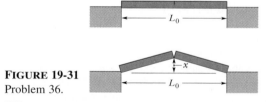

FIGURE 19-31
Problem 36.

37P. A composite bar of length $L = L_1 + L_2$ is made from a bar of material 1 and length L_1 attached to a bar of material 2 and length L_2, as shown in Fig. 19-32. (a) Show that the coefficient of linear expansion α for this composite bar is given by $\alpha = (\alpha_1 L_1 + \alpha_2 L_2)/L$. (b) Using steel and brass, design such a composite bar whose length is 52.4 cm and whose coefficient of linear expansion is $13.0 \times 10^{-6}/C°$.

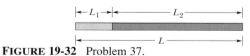

FIGURE 19-32 Problem 37.

SECTION 19-7 The Absorption of Heat by Solids and Liquids

38E. It is possible to melt ice by rubbing one block of it against another. How much work, in joules, would you have to do to get 1.00 g of ice to melt?

39E. A certain substance has a mass per mole of 50 g/mol. When 314 J of heat is added to a 30.0 g sample of this material, its temperature rises from 25.0°C to 45.0°C. (a) What is the specific heat of this substance? (b) How many moles of the substance are present? (c) What is the molar specific heat of the substance?

40E. In a certain solar house, energy from the Sun is stored in barrels filled with water. In a particular winter stretch of five cloudy days, 1.00×10^6 kcal is needed to maintain the inside of the house at 22.0°C. Assuming that the water in the barrels is at 50.0°C and that the water has a density of 1.00×10^3 kg/m³, what volume of water is required?

41E. A diet doctor encourages people to diet by drinking ice water. His theory is that the body must burn off enough fat to raise the temperature of the water from 0.00°C to the body temperature of 37.0°C. How many liters of ice water would have to be consumed to burn off 454 g (about 1 lb) of fat, assuming that this much fat burning requires 3500 Cal be transferred to the ice water? Why is it not advisable to follow this diet? One liter = 10^3 cm³. The density of water is 1.00 g/cm³.

42E. Icebergs in the North Atlantic present hazards to shipping, causing the lengths of shipping routes to increase by about 30% during the iceberg season. Attempts to destroy icebergs include planting explosives, bombing, torpedoing, shelling, ramming, and coating with black soot. Suppose that direct melting of the iceberg, by placing heat sources in the ice, is tried. How much heat is required to melt 10% of an iceberg with a mass of 200,000 metric tons? (Use 1 metric ton = 1000 kg.)

43E. How much water remains unfrozen after 50.2 kJ of heat has been extracted from 260 g of liquid water initially at its freezing point?

44E. Calculate the minimum amount of heat, in joules, required to completely melt 130 g of silver initially at 15.0°C.

45E. A room is lighted by four 100 W incandescent light bulbs. (The power of 100 W is the rate at which a bulb converts electrical energy to heat and visible light.) Assuming that 90% of the energy is converted to heat, how much heat is added to the room in 1.00 h?

46E. What quantity of butter (6.0 Cal/g = 6000 cal/g) would supply the energy needed for a 160 lb man to ascend to the top of Mt. Everest, elevation 29,000 ft, from sea level?

47E. An energetic athlete dissipates all the energy in a diet of 4000 Cal/day. If he were to release this energy at a steady rate, how would this conversion of energy compare with that of a 100 W bulb? (The power of 100 W is the rate at which the bulb converts electrical energy to heat and visible light.)

48E. If the heat necessary to warm a mass m of water from 68°F to 78°F were somehow converted to translational kinetic energy of that water, what would be the speed of the water?

49E. A power of 0.400 hp is required for 2.00 min to drill a hole in a 1.60-lb copper block. (a) How much thermal energy in Btu is generated? (b) What is the rise in temperature of the copper if only 75.0% of the power warms the copper? (Use 1 ft·lb = 1.285×10^{-3} Btu.)

50E. An object of mass 6.00 kg falls through a height of 50.0 m and, by means of a mechanical linkage, rotates a paddle wheel that stirs 0.600 kg of water. The water is initially at 15.0°C. What is the maximum possible temperature rise of the water?

51E. One way to keep the contents of a garage from becoming too cold on a night when a severe subfreezing temperature is forecast is to put a tub of water in the garage. If the mass of the water is 125 kg and its initial temperature is 20°C, (a) how much energy must the water transfer to its surroundings in order to freeze completely and (b) what is the lowest possible temperature of the water and its surroundings until that happens?

52E. A small electric immersion heater is used to heat 100 g of water for a cup of instant coffee. The heater is labeled ''200 watts,'' which means that it converts electrical energy to thermal energy at this rate. Calculate the time required to bring this water from 23°C to the boiling point, ignoring any heat losses.

53E. A pickup truck whose mass is 2200 kg is speeding along the highway at 65.0 mi/h. (a) If you could use all this kinetic energy to vaporize water already at 100°C, how much water could you vaporize? (b) If you had to buy this amount of energy from your local utility company at 12¢/kW·h, how much would it cost you? Guess at the answers before you figure them out; you may be surprised.

54E. A 150 g copper bowl contains 220 g of water, both at 20.0°C. A very hot 300 g copper cylinder is dropped into the water, causing the water to boil, with 5.00 g being converted to steam. The final temperature of the system is 100°C. (a) How much heat was transferred to the water? (b) How much to the bowl? (c) What was the original temperature of the cylinder?

55P. Calculate the specific heat of a metal from the following data. A container made of the metal has a mass of 3.6 kg and contains 14 kg of water. A 1.8 kg piece of the metal initially at a temperature of 180°C is dropped into the water. The container and water initially have a temperature of 16.0°C, and the final temperature of the entire system is 18.0°C.

56P. A thermometer of mass 0.0550 kg and of specific heat 0.837 kJ/kg·K reads 15.0°C. It is then completely immersed in 0.300 kg of water, and it comes to the same final temperature as the water. If the thermometer reads 44.4°C, what was the temperature of the water before insertion of the thermometer?

57P. How long does it take a 2.0×10^5 Btu/h water heater to raise the temperature of 40 gal of water from 70°F to 100°F?

58P. An athlete needs to lose weight and decides to do it by ''pumping iron.'' (a) How many times must an 80.0 kg weight be lifted a distance of 1.00 m in order to burn off 1 lb of fat, assuming that that much fat is equivalent to 3500 Cal? (b) If the weight is lifted once every 2.00 s, how long does the task take?

59P. A 1500 kg Buick moving at 90 km/h brakes to rest, at uniform deceleration and without skidding, over a distance of 80 m. At what average rate is mechanical energy transferred to thermal energy in the brake system?

60P. A chef, upon finding his stove out of order, decides to boil the water for his wife's coffee by shaking it in a thermos flask. Suppose that he uses 500 cm³ of tap water at 59°F, and that the water falls 1.0 ft each shake, the chef making 30 shakes each minute. Neglecting any loss of thermal energy by the flask, how long must he shake the flask before the water boils?

61P. A block of ice, at its melting point and of initial mass 50.0 kg, slides along a horizontal surface, starting at a speed of 5.38 m/s and finally coming to rest after traveling 28.3 m. Compute the mass of ice melted as a result of the friction between the block and the surface. (Assume that all the heat generated owing to friction goes into the block of ice.)

62P. The specific heat of a substance varies with temperature according to $c = 0.20 + 0.14T + 0.023T^2$, with T in °C and c in cal/g·K. Find the heat required to raise the temperature of 2.0 g of this substance from 5.0°C to 15°C.

63P. In a solar water heater, energy from the Sun is gathered by water that circulates through tubes in a rooftop collector. The solar radiation enters the collector through a transparent cover and warms the water in the tubes; this water is pumped into a holding tank. Assuming that the efficiency of the overall system is 20% (that is, 80% of the incident solar energy is lost from the system), what collector area is necessary to raise the temperature of 200 L of water in the tank from 20°C to 40°C in 1.0 h? The intensity of incident sunlight is 700 W/m².

64P. An insulated thermos contains 130 cm³ of hot coffee, at a temperature of 80.0°C. You put in a 12.0 g ice cube at its melting point to cool the coffee. By how many degrees will your coffee have cooled once the ice has melted? Treat the coffee as though it were pure water.

65P. What mass of steam at 100°C must be mixed with 150 g of ice at its melting point, in a thermally insulated container, to produce liquid water at 50°C?

66P. A person makes a quantity of iced tea by mixing 500 g of hot tea (essentially water) with an equal mass of ice at its melting point. If the initial hot tea is at a temperature of (a) 90°C and (b) 70°C, what are the temperature and mass of the remaining ice when the tea and ice reach a common temperature?

67P. (a) Two 50 g ice cubes are dropped into 200 g of water in a glass. If the water was initially at 25°C, and the ice came directly from a freezer at −15°C, what will be the temperature of the drink when the ice and water reach the same temperature? (b) Suppose that only one ice cube had been used in (a); what would be the final temperature of the drink? Neglect the heat capacity of the glass.

68P. A 20.0 g copper ring has a diameter of exactly 1.00000 in. at its temperature of 0.000°C. An aluminum sphere has a diameter of exactly 1.00200 in. at its temperature of 100.0°C. The sphere is placed on top of the ring (Fig. 19-33), and the two are allowed to

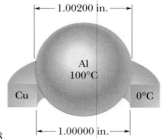

FIGURE 19-33 Problem 68.

come to thermal equilibrium, with no heat lost to the surroundings. The sphere just passes through the ring at the equilibrium temperature. What is the mass of the sphere?

69P. A *flow calorimeter* is a device used to measure the specific heat of a liquid. Heat is added at a known rate to a stream of the liquid as it passes through the calorimeter at a known rate. Measurement of the resulting temperature difference between the inflow and the outflow points of the liquid stream enables us to compute the specific heat of the liquid. Suppose a liquid of density 0.85 g/cm³ flows through a calorimeter at the rate of 8.0 cm³/s. Heat is added at the rate of 250 W by means of an electric heating coil, and a temperature difference of 15 C° is established in steady-state conditions between the inflow and the outflow points. What is the specific heat of the liquid?

70P. By means of a heating coil, energy is transferred at a constant rate to a substance in a thermally insulated container. The temperature of the substance is measured as a function of time. (a) Show how we can deduce from this information the way in which the heat capacity of the substance depends on the temperature. (b) Suppose that in a certain temperature range the temperature T is proportional to t^3, where t is the time. How does the heat capacity depend on T in this range?

SECTION 19-10 Some Special Cases of the First Law of Thermodynamics

71E. A sample of gas expands from 1.0 m³ to 4.0 m³ while its pressure decreases from 40 Pa to 10 Pa. How much work is done by the gas if its pressure changes with volume via each of the three paths shown in the *p-V* diagram in Fig. 19-34?

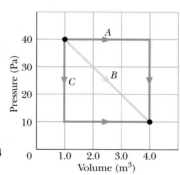

FIGURE 19-34
Exercise 71.

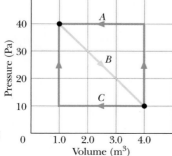

FIGURE 19-35
Exercise 72.

72E. A sample of gas expands from 1.0 m³ to 4.0 m³ along path B in the *p-V* diagram in Fig. 19-35. It is then compressed back to 1.0 m³ along either path A or path C. Compute the net work done by the gas for the complete cycle in each case.

73E. Consider that 200 J of work is done on a system and 70.0 cal of heat is extracted from the system. In the sense of the first law of thermodynamics, what are the values (including algebraic signs) of (a) W, (b) Q, and (c) ΔE_{int}?

74E. A thermodynamic system is taken from an initial state A to another state B and back again to A, via state C, as shown by path $ABCA$ in the *p-V* diagram of Fig. 19-36a. (a) Complete the table in Fig. 19-36b by filling in either + or − for the sign of each thermodynamic quantity associated with each process. (b) Calculate the numerical value of the work done by the system for the complete cycle $ABCA$.

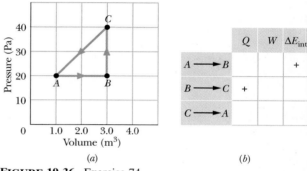

FIGURE 19-36 Exercise 74.

75E. Gas within a chamber passes through the cycle shown in Fig. 19-37. Determine the net heat added to the system during process CA if the heat Q_{AB} added during process AB is 20.0 J, no heat is transferred during process BC, and the net work done during the cycle is 15.0 J.

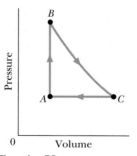

FIGURE 19-37 Exercise 75.

76E. Gas within a closed chamber undergoes the processes shown in the *p-V* diagram of Fig. 19-38. Calculate the net heat added to the system during one complete cycle.

77P. Figure 19-39a shows a cylinder containing gas and closed by a movable piston. The cylinder is kept submerged in an ice–water mixture. The piston is *quickly* pushed down from position 1 to position 2 and then held at position 2 until the gas is again at the temperature of the ice–water mixture; it then is *slowly* raised back to position 1. Figure 19-39b is a *p-V* diagram for the process.

If 100 g of ice is melted during the cycle, how much work has been done *on* the gas?

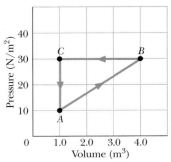

FIGURE 19-38   Exercise 76.

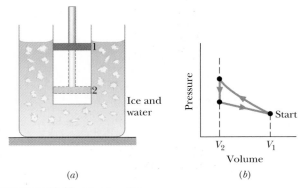

FIGURE 19-39   Problem 77.

78P. When a system is taken from state *i* to state *f* along path *iaf* in Fig. 19-40, $Q = 50$ cal and $W = 20$ cal. Along path *ibf*, $Q = 36$ cal. (a) What is *W* along path *ibf*? (b) If $W = -13$ cal for the curved return path *fi*, what is *Q* for this path? (c) Take $E_{int,i} = 10$ cal. What is $E_{int,f}$? (d) If $E_{int,b} = 22$ cal, what are the values of *Q* for path *ib* and path *bf*?

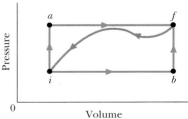

FIGURE 19-40   Problem 78.

SECTION 19-11   Heat Transfer Mechanisms

79E. The average rate at which heat is conducted through the ground surface in North America is 54.0 mW/m², and the average thermal conductivity of the near-surface rocks is 2.50 W/m·K. Assuming a surface temperature of 10.0°C, what should be the temperature at a depth of 35.0 km (near the base of the crust)? Ignore the heat generated by the presence of radioactive elements.

80E. The ceiling of a single-family dwelling in a cold climate should have an *R*-value of 30. To give such insulation, how thick

would a layer of (a) polyurethane foam and (b) silver have to be?

81E. The thermal conductivity of Pyrex glass at 0°C is 2.9×10^{-3} cal/cm·C°·s. (a) Express the quantity in W/m·K and in Btu/ft·F°·h. (b) What is the *R*-value for a 0.25 in. sheet of such glass?

82E. (a) Calculate the rate at which body heat is conducted through the clothing of a skier in a steady-state process, given the following data: the body surface area is 1.8 m² and the clothing is 1.0 cm thick; the skin surface temperature is 33°C, whereas the outer surface of the clothing is at 1.0°C; the thermal conductivity of the clothing is 0.040 W/m·K. (b) How would the answer to (a) change if, after a fall, the skier's clothes became soaked with water of thermal conductivity 0.60 W/m·K?

83E. Consider the slab shown in Fig. 19-17. Suppose that $L = 25.0$ cm, $A = 90.0$ cm², and the material is copper. If $T_H = 125$°C, $T_C = 10.0$°C, and a steady state is reached, find the rate of heat transfer through the slab.

84E. A cylindrical copper rod of length 1.2 m and cross-sectional area 4.8 cm² is insulated to prevent heat loss through its surface. The ends are maintained at a temperature difference of 100 C° by having one end in a water–ice mixture and the other in boiling water and steam. (a) Find the rate at which heat is conducted along the rod. (b) Find the rate at which ice melts at the cold end.

85E. Show that the temperature T_X at the interface of a compound slab (Fig. 19-18) is given by

$$T_X = \frac{R_1 T_H + R_2 T_C}{R_1 + R_2}.$$

86E. If you were to walk briefly in space without a spacesuit while far from the Sun (as an astronaut does in the movie *2001*), you would feel the cold of space: while you radiated thermal energy, you would absorb almost none from your environment. (a) At what rate would you lose energy? (b) How much energy would you lose in 30 s? Assume that your emissivity is 0.90, and estimate other data needed in the calculations.

87E. Four square pieces of insulation of two different materials, all with the same thickness and area *A*, are available to cover an opening of area 2*A*. This can be done in either of the two ways shown in Fig. 19-41. Which arrangement, (*a*) or (*b*), would give the lower heat flow if $k_2 \neq k_1$?

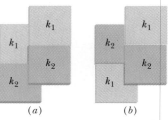

FIGURE 19-41   Exercise 87.

88P. Two identical rectangular rods of metal are welded end to end as shown in Fig. 19-42*a*, and 10 J of heat is conducted (in a steady-state process) through the rods in 2.0 min. How long would it take for 10 J to be conducted through the rods if they are welded together as shown in Fig. 19-42b?

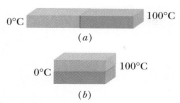

FIGURE 19-42 Problem 88.

89P. Compute the rate of heat conduction through two storm doors, both 2.0 m high and 0.75 m wide. (a) One door is made with aluminum panels 1.5 mm thick and a glass pane 3.0 mm thick; the glass covers 75% of the aluminum surface (the structural frame has a negligible area). (b) The second door is made entirely of white pine averaging 2.5 cm in thickness. Take the temperature drop across each door to be 33 C°; see Table 19-6.

90P. A large cylindrical water tank with a bottom 1.7 m in diameter is made of iron boilerplate 5.2 mm thick. Water in the tank is heated by a gas burner underneath that is able to maintain a temperature difference of 2.3 C° between the top and bottom surfaces of the bottom plate. How much heat is conducted through that plate in 5.0 min? (Iron has a thermal conductivity of 67 W/m·K.)

91P. (a) What is the rate of heat loss in watts per square meter through a glass window 3.0 mm thick if the outside temperature is −20°F and the inside temperature is +72°F? (b) A storm window is installed having the same thickness of glass but with an air gap of 7.5 cm between the two windows. What will be the corresponding rate of heat loss assuming that conduction is the only important heat-loss mechanism?

92P. A sphere of radius 0.500 m, temperature 27.0°C, and emissivity 0.850 is located in an environment of temperature 77.0°C. At what rate does the sphere (a) emit thermal radiation and (b) absorb thermal radiation? (c) What is the sphere's net rate of energy exchange?

93P. A cube of edge length 6.0×10^{-6} m, emissivity 0.75, and temperature −100°C floats in an environment of temperature −150°C. What is the cube's net thermal radiation transfer rate?

94P. A solid cylinder of radius $r_1 = 2.5$ cm, length $h_1 =$ 5.0 cm, emissivity 0.85, and temperature 30°C is suspended in an environment of temperature 50°C. (a) What is the cylinder's net thermal radiation transfer rate P_1? (b) If the cylinder is stretched until its radius is $r_2 = 0.50$ cm, its net thermal radiation transfer rate becomes P_2. What is the ratio P_2/P_1?

95P. A tank of water has been outdoors in cold weather and a slab of ice 5.0 cm thick has formed on its surface (Fig. 19-43). The air above the ice is at −10°C. Calculate the rate of formation of ice (in centimeters per hour) on the bottom surface of the ice slab. Take the thermal conductivity and density of ice to be 0.0040 cal/s·cm·C° and 0.92 g/cm³. Assume that heat is not transferred through the walls or bottom of the tank.

FIGURE 19-43 Problem 95.

96P. Ice has formed on a shallow pond and a steady state has been reached, with the air above the ice at −5.0°C and the bottom of the pond at 4.0°C. If the total depth of *ice + water* is 1.4 m, how thick is the ice? (Assume that the thermal conductivities of ice and water are 0.40 and 0.12 cal/m·C°·s, respectively.)

97P. Three metal rods, one copper, one aluminum, and one brass, are each 6.00 cm long and 1.00 cm in diameter. These rods are placed end to end, with the aluminum between the other two. The free ends of the copper and brass rods are maintained at the boiling point and the freezing point of water, respectively. Find the steady-state temperatures of the copper–aluminum junction and the aluminum–brass junction. The thermal conductivity of brass is 109 W/m·K.

20

The Kinetic Theory of Gases

Suppose that you return to your chilly dwelling after snowshoeing for miles on a cold winter day. Your first thought is to light a stove. But why, exactly, would you do that? Is it because the stove will increase the store of internal (thermal) energy of the air in the dwelling, until eventually the air will have enough of that internal energy to keep you comfortable? As logical as this reasoning sounds, it is flawed, because the air's store of internal energy will not be changed by the stove. How can that be? And if it is so, why would you bother to light the stove?

20-1 A NEW WAY TO LOOK AT GASES

Classical thermodynamics—the subject of the last chapter—has nothing to say about atoms. Its laws are concerned only with such macroscopic variables as pressure, volume, and temperature. However, we know that a gas is made up of atoms or molecules (groups of atoms bound together). The pressure exerted by a gas must surely be related to the steady drumbeat of its molecules on the walls of its container. The ability of a gas to take on the volume of its container must surely be due to the freedom of motion of its molecules. And the temperature and internal energy of a gas must surely be related to the kinetic energy of these molecules. Perhaps we can learn something about gases by approaching the subject from this direction. We call this molecular approach the **kinetic theory of gases.** It is the subject of this chapter.

20-2 AVOGADRO'S NUMBER

When our thinking is slanted toward molecules, it makes sense to measure the sizes of our samples in moles. If we do so, we can be certain that we are comparing samples that contain the same number of molecules. The *mole* is one of the seven SI base units and is defined as follows:

> One mole is the number of atoms in a 12 g sample of carbon-12.

We speak of a ''mole of helium'' or ''a mole of water,'' meaning a certain number of the elementary units of the substance. But we could just as well speak of a mole of tennis balls, where, of course, the elementary unit would be a tennis ball.

The obvious question now is: ''Just how many atoms or molecules are there in a mole?'' The answer is determined experimentally and, as you saw in Chapter 19, is

$$N_A = 6.02 \times 10^{23} \text{ mol}^{-1} \qquad \begin{matrix}\text{(Avogadro's} \\ \text{number).}\end{matrix} \qquad (20\text{-}1)$$

This number is called **Avogadro's number** after Italian scientist Amadeo Avogadro (1776–1856), who suggested that all gases contain the same number of molecules or atoms when they occupy the same volume under the same conditions of temperature and pressure.

The number of moles n contained in a sample of any substance can be found from

$$n = \frac{N}{N_A}, \qquad (20\text{-}2)$$

in which N is the number of molecules in the sample. The number of moles in a sample can also be found from the mass M_{sam} of the sample and either the *molar mass M* (the mass of 1 mole of that substance) or the mass m of one molecule:

$$n = \frac{M_{sam}}{M} = \frac{M_{sam}}{m N_A}. \qquad (20\text{-}3)$$

The enormously large value of Avogadro's number suggests how tiny and how numerous atoms must be. A mole of air, for example, can easily fit into a suitcase. Yet, if these molecules were spread uniformly over the surface of Earth, there would be about 120,000 of them per square centimeter. A second example: one mole of tennis balls would fill a volume equal to that of seven Moons!

PROBLEM SOLVING TACTICS

TACTIC 1: *Avogadro's Number of What?*

In Eq. 20-1, Avogadro's number is expressed in terms of mol^{-1}, which is the inverse mole, or $1/\text{mol}$. We could instead explicitly state the elementary unit involved in a given situation. For example, if the elementary unit is an atom, we might write $N_A = 6.02 \times 10^{23}$ atoms/mole. If the elementary unit is a molecule, then we might write $N_A = 6.02 \times 10^{23}$ molecules/mole. And if we are interested in a *great* many tennis balls, we can write $N_A = 6.02 \times 10^{23}$ tennis balls/mole.

20-3 IDEAL GASES

Our goal in this chapter is to explain the macroscopic properties of a gas—such as its pressure and its temperature—in terms of the behavior of the molecules that make it up. But there is an immediate problem: which gas? Should it be hydrogen or oxygen, or methane, or perhaps uranium hexafluoride? They are all different. However, experimenters have found that if we confine 1 mole samples of various gases in boxes of identical volume and hold the gases at the same temperature, then their measured pressures are nearly—though not exactly—the same. If we repeat the measurements at lower gas densities, then these small differences in the measured pressures tend to disappear. Further experiments show that, at low enough densities, all real gases tend to obey the relation

$$pV = nRT \qquad \text{(ideal gas law),} \qquad (20\text{-}4)$$

in which p is the absolute (not gauge) pressure, n is the number of moles of gas present, and R, the **gas constant,** has the same value for all gases, namely,

$$R = 8.31 \text{ J/mol} \cdot \text{K}. \qquad (20\text{-}5)$$

The temperature T in Eq. 20-4 must be expressed in kelvins. Equation 20-4 is called the **ideal gas law.** Provided the gas density is reasonably low, Eq. 20-4 holds for any type of gas, or a mixture of different types, with n being the total number of moles present.

You may well ask, "What is an *ideal gas* and what is so 'ideal' about one?" The answer lies in the simplicity of the law (Eq. 20-4) that governs its macroscopic properties. Using this law—as you will see —we can deduce many properties of the ideal gas in a simple way. Although there is no such thing in nature as a truly ideal gas, *all* gases approach the ideal state at low enough densities, that is, under conditions in which their molecules are far enough apart. Thus the ideal gas concept allows us to gain useful insights into the limiting behavior of real gases.

Work Done by an Ideal Gas at Constant Temperature

Suppose that a sample of n moles of an ideal gas, confined to a piston–cylinder arrangement, is allowed to expand from an initial volume V_i to a final volume V_f. Suppose further that the temperature T of the gas is held constant throughout the process. Such a process is called an **isothermal expansion** (and the reverse is called an **isothermal compression**).

On a p-V diagram, an *isotherm* is a curve that connects points that have the same temperature. It is then a graph of pressure versus volume for a gas whose temperature T is held constant, that is, a graph of

$$p = nRT \frac{1}{V} = \text{(a constant)} \frac{1}{V}. \qquad (20\text{-}6)$$

Figure 20-1 shows three isotherms, each corresponding to a different (constant) value of T. Note that the values of T increase upward to the right.

The p-V diagram for an isothermal expansion or compression is also a graph of pressure versus volume for a gas whose temperature is held constant. Thus, such a p-V diagram must follow along an isotherm, as does the expansion from state i to state f in Fig. 20-1.

Let us calculate the work done by an ideal gas during an isothermal expansion. Equation 19-23,

$$W = \int_{V_i}^{V_f} p \, dV, \qquad (20\text{-}7)$$

is a general expression for the work done during any change in volume of a gas. Because we are now dealing with an ideal gas, we can use Eq. 20-4 to substitute for p, obtaining

$$W = \int_{V_i}^{V_f} \frac{nRT}{V} \, dV. \qquad (20\text{-}8)$$

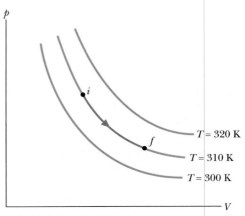

FIGURE 20-1   Three isotherms on a p-V diagram. The path shown along the middle isotherm represents an isothermal expansion of a gas from an initial state i to a final state f. The path from f to i along the isotherm would represent the reverse process, an isothermal compression.

And because we are considering an isothermal expansion, T is constant and we can move it in front of the integral sign to write

$$W = nRT \int_{V_i}^{V_f} \frac{dV}{V} = nRT \left[\ln V \right]_{V_i}^{V_f}. \qquad (20\text{-}9)$$

By evaluating at the limits and then using the relationship $\ln a - \ln b = \ln (a/b)$, we find that

$$W = nRT \ln \frac{V_f}{V_i} \qquad \begin{array}{l}\text{(ideal gas,}\\ \text{isothermal process).}\end{array} \qquad (20\text{-}10)$$

Recall that the symbol ln specifies a *natural* logarithm, that is, a logarithm to base e.

For an expansion, $V_f > V_i$ by definition, so the ratio V_f/V_i in Eq. 20-10 is greater than unity. The natural logarithm of a quantity greater than unity is positive, and so the work W done by an ideal gas during an isothermal expansion is positive, as we expect. For a compression, we have $V_f < V_i$, so the ratio of volumes in Eq. 20-10 is less than unity. The natural logarithm in that equation—hence the work W—is negative, again as we expect.

Equation 20-10 does not give the work W done by an ideal gas during *every* thermodynamic process. Instead, it gives the work only when the temperature is held constant. If the temperature varies, then the symbol T in Eq. 20-8 cannot be moved in front of the integral symbol as in Eq. 20-9, and thus we do not end up with Eq. 20-10.

However, we can go back to Eq. 20-7 to find the work W done by an ideal gas (or any other gas) during two more processes—a constant-volume process and a constant-pressure process. If the volume of the gas is constant, then Eq. 20-7 yields

$$W = 0 \quad \text{(constant-volume process).} \qquad (20\text{-}11)$$

If, instead, the volume changes while the pressure p of the gas is held constant, then Eq. 20-7 becomes

$$W = p(V_f - V_i) = p \, \Delta V \quad \begin{array}{l}\text{(constant-pressure}\\ \text{process).}\end{array} \quad (20\text{-}12)$$

CHECKPOINT 1: An ideal gas has an initial pressure of 3 pressure units and an initial volume of 4 volume units. The table gives the final pressure and volume of the gas (in those same units) in five processes. Which processes start and end on the same isotherm?

	a	b	c	d	e
p	12	6	5	4	1
V	1	2	7	3	12

SAMPLE PROBLEM 20-1

A cylinder contains 12 L of oxygen at 20°C and 15 atm. The temperature is raised to 35°C, and the volume reduced to 8.5 L. What is the final pressure of the gas in atmospheres? Assume that the gas is ideal.

SOLUTION: From Eq. 20-4 we may write

$$nR = \frac{p_i V_i}{T_i} = \frac{p_f V_f}{T_f}.$$

Solving for p_f yields

$$p_f = \frac{p_i T_f V_i}{T_i V_f}. \quad (20\text{-}13)$$

Note here that if we converted the given initial and final volumes from liters to the proper units of cubic meters, the multiplying conversion factors would cancel out of Eq. 20-13. The same would be true for conversion factors that convert the pressures from atmospheres to the proper pascals. However, to convert the given temperatures to kelvins requires the addition of an amount that would not cancel and thus must be included. Hence, we must write

$$T_i = (273 + 20) \text{ K} = 293 \text{ K}$$

and $\qquad T_f = (273 + 35) \text{ K} = 308 \text{ K}.$

Inserting the given data into Eq. 20-13 then yields

$$p_f = \frac{(15 \text{ atm})(308 \text{ K})(12 \text{ L})}{(293 \text{ K})(8.5 \text{ L})} = 22 \text{ atm.} \quad \text{(Answer)}$$

SAMPLE PROBLEM 20-2

One mole of oxygen (assume it to be an ideal gas) expands at a constant temperature T of 310 K from an initial volume V_i of 12 L to a final volume V_f of 19 L.

(a) How much work is done by the expanding gas?

SOLUTION: From Eq. 20-10 we have

$$W = nRT \ln \frac{V_f}{V_i}$$

$$= (1 \text{ mol})(8.31 \text{ J/mol} \cdot \text{K})(310 \text{ K}) \ln \frac{19 \text{ L}}{12 \text{ L}}$$

$$= 1180 \text{ J.} \quad \text{(Answer)}$$

The expansion is graphed in the p-V diagram of Fig. 20-2. The work done by the gas during the expansion is represented by the area beneath the curve if.

(b) How much work is done by the gas during an isothermal *compression* from $V_i = 19 \text{ L}$ to $V_f = 12 \text{ L}$?

SOLUTION: We proceed as in (a), finding

$$W = nRT \ln \frac{V_f}{V_i}$$

$$= (1 \text{ mol})(8.31 \text{ J/mol} \cdot \text{K})(310 \text{ K}) \ln \frac{12 \text{ L}}{19 \text{ L}}$$

$$= -1180 \text{ J.} \quad \text{(Answer)}$$

This result is equal in magnitude but opposite in sign to the result found in (a) for an isothermal expansion. The minus sign tells us that an external agent must do 1180 J of work *on* the gas to compress it.

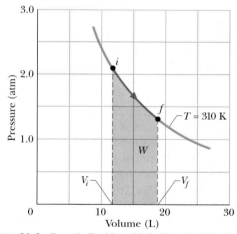

FIGURE 20-2 Sample Problem 20-2. The shaded area represents the work done by 1 mol of oxygen in expanding at a constant temperature T of 310 K.

20-4 PRESSURE, TEMPERATURE, AND RMS SPEED

Here is our first kinetic theory problem. Let n moles of an ideal gas be confined in a cubical box of volume V, as in Fig. 20-3. The walls of the box are held at temperature T. What is the connection between the pressure p exerted by the gas on the walls and the speeds of the molecules?

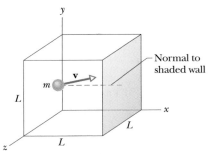

FIGURE 20-3 A cubical box of edge L, containing n moles of an ideal gas. A molecule of mass m and velocity $\mathbf{v}$ is about to collide with the shaded wall of area L^2. A normal to that wall is shown.

The molecules of gas in the box are moving in all directions and with various speeds, bumping into each other and bouncing from the walls of the box like balls in a racquetball court. We ignore (for the time being) collisions of the molecules with one another and consider only elastic collisions with the walls.

Figure 20-3 shows a typical gas molecule, of mass m and velocity $\mathbf{v}$, that is about to collide with the shaded wall. Because we assume that any collision of a molecule with a wall is elastic, when this molecule collides with the shaded wall, the only component of its velocity that is changed is the x component, and that component is reversed. This means that the only change in the particle's momentum is then along the x axis, and that change is

$$\Delta p_x = (-mv_x) - (mv_x) = -2mv_x.$$

Hence the momentum Δp_x delivered to the wall by the molecule during the collision is $+2mv_x$. (Because in this book the symbol p represents both momentum and pressure, we must be careful to note that here p represents momentum and is a vector quantity.)

The molecule of Fig. 20-3 will hit the shaded wall repeatedly. The time Δt between collisions is the time the molecule takes to travel to the opposite wall and back again (a distance of $2L$) at speed v_x. Thus Δt is equal to $2L/v_x$. (Note that this result holds even if the molecule bounces off any of the other walls along the way, because those walls are parallel to x and so cannot change v_x.) Thus the average rate at which momentum is delivered to the shaded wall by this single molecule is

$$\frac{\Delta p_x}{\Delta t} = \frac{2mv_x}{2L/v_x} = \frac{mv_x^2}{L}.$$

From Newton's second law ($\mathbf{F} = d\mathbf{p}/dt$), the rate at which momentum is delivered to the wall is the force acting on that wall. To find the total force, we must add up the contributions of all the molecules that strike the wall, al-

lowing for the possibility that they all have different speeds. Dividing the total force F_x by the area of the wall ($= L^2$) then gives the pressure p on that wall, where now and in the rest of this discussion, p represents pressure. Thus,

$$p = \frac{F_x}{L^2} = \frac{mv_{x1}^2/L + mv_{x2}^2/L + \cdots + mv_{xN}^2/L}{L^2}$$

$$= \left(\frac{m}{L^3}\right)(v_{x1}^2 + v_{x2}^2 + \cdots + v_{xN}^2), \qquad (20\text{-}14)$$

where N is the number of molecules in the box.

Since $N = nN_A$, there are nN_A terms in the second parentheses of Eq. 20-14. So we can replace that quantity by $nN_A \overline{v_x^2}$, where $\overline{v_x^2}$ is the average value of the square of the x components of all the molecular speeds. Equation 20-14 then becomes

$$p = \frac{nmN_A}{L^3}\overline{v_x^2}.$$

But mN_A is the molar mass M of the gas (that is, the mass of 1 mole of the gas). Also, L^3 is the volume of the box, so

$$p = \frac{nM\overline{v_x^2}}{V}. \qquad (20\text{-}15)$$

For any molecule, $v^2 = v_x^2 + v_y^2 + v_z^2$. Because there are many molecules and because they are all moving in random directions, the average values of the squares of their velocity components are equal, so that $\overline{v_x^2} = \frac{1}{3}\overline{v^2}$. Thus Eq. 20-15 becomes

$$p = \frac{nM\overline{v^2}}{3V}. \qquad (20\text{-}16)$$

The square root of $\overline{v^2}$ is a kind of average speed, called the **root-mean-square speed** of the molecules and symbolized by v_{rms}. Its name describes it rather well: you *square* each speed, you find the *mean* (that is, the average) of all these squared speeds, and then you take the square *root*. With $\sqrt{\overline{v^2}} = \overline{v}_{\text{rms}}$, we can then write Eq. 20-16 as

$$p = \frac{nMv_{\text{rms}}^2}{3V}. \qquad (20\text{-}17)$$

Equation 20-17 is very much in the spirit of kinetic theory. It tells us how the pressure of the gas (a purely macroscopic quantity) depends on the speed of the molecules (a purely microscopic quantity).

We can turn Eq. 20-17 around and use it to calculate v_{rms}. Combining Eq. 20-17 with the ideal gas law ($pV = nRT$) leads to

$$v_{\text{rms}} = \sqrt{\frac{3RT}{M}}. \qquad (20\text{-}18)$$

TABLE 20-1 **SOME MOLECULAR SPEEDS
AT ROOM TEMPERATURE ($T = 300$ K)**[a]

GAS	MOLAR MASS (10^{-3} kg/mol)	v_{rms} (m/s)
Hydrogen (H_2)	2.02	1920
Helium (He)	4.0	1370
Water vapor (H_2O)	18.0	645
Nitrogen (N_2)	28.0	517
Oxygen (O_2)	32.0	483
Carbon dioxide (CO_2)	44.0	412
Sulfur dioxide (SO_2)	64.1	342

[a]For convenience, we often set room temperature = 300 K even though (at 27°C or 81°F) that represents a fairly warm room.

Table 20-1 shows some rms speeds calculated from Eq. 20-18. The speeds are surprisingly high. For hydrogen molecules at room temperature (300 K), the rms speed is 1920 m/s or 4300 mi/h—faster than a speeding bullet! On the surface of the Sun, where the temperature is 2×10^6 K, the rms speed of hydrogen molecules would be 82 times larger than at room temperature were it not for the fact that at such high speeds, the molecules cannot survive the collisions among themselves. Remember too that the rms speed is only a kind of average speed; many molecules move much faster than this, and some much slower.

The speed of sound in a gas is closely related to the rms speed of the molecules of that gas. In a sound wave, the disturbance is passed on from molecule to molecule by means of collisions. The wave cannot move any faster than the "average" speed of the molecules. In fact, the speed of sound must be somewhat less than this "average" molecular speed because not all molecules are moving in exactly the same direction as the wave. As examples, at room temperature, the rms speeds of hydrogen and nitrogen molecules are 1920 m/s and 517 m/s, respectively. The speeds of sound in these two gases at this temperature are 1350 m/s and 350 m/s, respectively.

A question often arises: If molecules move so fast, why does it take as long as a minute or so before you can smell perfume if someone opens a bottle across a room? A glance ahead at Fig. 20-4 (Section 20-6) suggests that—although the molecules move very fast between collisions—a given molecule will wander only very slowly away from its release point.

SAMPLE PROBLEM 20-3

Here are five numbers: 5, 11, 32, 67, and 89.

(a) What is the average value $\bar{n}$ of these numbers?

SOLUTION: We find this from

$$\bar{n} = \frac{5 + 11 + 32 + 67 + 89}{5} = 40.8 \quad \text{(Answer)}$$

(b) What is the rms value n_{rms} of these numbers?

SOLUTION: We find this from

$$n_{rms} = \sqrt{\frac{5^2 + 11^2 + 32^2 + 67^2 + 89^2}{5}}$$

$$= 52.1. \quad \text{(Answer)}$$

The rms value is greater than the average value because the larger numbers—being squared—are relatively more important in forming the rms value. To test this, let us replace 89 in our set of five numbers by 300. The average value of the new set of five numbers (as you can easily show) is 2.0 times the previous average value. The rms value, however, is 2.7 times the previous rms value.

The rms values of variables occur in many branches of physics and engineering. The value 120 volts printed on an electric light bulb, for example, is an rms voltage.

20-5 TRANSLATIONAL KINETIC ENERGY

We again consider a single molecule of an ideal gas as it moves around in the box of Fig. 20-3, but we now assume that its speed changes when it collides with other molecules. Its translational kinetic energy at any instant is $\frac{1}{2}mv^2$. Its *average* translational kinetic energy over the time that we watch it is

$$\bar{K} = \overline{\tfrac{1}{2}mv^2} = \tfrac{1}{2}m\overline{v^2} = \tfrac{1}{2}mv_{rms}^2, \quad (20\text{-}19)$$

in which we make the assumption that the average speed of the molecule during our observation is the same as the average speed of all the molecules at any given time. (Provided the total energy of the gas is not changing and we observe our molecule for long enough, this assumption is appropriate.) Substituting for v_{rms} from Eq. 20-18 leads to

$$\bar{K} = (\tfrac{1}{2}m) \frac{3RT}{M}.$$

But M/m, the molar mass divided by the mass of a molecule, is simply Avogadro's number, so

$$\bar{K} = \frac{3RT}{2N_A},$$

which we can write as

$$\bar{K} = \tfrac{3}{2}kT. \quad (20\text{-}20)$$

The constant k, called the **Boltzmann constant,** is the ratio of the gas constant R to Avogadro's number N_A. It is sometimes called the gas constant for a single molecule (rather than for a mole), and its value is

$$k = \frac{R}{N_A} = \frac{8.31 \text{ J/mol} \cdot \text{K}}{6.02 \times 10^{23} \text{ mol}^{-1}} = 1.38 \times 10^{-23} \text{ J/K}$$

$$= 8.62 \times 10^{-5} \text{ eV/K}. \qquad (20\text{-}21)$$

Equation 20-20 tells us something unexpected:

At a given temperature T, all ideal gas molecules—no matter what their mass—have the same average translational kinetic energy, namely, $\frac{3}{2}kT$. When we measure the temperature of a gas, we are also measuring the average translational kinetic energy of its molecules.

CHECKPOINT 2: A gas mixture consists of molecules of types 1, 2, and 3, with molecular masses $m_1 > m_2 > m_3$. Rank the three types according to (a) average kinetic energy and (b) rms speed, greatest first.

SAMPLE PROBLEM 20-4

What is the average translational kinetic energy (in electron-volts) of the oxygen molecules in air at room temperature (= 300 K)? Of the nitrogen molecules?

SOLUTION: The average translational kinetic energy depends only on the temperature and not on the nature of the molecule. For both oxygen and nitrogen molecules it is given by Eq. 20-20 as

$$\overline{K} = \tfrac{3}{2}kT = (\tfrac{3}{2})(8.62 \times 10^{-5} \text{ eV/K})(300 \text{ K})$$

$$= 0.039 \text{ eV}. \qquad \text{(Answer)}$$

Physicists find it useful to remember that the mean translational kinetic energy of *any* molecule at room temperature is about $\frac{1}{25}$ eV, which is essentially the above result.

From Table 20-1 we see that the rms speed of the molecules of oxygen (for which $M = 32.0$ g/mol) is 483 m/s. That for the molecules of nitrogen ($M = 28.0$ g/mol) is 517 m/s. Thus the lighter molecule has the larger rms speed, consistent with the fact that the two kinds of molecules have the same average translational kinetic energy.

20-6 MEAN FREE PATH

We continue to examine the motion of molecules in an ideal gas. Figure 20-4 shows the path of a typical molecule as it moves through the gas, changing both speed and di-

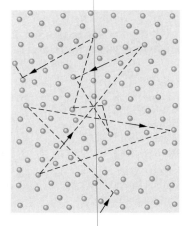

FIGURE 20-4 A molecule traveling through a gas, colliding with other gas molecules in its path. Although the other molecules are shown as stationary, they are also moving in a similar fashion.

rection abruptly as it collides elastically with other molecules. Between collisions, our typical molecule moves in a straight line at constant speed. Although the figure shows all the other molecules as stationary, they too are moving in much the same way.

One useful parameter to describe this random motion is the **mean free path** λ. As its name implies, λ is the average distance traversed by a molecule between collisions. We expect λ to vary inversely with N/V, the number of molecules per unit volume (or density of molecules). The larger N/V is, the more collisions there should be and the smaller the mean free path. We also expect λ to vary inversely with the size of the molecules. (If the molecules were true points, they would never collide and the mean free path would be infinite.) Thus the larger the molecules, the smaller the mean free path. We can even predict that λ should vary (inversely) as the *square* of the molecular diameter because the cross section of a molecule—not its diameter—determines its effective target area.

The expression for the mean free path does, in fact, turn out to be

$$\lambda = \frac{1}{\sqrt{2}\,\pi d^2 \, N/V} \qquad \text{(mean free path).} \quad (20\text{-}22)$$

To justify Eq. 20-22, we focus attention on a single molecule and assume—as Fig. 20-4 suggests—that our molecule is traveling with a constant speed v and that all the other molecules are at rest. Later, we shall relax this assumption.

We assume further that the molecules are spheres of diameter d. A collision will then take place if the centers of the molecules come within a distance d of each other, as in Fig. 20-5a. Another, more helpful way to look at the situation is to consider our single molecule to have a *radius* of d and all the other molecules to be *points,* as in Fig. 20-5b. This does not change our criterion for a collision.

As our single molecule zigzags through the gas, it sweeps out a short cylinder of cross-sectional area πd^2

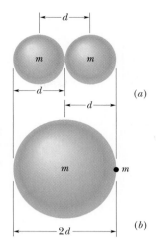

FIGURE 20-5 (*a*) A collision occurs when the centers of two molecules come within a distance *d* of each other, *d* being the molecular diameter. (*b*) An equivalent but more convenient representation is to think of the moving molecule as having a *radius d* and all other molecules as being points. The condition for a collision is unchanged.

speed of our single molecule *relative to the other molecules,* which are moving. It is this latter average speed that determines the number of collisions. A detailed calculation, taking into account the actual speed distribution of the molecules, gives $\overline{v_{rel}} = \sqrt{2}\,\overline{v}$ and thus the factor $\sqrt{2}$.

The mean free path of air molecules at sea level is about 0.1 μm. At an altitude of 100 km, the density of air has dropped to such an extent that the mean free path rises to about 16 cm. At 300 km, the mean free path is about 20 km. A problem faced by those who would study the physics and chemistry of the upper atmosphere in the laboratory is the unavailability of containers large enough to hold gas samples that simulate upper atmospheric conditions. Yet studies of the concentrations of Freon, carbon dioxide, and ozone in the upper atmosphere are of vital public concern.

between successive collisions. If we watch this molecule for a time interval Δt, it moves a distance $v\,\Delta t$, where v is its assumed speed. Thus, if we align all the short cylinders swept out in Δt, we form a composite cylinder (Fig. 20-6) of length $v\,\Delta t$ and volume $(\pi d^2)(v\,\Delta t)$. The number of collisions that occur is then equal to the number of (point) molecules that lie within this cylinder.

Since N/V is the number of molecules per unit volume, the number of collisions is N/V times the volume of the cylinder, or $(N/V)(\pi d^2 v\,\Delta t)$. The mean free path is the length of the path (and of the cylinder) divided by this number:

$$\lambda = \frac{\text{length of path}}{\text{number of collisions}} \approx \frac{v\,\Delta t}{\pi d^2 v\,\Delta t\, N/V}$$

$$= \frac{1}{\pi d^2\, N/V}. \qquad (20\text{-}23)$$

This equation is only approximate because it is based on the assumption that all the molecules except one are at rest. In fact, *all* the molecules are moving; when this is taken properly into account, Eq. 20-22 results. Note that it differs from the (approximate) Eq. 20-23 only by a factor of $1/\sqrt{2}$.

We can even get a glimpse of what is "approximate" about Eq. 20-23. The v in the numerator and that in the denominator are—strictly—not the same. The v in the numerator is $\overline{v}$, the mean speed of the molecule *relative to the container.* The v in the denominator is $\overline{v_{rel}}$, the mean

FIGURE 20-6 In time Δt the moving molecule effectively sweeps out a cylinder of length $v\,\Delta t$ and radius d.

SAMPLE PROBLEM 20-5

The molecular diameters of gas molecules of different kinds can be found experimentally by measuring the rates at which the different gases diffuse (spread) into each other. For oxygen, $d = 2.9 \times 10^{-10}$ m has been reported.

(a) What is the mean free path for oxygen at room temperature ($T = 300$ K) and at an atmospheric pressure of 1.0 atm? Assume it is an ideal gas under these conditions.

SOLUTION: Let us first find N/V, the number of molecules per unit volume under these conditions. From the ideal gas law, 1.0 mol of any ideal gas occupies a volume equal to

$$V = \frac{nRT}{p} = \frac{(1.0\ \text{mol})(8.31\ \text{J/mol}\cdot\text{K})(300\ \text{K})}{(1.0\ \text{atm})(1.01 \times 10^5\ \text{Pa/atm})}$$

$$= 2.47 \times 10^{-2}\ \text{m}^3.$$

Because we have 1.0 mol of the gas in that volume, the number of molecules per unit volume is

$$\frac{N}{V} = \frac{nN_A}{V} = \frac{(1.0\ \text{mol})(6.02 \times 10^{23}\ \text{molecules/mol})}{2.47 \times 10^{-2}\ \text{m}^3}$$

$$= 2.44 \times 10^{25}\ \text{molecules/m}^3.$$

Equation 20-22 then gives

$$\lambda = \frac{1}{\sqrt{2}\,\pi d^2\, N/V}$$

$$= \frac{1}{(\sqrt{2}\,\pi)(2.9 \times 10^{-10}\ \text{m})^2(2.44 \times 10^{25}\ \text{m}^{-3})}$$

$$= 1.1 \times 10^{-7}\ \text{m}. \qquad (\text{Answer})$$

This is about 380 molecular diameters. On average, the molecules in such a gas are only about 11 molecular diameters apart.

(b) If the average speed of an oxygen molecule is 450 m/s, what is the average collision rate?

SOLUTION: We find this rate by dividing the average speed by the mean free path to get

$$\text{rate} = \frac{v}{\lambda} = \frac{450 \text{ m/s}}{1.1 \times 10^{-7} \text{ m}} = 4.1 \times 10^9 \text{ s}^{-1}. \quad \text{(Answer)}$$

Thus—on average—every oxygen molecule makes more than 4 billion collisions per second!

CHECKPOINT **3:** One mole of gas A, with molecular radius $2d_0$ and average molecular speed v_0, is placed inside a certain container. One mole of gas B, with molecular radius d_0 and average molecular speed $2v_0$ (the molecules of B are smaller but faster), is placed in an identical container. Which gas has the greater average collision rate within its container?

20-7 THE DISTRIBUTION OF MOLECULAR SPEEDS (OPTIONAL)

The root-mean-square speed v_{rms} gives us a general idea of the molecular speed in a gas at a given temperature. We often want to know more. For example, what fraction of the molecules have speeds greater than the rms value? Greater than twice the rms value? To answer such questions, we need to know how the possible values of speed are distributed among the molecules. Figure 20-7a shows this distribution for oxygen molecules at room temperature ($T = 300$ K); Fig. 20-7b compares it with the distribution at $T = 80$ K.

In 1852, Scottish physicist James Clerk Maxwell first solved the problem of finding the speed distribution of gas molecules. His result, known as **Maxwell's speed distribution law,** is

$$P(v) = 4\pi \left(\frac{M}{2\pi RT} \right)^{3/2} v^2 e^{-Mv^2/2RT}. \quad (20\text{-}24)$$

Here v is the molecular speed, T is the gas temperature, M is the molar mass of the gas, and R is the gas constant. It is this equation that is plotted in Fig. 20-7a,b. The quantity $P(v)$ in Eq. 20-24 and Fig. 20-7 is a *probability distribution function*, defined as follows: The product $P(v)\,dv$ (which is a dimensionless quantity) is the fraction of molecules whose speeds lie in the range v to $v + dv$.

As Fig. 20-7a shows, this fraction is equal to the area of a strip whose height is $P(v)$ and whose width is dv. The total area under the distribution curve corresponds to the fraction of the molecules whose speeds lie between zero and infinity. All molecules fall into this category, so the value of this total area is unity.

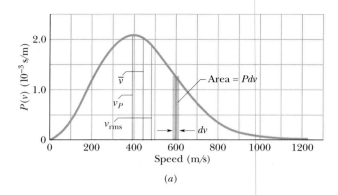

(a)

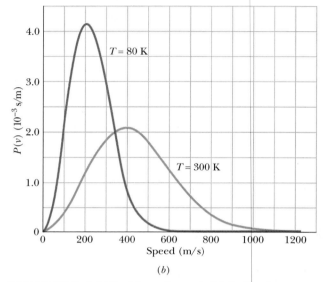

(b)

FIGURE 20-7 (a) The Maxwell speed distribution for oxygen molecules at $T = 300$ K. The three characteristic speeds are marked. (b) The curves for 300 K and 80 K. Note that the molecules move more slowly at the lower temperature. Because these are probability distributions, the area under each curve has a numerical value of unity.

Figure 20-7a also shows the root-mean-square speed v_{rms} (= 483 m/s) and two other measures of the speed distribution of the oxygen molecules. The *most probable speed* v_P (= 395 m/s) is the speed at which $P(v)$ is a maximum. The *average speed* $\bar{v}$ (= 445 m/s) is—as its name suggests—a simple average of the molecular speeds. A small number of molecules, lying in the extreme right-hand tail of the distribution curve, can have speeds that are several times the average speed. This simple fact, as we shall demonstrate, makes possible both rain and sunshine.

Rain: The speed distribution of water molecules in, say, a pond at summertime temperatures can be represented by a curve similar to that of Fig. 20-7a. Most of the molecules do not have nearly enough kinetic energy to escape from the water through its surface. However, small numbers of very fast molecules with speeds far out in the tail of the curve can do so. It is these water molecules that evaporate, making clouds and rain a possibility.

As the fast water molecules leave the surface, carrying energy with them, the temperature of the remaining water is maintained by heat transfer from the surroundings. Other fast molecules — produced in particularly favorable collisions — quickly take the place of those that have left, and the speed distribution is maintained.

Sunshine: Let the distribution curve of Fig. 20-7a now refer to protons in the core of the Sun. The Sun's energy is supplied by a nuclear fusion process that starts with the merging of two protons. However, protons repel each other because of their electrical charges, and protons of average speed do not have enough kinetic energy to overcome the repulsion and get close enough to merge. Very fast protons with speeds in the tail of the distribution curve can do so, however, and thus the Sun can shine.

SAMPLE PROBLEM 20-6

A container is filled with oxygen gas maintained at room temperature (300 K). What fraction of the molecules have speeds in the range 599–601 m/s? The molar mass M of oxygen is 0.0320 kg/mol.

SOLUTION: This speed interval Δv (= 2 m/s) is so small that we can treat it as a differential and say that the fraction frac that we seek is given very closely by $P(v)\,\Delta v$, where $P(v)$ is to be evaluated at $v = 600$ m/s, the midpoint of the interval; see the gold strip in Fig. 20-7a. Thus, using Eq. 20-24, we find

$$\text{frac} = P(v)\,\Delta v = 4\pi \left(\frac{M}{2\pi RT} \right)^{3/2} v^2 e^{-Mv^2/2RT}\,\Delta v.$$

For convenience in calculating, let us break this down into five factors, as

$$\text{frac} = (4\pi)(A)(v^2)(e^B)(\Delta v) \qquad (20\text{-}25)$$

in which A and B are

$$A = \left(\frac{M}{2\pi RT} \right)^{3/2} = \left(\frac{0.0320 \text{ kg/mol}}{(2\pi)(8.31 \text{ J/mol}\cdot\text{K})(300 \text{ K})} \right)^{3/2}$$

$$= 2.92 \times 10^{-9} \text{ s}^3/\text{m}^3$$

and $\quad B = -\dfrac{Mv^2}{2RT} = -\dfrac{(0.0320 \text{ kg/mol})(600 \text{ m/s})^2}{(2)(8.31 \text{ J/mol}\cdot\text{K})(300 \text{ K})}$

$$= -2.31.$$

Substituting A and B into Eq. 20-25 yields

frac $= (4\pi)(A)(v^2)(e^B)(\Delta v)$

$= (4\pi)(2.92 \times 10^{-9} \text{ s}^3/\text{m}^3)(600 \text{ m/s})^2(e^{-2.31})(2 \text{ m/s})$

$= 2.62 \times 10^{-3}.$ \hfill (Answer)

Thus, at room temperature, 0.262% of the oxygen molecules will have speeds that lie in the narrow range between 599 and 601 m/s. If the gold strip of Fig. 20-7a were drawn to the scale of this problem, it would be a very thin strip indeed.

SAMPLE PROBLEM 20-7

(a) What is the average speed v of oxygen gas molecules at $T = 300$ K? The molar mass M of oxygen is 0.0320 kg/mol.

SOLUTION: To find the average speed, we weight each speed v by $P(v)\,dv$, which is the fraction of the molecules whose speeds lie in the interval v to $v + dv$. We then add up (that is, integrate) these fractions over the entire range of speeds. Thus

$$\bar{v} = \int_0^\infty v P(v)\,dv. \qquad (20\text{-}26)$$

The next step is to substitute for $P(v)$ from Eq. 20-24 and evaluate the integral that results. From a table of integrals* we find

$$\bar{v} = \sqrt{\frac{8RT}{\pi M}} \qquad \text{(average speed).} \qquad (20\text{-}27)$$

Substituting numerical values yields

$$\bar{v} = \sqrt{\frac{(8)(8.31 \text{ J/mol}\cdot\text{K})(300 \text{ K})}{(\pi)(0.0320 \text{ kg/mol})}}$$

$$= 445 \text{ m/s.} \qquad \text{(Answer)}$$

(b) What is the root-mean-square speed v_{rms} of the oxygen molecules?

SOLUTION: We proceed as in (a) above except that we multiply v^2 (rather than simply v) by the weighting factor $P(v)\,dv$. This leads, after integration, to

$$\overline{v^2} = \int_0^\infty v^2 P(v)\,dv = \frac{3RT}{M}.$$

The rms speed is the square root of this quantity, or

$$v_{\text{rms}} = \sqrt{\overline{v^2}} = \sqrt{\frac{3RT}{M}} \qquad \text{(rms speed).} \qquad (20\text{-}28)$$

Equation 20-28 is identical to Eq. 20-18, which we derived earlier. The numerical calculation gives

$$v_{\text{rms}} = \sqrt{\frac{(3)(8.31 \text{ J/mol}\cdot\text{K})(300 \text{ K})}{(0.0320 \text{ kg/mol})}}$$

$$= 483 \text{ m/s.} \qquad \text{(Answer)}$$

(c) What is the most probable speed v_P?

SOLUTION: The most probable speed is the speed at which $P(v)$ of Eq. 20-24 has its maximum value. We find it by re-

---

*One such table is found in Section A of the familiar *CRC Handbook of Chemistry and Physics*; see integral 667 listed under "Definite Integrals."

TABLE 20-2 SPEED PARAMETERS FOR THE MAXWELL SPEED DISTRIBUTION

PARAMETER	SYMBOL	FORMULA	FOR OXYGEN AT 300 K
Most probable speed	v_P	$\sqrt{2RT/M}$	395 m/s
Average speed	$\bar{v}$	$\sqrt{8RT/\pi M}$	445 m/s
Root-mean-square speed	v_{rms}	$\sqrt{3RT/M}$	483 m/s

quiring that $dP/dv = 0$ and solving for v. Doing so yields (as you should show)

$$v_P = \sqrt{\frac{2RT}{M}} \qquad \text{(most probable speed).} \qquad (20\text{-}29)$$

Numerically, this yields

$$v_P = \sqrt{\frac{(2)(8.31\ \text{J/mol}\cdot\text{K})(300\ \text{K})}{(0.0320\ \text{kg/mol})}}$$

$$= 395\ \text{m/s.} \qquad \text{(Answer)}$$

Table 20-2 summarizes the three measures of the Maxwell speed distribution.

20-8 THE MOLAR SPECIFIC HEATS OF AN IDEAL GAS

In this section, we want to derive from atomic or molecular considerations an expression for the internal energy E_{int} of an ideal gas. We shall then use that result to derive an expression for the molar specific heats of an ideal gas.

Internal Energy E_{int}

Let us first assume that our ideal gas is a *monatomic gas* (which has individual atoms rather than molecules), such as helium, neon, or argon. Next, recall from Chapter 8 that internal energy is the energy associated with random motions of atoms and molecules. So let us assume that the internal energy E_{int} of our ideal gas is simply the sum of the translational kinetic energies of its atoms. (Individual atoms do not have rotational kinetic energy.) The average translational kinetic energy of a single atom depends only on the gas temperature and is given by Eq. 20-20 as $\bar{K} = \frac{3}{2}kT$. A sample of n moles of such a gas contains nN_A atoms. The internal energy E_{int} of the sample is then

$$E_{int} = (nN_A)\bar{K} = (nN_A)(\tfrac{3}{2}kT)$$

or, since $N_A k = R$, the gas constant,

$$E_{int} = \tfrac{3}{2}nRT \qquad \text{(monatomic ideal gas).} \qquad (20\text{-}30)$$

Thus,

> The internal energy E_{int} of an ideal gas is a function of the gas temperature *only*; it does not depend on any other variable.

With Eq. 20-30 in hand, we are now able to derive an expression for the molar specific heat of an ideal gas. Actually, we derive two expressions—one for the case in which the volume of the gas remains constant as heat is added to it, and one for the case in which the pressure of the gas remains constant as heat is added. The symbols for these two molar specific heats are C_V and C_p, respectively. (By convention, the capital letter C is used in both cases, even though C_V and C_P represent a type of specific heat.)

Molar Specific Heat at Constant Volume

Figure 20-8a shows n moles of an ideal gas at pressure p and temperature T, confined to a cylinder of fixed volume V. This *initial state i* of the gas is marked on the p-V diagram of Fig. 20-8b. Suppose now that you add a small amount of heat Q to the gas, by slowly turning up the temperature of the thermal reservoir on which the cylinder rests. The gas temperature rises a small amount to $T + \Delta T$, and its pressure to $p + \Delta p$, bringing the gas to *final state f*.

The defining equation for C_V, the molar specific heat at constant volume, is, in parallel with Eq. 19-15,

$$Q = nC_V \Delta T \qquad \text{(constant volume).} \qquad (20\text{-}31)$$

Substituting this expression for Q into the first law of thermodynamics (Eq. 19-24), we find

$$\Delta E_{int} = nC_V \Delta T - W.$$

With the volume held constant, the gas cannot do any work; so $W = 0$. Then solving for C_V we obtain

$$C_V = \frac{1}{n}\frac{\Delta E_{int}}{\Delta T}. \qquad (20\text{-}32)$$

From Eq. 20-30 we see that $\Delta E_{int}/\Delta T = \frac{3}{2}nR$. Substituting

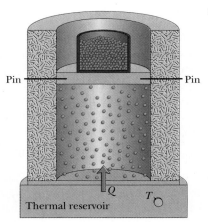

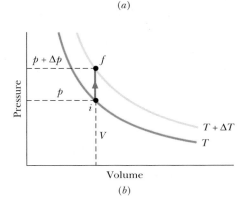

FIGURE 20-8
(a) The temperature of an ideal gas is raised from T to $T + \Delta T$ in a constant-volume process. Heat is added but no work is done. (b) The process on a p-V diagram.

atoms) are greater than those for monatomic gases for reasons that will be suggested in Section 20-9.

We can now generalize Eq. 20-30 for the internal energy of any ideal gas by substituting C_V for $\frac{3}{2}R$; we get

$$E_{int} = nC_V T \qquad \text{(any ideal gas).} \qquad (20\text{-}34)$$

This equation applies not only to an ideal monatomic gas but also to diatomic and polyatomic ideal gases, provided the appropriate value of C_V is used. Just as with Eq. 20-30, we see that the internal energy depends on the temperature of the gas but not on its pressure or density.

When an ideal gas that is confined to a container undergoes a temperature change ΔT, then from either Eq. 20-32 or Eq. 20-34 we can write the resulting change in its internal energy as

$$\Delta E_{int} = nC_V \, \Delta T \qquad \begin{array}{l}\text{(ideal gas,}\\\text{any process).}\end{array} \qquad (20\text{-}35)$$

This equation tells us:

> A change in the internal energy E_{int} of a confined ideal gas depends on the change in the gas temperature only; it does *not* depend on what type of process produces the change in the temperature.

this result into Eq. 20-32 yields

$$C_V = \tfrac{3}{2}R = 12.5 \text{ J/mol} \cdot \text{K} \qquad \begin{array}{l}\text{(monatomic}\\\text{gas).}\end{array} \qquad (20\text{-}33)$$

As Table 20-3 shows, this prediction of the kinetic theory (for ideal gases) agrees very well with experiment for real monatomic gases, the case that we have assumed. The (predicted and) experimental values of C_V for *diatomic gases* (which have molecules with two atoms) and *polyatomic gases* (which have molecules with more than two

As examples, consider the three paths between the two isotherms in the p-V diagram of Fig. 20-9. Path 1 represents a constant-volume process. Path 2 represents a constant-pressure process (that we are about to examine). And path 3 represents a process in which no heat is ex-

TABLE 20-3 MOLAR SPECIFIC HEATS

MOLECULE	EXAMPLE		C_V (J/mol·K)
Monatomic	Ideal		$\frac{3}{2}R = 12.5$
	Real	He	12.5
		Ar	12.6
Diatomic	Ideal		$\frac{5}{2}R = 20.8$
	Real	N_2	20.7
		O_2	20.8
Polyatomic	Ideal		$3R = 24.9$
	Real	NH_4	29.0
		CO_2	29.7

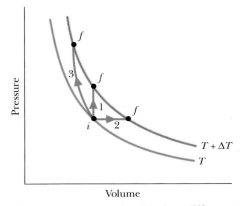

FIGURE 20-9 Three paths representing three different processes that take an ideal gas from an initial state i at temperature T to some final state f at temperature $T + \Delta T$. The change ΔE_{int} in the internal energy of the gas is the same for these three processes and for any others that result in the same change of temperature.

changed with the system's environment (we discuss this in Section 20-11). Although the values of heat Q and work W associated with these three paths differ, as do p_f and V_f, the values of ΔE_{int} associated with the three paths are identical and are all given by Eq. 20-35, because they all involve the same temperature change ΔT. So, no matter what path is actually traversed between T and $T + \Delta T$, we can always use path 1 and Eq. 20-35 to compute ΔE_{int} easily.

Molar Specific Heat at Constant Pressure

We now assume that the temperature of the ideal gas is increased by the same small amount ΔT as previously, but that the necessary heat Q is added with the gas under constant pressure. A mechanism for doing this is shown in Fig. 20-10a; the p-V diagram for the process is plotted in Fig. 20-10b. We can guess at once that the molar specific heat at constant pressure C_p, which we define with

$$Q = nC_p \, \Delta T \qquad \text{(constant pressure)}, \qquad (20\text{-}36)$$

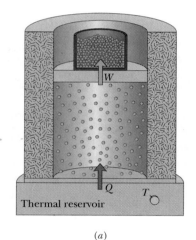

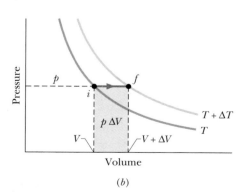

(a)

(b)

FIGURE 20-10   (a) The temperature of an ideal gas is raised from T to $T + \Delta T$ in a constant-pressure process. Heat is added and work is done in lifting the loaded piston. (b) The process on a p-V diagram. The work $p \, \Delta V$ is given by the shaded area.

is *greater* than the molar specific heat at constant volume, because energy must now be supplied not only to raise the temperature of the gas but also for the gas to do work, that is, to lift the weighted piston of Fig. 20-10a.

To relate C_p to C_V, we start with the first law of thermodynamics (Eq. 19-24):

$$\Delta E_{int} = Q - W. \qquad (20\text{-}37)$$

We next replace each term in Eq. 20-37. For ΔE_{int}, we substitute from Eq. 20-35. For Q, we substitute from Eq. 20-36. To replace W, we first note that since the pressure remains constant, Eq. 19-23 tells us that $W = p \, \Delta V$. Then we note that, using the ideal gas equation ($pV = nRT$), we can write

$$W = p \, \Delta V = nR \, \Delta T.$$

Making these substitutions in Eq. 20-37, and then dividing through by $n \, \Delta T$, we find

$$C_V = C_p - R,$$

and then

$$C_p = C_V + R. \qquad (20\text{-}38)$$

This prediction of kinetic theory agrees well with experiment, not only for monatomic gases but for gases in general, as long as their density is low enough so that we may treat them as ideal.

CHECKPOINT **4:** The figure shows five paths traversed by a gas on a p-V diagram. Rank the paths according to the change in internal energy of the gas, greatest first.

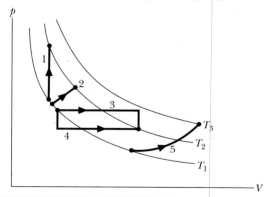

SAMPLE PROBLEM 20-8

A bubble of 5.00 mol of (monatomic) helium is submerged at a certain depth in liquid water when the water (and thus the helium) undergoes a temperature increase ΔT of 20.0 C° at constant pressure. As a result, the bubble expands.

(a) How much heat Q is added to the helium during the expansion and temperature increase?

SOLUTION: Treating the helium as an ideal gas, we use Eqs. 20-38 and 20-33 to write

$$C_p = C_V + R = \tfrac{5}{2}R.$$

We substitute this into Eq. 20-36 to obtain

$$
\begin{aligned}
Q = nC_p \, \Delta T &= n(\tfrac{5}{2}R) \, \Delta T \\
&= (5.00 \text{ mol})(2.5)(8.31 \text{ J/mol·K})(20.0 \text{ C°}) \\
&= 2077.5 \text{ J} \approx 2080 \text{ J}. \qquad \text{(Answer)}
\end{aligned}
$$

(b) What is the change ΔE_{int} in the internal energy of the helium during the temperature increase?

SOLUTION: Even though the temperature of the helium increases at constant pressure (and *not* at constant volume), we use Eq. 20-35 to calculate the change in internal energy:

$$
\begin{aligned}
\Delta E_{\text{int}} = nC_V \, \Delta T \\
&= (5.00 \text{ mol})(1.5)(8.31 \text{ J/mol·K})(20.0 \text{ C°}) \\
&= 1246.5 \text{ J} \approx 1250 \text{ J}. \qquad \text{(Answer)}
\end{aligned}
$$

(c) How much work W is done by the helium as it expands against the pressure of the surrounding water during the temperature increase?

SOLUTION: From the first law of thermodynamics,

$$
\begin{aligned}
W = Q - \Delta E_{\text{int}} &= 2077.5 \text{ J} - 1246.5 \text{ J} \\
&= 831 \text{ J}. \qquad \text{(Answer)}
\end{aligned}
$$

Note that during the temperature increase, only a portion (1250 J) of the heat (2080 J) that is transferred into the helium goes to increasing the internal energy of the helium and thus the temperature of the helium. The rest (831 J) is transferred out of the helium as work that the helium does during the expansion. If the water were frozen, it would not allow that expansion. Then the same temperature increase of 20.0 C° would require only 1250 J of heat, because no work would be done by the helium.

20-9 DEGREES OF FREEDOM AND MOLAR SPECIFIC HEATS

As Table 20-3 shows, the prediction that $C_V = \tfrac{3}{2}R$ agrees with experiment for monatomic gases but fails for diatomic and polyatomic gases. Let us try to explain the discrepancy by considering the possibility that molecules with more than one atom can store internal energy in forms other than translational motion.

Figure 20-11 shows kinetic theory models of helium (a monatomic gas), oxygen (diatomic), and methane (polyatomic). On the basis of their structure, it seems reasonable to assume that monatomic molecules—which are essentially pointlike and have only a very small rotational inertia

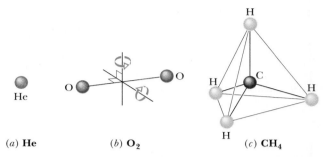

(a) He   (b) O₂   (c) CH₄

FIGURE 20-11 Models of molecules as used in kinetic theory: (*a*) helium, a typical monatomic molecule; (*b*) oxygen, a typical diatomic molecule (two rotation axes are shown); and (*c*) methane, a typical polyatomic molecule.

about any axis—can store energy only in their translational motion. Diatomic and polyatomic molecules, however, should be able to store substantial additional amounts of energy by rotating or oscillating.

To take these possibilities into account quantitatively, we use the theorem of the **equipartition of energy,** introduced by James Clerk Maxwell:

> Every kind of molecule has a certain number f of *degrees of freedom*, which are independent ways in which the molecule can store energy. Each such degree of freedom has associated with it—on average—an energy of $\tfrac{1}{2}kT$ per molecule (or $\tfrac{1}{2}RT$ per mole).

For translational motion, there are three degrees of freedom, corresponding to the three perpendicular axes along which such motion can occur. For rotational motion, a monatomic molecule has no degrees of freedom. A diatomic molecule—the rigid dumbbell of Fig. 20-11*b*—has two rotational degrees of freedom, corresponding to the two perpendicular axes about which it can store rotational energy. Such a molecule cannot store rotational energy about the axis connecting the nuclei of its two constituent atoms because its rotational inertia about this axis is approximately zero. A molecule with more than two atoms has six degrees of freedom, three rotational and three translational.

To extend the treatment of Section 20-8 to ideal diatomic and polyatomic gases, it is necessary to retrace the derivations of that section in detail, replacing Eq. 20-30 ($E_{\text{int}} = \tfrac{3}{2}nRT$) by $E_{\text{int}} = (f/2)nRT$, where f is the number of degrees of freedom listed in Table 20-4. Doing so leads to the prediction

$$C_V = \left(\frac{f}{2}\right) R = 4.16f \text{ J/mol·K}, \qquad (20\text{-}39)$$

TABLE 20-4 **DEGREES OF FREEDOM FOR VARIOUS MOLECULES**

		DEGREES OF FREEDOM			PREDICTED MOLAR SPECIFIC HEATS	
MOLECULE	EXAMPLE	TRANSLATIONAL	ROTATIONAL	TOTAL (f)	C_V (Eq. 20-39)	$C_p = C_V + R$
Monatomic	He	3	0	3	$\frac{3}{2}R$	$\frac{5}{2}R$
Diatomic	O_2	3	2	5	$\frac{5}{2}R$	$\frac{7}{2}R$
Polyatomic	CH_4	3	3	6	$3R$	$4R$

which agrees — as it must — with Eq. 20-33 for monatomic gases ($f = 3$). As Table 20-3 shows, this prediction also agrees with experiment for diatomic gases ($f = 5$), but it is too low for polyatomic gases.

SAMPLE PROBLEM 20-9

A cabin of volume V is filled with air (which we consider to be an ideal diatomic gas) at an initial low temperature T_1. After you light a wood stove, the air temperature increases to T_2. What is the resulting change in the store of internal energy of the air in the cabin?

SOLUTION: This situation differs from other situations we have examined in that the container (the cabin) is not sealed. If it were sealed, the ideal gas law ($pV = nRT$) would tell us that as the temperature of the air in the cabin increased, the air pressure would also. However, because the cabin is not airtight, air molecules leave through various openings as the air temperature increases, so that the air pressure inside the cabin always matches the air pressure outside it.

From Eq. 20-35, the change in the internal energy of the air in the room is

$$\Delta E_{int} = C_V \, \Delta(nT),$$

where both n and T can change. Using the ideal gas law, we may substitute $\Delta(pV)/R$ for $\Delta(nT)$, finding

$$\Delta E_{int} = \frac{C_V}{R} \, \Delta(pV).$$

From this we see that since neither the pressure p nor the volume V of the air within the cabin changes,

$$\Delta E_{int} = 0, \qquad \text{(Answer)}$$

even though the temperature changes.

So why does the cabin feel more comfortable at the higher temperature? There are at least two factors involved. You have a tendency to cool because (1) you emit electromagnetic radiation (thermal radiation) and (2) you exchange energy with air molecules that collide with you. If you increase the room temperature, (1) you increase the amount of thermal radiation you intercept from the surfaces within the room, replacing heat you have emitted; and (2) you increase the kinetic energy of the air molecules that collide with you, so you gain more energy from them.

20-10 A HINT OF QUANTUM THEORY

We can improve the agreement of kinetic theory with experiment by including the oscillations of the atoms in a gas of diatomic or polyatomic molecules. For example, the two atoms in the O_2 molecule of Fig. 20-11b can oscillate toward and away from each other, with the interconnecting bond acting like an oscillating spring. However, experiment shows that such oscillations occur only at relatively high temperatures of the gas — the motion is "turned on" only when the gas molecules have relatively large energies. Rotational motion is also subject to such "turning on," but at a lower gas temperature.

Figure 20-12 is of help in seeing the inception of rotational motion and oscillatory motion. The ratio C_V/R for diatomic hydrogen gas (H_2) is plotted there against temperature, with the temperature scale logarithmic to cover several orders of magnitude. Below about 80 K, we find that $C_V/R = 1.5$. This result implies that only the three translational degrees of freedom of hydrogen are involved in the specific heat. As the temperature increases, the value

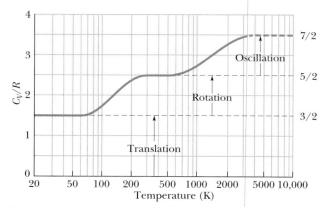

FIGURE 20-12 A plot of C_V/R versus temperature for (diatomic) hydrogen gas. Because rotational and oscillatory motions begin at certain energies, only translation is possible at very low temperatures. As the temperature increases, rotational motion can begin. At still higher temperatures, oscillatory motion can begin.

of C_V/R gradually increases to 2.5, implying that two additional degrees of freedom have become involved. Quantum theory shows that these two degrees of freedom are associated with the rotation of the hydrogen molecules and that this motion requires a certain minimum amount of energy. At very low temperatures (below 80 K), the molecules do not have enough energy to rotate. As the temperature increases from 80 K, first a few molecules and then more and more obtain enough energy to rotate, and C_V/R increases, until all of them are rotating and $C_V/R = 2.5$.

Similarly, quantum theory shows that oscillatory motion of the molecules requires a certain minimum amount of energy. This minimum amount is not met until the molecules reach a temperature of about 1000 K, as shown in Fig. 20-12. As the temperature increases beyond 1000 K, the number of molecules with enough energy to oscillate increases, and C_V/R increases, until all of them are oscillating and $C_V/R = 3.5$. (In Fig. 20-12, the plotted curve stops at 3200 K because at that temperature, the atoms of a hydrogen molecule oscillate so much that they overwhelm their common bond, and the molecule then *dissociates* into two separate atoms.)

20-11 THE ADIABATIC EXPANSION OF AN IDEAL GAS

We saw in Section 18-2 that sound waves are propagated through air and other gases as a series of compressions and expansions; these variations in the medium take place so rapidly that there is no time for heat transfer from one part of the medium to another. As we saw in Section 19-10, a process for which $Q = 0$ is an *adiabatic process*. We can ensure that $Q = 0$ either by carrying out the process very quickly (as in sound waves) or by doing it slowly in a well-insulated container. Let us see what the kinetic theory has to say about adiabatic processes.

Figure 20-13a shows an insulated cylinder containing an ideal gas and resting on an insulating stand. By removing weight from the piston, we can allow the gas to expand adiabatically. As the volume increases, both the pressure and the temperature drop. We shall prove below that the relation between the pressure and the volume during such an adiabatic process is

$$pV^\gamma = \text{a constant} \qquad \text{(adiabatic process)}, \qquad (20\text{-}40)$$

in which $\gamma = C_p/C_V$, the ratio of the molar specific heats for the gas. On a p-V diagram such as that in Fig. 20-13b, the process occurs along a line (called an *adiabat*) that has the equation $p = (\text{a constant})/V^\gamma$. When the gas goes from an initial state i to a final state f, we can rewrite Eq. 20-40 as

$$p_i V_i^\gamma = p_f V_f^\gamma \qquad \text{(adiabatic process)}. \qquad (20\text{-}41)$$

We can also write an equation for an adiabatic process in terms of T and V. To do so, we use the ideal gas equation ($pV = nRT$) to eliminate p from Eq. 20-40, finding

$$\left(\frac{nRT}{V}\right) V^\gamma = \text{a constant}.$$

Because n and R are constants, we can rewrite this in the alternative form

$$TV^{\gamma-1} = \text{a constant} \qquad \text{(adiabatic process)}, \qquad (20\text{-}42)$$

in which the constant is different from that in Eq. 20-40. When the gas goes from an initial state i to a final state f, we can rewrite Eq. 20-42 as

$$T_i V_i^{\gamma-1} = T_f V_f^{\gamma-1} \qquad \text{(adiabatic process)}. \qquad (20\text{-}43)$$

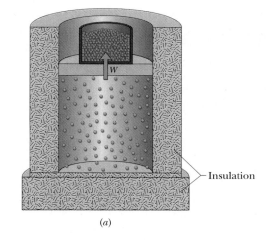

(a)

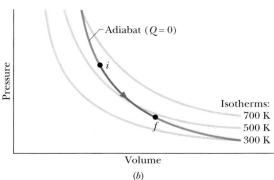

(b)

FIGURE 20-13 (a) The volume of an ideal gas is increased by removing weight from the piston. The process is adiabatic ($Q = 0$). (b) The process proceeds from i to f along an adiabat on a p-V diagram.

Proof of Eq. 20-40

Suppose that you remove some lead shot from the piston of Fig. 20-13a, allowing the ideal gas to push the piston and the remaining shot upward and thus to increase the volume by a differential amount dV. Since the volume change is tiny, we may assume that the pressure p of the gas on the piston is constant during the change. This assumption allows us to say that the work dW done by the gas during the volume increase is equal to $p\,dV$. From Eq. 19-25, the first law of thermodynamics can then be written as

$$dE_{int} = Q - p\,dV. \qquad (20\text{-}44)$$

Since the gas is thermally insulated (and thus the expansion is adiabatic), we substitute $Q = 0$. Then we use Eq. 20-35 to substitute $nC_V\,dT$ for dE_{int}. With these substitutions, and after some rearranging, we have

$$n\,dT = -\left(\frac{p}{C_V}\right)dV. \qquad (20\text{-}45)$$

Now from the ideal gas law ($pV = nRT$) we have

$$p\,dV + V\,dp = nR\,dT. \qquad (20\text{-}46)$$

Replacing R with its equal, $C_p - C_V$, in Eq. 20-46 yields

$$n\,dT = \frac{p\,dV + V\,dp}{C_p - C_V}. \qquad (20\text{-}47)$$

Equating Eqs. 20-45 and 20-47 and rearranging then give

$$\frac{dp}{p} + \left(\frac{C_p}{C_V}\right)\frac{dV}{V} = 0.$$

Replacing the ratio of the molar specific heats with γ and integrating (see Appendix E) yield

$$\ln p + \gamma \ln V = \text{a constant.}$$

Rewriting the left side as $\ln pV^\gamma$ and then taking the antilog of both sides, we find

$$pV^\gamma = \text{a constant}, \qquad (20\text{-}48)$$

which we set out to prove.

Free Expansion

Recall from Section 19-10 that a free expansion of a gas is an adiabatic process that involves no work done on or by the gas, and no change in the internal energy of the gas. A free expansion is thus quite different from the type of adiabatic process described by Eqs. 20-40 through 20-48, in which work is done and the internal energy changes. Those equations then do *not* apply to a free expansion, even though such an expansion is adiabatic.

Also recall that in a free expansion, a gas is in equilibrium only at its initial and final points; thus we can plot only those points, but not the expansion itself, on a p-V diagram. In addition, because $\Delta E_{int} = 0$, the temperature of the final state must be that of the initial state. Thus, the initial and final points on a p-V diagram must be on the same isotherm, and instead of Eq. 20-43 we have

$$T_i = T_f \qquad \text{(free expansion).} \qquad (20\text{-}49)$$

If we next assume that the gas is ideal ($pV = nRT$), then with no change in temperature, there can be no change in the product pV. Thus, instead of Eq. 20-40 a free expansion involves the relation

$$p_iV_i = p_fV_f \qquad \text{(free expansion).} \qquad (20\text{-}50)$$

SAMPLE PROBLEM 20-10

In Sample Problem 20-2, 1 mol of oxygen (assumed to be an ideal gas) expands isothermally (at 310 K) from an initial volume of 12 L to a final volume of 19 L.

(a) What would be the final temperature if the gas had expanded adiabatically to this same final volume? Oxygen (O_2) is diatomic and here has rotation but not oscillation.

SOLUTION: For a diatomic gas whose molecules have rotation but not oscillations, $C_p = \frac{7}{2}R$ and $C_V = \frac{5}{2}R$. So, $\gamma = C_p/C_V = \frac{7}{5} = 1.40$. From Eq. 20-43, we can then write

$$T_f = \frac{T_iV_i^{\gamma-1}}{V_f^{\gamma-1}} = \frac{(310\text{ K})(12\text{ L})^{1.40-1}}{(19\text{ L})^{1.40-1}}$$

$$= 258\text{ K.} \qquad \text{(Answer)}$$

The fact that the gas has cooled (from 310 K to 258 K) means that its internal energy has been correspondingly reduced. The

---

TABLE 20-5 FOUR SPECIAL PROCESSES

PATH IN FIG. 20-14	CONSTANT QUANTITY	PROCESS TYPE	SOME SPECIAL RESULTS ($\Delta E_{int} = Q - W$ and $\Delta E_{int} = nC_V \Delta T$ for all paths)
1	p	Isobaric	$Q = nC_p \Delta T; \quad W = p\Delta V$
2	T	Isothermal	$Q = W = nRT \ln(V_f/V_i); \quad \Delta E_{int} = 0$
3	$pV^\gamma, \quad TV^{\gamma-1}$	Adiabatic	$Q = 0; \quad W = -\Delta E_{int}$
4	V	Isochoric	$Q = \Delta E_{int} = nC_V \Delta T; \quad W = 0$

lost internal energy has gone to do the work of expansion (such as lifting the weighted piston in Fig. 20-13a).

(b) What would be the final temperature and pressure if, instead, the gas had expanded freely to the new volume, from an initial pressure of 2.0 Pa?

SOLUTION: Because the temperature does not change in a free expansion,

$$T_f = T_i = 310 \text{ K.} \qquad \text{(Answer)}$$

We find the new pressure using Eq. 20-50, which gives us

$$p_f = p_i \frac{V_i}{V_f} = (2.0 \text{ Pa}) \frac{12 \text{ L}}{19 \text{ L}} = 1.3 \text{ Pa.} \quad \text{(Answer)}$$

PROBLEM SOLVING TACTICS

TACTIC 2: *A Graphical Summary of Four Gas Processes*

In this chapter we have discussed four special processes that an ideal gas can undergo. An example of each is shown in Fig.

20-14, and some associated characteristics are given in Table 20-5, including two process names (isobaric and isochoric) that we have not used but which you might see in other courses.

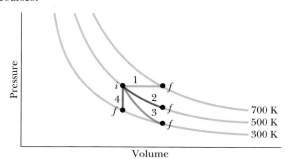

FIGURE 20-14 A *p-V* diagram representing four special processes for an ideal gas. See Table 20-5.

CHECKPOINT **5:** Rank paths 1, 2, and 3 in Fig. 20-14 according to the heat transfer to the gas, greatest first.

REVIEW & SUMMARY

Kinetic Theory of Gases

The *kinetic theory of gases* relates the *macroscopic* properties of gases (for example, pressure and temperature) to the *microscopic* properties of the gas molecules (for example, speed and kinetic energy).

Avogadro's Number

One mole of a substance contains N_A (*Avogadro's number*) elementary units (usually atoms or molecules), where N_A is found experimentally to be

$$N_A = 6.02 \times 10^{23} \text{ mol}^{-1} \qquad \begin{array}{l}\text{(Avogadro's} \\ \text{number).}\end{array} \quad \text{(20-1)}$$

One molar mass *M* of any substance is the mass of one mole of the substance.

Ideal Gas

An *ideal gas* is one for which the pressure *p*, volume *V*, and temperature *T* are related by

$$pV = nRT \qquad \text{(ideal gas law).} \qquad \text{(20-4)}$$

Here *n* is the number of moles of the gas present and *R* (= 8.31 J/mol·K) is the **gas constant.**

Work in an Isothermal Volume Change

The work done *by* an ideal gas during an **isothermal** (constant-temperature) change from volume V_i to volume V_f is

$$W = nRT \ln \frac{V_f}{V_i} \qquad \begin{array}{l}\text{(ideal gas,} \\ \text{isothermal process).}\end{array} \quad \text{(20-10)}$$

Pressure, Temperature, and Molecular Speed

The pressure exerted by *n* moles of an ideal gas, in terms of the speed of its molecules, is

$$p = \frac{nMv_{rms}^2}{3V}, \qquad \text{(20-17)}$$

where $v_{rms} = \sqrt{\overline{v^2}}$ is the **root-mean-square speed** of the molecules of the gas. With Eq. 20-4 this gives

$$v_{rms} = \sqrt{\frac{3RT}{M}}. \qquad \text{(20-18)}$$

Temperature and Kinetic Energy

The average translational kinetic energy $\overline{K}$ per molecule of an ideal gas is

$$\overline{K} = \tfrac{3}{2}kT, \qquad \text{(20-20)}$$

in which k (= R/N_A = 1.38×10^{-23} J/K) is the **Boltzmann constant.**

Mean Free Path

The *mean free path* λ of a gas molecule is its average path length between collisions and is given by

$$\lambda = \frac{1}{\sqrt{2}\pi d^2 \, N/V}, \qquad \text{(20-22)}$$

where N/V is the number of molecules per unit volume and *d* is the molecular diameter.

Maxwell Speed Distribution

The *Maxwell speed distribution* $P(v)$ is a function such that

$P(v)\, dv$ gives the *fraction* of molecules with speeds between v and $v + dv$:

$$P(v) = 4\pi \left(\frac{M}{2\pi RT}\right)^{3/2} v^2 e^{-Mv^2/2RT}. \qquad (20\text{-}24)$$

Three measures of the distribution of speeds among the molecules of a gas are

$$v_P = \sqrt{\frac{2RT}{M}} \qquad \text{(most probable speed),} \qquad (20\text{-}29)$$

$$\bar{v} = \sqrt{\frac{8RT}{\pi M}} \qquad \text{(average speed),} \qquad (20\text{-}27)$$

and the rms speed defined above in Eq. 20-18.

Molar Specific Heats

The molar specific heat C_V of a gas at constant volume is defined as

$$C_V = \frac{1}{n}\frac{Q}{\Delta T} = \frac{1}{n}\frac{\Delta E_{\text{int}}}{\Delta T}, \qquad (20\text{-}31, 20\text{-}32)$$

in which Q is the heat transferred to or from a sample of n moles of the gas, ΔT is the resulting temperature change of the gas, and ΔE_{int} is the resulting change in the internal energy of the gas. For an ideal monatomic gas,

$$C_V = \tfrac{3}{2}R = 12.5 \text{ J/mol} \cdot \text{K}. \qquad (20\text{-}33)$$

The molar specific heat C_p of a gas at constant pressure is defined to be

$$C_p = \frac{1}{n}\frac{Q}{\Delta T}, \qquad (20\text{-}36)$$

in which Q, n, and ΔT are defined as above. C_p is also given by

$$C_p = C_V + R. \qquad (20\text{-}38)$$

For n moles of an ideal gas,

$$E_{\text{int}} = nC_V T \qquad \text{(ideal gas).} \qquad (20\text{-}34)$$

If n moles of a confined ideal gas undergo a temperature change ΔT due to *any* process, the change in the internal energy of the gas is

$$\Delta E_{\text{int}} = nC_V \Delta T \qquad \text{(ideal gas, any process),} \qquad (20\text{-}35)$$

in which the appropriate value of C_V must be substituted, according to the type of ideal gas.

Degrees of Freedom and C_V

We find C_V itself by using the *equipartition of energy* theorem, which states that every *degree of freedom* of a molecule (that is, every independent way it can store energy) has associated with it—on average—an energy $\tfrac{1}{2}kT$ per molecule ($= \tfrac{1}{2}RT$ per mole). If f is the number of degrees of freedom, then $E_{\text{int}} = (f/2)nRT$ and

$$C_V = \left(\frac{f}{2}\right) R = 4.16f \text{ J/mol} \cdot \text{K}. \qquad (20\text{-}39)$$

For monatomic gases $f = 3$ (three translational degrees); for diatomic gases $f = 5$ (three translational and two rotational degrees).

Adiabatic Process

When an ideal gas undergoes a slow adiabatic volume change (a change for which $Q = 0$), its pressure and volume are related by

$$pV^\gamma = \text{a constant} \qquad \text{(adiabatic process),} \qquad (20\text{-}40)$$

in which $\gamma\,(= C_p/C_V)$ is the ratio of molar specific heats for the gas. For a free expansion, however, $pV = $ a constant.

QUESTIONS

1. If the temperature of an ideal gas is changed from 20°C to 40°C while the volume is unchanged, is the pressure of the gas doubled, increased but less than doubled, or increased but more than doubled?

2. Two rooms of equal volume are connected by an open passageway and are maintained at different temperatures. Which room has more air molecules?

3. The molar masses and Kelvin temperatures of three ideal gases are (a) M_0 and $2T_0$, (b) $2M_0$ and T_0, and (c) $6M_0$ and $6T_0$. Rank the gases according to the rms speeds of their molecules, greatest first.

4. The volume of a gas and the number of gas molecules within that volume for four situations are (a) $2V_0$ and N_0, (b) $3V_0$ and $3N_0$, (c) $8V_0$ and $4N_0$, and (d) $3V_0$ and $9N_0$. Rank the situations according to the mean free path of the molecules, greatest first.

5. In Sample Problem 20-2, how much heat is transferred during the expansion of part (a)?

6. Figure 20-15 shows the initial state of an ideal gas and an isotherm through that state. Which of the paths shown result in a decrease in the temperature of the gas?

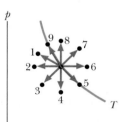

FIGURE 20-15
Question 6.

7. The table gives the heat transfer Q to or from an ideal gas and either the work W done by the gas or the work W_{on} done on the gas, all in joules, for four situations. Rank the four situations in terms of the temperature change of the gas, most positive first, most negative last.

	a	b	c	d
Q	-50	$+35$	-15	$+20$
W	-50	$+35$		
W_{on}			-40	$+40$

8. For a temperature increase of ΔT_1, a certain amount of an ideal gas requires 30 J when heated at constant volume and 50 J when heated at constant pressure. How much work is done by the gas in the second situation?

9. An ideal diatomic gas, with molecular rotation but not oscillation, loses heat Q. Is the resulting decrease in the internal energy of the gas greater if the loss occurs in a constant-volume process or in a constant-pressure process?

10. A certain amount of heat is to be transferred to 1 mol of a monatomic gas (a) at constant pressure and (b) at constant volume, and to 1 mol of a diatomic gas (c) at constant pressure and (d) at constant volume. Figure 20-16 shows four paths from an initial point to four final points on a p-V diagram. Which path goes with which process? (e) Are the molecules of the diatomic gas rotating?

FIGURE 20-16
Question 10.

11. Does the temperature of an ideal gas increase, decrease, or stay the same during (a) an isothermal expansion, (b) an expansion at constant pressure, (c) an adiabatic expansion, and (d) an increase in pressure at constant volume?

12. (a) Rank the four paths of Fig. 20-14 according to the work done by the gas, greatest first. (b) Rank paths 1, 2, and 3 according to the change in the internal energy of the gas, most positive first and most negative last.

13. In the p-V diagram of Fig. 20-17, the gas does 5 J of work along isotherm ab and 4 J along adiabat bc. What is the change in the internal energy of the gas if the gas traverses the straight path from a to c?

FIGURE 20-17
Question 13.

14. The p-V diagram of Fig. 20-18 (not to scale) shows eight separate transformations (lettered a to h) that an ideal gas undergoes in proceeding from an initial state to a final state. The table shows eight sets of values for Q, W, and ΔE_{int} (in joules), numbered 1 to 8. Which set of energy changes goes with which transformation? (*Hint:* Compare the paths in Fig. 20-18 to those in Fig. 20-14.)

	1	2	3	4	5	6	7	8
Q	10	5	-20		-10	-12		10
W	10		-10	10		-12	-10	
ΔE_{int}		2		-10	-10		10	10

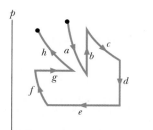

FIGURE 20-18
Question 14.

15. (a) The p-V diagram in Fig. 20-19 shows adiabatic expansions from a common point to volume V_f for a monatomic gas, a diatomic gas, and a polyatomic gas. Which path goes with which gas? (b) If the monatomic gas had, instead, freely expanded to the same final volume, would its final pressure be more than, less than, or the same as previously? (*Hint:* How would the initial and final states for the free expansion be plotted on Fig. 20-14?)

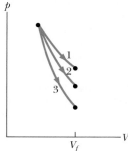

FIGURE 20-19
Question 15.

EXERCISES & PROBLEMS

SECTION 20-2 Avogadro's Number

1E. Gold has a molar mass of 197 g/mol. (a) How many moles of gold are in a 2.50 g sample of pure gold? (b) How many atoms are in the sample?

2E. Find the mass in kilograms of 7.50×10^{24} atoms of arsenic, which has a molar mass of 74.9 g/mol.

3P. If the water molecules in 1.00 g of water were distributed uniformly over the surface of Earth, how many such molecules would there be on 1.00 cm² of the surface?

4P. Consider this sentence: A _____ of water contains about as many molecules as there are _____ s of water in all the oceans. What single word best fits both blank spaces: drop, teaspoon, tablespoon, cup, quart, barrel, or ton? The oceans cover 75% of Earth's surface and have an average depth of about 5 km. (After Edward M. Purcell.)

5P. A distinguished scientist has written: ''There are enough molecules in the ink that makes one letter of this sentence to provide not only one for every inhabitant of Earth, but one for every creature if each star of our galaxy had a planet as populous as Earth.'' Check this statement. Assume the ink sample (molar mass = 18 g/mol) to have a mass of 1 μg, the population of Earth to be 5×10^9, and the number of stars in our galaxy to be 10^{11}.

SECTION 20-3 Ideal Gases

6E. (a) What is the volume occupied by 1.00 mol of an ideal gas at standard conditions, that is, at 1.00 atm ($= 1.01 \times 10^5$ Pa) and 0°C ($= 273$ K)? (b) Show that the number of molecules per cubic centimeter (the *Loschmidt number*) at standard conditions is 2.69×10^{19}.

7E. Compute (a) the number of moles and (b) the number of molecules in 1.00 cm³ of an ideal gas at a pressure of 100 Pa and a temperature of 220 K.

8E. The best vacuum that can be attained in the laboratory corresponds to a pressure of about 1.00×10^{-18} atm, or 1.01×10^{-13} Pa. How many gas molecules are there per cubic centimeter in such a vacuum at 293 K?

9E. A quantity of ideal gas at 10.0°C and a pressure of 100 kPa occupies a volume of 2.50 m³. (a) How many moles of the gas are present? (b) If the pressure is now raised to 300 kPa and the temperature is raised to 30.0°C, how much volume will the gas occupy? Assume no leaks.

10E. Oxygen gas having a volume of 1000 cm³ at 40.0°C and 1.01×10^5 Pa expands until its volume is 1500 cm³ and its pressure is 1.06×10^5 Pa. Find (a) the number of moles of oxygen present and (b) the final temperature of the sample.

11E. An automobile tire has a volume of 1000 in.³ and contains air at a gauge pressure of 24.0 lb/in.² when the temperature is 0.00°C. What is the gauge pressure of the air in the tires when its temperature rises to 27.0°C and its volume increases to 1020 in.³? (*Hint:* It is not necessary to convert from British units to SI units; why? Use $P_{atm} = 14.7$ lb/in.².)

12E. Calculate the work done by an external agent during an isothermal compression of 1.00 mol of oxygen from a volume of 22.4 L at 0°C and 1.00 atm pressure to 16.8 L.

13P. (a) What is the number of molecules per cubic meter in air at 20°C and at a pressure of 1.0 atm ($= 1.01 \times 10^5$ Pa)? (b) What is the mass of this 1 m³ of air? Assume that 75% of the molecules are nitrogen (N_2) and 25% oxygen (O_2).

14P. Pressure p, volume V, and temperature T for a certain material are related by

$$p = \frac{AT - BT^2}{V},$$

where A and B are constants. Find an expression for the work done by the material if the temperature changes from T_1 to T_2 while the pressure remains constant.

15P. Air that occupies 0.14 m³ at 1.03×10^5 Pa gauge pressure is expanded isothermally to atmospheric pressure and then cooled at constant pressure until it reaches its initial volume. Compute the work done by the air.

16P. Consider a given mass of an ideal gas. Compare curves representing constant-pressure, constant-volume, and isothermal processes on (a) a p-V diagram, (b) a p-T diagram, and (c) a V-T diagram. (d) How do these curves depend on the mass of gas?

17P. A container encloses two ideal gases. Two moles of the first gas are present, with molar mass M_1. The second gas has molar mass $M_2 = 3M_1$, and 0.5 mol of this gas is present. What fraction of the total pressure on the container wall is attributable to the second gas? (The kinetic theory explanation of pressure leads to the experimentally discovered law of partial pressures for a mixture of gases that do not react chemically: *the total pressure exerted by the mixture is equal to the sum of the pressures that the several gases would exert separately if each were to occupy the vessel alone.*)

18P. A sample of an ideal gas is taken through the cyclic process *abca* shown in Fig. 20-20; at point a, $T = 200$ K. (a) How many moles of gas are in the sample? What are (b) the temperature of the gas at point b, (c) the temperature of the gas at point c, and (d) the net heat added to the gas during the cycle?

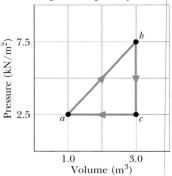

FIGURE 20-20 Problem 18.

19P. An ideal gas initially at 300 K is compressed at a constant pressure of 25 N/m² from a volume of 3.0 m³ to a volume of 1.8 m³. In the process, 75 J of heat is lost by the gas. What are (a) the change in internal energy of the gas and (b) the final temperature of the gas?

20P. A weather balloon is loosely inflated with helium at a pressure of 1.0 atm ($= 760$ torr) and a temperature of 20°C. The gas volume is 2.2 m³. At an elevation of 20,000 ft, the atmospheric pressure is down to 380 torr and the helium has expanded, being under no restraint from the confining bag. At this elevation the gas temperature is −48°C. What is the gas volume now?

21P. An air bubble of 20 cm³ volume is at the bottom of a lake 40 m deep where the temperature is 4.0°C. The bubble rises to the surface, which is at a temperature of 20°C. Take the temperature of the bubble to be the same as that of the surrounding water and find its volume just before it reaches the surface.

22P. A pipe of length $L = 25.0$ m that is open at one end contains air at atmospheric pressure. It is thrust vertically into a freshwater lake until the water rises halfway up in the pipe, as shown in Fig. 20-21. What is the depth h of the lower end of the pipe? Assume that the temperature is the same everywhere and does not change.

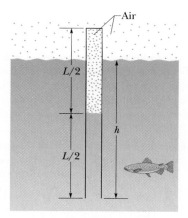

FIGURE 20-21
Problem 22.

23P. The envelope and basket of a hot-air balloon have a combined weight of 550 lb, and the envelope has a capacity of 77,000 ft³. When it is fully inflated, what should be the temperature of the enclosed air to give the balloon a lifting capacity of 600 lb (in addition to its own weight)? Assume that the surrounding air, at 20.0°C, has a weight density of 7.56×10^{-2} lb/ft³.

24P. A steel tank contains 300 g of ammonia gas (NH_3) at an absolute pressure of 1.35×10^6 Pa and temperature of 77°C. (a) What is the volume of the tank? (b) The tank is checked later when the temperature has dropped to 22°C, and the absolute pressure is found to be 8.7×10^5 Pa. How many grams of gas have leaked out of the tank?

25P. Container A in Fig. 20-22 holds an ideal gas at a pressure of 5.0×10^5 Pa and a temperature of 300 K. It is connected by a thin tube to container B with four times the volume of A. Container B holds the same ideal gas at a pressure of 1.0×10^5 Pa and a temperature of 400 K. The connecting valve is opened, and equilibrium is achieved at a common pressure while the temperature of each container is kept constant at its initial value. What is the final pressure in the system?

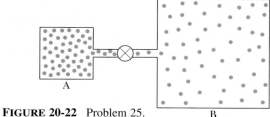

FIGURE 20-22 Problem 25.

A

B

SECTION 20-4 Pressure, Temperature, and rms Speed

26E. Calculate the root-mean-square speed of helium atoms at 1000 K. The molar mass of helium is 4.00 g/mol.

27E. The lowest possible temperature in outer space is 2.7 K. What is the root-mean-square speed of hydrogen molecules at this temperature? (Use Table 20-1.)

28E. Find the rms speed of argon atoms at 313 K. The molar mass of argon is 39.9 g/mol.

29E. The Sun is a huge ball of hot ideal gas. The temperature and pressure in the Sun's atmosphere are 2.00×10^6 K and 0.0300 Pa. Calculate the rms speed of free electrons (mass = 9.11×10^{-31} kg) there.

30E. (a) Compute the root-mean-square speed of a nitrogen molecule at 20.0°C. At what temperatures will the root-mean-square speed be (b) half that value and (c) twice that value?

31E. At what temperature do the atoms of helium gas have the same rms speed as molecules of hydrogen gas have at 20.0°C?

32P. At 273 K and 1.00×10^{-2} atm, the density of a gas is 1.24×10^{-5} g/cm³. (a) Find v_{rms} for the gas molecules. (b) Find the molar mass of the gas and identify it.

33P. The mass of the H_2 molecule is 3.3×10^{-24} g. If 10^{23} H_2 molecules per second strike 2.0 cm² of wall at an angle of 55° with the normal when moving with a speed of 1.0×10^5 cm/s, what pressure do they exert on the wall?

SECTION 20-5 Translational Kinetic Energy

34E. What is the average translational kinetic energy of nitrogen molecules at 1600 K, (a) in joules and (b) in electron-volts?

35E. (a) Determine the average value in electron-volts of the translational kinetic energy of the particles of an ideal gas at 0.00°C and at 100°C. (b) What is the translational kinetic energy per mole of an ideal gas at these temperatures, in joules?

36E. At what temperature is the average translational kinetic energy of a molecule equal to 1.00 eV?

37E. Oxygen (O_2) gas at 273 K and 1.0 atm pressure is confined to a cubical container 10 cm on a side. Calculate the ratio of (1) the change in gravitational potential energy of an oxygen molecule falling the height of the box to (2) the molecule's average translational kinetic energy.

38P. Show that the ideal gas equation, Eq. 20-4, can be written in these alternative forms: (a) $p = \rho RT/M$, where ρ is the mass density of the gas and M the molar mass; (b) $pV = NkT$, where N is the number of gas particles (atoms or molecules).

39P. Water standing in the open at 32.0°C evaporates because of the escape of some of the surface molecules. The heat of vaporization (539 cal/g) is approximately equal to ϵn, where ϵ is the average energy of the escaping molecules and n is the number of molecules per gram. (a) Find ϵ. (b) What is the ratio of ϵ to the average kinetic energy of H_2O molecules, assuming the latter is related to temperature in the same way as it is for gases?

40P. *Avogadro's law* states that under the same conditions of temperature and pressure, equal volumes of gas contain equal numbers of molecules. Is this law equivalent to the ideal gas law? Explain.

SECTION 20-6 Mean Free Path

41E. The mean free path of nitrogen molecules at 0.0°C and 1.0 atm is 0.80×10^{-5} cm. At this temperature and pressure there are 2.7×10^{19} molecules/cm³. What is the molecular diameter?

42E. At 2500 km above Earth's surface, the density of the atmosphere is about 1 molecule/cm³. (a) What mean free path is predicted by Eq. 20-22 and (b) what is its significance under these conditions? Assume a molecular diameter of 2.0×10^{-8} cm.

43E. What is the mean free path for 15 spherical jelly beans in a bag that is vigorously shaken? The volume of the bag is 1.0 L, and the diameter of a jelly bean is 1.0 cm.

44E. Derive an expression, in terms of N/V, $\bar{v}$, and d, for the collision frequency of a gas atom or molecule.

45P. In a certain particle accelerator, protons travel around a circular path of diameter 23.0 m in an evacuated chamber, whose residual gas is at 295 K and 1.00×10^{-6} torr pressure. (a) Calculate the number of gas molecules per cubic centimeter at this pressure. (b) What is the mean free path of the gas molecules if the molecular diameter is 2.00×10^{-8} cm?

46P. At what frequency would the wavelength of sound in air be equal to the mean free path of oxygen molecules at 1.0 atm pressure and 0.0°C? Take the diameter of the oxygen molecule to be 3.0×10^{-8} cm.

47P. (a) What is the molar volume (volume per mole) of an ideal gas at standard conditions (0.00°C, 1.00 atm)? (b) Calculate the ratio of the root-mean-square speed of helium atoms to that of neon atoms under these conditions. (c) What is the mean free path of helium atoms under these conditions? Assume the atomic diameter d to be 1.00×10^{-8} cm. (d) What is the mean free path of neon atoms under these conditions? Assume the same atomic diameter as for helium. (e) Comment on the results of (c) and (d) in view of the fact that the helium atoms are traveling faster than the neon atoms.

48P. The mean free path λ of the molecules of a gas may be determined from certain measurements (for example, from measurement of the viscosity of the gas). At 20°C and 750 torr pressure such measurements yield values of λ_{Ar} (argon) = 9.9×10^{-6} cm and λ_{N_2} (nitrogen) = 27.5×10^{-6} cm. (a) Find the ratio of the effective diameter of argon to that of nitrogen. What is the mean free path of argon at (b) 20°C and 150 torr, and (c) −40°C and 750 torr?

49P. Show that about 10^{13} air molecules are needed to cover the period that closes this sentence. Show that about 10^{21} air molecules collide with that period each second.

SECTION 20-7 The Distribution of Molecular Speeds

50E. The speeds of 10 molecules are 2.0, 3.0, 4.0,· · ·, 11 km/s. (a) What is their average speed? (b) What is their root-mean-square speed?

51E. Twenty-two particles have speeds as follows (N_i represents the number of particles that have speed v_i):

N_i	2	4	6	8	2
v_i (cm/s)	1.0	2.0	3.0	4.0	5.0

(a) Compute the average speed $\bar{v}$. (b) Compute the root-mean-square speed v_{rms}. (c) Of the five speeds shown, which is the most probable speed v_P?

52E. (a) Ten particles are moving with the following speeds: four at 200 m/s, two at 500 m/s, and four at 600 m/s. Calculate the average and root-mean-square speeds. Is $v_{rms} > \bar{v}$? (b) Make up your own speed distribution for the 10 particles and show that $v_{rms} \geq \bar{v}$ for your distribution. (c) Under what condition (if any) does $v_{rms} = \bar{v}$?

53E. Consider the distribution of speeds shown in Fig. 20-23. (a) Rank v_{rms}, $\bar{v}$, and v_P, greatest first. (b) How does the ranking in (a) compare with that for a Maxwellian distribution?

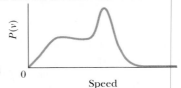

FIGURE 20-23
Exercise 53.

54E. It is found that the most probable speed of molecules in a gas at equilibrium temperature T_2 is the same as the rms speed of the molecules in this gas when its equilibrium temperature is T_1. Calculate T_2/T_1.

55P. (a) Compute the temperatures at which the rms speed is equal to the speed of escape from Earth's surface for molecular hydrogen and for molecular oxygen. (b) Do the same for the Moon, assuming the gravitational acceleration on its surface to be 0.16g. (c) The temperature high in Earth's upper atmosphere is about 1000 K. Would you expect to find much hydrogen there? Much oxygen?

56P. A molecule of hydrogen (diameter 1.0×10^{-8} cm), traveling with the rms speed, escapes from a furnace ($T = 4000$ K) into a chamber containing atoms of cold argon (diameter 3.0×10^{-8} cm) at a density of 4.0×10^{19} atoms/cm³. (a) What is the speed of the hydrogen molecule? (b) If the H_2 molecule and an argon atom collide, what is the closest their centers can be, considering each as spherical? (c) What is the initial number of collisions per second experienced by the hydrogen molecule?

57P. Two containers are at the same temperature. The first contains gas with pressure p_1, molecular mass m_1, and root-mean-square speed v_{rms1}. The second contains gas with pressure $2p_1$, molecular mass m_2, and average speed $\bar{v}_2 = 2v_{rms1}$. Find the mass ratio m_1/m_2.

58P. For the hypothetical speed distribution for N gas particles shown in Fig. 20-24 [$P(v) = Cv^2$ for $0 < v \leq v_0$; $P(v) = 0$ for $v > v_0$], find (a) an expression for C in terms of N and v_0, (b) the average speed of the particles, and (c) their rms speed.

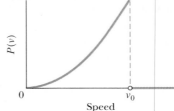

FIGURE 20-24
Problem 58.

59P. A hypothetical sample of N gas particles has the speed distribution shown in Fig. 20-25, where $P(v) = 0$ for $v > 2v_0$. (a) Express a in terms of N and v_0. (b) How many of the particles have speeds between $1.5v_0$ and $2.0v_0$? (c) Express the average speed of the particles in terms of v_0. (d) Find v_{rms}.

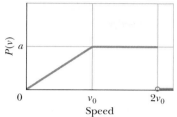

FIGURE 20-25
Problem 59.

SECTION 20-8 The Molar Specific Heats of an Ideal Gas

60E. What is the internal energy of 1.0 mol of an ideal monatomic gas at 273 K?

61E. One mole of an ideal gas undergoes an isothermal expansion. Find the heat added to the gas in terms of the initial and final volumes and the temperature. (*Hint:* Use the first law of thermodynamics.)

62E. The mass of a helium atom is 6.66×10^{-27} kg. Compute the specific heat at constant volume for (monatomic) helium gas (in J/kg·K) from the molar specific heat at constant volume.

63P. Let 20.9 J of heat be added to a particular ideal gas. As a result, its volume changes from 50.0 cm³ to 100 cm³ while the pressure remains constant at 1.00 atm. (a) By how much did the internal energy of the gas change? If the quantity of gas present is 2.00×10^{-3} mol, find the molar specific heat at (b) constant pressure and (c) constant volume.

64P. A quantity of ideal monatomic gas consists of n moles initially at temperature T_1. The pressure and volume are then slowly doubled in such a manner as to trace out a straight line on a p-V diagram. In terms of n, R, and T_1, what are (a) W, (b) ΔE_{int}, and (c) Q? (d) If one were to define a molar specific heat for this process, what would be its value?

65P. A container holds a mixture of three nonreacting gases: n_1 moles of the first gas with molar specific heat at constant volume C_1, and so on. Find the molar specific heat at constant volume of the mixture, in terms of the molar specific heats and quantities of the separate gases.

66P. The mass of a gas molecule can be computed from the specific heat at constant volume c_V. Take $c_V = 0.075$ cal/g·C° for argon and calculate (a) the mass of an argon atom and (b) the molar mass of argon.

67P. One mole of an ideal diatomic gas undergoes a transition from a to c along the diagonal path in Fig. 20-26. The temperature of the gas at point a is 1200 K. During the transition, (a) what is the change in internal energy of the gas, and (b) how much heat is added to the gas? (c) How much heat must be added to the gas if it goes from a to c along the indirect path abc?

SECTION 20-9 Degrees of Freedom and Molar Specific Heats

68E. One mole of oxygen (O_2) is heated at constant pressure starting at 0°C. How much heat must be added to the gas to double its volume? (The molecules rotate but do not oscillate.)

69E. Suppose 12.0 g of oxygen (O_2) is heated at constant atmospheric pressure from 25.0°C to 125°C. (a) How many moles of oxygen are present? (See Table 20-1.) (b) How much heat is transferred to the oxygen? (The molecules rotate but do not oscillate.) (c) What fraction of the heat is used to raise the internal energy of the oxygen?

70P. Suppose 4.00 mol of an ideal diatomic gas, with molecular rotation but not oscillation, experiences a temperature increase of 60.0 K under constant-pressure conditions. (a) How much heat was added to the gas? (b) How much did the internal energy of the gas increase? (c) How much work was done by the gas? (d) How much did the translational kinetic energy of the gas increase?

SECTION 20-11 The Adiabatic Expansion of an Ideal Gas

71E. A mass of gas occupies a volume of 4.3 L at a pressure of 1.2 atm and a temperature of 310 K. It is compressed adiabatically to a volume of 0.76 L. Determine (a) the final pressure and (b) the final temperature, assuming the gas to be an ideal gas for which $\gamma = 1.4$.

72E. (a) One liter of gas with $\gamma = 1.3$ is at 273 K and 1.0 atm pressure. It is suddenly compressed (adiabatically) to half its original volume. Find its final pressure and temperature. (b) The gas is now cooled back to 273 K at constant pressure. What is its final volume?

73E. Let n moles of an ideal gas expand adiabatically from an initial temperature T_1 to a final temperature T_2. Prove that the work done by the gas is $nC_V(T_1 - T_2)$, where C_V is the molar specific heat at constant volume. (*Hint:* Use the first law of thermodynamics.)

74E. We know that $pV^\gamma =$ a constant for an adiabatic process. Evaluate "a constant" for an adiabatic process involving exactly 2.0 mol of an ideal gas passing through the state having exactly $p = 1.0$ atm and $T = 300$ K. Assume a diatomic gas whose molecules have rotation but not oscillation.

75E. For adiabatic processes in an ideal gas, show that (a) the bulk modulus is given by

$$B = -V\frac{dp}{dV} = \gamma p,$$

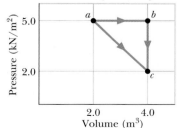

FIGURE 20-26
Problem 67.

and therefore (b) the speed of sound in the gas is

$$v_s = \sqrt{\frac{\gamma p}{\rho}} = \sqrt{\frac{\gamma RT}{M}}.$$

See Eqs. 18-2 and 18-3.

76E. Air at 0.000°C and 1.00 atm pressure has a density of 1.29×10^{-3} g/cm³, and the speed of sound in air is 331 m/s at that temperature. Compute the ratio γ of the molar specific heats of air. (*Hint:* See Exercise 75.)

77E. The speed of sound in different gases at the same temperature depends only on the molar mass of the gases. Show that $v_1/v_2 = \sqrt{M_2/M_1}$ (at constant T), where v_1 is the speed of sound in a gas of molar mass M_1 and v_2 is the speed of sound in a gas of molar mass M_2. (*Hint:* See Exercise 75.)

78P. The molar mass of iodine is 127 g/mol. A standing wave in a tube filled with iodine gas at 400 K has nodes that are 6.77 cm apart when the frequency is 1400 Hz. (a) What is γ? (b) Is iodine gas monatomic or diatomic? (*Hint:* See Exercise 75.)

79P. Knowing that C_V, the molar specific heat at constant volume, for a gas in a container is $5.0R$, calculate the ratio of the speed of sound in that gas to the rms speed of its molecules at temperature T. (*Hint:* See Exercise 75.)

80P. (a) An ideal gas initially at pressure p_0 undergoes a free expansion until its volume is 3.00 times its initial volume. What then is its pressure? (b) The gas is next slowly and adiabatically compressed back to its original volume. The pressure after compression is $(3.00)^{1/3}p_0$. Is the gas monatomic, diatomic, or polyatomic? (c) How does the average kinetic energy per molecule in this final state compare with that in the initial state?

81P. An ideal gas experiences an adiabatic compression from $p = 1.0$ atm, $V = 1.0 \times 10^6$ L, $T = 0.0°C$ to $p = 1.0 \times 10^5$ atm, $V = 1.0 \times 10^3$ L. (a) Is the gas monatomic, diatomic, or polyatomic? (b) What is its final temperature? (c) How many moles of gas are present? (d) What is the total translational kinetic energy per mole before and after the compression? (e) What is the ratio of the squares of the rms speeds before and after the compression?

82P. A sample of ideal gas expands from an initial pressure and volume of 32 atm and 1.0 L to a final volume of 4.0 L. The initial temperature of the gas is 300 K. What are the final pressure and temperature of the gas and how much work is done by the gas during the expansion, if the expansion is (a) isothermal, (b) adiabatic and the gas is monatomic, and (c) adiabatic and the gas is diatomic?

83P. An ideal gas, at initial temperature T_1 and initial volume 2 m³, is expanded adiabatically to a volume of 4 m³, then expanded isothermally to a volume of 10 m³, and then compressed adiabatically until its temperature is again T_1. What is its final volume?

84P. For a certain ideal gas C_V is 6.00 cal/mol·K. The temperature of 3.0 mol of the gas is raised 50 K by each of three different processes: at constant volume, at constant pressure, and by an

adiabatic compression. Complete the table, showing for each process the heat Q added (or subtracted), the work W done by the gas, the change ΔE_{int} in internal energy of the gas, and the change ΔK in total translational kinetic energy of the gas.

PROCESS	Q	W	ΔE_{int}	ΔK
Constant volume	_____	_____	_____	_____
Constant pressure	_____	_____	_____	_____
Adiabatic	_____	_____	_____	_____

85P. One mole of an ideal monatomic gas traverses the cycle shown in Fig. 20-27. Process $1 \rightarrow 2$ takes place at constant volume, process $2 \rightarrow 3$ is adiabatic, and process $3 \rightarrow 1$ takes place at constant pressure. (a) Compute the heat Q, the change in internal energy ΔE_{int}, and the work done W, for each of the three processes and for the cycle as a whole. (b) If the initial pressure at point 1 is 1.00 atm, find the pressure and the volume at points 2 and 3. Use 1.00 atm $= 1.013 \times 10^5$ Pa and $R = 8.314$ J/mol·K.

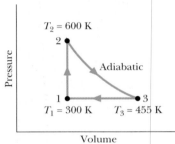

FIGURE 20-27
Problem 85.

Electronic Computation

86P. In an industrial process the volume of 25.0 moles of a monatomic ideal gas is reduced at a uniform rate from 0.616 m³ to 0.308 m³ in 2.00 h while its temperature is increased at a uniform rate from 27.0°C to 450°C. Throughout the process, the gas passes through thermodynamic equilibrium states. (a) Plot the cumulative work done on the gas and the cumulative heat absorbed by the gas as a function of time for the duration of the process. (b) What are the values of these quantities for the entire process? (c) What is the molar specific heat for the process? (*Hint:* To evaluate the integral for the work, you might use

$$\int \frac{a + bx}{A + Bx}\,dx = \frac{bx}{B} + \frac{aB - bA}{B^2}\ln(A + Bx),$$

an indefinite integral.) (d) Compare these values with those that would be obtained if the volume is first reduced at constant temperature, then the temperature is increased at constant volume, with the gas reaching the same final state.

21
Entropy and the Second Law of Thermodynamics

An anonymous graffito on a wall of the Pecan Street Cafe in Austin, Texas, read: "Time is God's way of keeping things from happening all at once." Time also has direction: some things happen in a certain sequence and could never happen on their own in a reverse sequence. As an example, an accidentally dropped egg splatters in a cup. The reverse process, a splattered egg re-forming into a whole egg and jumping up to an outstretched hand, will never happen on its own. But why not? Why can't that process be reversed, like a videotape run backward? What in the world gives direction to time?

21-1 SOME ONE-WAY PROCESSES

Suppose you come indoors on a very cold day and wrap your cold hands around a warm mug of cocoa. Then your hands get warmer and the mug gets cooler. But it never happens the other way around; that is, your cold hands never get still colder while the warm mug still warmer.

The system consisting of your hands and the mug is a *closed system*, one that is isolated from its environment. Here are some other one-way processes that occur in closed systems: (1) A crate sliding over an ordinary surface eventually stops. But you never see an initially stationary crate start to move all by itself. (2) If you drop a glob of putty, it falls to the floor. But an initially motionless glob of putty never leaps spontaneously into the air. (3) If you puncture a helium-filled balloon in a closed room, the helium gas spreads throughout the room. But the individual helium atoms will never clump up again into the shape of the balloon. We say that such one-way processes are **irreversible,** meaning that they cannot be reversed by means of only small changes in their environment.

The one-way character of such thermodynamic processes is so pervasive that we take it for granted. If these processes were to happen *spontaneously* (on their own) in the "wrong" direction, we would be astonished beyond belief. *Yet none of these "wrong-way" events would violate the law of conservation of energy.* In the cocoa mug example, that law would be obeyed even for a wrong-way transfer of heat between hands and mug. It would be obeyed even if a stationary crate or a stationary glob of putty suddenly were to transfer some of its thermal energy into kinetic energy and begin to move. And it would be obeyed even if the helium atoms released from a balloon were, on their own, to clump together again, which would require no change in their energy.

Thus changes in energy within a closed system do not set the direction of irreversible processes. Rather, that direction is set by another property that we shall discuss in this chapter—the **change in entropy** ΔS of the system. The change in entropy of a system is defined in the next section, but we can here state its central property, often called the *entropy postulate:*

> If an irreversible process occurs in a closed system, the entropy S of the system always increases; it never decreases.

Entropy differs from energy in that it does *not* obey a conservation law. The *energy* of a closed system is conserved; it always remains constant. For irreversible processes, the *entropy* of a closed system always increases. Because of this property, the change in entropy is sometimes called "the arrow of time." For example, we associate the egg of our opening photograph, breaking irreversibly as it drops into a cup, with the forward direction of time and with an increase in entropy. The backward direction of time (a videotape run backward) would correspond to the broken egg re-forming into a whole egg and rising into the air. This backward process, which would result in an entropy decrease, never happens.

There are two equivalent ways to define the change in entropy of a system: (1) in terms of the system's temperature and the energy it gains or loses as heat, and (2) by counting the ways in which the atoms or molecules that make up the system can be arranged. We use the first approach in the next section, and the second in Section 21-6.

21-2 CHANGE IN ENTROPY

Let's approach the definition of *change in entropy* by looking again at a process that we described in Section 19-10: the free expansion of an ideal gas. Figure 21-1a shows the gas in its initial equilibrium state i, confined by a closed stopcock to the left half of a thermally insulated container. If we open the stopcock, the gas rushes to fill the entire container, eventually reaching the final equilibrium state f shown in Fig. 21-1b. This is an irreversible process; all the molecules of the gas will never return, by themselves, to the left half of the container.

The p-V plot of the process in Fig. 21-2 shows the pressure and volume of the gas in its initial state i and the final state f. Pressure and volume are *state properties*, properties that depend only on the state of the gas and not on how it reached that state. Other state properties are temperature and energy. We now assume that the gas has still another state property—its entropy. Furthermore, we define the **change in entropy** $S_f - S_i$ of a system during a process that takes the system from an initial state i to a final state f as

$$\Delta S = S_f - S_i = \int_i^f \frac{dQ}{T} \qquad \text{(change in entropy defined).} \qquad (21\text{-}1)$$

Here Q is the energy transferred as heat to or from the system during the process, and T is the temperature of the system in kelvins. Thus an entropy change depends not only on the heat transferred but also on the temperature at which the transfer takes place. Because T is always positive, the sign of ΔS is the same as that of Q. We see from Eq. 21-1 that the SI unit for entropy and entropy change is the joule per kelvin.

There is a problem, however, in applying Eq. 21-1 to the free expansion of Fig. 21-1. As the gas rushes to fill the entire container, the pressure, temperature, and volume of

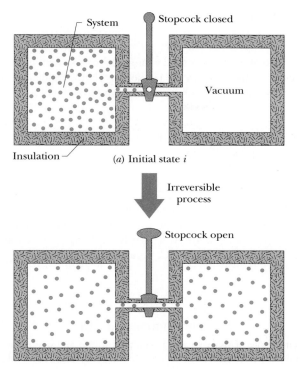

(a) Initial state i

Irreversible process

Stopcock open

(b) Final state f

FIGURE 21-1 The free expansion of an ideal gas. (a) The gas is confined to the left half of an insulated container by a closed stopcock. (b) When the stopcock is opened, the gas rushes to fill the entire container. This process is irreversible; that is, it does not occur in reverse, with the gas spontaneously collecting itself in the left half of the container.

the gas fluctuate unpredictably. That is, they do not have a sequence of well-defined equilibrium values during the intermediate stages of the change from initial equilibrium state i to final equilibrium state f. Thus we cannot trace a pressure–volume path for the free expansion on the p-V plot of Fig. 21-2 and, more important, we cannot find a relation between Q and T that allows us to integrate as Eq. 21-1 requires.

However, if entropy is truly a state property, the difference in entropy between states i and f must depend *only on those states* and not at all on the way the system went

FIGURE 21-2 A p-V diagram showing the initial state i and the final state f of the free expansion of Fig. 21-1. The intermediate states of the gas cannot be shown because they are not equilibrium states.

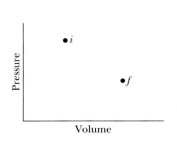

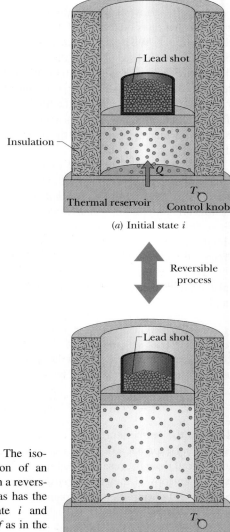

FIGURE 21-3 The isothermal expansion of an ideal gas, done in a reversible way. The gas has the same initial state i and same final state f as in the irreversible process of Figs. 21-1 and 21-2.

(a) Initial state i

Reversible process

(b) Final state f

from one state to the other. Suppose, then, that we replace the irreversible free expansion of Fig. 21-1 with a *reversible* process that connects states i and f. With a reversible process we can trace a pressure–volume path on a p-V plot, and we can find a relation between Q and T that allows us to use Eq. 21-1 to obtain the entropy change.

We saw in Section 19-10 that the temperature of an ideal gas does not change during a free expansion: $T_i = T_f = T$. So points i and f in Fig. 21-2 must be on the same isotherm. A convenient replacement process is then a reversible isothermal expansion from state i to state f, which actually proceeds *along* that isotherm. Furthermore, because T is constant throughout a reversible isothermal expansion, the integral of Eq. 21-1 is greatly simplified.

Figure 21-3 shows how to produce such a reversible isothermal expansion. We confine the gas to an insulated cylinder that rests on a thermal reservoir maintained at the temperature T. We begin by placing just enough lead shot on the movable piston so that the pressure and volume of the gas are those of the initial state i of Fig. 21-1a. We then remove shot slowly (piece by piece) until the pressure and volume of the gas are those of the final state f of Fig. 21-1b. The temperature of the gas does not change because the gas remains in thermal contact with the reservoir throughout the process.

The reversible isothermal expansion of Fig. 21-3 is physically quite different from the irreversible free expansion of Fig. 21-1. However, *both processes have the same initial state and the same final state and thus must have the same change in entropy*. Because we removed the lead shot slowly, the intermediate states of the gas are equilibrium states, so we can plot them on a p-V diagram (Fig. 21-4).

To apply Eq. 21-1 to the isothermal expansion, we take the constant temperature T outside the integral, obtaining

$$\Delta S = \frac{1}{T} \int_i^f dQ.$$

Because $\int dQ = Q$, where Q is the total energy transferred as heat during the process, we have

$$\Delta S = S_f - S_i = \frac{Q}{T} \qquad \text{(change in entropy, isothermal process).} \qquad (21\text{-}2)$$

To keep the temperature T of the gas constant during the isothermal expansion of Fig. 21-3, heat Q must have been transferred *from* the reservoir *to* the gas. Thus Q is positive and the entropy of the gas *increases* during the isothermal process and during the free expansion of Fig. 21-1.

To summarize:

To find the entropy change for an irreversible process occurring in a closed system, replace that process with any reversible process that connects the same initial and final states. Calculate the entropy change for this reversible process with Eq. 21-1.

When the temperature change ΔT of a system is small relative to the temperature (in kelvins) before and after the process, the entropy change can be approximated as

$$\Delta S = S_f - S_i \approx \frac{Q}{\overline{T}}, \qquad (21\text{-}3)$$

where $\overline{T}$ is the average temperature of the system during the process.

CHECKPOINT **1:** Water is heated on a stove. Rank the entropy changes of the water as its temperature rises (a) from 20°C to 30°C, (b) from 30°C to 35°C, and (c) from 80°C to 85°C, greatest first.

SAMPLE PROBLEM 21-1

One mole of nitrogen gas is confined to the left side of the container of Fig. 21-1. You open the stopcock and the volume of the gas doubles. What is the entropy change of the gas for this irreversible process? Treat the gas as ideal.

SOLUTION: As we did above, we replace the irreversible process of Fig. 21-1 with the isothermal expansion of Fig. 21-3 and calculate the entropy change for this latter process.

From Table 20-5, the heat Q added to an ideal gas as it expands isothermally at temperature T from an initial volume V_i to a final volume V_f is

$$Q = nRT \ln \frac{V_f}{V_i},$$

in which n is the number of moles of gas present.

From Eq. 21-2 the entropy change for this process is

$$\Delta S = \frac{Q}{T} = \frac{nRT \ln(V_f/V_i)}{T} = nR \ln \frac{V_f}{V_i}.$$

Substituting $n = 1.00$ mol and $V_f/V_i = 2$, we find

$$\Delta S = nR \ln \frac{V_f}{V_i} = (1.00 \text{ mol})(8.31 \text{ J/mol} \cdot \text{K})(\ln 2)$$

$$= +5.76 \text{ J/K}. \qquad \text{(Answer)}$$

This is also the entropy change for the free expansion—and for all other processes that connect the initial and final states shown in Fig. 21-2. ΔS is positive, so the entropy increases, in accordance with the entropy postulate of Section 21-1.

FIGURE 21-4 A p-V diagram for the reversible isothermal expansion of Fig. 21-3. The intermediate states, which are now equilibrium states, are shown.

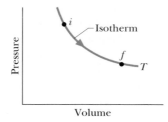

SAMPLE PROBLEM 21-2

Figure 21-5a shows two identical copper blocks of mass $m = 1.5$ kg: block L is at temperature $T_{iL} = 60$°C and block R is at temperature $T_{iR} = 20$°C. The blocks are in a thermally insulated box and are separated by an insulating shutter. When we lift the shutter, the blocks eventually come to the equilibrium temperature $T_f = 40°C$ (Fig. 21-5b). What is the net entropy change of the two-block system during this irreversible process? The specific heat of copper is 386 J/kg·K.

SOLUTION: To calculate the entropy change, we must find a reversible process that takes the system from the initial state of Fig. 21-5a to the final state of Fig. 21-5b. We then can calculate the net entropy change ΔS_{rev} of the reversible process using Eq. 21-1. For such a reversible process we need a thermal reservoir whose temperature can be changed slowly (say, by turning a knob). We then take the blocks through the following two steps, illustrated in Fig. 21-6:

STEP 1. With the reservoir's temperature set at 60°C, put block L on the reservoir. (Since block and reservoir are at the same temperature, they are already in thermal equilibrium.) Then slowly lower the temperature of the reservoir and the block to 40°C. As the block's temperature changes by each increment dT during this process, heat dQ is transferred *from* the block to the reservoir. Using Eq. 19-15, we can write this transferred energy as $dQ = mc \, dT$, where c is the specific heat of copper. According to Eq. 21-1, the entropy change ΔS_L of block L during the full temperature change from initial temperature T_{iL} (= 60°C = 333 K) to final temperature T_f (= 40°C = 313 K) is

$$\Delta S_L = \int_i^f \frac{dQ}{T} = \int_{T_{iL}}^{T_f} \frac{mc \, dT}{T} = mc \int_{T_{iL}}^{T_f} \frac{dT}{T}$$

$$= mc \ln \frac{T_f}{T_{iL}}.$$

Inserting the given data yields

$$\Delta S_L = (1.5 \text{ kg})(386 \text{ J/kg·K}) \ln \frac{313 \text{ K}}{333 \text{ K}}$$

$$= -35.86 \text{ J/K}.$$

STEP 2. With the reservoir's temperature now set at 20°C, put block R on the reservoir. Then slowly raise the temperature of the reservoir and the block to 40°C. With the same reasoning used to find ΔS_L, you can show that the entropy change ΔS_R of block R during this process is

$$\Delta S_R = (1.5 \text{ kg})(386 \text{ J/kg·K}) \ln \frac{313 \text{ K}}{293 \text{ K}}$$

$$= +38.23 \text{ J/K}.$$

The net entropy change ΔS_{rev} of the two-block system undergoing this two-step reversible process is then

$$\Delta S_{rev} = \Delta S_L + \Delta S_R$$

$$= -35.86 \text{ J/K} + 38.23 \text{ J/K} = 2.4 \text{ J/K}.$$

Thus, the net entropy change ΔS_{irrev} for the two-block system undergoing the actual irreversible process is

$$\Delta S_{irrev} = \Delta S_{rev} = 2.4 \text{ J/K}. \qquad \text{(Answer)}$$

This result is positive, in accordance with the entropy postulate of Section 21-1.

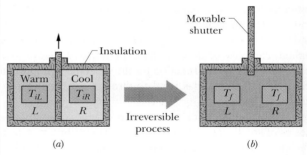

FIGURE 21-5 Sample Problem 21-2. (a) In the initial state, two copper blocks L and R, identical except for their temperatures, are in an insulating box and are separated by an insulating shutter. (b) When the shutter is removed, the blocks exchange heat and come to a final state, both with the same temperature T_f. The process is irreversible.

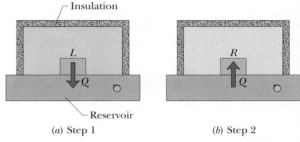

FIGURE 21-6 The blocks of Fig. 21-5 can proceed from their initial state to their final state in a reversible way if we use a reservoir with a controllable temperature (a) to extract heat reversibly from block L and (b) to add heat reversibly to block R.

CHECKPOINT 2: An ideal gas has temperature T_1 at the initial state i shown in the p-V diagram. The gas has a higher temperature T_2 at final states a and b, which it can reach along the paths shown. Is the entropy change along the path to state a larger than, smaller than, or the same as that along the path to state b?

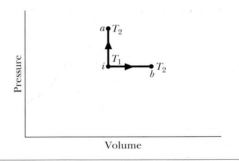

Entropy as a State Function

We have assumed that entropy, like pressure, energy, and temperature, is a property of the state of a system and independent of how that state is reached. That entropy is indeed a *state function* (as state properties are usually called) can only be deduced by experiment. However, we can prove it is a state function for the special and important case where an ideal gas is taken through a reversible process.

To make the process reversible, it is done slowly in a series of small steps, with the gas in an equilibrium state at the end of each step. For each small step, the heat transfer to or from the gas is dQ, the work done by the gas is dW, and the change in internal energy is dE_{int}. These are related by the first law of thermodynamics in differential form (Eq. 19-25):

$$dQ - dW = dE_{int}.$$

Because the steps are reversible, with the gas in equilibrium states, we can use Eq. 19-22 to replace dW with $p\, dV$ and Eq. 20-35 to replace dE_{int} with $nC_V\, dT$. Solving for dQ then leads to

$$dQ = p\, dV + nC_V\, dT.$$

Using the ideal gas law, we replace p in this equation with nRT/V. Then we divide each term in the resulting equation by T, obtaining

$$\frac{dQ}{T} = nR\frac{dV}{V} + nC_V\frac{dT}{T}.$$

Now let us integrate each term of this equation between an arbitrary initial state i and an arbitrary final state f to get

$$\int_i^f \frac{dQ}{T} = \int_i^f nR\frac{dV}{V} + \int_i^f nC_V\frac{dT}{T}.$$

The quantity on the left is the entropy change $\Delta S\ (= S_f - S_i)$ defined by Eq. 21-1. Substituting this and integrating the quantities on the right yield

$$\Delta S = S_f - S_i = nR \ln\frac{V_f}{V_i} + nC_V \ln\frac{T_f}{T_i}.$$

Note that we did not have to specify a particular reversible process when we integrated. So the integration must hold for all reversible processes that take the gas from state i to state f. Thus the change in entropy ΔS between the initial and final states of an ideal gas depends only on properties of the initial state (V_i and T_i) and properties of the final states (V_f and T_f); ΔS does not depend on how the gas changes between the two states.

21-3 THE SECOND LAW OF THERMODYNAMICS

Here is a puzzle. We saw in Sample Problem 21-1 that if we cause the reversible process of Fig. 21-3 to proceed from (*a*) to (*b*) in that figure, the change in entropy of the gas—which we take as our system—is positive. However, because the process is reversible, we can just as easily make it proceed from (*b*) to (*a*), simply by slowly adding lead shot to the piston of Fig. 21-3*b* until the original volume of the gas is restored. In this reverse process, heat must be extracted *from the gas* to keep its temperature from rising. Hence Q is negative and so, from Eq. 21-2, the entropy of the gas must decrease.

Doesn't this decrease in the entropy of the gas violate the entropy postulate of Section 21-1, which states that entropy always increases? No, because that postulate holds only for *irreversible* processes occurring in closed systems. The procedure suggested above does not meet these requirements: the process is *not* irreversible and (because energy is transferred as heat from the gas to the reservoir) the system—which is the gas alone—is *not* closed.

However, if we include the reservoir, along with the gas, as part of the system, then we do have a closed system. Let's check the change in entropy of the enlarged system *gas + reservoir* for the process that takes it from (*b*) to (*a*) in Fig. 21-3. During this reversible process, energy is transferred as heat from the gas to the reservoir, that is, from one part of the enlarged system to another. Let $|Q|$ represent the absolute value (or magnitude) of this heat. With Eq. 21-2, we can then calculate separately the entropy changes for the gas (which loses $|Q|$) and the reservoir (which gains $|Q|$). We get

$$\Delta S_{gas} = -\frac{|Q|}{T} \quad \text{and} \quad \Delta S_{res} = +\frac{|Q|}{T}.$$

The entropy change of the closed system is the sum of these two quantities, which is zero.

With this result, we can modify the entropy postulate of Section 21-1 to include both reversible and irreversible processes:

> If a process occurs in a closed system, the entropy of the system increases for irreversible processes and remains constant for reversible processes. It never decreases.

Although entropy may decrease in part of a closed system, there will always be an equal or larger entropy increase in another part of the system, so that the entropy of the system as a whole never decreases. This fact is one form of the

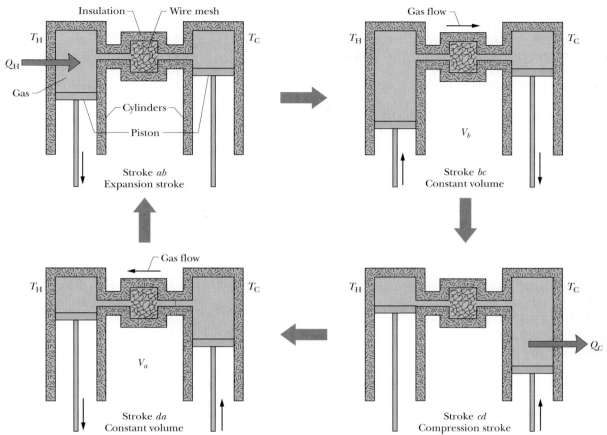

FIGURE 21-7 A schematic diagram of a Stirling engine, showing one cycle of operation, in the clockwise sequence of stroke *ab*, stroke *bc*, stroke *cd*, and then stroke *da*. Heat Q_H is added to the working substance of the engine during stroke *ab*, and heat Q_C is extracted from it during stroke *cd*. Heat is transferred to the wire mesh (to store the energy momentarily) during constant-volume stroke *bc* and then transferred back to the working substance during constant-volume stroke *da*.

second law of thermodynamics and can be written as

$$\Delta S \geq 0 \qquad \begin{array}{l}\text{(second law of}\\ \text{thermodynamics),}\end{array} \qquad (21\text{-}4)$$

where the greater-than sign applies to irreversible processes and the equals sign to reversible processes. Equation 21-4 applies only to closed systems.

In the real world almost all processes are irreversible to some extent because of friction, turbulence, and other factors. So the entropy of real closed systems undergoing real processes always increases. Processes in which the system's entropy remains constant are always idealizations.

21-4 ENTROPY IN THE REAL WORLD: ENGINES

A **heat engine,** or, more simply, an **engine,** is a device that exchanges heat with its environment and does work as it continuously repeats a set sequence of processes. As a prototype, we consider the **Stirling engine,** which was first proposed in 1816 by Robert Stirling. This engine, long neglected, is now being developed for use in automobiles and in spacecraft. A Stirling engine delivering 5000 hp (3.7 MW) has been built.

Figure 21-7 shows the elements of the Stirling engine. The left-hand cylinder is maintained at a temperature T_H by a high-temperature thermal reservoir, perhaps a propane burner. The right-hand cylinder is maintained at a lower temperature T_C by a low-temperature reservoir, perhaps circulating cooling water. (Neither reservoir is shown in Fig. 21-7.) The *working substance* of the engine is a gas. As the engine operates, the working substance moves between the cylinders, transferring heat between the reservoirs and doing work. (Do not confuse the working substance of an engine with the fuel. The function of the fuel, such as propane in a burner, is to maintain the temperature of the high-temperature reservoir.)

The two cylinders in Fig. 21-7 are connected by an insulated chamber, partially stuffed with wire mesh. The mesh stores thermal energy momentarily during the operation of the engine. The two pistons, one in each cylinder, are connected by a complex mechanical linkage that permits them to move up or down as shown by the vertical arrows in Fig. 21-7. They are also connected to a crankshaft, which is not shown.

The continuously repeated sequence of processes followed by an engine is called the *cycle* of the engine. The clockwise sequence of illustrations in Fig. 21-7 is a schematic diagram of the *Stirling cycle*. The cycle, which is also represented on the p-V plot in Fig. 21-8, consists of four processes, called *strokes:*

1. An isothermal expansion (stroke ab) at temperature T_H, produced when the left piston moves down. As a result of this expansion, heat Q_H is transferred to the gas from the left-cylinder walls, which are maintained at temperature T_H by the high-temperature reservoir.

2. A constant-volume process (stroke bc) in which the temperature of the gas decreases from T_H to T_C. To accomplish this, hot gas is pushed through the wire mesh as the moving pistons increase the volume of the right cylinder, and decrease that of the left cylinder, by equal amounts. As it is pushed through, the hot gas delivers heat to the wire mesh.

3. An isothermal compression (stroke cd) at temperature T_C back to the original volume, produced when the right piston moves up. As a result of this compression, heat Q_C is transferred from the gas to the right-cylinder walls, which are maintained at temperature T_C by the low-temperature reservoir.

4. A constant-volume process (stroke da) in which the temperature of the gas increases from T_C to T_H. To accomplish this, cool gas is pushed through the hot wire mesh as the volume of the left cylinder is increased and that of the right cylinder decreased, by equal amounts.

As the pistons move during a cycle, they cause the crankshaft to rotate. The continuous cycling of the engine causes a continuous rotation of the crankshaft, which can then be put to use. That is, the engine can do work.

If we strip the Stirling engine to its essentials, we can represent it with Fig. 21-9, which actually represents any engine that operates between a high-temperature reservoir at temperature T_H and a low-temperature reservoir at temperature T_C. Once every cycle of operation, heat Q_H moves from the high-temperature reservoir to the working substance, the engine does work W, and heat Q_C moves from the working substance to the low-temperature reservoir.

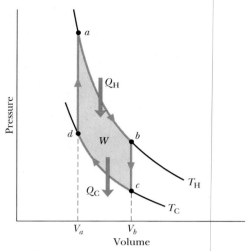

FIGURE 21-8 A p-V plot for a Stirling engine. The isothermal expansion ab, the isothermal compression cd, and the two constant-volume processes bc and da correspond to the four strokes of Fig. 21-7.

The **thermal efficiency** (or simply **efficiency**) of an engine is a measure of its ability to transform energy into work. For the engine of Fig. 21-9, heat Q_H is energy we pay for, perhaps as the cost of propane in a propane burner, and W is energy we get, to be used. So, it seems reasonable to define the efficiency of an engine as

$$\varepsilon = \frac{\text{energy we get}}{\text{energy we pay for}} = \frac{|W|}{|Q_H|} \quad \begin{array}{l}\text{(efficiency,} \\ \text{any engine).}\end{array} \quad (21\text{-}5)$$

(We use the absolute value symbols because we want a ratio of only magnitudes.)

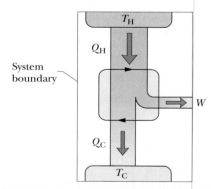

FIGURE 21-9 The elements of an engine. The two black arrowheads on the central loop suggest the working substance operating in a cycle, as if on a p-V plot. Heat Q_H is transferred from the high-temperature reservoir at temperature T_H to the working substance. Heat Q_C is transferred from the working substance to the low-temperature reservoir at temperature T_C. Work W is done by the engine (actually by the working substance) on something in the environment.

FIGURE 21-10 The North Anna nuclear power plant near Charlottesville, Virginia, which generates electrical energy at the rate of 900 MW. At the same time, by design, it discards energy into the nearby river at the rate of 2100 MW. This plant—and all others like it—throws away more energy than it delivers in useful form. It is a real counterpart to the ideal engine of Fig. 21-9.

Engines in the real world—*real engines*—that operate between the same high- and low-temperature reservoirs can vary in efficiency because they involve irreversible processes, with wasteful energy transfers. The wasteful transfers are often due to friction between moving parts and turbulence in the working substance. If we could make all the processes in an engine's operation reversible, eliminating wasteful energy transfers, we would have an *ideal engine.* The efficiency of this hypothetical (imaginary) engine is the upper limit for the efficiency of real engines. Let us find this limiting efficiency.

Any engine returns to its original state at the end of each cycle. This means that the internal energy of the working substance returns to its original value, and thus $\Delta E_{\text{int}} = 0$. For an ideal engine the only net energy transfers during a cycle are heat $|Q_H|$ from the high-temperature reservoir, heat $|Q_C|$ to the low-temperature reservoir, and the work $|W|$ done by the engine. The first law of thermodynamics ($\Delta E_{\text{int}} = Q - W$) then gives us

$$0 = (|Q_H| - |Q_C|) - |W|. \qquad (21\text{-}6)$$

In words, the difference between the net heat transferred per cycle $|Q_H| - |Q_C|$ and the work $|W|$ done by the engine must be zero. Solving Eq. 21-6 for $|W|$ and substituting the result into Eq. 21-5, we find that the efficiency of an ideal engine is

$$\varepsilon = \frac{|W|}{|Q_H|} = \frac{|Q_H| - |Q_C|}{|Q_H|} = 1 - \frac{|Q_C|}{|Q_H|} \qquad (21\text{-}7)$$
$$\text{(efficiency, ideal engine).}$$

We can find another expression for the efficiency of an ideal engine if we examine the entropy changes that take place during one cycle. Because all the engine processes are reversible, Eq. 21-4 tells us that the net entropy change per cycle for the entire closed system (the working substance of the engine plus the high- and low-temperature reservoirs) must be zero. The entropy change of the working substance is zero because the substance returns to its initial state at the end of each cycle. The entropy of the high-temperature reservoir *decreases* during each cycle, because heat is extracted from it. And the entropy of the low-temperature reservoir *increases,* because heat is added to it. So, for ΔS to be zero for the entire system, the magnitude of the entropy decrease of the high-temperature reservoir must equal the magnitude of the entropy increase of the low-temperature reservoir. Writing these two entropy changes in the form Q/T (via Eq. 21-2), we have

$$\frac{|Q_H|}{T_H} = \frac{|Q_C|}{T_C},$$

or
$$\frac{|Q_C|}{|Q_H|} = \frac{T_C}{T_H}. \qquad (21\text{-}8)$$

Then combining Eq. 21-8 with Eq. 21-7 yields, for the efficiency of an ideal engine operating between temperatures T_H and T_C,

$$\varepsilon = 1 - \frac{T_C}{T_H} \qquad \begin{array}{l}\text{(efficiency,}\\ \text{ideal engine).}\end{array} \qquad (21\text{-}9)$$

Note that we have not specified either the cycle or the working substance for the ideal engine. They do not matter: any *ideal* engine whose high- and low-temperature reservoirs are at T_H and T_C has the efficiency given by Eq. 21-9. Any *real* engine, with irreversible processes and wasteful energy transfers, is less efficient. If your car engine were an ideal engine, it would have an efficiency of about 55% according to Eq. 21-9; its actual efficiency is probably about 25%. A nuclear power plant (Fig. 21-10) is a much more complex engine, drawing heat from a reactor core, doing work by means of a turbine, and discharging heat to a nearby river. If the power plant were an ideal engine, its efficiency would be about 40%; its actual efficiency is about 30%.

Inventors continually try to improve engine efficiencies by reducing the energy $|Q_C|$ that is "thrown away" during each cycle. The inventor's dream is to produce the *perfect engine,* diagrammed in Fig. 21-11, in which $|Q_C|$ is reduced to zero and $|Q_H|$ is converted completely to work $|W|$. A perfect engine on an ocean liner, for example, could extract heat from the water and use it to drive the propellers, with no fuel cost. Alas, such a perfect engine is only a dream: inspection of Eq. 21-8 shows that $|Q_C|$ can be zero only if $T_C = 0$ K or $T_H \rightarrow \infty$, requirements that are impossible to meet. Instead, decades of practical en-

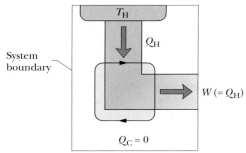

FIGURE 21-11 The elements of a perfect engine, that is, one that converts heat Q_H from a high-temperature reservoir directly to work W with 100% efficiency.

gineering experience have led to the following alternative version of the second law of thermodynamics:

> No series of processes is possible whose sole result is the absorption of heat from a thermal reservoir and the complete conversion of this energy to work.

In short, *there are no perfect engines.*

CHECKPOINT 3: Three ideal engines operate between reservoir temperatures of (a) 400 and 500 K; (b) 600 and 800 K; and (c) 400 and 600 K. Rank the engines according to their thermal efficiencies, greatest first.

SAMPLE PROBLEM 21-3

A model Stirling engine uses $n = 8.10 \times 10^{-3}$ mol of gas (assumed to be ideal) as a working substance. It operates between a high-temperature reservoir at $T_H = 95°C$ and a low-temperature reservoir at $T_C = 24°C$; the volume of its working substance doubles during the expansion stroke; and it runs at a rate of 0.70 cycle per second. Assume that the engine is ideal.

(a) How much work does the engine do per cycle?

SOLUTION: The net work per cycle is the algebraic sum of the work done by the gas during the four processes that make up the cycle shown in Figs. 21-7 and 21-8. The work done by the gas during stroke ab, an isothermal expansion from V_a to V_b, is, from Eq. 20-10,

$$W_{ab} = nRT_H \ln \frac{V_b}{V_a}.$$

The work done by the gas during stroke cd, an isothermal compression from V_b to V_a, is, similarly,

$$W_{cd} = nRT_C \ln \frac{V_a}{V_b}.$$

The work done during each of the two constant-volume processes (bc and da) is zero, so the net work for the entire cycle is

$$W = W_{ab} + W_{bc} + W_{cd} + W_{da}$$
$$= nRT_H \ln \frac{V_b}{V_a} + 0 + nRT_C \ln \frac{V_a}{V_b} + 0.$$

Because $\ln(V_a/V_b) = -\ln(V_b/V_a)$, we can write this as

$$W = nR(T_H - T_C) \ln \frac{V_b}{V_a}.$$

Substituting known data and $V_b/V_a = 2$ then yields

$$W = (8.10 \times 10^{-3} \text{ mol})(8.31 \text{ J/mol} \cdot \text{K})(95°C - 24°C)(\ln 2)$$
$$= 3.31 \text{ J} \approx 3.3 \text{ J}. \qquad \text{(Answer)}$$

Thus, during each cycle of operation, this engine does 3.3 J of work via its crankshaft.

(b) What is the power P of the engine?

SOLUTION: The power of the engine is the work W done per cycle divided by the duration t of each cycle. The engine operates at 0.70 cycle/s, so the duration of one cycle is $1/0.70 = 1.43$ s. The power is then

$$P = \frac{W}{t} = \frac{3.31 \text{ J}}{1.43 \text{ s}} \approx 2.3 \text{ W}. \qquad \text{(Answer)}$$

(c) How much heat is transferred *to* the gas *from* the high-temperature reservoir during each cycle?

SOLUTION: From Table 20-5, the magnitude $|Q_H|$ of the heat transferred to an ideal gas during an isothermal expansion (process ab in Fig. 21-8) is

$$|Q_H| = nRT_H \ln \frac{V_b}{V_a}.$$

Substituting given data and temperature $T_H = (95 + 273)$ K $= 368$ K yields

$$|Q_H| = (8.10 \times 10^{-3} \text{ mol})(8.31 \text{ J/mol} \cdot \text{K})(368 \text{ K})(\ln 2)$$
$$= 17.2 \text{ J} \approx 17 \text{ J}. \qquad \text{(Answer)}$$

(d) What is the efficiency of the engine?

SOLUTION: Substituting the temperatures of the two reservoirs into Eq. 21-9 gives us

$$\varepsilon = 1 - \frac{T_C}{T_H} = 1 - \frac{(24 + 273) \text{ K}}{(95 + 273) \text{ K}}$$
$$= 0.193 \approx 19\%. \qquad \text{(Answer)}$$

(Equations 21-5 and 21-7 give the same answer for ε.)

(e) What are the entropy changes per cycle for the gas, the high-temperature reservoir, the low-temperature reservoir, and the wire mesh?

SOLUTION: The gas operates in a cycle, so its net entropy change per cycle ΔS_{gas} is zero. For the thermal reservoirs, heat is transferred at constant temperature, so we can use Eq. 21-2 to calculate their entropy changes.

We first take the high-temperature reservoir as our system. In (c) we found that the magnitude of the heat transferred is $|Q_H| = 17.2$ J. Because the reservoir lost the heat, we can write the transfer as $-|Q_H|$. Then from Eq. 21-2, the entropy change of the reservoir is

$$\Delta S_H = \frac{-|Q_H|}{T_H} = \frac{-17.2 \text{ J}}{(95 + 273) \text{ K}} = -0.047 \text{ J/K.} \quad \text{(Answer)}$$

The entropy of the high-temperature reservoir decreases because heat is lost.

Heat enters the low-temperature reservoir from the gas, so the entropy of that reservoir increases. If you apply Eq. 21-6 or follow the example of our calculation in (c), you will find that this reservoir gains heat $|Q_C| = 13.9$ J and undergoes an entropy change of

$$\Delta S_C = \frac{+13.9 \text{ J}}{(24 + 273) \text{ K}} = +0.047 \text{ J/K.} \quad \text{(Answer)}$$

As for the mesh, the heat it gains in stroke *bc* of Fig. 21-7 it loses in stroke *da*. And these two heat transfers occur over the same range of temperatures. So from Eq. 21-2, the net entropy change per cycle ΔS_{mesh} is zero.

The entropy change per cycle for the entire system, including the reservoirs, is then

$$\Delta S = \Delta S_H + \Delta S_C + \Delta S_{gas} + \Delta S_{mesh}$$
$$= -0.047 \text{ J/K} + 0.047 \text{ J/K} + 0 + 0$$
$$= 0. \quad \text{(Answer)}$$

This is just what the second law of thermodynamics predicts for the entropy change associated with a set of reversible processes occurring in a closed system, such as the ideal engine here.

SAMPLE PROBLEM 21-4

An inventor claims to have constructed an engine that has an efficiency of 75% when operated between the boiling and freezing points of water. We shall evaluate the credibility of the inventor's claim in two (related) ways.

(a) Compare the claimed 75% efficiency to that of an ideal engine operating between the same two temperatures.

SOLUTION: With Eq. 21-9 we obtain

$$\varepsilon = 1 - \frac{T_C}{T_H} = 1 - \frac{(0 + 273) \text{ K}}{(100 + 273) \text{ K}} = 0.268 \approx 27\%.$$

An ideal engine operating between the given temperatures would have an efficiency of 27%; any real engine (with its inevitable irreversible processes and wasteful energy transfers) must be less efficient. Thus the claimed efficiency of 75% for the given temperatures is impossible.

(b) Now, starting over, assume that the engine is good enough to be approximately an ideal engine. Then determine the en-

tropy change ΔS that occurs within the closed system consisting of the engine and its reservoirs during one cycle. Compute ΔS in terms of $|Q_H|$, the magnitude of the heat transferred each cycle from the high-temperature reservoir to the working substance.

SOLUTION: Because the working substance returns to its initial state at the end of each cycle, its entropy change ΔS_{ws} during one cycle is zero. The high-temperature reservoir loses heat $|Q_H|$ at temperature T_H, and the low-temperature reservoir gains heat $|Q_C|$ at temperature T_C. Then from Eq. 21-2, their entropy changes are

$$\Delta S_H = \frac{-|Q_H|}{T_H} \quad \text{and} \quad \Delta S_C = \frac{+|Q_C|}{T_C}.$$

The total entropy change is

$$\Delta S = \Delta S_H + \Delta S_C + \Delta S_{ws}$$
$$= -\frac{|Q_H|}{T_H} + \frac{|Q_C|}{T_C} + 0. \quad (21\text{-}10)$$

We can find a substitute for $|Q_C|$ with Eq. 21-7:

$$|Q_C| = |Q_H| (1 - \varepsilon).$$

Now, combining this with Eq. 21-10, we have, after some algebra,

$$\Delta S = |Q_H| \left(\frac{1 - \varepsilon}{T_C} - \frac{1}{T_H} \right).$$

Substituting the data gives us

$$\Delta S = |Q_H| \left(\frac{1 - 0.75}{273 \text{ K}} - \frac{1}{373 \text{ K}} \right)$$
$$= -0.0018 |Q_H|.$$

Thus, if the inventor's claim were correct, the entropy of the system would decrease with each cycle. This is impossible. If every process that occurs in the engine were reversible, ΔS would be zero. If some were irreversible, ΔS would be greater than zero. No set of processes in a closed system can lead to ΔS less than zero.

PROBLEM SOLVING TACTICS

TACTIC 1: *The Language of Thermodynamics*
A rich, but sometimes misleading, language is used in scientific and engineering studies of thermodynamics. And regardless of your major, you will need to understand what the language means. You may see statements that say heat is absorbed, extracted, rejected, discharged, discarded, withdrawn, delivered, gained, lost, or expelled, or that it flows from one body to another (as if it were a liquid). You may also see statements that describe a body as *having* heat (as if heat can be held or possessed), or that its heat is increased or decreased. You should always keep in mind what is meant by the term *heat*:

Heat is energy that is transferred from one body to another body owing to a difference in the temperatures of the bodies.

When we identify one of the bodies as being our system of interest, any such transfer of energy into the system is positive heat Q, and any such transfer out of the system is negative Q.

The term *work* also requires close attention. You may see statements that say work is produced or generated, or combined with heat or changed from heat. Here is what is meant by the term *work*:

Work is energy that is transferred from one body to another body owing to a force that acts between them.

When we identify one of the bodies as being our system of interest, any such transfer of energy out of the system is either positive work W done *by* the system or negative work W done *on* the system. And any such transfer of energy into the system is negative work done *by* the system or positive work done *on* the system. (The preposition that is used is important.) Obviously, this can be confusing: whenever you see the term *work*, you should read carefully to determine the intent.

21-5 ENTROPY IN THE REAL WORLD: REFRIGERATORS

A **refrigerator** is a device that uses work to transfer thermal energy from a low-temperature reservoir to a high-temperature reservoir as it continuously repeats a set sequence of processes. In a household refrigerator, for example, work is done by an electrical compressor to transfer thermal energy from the food storage compartment (a low-temperature reservoir) to the room (a high-temperature reservoir). Figure 21-12 represents the basic elements of a refrigerator. If we assume that the processes involved in the refrigerator's operation are reversible, then we have an *ideal refrigerator*. Such a hypothetical refrigerator is just a (hypothetical) ideal engine operating in reverse.

Air conditioners and heat pumps are, in essence, refrigerators. The differences are only in the nature of the high- and low-temperature reservoirs. For an air conditioner, the low-temperature reservoir is the room that is to be cooled, and the high-temperature reservoir is the outdoors. A heat pump is an air conditioner that can be operated in reverse to heat a room; then the room is the high-temperature reservoir and heat is transferred to it from outdoors.

The designer of a refrigerator would like to extract as much heat $|Q_C|$ as possible from the low-temperature res-

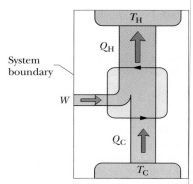

FIGURE 21-12 The elements of a refrigerator. The two black arrowheads on the central loop suggest the working substance operating in a cycle, as if on a p-V plot. Heat Q_C is transferred to the working substance from the low-temperature reservoir. Heat Q_H is transferred to the high-temperature reservoir from the working substance. Work W is done on the refrigerator (on the working substance) by something in the environment.

ervoir (what we want) for the least amount of work W (what we pay for). As a measure of the efficiency of a refrigerator, then, a reasonable definition is

$$K = \frac{\text{what we want}}{\text{what we pay for}} = \frac{|Q_C|}{|W|} \quad \text{(coefficient of performance),} \quad (21\text{-}11)$$

where K is called the *coefficient of performance*. If the refrigerator is ideal, the first law of thermodynamics gives $|W| = |Q_H| - |Q_C|$, where $|Q_H|$ is the magnitude of the heat transferred to the high-temperature reservoir. Equation 21-11 then becomes

$$K = \frac{|Q_C|}{|Q_H| - |Q_C|}. \quad (21\text{-}12)$$

Because an ideal refrigerator is an ideal engine operating in reverse, we can combine Eq. 21-8 with Eq. 21-12; after some algebra we find

$$K = \frac{T_C}{T_H - T_C} \quad \text{(coefficient of performance, ideal refrigerator).} \quad (21\text{-}13)$$

For typical room air conditioners, $K \approx 2.5$. For household refrigerators, $K \approx 5$. Perversely, the value of K is higher the closer the temperatures of the two reservoirs are to each other. That is why heat pumps are more effective in temperate climates than in climates where the outside temperature varies widely.

It would be nice to own a refrigerator that did not require some input of work, that is, one that would run without being plugged in. Figure 21-13 represents another "inventor's dream," a *perfect refrigerator* that transfers heat Q from a cold reservoir to a warm reservoir without

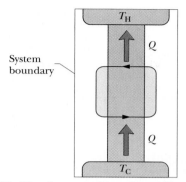

FIGURE 21-13   The elements of a perfect refrigerator, that is, one that transfers heat from a low-temperature reservoir to a high-temperature reservoir without any input of work.

the need for work. Because the unit operates in cycles, the entropy of the working substance does not change during a complete cycle. The entropies of the two reservoirs, however, do change: the entropy change for the cold reservoir is $-|Q|/T_C$, and that for the warm reservoir is $+|Q|/T_H$. So, the net entropy change for the entire system is

$$\Delta S = -\frac{|Q|}{T_C} + \frac{|Q|}{T_H}.$$

Because $T_H > T_C$, the right side of this equation is negative and thus the net change in entropy per cycle for the closed system *refrigerator + reservoirs* is also negative. Because such a decrease in entropy violates the second law of thermodynamics (Eq. 21-4), a perfect refrigerator does not exist. (If you want your refrigerator to operate, you must plug it in.)

This result leads us to another (equivalent) formulation of the second law of thermodynamics:

> No series of processes is possible whose sole result is the transfer of heat from a reservoir at a given temperature to a reservoir at a higher temperature.

In short, *there are no perfect refrigerators.*

CHECKPOINT 4: You wish to increase the coefficient of performance of an ideal refrigerator. You can do so by (a) running the cold chamber at a slightly higher temperature, (b) running it at a slightly lower temperature, (c) moving the unit to a slightly warmer room, or (d) moving it to a slightly cooler room. The temperature changes are to be the same in all four cases. List the changes according to the resulting coefficients of performance, greatest first.

SAMPLE PROBLEM 21-5

An ideal refrigerator, with coefficient of performance $K = 4.7$, extracts heat from the cold chamber at the rate of 250 J/cycle.

(a) How much work per cycle is required to operate the refrigerator?

SOLUTION: From Eq. 21-11, we find

$$|W| = \frac{|Q_C|}{K} = \frac{250 \text{ J}}{4.7} = 53 \text{ J}. \qquad \text{(Answer)}$$

(b) How much heat per cycle is discharged to the room?

SOLUTION: For an ideal refrigerator, with reversible processes and no energy transfers other than $|Q_H|$, $|Q_C|$, and $|W|$, the first law of thermodynamics tells us that

$$\Delta E_{int} = (|Q_H| - |Q_C|) - |W|.$$

Here $\Delta E_{int} = 0$ because the working substance operates in a cycle. Solving for $|Q_H|$ and inserting known data then yield

$$|Q_H| = |W| + |Q_C| = 53 \text{ J} + 250 \text{ J}$$
$$= 303 \text{ J} \approx 300 \text{ J}. \qquad \text{(Answer)}$$

21-6   A STATISTICAL VIEW OF ENTROPY

In Chapter 20 we saw that macroscopic properties of gases can be explained in terms of their microscopic, or molecular, behavior; such explanations are part of the study called *statistical mechanics*. Recall, for example, that we were able to account for the pressure exerted by a gas on the walls of its container in terms of the momentum transferred to those walls by rebounding gas molecules.

Here we shall focus our attention on a single problem involving the distribution of gas molecules between two halves of a container (a box). This problem is reasonably simple to analyze, and it allows us to use statistical mechanics to calculate the entropy change for a free expansion of an ideal gas. You will see in Sample Problem 21-6 that statistical mechanics leads to the same entropy change we obtained in Sample Problem 21-1 using thermodynamics.

Consider the box of Fig. 21-14, which contains four identical (and thus indistinguishable) molecules of a gas. At any instant, a given molecule will be in either the left half of the box or the right half; because the two halves of the box have equal volumes, the molecule has the same likelihood, or probability, of being in the left half as in the right half. Table 21-1, in which L stands for a molecule in the left half of the box and R for a molecule in the right half, shows the 16 possible ways in which the four identi-

TABLE 21-1   **FOUR MOLECULES IN A BOX**

LOCATION OF MOLECULE a b c d				CONFIGURATION LABEL	MULTIPLICITY W (NUMBER OF MICROSTATES)	CONFIGURATION PROBABILITY	CALCULATION OF W BY EQ. 21-14	ENTROPY $(10^{-23}$ J/K) FROM EQ. 21-15
L	L	L	L	I	1	1/16	$4!/(4!\,0!) = 1$	0.00
R	L	L	L					
L	R	L	L	II	4	4/16	$4!/(3!\,1!) = 4$	1.91
L	L	R	L					
L	L	L	R					
L	L	R	R					
L	R	L	R					
R	L	L	R	III	6	6/16	$4!/(2!\,2!) = 6$	2.47
L	R	R	L					
R	L	R	L					
R	R	L	L					
L	R	R	R					
R	L	R	R	IV	4	4/16	$4!/(1!\,3!) = 4$	1.91
R	R	L	R					
R	R	R	L					
R	R	R	R	V	1	1/16	$4!/(0!\,4!) = 1$	0.00

Total number of microstates 16

cal molecules, which we label a, b, c, and d, can be in the two halves of the box. We call each of these 16 arrangements a **microstate** of the system. (Although the four particles in Fig. 21-14 are truly indistinguishable, we assign letters to them temporarily as an aid in calculation. Our final result will not be affected by doing this.)

In Table 21-1, the 16 microstates are grouped into five **configurations,** each made up of *equivalent microstates* and labeled with a Roman numeral. Figure 21-14a shows configuration II, which we may describe as "one molecule in the right half of the box." Table 21-1 displays the four microstates that make up this configuration; they are equivalent because, since the molecules are identical, the situation in which molecule a is in the right half is physically no different from the situation in which molecule b (or c or d) is in the right half. Figure 21-14b shows configuration III, which is "two molecules in each half of the box"; Table 21-1 lists the six microstates that form this configuration. We call the number of microstates associated with a configuration the *multiplicity* W of that configuration. (We have used the same symbol for work; do not confuse these two uses.)

A basic assumption of statistical mechanics is that *all microstates are equally probable.* That is, if we were to take a great many snapshots of the molecules as they jostle around the box in random fashion, and then count the number of times each microstate occurred, we would find that all the microstates occur equally often. So, at any given instant, the probability of finding the molecules in any one microstate is equal to that of finding them in any

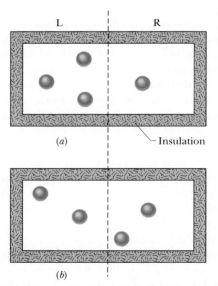

FIGURE 21-14   An insulated box contains four gas molecules. Each molecule has the same probability of being in the left half (L) of the box as in the right half (R). The arrangement in (a) corresponds to configuration II in Table 21-1, and that in (b) corresponds to configuration III.

other microstate. That is, the system will spend, on average, the same amount of time in each of the 16 microstates listed in Table 21-1.

Because the microstates are equally probable, but different configurations have *different numbers of microstates,* the configurations are *not* equally probable. The *probability (of occurrence) of a configuration* is the ratio of its multiplicity to the total number of microstates of the system. In Table 21-1, configuration III, with six microstates and thus a multiplicity of 6, is the most probable configuration, with a probability of 6/16 = 0.375. This means that the system is in configuration III 37.5% of the time. Configurations I and V, in which the gas occupies only half the box, are the least probable, with probabilities of 1/16 = 0.0625 or 6.25% each.

Now recall the free expansion process of Sample Problem 21-1; there we found that the state in which the gas fills its container has greater entropy than the state in which it occupies only half the container. Here we have found that the state (or configuration) in which the gas fills its container uniformly is more probable than a state in which the gas occupies only half the container. The state with higher entropy seems also to have the higher probability of occurrence.

Four molecules in a box (N = 4) are not very many on which to base a conclusion. Let us increase the number N of molecules to 100 and again compare the amount of time during which half the molecules are in each half of the box to that during which all the molecules are in the left half of the box. The ratio is not 6 to 1 (as it is for N = 4 in Table 21-1) but about 10^{29} to 1. Imagine what this ratio must be for the much more practical case of $N = 10^{22}$, which is

about the number of air molecules in a child's inflated balloon. The probability is then overwhelming for a uniform distribution of molecules throughout the container.

For large values of N, there are very large numbers of microstates. But nearly all the microstates correspond to an approximately equal division of the molecules between the two halves of the box, as Fig. 21-15 indicates. Even though the measured temperature and pressure of the gas remain constant, the gas is churning away endlessly as its molecules "visit" all the possible microstates with equal prob-

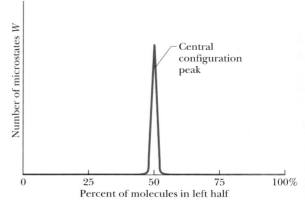

FIGURE 21-15 For a *large* number of molecules in a box, a plot of the number of microstates that require various percentages of the molecules to be in the left half of the box. Nearly all the microstates correspond to an approximately equal sharing of the molecules between the two halves of the box; those microstates form the central *configuration peak* on the plot. For $N \approx 10^{22}$, the central configuration peak would be much too narrow to be drawn on this plot.

Sometimes striking order appears naturally. (*a*) For example, Giant's Causeway in Northern Ireland consists of tall stone columns, many of which are hexagonal in cross section and appear to have been designed. The columns formed when hot magma seeped out of the ground and cooled. (*b*) Order can also be seen in

the *sorted circles* of gravel and stones that occur naturally on an island north of Norway. Just how the columns and circles were produced is still debated, but we can be certain that as the entropy of the magma and stones decreased, the entropy of their environment increased even more.

(*a*)

(*b*)

abilities. However, because so few microstates lie outside the narrow *central configuration peak* of Fig. 21-15, we might as well assume that the gas molecules are always evenly divided between the two halves of the box. And, as we saw, this is the configuration with the greatest entropy.

For the four molecules of Fig. 21-14, we can easily list the microstates associated with each of the five configurations and then find the multiplicities. For the general case of N molecules, where N can be large, we find the multiplicity of a configuration with

$$W = \frac{N!}{n_L! \, n_R!} \qquad \text{(multiplicity of a configuration).} \qquad (21\text{-}14)$$

Here N is the total number of molecules, n_L is the number in the left half of the box, and n_R is the number in the right half. The symbol "!" indicates a *factorial*, which is defined for any nonnegative integer n as

$$n! = n(n-1)(n-2) \cdots (1).$$

For example, $5! = 5 \times 4 \times 3 \times 2 \times 1 = 120$. By definition, $0! = 1$. Table 21-1 includes the calculation of W for each of its configurations, using Eq. 21-14.

We have seen that configurations with higher probability of occurrence also have higher entropy. Because the multiplicity of a configuration is a measure of its probability, we may say that states with greater multiplicity have greater entropy. Austrian physicist Ludwig Boltzmann first pointed out the following relationship between these two quantities in 1877:

$$S = k \ln W \qquad \text{(Boltzmann's entropy equation).} \qquad (21\text{-}15)$$

Here S is the entropy of a configuration, W is the multiplicity of the configuration, and k ($= 1.38 \times 10^{-23}$ J/K) is the Boltzmann constant, first encountered in Section 20-5. This famous formula is engraved on Boltzmann's tombstone.

The rightmost column of Table 21-1 displays the entropies of the configurations of the four-molecule system of Fig. 21-14, as computed with Eq. 21-15. Configuration III, which has the greatest multiplicity, also has the greatest entropy.

When you use Eq. 21-14 to calculate W, your calculator may signal an error if you try to find the factorial of a number much greater than a few hundred. Fortunately, there is a very good approximation, known as **Stirling's approximation,** not for $N!$ but for $\ln N!$—exactly what is needed in Eq. 21-15. Stirling's approximation is

$$\ln N! \approx N(\ln N) - N \qquad \text{(Stirling's approximation).} \qquad (21\text{-}16)$$

(The Stirling of this approximation is not the Stirling of the Stirling engine.)

CHECKPOINT **5:** A box contains one mole of a gas. Consider two configurations: (a) each half of the box contains half the molecules, and (b) each third of the box contains one-third of the molecules. Which configuration has more microstates?

SAMPLE PROBLEM 21-6

In Sample Problem 21-1 we showed that when n moles of an ideal gas doubles its volume in a free expansion, the entropy increase from the initial state i to the final state f is $S_f - S_i = nR \ln 2$. Derive this result with statistical mechanics.

SOLUTION: Let N be the number of molecules in n moles of the gas. From Eq. 21-14, the multiplicity of the initial state, in which all N molecules are in the left half of the container in Fig. 21-1, is

$$W_i = \frac{N!}{N! \, 0!} = 1.$$

The multiplicity of the final state, in which $N/2$ molecules are in each half of the container, is

$$W_f = \frac{N!}{(N/2)! \, (N/2)!}.$$

From Eq. 21-15, the initial and final entropies are

$$S_i = k \ln W_i = k \ln 1 = 0$$

and $\qquad S_f = k \ln W_f = k \ln(N!) - 2k \ln[(N/2)!].\qquad (21\text{-}17)$

In writing Eq. 21-17, we have used the relation

$$\ln \frac{a}{b^2} = \ln a - 2 \ln b.$$

Now, applying Eq. 21-16 to evaluate Eq. 21-17, we find that

$$\begin{aligned} S_f &= k \ln(N!) - 2k \ln[(N/2)!] \\ &= k[N(\ln N) - N] - 2k[(N/2) \ln(N/2) - (N/2)] \\ &= k[N(\ln N) - N - N \ln(N/2) + N] \\ &= k[N(\ln N) - N(\ln N - \ln 2)] = kN \ln 2. \quad (21\text{-}18) \end{aligned}$$

We can substitute for kN by using the relations

$$k = \frac{R}{N_A} \quad \text{and} \quad N = nN_A$$

from Eqs. 20-21 and 20-2, respectively. Here N_A is the Avogadro constant and R is the universal gas constant. With these two relations we see that $kN = nR$. So, Eq. 21-18 becomes

$$S_f = nR \ln 2.$$

The change in entropy from the initial state to the final is thus

$$S_f - S_i = nR \ln 2 - 0 = nR \ln 2, \qquad \text{(Answer)}$$

which is what we set out to show. In Sample Problem 21-1 we calculated this entropy increase for a free expansion with thermodynamics by finding an equivalent reversible process and calculating the entropy change for *that* process in terms of

temperature and heat transfer. Here we have calculated the same increase with statistical mechanics using the fact that the system consists of molecules.

REVIEW & SUMMARY

One-Way Processes

An **irreversible process** is one that cannot be reversed by means of small changes in the environment. The direction in which an irreversible process proceeds is set by the **change in entropy** ΔS of the system undergoing the process. Entropy S is a *state property* (or *state function*) of the system; that is, it depends only on the state of the system and not on how the system reached that state. The *entropy postulate* states (in part): *If an irreversible process occurs in a closed system, the entropy of the system always increases.*

Calculating Entropy Change

The **entropy change** ΔS for an irreversible process that takes a system from an initial state i to a final state f is exactly equal to the entropy change ΔS for *any reversible process* that takes the system between those same two states. We can compute the latter (but not the former) with

$$\Delta S = S_f - S_i = \int_i^f \frac{dQ}{T}. \tag{21-1}$$

Here Q is the energy transferred as heat to or from the system during the process, and T is the temperature of the system in kelvins.

For a reversible isothermal process, Eq. 21-1 reduces to

$$\Delta S = S_f - S_i = \frac{Q}{T}. \tag{21-2}$$

When the temperature change ΔT of a system is small relative to the temperature (in kelvins) before and after the process, the entropy change can be approximated as

$$\Delta S = S_f - S_i \approx \frac{Q}{\bar{T}}, \tag{21-3}$$

where $\bar{T}$ is the system's average temperature during the process.

The Second Law of Thermodynamics

This law, which is an extension of the entropy postulate, states: *If a process occurs in a closed system, the entropy of the system increases for irreversible processes and remains constant for reversible processes. It never decreases.* In equation form,

$$\Delta S \geq 0. \tag{21-4}$$

Engines

An **engine** is a device that, operating in a cycle, extracts heat $|Q_H|$ from a high-temperature reservoir, does a certain amount of work $|W|$, and discharges heat $|Q_C|$ to a low-temperature reservoir. The

efficiency ε of an engine is defined as

$$\varepsilon = \frac{\text{energy we get}}{\text{energy we pay for}} = \frac{|W|}{|Q_H|}. \tag{21-5}$$

An *ideal engine* is one in which all processes are reversible and the only net energy transfers in a cycle are $|Q_H|$, $|W|$, and $|Q_C|$. Then the first law of thermodynamics becomes

$$\Delta E_{int} = (|Q_H| - |Q_C|) - |W| = 0. \tag{21-6}$$

For an ideal engine, Eq. 21-5 can then be rewritten as

$$\varepsilon = 1 - \frac{|Q_C|}{|Q_H|} = 1 - \frac{T_C}{T_H}, \tag{21-7, 21-9}$$

in which T_H and T_C are the temperatures of the high- and low-temperature reservoirs, respectively. Real engines always have an efficiency lower than that given by Eqs. 21-7 and 21-9.

A *perfect engine* is an imaginary engine in which heat extracted from the high-temperature reservoir is converted completely to work. Because such a conversion would decrease the entropy of the system with each cycle, a perfect engine would violate the second law of thermodynamics. That law can then be restated as follows: No series of processes is possible whose sole result is the absorption of heat from a thermal reservoir and the complete conversion of this energy to work.

Refrigerators

A **refrigerator** (which may also be an air conditioner or a heat pump) is a device that, operating in a cycle, has work $|W|$ done on it, extracts heat $|Q_C|$ from a low-temperature reservoir, and discharges heat $|Q_H|$ to a high-temperature reservoir. The *coefficient of performance K* of a refrigerator is defined as

$$K = \frac{\text{what we want}}{\text{what we pay for}} = \frac{|Q_C|}{|W|}. \tag{21-11}$$

An *ideal refrigerator* is one in which all processes are reversible and the only net energy transfers in a cycle are $|Q_H|$, $|W|$, and $|Q_C|$. For an ideal refrigerator, Eq. 21-11 becomes

$$K = \frac{|Q_C|}{|Q_H| - |Q_C|} = \frac{T_C}{T_H - T_C}, \tag{21-12, 21-13}$$

in which T_H and T_C are the temperatures of the high- and low-temperature reservoirs, respectively.

A *perfect refrigerator* is an imaginary refrigerator in which heat extracted from the low-temperature reservoir is converted completely to heat discharged to the high-temperature reservoir, without any need for work. Because this conversion would decrease the entropy of the system with each cycle, a perfect refrigerator would violate the second law of thermodynamics. That law can then be restated as follows: No series of processes is possible whose sole result is the transfer of heat from a reservoir at a given temperature to a reservoir at a higher temperature.

Entropy from a Statistical View

The entropy of a system can be defined in terms of the possible distributions of its molecules. For identical molecules, each possible distribution of molecules is called a **microstate** of the system. All equivalent microstates are grouped into a **configuration** of the system. The number of microstates in a configuration is the **multiplicity** W of the configuration.

For a system of N molecules that may be distributed between the two halves of a box, the multiplicity is given by

$$W = \frac{N!}{n_L! \, n_R!}, \qquad (21\text{-}14)$$

in which n_L is the number of molecules in the left half of the box and n_R is the number in the right half. A basic assumption of *statistical mechanics* is that all the microstates are equally probable. Thus, configurations with a large multiplicity occur most often. When N is very large (say, $N = 10^{22}$ molecules or more), the molecules are nearly always in the configuration in which $n_L = n_R$.

The multiplicity W of a configuration of a system and the entropy S of the system in that configuration are related by Boltzmann's entropy equation:

$$S = k \ln W, \qquad (21\text{-}15)$$

where $k = 1.38 \times 10^{-23}$ J/K is the Boltzmann constant.

When N is very large (the usual case), we can approximate $\ln N!$ with *Stirling's approximation*:

$$\ln N! \approx N(\ln N) - N. \qquad (21\text{-}16)$$

QUESTIONS

1. The net heat transfer between the Sun and Earth is from Sun to Earth. Does the entropy of the Earth–Sun system increase, decrease, or remain the same during this process?

2. A gas, confined to an insulated cylinder, is compressed adiabatically to half its volume. Does the entropy of the gas increase, decrease, or remain unchanged during this process?

3. A drop of water boils away *slowly* on a hot plate whose temperature is slightly above the boiling point. (a) Does the entropy of the water increase, decrease, or remain the same during this process? (b) Does the entropy of the system *water + hot plate* increase, decrease, or remain the same?

4. Point i in Fig. 21-16 represents the initial state of an ideal gas at temperature T. Taking algebraic signs into account, rank the entropy changes as the gas moves, successively and reversibly, from point i to points a, b, c, and d, greatest first.

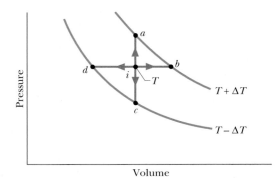

FIGURE 21-16   Question 4.

5. You allow a gas to expand freely from volume V to volume $2V$. Later you allow that gas to expand freely from volume $2V$ to volume $3V$. Is the net entropy change for these two expansions

greater than, less than, or equal to the entropy change that would occur if you allowed the gas to expand freely from volume V directly to volume $3V$?

6. Three ideal engines operate between temperature limits of (a) 400 and 500 K; (b) 500 and 600 K; and (c) 400 and 600 K. Each engine extracts the same amount of energy per cycle from the high-temperature reservoir. Rank the magnitudes of the work done by the engines per cycle, greatest first.

7. You wish to increase the efficiency of an ideal engine. Is it better to raise the temperature of the high-temperature reservoir by a certain amount or lower the temperature of the low-temperature reservoir by the same amount?

8. The work done by an ideal engine is used to lift a heavy safe, as shown in Fig. 21-17. Does the entropy of the *engine + Earth + safe* system increase, decrease, or remain the same during this process?

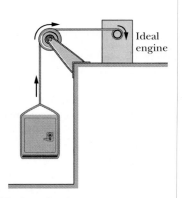

FIGURE 21-17   Question 8.

9. Does the entropy per cycle increase, decrease, or remain the same for (a) an ideal engine, (b) a real engine, and (c) a perfect engine (which is, of course, impossible to build)?

10. If you leave the door of your kitchen refrigerator open for several hours, does the temperature of the kitchen increase, decrease, or remain the same? Assume that the kitchen is closed and well insulated.

11. Does the entropy per cycle increase, decrease, or remain the same for (a) an ideal refrigerator, (b) a real refrigerator, and (c) a perfect refrigerator (which is, of course, impossible to build)?

12. In an ideal refrigerator, is the magnitude $|Q_H|$ of the heat transferred per cycle to the high-temperature reservoir always less than, always greater than, or always equal to the magnitude of the work W done per cycle?

13. A box contains 100 atoms in a configuration, with 50 atoms in each half of the box. Suppose that you could count the different microstates associated with this configuration at the rate of 100 billion states per second, using a supercomputer. Without written calculation, guess how much computing time you would need: a day, a year, or much more than a year.

14. A basic assumption of statistical mechanics is that all microstates are equally probable. How would the air in your bedroom behave if all configurations were equally probable (they aren't)?

<h1 style="text-align:center">EXERCISES & PROBLEMS</h1>

SECTION 21-2 Change in Entropy

1E. An ideal gas undergoes a reversible isothermal expansion at 132°C. The entropy of the gas increases by 46.0 J/K. How much heat was absorbed?

2E. A 2.50 mol sample of an ideal gas expands reversibly and isothermally at 360 K until its volume is doubled. What is the increase in entropy of the gas?

3E. An ideal gas undergoes a reversible isothermal expansion at 77.0°C, increasing its volume from 1.30 L to 3.40 L. The entropy change of the gas is 22.0 J/K. How many moles of gas are present?

4E. Construct plots of T versus S, S versus E_{int}, and S versus V for the isothermal expansion process of Fig. 21-4.

5E. An ideal gas in contact with a constant-temperature reservoir undergoes a reversible isothermal expansion to twice its initial volume. Show that the reservoir's change in entropy is independent of its temperature.

6E. Four moles of an ideal gas are expanded from volume V_1 to volume $V_2 = 2V_1$. If the expansion is isothermal at temperature $T = 400$ K, find (a) the work done by the expanding gas and (b) the change in its entropy. (c) If the expansion is reversibly adiabatic instead of isothermal, what is the entropy change of the gas?

7E. Find (a) the heat absorbed and (b) the change in entropy of a 2.00 kg block of copper whose temperature is increased reversibly from 25°C to 100°C. The specific heat of copper is 386 J/kg·K.

8E. An ideal monatomic gas at initial temperature T_0 (in kelvins) expands from initial volume V_0 to volume $2V_0$ by each of the five processes indicated in the T-V diagram of Fig. 21-18. In which process is the expansion (a) isothermal, (b) isobaric (constant pressure), and (c) adiabatic? Explain your answers. (d) In which processes does the entropy of the gas decrease?

9E. The temperature of 1.0 mol of a monatomic ideal gas is raised reversibly from 300 K to 400 K, with its volume kept constant. What is the entropy change of the gas? (*Hint:* Recall that $E_{int} = \frac{3}{2}nRT$ for such a gas.)

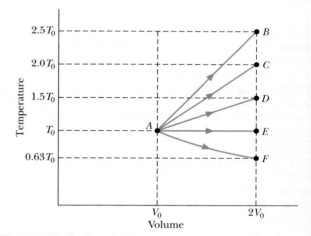

FIGURE 21-18 Exercise 8.

10E. Repeat Exercise 9, with the pressure now kept constant.

11E. (a) What is the entropy change of a 12.0 g ice cube that melts completely in a bucket of water whose temperature is just above the freezing point of water? (b) What is the entropy change of a 5.00 g spoonful of water that evaporates completely on a hot plate whose temperature is slightly above the boiling point of water?

12E. Suppose that 260 J of heat is conducted from a constant-temperature reservoir at 400 K to another reservoir at (a) 100 K, (b) 200 K, (c) 300 K, and (d) 360 K. What are the net changes in entropy of the reservoirs for those four cases? (e) What is the trend in those changes?

13E. A brass rod is in thermal contact with a constant-temperature reservoir at 130°C at one end and another reservoir at 24.0°C at the other end. (a) Compute the total change in entropy of the rod–reservoirs system that occurs when 5030 J of heat is conducted through the rod, from one reservoir to the other. (b) Does the entropy of the rod change in the process?

14E. At very low temperatures, the molar specific heat C_V of many solids is approximately $C_V = AT^3$, where A depends on the particular substance. For aluminum, $A = 3.15 \times 10^{-5}$ J/mol·K⁴. Find the entropy change for 4.00 mol of aluminum when its temperature is raised from 5.00 K to 10.0 K.

15P. A 2.0 mol sample of an ideal monatomic gas undergoes the reversible process shown in Fig. 21-19. (a) How much heat is absorbed by the gas? (b) What is the change in the internal energy of the gas? (c) How much work is done by the gas?

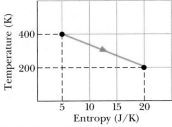

FIGURE 21-19   Problem 15.

16P. Heat can actually be removed from water at and then below the normal freezing point (0.0°C at atmospheric pressure) without causing the water to freeze; the water is said to be *supercooled*. Suppose a 1.00 g water drop is supercooled until its temperature is that of the surrounding air, which is at −5.00°C. The drop then suddenly and irreversibly freezes, transferring heat to the air until the drop is again at −5.00°C. What is the entropy change for the drop? (*Hint:* Use a three-step reversible process to take the water through the normal freezing point.) The specific heat of ice is 2220 J/kg·K.

17P. In an experiment, 200 g of aluminum (with a specific heat of 900 J/kg·K) at 100°C is mixed with 50.0 g of water at 20.0°C. (a) What is the equilibrium temperature? What are the entropy changes of (b) the aluminum, (c) the water, and (d) the aluminum−water system?

18P. An object of (constant) heat capacity C is heated from an initial temperature T_i to a final temperature T_f by a constant-temperature reservoir at T_f. (a) Represent the process on a graph of C/T versus T, and show graphically that the total change in entropy ΔS of the object−reservoir system is positive. (b) Explain how the use of reservoirs at intermediate temperatures would allow the process to be carried out in a way that makes ΔS as small as desired.

19P. In the irreversible process of Fig. 21-5, let the initial temperatures of identical blocks L and R be 305.5 and 294.5 K, respectively, and let 215 J be the heat transfer between the blocks required to reach equilibrium. Then for the reversible processes of Fig. 21-6, what are the entropy changes of (a) block L, (b) its reservoir, (c) block R, (d) its reservoir, (e) the two-block system, and (f) the system of the two blocks and the two reservoirs?

20P. Use the reversible apparatus of Fig. 21-6 to show that, if the process of Fig. 21-5 happened in reverse, the entropy of the system would decrease, a violation of the second law of thermodynamics.

21P. The entropy change ΔS between two states of a system, which is $\int (dQ/T)$, depends only on the properties of the initial and final states. That is, this integral is path independent, as is

shown for an ideal gas in Section 21-2. Show similarly that the integrals $\int dQ$, $\int (T\, dQ)$, and $\int (dQ/T^2)$ are not path independent and thus do not define new state properties for an ideal gas.

22P. An ideal diatomic gas, whose molecules are rotating but not oscillating, is taken through the cycle in Fig. 21-20. Determine for all three processes, in terms of p_1, V_1, T_1, and R: (a) p_2, p_3, and T_3 and (b) W, Q, ΔE_{int}, and ΔS per mole.

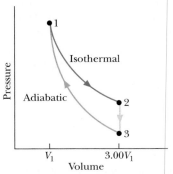

FIGURE 21-20   Problem 22.

23P. A mole of a monatomic ideal gas is taken from an initial pressure p and volume V to a final pressure $2p$ and volume $2V$ by two different processes: (I) It expands isothermally until its volume is doubled, and then its pressure is increased at constant volume to the final pressure. (II) It is compressed isothermally until its pressure is doubled, and then its volume is increased at constant pressure to the final volume. Show the path of each process on a p-V diagram. For each process calculate, in terms of p and V, (a) the heat absorbed by the gas in each part of the process, (b) the work done by the gas in each part of the process, (c) the change in internal energy of the gas, $E_{int,f} - E_{int,i}$, and (d) the change in entropy of the gas, $S_f - S_i$.

24P. A 50.0 g block of copper whose temperature is 400 K is placed in an insulating box with a 100 g block of lead whose temperature is 200 K. (a) What is the equilibrium temperature of the two-block system? (b) What is the change in the internal energy of the two-block system between the initial state and the equilibrium state? (c) What is the change in the entropy of the two-block system? See Table 19-3.

25P. A 10 g ice cube at −10°C is placed in a lake whose temperature is 15°C. Calculate the change in entropy of the cube−lake system as the ice cube comes to thermal equilibrium with the lake. The specific heat of ice is 2220 J/kg·K. (*Hint:* Will the ice cube affect the temperature of the lake?)

26P. An 8.0 g ice cube at −10°C is put into a Thermos flask containing 100 cm³ of water at 20°C. What is the change in entropy of the cube−water system when a final equilibrium state is reached? The specific heat of ice is 2220 J/kg·K.

27P. A mixture of 1773 g of water and 227 g of ice is in an initial equilibrium state at 0.00°C. The mixture is then, in a reversible process, brought to a second equilibrium state where the water−ice ratio, by mass, is 1:1 at 0.00°C. (a) Calculate the entropy change of the system during this process. (The heat of fusion for water is 333 kJ/kg.) (b) The system is then returned to the initial equilibrium state in an irreversible process (say, by using a Bunsen burner). Calculate the entropy change of the system during this process. (c) Are your answers consistent with the second law of thermodynamics?

28P. A cylinder contains n moles of a monatomic ideal gas. If the gas undergoes a reversible isothermal expansion from initial volume V_i to final volume V_f along path I in Fig. 21-21, its change in entropy is $\Delta S = nR \ln (V_f/V_i)$. (See Sample Problem 21-1.) Now consider path II in Fig. 21-21, which takes the gas from the same initial state i to state x by a reversible adiabatic expansion, and then from state x to the same final state f by a reversible constant-volume process. (a) Describe how you would carry out the two reversible processes for path II. (b) Show that the temperature of the gas in state x is

$$T_x = T_i(V_i/V_f)^{2/3}.$$

(c) What are the heat Q_I transferred along path I and the heat Q_{II} transferred along path II? Are they equal? (d) What is the entropy change ΔS for path II? Is the entropy change for path I equal to it? (e) Evaluate T_x, Q_I, Q_{II}, and ΔS for $n = 1$, $T_i = 500$ K, and $V_f/V_i = 2$.

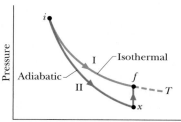

FIGURE 21-21 Problem 28.

29P. One mole of an ideal monatomic gas is taken through the cycle in Fig. 21-22. (a) How much work is done by the gas in going from state a to state c along path abc? What are the changes in internal energy and entropy in going (b) from b to c and (c) through one complete cycle? Express all answers in terms of the pressure p_0, volume V_0, and temperature T_0 of state a.

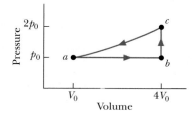

FIGURE 21-22
Problem 29.

30P. One mole of an ideal monatomic gas, at an initial pressure of 5.00 kPa and initial temperature of 600 K, expands from initial volume $V_i = 1.00$ m³ to final volume $V_f = 2.00$ m³. During the expansion, the pressure p and volume V of the gas are related by $p = 5.00 \exp[(V_i - V)/a]$, where p is in kilopascals, V_i and V are in cubic meters, and $a = 1.00$ m³. What are (a) the final pressure and (b) the final temperature of the gas? (c) How much work is done by the gas during the expansion? (d) What is the change in entropy of the gas during the expansion? (*Hint:* Use two simple reversible processes to find the entropy change.)

SECTION 21-4 Entropy in the Real World: Engines

31E. An athlete dives from a platform and comes to rest in the water; this process is irreversible, and the diver–water–Earth system is closed. Devise a reversible way of transforming the system from its initial state (diver on platform) to its final state (diver in water, with the temperature of the diver and water slightly increased due to frictional heating). You can use an ideal engine (as in Fig. 21-17) and a reservoir whose temperature can be controlled.

32E. A moving block is slowed to a stop by friction. Both the block and the surface along which it slides are in an insulated enclosure. (a) Devise a reversible process to change the system from its initial state (block moving) to its final state (block stationary; temperature of the block and surface slightly increased). Show that this reversible process results in an entropy increase for the closed block–surface system. For the process, you can use an ideal engine and a constant-temperature reservoir. (b) Show, using the same process, but in reverse, that if the temperature of the system were to decrease spontaneously and the block were to start moving again, the entropy of the system would decrease (a violation of the second law of thermodynamics).

33E. An ideal heat engine absorbs 52 kJ of heat and exhausts 36 kJ of heat in each cycle. Calculate (a) the efficiency and (b) the work done per cycle in kilojoules.

34E. Calculate the efficiency of a fossil-fuel power plant that consumes 380 metric tons of coal each hour to produce useful work at the rate of 750 MW. The heat of combustion of (the heat released by) 1.0 kg of coal is 28 MJ.

35E. An inventor claims to have invented four engines, each of which operates between constant-temperature reservoirs at 400 and 300 K. Data on each engine, per cycle of operation, are: engine A, $Q_H = 200$ J, $Q_C = -175$ J, and $W = 40$ J; engine B, $Q_H = 500$ J, $Q_C = -200$ J, and $W = 400$ J; engine C, $Q_H = 600$ J, $Q_C = -200$ J, and $W = 400$ J; engine D, $Q_H = 100$ J, $Q_C = -90$ J, and $W = 10$ J. Of the first and second laws of thermodynamics, which (if either) does each engine violate?

36E. A car engine delivers 8.2 kJ of work per cycle. (a) Before a tune-up, the efficiency is 25%. Calculate, per cycle, the heat absorbed from the combustion of fuel and the energy lost by the engine. (b) After a tune-up, the efficiency is 31%. What are the new values of the quantities calculated in (a) when 8.2 kJ of work is delivered per cycle?

37E. An ideal engine whose low-temperature reservoir is at 17°C has an efficiency of 40%. By how much should the temperature of the high-temperature reservoir be increased to increase the efficiency to 50%?

38E. An ideal heat engine operates in a Stirling cycle between 235°C and 115°C. It absorbs 6.30×10^4 J per cycle at the higher temperature. (a) What is the efficiency of the engine? (b) How much work per cycle is this engine capable of performing?

39E. In a hypothetical nuclear fusion reactor, the fuel is deuterium gas at a temperature of about 7×10^8 K. If this gas could be used to operate an ideal heat engine with $T_C = 100$°C, what would be the engine's efficiency?

40E. An ideal Stirling engine has an efficiency of 22.0%. It operates between constant-temperature reservoirs differing in temperature by 75.0 C°. What are the temperatures of the two reservoirs?

41E. Show that the heat extracted from the gas during stroke *bc* of the ideal Stirling engine (Fig. 21-8) is equal in magnitude to the heat added to the gas during stroke *da*. What is the magnitude of this heat transfer for the data given in Sample Problem 21-3? Assume that the working substance is a monatomic ideal gas.

42P. One mole of a monatomic ideal gas is taken through the reversible cycle shown in Fig. 21-23. Process *bc* is an adiabatic expansion, with $p_b = 10.0$ atm and $V_b = 1.00 \times 10^{-3}$ m³. Find (a) the heat added to the gas, (b) the heat leaving the gas, (c) the net work done by the gas, and (d) the efficiency of the cycle.

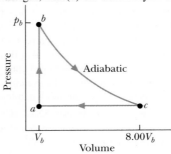

FIGURE 21-23
Problem 42.

43P. One mole of a monatomic ideal gas initially at a volume of 10 L and a temperature of 300 K is heated at constant volume to a temperature of 600 K, allowed to expand isothermally to its initial pressure, and finally compressed at constant pressure to its original volume, pressure, and temperature. (a) During the cycle, how much heat enters the system (the gas)? (b) What is the net work done by the gas? (c) What is the efficiency of the cycle?

44P. An ideal engine has a power of 500 W. It operates between constant-temperature reservoirs at 100°C and 60.0°C. What are (a) the rate of heat input and (b) the rate of exhaust heat output, in kilojoules per second?

45P. One mole of an ideal monatomic gas is taken through the cycle shown in Fig. 21-24. Assume that $p = 2p_0$, $V = 2V_0$, $p_0 = 1.01 \times 10^5$ Pa, and $V_0 = 0.0225$ m³. Calculate (a) the work done during the cycle, (b) the heat added during stroke *abc*, and (c) the efficiency of the cycle. (d) What is the efficiency of an ideal engine operating between the highest and lowest temperatures that occur in the cycle? How does this compare to the efficiency calculated in (c)?

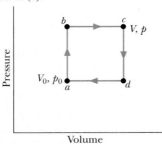

FIGURE 21-24   Problem 45.

46P. A combination mercury–steam turbine takes saturated mercury vapor from a boiler at 876°F and exhausts it to heat a steam boiler at 460°F. The steam turbine receives steam at this temperature and exhausts it to a condenser that is at 100°F. What is the maximum theoretical efficiency of the combination?

47P. In the first stage of a two-stage ideal heat engine, heat Q_1 is absorbed at temperature T_1, work W_1 is done, and heat Q_2 is expelled at a lower temperature T_2. The second stage absorbs that heat Q_2, does work W_2, and expels heat Q_3 at a still lower temperature T_3. Prove that the efficiency of the two-stage engine is $(T_1 - T_3)/T_1$.

48P. One mole of an ideal gas is used as the working substance of an engine that operates on the cycle shown in Fig. 21-25. *BC* and *DA* are reversible adiabatic processes. (a) Is the gas monatomic, diatomic, or polyatomic? (b) What is the efficiency of the engine?

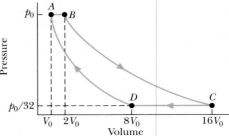

FIGURE 21-25
Problem 48.

49P. Suppose that a deep shaft were drilled in Earth's crust near one of the poles, where the surface temperature is −40°C, to a depth where the temperature is 800°C. (a) What is the theoretical limit to the efficiency of an engine operating between these temperatures? (b) If all the heat released into the low-temperature reservoir were used to melt ice that was initially at −40°C, at what rate could liquid water at 0°C be produced by a 100 MW power plant? The specific heat of ice is 2220 J/kg·K; water's heat of fusion is 333 kJ/kg. (Note that the engine can operate only between 0°C and 800°C in this case. Energy exhausted at −40°C cannot be used to raise the temperature of anything above −40°C.)

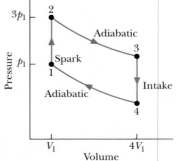

FIGURE 21-26
Problem 50.

50P. The operation of a gasoline internal combustion engine is represented by the cycle in Fig. 21-26. Assume an ideal gas and use a compression ratio of 4:1 ($V_4 = 4V_1$). Assume that $p_2 = 3p_1$. (a) Determine the pressure and temperature at each of the vertex points of the *p-V* diagram in terms of p_1, T_1, and the ratio γ of the molar specific heats of the gas. (b) What is the efficiency of the cycle?

SECTION 21-5 Entropy in the Real World: Refrigerators

51E. An inventor has built an engine (engine X) and claims that its efficiency ε_X is greater than the efficiency ε of an ideal engine operating between the same two temperatures. Suppose that you couple engine X to an ideal refrigerator (Fig. 21-27a) and adjust the cycle of engine X so that the work per cycle that it provides equals the work per cycle required by the ideal refrigerator. Treat this combination as a single unit and show that if the inventor's claim were true (if $\varepsilon_X > \varepsilon$), the combined unit would act as a perfect refrigerator (Fig. 21-27b), transferring heat from the low-temperature reservoir to the high-temperature reservoir without the need of work.

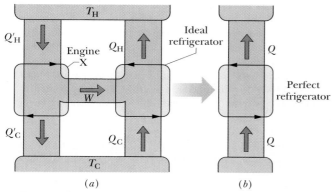

FIGURE 21-27 Exercise 51.

52E. An apparatus that liquefies helium is in a room maintained at 300 K. If the helium in the apparatus is at 4.0 K, what is the minimum ratio of heat delivered to the room to heat removed from the helium?

53E. An ideal engine has efficiency ε. Show that if you run it backward as an ideal refrigerator, the coefficient of performance will be $K = (1 - \varepsilon)/\varepsilon$.

54E. An ideal refrigerator does 150 J of work to remove 560 J of heat from its cold compartment. (a) What is the refrigerator's coefficient of performance? (b) How much heat per cycle is exhausted to the kitchen?

55E. To make ice, an ideal freezer extracts 42 kJ of heat at $-12°C$ during each cycle. The freezer has a coefficient of performance of 5.7. The room temperature is 26°C. (a) How much heat per cycle is delivered to the room? (b) How much work per cycle is required to run the freezer?

56E. (a) An ideal engine operates between a hot reservoir at 320 K and a cold reservoir at 260 K. If it absorbs 500 J of heat per cycle at the hot reservoir, how much work per cycle does it deliver? (b) If the same engine, working in reverse, functions as a refrigerator between the same two reservoirs, how much work per cycle must be supplied to remove 1000 J of heat from the cold reservoir?

57E. An ideal air conditioner takes heat from a room at 70°F and transfers it to the outdoors, which is at 96°F. For each joule of

electrical energy required to operate the air conditioner, how many joules of heat are removed from the room?

58E. Heat from the outdoors at $-5.0°C$ is transferred to a room at 17°C by the electric motor of a heat pump. If the heat pump is ideal, how many joules of heat will be delivered to the room for each joule of electric energy consumed?

59E. How much work must be done by an ideal refrigerator to transfer 1.0 J of heat (a) from a reservoir at 7.0°C to one at 27°C; (b) from a reservoir at $-73°C$ to one at 27°C; (c) from a reservoir at $-173°C$ to one at 27°C; and (d) from a reservoir at $-223°C$ to one at 27°C?

60P. An ideal heat pump is used to heat a building. The outside temperature is $-5.0°C$ and the temperature inside the building is to be maintained at 22°C. The coefficient of performance is 3.8, and the heat pump delivers 7.54 MJ of heat to the building each hour. At what rate must work be done to run the heat pump?

61P. The motor in a refrigerator has a power of 200 W. If the freezing compartment is at 270 K and the outside air is at 300 K, assuming ideal efficiency, what is the maximum amount of heat that can be extracted from the freezing compartment in 10.0 min?

62P. An air conditioner operating between 93°F and 70°F is rated at 4000 Btu/h cooling capacity. Its coefficient of performance is 27% of that of an ideal refrigerator operating between the same two temperatures. What horsepower is required of the air conditioner motor?

63P. An ideal engine works between temperatures T_1 and T_2. It drives an ideal refrigerator that works between temperatures T_3 and T_4 (Fig. 21-28). Find the ratio Q_3/Q_1 in terms of T_1, T_2, T_3, and T_4.

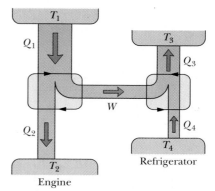

FIGURE 21-28 Problem 63.

SECTION 21-6 A Statistical View of Entropy

64E. Construct a table like Table 21-1 for six molecules rather than four.

65E. Show that for N molecules in a box, the number of possible microstates is 2^N when microstates are defined by whether a given molecule is in the left half of the box or the right half. Check this for the situation of Table 21-1.

66E. Check your calculator against Stirling's approximation, with N equal to (a) 50, (b) 100, and (c) 250. Is there a trend?

67P. Suppose you put 100 pennies in a cup, shake the cup, and then toss the pennies onto the floor. (a) How many different head–tail arrangements (microstates) are possible for the 100 pennies? (*Hint:* See Exercise 65.) What is the probability of finding exactly (b) 50 heads, (c) 48 heads, (d) 48 tails, (e) 40 heads, and (f) 30 heads?

68P. A box contains N gas molecules, equally divided between its two halves. For $N = 50$: (a) What is the multiplicity of this central configuration? (b) What is the total number of microstates for the system? (*Hint:* See Exercise 65.) (c) What percentage of the time does the system spend in its central configuration? (d) Repeat (a) through (c) for $N = 100$. (e) Repeat (a) through (c) for $N = 200$. (f) As N increases, you will find that the system spends *less* time (not more) in its central configuration. Explain why this is so.

69P. A box contains N gas molecules. Consider the box to be divided into three equal parts. (a) By extension of Eq. 21-14, write a formula for the multiplicity of any given configuration. (b) Consider two configurations: configuration A with equal numbers of molecules in all three thirds of the box, and configuration B with equal numbers of molecules in both halves of the box. What is the ratio W_A/W_B of the multiplicity of configuration A to that of configuration B? (c) Evaluate W_A/W_B for $N = 100$. (Because 100 is not evenly divisible by 3, put 34 molecules into one of the three box parts and 33 in each of the other parts for configuration A.)

70P. A box contains N molecules. Consider two configurations: configuration A with an equal division of the molecules between the two halves of the box, and configuration B with 60.0% of the molecules in the left half of the box and 40.0% in the right half. For $N = 50$, what are the multiplicities of (a) configuration A and (b) configuration B? (c) What is the ratio f of the time the system spends in configuration B to that in configuration A? (d) Repeat (a) through (c) for $N = 100$. (e) Repeat (a) through (c) for $N = 200$. (f) With an increase in N, does f increase, decrease, or remain the same?

Electronic Computation

71. The temperature of 25 mol of an ideal diatomic gas, originally 300 K, is increased by placing the gas in contact with a thermal reservoir at 800 K. The pressure remains constant throughout the increase. (a) Calculate the change in the entropy of the gas, the change in the entropy of the reservoir, and the change in the total entropy of the gas-reservoir system. (b) Suppose the same change in state is carried out by placing the gas in contact successively with n intermediate reservoirs, equally spaced in temperature between 300 K and 800 K. Calculate the total change in the entropy of the reservoirs and the change in the entropy of the closed system consisting of the gas and the reservoirs for (b) $n = 1$, (c) $n = 10$, (d) $n = 50$, and (e) $n = 100$. With each increase in n, the process more closely approximates a reversible process.

Boiling and the Leidenfrost Effect

JEARL WALKER
Cleveland State University

How does water boil? As commonplace as the event is, you may not have noticed all of its curious features. Some of the features are important in industrial applications, while others appear to be the basis for certain dangerous stunts once performed by daredevils in carnival sideshows.

Arrange for a pan of tap water to be heated from below by a flame or electric heat source. As the water warms, air molecules are driven out of solution in the water, collecting as tiny bubbles in crevices along the bottom of the pan (Fig. 1a). The air bubbles gradually inflate, and then they begin to pinch off from the crevices and rise to the top surface of the water (Fig. 1b–f). As they leave, more air bubbles form in the crevices and pinch off, until the supply of air in the water is depleted. The formation of air bubbles is a sign that the water is heating but has nothing to do with boiling.

Water that is directly exposed to the atmosphere boils at what is sometimes called its normal boiling temperature T_s. For example, T_s is about 100°C when the air pressure is 1 atm. Since the water at the bottom of your pan is not directly exposed to the atmosphere, it remains liquid even when it *superheats* above T_s by as much as a few degrees. During this process, the water is constantly mixed by convection as hot water rises and cooler water descends.

If you continue to increase the pan's temperature, the bottom layer of water begins to vaporize, with water mole-cules gathering in small vapor bubbles in the now dry crevices, as the air bubbles do in Fig. 1. This phase of boiling is signaled by pops, pings, and eventually buzzing. The water almost sings its displeasure at being heated. Every time a vapor bubble expands upward into slightly cooler water, the bubble suddenly collapses because the vapor within it condenses. Each collapse sends out a sound wave, the ping you hear. Once the temperature of the bulk water increases, the bubbles may not collapse until after they pinch off from the crevices and ascend part of the way to the top surface of the water. This phase of boiling is labeled "isolated vapor bubbles" in Fig. 2.

If you still increase the pan's temperature, the clamor of collapsing bubbles first grows louder and then disappears. The noise begins to soften when the bulk liquid is sufficiently hot that the vapor bubbles reach the top surface of the water. There they pop open with a light splash. The water is now in full boil.

If your heat source is a kitchen stove, the story stops at this point. However, with a laboratory burner you can continue to increase the pan's temperature. The vapor bubbles

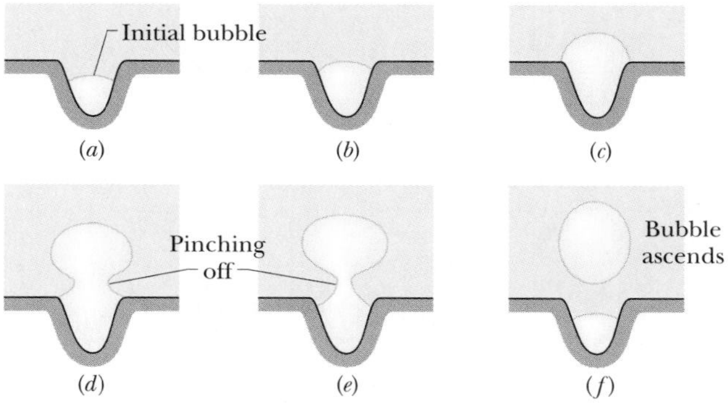

FIGURE 1 (*a*) A bubble forms in the crevice of a scratch along the bottom of a pan of water. (*b-f*) The bubble grows, pinches off, and then ascends through the water.

(*a*) Initial bubble
(*b*)
(*c*)
(*d*) Pinching off
(*e*)
(*f*) Bubble ascends

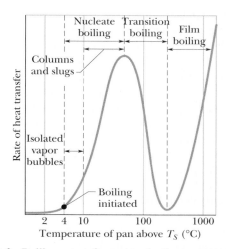

FIGURE 2 Boiling curve for water. As the temperature at the bottom of the pan is increased above the normal boiling point, the rate at which energy is transferred from the pan bottom to the water increases at first. However, above a certain temperature, the transfer almost disappears. At even higher temperatures, the transfer reappears.

533

next become so abundant and pinch off from their crevices so frequently that they coalesce, forming columns of vapor that violently and chaotically churn upward, sometimes meeting previously detached "slugs" of vapor.

The production of vapor bubbles and columns is called *nucleate boiling* because the formation and growth of the bubbles depend on crevices serving as *nucleating sites* (sites of formation). Whenever you increase the pan's temperature, the rate at which heat is transferred to the water increases. If you continue to raise the pan's temperature past the stage of columns and slugs, the boiling enters a new phase called the *transition regime*. Then each increase in the pan's temperature reduces the rate at which heat is transferred to the water. The decrease is not paradoxical. In the transition regime, much of the bottom of the pan is covered by a layer of vapor. Since water vapor conducts heat about an order of magnitude more poorly than does liquid water, the transfer of heat to the water is diminished. The hotter the pan becomes, the less direct contact the water has with it and the worse the transfer of heat becomes. This situation can be dangerous in a *heat exchanger,* whose purpose is to transfer heat from a heated object. If the water in the heat exchanger is allowed to enter the transition regime, the object may destructively overheat because of diminished transfer of heat from it.

Suppose you continue to increase the temperature of the pan. Eventually, the whole of the bottom surface is covered with vapor. Then heat is slowly transferred to the liquid above the vapor by radiation and gradual conduction. This phase is called *film boiling.*

Although you cannot obtain film boiling in a pan of water on a kitchen stove, it is still commonplace in the kitchen. My grandmother once demonstrated how it serves to indicate when her skillet is hot enough for pancake batter. After she heated the empty skillet for a while, she sprinkled a few drops of water into it. The drops sizzled away within seconds. Their rapid disappearance warned her that the skillet was insufficiently hot for the batter. After further heating the skillet, she repeated her test with a few more sprinkled water drops. This time they beaded up and danced over the metal, lasting well over a minute before they disappeared. The skillet was then hot enough for my grandmother's batter.

To study her demonstration, I arranged for a flat metal plate to be heated by a laboratory burner. While monitoring the temperature of the plate with a thermocouple, I carefully released a drop of distilled water from a syringe held just above the plate. The drop fell into a dent I had made in the plate with a ball-peen hammer. The syringe allowed me to release drops of uniform size. Once a drop was released, I timed how long it survived on the plate. Afterward, I plotted the survival times of the drops versus the plate

temperature (Fig. 3). The graph has a curious peak. When the plate temperature was between 100 and about 200°C, each drop spread over the plate in a thin layer and rapidly vaporized. When the plate temperature was about 200°C, a drop deposited on the plate beaded up and survived for over a minute. At even higher plate temperatures, the water beads did not survive quite as long. Similar experiments with tap water generated a graph with a flatter peak, probably because suspended particles of impurities in the drops breached the vapor layer, conducting heat into the drops.

The fact that a water drop is long lived when deposited on metal that is much hotter than the boiling temperature of water was first reported by Hermann Boerhaave in 1732. It was not investigated extensively until 1756 when Johann Gottlob Leidenfrost published "A Tract About Some Qualities of Common Water." Because Leidenfrost's work was not translated from the Latin until 1965, it was not widely read. Still, his name is now associated with the phenomenon. In addition, the temperature corresponding to the peak in a graph such as I made is called the Leidenfrost point.

Leidenfrost conducted his experiments with an iron spoon that was heated red-hot in a fireplace. After placing a drop of water into the spoon, he timed its duration by the swings of a pendulum. He noted that the drop seemed to suck the light and heat form the spoon, leaving a spot duller than the rest of the spoon. The first drop deposited in the spoon lasted 30 s while the next drop lasted only 10 s. Additional drops lasted only a few seconds.

Leidenfrost misunderstood his demonstrations because he did not realize that the longer-lasting drops were actually boiling. Let me explain in terms of my experi-

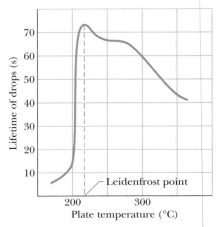

FIGURE 3 Drop lifetimes on a hot plate. Strangely, in a certain temperature range, the drops last longer when the hot plate is hotter.

ments. When the temperature of the plate is less than the Leidenfrost point, the water spreads over the plate and rapidly conducts heat from it, resulting in complete vaporization within seconds. When the temperature is at or above the Leidenfrost point, the bottom surface of a drop deposited on the plate almost immediately vaporizes. The gas pressure from this vapor layer prevents the rest of the drop from touching the plate (Fig. 4). The layer thus protects and supports the drop for the next minute or so. The layer is constantly replenished as additional water vaporizes from the bottom surface of the drop because of heat radiated and conducted through the layer from the plate. Although the layer is less than 0.1 mm thick near its outer boundary and only about 0.2 mm thick at its center, it dramatically slows the vaporization of the drop.

After reading the translation of Leidenfrost's research, I happened upon a description of a curious stunt that was performed in the sideshows of carnivals around the turn of the century. Reportedly, a performer was able to dip wet fingers into molten lead. Assuming that the stunt involved no trickery, I conjectured that it must depend on the Leidenfrost effect. As soon as the performer's wet flesh touched the hot liquid metal, part of the water vaporized, coating the fingers with a vapor layer. If the dip was brief, the flesh would not be heated significantly.

I could not resist the temptation to test my explanation. With a laboratory burner, I melted down a sizable slab of lead in a crucible. I heated the lead until its temperature was over 400°C, well above its melting temperature of 328°C. After wetting a finger in tap water, I prepared to touch the top surface of the molten lead. I must confess that I had an assistant standing ready with first-aid materials. I must also confess that my first several attempts failed because my brain refused to allow this ridiculous experiment, always directing my finger to miss the lead.

When I finally overcame my fears and briefly touched the lead, I was amazed. I felt no heat. Just as I had guessed, part of the water on the finger vaporized, forming a protective layer. Since the contact was brief, radiation and conduction of heat through the vapor layer were insufficient to raise perceptibly the temperature of my flesh. I grew braver. After wetting my hand, I dipped all my fingers into

FIGURE 5 Walker demonstrating the Leidenfrost effect with molten lead. He has just plunged his fingers into the lead, touching the bottom of the pan. The temperature of the lead is given in degrees Fahrenheit on the industrial thermometer.

the lead, touching the bottom of the container (Fig. 5). The contact with the lead was still too brief to result in a burn. Apparently, the Leidenfrost effect, or more exactly, the immediate presence of film boiling, protected my fingers.

I still questioned my explanation. Could I possibly touch the lead with a dry finger without suffering a burn? Leaving aside all rational thought, I tried it, immediately realizing my folly when pain raced through the finger. Later, I tested a dry weiner, forcing it into the molten lead for several seconds. The skin of the weiner quickly blackened. It lacked the protection of film boiling just as my dry finger had.

I must caution that dipping fingers into molten lead presents several serious dangers. If the lead is only slightly above its melting point, the loss of heat from it when the water is vaporized may solidify the lead around the fingers. If I were to pull the resulting glove of hot, solid lead up from the container, it will be in contact with my fingers so long that my fingers are certain to be badly burned. I must also contend with the possibility of splashing and spillage. In addition, there is the acute danger of having too much water on the fingers. When the surplus water rapidly va-

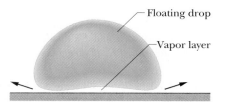

FIGURE 4 A Leidenfrost drop in cross section.

Floating drop

Vapor layer

porizes, it can blow molten lead over the surroundings and, most seriously, into the eyes. I have been scarred on my arms and face from such explosive vaporizations. *You should never repeat this demonstration.*

Film boiling can also be seen when liquid nitrogen is spilled. The drops and globs bead up as they skate over the floor. The liquid is at a temperature of about $-200°C$. When the spilled liquid nears the floor, its bottom surface vaporizes. The vapor layer then provides support for the rest of the liquid, allowing the liquid to survive for a surprisingly long time.

I was told of a stunt where a performer poured liquid nitrogen into his mouth without being hurt by its extreme cold. The liquid immediately underwent film boiling on its bottom surface and thus did not directly touch the tongue. Foolishly, I repeated this demonstration. For several dozen times the stunt went smoothly and dramatically. With a large glob of liquid nitrogen in my mouth, I concentrated on not swallowing while I breathed outward. The moisture in my cold breath condensed, creating a terrific plume that extended about a meter from my mouth. However, on my last attempt the liquid thermally contracted two of my front teeth so severely that the enamel ruptured into a ''road map'' of fissures. My dentist convinced me to drop the demonstration.

The Leidenfrost effect may also play a role in another foolhardy demonstration: walking over hot coals. At times the news media have carried reports of a performer striding over red-hot coals with much hoopla and mystic nonsense, perhaps claiming that protection from a bad burn is afforded by ''mind over matter.'' Actually, physics protects the feet when the walk is successful. Particularly important is the fact that although the surface of the coals is quite hot, it contains surprisingly little energy. If the performer walks at a moderate pace, a footfall is so brief that the foot conducts little energy from the coals. Of course, a slower walk invites a burn because the longer contact allows heat to be conducted to the foot from the interior of the coals.

If the feet are wet prior to the walk, the liquid might also help protect them. To wet the feet a performer might walk over wet grass just before reaching the hot coals. Instead, the feet might just be sweaty because of the heat from the coals or the excitement of the performance. Once the performer is on the coals, some of the heat vaporizes the liquid on the feet, leaving less heat to be conducted to the flesh. In addition, there may be points of contact where the liquid undergoes film boiling, thereby providing brief protection from the coals.

I have walked over hot coals on five occasions. For four of the walks I was fearful enough that my feet were sweaty. However, on the fifth walk I took my safety so much for granted that my feet were dry. The burns I suf-

fered then were extensive and terribly painful. My feet did not heal for weeks.

My failure may have been due to a lack of film boiling on the feet, but I had also neglected an additional safety factor. On the other days I had taken the precaution of clutching an earlier edition of this book to my chest during the walks so as to bolster my belief in physics. Alas, I forgot the book on the day when I was so badly burned.

I have long argued that degree-granting programs should employ ''fire-walking'' as a last exam. The chairperson of the program should wait on the far side of a bed of red-hot coals while a degree candidate is forced to walk over the coals. If the candidate's belief in physics is strong enough that the feet are left undamaged, the chairperson hands the candidate a graduation certificate. The test would be more revealing than traditional final exams.

REFERENCES

Leidenfrost, Johann Gottlob, ''On the Fixation of Water in Diverse Fire,'' *International Journal of Heat and Mass Transfer,* Vol. 9, pages 1153–1166 (1966).

Gottfried, B. S., C. J. Lee, and K. J. Bell, ''The Leidenfrost Phenomenon: Film Boiling of Liquid Droplets on a Flat Plate,'' *International Journal of Heat and Mass Transfer,* Vol. 9, pages 1167–1187 (1966).

Hall, R. S., S. J. Board, A. J. Clare, R. B. Duffey, T. S. Playle, and D. H. Poole, ''Inverse Leidenfrost Phenomenon,'' *Nature,* Vol. 224, pages 266–267 (1969).

Walker, Jearl, ''The Amateur Scientist,'' *Scientific American,* Vol. 237, pages 126–131 + 140 (August 1977).

Curzon, F. L. ''The Leidenfrost Phenomenon,'' *American Journal of Physics,* Vol. 46, pages 825–828 (1978).

Leikind, Bernard J., and William J. McCarthy, ''An Investigation of Firewalking,'' *Skeptical Inquirer,* Vol. 10, No. 1, pages 23–34 (Fall 1985).

Bent, Henry A., ''Droplet on a Hot Metal Spoon,'' *American Journal of Physics,* Vol. 54, page 967 (1986).

Leikind, B. J., and W. J. McCarthy, ''Firewalking,'' *Experientia,* Vol. 44, pages 310–315 (1988).

Thimbleby, Harold, ''The Leidenfrost Phenomenon,'' *Physics Education,* Vol. 24, pages 300–303 (1989).

Taylor, John R., ''Firewalking: A Lesson in Physics,'' *The Physics Teacher,* Vol. 27, pages 166–168 (March 1989).

Zhang, S., and G. Gogos, ''Film Evaporation of a Spherical Droplet over a Hot Surface: Fluid Mechanics and Heat/Mass Transfer Analysis,'' *Journal of Fluid Mechanics,* Vol. 222, pages 543–563 (1991).

Agrawal, D. C., and V. J. Menon, ''Boiling and the Leidenfrost Effect in a Gravity-free Zone: A Speculation,'' *Physics Education,* Vol. 29, pages 39–42 (1994).

Appendix A
*The International System of Units (SI)**

1. THE SI BASE UNITS

QUANTITY	NAME	SYMBOL	DEFINITION
length	meter	m	". . . the length of the path traveled by light in vacuum in 1/299,792,458 of a second." (1983)
mass	kilogram	kg	". . . this prototype [a certain platinum–iridium cylinder] shall henceforth be considered to be the unit of mass." (1889)
time	second	s	". . . the duration of 9,192,631,770 periods of the radiation corresponding to the transition between the two hyperfine levels of the ground state of the cesium-133 atom." (1967)
electric current	ampere	A	". . . that constant current which, if maintained in two straight parallel conductors of infinite length, of negligible circular cross section, and placed 1 meter apart in vacuum, would produce between these conductors a force equal to 2×10^{-7} newton per meter of length." (1946)
thermodynamic temperature	kelvin	K	". . . the fraction 1/273.16 of the thermodynamic temperature of the triple point of water." (1967)
amount of substance	mole	mol	". . . the amount of substance of a system which contains as many elementary entities as there are atoms in 0.012 kilogram of carbon-12." (1971)
luminous intensity	candela	cd	". . . the luminous intensity, in the perpendicular direction, of a surface of 1/600,000 square meter of a blackbody at the temperature of freezing platinum under a pressure of 101.325 newtons per square meter." (1967)

*Adapted from "The International System of Units (SI)," National Bureau of Standards Special Publication 330, 1972 edition. The definitions above were adopted by the General Conference of Weights and Measures, an international body, on the dates shown. In this book we do not use the candela.

2. SOME SI DERIVED UNITS

QUANTITY	NAME OF UNIT	SYMBOL	
area	square meter	m^2	
volume	cubic meter	m^3	
frequency	hertz	Hz	s^{-1}
mass density (density)	kilogram per cubic meter	kg/m^3	
speed, velocity	meter per second	m/s	
angular velocity	radian per second	rad/s	
acceleration	meter per second per second	m/s^2	
angular acceleration	radian per second per second	rad/s^2	
force	newton	N	$kg \cdot m/s^2$
pressure	pascal	Pa	N/m^2
work, energy, quantity of heat	joule	J	$N \cdot m$
power	watt	W	J/s
quantity of electric charge	coulomb	C	$A \cdot s$
potential difference, electromotive force	volt	V	W/A
electric field strength	volt per meter (or newton per coulomb)	V/m	N/C
electric resistance	ohm	Ω	V/A
capacitance	farad	F	$A \cdot s/V$
magnetic flux	weber	Wb	$V \cdot s$
inductance	henry	H	$V \cdot s/A$
magnetic flux density	tesla	T	Wb/m^2
magnetic field strength	ampere per meter	A/m	
entropy	joule per kelvin	J/K	
specific heat	joule per kilogram kelvin	$J/(kg \cdot K)$	
thermal conductivity	watt per meter kelvin	$W/(m \cdot K)$	
radiant intensity	watt per steradian	W/sr	

3. THE SI SUPPLEMENTARY UNITS

QUANTITY	NAME OF UNIT	SYMBOL
plane angle	radian	rad
solid angle	steradian	sr

Appendix **B**
*Some Fundamental Constants of Physics**

CONSTANT	SYMBOL	COMPUTATIONAL VALUE	BEST (1986) VALUE VALUE[a]	UNCERTAINTY[b]
Speed of light in a vacuum	c	3.00×10^8 m/s	2.99792458	exact
Elementary charge	e	1.60×10^{-19} C	1.60217733	0.30
Gravitational constant	G	6.67×10^{-11} m³/s²·kg	6.67259	128
Universal gas constant	R	8.31 J/mol·K	8.314510	8.4
Avogadro constant	N_A	6.02×10^{23} mol⁻¹	6.0221367	0.59
Boltzmann constant	k	1.38×10^{-23} J/K	1.380658	8.5
Stefan-Boltzmann constant	σ	5.67×10^{-8} W/m²·K⁴	5.67051	34
Molar volume of ideal gas at STP[d]	V_m	2.24×10^{-2} m³/mol	2.241409	8.4
Permittivity constant	ϵ_0	8.85×10^{-12} F/m	8.85418781762	exact
Permeability constant	μ_0	1.26×10^{-6} H/m	1.25663706143	exact
Planck constant	h	6.63×10^{-34} J·s	6.6260755	0.60
Electron mass[c]	m_e	9.11×10^{-31} kg	9.1093897	0.59
		5.49×10^{-4} u	5.48579903	0.023
Proton mass[c]	m_p	1.67×10^{-27} kg	1.6726231	0.59
		1.0073 u	1.0072764660	0.005
Ratio of proton mass to electron mass	m_p/m_e	1840	1836.152701	0.020
Electron charge-to-mass ratio	e/m_e	1.76×10^{11} C/kg	1.75881961	0.30
Neutron mass[c]	m_n	1.68×10^{-27} kg	1.6749286	0.59
		1.0087 u	1.0086649235	0.0023
Hydrogen atom mass[c]	m_{1_H}	1.0078 u	1.0078250316	0.0005
Deuterium atom mass[c]	m_{2_H}	2.0141 u	2.0141017779	0.0005
Helium atom mass[c]	$m_{4_{He}}$	4.0026 u	4.0026032	0.067
Muon mass	m_μ	1.88×10^{-28} kg	1.8835326	0.61
Electron magnetic moment	μ_e	9.28×10^{-24} J/T	9.2847701	0.34
Proton magnetic moment	μ_p	1.41×10^{-26} J/T	1.41060761	0.34
Bohr magneton	μ_B	9.27×10^{-24} J/T	9.2740154	0.34
Nuclear magneton	μ_N	5.05×10^{-27} J/T	5.0507866	0.34
Bohr radius	r_B	5.29×10^{-11} m	5.29177249	0.045
Rydberg constant	R	1.10×10^7 m⁻¹	1.0973731534	0.0012
Electron Compton wavelength	λ_C	2.43×10^{-12} m	2.42631058	0.089

[a]Values given in this column should be given the same unit and power of 10 as the computational value.   [b]Parts per million.   [c]Masses given in u are in unified atomic mass units, where 1 u = $1.6605402 \times 10^{-27}$ kg.   [d]STP means standard temperature and pressure: 0°C and 1.0 atm (0.1 MPa).

*The values in this table were largely selected from a longer list in *Symbols, Units and Nomenclature in Physics* (IUPAP), prepared by E. Richard Cohen and Pierre Giacomo, 1986.

Appendix C
Some Astronomical Data

SOME DISTANCES FROM THE EARTH

To the moon*	3.82×10^8 m
To the sun*	1.50×10^{11} m
To the nearest star (Proxima Centauri)	4.04×10^{16} m
To the center of our galaxy	2.2×10^{20} m
To the Andromeda Galaxy	2.1×10^{22} m
To the edge of the observable universe	$\sim 10^{26}$ m

*Mean distance.

THE SUN, THE EARTH, AND THE MOON

PROPERTY	UNIT	SUN	EARTH	MOON
Mass	kg	1.99×10^{30}	5.98×10^{24}	7.36×10^{22}
Mean radius	m	6.96×10^8	6.37×10^6	1.74×10^6
Mean density	kg/m^3	1410	5520	3340
Free-fall acceleration at the surface	m/s^2	274	9.81	1.67
Escape velocity	km/s	618	11.2	2.38
Period of rotation[a]	—	37 d at poles[b] 26 d at equator[b]	23 h 56 min	27.3 d
Radiation power[c]	W	3.90×10^{26}		

[a]Measured with respect to the distant stars.

[b]The sun, a ball of gas, does not rotate as a rigid body.

[c]Just outside the Earth's atmosphere solar energy is received, assuming normal incidence, at the rate of 1340 W/m^2.

SOME PROPERTIES OF THE PLANETS

	MERCURY	VENUS	EARTH	MARS	JUPITER	SATURN	URANUS	NEPTUNE	PLUTO
Mean distance from sun, 10^6 km	57.9	108	150	228	778	1430	2870	4500	5900
Period of revolution, y	0.241	0.615	1.00	1.88	11.9	29.5	84.0	165	248
Period of rotation,[a] d	58.7	−243[b]	0.997	1.03	0.409	0.426	−0.451[b]	0.658	6.39
Orbital speed, km/s	47.9	35.0	29.8	24.1	13.1	9.64	6.81	5.43	4.74
Inclination of axis to orbit	<28°	≈3°	23.4°	25.0°	3.08°	26.7°	97.9°	29.6°	57.5°
Inclination of orbit to Earth's orbit	7.00°	3.39°		1.85°	1.30°	2.49°	0.77°	1.77°	17.2°
Eccentricity of orbit	0.206	0.0068	0.0167	0.0934	0.0485	0.0556	0.0472	0.0086	0.250
Equatorial diameter, km	4880	12,100	12,800	6790	143,000	120,000	51,800	49,500	2300
Mass (Earth = 1)	0.0558	0.815	1.000	0.107	318	95.1	14.5	17.2	0.002
Density (water = 1)	5.60	5.20	5.52	3.95	1.31	0.704	1.21	1.67	2.03
Surface value of g,[c] m/s^2	3.78	8.60	9.78	3.72	22.9	9.05	7.77	11.0	0.5
Escape velocity,[c] km/s	4.3	10.3	11.2	5.0	59.5	35.6	21.2	23.6	1.1
Known satellites	0	0	1	2	16 + ring	18 + rings	15 + rings	8 + rings	1

[a]Measured with respect to the distant stars.

[b]Venus and Uranus rotate opposite their orbital motion.

[c]Gravitational acceleration measured at the planet's equator.

Appendix D
Conversion Factors

Conversion factors may be read directly from these tables. For example, 1 degree = 2.778×10^{-3} revolutions, so $16.7° = 16.7 \times 2.778 \times 10^{-3}$ rev. The SI quantities are fully capitalized.

Adapted in part from G. Shortley and D. Williams, *Elements of Physics,* Prentice-Hall, Englewood Cliffs, NJ, 1971.

PLANE ANGLE

	°	′	″	RADIAN	rev
1 degree =	1	60	3600	1.745×10^{-2}	2.778×10^{-3}
1 minute =	1.667×10^{-2}	1	60	2.909×10^{-4}	4.630×10^{-5}
1 second =	2.778×10^{-4}	1.667×10^{-2}	1	4.848×10^{-6}	7.716×10^{-7}
1 RADIAN =	57.30	3438	2.063×10^{5}	1	0.1592
1 revolution =	360	2.16×10^{4}	1.296×10^{6}	6.283	1

SOLID ANGLE

1 sphere = 4π steradians = 12.57 steradians

LENGTH

	cm	METER	km	in.	ft	mi
1 centimeter =	1	10^{-2}	10^{-5}	0.3937	3.281×10^{-2}	6.214×10^{-6}
1 METER =	100	1	10^{-3}	39.37	3.281	6.214×10^{-4}
1 kilometer =	10^{5}	1000	1	3.937×10^{4}	3281	0.6214
1 inch =	2.540	2.540×10^{-2}	2.540×10^{-5}	1	8.333×10^{-2}	1.578×10^{-5}
1 foot =	30.48	0.3048	3.048×10^{-4}	12	1	1.894×10^{-4}
1 mile =	1.609×10^{5}	1609	1.609	6.336×10^{4}	5280	1

1 angström = 10^{-10} m

1 nautical mile = 1852 m
  = 1.151 miles = 6076 ft

1 fermi = 10^{-15} m

1 light-year = 9.460×10^{12} km

1 parsec = 3.084×10^{13} km

1 fathom = 6 ft

1 Bohr radius = 5.292×10^{-11} m

1 yard = 3 ft

1 rod = 16.5 ft

1 mil = 10^{-3} in.

1 nm = 10^{-9} m

AREA

	METER^2	cm^2	ft^2	in.^{2}
1 SQUARE METER =	1	10^{4}	10.76	1550
1 square centimeter =	10^{-4}	1	1.076×10^{-3}	0.1550
1 square foot =	9.290×10^{-2}	929.0	1	144
1 square inch =	6.452×10^{-4}	6.452	6.944×10^{-3}	1

1 square mile = 2.788×10^{7} ft^2
  = 640 acres

1 barn = 10^{-28} m^2

1 acre = 43,560 ft^2

1 hectare = 10^{4} m^2 = 2.471 acres

VOLUME

	METER^3	cm^3	L	ft^3	in.^{3}
1 CUBIC METER = 1		10^6	1000	35.31	6.102×10^4
1 cubic centimeter = 10^{-6}		1	1.000×10^{-3}	3.531×10^{-5}	6.102×10^{-2}
1 liter = 1.000×10^{-3}		1000	1	3.531×10^{-2}	61.02
1 cubic foot = 2.832×10^{-2}		2.832×10^4	28.32	1	1728
1 cubic inch = 1.639×10^{-5}		16.39	1.639×10^{-2}	5.787×10^{-4}	1

1 U.S. fluid gallon = 4 U.S. fluid quarts = 8 U.S. pints = 128 U.S. fluid ounces = 231 in.^{3}

1 British imperial gallon = 277.4 in.^{3} = 1.201 U.S. fluid gallons

MASS

Quantities in the colored areas are not mass units but are often used as such. When we write, for example, 1 kg "=" 2.205 lb, this means that a kilogram is a *mass* that *weighs* 2.205 pounds at a location where g has the standard value of 9.80665 m/s^2.

	g	KILOGRAM	slug	u	oz	lb	ton
1 gram = 1		0.001	6.852×10^{-5}	6.022×10^{23}	3.527×10^{-2}	2.205×10^{-3}	1.102×10^{-6}
1 KILOGRAM = 1000		1	6.852×10^{-2}	6.022×10^{26}	35.27	2.205	1.102×10^{-3}
1 slug = 1.459×10^4		14.59	1	8.786×10^{27}	514.8	32.17	1.609×10^{-2}
1 atomic mass unit = 1.661×10^{-24}		1.661×10^{-27}	1.138×10^{-28}	1	5.857×10^{-26}	3.662×10^{-27}	1.830×10^{-30}
1 ounce = 28.35		2.835×10^{-2}	1.943×10^{-3}	1.718×10^{25}	1	6.250×10^{-2}	3.125×10^{-5}
1 pound = 453.6		0.4536	3.108×10^{-2}	2.732×10^{26}	16	1	0.0005
1 ton = 9.072×10^5		907.2	62.16	5.463×10^{29}	3.2×10^4	2000	1

1 metric ton = 1000 kg

DENSITY

Quantities in the colored areas are weight densities and, as such, are dimensionally different from mass densities. See note for mass table.

	slug/ft^3	KILOGRAM/ METER^3	g/cm^3	lb/ft^3	lb/in.^{3}
1 slug per foot^3 = 1		515.4	0.5154	32.17	1.862×10^{-2}
1 KILOGRAM per METER^3 = 1.940×10^{-3}		1	0.001	6.243×10^{-2}	3.613×10^{-5}
1 gram per centimeter^3 = 1.940		1000	1	62.43	3.613×10^{-2}
1 pound per foot^3 = 3.108×10^{-2}		16.02	1.602×10^{-2}	1	5.787×10^{-4}
1 pound per inch^3 = 53.71		2.768×10^4	27.68	1728	1

TIME

	y	d	h	min	SECOND
1 year = 1		365.25	8.766×10^3	5.259×10^5	3.156×10^7
1 day = 2.738×10^{-3}		1	24	1440	8.640×10^4
1 hour = 1.141×10^{-4}		4.167×10^{-2}	1	60	3600
1 minute = 1.901×10^{-6}		6.944×10^{-4}	1.667×10^{-2}	1	60
1 SECOND = 3.169×10^{-8}		1.157×10^{-5}	2.778×10^{-4}	1.667×10^{-2}	1

SPEED

	ft/s	km/h	METER/SECOND	mi/h	cm/s
1 foot per second = 1		1.097	0.3048	0.6818	30.48
1 kilometer per hour = 0.9113		1	0.2778	0.6214	27.78
1 METER per SECOND = 3.281		3.6	1	2.237	100
1 mile per hour = 1.467		1.609	0.4470	1	44.70
1 centimeter per second = 3.281×10^{-2}		3.6×10^{-2}	0.01	2.237×10^{-2}	1

1 knot = 1 nautical mi/h = 1.688 ft/s   1 mi/min = 88.00 ft/s = 60.00 mi/h

FORCE

Force units in the colored areas are now little used. To clarify: 1 gram-force (= 1 gf) is the force of gravity that would act on an object whose mass is 1 gram at a location where g has the standard value of 9.80665 m/s^2.

	dyne	NEWTON	lb	pdl	gf	kgf
1 dyne = 1		10^{-5}	2.248×10^{-6}	7.233×10^{-5}	1.020×10^{-3}	1.020×10^{-6}
1 NEWTON = 10^5		1	0.2248	7.233	102.0	0.1020
1 pound = 4.448×10^5		4.448	1	32.17	453.6	0.4536
1 poundal = 1.383×10^4		0.1383	3.108×10^{-2}	1	14.10	1.410×10^{-2}
1 gram-force = 980.7		9.807×10^{-3}	2.205×10^{-3}	7.093×10^{-2}	1	0.001
1 kilogram-force = 9.807×10^5		9.807	2.205	70.93	1000	1

PRESSURE

	atm	dyne/cm^2	inch of water	cm Hg	PASCAL	lb/in.^{2}	lb/ft^2
1 atmosphere = 1		1.013×10^6	406.8	76	1.013×10^5	14.70	2116
1 dyne per centimeter^2 = 9.869×10^{-7}		1	4.015×10^{-4}	7.501×10^{-5}	0.1	1.405×10^{-5}	2.089×10^{-3}
1 inch of water^a at 4°C = 2.458×10^{-3}		2491	1	0.1868	249.1	3.613×10^{-2}	5.202
1 centimeter of mercury^a at 0°C = 1.316×10^{-2}		1.333×10^4	5.353	1	1333	0.1934	27.85
1 PASCAL = 9.869×10^{-6}		10	4.015×10^{-3}	7.501×10^{-4}	1	1.450×10^{-4}	2.089×10^{-2}
1 pound per inch^2 = 6.805×10^{-2}		6.895×10^4	27.68	5.171	6.895×10^3	1	144
1 pound per foot^2 = 4.725×10^{-4}		478.8	0.1922	3.591×10^{-2}	47.88	6.944×10^{-3}	1

^{a} Where the acceleration of gravity has the standard value of 9.80665 m/s^2.

1 bar = 10^6 dyne/cm^2 = 0.1 MPa   1 millibar = 10^3 dyne/cm^2 = 10^2 Pa   1 torr = 1 mm Hg

ENERGY, WORK, HEAT

Quantities in the colored areas are not properly energy units but are included for convenience. They arise from the relativistic mass–energy equivalence formula $E = mc^2$ and represent the energy released if a kilogram or unified atomic mass unit (u) is completely converted to energy (bottom two rows) or the mass that would be completely converted to one unit of energy (rightmost two columns).

	Btu	erg	ft·lb	hp·h	JOULE	cal	kW·h	eV	MeV	kg	u
1 British thermal unit =	1	1.055×10^{10}	777.9	3.929×10^{-4}	1055	252.0	2.930×10^{-4}	6.585×10^{21}	6.585×10^{15}	1.174×10^{-14}	7.070×10^{12}
1 erg =	9.481×10^{-11}	1	7.376×10^{-8}	3.725×10^{-14}	10^{-7}	2.389×10^{-8}	2.778×10^{-14}	6.242×10^{11}	6.242×10^{5}	1.113×10^{-24}	670.2
1 foot-pound =	1.285×10^{-3}	1.356×10^{7}	1	5.051×10^{-7}	1.356	0.3238	3.766×10^{-7}	8.464×10^{18}	8.464×10^{12}	1.509×10^{-17}	9.037×10^{9}
1 horsepower-hour =	2545	2.685×10^{13}	1.980×10^{6}	1	2.685×10^{6}	6.413×10^{5}	0.7457	1.676×10^{25}	1.676×10^{19}	2.988×10^{-11}	1.799×10^{16}
1 JOULE =	9.481×10^{-4}	10^{7}	0.7376	3.725×10^{-7}	1	0.2389	2.778×10^{-7}	6.242×10^{18}	6.242×10^{12}	1.113×10^{-17}	6.702×10^{9}
1 calorie =	3.969×10^{-3}	4.186×10^{7}	3.088	1.560×10^{-6}	4.186	1	1.163×10^{-6}	2.613×10^{19}	2.613×10^{13}	4.660×10^{-17}	2.806×10^{10}
1 kilowatt-hour =	3413	3.600×10^{13}	2.655×10^{6}	1.341	3.600×10^{6}	8.600×10^{5}	1	2.247×10^{25}	2.247×10^{19}	4.007×10^{-11}	2.413×10^{16}
1 electron-volt =	1.519×10^{-22}	1.602×10^{-12}	1.182×10^{-19}	5.967×10^{-26}	1.602×10^{-19}	3.827×10^{-20}	4.450×10^{-26}	1	10^{-6}	1.783×10^{-36}	1.074×10^{-9}
1 million electron-volts =	1.519×10^{-16}	1.602×10^{-6}	1.182×10^{-13}	5.967×10^{-20}	1.602×10^{-13}	3.827×10^{-14}	4.450×10^{-20}	10^{-6}	1	1.783×10^{-30}	1.074×10^{-3}
1 kilogram =	8.521×10^{13}	8.987×10^{23}	6.629×10^{16}	3.348×10^{10}	8.987×10^{16}	2.146×10^{16}	2.497×10^{10}	5.610×10^{35}	5.610×10^{29}	1	6.022×10^{26}
1 unified atomic mass unit =	1.415×10^{-13}	1.492×10^{-3}	1.101×10^{-10}	5.559×10^{-17}	1.492×10^{-10}	3.564×10^{-11}	4.146×10^{-17}	9.320×10^{8}	932.0	1.661×10^{-27}	1

POWER

	Btu/h	ft·lb/s	hp	cal/s	kW	WATT
1 British thermal unit per hour =	1	0.2161	3.929×10^{-4}	6.998×10^{-2}	2.930×10^{-4}	0.2930
1 foot-pound per second =	4.628	1	1.818×10^{-3}	0.3239	1.356×10^{-3}	1.356
1 horsepower =	2545	550	1	178.1	0.7457	745.7
1 calorie per second =	14.29	3.088	5.615×10^{-3}	1	4.186×10^{-3}	4.186
1 kilowatt =	3413	737.6	1.341	238.9	1	1000
1 WATT =	3.413	0.7376	1.341×10^{-3}	0.2389	0.001	1

MAGNETIC FIELD

	gauss	TESLA	milligauss
1 gauss =	1	10^{-4}	1000
1 TESLA =	10^{4}	1	10^{7}
1 milligauss =	0.001	10^{-7}	1

1 tesla = 1 weber/meter^2

MAGNETIC FLUX

	maxwell	WEBER
1 maxwell =	1	10^{-8}
1 WEBER =	10^{8}	1

GEOMETRY

Circle of radius r: circumference $= 2\pi r$; area $= \pi r^2$.

Sphere of radius r: area $= 4\pi r^2$; volume $= \frac{4}{3}\pi r^3$.

Right circular cylinder of radius r and height h:
area $= 2\pi r^2 + 2\pi rh$; volume $= \pi r^2 h$.

Triangle of base a and altitude h: area $= \frac{1}{2}ah$.

QUADRATIC FORMULA

If $ax^2 + bx + c = 0$, then $x = \dfrac{-b \pm \sqrt{b^2 - 4ac}}{2a}$.

TRIGONOMETRIC FUNCTIONS OF ANGLE θ

$\sin \theta = \dfrac{y}{r}$ $\cos \theta = \dfrac{x}{r}$

$\tan \theta = \dfrac{y}{x}$ $\cot \theta = \dfrac{x}{y}$

$\sec \theta = \dfrac{r}{x}$ $\csc \theta = \dfrac{r}{y}$

PYTHAGOREAN THEOREM

In this right triangle,
$$a^2 + b^2 = c^2$$

TRIANGLES

Angles are A, B, C

Opposite sides are a, b, c

Angles $A + B + C = 180°$

$\dfrac{\sin A}{a} = \dfrac{\sin B}{b} = \dfrac{\sin C}{c}$

$c^2 = a^2 + b^2 - 2ab \cos C$

Exterior angle $D = A + C$

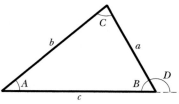

MATHEMATICAL SIGNS AND SYMBOLS

$=$ equals

$\approx$ equals approximately

$\sim$ is the order of magnitude of

$\neq$ is not equal to

$\equiv$ is identical to, is defined as

$>$ is greater than ($\gg$ is much greater than)

$<$ is less than ($\ll$ is much less than)

$\geq$ is greater than or equal to (or, is no less than)

$\leq$ is less than or equal to (or, is no more than)

$\pm$ plus or minus

$\propto$ is proportional to

Σ the sum of

$\bar{x}$ the average value of x

TRIGONOMETRIC IDENTITIES

$\sin(90° - \theta) = \cos \theta$

$\cos(90° - \theta) = \sin \theta$

$\sin \theta / \cos \theta = \tan \theta$

$\sin^2 \theta + \cos^2 \theta = 1$

$\sec^2 \theta - \tan^2 \theta = 1$

$\csc^2 \theta - \cot^2 \theta = 1$

$\sin 2\theta = 2 \sin \theta \cos \theta$

$\cos 2\theta = \cos^2 \theta - \sin^2 \theta = 2 \cos^2 \theta - 1 = 1 - 2 \sin^2 \theta$

$\sin(\alpha \pm \beta) = \sin \alpha \cos \beta \pm \cos \alpha \sin \beta$

$\cos(\alpha \pm \beta) = \cos \alpha \cos \beta \mp \sin \alpha \sin \beta$

$\tan(\alpha \pm \beta) = \dfrac{\tan \alpha \pm \tan \beta}{1 \mp \tan \alpha \tan \beta}$

$\sin \alpha \pm \sin \beta = 2 \sin \frac{1}{2}(\alpha \pm \beta) \cos \frac{1}{2}(\alpha \mp \beta)$

$\cos \alpha + \cos \beta = 2 \cos \frac{1}{2}(\alpha + \beta) \cos \frac{1}{2}(\alpha - \beta)$

$\cos \alpha - \cos \beta = -2 \sin \frac{1}{2}(\alpha + \beta) \sin \frac{1}{2}(\alpha - \beta)$

BINOMIAL THEOREM

$$(1 + x)^n = 1 + \frac{nx}{1!} + \frac{n(n-1)x^2}{2!} + \cdots \qquad (x^2 < 1)$$

EXPONENTIAL EXPANSION

$$e^x = 1 + x + \frac{x^2}{2!} + \frac{x^3}{3!} + \cdots$$

LOGARITHMIC EXPANSION

$$\ln(1 + x) = x - \tfrac{1}{2}x^2 + \tfrac{1}{3}x^3 - \cdots \qquad (|x| < 1)$$

TRIGONOMETRIC EXPANSIONS
(θ in radians)

$$\sin \theta = \theta - \frac{\theta^3}{3!} + \frac{\theta^5}{5!} - \cdots$$

$$\cos \theta = 1 - \frac{\theta^2}{2!} + \frac{\theta^4}{4!} - \cdots$$

$$\tan \theta = \theta + \frac{\theta^3}{3} + \frac{2\theta^5}{15} + \cdots$$

CRAMER'S RULE

Two simultaneous equations in unknowns x and y,

$$a_1 x + b_1 y = c_1 \qquad \text{and} \qquad a_2 x + b_2 y = c_2,$$

have the solutions

$$x = \frac{\begin{vmatrix} c_1 & b_1 \\ c_2 & b_2 \end{vmatrix}}{\begin{vmatrix} a_1 & b_1 \\ a_2 & b_2 \end{vmatrix}} = \frac{c_1 b_2 - c_2 b_1}{a_1 b_2 - a_2 b_1}$$

and

$$y = \frac{\begin{vmatrix} a_1 & c_1 \\ a_2 & c_2 \end{vmatrix}}{\begin{vmatrix} a_1 & b_1 \\ a_2 & b_2 \end{vmatrix}} = \frac{a_1 c_2 - a_2 c_1}{a_1 b_2 - a_2 b_1}.$$

PRODUCTS OF VECTORS

Let $\mathbf{i}$, $\mathbf{j}$, and $\mathbf{k}$ be unit vectors in the x, y, and z directions. Then

$$\mathbf{i} \cdot \mathbf{i} = \mathbf{j} \cdot \mathbf{j} = \mathbf{k} \cdot \mathbf{k} = 1, \qquad \mathbf{i} \cdot \mathbf{j} = \mathbf{j} \cdot \mathbf{k} = \mathbf{k} \cdot \mathbf{i} = 0,$$

$$\mathbf{i} \times \mathbf{i} = \mathbf{j} \times \mathbf{j} = \mathbf{k} \times \mathbf{k} = 0,$$

$$\mathbf{i} \times \mathbf{j} = \mathbf{k}, \qquad \mathbf{j} \times \mathbf{k} = \mathbf{i}, \qquad \mathbf{k} \times \mathbf{i} = \mathbf{j},$$

Any vector $\mathbf{a}$ with components a_x, a_y, and a_z along the x, y, and z axes can be written

$$\mathbf{a} = a_x \mathbf{i} + a_y \mathbf{j} + a_z \mathbf{k}.$$

Let $\mathbf{a}$, $\mathbf{b}$, and $\mathbf{c}$ be arbitrary vectors with magnitudes a, b, and c. Then

$$\mathbf{a} \times (\mathbf{b} + \mathbf{c}) = (\mathbf{a} \times \mathbf{b}) + (\mathbf{a} \times \mathbf{c})$$

$$(s\mathbf{a}) \times \mathbf{b} = \mathbf{a} \times (s\mathbf{b}) = s(\mathbf{a} \times \mathbf{b}) \qquad (s = \text{a scalar}).$$

Let θ be the smaller of the two angles between $\mathbf{a}$ and $\mathbf{b}$. Then

$$\mathbf{a} \cdot \mathbf{b} = \mathbf{b} \cdot \mathbf{a} = a_x b_x + a_y b_y + a_z b_z = ab \cos \theta$$

$$\mathbf{a} \times \mathbf{b} = -\mathbf{b} \times \mathbf{a} = \begin{vmatrix} \mathbf{i} & \mathbf{j} & \mathbf{k} \\ a_x & a_y & a_z \\ b_x & b_y & b_z \end{vmatrix}$$

$$= \mathbf{i} \begin{vmatrix} a_y & a_z \\ b_y & b_z \end{vmatrix} - \mathbf{j} \begin{vmatrix} a_x & a_z \\ b_x & b_z \end{vmatrix} + \mathbf{k} \begin{vmatrix} a_x & a_y \\ b_x & b_y \end{vmatrix}$$

$$= (a_y b_z - b_y a_z)\mathbf{i}$$

$$+ (a_z b_x - b_z a_x)\mathbf{j} + (a_x b_y - b_x a_y)\mathbf{k}$$

$$|\mathbf{a} \times \mathbf{b}| = ab \sin \theta$$

$$\mathbf{a} \cdot (\mathbf{b} \times \mathbf{c}) = \mathbf{b} \cdot (\mathbf{c} \times \mathbf{a}) = \mathbf{c} \cdot (\mathbf{a} \times \mathbf{b})$$

$$\mathbf{a} \times (\mathbf{b} \times \mathbf{c}) = (\mathbf{a} \cdot \mathbf{c})\mathbf{b} - (\mathbf{a} \cdot \mathbf{b})\mathbf{c}$$

DERIVATIVES AND INTEGRALS

In what follows, the letters u and v stand for any functions of x, and a and m are constants. To each of the indefinite integrals should be added an arbitrary constant of integration. The *Handbook of Chemistry and Physics* (CRC Press Inc.) gives a more extensive tabulation.

1. $\dfrac{dx}{dx} = 1$

2. $\dfrac{d}{dx}(au) = a\dfrac{du}{dx}$

3. $\dfrac{d}{dx}(u + v) = \dfrac{du}{dx} + \dfrac{dv}{dx}$

4. $\dfrac{d}{dx}x^m = mx^{m-1}$

5. $\dfrac{d}{dx}\ln x = \dfrac{1}{x}$

6. $\dfrac{d}{dx}(uv) = u\dfrac{dv}{dx} + v\dfrac{du}{dx}$

7. $\dfrac{d}{dx}e^x = e^x$

8. $\dfrac{d}{dx}\sin x = \cos x$

9. $\dfrac{d}{dx}\cos x = -\sin x$

10. $\dfrac{d}{dx}\tan x = \sec^2 x$

11. $\dfrac{d}{dx}\cot x = -\csc^2 x$

12. $\dfrac{d}{dx}\sec x = \tan x \sec x$

13. $\dfrac{d}{dx}\csc x = -\cot x \csc x$

14. $\dfrac{d}{dx}e^u = e^u\dfrac{du}{dx}$

15. $\dfrac{d}{dx}\sin u = \cos u\dfrac{du}{dx}$

16. $\dfrac{d}{dx}\cos u = -\sin u\dfrac{du}{dx}$

1. $\displaystyle\int dx = x$

2. $\displaystyle\int au\, dx = a\int u\, dx$

3. $\displaystyle\int (u + v)\, dx = \int u\, dx + \int v\, dx$

4. $\displaystyle\int x^m\, dx = \dfrac{x^{m+1}}{m+1} \quad (m \neq -1)$

5. $\displaystyle\int \dfrac{dx}{x} = \ln |x|$

6. $\displaystyle\int u\dfrac{dv}{dx}\, dx = uv - \int v\dfrac{du}{dx}\, dx$

7. $\displaystyle\int e^x\, dx = e^x$

8. $\displaystyle\int \sin x\, dx = -\cos x$

9. $\displaystyle\int \cos x\, dx = \sin x$

10. $\displaystyle\int \tan x\, dx = \ln |\sec x|$

11. $\displaystyle\int \sin^2 x\, dx = \tfrac{1}{2}x - \tfrac{1}{4}\sin 2x$

12. $\displaystyle\int e^{-ax}\, dx = -\dfrac{1}{a}e^{-ax}$

13. $\displaystyle\int xe^{-ax}\, dx = -\dfrac{1}{a^2}(ax + 1)e^{-ax}$

14. $\displaystyle\int x^2 e^{-ax}\, dx = -\dfrac{1}{a^3}(a^2x^2 + 2ax + 2)e^{-ax}$

15. $\displaystyle\int_0^{\infty} x^n e^{-ax}\, dx = \dfrac{n!}{a^{n+1}}$

16. $\displaystyle\int_0^{\infty} x^{2n} e^{-ax^2}\, dx = \dfrac{1\cdot 3\cdot 5\,\cdots\,(2n-1)}{2^{n+1}a^n}\sqrt{\dfrac{\pi}{a}}$

17. $\displaystyle\int \dfrac{dx}{\sqrt{x^2 + a^2}} = \ln(x + \sqrt{x^2 + a^2})$

18. $\displaystyle\int \dfrac{x\, dx}{(x^2 + a^2)^{3/2}} = -\dfrac{1}{(x^2 + a^2)^{1/2}}$

19. $\displaystyle\int \dfrac{dx}{(x^2 + a^2)^{3/2}} = \dfrac{x}{a^2(x^2 + a^2)^{1/2}}$

Appendix F
Properties of the Elements

All physical properties are for a pressure of 1 atm unless otherwise specified.

ELEMENT	SYMBOL	ATOMIC NUMBER, Z	MOLAR MASS, g/mol	DENSITY, g/cm³ AT 20°C	MELTING POINT, °C	BOILING POINT, °C	SPECIFIC HEAT, J/(g·°C) AT 25°C
Actinium	Ac	89	(227)	10.06	1323	(3473)	0.092
Aluminum	Al	13	26.9815	2.699	660	2450	0.900
Americium	Am	95	(243)	13.67	1541	—	—
Antimony	Sb	51	121.75	6.691	630.5	1380	0.205
Argon	Ar	18	39.948	1.6626×10^{-3}	−189.4	−185.8	0.523
Arsenic	As	33	74.9216	5.78	817 (28 atm)	613	0.331
Astatine	At	85	(210)	—	(302)	—	—
Barium	Ba	56	137.34	3.594	729	1640	0.205
Berkelium	Bk	97	(247)	14.79	—	—	—
Beryllium	Be	4	9.0122	1.848	1287	2770	1.83
Bismuth	Bi	83	208.980	9.747	271.37	1560	0.122
Boron	B	5	10.811	2.34	2030	—	1.11
Bromine	Br	35	79.909	3.12 (liquid)	−7.2	58	0.293
Cadmium	Cd	48	112.40	8.65	321.03	765	0.226
Calcium	Ca	20	40.08	1.55	838	1440	0.624
Californium	Cf	98	(251)	—	—	—	—
Carbon	C	6	12.01115	2.26	3727	4830	0.691
Cerium	Ce	58	140.12	6.768	804	3470	0.188
Cesium	Cs	55	132.905	1.873	28.40	690	0.243
Chlorine	Cl	17	35.453	3.214×10^{-3} (0°C)	−101	−34.7	0.486
Chromium	Cr	24	51.996	7.19	1857	2665	0.448
Cobalt	Co	27	58.9332	8.85	1495	2900	0.423
Copper	Cu	29	63.54	8.96	1083.40	2595	0.385
Curium	Cm	96	(247)	13.3	—	—	—
Dysprosium	Dy	66	162.50	8.55	1409	2330	0.172
Einsteinium	Es	99	(254)	—	—	—	—
Erbium	Er	68	167.26	9.15	1522	2630	0.167
Europium	Eu	63	151.96	5.243	817	1490	0.163
Fermium	Fm	100	(237)	—	—	—	—
Fluorine	F	9	18.9984	1.696×10^{-3} (0°C)	−219.6	−188.2	0.753
Francium	Fr	87	(223)	—	(27)	—	—
Gadolinium	Gd	64	157.25	7.90	1312	2730	0.234
Gallium	Ga	31	69.72	5.907	29.75	2237	0.377
Germanium	Ge	32	72.59	5.323	937.25	2830	0.322
Gold	Au	79	196.967	19.32	1064.43	2970	0.131
Hafnium	Hf	72	178.49	13.31	2227	5400	0.144
Hahnium	Ha	105	—	—	—	—	—
Hassium	Hs	108	—	—	—	—	—

continued on next page

ELEMENT	SYMBOL	ATOMIC NUMBER, Z	MOLAR MASS, g/mol	DENSITY, g/cm^3 AT 20°C	MELTING POINT, °C	BOILING POINT, °C	SPECIFIC HEAT, J/(g·°C) AT 25°C
Helium	He	2	4.0026	0.1664×10^{-3}	−269.7	−268.9	5.23
Holmium	Ho	67	164.930	8.79	1470	2330	0.165
Hydrogen	H	1	1.00797	0.08375×10^{-3}	−259.19	−252.7	14.4
Indium	In	49	114.82	7.31	156.634	2000	0.233
Iodine	I	53	126.9044	4.93	113.7	183	0.218
Iridium	Ir	77	192.2	22.5	2447	(5300)	0.130
Iron	Fe	26	55.847	7.874	1536.5	3000	0.447
Krypton	Kr	36	83.80	3.488×10^{-3}	−157.37	−152	0.247
Lanthanum	La	57	138.91	6.189	920	3470	0.195
Lawrencium	Lr	103	(257)		—	—	
Lead	Pb	82	207.19	11.35	327.45	1725	0.129
Lithium	Li	3	6.939	0.534	180.55	1300	3.58
Lutetium	Lu	71	174.97	9.849	1663	1930	0.155
Magnesium	Mg	12	24.312	1.738	650	1107	1.03
Manganese	Mn	25	54.9380	7.44	1244	2150	0.481
Meitnerium	Mt	109	—	—	—	—	—
Mendelevium	Md	101	(256)	—	—	—	—
Mercury	Hg	80	200.59	13.55	−38.87	357	0.138
Molybdenum	Mo	42	95.94	10.22	2617	5560	0.251
Neodymium	Nd	60	144.24	7.007	1016	3180	0.188
Neon	Ne	10	20.183	0.8387×10^{-3}	−248.597	−246.0	1.03
Neptunium	Np	93	(237)	20.25	637	—	1.26
Nickel	Ni	28	58.71	8.902	1453	2730	0.444
Nielsbohrium	Ns	107	—	—	—	—	—
Niobium	Nb	41	92.906	8.57	2468	4927	0.264
Nitrogen	N	7	14.0067	1.1649×10^{-3}	−210	−195.8	1.03
Nobelium	No	102	(255)	—	—	—	—
Osmium	Os	76	190.2	22.59	3027	5500	0.130
Oxygen	O	8	15.9994	1.3318×10^{-3}	−218.80	−183.0	0.913
Palladium	Pd	46	106.4	12.02	1552	3980	0.243
Phosphorus	P	15	30.9738	1.83	44.25	280	0.741
Platinum	Pt	78	195.09	21.45	1769	4530	0.134
Plutonium	Pu	94	(244)	19.8	640	3235	0.130
Polonium	Po	84	(210)	9.32	254	—	—
Potassium	K	19	39.102	0.862	63.20	760	0.758
Praseodymium	Pr	59	140.907	6.773	931	3020	0.197
Promethium	Pm	61	(145)	7.22	(1027)	—	—
Protactinium	Pa	91	(231)	15.37 (estimated)	(1230)	—	—
Radium	Ra	88	(226)	5.0	700	—	—
Radon	Rn	86	(222)	9.96×10^{-3} (0°C)	(−71)	−61.8	0.092
Rhenium	Re	75	186.2	21.02	3180	5900	0.134
Rhodium	Rh	45	102.905	12.41	1963	4500	0.243
Rubidium	Rb	37	85.47	1.532	39.49	688	0.364
Ruthenium	Ru	44	101.107	12.37	2250	4900	0.239
Rutherfordium	Rf	104	—	—	—	—	—
Samarium	Sm	62	150.35	7.52	1072	1630	0.197

continued on next page

ELEMENT	SYMBOL	ATOMIC NUMBER, Z	MOLAR MASS, g/mol	DENSITY, g/cm³ AT 20°C	MELTING POINT, °C	BOILING POINT, °C	SPECIFIC HEAT, J/(g·°C) AT 25°C
Scandium	Sc	21	44.956	2.99	1539	2730	0.569
Seaborgium	Sg	106	—	—	—	—	—
Selenium	Se	34	78.96	4.79	221	685	0.318
Silicon	Si	14	28.086	2.33	1412	2680	0.712
Silver	Ag	47	107.870	10.49	960.8	2210	0.234
Sodium	Na	11	22.9898	0.9712	97.85	892	1.23
Strontium	Sr	38	87.62	2.54	768	1380	0.737
Sulfur	S	16	32.064	2.07	119.0	444.6	0.707
Tantalum	Ta	73	180.948	16.6	3014	5425	0.138
Technetium	Tc	43	(99)	11.46	2200	—	0.209
Tellurium	Te	52	127.60	6.24	449.5	990	0.201
Terbium	Tb	65	158.924	8.229	1357	2530	0.180
Thallium	Tl	81	204.37	11.85	304	1457	0.130
Thorium	Th	90	(232)	11.72	1755	(3850)	0.117
Thulium	Tm	69	168.934	9.32	1545	1720	0.159
Tin	Sn	50	118.69	7.2984	231.868	2270	0.226
Titanium	Ti	22	47.90	4.54	1670	3260	0.523
Tungsten	W	74	183.85	19.3	3380	5930	0.134
Uranium	U	92	(238)	18.95	1132	3818	0.117
Vanadium	V	23	50.942	6.11	1902	3400	0.490
Xenon	Xe	54	131.30	5.495×10^{-3}	-111.79	-108	0.159
Ytterbium	Yb	70	173.04	6.965	824	1530	0.155
Yttrium	Y	39	88.905	4.469	1526	3030	0.297
Zinc	Zn	30	65.37	7.133	419.58	906	0.389
Zirconium	Zr	40	91.22	6.506	1852	3580	0.276

The values in parentheses in the column of molar masses are the mass numbers of the longest-lived isotopes of those elements that are radioactive. Melting points and boiling points in parentheses are uncertain.

The data for gases are valid only when these are in their usual molecular state, such as H_2, He, O_2, Ne, etc. The specific heats of the gases are the values at constant pressure.

Source: Adapted from Wehr, Richards, Adair, *Physics of the Atom,* 4th ed., Addison-Wesley, Reading, MA, 1984, and from J. Emsley, *The Elements,* 2nd ed., Clarendon Press, Oxford, 1991.

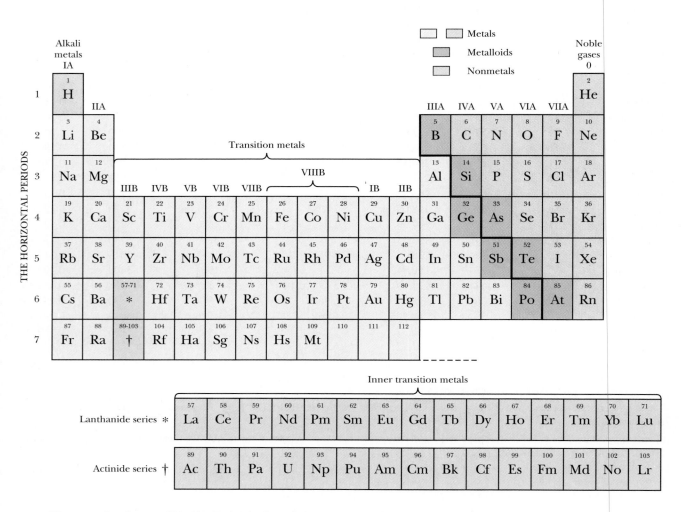

The names for elements 104–109 (Rutherfordium, Hahnium, Seaborgium, Nielsbohrium, Hassium, and Meitnerium, respectively) are those recommended by the American Chemical Society Nomenclature Committee. As of 1996, the names and symbols for elements 104–108 have not yet been approved by the appropriate international body. Elements 110, 111 and 112 have been discovered but, as of 1996, have not been provisionally named.

Appendix **H**
*Winners of the Nobel Prize in Physics**

1901 Wilhelm Konrad Röntgen *(1845–1923)* for the discovery of x rays

1902 Hendrik Antoon Lorentz *(1853–1928)* and Pieter Zeeman *(1865–1943)* for their researches into the influence of magnetism upon radiation phenomena

1903 Antoine Henri Becquerel *(1852–1908)* for his discovery of spontaneous radioactivity

Pierre Curie *(1859–1906)* and Marie Sklowdowska-Curie *(1867–1934)* for their joint researches on the radiation phenomena discovered by Becquerel

1904 Lord Rayleigh (John William Strutt) *(1842–1919)* for his investigations of the densities of the most important gases and for his discovery of argon

1905 Philipp Eduard Anton von Lenard *(1862–1947)* for his work on cathode rays

1906 Joseph John Thomson *(1856–1940)* for his theoretical and experimental investigations on the conduction of electricity by gases

1907 Albert Abraham Michelson *(1852–1931)* for his optical precision instruments and metrological investigations carried out with their aid

1908 Gabriel Lippmann *(1845–1921)* for his method of reproducing colors photographically based on the phenomena of interference

1909 Guglielmo Marconi *(1874–1937)* and Carl Ferdinand Braun *(1850–1918)* for their contributions to the development of wireless telegraphy

1910 Johannes Diderik van der Waals *(1837–1923)* for his work on the equation of state for gases and liquids

1911 Wilhelm Wien *(1864–1928)* for his discoveries regarding the laws governing the radiation of heat

1912 Nils Gustaf Dalén *(1869–1937)* for his invention of automatic regulators for use in conjunction with gas accumulators for illuminating lighthouses and buoys

1913 Heike Kamerlingh Onnes *(1853–1926)* for his investigations of the properties of matter at low temperatures which led, among other things, to the production of liquid helium

1914 Max von Laue *(1879–1960)* for his discovery of the diffraction of Röntgen rays by crystals

1915 William Henry Bragg *(1862–1942)* and William Lawrence Bragg *(1890–1971)* for their services in the analysis of crystal structure by means of x rays

1917 Charles Glover Barkla *(1877–1944)* for his discovery of the characteristic x rays of the elements

1918 Max Planck *(1858–1947)* for his discovery of energy quanta

1919 Johannes Stark *(1874–1957)* for his discovery of the Doppler effect in canal rays and the splitting of spectral lines in electric fields

1920 Charles-Édouard Guillaume *(1861–1938)* for the service he rendered to precision measurements in physics by his discovery of anomalies in nickel steel alloys

1921 Albert Einstein *(1879–1955)* for his services to theoretical physics, and especially for his discovery of the law of the photoelectric effect

1922 Niels Bohr *(1885–1962)* for the investigation of the structure of atoms, and of the radiation emanating from them

1923 Robert Andrews Millikan *(1868–1953)* for his work on the elementary charge of electricity and on the photoelectric effect

1924 Karl Manne Georg Siegbahn *(1886–1978)* for his discoveries and research in the field of x-ray spectroscopy

1925 James Franck *(1882–1964)* and Gustav Hertz *(1887–1975)* for their discovery of the laws governing the impact of an electron upon an atom

1926 Jean Baptiste Perrin *(1870–1942)* for his work on the discontinuous structure of matter, and especially for his discovery of sedimentation equilibrium

1927 Arthur Holly Compton *(1892–1962)* for his discovery of the effect named after him

Charles Thomson Rees Wilson (1869–1959) for his method of making the paths of electrically charged particles visible by condensation of vapor

1928 Owen Willans Richardson *(1879–1959)* for his work on the thermionic phenomenon and especially for the discovery of the law named after him

1929 Prince Louis Victor de Broglie *(1892–1987)* for his discovery of the wave nature of electrons

*See *Nobel Lectures, Physics,* 1901–1970, Elsevier Publishing Company, for biographies of the awardees and for lectures given by them on receiving the prize.

1930 Sir Chandrasekhara Venkata Raman *(1888–1970)* for his work on the scattering of light and for the discovery of the effect named after him

1932 Werner Heisenberg *(1901–1976)* for the creation of quantum mechanics, the application of which has, among other things, led to the discovery of the allotropic forms of hydrogen

1933 Erwin Schrödinger *(1887–1961)* and Paul Adrien Maurice Dirac *(1902–1984)* for the discovery of new productive forms of atomic theory

1935 James Chadwick *(1891–1974)* for his discovery of the neutron

1936 Victor Franz Hess *(1883–1964)* for the discovery of cosmic radiation

Carl David Anderson *(1905–1991)* for his discovery of the positron

1937 Clinton Joseph Davisson *(1881–1958)* and George Paget Thomson *(1892–1975)* for their experimental discovery of the diffraction of electrons by crystals

1938 Enrico Fermi *(1901–1954)* for his demonstrations of the existence of new radioactive elements produced by neutron irradiation, and for his related discovery of nuclear reactions brought about by slow neutrons

1939 Ernest Orlando Lawrence *(1901–1958)* for the invention and development of the cyclotron and for results obtained with it, especially for artificial radioactive elements

1943 Otto Stern *(1888–1969)* for his contribution to the development of the molecular-ray method and his discovery of the magnetic moment of the proton

1944 Isidor Isaac Rabi *(1898–1988)* for his resonance method for recording the magnetic properties of atomic nuclei

1945 Wolfgang Pauli *(1900–1958)* for the discovery of the Exclusion Principle (also called Pauli Principle)

1946 Percy Williams Bridgman *(1882–1961)* for the invention of an apparatus to produce extremely high pressures and for the discoveries he made therewith in the field of high-pressure physics

1947 Sir Edward Victor Appleton *(1892–1965)* for his investigations of the physics of the upper atmosphere, especially for the discovery of the so-called Appleton layer

1948 Patrick Maynard Stuart Blackett *(1897–1974)* for his development of the Wilson cloud-chamber method, and his discoveries therewith in nuclear physics and cosmic radiation

1949 Hideki Yukawa *(1907–1981)* for his prediction of the existence of mesons on the basis of theoretical work on nuclear forces

1950 Cecil Frank Powell *(1903–1969)* for his development of the photographic method of studying nuclear processes and his discoveries regarding mesons made with this method

1951 Sir John Douglas Cockcroft *(1897–1967)* and Ernest Thomas Sinton Walton *(1903–1995)* for their pioneer work on the transmutation of atomic nuclei by artificially accelerated atomic particles

1952 Felix Bloch *(1905–1983)* and Edward Mills Purcell *(1912–   )* for their development of new nuclear-magnetic precision methods and discoveries in connection therewith

1953 Frits Zernike *(1888–1966)* for his demonstration of the phase-contrast method, especially for his invention of the phase-contrast microscope

1954 Max Born *(1882–1970)* for his fundamental research in quantum mechanics, especially for his statistical interpretation of the wave function

Walther Bothe *(1891–1957)* for the coincidence method and his discoveries made therewith

1955 Willis Eugene Lamb *(1913–   )* for his discoveries concerning the fine structure of the hydrogen spectrum

Polykarp Kusch *(1911–1993)* for his precision determination of the magnetic moment of the electron

1956 William Shockley *(1910–1989)*, John Bardeen *(1908–1991)* and Walter Houser Brattain *(1902–1987)* for their researches on semiconductors and their discovery of the transistor effect

1957 Chen Ning Yang *(1922–   )* and Tsung Dao Lee *(1926–   )* for their penetrating investigation of the parity laws which has led to important discoveries regarding the elementary particles

1958 Pavel Aleksejevič Čerenkov *(1904–   )*, Il' ja Michajlovič Frank *(1908–1990)* and Igor' Evgen' evič Tamm *(1895–1971)* for the discovery and interpretation of the Cerenkov effect

1959 Emilio Gino Segrè *(1905–1989)* and Owen Chamberlain *(1920–   )* for their discovery of the antiproton

1960 Donald Arthur Glaser *(1926–   )* for the invention of the bubble chamber

1961 Robert Hofstadter *(1915–1990)* for his pioneering studies of electron scattering in atomic nuclei and for his thereby achieved discoveries concerning the structure of the nucleons

Rudolf Ludwig Mössbauer *(1929–   )* for his researches concerning the resonance absorption of γ rays and his discovery in this connection of the effect which bears his name

1962 Lev Davidovič Landau *(1908–1968)* for his pioneering theories of condensed matter, especially liquid helium

1963 Eugene P. Wigner *(1902–1995)* for his contributions to the theory of the atomic nucleus and the elementary particles, particularly through the discovery and application of fundamental symmetry principles

Maria Goeppert Mayer *(1906–1972)* and J. Hans D. Jensen *(1907–1973)* for their discoveries concerning nuclear shell structure

1964 Charles H. Townes *(1915–   )*, Nikolai G. Basov *(1922–   )* and Alexander M. Prochorov *(1916–   )* for fundamental work in the field of quantum electronics which has led to the construction of oscillators and amplifiers based on the maser–laser principle

1965 Sin-itiro Tomonaga *(1906–1979)*, Julian Schwinger *(1918–1994)* and Richard P. Feynman *(1918–1988)* for their fundamental work in quantum electrodynamics, with deep-ploughing consequences for the physics of elementary particles

1966 Alfred Kastler *(1902–1984)* for the discovery and development of optical methods for studying Hertzian resonance in atoms

1967 Hans Albrecht Bethe *(1906–   )* for his contributions to the theory of nuclear reactions, especially his discoveries concerning the energy production in stars

1968 Luis W. Alvarez *(1911–1988)* for his decisive contribution to elementary particle physics, in particular the discovery of a large number of resonance states, made possible through his development of the techniques of using the hydrogen bubble chamber and its data analysis

1969 Murray Gell-Mann *(1929–   )* for his contributions and discoveries concerning the classification of elementary particles and their interactions

1970 Hannes Alfvén *(1908–1995)* for fundamental work and discoveries in magneto-hydrodynamics with fruitful applications in different parts of plasma physics

Louis Néel *(1904–   )* for fundamental work and discoveries concerning antiferromagnetism and ferrimagnetism which have led to important applications in solid state physics

1971 Dennis Gabor *(1900–1979)* for his discovery of the principles of holography

1972 John Bardeen *(1908–1991)*, Leon N. Cooper *(1930–   )* and J. Robert Schrieffer *(1931–   )* for their development of a theory of superconductivity

1973 Leo Esaki *(1925–   )* for his discovery of tunneling in semiconductors

Ivar Giaever *(1929–   )* for his discovery of tunneling in superconductors

Brian D. Josephson *(1940–   )* for his theoretical prediction of the properties of a supercurrent through a tunnel barrier

1974 Antony Hewish *(1924–   )* for the discovery of pulsars

Sir Martin Ryle *(1918–1984)* for his pioneering work in radioastronomy

1975 Aage Bohr *(1922–   )*, Ben Mottelson *(1926–   )* and James Rainwater *(1917–1986)* for the discovery of the connection between collective motion and particle motion and the development of the theory of the structure of the atomic nucleus based on this connection

1976 Burton Richter *(1931–   )* and Samuel Chao Chung Ting *(1936–   )* for their (independent) discovery of an important fundamental particle

1977 Philip Warren Anderson *(1923–   )*, Nevill Francis Mott *(1905–1996)* and John Hasbrouck Van Vleck *(1899–1980)* for their fundamental theoretical investigations of the electronic structure of magnetic and disordered systems

1978 Peter L. Kapitza *(1894–1984)* for his basic inventions and discoveries in low-temperature physics

Arno A. Penzias *(1933–   )* and Robert Woodrow Wilson *(1936–   )* for their discovery of cosmic microwave background radiation

1979 Sheldon Lee Glashow *(1932–   )*, Abdus Salam *(1926–1996)*, and Steven Weinberg *(1933–   )* for their unified model of the action of the weak and electromagnetic forces and for their prediction of the existence of neutral currents

1980 James W. Cronin *(1931–   )* and Val L. Fitch *(1923–   )* for the discovery of violations of fundamental symmetry principles in the decay of neutral K mesons

1981 Nicolaas Bloembergen *(1920–   )* and Arthur Leonard Schawlow *(1921–   )* for their contribution to the development of laser spectroscopy

Kai M. Siegbahn *(1918–   )* for his contribution to high-resolution electron spectroscopy

1982 Kenneth Geddes Wilson *(1936–   )* for his method of analyzing the critical phenomena inherent in the changes of matter under the influence of pressure and temperature

1983 Subrehmanyan Chandrasekhar *(1910–1995)* for his theoretical studies of the structure and evolution of stars

William A. Fowler *(1911–1995)* for his studies of the formation of the chemical elements in the universe

1984 Carlo Rubbia *(1934–   )* and Simon van der Meer *(1925–   )* for their decisive contributions to the Large Project, which led to the discovery of the field particles W and Z, communicators of the weak interaction

1985 Klaus von Klitzing *(1943–   )* for his discovery of the quantized Hall resistance

1986 Ernst Ruska *(1906–1988)* for his invention of the electron microscope

Gerd Binnig *(1947–   )*, Heinrich Rohrer *(1933–   )* for their invention of the scanning tunneling microscope

1987 Karl Alex Müller *(1927–   )* and J. George Bednorz *(1950–   )* for their discovery of a new class of superconductors

1988 Leon M. Lederman *(1922–   )*, Melvin Schwartz *(1932–   )* and Jack Steinberger *(1921–   )* for the first use of a neutrino beam and the discovery of the muon neutrino

1989 Norman Ramsey *(1915–   )*, Hans Dehmelt *(1922–   )* and Wolfgang Paul *(1913–1993)* for their work that led to the development of atomic clocks and precision timing

1990 Jerome I. Friedman *(1930–   )*, Henry W. Kendall *(1926–   )* and Richard E. Taylor *(1929–   )* for demonstrating that protons and neutrons consist of quarks

1991 Pierre de Gennes *(1932–   )* for studies of order phenomena, such as in liquid crystals and polymers

1992 George Charpak *(1924–   )* for his invention of fast electronic detectors for high-energy particles

1993 Joseph H. Taylor *(1941–   )* and Russell A. Hulse *(1950–   )* for the discovery and interpretation of the first binary pulsar.

1994 Bertram N. Brockhouse *(1918–   )* and Clifford G. Shull *(1915–   )* for the development of neutron scattering techniques

1995 Martin L. Perl *(1927–   )* for the discovery of the tau lepton

Frederick Reines *(1918–   )* for the detection of the neutrino

1996 David M. Lee *(1931–   )*, Robert C. Richardson *(1937–   )* and Douglas D. Oscheroff *(1944–   )* for the discovery of superfluidity in helium-3

Answers to Checkpoints, Odd-Numbered Questions, Exercises, and Problems

Chapter 1

EP **3.** (a) 186 mi; (b) 3.0×10^8 mm **5.** (a) 10^9; (b) 10^{-4}; (c) 9.1×10^5 **7.** 32.2 km **9.** 0.020 km^3 **11.** (a) 250 ft^2; (b) 23.3 m^2; (c) 3060 ft^3; (d) 86.6 m^3 **13.** 8×10^2 km **15.** (a) 11.3 m^2/L; (b) 1.13×10^4 m^{-1}; (c) 2.17×10^{-3} gal/ft^2 **17.** (a) $d_{Sun}/d_{Moon} = 400$; (b) $V_{Sun}/V_{Moon} = 6.4 \times 10^7$; (c) 3.5×10^3 km **19.** (a) 0.98 ft/ns; (b) 0.30 mm/ps **21.** 3.156×10^7 s **23.** 5.79×10^{12} days **25.** (a) 0.013; (b) 0.54; (c) 10.3; (d) 31 m/s **27.** 15° **29.** 3.3 ft **31.** 2 days 5 hours **33.** (a) 2.99×10^{-26} kg; (b) 4.68×10^{46} **35.** 1.3×10^9 kg **37.** (a) 10^3 kg/m^3; (b) 158 kg/s **39.** (a) 1.18×10^{-29} m^3; (b) 0.282 nm

Chapter 2

CP **1.** b and c **2.** zero **3.** (a) 1 and 4; (b) 2 and 3; (c) 3 **4.** (a) plus; (b) minus; (c) minus; (d) plus **5.** 1 and 4 **6.** (a) plus; (b) minus; (c) $a = -g = -9.8$ m/s^2 **Q** **1.** (a) yes; (b) no; (c) yes; (d) yes **3.** (a) 2, 3; (b) 1, 3; (c) 4 **5.** all tie (see Eq. 2-16) **7.** (a) $-g$; (b) 2 m/s upward **9.** same **11.** $x = t^2$ and $x = 8(t - 2) + (1.5)(t - 2)^2$ **13.** increase **EP** **1.** (a) Lewis: 10.0 m/s, Rodgers: 5.41 m/s; (b) 1 h 10 min **3.** 309 ft **5.** 2 cm/y **7.** 6.71×10^8 mi/h, 9.84×10^8 ft/s, 1.00 ly/y **9.** (a) 5.7 ft/s; (b) 7.0 ft/s **11.** (a) 45 mi/h (72 km/h); (b) 43 mi/h (69 km/h); (c) 44 mi/h (71 km/h); (d) 0 **13.** (a) 28.5 cm/s; (b) 18.0 cm/s; (c) 40.5 cm/s; (d) 28.1 cm/s; (e) 30.3 cm/s **15.** (a) mathematically, an infinite number; (b) 60 km **17.** (a) 4 s > t > 2 s; (b) 3 s > t > 0; (c) 7 s > t > 3 s; (d) $t = 3$ s **19.** 100 m **23.** (a) The signs of v and a are: *AB*: +, −; *BC*: 0, 0; *CD*: +, +; *DE*: +, 0; (b) no; (c) no **25.** (e) situations (a), (b), and (d) **27.** (a) 80 m/s; (b) 110 m/s; (c) 20 m/s^2 **29.** (a) 1.10 m/s, 6.11 mm/s^2; (b) 1.47 m/s, 6.11 mm/s^2 **31.** (a) 2.00 s; (b) 12 cm from left edge of screen; (c) 9.00 cm/s^2, to the left; (d) to the right; (e) to the left; (f) 3.46 s **33.** 0.556 s **35.** each, 0.28 m/s^2 **37.** 2.8 m/s^2 **39.** 1.62×10^{15} m/s^2 **41.** 21g **43.** (a) 25g; (b) 400 m **45.** 90 m **47.** (a) 5.0 m/s^2; (b) 4.0 s; (c) 6.0 s; (d) 90 m **49.** (a) 5.00 m/s; (b) 1.67 m/s^2; (c) 7.50 m **51.** (a) 0.74 s; (b) −20 ft/s^2 **53.** (a) 0.75 s; (b) 50 m **55.** (a) 34.7 ft; (b) 41.6 s **57.** (a) 3.26 ft/s^2 **61.** (a) 31 m/s; (b) 6.4 s **63.** (a) 48.5 m/s; (b) 4.95 s; (c) 34.3 m/s; (d) 3.50 s **65.** (a) 5.44 s; (b) 53.3 m/s; (d) 5.80 m **67.** (a) 3.2 s; (b) 1.3 s **69.** 4.0 m/s **71.** (a) 350 ms; (b) 82 ms (each is for ascent and descent through the 15 cm) **73.** 857 m/s^2, upward **75.** (a) 1.23 cm; (b) 4 times, 9 times, 16 times, 25 times **77.** (a) 8.85 m/s; (b) 1.00 m **79.** 22 cm and 89 cm below the nozzle **81.** (a) 3.41 s; (b) 57 m **83.** (a) 40.0 ft/s **85.** 1.5 s **87.** (a) 5.4 s; (b) 41 m/s **89.** 20.4 m

91. (a) $d = v_f^2/2a' + T_R v_i$; (b) 9.0 m/s^2; (c) 0.66 s. **93.** (a) $v_j^2 = 2a'd_0(j - 1) + v_1^2$; (c) 7.0 m/s^2; (d) 14 m.

Chapter 3

CP **1.** (a) 7 m; (b) 1 m **2.** *c, d, f* **3.** (a) +, +; (b) +, −; (c) +, + **4.** (a) 90°; (b) 0 (vectors are parallel); (c) 180° (vectors are antiparallel) **5.** (a) 0° or 180°; (b) 90° **Q** **1.** **A** and **B** **3.** No, but **a** and −**b** are commutative: **a** + (−**b**) = (−**b**) + **a**. **5.** (a) **a** and **b** are parallel; (b) **b** = 0; (c) **a** and **b** are perpendicular **7.** (a)−(c) yes (example: 5**i** and −2**i**) **9.** all but *e* **11.** (a) minus, minus; (b) minus, minus **13.** (a) **B** and **C**, **D** and **E**; (b) **D** and **E** **15.** no (their orientations can differ) **17.** (a) 0 (vectors are parallel); (b) 0 (vectors are antiparallel) **EP** **1.** The displacements should be (a) parallel, (b) antiparallel, (c) perpendicular **3.** (b) 3.2 km, 41° south of west **5.** **a** + **b**: 4.2, 40° east of north; **b** − **a**: 8.0, 24° north of west **7.** (a) 38 units at 320°; (b) 130 units at 1.2°; (c) 62 units at 130° **9.** $a_x = -2.5$, $a_y = -6.9$ **11.** $r_x = 13$ m, $r_y = 7.5$ m **13.** (a) 14 cm, 45° left of straight down; (b) 20 cm, vertically up; (c) zero **15.** 4.74 km **17.** 168 cm, 32.5° above the floor **19.** $r_x = 12$, $r_y = -5.8$, $r_z = -2.8$ **21.** (a) 8**i** + 2**j**, 8.2, 14°; (b) 2**i** − 6**j**, 6.3, −72° relative to **i** **23.** (a) 5.0, −37°; (b) 10, 53°; (c) 11, 27°; (d) 11, 80°; (e) 11, 260°; the angles are relative to $+x$, the last two vectors are in opposite directions **25.** 4.1 **27.** (a) $r_x = 1.59$, $r_y = 12.1$; (b) 12.2; (c) 82.5° **29.** 3390 ft, horizontally **31.** (a) −2.83 m, −2.83 m, +5.00 m, 0 m, 3.00 m, 5.20 m; (b) 5.17 m, 2.37 m; (c) 5.69 m, 24.6° north of east; (d) 5.69 m, 24.6° south of west **35.** (a) $a_x = 9.51$ m, $a_y = 14.1$ m; (b) $a_x' = 13.4$ m, $a_y' = 10.5$ m **37.** (a) +y; (b) −y; (c) 0; (d) 0; (e) +z; (f) −z; (g) ab, both; (h) ab/d, +z **39.** yes **41.** (a) up, unit magnitude; (b) zero; (c) south, unit magnitude; (d) 1.00; (e) 0 **43.** (a) −18.8; (b) 26.9, +z direction **45.** (a) 12, out of page; (b) 12, into page; (c) 12, out of page **47.** (a) 11**i** + 5**j** − 7**k**; (b) 120° **51.** (a) 57°; (b) $c_x = \pm 2.2$, $c_y = \mp 4.5$ **53.** (a) −21; (b) −9; (c) 5**i** − 11**j** − 9**k**

Chapter 4

CP **1.** (a) $(8\mathbf{i} - 6\mathbf{j})$ m; (b) yes, the xy plane **2.** (a) first; (b) third **3.** (1) and (3) a_x and a_y are both constant and thus **a** is constant; (2) and (4) a_y is constant but a_x is not, thus **a** is not **4.** 4 m/s^3, −2 m/s, 3 m **5.** (a) v_x constant; (b) v_y initially positive, decreases to zero, and then becomes progressively more negative; (c) $a_x = 0$ throughout; (d) $a_y = -g$ throughout **6.** (a) −(4 m/s)**i**; (b) −(8 m/s^2)**j** **7.** (1) 0, distance not changing; (2) +70 km/h, distance increasing; (3) +80 km/h, distance decreasing **Q** **1.** (1) and (3) a_y is constant but a_x is not and thus **a** is not; (2) a_x is constant but a_y

is not and thus **a** is not; (4) a_x and a_y are both constant and thus **a** is constant; -2 m/s^2, 3 m/s **3.** (a) highest point; (b) lowest point **5.** (a) all tie; (b) 1 and 2 tie (the rocket is shot upward), then 3 and 4 tie (it is shot into the ground!) **7.** $(2\mathbf{i} - 4\mathbf{j})$ m/s **9.** (a) all tie; (b) all tie; (c) c, b, a; (d) c, b, a **11.** (a) no; (b) same **13.** (a) in your hands; (b) behind you; (c) in front of you **15.** (a) straight down; (b) curved; (c) more curved **17.** (a) 3; (b) 4. **EP** **1.** (a) $(-5.0\mathbf{i} + 8.0\mathbf{j})$ m; (b) 9.4 m, 122° from $+x$; (d) $(8\mathbf{i} - 8\mathbf{j})$ m; (e) 11 m, $-45°$ from $+x$ **3.** (a) $(-7.0\mathbf{i} + 12\mathbf{j})$ m; (b) xy plane **5.** (a) 671 mi, 63.4° south of east; (b) 298 mi/h, 63.4° south of east; (c) 400 mi/h **7.** (a) 6.79 km/h; (b) 6.96° **9.** (a) $(3\mathbf{i} - 8t\mathbf{j})$ m/s; (b) $(3\mathbf{i} - 16\mathbf{j})$ m/s; (c) 16 m/s, $-79°$ to $+x$ **11.** (a) $(8t\mathbf{j} + \mathbf{k})$ m/s; (b) $8\mathbf{j}$ m/s^2 **13.** $(-2.10\mathbf{i} + 2.81\mathbf{j})$ m/s^2 **15.** (a) $-1.5\mathbf{j}$ m/s; (b) $(4.5\mathbf{i} - 2.25\mathbf{j})$ m **17.** 60.0° **19.** (a) 63 ms; (b) 1.6×10^3 ft/s **21.** (a) 2.0 ns; (b) 2.0 mm; (c) $(1.0 \times 10^9\mathbf{i} - 2.0 \times 10^8\mathbf{j})$ cm/s **23.** (a) 3.03 s; (b) 758 m; (c) 29.7 m/s **25.** (a) 16 m/s, 23° above the horizontal; (b) 27 m/s, 57° below the horizontal **27.** (a) 32.4 m; (b) -37.7 m **29.** (b) 76° **31.** (a) 51.8 m; (b) 27.4 m/s; (c) 67.5 m **33.** (a) 194 m/s; (b) 38° **35.** 1.9 in. **37.** (a) 11 m; (b) 23 m; (c) 17 m/s, 63° below horizontal **41.** (a) 73 ft; (b) 7.6°; (c) 1.0 s **43.** 23 ft/s **45.** (a) 11 m; (b) 45 m/s **47.** 30 m above the release point **49.** 19 ft/s **51.** (a) 202 m/s; (b) 806 m; (c) 161 m/s, -171 m/s **53.** (a) 20 cm; (b) no, the ball hits the net only 4.4 cm above the ground **55.** yes; its center passes about 4.1 ft above the fence **57.** (a) 9.00×10^{22} m/s^2, toward the center; (b) 1.52×10^{-16} s **59.** (a) 6.7×10^6 m/s; (b) 1.4×10^{-7} s **61.** (a) 7.49 km/s; (b) 8.00 m/s^2 **63.** (a) 0.94 m; (b) 19 m/s; (c) 2400 m/s^2, toward center; (d) 0.05 s **65.** (a) 1.3×10^5 m/s; (b) 7.9×10^5 m/s^2 or $(8.0 \times 10^4)g$, toward the center; (c) both answers increase **67.** (a) 0.034 m/s^2; (b) 84 min **69.** 2.58 cm/s^2 **71.** 160 m/s **73.** 36 s, no **75.** 0.018 mi/s^2 from either frame **77.** 130° **79.** 60° **81.** (a) 5.8 m/s; (b) 16.7 m; (c) 67° **83.** 185 km/h, 22° south of west **85.** (a) from 75° east of south; (b) 30° east of north; substitute west for east to get second solution **87.** (a) 30° upstream; (b) 69 min; (c) 80 min; (d) 80 min; (e) perpendicular to the current, the shortest possible time is 60 min **89.** $0.83c$ **91.** (a) $0.35c$; (b) $0.62c$ **93.** For launch angles from 5° to 70°, it always moves away from the launch site. For a 75° launch angle, it moves toward the site from 11.5 s to 18.5 s after launch. For an 80° launch angle, it moves toward the site from 10.5 s to 20.5 s after launch. For an 85° launch angle, it moves toward the site from 10.5 s to 20.5 s after launch. For a 90° launch angle, it moves toward the site from 10 s to 20.5 s after launch. **95.** (a) 1.6 s; (b) no; (c) 14 m/s; (d) yes **97.** (a) $\Delta\mathbf{D} = (1.0\text{ m})\mathbf{i} - (2.0\text{ m})\mathbf{j} + (1.0\text{ m})\mathbf{k}$; (b) 2.4 m; (c) $\bar{\mathbf{v}} = (0.025\text{ m/s})\mathbf{i} - (0.050\text{ m/s})\mathbf{j} + (0.025\text{ m/s})\mathbf{k}$ (d) cannot be determined without additional information

Chapter 5

CP **1.** $c, d,$ and e **2.** (a) and (b) 2 N, leftward (acceleration is zero in each situation) **3.** (a) and (b) 1, 4, 3, 2 **4.** (a) equal; (b) greater (acceleration is upward, thus net force on body must be upward) **5.** (a) equal; (b) greater; (c) less **6.** (a) increase; (b) yes; (c) same; (d) yes **7.** (a) $F \sin\theta$; (b) increase **8.** 0 **Q** **1.** (a) yes; (b) yes; (c) yes; (d) yes **3.** (a) 2 and 4; (b) 2 and 4 **5.** (a) 50 N, upward; (b) 150 N, upward **7.** (a) less; (b) greater **9.** (a) no; (b) no; (c) no **11.** (a) increases; (b) increases; (c) decreases; (d) decreases **13.** (a) 20 kg; (b) 18 kg; (c) 10 kg; (d) all tie; (e) 3, 2, 1 **15.** d, c, a, b **EP** **1.** (a) $F_x = 1.88$ N, $F_y = 0.684$ N; (b) $(1.88\mathbf{i} + 0.684\mathbf{j})$ N **3.** (a) $(-6.26\mathbf{i} - 3.23\mathbf{j})$ N; (b) 7.0 N, 207° relative to $+x$ **5.** $(-2\mathbf{i} + 6\mathbf{j})$ N **7.** (a) 0; (b) $+20$ N; (c) -20 N; (d) -40 N; (e) -60 N **9.** (a) $(1\mathbf{i} - 1.3\mathbf{j})$ m/s^2; (b) 1.6 m/s^2 at $-50°$ from $+x$ **11.** (a) $\mathbf{F}_2$ and $\mathbf{F}_3$ are in the $-x$ direction, $\mathbf{a} = 0$; (b) $\mathbf{F}_2$ and $\mathbf{F}_3$ are in the $-x$ direction, $\mathbf{a}$ is on the x axis, $a = 0.83$ m/s^2; (c) $\mathbf{F}_2$ and $\mathbf{F}_3$ are at 34° from $-x$ direction; $\mathbf{a} = 0$ **13.** (a) 22 N, 2.3 kg; (b) 1100 N, 110 kg; (c) 1.6×10^4 N, 1.6×10^3 kg **15.** (a) 11 N, 2.2 kg; (b) 0, 2.2 kg **17.** (a) 44 N; (b) 78 N; (c) 54 N; (d) 152 N **19.** 1.18×10^4 N **21.** 1.2×10^5 N **23.** 16 N **25.** (a) 13 ft/s^2; (b) 190 lb **27.** (a) 42 N; (b) 72 N; (c) 4.9 m/s^2 **29.** (a) 0.02 m/s^2; (b) 8×10^4 km; (c) 2×10^3 m/s **31.** (a) 1.1×10^{-15} N; (b) 8.9×10^{-30} N **33.** (a) 5500 N; (b) 2.7 s; (c) 4 times as far; (d) twice the time **35.** (a) 4.9×10^5 N; (b) 1.5×10^6 N **37.** (a) 110 lb, up; (b) 110 lb, down **39.** (a) 0.74 m/s^2; (b) 7.3 m/s^2 **41.** (a) $\cos\theta$; (b) $\sqrt{\cos\theta}$ **43.** 1.8×10^4 N **45.** (a) 4.6×10^3 N; (b) 5.8×10^3 N **47.** (a) 250 m/s^2; (b) 2.0×10^4 N **49.** 23 kg **51.** (a) 620 N; (b) 580 N **53.** 1.9×10^5 lb **55.** (a) rope breaks; (b) 1.6 m/s^2 **57.** 4.6 N **59.** (a) allow a downward acceleration with magnitude ≥ 4.2 ft/s^2; (b) 13 ft/s or greater **61.** 195 N, up **63.** (a) 566 N; (b) 1130 N **65.** 18,000 N **67.** (a) 1.4×10^4 N; (b) 1.1×10^4 N; (c) 2700 N, toward the counterweight **69.** 6800 N, at 21° to the line of motion of the barge **71.** (a) 4.6 m/s^2; (b) 2.6 m/s^2 **73.** (b) $Fl/(m + M)$; (c) $MFl/(m + M)$; (d) $F(m + 2M)/2(m + M)$ **75.** $T_1 = 13$ N, $T_2 = 20$ N, $a = 3.2$ m/s^2

Chapter 6

CP **1.** (a) zero (because there is no attempt at sliding); (b) 5 N; (c) no; (d) yes **2.** (a) same (10 N); (b) decreases; (c) decreases **3.** greater **4.** (a) **a** downward; **N** upward; (b) **a** and **N** upward **5.** (a) $4R_1$; (b) $4R_1$ **6.** (a) same; (b) increases; (c) increases **Q** **1.** They slide at the same angle for all orders. **3.** (a) upward; (b) horizontal, toward you; (c) no change; (d) increases; (e) increases **5.** The frictional force $\mathbf{f}_s$ is initially directed up the ramp, decreases in magnitude to zero, and then is directed down the ramp, increasing in magnitude until the magnitude reaches $f_{s,\text{max}}$; thereafter, the magnitude of the frictional force is f_k, which is a constant smaller value. **7.** (a) decreases; (b) decreases; (c) increases; (d) increases **9.** (a) zero; (b) infinite **11.** 4, 3; then 1, 2, and 5 tie **13.** (a) less; (b) greater **EP** **1.** (a) 200 N; (b) 120 N **3.** 2° **5.** 440 N

7. (a) 110 N; (b) 130 N; (c) no; (d) 46 N; (e) 17 N
9. (a) 90 N; (b) 70 N; (c) 0.89 m/s² **11.** (a) no; (b) $(-12\mathbf{i} + 5\mathbf{j})$ N **13.** 20° **15.** (a) 0.13 N; (b) 0.12 **17.** $\mu_s = 0.58$, $\mu_k = 0.54$ **19.** (a) 0.11 m/s², 0.23 m/s²; (b) 0.041, 0.029
21. 36 m **23.** (a) 300 N; (b) 1.3 m/s² **25.** (a) 66 N; (b) 2.3 m/s² **27.** (a) $\mu_k mg/(\sin\theta - \mu_k \cos\theta)$; (b) $\theta_0 = \tan^{-1}\mu_s$
29. (b) 3.0×10^7 N **31.** 100 N **33.** 3.3 kg **35.** (a) 11 ft/s²; (b) 0.46 lb; (c) blocks move independently
37. (a) 27 N; (b) 3.0 m/s² **39.** (a) 6.1 m/s², leftward; (b) 0.98 m/s², leftward **41.** (a) 3.0×10^5 N; (b) 1.2°
43. 9.9 s **45.** 3.75 **47.** 12 cm **49.** 68 ft
51. (a) 3210 N; (b) yes **53.** 0.078 **55.** (a) 0.72 m/s; (b) 2.1 m/s²; (c) 0.50 N **57.** $\sqrt{Mgr/m}$ **59.** (a) 30 cm/s; (b) 180 cm/s², radially inward; (c) 3.6×10^{-3} N, radially inward; (d) 0.37 **61.** (a) 275 N; (b) 877 N **63.** 874 N
65. (a) at the bottom of the circle; (b) 31 ft/s **67.** (a) 9.5 m/s; (b) 20 m **69.** 13° **71.** (a) 0.0338 N; (b) 9.77 N

Chapter 7
CP **1.** (a) decrease; (b) same; (c) negative, zero **2.** d, c, b, a **3.** (a) same; (b) smaller **4.** (a) positive; (b) negative; (c) zero **5.** zero **Q** **1.** all tie **3.** (a) increasing; (b) same; (c) same; (d) increasing **5.** (a) positive; (b) negative; (c) negative **7.** (a) positive; (b) zero; (c) negative; (d) negative; (e) zero; (f) positive **9.** all tie **11.** c, d, a and b tie; then f, e. **13.** (a) 3 m; (b) 3 m; (c) 0 and 6 m; (d) negative direction of x **15.** (a) A; (b) B **17.** twice **EP** **1.** 1.8×10^{13} J; **3.** (a) 3610 J; (b) 1900 J; (c) 1.1×10^{10} J **5.** (a) 1×10^5 megatons TNT; (b) 1×10^7 bombs **7.** father, 2.4 m/s; son, 4.8 m/s
9. (a) 200 N; (b) 700 m; (c) -1.4×10^5 J; (d) 400 N, 350 m, -1.4×10^5 J **11.** 5000 J **13.** 47 keV **15.** 7.9 J
17. 530 J **19.** -37 J **21.** (a) 314 J; (b) -155 J; (c) 0; (d) 158 J **23.** (a) 98 N; (b) 4.0 cm; (c) 3.9 J; (d) -3.9 J
25. (a) $-3Mgd/4$; (b) Mgd; (c) $Mgd/4$ (d) $\sqrt{gd/2}$ **27.** 25 J
31. -6 J **33.** (a) 12 J; (b) 4.0 m; (c) 18 J
35. (a) -0.043 J; (b) -0.13 J **37.** (a) 6.6 m/s; (b) 4.7 m
39. (a) up; (b) 5.0 cm; (c) 5.0 J **41.** 270 kW
43. 235 kW **45.** 490 W **47.** (a) 100 J; (b) 67 W; (c) 33 W **49.** 0.99 hp **51.** (a) 0; (b) -350 W
53. (a) 79.4 keV; (b) 3.12 MeV; (c) 10.9 MeV **55.** (a) 32 J; (b) 8 W; (c) 78°

Chapter 8
CP **1.** no **2.** 3, 1, 2 **3.** (a) all tie; (b) all tie **4.** (a) CD, AB, BC (zero); (b) positive direction of x **5.** 2, 1, 3
6. decrease **7.** (a) seventh excited state, with energy E_7; (b) 1.3 eV **Q** **1.** -40 J **3.** (c) and (d) tie; then (a) and (b) tie **5.** (a) all tie; (b) all tie **7.** (a) 3, 2, 1; (b) 1, 2, 3 **9.** less than (smaller decrease in potential energy)
11. (a) $E < 3$ J, $K < 2$ J; (b) $E < 5$ J, $K < 4$ J
13. (a) increasing; (b) decreasing; (c) decreasing; (d) constant in AB and BC, decreasing in CD **EP** **1.** 15 J
3. (a) 167 J; (b) -167 J; (c) 196 J; (d) 29 J **5.** (a) 0; (b) $mgh/2$; (c) mgh; (d) $mgh/2$; (e) mgh **7.** (a) -0.80 J; (b) -0.80 J; (c) $+1.1$ J **9.** (a) $mgL(1 - \cos\theta)$; (b) $-mgL(1 - \cos\theta)$; (c) $mgL(1 - \cos\theta)$ **11.** (a) 18 J;

(b) 0; (c) 30 J; (d) 0; (e) parts b and d **13.** (a) 2.08 m/s; (b) 2.08 m/s **15.** (a) $\sqrt{2gL}$; (b) $2\sqrt{gL}$; (c) $\sqrt{2gL}$
17. 830 ft **19.** (a) 6.75 J; (b) -6.75 J; (c) 6.75 J; (d) 6.75 J; (e) -6.75 J; (f) 0.459 m **21.** (a) 21.0 m/s; (b) 21.0 m/s
23. (a) 0.98 J; (b) -0.98 J; (c) 3.1 N/cm **25.** (a) 39.2 J; (b) 39.2 J; (c) 4.00 m **27.** (a) 54 m/s; (b) 52 m/s; (c) 76 m, below **29.** (a) 39 ft/s; (b) 4.3 in. **31.** (a) 300 J; (b) 93.8 J; (c) 6.38 m **33.** (a) 4.8 m/s; (b) 2.4 m/s
35. (a) $[v_0^2 + 2gL(1 - \cos\theta_0)]^{1/2}$; (b) $(2gL \cos\theta_0)^{1/2}$; (c) $[gL(3 + 2\cos\theta_0)]^{1/2}$ **37.** (a) $U(x) = -Gm_1m_2/x$; (b) $Gm_1m_2d/x_1(x_1 + d)$ **39.** (a) $8mg$ leftward and mg downward; (b) $2.5R$ **43.** $mgL/32$ **47.** (a) $1.12(A/B)^{1/6}$; (b) repulsive; (c) attractive **49.** (a) turning point on left, none on right; molecule breaks apart; (b) turning points on both left and right; molecule does not break apart; (c) -1.2×10^{-19} J; (d) 2.2×10^{-19} J; (e) $\approx 1 \times 10^{-9}$ on each, directed toward the other; (f) $r < 0.2$ nm; (g) $r > 0.2$ nm; (h) $r = 0.2$ nm
51. -25 J **53.** (a) 2200 J; (b) -1500 J; (c) 700 J
55. 17 kW **57.** (a) -0.74 J; (b) -0.53 J **59.** -12 J
61. 54% **63.** 880 MW **65.** (a) 39 kW; (b) 39 kW
67. (a) 1.5 MJ; (b) 0.51 MJ; (c) 1.0 MJ; (d) 63 m/s
69. (a) 67 J; (b) 67 J; (c) 46 cm **71.** Your force on the cabbage does work. **73.** (a) -0.90 J; (b) 0.46 J; (c) 1.0 m/s
75. (a) 18 ft/s; (b) 18 ft **77.** 4.3 m **79.** (a) 31.0 J; (b) 5.35 m/s; (c) conservative **81.** 1.2 m **85.** in the center of the flat part **87.** (a) 24 ft/s; (b) 3.0 ft; (c) 9.0 ft; (d) 49 ft **89.** (a) 216 J; (b) 1180 N; (c) 432 J; (d) motor also supplies thermal energy to crate and belt **91.** (a) 1.1×10^{17} J; (b) 1.2 kg **93.** 7.28 MeV **95.** (a) release; (b) 17.6 MeV **97.** (a) 5.3 eV; (b) 0.9 eV **99.** (a) 7.2 J; (b) -7.2 J; (c) 86 cm; (d) 26 cm

Chapter 9
CP **1.** (a) origin; (b) fourth quadrant; (c) on y axis below origin; (d) origin; (e) third quadrant; (f) origin **2.** (a) to (c) at the center of mass, still at the origin (their forces are internal to the system and cannot move the center of mass) **3.** (a) 1, 3, and then 2 and 4 tie (zero force); (b) 3 **4.** (a) 0; (b) no; (c) negative x **5.** (a) 500 km/h; (b) 2600 km/h; (c) 1600 km/h **6.** (a) yes; (b) no **Q** **1.** point 4 **3.** (a) at the center of the sled; (b) $L/4$, to the right; (c) not at all (no net external force); (d) $L/4$, to the left; (e) L; (f) $L/2$; (g) $L/2$
5. (a) ac, cd, and bc; (b) bc; (c) bd and ad **7.** (a) 2 N, rightward; (b) 2 N, rightward; (c) greater than 2 N, rightward
9. b, c, a **11.** (a) yes; (b) 6 kg·m/s in $-x$ direction; (c) can't tell **EP** **1.** (a) 4600 km; (b) $0.73R_e$ **3.** (a) $x_{cm} = 1.1$ m, $y_{cm} = 1.3$ m; (b) shifts toward topmost particle **5.** $x_{cm} = -0.25$ m, $y_{cm} = 0$ **7.** in the iron, at midheight and midwidth, 2.7 cm from midlength **9.** $x_{cm} = y_{cm} = 20$ cm, $z_{cm} = 16$ cm **11.** (a) $H/2$; (b) $H/2$; (c) descends to lowest point and then ascends to $H/2$; (d) $(HM/m)(\sqrt{1 + m/M} - 1)$
13. 72 km/h **15.** (a) center of mass does not move; (b) 0.75 m **17.** 4.8 m/s **19.** (a) 22 m; (b) 9.3 m/s
21. 53 m **23.** 13.6 ft **25.** (a) 52.0 km/h; (b) 28.8 km/h
27. a proton **29.** (a) 30°; (b) $-0.572\mathbf{j}$ kg·m/s
31. (a) $(-4.0 \times 10^4\,\mathbf{i})$ kg·m/s; (b) west; (c) 0 **33.** $0.707c$
35. 0.57 m/s, toward center of mass **37.** it increases by

4.4 m/s   **39.** (a) rocket case: 7290 m/s, payload: 8200 m/s; (b) before: 1.271×10^{10} J, after: 1.275×10^{10} J   **41.** (a) -1; (b) 1830; (c) 1830; (d) same   **43.** 14 m/s, 135° from the other pieces   **45.** 190 m/s   **47.** (a) $0.200v_{rel}$; (b) $0.210v_{rel}$; (c) $0.209v_{rel}$ **49.** (a) 1.57×10^6 N; (b) 1.35×10^5 kg; (c) 2.08 km/s   **51.** 108 m/s   **53.** 2.2×10^{-3} **57.** fast barge: 46 N more; slow barge: no change   **59.** (a) 7.8 MJ; (b) 6.2   **61.** 690 W   **63.** 5.5×10^6 N   **65.** 24 W   **67.** 100 m   **69.** (a) 860 N; (b) 2.4 m/s   **71.** (a) 3.0×10^5 J; (b) 10 kW; (c) 20 kW   **73.** (a) 2.1×10^6 kg; (b) $\sqrt{100 + 1.5t}$ m/s; (c) $(1.5 \times 10^6)/\sqrt{100 + 1.5t}$ N; (d) 6.7 km   **75.** $t = (3d/2)^{2/3}(m/2P)^{1/3}$

Chapter 10

CP   **1.** (a) unchanged; (b) unchanged; (c) decreased   **2.** (a) zero; (b) positive; (c) positive direction of y **3.** (a) 4 kg·m/s; (b) 8 kg·m/s; (c) 3 J   **4.** (a) 0; (b) 4 kg·m/s   **5.** (a) 10 kg·m/s; (b) 14 kg·m/s; (c) 6 kg·m/s   **6.** (a) 2 kg·m/s; (b) 3 kg·m/s   **7.** (a) increases; (b) increases   **Q**   **1.** all tie   **3.** b and c   **5.** (a) one stationary; (b) 2; (c) 5; (d) equal (pool player's result)   **7.** (a) 1 and 4 tie; then 2 and 3 tie; (b) 1; 3 and 4 tie; then 2   **9.** (a) rightward; (b) rightward; (c) smaller   **11.** positive direction of x axis   **EP**   **1.** (a) 750 N; (b) 6.0 m/s   **3.** 6.2×10^4 N   **5.** 3000 N ($= 660$ lb)   **7.** 1.1 m   **9.** (a) 42 N·s; (b) 2100 N   **11.** (a) $(7.4 \times 10^3 \,\mathbf{i} - 7.4 \times 10^3 \,\mathbf{j})$ N·s; (b) $(-7.4 \times 10^3 \,\mathbf{i})$ N·s; (c) 2.3×10^3 N; (d) 2.1×10^4 N; (e) $-45°$ **13.** (a) 1.0 kg·m/s; (b) 250 J; (c) 10 N; (d) 1700 N   **15.** 5 N   **17.** $2\mu v$ **19.** 990 N   **21.** (a) 1.8 N·s, upward; (b) 180 N, downward   **25.** 8 m/s   **27.** 38 km/s   **29.** 4.2 m/s   **31.** (a) 99 g; (b) 1.9 m/s; (c) 0.93 m/s   **33.** (a) 1.2 kg; (b) 2.5 m/s   **35.** 7.8 kg   **37.** (a) 1/3; (b) $4h$ **39.** 35 cm   **41.** 3.0 m/s   **43.** (a) $(10\mathbf{i} + 15\mathbf{j})$ m/s; (b) 500 J lost   **45.** (a) 2.7 m/s; (b) 1400 m/s   **47.** (a) A: 4.6 m/s, B: 3.9 m/s; (b) 7.5 m/s   **49.** 20 J for the heavy particle, 40 J for the light particle   **51.** $mv^2/6$ **53.** 13 tons   **55.** 25 cm   **57.** 0.975 m/s, 0.841 m/s   **59.** (a) 4.1 ft/s; (b) 1700 ft·lb; (c) $v_{24} = 5.3$ ft/s, $v_{32} = 3.3$ ft/s   **61.** (a) 30° from the incoming proton's direction; (b) 250 m/s and 430 m/s   **63.** (a) 41°; (b) 4.76 m/s; (c) no   **65.** $v = V/4$ **67.** (a) 117° from the final direction of B; (b) no   **69.** 120°   **71.** (a) 1.9 m/s, 30° to initial direction; (b) no   **73.** (a) 3.4 m/s, deflected by 17° to the right; (b) 0.95 MJ   **75.** (a) 117 MeV; (b) equal and opposite momenta; (c) π^- **77.** (a) 4.94 MeV; (b) 0; (c) 4.85 MeV; (d) 0.09 MeV

Chapter 11

CP   **1.** (b) and (c)   **2.** (a) and (d)   **3.** (a) yes; (b) no; (c) yes; (d) yes   **4.** all tie   **5.** 1, 2, 4, 3   **6.** (a) 1 and 3 tie, 4; then 2 and 5 tie (zero)   **7.** (a) downward in the figure; (b) less   **Q**   **1.** (a) positive; (b) zero; (c) negative; (d) negative   **3.** (a) 2 and 3; (b) 1 and 3; (c) 4   **5.** (a) and (c)   **7.** (a) all tie; (b) 2, 3; then 1 and 4 tie   **9.** b, c, a   **11.** less   **13.** 90°; then 70° and 110° tie   **15.** Finite angular

displacements are not commutative.   **EP**   **1.** (a) 1.50 rad; (b) 85.9°; (c) 1.49 m   **3.** (a) 0.105 rad/s; (b) 1.75×10^{-3} rad/s; (c) 1.45×10^{-4} rad/s   **5.** (a) $\omega(2) = 4.0$ rad/s, $\omega(4) = 28$ rad/s; (b) 12 rad/s²; (c) $\alpha(2) = 6.0$ rad/s², $\alpha(4) = 18$ rad/s²   **7.** (a) $\omega_0 + at^4 - bt^3$; (b) $\theta_0 + \omega_0 t + at^5/5 - bt^4/4$ **9.** 11 rad/s   **11.** (a) 9000 rev/min²; (b) 420 rev   **13.** (a) 30 s; (b) 1800 rad   **15.** 200 rev/min   **17.** (a) 2.0 rad/s²; (b) 5.0 rad/s; (c) 10 rad/s; (d) 75 rad   **19.** (a) 13.5 s; (b) 27.0 rad/s   **21.** (a) 340 s; (b) -4.5×10^{-3} rad/s²; (c) 98 s   **23.** (a) 1.0 rev/s²; (b) 4.8 s; (c) 9.6 s; (d) 48 rev   **25.** 6.1 ft/s² (1.8 m/s²), toward the center   **27.** 0.13 rad/s   **29.** 5.6 rad/s²   **31.** (a) 5.1 h; (b) 8.1 h   **33.** (a) 2.50×10^{-3} rad/s; (b) 20.2 m/s²; (c) 0   **35.** (a) -1.1 rev/min²; (b) 9900 rev; (c) -0.99 mm/s²; (d) 31 m/s²   **37.** (a) 310 m/s; (b) 340 m/s   **39.** (a) 1.94 m/s²; (b) 75.1°, toward the center of the track   **41.** 16 s   **43.** (a) 73 cm/s²; (b) 0.075; (c) 0.11   **45.** 12.3 kg·m²   **47.** first cylinder: 1100 J; second cylinder: 9700 J   **49.** (a) 221 kg·m²; (b) 1.10×10^4 J   **51.** (a) 6490 kg·m²; (b) 4.36 MJ   **53.** 0.097 kg·m²   **57.** (a) 1300 g·cm²; (b) 550 g·cm²; (c) 1900 g·cm²; (d) $A + B$ **59.** (a) 49 MJ; (b) 100 min   **61.** 4.6 N·m   **63.** (a) $r_1 F_1 \sin \theta_1 - r_2 F_2 \sin \theta_2$; (b) -3.8 N·m   **65.** 1.28 kg·m²   **67.** 9.7 rad/s², counterclockwise   **69.** (a) 155 kg·m²; (b) 64.4 kg   **71.** (a) 420 rad/s²; (b) 500 rad/s   **73.** small sphere: (a) 0.689 N·m and (b) 3.05 N; large sphere: (a) 9.84 N·m and (b) 11.5 N   **75.** 1.73 m/s²; 6.92 m/s²   **77.** (a) 1.4 m/s; (b) 1.4 m/s   **79.** (a) 19.8 kJ; (b) 1.32 kW   **81.** (a) 8.2×10^{28} N·m; (b) 2.6×10^{29} J; (c) 3.0×10^{21} kW   **83.** $\sqrt{9g/4\ell}$ **85.** (a) 4.8×10^5 N; (b) 1.1×10^4 N·m; (c) 1.3×10^6 J   **87.** (a) $3g(1 - \cos \theta)$; (b) $\frac{3}{2}g \sin \theta$; (c) 41.8°   **89.** (a) 5.6 rad/s²; (b) 3.1 rad/s   **91.** (a) 42.1 km/h; (b) 3.09 rad/s²; (c) 7.57 kW   **93.** (a) 3.4×10^5 g·cm²; (b) 2.9×10^5 g·cm²; (c) 6.3×10^5 g·cm²; (d) (1.2 cm) $\mathbf{i}$ + (5.9 cm) $\mathbf{j}$

Chapter 12

CP   **1.** (a) same; (b) less   **2.** less   **3.** (a) $\pm z$; (b) $+y$; (c) $-x$ **4.** (a) 1 and 3 tie, then 2 and 4 tie, then 5 (zero); (b) 2 and 3   **5.** (a) 3, 1; then 2 and 4 tie (zero); (b) 3   **6.** (a) all tie (same τ, same t, thus same ΔL); (b) sphere, disk, hoop (reverse order of I)   **7.** (a) decreases; (b) same; (c) increases   **Q**   **1.** (a) same; (b) block; (c) block   **3.** (a) greater; (b) same   **5.** (a) L; (b) $1.5L$ **7.** b, then c and d tie; then a and e tie (zero)   **9.** a, then b and c tie; then e, d (zero)   **11.** (a) same; (b) increases, because of decrease in rotational inertia   **13.** (a) 30 units clockwise; (b) 2 then 4, then the others; or 4 then 2, then the others   **15.** (a) spins in place; (b) rolls toward you; (c) rolls away from you   **EP**   **1.** 1.00   **3.** (a) 59.3 rad/s; (b) -9.31 rad/s²; (c) 70.7 m   **5.** (a) -4.11 m/s²; (b) -16.4 rad/s²; (c) -2.54 N·m   **7.** (a) 8.0°; (b) $0.14g$ **9.** (a) 4.0 N, to the left; (b) 0.60 kg·m²   **11.** (a) $\frac{1}{2}mR^2$; (b) a solid circular cylinder   **13.** (a) $mg(R - r)$; (b) 2/7; (c) $(17/7)mg$ **15.** (a) $2.7R$; (b) $(50/7)mg$ **17.** (a) 13 cm/s²; (b) 4.4 s; (c) 55 cm/s; (d) 1.8×10^{-2} J; (e) 1.4 J; (f) 27 rev/s   **21.** (a) 24 N·m, in $+y$ direction; (b) 24 N·m, $-y$; (c) 12 N·m, $+y$;

(d) 12 N·m, $-y$ **23.** (a) $(-1.5\mathbf{i} - 4.0\mathbf{j} - \mathbf{k})$ N·m;
(b) $(-1.5\mathbf{i} - 4.0\mathbf{j} - \mathbf{k})$ N·m **25.** $-2.0\mathbf{i}$ N·m **27.** 9.8
kg·m²/s **29.** (a) 12 kg·m²/s, out of page; (b) 3.0 N·m,
out of page **31.** (a) 0; (b) $(8.0\mathbf{i} + 8.0\mathbf{k})$ N·m **33.** (a) mvd;
(b) no; (c) 0, yes **35.** (a) 3.15×10^{43} kg·m²/s; (b) 0.616
37. 4.5 N·m, parallel to xy plane at $-63°$ from $+x$
39. (a) 0; (b) 0; (c) $30t^3$ kg·m²/s, $90t^2$ N·m, both in $-z$
direction; (d) $30t^3$ kg·m²/s, $90t^2$ N·m, both in $+z$ direction
41. (a) $\frac{1}{2}mgt^2v_0 \cos \theta_0$; (b) $mgtv_0 \cos \theta_0$; (c) $mgtv_0 \cos \theta_0$
43. (a) -1.47 N·m; (b) 20.4 rad; (c) -29.9 J; (d) 19.9 W
45. (a) 12.2 kg·m²; (b) 308 kg·m²/s, down **47.** (a) 1/3;
(b) 1/9 **49.** $\omega_0 R_1 R_2 I_1 / (I_1 R_2^2 + I_2 R_1^2)$ **51.** (a) 3.6 rev/s;
(b) 3.0; (c) work done by man in moving weights inward
53. (a) 267 rev/min; (b) 2/3 **55.** 3.0 min **57.** 2.6 rad/s
59. (a) they revolve in a circle of 1.5 m radius at 0.93 rad/s;
(b) 8.4 rad/s; (c) $K_a = 98$ J, $K_b = 880$ J; (d) from the work
done in pulling inward **61.** $m/(M + m)(v/R)$
63. (a) $mvR/(I + MR^2)$; (b) $mvR^2/(I + MR^2)$ **65.** 1300 m/s
67. (a) 18 rad/s; (b) 0.92

69. $\theta = \cos^{-1}\left[1 - \dfrac{6m^2h}{\ell(2m + M)(3m + M)}\right]$

71. 5.28×10^{-35} J·s **73.** Any three are spin up; the other is
spin down. **75.** (a) The magnitude of the angular momentum
increases in proportion to t^2 and the magnitude of the torque
increases in proportion to t, in agreement with the second law
for rotation. (b) The magnitudes of the angular momentum and
torque again increase with time. But the change in the
magnitude of the angular momentum in any interval is less than
is predicted by proportionality to t^2 law and the change in the
torque is less than is predicted by proportionality to t. At any
position of the projectile the torque is less when drag is present
than when it is not.

Chapter 13

CP **1.** c, e, f **2.** (a) no; (b) at site of $\mathbf{F}_1$, perpendicular to
plane of figure; (c) 45 N **3.** (a) at C (to eliminate forces there
from a torque equation); (b) plus; (c) minus; (d) equal **4.** d
5. (a) equal; (b) B; (c) B **Q** **1.** (a) yes; (b) yes; (c) yes;
(d) no **3.** b **5.** (a) yes; (b) no; (c) no (it could balance the
torques but the forces would then be unbalanced) **7.** (a) a,
then b and c tie, then d **9.** (a) 20 N (the key is the pulley
with the 20 N weight); (b) 25 N **11.** (a) $\sin \theta$; (b) same;
(c) larger **13.** tie of A and B, then C **EP** **1.** (a) two;
(b) seven **3.** (a) 2.5 m; (b) 7.3° **5.** 120° **7.** 7920 N
9. (a) 840 N; (b) 530 N **11.** 0.536 m **13.** (a) 2770 N;
(b) 3890 N **15.** (a) 1160 N, down; (b) 1740 N, up; (c) left,
stretched; (d) right, compressed **17.** (a) 280 N; (b) 880 N,
71° above the horizontal **19.** bars BC, CD, and DA are under
tension due to forces T, diagonals AC and BD are compressed
due to forces $\sqrt{2}T$ **21.** (a) 1800 lb; (b) 822 lb; (c) 1270 lb
23. (a) 49 N; (b) 28 N; (c) 57 N; (d) 29° **25.** (a) 1900 N, up;
(b) 2100 N, down **27.** (a) 340 N; (b) 0.88 m; (c) increases,
decreases **29.** $W\sqrt{2rh - h^2}/(r - h)$ **31.** (a) $L/2$; (b) $L/4$;
(c) $L/6$; (d) $L/8$; (e) $25L/24$ **33.** (a) 6630 N; (b) $F_h = 5740$ N;
(c) $F_v = 5960$ N **35.** 2.20 m **37.** (a) 1.50 m; (b) 433 N;

(c) 250 N **39.** (a) $a_1 = L/2$, $a_2 = 5L/8$, $h = 9L/8$;
(b) $b_1 = 2L/3$, $b_2 = L/2$, $h = 7L/6$ **41.** (a) 47 lb; (b) 120 lb;
(c) 72 lb **43.** (a) 445 N; (b) 0.50; (c) 315 N
45. (a) 3.9 m/s²; (b) 2000 N on each rear wheel, 3500 N on
each front wheel; (c) 790 N on each rear wheel, 1410 N on
each front wheel **47.** (a) 1.9×10^{-3}; (b) 1.3×10^7 N/m²;
(c) 6.9×10^9 N/m² **49.** 3.1 cm **51.** 2.4×10^9 N/m²
53. (a) 1.8×10^7 N; (b) 1.4×10^7 N; (c) 16 **55.** (a) 867 N;
(b) 143 N; (c) 0.165

Chapter 14

CP **1.** all tie **2.** (a) 1, tie of 2 and 4, then 3; (b) line d
3. negative y direction **4.** (a) increase; (b) negative
5. (a) 2; (b) 1 **6.** (a) path 1 (decreased E (more negative)
gives decreased a); (b) less than (decreased a gives decreased T)
Q **1.** (a) between, closer to less massive particle; (b) no;
(c) no (other than infinity) **3.** $3GM^2/d^2$, leftward **5.** b, tie
of a and c, then d **7.** b, a, c **9.** (a) negative; (b) negative;
(c) postive; (d) all tie **11.** (a) all tie; (b) all tie
13. (a) same; (b) greater **EP** **1.** 19 m **3.** 2.16
5. 1/2 **7.** 3.4×10^5 km **9.** (a) 3.7×10^{-5} N, increasing y **11.** $M = m$ **13.** 3.2×10^{-7} N **15.** $(GmM/d^2) \times$
$\left[1 - \dfrac{1}{8(1 - R/2d)^2}\right]$ **17.** 2.6×10^6 m **19.** (a) $1.3 \times$
10^{12} m/s²; (b) 1.6×10^6 m/s **21.** (a) 17 N; (b) 2.5
23. (b) 1.9 h **27.** (a) $a_g = (3.03 \times 10^{43}$ kg·m/s²$)/M_h$;
(b) decrease; (c) 9.82 m/s²; (d) 7.30×10^{-15} m/s²; (e) no
29. 7.91 km/s **31.** (a) $(3.0 \times 10^{-7}m)$ N; (b) $(3.3 \times 10^{-7}m)$
N; (c) $(6.7 \times 10^{-7}mr)$ N **33.** (a) 9.83 m/s²; (b) 9.84 m/s²;
(c) 9.79 m/s² **35.** (a) -1.4×10^{-4} J; (b) less; (c) positive;
(d) negative **37.** (a) 0.74; (b) 3.7 m/s²; (c) 5.0 km/s
39. (a) 0.0451; (b) 28.5 **41.** $-Gm(M_E/R + M_M/r)$
43. (a) 5.0×10^{-11} J; (b) -5.0×10^{-11} J **45.** (a) 1700 m/s;
(b) 250 km; (c) 1400 m/s **47.** (a) 2.2×10^{-7} rad/s;
(b) 90 km/s **51.** (a) -1.67×10^{-8} J; (b) 0.56×10^{-8} J
55. 6.5×10^{23} kg **57.** 5×10^{10} **59.** (a) 7.82 km/s;
(b) 87.5 min **61.** (a) 6640 km; (b) 0.0136 **63.** (a) 39.5
AU³/M_S·y²; (b) $T^2 = r^3/M$ **65.** (a) 1.9×10^{13} m;
(b) $3.5R_P$ **67.** south, at 35.4° above the horizon
71. $2\pi r^{3/2}/\sqrt{G(M + m/4)}$ **73.** $\sqrt{GM/L}$ **75.** (a) 2.8 y;
(b) 1.0×10^{-4} **77.** (a) 1/2; (b) 1/2; (c) B, by 1.1×10^8 J
79. (a) 54 km/s; (b) 960 m/s; (c) $R_p/R_a = v_a/v_p$ **81.** (a) $4.6 \times$
10^5 J; (b) 260 **83.** (a) 7.5 km/s; (b) 97 min; (c) 410 km;
(d) 7.7 km/s; (e) 92 min; (f) 3.2×10^{-3} N; (g) if the satellite–
Earth system is considered isolated, its $\mathbf{L}$ is conserved
85. (a) 5540 s; (b) 7.68 km/s; (c) 7.60 km/s; (d) 5.78×10^{10} J;
(e) -11.8×10^{10} J; (f) -6.02×10^{10} J; (g) $6.63 \times$
10^6 m; (h) 170 s, new orbit **87.** (a) $(-7.0$ mm$)\mathbf{i} +$
$(3.0$ cm$)\mathbf{j}$; (b) $(-0.19$ m/s$)\mathbf{i} + (0.40$ m/s$)\mathbf{j}$ **89.** (a) $1.98 \times$
10^{30} kg; (b) 1.96×10^{30} kg

Chapter 15

CP **1.** all tie **2.** (a) all tie; (b) $0.95\rho_0$, ρ_0, $1.1\rho_0$
3. 13 cm³/s, outward **4.** (a) all tie; (b) 1, then 2 and 3 tie, 4;

(c) 4, 3, 2, 1   **Q   1.** *e*, then *b* and *d* tie, then *a* and *c* tie
3. (a) 1, 3, 2; (b) all tie; (c) no (you must consider the weight exerted on the scale via the walls)   **5.** 3, 4, 1, 2
7. (a) downward; (b) downward; (c) same   **9.** (a) same;
(b) same; (c) lower; (d) higher   **11.** (a) block 1,
counterclockwise; block 2, clockwise; (b) block 1, tip more;
block 2, right itself   **EP   1.** 1000 kg/m^3 **3.** 1.1×10^5 Pa
or 1.1 atm   **5.** 2.9×10^4 N   **7.** 6.0 lb/in.^2 **9.** 1.90×10^4
Pa   **11.** 5.4×10^4 Pa   **13.** 0.52 m   **15.** (a) 6.06×10^9 N;
(b) 20 atm   **17.** 0.412 cm   **19.** $\frac{1}{4}\rho g A(h_2 - h_1)^2$
21. 44 km   **23.** (a) $\rho g W D^2/2$; (b) $\rho g W D^3/6$; (c) $D/3$
25. (a) 2.2; (b) 2.4   **27.** -3.9×10^{-3} atm   **29.** (a) fA/a;
(b) 20 lb   **31.** 1070 g   **33.** 1.5 g/cm^3 **35.** 600 kg/m^3
37. (a) 670 kg/m^3; (b) 740 kg/m^3 **39.** 390 kg
41. (a) 1.2 kg; (b) 1300 kg/m^3 **43.** 0.126 m^3 **45.** five
47. (a) 1.80 m^3; (b) 4.75 m^3 **49.** 2.79 g/cm^3
51. (a) 9.4 N; (b) 1.6 N   **53.** 4.0 m   **55.** 28 ft/s   **57.** 43
cm/s   **59.** (a) 2.40 m/s; (b) 245 Pa   **61.** (a) 12 ft/s; (b) 13
lb/in.^2 **63.** $0.72 \text{ ft} \cdot \text{lb/ft}^3$ **65.** (a) 2; (b) $R_1/R_2 = \frac{1}{2}$; (c) drain
it until $h_2 = h_1/4$ **67.** 116 m/s   **69.** (a) 6.4 m^3; (b) 5.4 m/s;
(c) 9.8×10^4 Pa   **71.** (a) 560 Pa; (b) 5.0×10^4 N
73. 40 m/s   **75.** (b) $H - h$; (c) $H/2$ **77.** (b) $0.69 \text{ ft}^3/s$
79. (b) 63.3 m/s

Chapter 16

CP   1. (a) $-x_m$; (b) $+x_m$; (c) 0   **2.** a   **3.** (a) 5 J; (b) 2 J;
(c) 5 J   **4.** all tie (in Eq. 16-32, *m* is included in *I*)
5. 1, 2, 3 (the ratio *m/b* matters; *k* does not)   **Q   1.** c
3. (a) 0; (b) between 0 and $+x_m$; (c) between $-x_m$ and 0;
(d) between $-x_m$ and 0   **5.** (a) toward $-x_m$; (b) toward $+x_m$;
(c) between $-x_m$ and 0; (d) between $-x_m$ and 0; (e) decreasing;
(f) increasing   **7.** (a) 3, 2, 1; (b) all tie   **9.** 3, 2, 1
11. system with spring *A*   **13.** *b* (infinite period; does not
oscillate), *c*, *a*   **15.** (a) same; (b) same; (c) same; (d) smaller;
(e) smaller; (f) and (g) larger ($T = \infty$)   **EP   1.** (a) 0.50 s;
(b) 2.0 Hz; (c) 18 cm   **3.** (a) 245 N/m; (b) 0.284 s
5. 708 N/m   **7.** $f > 500$ Hz   **9.** (a) 100 N/m; (b) 0.45 s
11. (a) 6.28×10^5 rad/s; (b) 1.59 mm   **13.** (a) 1.0 mm;
(b) 0.75 m/s; (c) 570 m/s^2 **15.** (a) 1.29×10^5 N/m;
(b) 2.68 Hz   **17.** (a) 4.0 s; (b) $\pi/2$ rad/s; (c) 0.37 cm;
(d) $(0.37 \text{ cm}) \cos \frac{\pi}{2}t$; (e) $(-0.58 \text{ cm/s}) \sin \frac{\pi}{2}t$; (f) 0.58 cm/s;
(g) 0.91 cm/s^2; (h) 0; (i) 0.58 cm/s   **19.** (b) 12.47 kg;
(c) 54.43 kg   **21.** 1.6 kg   **23.** (a) 1.6 Hz; (b) 1.0 m/s, 0;
(c) 10 m/s^2, ± 10 cm; (d) $(-10 \text{ N/m})x$ **25.** 22 cm
27. (a) 25 cm; (b) 2.2 Hz   **29.** (a) 0.500 m; (b) -0.251 m;
(c) 3.06 m/s   **31.** (a) 0.183*A*; (b) same direction
37. (a) $k_1 = (n + 1)k/n$, $k_2 = (n + 1)k$; (b) $f_1 = \sqrt{(n + 1)/n}f$,
$f_2 = \sqrt{n + 1}f$ **39.** (b) 42 min   **41.** (a) 200 N/m;
(b) 1.39 kg; (c) 1.91 Hz   **43.** (a) 130 N/m; (b) 0.62 s; (c) 1.6
Hz; (d) 5.0 cm; (e) 0.51 m/s   **45.** (a) 3/4; (b) 1/4; (c) $x_m/\sqrt{2}$
47. (a) 3.5 m; (b) 0.75 s   **49.** (a) 0.21 m; (b) 1.6 Hz;
(c) 0.10 m   **51.** (a) 0.0625 J; (b) 0.03125 J   **53.** 12 s
55. (a) 39.5 rad/s; (b) 34.2 rad/s; (c) 124 rad/s^2 **57.** (a) 8.3 s;
(b) no   **59.** 9.47 m/s^2 **61.** 8.77 s   **63.** 5.6 cm
65. $2\pi\sqrt{(R^2 + 2d^2)/2gd}$ **67.** (a) $0.205 \text{ kg} \cdot \text{m}^2$; (b) 47.7 cm;
(c) 1.50 s   **71.** (a) $2\pi\sqrt{(L^2 + 12x^2)/12gx}$; (b) 0.289 m

73. 9.78 m/s^2 **75.** $2\pi\sqrt{m/3k}$ **77.** $(1/2\pi)(\sqrt{g^2 + v^4/R^2}/L)^{1/2}$
79. (b) smaller   **81.** (a) 2.0 s; (b) $18.5 \text{ N} \cdot \text{m/rad}$
83. 0.29*L*   **85.** 0.39   **87.** (a) 0.102 kg/s; (b) 0.137 J
89. $k = 490$ N/cm, $b = 1100$ kg/s   **91.** 1.9 in.
93. (a) $y_m = 8.8 \times 10^{-4}$ m, $T = 0.18$ s, $\omega = 35$ rad/s;
(b) $y_m = 5.6 \times 10^{-2}$ m, $T = 0.48$ s, $\omega = 13$ rad/s;
(c) $y_m = 3.3 \times 10^{-2}$ m, $T = 0.31$ s, $\omega = 20$ rad/s

Chapter 17

CP   1. a, 2; b, 3; c, 1   **2.** (a) 2, 3, 1; (b) 3, then 1 and 2 tie
3. *a*   **4.** 0.20 and 0.80 tie, then 0.60, 0.45   **5.** (a) 1; (b) 3;
(c) 2   **6.** (a) 75 Hz; (b) 525 Hz   **Q   1.** 7*d*   **3.** tie of *A*
and *B*, then *C*, *D*   **5.** intermediate (closer to fully destructive
interference)   **7.** a and d tie, then b and c tie   **9.** (a) 8;
(b) antinode; (c) longer; (d) lower   **11.** (a) integer multiples
of 3; (b) node; (c) node   **13.** string *A*   **15.** decrease
EP   1. (a) 75 Hz; (b) 13 ms   **3.** (a) 7.5×10^{14} to 4.3×10^{14}
Hz; (b) 1.0 to 200 m; (c) 6.0×10^{16} to 3.0×10^{19} Hz
5. $y = 0.010 \sin \pi(3.33x + 1100t)$, with *x* and *y* in m and *t*
in s   **11.** (a) $z = 3.0 \sin(60y - 10\pi t)$, with *z* in mm, *y* in cm,
and *t* in s; (b) 9.4 cm/s   **13.** (a) $y = 2.0 \sin 2\pi(0.10x - 400t)$, with *x* and *y* in cm and *t* in s; (b) 50 m/s; (c) 40 m/s
15. (b) 2.0 cm/s; (c) $y = (4.0 \text{ cm}) \sin(\pi x/10 - \pi t/5 + \pi)$,
where *x* is in cm and *t* is in s; (d) -2.5 cm/s   **17.** 129 m/s
19. 135 N   **23.** (a) 15 m/s; (b) 0.036 N   **25.** $y = 0.12 \sin(141x + 628t)$, with *y* in mm, *x* in m, and *t* in s
27. (a) 5.0 cm; (b) 40 cm; (c) 12 m/s; (d) 0.033 s; (e) 9.4 m/s;
(f) $5.0 \sin(16x + 190t + 0.79)$, with *x* in m, *y* in cm, and *t* in s
29. (a) $v_1 = 28.6$ m/s, $v_2 = 22.1$ m/s; (b) $M_1 = 188$ g, $M_2 = 313$ g   **31.** (a) $\sqrt{k(\Delta l)(l + \Delta l)/m}$ **33.** (a) $P_2 = 2P_1$;
(b) $P_2 = P_1/4$ **35.** (a) 3.77 m/s; (b) 12.3 N; (c) zero; (d) 46.3
W; (e) zero; (f) zero; (g) ± 0.50 cm   **37.** 82.8°, 1.45 rad,
0.23 wavelength   **39.** 5.0 cm   **41.** (a) 4.4 mm; (b) 112°
43. (a) $0.83y_1$; (b) 37°   **45.** (a) $2f_3$; (b) λ_3 **47.** 10 cm
49. (a) 82.0 m/s; (b) 16.8 m; (c) 4.88 Hz   **51.** 240 cm, 120 cm,
80 cm   **53.** 7.91 Hz, 15.8 Hz, 23.7 Hz   **55.** $f_{1A} = f_{4B}$,
$f_{2A} = f_{8B}$ **57.** (a) 2.0 Hz, 200 cm, 400 cm/s; (b) $x = 50$ cm,
150 cm, 250 cm, etc.; (c) $x = 0$, 100 cm, 200 cm, etc.
63. (a) 1.3 m; (b) $y' = 0.002 \sin(9.4x) \cos(3800t)$, with *x* and
y in m and *t* in s   **67.** (b) in the positive *x* direction;
interchange the amplitudes of the original two traveling waves;
(c) largest at $x = \lambda/4 = 6.26$ cm; smallest at $x = 0$ and
$x = \lambda/2 = 12.5$ cm; (d) the largest amplitude is 4.0 mm, which
is the sum of the amplitudes of the original traveling waves; the
smallest amplitude is 1.0 mm, which is the difference of the
amplitudes of the original traveling waves

Chapter 18

CP   1. beginning to decrease (example: mentally move the
curves of Fig. 18-6 rightward past the point at $x = 42$ m)
2. (a) fully constructive, 0; (b) fully constructive, 4   **3.** (a) 1
and 2 tie, then 3; (b) 3, then 1 and 2 tie   **4.** second
5. loosen   **6.** *a*, greater; *b*, less; *c*, can't tell; *d*, can't tell;
e, greater; *f*, less   **7.** (a) 222 m/s; (b) +20 m/s   **Q   1.** pulse

along path 2 **3.** (a) 2.0 wavelengths; (b) 1.5 wavelengths; (c) fully constructive, fully destructive **5.** (a) exactly out of phase; (b) exactly out of phase **7.** 70 dB **9.** (a) two; (b) antinode **11.** all odd harmonics **13.** 501, 503, and 508 Hz; or 505, 507, and 508 Hz **EP 1.** (a) $\approx 6\%$ **3.** the radio listener by about 0.85 s **5.** 7.9×10^{10} Pa **7.** If only the length is uncertain, it must be known to within 10^{-4} m. If only the time is imprecise, the uncertainty must be no more than one part in 10^8. **9.** 43.5 m **11.** 40.7 m **13.** 100 kHz **15.** (a) 2.29, 0.229, 22.9 kHz; (b) 1.14, 0.114, 11.4 kHz **17.** (a) 6.0 m/s; (b) $y = 0.30 \sin(\pi x/12 + 50\pi t)$, with x and y in cm and t in s **19.** 4.12 rad **21.** (a) $343 \times (1 + 2m)$ Hz, with m being an integer from 0 to 28; (b) $686m$ Hz, with m being an integer from 1 to 29 **23.** (a) eight; (b) eight **25.** 64.7 Hz, 129 Hz **27.** (a) 0.080 W/m^2; (b) 0.013 W/m^2 **29.** 36.8 nm **31.** (a) 1000; (b) 32 **33.** (a) 39.7 μW/m^2; (b) 171 nm; (c) 0.893 Pa **35.** (a) 59.7; (b) 2.81×10^{-4} **37.** $s_m \propto r^{-1/2}$ **39.** (a) 5000; (b) 71; (c) 71 **41.** 171 m **43.** 3.16 km **45.** (a) 5200 Hz; (b) amplitude$_{SAD}$/amplitude$_{SBD}$ = 2 **47.** 20 kHz **49.** by a factor of 4 **51.** water filled to a height of $\frac{7}{8}$, $\frac{5}{8}$, $\frac{3}{8}$, $\frac{1}{8}$ m **53.** (a) 5.0 cm from one end; (b) 1.2; (c) 1.2 **55.** (a) 1130, 1500, and 1880 Hz **57.** (a) 230 Hz; (b) higher **59.** (a) node; (c) 22 s **61.** 387 Hz **63.** 0.02 **65.** 3.8 Hz **67.** (a) 380 mi/h, away from owner; (b) 77 mi/h, away from owner **69.** 15.1 ft/s **71.** 2.6×10^8 m/s **73.** (a) 77.6 Hz; (b) 77.0 Hz **75.** 33.0 km **79.** (a) 970 Hz; (b) 1030 Hz; (c) 60 Hz, no **81.** (a) 1.02 kHz; (b) 1.04 kHz **83.** 1540 m/s **85.** 41 kHz **87.** (a) 2.0 kHz; (b) 2.0 kHz **89.** (a) 485.8 Hz; (b) 500.0 Hz; (c) 486.2 Hz; (d) 500.0 Hz **91.** 1×10^6 m/s, receding **93.** $0.13c$

Chapter 19

CP 1. (a) all tie; (b) 50°X, 50°Y, 50°W **2.** (a) 2 and 3 tie, then 1, then 4; (b) 3, 2, then 1 and 4 tie **3.** A **4.** c and e **5.** (a) zero; (b) negative **6.** b and d tie, then a, c **Q 1.** 25 S°, 25 U°, 25 R° **3.** c, then the rest tie **5.** B, then A and C tie **7.** (a) both clockwise; (b) both clockwise **9.** c, a, b **11.** upward (with liquid water on the exterior and at the bottom, $\Delta T = 0$ horizontally and downward) **13.** at the temperature of your fingers **15.** 3, 2, 1 **EP 1.** 2.71 K **3.** 0.05 kPa, nitrogen **5.** (a) 320°F; (b) -12.3°F **7.** (a) -96°F; (b) 56.7°C **9.** (a) -40°; (b) 575°; (c) Celsius and Kelvin cannot give the same reading **11.** (a) Dimensions are inverse time **13.** 4.4×10^{-3} cm **15.** 0.038 in. **17.** (a) 9.996 cm; (b) 68°C **19.** 170 km **21.** 0.32 cm^2 **23.** 29 cm^3 **25.** 0.432 cm^3 **27.** -157°C **29.** 360°C **35.** $+0.68$ s/h **37.** (b) use 39.3 cm of steel and 13.1 cm of brass **39.** (a) 523 J/kg · K; (b) 0.600; (c) 26.2 J/mol · K **41.** 94.6 L **43.** 109 g **45.** 1.30 MJ **47.** 1.9 times as great **49.** (a) 33.9 Btu; (b) 172 F° **51.** (a) 52 MJ; (b) 0°C **53.** (a) 411 g; (b) 3.1¢ **55.** 0.41 kJ/kg · K **57.** 3.0 min **59.** 73 kW **61.** 2.17 g **63.** 33 m^2 **65.** 33 g **67.** (a) 0°C; (b) 2.5°C **69.** 2500 J/kg · K **71.** A: 120 J, B: 75 J, C: 30 J **73.** (a) -200 J; (b) -293 J;

(c) -93 J **75.** -5.0 J **77.** 33.3 kJ **79.** 766°C **81.** (a) 1.2 W/m · K, 0.70 Btu/ft · F° · h; (b) 0.030 ft^2 · F° · h/Btu **83.** 1660 J/s **87.** arrangement b **89.** (a) 2.0 MW; (b) 220 W **91.** (a) 17 kW/m^2; (b) 18 W/m^2 **93.** -6.1 nW **95.** 0.40 cm/h **97.** Cu-Al, 84.3°C; Al-brass, 57.6°C

Chapter 20

CP 1. all but c **2.** (a) all tie; (b) 3, 2, 1 **3.** gas A **4.** 5 (greatest change in T), then tie of 1, 2, 3, and 4 **5.** 1, 2, 3 ($Q_3 = 0$, Q_2 goes into work W_2, but Q_1 goes into greater work W_1 and increases gas temperature) **Q 1.** increased but less than doubled **3.** a, c, b **5.** 1180 J **7.** d, tie of a and b, then c **9.** constant-volume process **11.** (a) same; (b) increases; (c) decreases; (d) increases **13.** -4 J **15.** (a) 1, polyatomic; 2, diatomic; 3, monatomic; (b) more **EP 1.** (a) 0.0127; (b) 7.65×10^{21} **3.** 6560 **5.** number of molecules in the ink $\approx 3 \times 10^{16}$; number of people $\approx 5 \times 10^{20}$; statement is wrong, by a factor of about 20,000 **7.** (a) 5.47×10^{-8} mol; (b) 3.29×10^{16} **9.** (a) 106; (b) 0.892 m^3 **11.** 27.0 lb/in.^{2} **13.** (a) 2.5×10^{25}; (b) 1.2 kg **15.** 5600 J **17.** 1/5 **19.** (a) -45 J; (b) 180 K **21.** 100 cm^3 **23.** 198°F **25.** 2.0×10^5 Pa **27.** 180 m/s **29.** 9.53×10^6 m/s **31.** 313°C **33.** 1.9 kPa **35.** (a) 0.0353 eV, 0.0483 eV; (b) 3400 J, 4650 J **37.** 9.1×10^{-6} **39.** (a) 6.75×10^{-20} J; (b) 10.7 **41.** 0.32 nm **43.** 15 cm **45.** (a) 3.27×10^{10}; (b) 172 m **47.** (a) 22.5 L; (b) 2.25; (c) 8.4×10^{-5} cm; (d) same as (c) **51.** (a) 3.2 cm/s; (b) 3.4 cm/s; (c) 4.0 cm/s **53.** (a) v_P, v_{rms}, $\bar{v}$ (b) reverse ranking **55.** (a) 1.0×10^4 K, 1.6×10^5 K; (b) 440 K, 7000 K **57.** 4.7 **59.** (a) $2N/3v_0$; (b) $N/3$; (c) $1.22v_0$; (d) $1.31v_0$ **61.** $RT \ln(V_f/V_i)$ **63.** (a) 15.9 J; (b) 34.4 J/mol · K; (c) 26.1 J/mol · K **65.** $(n_1C_1 + n_2C_2 + n_3C_3)/(n_1 + n_2 + n_3)$ **67.** (a) -5.0 kJ; (b) 2.0 kJ; (c) 5.0 kJ **69.** (a) 0.375 mol; (b) 1090 J; (c) 0.714 **71.** (a) 14 atm; (b) 620 K **79.** 0.63 **81.** (a) monatomic; (b) 2.7×10^4 K; (c) 4.5×10^4 mol; (d) 3.4 kJ, 340 kJ; (e) 0.01 **83.** 5 m^3 **85.** (a) in joules, in the order Q, ΔE_{int}, W: $1 \rightarrow 2$: 3740, 3740, 0; $2 \rightarrow 3$: 0, -1810, 1810; $3 \rightarrow 1$: -3220, -1930, -1290; cycle: 520, 0, 520; (b) $V_2 = 0.0246$ m^3, $p_2 = 2.00$ atm, $V_3 = 0.0373$ m^3, $p_3 = 1.00$ atm

Chapter 21

CP 1. a, b, c **2.** smaller **3.** c, b, a **4.** a, d, c, b **5.** b **Q 1.** increase **3.** (a) increase; (b) same **5.** equal **7.** lower the temperature of the low temperature reservoir **9.** (a) same; (b) increase; (c) decrease **11.** (a) same; (b) increase; (c) decrease **13.** more than the age of the universe **EP 1.** 1.86×10^4 J **3.** 2.75 mol **7.** (a) 5.79×10^4 J; (b) 173 J/K **9.** $+3.59$ J/K **11.** (a) 14.6 J/K; (b) 30.2 J/K **13.** (a) 4.45 J/K; (b) no **15.** (a) 4500 J; (b) -5000 J; (c) 9500 J **17.** (a) 57.0°C; (b) -22.1 J/K; (c) $+24.9$ J/K; (d) $+2.8$ J/K **19.** (a) -710 mJ/K; (b) $+710$ mJ/K; (c) $+723$ mJ/K;

(d) -723 mJ/K; (e) $+13$ mJ/K; (f) 0   **23.** (a) (I) constant T, $Q = pV \ln 2$; constant V, $Q = 4.5pV$; (II) constant T, $Q = -pV \ln 2$; constant p, $Q = 7.5pV$; (b) (I) constant T, $W = pV \ln 2$; constant V, $W = 0$; (II) constant T, $W = -pV \ln 2$; constant p, $W = 3pV$; (c) $4.5pV$ for either case; (d) $4R \ln 2$ for either case   **25.** 0.75 J/K   **27.** (a) -943 J/K; (b) $+943$ J/K; (c) yes   **29.** (a) $3p_0V_0$; (b) $6RT_0$, $(3/2)R \ln 2$; (c) both are zero   **33.** (a) 31%; (b) 16 kJ   **35.** engine A, first; engine B, first and second; engine C, second; engine D, neither   **37.** 97 K   **39.** 99.99995%   **41.** 7.2 J/cycle   **43.** (a) 7200 J; (b) 960 J; (c) 13%   **45.** (a) 2270 J; (b) 14,800 J; (c) 15.4%; (d) 75.0%, greater   **49.** (a) 78%; (b) 81 kg/s   **55.** (a) 49 kJ; (b) 7.4 kJ   **57.** 21 J   **59.** (a) 0.071 J; (b) 0.50 J; (c) 2.0 J; (d) 5.0 J   **61.** 1.08 MJ   **63.** $[1 - (T_2/T_1)] \div [1 - (T_4/T_3)]$ **67.** (a) 1.27×10^{30}; (b) 7.9%; (c) 7.3%; (d) 7.3%; (e) 1.1%; (f) 0.0023%   **69.** (a) $W = N!/(n_1! \, n_2! \, n_3!)$; (b) $[(N/2)! \, (N/2)!]/[(N/3)! \, (N/3)! \, (N/3)!]$; (c) 4.2×10^{16}

Chapter 22

CP **1.** C and D attract; B and D attract   **2.** (a) leftward; (b) leftward; (c) leftward   **3.** (a) a, c, b; (b) less than   **4.** $-15e$ (net charge of $-30e$ is equally shared)   **Q** **1.** No, only for charged particles, charged particle-like objects, and spherical shells (including solid spheres) of uniform charge   **3.** a and b **5.** two points: one to the left of the particles and one between the protons   **7.** $6q^2/4\pi\epsilon_0 d^2$, leftward   **9.** (a) same; (b) less than; (c) cancel; (d) add; (e) the adding components; (f) positive direction of y; (g) negative direction of y; (h) positive direction of x; (i) negative direction of x **11.** (a) A, B, and D; (b) all four; (c) Connect A and D; disconnect them; then connect one of them to B. (There are two more solutions.)   **13.** (a) possibly; (b) definitely   **15.** same   **17.** D **EP** **1.** 0.50 C   **3.** 2.81 N on each   **5.** (a) 4.9×10^{-7} kg; (b) 7.1×10^{-11} C   **7.** $3F/8$ **9.** (a) 1.60 N; (b) 2.77 N   **11.** (a) $q_1 = 9q_2$; (b) $q_1 = -25q_2$ **13.** either -1.00 μC and $+3.00$ μC or $+1.00$ μC and -3.00 μC   **15.** (a) 36 N, $-10°$ from the x axis; (b) $x = -8.3$ cm, $y = +2.7$ cm   **17.** (a) 5.7×10^{13} C, no; (b) 6.0×10^5 kg   **19.** (a) $Q = -2\sqrt{2}q$; (b) no   **21.** 3.1 cm   **23.** 2.89×10^{-9} N   **25.** -1.32×10^{13} C   **27.** (a) 3.2×10^{-19} C; (b) two   **29.** (a) 8.99×10^{-19} N; (b) 625   **31.** 5.1 m below the electron   **33.** 1.3 days   **35.** (a) 0; (b) 1.9×10^{-9} N   **37.** 10^{18} N   **39.** (a) ^{9}B; (b) ^{13}N; (c) ^{12}C   **41.** (a) $F = (Q^2/4\pi\epsilon_0 d^2)\alpha(1 - \alpha)$; (c) 0.5; (d) 0.15 and 0.85

Chapter 23

CP **1.** (a) rightward; (b) leftward; (c) leftward; (d) rightward (p and e have same charge magnitude, and p is farther)   **2.** all tie   **3.** (a) toward positive y; (b) toward positive x; (c) toward negative y **4.** (a) leftward; (b) leftward; (c) decrease   **5.** (a) all tie; (b) 1 and 3 tie, then 2 and 4 tie   **Q** **1.** (a) toward positive x; (b) downward and to the right;

(c) A **3.** two points: one to the left of the particles, the other between the protons   **5.** (a) yes; (b) toward; (c) no (the field vectors are not along the same line); (d) cancel; (e) add; (f) adding components; (g) toward negative y **7.** (a) 3, then 1 and 2 tie (zero); (b) all tie; (c) 1 and 2 tie, then 3   **9.** (a) rightward; (b) $+q_1$ and $-q_3$, increase; q_2, decrease; n, same   **11.** a, b, c **13.** (a) 4, 3, 1, 2; (b) 3, then 1 and 4 tie, then 2   **EP** **1.** (a) 6.4×10^{-18} N; (b) 20 N/C   **3.** to the right in the figure   **7.** 56 pC   **9.** 3.07×10^{21} N/C, radially outward   **13.** (a) $q/8\pi\epsilon_0 d^2$, to the left; $3q/\pi\epsilon_0 d^2$, to the right; $7q/16\pi\epsilon_0 d^2$, to the left   **15.** 0   **17.** 9:30   **19.** $E = q/\pi\epsilon_0 a^2$, along bisector, away from triangle   **21.** $7.4q/4\pi\epsilon_0 d^2$, 28° counterclockwise to $+x$ **23.** 6.88×10^{-28} C·m   **25.** $(1/4\pi\epsilon_0)(p/r^3)$, antiparallel to $\mathbf{p}$ **29.** $R/\sqrt{2}$ **31.** $(1/4\pi\epsilon_0)(4q/\pi R^2)$, toward decreasing y **37.** (a) 0.10 μC; (b) 1.3×10^{17}; (c) 5.0×10^{-6} **39.** 3.51×10^{15} m/s²   **41.** 6.6×10^{-15} N   **43.** 2.03×10^{-7} N/C, up   **45.** (a) -0.029 C; (b) repulsive forces would explode the sphere   **47.** (a) 1.92×10^{12} m/s²; (b) 1.96×10^5 m/s   **49.** (a) 8.87×10^{-15} N; (b) 120   **51.** 1.64×10^{-19} C ($\approx$3% high)   **53.** (a) 0.245 N, 11.3° clockwise from the $+x$ axis; (b) $x = 108$ m, $y = -21.6$ m   **55.** 27μm   **57.** (a) yes; (b) upper plate, 2.73 cm   **59.** (a) 0; (b) 8.5×10^{-22} N·m; (c) 0   **61.** $(1/2\pi)\sqrt{pE/I}$ **63.** (a) $E = (2q/4\pi\epsilon_0 d^2)(\alpha/(1 + \alpha^2)^{3/2})$; (c) 0.707; (d) 0.21 and 1.9

Chapter 24

CP **1.** (a) $+EA$; (b) $-EA$; (c) 0; (d) 0   **2.** (a) 2; (b) 3; (c) 1   **3.** (a) equal; (b) equal; (c) equal   **4.** (a) $+50e$; (b) $-150e$ **5.** 3 and 4 tie, then 2, 1   **Q** **1.** (a) 8 N·m²/C; (b) 0   **3.** (a) all tie (zero); (b) all tie   **5.** $+13q/\epsilon_0$ **7.** all tie   **9.** all tie   **11.** $2\sigma, \sigma, 3\sigma$; or $3\sigma, \sigma, 2\sigma$ **13.** (a) all tie ($E = 0$); (b) all tie   **15.** (a) same ($E = 0$); (b) decrease; (c) decrease (to zero); (d) same   **EP** **1.** (a) 693 kg/s; (b) 693 kg/s; (c) 347 kg/s; (d) 347 kg/s; (e) 575 kg/s   **3.** (a) 0; (b) -3.92 N·m²/C; (c) 0; (d) 0 for each field   **5.** (a) enclose $2q$ and $-2q$, or enclose all four charges; (b) enclose $2q$ and q; (c) not possible   **7.** 2.0×10^5 N·m²/C   **9.** $q/6\epsilon_0$ **11.** (a) $-\pi R^2 E$; (b) $\pi R^2 E$ **13.** -4.2×10^{-10} C   **15.** 0 through each of the three faces meeting at q, $q/24\epsilon_0$ through each of the other faces   **17.** 2.0 μC/m²   **19.** (a) 4.5×10^{-7} C/m²; (b) 5.1×10^4 N/C   **21.** (a) -3.0×10^{-6} C; (b) $+1.3 \times 10^{-5}$ C   **23.** (a) 0.32 μC; (b) 0.14 μC   **27.** (a) $E = q/2\pi\epsilon_0 Lr$, radially inward; (b) $-q$ on both inner and outer surfaces; (c) $E = q/2\pi\epsilon_0 Lr$, radially outward   **29.** 3.6 nC   **31.** (b) $\rho R^2/2\epsilon_0 r$ **33.** (a) 5.3×10^7 N/C; (b) 60 N/C   **35.** 5.0 nC/m²   **37.** 0.44 mm   **39.** (a) 4.9×10^{-22} C/m²; (b) downward   **41.** (a) $\rho x/\epsilon_0$; (b) $\rho d/2\epsilon_0$ **43.** (a) -750 N·m²/C; (b) -6.64 nC   **45.** (a) 4.0×10^6 N/C; (b) 0   **47.** (a) 0; (b) $q_a/4\pi\epsilon_0 r^2$; (c) $(q_a + q_b)/4\pi\epsilon_0 r^2$ **51.** (a) $-q$; (b) $+q$; (c) $E = q/4\pi\epsilon_0 r^2$, radially outward; (d) $E = 0$; (e) $E = q/4\pi\epsilon_0 r^2$, radially outward; (f) 0; (g) $E = q/4\pi\epsilon_0 r^2$, radially outward; (h) yes, charge is induced; (i) no; (j) yes; (k) no; (l) no   **53.** (a) $E = (q/4\pi\epsilon_0 a^3)r$; (b) $E = q/4\pi\epsilon_0 r^2$; (c) 0; (d) 0; (e) inner, $-q$; outer, 0   **55.** $q/2\pi a^2$ **59.** $\alpha = 0.80$

Chapter 25

CP **1.** (a) negative; (b) increase **2.** (a) positive;
(b) higher **3.** (a) rightward; (b) 1, 2, 3, 5: positive; 4,
negative; (c) 3, then 1, 2, and 5 tie, then 4 **4.** all tie **5.** a,
c (zero), b **6.** (a) 2, then 1 and 3 tie; (b) 3; (c) accelerate
leftward **7.** closer (half of 9.23 fm) **Q** **1.** (a) higher;
(b) positive; (c) negative; (d) all tie **3.** (a) 1 and 2; (b) none;
(c) no; (d) 1 and 2 yes, 3 and 4 no **5.** b, then a, c, and d tie
7. (a) negative; (b) zero **9.** (a) 1, then 2 and 3 tie; (b) 3
11. left **13.** a, b, c **15.** (a) c, b, a; (b) zero
17. (a) positive; (b) positive; (c) negative; (d) all tie
19. (a) no; (b) yes **21.** no (a particle at the intersection
would have two different potential energies) **23.** (a)–(b) all
tie; (c) C, B, A; (d) all tie **EP** **1.** 1.2 GeV **3.** (a) $3.0 \times$
10^{10} J; (b) 7.7 km/s; (c) 9.0×10^4 kg **7.** 2.90 kV **9.** 8.8
mm **11.** (a) 136 MV/m; (b) 8.82 kV/m **13.** (b) because
$V = 0$ point is chosen differently; (c) $q/(8\pi\epsilon_0 R)$; (d) potential
differences are independent of the choice for the $V = 0$ point
15. (a) -4500 V; (b) -4500 V **17.** 843 V **19.** 2.8×10^5
21. $x = d/4$ and $x = -d/2$ **23.** none **25.** (a) 3.3 nC;
(b) 12 nC/m^2 **27.** 6.4×10^8 V **29.** 190 MV
31. (a) -4.8 nm; (b) 8.1 nm; (c) no **33.** 16.3 μV
35. (a) $\dfrac{2\lambda}{4\pi\epsilon_0} \ln\left[\dfrac{L/2 + (L^2/4 + d^2)^{1/2}}{d}\right]$; (b) 0
37. (a) $-5Q/4\pi\epsilon_0 R$; (b) $-5Q/4\pi\epsilon_0(z^2 + R^2)^{1/2}$
39. $0.113\sigma R/\epsilon_0$ **41.** $(Q/4\pi\epsilon_0 L)\ln(1 + L/d)$ **43.** 670 V/m
45. $p/2\pi\epsilon_0 r^3$ **47.** 39 V/m, $-x$ direction
51. (a) $\dfrac{c}{4\pi\epsilon_0}[\sqrt{L^2 + y^2} - y]$; (b) $\dfrac{c}{4\pi\epsilon_0}\left[1 - \dfrac{y}{\sqrt{L^2 + y^2}}\right]$
53. (a) 2.5 MV; (b) 5.1 J; (c) 6.9 J **55.** (a) 2.72×10^{-14} J;
(b) 3.02×10^{-31} kg, about $\frac{1}{3}$ of accepted value **57.** (a) 0.484
MeV; (b) 0 **59.** 2.1 d **61.** 0 **63.** (a) 27.2 V; (b) -27.2
eV; (c) 13.6 eV; (d) 13.6 eV **65.** 1.8×10^{-10} J
67. 1.48×10^7 m/s **69.** $qQ/4\pi\epsilon_0 K$ **71.** 0.32 km/s
73. 1.6×10^{-9} m **77.** (a) $V_1 = V_2$; (b) $q_1 = q/3$, $q_2 =$
$2q/3$; (c) 2 **79.** (a) -0.12 V; (b) 1.8×10^{-8} N/C, radially
inward **81.** (a) 12,000 N/C; (b) 1800 V; (c) 5.8 cm
83. (c) 4.24 V

Chapter 26

CP **1.** (a) same; (b) same **2.** (a) decreases; (b) increases;
(c) decreases **3.** (a) V, $q/2$; (b) $V/2$, q **4.** (a) $q_0 = q_1 +$
q_{34}; (b) equal (C_3 and C_4 are in series) **5.** (a) same;
(b)–(d) increase; (e) same (same potential difference across
same plate separation) **6.** (a) same; (b) decrease;
(c) increase **Q** **1.** a, 2; b, 1; c, 3 **3.** (a) increase;
(b) same **5.** (a) parallel; (b) series **7.** (a) $C/3$; (b) $3C$;
(c) parallel **9.** (a) equal; (b) less **11.** (a)–(d) less
13. (a) 2; (b) 3; (c) 1 **15.** Increase plate separation d, but
also plate area A, keeping A/d constant. **EP** **1.** 7.5 pC
3. 3.0 mC **5.** (a) 140 pF; (b) 17 nC **7.** (a) 84.5 pF;
(b) 191 cm^2 **9.** (a) 11 cm^2; (b) 11 pF; (c) 1.2 V
13. (b) 4.6×10^{-5}/K **15.** 7.33 μF **17.** 315 mC
19. (a) 10.0 μF; (b) $q_2 = 0.800$ mC, $q_1 = 1.20$ mC; (c) 200 V
for both **21.** (a) $d/3$; (b) $3d$ **25.** (a) five in series; (b) three

arrays as in (a) in parallel (and other possibilities) **27.** 43 pF
29. (a) 50 V; (b) 5.0×10^{-5} C; (c) 1.5×10^{-4} C
31. (a) $q_1 = 9.0$ μC, $q_2 = 16$ μC, $q_3 = 9.0$ μC, $q_4 = 16$ μC;
(b) $q_1 = 8.4$ μC, $q_2 = 17$ μC, $q_3 = 11$ μC, $q_4 = 14$ μC
33. 99.6 nJ **35.** 72 F **37.** 4.9% **39.** 0.27 J **41.** 0.11
J/m^3 **43.** (a) 2.0 J **45.** (a) $q_1 = 0.21$ mC, $q_2 = 0.11$ mC,
$q_3 = 0.32$ mC; (b) $V_1 = V_2 = 21$ V, $V_3 = 79$ V; (c) $U_1 = 2.2$
mJ, $U_2 = 1.1$ mJ, $U_3 = 13$ mJ **47.** (a) $q_1 = q_2 = 0.33$ mC,
$q_3 = 0.40$ mC; (b) $V_1 = 33$ V, $V_2 = 67$ V, $V_3 = 100$ V;
(c) $U_1 = 5.6$ mJ, $U_2 = 11$ mJ, $U_3 = 20$ mJ **53.** Pyrex
55. (a) 6.2 cm; (b) 280 pF **57.** 0.63 m^2 **59.** (a) 2.85 m^3;
(b) 1.01×10^4 **61.** (a) $\epsilon_0 A/(d-b)$; (b) $d/(d-b)$;
(c) $-q^2 b/2\epsilon_0 A$, sucked in **65.** $\dfrac{\epsilon_0 A}{4d}\left(\kappa_1 + \dfrac{2\kappa_2\kappa_3}{\kappa_2 + \kappa_3}\right)$
67. (a) 13.4 pF; (b) 1.15 nC; (c) 1.13×10^4 N/C; (d) $4.33 \times$
10^3 N/C **69.** (a) 7.1; (b) 0.77 μC **71.** (a) 0.606; (b) 0.394

Chapter 27

CP **1.** 8 A, rightward **2.** (a)–(c) rightward **3.** a and c
tie, then b **4.** Device 2 **5.** (a) and (b) tie, then (d), then
(c) **Q** **1.** a, b, and c tie, then d (zero) **3.** b, a, c **5.** tie
of A, B, and C, then a tie of $A + B$ and $B + C$, then $A +$
$B + C$ **7.** (a)–(c) 1 and 2 tie, then 3 **9.** C, A, B
11. b, a, c **13.** (a) conductors: 1 and 4; semiconductors: 2
and 3; (b) 2 and 3; (c) all four **EP** **1.** 1.25×10^{15} **3.** 6.7
μC/m^2 **5.** 14-gauge **7.** (a) 2.4×10^{-5} A/m^2; (b) $1.8 \times$
10^{-15} m/s **9.** 0.67 A, toward the negative terminal
11. (a) 0.654 μA/m^2; (b) 83.4 MA **13.** 13 min
15. (a) $J_0 A/3$; (b) $2J_0 A/3$ **17.** 2.0×10^{-8} $\Omega \cdot$m
19. 100 V **21.** (a) 1.53 kA; (b) 54.1 MA/m^2; (c) $10.6 \times$
10^{-8} $\Omega \cdot$m, platinum **23.** (a) 253°C; (b) yes **25.** (a) 0.38 mV;
(b) negative; (c) 3 min 58 s **27.** 54 Ω **29.** 3.0
31. (a) 1.3 mΩ; (b) 4.6 mm **33.** (a) 6.00 mA; (b) $1.59 \times$
10^{-8} V; (c) 21.2 nΩ **35.** 2000 K **37.** (a) copper: $5.32 \times$
10^5 A/m^2, aluminum: 3.27×10^5 A/m^2; (b) copper: 1.01 kg/m,
aluminum: 0.495 kg/m **39.** 0.40 Ω **41.** (a) $R = \rho L/\pi ab$
43. 14 kC **45.** 11.1 Ω **47.** (a) 1.0 kW; (b) 25¢
49. 0.135 W **51.** (a) 1.74 A; (b) 2.15 MA/m^2; (c) 36.3
mV/m; (d) 2.09 W **53.** (a) 1.3×10^5 A/m^2; (b) 94 mV
55. (a) \$4.46 for a 31-day month; (b) 144 Ω; (c) 0.833 A
57. 660 W **59.** (a) 3.1×10^{11}; (b) 25 μA; (c) 1300 W,
25 MW **61.** 27 cm/s **63.** (a) 120 Ω; (b) 107 Ω; (c) $5.3 \times$
10^{-3}/C°; (d) 5.9×10^{-3}/C°; (e) 276 Ω

Chapter 28

CP **1.** (a) rightward; (b) all tie; (c) b, then a and c tie; (d) b,
then a and c tie **2.** (a) all tie; (b) R_1, R_2, R_3 **3.** (a) less;
(b) greater; (c) equal **4.** (a) $V/2$, i; (b) V, $i/2$ **5.** (a) 1, 2, 4,
3; (b) 4, tie of 1 and 2; then 3 **Q** **1.** 3, 4, 1, 2 **3.** (a) no;
(b) yes; (c) all tie (the circuits are the same) **5.** parallel, R_2,
R_1, series **7.** (a) equal; (b) more **9.** (a) less; (b) less;
(c) more **11.** C_1, 15 V; C_2, 35 V; C_3, 20 V; C_4, 20 V; C_5,
30 V **13.** 60 μC **15.** c, b, a **17.** (a) all tie; (b) 1, 3, 2
19. 1, 3, and 4 tie (8 V on each resistor), then 2 and 5 tie (4 V
on each resistor) **EP** **1.** (a) \$320; (b) 4.8¢ **3.** 11 kJ

5. (a) counterclockwise; (b) battery 1; (c) B   **7.** (a) 80 J;
(b) 67 J; (c) 13 J converted to thermal energy within battery
9. (a) 14 V; (b) 100 W; (c) 600 W; (d) 10 V, 100 W
11. (a) 50 V; (b) 48 V; (c) B is connected to the negative
terminal   **13.** 2.5 V   **15.** (a) 6.9 km; (b) 20 Ω
17. 8.0 Ω **19.** 10^{-6} **21.** the cable   **23.** (a) 1000 Ω;
(b) 300 mV; (c) 2.3×10^{-3} **25.** (a) 3.41 A or 0.586 A;
(b) 0.293 V or 1.71 V   **27.** 5.56 A   **29.** 4.0 Ω and 12 Ω
31. 4.50 Ω **33.** 0.00, 2.00, 2.40, 2.86, 3.00, 3.60, 3.75,
3.94 A   **35.** $V_d - V_c = +0.25$ V, by all paths   **37.** three
39. (a) 2.50 Ω; (b) 3.13 Ω **41.** nine   **43.** (a) left branch:
0.67 A down; center branch: 0.33 A up; right branch: 0.33 A
up; (b) 3.3 V   **47.** (a) 120 Ω; (b) $i_1 = 51$ mA, $i_2 = i_3 =$
19 mA, $i_4 = 13$ mA   **49.** (a) 19.5 Ω; (b) 0; (c) ∞; (d) 82.3 W,
57.6 W   **51.** (a) Cu: 1.11 A, Al: 0.893 A; (b) 126 m
53. (a) 13.5 kΩ; (b) 1500 Ω; (c) 167 Ω; (d) 1480 Ω
55. 0.45 A   **57.** (a) 12.5 V; (b) 50 A   **59.** -0.9%
65. (a) 0.41τ; (b) 1.1τ **67.** 4.6   **69.** (a) 0.955 μC/s;
(b) 1.08 μW; (c) 2.74 μW; (d) 3.82 μW   **71.** 2.35 MΩ
73. 0.72 MΩ **75.** 24.8 Ω to 14.9 kΩ **77.** (a) at $t = 0$,
$i_1 = 1.1$ mA, $i_2 = i_3 = 0.55$ mA; at $t = \infty$, $i_1 = i_2 = 0.82$ mA,
$i_3 = 0$; (c) at $t = 0$, $V_2 = 400$ V; at $t = \infty$, $V_2 = 600$ V;
(d) after several time constants ($\tau = 7.1$ s) have elapsed
79. (a) $V_T = -ir + \mathcal{E}$; (b) 13.6 V; (c) 0.060 Ω
81. (a) 6.4 V; (b) 3.6 W; (c) 17 W; (d) -5.6 W; (e) a

Chapter 29

CP **1.** $a, +z; b, -x; c, F_B = 0$ **2.** 2, then tie of 1 and 3
(zero); (b) 4   **3.** (a) $+z$ and $-z$ tie, then $+y$ and $-y$ tie, then
$+x$ and $-x$ tie (zero); (b) $+y$ **4.** (a) electron;
(b) clockwise   **5.** $-y$ **6.** (a) all tie; (b) 1 and 4 tie, then 2
and 3 tie   **Q** **1.** (a) all tie; (b) 1 and 2 (charge is negative)
3. a, no, $\mathbf{v}$ and $\mathbf{F}_B$ must be perpendicular; b, yes; c, no, $\mathbf{B}$ and
$\mathbf{F}_B$ must be perpendicular   **5.** (a) $\mathbf{F}_E$; (b) $\mathbf{F}_B$ **7.** (a) right;
(b) right   **9.** (a) negative; (b) equal; (c) equal; (d) half a
circle   **11.** (a) $\mathbf{B}_1$; (b) $\mathbf{B}_1$ into page; $\mathbf{B}_2$ out of page; (c) less
13. all   **15.** all tie   **17.** (a) positive; (b) (1) and (2) tie, then
(3) which is zero   **EP** **1.** M/QT **3.** (a) 9.56×10^{-14} N, 0;
(b) $0.267°$ **5.** (a) east; (b) 6.28×10^{14} m/s²; (c) 2.98 mm
7. 0.75$\mathbf{k}$ T   **9.** (a) 3.4×10^{-4} T, horizontal and to the left
as viewed along $\mathbf{v}_0$; (b) yes, if velocity is the same as the
electron's velocity   **11.** $(-11.4\mathbf{i} - 6.00\mathbf{j} + 4.80\mathbf{k})$ V/m
13. 680 kV/m   **17.** (b) 2.84×10^{-3} **19.** (a) 1.11×10^7 m/s;
(b) 0.316 mm   **21.** (a) 0.34 mm; (b) 2.6 keV
23. (a) 2.05×10^7 m/s; (b) 467 μT; (c) 13.1 MHz; (d) 76.3 ns
25. (a) 2.60×10^6 m/s; (b) 0.109 μs; (c) 0.140 MeV;
(d) 70 kV   **29.** (a) 1.0 MeV; (b) 0.5 MeV   **31.** $R_d = \sqrt{2}R_p$;
$R_\alpha = R_p$.   **33.** (a) $B\sqrt{mq/2V}\,\Delta x$; (b) 8.2 mm   **37.** (a) $-q$;
(b) $\pi m/qB$ **39.** $B_{min} = \sqrt{mV/2ed^2}$ **41.** (a) 22 cm;
(b) 21 MHz   **43.** neutron moves tangent to original path,
proton moves in a circular orbit of radius 25 cm   **45.** 28.2 N,
horizontally west   **47.** 20.1 N   **49.** $Bitd/m$, away from
generator   **51.** $-0.35\mathbf{k}$ N   **53.** 0.10 T, at $31°$ from the
vertical   **55.** 4.3×10^{-3} N$\cdot$m, negative y **59.** $qvaB/2$
61. (a) 540 Ω, in series; (b) 2.52 Ω, in parallel
63. (a) 12.7 A; (b) 0.0805 N$\cdot$m   **65.** (a) 0.184 A$\cdot$m²;

(b) 1.45 N$\cdot$m   **67.** (a) 20 min; (b) 5.9×10^{-2} N$\cdot$m
69. (a) $(8.0 \times 10^{-4}$ N$\cdot$m$)(-1.2\mathbf{i} - 0.90\mathbf{j} + 1.0\mathbf{k})$;
(b) -6.0×10^{-4} J

Chapter 30

CP **1.** a, c, b **2.** b, c, a **3.** d, tie of a and c, then b
4. d, a, tie of b and c (zero)   **Q** **1.** c, d, then a and b tie
3. 2 and 4   **5.** a, b, c **7.** b, d, c, a (zero)   **9.** (a) 1, $+x$;
2, $-y$; (b) 1, $+y$; 2, $+x$ **11.** outward   **13.** c and
d tie, then b, a **15.** d, then tie of a and e, then b, c
17. 0 (dot product is zero)   **EP** **1.** 32.1 A
3. (a) 3.3 μT; (b) yes   **5.** (a) $(0.24\mathbf{i})$ nT; (b) 0; (c) $(-43\mathbf{k})$ pT;
(d) $(0.14\mathbf{k})$ nT   **7.** (a) 16 A; (b) west to east   **9.** 0
11. (a) 0; (b) $\mu_0 i/4R$, into the page; (c) same as (b)
13. $\mu_0 i\theta\,(1/b - 1/a)/4\pi$, out of page   **15.** (a) 1.0 mT, out of
the figure; (b) 0.80 mT, out of the figure   **25.** 200 μT, into
page   **27.** (a) it is impossible to have other than $B = 0$
midway between them; (b) 30 A   **29.** 4.3 A, out of page
35. $0.338\mu_0 i^2/a$, toward the center of the square
37. (b) to the right   **39.** (b) 2.3 km/s   **41.** $+5\mu_0 i$

47. (a) $\mu_0 ir/2\pi c^2$; (b) $\mu_0 i/2\pi r$; (c) $\dfrac{\mu_0 i}{2\pi(a^2 - b^2)}\left|\dfrac{a^2 - r^2}{r}\right|$;

(d) 0   **49.** $3i/8$, into page   **53.** 0.30 mT   **55.** 108 m
61. 0.272 A   **63.** (a) 4; (b) 1/2   **65.** (a) 2.4 A$\cdot$m²;
(b) 46 cm   **67.** (a) $\mu_0 i(1/a + 1/b)/4$, into page; (b) $\frac{1}{2}i\pi(a^2 + b^2)$,
into page   **69.** (a) 79 μT; (b) 1.1×10^{-6} N$\cdot$m
71. (b) $(0.060\,\mathbf{j})$ A$\cdot$m²; (c) $(9.6 \times 10^{-11}\mathbf{j})$ T, $(-4.8 \times 10^{-11}\mathbf{j})$ T   **73.** (a) B from sum: 7.069×10^{-5} T; $\mu_0 in =$
5.027×10^{-5} T; 40% difference; (b) B from sum: 1.043×10^{-4} T; $\mu_0 in = 1.005 \times 10^{-4}$ T; 4% difference; (c) B from
sum: 2.506×10^{-4} T; $\mu_0 in = 2.513 \times 10^{-4}$ T; 0.3%
difference   **75.** (a) $\mathbf{B} = (\mu_0/2\pi)\,[i_1/(x - a) + i_2/x]\mathbf{j}$;
(b) $\mathbf{B} = (\mu_0/2\pi)\,(i_1/a)\,(1 + b/2)\mathbf{j}$

Chapter 31

CP **1.** b, then d and e tie, and then a and c tie (zero)   **2.** a
and b tie, then c (zero)   **3.** c and d tie, then a and b tie
4. b, out; c, out; d, into; e, into   **5.** d and e   **6.** (a) 2, 3, 1
(zero); (b) 2, 3, 1   **7.** a and b tie, then c **Q** **1.** (a) all tie
(zero); (b) all tie (nonzero); (c) 3, then tie of 1 and 2 (zero)
3. out   **5.** (a) into; (b) counterclockwise; (c) larger
7. (a) leftward; (b) rightward   **9.** c, a, b **11.** (a) 1, 3, 2;
(b) 1 and 3 tie, then 2   **13.** a, tie of b and c   **15.** (a) more;
(b) same; (c) same; (d) same (zero)   **17.** a, 2; b, 4; c, 1; d, 3
EP **1.** 57 μWb   **3.** 1.5 mV   **5.** (a) 31 mV; (b) right to
left   **7.** (a) 0.40 V; (b) 20 A   **9.** (b) 58 mA   **11.** 1.2 mV
13. 1.15 μWb   **15.** 51 mV, clockwise when viewed along the
direction of $\mathbf{B}$ **17.** (a) 1.26×10^{-4} T, 0, -1.26×10^{-4} T;
(b) 5.04×10^{-8} V   **19.** (b) no   **21.** 15.5 μC
23. (a) 24 μV; (b) from c to b **25.** (b) design it so that
$Nab = (5/2\pi)$ m²   **27.** (a) 0.598 μV; (b) counterclockwise

29. (a) $\dfrac{\mu_0 ia}{2\pi}\left(\dfrac{2r + b}{2r - b}\right)$; (b) $2\mu_0 iabv/\pi R(4r^2 - b^2)$

31. $A^2 B^2/R\Delta t$ **33.** (a) 48.1 mV; (b) 2.67 mA; (c) 0.128 mW
35. $v_t = mgR/B^2L^2$ **37.** 268 W   **39.** (a) 240 μV; (b) 0.600

mA; (c) 0.144 μW; (d) 2.88 $\times$ 10^{-8} N; (e) same as (c)
41. 1, -1.07 mV; 2, -2.40 mV; 3, 1.33 mV   **43.** at a:
4.4 $\times$ 10^7 m/s^2, to the right; at b: 0; at c: 4.4 $\times$ 10^7 m/s^2, to the
left   **45.** 0.10 μWb   **47.** (a) 800; (b) 2.5 $\times$ 10^{-4} H/m
49. (a) $\mu_0 i/W$; (b) $\pi \mu_0 R^2/W$ **51.** (a) decreasing;
(b) 0.68 mH   **53.** (a) 0.10 H; (b) 1.3 V   **55.** (a) 16 kV;
(b) 3.1 kV; (c) 23 kV   **57.** (b) so that the changing magnetic
field of one does not induce current in the other;
(c) $1/L_{eq} = \sum\limits_{j=1}^{N} (1/L_j)$ **59.** 6.91   **61.** 1.54 s   **63.** (a) 8.45
ns; (b) 7.37 mA   **65.** $(42 + 20t)$ V   **67.** 12.0 A/s
69. (a) $i_1 = i_2 = 3.33$ A; (b) $i_1 = 4.55$ A, $i_2 = 2.73$ A;
(c) $i_1 = 0$, $i_2 = 1.82$ A; (d) $i_1 = i_2 = 0$ **71.** $\mathscr{E}L_1/R(L_1 + L_2)$
73. (a) $i(1 - e^{-Rt/L})$ **75.** $1.23\tau_L$ **77.** (a) 240 W;
(b) 150 W; (c) 390 W   **79.** (a) 97.9 H; (b) 0.196 mJ
81. (a) 10.5 mJ; (b) 14.1 mJ   **83.** (a) 34.2 J/m^3; (b) 49.4
mJ   **85.** 1.5 $\times$ 10^8 V/m   **87.** $(\mu_0 l/2\pi)\ln(b/a)$ **89.** (a) 1.3
mT; (b) 0.63 J/m^3 **91.** (a) 1.0 J/m^3; (b) 4.8 $\times$ 10^{-15} J/m^3
93. (a) 1.67 mH; (b) 6.00 mWb   **95.** 13 H   **99.** magnetic
field exists only within the cross section of solenoid 1

Chapter 32

CP   **1.** d, b, c, a (zero)   **2.** (a) 2; (b) 1   **3.** (a) away;
(b) away; (c) less   **4.** (a) toward; (b) toward; (c) less
5. a, c, b, d (zero)   **6.** tie of b, c, and d, then a
Q   **1.** (a) a, c, f; (b) bar gh **3.** supplied   **5.** (a) all down;
(b) 1 up, 2 down, 3 zero   **7.** (a) 1 up, 2 up, 3 down;
(b) 1 down, 2 up, 3 zero   **9.** (a) rightward; (b) leftward
11. (a) decreasing; (b) decreasing   **13.** (a) tie of a and b, then
c, d; (b) none (plate lacks circular symmetry, so **B** is not
tangent to a circular loop); (c) none   **15.** 1/4   **EP**   **1.** (b)
sign is minus; (c) no, compensating positive flux through open
end near magnet   **3.** 47 μWb, inward   **5.** 55 μT   **7.** (a)
600 MA; (b) yes; (c) no   **9.** (a) 31.0 μT, 0°; (b) 55.9 μT,
73.9°; (c) 62.0 μT, 90°   **11.** 4.6 $\times$ 10^{-24} J   **13.** (a) 5.3 $\times$
10^{11} V/m; (b) 20 mT; (c) 660   **15.** (a) 7; (b) 7; (c) $3h/2\pi$, 0;
(d) $3eh/4\pi m$, 0; (e) $3.5h/2\pi$; (f) 8   **17.** (b) in the direction
of the angular momentum vector   **19.** $\Delta\mu = e^2 r^2 B/4m$
21. 20.8 mJ/T   **23.** yes   **25.** (a) 4 K; (b) 1 K
29. (a) 3.0 μT; (b) 5.6 $\times$ 10^{-10} eV   **31.** (a) 8.9 A $\cdot$ m^2;
(b) 13 N $\cdot$ m   **35.** (a) 0.14 A; (b) 79 μC   **37.** 2.4 $\times$
10^{13} V/m $\cdot$ s   **39.** 1.9 pT   **41.** 7.5 $\times$ 10^5 V/s
43. 7.2 $\times$ 10^{12} V/m $\cdot$ s   **45.** (a) 2.1 $\times$ 10^{-8} A, downward;
(b) clockwise   **47.** (a) 0.63 μT; (b) 2.3 $\times$ 10^{12} V/m $\cdot$ s
49. (a) 2.0 A; (b) 2.3 $\times$ 10^{11} V/m $\cdot$ s; (c) 0.50 A;
(d) 0.63 μT $\cdot$ m   **51.** (a) 7.60 μA; (b) 859 kV $\cdot$ m/s;
(c) 3.39 mm; (d) 5.16 pT

Chapter 33

CP   **1.** (a) $T/2$, (b) T, (c) $T/2$, (d) $T/4$ **2.** (a) 5 V;
(b) 150 μJ   **3.** (a) 1; (b) 2   **4.** (a) C, B, A; (b) 1, A;
2, B; 3, S; 4, C; (c) A **5.** (a) increases; (b) decreases
6. (a) 1, lags; 2, leads; 3, in phase; (b) 3 ($\omega_d = \omega$ when
$X_L = X_C$)   **7.** (a) increase (circuit is mainly capacitive;
increase C to decrease X_C to be closer to resonance for

maximum P_{av}); (b) closer   **8.** step-up   **Q**   **1.** (a) $T/4$,
(b) $T/4$, (c) $T/2$ (see Fig. 33-2), (d) $T/2$ (see Eq. 31-40)
3. b, a, c **5.** (a) 3, 1, 2; (b) 2, tie of 1 and 3
7. slower   **9.** (a) 1 and 4; (b) 2 and 3   **11.** (a) 3,
then 1 and 2 tie; (b) 2, 1, 3   **13.** (a) negative; (b) lead
15. (a)–(c) rightward, increase   **EP**   **1.** 9.14 nF
3. 45.2 mA   **5.** (a) 6.00 μs; (b) 167 kHz; (c) 3.00 μs
7. (a) 89 rad/s; (b) 70 ms; (c) 25 μF   **9.** 38 μH
11. 7.0 $\times$ 10^{-4} s   **15.** (a) 3.0 nC; (b) 1.7 mA; (c) 4.5 nJ
17. (a) 3.60 mH; (b) 1.33 kHz; (c) 0.188 ms   **19.** 600, 710,
1100, 1300 Hz   **21.** (a) $Q/\sqrt{3}$; (b) 0.152   **25.** (a) 1.98 μJ;
(b) 5.56 μC; (c) 12.6 mA; (d) $-46.9°$; (e) $+46.9°$ **27.** (a) 0;
(b) $2i(t)$ **29.** (a) 356 μs; (b) 2.50 mH; (c) 3.20 mJ
31. 8.66 mΩ **33.** $(L/R)\ln 2$ **35.** (a) $\pi/2$ rad; (b) $q =$
$(I/\omega') e^{-Rt/2L} \sin \omega' t$ **39.** (a) 0.0955 A; (b) 0.0119 A
41. (a) 4.60 kHz; (b) 26.6 nF; (c) $X_L = 2.60$ kΩ, $X_C =$
0.650 kΩ **43.** (a) 0.65 kHz; (b) 24 Ω **45.** (a) 39.1 mA;
(b) 0; (c) 33.9 mA   **47.** (a) 6.73 ms; (b) 2.24 ms;
(c) capacitor; (d) 59.0 μF   **49.** (a) $X_C = 0$, $X_L = 86.7$ Ω,
$Z = 182$ Ω, $I = 198$ mA, $\phi = 28.5°$ **51.** (a) $X_C = 37.9$ Ω,
$X_L = 86.7$ Ω, $Z = 167$ Ω, $I = 216$ mA, $\phi = 17.1°$
53. (a) 2.35 mH; (b) they move away from 1.40 kHz
55. 1000V   **57.** (a) 36.0 V; (b) 27.3 V; (c) 17.0 V;
(d) -8.34 V   **59.** (a) 224 rad/s; (b) 6.00 A; (c) 228 rad/s,
219 rad/s; (d) 0.040   **61.** (a) 707 Ω; (b) 32.2 mH;
(c) 21.9 nF   **63.** (a) resonance at $f = 1/2\pi\sqrt{LC} = 85.7$ Hz;
(b) 15.6 μF; (c) 225 mA   **65.** (a) 796 Hz; (b) no change;
(c) decreased; (d) increased   **69.** 141 V   **71.** (a) taking;
(b) supplying   **73.** 0, 9.00 W, 3.14 W, 1.82 W
75. 177 Ω, no   **77.** 7.61 A   **83.** (a) 117 μF; (b) 0;
(c) 90.0 W, 0; (d) 0°, 90°; (e) 1, 0   **85.** (a) 2.59 A;
(b) 38.8 V, 159 V, 224 V, 64.2 V, 75.0 V; (c) 100 W for R,
0 for L and C.   **87.** (a) 2.4 V; (b) 3.2 mA, 0.16 A
89. (a) 1.9 V, 5.9 W; (b) 19 V, 590 W; (c) 0.19 kV, 59 kW
91. (a) $X_C = [(2\pi)(45 \times 10^{-6}$ F$)f]^{-1}$; (c) 17.7 Hz
93. (a) $X_L = (2\pi)(40 \times 10^{-3}$ H$)f$; (c) 796 Hz
95. (b) 61 Hz; (c) 90 Ω and 61 Hz

Chapter 34

CP   **1.** (a) (Use Fig. 34-5.) On right side of rectangle, **E** is in
negative y direction; on left side, **E** + d**E** is greater and in
same direction; (b) **E** is downward. On right side, **B** is in
negative z direction; on left side, **B** + d**B** is greater and in
same direction.   **2.** positive direction of x **3.** (a) same;
(b) decrease   **4.** a, d, b, c (zero)   **5.** a **6.** (a) yes; (b) no
Q   **1.** (a) positive direction of z; (b) x **3.** (a) same;
(b) increase; (c) decrease   **5.** both 20° clockwise from the y
axis   **7.** two   **9.** b, 30°; c, 60°; d, 60°; e, 30°; f, 60°
11. d, b, a, c **13.** (a) b; (b) blue; (c) c **15.** 1.5
EP   **1.** (a) 4.7 $\times$ 10^{-3} Hz; (b) 3 min 32 s   **3.** (a) 4.5 $\times$ 10^{24}
Hz; (b) 1.0 $\times$ 10^4 km or 1.6 Earth radii   **7.** it would steadily
increase; (b) the summed discrepancies between the apparent
time of eclipse and those observed from x; the radius of Earth's
orbit   **9.** 5.0 $\times$ 10^{-21} H   **11.** 1.07 pT   **17.** 4.8 $\times$ 10^{-29}
W/m^2 **19.** 4.51 $\times$ 10^{-10} **21.** 89 cm   **23.** 1.2 MW/m^2
25. 820 m   **27.** (a) 1.03 kV/m; 3.43 μT   **29.** (a) 1.4 $\times$

10^{-22} W; (b) 1.1×10^{15} W **31.** (a) 87 mV/m; (b) 0.30 nT; (c) 13 kW **33.** 3.3×10^{-8} Pa **35.** (a) 4.7×10^{-6} Pa; (b) 2.1×10^{10} times smaller **37.** 5.9×10^{-8} Pa **39.** (a) 3.97 GW/m²; (b) 13.2 Pa; (c) 1.67×10^{-11} N; (d) 3.14×10^3 m/s² **41.** $I(2 - frac)/c$ **43.** $p_{r\perp} \cos^2 \theta$ **45.** 1.9 mm/s **47.** (b) 580 nm **49.** (a) 1.9 V/m; (b) 1.7×10^{-11} Pa **51.** 1/8 **53.** 3.1% **55.** 20° or 70° **57.** 19 W/m² **59.** (a) 2 sheets; (b) 5 sheets **61.** 180° **63.** 1.26 **65.** 1.07 m **69.** (a) 0; (b) 20°; (c) still 0 and 20° **73.** 1.41 **75.** 1.22 **77.** 182 cm **79.** (a) no; (b) yes; (c) about 43° **81.** (a) 35.6°; (b) 53.1° **83.** (b) 23.2° **85.** (a) 53°; (b) yes **87.** (a) 55.5°; (b) 55.8°

Chapter 35

CP Kaleidoscope answer: two mirrors that form a V with an angle of 60° **1.** $0.2d$, $1.8d$, $2.2d$ **2.** (a) real; (b) inverted; (c) same **3.** (a) e; (b) virtual, same **4.** virtual, same as object, diverging **Q** **1.** c **3.** (a) a; (b) c **5.** (a) no; (b) yes (fourth is off mirror ed) **7.** (a) from infinity to the focal point; (b) decrease continually **9.** d (infinite), tie of a and b, then c **11.** mirror, equal; lens, greater **13.** (a) all but variation 2; (b) for 1, 3, and 4: right, inverted; for 5 and 6: left, same **15.** (a) less; (b) less **EP** **1.** (a) virtual; (b) same; (c) same; (d) $D + L$ **3.** 40 cm **7.** (a) 7; (b) 5; (c) 1 to 3; (d) depends on the position of O and your perspective **11.** new illumination is 10/9 of the old **15.** 10.5 cm **19.** (a) 2.00; (b) none **23.** 1.14 **25.** (b) separate the lenses by a distance $f_2 - |f_1|$, where f_2 is the focal length of the converging lens **27.** 45 mm, 90 mm **29.** (a) $+40$ cm; (b) at infinity **33.** (a) 40 cm, real; (b) 80 cm, real; (c) 240 cm, real; (d) -40 cm, virtual; (e) -80 cm, virtual; (f) -240 cm, virtual **35.** same orientation, virtual, 30 cm to left of second lens, $m = 1$ **37.** (a) final image coincides in location with the object; it is real, inverted, and $m = -1.0$ **39.** (a) coincides in location with the original object and is enlarged 5.0 times; (c) virtual; (d) yes

45. $i = \dfrac{(2 - n)r}{2(n - 1)}$, to the right of the right side of the sphere

47. 2.1 mm **49.** (b) when image is at near point **51.** (b) farsighted **53.** -125

Chapter 36

CP **1.** b (least n), c, a **2.** (a) top; (b) bright intermediate illumination (phase difference is 2.1 wavelengths) **3.** (a) 3λ, 3; (b) 2.5λ, 2.5 **4.** a and d tie (amplitude of resultant wave is $4E_0$), then b and c tie (amplitude of resultant wave is $2E_0$) **5.** (a) 1 and 4; (b) 1 and 4 **Q** **1.** a, c, b **3.** (a) 300 nm; (b) exactly out of phase **5.** c **7.** (a) increase; (b) 1λ **9.** down **11.** (a) maximum; (b) minimum; (c) alternates **13.** d **15.** (a) 0.5 wavelength; (b) 1 wavelength **17.** bright **19.** all **EP** **1.** (a) 5.09×10^{14} Hz; (b) 388 nm; (c) 1.97×10^8 m/s **5.** 2.1×10^8 m/s **7.** the time is longer for the pipeline containing air, by about 1.55 ns **9.** 22°, refraction reduces θ **11.** (a) pulse 2; (b) $0.03L/c$ **13.** (a) 1.70 (or 0.70); (b) 1.70 (or 0.70); (c) 1.30

(or 0.30); (d) brightness is identical, close to fully destructive interference **15.** (a) 0.833; (b) intermediate, closer to fully constructive interference **17.** $(2m + 1)\pi$ **19.** 2.25 mm **21.** 648 nm **23.** 1.6 mm **25.** 16 **27.** 0.072 mm **29.** 8.75λ **31.** 0.03% **33.** 6.64 μm **35.** $y = 17 \sin(\omega t + 13°)$ **39.** (a) 1.17 m, 3.00 m, 7.50 m; (b) no **41.** $I = \frac{1}{9}I_m[1 + 8 \cos^2(\pi d \sin \theta/\lambda)]$, $I_m =$ intensity of central maximum **43.** $L = (m + \frac{1}{2})\lambda/2$, for $m = 0, 1, 2, \ldots$ **45.** 0.117 μm, 0.352 μm **47.** $\lambda/5$ **49.** 70.0 nm **51.** none **53.** (a) 552 nm; (b) 442 nm **55.** 338 nm **59.** $2n_2L/\cos \theta_r = (m + \frac{1}{2})\lambda$, for $m = 0, 1, 2, \ldots$, where $\theta_r = \sin^{-1}[(\sin \theta_i)/n_2]$ **61.** intensity is diminished by 88% at 450 nm and by 94% at 650 nm **63.** (a) dark; (b) blue end **65.** 1.89 μm **67.** 1.00025 **69.** (a) 34; (b) 46 **73.** 588 nm **75.** 1.0003 **77.** $I = I_m \cos^2(2\pi x/\lambda)$

Chapter 37

CP **1.** (a) expand; (b) expand **2.** (a) second side maximum; (b) 2.5 **3.** (a) red; (b) violet **4.** diminish **5.** (a) increase; (b) same **6.** (a) left; (b) less **Q** **1.** (a) contract; (b) contract **3.** with megaphone (larger opening, less diffraction) **5.** four **7.** (a) larger; (b) red **9.** (a) decrease; (b) same; (c) in place **11.** (a) A; (b) left; (c) left; (d) right **EP** **1.** 690 nm **3.** 60.4 μm **5.** (a) 2.5 mm; (b) 2.2×10^{-4} rad **7.** (a) 70 cm; (b) 1.0 mm **9.** 41.2 m from the central axis **11.** 160° **15.** (d) 53°, 10°, 5.1° **19.** (a) 1.3×10^{-4} rad; (b) 10 km **21.** 50 m **23.** 30.5 μm **25.** 1600 km **27.** (a) 17.1 m; (b) 1.37×10^{-10} **29.** 27 cm **31.** 4.7 cm **33.** (a) 0.347°; (b) 0.97° **35.** (a) red; (b) 130 μm **37.** five **41.** $\lambda D/d$ **43.** (a) 5.05 μm; (b) 20.2 μm **45.** (a) 3.33 μm; (b) 0, $\pm 10.2°$, $\pm 20.7°$, $\pm 32.0°$, $\pm 45.0°$, $\pm 62.2°$ **47.** all wavelengths shorter than 635 nm **49.** 13,600 **51.** 500 nm **53.** (a) three; (b) 0.051° **55.** 523 nm **61.** 470 nm to 560 nm **63.** 491 **65.** 3650 **67.** (a) 1.0×10^4 nm; (b) 3.3 mm **69.** (a) 0.032°/nm, 0.076°/nm, 0.24°/nm; (b) 40,000, 80,000, 120,000 **71.** (a) $\tan \theta$; (b) 0.89 **73.** 0.26 nm **75.** 6.8° **77.** (a) 170 pm; (b) 130 pm **81.** 0.570 nm **83.** 30.6°, 15.3° (clockwise); 3.08°, 37.8° (counterclockwise)

Chapter 38

CP **1.** (a) same (speed of light postulate); (b) no (the start and end of the flight are spatially separated); (c) no (again, because of the spatial separation) **2.** (a) Sally's; (b) Sally's **3.** a, negative; b, positive; c, negative **4.** (a) right; (b) more **5.** (a) equal; (b) less **Q** **1.** all tie (pulse speed is c) **3.** (a) C_1; (b) C_1 **5.** (a) 3, 2, 1; (b) 1 and 3 tie, then 2 **7.** (a) negative; (b) positive **9.** c, then b and d tie, then a **11.** (a) 3, tie of 1 and 2, then 4; (b) 4, tie of 1 and 2, then 3; (c) 1, 4, 2, 3 **13.** greater than f_1 **EP** **1.** (a) 3×10^{-18}; (b) 8.2×10^{-8}; (c) 1.1×10^{-6}; (d) 3.7×10^{-5}; (e) 0.10 **3.** $0.75c$ **5.** $0.99c$ **7.** 55 m **9.** 1.32 m **11.** 0.63 m **13.** 6.4 cm **15.** (a) 26 y; (b) 52 y; (c) 3.7 y **17.** (b) $0.999\,999\,15c$ **19.** (a) $x' = 0$, $t' = 2.29$ s; (b) $x' =$

6.55×10^8 m, $t' = 3.16$ s   **21.** (a) 25.8 μs; (b) small flash
23. (a) 1.25; (b) 0.800 μs   **25.** 2.40 μs   **27.** (a) 0.84c, in
the direction of increasing x; (b) 0.21c, in the direction of
increasing x; the classical predictions are 1.1c and 0.15c
29. (a) 0.35c; (b) 0.62c **31.** 1.2 μs   **33.** seven
35. 22.9 MHz   **37.** +2.97 nm   **39.** (a) $\tau_0/\sqrt{1 - v^2/c^2}$
41. (a) 0.134c; (b) 4.65 keV; (c) 1.1%   **43.** (a) 0.9988, 20.6;
(b) 0.145, 1.01; (c) 0.073, 1.0027   **45.** (a) 5.71 GeV,
6.65 GeV, 6.58 GeV/c; (b) 3.11 MeV, 3.62 MeV,
3.59 MeV/c **47.** 18 smu/y   **49.** (a) 0.943c; (b) 0.866c
51. (a) 256 kV; (b) 0.746c **53.** $\sqrt{8}mc$ **55.** 6.65×10^6 mi,
or 270 earth circumferences   **57.** 110 km   **59.** (a) 2.7 $\times$

10^{14} J; (b) 1.8×10^7 kg; (c) 6.0×10^6 **61.** 4.00 u, probably
a helium nucleus   **63.** 330 mT
65.

SIGNAL	TIME SENT (h)	TIME REPLY RECEIVED (h)	TIME REPORTED	DISTANCE (m)
1	6.0	400	11.8	2.10×10^{14}
2	12.0	800	23.6	4.19×10^{14}
3	18.0	1200	35.5	6.29×10^{14}
4	24.0	1600	47.3	8.38×10^{14}
5	30.0	2000	59.1	1.05×10^{15}

67. (a) $vt \sin \theta$; (b) $t[1 - (v/c) \cos \theta]$; (c) 3.24c

Index

Page references followed by lowercase roman t indicate material in tables. Page references followed by lowercase italic *n* indicate material in footnotes.

Photo Credits

Page 884: Courtesy Matthew J. Wheeler. Page 894: Piergiorgio Scharandis/Black Star.

Chapter 36

Page 901: E.R. Degginger. Page 905: Runk Schoenberger/ Grant Heilman Photography. Page 906: From *Atlas of Optical Phenomena* by M. Cagnet et al., Springer-Verlag, Prentice Hall, 1962. Page 914: Richard Megna/Fundamental Photographs. Page 917: Courtesy Dr. Helen Ghiradella, Department of Biological Sciences, SUNY, Albany. Page 926: Courtesy Bausch & Lomb.

Chapter 37

Page 929: Georges Seurat, French, 1859–1891, *A Sunday on La Grande Jatte,* 1884. Oil on canvas; 1884–86, 207.5 × 308 cm; Helen Birch Bartlett Memorial Collection, 1926. Photograph

©1996, The Art Institute of Chicago. All Rights Reserved. Page 930: Ken Kay/Fundamental Photographs. Pages 931, 937 (bottom left) and 940: From *Atlas of Optical Phenomena* by Cagnet, Francon, Thierr, Springer-Verlag, Berlin, 1962. Page 937 (bottom right): AP/Wide World Photos, Inc. Page 938: Cath Ellis/ Science Photo Library/Photo Researchers. Page 945 (left): Department of Physics, Imperial College/Science Photo Library/ Photo Researchers. Page 945 (top right): Peter L. Chapman/Stock, Boston. Page 953: Kjell B. Sandved/Bruce Coleman, Inc. Page 954: Courtesy Professor Robert Greenler, Physics Department, University of Wisconsin.

Chapter 38

Page 958: T. Tracy/FPG International. Page 959: Courtesy Hebrew University of Jerusalem, Israel.